Encyclopaedia of Mathematical Sciences
Volume 125

Operator Algebras and Non-Commutative Geometry

Springer
Berlin
Heidelberg
New York
Hong Kong
London
Milan
Paris
Tokyo

M. Takesaki

Theory of Operator Algebras II

Springer

Author

Masamichi Takesaki
University of California
Department of Mathematics
Los Angeles, CA 90095-1555
USA

e-mail: mt@math.ucla.edu

Founding editor of the Encyclopaedia of Mathematical Sciences:
R. V. Gamkrelidze

Mathematics Subject Classification (2000):
22D25, 46LXX, 47CXX, 47DXX

Theory of Operator Algebras I by M. Takesaki was published as Vol. 124
of the Encyclopaedia of Mathematical Sciences, ISBN 3-540-42248-X,
Theory of Operator Algebras III by M. Takesaki was published as Vol. 127
of the Encyclopaedia of Mathematical Sciences, ISBN 3-540-42913-1

ISSN 0938-0396
ISBN 3-540-42914-X Springer-Verlag Berlin Heidelberg New York

Springer-Verlag Berlin Heidelberg New York
a member of BertelsmannSpringer Science+Business Media GmbH
http://www.springer.de

Printed in Germany

Typesetting: Typeset in LaTeX
by Johannes Küster · typoma · www.typoma.com,
based on the author's plainTeX files.
Cover Design: E. Kirchner, Heidelberg, Germany
Printed on acid-free paper SPIN: 10833243 46/3142 LK 5 4 3 2 1 0

Preface to the Encyclopaedia Subseries on Operator Algebras and Non-Commutative Geometry

The theory of von Neumann algebras was initiated in a series of papers by Murray and von Neumann in the 1930's and 1940's. A von Neumann algebra is a self-adjoint unital subalgebra M of the algebra of bounded operators of a Hilbert space which is closed in the weak operator topology. According to von Neumann's bicommutant theorem, M is closed in the weak operator topology if and only if it is equal to the commutant of its commutant. A *factor* is a von Neumann algebra with trivial centre and the work of Murray and von Neumann contained a reduction of all von Neumann algebras to factors and a classification of factors into types I, II and III.

C^*-algebras are self-adjoint operator algebras on Hilbert space which are closed in the norm topology. Their study was begun in the work of Gelfand and Naimark who showed that such algebras can be characterized abstractly as involutive Banach algebras, satisfying an algebraic relation connecting the norm and the involution. They also obtained the fundamental result that a commutative unital C^*-algebra is isomorphic to the algebra of complex valued continuous functions on a compact space – its spectrum.

Since then the subject of operator algebras has evolved into a huge mathematical endeavour interacting with almost every branch of mathematics and several areas of theoretical physics.

Up into the sixties much of the work on C^*-algebras was centered around representation theory and the study of C^*-algebras of type I (these algebras are characterized by the fact that they have a well behaved representation theory). Finite dimensional C^*-algebras are easily seen to be just direct sums of matrix algebras. However, by taking algebras which are closures in norm of finite dimensional algebras one obtains already a rich class of C^*-algebras – the so-called AF-algebras – which are not of type I. The idea of taking the closure of an inductive limit of finite-dimensional algebras had already appeared in the work of Murray-von Neumann who used it to construct a fundamental example of a factor of type II – the "hyperfinite" (nowadays also called approximately finite dimensional) factor.

One key to an understanding of the class of AF-algebras turned out to be K-theory. The techniques of K-theory, along with its dual, *Ext*-theory, also found immediate applications in the study of many new examples of C^*-algebras that arose in the end of the seventies. These examples include for instance "the noncommutative tori" or other crossed products of abelian C^*-algebras by groups of homeomorphisms and abstract C^*-algebras generated by isometries with certain relations, now known as the algebras $\mathcal{O}_n$. At the same time, examples of algebras were increasingly studied that codify data from differential geometry or from topological dynamical systems.

On the other hand, a little earlier in the seventies, the theory of von Neumann algebras underwent a vigorous growth after the discovery of a natural infinite family of pairwise nonisomorphic factors of type III and the advent of Tomita-Takesaki theory. This development culminated in Connes' great classification theorems for approximately finite dimensional ("injective") von Neumann algebras.

Perhaps the most significant area in which operator algebras have been used is mathematical physics, especially in quantum statistical mechanics and in the foundations of quantum field theory. Von Neumann explicitly mentioned quantum theory as one of his motivations for developing the theory of rings of operators and his foresight was confirmed in the algebraic quantum field theory proposed by Haag and Kastler. In this theory a von Neumann algebra is associated with each region of space-time, obeying certain axioms. The inductive limit of these von Neumann algebras is a C^*-algebra which contains a lot of information on the quantum field theory in question. This point of view was particularly successful in the analysis of superselection sectors.

In 1980 the subject of operator algebras was entirely covered in a single big three weeks meeting in Kingston Ontario. This meeting served as a review of the classification theorems for von Neumann algebras and the success of K-theory as a tool in C^*-algebras. But the meeting also contained a preview of what was to be an explosive growth in the field. The study of the von Neumann algebra of a foliation was being developed in the far more precise C^*-framework which would lead to index theorems for foliations incorporating techniques and ideas from many branches of mathematics hitherto unconnected with operator algebras.

Many of the new developments began in the decade following the Kingston meeting. On the C^*-side was Kasparov's KK-theory – the bivariant form of K-theory for which operator algebraic methods are absolutely essential. Cyclic cohomology was discovered through an analysis of the fine structure of extensions of C^*-algebras These ideas and many others were integrated into Connes' vast *Noncommutative Geometry* program. In cyclic theory and in connection with many other aspects of noncommutative geometry, the need for going beyond the class of C^*-algebras became apparent. Thanks to recent progress, both on the cyclic homology side as well as on the K-theory side, there is now a well developed bivariant K-theory and cyclic theory for a natural class of topological algebras as well as a bivariant character taking K-theory to cyclic theory. The 1990's also saw huge progress in the classification theory of nuclear C^*-algebras in terms of K-theoretic invariants, based on new insight into the structure of exact C^*-algebras.

On the von Neumann algebra side, the study of subfactors began in 1982 with the definition of the *index* of a subfactor in terms of the Murray-von Neumann theory and a result showing that the index was surprisingly restricted in its possible values. A rich theory was developed refining and clarifying the index. Surprising connections with knot theory, statistical mechanics and quantum field theory have been found. The superselection theory mentioned above turned out to have fascinating links to subfactor theory. The subfactors themselves were constructed in the representation theory of loop groups.

Beginning in the early 1980's Voiculescu initiated the theory of free probability and showed how to understand the free group von Neumann algebras in terms of random matrices, leading to the extraordinary result that the von Neumann algebra M of the free group on infinitely many generators has full fundamental group, i.e. pMp is isomorphic to M for every non-zero projection $p \in M$. The subsequent introduction of free entropy led to the solution of more old problems in von Neumann algebras such as the lack of a Cartan subalgebra in the free group von Neumann algebras.

Many of the topics mentioned in the (obviously incomplete) list above have become large industries in their own right. So it is clear that a conference like the one in Kingston is no longer possible. Nevertheless the subject does retain a certain unity and sense of identity so we felt it appropriate and useful to create a series of encylopaedia volumes documenting the fundamentals of the theory and defining the current state of the subject.

In particular, our series will include volumes treating the essential technical results of C^*-algebra theory and von Neumann algebra theory including sections on noncommutative dynamical systems, entropy and derivations. It will include an account of K-theory and bivariant K-theory with applications and in particular the index theorem for foliations. Another volume will be devoted to cyclic homology and bivariant K-theory for topological algebras with applications to index theorems. On the von Neumann algebra side, we plan volumes on the structure of subfactors and on free probability and free entropy. Another volume shall be dedicated to the connections between operator algebras and quantum field theory.

October 2001

subseries editors:
Joachim Cuntz
Vaughan Jones

In loving memory of our daughter

Yuki

whose childhood was greatly influenced by events and developments in Operator Algebras.

Contents
of Theory of Operator Algebras I, II and III

Theory of Operator Algebras I

Theory of Operator Algebras II

Theory of Operator Algebras III

Contents
Theory of Operator Algebras II

Contents
Theory of Operator Algebras I

Contents
Theory of Operator Algebras III

Preface

The author believes that the theory of operator algebras should be viewed as a number theory in analysis. Number theory has been attracting the interest of humans ever since civilization began. Every culture in the world throughout history has given special meanings to certain numbers.

For example, a number may represent a position, quantity and/or quality. Today's civilization would be just impossible without numbers. People have been attracted to the mysteries of numbers throughout history. Accordingly, number theory is the oldest and most developed area of mathematics. Throughout the mathematical path to the present day, people have gradually learned properties of numbers. It is surprising to find that the number zero was not recognized until Hindus found it about one thousand years ago (although it is recognized that Mayans found it as well). Compared to this old field of mathematics, the theory of operator algebras is very new; its foundation was given by the pioneering work of J. von Neumann and his collaborator F. J. Murray in the early part of the twentieth century, i.e. in the thirties. Subsequent major development occurred only a decade later in the late forties and the early fifties. But since then it has marked steady progress reaching new heights today. The theory handles self-adjoint algebras of bounded operators on a Hilbert space. The advent of quantum physics at the turn of century forced one to consider non-commutative variables. One needed to broaden the concept of numbers. Integers, rational numbers, real numbers and complex numbers are all commutative. Among the few noncommutative mathematical systems available at the beginning of quantum mechanics were matrix algebras, which did not accommodate the needs of quantum physics because the Heisenberg uncertainty principle and/or Heisenberg commutation relation do not allow one to stay in the realm of finite matrices. One needs to consider algebras of operators on a Hilbert space of infinite dimension. Some of these operators correspond to important physical quantities. One has to include operators in the list of "numbers". Number theory tells us to put numbers in a field to study them more efficiently. Similarly, the theory of operator algebras puts operators of interest in an algebra and we study the algebra and its structure first. The infinite dimensionality of the underlying Hilbert space poses big challenges and also presents interesting new phenomenon which do not occur in the classical frame work. We have already seen some of them in the first volume. For example, the continuity of dimensions in a factor of type II_1 is one of them. The infinite dimensionality of our objects forces us to create sophisticated methods to handle approximations. Simple minded counting does not lead to the heart of the matter. For example, it is impossible to introduce a simple minded coordinate system in an infinite dimensional operator algebra, thus mathematical induction based on a basis does not fly. The early part of the theory, in the period of the forties through the early sixties were spent on this issue. Luckily there is a remarkable similarity between the theory of measures on a locally compact space and the theory of operator algebras. The first volume was devoted to the pursuit of this similarity.

The second volume of "Theory of Operator Algebras" is devoted to the study of the structure of von Neumann algebras of type III and their automorphism groups, cf. Chapter VI through Chapter XII; and the third volume is devoted to the study of the fine structure analysis of approximately finite dimensional factors and their automorphism groups, cf. Chapter XIII through Chapter XVIII. The last chapter, Chapter XIX, is an introduction to the theory of subfactors and their symmetries. One should note that the class of von Neumann algebras of type III is given by exclusion, i.e., by the absence of a non-trivial trace or a non-zero finite projection. This situation presented the major obstruction for the study of von Neumann algebras from the beginning of the subject until the advent of Tomita-Takesaki theory in the late sixties whilst many examples had been found to be of type III: the infinity of non-isomorphic factors were first established for factors of type III by Powers in 1967, [670], before the discovery of infinitely many non-isomorphic factors of type II_1 or II_∞, [635, 686], and most examples from quantum physics were shown to be of type III, [430]. It was the Tomita-Takesaki theory which broke the ice. It is still amazing that the subject defined by exclusion admits such a fine structural analysis since usually exclusion does not allow one to find any alternative and is viewed as pathological. Of course, a von Neumann algebra of type III had been pathological until we discover their fine structure. We will explore this in full detail through the second volume.

Each chapter has its own introduction which describes the content of that chapter and the basic strategy so that the reader can get a quick overview of the chapter.

In the second and third volume, we present two major items in the theory of von Neumann algebras: one is the analogy with integration theory on an abstract measure space and the other is the emphatic importance of automorphisms of algebras, i.e. we emphasize the symmetries of our objects following the modern point of view of E. Galois.

In general, the theory of von Neumann algebras is considered to be non-commutative integration. In Volume I, the similarity between von Neumann algebras and measure spaces are examined from the point of view of Banach space duality. In the second and third volume, non-commutative integration goes far beyond the analogy with ordinary integration. Since it is not our main interest to examine how ordinary integration should be formulated based on commutative von Neumann algebras, it is not discussed here in detail beyond a few comments. Still it is possible to develop a theory which covers the ordinary integration theory based on the operator algebra approach. In fact, such a theory has been explored by G. K. Pedersen, [653, Chapter 6], and it does eliminate pathological uninteresting measure spaces easily. The main difference between the operator algebra approach and the conventional approach to integration theory relies on the fact that in operator algebras one considers functions first, or equivalently variables, and then one views the underlying points as the spectrum of the variables; whilst in the ordinary approach one considers points first and views variables as functions on the set of points. We would like to point out here, however, that in practice we never observe points directly only approximately by successive evaluations of coordinates. Besides this philosophical difference, there is another major difference between the ordinary integration theory and the non-commutative integration theory which rests on the fact that a weight, a non-commutative counterpart of a σ-finite measure, gives rise to a one-parameter automorphism group, called the modular automorphism group, of the von Neumann algebra in question. This modular automorphism group can be considered as the time evolution of the system, i.e., in the non-commutative world a state determines the associated dynamics. The appearance of the modular automorphism group distinguishes our theory sharply from the classical theory. The modular automorphism group gives us abundant non-trivial information precisely when there is no trace on the algebra in question. Since the ordinary integration is a trace, the modular automorphism group is trivial in that case and cannot be appreciated. Furthermore, thanks to the Connes cocycle derivative theorem, Theorem VIII.3.3, the modular automorphism group is unique up to perturbation by a one unitary cocycle, which allows us to relate the structure of a von Neumann algebra of type III to that of the associated von Neumann algebra of type II_∞ equipped with a trace scaling one parameter automorphism group, cf. Chapter XII. As a byproduct of our non-commutative integration theory, a duality theorem attributed to Pontrjagin, van Kampen, Tannaka, Stinespring, Eymard, Saito and Tatsuuma, is presented in §3, Chapter VII. With this exception, no discussion of examples is presented in the second volume, Chapter VI through Chapter XII. Extensive discussions of examples and constructions of factors occupy the third volume starting in Chapter XIII and through Chapter XVIII.

The so-called Murray-von Neumann measure space construction of factors is closely investigated first in Chapter XIII yielding the Krieger construction of factors and the theory of measured groupoids. Systematic study of approximately finite dimensional factors occupies most of the third volume, cf. Chapter XIV through Chapter XIX. The theory is highlighted by the celebrated classification theorem of Alain Connes in the form of Theorems XVI.1.9, XVIII.1.1, XVIII.2.1 to which W. Krieger made a substantial contribution also, and XVIII.4.16 which requires one full section of preparation given by U. Haagerup, [550]. The last chapter, Chapter XIX, is devoted to an introduction to the theory of subfactors of an AFD factor created by V. F. R. Jones, and concludes with a classification theorem of Popa, Theorem XIX.4.16, for subfactors of an AFD factor of type II_1 with small indices.

The three volume book, "Theory of Operator Algebras", is a product of the author's research and teaching activities at the Department of Mathematics at University of California, Los Angeles, spanning the years from 1969 through the present time. It is important to mention the following: the author's visit to the University of Pennsylvania from 1968 through 1969 where the foundation of Tomita-Takesaki theory was established; the author's participation in various research activities which include several short and long visits to the University of Marseille-Aix-Luminy;

several short visits to RIMS of Kyoto University; one year participation in the Mathematical Physics Project of 1975–1976 at ZiF, University of Bielefeld; a full year participation in the operator algebra project of MSRI for 1984–1985; a one year visit to IHES, 1988–1989; two one month long participations in the one year project (1988–1989) on operator algebras at the Mittag-Leffler Institute; several visits to the University of New South Wales; and several month long visits to the Mathematics Institute of University of Warwick. The author would like to express here his sincere gratitude to these institutions and to the mathematicians who hosted him warmly and worked with him. Special thanks are due to Professor Richard V. Kadison with whom the author discussed the philosophy of the subject at length so many times, and to Professor Daniel Kastler who encouraged him in many ways and provided the opportunity to work with him and others including Alain Connes. Throughout the period of the preparation of the book, the author has been continuously supported by the National Science Foundation. Here he would like to record his appreciation of that support. The Guggenheim Foundation also gave the author support at a critical period of his career, for which the author is very grateful. The author also would like to express his gratitude to Professor Masahiro Nakamura who has constantly given his moral support to the author, to Professor Takashi Turumaru whose beautiful lectures inspired the author to be a functional analyst and to the late Professor Yoshinao Misonou under whose leadership the author started his career as a functional analyst. At the final stage of the preparation of the manuscript, Dr. Un Kit Hui and Dr. Toshihiko Masuda took pains to help the author to edit the manuscript. Although any misprints and mistakes are the author's responsibility, the author would like to thank them here.

Guidance to the Reader

Each chapter has its own introduction so that one can quickly get an overview of the content of the chapter. Theorems, Propositions, Lemmas and Definitions are numbered in one sequence, whilst formulas and equations are numbered in each section separately without reference to the section. Formulas (respectively, equations) are referred to by the formula number (respectively, equation number) alone if it is quoted in the same section, and by the section number followed by the formula number if it is quoted in a different section but in the same chapter, and finally by the chapter number, the section number and the formula number (respectively, equation number) if it is quoted in a different chapter. Some exercises are selected to help the reader to get information and techniques not covered in the main text, so they can be viewed as a supplement to the text. Those exercises taken directly literatures are marked by a †-sign, and the references are cited there.

To keep the book within a reasonable size, this three volume book does not include the materials related to the following important areas of operator algebras: K-theory for C^*-algebras, geometric theory of operator algebras such as cyclic cohomology, the classification theory of nuclear C^*-algebras, free probability theory and the advanced theory of subfactors. The interested readers are referred to the forthcoming books in this operator algebra series of encyclopedia.

Chapter VI

Left Hilbert Algebras

§ 0 Introduction

This chapter is devoted to the foundation of non-commutative integration theory. In the first volume of this book, we have seen the strong similarity between the theory of operator algebras and the integration theory. To explore this similarity further, it is necessary to work on the theory of left Hilbert algebras. It is the non-commutative counter part of the algebra of all bounded square integrable functions on a measure space.

Although it is customary to begin the real analysis from a space X equipped with a σ-algebra of subsets and a countably additive positive real valued set function, i.e. a measure, on the σ-algebra, one should note that to rule out several non-sense, one has to impose several conditions on the measure space which reduces essentially the theory down to the Lebesgue integration on the real line. If one pays more attention on the algebra $L^2(X, \mu) \cap L^\infty(X, \mu)$, then one can develop a much more streamlined theory of integration. Namely, the usual measure theory is nothing but the theory of commutative (left) Hilbert algebra. The measure comes from the inner product of the underlying Hilbert space. Namely, the usual measure theory postulates are not strong enough to rule out several triviality such as the existence of measure spaces which has one dimensional function spaces yet fails to be a one point set. As long as one sticks to the space of points as the starting point, one can not avoid this problem.

The theory of operator algebras takes different approach. We start from an algebra and then develop the theory. Here we begin with an involutive algebra equipped with an inner product which behaves nicely. Theorem 1.19 is the key to the entire theory. Unlike the classical case, i.e. the commutative case or measure theory, it comes with a time evolution as an intrinsic dynamics of the system which makes the entire theory very interesting.

Section 1 is devoted to the fundamentals of the theory. Section 2 is devoted to the algebra of analytic elements of the dynamics which enables one to generate the left and right von Neumann algebras from one algebra. We relate then the theory to the disintegration theory in Section 3.

§ 1 Left Hilbert Algebras and Right Hilbert Algebras

In order to develop the theory of weights and states beyond their formal aspect given in the Section VII.1, we need to introduce new algebras, called left Hilbert algebras.

Definition 1.1. An involutive algebra $\mathfrak{A}$ over **C** with involution $\xi \in \mathfrak{A} \mapsto \xi^\sharp \in \mathfrak{A}$ (resp. $\xi \mapsto \xi^\flat$) is called a *left* (resp. *right*) *Hilbert algebra* if $\mathfrak{A}$ admits an inner product satisfying the following postulates:

a) Each fixed $\xi \in \mathfrak{A}$ gives rise to a bounded operator $\pi_\ell(\xi)\colon \eta \in \mathfrak{A} \mapsto \xi\eta \in \mathfrak{A}$ (resp. $\pi_r(\xi)\colon \eta \in \mathfrak{A} \mapsto \eta\xi \in \mathfrak{A}$) by multiplying from the left (resp. right);
b) $(\xi\eta \mid \zeta) = (\eta \mid \xi^\sharp\zeta)$ (resp. $(\xi\eta \mid \zeta) = (\xi \mid \zeta\eta^\flat)$);
c) The involution: $\xi \in \mathfrak{A} \mapsto \xi^\sharp \in \mathfrak{A}$ (resp. $\xi \in \mathfrak{A} \mapsto \xi^\flat \in \mathfrak{A}$) is preclosed;
d) The subalgebra, denoted $\mathfrak{A}^2$, spanned linearly by all possible products $\xi\eta$, $\xi, \eta \in \mathfrak{A}$, is dense in $\mathfrak{A}$ with respect to the inner product.

If the involution of a left Hilbert algebra $\mathfrak{A}$ is an isometry, then it is also a right Hilbert algebra. In this case, we say that $\mathfrak{A}$ is a (*unimodular*) Hilbert algebra and the involution of $\mathfrak{A}$ is denoted by $\xi \in \mathfrak{A} \mapsto \xi^* \in \mathfrak{A}$.

Example 1.2. Let G be a locally compact group with left Haar measure ds and modular function $\delta_G(s)$. The vector space $\mathcal{K}(G)$ of all continuous functions on G with compact support is a left Hilbert algebra with respect to the following product, involution and inner product:

$$\left.\begin{aligned} (\xi\eta)(s) &= \int_G \xi(t)\eta(t^{-1}s)\,\mathrm{d}t; \\ \xi^\sharp(s) &= \delta_G(s^{-1})\overline{\xi(s^{-1})}, \qquad s \in G; \\ (\xi \mid \eta) &= \int_G \xi(s)\overline{\eta(s)}\,\mathrm{d}s\,. \end{aligned}\right\} \tag{1}$$

If we define the involution by

$$\xi^\flat(s) = \overline{\xi(s^{-1})} \tag{2}$$

then $\mathcal{K}(G)$ is a right Hilbert algebra.

Example 1.3. Suppose that a von Neumann algebra $\{\mathcal{M}, \mathfrak{H}\}$ admits a cyclic and separating vector ξ_0, i.e. $[\mathcal{M}\xi_0] = [\mathcal{M}'\xi_0] = \mathfrak{H}$. Set $\mathfrak{A} = \mathcal{M}\xi_0$ and

$$\left.\begin{aligned} (x\xi_0)(y\xi_0) &= xy\xi_0, \quad x, y \in \mathcal{M}, \\ (x\xi_0)^\sharp &= x^*\xi_0, \end{aligned}\right\} \tag{3}$$

and consider the inner product in $\mathfrak{A}$ inherited from $\mathfrak{H}$. It follows that $\mathfrak{A}$ is a left Hilbert algebra. The vector ξ_0 is the identity of $\mathfrak{A}$. So in this case, $\mathfrak{A}$ is a unital left Hilbert algebra.

It will be shown that if φ is a faithful (semi-finite normal) weight[1] on $\mathcal{M}$, then $\mathfrak{n}_\varphi \cap \mathfrak{n}_\varphi^* = \mathfrak{A}_\varphi$ is a left Hilbert algebra.

We now concentrate ourselves to the study of a fixed left Hilbert algebra $\mathfrak{A}$. Let $\mathfrak{H}$ denote the completion of $\mathfrak{A}$. To each $\xi \in \mathfrak{A}$, there corresponds a unique bounded operator $\pi_\ell(\xi)$ by (a) and the map $\pi_\ell \colon \xi \in \mathfrak{A} \mapsto \pi_\ell(\xi) \in \mathcal{L}(\mathfrak{H})$ is a *-representation of $\mathfrak{A}$. The postulate (d) guarantees that the *-algebra $\pi_\ell(\mathfrak{A})$ is non-degenerate. The von Neumann algebra generated by $\pi_\ell(\mathfrak{A})$ is called the *left von Neumann algebra* and we denote it by $\mathcal{R}_\ell(\mathfrak{A})$, i.e.

$$\mathcal{R}_\ell(\mathfrak{A}) = \pi_\ell(\mathfrak{A})''. \tag{4}$$

If $\mathfrak{A}$ is a right Hilbert algebra instead, then to each $\eta \in \mathfrak{A}$ there corresponds a bounded operator $\pi_r(\eta) \in \mathcal{L}(\mathfrak{H})$ such that

$$\pi_r(\eta)\xi = \xi\eta, \quad \xi \in \mathfrak{A}.$$

It then follows that π_r is an *anti* *-representation of $\mathfrak{A}$ in the sense that π_r reverses the order of product, i.e.

$$\pi_r(\eta_1\eta_2) = \pi_r(\eta_2)\pi_r(\eta_1),$$

and that $\pi_r(\mathfrak{A})$ is a non degenerate *-algebra of operators on $\mathfrak{H}$. The von Neumann algebra generated by $\pi_r(\mathfrak{A})$ is called the *right von Neumann algebra* of $\mathfrak{A}$ and we write it $\mathcal{R}_r(\mathfrak{A})$, i.e.

$$\mathcal{R}_r(\mathfrak{A}) = \pi_r(\mathfrak{A})''. \tag{4'}$$

We now return to the left Hilbert algebra $\mathfrak{A}$. We first consider the involution: $\xi \in \mathfrak{A} \mapsto \xi^\sharp \in \mathfrak{A}$ and denote it by S_0. By postulate (c), S_0 is preclosed. We shall denote the closure of S_0 and its domain by S and $\mathfrak{D}^\sharp$ respectively. We often write $\xi^\sharp$ for $S\xi \in \mathfrak{D}^\sharp$. We equip $\mathfrak{D}^\sharp$ with a new inner product and a norm:

$$\left.\begin{aligned} (\xi \mid \eta)_\sharp &= (\xi \mid \eta) + (S\eta \mid S\xi), \quad \xi, \eta \in \mathfrak{D}^\sharp; \\ \|\xi\|_\sharp &= \sqrt{\|\xi\|^2 + \|S\xi\|^2}. \end{aligned}\right\} \tag{5}$$

We note here that the order of η and ξ in the second term of the right hand side is reversed due to the antilinearity of S.

Lemma 1.4.

(i) *For a vector $\xi \in \mathfrak{H}$ to belong to $\mathfrak{D}^\sharp$ it is necessary and sufficient that there exists a sequence $\{\xi_n\}$ in $\mathfrak{A}$ such that $\lim_{n\to\infty}\|\xi_n - \xi\| = 0$ and $\{\xi_n^\sharp\}$ is a Cauchy sequence in $\mathfrak{H}$. If this is the case, then*

$$\xi^\sharp = \lim_{n\to\infty} \xi_n^\sharp.$$

(ii) *$\mathfrak{D}^\sharp$ is complete with respect to the new norm* (5), *and $\mathfrak{A}$ is dense in $\mathfrak{D}^\sharp$.*

We leave the proof to the reader.

[1] See Definition VII.1.1.

Lemma 1.5.

(i) $S = S^{-1}$.

(ii) *There exists an antilinear densely defined closed operator* F *with domain* $\mathfrak{D}^\flat$ *such that*

a) $\mathfrak{D}^\flat = \{\eta \in \mathfrak{H} : \xi \in \mathfrak{D}^\sharp \mapsto (\eta \mid S\xi)$ *is bounded*$\}$;

b) $(S\xi \mid \eta) = (F\eta \mid \xi), \quad \xi \in \mathfrak{D}^\sharp, \ \eta \in \mathfrak{D}^\flat$. (6)

(iii) $F = F^{-1}$.

(iv) $\Delta = FS$ *is a linear positive non-singular self-adjoint operator such that* $\mathfrak{D}(\Delta^{1/2}) = \mathfrak{D}^\sharp$.

(v) *There exists an antilinear isometry* J *of* $\mathfrak{H}$ *onto itself;*

a) $(J\xi \mid J\eta) = (\eta \mid \xi), \quad \xi, \eta \in \mathfrak{H}$, (7)

b) $J = J^{-1}$, *equivalently* $J^2 = 1$, (8)

c) $J\Delta J = \Delta^{-1}$, (9)

d) $S = J\Delta^{1/2} = \Delta^{-1/2}J$;

e) $F = J\Delta^{-1/2} = \Delta^{1/2}J$. (10)

(vi) J *and* Δ *are uniquely determined by the property* (v-d) *and* $\mathfrak{D}(\Delta^{1/2}) = \mathfrak{D}^\sharp$.

PROOF: Let $\mathfrak{H}^*$ be the Banach space dual of $\mathfrak{H}$. The Riesz representation theorem means the existence of an antilinear isometry C of $\mathfrak{H}$ onto $\mathfrak{H}^*$ such that

$$(\xi \mid \eta) = \langle \xi, C\eta \rangle, \quad \xi, \eta \in \mathfrak{H},$$

where $\langle \ , \ \rangle$ denotes the bilinear form on $\mathfrak{H} \times \mathfrak{H}^*$. If we set

$$(\xi^* \mid \eta^*) = (C^{-1}\eta^* \mid C^{-1}\xi^*), \quad \xi^*, \eta^* \in \mathfrak{H}^*,$$

$\mathfrak{H}^*$ becomes a Hilbert space. We put $S_1 = CS_0$. Then S_1 is a preclosed linear operator with domain $\mathfrak{D}(S_0) = \mathfrak{A}$ and range $C\mathfrak{D}(S_0) = C\mathfrak{A}$. Let $\overline{S}_1$ denote the closure of S_1. It is clear that $\mathfrak{D}(\overline{S}_1) = \mathfrak{D}^\sharp$ and $\overline{S}_1 = CS$. Since $S_0 = S_0^{-1}$, S maps $\mathfrak{D}^\sharp$ onto $\mathfrak{D}^\sharp$ bijectively, so that S^{-1} exists and $S = S^{-1}$. Hence $\overline{S}_1$ is a non-singular closed operator from $\mathfrak{D}^\sharp$ onto $C\mathfrak{D}^\sharp$. We set here $\Delta = \overline{S}_1^* \overline{S}_1$ and let $\overline{S}_1 = U\Delta^{1/2}$ be the polar decomposition of $\overline{S}_1$. The non-singularity of $\overline{S}_1$ means that Δ is non-singular. We then put $J = C^{-1}U$. It follows that

$$S = C^{-1}\overline{S}_1 = C^{-1}U\Delta^{\frac{1}{2}} = J\Delta^{\frac{1}{2}}.$$

Now $\overline{S}_1^{-1} = SC^{-1}$, so the right polar decomposition of $\overline{S}_1^{-1}$:

$$\overline{S}_1^{-1} = \Delta^{-\frac{1}{2}}U^* = U^*(U\Delta^{-\frac{1}{2}}U^*)$$

means the following:

$$S = \Delta^{-\frac{1}{2}}U^*C = \Delta^{-\frac{1}{2}}(C^{-1}U)^{-1} = \Delta^{-\frac{1}{2}}J^{-1} = J^{-1}(J\Delta^{-\frac{1}{2}}J^{-1}).$$

Hence we get

$$J\Delta^{-\frac{1}{2}}J^{-1} = JS = J^2\Delta^{\frac{1}{2}}.$$

Clearly, $J\Delta^{-1/2}J^{-1}$ and $\Delta^{1/2}$ are both positive self-adjoint operators, and J^2 is a unitary. The uniqueness of the polar decomposition yields that $J^2 = 1$ and $J\Delta^{-1/2}J^{-1} = \Delta^{1/2}$, so that $J = J^{-1}$ and we obtain (v-b).

We now define F to be $\overline{S}_1^*C$ with domain $\mathfrak{D}^\flat = C^{-1}\mathfrak{D}(\overline{S}_1^*)$. It follows that $\eta \in \mathfrak{H}$ belongs to $\mathfrak{D}^\flat$ if and only if $C\eta \in \mathfrak{D}(\overline{S}_1^*)$, if and only if $\xi \in \mathfrak{D}(S_1) \mapsto (S_1\xi \mid C\eta)$ is bounded, but we have $(S_1\xi \mid C\eta) = (\eta \mid S\xi)$, and that if $\xi \in \mathfrak{D}$ and $\eta \in \mathfrak{D}^\flat$, then

$$(\eta \mid S\xi) = (S_1\xi \mid C\eta) = (\xi \mid S_1^*C\eta) = (\xi \mid F\eta).$$

Therefore (ii) follows. The rest of our claim follows from the computation:

$$\begin{aligned} F &= \overline{S}_1^*C = (U\Delta^{\frac{1}{2}})^*C = \Delta^{\frac{1}{2}}U^*C = \Delta^{\frac{1}{2}}(C^{-1}U)^{-1} \\ &= \Delta^{\frac{1}{2}}J^{-1} = \Delta^{\frac{1}{2}}J = J\Delta^{-\frac{1}{2}}. \end{aligned}$$

Q.E.D.

Definition 1.6. The operators Δ and J are called the *modular operator* and the *modular conjugation* of the left Hilbert algebra $\mathfrak{A}$ respectively.

The principal objective of the section is to show that

$$\left.\begin{aligned} \Delta^{it}\mathcal{R}_\ell(\mathfrak{A})\Delta^{-it} &= \mathcal{R}_\ell(\mathfrak{A}), \quad t \in \mathbf{R}, \\ J\mathcal{R}_\ell(\mathfrak{A})J &= \mathcal{R}_\ell(\mathfrak{A})'. \end{aligned}\right\} \tag{11}$$

To this end, we need several preparations.

Definition 1.7. A vector $\eta \in \mathfrak{H}$ is said to be *right bounded* if

$$\sup\left\{ \|\pi_\ell(\xi)\eta\| : \xi \in \mathfrak{A},\ \|\xi\| \leq 1 \right\} = c < +\infty. \tag{12}$$

We denote the set of all right bounded vectors by $\mathfrak{B}'$.

A vector $\eta \in \mathfrak{H}$ is right bounded if and only if there exists an $a \in \mathcal{L}(\mathfrak{H})$ such that

$$a\xi = \pi_\ell(\xi)\eta, \quad \xi \in \mathfrak{A}. \tag{13}$$

The operator a is uniquely determined by η, so write $\pi_r(\eta)$ for this a. It further follows that $\mathfrak{B}'$ is a linear subspace, not necessarily closed, of $\mathfrak{H}$ and

$$\pi_r(\lambda\eta + \mu\zeta) = \lambda\pi_r(\eta) + \mu\pi_r(\zeta), \quad \eta, \zeta \in \mathfrak{B}'. \tag{14}$$

Lemma 1.8.

(i) $\mathfrak{B}'$ *is invariant under* $\mathcal{R}_\ell(\mathfrak{A})'$.

(ii) $\mathfrak{n}_r = \pi_r(\mathfrak{B}')$ *is a left ideal of* $\mathcal{R}_\ell(\mathfrak{A})'$ *and*

$$\pi_r(a\eta) = a\pi_r(\eta), \quad a \in \mathcal{R}_\ell(\mathfrak{A})', \quad \eta \in \mathfrak{B}'. \tag{15}$$

PROOF: If $a \in \mathcal{R}_\ell(\mathfrak{A})'$, $\xi \in \mathfrak{A}$ and $\eta \in \mathfrak{B}'$, we have

$$\pi_\ell(\xi)(a\eta) = a\pi_\ell(\xi)\eta = a\pi_r(\eta)\xi,$$

so that $a\eta \in \mathfrak{B}'$ and $\pi_r(a\eta) = a\pi_r(\eta)$. If $\zeta \in \mathfrak{A}$, then

$$\pi_r(\eta)\pi_\ell(\xi)\zeta = \pi_r(\eta)(\xi\zeta) = \pi_\ell(\xi\zeta)\eta = \pi_\ell(\xi)\pi_\ell(\zeta)\eta = \pi_\ell(\xi)\pi_r(\eta)\zeta,$$

so that $\pi_r(\eta)$ and $\pi_\ell(\xi)$ commute, which means that $\pi_r(\eta) \in \mathcal{R}_\ell(\mathfrak{A})'$. Q.E.D.

We now want to extend multiplication from $\mathfrak{A}$ to a wider class of pairs of vectors. First, we set

$$\xi\eta = \pi_\ell(\xi)\eta, \quad \xi \in \mathfrak{A} \quad \text{and} \quad \eta \in \mathfrak{H}, \tag{16}$$

$$\xi\eta = \pi_r(\eta)\xi, \quad \xi \in \mathfrak{H} \quad \text{and} \quad \eta \in \mathfrak{B}'. \tag{17}$$

The commutativity of $\pi_\ell(\mathfrak{A})$ and $\pi_r(\mathfrak{B}')$ implies that (16) and (17) are mutually consistent. We shall often write $\eta^\flat$ for $F\eta$. For the later convenience, we record this here:

$$\xi^\sharp = S\xi, \qquad \xi \in \mathfrak{D}^\sharp, \qquad \eta^\flat = F\eta, \qquad \eta \in \mathfrak{D}^\flat. \tag{18}$$

We then define

$$\mathfrak{A}' = \mathfrak{B}' \cap \mathfrak{D}^\flat. \tag{19}$$

Lemma 1.9.

(i) $\pi_r(\mathfrak{B}')^*\mathfrak{B}' \subset \mathfrak{A}'$.

(ii) $(\pi_r(\eta_1)^*\eta_2)^\flat = \pi_r(\eta_2)^*\eta_1, \quad \eta_1, \eta_2 \in \mathfrak{B}',$ (20)

(iii) $\mathfrak{A}'$ *satisfies postulates* (a), (b) *and* (c) *for a right Hilbert algebra.*

PROOF: Let $\eta_1, \eta_2 \in \mathfrak{B}'$ and set $\eta = \pi_r(\eta_1)^*\eta_2$. By the last lemma, η is right bounded, i.e. $\eta \in \mathfrak{B}'$. If $\xi \in \mathfrak{A}$, then

$$\begin{aligned}(\xi^\sharp \mid \eta) &= \left(\xi^\sharp \mid \pi_r(\eta_1)^*\eta_2\right) = \left(\pi_r(\eta_1)\xi^\sharp \mid \eta_2\right) = \left(\pi_\ell(\xi^\sharp)\eta_1 \mid \eta_2\right)\\ &= \left(\pi_\ell(\xi)^*\eta_1 \mid \eta_2\right) = \left(\eta_1 \mid \pi_\ell(\xi)\eta_2\right)\\ &= \left(\eta_1 \mid \pi_r(\eta_2)\xi\right) = \left(\pi_r(\eta_2)^*\eta_1 \mid \xi\right),\end{aligned}$$

so that $\eta \in \mathfrak{D}^\flat$ and $\eta^\flat = \pi_r(\eta_2)^*\eta_1$. The rest of the lemma is clear. Q.E.D.

Lemma 1.10. *Let $\eta \in \mathfrak{D}^\flat$ and define operators a_0 and b_0 with domain $\mathfrak{A}$ by the following:*

$$a_0\xi = \pi_\ell(\xi)\eta, \quad b_0\xi = \pi_\ell(\xi)\eta^\flat, \quad \xi \in \mathfrak{A}. \tag{21}$$

Then we conclude:

(i) *a_0 and b_0 are both preclosed, $a_0 \subset b_0^*$ and $b_0 \subset a_0^*$;*

(ii) *if $\pi_r(\eta) = a_0^{**}$ and $\pi_r(\eta^\flat) = b_0^{**}$, then $\pi_r(\eta)$ and $\pi_r(\eta^\flat)$ are affiliated with $\mathcal{R}_\ell(\mathfrak{A})'$ in the sense that every unitary in $\mathcal{R}_\ell(\mathfrak{A})$ commutes with $\pi_r(\eta)$ and $\pi_r(\eta^\flat)$.*

PROOF:

(i) If $\xi, \zeta \in \mathfrak{A}$, then

$$\begin{aligned}(a_0\xi \mid \zeta) &= (\pi_\ell(\xi)\eta \mid \zeta) = (\eta \mid \pi_\ell(\xi)^*\zeta) = (\eta \mid \xi^\sharp\zeta) \\ &= ((\xi^\sharp\zeta)^\sharp \mid \eta^\flat) = (\zeta^\sharp\xi \mid \eta^\flat) = (\xi \mid \pi_\ell(\zeta)\eta^\flat) = (\xi \mid b_0\zeta),\end{aligned}$$

thus the first claim follows.

(ii) The affiliation of $\pi_r(\eta)$ and $\pi_r(\eta^\flat)$ to $\mathcal{R}_\ell(\mathfrak{A})'$ is equivalent to that of a_0^* and b_0^*. By symmetry, it suffices to prove the claim for a_0^*. Suppose that $\zeta \in \mathfrak{D}(a_0^*)$ and $\xi_1, \xi_2 \in \mathfrak{A}$. Then

$$\begin{aligned}(a_0\xi_1 \mid \pi_\ell(\xi_2)\zeta) &= (\pi_\ell(\xi_2)^*\pi_\ell(\xi_1)\eta \mid \zeta) = (\pi_\ell(\xi_2^\sharp\xi_1)\eta \mid \zeta) \\ &= (a_0(\xi_2^\sharp\xi_1) \mid \zeta) = (\xi_2^\sharp\xi_1 \mid a_0^*\zeta) = (\xi_1 \mid \pi_\ell(\xi_2)a_0^*\zeta),\end{aligned}$$

so that $\pi_\ell(\xi_2)\zeta \in \mathfrak{D}(a_0^*)$ and $a_0^*\pi_\ell(\xi_2)\zeta = \pi_\ell(\xi_2)a_0^*\zeta$. Therefore, a_0^* affiliates to $\mathcal{R}_\ell(\mathfrak{A})'$ by the following lemma with $A = \pi_\ell(\mathfrak{A})$. Q.E.D.

Lemma 1.11. *Let $\{\mathcal{M}, \mathfrak{H}\}$ be a von Neumann algebra. If a is a closed operator on $\mathfrak{H}$ with domain $\mathfrak{D}(a)$, then a necessary and sufficient condition for a to affiliate to $\mathcal{M}$ is the existence of a σ-weakly dense $*$-subalgebra A of $\mathcal{M}'$ such that*

$$x\mathfrak{D}(a) \subset \mathfrak{D}(a) \quad \text{and} \quad xa\xi = ax\xi, \quad x \in A, \quad \xi \in \mathfrak{D}(a).$$

PROOF: The necessity of the condition is a triviality. Suppose that the condition holds. If $x \in \mathcal{M}'$, then there exists a net $\{x_i\}$ in A converging to x strongly. For each $\xi \in \mathfrak{D}(a)$, we have

$$xa\xi = \lim_i x_i a\xi = \lim_i ax_i\xi,$$

so that $x\xi$ belongs to $\mathfrak{D}(a)$ and $xa\xi = ax\xi$. The means precisely the affiliation of a to $\mathcal{M}'$. Q.E.D.

We return to the study of the original $\mathfrak{A}$ and $\mathfrak{A}'$.

Lemma 1.12. *Let $\mathcal{K}(0,\infty)$ denote the algebra of continuous functions on the open half line $]0,\infty[$ with compact support. For a fixed $\eta \in \mathfrak{D}^\flat$, let*

$$\pi_r(\eta) = uh = ku \tag{22}$$

be the left and right polar decomposition. If $f \in \mathcal{K}(0,\infty)$, then $f(h)\eta^\flat$ and $f(k)\eta$ are both right bounded and

$$\begin{aligned}\pi_r(f(h)\eta^\flat) &= hf(h)u^* \in \mathcal{R}_\ell(\mathfrak{A})', \\ \pi_r(f(k)\eta) &= kf(k)u \in \mathcal{R}_\ell(\mathfrak{A})'.\end{aligned} \tag{23}$$

PROOF: If $\xi \in \mathfrak{A}$, then we have

$$\pi_\ell(\xi)f(h)\eta^\flat = f(h)\pi_\ell(\xi)\eta^\flat = f(h)\pi_r(\eta)^*\xi = f(h)hu^*\xi = hf(h)u^*\xi;$$
$$\pi_\ell(\xi)f(k)\eta = f(k)\pi_\ell(\xi)\eta = f(k)ku\xi = kf(k)u\xi.$$

Since $hf(h)u^*$ and $kf(k)u$ are both bounded and belong to $\mathcal{R}_\ell(\mathfrak{A})'$, $f(h)\eta^\flat$ and $f(k)\eta$ are right bounded and (23) holds. Q.E.D.

Lemma 1.13. *$\mathfrak{A}'$ and $(\mathfrak{A}')^2$ are both dense in $\mathfrak{D}^\flat$ with respect to the $\|\ \|_\flat$ given by*

$$\|\eta\|_\flat = \sqrt{\|\eta\|^2 + \|\eta^\flat\|^2}, \quad \eta \in \mathfrak{D}^\flat. \tag{5'}$$

In particular, they are both dense in $\mathfrak{H}$; thus $\mathfrak{A}'$ is a right Hilbert algebra.

PROOF: We set

$$\mathfrak{n}_r = \pi_r(\mathfrak{B}'). \tag{24}$$

Let $\eta \in \mathfrak{D}^\flat$ and keep the notations in the last lemma. By the last lemma, we have, for each $f \in \mathcal{K}(0, \infty)$,

$$hf(h) = u^*kf(k)u = \pi_r\big(u^*f(k)\eta\big) \in \mathfrak{n}_r,$$
$$kf(k) = uhf(h)u^* = \pi_r\big(uf(h)\eta^\flat\big) \in \mathfrak{n}_r.$$

Choose a $g \in \mathcal{K}(0, \infty)$ such that $f(\lambda) = \lambda g(\lambda)f(\lambda)$, $\lambda \geq 0$. Then we get $f(h) = g(h)hf(h) \in \mathfrak{n}_r$ and $f(k) = g(k)kf(k) \in \mathfrak{n}_r$. Choosing $f_1, f_2 \in \mathcal{K}(0, \infty)$ so that $f(\lambda) = \overline{f}_1(\lambda)f_2(\lambda)$, $\lambda > 0$, we get

$$f(h) = f_1(h)^* f_2(h) \in \mathfrak{n}_r^* \mathfrak{n}_r \quad \text{and} \quad f(k) = f_1(k)^* f_2(k) \in \mathfrak{n}_r^* \mathfrak{n}_r.$$

By Lemma 1.9, we have $\mathfrak{n}_r^*\mathfrak{n}_r \subset \pi_r(\mathfrak{A}')$, so that $f(h)$ and $f(k)$ both belong to $\pi_r(\mathfrak{A}')$. Hence $f(k)\eta$ and $f(h)\eta^\flat$ both belong to $\mathfrak{A}'$ and (23) implies

$$\big(f(k)\eta\big)^\flat = f(h)^*\eta^\flat, \quad f \in \mathcal{K}(0, \infty). \tag{25}$$

We note that $f(k)\eta$ and $f(h)\eta^\flat$ in fact belong to $(\mathfrak{A}')^2$ by the above argument.

We now choose an increasing positive sequence $\{f_n\}$ in $\mathcal{K}(0, \infty)$ such that $\lim_{n\to\infty} f_n(\lambda) = 1$, $\lambda > 0$. It follows that $\{f_n(h)\}$ and $\{f_n(k)\}$ converge strongly to the range projection p and q of h and k respectively. But the range projection of $\pi_r(\eta)^*$ coincides with p and so does the range projection of $\pi_r(\eta)$ with q. Hence if $q\eta = \eta$ and $p\eta^\flat = \eta^\flat$, then $\{f_n(k)\eta\}$ converges to η and $\{f_n(h)\eta^\flat\}$ converges to $\eta^\flat$, which means the convergence of $\{f_n(k)\eta\}$ to η in $\mathfrak{D}^\flat$. To prove $q\eta = \eta$ and $p\eta^\flat = \eta^\flat$, we choose a net $\{\xi_i\}$ in $\mathfrak{A}$ such that $\{\pi_\ell(\xi_i)\}$ converges strongly to the identity. It then follows that

$$\eta = \lim \pi_\ell(\xi_i)\eta = \lim \pi_r(\eta)\xi_i \in q\mathfrak{H};$$
$$\eta^\flat = \lim \pi_\ell(\xi_i)\eta^\flat = \lim \pi_r(\eta)^*\xi_i \in p\mathfrak{H}.$$

Q.E.D.

Lemma 1.14. $\mathcal{R}_\ell(\mathfrak{A})' = \mathcal{R}_r(\mathfrak{A}')$.

PROOF: By the last lemma, $\pi_r(\mathfrak{A}')$ is a non-degenerate *-subalgebra of $\pi_\ell(\mathfrak{A})'$. Hence $\pi_r(\mathfrak{A}')$ contains a bounded net $\{a_i\}$ converging to 1 σ-strongly*.

So if $x \in \mathcal{R}_\ell(\mathfrak{A})'$, then $x = \lim a_i^* x a_i$, but $a_i^* x a_i \in \mathfrak{n}_r^* \mathfrak{n}_r \subset \pi_r(\mathfrak{A}')$. Thus $\mathcal{R}_\ell(\mathfrak{A})' = \pi_r(\mathfrak{A}')''$. Q.E.D.

Lemma 1.15.

(i) $\mathfrak{A}^2$ *is dense in* $\{\mathfrak{D}^\sharp, \| \ \|_\sharp\}$, *i.e. if* $\eta_1, \eta_2 \in \mathfrak{H}$ *satisfy the equation:*

$$(\xi_1^\sharp \xi_2 \mid \eta_1) = (\eta_2 \mid \xi_2^\sharp \xi_1), \quad \xi_1, \xi_2 \in \mathfrak{A}, \tag{26}$$

then $\eta_1 \in \mathfrak{D}^\flat$ *and* $\eta_1^\flat = \eta_2$.

(ii) $\pi_r(\mathfrak{A}') = \mathfrak{n}_r \cap \mathfrak{n}_r^*$.

PROOF:

(i) Fix an arbitrary $\xi \in \mathfrak{A}$. Since $\pi_r(\mathfrak{A}')$ is non-degenerate, $\xi \in [\pi_r(\mathfrak{A}')\mathfrak{H}]$. Multiplying by a scalar, we assume that $\|\pi_\ell(\xi)\| \le 1$. We next set

$$p_n(t) = 1 - (1-t)^n. \tag{27}$$

For any operator a with $\|a\| \le 1$, $p_n(aa^*)$ converges to the range projection $s_\ell(a)$. We now compute

$$\begin{aligned}
\xi &= \lim_{n\to\infty} p_n\big(\pi_\ell(\xi)\pi_\ell(\xi)^*\big)\xi = \lim_{n\to\infty} p_n(\xi\xi^\sharp)\xi;\\
\xi^\sharp &= \lim_{n\to\infty} p_n\big(\pi_\ell(\xi^\sharp)\pi_\ell(\xi^\sharp)^*\big)\xi^\sharp = \lim_{n\to\infty} p_n(\xi^\sharp\xi)\xi^\sharp\\
&= \lim_{n\to\infty} \xi^\sharp p_n(\xi\xi^\sharp) = \lim_{n\to\infty}\big(p_n(\xi\xi^\sharp)\xi\big)^\sharp.
\end{aligned}$$

(ii) Suppose that $\eta_1 \in \mathfrak{B}'$ and $\eta_2 \in \mathfrak{B}'$ satisfy the equation $\pi_r(\eta_1)^* = \pi_r(\eta_2)$. For every $\xi_1, \xi_2 \in \mathfrak{A}$, we have

$$\begin{aligned}
(\xi_1^\sharp\xi_2 \mid \eta_1) &= (\xi_2 \mid \pi_\ell(\xi_1)\eta_1) = (\xi_2 \mid \pi_r(\eta_1)\xi_1) = (\pi_r(\eta_1)^*\xi_2 \mid \xi_1)\\
&= (\pi_r(\eta_2)\xi_2 \mid \xi_1) = (\pi_\ell(\xi_2)\eta_2 \mid \xi_1) = (\eta_2 \mid \xi_2^\sharp\xi_1),
\end{aligned}$$

so that $\eta_1 \in \mathfrak{D}^\flat$ and $\eta_1^\flat = \eta_2$ by (i). Hence $\eta_1 \in \mathfrak{D}^\flat \cap \mathfrak{B}' = \mathfrak{A}'$. Therefore $\mathfrak{n}_r \cap \mathfrak{n}_r^* \subset \pi_r(\mathfrak{A}')$. The converse inclusion is clear. Thus we are through. Q.E.D.

We now start from the right Hilbert algebra $\mathfrak{A}'$, and then dualize the above discussion entirely. However, the adjoint operator of the involution: $\eta \in \mathfrak{A}' \to \eta^\flat \in \mathfrak{A}'$ coincides precisely with the original one: $\xi \in \mathfrak{D}^\sharp \to \xi^\sharp \in \mathfrak{D}^\sharp$ by Lemma 1.13.

Definition 1.7′. A vector $\xi \in \mathfrak{H}$ is called *left bounded* if

$$\sup\{\, \|\pi_r(\eta)\xi\| : \eta \in \mathfrak{A}', \ \|\eta\| \le 1\,\} = c < +\infty. \tag{12′}$$

We shall denote by $\mathfrak{B}$ the set of all left bounded vectors.

Clearly, $\mathfrak{B}$ contains $\mathfrak{A}$ and to each $\xi \in \mathfrak{B}$ there corresponds a bounded operator $\pi_\ell(\xi)$ on $\mathfrak{H}$ determined by:

$$\pi_\ell(\xi)\eta = \pi_r(\eta)\xi, \quad \xi \in \mathfrak{B}, \quad \eta \in \mathfrak{A}'. \tag{13'}$$

Dualizing the argument of Lemma 1.8, we obtain the next result. We shall leave the proof to the reader.

Lemma 1.8′.
(i) $\mathfrak{B}$ *is invariant under* $\mathcal{R}_\ell(\mathfrak{A})$.
(ii) $\mathfrak{n}_\ell = \pi_\ell(\mathfrak{B})$ *is a left ideal of* $\mathcal{R}_\ell(\mathfrak{A})$*, and*

$$\pi_\ell(a\xi) = a\pi_\ell(\xi), \quad a \in \mathcal{R}_\ell(\mathfrak{A}), \quad \xi \in \mathfrak{B}. \tag{15'}$$

As before, we further extend products of vectors of $\mathfrak{H}$ as follows:

$$\xi\eta = \pi_\ell(\xi)\eta, \quad \xi \in \mathfrak{B}, \quad \eta \in \mathfrak{H}.$$

The reader should check the above definition of products is consistent with (16) and (17). We then set

$$\mathfrak{A}'' = \mathfrak{B} \cap \mathfrak{D}^\sharp. \tag{28}$$

It then follows that $\mathfrak{A}''$ is a left Hilbert algebra and $\mathfrak{A} \subset \mathfrak{A}''$, so that $\mathcal{R}_\ell(\mathfrak{A}'') = \mathcal{R}_\ell(\mathfrak{A})$. As before, we obtain the following:

Lemma 1.15′.
(ii′) $\pi_\ell(\mathfrak{A}'') = \mathfrak{n}_\ell \cap \mathfrak{n}_\ell^*$.

If we continue the dualization, then we come to the following situation:

$$\mathfrak{A} \subset \mathfrak{A}'' = \mathfrak{A}^{(\mathrm{iv})} = \cdots,$$
$$\mathfrak{A}' = \mathfrak{A}''' = \cdots.$$

Definition 1.16. A left Hilbert algebra $\mathfrak{A}$ is said to be *full* if $\mathfrak{A} = \mathfrak{A}''$. Two left Hilbert algebras $\mathfrak{A}_1$ and $\mathfrak{A}_2$ are said to be *equivalent* if $\mathfrak{A}_1''$ and $\mathfrak{A}_2''$ are isometrically $*$-isomorphic.

Since the replacement of $\mathfrak{A}$ by $\mathfrak{A}''$ does not affect on the von Neumann algebras $\mathcal{R}_\ell(\mathfrak{A})$ and $\mathcal{R}_\ell(\mathfrak{A})'$, we shall assume that *the left Hilbert algebra* $\mathfrak{A}$ under consideration *is full.*

Lemma 1.10′. *Let* $\xi \in \mathfrak{D}^\sharp$ *and define operators* a_0 *and* b_0 *with domain* $\mathfrak{A}'$ *by the following:*

$$a_0\eta = \pi_r(\eta)\xi, \quad b_0\eta = \pi_r(\eta)\xi^\sharp, \quad \eta \in \mathfrak{A}'. \tag{21'}$$

We then conclude:

(i) a_0 *and* b_0 *are both preclosed,* $a_0 \subset b_0^*$ *and* $b_0 \subset a_0^*$;

(ii) *If* $\pi_\ell(\xi) = a_0^{**}$ *and* $\pi_\ell(\xi^\sharp) = b_0^{**}$, *then* $\pi_\ell(\xi)$ *and* $\pi_\ell(\xi^\sharp)$ *are both affiliated with* $\mathcal{R}_\ell(\mathfrak{A})$.

We leave the proof to the reader.

We now come to the first result which links $\mathcal{R}_\ell(\mathfrak{A})$ and $\mathcal{R}_\ell(\mathfrak{A})'$. We maintain the previous notations and assumptions, in particular we assume that $\mathfrak{A}$ is a full left Hilbert algebra.

Lemma 1.17. *For each* $\omega \in \mathbf{C}$, *but* $\omega \notin \mathbf{R}_+$, *we set*

$$\gamma(\omega) = \frac{1}{\sqrt{2(|\omega| - \operatorname{Re}\omega)}}. \tag{29}$$

(i) $(\Delta - \omega)^{-1}\mathfrak{A}' \subset \mathfrak{A}$ *and*

$$\left\|\pi_\ell\big((\Delta - \omega)^{-1}\eta\big)\right\| \le \gamma(\omega)\left\|\pi_r(\eta)\right\|, \quad \eta \in \mathfrak{A}'. \tag{30}$$

(ii) $(\Delta^{-1} - \omega)^{-1}\mathfrak{A} \subset \mathfrak{A}'$ *and*

$$\left\|\pi_r\big((\Delta^{-1} - \omega)^{-1}\xi\big)\right\| \le \gamma(\omega)\left\|\pi_\ell(\xi)\right\|, \quad \xi \in \mathfrak{A}. \tag{30'}$$

PROOF: By symmetry, we prove only (i). We fix $\eta \in \mathfrak{A}'$ and set $\xi = (\Delta - \omega)^{-1}\eta$. Clearly, $\xi \in \mathfrak{D}(\Delta) \subset \mathfrak{D}^\sharp$. Let $\pi_\ell(\xi) = uh = ku$ be the left and right polar decomposition of $\pi_\ell(\xi)$. By the dual form of Lemma 1.12, we have $f(k)\xi \in \mathfrak{A}$ for every $f \in \mathcal{K}(0, \infty)$ and

$$\big(f(k)\xi\big)^\sharp = \overline{f}(h)\xi^\sharp, \quad f \in \mathcal{K}(0, \infty). \tag{25'}$$

Furthermore, we have

$$\begin{aligned}
2\big(|\omega| - \operatorname{Re}\omega\big)\left\|hf(h)\xi^\sharp\right\|^2 &= 2\big(|\omega| - \operatorname{Re}\omega\big)\big(h\overline{f}(h)hf(h)\xi^\sharp \mid \xi^\sharp\big) \\
&= 2\big(|\omega| - \operatorname{Re}\omega\big)\big(\{k\overline{f}(k)kf(k)\xi\}^\sharp \mid \xi^\sharp\big) \\
&= 2\big(|\omega| - \operatorname{Re}\omega\big)\big(\Delta\xi \mid kf(k)^*kf(k)\xi\big) \\
&\le 2|\omega|\left\|kf(k)\Delta\xi\right\|\left\|kf(k)\xi\right\| - 2\operatorname{Re}\omega\big(kf(k)\Delta\xi \mid kf(k)\xi\big) \\
&\le \left\|kf(k)(\Delta - \omega)\xi\right\|^2 = \left\|kf(k)\eta\right\|^2 = \left\|f(k)k\eta\right\|^2 \\
&= \left\|f(k)u\pi_\ell(\xi)^*\eta\right\|^2 = \left\|f(k)u\pi_\ell(\xi^\sharp)\eta\right\|^2 = \|f(k)u\pi_r(\eta)\xi^\sharp\|^2 \\
&= \left\|\pi_r(\eta)uf(h)\xi^\sharp\right\|^2 \le \left\|\pi_r(\eta)\right\|^2\left\|f(h)\xi^\sharp\right\|^2.
\end{aligned}$$

Hence we obtain an inequality:

$$\left\|hf(h)\xi^\sharp\right\|^2 \le \gamma(\omega)^2\left\|\pi_r(\eta)\right\|^2\left\|f(h)\xi^\sharp\right\|^2. \tag{31}$$

We now let

$$h = \int_0^\infty \lambda \, de(\lambda)$$

be the spectral decomposition of h. The inequality (31) means then

$$\int_0^\infty \lambda^2 |f(\lambda)|^2 \, d\|e(\lambda)\xi^\sharp\|^2 \le c^2 \int_0^\infty |f(\lambda)|^2 \, d\|e(\lambda)\xi^\sharp\|^2, \quad f \in \mathcal{K}(0,\infty),$$

with $c = \gamma(\omega)\|\pi_r(\eta)\|$. This inequality means that the measure $d\|e(\lambda)\xi^\sharp\|^2$ on $[0, \infty[$ is supported by the interval $[0, c]$. Hence $e(c)\xi^\sharp = \xi^\sharp$. Therefore, we have, for each $\zeta \in \mathfrak{A}'$.

$$e(c)\pi_\ell(\xi)^*\zeta = e(c)\pi_\ell(\xi^\sharp)\zeta = e(c)\pi_r(\zeta)\xi^\sharp = \pi_r(\zeta)e(c)\xi^\sharp = \pi_r(\zeta)\xi^\sharp = \pi_\ell(\xi)^*\zeta,$$

so that

$$\|\pi_\ell(\xi^\sharp)\zeta\| = \|e(c)\pi_\ell(\xi)^*\zeta\| = \|e(c)hu^*\zeta\| \le c\|\zeta\|.$$

Hence $\xi^\sharp$ is left bounded, equivalently ξ is left bounded, and we obtain $\|\pi_\ell(\xi)\| = \|\pi_\ell(\xi)^*\| \le c$. Q.E.D.

Lemma 1.18. *If $\xi = (\Delta + s)^{-1}\eta$ for $\eta \in \mathfrak{A}'$ and $s > 0$, then for each $\zeta_1, \zeta_2 \in \mathfrak{D}(\Delta^{-1/2}) \cap \mathfrak{D}(\Delta^{-1/2})$ we have*

$$\left(\pi_r(\eta)\zeta_1 \mid \zeta_2\right) = \left(J\pi_\ell(\xi)^*J\Delta^{-\frac{1}{2}}\zeta_1 \mid \Delta^{\frac{1}{2}}\zeta_2\right) + s\left(J\pi_\ell(\xi)^*J\Delta^{\frac{1}{2}}\zeta_1 \mid \Delta^{-\frac{1}{2}}\zeta_2\right). \quad (32)$$

PROOF: Suppose $\zeta_1, \zeta_2 \in \mathfrak{A} \cap \mathfrak{D}(\Delta^{-1/2})$. We then compute:

$$\begin{aligned}
\left(\pi_r(\eta)\zeta_1 \mid \zeta_2\right) &= \left(\pi_\ell(\zeta_1)\eta \mid \zeta_2\right) = \left(\eta \mid \zeta_1^\sharp\zeta_2\right) = \left((\Delta + s)\xi \mid \zeta_1^\sharp\zeta_2\right) \\
&= \left(FS\xi \mid \zeta_1^\sharp\zeta_2\right) + s\left(\xi \mid \zeta_1^\sharp\zeta_2\right) = \left(\zeta_2^\sharp\zeta_1 \mid S\xi\right) + s\left(\zeta_1\xi \mid \zeta_2\right) \\
&= \left(\zeta_1 \mid \zeta_2\xi^\sharp\right) + s\left((\xi^\sharp\zeta_1^\sharp)^\sharp \mid \zeta_2\right) = \left(\zeta_1 \mid (\xi\zeta_2^\sharp)^\sharp\right) + s\left((\xi^\sharp\zeta_1^\sharp)^\sharp \mid \zeta_2\right) \\
&= \left(\zeta_1 \mid \Delta^{-\frac{1}{2}}J\pi_\ell(\xi)J\Delta^{\frac{1}{2}}\zeta_2\right) + s\left(\Delta^{-\frac{1}{2}}J\pi_\ell(\xi)^*J\Delta^{\frac{1}{2}}\zeta_1 \mid \zeta_2\right) \\
&= \left(J\pi_\ell(\xi)^*J\Delta^{-\frac{1}{2}}\zeta_1 \mid \Delta^{\frac{1}{2}}\zeta_2\right) + s\left(J\pi_\ell(\xi)^*J\Delta^{\frac{1}{2}}\zeta_1 \mid \Delta^{-\frac{1}{2}}\zeta_2\right).
\end{aligned}$$

Therefore, the equality (32) holds for $\zeta_1, \zeta_2 \in \mathfrak{A} \cap \mathfrak{D}(\Delta^{-1/2})$. Since the both sides of (32) are sesquilinear forms on $\mathfrak{D}(\Delta^{1/2}) \cap \mathfrak{D}(\Delta^{-1/2})$, to complete the proof, it suffices to show that every vector $\zeta \in \mathfrak{D}(\Delta^{1/2}) \cap \mathfrak{D}(\Delta^{-1/2})$ is approximated by a sequence $\{\zeta_n\}$ in $\mathfrak{A} \cap \mathfrak{D}(\Delta^{-1/2})$ in the sense that

$$\lim\|\zeta - \zeta_n\| = 0, \qquad \lim\|\Delta^{\frac{1}{2}}(\zeta - \zeta_n)\| = 0, \qquad \lim\|\Delta^{-\frac{1}{2}}(\zeta - \zeta_n)\| = 0.$$

Now, we know that $\Delta^{-1/2}\mathfrak{A}' = JF\mathfrak{A}' = J\mathfrak{A}'$ is dense in $\mathfrak{H}$, so given a $\zeta \in \mathfrak{D}(\Delta^{1/2}) \cap \mathfrak{D}(\Delta^{-1/2})$ we choose a sequence $\{\eta_n\} \subset \mathfrak{A}'$ such that

$$(\Delta^{\frac{1}{2}} + \Delta^{-\frac{1}{2}})\zeta = \lim_{n\to\infty} \Delta^{-\frac{1}{2}}\eta_n.$$

Since $\Delta^{-1/2}\big(\Delta^{1/2} + \Delta^{-1/2}\big)^{-1} = (1+\Delta)^{-1}$, we set

$$\zeta_n = (1+\Delta)^{-1}\eta_n,$$

and conclude by Lemma 1.17 that $\zeta_n \in \mathfrak{A} \cap \mathfrak{D}(\Delta^{-1/2})$ and

$$\begin{aligned}\zeta &= \big(\Delta^{\frac{1}{2}} + \Delta^{-\frac{1}{2}}\big)^{-1} \lim_{n\to\infty} \Delta^{-\frac{1}{2}}\eta_n = \lim_{n\to\infty} \Delta^{-\frac{1}{2}}\big(\Delta^{\frac{1}{2}} + \Delta^{-\frac{1}{2}}\big)^{-1}\eta_n \\ &= \lim_{n\to\infty} (1+\Delta)^{-1}\eta_n = \lim_{n\to\infty} \zeta_n.\end{aligned}$$

We now compute:

$$\begin{aligned}\Delta^{\frac{1}{2}}\zeta &= \Delta^{\frac{1}{2}}\big(\Delta^{\frac{1}{2}} + \Delta^{-\frac{1}{2}}\big)^{-1}\big(\Delta^{\frac{1}{2}} + \Delta^{-\frac{1}{2}}\big)\zeta \\ &= \Delta(1+\Delta)^{-1} \lim_{n\to\infty} \Delta^{-\frac{1}{2}}\eta_n = \lim_{n\to\infty} \Delta^{\frac{1}{2}}(1+\Delta)^{-1}\eta_n = \lim_{n\to\infty} \Delta^{\frac{1}{2}}\zeta_n;\end{aligned}$$

$$\begin{aligned}\Delta^{-\frac{1}{2}}\zeta &= \Delta^{-\frac{1}{2}}\big(\Delta^{\frac{1}{2}} + \Delta^{-\frac{1}{2}}\big)^{-1}\big(\Delta^{\frac{1}{2}} + \Delta^{-\frac{1}{2}}\big)\zeta \\ &= (1+\Delta)^{-1} \lim_{n\to\infty} \Delta^{-\frac{1}{2}}\eta_n = \lim_{n\to\infty} \Delta^{-\frac{1}{2}}(1+\Delta)^{-1}\eta_n = \lim_{n\to\infty} \Delta^{-\frac{1}{2}}\zeta_n.\end{aligned}$$

Q.E.D.

Before making a small excursion into Fourier analysis in an abstract Banach algebra, we state the main theorem of the section:

Theorem 1.19. *Let $\mathfrak{A}$ be a left Hilbert algebra with the associated modular operator Δ and modular conjugation J.*

(i)
$$\begin{aligned} J\mathcal{R}_\ell(\mathfrak{A})J &= \mathcal{R}_\ell(\mathfrak{A})'; \\ J\mathcal{R}_\ell(\mathfrak{A})'J &= \mathcal{R}_\ell(\mathfrak{A}); \end{aligned} \tag{33}$$

$$\begin{aligned} \Delta^{it}\mathcal{R}_\ell(\mathfrak{A})\Delta^{-it} &= \mathcal{R}_\ell(\mathfrak{A}); \\ \Delta^{it}\mathcal{R}_\ell(\mathfrak{A})'\Delta^{-it} &= \mathcal{R}_\ell(\mathfrak{A})', \quad t \in \mathbf{R}. \end{aligned} \tag{34}$$

(ii) *The one parameter unitary group $\{\Delta^{it} : t \in \mathbf{R}\}$ acts on $\mathfrak{A}''$ and $\mathfrak{A}'$ as automorphisms and the modular conjugation J maps $\mathfrak{A}''$ (resp. $\mathfrak{A}'$) onto $\mathfrak{A}'$ (resp. $\mathfrak{A}''$) anti-isomorphically in the sense that*

$$J(\xi\eta) = (J\eta)(J\xi), \quad \xi, \eta \in \mathfrak{A}''. \tag{35}$$

In order to prove the theorem, we need the following result:

Lemma 1.20. *Let A be a unital Banach algebra. Suppose $\{u(\alpha) : \alpha \in \mathbf{C}\}$ is a complex one parameter subgroup of the group $\mathrm{GL}(A)$ of invertible elements, i.e.*

$$u(\alpha+\beta) = u(\alpha)u(\beta), \quad \alpha, \beta \in \mathbf{C}.$$

Furthermore, assume that $\alpha \in \mathbf{C} \to u(\alpha) \in A$ is holomorphic and

$$\sup\{\|u(t)\| : t \in \mathbf{R}\} = M < +\infty.$$

Then, for any $s \in \mathbf{R}$, $\mathrm{e}^{-\frac{s}{2}}u\left(-\frac{\mathrm{i}}{2}\right) + \mathrm{e}^{\frac{s}{2}}u\left(\frac{\mathrm{i}}{2}\right)$ is invertible and

$$\left[\mathrm{e}^{-\frac{s}{2}}u\left(-\frac{\mathrm{i}}{2}\right) + \mathrm{e}^{\frac{s}{2}}u\left(\frac{\mathrm{i}}{2}\right)\right]^{-1} = \int_{-\infty}^{\infty} \frac{\mathrm{e}^{-\mathrm{i}st}}{\mathrm{e}^{\pi t} + \mathrm{e}^{-\pi t}} u(t)\, \mathrm{d}t\,. \tag{36}$$

PROOF: Set for a fixed $s \in \mathbf{R}$,

$$f(\alpha) = \frac{\mathrm{e}^{\mathrm{i}s\alpha}}{\mathrm{e}^{\pi\alpha} - \mathrm{e}^{-\pi\alpha}} u(\alpha), \quad \alpha \in \mathbf{C}.$$

It follows that f is a meromorphic A-valued function with simple poles at $\alpha = \mathrm{i}n$, $n \in \mathbf{Z}$. If $\alpha = t + \mathrm{i}r,\ t, r \in \mathbf{R}$, then we have the estimate:

$$\|f(\alpha)\| \leq M\mathrm{e}^{-sr}\left|\frac{1}{\mathrm{e}^{\pi\alpha} - \mathrm{e}^{-\pi\alpha}}\right| \|u(\mathrm{i}r)\|.$$

We now consider the simple closed curve c_R for $R > 0$ indicated below:

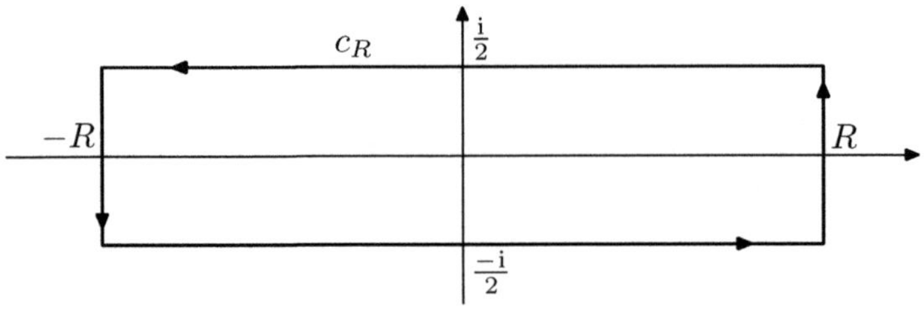

We then integrate f along the curve c_R:

$$I_R = \oint_{c_R} f(\alpha)\, \mathrm{d}\alpha\,.$$

The above estimate of $\|f(\alpha)\|$ implies the convergence:

$$I = \lim_{R\to\infty} I_R = \int_{-\infty}^{\infty} f\left(t - \frac{\mathrm{i}}{2}\right) \mathrm{d}t - \int_{-\infty}^{\infty} f\left(t + \frac{\mathrm{i}}{2}\right) \mathrm{d}t\,.$$

By the residue theorem, we obtain

$$I = 2\pi \mathrm{i} \lim_{\alpha \to 0} \alpha f(\alpha) = \mathrm{i},$$

so that we come to the following computation:

$$\begin{aligned}
\mathrm{i} &= \int_{-\infty}^{\infty} \frac{\mathrm{e}^{-\mathrm{i}s\left(t-\frac{\mathrm{i}}{2}\right)}}{\mathrm{e}^{\pi\left(t-\frac{\mathrm{i}}{2}\right)} - \mathrm{e}^{-\pi\left(t-\frac{\mathrm{i}}{2}\right)}} u\left(t - \frac{\mathrm{i}}{2}\right) \mathrm{d}t \\
&\quad - \int_{-\infty}^{\infty} \frac{\mathrm{e}^{-\mathrm{i}s\left(t+\frac{\mathrm{i}}{2}\right)}}{\mathrm{e}^{\pi\left(t+\frac{\mathrm{i}}{2}\right)} - \mathrm{e}^{-\pi\left(t+\frac{\mathrm{i}}{2}\right)}} u\left(t + \frac{\mathrm{i}}{2}\right) \mathrm{d}t \\
&= \mathrm{i}\left[\mathrm{e}^{-\frac{s}{2}} u\left(-\frac{\mathrm{i}}{2}\right) + \mathrm{e}^{\frac{s}{2}} u\left(\frac{\mathrm{i}}{2}\right)\right] \int_{-\infty}^{\infty} \frac{\mathrm{e}^{-\mathrm{i}st}}{\mathrm{e}^{\pi t} + \mathrm{e}^{-\pi t}} u(t)\, \mathrm{d}t\,.
\end{aligned}$$

Thus (36) follows. Q.E.D.

For the convenience of later use, we rewrite (36) in the following form:

$$\mathrm{e}^{\frac{s}{2}} u\left(-\frac{\mathrm{i}}{2}\right)\left(u(-\mathrm{i}) + \mathrm{e}^{s}\right)^{-1} = \int_{-\infty}^{\infty} \frac{\mathrm{e}^{-\mathrm{i}st}}{\mathrm{e}^{\pi t} + \mathrm{e}^{-\pi t}} u(t)\, \mathrm{d}t, \quad s \in \mathbf{R}. \tag{37}$$

We apply the above integral formula to the one parameter unitary group $\Delta^{\mathrm{i}t}$ given by the modular operator Δ.

Lemma 1.21. *Under the assumption of Theorem* 1.19, *we have*

$$\mathrm{e}^{\frac{s}{2}} \Delta^{\frac{1}{2}} (\Delta + \mathrm{e}^{s})^{-1} = \int_{-\infty}^{\infty} \frac{\mathrm{e}^{-\mathrm{i}st}}{\mathrm{e}^{\pi t} + \mathrm{e}^{-\pi t}} \Delta^{\mathrm{i}t}\, \mathrm{d}t\,. \tag{38}$$

PROOF: With the spectral decomposition of Δ:

$$\Delta = \int_0^{\infty} \lambda\, \mathrm{d}E(\lambda),$$

we set $E_r = E(r) - E(1/r)$ for $r > 1$. We apply (37) to $A = \mathcal{L}(E_r\mathfrak{H})$ and $u(\alpha) = (\Delta E_r)^{\mathrm{i}\alpha}$, $\alpha \in \mathbf{C}$, to obtain

$$\mathrm{e}^{\frac{s}{2}} \Delta^{\frac{1}{2}} (\Delta + \mathrm{e}^{s})^{-1} E_r = \int_{-\infty}^{\infty} \frac{\mathrm{e}^{-\mathrm{i}st}}{\mathrm{e}^{\pi t} + \mathrm{e}^{-\pi t}} \Delta^{\mathrm{i}t} E_r\, \mathrm{d}t\,.$$

We then let $r \to +\infty$ to conclude the lemma. Q.E.D.

Another application of formula (37) is the following:

Lemma 1.22. *Under the assumption of Theorem* 1.19, *if* $x, y \in \mathcal{L}(\mathfrak{H})$ *and* $s \in \mathbf{R}$ *satisfy the following equation for every* $\zeta_1, \zeta_2 \in \mathfrak{D}(\Delta^{1/2}) \cap \mathfrak{D}(\Delta^{-1/2})$:

$$(x\zeta_1 \mid \zeta_2) = (y\Delta^{-\frac{1}{2}}\zeta_1 \mid \Delta^{\frac{1}{2}}\zeta_2) + \mathrm{e}^s(y\Delta^{\frac{1}{2}}\zeta_1 \mid \Delta^{-\frac{1}{2}}\zeta_2),$$

then

$$\mathrm{e}^{\frac{s}{2}} y = \int_{-\infty}^{\infty} \frac{\mathrm{e}^{-\mathrm{i}st}}{\mathrm{e}^{\pi t} + \mathrm{e}^{-\pi t}} \Delta^{\mathrm{i}t} x \Delta^{-\mathrm{i}t}\, \mathrm{d}t\,. \tag{39}$$

PROOF: Let $\{E_r : r > 1\}$ be as above and $A = \mathcal{L}(\mathcal{L}(E_r\mathfrak{H}))$ and $\{u(\alpha) : \alpha \in \mathbf{C}\}$ be the group:

$$\sigma_\alpha(x) = \Delta^{\mathrm{i}\alpha} x \Delta^{-\mathrm{i}\alpha}, \quad x \in \mathcal{L}(E_r\mathfrak{H}).$$

By assumption, we have, for each $\zeta_1, \zeta_2 \in \mathfrak{H}$,

$$\begin{aligned}(E_r x E_r \zeta_1 \mid \zeta_2) &= \left(\Delta^{\frac{1}{2}} E_r y \Delta^{-\frac{1}{2}} E_r \zeta_1 \,\middle|\, \zeta_2\right) + \mathrm{e}^s\left(\Delta^{-\frac{1}{2}} E_r y \Delta^{\frac{1}{2}} E_r \zeta_1 \,\middle|\, \zeta_2\right)\\ &= \left(\left[\sigma_{-\frac{\mathrm{i}}{2}}(E_r y E_r) + \mathrm{e}^s \sigma_{\frac{\mathrm{i}}{2}}(E_r y E_r)\right]\zeta_1 \,\middle|\, \zeta_2\right).\end{aligned}$$

Hence we get

$$E_r x E_r = (\sigma_{-\frac{\mathrm{i}}{2}} + \mathrm{e}^s \sigma_{\frac{\mathrm{i}}{2}})(E_r y E_r) = \mathrm{e}^{\frac{s}{2}}\,(\mathrm{e}^{-\frac{s}{2}}\sigma_{-\frac{\mathrm{i}}{2}} + \mathrm{e}^{\frac{s}{2}}\sigma_{\frac{\mathrm{i}}{2}})(E_r y E_r).$$

By Lemma 1.20, we get

$$\mathrm{e}^{\frac{s}{2}} E_r y E_r = \int_{-\infty}^{\infty} \frac{\mathrm{e}^{-\mathrm{i}st}}{\mathrm{e}^{\pi t} + \mathrm{e}^{-\pi t}} \Delta^{\mathrm{i}t} E_r x E_r \Delta^{-\mathrm{i}t}\, \mathrm{d}t\,.$$

Letting $r \to \infty$, we obtain the formula (39). Q.E.D.

We define the following transformation in $\mathcal{L}(\mathfrak{H})$:

$$\rho_s(x) = \int_{-\infty}^{\infty} \frac{\mathrm{e}^{-\mathrm{i}st}}{\mathrm{e}^{\pi t} + \mathrm{e}^{-\pi t}} \Delta^{\mathrm{i}t} x \Delta^{-\mathrm{i}t}\, \mathrm{d}t, \quad x \in \mathcal{L}(\mathfrak{H}), \quad s \in \mathbf{R}. \tag{40}$$

PROOF OF THEOREM 1.19: Lemma 1.18 and 1.22 together imply the formula:

$$\mathrm{e}^{\frac{s}{2}} J\pi_\ell((\Delta + \mathrm{e}^s)^{-1}\eta)^* J = \rho_s(\pi_r(\eta)), \quad \eta \in \mathfrak{A}', \quad s \in \mathbf{R}.$$

For each $\zeta \in \mathfrak{A}'$, we get, by (38),

$$\begin{aligned}J\rho_s(\pi_r(\eta))J\zeta &= \mathrm{e}^{\frac{s}{2}}\pi_\ell((\Delta + e^s)^{-1}\eta)^*\zeta = \mathrm{e}^{\frac{s}{2}}(S(\Delta + \mathrm{e}^s)^{-1}\eta)\zeta\\ &= \mathrm{e}^{\frac{s}{2}}\pi_r(\zeta) J \Delta^{\frac{1}{2}}(\Delta + \mathrm{e}^s)^{-1}\eta\\ &= \pi_r(\zeta) J \int_{-\infty}^{\infty} \frac{\mathrm{e}^{-\mathrm{i}st}}{\mathrm{e}^{\pi t} + \mathrm{e}^{-\pi t}} \Delta^{\mathrm{i}t}\eta\, \mathrm{d}t,\end{aligned}$$

so that we obtain

$$\int_{-\infty}^{\infty} \frac{e^{-ist}}{e^{\pi t} + e^{-\pi t}} \left(J\Delta^{it}\pi_r(\eta)\Delta^{-it} J\zeta - \pi_r(\zeta) J\Delta^{it}\eta \right) dt = 0.$$

By the uniqueness of the Fourier transform we conclude

$$\frac{1}{e^{\pi t} + e^{-\pi t}} \left(J\Delta^{it}\pi_r(\eta)\Delta^{-it} J\zeta - \pi_r(\zeta) J\Delta^{it}\eta \right) = 0,$$

hence

$$J\Delta^{it}\pi_r(\eta)\Delta^{-it} J\zeta = \pi_r(\zeta) J\Delta^{it}\eta.$$

This is true for all $\zeta \in \mathfrak{A}'$; so $J\Delta^{it}\eta$ is left bounded and

$$\pi_\ell(J\Delta^{it}\eta) = J\Delta^{it}\pi_r(\eta)\Delta^{-it} J, \quad \eta \in \mathfrak{A}', \quad t \in \mathbf{R}. \tag{41}$$

Since $J\mathfrak{D}^\flat = J\mathfrak{D}(\Delta^{-1/2}) = \mathfrak{D}(\Delta^{1/2}) = \mathfrak{D}^\sharp$, $J\Delta^{it}\eta$ belongs to $\mathfrak{A}$. Setting $t = 0$, we obtain $J\mathfrak{A}' \subset \mathfrak{A}''$ and

$$\pi_\ell(J\eta) = J\pi_r(\eta)J, \quad \eta \in \mathfrak{A}'. \tag{42}$$

This implies immediately (35) for $\mathfrak{A}'$. By symmetry, we obtain also $J\mathfrak{A}'' \subset \mathfrak{A}'$ and $J\pi_\ell(\xi)J = \pi_r(J\xi)$ for $\xi \in \mathfrak{A}''$. Thus, we conclude that

$$J\mathfrak{A}'' = \mathfrak{A}', \qquad J\mathfrak{A}' = \mathfrak{A}'',$$

$$\Delta^{it}\mathfrak{A}'' = \mathfrak{A}'', \qquad \Delta^{it}\mathfrak{A}' = \mathfrak{A}', \quad t \in \mathbf{R}.$$

The formula (41) means precisely that $\{\Delta^{it}\}$ is a one parameter group of automorphisms of $\mathfrak{A}''$ and $\mathfrak{A}'$. Thus we conclude the theorem. Q.E.D.

Proposition 1.23. *If $\mathfrak{A}$ is a left Hilbert algebra, then every central element $a \in \mathcal{R}_\ell(\mathfrak{A})$ leaves $\mathfrak{D}^\sharp$ and $\mathfrak{D}^\flat$ invariant and*

$$(a\xi)^\sharp = a^*\xi^\sharp, \quad \xi \in \mathfrak{D}^\sharp,$$

$$(a\eta)^\flat = a^*\eta^\flat, \quad \eta \in \mathfrak{D}^\flat.$$

Furthermore, we have

$$JaJ = a^*, \qquad \Delta^{it}a\Delta^{-it} = a, \quad t \in \mathbf{R}.$$

PROOF: We may assume that $\mathfrak{A}$ is full. Recall the left ideals $\mathfrak{n}_r$ of $\mathcal{R}_\ell(\mathfrak{A})'$ and $\mathfrak{n}_\ell$ of $\mathcal{R}_\ell(\mathfrak{A})$ of Lemma 1.8 and Lemma 1.8′. We then have $a(\mathfrak{n}_r \cap \mathfrak{n}_r^*) \subset \mathfrak{n}_r \cap \mathfrak{n}_r^*$ and $a(\mathfrak{n}_\ell \cap \mathfrak{n}_\ell^*) \subset \mathfrak{n}_\ell \cap \mathfrak{n}_\ell^*$, so that $a\pi_r(\mathfrak{A}') \subset \pi_r(\mathfrak{A}')$ and $a\pi_\ell(\mathfrak{A}) \subset \pi_\ell(\mathfrak{A})$, which means that $a\mathfrak{A}' \subset \mathfrak{A}'$ and $a\mathfrak{A} \subset \mathfrak{A}$. Furthermore, if $\eta \in \mathfrak{A}'$, then

$$\pi_r\big((a\eta)^\flat\big) = \pi_r(a\eta)^* = \big(a\pi_r(\eta)\big)^* = a^*\pi_r(\eta)^* = a^*\pi_r(\eta^\flat) = \pi_r(a^*\eta^\flat),$$

so that $(a\eta)^\flat = a^*\eta^\flat$. By symmetry, $(a\xi)^* = a^*\xi^\sharp$ for $a \in \mathfrak{A}$.

If a is unitary, then $a\mathfrak{A} = \mathfrak{A}$ and $a\mathfrak{A}' = \mathfrak{A}'$. As seen above, $Su\xi = u^*S\xi,\ \xi \in \mathfrak{A}$, so that $uSu = S$, which means that

$$J\Delta^{\frac{1}{2}} = uJ\Delta^{\frac{1}{2}}u = uJuu^*\Delta^{\frac{1}{2}}u.$$

The uniqueness of the polar decomposition implies that

$$J = uJu \qquad \text{and} \qquad \Delta^{\frac{1}{2}} = u^*\Delta^{\frac{1}{2}}u.$$

Hence we get $Ju = u^*J$ and $\Delta^{it} = u^*\Delta^{it}u$. Since $\mathcal{R}_\ell(\mathfrak{A}) \cap \mathcal{R}_\ell(\mathfrak{A})'$ is spanned linearly by the unitary group, the last assertion of the proposition follows. Q.E.D.

We want now to study the topological properties of the mapping π_ℓ.

Proposition 1.24. *Let $\mathfrak{A}$ be a full left Hilbert algebra with completion $\mathfrak{H}$ and the associated algebra $\mathfrak{B}$ of all left bounded vectors.*

(i) *The map $\pi_\ell\colon \xi \in \mathfrak{A} \mapsto \pi_\ell(\xi) \in \mathcal{R}_\ell(\mathfrak{A})$ is closed with respect to the Hilbert space norm topology in $\mathfrak{D}^\sharp$ and the σ-strong* topology in $\mathcal{R}_\ell(\mathfrak{A})$.*
(ii) *The map $\pi_\ell\colon \xi \in \mathfrak{B} \mapsto \pi_\ell(\xi) \in \mathcal{R}_\ell(\mathfrak{A})$ is closed with respect to the Hilbert space norm topology in $\mathfrak{H}$ and the σ-strong topology in $\mathcal{R}_\ell(\mathfrak{A})$.*

PROOF:
(i) Let $\{\xi_i\}$ be a net in $\mathfrak{A}$ such that $\{\xi_i\}$ converges to ξ in $\mathfrak{D}^\sharp$ and $\{\pi_\ell(\xi_i)\}$ converges to $x \in \mathcal{R}_\ell(\mathfrak{A})$ σ-strongly*. Let $\xi = \lim \xi_i$. For any $\eta \in \mathfrak{A}'$, we have

$$\begin{aligned}\pi_r(\eta)\xi &= \lim \pi_r(\eta)\xi_i = \lim \pi_\ell(\xi_i)\eta = x\eta;\\ \pi_r(\eta)\xi^\sharp &= \lim \pi_r\xi_i^\sharp = \lim \pi_\ell(\xi_i)^*\eta = x^*\eta,\end{aligned}$$

so that ξ is left bounded and $x = \pi_\ell(\xi)$.
(ii) This follows from the above same arguments. Q.E.D.

Lemma 1.25. *Let $\{x_i\}$ be a net in $\mathcal{L}(\mathfrak{H})$ and $x \in \mathcal{L}(\mathfrak{H})$. If there exist dense subset $\mathfrak{M}$ and $\mathfrak{N}$ of $\mathfrak{H}$ such that $\lim_i x_i\xi = x\xi$, $\xi \in \mathfrak{M}$, and $\lim_i x_i^*\eta = x^*\eta$, $\eta \in \mathfrak{N}$, in norm, then for any continuous bounded function f on $[0, \infty)$ we have the convergence:*

$$\lim_i f(x_i^*x_i) = f(x^*x), \quad \lim_i f(x_ix_i^*) = f(xx^*)$$

in the strong operator topology.

PROOF: Let $g(t) = f(t^2)$, $t \in \mathbf{R}$, and h_i and h be the self-adjoint operators on $\mathfrak{H} \oplus \mathfrak{H}$ given by the matrices:

$$h_i = \begin{pmatrix} 0 & x_i^* \\ x_i & 0 \end{pmatrix} \quad \text{and} \quad h = \begin{pmatrix} 0 & x^* \\ x & 0 \end{pmatrix}.$$

It then follows that $\lim_i h_i\xi = h\xi$ for any $\xi \in \mathfrak{M} \oplus \mathfrak{N}$. Hence by A.6 $g(h_i)$ converges to $\{g(h)\}$ in the strong operator topology. But we have

$$g\left[\begin{pmatrix} 0 & y^* \\ y & 0 \end{pmatrix}\right] = \begin{pmatrix} f(y^*y) & 0 \\ 0 & f(yy^*) \end{pmatrix}, \quad y \in \mathcal{L}(\mathfrak{H}). \qquad \text{Q.E.D.}$$

Theorem 1.26. *Let $\mathfrak{A}$ be a left Hilbert algebra with completion $\mathfrak{H}$.*

(i) *If $\xi \in \mathfrak{A}''$, then there exists a sequence $\{\xi_n\}$ in $\mathfrak{A}$ such that*

$$\lim_{n\to\infty} \|\xi - \xi_n\|_\sharp = 0 \quad \textit{and} \quad \|\pi_\ell(\xi_n)\| \le \|\pi_\ell(\xi)\|.$$

Hence $\{\pi_\ell(\xi_n)\}$ converges to $\pi_\ell(\xi)$ in the strong operator topology.*

(ii) *If $\xi \in \mathfrak{H}$ is left bounded, then there exists a sequence $\{\xi_n\}$ in $\mathfrak{A}$ such that*

$$\lim_{n\to\infty} \|\xi - \xi_n\| = 0 \quad \textit{and} \quad \|\pi_\ell(\xi_n)\| \le \|\pi_\ell(\xi)\|.$$

Hence $\{\pi_\ell(\xi_n)\}$ converges to $\pi_\ell(\xi)$ in the strong operator topology.

PROOF:

(i) We may assume that $\|\pi_\ell(\xi)\| = 1$. Since $\xi \in \mathfrak{D}^\sharp$, we can find a sequence $\{\zeta_n\}$ in $\mathfrak{A}$ such that $\|\zeta_n - \xi\|_\sharp \to 0$ as $n \to \infty$. We set $x = \pi_\ell(\zeta)$ and $x_n = \pi_\ell(\xi_n)$. For every $\eta \in \mathfrak{A}'$, we have the convergence:

$$\|x\eta - x_n\eta\| \to 0 \quad \text{and} \quad \|x^*\eta - x_n^*\eta\| \to 0$$

as $n \to \infty$. Hence Lemma 1.25 applies to $\{x_n\}$ and $\{x\}$. We consider the function f on $[0, \infty[$:

$$f(t) = \begin{cases} 1, & 0 \le t \le 1 \\ t^{-\frac{1}{2}}, & t > 1. \end{cases}$$

By the lemma, $f(x_n^*x_n) \to f(x^*x) = 1$ and $f(x_nx_n^*) \to f(xx^*) = 1$ strongly. Set $\xi_n = f(x_nx_n^*)\zeta_n$. The dualized version of (25), say (25′), shows that $\xi_n \in \mathfrak{A}''$ and $\xi_n^\sharp = f(x_n^*x_n)\zeta_n^\sharp$. Furthermore, we have

$$\pi_\ell(\xi_n) = f(x_nx_n^*)x_n, \qquad \pi_\ell(\xi_n)^* = f(x_n^*x_n)x_n^*.$$

Now, we have

$$\begin{aligned} \|\xi_n - \xi\| &\le \|f(x_nx_n^*)(\zeta_n - \xi)\| + \|f(x_nx_n^*)\xi - \xi\| \\ &\le \|\zeta_n - \xi\| + \|((f(x_nx_n^*) - 1)\xi\| \to 0 \quad \text{as } n \to \infty. \end{aligned}$$

Similarly, we get $\|\xi_n^\sharp - \xi\| \to 0$ as $n \to \infty$. Moreover, we have

$$\|\pi_\ell(\xi_n)\|^2 = \|\pi_\ell(\xi_n)\pi_\ell(\xi_n)^*\| = \|f(x_nx_n^*)x_nx_n^*f(x_nx_n^*)\| \le \sup_{t\ge 1}|tf(t)^2| = 1.$$

We now approximate f by polynomials $\{p_n\}$ on $[0, \infty[$ so that

$$\left|p_n(t) - f(t)\right| \leq \frac{1}{n} \min\left\{\frac{1}{\|x_n\|}, \frac{1}{\|\zeta_n\|}, \frac{1}{\|\zeta_n^\sharp\|}\right\}, \qquad 0 \leq t \leq \|x_n\|^2.$$

Set $\xi_n' = p_n(\zeta_n \zeta_n^\sharp)\zeta_n$. We then have $\xi_n \in \mathfrak{A}$ and $\xi_n^\sharp = p_n(\zeta_n^\sharp \zeta_n)\zeta_n^\sharp$. Now we get

$$\left\|\xi_n' - \xi_n\right\| \leq \left\|p_n(x_n x_n^*) - f(x_n x_n^*)\right\| \|\zeta_n\| \leq 1/n;$$

$$\left\|(\xi_n')^\sharp - \xi_n^\sharp\right\| \leq \left\|p_n(x_n^* x_n)\right\| \|\zeta_n^\sharp\| \leq 1/n.$$

Thus, $\|\xi_n' - \xi_n\|_\sharp \to 0$ as $n \to \infty$. Furthermore, we compute:

$$\begin{aligned}\left\|\pi_\ell(\xi_n')\right\| &= \left\|p_n(x_n x_n^*)x_n\right\| \leq \left\|p_n(x_n^* x_n) - f(x_n x_n^*)\right\| \|x_n\| + \left\|f(x_n x_n^*)x_n\right\| \\ &\leq \frac{1}{n} + 1.\end{aligned}$$

Thus, setting $\xi_n'' = \left(1 + \frac{1}{n}\right)^{-1} \xi_n'$, we obtain a sequence $\{\xi_n''\}$ such that

$$\|\xi - \xi_n''\|_\sharp \to 0 \quad \text{and} \quad \left\|\pi_\ell(\xi_n'')\right\| \leq 1.$$

Now, if $\eta \in \mathfrak{A}'$, then we have, as $n \to \infty$,

$$\pi_\ell(\xi_n'')\eta = \pi_r(\eta)\xi_n'' \;\to\; \pi_r(\eta)\xi = \pi_\ell(\xi)\eta;$$

$$\pi_\ell(\xi_n'')^*\eta = \pi_r(\eta)(\xi_n'')^\sharp \;\to\; \pi_r(\eta)\xi^\sharp = \pi_\ell(\xi)^*\eta.$$

Since $\left\{\pi_\ell(\xi_n'')\right\}$ is bounded, it converges to $\pi_\ell(\xi)$ strongly*.

(ii) Suppose that $\xi \in \mathfrak{B}$. We may assume again that $\left\|\pi_\ell(\xi_n'')\right\| = 1$. Let $\pi_\ell(\xi) = uh$ be the polar decomposition and set $\zeta = u^*\xi$. Then $\zeta \in \mathfrak{B}$ and $\pi_\ell(\zeta) = h$ is self-adjoint, so that ζ belong to $\mathfrak{A}''$ and $\zeta = \zeta^\sharp$. By (i), we choose a sequence $\{\zeta_n\}$ in $\mathfrak{A}$ such that $\|\zeta - \zeta_n\| < \frac{n}{2}$ and $\left\|\pi_\ell(\zeta_n)\right\| \leq 1$. Since $u \in \mathcal{R}_\ell(\mathfrak{A})$, there exists, by Kaplansky's density theorem, Theorem IV.4.8, a sequence $\{\eta_n\}$ in $\mathfrak{A}$ such that

$$\left\|\pi_\ell(\eta_n)\right\| \leq 1, \qquad \left\|\eta_n \zeta_n - u\zeta_n\right\| < \frac{n}{2}.$$

We then set $\xi_n = \eta_n \zeta_n \in \mathfrak{A}$ and get

$$\|\xi_n - \xi\| \leq \left\|\eta_n \zeta_n - u\zeta_n\right\| + \left\|u(\zeta_n - \zeta)\right\| < \frac{1}{n};$$

$$\left\|\pi_\ell(\xi_n)\right\| = \left\|\pi_\ell(\eta_n)\pi_\ell(\zeta_n)\right\| \leq 1.$$

The same arguments as in the last part of (i) show that $\left\{\pi_\ell(\xi_n)\right\}$ converges to $\pi_\ell(\xi)$ strongly. Since we have no control about $\xi_n^\sharp$, we get only the strong convergence.

Q.E.D.

Exercise VI.1

1) Let $\mathfrak{A}$ be a full left Hilbert algebra with completion $\mathfrak{H}$. Define a new norm $|||\cdot|||$ in $\mathfrak{A}$ by $|||\xi||| = \max\{\|\pi_\ell(\xi)\|, \|\xi\|, \|\xi^\sharp\|\}$. Prove that $\{\mathfrak{A}, |||\cdot|||\}$ is an involutive Banach algebra.

2) Prove that a left Hilbert algebra $\mathfrak{A}$ is full if and only if the set $\{\xi \in \mathfrak{A} : \|\pi(\xi)\| \le 1\}$ is closed in $\mathfrak{D}^\sharp$.

3) Let $\mathfrak{A}$ be a complete left Hilbert algebra, so that $\mathfrak{H} = \mathfrak{A}$.

(**a**) Show that the set $\mathrm{Proj}(\mathfrak{A})$ of non-zero projections (i.e. self-adjoint idempotents) in $\mathfrak{A}$ is non-empty. (Hint: If $\xi = \xi^\sharp \neq 0$, then there exists $a \in \mathcal{R}_\ell(\mathfrak{A})$ such that $a\pi_\ell(\xi)$ is a non-zero spectral projection of $\pi_\ell(\xi)$, but $a\pi_\ell(\xi) = \pi_\ell(a\xi)$.)

(**b**) Show that there exists $M > 0$ such that $\|\xi\eta\| \le M\|\xi\|\|\eta\|$, $\xi, \eta \in \mathfrak{A}$. (Hint: Apply the uniform boundedness theorem.)

(**c**) Observe that if $e \in \mathrm{Proj}(\mathfrak{A})$, then $\|e\| = \|e^2\| \le M\|e\|^2$, so that $\|e\| \ge 1/M$.

(**d**) Let $\lambda = \inf\{\|e\|^2 : e \in \mathrm{Proj}(\mathfrak{A})\} \ge 1/M$. Show that if $\|e\|^2 < 2\lambda$, $e \in \mathrm{Proj}(\mathfrak{A})$, then $\pi_\ell(e)$ is a minimal projection of $\mathcal{R}_\ell(\mathfrak{A})$. (Hint: If $\pi(e) = p_1 + p_2$, $p_1p_2 = 0$, with $p_1, p_2 \in \mathrm{Proj}(\mathcal{R}_\ell(\mathfrak{A}))$, then there exist $e_1, e_2 \in \mathfrak{A}$ such that $\pi_\ell(e_1) = p_1$, $\pi_\ell(e_2) = p_2$ and $e = e_1 + e_2$. Then $e_1e_2 = 0$ and

$$(e_1 \mid e_2) = (e_1^2 \mid e_2) = (e_1 \mid e_1e_2) = 0;$$

$$2\lambda \le \|e_1\|^2 + \|e_2\|^2 = \|e\|^2 < 2\lambda,$$

which is a contradiction.)

(**e**) Prove that $\mathcal{R}_\ell(\mathfrak{A})$ is a direct sum of factors of type I. (Hint: Prove that $\mathcal{R}_\ell(\mathfrak{A})$ is generated by minimal projections.)

4) Let $\mathfrak{K}$ be a Hilbert space and $\overline{\mathfrak{K}}$ the Banach space dual of $\mathfrak{K}$. For each $\eta \in \mathfrak{K}$, let $\overline{\eta}$ be the vector of $\overline{\mathfrak{K}}$ corresponding to η, i.e. $\langle \xi, \overline{\eta}\rangle = (\xi \mid \eta)$, $\xi, \eta \in \mathfrak{K}$. If $\eta \in \overline{\mathfrak{K}}$, then let $\overline{\eta}$ be the vector of $\mathfrak{K}$ corresponding to η, i.e. $\overline{\overline{\eta}} = \eta$ for $\eta \in \overline{\mathfrak{K}}$ or $\eta \in \overline{\mathfrak{K}}$. Let h be a positive non-singular self-adjoint operator on $\mathfrak{K}$ with domain $\mathfrak{D}(h)$. Let $\mathfrak{A} = \mathfrak{D}(h) \otimes \mathfrak{D}(h^{-1})$ be the algebraic tensor product. Define a multiplication and a $\sharp$-operation in $\mathfrak{A}$ by:

$$(\xi_1 \otimes \overline{\eta}_1)(\xi_2 \otimes \overline{\eta}_2) = (\xi_2 \mid \overline{h^{-1}\eta_1})\xi_1 \otimes \overline{\eta}_2;$$

$$(\xi \otimes \overline{\eta})^\sharp = h^{-1}\eta \otimes \overline{h\xi}, \qquad \xi_1, \xi_2, \xi \in \mathfrak{D}(h), \quad \eta_1, \eta_2, \eta \in \mathfrak{D}(h^{-1}).$$

Prove that $\mathfrak{A}$ is a left Hilbert algebra with completion $\mathfrak{K} \otimes \overline{\mathfrak{K}}$ such that $\mathcal{R}_\ell(\mathfrak{A}) = \mathcal{L}(\mathfrak{K}) \otimes \mathbf{C}$, $J(\xi \otimes \overline{\eta}) = \eta \otimes \overline{\xi}$ and $\Delta^{1/2}(\xi \otimes \overline{\eta}) = h\xi \otimes \overline{h^{-1}\eta}$.

§ 2 Tomita Algebras

Given a left Hilbert algebra $\mathfrak{A}$, $\mathfrak{A}''$ and $\mathfrak{A}'$ do not agree in general. It is desirable to have a "self-adjoint" subalgebra $\mathfrak{A}_0$ of $\mathfrak{A}''$ and $\mathfrak{A}'$ such that $\mathfrak{A}'' = \mathfrak{A}_0''$ and $\mathfrak{A}' = \mathfrak{A}_0'$. For this purpose, the following Tomita algebras are important:

Definition 2.1. A left Hilbert algebra $\mathfrak{A}$ is called a *Tomita algebra* if $\mathfrak{A}$ admits a complex one parameter group $\{U(\alpha) : \alpha \in \mathbf{C}\}$ of automorphisms, not *-preserving, with the following properties:

a) The function: $\alpha \in \mathbf{C} \to \big(U(\alpha)\xi \mid \eta\big)$ is entire;

b) $\big(U(\alpha)\xi\big)^\sharp = U(\overline{\alpha})\xi^\sharp$;

c) $\big(U(\alpha)\xi \mid \eta\big) = \big(\xi \mid U(-\overline{\alpha})\eta\big)$, $\alpha \in \mathbf{C}$, $\xi, \eta \in \mathfrak{A}$;

d) $(\xi^\sharp \mid \eta^\sharp) = \big(U(-i)\eta \mid \xi\big)$.

The group $\{U(\alpha) : \alpha \in \mathbf{C}\}$ is called the *modular automorphism group* of $\mathfrak{A}$.

Theorem 2.2.

(i) *Given a full left Hilbert algebra $\mathfrak{A}$ with the modular operator Δ, if we set*

$$\mathfrak{A}_0 = \Big\{\xi \in \bigcap_{n\in\mathbf{Z}} \mathfrak{D}(\Delta^n) : \Delta^n\xi \in \mathfrak{A},\ n \in \mathbf{Z}\Big\}, \tag{1}$$

then $\mathfrak{A}_0$ is a Tomita algebra with respect to $\{\Delta^{\mathrm{i}\alpha} : \alpha \in \mathbf{C}\}$ such that

$$\mathfrak{A} = \mathfrak{A}_0'', \qquad \mathfrak{A}_0' = \mathfrak{A}' \qquad \text{and} \qquad J\mathfrak{A}_0 = \mathfrak{A}_0.$$

Hence, in particular we have

$$\mathcal{R}_\ell(\mathfrak{A}_0) = \mathcal{R}_\ell(\mathfrak{A}) \qquad \text{and} \qquad \mathcal{R}_r(\mathfrak{A}_0) = \mathcal{R}_\ell(\mathfrak{A})'.$$

(ii) *If $\mathfrak{A}$ is a Tomita algebra, then with new involution:*

$$\xi^\flat = U(-\mathrm{i})\xi^\sharp, \quad \xi \in \mathfrak{A}, \tag{2}$$

$\mathfrak{A}$ is a right Hilbert algebra and

$$\mathcal{R}_\ell(\mathfrak{A})' = \mathcal{R}_r(\mathfrak{A}).$$

Furthermore, the modular operator Δ is the closure of $U(-\mathrm{i})$.

To prove the theorem, we need a couple of lemmas. The first one provides a useful criterion for a vector to belong to the domain of a power of a self-adjoint operator.

Lemma 2.3. *Let H be a non-singular self-adjoint positive operator on a Hilbert space $\mathfrak{H}$. For a fixed $\alpha \in \mathbf{R}$ and $\xi \in \mathfrak{H}$, the following two conditions are equivalent:*

(i) *ξ belongs to the domain $\mathfrak{D}(H^\alpha)$ of H^α;*

(ii) *The* $\mathfrak{H}$ *-valued function:* $t \in \mathbf{R} \mapsto H^{\mathrm{i}t}\xi \in \mathfrak{H}$ *can be extended to an* $\mathfrak{H}$ *-valued function:* $\omega \in \mathbf{D}_\alpha \mapsto \xi(\omega) \in \mathfrak{H}$ *such that* ξ *is continuous and bounded on the closure* $\overline{\mathbf{D}}_\alpha$ *and holomorphic in* $\mathbf{D}_\alpha$*, where* $\mathbf{D}_\alpha$ *is the horizontal strip bounded by* $\mathbf{R}$ *and* $\mathbf{R} - \mathrm{i}\alpha$.

PROOF: Considering H^{-1} if necessary, we may assume $\alpha > 0$.

(i) $\Longrightarrow$ (ii): Assume (i). If $\omega = t - \mathrm{i}s \in \mathbf{D}_\alpha$, then $\mathfrak{D}(H^{i\omega}) = \mathfrak{D}(H^s)$. The inequality,

$$\|H^{\mathrm{i}\omega}\xi\| = \|H^s\xi\| \leq \left\|(1+H)^s\xi\right\| \leq \left\|(1+H)^\alpha\xi\right\|,$$

shows that $\xi(\omega) = H^{\mathrm{i}\omega}\xi,\ \omega \in \mathbf{D}_\alpha$, is bounded and continuous on $\mathbf{D}_\alpha$. Let

$$H = \int_0^\infty \lambda \,\mathrm{d}E(\lambda)$$

be the spectral decomposition and set

$$\mathfrak{M} = \bigcup_{n=1}^\infty \big[E(n) - E(1/n)\big]\mathfrak{H}.$$

For each $\eta \in \mathfrak{M}$, setting $\eta(\omega) = H^{\mathrm{i}\omega}\eta,\ \omega \in \mathbf{C}$, we obtain an $\mathfrak{H}$-valued function $\eta(\,\cdot\,)$. We can see from the integral representation,

$$\left(H^{i\omega}\eta \mid \zeta\right) = \int_{1/n}^{n} \lambda^{\mathrm{i}\omega}\,\mathrm{d}\big(E(\lambda)\eta \mid \zeta\big), \qquad \zeta \in \mathfrak{H},$$

with a sufficiently large n, that $\eta(\,\cdot\,)$ is entire. Now, we have, for every $\eta \in \mathfrak{M}$,

$$\big(\xi(\omega) \mid \eta\big) = \big(H^{\mathrm{i}\omega}\xi \mid \eta\big) = \big(\xi \mid H^{-\mathrm{i}\omega}\eta\big) = \overline{\big(\eta(-\mathrm{i}\omega) \mid \xi\big)}.$$

Hence the function: $\omega \in \mathbf{D}_\alpha \mapsto \big(\xi(\omega) \mid \eta\big)$ is holomorphic for every $\eta \in \mathfrak{M}$. Since $\mathfrak{M}$ is dense in $\mathfrak{H}$, $\xi(\,\cdot\,)$ is holomorphic in $\mathbf{D}_\alpha$ by A.1. Thus we obtain (ii).

(ii) $\Longrightarrow$ (i): With $\xi(t) = H^{\mathrm{i}t}\xi$, suppose that $\xi(\,\cdot\,)$ is extended to a function on $\mathbf{D}_\alpha$ as in (ii). By the previous argument, each $\eta \in \mathfrak{D}(H^\alpha)$ gives rise to a bounded continuous function $\eta(\omega) = H^{\mathrm{i}\omega}\eta$ on $\overline{\mathbf{D}}_\alpha$ which is holomorphic in $\mathbf{D}_\alpha$. We consider two functions on $\overline{\mathbf{D}}_\alpha$:

$$\omega \in \overline{\mathbf{D}}_\alpha \mapsto \big(\xi(\omega) \mid \eta\big) \qquad \text{and} \qquad \omega \in \overline{\mathbf{D}}_\alpha \to \big(\xi \mid H^{-\mathrm{i}\overline{\omega}}\eta\big).$$

Then they agree on the real line $\mathbf{R}$. The analyticity implies that they agree on the entire strip $\overline{\mathbf{D}}_\alpha$; hence

$$\big(\xi(\omega) \mid \eta\big) = \big(\xi \mid H^{-\mathrm{i}\overline{\omega}}\eta\big), \qquad \eta \in \mathfrak{D}(H^\alpha),$$

Setting $\omega = -\mathrm{i}\alpha$, we get

$$\big(\xi(-\mathrm{i}\alpha) \mid \eta\big) = \big(\xi \mid H^\alpha\eta\big), \qquad \eta \in \mathfrak{D}(H^\alpha),$$

which shows that $\xi \in \mathfrak{D}(H^\alpha)$ and $H^\alpha\xi = \xi(-\mathrm{i}\alpha)$. Q.E.D.

Lemma 2.4. *Let K be a compact convex subset of a locally convex vector space E. If a function $x : t \in \mathbf{R} \mapsto x(t) \in E$ is continuous and takes values in K, then the Bochner integral for each $r > 0$*

$$x_r = \sqrt{\frac{r}{\pi}} \int_{\mathbf{R}} \mathrm{e}^{-rt^2} x(t)\,\mathrm{d}t$$

belongs to K and $\lim_{r\to\infty} x_r = x(0)$.

PROOF: Since $\mathrm{e}^{-rt^2} > 0$ and $(r/\pi)^{1/2} \int_{-\infty}^{\infty} \mathrm{e}^{-rt^2}\,\mathrm{d}t = 1$, x_r belongs to K by the compactness and the convexity of K. If p is a continuous semi-norm on E, then

$$\begin{aligned} p\left(\sqrt{\frac{r}{\pi}} \int_{\mathbf{R}} \mathrm{e}^{-rt^2} x(t)\,\mathrm{d}t - x(0)\right) &= p\left(\sqrt{\frac{r}{\pi}} \int_{\mathbf{R}} \mathrm{e}^{-rt^2} \big[x(t) - x(0)\big]\,\mathrm{d}t\right) \\ &\leq \sqrt{\frac{r}{\pi}} \int_{\mathbf{R}} \mathrm{e}^{-rt^2} p\big[x(t) - x(0)\big]\,\mathrm{d}t\,. \end{aligned}$$

On the other hand, for any continuous bounded function f on $\mathbf{R}$, we have

$$\lim_{r\to\infty} \sqrt{\frac{r}{\pi}} \int_{\mathbf{R}} \mathrm{e}^{-rt^2} f(t)\,\mathrm{d}t = f(0).$$

Therefore, we conclude that the last expression in the above inequality converges to zero as $r \to \infty$. The convergence holds for every continuous seminorm p, we conclude that x_r converges to $x(0)$ and $r \to \infty$. Q.E.D.

PROOF OF THEOREM 2.2:

(i) Suppose that $\mathfrak{A}$ is a left Hilbert algebra and $\mathfrak{A}_0$ is defined by (1). If $\xi \in \mathfrak{A}_0$, then $\xi = (1 + \Delta^{-1})^{-1}(1 + \Delta^{-1})\xi \in (1 + \Delta^{-1})^{-1}\mathfrak{A}$, so that ξ belongs to $\mathfrak{A}'$ by Lemma 1.17. Hence $\mathfrak{A}_0 \subset \mathfrak{A} \cap \mathfrak{A}'$.

Suppose $\xi \in \mathfrak{A}_0$. We know that the function: $\alpha \in \mathbf{C} \mapsto \Delta^{\mathrm{i}\alpha}\xi \in \mathfrak{H}$ is entire by Lemma 2.3. We want to show that $\Delta^{i\alpha}\xi \in \mathfrak{A}_0$, $\alpha \in \mathbf{C}$. Let $\alpha = r + is$ and $n = [s]$, the integral part of s. We know that $\Delta^{\mathrm{i}r}\xi \in \mathfrak{A}_0$ by Theorem 1.19 and $\Delta^m\xi \in \mathfrak{A}_0$, $m \in \mathbf{Z}$, by definition. For any $\eta \in \mathfrak{A}'$, we have for each $m \in \mathbf{Z}$.

$$\sup_{t\in\mathbf{R}} \big\|\pi_r(\eta)\Delta^{\mathrm{i}t+m}\xi\big\| \leq \big\|\pi_\ell(\Delta^m\xi)\big\|\,\|\eta\|.$$

By the Phragmen-Lindelöf theorem, we have

$$\big\|\pi_r(\eta)\Delta^{\mathrm{i}\alpha}\xi\big\| \leq \max\Big\{\big\|\pi_\ell(\Delta^{-n}\xi)\big\|,\ \big\|\pi_\ell(\Delta^{-n-1}\xi)\big\|\Big\}\|\eta\|,$$

so that $\Delta^{\mathrm{i}\alpha}\xi$ is left bounded. Since $\Delta^{\mathrm{i}\alpha}\xi \in \mathfrak{D}(\Delta^m)$, $m \in \mathbf{Z}$, we conclude that $\Delta^{\mathrm{i}\alpha}\xi \in \mathfrak{A}$, which means $\Delta^{\mathrm{i}\alpha}\xi \in \mathfrak{A}_0$ because α is arbitrary. Therefore, $\{\Delta^{\mathrm{i}\alpha} : \alpha \in \mathbf{C}\}$ leaves $\mathfrak{A}_0$ globally invariant.

By Theorem 1.19, $\{\Delta^{it} : t \in \mathbf{R}\}$ is a one parameter automorphism group of $\mathfrak{A}$. If ξ and η are vectors of $\mathfrak{A}_0$, then the uniqueness of holomorphic extension of a function over the reals: $t \mapsto \Delta^{it}(\xi\eta) = (\Delta^{it}\xi)(\Delta^{it}\eta)$ implies that $\Delta^{i\alpha}$, $\alpha \in \mathbf{C}$, is multiplicative. Hence $\{\Delta^{i\alpha} : \alpha \in \mathbf{C}\}$ is a complex one parameter group of automorphisms of $\mathfrak{A}_0$. By now, it is easy to see that $\{\mathfrak{A}_0, \Delta^{i\alpha} : \alpha \in \mathbf{C}\}$ satisfies postulates (a)–(d) in Definition 2.1.

We now show that $\mathfrak{A}_0'' = \mathfrak{A}$. For a fixed $\xi \in \mathfrak{A}$ and each $r > 0$ we set

$$\xi_r = \sqrt{\frac{r}{\pi}} \int_{\mathbf{R}} \mathrm{e}^{-rt^2} \Delta^{it}\xi \, \mathrm{d}t .$$

It follows that $\xi_r \in \mathfrak{D}(\Delta^{i\alpha})$, $\alpha \in \mathbf{C}$, and

$$\Delta^{i\alpha}\xi_r = \sqrt{\frac{r}{\pi}} \int_{\mathbf{R}} \mathrm{e}^{-r\,(t-\alpha)^2} \Delta^{it}\xi \, \mathrm{d}t .$$

We then have for each $\eta \in \mathfrak{A}'$

$$\begin{aligned}
\pi_r(\eta)\Delta^{i\alpha}\xi_r &= \sqrt{\frac{r}{\pi}} \int_{\mathbf{R}} \mathrm{e}^{-r\,(t-\alpha)^2} \pi_r(\eta)\Delta^{it}\xi \, \mathrm{d}t \\
&= \sqrt{\frac{r}{\pi}} \int_{\mathbf{R}} \mathrm{e}^{-r\,(t-\alpha)^2} \pi_\ell(\Delta^{it}\xi)\eta \, \mathrm{d}t \\
&= \sqrt{\frac{r}{\pi}} \left(\int_{\mathbf{R}} \mathrm{e}^{-r\,(t-\alpha)^2} \Delta^{it}\pi_\ell(\xi)\Delta^{-it} \, \mathrm{d}t \right)\eta,
\end{aligned}$$

so that $\Delta^{i\alpha}\xi_r$ is left bounded by Lemma 2.4. Hence $\Delta^{i\alpha}\xi_r \in \mathfrak{A}$ for every $\alpha \in \mathbf{C}$, consequently ξ_r belongs to $\mathfrak{A}_0$. By Lemma 2.4, we get

$$\lim_{r\to\infty} \xi_r = \xi, \qquad \lim_{r\to\infty} \Delta^{\frac{1}{2}}\xi_r = \lim_{r\to\infty} (\Delta^{\frac{1}{2}}\xi)_r = \Delta^{\frac{1}{2}}\xi_r,$$

so that ξ is approximated by ξ_r in $\sharp$-norm. Therefore, $\mathfrak{A}_0$ is a core of $\Delta^{1/2}$. Furthermore, $\mathfrak{A}_0 = J\mathfrak{A}_0$ is dense in $J\mathfrak{D}^\sharp = \mathfrak{D}^\flat$, so that $\mathfrak{A}_0$ is also a core of $\Delta^{-1/2}$. Therefore, the closure of the $\sharp$-operation in $\mathfrak{A}_0$ agrees with S and that of the $\flat$-operation in $\mathfrak{A}_0$ agrees with F.

If $\xi \in \mathfrak{A}$, then $\pi_\ell(\xi_r)$ converges to σ-strongly as $r \to +\infty$, so that if $\eta \in \mathfrak{H}$ is right bounded with respect to $\mathfrak{A}_0$ then

$$\|\pi_\ell(\xi)\eta\| = \lim_{r\to\infty} \|\pi_\ell(\xi_r)\eta\| \le \lim_{r\to\infty} c\|\xi_r\| \le c\|\xi\|,$$

which implies the right boundedness of η with respect to $\mathfrak{A}$. Therefore, we get $\mathfrak{A}_0' = \mathfrak{A}'$. Thus we completed the proof of the first half.

(ii) Suppose that $\{\mathfrak{A}, U(\alpha) : \alpha \in \mathbf{C}\}$ is a Tomita algebra. Let Δ and J be the associated modular operator and the modular conjugation, and let $\mathfrak{H}$ be the completion of $\mathfrak{A}$. By the group property and Definition 2.1.(c) of $U(\alpha)$, if $t \in \mathbf{R}$, we have, for each $\xi \in \mathfrak{A}$,

$$\begin{aligned}\|U(t)\xi\|^2 &= \big(U(t)\xi \mid U(t)\xi\big) = \big(\xi \mid U(-t)U(t)\xi\big)\\ &= \big(\xi \mid U(0)\xi\big) = \big(\xi \mid \xi\big) = \|\xi\|^2.\end{aligned}$$

Hence $U(t)$ can be extended to a unitary on $\mathfrak{H}$, which is denoted by $U(t)$ again. By condition (a), the map: $\alpha \in \mathbf{C} \mapsto U(\alpha)\xi$ is entire in norm, see A.1, for each $\xi \in \mathfrak{A}$, so that $\{U(t)\}$ is a one parameter unitary group on $\mathfrak{A}$ which leaves $\mathfrak{A}$ invariant. Let H be the infinitesimal generator of $U(t)$, see A.3 and A.4. It then follows from A.5 that H is essentially self-adjoint on $\mathfrak{A}$, that is, H is the closure of its restriction to $\mathfrak{A}$. Furthermore, we have $U(t) = \exp(\mathrm{i}tH),\ t \in \mathbf{R}$, by Stone's Theorem, A.3. Making use of the spectral decomposition of $H = \int_{\mathbf{R}} \lambda\, \mathrm{d}E(\lambda)$, we consider

$$\exp(\mathrm{i}\alpha H) = \int_{\mathbf{R}} \mathrm{e}^{\mathrm{i}\lambda\alpha}\, \mathrm{d}E(\lambda), \qquad \alpha \in \mathbf{C}.$$

Then $\exp(\mathrm{i}\alpha H)$ is closed and extends $U(\alpha)$ for each $\alpha \in \mathbf{C}$, so that $\mathfrak{A} \subset \mathfrak{D}\big(\exp(nH)\big)$ for every $n \in \mathbf{Z}$. We want to show that $\Delta = \exp H$.

First, we observe that condition 2.1.(b) means

$$SU(\alpha)\xi = U(\overline{\alpha})S\xi, \quad \xi \in \mathfrak{A}, \quad \alpha \in \mathbf{C},$$

which means that

$$J\Delta^{\frac{1}{2}}U(\alpha)\xi = U\big(\overline{\alpha}\big)J\Delta^{\frac{1}{2}}\xi, \quad \xi \in \mathfrak{A}, \quad \alpha \in \mathbf{C}.$$

Hence, if $t \in \mathbf{R}$, then the uniqueness of the polar decomposition entails

$$JU(t) = U(t)J \quad \text{and} \quad U(t)\Delta^{\frac{1}{2}} = \Delta^{\frac{1}{2}}U(t), \quad t \in \mathbf{R}.$$

Therefore, the spectral projections of H and Δ commute, which is in turn equivalent to the commutativity of $\{U(t)\}$ and $\{\Delta^{\mathrm{i}s}\}$. Condition 2.1.(d) for $\mathfrak{A}$ then implies that $\mathfrak{A} \subset \mathfrak{D}^\flat$ and $\eta^\flat = U(-\mathrm{i})\eta^\sharp$ for every $\eta \in \mathfrak{A}$, and that

$$\Delta\xi = FS\xi = U(-\mathrm{i})\xi = \exp(H)\xi, \quad \xi \in \mathfrak{A}.$$

We know that $\Delta^{1/2}$ is essentially self-adjoint on $\mathfrak{A}$, so we want to prove that $\mathfrak{A}$ is a core for $\exp(sH),\ s \in \mathbf{R}$. Considering the graph norm in $\mathfrak{D}\big(\exp(sH)\big)$, i.e. with $K = \exp(sH)$

$$\|\xi\|_K = \sqrt{\|\xi\|^2 + \|K\xi\|^2}, \quad \xi \in \mathfrak{D}(K),$$

the density of $(1 + K)\mathfrak{A}$ in $\mathfrak{H}$ is equivalent to the essential self-adjointness of K on $\mathfrak{A}$. Applying Lemma 1.21 to $\{K^{\mathrm{i}t}\}$ and $s = 0$, we have

$$\big(K^{\frac{1}{2}} + K^{-\frac{1}{2}}\big)^{-1} = \int_{\mathbf{R}} \frac{1}{\mathrm{e}^{\pi t} + \mathrm{e}^{-\pi t}} K^{\mathrm{i}t}\, \mathrm{d}t = \int_{\mathbf{R}} \frac{1}{\mathrm{e}^{\pi t} + \mathrm{e}^{-\pi t}} U(st)\, \mathrm{d}t\,.$$

We then compute for each $\xi \in \mathfrak{A}$, keeping in mind that $K = U(-\mathrm{i}s)$:

$$\begin{aligned}
\xi &= (1+K)^{-1}(1+K)\xi = \left(K^{-\frac{1}{2}} + K^{\frac{1}{2}}\right)^{-1} K^{-\frac{1}{2}}(1+K)\xi \\
&= \int_{\mathbf{R}} \frac{1}{\mathrm{e}^{\pi t} + \mathrm{e}^{-\pi t}} U(st) K^{-\frac{1}{2}}(1+K)\xi \,\mathrm{d}t \\
&= \int_{\mathbf{R}} (1+K)\frac{1}{\mathrm{e}^{\pi t} + \mathrm{e}^{-\pi t}} U\left(s\left(t - \frac{\mathrm{i}}{2}\right)\right)\xi \,\mathrm{d}t \,.
\end{aligned}$$

Approximating the above integral by Riemann sum, we conclude that ξ is arbitrarily well approximated by $(1+K)\mathfrak{A}$. Therefore, $\exp(sH)$ is essentially self-adjoint on $\mathfrak{A}$, thus we conclude now that

$$\Delta^{\frac{1}{2}} = \exp\left(\frac{1}{2}H\right),$$

so that $\Delta^{\mathrm{i}t} = U(t)$, $t \in \mathbf{R}$, as required.

Now, we can show easily that $\mathfrak{A}$ is a right Hilbert algebra with involution $\flat$. Namely, we have, for each $\xi, \eta, \zeta \in \mathfrak{A}$,

$$\begin{aligned}
J(\xi\eta) &= \Delta^{\frac{1}{2}} S(\xi\eta) = U\left(-\frac{\mathrm{i}}{2}\right)(\eta^\sharp \xi^\sharp) \\
&= \left(U\left(-\frac{\mathrm{i}}{2}\right)\eta^\sharp\right)\left(U\left(-\frac{\mathrm{i}}{2}\right)\xi^\sharp\right) = (J\eta)(J\xi);
\end{aligned}$$

$$\begin{aligned}
(\xi\eta \mid \zeta) &= (J\zeta \mid J(\xi\eta)) = (J\zeta \mid (J\eta)(J\xi)) = ((J\eta)^\sharp J\zeta \mid J\xi) \\
&= \left(\xi \,\middle|\, J[(\Delta^{-\frac{1}{2}}\eta)J\zeta]\right) = (\xi \mid \zeta\eta^\flat).
\end{aligned}$$

Thus we conclude that $J\mathfrak{A} = \mathfrak{A}$ and

$$\begin{aligned}
\pi_r(\eta) &= J\pi_\ell(J\eta)J, \quad \eta \in \mathfrak{A}, \\
\pi_r(\eta)^* &= \pi_r(\eta^\flat).
\end{aligned}$$

Since $\mathfrak{A}$ is a core of $\Delta^{1/2}$, $J\mathfrak{A} = \mathfrak{A}$ is also a core of $\Delta^{-1/2}$. Therefore, we conclude that $\pi_r(\mathfrak{A})$ generates $\mathcal{R}_\ell(\mathfrak{A})'$, i.e.

$$\mathcal{R}_r(\mathfrak{A}) = \mathcal{R}_\ell(\mathfrak{A})',$$

and that $\mathfrak{A}$ is equivalent to $\mathfrak{A}'$ as a right Hilbert algebra. Q.E.D.

Exercise VI.2

1) Show that the Tomita algebra $\mathfrak{A}$ constructed in Theorem 2.2 is complete with respect to the locally convex topology determined by the system of norms:

$$p_n(\xi) = \max\left\{\left\|\pi_\ell(\Delta^n\xi)\right\|, \|\Delta^n\xi\|\right\}, \quad n \in \mathbf{Z}.$$

§ 3 Direct Integral of Left Hilbert Algebras and Tomita Algebras

As indicated in the section title, we consider direct integrals of left Hilbert algebras and Tomita algebras. We begin by the following:

Definition 3.1. Let $\{\Gamma, \mu\}$ be a σ-finite standard measure space in the sense that μ is a σ-finite measure defined in a standard Borel space Γ. A field $\left\{\mathfrak{A}(\gamma) : \gamma \in \Gamma\right\}$ of left Hilbert algebras over $\{\Gamma, \mu\}$ is said to be *measurable* if there are countably many vector fields $\left\{\xi_j(\gamma) : \gamma \in \Gamma,\ j \in \mathbf{N}\right\}$ with the following properties:

(i) $\xi_j(\gamma) \in \mathfrak{A}(\gamma)$ for μ-almost every γ and the fields: $\gamma \mapsto \xi_j(\gamma)$ are fundamental for the field $\mathfrak{H}(\gamma)$ of Hilbert spaces where $\mathfrak{H}(\gamma)$ is the completion of $\mathfrak{A}(\gamma)$, cf. Definition IV.8.9;
(ii) The fields $\gamma \mapsto \xi_j(\gamma)^\sharp$ are μ-measurable;
(iii) $\left\{\xi_j(\gamma) : j \in \mathbf{N}\right\}$ is dense in $\mathfrak{A}(\gamma)$ with respect to the $\sharp$-norm for μ-almost every $\gamma \in \Gamma$;
(iv) The fields: $\gamma \mapsto \xi_j(\gamma)\xi_k(\gamma)$ are μ-measurable.

Such a set of vectors fields $\gamma \mapsto \xi_j(\gamma),\ j = 1, 2, \dots$ is called *fundamental* for the field: $\gamma \mapsto \mathfrak{A}(\gamma)$ of left Hilbert algebras.

A *measurable field of right* Hilbert algebras is defined similarly.

Suppose that $\gamma \in \Gamma \mapsto \mathfrak{A}(\gamma)$ is a measurable field of left Hilbert algebras over a standard measure space $\{\Gamma, \mu\}$. Let $\mathfrak{H}(\gamma)$ be the completion of $\mathfrak{A}(\gamma)$, and $\mathfrak{D}^\sharp(\gamma)$ the definition domain of the closure $S(\gamma)$ of the $\sharp$-operation of $\mathfrak{A}(\gamma)$. We use the notations: $\Delta(\gamma),\ J(\gamma),\ \mathfrak{D}^\flat(\gamma),\ \mathfrak{A}'(\gamma)$ and so on in the obvious way. It follows that $\gamma \in \Gamma \mapsto \mathfrak{D}^\sharp(\gamma)$ and $\gamma \in \Gamma \mapsto \mathfrak{D}^\flat(\gamma)$ are both measurable fields of Hilbert spaces.

Lemma 3.2. *In the above setting, the fields:* $\gamma \mapsto J(\gamma)\xi_j(\gamma),\quad j \in \mathbf{N}$, *are measurable.*

Instead of the lemma, we discuss a little more general result concerning fields of closed operators (but they are essentially the same result). For each densely defined closed operator T on a Hilbert space $\mathfrak{H}$, we consider the new inner product in the domain $\mathfrak{D}(T)$ defined as follows:

$$(\xi \mid \eta)_T = (\xi \mid \eta) + (T\xi \mid T\eta), \qquad \xi, \eta \in \mathfrak{D}(T). \tag{1}$$

We call the norm $\|\cdot\|_T$ of $\mathfrak{D}(T)$ the *graph norm* of T.

Lemma 3.3. *Let* $\gamma \in \Gamma \mapsto \mathfrak{H}(\gamma)$ *be a measurable field of Hilbert spaces on a standard measure space* $\{\Gamma, \mu\}$. *For a field:* $\gamma \in \Gamma \mapsto T(\gamma)$ *of densely defined closed operators on* $\mathfrak{H}(\gamma)$, *the following conditions are equivalent:*

(i) *The field:* $\gamma \in \Gamma \mapsto \mathfrak{D}(T(\gamma))$ *of domains is measurable as a field of Hilbert spaces equipped with the graph norms. Namely, there exists a sequence* $\{\xi_n(\gamma)\}$ *of measurable vector fields such that* $\{\xi_n(\gamma)\}$ *is dense in* $\mathfrak{D}(T(\gamma))$ *with respect to the graph norm* (1) *for almost every* $\gamma \in \Gamma$. *Furthermore,* $\{T(\gamma)\xi(\gamma)\}$ *is measurable if* $\{\xi(\gamma)\}$ *is measurable in the field* $\{\mathfrak{D}(T(\gamma))\}$.

(ii) *The field:* $\gamma \in \Gamma \mapsto \mathfrak{G}(T(\gamma)) \subset \mathfrak{H}(\gamma) \oplus \mathfrak{H}(\gamma)$ *of the graphs of* $\{T(\gamma)\}$ *is a measurable field of closed subspaces of* $\mathfrak{H}(\gamma) \oplus \mathfrak{H}(\gamma)$;

(iii) *With the polar decomposition:*

$$T(\gamma) = U(\gamma)H(\gamma),$$

the fields: $\gamma \in T \mapsto U(\gamma) \in \mathcal{L}(\mathfrak{H}(\gamma))$ *and* $\gamma \in \Gamma \mapsto (1 + H(\gamma))^{-1} \in \mathcal{L}(\mathfrak{H}(\gamma))$ *are both measurable operator fields.*

PROOF: As in Chapter IV, §8, we may assume that $\mathfrak{H}(\gamma) = \mathfrak{H}_0$ with a fixed separable Hilbert space $\mathfrak{H}_0$.

(i) $\Longleftrightarrow$ (i): This equivalence follows from Theorem IV.8.13.

(ii) $\Longrightarrow$ (iii): Let $P(\gamma)$ be the projection of $\mathfrak{H}(\gamma) \oplus \mathfrak{H}(\gamma)$ onto $\mathfrak{G}(T(\gamma))$. Writing $P(\gamma)$ as a 2×2-matrix with operator entries, we set

$$P(\gamma) = \begin{pmatrix} a(\gamma) & b(\gamma) \\ b(\gamma)^* & d(\gamma) \end{pmatrix}. \tag{2}$$

The fact that $P(\gamma)(\mathfrak{H}(\gamma) \oplus \mathfrak{H}(\gamma)) = \mathfrak{G}(T(\gamma))$, implies that

$$T(\gamma)a(\gamma) = b(\gamma)^*, \qquad T(\gamma)b(\gamma) = d(\gamma). \tag{3}$$

Since $P(\gamma) = P(\gamma)^2$, we have

$$\begin{gathered} a(\gamma) = a(\gamma)^2 + b(\gamma)b(\gamma)^*, \qquad b(\gamma) = a(\gamma)b(\gamma) + b(\gamma)d(\gamma), \\ d(\gamma) = b(\gamma)^*b(\gamma) + d(\gamma)^2. \end{gathered} \tag{4}$$

The equations (3) and (4) together imply that

$$\begin{gathered} a(\gamma) = (1 + T(\gamma)^*T(\gamma))^{-1}, \qquad b(\gamma) = T(\gamma)^*(1 + T(\gamma)T(\gamma)^*)^{-1}, \\ d(\gamma) = T(\gamma)T(\gamma)^*(1 + T(\gamma)T(\gamma)^*)^{-1}. \end{gathered} \tag{5}$$

Thus, we obtain

$$P(\gamma) = \begin{pmatrix} [1 + T(\gamma)^*T(\gamma)]^{-1} & T(\gamma)^*(1 + T(\gamma)T(\gamma)^*)^{-1} \\ T(\gamma)(1 + T(\gamma)^*T(\gamma))^{-1} & T(\gamma)T(\gamma)^*(1 + T(\gamma)T(\gamma)^*)^{-1} \end{pmatrix}$$

Therefore, the fields: $\gamma \in \Gamma \mapsto (1 + T(\gamma)^*T(\gamma))^{-1} \in \mathcal{L}(\mathfrak{H}(\gamma))$ and $\gamma \in \Gamma \mapsto T(\gamma)(1 + T(\gamma)^*T(\gamma))^{-1} \in \mathcal{L}(\mathfrak{H}(\gamma))$ are both measurable. Therefore, if χ_λ is the characteristic function of $]-\infty, \lambda]$, then the field: $\gamma \in \Gamma \mapsto \chi_\lambda((1 + T(\gamma)^*T(\gamma))^{-1}) \in \mathcal{L}(\mathfrak{H}(\gamma))$ is measurable. It then follows that the field: $\gamma \mapsto \chi_\lambda(H(\gamma)) = 1 - \chi_\mu((1 + T(\gamma)^*T(\gamma))^{-1})$ is measurable where $\mu = \sqrt{\frac{1}{\lambda} - 1}$, so that $\{(\varepsilon + H(\gamma))^{-1}\}$ is a measurable field for any $\varepsilon > 0$.

Now, if $\{\xi(\gamma)\}$ is a measurable vector field of $\{\mathfrak{H}(\gamma)\}$, then

$$U(\gamma)\xi(\xi) = \lim_{\varepsilon\to 0} T(\gamma)\big(\varepsilon + H(\gamma)\big)^{-1}\xi(\gamma), \quad \gamma \in \Gamma,$$

so that $\{U(\gamma)\xi(\gamma)\}$ is measurable. Hence $U(\gamma)$ is also measurable.

(iii) $\Longrightarrow$ (i): Suppose that (iii) holds. Since $\mathfrak{D}\big(T(\gamma)\big) = \big(1 + H(\gamma)\big)^{-1}\mathfrak{H}(\gamma)$, we conclude that $\{(1+H(\gamma))^{-1}\xi_n(\gamma)\}$, when $\{\xi_n(\gamma)\}$ runs a fundamental sequence of measurable vector fields of $\{\mathfrak{H}(\gamma)\}$, gives the measurable field $\{\mathfrak{D}(T(\gamma))\}$ of Hilbert spaces, and that $\{T(\gamma)(1+H(\gamma))^{-1}\xi_n(\gamma)\} = \{U(\gamma)H(\gamma)(1+H(\gamma))^{-1}\xi_n(\gamma)\}$ are measurable vector fields. Thus condition (i) is fulfilled. Q.E.D.

Definition 3.4. A field $\{T(\gamma)\}$ of densely defined closed operators defined on the measurable field $\{\mathfrak{H}(\gamma)\}$ of Hilbert spaces is said to be *measurable* if any of the three conditions in Lemma 3.3 holds. The direct integral:

$$T = \int_\Gamma^\oplus T(\gamma)\,\mathrm{d}\mu(\gamma) \quad \text{on} \quad \mathfrak{H} = \int_\Gamma^\oplus \mathfrak{H}(\gamma)\,\mathrm{d}\mu(\gamma) \tag{6}$$

is naturally defined as an operator with domain $\mathfrak{D}$ consisting of those vectors $\xi = \int_\Gamma^\oplus \xi(\gamma)\,\mathrm{d}\mu(\gamma)$ such that $\xi(\gamma) \in \mathfrak{D}\big(T(\gamma)\big)$ for almost every $\gamma \in \Gamma$ and

$$\int_\Gamma^\oplus \|T(\gamma)\xi(\gamma)\|^2\,\mathrm{d}\mu(\gamma) < +\infty. \tag{7}$$

It is easy to see that the graph $\mathfrak{G}(T)$ is given by:

$$\mathfrak{G}(T) = \int_\Gamma^\oplus \mathfrak{G}\big(T(\gamma)\big)\,\mathrm{d}\mu(\gamma)\,. \tag{8}$$

If each $T(\gamma)$ is normal in the sense that $T(\gamma)^*T(\gamma) = T(\gamma)T(\gamma)^*$, then for any continuous bounded function f on **C**, $\{f(T(\gamma))\}$ is a bounded measurable field of bounded operators because $f\big(T(\gamma)\big) = g\big(U(\gamma)H(\gamma)(1 + H(\gamma))^{-1}\big)$ where g is a bounded continuous function on the unit disk such that $g\big(z/(1 + |z|)\big) = f(z)$. Since the class of functions f such that $\{f(T(\gamma))\}$ is a measurable field of closed operators is closed under pointwise limit, $\{f(T(\gamma))\}$ is measurable for any Borel function over **C**.

Now, the conclusion of Lemma 3.2 follows immediately from Lemma 3.3, even though the involved operators are conjugately linear.

Returning to the original setting, we see that the field $\{\Delta(\gamma)\}$ of modular operators is measurable in the above sense. Making use of the Gram-Schmidt orthogonalization, we may take a fundamental sequence $\{\xi_n(\gamma)\}$ of measurable fields in $\{\mathfrak{A}(\gamma)\}$ so that $\{\xi_n(\gamma)\}$ are a field of orthonormal bases for $\{\mathfrak{H}(\gamma)\}$ in the sense that

$$\eta(\gamma) = \sum_{n=1}^{\infty} \big(\eta(\gamma) \mid \xi_n(\gamma)\big)\xi_n(\gamma) \tag{9}$$

for every $\eta(\gamma) \in \mathfrak{H}(\gamma)$ and $\big(\xi_j(\gamma) \mid \xi_k(\gamma)\big) = 0,\ j \neq k$, and $\|\xi_n(\gamma)\| = 1$ if $n \leq \dim \mathfrak{H}(\gamma)$, $\|\xi_n(\gamma)\| = 0$ if $n > \dim \mathfrak{H}(\gamma)$. We call such a sequence $\{\xi_n(\gamma)\}$ of fields an *orthonormal fundamental* sequence of measurable fields. With this type of fundamental sequence, we can conclude the following result easily.

Lemma 3.5. *If $\gamma \mapsto \xi(\gamma) \in \mathfrak{A}(\gamma)$ is measurable, then the field $\{\pi_\ell(\xi(\gamma))\}$ of operators in $\mathfrak{H}(\gamma)$ is measurable.*

Theorem 3.6. *If $\{\mathfrak{A}(\gamma)\}$ is a measurable field of left Hilbert algebras on a standard measure space $\{\Gamma(\gamma), \mu\}$ then $\{\mathfrak{A}(\gamma)'\}$ is also a measurable field of right Hilbert algebras.*

PROOF: We know already that the field $\{\Delta(\gamma)\}$ of modular operators is measurable. Lemma 3.3 assures that any power $\{\Delta(\gamma)^\alpha\}$ of $\{\Delta(\gamma)\}$ is measurable. In particular, $\{\Delta(\gamma)^{-1/2}\}$ is measurable. Hence, Lemma 3.2 yields that if $\{\eta(\gamma)\}$ is a measurable field in $\{\mathfrak{D}(\Delta(\gamma)^{-1/2})\}$ then $\{\eta(\gamma)^\flat\} = \{J(\gamma)\Delta(\gamma)^{-1/2}\eta(\gamma)\}$ is measurable. We shall use the notations: $\mathfrak{D}^\sharp(\gamma)$ and $\mathfrak{D}^\flat(\gamma)$ for $\mathfrak{D}(\Delta(\gamma)^{1/2})$ and $\mathfrak{D}(\Delta(\gamma)^{-1/2})$ respectively.

For each measurable field $\{\eta(\gamma)\}$ in $\{\mathfrak{D}^\flat(\gamma)\}$, the field $\{\pi_r(\eta(\gamma))\}$ of closed operators defined in Lemma 1.10 is measurable by Lemma 3.5 and Lemma 3.3. We fix such a field $\{\eta(\gamma)\}$ and set $a(\gamma) = \pi_r(\eta(\gamma))$. Let $a(\gamma) = u(\gamma)h(\gamma) = k(\gamma)u(\gamma)$ be the two sided polar decomposition. By Lemma 1.12, $f(k(\gamma))\eta(\gamma) \in \mathfrak{A}(\gamma)'$ for every $f \in \mathcal{K}(0,\infty)$ and $\big(f(k(\gamma))\eta(\gamma)\big)^\flat = \overline{f}(h(\gamma))\eta(\gamma)^\flat$. Lemma 3.3 says that the field $\{f(k(\gamma))\eta(\gamma)\}$ is measurable in the field $\{\mathfrak{H}(\gamma)\}$. If $\{f_n\}$ is a sequence in $\mathcal{K}_+(0,\infty)$ such that $f_n(\lambda) \nearrow 1$, then $f_n(k(\gamma))\eta(\gamma) \to \eta(\gamma)$ and $\big[f_n(k(\gamma))\eta(\gamma)\big]^\flat \to \eta(\gamma)^\flat$ as $n \to \infty$. Thus, choosing a fundamental sequence $\{\eta_j(\gamma)\}$ in $\{\mathfrak{D}^\flat(\gamma)\}$ and a sequence $\{f_n\}$ in $\mathcal{K}_+(0,\infty)$ with $f_n(\lambda) \nearrow 1$, the sequence $\{f_n\big([\pi_r(\eta_j(\gamma))\pi_r(\eta_j(\gamma))^*]^{1/2}\big)\eta_j(\gamma)\}$ gives a fundamental sequence of measurable fields for $\{\mathfrak{A}(\gamma)'\}$. Therefore, $\{\mathfrak{A}(\gamma)'\}$ is a measurable field of right Hilbert algebras. Q.E.D.

Definition 3.7. Given a measurable field $\{\mathfrak{A}(\gamma)\}$ of left Hilbert algebras over a standard measure space $\{\Gamma, \mu\}$ with $\mathfrak{H}(\gamma)$ the completion of $\mathfrak{A}(\gamma)$, we consider the direct integral:

$$\mathfrak{H} = \int_\Gamma^\oplus \mathfrak{H}(\gamma)\, d\mu(\gamma)\,. \tag{10}$$

Let $\mathfrak{A}$ be the set of $\xi = \int_\Gamma^\oplus \xi(\gamma)\, d\mu(\gamma) \in \mathfrak{H}$ with the properties:

(i) $\xi(\gamma) \in \mathfrak{A}(\gamma)$ for $\mu - a \cdot e \cdot \gamma \in \Gamma$;

(ii) $\int_\Gamma \|\xi(\gamma)^\sharp\|^2 \, \mathrm{d}\mu(\gamma) < +\infty$;

(iii) $\operatorname{ess\,sup}\|\pi_\ell(\xi(\gamma))\| < +\infty$.

We call $\mathfrak{A}$ the *direct integral* of $\{\mathfrak{A}(\gamma)\}$ and write

$$\mathfrak{A} = \int_\Gamma^\oplus \mathfrak{A}(\gamma) \, \mathrm{d}\mu(\gamma) \,. \tag{11}$$

The *direct integral* of *right* Hilbert algebras is similarly defined.

Theorem 3.8.

(i) *The direct integral $\mathfrak{A}$ of a measurable field of left Hilbert algebras $\{\mathfrak{H}(\gamma)\}$ is a left Hilbert algebra equipped with the obvious algebraic structure such that the completion of $\mathfrak{A}$ is the direct integral $\mathfrak{H}$ given by* (10).

(ii)
$$\left\{ \int_\Gamma^\oplus \mathfrak{A}(\gamma) \, \mathrm{d}\mu(\gamma) \right\}' = \int_\Gamma^\oplus \mathfrak{A}(\gamma)' \, \mathrm{d}\mu(\gamma) \,. \tag{12}$$

PROOF: The proof of assertion (i) is routine, so we leave it to the reader.

(ii) By Theorem 3.6, the field $\{\mathfrak{A}(\gamma)'\}$ is measurable, so the both sides of (12) have meaning. Let $\mathfrak{A}_1$ denote the right hand side of (12). It is clear that $\mathfrak{A}_1 \subset \mathfrak{A}'$. Let $\{\xi_j(\gamma)\}$ (resp. $\{\eta_j(\gamma)\}$) be a fundamental sequence in $\{\mathfrak{A}(\gamma)\}$ (resp. $\{\mathfrak{A}(\gamma)'\}$) such that $\xi_j = \int_\Gamma^\oplus \xi_j(\gamma) \, \mathrm{d}\mu(\gamma) \in \mathfrak{A}$ (resp. $\eta_j = \int_\gamma^\oplus \eta_j(\gamma) \, \mathrm{d}\mu(\gamma) \in \mathfrak{A}_1$). For short, we shall often write $\xi \sim \{\xi(\gamma)\}$ instead of $\xi = \int_\Gamma^\oplus \xi(\gamma) \, \mathrm{d}\mu(\gamma)$.

Suppose that $\eta \sim \{\eta(\gamma)\} \in \mathfrak{A}'$ and $\eta^\flat \sim \{\eta^\flat(\gamma)\}$. Then for all j and a bounded measurable function f on Γ,

$$\begin{aligned}\int_\Gamma f(\gamma)(\xi_j(\gamma) \mid \eta^\flat(\gamma)) \, \mathrm{d}\mu(\gamma) &= (\xi_j \mid (f\eta)^\flat) = (f\eta \mid \xi_j^\sharp) \\ &= \int_\Gamma f(\gamma)(\eta(\gamma) \mid \xi_j(\gamma)^\sharp) \, \mathrm{d}\mu(\gamma),\end{aligned}$$

so that $\eta(\gamma) \in \mathfrak{D}^\flat(\gamma)$ for μ-almost every, and $\eta^\flat(\gamma) = \eta(\gamma)^\flat$ for μ-almost every $\gamma \in \Gamma$. Since $\eta^\flat \in \mathfrak{H}$, $\gamma \mapsto \|\eta(\gamma)^\flat\|^2$ is integrable.

We now check the essential boundedness of $\|\pi_r(\eta(\gamma))\|$ and also prove that $\pi_r(\eta) \sim \{\pi_r(\eta(\gamma))\}$. For any $f \in L^\infty(\Gamma, \mu)$, we have

$$\begin{aligned}\int_\Gamma f(\gamma)\Big((\pi_r(\eta)\xi_j)(\gamma) \,\Big|\, \xi_k(\gamma)\Big) \, \mathrm{d}\mu(\gamma) &= \int_\Gamma f(\gamma)\Big(\pi_\ell(\xi_j)\eta)(\gamma) \,\Big|\, \xi_k(\gamma)\Big) \, \mathrm{d}\mu(\gamma) \\ &= \int_\Gamma f(\gamma)\Big(\pi_\ell(\xi_j(\gamma))\eta(\gamma) \,\Big|\, \xi_k(\gamma)\Big) \, \mathrm{d}\mu(\gamma) \\ &= \int_\Gamma f(\gamma)\Big(\pi_r(\eta(\gamma))\xi_j(\gamma) \,\Big|\, \xi_k(\gamma)\Big) \, \mathrm{d}\mu(\gamma),\end{aligned}$$

so that $(\pi_r(\eta)\xi_j)(\gamma) = \pi_r(\eta)\xi_j(\gamma)$ for μ-almost every. Hence $\eta(\gamma) \in \mathfrak{A}(\gamma)'$ for μ-almost every γ and $\pi_r(\eta) \sim \{\pi_r(\eta(\gamma))\}$. Thus $\{\|\pi_r(\eta(\gamma))\|\}$ is essentially bounded. We then conclude that $\eta \in \mathfrak{A}_1$. Q.E.D.

Corollary 3.9. *The direct integral of left Hilbert algebras* $\{\mathfrak{A}(\gamma)\}$ *over a standard measure space* $\{\Gamma, \mu\}$ *is full if and only if the component algebras* $\mathfrak{A}(\gamma)$ *are full for* μ *-almost every* $\gamma \in \Gamma$.

The following result follows now immediately.

Proposition 3.10. *If* $\mathfrak{A}$ *is the direct integral of a measurable field* $\mathfrak{A}(\gamma)$ *of left Hilbert algebras over a standard measure space* $\{\Gamma, \mu\}$, *then the field* $\{\mathcal{R}_\ell(\mathfrak{A}(\gamma))\}$ *of the left von Neumann algebras is measurable and*

$$\mathcal{R}_\ell(\mathfrak{A}) = \int_\Gamma^\oplus \mathcal{R}_\ell(\mathfrak{A}(\gamma))\, \mathrm{d}\mu(\gamma)\,. \tag{13}$$

We now consider direct integrals of Tomita algebras. Let us begin by formulating the measurability of a field of Tomita algebras.

Definition 3.11. A field $\{\mathfrak{T}(\gamma)\}$ of Tomita algebras over $\{\Gamma, \mu\}$ is called *measurable* if it is measurable as a field of left Hilbert algebras. The *direct integral* $\mathfrak{T}$ of a measurable field $\{\mathfrak{T}(\gamma)\}$ of Tomita algebras is defined to be the subset of $\mathfrak{A} = \int_\Gamma^\oplus \mathfrak{T}(\gamma)\, \mathrm{d}\mu(\gamma)$, the left Hilbert algebra direct integral, consisting of those vectors $\xi \sim \{\xi(\gamma)\} \in \mathfrak{A}$ such that for each $n \in \mathbf{Z}$, $\Delta^n\xi \in \mathfrak{A}$,[2] where

$$\Delta = \int_\Gamma^\oplus \Delta(\gamma)\, \mathrm{d}\mu(\gamma) \tag{14}$$

in the sense of (6). We write

$$\mathfrak{T} = \int_\Gamma^\oplus \mathfrak{T}(\gamma)\, \mathrm{d}\mu(\gamma)\,. \tag{15}$$

and call it the *Tomita algebra direct integral* when we need to distinguish it from the left Hilbert algebra direct integral. We define the obvious algebraic structure in $\mathfrak{T}$.

Proposition 3.12. *The Tomita algebra direct integral* $\mathfrak{T}$ *of a measurable field* $\{\mathfrak{T}(\gamma)\}$ *of Tomita algebras over* $\{\Gamma, \mu\}$ *is indeed a Tomita algebra which is equivalent to the left Hilbert algebra direct integral* $\mathfrak{A}$ *of* $\{\mathfrak{T}(\gamma)\}$.

PROOF: The proof for the fact that $\mathfrak{T}$ is a Tomita algebra is routine, so we leave it to the reader. We prove here only the equivalence of $\mathfrak{T}$ and $\mathfrak{A}$.

[2] Of course, it should be understood that ξ belongs to $\mathfrak{D}(\Delta^n)$.

Writing Γ as a union of countable disjoint Borel subsets with finite measure, we may assume that Γ has finite measure. Let $\xi \sim \{\xi(\gamma)\}$ be a vector of $\mathfrak{A}$ and $\{\eta_j(\gamma)\}$ be a sequence of measurable vector fields in $\{\mathfrak{T}(\gamma)\}$ such that for each $\gamma \in \Gamma$, $\{\eta_j(\gamma) : j \in \mathbf{N}\}$ is dense in the unit ball of the completion $\mathfrak{H}(\gamma)$ of $\mathfrak{T}(\gamma)$. We then observe that

$$g_n(\gamma) = \|\Delta(\gamma)^n \xi(\gamma)\| = \sup_j \left|\left(\Delta(\gamma)^n \xi(\gamma) \mid \eta_j(\gamma)\right)\right|$$

$$f_n(\gamma) = \left\|\pi_\ell\left(\Delta(\gamma)^n \xi(\gamma)\right)\right\| = \sup_j \left\|\pi_\ell\left(\Delta(\gamma)^n \xi(\gamma)\right)\eta_j(\gamma)\right\|$$

are both Borel functions of $\gamma \in \Gamma$ for each $n \in \mathbf{Z}$. We set $h_n(\gamma)=\max\{f_j(\gamma), g_k(\gamma) : 1 \leq j,\ k \leq n\}$. Then $\{h_n\}$ is an increasing sequence of Borel functions on Γ. Since $\{\Gamma, \mu\}$ is standard, we may assume that Γ is a closed subset of the unit interval $[0, 1]$. By Lusin's theorem, for any $\varepsilon > 0$ there exists a compact set Γ_ε such that h_n are all continuous on Γ_ε, in particular they are bounded, and $\mu(\Gamma_\varepsilon^c) < \varepsilon$. It is possible to arrange that $\{\Gamma_n\}$ is an increasing sequence of compact sets in Γ such that, $\{h_k\}$ are all continuous on Γ_n and $\mu(\Gamma_n^c) < 1/n$. We then set $\xi_n(\gamma) = \xi(\gamma)$ if $\gamma \in \Gamma_n$ and zero otherwise. Since each h_k is bounded on Γ_n, $\xi_n \sim \{\xi_n(\gamma)\}$ belongs to $\mathfrak{T}$, and $\{\xi_n\}$ converges to ξ in the graph norm of $\Delta^{1/2}$. This proves that $\mathfrak{T}$ is dense in $\mathfrak{A}$ with respect to the graph norm of $\Delta^{1/2}$, i.e., the $\sharp$-norm of $\mathfrak{A}$. This shows that $\mathfrak{T}$ is equivalent to $\mathfrak{A}$. Q.E.D.

Theorem 3.13. *Let $\{\mathfrak{A}(\gamma)\}$ be a measurable field of full left Hilbert algebras over a standard measure space $\{\Gamma, \mu\}$ with $\mathfrak{A} = \int_\Gamma^\oplus \mathfrak{A}(\gamma)\, \mathrm{d}\mu(\gamma)$, the left Hilbert algebra direct integral, and let $\mathfrak{A}_0$, $\mathfrak{A}(\gamma)_0$ be the maximal Tomita algebras associated with $\mathfrak{A}$ and $\mathfrak{A}(\gamma)$ respectively. Then $\{\mathfrak{A}(\gamma)_0\}$ is a measurable field of Tomita algebras and we have*

$$\mathfrak{A}_0 = \left(\int_\Gamma^\oplus \mathfrak{A}(\gamma)\, \mathrm{d}\mu(\gamma)\right)_0 = \int_\Gamma^\oplus \mathfrak{A}(\gamma)_0\, \mathrm{d}\mu(\gamma) \tag{16}$$

as the Tomita algebra direct integral.

PROOF: Let

$$\mathfrak{H} = \int_\Gamma^\oplus \mathfrak{H}(\gamma)\, \mathrm{d}\mu(\gamma), \qquad \Delta = \int_\Gamma^\oplus \Delta(\gamma)\, \mathrm{d}\mu(\gamma) \tag{17}$$

with $\mathfrak{H}(\gamma)$ the completion of $\mathfrak{A}(\gamma)$ and $\Delta(\gamma)$ the modular operator of $\mathfrak{A}(\gamma)$. For each $r > 0$, we set

$$f_r(\lambda) = \sqrt{\frac{r}{\pi}} \int_{\mathbf{R}} \mathrm{e}^{-rt^2} \lambda^{it}\, \mathrm{d}t = \exp\left(-(\log\lambda)^2/4r\right), \quad \lambda > 0. \tag{18}$$

The discussion in the previous section shows that

$$f_r(\Delta)\mathfrak{A} \subset \mathfrak{A}_0, \qquad f_r\big(\Delta(\gamma)\big)\mathfrak{A}(\gamma) \subset \mathfrak{A}(\gamma)_0$$

and that

$$\lim_{r\to\infty} f_r(\Delta)\xi = \xi, \qquad \lim_{r\to\infty} \Delta^{\frac{1}{2}} f_r(\Delta)\xi = \Delta^{\frac{1}{2}}\xi$$

if $\xi \in \mathfrak{D}(\Delta^{1/2})$. This means that if $\{\xi_k(\gamma)\}$ is a fundamental sequence of measurable vector fields in $\{\mathfrak{A}(\gamma)\}$, then $\{f_n\big(\Delta(\gamma)\big)\xi_k(\gamma)\}_{k,n}$ gives a fundamental sequence of measurable fields in $\{\mathfrak{A}(\gamma)_0\}$. Hence $\{\mathfrak{A}(\gamma)_0\}$ is measurable.

We now let $\mathfrak{T}$ be the Tomita algebra direct integral of $\{\mathfrak{A}(\gamma)_0\}$. We then have $\mathfrak{T} \subset \mathfrak{A}_0$. Now, let $\xi \sim \{\xi(\gamma)\}$ be a vector in $\mathfrak{A}_0$. Since $\Delta^n\xi \in \mathfrak{A}$, $\Delta(\gamma)^n\xi(\gamma)$ belongs to $\mathfrak{A}(\gamma)$ for almost every $\gamma \in \Gamma$. Hence $\xi(\gamma) \in \mathfrak{A}(\gamma)_0$ for almost every $\gamma \in \Gamma$. This means that $\Delta^n\xi$ belongs to the left Hilbert algebra direct integral of $\{\mathfrak{A}(\gamma)_0\}$, so that $\xi \in \mathfrak{T}$ by definition. Thus we obtain $\mathfrak{A}_0 = \mathfrak{T}$. Q.E.D.

We now discuss the existence and the uniqueness of direct integral decomposition.

Theorem 3.14 (Existence of Disintegration). *Let $\mathfrak{A}$ be a full left Hilbert algebra with separable completion $\mathfrak{H}$. Let $\mathcal{M} = \mathcal{R}_\ell(\mathfrak{A})$. If $\mathcal{A}$ is a von Neumann subalgebra of the center $\mathcal{Z} = \mathcal{M} \cap \mathcal{M}'$, then there exists a measurable field $\{\mathfrak{A}(\gamma)\}$ of full left Hilbert algebras on a standard measure space $\{\Gamma, \mu\}$ such that*

(i) $\mathfrak{A} = \int_\Gamma^\oplus \mathfrak{A}(\gamma)\, d\mu(\gamma)$; (19)

(ii) *$\mathcal{A}$ is the diagonal algebra.*

PROOF: By Theorem IV.8.21, we have that

$$\{\mathcal{M}, \mathfrak{H}\} = \int_\Gamma^\oplus \{\mathcal{M}(\gamma), \mathfrak{H}(\gamma)\}\, d\mu(\gamma) \tag{20}$$

and $\mathcal{A}$ is the diagonal algebra with a standard measure space $\{\Gamma, \mu\}$.

We first choose a countable subset $\mathfrak{A}_1$ of $\mathfrak{A}$ such that

a) $\mathfrak{A}_1$ is dense in $\mathfrak{A}$ with respect to the $\sharp$-norm;
b) $\mathfrak{A}_1$ is an algebra over the complex rational number field $\mathbf{Q}(\mathrm{i})$;
c) $\mathfrak{A}_1$ is closed under the $\sharp$-operation.

We then write $\mathfrak{A}_1 = \{\xi_j : j = 1, 2, \dots\}$. For each j, we choose a measurable vector field $\{\xi_j(\gamma)\}$ so that $\xi_j \sim \{\xi_j(\gamma)\}$ with respect to (20). We set $\mathfrak{A}_1(\gamma) = \{\xi_j(\gamma) : j = 1, 2, \dots\}$ and want to endow $\mathfrak{A}_1(\gamma)$ with the structure of an involutive algebra over $\mathbf{Q}(\mathrm{i})$. We observe that

$$\xi_j(\gamma) + \xi_k(\gamma) = (\xi_j + \xi_k)(\gamma) \qquad \text{for almost every } \gamma \in \Gamma$$
$$\lambda\xi_j(\gamma) = (\lambda\xi_j)(\gamma), \quad \lambda \in \mathbf{Q}(\mathrm{i}), \quad \text{for almost every } \gamma \in \Gamma.$$

Deleting a null set from Γ, we may assume that the above relations hold for every $\gamma \in \Gamma$. We then set

$$\xi_j(\gamma)\xi_k(\gamma) = (\xi_j\xi_k)(\gamma), \qquad \xi_j(\gamma)^\sharp = \xi_j^\sharp(\gamma). \tag{21}$$

To see that this definition makes sense for almost every $\gamma \in \Gamma$, for each i, j, k, ℓ we set

$$N_{i,j,k,\ell} = \Big\{\gamma \in \Gamma : \xi_i(\gamma) = \xi_k(\gamma),\ \xi_j(\gamma) = \xi_\ell(\gamma),\ (\xi_i\xi_j)(\gamma) \neq (\xi_k\xi_\ell)(\gamma)\Big\}.$$

Then $N_{i,j,k,\ell}$ is a measurable subset of Γ. We claim that $\mu(N_{i,j,k,\ell}) = 0$. Otherwise, let e be the projection of $\mathcal{A}$ corresponding to $N_{i,j,k,\ell}$. We then have

$$e\xi_i = e\xi_k, \qquad e\xi_j = e\xi_\ell \qquad \text{and} \qquad e(\xi_i\xi_j) \neq e(\xi_k\xi_\ell).$$

Since e is a central projection of $\mathcal{M} = \mathcal{R}_\ell(\mathfrak{A})$, we have

$$e(\xi_i\xi_j) = (e\xi_i)(e\xi_j) = (e\xi_k)(e\xi_\ell) = e(\xi_k\xi_\ell),$$

which is absurd. Thus we have $e = 0$. Namely, $\mu(N_{i,j,k,\ell}) = 0$. Deleting the union of $N_{i,j,k,\ell}$ which is still negligible, we observe that the first equation of (21) makes sense. Similarly, the second half of (21) makes sense for almost every $\gamma \in \Gamma$. Therefore, deleting a null set from Γ, we can conclude that $\mathfrak{A}_1(\gamma)$ is an involutive algebra over $\mathbf{Q}(\mathrm{i})$ for every $\gamma \in \Gamma$ such that

(i) $\big(\xi_j(\gamma)\xi_k(\gamma) \,\big|\, \xi_\ell(\gamma)\big) = \big(\xi_k(\gamma) \,\big|\, \xi_j(\gamma)^\sharp\xi_\ell(\gamma)\big)$,
(ii) the map: $\xi_k(\gamma) \to (\xi_j\xi_k)(\gamma)$ is continuous,
(iii) the set $\mathfrak{A}_1(\gamma)^2$ is dense in $\mathfrak{A}(\gamma)^2$.

To prove the preclosedness of the involution in $\mathfrak{A}_1(\gamma)$, we repeat the same arguments for $\mathfrak{A}'$ so that we get a countable involutive subalgebra $\mathfrak{A}_1'$ of $\mathfrak{A}'$ over $\mathbf{Q}(\mathrm{i})$ and γ-component $\mathfrak{A}_1'(\gamma)$. We then observe that

$$\big(\xi_j^\sharp(\gamma) \,\big|\, \eta_k(\gamma)\big) = \big(\eta_k^\flat(\gamma) \,\big|\, \xi_j(\gamma)\big), \quad j, k = 1, 2, \ldots$$

for almost every $\gamma \in \Gamma$. Deleting a null set from Γ, we conclude that the $\sharp$-operation in $\mathfrak{A}_1(\gamma)$ is preclosed. Let $\mathfrak{A}(\gamma) = \big(\mathbf{C}\mathfrak{A}_1(\gamma)\big)''$. We then observe that $\{\xi_j(\gamma)\}$ gives rise to a measurable field structure in $\{\mathfrak{A}(\gamma)\}$. By construction, we have $\int_\Gamma^\oplus \mathfrak{A}(\gamma)\,d\mu(\gamma) \supset \mathfrak{A}_1$. By Corollary 3.9, we conclude that $\mathfrak{A} = \int_\Gamma^\oplus \mathfrak{A}(\gamma)\,d\mu(\gamma)$.

Q.E.D.

Theorem 3.15 (Uniqueness). *Let $\{\mathfrak{A}_1(\gamma)\}$ be two measurable fields of full left Hilbert algebras over a standard measure space $\{\Gamma, \mu\}$.*

(i) *If*

$$\mathfrak{A} = \int_\Gamma^\oplus \mathfrak{A}_1(\gamma)\,d\mu(\gamma) = \int_\Gamma^\oplus \mathfrak{A}_2(\gamma)\,d\mu(\gamma),$$

then $\mathfrak{A}_1(\gamma) = \mathfrak{A}_2(\gamma)$ for almost every $\gamma \in \Gamma$.

(ii) *If for almost every $\gamma \in \Gamma$ there exists a unitary isomorphism $U(\gamma)$ of $\mathfrak{A}_1(\gamma)$ onto $\mathfrak{A}_2(\gamma)$, then one can choose a measurable field $V(\gamma)$ of unitary isomorphisms of $\{\mathfrak{A}_1(\gamma)\}$ onto $\{\mathfrak{A}_2(\gamma)\}$, so that*

$$\int_\Gamma^\oplus \mathfrak{A}_1(\gamma)\, d\mu(\gamma) \cong \int_\Gamma^\oplus \mathfrak{A}_2(\gamma)\, d\mu(\gamma)$$

under the unitary isomorphism V:

$$V = \int_\Gamma^\oplus V(\gamma)\, d\mu(\gamma)\,.$$

PROOF: By Theorem IV.8.23, the completion of $\mathfrak{A}_1(\gamma)$ and $\mathfrak{A}_2(\gamma)$ agree, say $\mathfrak{H}(\gamma)$, and $\mathcal{R}_\ell\big(\mathfrak{A}_1(\gamma)\big) = \mathcal{R}_\ell\big(\mathfrak{A}_2(\gamma)\big)$ for almost every $\gamma \in \Gamma$.

Let $\{\xi_j(\gamma)\}$ be a fundamental sequence of measurable vector fields for $\{\mathfrak{A}_1(\gamma)\}$ such that

a) the functions: $\gamma \to \|\xi_j(\gamma)\|$, $\|\xi_j(\gamma)^\sharp\|$ and $\|\pi_\ell(\xi_j(\gamma))\|$ are all bounded;
b) the set of fields $\{\, \xi_j(\gamma) : j = 1, 2, \dots \}$ forms a $\sharp$-algebra over $\mathbf{Q}(\mathrm{i})$.

Let Γ_1 be a subset of Γ with $\mu(\Gamma_1) < +\infty$. Let

$$\xi_{\Gamma_1, j}(\gamma) = \begin{cases} \xi_j(\gamma), & \gamma \in \Gamma_1, \\ 0, & \gamma \notin \Gamma_1. \end{cases}$$

Then $\{\xi_{\Gamma_1,j}(\gamma)\} \sim \xi_{\Gamma_1,j}$ is a vector in $\mathfrak{A}$ and $\{\xi_{\Gamma_1,j}\}$ is a $\sharp$-subalgebra of $\mathfrak{A}$ over $\mathbf{Q}(\mathrm{i})$. It follows that after the deletion of a null set in Γ_1, we have $\xi_{\Gamma_1,j}(\gamma) \in \mathfrak{A}_2(\gamma)$ by Corollary 3.9. Hence $\xi_j(\gamma) \in \mathfrak{A}_2(\gamma)$ for almost every $\gamma \in \Gamma$ due to the σ-finiteness of μ. Furthermore, $(\xi_j\xi_k)(\gamma) = \xi_j(\gamma)\xi_k(\gamma)$ and $\xi_j^\sharp(\gamma) = \xi_j(\gamma)^\sharp$ for almost every $\gamma \in \Gamma$, which do not depend on whether we calculate in $\mathfrak{A}_1(\gamma)$ or $\mathfrak{A}_2(\gamma)$.

But $\{\xi_j(\gamma)\}$ is a fundamental sequence for $\{\mathfrak{A}_1(\gamma)\}$, so $\mathfrak{A}_1(\gamma)$ is the $\sharp$-closure of $\{\, \xi_j(\gamma) : j = 1, 2, \dots \}$. Hence we have $\mathfrak{A}_1(\gamma) \subset \mathfrak{A}_2(\gamma)$ for almost every $\gamma \in \Gamma$, and by symmetry $\mathfrak{A}_1(\gamma) = \mathfrak{A}_2(\gamma)$ for almost every $\gamma \in \Gamma$.

(ii) Considering the partition of Γ according to the dimension of $\mathfrak{H}(\gamma)$: $\Gamma_n = \{\, \gamma \in \Gamma : \dim \mathfrak{H}(\gamma) = n \,\}$, we may assume that $\dim \mathfrak{H}(\gamma) = \dim \mathfrak{H}_0$ for some fixed separable Hilbert space. By Theorem IV.8.13, we may assume that the measurable fields $\{\mathfrak{A}_1(\gamma)\}$ and $\{\mathfrak{A}_2(\gamma)\}$ are fields of dense linear subspaces of $\mathfrak{H}_0$ such that a vector field: $\gamma \in \Gamma \mapsto \xi(\gamma) \in \mathfrak{A}_i(\gamma)$, $i = 1, 2$, is measurable if and only if the map: $\gamma \in \Gamma \mapsto \xi(\gamma) \in \mathfrak{H}_0$ is a measurable map of Γ into $\mathfrak{H}_0$. Let $\mathcal{U}(\mathfrak{H}_0)$ be the unitary group of $\mathfrak{H}_0$. Let Σ be the set of all pair $(\gamma, U) \in \Gamma \times \mathcal{U}(\mathfrak{H}_0)$ such that U is an isomorphism of $\mathfrak{A}_1(\gamma)$ onto $\mathfrak{A}_2(\gamma)$ as left Hilbert algebras.

We claim that Σ is a Borel subset of the standard space $\Gamma \times \mathcal{U}(\mathfrak{H}_0)$. Let $\{\, \xi_j(\gamma) : j \in \mathbf{N} \}$ (resp. $\{\, \eta_j(\gamma) : j \in \mathbf{N} \}$) be a fundamental sequence of $\{\mathfrak{A}_1(\gamma)\}$ (resp. $\mathfrak{A}_2(\gamma)'$). Then the fact that $(\gamma, U) \in \Sigma$ is characterized by the following conditions:

a) $\left(U\xi_i(\gamma) \mid \eta_j(\gamma)^\flat\right) = \left(\eta_j(\gamma) \mid U(\xi_i(\gamma)^\sharp)\right), \quad i, j \in \mathbf{N};$

b) $\pi_r\left(\eta_j(\gamma)\right)u\xi_i(\gamma) = u\pi_\ell\left(\xi_i(\gamma)\right)u^*\eta_j(\gamma);$

c) $U\Delta_1(\gamma)^{it}U^* = \Delta_2(\gamma)^{it}, \quad t \in \mathbf{Q}.$

Therefore, Σ is a Borel subset of $\Gamma \times \mathcal{U}(\mathfrak{H}_0)$.

The assumption of (ii) entails that the projection $\mathrm{pr}_\mathrm{I}(\Sigma)$ of Σ to Γ covers Γ except for a null set. Therefore, the measurable cross-section theorem, Theorem A.16, assures the existence of a measurable field: $\gamma \in \Gamma \mapsto V(\gamma) \in \mathcal{U}(\mathfrak{H}_0)$ such that $\left(\gamma, V(\gamma)\right) \in \Sigma$ for almost every $\gamma \in \Gamma$. Now, the rest of the proof is an easy exercise for the reader. Q.E.D.

Notes on Chapter VI

The origin of the concept of left Hilbert algebras can be traced to the work of Murray and von Neumann, [240], which states that the set of all densely defined closed operators affiliated to a finite factor forms an algebra under the natural operations. A surprising result even from today's point of view that a symmetric operator affiliated with a finite von Neumann algebra is automatically self-adjoint seems to make Murray and von Neumann believe that the theory of finite von Neumann algebras would play essential role in the theory of quantum mechanics as well as in the theory of differential equations. The history did not develop along their dream. It was proved later in 1963 by Araki and Woods, [430], that most of factors arising from physics are indeed of type III. At the moment, the usual theory of differential equations does not use directly the theory of operator algebras. However, the modern theory of elliptic differential operators on manifolds does make use of the theory of operator algebras. As seen later, the theory of von Neumann algebras of type III reduces the entire theory to the theory of von Neumann algebras of type II equipped with special kind of one parameter automorphism groups. Thus, the entire theory has indeed returned back to the theory of von Neumann algebras of type II_1 in a very round about way.

The definition of Hilbert algebras was given by Nakano, [646], following the idea of H-system due to Ambrose, [420], which are unimodular Hilbert algebras in our terminology. But the commutation theorem for the left and right von Neumann algebras and the existence of the conjugation were proven by I. E. Segal and R. Godement independently, [326, 528, 529]. They used Hilbert algebras to prove the commutation theorem for the regular representation of an unimodular locally compact group. In order handle non-unimodular groups, Dixmier further extended the theory to quasi-Hilbert algebras, [502]. This is indeed the predecessor of left Hilbert algebras. However, in Dixmier's theory the modular operator, or more precisely $\Delta^{1/2}$ and $\Delta^{-1/2}$, was not considered as the key element of the theory, it is rather viewed as an adjusting term of the conjugation. Pukanszky later treated this more carefully, [672], to characterize the semi-finiteness of the left von Neumann algebra. But the recognition of the modular operator as the absolute value part of the polar decomposition of the involution was still missing. It was Tomita who realized that the involution of a von Neumann algebra is preclosed in the Hilbert space structure given by a faithful normal state, and thus its polar decomposition gives rise to the modular operator. It was as early as 1959, [720]. Unfortunately, this work did not receive much attention. With the above polar decomposition of the involution, the tensor product commutation theorem, $(\mathcal{M} \overline{\otimes} \mathcal{N})' = \mathcal{M}' \overline{\otimes} \mathcal{N}'$, could be easily proved at that time. He did not pay much attention on this pressing problem at that time and the world did not pay much attention to his work either. Several years elapsed before every thing came under light.

In 1967, Tomita finally proposed a theory which covers the tensor product commutation theorem, [374], and asserts the fundamental theorem, Theorem VI.1.19. But the paper was not carefully written. As a result, once again he failed to convince the world despite the importance of his work. But this time, luckily his claim drew considerable attention of the world experts.

The definition of left Hilbert algebras was given in the present form under the term, "generalized Hilbert algebras", in [708] and the entire theory of Tomita was justified in this work. Tomita algebras were introduced by Tomita in the work cited above under the terminology "modular Hilbert algebras". To honor his contribution, we propose to call it *Tomita* algebras as in the text.

Originally, the direct (dis)integration of Hilbert algebras was studied in conjunction with the study of group algebras as in the case of the theory of Hilbert algebras itself. In particular, it is directly related to the Plancherel formula for locally compact non-abelian groups. In the early fifties, there were quite a few articles on the subject published, following the works of Dixmier, Godement and Segal cited above and [689]. The two books of Dixmier, [4] and [5], are good source of the references of the theory. As the theory of left Hilbert algebras was established after the tidal waves on the non-commutative Plancherel formula had passed, there are only relatively small number of articles published on the subject. As it involves the direct integral of unbounded operators, it requires a bit of care of the details. First of such was done by Sutherland, [696], and by Lance, [623].

Chapter VII
Weights

§ 0 Introduction

Weights correspond precisely to measures on a space equipped with a σ-algebra. Not all von Neumann algebras admit a faithful normal state. In fact, even in the separable case, to restrict ourselves to bounded functionals prevents us from developing the theory naturally, which will become clearer in the later chapters such as in Chapter XII.

In Section 1, we adapt the Gelfand-Naimark-Segal construction of the cyclic representation from a state to that of the semi-cyclic representation from a weight. In Section 2, we will see that there is one to one correspondence between left Hilbert algebras and faithful semi-finite normal weights. Here one observes that the inner product of the underlying Hilbert space is the source of non-commutative integration theory. As mentioned in the introduction of Chapter VI, one can develop an integration theory entirely based on the theory of weights on a commutative von Neumann algebra.

Section 3 is devoted to the Plancherel theorem. As a locally compact group gives rise to two algebras of operators: one is the multiplication operators on $L^2(G)$ by bounded measurable functions and the other is the convolution operators by integrable functions, the Hilbert space $L^2(G)$ gives rise to two "integrations": one is the usual integration over G relative to the Haar measure and the second is the subject of Section 3, i.e. the Plancherel weight. In this section, we prove a duality theorem which has been gradually polished up to the present form started with the Pontrjagin-Kampen duality theorem. Then Section 4 discusses briefly weights on a C^*-algebra.

§ 1 Weights and Semi-Cyclic Representations

In this section, we first give a simultaneous generalization of positive linear functionals and traces, which corresponds to infinite measures in the commutative case. We then give a construction of the associated representation.

Definition 1.1. A *weight* on a von Neumann algebra $\mathcal{M}$ is a map $\varphi\colon \mathcal{M}_+ \mapsto [0, \infty]$, the extended positive real numbers, satisfying the following conditions:

$$\varphi(x+y) = \varphi(x) + \varphi(y), \quad x, y, \in \mathcal{M}_+; \qquad \varphi(\lambda x) = \lambda\varphi(x), \quad \lambda \geq 0,$$

where we use the convention: $0\,(+\infty) = 0$. It is said to be *semi-finite* if

$$\mathfrak{p}_\varphi = \left\{ x \in \mathcal{M}_+ : \varphi(x) < +\infty \right\} \tag{1}$$

generates $\mathcal{M}$; *faithful* if $\varphi(x) \neq 0$ for every non-zero $x \in \mathcal{M}_+$; *normal* if

$$\varphi(\sup x_i) = \sup \varphi(x_i) \tag{2}$$

for every bounded increasing net $\{x_i\}$ in $\mathcal{M}_+$.

Lemma 1.2. *If $\mathfrak{p}$ is a hereditary convex subcone of $\mathcal{M}_+$ in the sense that*

$$\mathfrak{p} + \mathfrak{p} \subset \mathfrak{p}, \quad \lambda\mathfrak{p} \subset \mathfrak{p}, \quad \lambda \geq 0;$$
$$0 \leq y \leq x, \quad x \in \mathfrak{p}, \quad \Longrightarrow y \in \mathfrak{p},$$

then we conclude the following:

(i) $\mathfrak{n} = \left\{ x \in \mathcal{M} : x^*x \in \mathfrak{p} \right\}$ *is a left ideal of* $\mathcal{M}$;

(ii) $\mathfrak{m} = \left\{ \sum_{i=1}^n y_i^* x_i : x_1, \dots, x_n, y_1, \dots, y_n \in \mathfrak{n} \right\}$ *is a hereditary* **-subalgebra such that* $\mathfrak{m} \cap \mathcal{M}_+ = \mathfrak{p}$ *and every element of* $\mathfrak{m}$ *is a linear combination of four elements of* $\mathfrak{p}$.

PROOF:

(i) The inequality, $(ax)^*ax \leq \|a\|^2 x^*x$, yields that $a\mathfrak{n} \subset \mathfrak{n}$. Another inequality,

$$(x \pm y)^*(x \pm y) \leq 2(x^*x + y^*y),$$

shows that $\mathfrak{n}$ is additive. Therefore, $\mathfrak{n}$ is a left ideal.

(ii) The fact that $\mathfrak{m}$ is a *-subalgebra of $\mathcal{M}$ follows from (i). The polarization identity,

$$4y^*x = \sum_{k=0}^{3} \mathrm{i}^k\,(x + \mathrm{i}^k y)^*(x + \mathrm{i}^k y), \tag{3}$$

implies that $\mathfrak{m}$ is spanned linearly by $\mathfrak{p}$. Suppose

$$a = \sum_{k=1}^{n} y_k^* x_k \in \mathfrak{m} \cap \mathcal{M}_+, \quad x_1, \dots, x_n, y_1, \dots, y_n \in \mathfrak{n}.$$

Since $a = a^*$, the above (3) implies that

$$a = \frac{1}{2} \sum_{k=1}^{n} \left\{ (x_k + y_k)^*(x_k + y_k) - (x_k - y_k)^*(x_k - y_k) \right\}.$$

So, if we set

$$b = \frac{1}{2} \sum_{k=1}^{n} (x_k + y_k)^*(x_k + y_k) \in \mathfrak{p},$$

then we have $0 \leq a \leq b$, thus $a \in \mathfrak{p}$. Q.E.D.

We apply this lemma to $\mathfrak{p}_\varphi$ with φ a weight on $\mathcal{M}$. In this case, we denote $\mathfrak{n}$ and $\mathfrak{m}$ by $\mathfrak{n}_\varphi$ and $\mathfrak{m}_\varphi$ respectively. It follows that the weight φ is extended to a linear functional $\dot{\varphi}$ on $\mathfrak{m}_\varphi$.

Definition 1.3. The hereditary *-subalgebra $\mathfrak{m}_\varphi$ is called the *definition domain* of the weight φ or the *definition subalgebra* of φ.

We fix a weight φ on $\mathcal{M}$. The set

$$N_\varphi = \left\{ x \in \mathcal{M} : \varphi(x^*x) = 0 \right\} \tag{4}$$

is a left ideal of $\mathcal{M}$ contained in $\mathfrak{n}_\varphi$. We then consider the quotient space $\mathfrak{n}_\varphi / N_\varphi$ and denote the canonical map: $x \in \mathfrak{n}_\varphi \mapsto x + N_\varphi \in \mathfrak{n}_\varphi / N_\varphi$, i.e.,

$$\eta_\varphi(x) = x + N_\varphi \in \mathfrak{n}_\varphi / N_\varphi. \tag{5}$$

The sesquilinear functional:

$$\left(\eta_\varphi(x) \mid \eta_\varphi(y)\right) = \varphi(y^*x), \qquad x, y \in \mathfrak{n}_\varphi, \tag{6}$$

makes $\mathfrak{n}_\varphi / N_\varphi$ a pre-Hilbert space. Its completion will be denoted by $\mathfrak{H}_\varphi$. Since both N_φ and $\mathfrak{n}_\varphi$ are left ideals, the quotient space $\mathfrak{n}_\varphi / N_\varphi$ is a left $\mathcal{M}$-module. It is easy to see that the action of $\mathcal{M}$ on $\mathfrak{n}_\varphi / N_\varphi$ is extended to that on the Hilbert space $\mathfrak{H}_\varphi$ by the inequality, $(ax)^*ax \leq \|a\|^2 x^*x$. Thus, we obtain a representation π_φ of $\mathcal{M}$ on $\mathfrak{H}_\varphi$.

Proposition 1.4. *If φ is a semi-finite normal weight, then the representation $\left\{\pi_\varphi, \mathfrak{H}_\varphi\right\}$ is a non-degenerate normal *-representation. In addition, if φ is faithful, then so is π_φ.*

PROOF: For each $a \in \mathcal{M}$, $x, y \in \mathfrak{n}_\varphi$, we have

$$\begin{aligned}\left(\pi_\varphi(a)\eta_\varphi(x) \mid \eta_\varphi(y)\right) &= \left(\eta_\varphi(ax) \mid (\eta_\varphi(y)\right) = \dot{\varphi}(y^*ax) \\ &= \dot{\varphi}\left((a^*y)^*x\right) = \left(\eta_\varphi(x) \mid \pi_\varphi(a^*)\eta_\varphi(y)\right),\end{aligned}$$

so that π_φ is a *-representation.

Since $\mathcal{M}$ is unital, π_φ is non-degenerate.

The normality of π_φ follows immediately from that of φ. Q.E.D.

Definition 1.5. The triplet $\left\{\pi_\varphi, \mathfrak{H}_\varphi, \eta_\varphi\right\}$ is called *the semi-cyclic* representation of $\mathcal{M}$. In general, *a semi-cyclic* representation of $\mathcal{M}$ is a triplet $\left\{\pi, \mathfrak{H}, \eta\right\}$ consisting of a representation $\{\pi, \mathfrak{H}\}$ and a linear map $\eta: \mathfrak{n} \mapsto \mathfrak{H}$ from a left ideal $\mathfrak{n}$ of $\mathcal{M}$ into $\mathfrak{H}$ with dense range such that

$$\pi(a)\eta(x) = \eta(ax), \qquad a \in \mathcal{M}, \qquad x \in \mathfrak{n}.$$

In order to study semi-cyclic representations, we need the following:

Lemma 1.6. *Let* $\{\mathcal{M}, \mathfrak{H}\}$ *be a von Neumann algebra.*

(i) *If* $x, y \in \mathcal{M}$ *satisfy the inequality* $y^*y \leq x^*x$, *then there exists uniquely an* $s \in \mathcal{M}$ *such that*

$$y = sx \quad \text{and} \quad s[x\mathfrak{H}]^\perp = \{0\}. \tag{7}$$

Furthermore, the s *is at most of norm one.*

(ii) *If* $\{x_i\}_{i\in I}$ *is a family in* $\mathcal{M}$ *such that* $a = \sum x_i^* x_i$ *converges strongly, and if the operators* $\{s_i\}$ *determined by* $x_i = s_i a^{1/2}$ *and* $s_i[a\mathfrak{H}]^\perp = \{0\}$, *then the sum*

$$p = \sum_{i\in I} s_i^* s_i$$

is strongly convergent and p *is the range projection of* a, *the projection to the closure of the range of* a.

(iii) *If* $\{x_i\}$ *is a bounded increasing net in* $\mathcal{M}_+$ *with* $x = \sup x_i$, *and if* $\{s_i\}$ *are the operators such that* $x_i^{1/2} = s_i x^{1/2}$ *and* $s_i[x\mathfrak{H}]^\perp = \{0\}$, *then* $\{s_i^* s_i\}$ *is increasing and* $p = \sup s_i^* s_i$ *is the range projection of* x, *and also* $\{s_i\}$ *converges to* p *strongly.*

PROOF:

(i) For each $\xi \in \mathfrak{H}$, we have

$$\|y\xi\|^2 = (y^*y\xi \mid \xi) \leq (x^*x\xi \mid \xi) = \|x\xi\|^2,$$

so that the map $s_0 \colon x\xi \in x\mathfrak{H} \mapsto y\xi \in y\mathfrak{H}$ is a well-defined linear map with $\|s_0\| \leq 1$; hence we can extend it to a bounded operator, denoted by s_0 again, on the closure $[x\mathfrak{H}]$ of $x\mathfrak{H}$. With p the projection to $[x\mathfrak{H}]$, we set $s = s_0 p$. Then s is the required operator. The uniqueness of s follows from its construction. If $u \in \mathcal{U}(\mathcal{M}')$, the unitary group of $\mathcal{M}'$, then the uniqueness of s together with $x = uxu^*$ and $y = uyu^*$ implies $s = usu^*$. Therefore s belongs to $\mathcal{M}$.

(ii) For each $\xi \in \mathfrak{H}$, set $\eta = a^{1/2}\xi$. If J is a finite subset of I, then

$$\begin{aligned}\Big(\sum_{i\in J} s_i^* s_i \eta \;\Big|\; \eta\Big) &= \Big(\sum_{i\in J} s_i^* s_i a^{\frac12}\xi \;\Big|\; a^{\frac12}\xi\Big) = \Big(\sum_{i\in J} a^{\frac12} s_i^* s_i a^{\frac12}\xi \;\Big|\; \xi\Big)\\ &= \Big(\sum_{i\in J} x_i^* x_i \xi \;\Big|\; \xi\Big) \leq (a\xi \mid \xi) = \|a^{\frac12}\xi\|^2 = \|\eta\|^2.\end{aligned}$$

Setting $p_J = \sum_{i\in J} s_i^* s_i$, we have $(p_J\eta \mid \eta) \leq (p\eta \mid \eta)$ for every η in the algebraic direct sum $a^{1/2}\mathfrak{H} + (1-p)\mathfrak{H}$. Hence we have $p_J \leq p$ by continuity. Since $\{p_J\}$ is an increasing net it converges strongly to an operator, say p_0 in $\mathcal{M}$, and $p_0 \leq p$. But, if $\eta = a^{1/2}\xi$, $\xi \in \mathfrak{H}$, then

$$\begin{aligned}(p_0\eta \mid \eta) &= \sum_{i\in I}(s_i^* s_i a^{\frac12}\xi \mid a^{\frac12}\xi) = \Big(\sum_{i\in I} x_i^* x_i \xi \;\Big|\; \xi\Big)\\ &= (a\xi \mid \xi) = \|a^{\frac12}\xi\|^2 = \|\eta\|^2,\end{aligned}$$

so that $(p_0\eta \mid \eta) = \|\eta\|^2$ for every $\eta \in p\mathfrak{H} = [a^{1/2}\mathfrak{H}]$ by continuity, which means $p_0 = p$.

(iii) If $\eta = x^{1/2}\xi,\ \xi \in \mathfrak{H}$, and $i \leq j$, then

$$\begin{aligned}\left(s_i^* s_i \eta \mid \eta\right) &= \left(s_i^* s_i a^{\frac{1}{2}} \xi \mid a^{\frac{1}{2}} \xi\right) = (x_i \xi \mid \xi) \leq (x_j \xi \mid \xi) \\ &= \left(s_j^* s_j \eta \mid \eta\right) \leq (x \xi \mid \xi) = \|\eta\|^2,\end{aligned}$$

so that $\{s_i^* s_i\}$ is increasing and majorized by p. Let $p_0 = \sup s_i^* s_i$. The argument similar to that in (ii) shows $p = p_0$ and the strong convergence of $\{s_i\}$. Q.E.D.

Lemma 1.7. *If $\mathcal{M}$ is σ-finite, then every completely additive weight φ on $\mathcal{M}$ in the sense that $\varphi\left(\sum_{i \in I} x_i\right) = \sum_{i \in I} \varphi(x_i),\ \{x_i\} \subset \mathcal{M}_+$, is normal.*

PROOF: By assumption, $\mathcal{M}$ admits a faithful normal state ω. Suppose that $\{x_i\}$ is a bounded increasing net in $\mathcal{M}_+$ with $x = \sup x_i$. By induction, we choose an increasing sequence $\{x_n\}$ from $\{x_i\}$ such that $\omega(x_n) > \omega(x) - 1/n$ for each $n \in \mathbf{N}$. Since $x_n \leq x,\ y = \lim x_n \leq x$ converges σ-strongly. But we have

$$\omega(x) - \frac{1}{n} < \omega(x_n) \leq \omega(y) \leq \omega(x), \quad n = 1, 2, \ldots,$$

so that $\omega(x) = \omega(y)$, yielding $x = y$ by the faithfulness of ω.

Now, setting $x_0 = 0$ and $y_n = x_n - x_{n-1},\ n = 1, 2, \ldots$, we have $x = \sum y_n$ and the complete additivity of φ implies

$$\varphi(x) = \sum_{n=1}^{\infty} \varphi(y_n) = \lim \varphi(x_n) \leq \lim \varphi(x_i) \leq \varphi(x),$$

so that $\varphi(x) = \sup \varphi(x_i)$. Thus φ is normal. Q.E.D.

We now fix a normal weight φ on $\mathcal{M}$ together with the associated semicyclic representation $\left\{\pi_\varphi,\ \mathfrak{H}_\varphi,\ \eta_\varphi\right\}$. The following result is the unbounded analogue of Proposition IV.3.10:

Lemma 1.8. *There exists uniquely a completely positive map θ_φ from the definition subalgebra $\mathfrak{m}_\varphi$ of φ into $\pi_\varphi(\mathcal{M})'_*$ determined by the formula:*

$$\left\langle a, \theta_\varphi(y^* x)\right\rangle = \left(a\eta_\varphi(x) \mid \eta_\varphi(y)\right), \quad a \in \pi_\varphi(\mathcal{M})', \quad x, y \in \mathfrak{n}_\varphi. \tag{8}$$

PROOF: The uniqueness follows from the fact that every element of $\mathfrak{m}_\varphi$ is of the form $y^* x,\ x, y \in \mathfrak{n}_\varphi$, the fact driven by the polar decomposition.

If $x^* x = y^* y,\ x, y \in \mathfrak{n}_\varphi$, then $y = sx$ by Lemma 1.6.(i); but in this special case, s is a partial isometry and $x = s^* y$. For each $a \in \pi_\varphi(\mathcal{M})'$, we have

$$\begin{aligned}\left(a\eta_\varphi(x) \mid \eta_\varphi(x)\right) &= \left(a\eta_\varphi(s^* y) \mid \eta_\varphi(s^* y)\right) = \left(\pi_\varphi(s) a \pi_\varphi(s^*) \eta_\varphi(y) \mid \eta_\varphi(y)\right) \\ &= \left(a\eta_\varphi(y) \mid \eta_\varphi(y)\right).\end{aligned}$$

Hence a map: $x^* x \in \mathfrak{m}_\varphi^+ \mapsto \theta_\varphi(x^* x) \in \pi_\varphi(\mathcal{M})'_*$ is well-defined. Clearly, $\theta_\varphi(\lambda x) = \lambda \theta_\varphi(x),\ \lambda > 0$.

We shall show the additivity. Suppose $z = x + y,\ x, y \in \mathfrak{m}_\varphi^+$. Choose s and t in $\mathcal{M}$ by Lemma 1.6 so that $x^{1/2} = sz^{1/2},\ y^{1/2} = tz^{1/2}$ and $p = s^*s + t^*t$ is the range projection $s_\ell(z)$ of z. Now we have, with $a \in \pi_\varphi(\mathcal{M})'$,

$$\begin{aligned}\left(a\eta_\varphi(z^{\frac{1}{2}}) \mid \eta_\varphi(z^{\frac{1}{2}})\right) &= \left(a\pi_\varphi(s^*s + t^*t)\eta_\varphi(z^{\frac{1}{2}}) \mid \eta_\varphi(z^{\frac{1}{2}})\right) \\ &= \left(a\pi_\varphi(s^*s)\eta_\varphi(z^{\frac{1}{2}}) \mid \eta_\varphi(z^{\frac{1}{2}})\right) + \left(a\pi_\varphi(t^*t)\eta_\varphi(z^{\frac{1}{2}}) \mid \eta_\varphi(z^{\frac{1}{2}})\right) \\ &= \left(a\eta_\varphi(x^{\frac{1}{2}}) \mid \eta_\varphi(x^{\frac{1}{2}})\right) + \left(a\eta_\varphi(y^{\frac{1}{2}}) \mid \eta_\varphi(y^{\frac{1}{2}})\right) \\ &= \langle a, \theta_\varphi(x)\rangle + \langle a, \theta_\varphi(y)\rangle = \langle a, \theta_\varphi(x) + \theta_\varphi(y)\rangle,\end{aligned}$$

so that $\theta_\varphi(z) = \theta_\varphi(x) + \theta_\varphi(y)$. Thus θ_φ is additive on $\mathfrak{m}_\varphi^+$. It is then easy to see that θ_φ is linearly extendable to the whole of $\mathfrak{m}_\varphi$.

If $x_1, \dots, x_n \in \mathfrak{n}_\varphi$ and $a_1, \dots, a_n \in \pi_\varphi(\mathcal{M})'$, then we have

$$\begin{aligned}\left\langle [a_i^* a_j], (\theta_\varphi)_n [x_i^* x_j]\right\rangle &= \left\langle [a_i^* a_j], [\theta_\varphi(x_i^* x_j)]\right\rangle = \sum_{i,j=1}^n \langle a_i^* a_j, \theta_\varphi(x_i^* x_j)\rangle \\ &= \sum_{i,j=1}^n \left(a_i^* a_j \eta_\varphi(x_j) \mid \eta_\varphi(x_i)\right) = \left\| \sum_{j=1}^n a_j \eta_\varphi(x_j)\right\|^2 \geq 0,\end{aligned}$$

so that θ_φ is completely positive. Q.E.D.

Lemma 1.9. *With the notations as before, if $h \in \mathfrak{m}_\varphi$ is self-adjoint, then*

$$\|\theta_\varphi(h)\| = \inf\{\, \varphi(a) + \varphi(b) : h = a - b,\ a, b \in \mathfrak{m}_\varphi^+ \,\}. \tag{9}$$

PROOF: Let $\rho(h)$ be the number on the right hand side of (9). Then ρ is a function on the self-adjoint part $\mathfrak{m}_{\varphi,h}$ with the property:

$$\rho(\lambda x) = |\lambda| \rho(x) \geq 0, \quad \lambda \in \mathbf{R}.$$

We shall show that ρ is subadditive. Fix x_1 and x_2 in $\mathfrak{m}_{\varphi,h}$ and $\varepsilon > 0$. Choose y_1, y_2, z_1 and z_2 in $\mathfrak{m}_\varphi^+$ so that

$$x_i = y_i - z_i, \qquad \varphi(y_i) + \varphi(z_i) < \rho(x_i) + \varepsilon, \qquad i = 1, 2.$$

Setting $x = x_1 + x_2,\ y = y_1 + y_2$ and $z = z_1 + z_2$, we get $x = y - z$ and

$$\rho(x) \leq \varphi(y) + \varphi(z) \leq \rho(x_1) + \rho(x_2) + 2\varepsilon.$$

Thus we conclude $\rho(x) \leq \rho(x_1) + \rho(x_2)$. Therefore, ρ is semi-norm on $\mathfrak{m}_{\varphi,h}$. Furthermore, ρ and φ agree on the positive part m_φ^+. If $x \in \mathfrak{m}_\varphi^+$, then $\theta_\varphi(x) \geq 0$ and so

$$\|\theta_\varphi(x)\| = \langle 1, \theta_\varphi(x)\rangle = \|\eta_\varphi(x^{\frac{1}{2}})\|^2 = \varphi(x) = \rho(x).$$

If $x = y - z$ with $y, z \in \mathfrak{m}_\varphi^+$, then

$$\|\theta_\varphi(x)\| \le \|\theta_\varphi(y)\| + \|\theta_\varphi(z)\| = \varphi(y) + \varphi(z).$$

Thus, we get $\|\theta_\varphi(x)\| \le \rho(x),\ x \in \mathfrak{m}_{\varphi,h}$.

We now show the reversed inequality. We fix an $x_0 \in \mathfrak{m}_{\varphi,h}$. By the Hahn-Banach extension theorem, we find a real valued linear functional ψ on $\mathfrak{m}_{\varphi,h}$ such that

$$\psi(x_0) = \rho(x_0), \qquad |\psi(x)| \le \rho(x), \qquad x \in \mathfrak{m}_{\varphi,h}.$$

We extend ψ linearly to the entire $\mathfrak{m}_\varphi$ as a self-adjoint linear functional which we shall denote by ψ again. The inequality:

$$-\varphi(x^*x) = -\rho(x^*x) \le \psi(x^*x) \le \rho(x^*x) = \varphi(x^*x), \quad x \in \mathfrak{n}_\varphi$$

guarantees that the sequilinear form: $(x, y) \in \mathfrak{n}_\varphi \times \mathfrak{n}_\varphi \mapsto \psi(y^*x)$ can be extended to the entire Hilbert space $\mathfrak{H}_\varphi$ and gives rise to a bounded operator a with $\|a\| \le 1$ such that

$$\big(a\eta_\varphi(x) \mid \eta_\varphi(y)\big) = \psi(y^*x), \quad x, y \in \mathfrak{n}_\varphi.$$

It is routine to check that a belongs to $\pi_\varphi(\mathcal{M})'$. Since ψ is self-adjoint, a is self-adjoint. Now, let $x_0 = uh$ be the polar decomposition of x_0. Since $x_0 = x_0^*$, u is the difference of two orthogonal projections, and u and h commute. Note that $h^{1/2} \in \mathfrak{n}_\varphi$. We then compute:

$$\begin{aligned}\rho(x_0) = \psi(x_0) = \psi\big(h^{\frac12}h^{\frac12}u\big) &= \Big(a\eta_\varphi\big(h^{\frac12}u\big) \,\Big|\, \eta_\varphi\big(h^{\frac12}\big)\Big)\\ &= \big\langle a, \theta_\varphi(hu)\big\rangle \le \|\theta_\varphi(hu)\| = \|\theta_\varphi(x_0)\|,\end{aligned}$$

establishing the desired inequality. Q.E.D.

Lemma 1.10. *We keep the same notation. Suppose that $\{x_n\}$ is a bounded sequence of $\mathfrak{m}_\varphi^+$.*

(i) *If $\{x_n\}$ converges to an $x \in \mathcal{M}$ σ-strongly and if $\{\theta_\varphi(x_n)\}$ converges in norm, then x belongs to $\mathfrak{m}_\varphi^+$.*
(ii) *If $\{x_n\}$ converges to zero σ-strongly and if $\{\theta_\varphi(x_n)\}$ converges in norm, then the limit of $\{\theta_\varphi(x_n)\}$ must be zero.*

This lemma asserts that the map θ_φ is closed when it is restricted to $\mathfrak{m}_\varphi^+$. In fact, we shall see later that θ_φ is closed.

PROOF:

(i) Fix $\varepsilon > 0$ and $\psi = \lim_{n\to\infty} \theta_\varphi(x_n)$. Choose a subsequence $\{y_n\}$ from $\{x_n\}$ so that $\|\psi - \theta_\varphi(y_n)\| < \varepsilon/2^{n+1}$. Namely, we have

$$\|\theta_\varphi(y_{n+1}) - \theta_\varphi(y_n)\| < \varepsilon/2^n.$$

By the previous lemma, we choose $a_n, b_n \in \mathfrak{m}_\varphi^+$ such that

$$y_{n+1} - y_n = a_n - b_n, \qquad \varphi(a_n) + \varphi(b_n) < \varepsilon/2^n.$$

It follows that

$$y_{n+1} \leq y_1 + \sum_{k=1}^{n} a_k.$$

For each $\alpha > 0$, we define a function f_α as follows:

$$f_\alpha(t) = t/(1+\alpha t), \quad t \in \left]-1/\alpha, +\infty\right[. \tag{10}$$

It follows that f_α is operator monotone in the sense that

$$-1/\alpha < x \leq y \Longrightarrow f_\alpha(x) \leq f_\alpha(y).$$

Hence we get

$$f_\alpha(y_{n+1}) \leq f_\alpha\left(y_1 + \sum_{k=1}^{n} a_k\right) \leq 1/\alpha.$$

Since $\left\{f_\alpha\left(y_1 + \sum_{k=1}^{n} a_k\right)\right\}$ is now bounded and increasing, it converges σ-strongly. Let c_α be its limit in $\mathcal{M}_+$. Then we have

$$c_\alpha = \lim_{n\to\infty} f_\alpha\left(y_1 + \sum_{k=1}^{n} a_k\right) \geq \lim_{n\to\infty} f_\alpha(y_{n+1}) = f_\alpha(x).$$

By the normality of φ, we get

$$\begin{aligned}
\varphi\big(f_\alpha(x)\big) &\leq \lim_{n\to\infty} \varphi\left(f_\alpha\left(y_1 + \sum_{k=1}^{n} a_i\right)\right) \\
&\leq \lim_{n\to\infty} \varphi\left(y_1 + \sum_{k=1}^{n} a_k\right) \qquad (\text{since } f_\alpha(a) \leq a, \ a \in \mathcal{M}_+) \\
&= \varphi(y_1) + \sum_{k=1}^{\infty} \varphi(a_k) \leq \varphi(y_1) + \sum_{n=1} \varepsilon/2^n = \varphi(y_1) + \varepsilon.
\end{aligned}$$

Now, the fact that

$$x = \lim_{\alpha\to 0} f_\alpha(x), \qquad f_\alpha(x) \leq f_\beta(x) \quad \text{if} \quad 0 < \beta < \alpha,$$

together with the normality of φ implies

$$\varphi(x) = \lim_{\alpha\to 0} \varphi\big(f_\alpha(x)\big) \leq \varphi(y_1) + \varepsilon < +\infty,$$

hence the conclusion $x \in \mathfrak{m}_\varphi^+$.

(ii) We fix $\varepsilon > 0$, ψ, $\{y_n\}$, $\{a_n\}$, $\{b_n\}$ and f_α as before. We use, however, the inequality:

$$y_1 - y_{n+1} \leq \sum_{k=1}^{n} b_k.$$

By the uniform boundedness theorem, $\{y_n\}$ is bounded. Set $K = \sup\|y_n\|$. Then $y_1 - y_{n+1} \geq -K$, so that if $0 < \alpha < 1/K$, the following inequality makes sense:

$$f_\alpha(y_1 - y_{n+1}) \leq f_\alpha\Big(\sum_{k=1}^n b_k\Big).$$

Let $d_\alpha = \lim_{n\to\infty} f_\alpha\big(\sum_{k=1}^n b_k\big)$. By assumption, $\lim_{n\to\infty} y_n = 0$, so

$$f_\alpha(y_1) = \lim_{n\to\infty} f_\alpha(y_1 - y_{n+1}) \leq \lim_{n\to\infty} f_\alpha\Big(\sum_{k=1}^n b_k\Big) = d_\alpha.$$

By the normality of φ, we get an estimate:

$$\varphi\big(f_\alpha(y_1)\big) \leq \varphi(d_\alpha) = \lim_{n\to\infty} \varphi\Big(f_\alpha\Big(\sum_{k=1}^n b_k\Big)\Big).$$

Thus, the inequality:

$$\varphi\Big(f_\alpha\Big(\sum_{k=1}^n b_k\Big)\Big) \leq \varphi\Big(\sum_{k=1}^n b_k\Big) < \sum_{k=1}^n \varepsilon/2^k < \varepsilon$$

implies that $\varphi\big(f_\alpha(y_1)\big) \leq \varepsilon$. As $\alpha \to 0$, $f_\alpha(y_1)$ converges to y_1 from below and we get

$$\varphi(y_1) = \lim_{\alpha\to 0} \varphi\big(f_\alpha(y_1)\big) \leq \varepsilon.$$

Hence,

$$\|\psi\| \leq \big\|\psi - \theta_\varphi(y_1)\big\| + \big\|\varphi(y_1)\big\| < \varepsilon/2 + \varphi(y_1) < 2\varepsilon,$$

so we conclude $\psi = 0$. Q.E.D.

Although we still need preparation, it may be the place to state the result to motivate further discussion.[1]

Theorem 1.11. *If φ is a weight on a von Neumann algebra $\mathcal{M}$, then the following conditions are mutually equivalent:*

(i) *φ is completely additive in the sense that*

$$\varphi\Big(\sum_{i\in I} x_i\Big) = \sum_{i\in I} \varphi(x_i)$$

for every σ-strongly summable family $\{x_i\}_{i\in I}$ in $\mathcal{M}_+$;

(ii) *φ is normal;*

[1] Those readers who want to proceed quickly can skip the proof of the theorem and jump to the end of the proof of the theorem on page 55. In this case, take (iv) as the definition of the normality for φ instead of (2).

(iii) *φ is σ-weakly lower semi continuous;*
(iv) *If we define*

$$\Phi_\varphi = \left\{ \omega \in \mathcal{M}_*^+ : \omega(x) \le \varphi(x),\ x \in \mathcal{M}_+ \right\}, \tag{11}$$

then

$$\varphi(x) = \sup\left\{ \omega(x) : \omega \in \Phi_\varphi \right\}, \quad x \in \mathcal{M}_+, \tag{12}$$

The implications: (iv) $\Longrightarrow$ (iii) $\Longrightarrow$ (ii) $\Longrightarrow$ (i) are trivial. The hard part is the implication (i) $\Longrightarrow$ (iv). Since (iv) described the form of φ in terms of $\mathcal{M}_*^+$ it is important in the later discussion. We need further preparation.

We maintain the notation established above. Consider the graph of η_φ:

$$G = \left\{ \left(x, \eta_\varphi(x)\right) \in \mathcal{M} \oplus \mathfrak{H}_\varphi : x \in \mathfrak{n}_\varphi \right\}. \tag{13}$$

In the direct sum $\mathcal{M} \oplus \mathfrak{H}_\varphi$, we consider the norm:

$$\left\|(x, \xi)\right\| = \max\left\{ \|x\|, \|\xi\| \right\}, \quad x \in \mathcal{M}, \quad \xi \in \mathfrak{H}_\varphi. \tag{14}$$

Then it is the Banach space dual of the direct sum Banach space $\mathcal{M}_* \oplus \mathfrak{H}_\varphi^*$ equipped with norm:

$$\left\|(w, \xi^*)\right\| = \|w\| + \|\xi^*\|, \quad w \in \mathcal{M}_*, \quad \xi^* \in \mathfrak{H}^*. \tag{15}$$

Lemma 1.12. *If $\mathcal{M}$ is σ-finite, then the unit ball of G is weak* compact in the above Banach space $\mathcal{M} \oplus \mathfrak{H}_\varphi$.*

PROOF: Let B be the closed unit ball of $\mathcal{M} \oplus \mathfrak{H}_\varphi$. Since $B \cap G$ is convex, the weak* closedness of $B \cap G$ is equivalent to its closedness under any locally convex topology in $\mathcal{M} \oplus \mathfrak{H}_\varphi$ having $\mathcal{M}_* \oplus \mathfrak{H}_\varphi^*$ as the dual space. We consider therefore the product topology $\mathcal{T}$ of σ-strong* topology in $\mathcal{M}$ and the topology in $\mathfrak{H}_\varphi$, and prove the $\mathcal{T}$-closedness of $B \cap G$.

The σ-finiteness of $\mathcal{M}$ implies the metrizability of the unit ball of $\mathcal{M}$ with respect to the σ-strong* topology. Therefore, if (x, ξ) is a limit point of $B \cap G$, then there exists a sequence $\{x_n\}$ in $\mathfrak{n}_\varphi$ such that $\{x_n\}$ converges to x σ-strongly* and $\left\|\eta_\varphi(x_n) - \xi\right\| \to 0$. Since $\left(x_n, \eta_\varphi(x_n)\right) \in B$, we have $\left\|x_n\right\| \le 1$ and $\left\|\eta_\varphi(x_n)\right\| \le 1$, so that $\{x_n^* x_n\}$ converges to x^*x σ-strongly and $\theta_\varphi(x_n^* x_n) = \omega_{\eta_\varphi}(x_n)$ converges to ω_ξ in norm, where ω_ζ, $\zeta \in \mathfrak{H}_\varphi$, means the vectorial functional: $y \in \mathcal{M} \to (y\zeta \mid \zeta)$. By Lemma 1.10.(i), x belongs to $\mathfrak{n}_\varphi$. It then follows that $\left\{(x_n - x)^*(x_n - x)\right\}$ converges to zero σ-strongly and $\theta_\varphi\left((x_n - x)^*(x_n - x)\right) = \omega_{\eta_\varphi(x_n - x)}$ converges to $\omega_{\xi - \eta_\varphi(x)}$ in norm. Thus Lemma 1.10.(ii), says that $\omega_{\xi - \eta_\varphi(x)} = 0$ in $\pi_\varphi(\mathcal{M})'_*$, so that

$$\left\|\xi - \eta_\varphi(x)\right\|^2 = \omega_{\xi - \eta_\varphi(x)}(1) = 0.$$

Hence $\xi = \eta_\varphi(x)$, which concludes $(x, \xi) \in B \cap G$. Q.E.D.

Lemma 1.13. *If $\mathcal{M}$ is σ-finite, then conditions* (i), (ii) *and* (iii) *in Theorem* 1.11 *are equivalent.*

PROOF: The equivalence (i) $\Longleftrightarrow$ (ii) has already been shown in Lemma 1.7. As mentioned above, the implication (iii) $\Longrightarrow$ (ii) is trivial; so we shall show only the implication (ii) $\Longrightarrow$ (iii).

By the previous lemma, the graph G of η_φ is weak* closed. We set for each $r, s > 0$,

$$B_{r,s} = \left\{ (x, \xi) \in \mathcal{M} \oplus \mathfrak{H}_\varphi : \|x\| \leq r,\ \|\xi\| \leq s \right\}.$$

Then $B_{r,s} \cap G$ is weak* compact, so that the projection image $C_{r,s}$ of $B_{r,s} \cap G$ in $\mathcal{M}$:

$$C_{r,s} = \left\{ x \in \mathcal{M} : \|x\| \leq r,\ \varphi(x^*x) \leq s^2 \right\}$$

is σ-weakly compact. We then set

$$E_s = \left\{ x \in \mathcal{M} : \varphi(x^*x) \leq s^2 \right\}, \quad s > 0.$$

If $\mathcal{S}$ is the closed unit ball of $\mathcal{M}$, then $C_{r,s} = E_s \cap r\mathcal{S}$. Thus, the convexity of E_s implies the σ-weak closedness of E_s by a theorem of Banach. We now set

$$F_s = \left\{ x \in \mathcal{M}_+ : \varphi(x) \leq s^2 \right\}$$

and note that the σ-weak lower semi-continuity of φ is, by definition, equivalent to the σ-weak closedness of F_s. By the convexity of F_s, it suffices to prove the σ-strong closedness of $F_s \cap r\mathcal{S}$. Suppose $\{x_i\}$ is a net in $F_s \cap r\mathcal{S}$ converging σ-strongly to an $x \in \mathcal{M}_+$. The square root operation is σ-strongly continuous on $F_s \cap r\mathcal{S}$, so that $\left\{x_i^{1/2}\right\}$ converges σ-strongly to $x^{1/2}$. But $\left\{x_i^{1/2}\right\}$ is a net in E_s, so that its σ-strong limit $x^{1/2}$ belong to E_s. Thus x belongs to F_s. Q.E.D.

Lemma 1.14. *Let Σ denote the set of σ-finite projection of $\mathcal{M}$ and $\mathcal{M}_0 = \bigcup\left\{ p\mathcal{M}p : p \in \Sigma \right\}$. Then $\mathcal{M}_0$ is an ideal of $\mathcal{M}$ and the limit of each σ-weakly convergent sequence belongs to $\mathcal{M}_0$.*

PROOF: Since $x \in \mathcal{M}_0$ is equivalent to the fact that $s_\ell(x) \in \Sigma$ and $s_r(x) \in \Sigma$, $\mathcal{M}_0$ is an ideal of $\mathcal{M}$, where $s_\ell(x)$ and $s_r(x)$ denote the range projections of x and x^* respectively.

Suppose that $\{x_n\}$ is a sequence in $\mathcal{M}_0$ converging to x σ-weakly. By assumption, there exists a sequence $\{p_n\}$ in Σ such that $x_n = p_n x_n p_n$. Set $p = \bigvee_{n=1}^\infty p_n$. It follows that p belongs to Σ and $x_n = p x_n p$, so that $x = pxp$. Thus x belongs to $\mathcal{M}_0$. Q.E.D.

Lemma 1.15. *We keep the previously introduced notation. Suppose that a convex set $F \subseteq \mathcal{M}_0^+$ is hereditary in the sense that $0 \leq y \leq x \in F$ implies $y \in F$. A necessary and sufficient condition for F to be σ-weakly closed in $\mathcal{M}_0$ is that $F \cap p\mathcal{M}p$ is σ-weakly closed for every $p \in \Sigma$.*

PROOF: The necessity of the condition is clear, so that we shall show only the sufficiency.

Assume that $F \cap p\mathcal{M}p,\ p \in \Sigma$, is σ-weakly closed. Let ω be a normal state on $\mathcal{M}$ with $p = s(\omega)$. Then $d(x, y) = \omega\big((x-y)^*(x-y)\big)^{1/2},\ x, y \in \mathcal{M}_p \cap \mathcal{S}$, gives a metric on $\mathcal{M}_p \cap \mathcal{S}$ for the σ-strong topology there. We set $E = \{x \in \mathcal{M} : x^*x \in F\}$. Then we have, for $x, y \in E$ and $0 \le \lambda \le 1$,

$$
\begin{aligned}
0 &\le \big(\lambda x + (1-\lambda)y\big)^*\big(\lambda x + (1-\lambda)y\big)\\
&= \lambda^2 x^*x + (1-\lambda)^2 y^*y + \lambda(1-\lambda)(x^*y + y^*x)\\
&\le \lambda^2 x^*x + (1-\lambda)^2 y^*y + \lambda(1-\lambda)(x^*x + y^*y)\\
&= \lambda x^*x + (1-\lambda)y^*y \in F
\end{aligned}
$$

so that $\lambda x + (1-\lambda)y \in E$. Hence E is convex. Furthermore, if $a \in \mathcal{S}$, then $aE \subset E$ by the inequality $(ax)^*ax \le x^*x$.

Now, we show the σ-weak closedness of $E \cap p\mathcal{M}p,\ p \in \Sigma$. By convexity, it is enough to prove the σ-strong* closedness of $E \cap p\mathcal{M}p \cap r\mathcal{S},\ r > 0$. But this set is the inverse image of $f \cap p\mathcal{M}p$ under the map: $x \in r\mathcal{S} \cap p\mathcal{M}p \mapsto x^*x \in r^2\mathcal{S} \cap p\mathcal{M}p$ which is σ-strong* continuous. By assumption, $F \cap p\mathcal{M}p$ is σ-weakly closed as required.

We then show the σ-weak closedness of $pE,\ p \in \Sigma$, which is equivalent to that of E^*p. It is again enough to prove the σ-strong closedness of $r\mathcal{S} \cap E^*p$. If x is a point in the σ-strong closure of $E^*p \cap r\mathcal{S}$, then x is approximated σ-strongly by a sequence $\{x_n^*\}$ in $E^*p \cap r\mathcal{S}$ as shown above. Thus, $\{x_n^*\}$ converges σ-weakly to x^* and there exists by the previous lemma a projection $q \in \Sigma$ such that $x_n = qx_nq$, so that $x_n \in E^* \cap q\mathcal{M}q$. Hence x^* belongs to $E \cap q\mathcal{M}q$ because $E \cap q\mathcal{M}q$ is σ-weakly closed as just shown. Therefore, $x \in E^* \cap \mathcal{M}p = E^*p$. Namely, pE is σ-weakly closed.

Let $\tilde{F}$ be the σ-strong closure of F and $y \in \tilde{F} \cap \mathcal{M}_0$. Set $p = s_\ell(y) \in \Sigma$ and let $\{y_i\}$ be a net in F converging σ-strongly to y. By Theorem II.4.7, $\{y_i^{1/2}\}$ converges to $y^{1/2}$ σ-strongly, and thus $\{y_i^{1/2}\}$ converges to $py^{1/2}$. The closedness of pE implies that $py^{1/2} = y^{1/2} \in pE \subset E$. Therefore, y belongs to F, concluding $\tilde{F} \cap \mathcal{M}_0 = F$. Q.E.D.

PROOF OF THE EQUIVALENCE (i) $\Longleftrightarrow$ (iii) IN THEOREM 1.11: Suppose φ is a weight on $\mathcal{M}$ satisfying (i), i.e. a completely additive weight. By Lemma 1.13, φ is σ-weakly lower semi-continuous on $(p\mathcal{M}p)_+$ for each $p \in \Sigma$. Hence, if we set $F = \{x \in \mathcal{M}_+ : \varphi(x) \le 1\}$, then $F \cap p\mathcal{M}p$ is σ-weakly closed. Of course, F is convex and hereditary. Thus, the previous lemma shows that $F \cap \mathcal{M}_0$ is relatively σ-weakly closed in $\mathcal{M}_0$. Let $\{p_i\}_{i \in I}$ be a maximal orthogonal family on Σ. Then $\Sigma p_i = 1$. For each finite subset J of I, set $q_J = \sum_{i \in J} p_i$. It follows that $q_J \in \Sigma$ and $\{q_J\}$ increases up to 1. To show the σ-weak closedness of F, it suffices to prove the σ-strong closedness of $F \cap r\mathcal{S},\ r > 0$, by the convexity of F. Let $\{x_\lambda : \lambda \in \Lambda\}$ be a net in $F \cap r\mathcal{S}$ converging σ-strongly to x. For each finite subset J of I,

$\{x_\lambda^{1/2} q_J x_\lambda^{1/2} : \lambda \in \Lambda\}$ converges to $x^{1/2} q_J x^{1/2}$ σ-strongly. Since $\mathcal{M}_0$ is an ideal of $\mathcal{M}$, both $\{x_\lambda^{1/2} q_J x_\lambda^{1/2}\}$ and $x^{1/2} q_J x^{1/2}$ belong to $\mathcal{M}_0$. Since $x_\lambda^{1/2} q_J x_\lambda^{1/2} \in F$, $x^{1/2} q_J x^{1/2}$ belongs to $F \cap \mathcal{M}_0$ as shown above. Namely, we have $\varphi(x^{1/2} q_J x^{1/2}) \leq 1$, thus we conclude, using the complete additivity of φ,

$$\begin{aligned}\varphi(x) &= \varphi\Big(\sum_{i \in I} x^{\frac{1}{2}} p_i x^{\frac{1}{2}}\Big) = \sum_{i \in I} \varphi\big(x^{\frac{1}{2}} p_i x^{\frac{1}{2}}\big) \\ &= \lim_J \sum_{i \in J} \varphi\big(x^{\frac{1}{2}} p_i x^{\frac{1}{2}}\big) = \lim_J \varphi\big(x^{\frac{1}{2}} q_J x^{\frac{1}{2}}\big) \\ &\leq 1.\end{aligned}$$

Hence $x \in F$. Therefore, F is σ-weakly closed. Now, for each $s > 0$, we have

$$F_s = \{x \in \mathcal{M}_+ : \varphi(x) \leq s^2\} = s^2 F,$$

concluding the σ-weak lower semi-continuity of φ. Q.E.D.

Our next task is to show that a normal weight φ is of the form (12). We need a few results in ordered locally convex vector spaces.

Let A be an ordered locally convex topological vector space over $\mathbf{R}$. With A_+ the positive part, we assume

$$A = A_+ - A_+. \tag{16}$$

In the dual space A^*, we define the dual positive cone:

$$A_+^* = \{\omega \in A^* : \omega(x) \geq 0,\ x \in A_+\}. \tag{17}$$

It follows that A_+^* gives rise to an ordered vector structure in A^*. We say that a subset F of A_+ is *hereditary* if $y \in F$ whenever $0 \leq y \leq x$ and $x \in F$. The polar F° of a subset F of A is, by definition, given as follows:

$$F^\circ = \{\omega \in A^* : \omega(x) \leq 1,\ x \in F\}. \tag{18}$$

We shall use the notation:

$$F^\wedge = F^\circ \cap A_+^* = \{\omega \in A_+^* : \omega(x) \leq 1,\ x \in F\}. \tag{19}$$

For a subset E of A_+^*, we define the polar E° and $E^\wedge$ in A similarly.

Lemma 1.16. *Under the above setting, the following conditions are mutually equivalent:*

(i) *For every hereditary convex closed subset F of A_+*

$$F = (F - A_+)^- \cap A_+, \tag{20}$$

where the bar means the closure.

(ii) *For every hereditary convex closed subset* F *of* A_+

$$F = F^{\wedge\wedge}. \tag{21}$$

(iii) *If* φ *is an extended real* $[0, \infty]$ *valued lower semi-continuous function on* A_+ *such that*

$$\left.\begin{aligned} &\varphi(x) \le \varphi(y), && \text{if } 0 \le x \le y; \\ &\varphi(x+y) \le \varphi(x) + \varphi(y), && \text{if } x, y \in A_+; \\ &\varphi(\lambda x) = \lambda\varphi(x), && \text{if } \lambda \in \mathbf{R}_+, \end{aligned}\right\} \tag{22}$$

then φ *is of the form:*

$$\varphi(x) = \sup\{\, \omega(x) : \omega \in \Phi \,\}, \quad x \in A_+, \tag{23}$$

where

$$\Phi = \{\, \omega \in A_+^* : \omega(x) \le \varphi(x),\ x \in A_+ \,\}. \tag{24}$$

PROOF:
(i) $\Longrightarrow$ (ii): Since $F - A_+ = \operatorname{conv}(F \cup (-A_+))$, the convex hull, we have

$$(F - A_+)^\circ = (F \cup (-A_+))^\circ = F^\circ \cap A_+^* = F^\wedge,$$

so that the Hahn-Banach separation theorem yields

$$(F - A_+)^- = (F - A_+)^{\circ\circ} = (F^\wedge)^\circ \cap A_+ = F^{\wedge\wedge}.$$

(ii) $\Longrightarrow$ (iii): Suppose that φ is a function on A_+ satisfying the assumption in (iii) and set $F = \{\, x \in A_+ : \varphi(x) \le 1 \,\}$. Define $\Phi = F^\wedge$. It follows that Φ is precisely the set given by (23). We then set

$$\psi(x) = \sup\{\, \omega(x) : \omega \in \Phi \,\}, \quad x \in A_+.$$

Obviously, we have $\psi(x) \le \varphi(x),\ x \in A_+$. Suppose that there exists an $x_0 \in A_+$ such that $\psi(x_0) < \varphi(x_0)$. By multiplying a scalar, we may assume that $\psi(x_0) < 1 < \varphi(x_0)$. By the definition of F, $x_0 \notin F$. But, on the other hand condition (ii) says, together with the lower semi-continuity of φ, that

$$F = \bar{F} = F^{\wedge\wedge} = \{\, x \in A_+ : \omega(x) \le 1,\ \omega \in \Phi \,\} = \{\, x \in A_+ : \psi(x) \le 1 \,\},$$

so that x_0 must belong to F, which is a contradiction. Thus $\psi = \varphi$.
(iii) $\Longrightarrow$ (i): Suppose that F is a closed hereditary convex subset of A_+. We set

$$\varphi(x) = \inf\{\, r > 0 : x \in rF \,\}, \quad x \in A_+.$$

It follows that φ satisfies the assumption of (iii). The closedness of F yields that $F = \{\, x \in A_+ : \varphi(x) \le 1 \,\}$. Hence

$$F = \{\, x \in A_+ : \omega(x) \le 1,\ \omega \in \Phi \,\} = \Phi^\circ \cap A_+.$$

But we have

$$\begin{aligned}\Phi &= \{\omega \in A_+^* : \omega(x) \le \varphi(x),\ x \in A_+\} \\ &= \{\omega \in A_+^* : \omega(x) \le 1,\ x \in F\} = F^\circ \cap A_+^* \\ &= \{F \cup (-A_+)\}^\circ = (F - A_+)^\circ,\end{aligned}$$

so that $\Phi^\circ = (F - A_+)^{\circ\circ} = (F - A_+)^-$ by the separation theorem. Q.E.D.

We are now ready to complete the proof of the theorem.

PROOF OF THE IMPLICATION (iii) $\Longrightarrow$ (iv) IN THEOREM 1.11: By the last lemma, the only thing we have to do is to show that the self-adjoint part $\mathcal{M}_h$ of $\mathcal{M}$ satisfies condition (i) of the lemma.

Let E be a σ-strongly closed hereditary convex subset of $\mathcal{M}_+$. For each $x \in \mathcal{M}_h$, we set $\alpha_x = \sup\{\alpha > 0 : x \ge -1/\alpha\}$. We consider the family $\{f_\alpha : \alpha > 0\}$ of functions given by (10) and let

$$G = \{x \in \mathcal{M}_h : f_\alpha(x) \in E - \mathcal{M}_+,\ \alpha \in]0, \alpha_x[\}. \tag{25}$$

We first show that $G \cap r\mathcal{S}$, $r > 0$, is σ-strongly closed. Suppose $\{x_i : i \in I\}$ is a net in $G \cap r\mathcal{S}$ converging σ-strongly to an $x \in \mathcal{M}_h$. By assumption, $f_\alpha(x_i) \in E - \mathcal{M}_+$ for $0 < \alpha < 1/2r$, so that we can find a $y_i \in E$ such that $f_\alpha(x_i) \le y_i$. Hence

$$f_{2\alpha}(x_i) = f_\alpha\big(f_\alpha(x_i)\big) \le f_\alpha(y_i) \le 1/\alpha.$$

If $0 < \alpha < 1/2r$, then $f_{2\alpha}$ is continuous on $[-r, r]$, $\{f_{2\alpha}(x_i) : i \in I\}$ converges to $f_{2\alpha}(x)$ σ-strongly. By the boundedness of $\{f_\alpha(y_i) : i \in I\}$, there exists a subnet $\{y_j : j \in J\}$ of $\{y_i\}$ such that $\{f_\alpha(y_j) : j \in J\}$ converges to a $y_\alpha \in \mathcal{M}_h$ σ-weakly. Since $0 \le f_\alpha(y_j) \le y_j \in E$, $f_\alpha(y_j)$ belongs to E, so that $y_\alpha \in E$ by the σ-weak closedness of E. Now, the inequality:

$$y_\alpha - f_{2\alpha}(x) = \lim_J \big(f_\alpha(y_j) - f_{2\alpha}(x_j)\big) \ge 0$$

entails $f_{2\alpha}(x) \in E - \mathcal{M}_+$ for $0 < \alpha < 1/2r$, that is, $f_\alpha(x) \in E - \mathcal{M}_+$ if $0 < \alpha < 1/r$. Since $f_\alpha(t) \ge f_\beta(t)$ for $0 < \alpha < \beta$ and $t \in \big]-1/\alpha, \infty\big[$, we have $f_\beta(x) \le f_\alpha(x)$, which means that

$$f_\beta(x) \in (E - \mathcal{M}_+) - \mathcal{M}_+ = E - \mathcal{M}_+, \quad 0 < \beta < \alpha_x.$$

Therefore, we conclude $x \in G$ and so the σ-strong closedness of G.

We next show

$$G \cap r\mathcal{S} = \big((E - \mathcal{M}_+) \cap s\mathcal{S}\big)^- \cap r\mathcal{S}, \quad s > r, \tag{26}$$

where the bar means the σ-strong closure. It is clear that this would imply the convexity of $G \cap r\mathcal{S}$. Now suppose $x \in G \cap r\mathcal{S}$. Then $f_\alpha(x) \in E - \mathcal{M}_+$ for $0 < \alpha < \alpha_x$.

For a sufficiently small $\alpha > 0$, $f_\alpha(x) \in s\mathcal{S}$ and $f_\alpha(x) \nearrow x$ as $\alpha \searrow 0$, so that $x \in \{(E - \mathcal{M}_+) \cap s\mathcal{S}\}^-$ for each $s > r$. On the other hand, $G \supset E - \mathcal{M}_+$ because $f_\alpha(x) \le x$ for $0 < \alpha < \alpha_x$, so that $G \cap s\mathcal{S} \supset (E - \mathcal{M}_+) \cap s\mathcal{S}$. Hence $G \cap s\mathcal{S} \supset ((E - \mathcal{M}_+) \cap s\mathcal{S})^-$ by the closedness of G as just shown. Therefore, we get

$$G \cap r\mathcal{S} \supset [E - \mathcal{M}_+) \cap s\mathcal{S}]^- \cap s\mathcal{S}, \quad s > r.$$

Thus we obtain (26) which readily implies the convexity of $G \cap r\mathcal{S}$. Therefore, G itself is convex, which in turn entails the σ-weak closedness of G. Since $E - \mathcal{M}_+ \subset G \subset (E - \mathcal{M}_+)^-$, we conclude

$$G = (E - \mathcal{M}_+)^-.$$

Now, if $x \in (E - \mathcal{M}_+)^- \cap \mathcal{M}_+ = G \cap \mathcal{M}_+$, then $0 \le f_\alpha(x) \le x$ and $f_\alpha(x) \in E - \mathcal{M}_+$, so that $f_\alpha(x) \in E$ by the heredity of E. Hence $x = \lim_{\alpha \to 0} f_\alpha(x) \in E$ by the closedness assumption on E. Namely, we have $E \supset (E - \mathcal{M}_+)^- \cap \mathcal{M}_+$. Since the reversed inclusion is clear, we get $E = \overline{(E - \mathcal{M}_+)} \cap \mathcal{M}_+$ as desired. Q.E.D.

From now on, we mean by a normal weight a weight φ on $\mathcal{M}$ of the form (11) and (12). We fix φ and $\{\pi_\varphi, \mathfrak{H}_\varphi, \eta_\varphi\}$, the semi-cyclic representation of $\mathcal{M}$. Set

$$E_\varphi = \bigcup_{\lambda \ge 0} \lambda \Phi_\varphi. \tag{27}$$

For each $\omega \in E_\varphi$, we define a sesquilinear form B_ω on $\eta_\varphi(\mathfrak{n}_\varphi)$ as follows:

$$B_\omega(\eta_\varphi(x), \eta_\varphi(y)) = \omega(y^*x), \quad x, y \in \mathfrak{n}_\varphi. \tag{28}$$

Since $\omega \le \lambda\varphi$ for some $\lambda > 0$, B_ω is bounded, so that there exists a unique $h_\omega \ge 0$ on $\mathfrak{H}_\varphi$ such that

$$(h_\omega \eta_\varphi(x) \mid \eta_\varphi(y)) = \omega(y^*x), \quad x, y \in \mathfrak{n}_\varphi. \tag{28'}$$

It is easy to see that h_ω belongs to $\pi_\varphi(\mathcal{M})'$. Now we consider the cyclic representation $\{\pi_\omega, \mathfrak{H}_\omega, \xi_\omega\}$ of $\mathcal{M}$ determined by ω. The inequality $\omega \le \lambda\varphi$ assures that the map t_ω defined by

$$t_\omega \eta_\varphi(x) = \pi_\omega(x)\xi_\omega, \quad x \in \mathfrak{n}_\varphi, \tag{29}$$

can be extended to a bounded operator from $\mathfrak{H}_\varphi$ into $\mathfrak{H}_\omega$. We use the same symbol t_ω to denote this extended operator. It follows easily that

$$\left.\begin{aligned} t_\omega \pi_\varphi(a) &= \pi_\omega(a) t_\omega, \quad a \in \mathcal{M}, \\ h_\omega &= t_\omega^* t_\omega. \end{aligned}\right\} \tag{30}$$

We then let

$$t_\omega = u_\omega h_\omega^{\frac{1}{2}} \tag{31}$$

be the polar decomposition of t_ω. The intertwining relation (30) entails the following:

$$u_\omega \pi_\varphi(a) = \pi_\omega(a) u_\omega, \quad a \in \mathcal{M}. \tag{31'}$$

We set here

$$\eta_\omega = u_\omega^* \xi_\omega \tag{32}$$

and obtain, for each $x \in \mathfrak{n}_\varphi$,

$$\pi_\varphi(x)\eta_\omega = \pi_\varphi(x)u_\omega^*\xi_\omega = u_\omega^*\pi_\omega(x)\xi_\omega = u_\omega^* t_\omega \eta_\varphi(x) = h_\omega^{\frac{1}{2}}\eta_\varphi(x),$$

that is

$$h_\omega^{\frac{1}{2}}\eta_\varphi(x) = \pi_\varphi(x)\eta_\omega, \quad x \in \mathfrak{n}_\varphi. \tag{33}$$

Now put

$$\mathfrak{p}_{\varphi'} = \{ h_\omega : \omega \in E_\varphi \} \subset \pi_\varphi(\mathcal{M})'_+. \tag{34}$$

Theorem 1.17. *In the above setting, if we set*

$$\varphi'(x) = \begin{cases} \|\omega\| & \text{for } x = h_\omega \in \mathfrak{p}_{\varphi'}, \\ +\infty & \text{otherwise,} \end{cases} \tag{35}$$

then φ' is a faithful semi-finite normal weight on $\pi_\varphi(\mathcal{M})'$.

PROOF: By its definition, (27), E_φ is a convex subcone of $\mathcal{M}_*^+$ and the correspondence: $\omega \in E_\varphi \mapsto h_\omega \in \pi_\varphi(\mathcal{M})'_+$ is homogeneous and additive, so that $\mathfrak{p}_{\varphi'}$ is a convex subcone of $\pi_\varphi(\mathcal{M})'_+$. If $0 \le x \le h_\omega$, $x \in \mathcal{M}'$, and $\omega \in E_\varphi$, then $x^{1/2} = s\, h_\omega^{1/2}$, $\|s\| \le 1$, by Lemma 1.6, and we define for each $a \in \mathcal{M}$, $\rho(a) = \big(\pi_\varphi(a)s\eta_\omega \mid s\eta_\omega\big)$. We then get, for $a, b \in \mathfrak{n}_\varphi$,

$$\begin{aligned} \rho(b^*a) &= \big(\pi_\varphi(b^*a)s\eta_\omega \mid s\eta_\omega\big) = \big(\pi_\varphi(a)s\eta_\omega \mid \pi_\varphi(b)s\eta_\omega\big) \\ &= \big(s\pi_\varphi(a)\eta_\omega \mid s\pi_\omega(b)\eta_\omega\big) = \big(sh_\omega^{\frac{1}{2}}\eta_\varphi(a) \mid sh_\omega^{\frac{1}{2}}(b)\big) \qquad \text{by (33)} \\ &= \big(x^{\frac{1}{2}}\eta_\varphi(a) \mid x^{\frac{1}{2}}\eta_\varphi(b)\big) = \big(x\eta_\varphi(a) \mid \eta_\varphi(b)\big), \end{aligned}$$

so that $0 \le \rho \le \omega$; hence $\rho \in E_\varphi$ and $h_\rho = x \in \mathfrak{p}_{\varphi'}$. Therefore, $\mathfrak{p}_{\varphi'}$ is hereditary and

$$\varphi'(x) = \|\rho\| \le \|\omega\| = \varphi'(h_\omega).$$

Thus φ' is well-defined and monotone increasing.

We now show the complete additivity of φ'. Suppose that $h = \sum_{i\in I} h_i$ converges σ-strongly in $\pi_\varphi(\mathcal{M})'_+$. Assume first $\sum_{i\in E} \varphi'(h_i) < +\infty$. By the construction of φ, each h_i is of the form: $h_i = h_{\omega_i}$, $\omega_i \in E_\varphi$, and the assumption means the convergence:

$$\sum_{i\in I} \|\omega_i\| = \sum_{i\in I} \varphi'(h_i) < +\infty,$$

so that $\omega = \sum_{i\in I} \omega_i \in \mathcal{M}_*^+$ converges in norm. For each $x \in \mathfrak{n}_\varphi$, we have

$$\begin{aligned} \omega(x^*x) &= \sum_{i\in I} \omega_i(x^*x) = \sum_{i\in I} \big(h_i\eta_\varphi(x) \mid \eta_\varphi(x)\big) = \big(h\eta_\varphi(x) \mid \eta_\varphi(x)\big) \\ &= \big\|h^{\frac{1}{2}}\eta_\varphi(x)\big\|^2 \le \|h\| \big\|\eta_\varphi(x)\big\|^2 = \|h\|\varphi(x^*x), \end{aligned}$$

so that $\omega \in E_\varphi$ and $h = h_\omega$, and hence

$$\varphi'(h) = \|\omega\| = \sum \|\omega_i\| = \sum \varphi'(h_i).$$

Conversely, suppose $\varphi'(h) < +\infty$. In this case, $h = h_\omega$ for some $\omega \in E_\varphi$. Since $\mathfrak{p}'_\varphi$ is hereditary, $h_i \in \mathfrak{p}_{\varphi'}$, $i \in I$, and for any finite $J \subset I$,

$$\sum_{i \in J} \varphi'(h_i) = \varphi'\Bigl(\sum_{i \in J} h_i\Bigr) \le \varphi'(h) < +\infty.$$

Thus $\sum_{i \in I} \varphi'(h_i) < +\infty$. By the above argument, we conclude $\varphi'(h) = \sum_{i \in I} \varphi'(h_i)$. Thus, φ' is completely additive.

By its construction, φ' is faithful. To complete the proof, we prove the semi-finiteness of φ'. To this end, let

$$\mathfrak{n}_{\varphi'} = \bigl\{ x \in \pi_\varphi(\mathcal{M})' : \varphi'(x^*x) < +\infty \bigr\};$$

$$\mathfrak{m}_{\varphi'} = \Bigl\{ \sum_{i=1}^{n} y_i^* x_i : x_i, \dots, x_n, y_1, \dots, y_n \in \mathfrak{n}_{\varphi'} \Bigr\} = \mathfrak{n}_{\varphi'}^* \mathfrak{n}_{\varphi'}.$$

By Lemma 1.2, $\mathfrak{p}_{\varphi'} = \mathfrak{m}_{\varphi'}^+$ and it follows that

$$\bigl\{ h_\omega : \omega \in \Phi_\varphi \bigr\} = \bigl\{ x \in \mathfrak{m}_{\varphi'}^+ : \|x\| \le 1 \bigr\}.$$

For each $x \in \mathfrak{n}_\varphi$, we have

$$\begin{aligned}
\|\eta_\varphi(x)\|^2 &= \varphi(x^*x) = \sup\{ \omega(x^*x) : \omega \in \Phi_\varphi \} \\
&= \sup\Bigl\{ \|h_\omega^{\frac{1}{2}} \eta_\varphi(x)\|^2 : \omega \in \Phi_\varphi \Bigr\} \\
&= \sup\Bigl\{ \|h \eta_\varphi(x)\|^2 : h \in \mathfrak{n}_{\varphi'},\ 0 \le h \le 1 \Bigr\},
\end{aligned}$$

so that the *-algebra $\mathfrak{n}_{\varphi'} \cap \mathfrak{n}_{\varphi'}^*$ is non-degenerate on $\mathfrak{H}_\varphi$. Hence the open unit ball of $\bigl(\mathfrak{n}_{\varphi'} \cap \mathfrak{n}_{\varphi'}^*\bigr)_+$ is upward directed and converges to the identity σ-strongly. Since $a^* \pi_\varphi(\mathcal{M})' a \subset \mathfrak{n}_{\varphi'} \cap \mathfrak{n}_{\varphi'}^*$ for every $a \in \mathfrak{n}_{\varphi'}$, $\pi_\varphi(\mathcal{M})'$ is the σ-strong closure of $\mathfrak{n}_{\varphi'} \cap \mathfrak{n}_{\varphi'}^*$. Thus φ' is semi-finite on $\pi_\varphi(\mathcal{M})'$. Q.E.D.

Definition 1.18. The weight φ' on $\pi_\varphi(\mathcal{M})'$ given by (35) is called the *opposite* weight of φ.

Let φ be a normal weight on $\mathcal{M}$. Let e be the projection of $\mathcal{M}$ such that $\mathcal{M}e = \mathfrak{n}_\varphi^-$, the σ-strong closure, and let f the projection of $\mathcal{M}$ such that $\mathcal{M}f = N_\varphi = \bigl\{ x \in \mathcal{M} : \varphi(x^*x) = 0 \bigr\}$. It is clear that φ is semi-finite on $e\mathcal{M}e$ and faithful on $(1-f)\mathcal{M}(1-f)$. indexWeight!, support of a ~We call the projection $e - f$ the *support* of φ and denote it by $s(\varphi)$.

In the rest of this book, we mean by a weight on a von Neumann algebra a *semi-finite normal* one unless otherwise is explicitly mentioned.

§ 2 Left Hilbert Algebras and Weights

We discuss in this section the relation between full left Hilbert algebras and faithful semi-finite normal weights on von Neumann algebras.

Let $\mathfrak{A}$ be a full left Hilbert algebra with completion $\mathfrak{H}$ and left von Neumann algebra $\mathcal{M} = \mathcal{R}_\ell(\mathfrak{A})$. We shall use the notations in the last chapter and the last section freely. By Lemma VI.1.15, we have

$$\pi_\ell(\mathfrak{A}) = \mathfrak{n}_\ell \cap \mathfrak{n}_\ell^*, \qquad \pi_r(\mathfrak{A}') = \mathfrak{n}_r \cap \mathfrak{n}_r^*. \tag{1}$$

Set

$$\left.\begin{aligned} \mathfrak{m}_\ell = \mathfrak{n}_\ell^*\mathfrak{n}_\ell &= \left\{\sum_{i=1}^n x_i^* y_i \ : x_1, \dots, x_n, \ y_1, \dots, y_n \in \mathfrak{n}_\ell\right\} \\ \mathfrak{m}_r = \mathfrak{n}_r^*\mathfrak{n}_r &= \left\{\sum_{i=1}^m u_i^* v_i \ : u_1, \dots, u_m, \ v_1, \dots, v_m \in \mathfrak{n}_r\right\} \end{aligned}\right\} \tag{2}$$

We then define an extended positive real valued function φ_ℓ on $\mathcal{M}_+$ as follows:

$$\varphi_\ell(x) = \begin{cases} \|\xi\|^2 & \text{if } x^{\frac{1}{2}} = \pi_\ell(\xi), \ \xi \in \mathfrak{A}; \\ +\infty & \text{otherwise.} \end{cases} \tag{3}$$

Similarly, we define φ_r on $\mathcal{M}'_+$:

$$\varphi_r(y) = \begin{cases} \|\eta\|^2 & \text{if } y^{\frac{1}{2}} = \pi_r(\eta), \ \eta \in \mathfrak{A}'; \\ +\infty & \text{otherwise..} \end{cases} \tag{3'}$$

Lemma 2.1. *In the above situation,* $\mathfrak{m}_\ell^+$ $(= \mathfrak{m}_\ell \cap \mathcal{M}_+)$ *(resp.* $\mathfrak{m}_r^+$ *) is a hereditary convex subcone of* $\mathcal{M}_+$ *(resp.* $\mathcal{M}'_+$ *). Furthermore,* $\mathfrak{n}_\ell$ *and* $\mathfrak{n}_r$ *are given by the following:*

$$\mathfrak{n}_\ell = \left\{x \in \mathcal{M} : x^*x \in \mathfrak{m}_\ell^+\right\}, \qquad \mathfrak{n}_r = \left\{y \in \mathcal{M}' : y^*y \in \mathfrak{m}_r^+\right\}. \tag{4}$$

PROOF: Since $\mathfrak{m}_\ell$ is a self-adjoint subalgebra of $\mathcal{M}$, $\mathfrak{m}_\ell^+$ is a convex subcone of $\mathcal{M}_+$. Suppose that $a \in \mathfrak{m}_\ell^+$ dominates $b \in \mathcal{M}_+$, i.e. $0 \leq b \leq a$. By (2), a is of the form $a = \sum_{i=1}^n x_i^* y_i$, $x_1, \dots, x_n$, $y_1, \dots, y_n \in \mathfrak{n}_\ell$, so we have

$$a = \frac{1}{2}(a + a^*) = \sum_{i=1}^n \frac{1}{2}(x_i^* y_i + y_i^* x_i) \leq \sum_{i=1}^n (x_i^* x_i + y_i^* y_i) \in \mathfrak{m}_\ell^+.$$

Hence replacing a by the last expression in the above inequality, b may be assumed to be dominated by $a = \sum_{i=1}^n x_i^* x_i$, $x_1 \dots, x_n \in \mathfrak{n}_\ell$. We then want to show $b \in \mathfrak{m}_\ell^+$. By definition, x_j's are of the form $x_j = \pi_\ell(\xi_j)$, $\xi_j \in \mathfrak{B}$, $1 \leq j \leq n$. By

Lemma 1.6.(ii), we choose $s_1, \dots, s_n$ from the unit ball of $\mathcal{M}$ so that $x_i^{1/2} = s_i a^{1/2}$ and $p = \sum_{i=1}^n s_i^* s_i$ is the range projection $s(a)$ of a. We set here $\xi = \sum_{i=1}^n s_i^* \xi_i$. We then have $\xi \in \mathfrak{B}$ and

$$\pi_\ell(\xi) = \sum_{i=1}^n s_i^* \pi_\ell(\xi_i) = \sum_{i=1}^n s_i^* x_i = \sum_{i=1}^n s_i^* s_i a^{\frac{1}{2}} = pa^{\frac{1}{2}} = a^{\frac{1}{2}}.$$

Hence $a^{1/2} = \pi_\ell(\xi)$. By Lemma 1.6.(i), we choose an $s \in \mathcal{M}$ so that $b^{1/2} = sa^{1/2}$. Then we have

$$b^{\frac{1}{2}} = s\pi_\ell(\xi) = \pi_\ell(s\xi) \in \mathfrak{n}_\ell,$$

so that $b = b^{1/2}b^{1/2} \in \mathfrak{m}_\ell^+$. Thus $\mathfrak{m}_\ell^+$ is hereditary.

If $x \in \mathfrak{n}_\ell$, then $x^*x \in \mathfrak{m}_\ell^+$ by (2). Conversely, if $x^*x \in \mathfrak{m}_\ell^+$, then $|x| = (x^*x)^{1/2}$ is of the form $|x| = \pi_\ell(\xi)$ for some $\xi \in \mathfrak{B}$ as seen above. Writing $x = u|x|$ in the polar decomposition, we have $x = u\pi_\ell(\xi) = \pi_\ell(u\xi) \in \mathfrak{n}_\ell$.

By symmetry, the assertions for $\mathfrak{n}_r$ follow. Q.E.D.

For each $\xi \in \mathfrak{H}$, we define $\omega_\xi^\ell \in \mathcal{M}_*^+$ and $\omega_\xi^r \in (\mathcal{M})_*^+$ as follows:

$$\omega_\xi^\ell(x) = (x\xi \mid \xi), \quad x \in \mathcal{M}; \qquad \omega_\xi^r(y) = (y\xi \mid \xi), \quad x \in \mathcal{M}';$$

We then consider the sets defined by:

$$\left.\begin{aligned} \Phi_\ell &= \left\{\omega_\eta^\ell : \eta \in \mathfrak{B}', \ \left\|\pi_r(\eta)\right\| \le 1\right\}, \\ \Phi_{\ell,0} &= \left\{\omega_\eta^\ell : \eta \in \mathfrak{B}', \ \left\|\pi_r(\eta)\right\| < 1\right\}, \\ \Phi_r &= \left\{\omega_\xi^r : \xi \in \mathfrak{B}, \ \left\|\pi_\ell(\xi)\right\| \le 1\right\}, \\ \Phi_{r,0} &= \left\{\omega_\xi^r : \xi \in \mathfrak{B}, \ \left\|\pi_\ell(\xi)\right\| < 1\right\}. \end{aligned}\right\} \tag{5}$$

Lemma 2.2. *Let $\mathcal{S}$ and $\mathcal{S}_0$ (resp. $\mathcal{S}'$ and $\mathcal{S}_0'$) denote the closed unit ball and the open unit ball of $\mathcal{M}$ (resp. $\mathcal{M}'$). There exists a completely positive map θ (resp. θ') from $\mathfrak{m}_\ell$ into $\mathcal{M}_*'$ (resp. $\mathfrak{m}_r$ into $\mathcal{M}_*$) such that*

$$\left.\begin{aligned} \theta\big(\pi_\ell(\xi)^*\pi_\ell(\xi)\big) &= \omega_\xi^r, \quad \xi \in \mathfrak{B}, \\ \theta'\big(\pi_r(\eta)^*\pi_r(\eta)\big) &= \omega_\eta^\ell, \quad \eta \in \mathfrak{B}', \end{aligned}\right\} \tag{6}$$

$$\left.\begin{aligned} \theta\big(\mathfrak{m}_\ell^+ \cap \mathcal{S}\big) &= \Phi_r, \qquad \theta\big(\mathfrak{m}_r^+ \cap \mathcal{S}_0\big) = \Phi_{r,0}. \\ \theta'\big(\mathfrak{m}_r^+ \cap \mathcal{S}'\big) &= \Phi_\ell, \qquad \theta'\big(\mathfrak{m}_\ell^+ \cap \mathcal{S}_0'\big) = \Phi_{\ell,0}. \end{aligned}\right\} \tag{7}$$

The proof goes exactly the same line as Lemma 1.8. We leave the proof to the reader as an exercise.

Corollary 2.3. *The sets $\Phi_{\ell,0}$ and $\Phi_{r,0}$ are hereditary convex subsets of $\mathcal{M}_*^+$ and $(\mathcal{M}')_*^+$ respectively.*

Lemma 2.4. *The functions φ_ℓ and φ_r in (3) and (3′) are given as follows:*

$$\left.\begin{aligned} \varphi_\ell(x) &= \sup\{\,\omega(x) : \omega \in \Phi_{\ell,0}\,\}, \quad x \in \mathcal{M}_+, \\ \varphi_r(y) &= \sup\{\,\omega(y) : \omega \in \Phi_{r,0}\,\}, \quad y \in \mathcal{M}'_+. \end{aligned}\right\} \tag{8}$$

PROOF: By symmetry, we shall prove only the first equality. Let $\psi(x),\ x \in \mathcal{M}_+$, be the right hand side of the first equation. We want to show $\varphi_\ell = \psi$.

Since $\Phi_{\ell,0}$ is directed upward by (7), ψ is a normal weight on $\mathcal{M}_+$. If $a \in \mathfrak{m}_\ell^+$, then $a^{1/2} = \pi_\ell(\xi)$ with $\xi \in \mathfrak{A}$. Keeping in mind the fact that $\mathcal{S}'_0 \cap \mathfrak{n}_r^+$ is upward directed and converges strongly to the identity, we compute:

$$\begin{aligned} \psi(a) = \sup\{\,\omega(a) : \omega \in \Phi_{\ell,0}\,\} &= \sup\Big\{\,\big(a\eta \mid \eta\big) : \eta \in \mathfrak{B}',\ \big\|\pi_r(\eta)\big\| < 1\,\Big\} \\ &= \sup\Big\{\,\big\|a^{\frac{1}{2}}\eta\big\|^2 : \eta \in \mathfrak{B}',\ \big\|\pi_r(\eta)\big\| < 1\,\Big\} \\ &= \sup\Big\{\,\big\|\pi_r(\eta)\xi\,\big\|^2 : \eta \in \mathfrak{B}',\ \big\|\pi_r(\eta)\big\| < 1\,\Big\} \\ &= \sup\Big\{\,\big\|b\xi\,\big\|^2 : b \in \mathcal{S}'_0 \cap \mathfrak{n}_r\,\Big\} = \|\xi\|^2 = \varphi_\ell(a). \end{aligned}$$

Conversely, suppose $\psi(a) = \lambda < +\infty,\ a \in \mathcal{M}_+$. We define a functional ω'_a on $\mathfrak{m}_r$ as follows:

$$\omega'_a(y) = \big\langle a, \theta'(y)\big\rangle, \quad y \in \mathfrak{m}_r.$$

Since $\theta'(\mathfrak{m}_r^+) \subset \mathcal{M}_*^+$, ω'_a is positive and

$$\begin{aligned} \|\omega'_a\| &= \sup\{\,\omega'_a(y) : y \in \mathfrak{m}_r^+ \cap \mathcal{S}'_0\,\} = \sup\Big\{\,\big\langle a, \theta'(y)\big\rangle : y \in \mathfrak{m}_r^+ \cap \mathcal{S}'_0\,\Big\} \\ &= \sup\{\,\langle a, \omega\rangle : \omega \in \Phi_{\ell,0}\,\} = \psi(a) = \lambda < +\infty. \end{aligned}$$

Hence ω'_a is bounded, so that it can be extended to the norm closure A_r of $\mathfrak{m}_r$ as a positive linear functional, which will be denoted by ω'_a again. On the C^*-algebra A_r, we have the inequality:

$$\big|\omega'_a(y)\big|^2 \le \|\omega'_a\|\omega'_a(y^*y), \quad y \in A_r.$$

Hence, we have for every $\eta \in \mathfrak{B}'$,

$$\Big|\omega'_a\big(\pi_r(\eta)\big)\Big| \le \sqrt{\lambda}\,\omega'_a\big(\pi_r(\eta)^*\pi_r(\eta)\big)^{\frac{1}{2}} = \sqrt{\lambda}\,\big(a\eta \mid \eta\big)^{\frac{1}{2}} = \sqrt{\lambda}\,\big\|a^{\frac{1}{2}}\eta\big\|.$$

By the Riesz representation theorem, there exists a vector $\xi \in \big[a^{1/2}\mathfrak{H}\big]$ such that

$$\omega'_a\big(\pi_r(\eta)\big) = \big(a^{\frac{1}{2}}\eta \mid \xi\big), \quad \eta \in \mathfrak{B}'.$$

Therefore, for each $\eta, \zeta \in \mathfrak{B}'$, we have

$$\begin{aligned}\left(a^{\frac{1}{2}}\eta \mid a^{\frac{1}{2}}\zeta\right) &= \left(a\eta \mid \zeta\right) = \omega'_a\left(\pi_r(\zeta)^*\pi_r(\eta)\right)\\ &= \left(a^{\frac{1}{2}}\pi_r(\zeta)^*\eta \mid \xi\right) = \left(a^{\frac{1}{2}}\eta \mid \pi_r(\zeta)\xi\right).\end{aligned}$$

Since we have

$$\pi_r(\zeta)\xi \in \pi_r(\zeta)\left[a^{\frac{1}{2}}\mathfrak{H}\right] \subset \left[a^{\frac{1}{2}}\pi_r(\zeta)\mathfrak{H}\right] \subset \left[a^{\frac{1}{2}}\mathfrak{H}\right],$$

we conclude that

$$a^{\frac{1}{2}}\zeta = \pi_r(\zeta)\xi, \quad \zeta \in \mathfrak{B}'.$$

Thus ξ is left bounded and $a^{1/2} = \pi_\ell(\xi) \in \mathfrak{n}_\ell^*$. Hence $a = (a^{1/2})^2 \in \mathfrak{m}_\ell^+$ and $\varphi_\ell(a) = \|\xi\|^2 < +\infty$. Therefore, the first half of the proof applies to a, concluding $\varphi_\ell(a) = \psi(a)$. Thus (8) holds. Q.E.D.

We are now ready to show that a full left Hilbert algebra gives rise to a faithful semi-finite normal weight on the left von Neumann algebra.

Theorem 2.5. *If $\mathfrak{A}$ is a full left Hilbert algebra with left von Neumann algebra $\mathcal{M} = \mathcal{R}_\ell(\mathfrak{A})$, then the formulas (3) and (3′) give faithful semi-finite normal weights φ_ℓ on $\mathcal{M}$ and φ_r on $\mathcal{M}'$ with the following properties:*

(i) *The action of $\mathcal{M}$ on $\mathfrak{H}$, the completion of $\mathfrak{A}$, is unitarily equivalent to the semi-cyclic representation π_{φ_ℓ} of $\mathcal{M}$ on $\mathfrak{H}_{\varphi_\ell}$ under the correspondence:*

$$U\xi = \eta_{\varphi_\ell}\left(\pi_\ell(\xi)\right), \quad \xi \in \mathfrak{B}. \tag{9}$$

(ii) *Identifying $\mathfrak{H}$ and $\mathfrak{H}_{\varphi_\ell}$ under the unitary U of (9), φ_r is the opposite weight of φ_ℓ.*

(iii)

$$\left.\begin{aligned}\mathfrak{m}_{\varphi_\ell} &= \mathfrak{m}_\ell, \qquad \mathfrak{n}_{\varphi_\ell} = \mathfrak{n}_\ell,\\ \mathfrak{m}_{\varphi_r} &= \mathfrak{m}_r \qquad \mathfrak{n}_{\varphi_r} = \mathfrak{n}_r,\\ \dot\varphi_\ell\left(\pi_\ell(\eta)^*\pi_\ell(\xi)\right) &= (\xi \mid \eta), \quad \xi, \eta \in \mathfrak{B},\\ \dot\varphi_r\left(\pi_r(\eta)^*\pi_r(\xi)\right) &= (\xi \mid \eta), \quad \xi, \eta \in \mathfrak{B}'.\end{aligned}\right\} \tag{10}$$

PROOF: We continue to use the notations established above. The last lemma shows that φ_ℓ and φ_r are both normal weights. Furthermore, the semi-finiteness of φ_ℓ follows from the fact that $\mathfrak{m}_\ell^+$ generates $\mathcal{M}$. To simplify the notation, we write φ for φ_ℓ and consider the semi-cyclic representation $\left\{\pi_\varphi, \mathfrak{H}_\varphi, \eta_\varphi\right\}$. If $\xi_1, \xi_2 \in \mathfrak{B}$, then

$$(\xi_1 \mid \xi_2) = \dot\varphi\left(\pi_\ell(\xi_2)^*\pi_\ell(\xi_1)\right) = \left(\eta_\varphi\left(\pi_\ell(\xi_1)\right) \,\middle|\, \eta_\varphi\left(\pi_\ell(\xi_2)\right)\right)$$

by the polarization identity applied to (3). Therefore, the map U of (9) can be extended to an isometry from of $\mathfrak{H}$ onto $\mathfrak{H}_\varphi$ which we denote by U again. Furthermore, it follows immediately that $UaU^* = \pi_\varphi(a)$ for $a \in \mathcal{M}$. If we identify $\mathfrak{H}_\varphi$ with $\mathfrak{H}$ via U, then the maps π_ℓ and η_φ are inverse of each other.

The assertion (ii) follows from Lemma 2.2 and Corollary 2.3. The rest of the theorem is now obvious. Q.E.D.

The next result states the converse of the above theorem:

Theorem 2.6. *Let φ be a faithful semi-finite normal weight on a von Neumann algebra $\mathcal{M}$. Let*

$$\mathfrak{A}_\varphi = \eta_\varphi\big(\mathfrak{n}_\varphi \cap \mathfrak{n}_\varphi^*\big). \tag{11}$$

If we define an involutive algebra structure in $\mathfrak{A}_\varphi$ in the following fashion:

$$\left.\begin{aligned} \eta_\varphi(x)\eta_\varphi(y) &= \eta_\varphi(xy), \quad x, y \in \mathfrak{n}_\varphi \cap \mathfrak{n}_\varphi^*, \\ \eta_\varphi(x)^\sharp &= \eta_\varphi(x^*), \end{aligned}\right\} \tag{12}$$

then $\mathfrak{A}_\varphi$ is a full left Hilbert algebra such that

(i) $\pi_\varphi(\mathcal{M}) = \mathcal{R}_\ell(\mathfrak{A}_\varphi)$;
(ii) *if we identify $\mathcal{M}$ and $\mathcal{R}_\ell(\mathfrak{A}_\varphi)$ via π_φ, then the weight φ_ℓ associated with $\mathfrak{A}_\varphi$ agrees with the original weight φ.*

PROOF: If $x, y \in \mathfrak{n}_\varphi \cap \mathfrak{n}_\varphi^*$, then $\pi_\varphi(x)\eta_\varphi(y) = \eta_\varphi(x)\eta_\varphi(y)$, so that the multiplication in $\mathfrak{A}_\varphi$ from the left is continuous and $\pi_\ell\big(\eta_\varphi(x)\big) = \pi_\varphi(x)$. It also follows that $\pi_\ell(\mathfrak{A}_\varphi) = \pi_\varphi(\mathfrak{n}_\varphi \cap \mathfrak{n}_\varphi^*)$. Hence $\pi_\ell(\mathfrak{A}_\varphi)$ generates $\pi_\varphi(\mathcal{M})$ as a von Neumann algebra; thus $\pi_\ell(\mathfrak{A}_\varphi)$ is non-degenerate on $\mathfrak{H}_\varphi$, which means that $\mathfrak{A}_\varphi^2$ is dense in $\mathfrak{H}_\varphi$ because $\pi_\varphi(\mathfrak{n}_\varphi \cap \mathfrak{n}_\varphi^*)$ contains an increasing net converging strongly to 1. The condition (b) in Definition VI.1.1 follows from the fact that π_φ is a $*$-representation. Therefore, the condition (c) remains to be shown. Namely, we must prove the preclosedness of the $\sharp$-operation in $\mathfrak{A}_\varphi$. This is the place where Theorem 1.11.(iv) plays a crucial role.

We set

$$\Phi_{\varphi,0} = \big\{\omega \in \mathcal{M}_*^+ : (1+\varepsilon)\omega \le \varphi \text{ for some } \varepsilon > 0\big\}. \tag{13}$$

To each $\omega \in \Phi_{\varphi,0}$ there correspond an $h_\omega \in \pi_\varphi(\mathcal{M})'_+$ with $\|h_\omega\| < 1$ and an $\eta_\omega \in \mathfrak{H}_\varphi$ such that

$$\omega(x) = \big(x\eta_\omega \mid \eta_\omega\big), \quad x \in \mathcal{M}, \qquad h_\omega^{\frac{1}{2}}\eta_\varphi(x) = \pi_\varphi(x)\eta_\omega, \quad x \in \mathfrak{n}_\varphi. \tag{14}$$

Now, for each $\omega_1, \omega_2 \in \Phi_{\varphi,0}$, $b \in \pi_\varphi(\mathcal{M})'$ and $x \in \mathfrak{n}_\varphi \cap \mathfrak{n}_\varphi^*$, we compute:

$$\begin{aligned} \big(\eta_\varphi(x^*) \mid h_{\omega_1}^{\frac{1}{2}} b\eta_{\omega_2}\big) &= \big(h_{\omega_1}^{\frac{1}{2}}\eta_\varphi(x^*) \mid b\eta_{\omega_2}\big) = \big(\pi_\varphi(x)^*\eta_{\omega_1} \mid \eta_{\omega_2})\big) \\ &= \big(\eta_{\omega_1} \mid \pi_\varphi(x)b\eta_{\omega_2}\big) = \big(\eta_{\omega_1} \mid b\pi_\varphi(x)\eta_{\omega_2}\big) \\ &= \big(\eta_{\omega_1} \mid bh_{\omega_2}^{\frac{1}{2}}\eta_\varphi(x)\big) = \big(h_{\omega_2}^{\frac{1}{2}}b^*\eta_{\omega_1} \mid \eta_\varphi(x)\big). \end{aligned}$$

If $\{x_n\}$ is a sequence in $\mathfrak{n}_\varphi \cap \mathfrak{n}^*$ such that

$$\lim_{n\to\infty} \eta_\varphi(x_n) = 0 \qquad \text{and} \qquad \lim_{n\to\infty} \eta_\varphi(x_n^*) = \xi \in \mathfrak{H}_\varphi$$

in norm, then the above computation shows that

$$\left(\xi \mid h_{\omega_1}^{\frac{1}{2}} b \eta_{\omega_2}\right) = \lim_{n\to\infty} \left(h_{\omega_2}^{\frac{1}{2}} b^* \eta_{\omega_1} \mid \eta_\varphi(x_n)\right) = 0$$

for every $b \in \pi_\varphi(\mathcal{M})'$ and $\omega_1, \omega_2 \in \Phi_{\varphi,0}$. To show the preclosedness of the $\sharp$-operation of $\mathfrak{A}_\varphi$, it suffices to prove that the union:

$$\bigcup \left\{ h_{\omega_1}^{\frac{1}{2}} \pi_\varphi(\mathcal{M})' \eta_{\omega_2} : \omega_1, \omega_2 \in \Phi_{\varphi,0} \right\}$$

is total in $\mathfrak{H}_\varphi$, because this implies immediately that $\xi = 0$. Now, we compute for $x \in \mathfrak{n}_\varphi \cap \mathfrak{n}_\varphi^*$:

$$\begin{aligned}
\|\eta_\varphi(x)\|^2 &= \sup\{\omega(x^*x) : \omega \in \Phi_{\varphi,0}\} \qquad \text{by (12)} \\
&= \sup\{\|x\eta_\omega\|^2 : \omega \in \Phi_{\varphi,0}\} = \sup\left\{\left\|h_\omega^{\frac{1}{2}} \eta_\varphi(x)\right\|^2 : \omega \in \Phi_{\varphi,0}\right\} \\
&= \sup\left\{\left(h_\omega \eta_\varphi(x) \mid \eta_\varphi(x)\right) : \omega \in \Phi_{\varphi,0}\right\}
\end{aligned}$$

Since $\Phi_{\varphi,0}$ is convex and the map: $\omega \in \Phi_{\varphi,0} \mapsto h_\omega \in \mathcal{M}'_+$ is affine, 1 is the strong closure of $\{h_\omega : \omega \in \Phi_{\varphi,0}\}$. Hence $\bigcup\{h_{\omega_1}^{1/2} \pi_\varphi(\mathcal{M})' \eta_{\omega_2} : \omega_1 \in \Phi_{\varphi,0}\}$ is dense in $\pi_\varphi(\mathcal{M})' \eta_{\omega_2}$.

Now we prove that $\bigcup\{\pi_\varphi(\mathcal{M})' \eta_\omega : \omega \in \Phi_{\varphi,0}\}$ is total in $\mathfrak{H}_\varphi$. Set $\mathfrak{K} = \left[\bigcup \pi_\varphi(\mathcal{M})' \eta_\omega : \omega \in \Phi_{\varphi,0}\right]$, the closed subspace spanned by the union, and let e be the projection of $\mathfrak{H}_\varphi$ onto $\mathfrak{K}$. Since $\mathfrak{K}$ is invariant under $\pi_\varphi(\mathcal{M})'$, e belongs to $\pi_\varphi(\mathcal{M})$ and $(1-e)\eta_\omega = 0$ for every $\omega \in \Phi_{\varphi,0}$. If $f \in \mathrm{Proj}(\mathcal{M})$ with $\pi_\varphi(f) = 1 - e$, then we have

$$\begin{aligned}
\varphi(f) &= \sup\{\omega(f) : \omega \in \Phi_{\varphi,0}\} = \sup\left\{\left(\pi_\varphi(f)\eta_\omega \mid \eta_\omega\right) : \omega \in \Phi_{\varphi,0}\right\} \\
&= \sup\left\{\left((1-e)\eta_\omega \mid \eta_\omega\right) : \omega \in \Phi_{\varphi,0}\right\} = 0.
\end{aligned}$$

Since φ is faithful, we have $f = 0$ and $e = 1$, which means precisely that $\mathfrak{K} = \mathfrak{H}_\varphi$. Thus, we conclude now that the $\sharp$-operation $\mathfrak{A}_\varphi$ is preclosed, and that $\mathfrak{A}_\varphi$ is a left Hilbert algebra.

Next, we want to prove that $\mathfrak{A}_\varphi$ is full. From the above arguments, it follows that the vector η_ω corresponding to $\omega \in \Phi_{\varphi,0}$ is right bounded and $\pi_r(\eta_\omega) = h_\omega^{1/2}$. The self-adjointness of $h_\omega^{1/2}$ implies that $\eta_\omega \in \mathfrak{A}'_\varphi$ by Lemma VI.1.15. Now, set $x = \pi_\ell(\xi) \in \pi_\varphi(\mathcal{M})$ for a left bounded vector $\xi \in \mathfrak{H}_\varphi$ and compute:

$$\begin{aligned}
\varphi(x^*x) &= \sup\{\omega(x^*x) : \omega \in \Phi_{\varphi,0}\} = \sup\{\|x\eta_\omega\|^2 : \omega \in \Phi_{\varphi,0}\} \\
&= \sup\left\{\left\|h_\omega^{\frac{1}{2}} \eta_\varphi(x)\right\|^2 : \omega \in \Phi_{\varphi,0}\right\} = \|\xi\|^2 < +\infty.
\end{aligned}$$

Hence $x \in \mathfrak{n}_\varphi$ and $\varphi(x^*x) = \|\xi\|^2$. Thus $\pi_\ell(\mathfrak{B}_\varphi) \subset \pi_\varphi(\mathfrak{n}_\varphi)$, where $\mathfrak{B}_\varphi$ means the set of all left bounded vectors in $\mathfrak{H}_\varphi$. By Lemma VI.1.15′.(ii′) we have

$$\pi_\ell(\mathfrak{A}_\varphi) \subset \pi_\ell(\mathfrak{A}''_\varphi) = \pi_\ell(\mathfrak{B}_\varphi) \cap \pi_\ell(\mathfrak{B}_\varphi)^* \subset \pi_\varphi\big(\mathfrak{n}_\varphi \cap \mathfrak{n}_\varphi^*\big) = \pi_\ell(\mathfrak{A}_\varphi),$$

which means that $\mathfrak{A}_\varphi = \mathfrak{A}''_\varphi$. Hence $\mathfrak{A}_\varphi$ is full.

Finally, if $x = uh \in \mathfrak{n}_\varphi$ is the polar decomposition, then $h = u^*x \in \mathfrak{n}_\varphi \cap \mathfrak{n}_\varphi^*$, so that $\eta_\varphi(x) = \pi_\varphi(h) \in \mathfrak{B}_\varphi$; hence $\eta_\varphi(x)$ is left bounded and $\pi_\ell\big(\eta_\varphi(x)\big) = \pi_\varphi(x)$. Therefore, we obtain $\pi_\ell(\mathfrak{B}_\varphi) = \pi_\varphi(\mathfrak{n}_\varphi)$. Now, we have, for every $x \in \mathfrak{n}_\varphi = \pi_\varphi^{-1}(\mathfrak{n}_\ell)$,

$$\varphi(x^*x) = \big\|\eta_\varphi(x)\big\|^2 = \varphi_\ell\Big(\pi_\ell\big(\eta_\varphi(x)\big)^*\pi_\ell\big(\eta_\varphi(x)\big)\Big) = \varphi_\ell\big(\pi_\varphi(x)^*\pi_\varphi(x)\big),$$

which entails that $\varphi = \varphi_\ell \circ \pi_\varphi$. Hence φ is identified with φ_ℓ via π_φ. Q.E.D.

We close this section with the following result which guarantees the previous theory of weights and left Hilbert algebras is applicable to any von Neumann algebra.

Theorem 2.7. *Every von Neumann algebra admits a faithful semi-finite normal weight.*

PROOF: Let $\{\omega_i : i \in I\}$ be a maximal family of normal positive linear functionals on a given von Neumann algebra $\mathcal{M}$ with orthogonal support $\{s(\omega_i)\}$. By maximality, we have $\sum_{i\in I} s(\omega_i) = 1$. We then set

$$\varphi(x) = \sum_{i\in I} \omega_i(x), \quad x \in \mathcal{M}_+.$$

Rewriting φ in the form: [2]

$$\varphi(x) = \sup\Big\{\sum_{i\in J} \omega_i(x) : J \subset\subset I\Big\}$$

we conclude that φ is a normal weight. For each $J \subset\subset I$, set $p_J = \sum_{i\in J} s(\omega_i) \in \mathrm{Proj}(\mathcal{M})$. We then have $p_J \in \mathfrak{m}_\varphi^+$ and $p_J \nearrow 1$, so that φ is semi-finite. If $\varphi(x) = 0$, $x \in \mathcal{M}_+$, then $\omega_i(x) = 0$ for every $i \in I$, so that $x^{1/2}s(\omega_i) = 0$, $i \in I$; hence $x^{1/2} = x^{1/2}\sum s(\omega_i) = \sum x^{1/2}s(\omega_i) = 0$. Thus, φ must be faithful. Q.E.D.

Exercise VII.2

1) Let φ be a faithful semi-finite normal weight on a von Neumann algebra $\mathcal{M}$. Let $\mathfrak{A}$ be the associated left Hilbert algebra, i.e. $\mathfrak{A} = \eta_\varphi\big(\mathfrak{n}_\varphi \cap \mathfrak{n}_\varphi^*\big)$ and $\mathfrak{B}$ be the algebra of all left bounded elements in $\mathfrak{H}_\varphi$. Identify $\mathcal{M}$ with $\pi_\varphi(\mathcal{M})$.

2 $J \subset\subset I$ means that J is a finite subset of I.

(**a**) Show that the map $\eta_\varphi : x \in \mathfrak{n}_\varphi \cap \mathfrak{n}_\varphi^* \mapsto \eta_\varphi(x) \in \mathfrak{A}$ is a closed linear map relative to the σ-strong* topology in $\mathcal{M}$ and the norm topology in $\mathfrak{H}_\varphi$, whose inverse is in fact the map $\pi_\ell : \xi \in \mathfrak{A} \mapsto \pi_\ell(\xi) \in \mathcal{M}$.

(**b**) Show that the map $\pi_\ell : \xi \in \mathfrak{B} \mapsto \pi_\ell(\xi) \in \mathcal{M}$ is a closed linear map relative to the norm topology in $\mathfrak{H}_\varphi$ and the σ strong topology in $\mathcal{M}$, whose inverse is $\eta_\varphi : x \in \mathfrak{n}_\varphi \mapsto \eta_\varphi(x) \in \mathfrak{B}$.

§ 3 The Plancherel Weight and the Fourier Algebra

In this section, we discuss the von Neumann algebra $\mathcal{R}_\ell(G)$ generated by the left regular representation λ of a locally compact group G. In particular, we study the relation between $\mathcal{R}_\ell(G)$ and $\mathcal{R}_r(G)$, the von Neumann algebra generated by the right regular representation ρ of G, and show that they are commutant of each other. Furthermore, we construct a faithful semi-finite normal weight ψ on $\mathcal{R}_\ell(G)$ which may be regarded as dual to the Haar integral over G.

Let G be a locally compact group with a left invariant Haar measure dg, i.e.

$$\int_G f(hg)\,dg = \int_G f(g)\,dg \tag{1}$$

for any continuous function f on G with compact support. We denote by $\mathcal{K}(G)$ the vector space of all continuous functions on G with compact support and by δ_G the modular function of G, i.e.

$$\int_G f(gh)\,dg = \delta_G(h)^{-1} \int_G f(g)\,dg, \tag{2}$$

or

$$\int_G f(g^{-1})\delta_G(g)^{-1}\,dg = \int_G f(g)\,dg\,. \tag{2'}$$

On the Hilbert space $L^2(G)$ of all square integrable functions on G, we define two representations λ and ρ of G as follows:

$$\left.\begin{aligned} \big(\lambda(h)\xi\big)(g) &= \xi\big(h^{-1}g\big), \quad g, h \in G,\\ \big(\rho(h)\xi\big)(g) &= \delta_G(h)^{\frac{1}{2}}\xi(gh), \quad \xi \in L^2(G). \end{aligned}\right\} \tag{3}$$

It follows that λ and ρ are both unitary representations of G and they commute with each other. The representations λ and ρ are respectively called the *left* and *right regular* representations of G. The von Neumann algebra generated by $\lambda(G)$ (resp. $\rho(G)$) is denoted by $\mathcal{R}_\ell(G)$ (resp. $\mathcal{R}_r(G)$) and called the *left* (resp. *right*) *group von Neumann algebra* of G. We have discussed this algebra already in Chapter V for a discrete group in connection with the construction of factors of type II_1.

In Example VI.1.2, we observed that $\mathcal{K}(G)$ equipped with the algebraic structure given by (1), Chap. VI, §1, is a left Hilbert algebra. We define a non-singular self-adjoint operator Δ_G on $L^2(G)$ as follows:

$$\big(\Delta_G\xi\big)(g) = \delta_G(g)\xi(g), \quad \xi \in \mathfrak{D}(\Delta_G), \tag{4}$$

where

$$\mathfrak{D}\big(\Delta_G\big) = \left\{\xi \in L^2(G) : \int_G \delta_G(g)^2 \big|\xi(g)\big|^2 \, dg < +\infty\right\}. \tag{5}$$

The next result follows immediately:

Proposition 3.1. *Equipped with the complex one parameter group* Δ_G^α, $\mathcal{K}(G)$ *is a Tomita algebra such that*

$$\mathcal{R}_\ell\big(\mathcal{K}(G)\big) = \mathcal{R}_\ell(G) \qquad \textit{and} \qquad \mathcal{R}_r\big(\mathcal{K}(G)\big) = \mathcal{R}_r(G). \tag{6}$$

In particular, the left and the right group von Neumann algebras are commutant of each other.

PROOF: The fact that $\mathcal{K}(G)$ is a Tomita algebra can be seen by a routine inspection of the postulates. Hence we only check the relation (6). We observe quickly that $\lambda(G)$ and $\pi_r\big(\mathcal{K}(G)\big)$ (resp. $\rho(G)$ and $\pi_\ell(\mathcal{K}(G))$) commutes, so that $\mathcal{R}_\ell(G)' \supset \mathcal{R}_r\big(\mathcal{K}(G)\big)$ and $\mathcal{R}_r(G)' \supset \mathcal{R}_\ell\big(\mathcal{K}(G)\big)$. Hence we have

$$\mathcal{R}_\ell(G) \subset \mathcal{R}_\ell\big(\mathcal{K}(G)\big), \qquad \mathcal{R}_r(G) \subset \mathcal{R}_r\big(\mathcal{K}(G)\big)$$

by Theorem VI.2.2.(ii). On the other hand, if $\xi \in \mathcal{K}(G)$, then for each $\eta \in L^2(G)$ and $\zeta \in \mathcal{K}(G)$, we have

$$\begin{aligned}
\big(\pi(\xi)\eta \mid \xi\big) &= \big(\eta \mid \pi_\ell(\xi)^*\zeta\big) = \big(\eta \mid \pi_\ell(\xi^\sharp)\zeta\big) \\
&= \int \eta(g)\overline{\big(\pi_\ell(\xi^\sharp)\zeta\big)(g)}\, dg = \iint_{G\times G} \eta(g)\overline{\xi^\sharp(h)\zeta(h^{-1}g)}\, dh\, dg \\
&= \iint_{G\times G} \eta(g)\delta_G(h)^{-1}\xi(h)^{-1}\overline{\zeta(h^{-1}g)}\, dh\, dg \\
&= \iint_{G\times G} \eta(g)\xi(h)\overline{\zeta(hg)}\, dh\, dg = \iint_{G\times G} \eta(g)\xi(h)\overline{\zeta(hg)}\, dg\, dh \\
&= \iint_{G\times G} \xi(h)\eta(h^{-1}g)\overline{\zeta(g)}\, dg\, dh \\
&= \int_G \xi(h)\big(\lambda(h)\eta \mid \zeta\big)\, dh = \left(\int_G \xi(h)\lambda(h)\, dh\, \eta \,\middle|\, \zeta\right).
\end{aligned}$$

Since the both ends of the above calculation are continuous functionals of ζ in the L^2-norm, we have

$$\pi(\xi) = \int_G \xi(g)\lambda(g)\, dg\,. \tag{7}$$

Therefore, $\pi_\ell(\xi)$ is approximated by linear combinations of $\{\lambda(g)\}$ in the weak operator topology, so that we have $\pi_\ell(\xi) \in \mathcal{R}_\ell(G)$; thus $\pi_\ell(\mathcal{K}(G)) \subset \mathcal{R}_\ell(G)$ and similarly $\mathcal{R}_r(\mathcal{K}(G)) \subset \mathcal{R}_r(G)$. Hence we conclude (6). Q.E.D.

At this point, we introduce two operations on functions f on G:

$$\left.\begin{aligned} f^\wedge(g) &= \delta_G(g)^{-1} f(g^{-1}), \\ f^\vee(g) &= f(g^{-1}). \end{aligned}\right\} \tag{8}$$

We denote the associated full left Hilbert algebra by $\mathfrak{A}_\ell(G)$. We also use the notations $\mathfrak{A}_r(G)$, $\mathfrak{B}_\ell(G)$ and $\mathfrak{B}_r(G)$ accordingly.

Definition 3.2. The weight on $\mathcal{R}_\ell(G)$ associated with the full left Hilbert algebra $\mathfrak{A}_\ell(G)$ is called the *Plancherel* weight and denoted by ψ_G.

We now define the convolution of two measurable functions ξ and η on G by the integral:

$$(\xi * \eta)(g) = \int \xi(h)\eta(h^{-1}g)\, dh \tag{9}$$

whenever the above integral makes sense.

Lemma 3.3. *If $\xi, \eta \in L^2(G)$, then $\xi * \eta^\vee$ is a continuous function on G vanishing at infinity and*

$$\big(\lambda(g)\xi \mid \eta\big) = \big(\bar\eta * \xi^\vee\big)(g), \quad g \in G. \tag{10}$$

PROOF: Suppose first $\xi, \eta \in \mathcal{K}(G)$ and compute:

$$\big(\lambda(g)\xi \mid \eta\big) = \int_G \xi(g^{-1}h)\overline{\eta(h)}\, dh = \int_G \bar\eta(h)\xi^\vee(h^{-1}g)\, dh = \bar\eta * \xi^\vee(g).$$

We know that $\bar\eta * \xi^\vee$ belongs to $\mathcal{K}(G)$. If ξ and η are in $L^2(G)$, then we choose sequences $\{\xi_n\}$ and $\{\eta_n\}$ from $\mathcal{K}(G)$ with $\lim\|\xi - \xi_n\| = 0$ and $\lim\|\eta - \eta_n\| = 0$. Then we have

$$\begin{aligned} \Big|\big(\lambda(g)\xi \mid \eta\big) - \big(\lambda(g)\xi_n \mid \eta_n\big)\Big| &\le \big\|\lambda(g)(\xi - \xi_n)\big\|\,\|\eta\| + \big\|\lambda(g)\xi_n\big\|\,\big\|\eta - \eta_n\big\| \\ &= \big\|\xi - \xi_n\big\|\,\|\eta\| + \|\xi_n\|\,\big\|\eta - \eta_n\big\| \to 0 \end{aligned}$$

uniformly in g as $n \to \infty$. Hence the left hand side of (10) is a continuous function on G vanishing at infinity. On the other hand, the right hand side of (10) is precisely the inner product between the L^2-function $\lambda(g)\xi$ and η, so that the integral converges. Q.E.D.

Considering the right regular representation ρ of G, we have the following result whose proof will be left to the reader.

Lemma 3.3′. *If $\xi, \eta \in L^2(G)$, then $\eta^\sharp * \xi$ is a continuous function on G vanishing at infinity and*

$$\big(\rho(g)\xi \mid \eta\big) = \big(\eta^\sharp * \xi\big)(g), \quad g \in G.$$

Theorem 3.4. *If $\xi \in L^2(G)$ is left bounded with respect to the Tomita algebra $\mathcal{K}(G)$, then*

$$\psi_G\big(\pi_\ell(\xi) * \pi_\ell(\xi)\big) = \xi^\sharp * \xi(e) = \|\xi\|^2. \tag{11}$$

The theorem follows naturally from the definition of the Plancherel weight and Lemma 3.3′.

We now define a unitary W_G on $L^2(G \times G)$ as follows:

$$\big(W_G\xi\big)(g,h) = \xi(g, gh), \quad \xi \in L^2(G \times G), \quad g, h \in G. \tag{12}$$

Definition 3.5. The operator W_G will be called the *structure operator* of G.

We shall see that the structure operator W_G, together with the tensor product $L^2(G) \otimes L^2(G) = L^2(G) = L^2(G \times G)$, determines the group G completely, Theorem 3.9.

We represent the abelian von Neumann algebra $L^\infty(G)$ on $L^2(G)$ by multiplication. We have immediately the next fact:

Lemma 3.6.

(i) *The structure operator W_G belongs to $L^\infty(G) \overline{\otimes} \mathcal{R}_\ell(G)$ by identifying $L^2(G \times G)$ with $L^2(G) \otimes L^2(G)$.*

(ii) $$W_G^*\big(\lambda(g) \otimes 1\big)W_G = \lambda(g) \otimes \lambda(g), \quad g \in G. \tag{13}$$

Lemma 3.7. *Set*

$$A(G) = \big\{\, \xi * \eta^\vee : \xi, \eta \in L^2(G) \,\big\}. \tag{14}$$

Then $A(G)$ is a dense *-subalgebra of the abelian C^*-algebra $C_\infty(G)$ of all continuous functions on G vanishing at infinity.

PROOF: We recall first that every element of $\mathcal{R}_\ell(G)_*$ is of the form $\omega_{\xi,\eta}$, $\xi, \eta \in L^2(G)$, by Theorem V.3.15. Since linear combinations of $\lambda(G)$ are σ-weakly dense in $\mathcal{R}_\ell(G)$, the correspondence $\bar{\eta} * \xi^\vee \in A(G) \longleftrightarrow \omega_{\xi,\eta} \in \mathcal{R}_\ell(G)_*$ is bijective, so that $A(G)$ is a linear subspace of $C_\infty(G)$ by the above remark. The self-adjointness of $A(G)$ is clear since $L^2(G)$ is invariant under the complex conjugation.

Now, we consider an isomorphism π of $\mathcal{R}_\ell(G)$ into $\mathcal{L}\big(L^2(G \times G)\big)$ given by the following:

$$\pi(x) = W_G^*(x \otimes 1)W_G, \quad x \in \mathcal{R}_\ell(G). \tag{15}$$

By Lemma 3.6, π maps $\mathcal{R}_\ell(G)$ into $\mathcal{R}_\ell(G)\,\overline{\otimes}\,\mathcal{R}_\ell(G)$. If $\xi_1, \xi_2, \eta_1, \eta_2 \in L^2(G)$, ${}^t\pi\left(\omega_{\xi_1,\eta_1} \otimes \omega_{\xi_2,\eta_2}\right)$ belongs to $\mathcal{R}_\ell(G)_*$, so that there exist $\xi, \eta \in L^2(G)$ such that

$$\omega_{\xi,\eta} = {}^t\pi\left(\omega_{\xi_1,\eta_1} \otimes \omega_{\xi_2,\eta_2}\right).$$

We then compute:

$$\begin{aligned}
\bar\eta_1 * \xi_1^\vee(g)\bar\eta_2 * \xi_2^\vee(g) &= \left(\lambda(g)\xi_1 \mid \eta_1\right)\left(\lambda(g)\xi_2 \mid \eta_2\right)\\
&= \left(\left(\lambda(g)\otimes\lambda(g)\right)(\xi_1 \otimes \xi_2) \;\middle|\; \eta_1 \otimes \eta_2\right)\\
&= \left(W_G^*\left(\lambda(g)\otimes 1\right)W_G\left(\xi_1\otimes\xi_2\right) \;\middle|\; \eta_1\otimes\eta_2\right)\\
&= \left\langle \pi\left(\lambda(g)\right), \omega_{\xi_1,\eta_1}\otimes\omega_{\xi_2,\eta_2}\right\rangle\\
&= \left\langle \lambda(g), \omega_{\xi,\eta}\right\rangle = \bar\eta * \xi^\vee(g).
\end{aligned}$$

Hence we conclude $\left(\bar\eta_1 * \xi_1^\vee\right)\left(\bar\eta_2 * \xi_2^\vee\right) = \bar\eta * \xi$. Thus $A(G)$ is a subalgebra of $C_\infty(G)$.

Since $A(G)$ separates points of G because it contains $\mathcal{K}(G) * \mathcal{K}(G)$, $A(G)$ is dense in $C_\infty(G)$. Q.E.D.

Definition 3.8. Identifying $A(G)$ with the predual $\mathcal{R}_\ell(G)_*$ under the correspondence: $\bar\eta * \xi^\vee \longleftrightarrow \omega_{\xi,\eta}$, $A(G)$ is a commutative Banach algebra. We call this Banach algebra $A(G)$ the *Fourier algebra* of G.

Theorem 3.9. *For a non-zero $x \in \mathcal{L}\left(L^2(G)\right)$, the following conditions are equivalent:*

(i) $W_G^*(x\otimes 1)W_G = x \otimes x\,;$ (16)

(ii) $x \in \mathcal{R}_\ell(G)$ *and* $\varphi \in A(G) \mapsto \langle x, \varphi\rangle \in \mathbf{C}$ *is a character;*

(iii) $x = \lambda(g)$ *for some* $g \in G$.

PROOF: The implications: (iii) $\Longrightarrow$ (i) $\Longleftrightarrow$ (ii) are easy. The hard part is the last step (ii) $\Longrightarrow$ (iii), which needs some preparation. Let us finish first the easy part.

The implication (iii) $\Longrightarrow$ (i) is exactly the equality (13).

(i) $\Longrightarrow$ (ii): Suppose x satisfies (16). If $g \in G$, then

$$\begin{aligned}
x\otimes\rho(g)x &= \left(1\otimes\rho(g)\right)W_G^*(x\otimes 1)W_G\\
&= W_G^*\left(1\otimes\rho(g)\right)(x\otimes 1)W_g \qquad \text{(by Lemma 3.6.(i))}\\
&= W_G^*(x\otimes 1)\left(1\otimes\rho(g)\right)W_G = W_G^*(x\otimes 1)W_G\left(1\otimes\rho(g)\right)\\
&= x\otimes x\rho(g);
\end{aligned}$$

hence $\rho(g)x = x\rho(g)$. Therefore, x belongs to $\mathcal{R}_\ell(G)$. If $\varphi, \psi \in A(G)$, then we have, as in Lemma 3.7,

$$\left\langle x, \varphi\cdot\psi\right\rangle = \left\langle\pi(x), \varphi\otimes\psi\right\rangle = \left\langle x\otimes x, \varphi\otimes\psi\right\rangle = \langle x,\varphi\rangle\langle x,\psi\rangle, \tag{17}$$

so that x is a character as a functional on $A(G)$.

To prove the implication (ii) $\Longrightarrow$ (i), assume that $x \in \mathcal{R}_\ell(G)$ satisfies the condition (ii). For every $\varphi, \psi \in A(G)$, we have, as in Lemma 3.7,

$$\begin{aligned}\langle W_G^*(x \otimes 1)W_G, \varphi \otimes \psi\rangle &= \langle \pi(x), \varphi \otimes \psi\rangle = \langle x, \varphi \cdot \psi\rangle \\ &= \langle x, \varphi\rangle\langle x, \psi\rangle \qquad \text{(by assumption)} \\ &= \langle x \otimes x, \varphi \otimes \psi\rangle,\end{aligned}$$

so that (16) follows.

We need some preparations for the proof of the implication (i) $\Longrightarrow$ (iii).

Q.E.D.

First, we set

$$\mathcal{G} = \left\{ x \in \mathcal{L}(L^2(G)) : W_G^*(x \otimes 1)W_G = x \otimes x,\ x \neq 0 \right\}. \tag{18}$$

We want to prove that $\mathcal{G}$ is precisely $\lambda(G)$. We begin this project by the next result concerning W_G:

Lemma 3.10. *On $L^2(G)$, we set*

$$(C\xi)(g) = \overline{\xi(g)}, \quad \xi \in L^2(G), \tag{19}$$

$$\big(\lambda(f)\xi\big)(g) = f * \xi(g), \quad f \in L^1(G). \tag{20}$$

Then W_G commutes with C and satisfies the following equation:

$$\big(W_G(f \otimes \xi) \mid \eta \otimes \zeta\big) = \langle \xi * \zeta^\flat, f \cdot \bar\eta\rangle = \big(\xi \mid \lambda(\bar f \eta)\zeta\big), \quad f, \xi, \eta, \zeta \in L^2(G), \tag{21}$$

where $\xi * \zeta^\flat \in A(G)$ and $f \cdot \bar\eta \in L^1(G)$.

PROOF: The commutativity of W_G and C follows from their definitions, (12) and (19). We next compute:

$$\begin{aligned}\big(W_G(f \otimes \xi) \mid \eta \otimes \zeta\big) &= \iint f(g)\xi(gh)\overline{\eta(g)\zeta(h)}\, dg\, dh \\ &= \int_G f(g)\overline{\eta(g)}\left(\int \xi(h)\overline{\zeta(g^{-1}h)}\, dh\right) dg \\ &= \int_G f(g)\overline{\eta(g)} \int \xi(h)\zeta^\flat(h^{-1}g)\, dh)\, dg \\ &= \langle \xi * \zeta^\flat, f\bar\eta\rangle = \big(\xi \mid \lambda(\bar f \eta)\zeta\big)\end{aligned}$$

for every $f, \xi, \eta, \zeta \in L^2(G)$. Q.E.D.

Remark 3.11. We interpret the equation (21) as follows: Being the predual $\mathcal{R}_\ell(G)_*$, $A(G)$ can be viewed as "L^1-space". If $m(f)$ denotes the multiplication operator given by $f \in L^\infty(G)$, then (21) is reformulated as follows:

$$\begin{aligned}\langle m(\bar{\xi} * \zeta^\vee), \omega_{\eta, f}\rangle &= \langle \bar{\xi} * \zeta^\vee, \eta \bar{f}\rangle = \langle \xi * \zeta^\flat, f\bar{\eta}\rangle^- \\ &= \big(W_G(f \otimes \xi) \mid \eta \otimes \zeta\big)^- = \big(\xi \mid \lambda(\bar{f}\eta)\zeta\big)^- \\ &= \langle \lambda(\bar{f}\eta), \omega_{\xi,\zeta}\rangle.\end{aligned} \tag{21'}$$

Hence we have

$$\langle m(\varphi), f\rangle = \langle \lambda(f), \varphi\rangle, \quad f \in L^1(G), \quad \varphi \in A(G). \tag{22}$$

This shows that m and λ are in duality, and that W_G controls the entire duality structure involving $L^2(G)$, the neutral space, $L^1(G)$, $L^\infty(G)$, $A(G)$ and $\mathcal{R}_\ell(G)$.

Lemma 3.12. *The set $\mathcal{G}$, defined by* (18), *enjoys the following properties:*

(i) *$\mathcal{G}$ is closed under the operations: $x \mapsto x^*$, $x \mapsto \bar{x} = CxC$ and $x \mapsto x^\wedge = \bar{x}^*$;*
(ii) *If $x_1, x_2 \in \mathcal{G}$, then $x_1x_2 \in \mathcal{G}$ provided $x_1x_2 \neq 0$;*
(iii) *If $x = u|x|$ is the polar decomposition of an $x \in \mathcal{G}$, then u and $|x|$ both fall in $\mathcal{G}$;*
(iv) *If $x \in \mathcal{G}$, then*

$$m(f)x\xi = xm(\xi)x^\wedge f, \quad f, \xi \in L^2(G) \cap L^\infty(G); \tag{23}$$

(v) *If $e \in \mathcal{G}$ is a projection, then $e = 1$;*
(vi) *$\mathcal{G}$ is a group of unitaries in $\mathcal{R}_\ell(G)$ commuting with C;*
(vii) *Each $u \in \mathcal{G}$ leaves invariant the *-subalgebra $L^2(G) \cap L^\infty(G)$ of $L^\infty(G)$ on which it acts multiplicatively:*

$$u(\xi \cdot \eta) = (u\xi) \cdot (u\eta), \quad \xi, \eta \in L^2(G) \cap L^\infty(G). \tag{24}$$

PROOF: The assertions (i) and (ii) follow from the multiplicativity in x of the both sides of the equation (16).

(iii) This follows from the uniqueness of the polar decomposition.

(iv) For $f, \xi \in L^2(G) \cap L^\infty(G)$ and $\eta, \zeta \in L^2(G)$, we have, by (21′) and (16),

$$\begin{aligned}\Big(\eta \,\Big|\, \lambda\big(m(f)x\xi\big)\zeta\Big) &= \big(W_G(\bar{f} \otimes \eta) \mid x\xi \otimes \zeta\big) = \big(\bar{f} \otimes \eta \mid W_G^*(x \otimes 1)\xi \otimes \zeta\big) \\ &= \big(\bar{f} \otimes \eta \mid (x \otimes x)W_G^*(\xi \otimes \zeta)\big) = \big((x^* \otimes x^*)(\bar{f} \otimes \eta) \mid W_G^*(\xi \otimes \zeta)\big) \\ &= \big(x^*\bar{f} \otimes x^*\eta \mid W_G^*(\xi \otimes \zeta)\big) = \big(W_G(x^*\bar{f} \otimes x^*\eta) \mid \xi \otimes \zeta\big) \\ &= \Big(x^*\eta \,\Big|\, \lambda\big((x^*\bar{f})^-\xi\big)\zeta\Big) = \Big(x^*\eta \,\Big|\, \lambda\big((x^\wedge f) \cdot \xi\big)\zeta\Big) \\ &= \Big(\eta \,\Big|\, x\lambda\big((\hat{x}f)\xi\big)\zeta\Big) = \Big(\eta \,\Big|\, x\lambda\big(m(\xi)\hat{x}f\big)\zeta\Big).\end{aligned}$$

Since $m(\xi)xf \in L^2(G)$ is left bounded with respect to the left Hilbert algebra $\mathcal{K}(G)$, $\lambda\big(m(\xi)\hat{x}f\big) = \pi_\ell\big(m(\xi)\hat{x}f\big)$; hence we have $x\lambda\big(m(\xi)\hat{x}f\big) = \lambda\big(xm(\xi)xf\big)$. Thus (23) follows.

(v) Let $e \in \mathcal{G}$ be a projection. Then we have, using (23) twice

$$m(f)e\xi = em(\xi)\hat{e}f = e\hat{e}m(f)e\xi, \quad f, \xi \in L^\infty(G) \cap L^2(G).$$

Hence we get $m(f)e = e\hat{e}m(f)e$, so that $m(f)e = em(f)e$; hence e and $m(f)$ commute. By density, e commutes with $L^\infty(G)$. By the maximal commutativity of $L^\infty(G)$, we have

$$e \otimes e = W_G^*(e \otimes 1)W_G = e \otimes 1,$$

thus $e = 1$.

(vi) If $x = u|x|$ is the polar decomposition of an $x \in \mathcal{G}$, then $u^*u = uu^* = 1$ by (v). Hence x and x^* both have dense range. We now look at (22) again and have

$$m(f)x\xi = xm(\xi)\hat{x}f = x\hat{x}m(f)x\xi$$

for every $f, \xi \in L^2(G) \cap L^\infty(G)$. The density of $\{m(f)x\xi\}$ in $L^2(G)$ implies that $x\hat{x} = 1$. Similarly $\hat{x}x = 1$. Applying this to $|x|$, we get $1 = |x|^\wedge|x| = |x||x|^\wedge$. Hence $|x|^\wedge = |x|^{-1}$. But $\mathcal{G}$ is in the unit ball of $\mathcal{R}_\ell(G)$, being the set of characters of the Banach algebra $A(G)$. Hence $\||x|\| \leq 1$ and $\||x|^{-1}\| \leq 1$. This is possible only when $|x| = 1$. Therefore, $x = u$.

(vii) The equality (23), together with the above arguments, shows that

$$m(uf) = um(f)u^*, \quad f \in L^2(G) \cap L^\infty(G), \quad u \in \mathcal{G}. \tag{23'}$$

This means that the multiplication operator $m(uf)$ is bounded, so that uf must be bounded. Hence u leaves $L^2(G)\cap L^\infty(G)$ invariant. Furthermore, (23′) entails (24).

Q.E.D.

PROOF OF THEOREM 3.9 (i) $\Longrightarrow$ (iii): Since $\mathcal{G}$ is the set of all nonzero solutions of the equation (16), $\mathcal{G}$ is locally compact with respect to the weak operator topology. Being a subgroup of $\mathcal{U}(\mathcal{R}_\ell(G))$, the strong operator topology and the weak operator topology agree on $\mathcal{G}$ and the both of them give a topological group structure on $\mathcal{G}$. Hence $\mathcal{G}$ is a locally compact group. The left regular representation λ maps G into $\mathcal{G}$ injectively and of course continuously.

We have identified $A(G)$ with $\mathcal{R}_\ell(G)_*$, so that each $\varphi \in A(G)$ is a normal linear functional on $\mathcal{R}_\ell(G)$ and in particular a continuous function on $\mathcal{G}$. Since the zero of $\mathcal{R}_\ell(G)$ is the point at infinity of $\mathcal{G}$, $A(G)$ is a subalgebra of $C_\infty(\mathcal{G})$. Since $\overline{\varphi(u)} = \varphi^*(u^*)$, $u \in \mathcal{G}$ and $\varphi \in A(G)$, $A(G)$ is self-adjoint in $C_\infty(\mathcal{G})$, i.e. closed under the conjugation. Thus, $C_\infty(\mathcal{G})$ is the enveloping C^*-algebra of $A(G)$, which means that every representation of $A(G)$ as an involutive Banach algebra, is extended to a representation of $C_\infty(\mathcal{G})$. On the other hand, $A(G)$ is a self-adjoint dense subalgebra of $C_\infty(G)$. Therefore, the multiplication representation: $\varphi \in A(G) \mapsto m(\varphi) \in \mathcal{L}(L^2(G))$ can be extended to a representation of $C_\infty(\mathcal{G})$.

Now, by the Riesz-Markov theorem, each pair $\xi, \eta, \in L^2(G)$ gives rise to a Radon measure $\mu_{\xi,\eta}$ on $\mathcal{G}$ such that

$$\int_{\mathcal{G}} \varphi(u)\, \mathrm{d}\mu_{\xi,\eta}(u) = (m(\varphi)\xi \mid \eta). \tag{25}$$

By (22) we have

$$\int_G \varphi(u)\, \mathrm{d}\mu_{\xi,\eta}(u) = \langle \lambda(\xi \cdot \bar{\eta}), \varphi \rangle, \quad \xi, \eta \in L^2(G), \quad \varphi \in A(G).$$

If $\varphi \in A(G)$, then we have

$$d\mu_{\varphi\xi,\eta}(u) = \varphi(u)\, d\mu_{\xi,\eta}(u),$$

so that we have

$$\varphi(u)\overline{\psi(u)}\, d\mu_{\xi,\eta}(u) = d\mu_{\varphi\xi,\psi\eta}(u) = \xi(u)\overline{\eta(u)}\, d\mu_{\varphi,\psi}(u)\,. \tag{26}$$

for every $\varphi, \psi, \xi, \eta \in L^2(G) \cap A(G)$. Therefore, $\mathrm{supp}(\mu_{\varphi,\varphi})$, the support of the measure $\mu_{\varphi,\varphi}$, is contained in that of φ for any $\varphi \in L^2(G) \cap A(G)$. Thus we can define a positive Radon measure μ on G by

$$d\mu(u) = \frac{1}{|\varphi(u)|^2}\, d\mu_{\varphi,\varphi}(u), \quad \varphi \in L^2(G) \cap A(G), \tag{27}$$

which satisfies the following

$$d\mu_{\xi,\eta}(u) = \xi(u)\eta(u)\, d\mu(u), \quad \xi, \eta \in L^2(G) \cap A(G) \tag{27'}$$

by the polarization identity. If $u_0 \in \mathcal{G}$ is fixed, then we have, for each pair $\xi, \eta \in A(G) \cap L^2(G)$,

$$\begin{aligned}\int_{\mathcal{G}} (\xi\bar{\eta})(uu_0)\, d\mu(u) &= \int_{\mathcal{G}} \big[u_0(\xi\bar{\eta})\big](u)\, d\mu(u) = \int_{\mathcal{G}} (u_0(\xi)(u_0\bar{\eta})(u)\, d\mu(u) \\ &= \int_{\mathcal{G}} d\mu_{u_0\xi,u_0\eta}(u) = \big(u_0\xi \mid u_0\eta\big) \\ &= (\xi \mid \eta) = \int_{\mathcal{G}} (\xi\bar{\eta})(u)\, d\mu(u)\,.\end{aligned}$$

Therefore, μ is a right invariant Haar measure on $\mathcal{G}$. Hence, the support of μ must be the entire $\mathcal{G}$. On the other hand, we have, for each $\varphi \in A(G),\ \xi, \eta \in L^2(G) \cap A(G)$,

$$\int_G \varphi(g)\xi(g)\eta(g)\, dg = \big(m(\varphi)\xi \mid \eta\big) = \int_{\mathcal{G}} \varphi(u)\xi(u)\bar{\eta}(u)\, d\mu(u)\,.$$

Therefore, the essential supremums of $|\varphi(\,\cdot\,)|$ of any $\varphi \in A(G)$ over G and over $\mathcal{G}$ agree, which means that

$$\sup\big\{\, |\varphi(g)| : g \in G \,\big\} = \sup\big\{\, |\varphi(u)| : u \in \mathcal{G} \,\big\}.$$

Hence the restriction map of $C_\infty(\mathcal{G})$ onto $C_\infty(G)$ must be an isomorphism. Therefore, we conclude that $G = \mathcal{G}$. Q.E.D.

We now assume that G is commutative and write the group operation additively. We shall show the Plancherel formula and the van Kampen- Pontrjagin duality theorem.

We recall first that the group C^*-algebra $C^*(G)$ is the completion of $L^1(G)$ under the new norm $\|f\|' = \sup\{\|\pi(f)\| : \pi$ runs over all *-representations of $L^1(G)\}$, or equivalently the closure of $L^1(G)$ under the universal *-representation π of $L^1(G)$. We also recall that *-representations of $L^1(G)$ are in one-to-one correspondence with unitary representations of G. We use the same notation for the corresponding representations of $L^1(G)$ and G. It then follows that

$$u(g)u(f)\xi = u\big(\lambda(g)f\big)\xi, \quad g \in G, \quad f \in L^1(G) \tag{28}$$

$$u(f) = \int_G f(g)u(g)\,\mathrm{d}g, \quad \xi \in \mathfrak{H}_u.$$

Let $\widehat{G}$ denote the spectrum of the abelian C^*-algebra $C^*(G)$. By the Gelfand representation theorem, Theorem I.4.4, we have $C^*(G) \simeq C_\infty(\widehat{G})$. Each character $p \in \widehat{G}$ is then a continuous homomorphism of G into the unit circle $\mathbf{T} = \{\lambda \in \mathbf{C} : |\lambda| = 1\}$, because p is a one dimensional unitary representation and $\mathbf{T}$ is the unitary group of $\mathbf{C}$. We now define the coupling of G and $\widehat{G}$ as follows:

$$\langle g, p\rangle = \overline{p(g)}, \quad g \in G, \quad p \in \widehat{G}. \tag{29}$$

Definition 3.13. The Gelfand transform $\mathcal{F}$ of $L^1(G)$ into $C_\infty(\widehat{G})$ is called the *Fourier transform*. In view of (28), the Fourier transform $\mathcal{F}$ is given by the following integral:

$$(\mathcal{F}f)(p) = \int_G \overline{\langle g, p\rangle} f(g)\,\mathrm{d}g, \quad f \in L^1(G), \quad p \in \widehat{G}. \tag{30}$$

Since $\rho(g) = \lambda(g)^{-1}$, there is no distinction between the left and right group von Neumann algebras, so that we write it simply $\mathcal{R}(G)$.

The next result is often called the *Plancherel formula* and in fact the term Plancherel weight came from it.

Theorem 3.14. *Let G be a locally compact abelian group. If we identify $C^*(G)$ with $C_\infty(\widehat{G})$ via the Gelfand transform, then the Plancherel weight ψ_G on $\mathcal{R}(G)$ gives rise to a Radon measure $\mathrm{d}p$ on $\widehat{G}$ such that*

(i) *The Fourier transform $\mathcal{F}$ restricted to $L^1(G) \cap L^2(G)$ can be extended to a unitary from $L^2(G)$ onto $L^2(\widehat{G})$, in particular*

$$\int_G |f(g)|^2\,\mathrm{d}g = \int_{\widehat{G}} |\mathcal{F}f(p)|^2\,\mathrm{d}p, \quad f \in L^1(G) \cap L^2(G). \tag{31}$$

(ii) *If we denote by $\mathcal{F}$ the extended Fourier transform, then $\mathcal{F}\mathcal{R}(G)\mathcal{F}^{-1}$ is precisely the multiplication von Neumann algebra $L^\infty(\widehat{G})$ on $L^2(\widehat{G})$.*

(iii) *The inverse transform $\widehat{\mathcal{F}}$ of $\mathcal{F}$ on $L^1(\widehat{G}) \cap L^2(\widehat{G})$ is given by the integral:*

$$(\widehat{\mathcal{F}}\varphi)(g) = \int_G \langle g, p\rangle \varphi(p)\,\mathrm{d}p, \quad \varphi \in L^1(\widehat{G}) \cap L^2(\widehat{G}). \tag{32}$$

Here, the function spaces over $\widehat{G}$, such as $L^1(\widehat{G})$, $L^2(\widehat{G})$, ..., are all with respect to the measure $\mathrm{d}p$.

PROOF:

(i) and (ii): We consider the Plancherel weight ψ_G. We identify again $C^*(G)$ and $C_\infty(\widehat{G})$, and view the regular representation λ of $C^*(G)$ as a representation of $C_\infty(\widehat{G})$ on $L^2(G)$. The (left) Hilbert algebra $L^1(G) \cap L^2(G)$ is dense in $L^1(G)$ with respect to the L^1-norm. Hence its image $\mathfrak{a} = \mathcal{F}(L^1(G) \cap L^2(G))$ is dense in $C_\infty(\widehat{G})$. Furthermore, $\lambda^{-1}(\mathfrak{m}_{\psi_G})$ contains the subalgebra $\mathfrak{a}^2$. Thus $\psi_G \circ \lambda$ is a weight on $C_\infty(\widehat{G})$ with dense definition domain. We want to show that $\psi_G \circ \lambda$ takes finite values on $\mathcal{K}(\widehat{G})$.

Let $\mathcal{I} = \lambda^{-1}(\mathfrak{m}_{\psi_G})$ and K a compact subset of $\widehat{G}$. Set $\mathcal{I}_K = \{\varphi \in C_\infty(\widehat{G}) : \varphi(p) = 0 \text{ for every } p \in K\}$, and consider the restriction map $\pi_K : C_\infty(\widehat{G}) \mapsto C(K) = C_\infty(\widehat{G})/\mathcal{I}_K$. Since $\mathcal{I}$ is dense in $C_\infty(\widehat{G})$, the image $\pi_K(\mathcal{I})$ is dense in $C(K)$, so that $\pi_K(\mathcal{I})$ must contain an invertible element of $C(K)$, and so $\pi_K(\mathcal{I}) = C(K)$. This means that $\mathcal{I} + \mathcal{I}_K = C_\infty(\widehat{G})$. Let L be a compact subset of $\widehat{G}$ and that K is contained in the interior of L. Choose $\varphi \in \mathcal{K}(G)$ such that $\varphi = 1$ on K and $\varphi = 0$ outside L. There exist $\varphi' \in \mathcal{I}$ and $\varphi'' \in \mathcal{I}_K$ such that $\varphi = \varphi' + \varphi''$. If $\psi \in \mathcal{K}(\widehat{G})$ with $\operatorname{supp} \psi \subset K$, then $\psi = \varphi\psi = (\varphi' + \varphi'')\psi = \varphi'\psi$ because $\varphi''\psi = 0$. Thus ψ belongs to $\mathcal{I}$. Namely, $\mathcal{K}(\widehat{G}) \subset \mathcal{I}$.

Therefore, $\psi_G \cdot \lambda$ takes finite values on $\mathcal{K}(\widehat{G})$, so that it gives rise to a Radon measure $\mathrm{d}p$ on $\widehat{G}$, and

$$\psi_G(\lambda(\varphi)) = \int_{\widehat{G}} \varphi(p)\,\mathrm{d}p, \quad \varphi \in C_\infty(\widehat{G}).$$

Thus, the definition of ψ_G implies the formula (31). Since $L^1(G) \cap L^2(G)$ is dense in $C^*(G) = C_\infty(\widehat{G})$, $\mathcal{F}(L^1(G) \cap L^2(G))$ is dense in $L^2(\widehat{G})$. Hence the extended Fourier transform maps $L^2(G)$ onto $L^2(\widehat{G})$ isometrically.

Now, $C^*(G) = C_\infty(\widehat{G})$ is represented on $L^2(\widehat{G})$ by multiplication, which is identified with λ. The von Neumann algebra generated by $C_\infty(\widehat{G})$ is therefore the algebra $L^\infty(\widehat{G})$ acting by multiplication. This means precisely that $\mathcal{F}\mathcal{R}(G)\mathcal{F}^{-1} = L^\infty(\widehat{G})$.

(iii) Suppose that $f \in L^1(G) \cap L^2(G)$ and $\varphi \in L^1(\widehat{G}) \cap L^2(\widehat{G})$. We then have

$$\begin{aligned}(\mathcal{F}f \mid \varphi) &= \int_{\widehat{G}} (\mathcal{F}f)(p)\overline{\varphi(p)}\,\mathrm{d}p = \int_{\widehat{G}} \Big(\int_G \langle g, p\rangle f(g)\,\mathrm{d}g \Big) \overline{\varphi(p)}\,\mathrm{d}p \\ &= \iint_{\widehat{G}\times G} f(g)\overline{\langle g, p\rangle \varphi(p)}\,\mathrm{d}g\,\mathrm{d}p \\ &= \int_G \overline{\Big(\int_{\widehat{G}} \langle g, p\rangle \varphi(p)\,\mathrm{d}p \Big)} f(g)\,\mathrm{d}g \qquad \text{(by Fubini's theorem).}\end{aligned}$$

Therefore, (32) follows. Q.E.D.

Definition 3.15. The measure dp on $\widehat{G}$ in Theorem 3.14 is called the *Plancherel measure.*

Since $\widehat{G}$ is the set of all continuous homomorphisms of G into $\mathbf{T}$, it forms a group under the pointwise multiplication. Namely, we define the additive group structure in $\widehat{G}$ by the following:

$$\left.\begin{aligned} \langle g, p+q\rangle &= \langle g, p\rangle\langle g, q\rangle, \quad g \in G, \\ \langle g, -p\rangle &= \overline{\langle g, p\rangle}, \quad p, q \in \widehat{G}. \end{aligned}\right\} \tag{33}$$

We can now state the van Kampen-Pontrjagin duality theorem.

Theorem 3.16. *Let G be a locally compact abelian group and $\widehat{G}$ the group of all continuous homomorphism of G into the unit circle $\mathbf{T}$ with the group structure given by* (33).

(i) *With respect to the uniform convergence topology on compact subsets of G, $\widehat{G}$ is a locally compact abelian group. Namely, a neighborhood $U(p, K, \varepsilon)$ of $p \in \widehat{G}$ determined by a compact set K in G and $\varepsilon > 0$ is given as follows:*

$$U(p, K, \varepsilon) = \left\{q \in \widehat{G} : \left|\langle g, p\rangle - \langle g, q\rangle\right| < \varepsilon,\ g \in K\right\}. \tag{34}$$

Furthermore, this topology coincides with the weak topology on $C^*(G)$, or equivalently on $L^1(G)$.*

(ii) *The Plancherel measure on $\widehat{G}$ is indeed a Haar measure of $\widehat{G}$.*

(iii) *Each $g \in G$ gives rise to a continuous homomorphism of $\widehat{G}$ into $\mathbf{T}$: $p \in \widehat{G} \mapsto \langle g, p\rangle \in \mathbf{T}$. Every continuous homomorphism of $\widehat{G}$ into $\mathbf{T}$ is of this form. Therefore, G is naturally isomorphic to $\widehat{\widehat{G}}$.*

PROOF:

(i) The topology of $\widehat{G}$ as $\mathrm{Sp}\left(C^*(G)\right)$ is given as the weak* topology. Since $\widehat{G}$ is in the unit ball of the conjugate space $C^*(G)^*$, the topology of $\widehat{G}$ is also given by the weak* topology on $L^1(G)$. Since each $p \in \widehat{G}$ is a function on G of absolute value one, it defines the unitary on $L^2(G)$ given by:

$$\left(\mu(p)\xi\right)(g) = p(g)\xi(g) = \overline{\langle g, p\rangle}\xi(g). \tag{35}$$

Then the topology of G is carried into the σ-weak topology of $\mu(G)$. Since the strong operator topology and the weak operator topology coincide on the unitary group of $L^\infty(G)$, $\mu(G)$ is a topological group. This means that the group operations in $\widehat{G}$ are continuous. Therefore, $\widehat{G}$ is a locally compact topological abelian group.

Let K be a compact subset of G. For a fixed $f \in L^1(G)$, the map: $g \in G \mapsto \lambda(g)f \in L^1(G)$ is continuous in norm. Hence $\lambda(K)f$ is a compact subset of $L^1(G)$, so that $\mathcal{F}\left(\lambda(K)f\right)$ is a compact subset of $C^*(G) = C_\infty(\widehat{G})$ in norm.

Hence, if $\{p_i\} \subset \widehat{G}$ converges to $p \in \widehat{G}$, then $\mathcal{F}(\lambda(g)f)(p_i) = \overline{\langle g, p_i\rangle}(\mathcal{F}f)(p_i)$ converges to $\overline{\langle g, p\rangle}(\mathcal{F}f)(p)$ uniformly in $g \in K$. Choosing $f \in L^1(G)$ with $(\mathcal{F}f)(p) \neq 0$, we conclude that $\langle g, p_i\rangle$ converges to $\langle g, p\rangle$ uniformly in $g \in K$. Conversely, if $\{p_i\}$ converges to $p \in \widehat{G}$ in the uniform convergence topology on compact sets, then for every $f \in \mathcal{K}(G)$, $\langle p_i, f\rangle = \int_G p_i(g)f(g)\,dg$ converges to $\langle p, f\rangle = \int_G p(g)f(g)\,dg$. Since $\mathcal{K}(G)$ is norm dense in $L^1(G)$ and $\{p_i\}$ is bounded in $L^\infty(G)$, $\{p_i\}$ converges to p in the $\sigma(L^\infty, L^1)$-topology. Hence $\{p_i\}$ converges to p in $\widehat{G} = \mathrm{Sp}(C^*(G))$. Therefore, the two topologies coincide on $\widehat{G}$.

(ii) For each $f \in L^1(G) \cap L^2(G)$ and $p, q \in \widehat{G}$, we have

$$\mathcal{F}(\mu(p)f)(q) = \int_G \overline{\langle g, q\rangle}(\mu(p)f)(g)\,dg = \int_G \overline{\langle g, p\rangle\langle g, p\rangle} f(g)\,dg$$
$$= (\mathcal{F}f)(p+q).$$

Therefore, $\mathcal{F}\mu(p)\mathcal{F}^{-1}$ is the translation by p. Hence, if φ and ψ are in $L^2(G)$, then we have, since $\mathcal{F}\mu(p)\mathcal{F}^{-1}$ is a unitary,

$$\int_{\widehat{G}} \varphi(p+q)\overline{\psi(p+q)}\,dq = (\mathcal{F}\mu(p)\mathcal{F}^{-1}\varphi \mid \mathcal{F}\mu(p)\mathcal{F}^{-1}\psi)$$
$$= (\varphi \mid \psi) = \int_{\widehat{G}} \varphi(q)\overline{\psi(q)}\,dq\,.$$

Hence the Plancherel measure is invariant under translations. Thus, the Plancherel measure must be a Haar measure of $\widehat{G}$.

(iii) The identification of $C^*(G)$ with $C_\infty(\widehat{G})$ extends to that of $\mathcal{R}(G)$ with $L^\infty(\widehat{G})$. Then predual $\mathcal{R}(G)_* = A(G)$ is identified with $L^1(\widehat{G})$. Now, set

$$\widehat{W}_G = (\mathcal{F} \otimes \mathcal{F})W_G(\mathcal{F} \otimes \mathcal{F})^{-1}. \tag{36}$$

It then follows that

$$(\widehat{W}_G\xi)(p, q) = \xi(p+q, q), \quad \xi \in L^2(\widehat{G} \times \widehat{G}). \tag{37}$$

Therefore, if $x \in \mathcal{R}(G)$, then we have

$$(\mathcal{F} \otimes \mathcal{F})(W^*(x \otimes 1)W)(\mathcal{F} \otimes \mathcal{F})^{-1}(p, q) = (\mathcal{F}x\mathcal{F}^{-1})(p+q).$$

Hence, if $\varphi, \psi \in L^1(\widehat{G})$, then we have, with π defined by (15),

$$\pi_*(\varphi \otimes \psi)(p) = \int_{\widehat{G}} \varphi(p-q)\psi(q)\,dq\,.$$

Thus, $A(G)$ is identified with the group algebra $L^1(\widehat{G})$. Therefore, Theorem 3.9 implies our assertion. Q.E.D.

Definition 3.17. The group $\widehat{G}$ is called the *dual group* of G. Each element of $\widehat{G}$ is called a *character* of G.

Remark 3.18. Since the support of the Plancherel measure is the entire dual group $\widehat{G}$, being a Haar measure, the regular representation of $C^*(G)$ is faithful. Hence the reduced group C^*-algebra $C_r^*(G)$ coincides with $C^*(G)$. Of particular interest is the case that $G = \mathbf{R}$. The dual group $\widehat{\mathbf{R}}$ is then identified with $\mathbf{R}$ itself under the pairing:

$$\langle s, t\rangle = \mathrm{e}^{2\pi i s t}, \quad s, t \in \mathbf{R}. \tag{38}$$

One checks that if the Lebesgue measure is taken as a Haar measure on $\mathbf{R}$, then the associated Plancherel measure on $\widehat{\mathbf{R}} = \mathbf{R}$ is also the Lebesgue measure, i.e.

$$\int_{\mathbf{R}} |f(s)|^2 \,\mathrm{d}s = \int_{\mathbf{R}} \left| \int_{\mathbf{R}} \mathrm{e}^{-2\pi i s t} f(s)\,\mathrm{d}s \right|^2 \mathrm{d}t, \quad f \in L^1(G) \cap L^2(G). \tag{39}$$

Now we return to the case of a general locally compact group G with left invariant measure $\mathrm{d}g$ as in the beginning of the section. The group C^*-algebra $C^*(G)$ is defined to be the universal enveloping C^*-algebra of the group algebra $L^1(G)$, i.e. the completion of $L^1(G)$ under the new norm:

$$\|f\|' = \sup\Big\{ \|\pi(f)\| : \pi \text{ runs over all }{}^*\text{-representations of } L^1(G) \Big\}.$$

The reduced group C^*-algebra $C_r^*(G)$ is the norm closure of $\lambda\big(L^1(G)\big)$ on $L^2(G)$.

Proposition 3.19.

(i) *If $\{\pi, \mathfrak{H}_\pi\}$ is a *-representation of $L^1(G)$, then there exists a unitary representation $\{u_\pi, \mathfrak{H}_\pi\}$ of G on the same Hilbert space $\mathfrak{H}_\pi$ such that*

$$\pi(f) = \int_G f(g) u_\pi(g)\,\mathrm{d}g, \quad f \in L^1(G). \tag{40}$$

(ii) *If $\{u, \mathfrak{H}_u\}$ is a unitary representation of G, then there exists a *-representation $\{\pi, \mathfrak{H}_u\}$ of $L^1(G)$ on the same Hilbert space $\mathfrak{H}_u$ such that $u = u_\pi$, i.e.,*

$$\pi(f) = \int_G f(g) u(g)\,\mathrm{d}g, \quad g \in L^1(G).$$

PROOF:

(i) As we mean by a *-representation of $L^1(G)$ a non-degenerated one, the linear combinations of products $\pi(f)\xi$, $f \in L^1(G)$, $\xi \in \mathfrak{H}_\pi$, is dense in $\mathfrak{H}_\pi$. The identity:

$$\begin{aligned}\Big(\pi\big(\lambda(g)f_1\big)\xi \,\Big|\, \pi\big(\lambda(g)f_2\big)\eta\Big) &= \Big(\pi\big(\lambda(g)f_2\big)^*\pi\big(\lambda(g)f_1\big)\xi \,\Big|\, \eta\Big)\\ &= \Big(\pi\Big(\big(\lambda(g)f_2\big)^* * \big(\lambda(g)f_1\big)\Big)\xi \,\Big|\, \eta\Big)\\ &= \Big(\pi\big(f_2^* f_1\big)\xi \,\Big|\, \eta\Big) = \Big(\pi\big(f_1\big)\xi \,\Big|\, \pi\big(f_2\big)\eta\Big),\end{aligned}$$

which follows from the fact that

$$\big(\lambda(g)f_1\big) * \big(\lambda(g)f_2\big) = f_1 * f_2, \quad f_1, f_2 \in L^1(G), \quad g \in G,$$

shows that the correspondence: $\pi(f)\xi \mapsto \pi\big(\lambda(g)f\big)\xi$ extends to a unitary operator $u_\pi(g)$ on $\mathfrak{H}$. The norm continuity of the map: $g \in G \mapsto \lambda(g)f \in L^1(G)$ for each $f \in L^1(G)$ guarantees the strong continuity of the map $g \in G \mapsto u_\pi(g) \in \mathcal{U}(\mathfrak{H}_\pi)$. It is easy to see that $u_\pi(gh) = u_\pi(g)u_\pi(h),\ g, h \in G$. Thus $\{u_\pi, \mathfrak{H}_\pi\}$ is a unitary representation of G. From the identity:

$$f_1 * f_2 = \int_G f_1(g)\big(\lambda(g)f_2\big)\, dg \quad \text{in } L^1(G)$$

as an $L^1(G)$-valued integral it follows that

$$\pi(f) = \int_G f(g)u_\pi(g)\, dg, \quad f \in L^1(G).$$

(ii) Given a unitary representation $\{u, \mathfrak{H}_u\}$ of G and an element $f \in L^1(G)$, we define an operator by

$$\pi(f) = \int_G f(g)u(g)\, dg\ .$$

Then a direct calculation shows that π is a *-representation of $L^1(G)$ such that $u = u_\pi$. Q.E.D.

Definition 3.20. A function $\omega(\,\cdot\,)$ in $L^\infty(G)$ is said to be *positive definite* if every $f \in L^1(G)$ gives positive integral:

$$\langle f^* * f, \omega\rangle = \iint_{G\times G} \overline{f(g)} f(h)\omega(g^{-1}h)\, dg\, dh \geq 0.$$

Proposition 3.21. *If $\omega(\,\cdot\,) \in L^\infty(G)$, then the following two conditions are equivalent:*

(i) *The function $\omega(\,\cdot\,)$ is positive definite;*
(ii) *There exists a unitary representation $\{u, \mathfrak{H}\}$ such that*

$$\omega(g) = \big(u(g)\xi \mid \xi\big) \quad \textit{for almost every } g \in G. \tag{41}$$

PROOF:
(i) $\Longrightarrow$ (ii): Suppose $\omega(\,\cdot\,)$ is positive definite. Then ω is a positive linear functional of the involutive Banach algebra $L^1(G)$. Hence we get a cyclic representation $\{\pi_\omega, \mathfrak{H}_\omega, \xi_\omega\}$ of $L^1(G)$ such that

$$\langle f, \omega\rangle = \big(\pi_\omega(f)\xi_\omega \mid \xi_\omega\big).$$

Proposition 3.19 then yields a unitary representation $\{u, \mathfrak{H}_\omega\}$ such that

$$\pi_\omega(f) = \int_G f(g)u(g)\,dg, \quad g \in L^1(G).$$

Therefore, we get

$$\begin{aligned}\int_G f(g)\omega(g)\,dg &= \langle f, \omega\rangle = \big(\pi_\omega(f)\xi \mid \xi\big)\\ &= \int_G f(g)\big(u(g)\xi \mid \xi\big)\,dg\end{aligned}$$

for every $f \in L^1(G)$. Therefore, the assertion (ii) follows.
(ii) $\Longrightarrow$ (i): Suppose that ω is of the form:

$$\omega(g) = \big(u(g)\xi \mid \xi\big) \quad \text{for almost every } g \in G.$$

Let π be the *-representation of $L^1(G)$ given by Proposition 3.19. Then for any $f \in L^1(G)$, we have

$$\begin{aligned}\big\langle f^* * f, \omega\big\rangle &= \int_G f^* * f(g)\big(u(g)\xi \mid \xi\big)\,dh\,dg = \big(\pi(f^* * f)\xi \mid \xi\big)\\ &= \big(\pi(f)\xi \mid \pi(f)\xi\big) = \|\pi(f)\xi\|^2 \geq 0.\end{aligned}$$

Q.E.D.

Since the function given by (41) is continuous, we get the following result immediately:

Corollary 3.22. *If $\omega \in L^\infty(G)$ is positive definite, then it agrees with a continuous positive definite function almost everywhere.*

We therefore consider only continuous positive definite functions.

Proposition 3.23. *For a continuous function ω on G the following conditions are equivalent:*

(i) *The function ω is positive definite;*
(ii) *For any finite set $\{g_1, \dots, g_n\} \subset G$, the matrix:*

$$\begin{pmatrix} \omega(e) & \omega(g_1^{-1}g_2) & \cdots & \omega(g_1^{-1}g_n)\\ \omega(g_2^{-1}g_1) & \omega(e) & \cdots & \omega(g_2^{-1}g_n)\\ \vdots & \vdots & \ddots & \vdots\\ \omega(g_n^{-1}g_1) & \omega(g_n^{-1}g_2) & \cdots & \omega(e)\end{pmatrix}$$

is a positive matrix.

If ω is positive definite, then $|\omega(g)| \leq \omega(e)$, $g \in G$.

PROOF:

(i) $\Longrightarrow$ (ii): Fix $\{g_1, \dots, g_n\} \subset G$ and $(\alpha_1, \dots, \alpha_n) \in \mathbf{C}^n$. We assume that ω is of the form (41). Then we get

$$\sum_{i,j=1}^{n} \bar{\alpha}_i \alpha_j \omega(g_i^{-1} g_j) = \sum_{i,j=1}^{n} \bar{\alpha}_i \alpha_j \Big(u(g_i^{-1} g_j)\xi \;\Big|\; \xi \Big)$$

$$= \Big(\sum_{j=1}^{n} \alpha_j u(g_j)\xi \;\Big|\; \sum_{i=1}^{n} \alpha_i u(g_i)\xi \Big) = \Big\| \sum_{i=1}^{n} \alpha_i u(g_i)\xi \Big\|^2 \geq 0.$$

(ii) $\Longrightarrow$ (i): For each $f \in \mathcal{K}(G)$ we have

$$\langle f^* * f, \omega\rangle = \int_G f^* * f(g)\omega(g)\, dg = \iint_{G\times G} f^*(h) f(h^{-1}g)\omega(g)\, dh\, dg$$

$$= \iint_{G\times G} \Delta\Big(h^{-1}) \overline{f(h^{-1})} f(h^{-1})g\Big)\omega(g)\, dh\, dg$$

$$= \iint_{G\times G} \Delta(h^{-1}) \overline{f(h^{-1})} f(g)\omega(hg)\, dg\, dh$$

$$= \iint_{G\times G} \overline{f(h)} f(g)\omega(h^{-1}g)\, dg\, dh\, .$$

Since $f \in \mathcal{K}(G)$, the above integral can be approximated arbitrarily well by Riemann sum:

$$\sum_{i,j=1}^{n} \overline{f(g_i)} f(g_k)\omega(g_i^{-1}g_k)|U_i||U_k| \geq 0$$

where U_i is a neighborhood of g_i for $i = 1, \dots, n$. Hence $\langle f^* * f, \omega\rangle \geq 0$ for every $f \in \mathcal{K}(G)$ follows. As $\mathcal{K}(G)$ is dense in $L^1(G)$, our assertion follows.

Suppose ω is positive definite. Then the 2×2-matrix $\begin{pmatrix} \omega(e) & \omega(g^{-1}) \\ \omega(g) & \omega(e) \end{pmatrix}$ is positive for every $g \in G$. Hence its determinant is positive:

$$\omega(e)^2 - |\omega(g)|^2 = \det \begin{pmatrix} \omega(e) & \omega(g^{-1}) \\ \omega(g) & \omega(e) \end{pmatrix} \geq 0. \qquad \text{Q.E.D.}$$

Now we consider the Banach space conjugate $B(G) = C^*(G)^*$ of the group C^*-algebra $C^*(G)$ of a locally compact group G. The inequality:

$$\|\pi(f)\| \leq \|f\|, \quad f \in L^1(G),$$

for every *-representation π of $L^1(G)$ by Proposition I.5.2 implies that every element $\omega \in B(G)$ is given by a function $\omega(\,\cdot\,) \in L^\infty(G)$:

$$\langle f, \omega\rangle = \int_G f(g)\omega(g)\, dg\, .$$

Moreover, the conjugate space $C^*(G)^*$ is spanned by its positive part as seen in Chapter III, Corollary 3.19 yields that the Banach space $B(G)$ is identified with a subspace of the algebra $C_b(G)$ of all bounded continuous functions on G. Of course the norm in $B(G)$ is different from the norm inherited from $C_b(G)$.

Proposition 3.24.
(i) *The Banach space $B(G)$ is a self-adjoint Banach algebra relative to the pointwise product inherited from $C_b(G)$ and its own norm given as the conjugate space norm on $C^*(G)^*$.*
(ii) *The Fourier algebra $A(G)$ is an ideal of $B(G)$.*

PROOF:
(i) Suppose that φ and ψ are two elements of $B(G)$. By Proposition 3.21 there exist unitary representations $\{u, \mathfrak{H}\}$ and $\{v, \mathfrak{K}\}$ and vectors $\xi_1, \xi_2 \in \mathfrak{H}$ and $\eta_1, \eta_2 \in \mathfrak{K}$ such that

$$\varphi(g) = \big(u(g)\xi_1 \mid \xi_2\big) \qquad \text{and} \qquad \psi(g) = \big(v(g)\eta_1 \mid \eta_2\big), \qquad g \in G.$$

Then we have

$$\varphi(g)\psi(g) = \Big(\big(u \otimes v\big)(g)\big(\xi_1 \otimes \eta_1\big) \Big| \big(\xi_2 \otimes \eta_2\big)\Big).$$

Hence the polarization identity implies that the product $\varphi\psi$ is a linear combination of four positive definite continuous functions. Hence $\varphi\psi \in B(G)$.

Let π_u and π_v be the *-representations of $L^1(G)$, hence the representations of $C^*(G)$, corresponding to u and v respectively. Consider the tensor product representation $\pi_u \otimes \pi_v$ of $C^*(G \times G)$. Then we have

$$\|\varphi\|_{B(G)}\|\psi\|_{B(G)} = \big\|\varphi \otimes \psi\big\|_{B(G\times G)}.$$

The diagonal embedding map $\delta\colon g \in G \mapsto (g, g) \in G \times G$ gives rise to an isomorphism δ of $C^*(G)$ into the multiplier C^*-algebra $M\big(C^*(G \times G)\big)$ of $C^*(G \times G)$ such that

$$\big\langle f, \varphi\psi\big\rangle = \big\langle \delta(f), (\varphi \otimes \psi)\big\rangle, \quad f \in L^1(G).$$

Consequently we get

$$\big\|\varphi\psi\big\|_{B(G)} \le \|\varphi\|_{B(G)}\|\psi\|_{B(G)}, \quad \varphi, \psi \in B(G).$$

Fix an arbitrary $\varphi \in B(G)$ and choose a unitary representation $\{u, \mathfrak{H}\}$ and vectors $\xi, \eta \in \mathfrak{H}$ such that

$$\varphi(g) = \big(u(g)\xi \mid \eta\big), \quad g \in G.$$

Let $\overline{\mathfrak{H}}$ be the Banach space conjugate of $\mathfrak{H}$, which is anti-isomorphic to $\mathfrak{H}$ under the correspondence: $\zeta \in \mathfrak{H} \longleftrightarrow \bar{\zeta} \in \overline{\mathfrak{H}}$ and $\langle \xi, \bar{\zeta}\rangle = (\xi \mid \zeta)$. Let $\bar{u}$ be the unitary representation of G on $\overline{\mathfrak{H}}$ given by:

$$\bar{u}(g)\bar{\zeta} = \overline{u(g)\zeta}, \quad g \in G,$$

i.e., $\bar{u}(g) = u(g^{-1})^{\mathrm{t}}$, $g \in G$, where the exponent t means the transposing operation.

Then we get

$$\left(\bar{u}(g)\bar{\xi} \mid \bar{\eta}\right) = \overline{\left(u(g)\xi \mid \bar{\eta}\right)} = \left(\eta \mid u(g)\xi\right) = \overline{\varphi(g)} = \bar{\varphi}(g), \quad g \in G.$$

Hence we get $\bar{\varphi} \in B(G)$, i.e., $B(G)$ is self-adjoint.

(ii) Let $\{u, \mathfrak{H}\}$ be a unitary representation of G. Identifying the tensor product $L^2(G) \otimes \mathfrak{H}$ with $L^2(\mathfrak{H}, G)$ the Hilbert space of all square integrable $\mathfrak{H}$-valued functions on G relative to the left Haar measure. Set

$$(W\xi)(g) = u(g)\xi(g), \quad \xi \in L^2(\mathfrak{H}, G), \quad g \in G,$$

to obtain a unitary W on $L^2(\mathfrak{H}, G)$ and compute:

$$\begin{aligned}\left(W\left(\lambda(g) \otimes 1\right)W^*\xi\right)(h) &= u(h)\left(\left(\lambda(g) \otimes 1\right)W^*\xi\right)(h)\\ &= u(h)\left(W^*\xi\right)\left(g^{-1}h\right) = u(h)u\left(g^{-1}h\right)^*\xi\left(g^{-1}h\right)\\ &= u(g)\xi\left(g^{-1}h\right) = \left(\left(\lambda(g) \otimes u(g)\right)\xi\right)(h).\end{aligned}$$

Therefore, we conclude that $W\left(\lambda(g) \otimes 1\right)W^* = \lambda(g) \otimes u(g),\ g \in G$. If $\varphi \in A(G)$ and $\psi \in B(G)$ are of the form:

$$\begin{aligned}\varphi(g) &= \left(\lambda(g)\xi_1 \mid \xi_2\right), \quad \xi_1, \xi_2 \in L^2(G);\\ \psi(g) &= \left(u(g)\eta_1 \mid \eta_2\right), \quad \eta_1, \eta_2 \in \mathfrak{H}, \quad g \in G,\end{aligned}$$

then we get

$$\begin{aligned}\varphi(g)\psi(g) &= \left(\left(\lambda(g) \otimes u(g)\right)\xi_1 \otimes \eta_1 \,\middle|\, \xi_2 \otimes \eta_2\right)\\ &= \left(W\left(\lambda(g) \otimes 1\right)W^*\left(\xi_1 \otimes \eta_1\right) \,\middle|\, \left(\xi_2 \otimes \eta_2\right)\right)\\ &= \left(\left(\lambda(g) \otimes 1\right)\zeta_1 \,\middle|\, \zeta_2\right)\\ &= \int_G \left(\zeta_1\left(g^{-1}h\right) \,\middle|\, \zeta_2(h)\right) \mathrm{d}h\end{aligned}$$

where $\zeta_1 = W^*(\xi_1 \otimes \eta_1)$ and $\zeta_2 = W^*(\xi_2 \otimes \eta_2)$. With an orthonormal basis $\left\{\varepsilon_i : i \in I\right\}$ of $\mathfrak{H}$, we get

$$\varphi(g)\psi(g) = \int_G \left(\sum_{i \in I}\left(\zeta_1(g^{-1}h) \mid \varepsilon_i\right)\left(\varepsilon_i \mid \zeta_2(h)\right)\right) \mathrm{d}h\,.$$

As we have

$$\int_G \left(\sum_{i \in I}\left|\left(\zeta_k(h) \mid \varepsilon_i\right)\right|^2\right) \mathrm{d}h = \int_G \left\|\zeta_k(h)\right\|^2 \mathrm{d}h < \infty, \quad k = 1, 2,$$

the above summation is absolutely summable and with

$$f_{k,i}(g) = \left(\zeta_k(g) \mid \varepsilon_i\right), \quad k = 1, 2,$$

we get

$$\varphi(g)\psi(g) = \sum_{i\in I}\int_G f_{1,i}(g^{-1}h)\overline{f_{2,i}(h)}\,dh = \sum_{i\in I}\bar{f}_{2,i} * f_{1,i}^{\vee}(g), \quad g \in G.$$

Now from the computation that

$$\infty > \int_G \|\zeta_k(g)\|^2\,dg = \int_G\Big(\sum_{i\in I}\big|(\zeta_k(g)\mid\varepsilon_i)\big|^2\Big)\,dg$$

$$= \sum_{i\in I}\int_G \big|(\zeta_k(g)\mid\varepsilon_i)\big|^2\,dg = \sum_{i\in I}\|f_{k,i}\|_2^2, \quad k = 1, 2,$$

it follows that the systems $\{f_{k,i} : i \in I\}$, $k = 1, 2$, are both square summable. Consequently the system $\varphi_i = \bar{\varphi}_{2,i} * f_{1,i}^{\vee} \in A(G)$ is norm summable in $A(G)$ because

$$\sum_{i\in G}\|\varphi_i\|_{A(G)} = \sum_{i\in I}\|\bar{f}_{2,i} * f_{1,i}^{\vee}\|_{A(G)} \le \sum_{i\in I}\|f_{1,i}\|_2\|f_{2,i}\|_2$$

$$\le \Big(\sum_{i\in I}\|f_{1,i}\|_2^2\Big)^{\frac{1}{2}}\Big(\sum_{i\in I}\|f_{2,i}\|_2^2\Big)^{\frac{1}{2}} < \infty.$$

Therefore we obtain $\varphi\psi = \sum_{i\in I}\varphi_i \in A(G)$. This shows that $A(G)$ is an ideal of $B(G)$. Q.E.D.

Definition 3.25. The Banach algebra $B(G)$ is called the *Fourier-Stieltjes algebra* of G.

Proposition 3.26. *Let $\mathfrak{S}(G)$ be the set of all states on the group C^*-algebra $C^*(G)$ of a locally compact group G.*

(i) *If $\varphi \in \mathfrak{S}(G)$, then we have the inequality:*

$$|\varphi(g)-\varphi(h)| \le \min\Big\{\sqrt{2-2\operatorname{Re}\varphi(g^{-1}h)}, \sqrt{2-2\operatorname{Re}\varphi(gh^{-1})}\Big\} \quad g, h \in G.$$

(ii) *The weak* topology on the state space $\mathfrak{S}(G)$ is precisely the uniform convergence topology on compact subsets of G.*

PROOF:

(i) We may assume that φ is of the form:

$$\varphi(g) = (u(g)\xi \mid \xi), \quad g \in G,$$

for some unitary representation $\{u, \mathfrak{H}\}$ and a unit vector $\xi \in \mathfrak{H}$. We then compute

$$|\varphi(g)-\varphi(h)| = \big|(u(g)\xi\mid\xi) - (u(h)\xi\mid\xi)\big| = \Big|\big((u(g)-u(h))\xi \,\big|\, \xi\big)\Big|$$

$$\begin{aligned}
&\le \left\|(u(g) - u(h))\xi\right\| = \left\|(1 - u(g^{-1}h))\xi\right\| \\
&= \left((1 - u(g^{-1}h))\xi \;\middle|\; (1 - u(g^{-1}h))\xi\right)^{\frac{1}{2}} \\
&= \left(2 - 2\operatorname{Re}\left(u(g^{-1}h)\xi \mid \xi\right)\right)^{\frac{1}{2}} = \left(2 - 2\varphi\left(g^{-1}h\right)\right)^{\frac{1}{2}},
\end{aligned}$$

and also

$$\left|\varphi(g) - \varphi(h)\right| = \left|\varphi\left(g^{-1}\right) - \varphi\left(h^{-1}\right)\right| \le \sqrt{2 - 2\operatorname{Re}\varphi\left(gh^{-1}\right)}.$$

(ii) Suppose $\left\{\varphi_i : i \in I\right\}$ is a net in $\mathfrak{S}(G)$ converging to $\varphi_0 \in \mathfrak{S}(G)$ uniformly on every compact subset of G. Fix $f \in L^1(G)$ and $\varepsilon > 0$. Choose a compact subset K of G and $i_0 \in I$ such that $\left\|(1 - \chi_K)f\right\|_1 < \varepsilon/4$ and $\left\|(\varphi_i - \varphi_0)\chi_K\right\|_\infty < \varepsilon/\left(2\|f\|_1\right)$ for $i \ge i_0$. Then we have for every $i \ge i_0$

$$\begin{aligned}
\left|\langle f, \varphi_i - \varphi_0\rangle\right| &\le \left|\langle f, \chi_K(\varphi_i - \varphi_0)\rangle\right| + \left|\langle f, (1 - \chi_K)(\varphi_i - \varphi_0)\rangle\right| \\
&\le \|f\|_1 \left\|\chi_K(\varphi_i - \varphi_0)\right\|_\infty + \left\|f(1 - \chi_K)\right\|_{C^*(G)} \left\|\varphi_i - \varphi_0\right\|_{B(G)} \\
&\le \frac{\varepsilon}{2} + 2\left\|f(1 - \chi_K)\right\|_1 \le \frac{\varepsilon}{2} + \frac{\varepsilon}{2} = \varepsilon.
\end{aligned}$$

Therefore the net $\{\varphi_i\}$ converges to φ_0 in the $\sigma\left(B(G), L^1(G)\right)$-topology. As the net $\{f_i\}$ is contained in the unit ball of $B(G)$ and $L^1(G)$ is norm dense in $C^*(G)$, the net $\{\varphi_i\}$ converges to φ_0 in the weak* topology.

Conversely, suppose $\left\{ f_i : i \in I \right\}$ is a net in $\mathfrak{S}(G)$ converging to φ_0 in the weak* topology. Fix a compact subset K of G. Let V be a relatively compact neighborhood of the identity $e \in G$ such that

$$V \subset \left\{g \in G : 1 - \varphi_0(g) \le \varepsilon\right\}.$$

Recall that $\varphi_0(e) = 1$ and $\varphi_0(g) \le \varphi_0(e),\ g \in G$. Set $f_V = \chi_V/|V| \in L^1(G)$. Let $\delta > 0$ be a small number which is to be determined later. Choose $i_1 \in I$ such that $\left|\langle f_V, \varphi_i - \varphi_0\rangle\right| < \delta$ for every $i \ge i_1$, i.e.,

$$\int_V \left|\varphi_i(g) - \varphi_0(g)\right| \,\mathrm{d}g < \delta|V|, \quad i \ge i_1.$$

For every $i \ge i_1$, we have

$$\left|\int_V (1 - \varphi_i(g)) \,\mathrm{d}g\right| \le \left|\int_V (1 - \varphi_0(g)) \,\mathrm{d}g\right| + \left|\int_V (\varphi_i(g) - \varphi_0(g)) \,\mathrm{d}g\right| < 2\delta|V|.$$

Therefore we get, for every $i \ge i_1$ and $g \in G$,

$$\left|f_V * \varphi_i(g) - \varphi_i(g)\right| = \left|\int_V f(h)\left(\varphi_i(h^{-1}g) - \varphi_i(g)\right) \,\mathrm{d}h\right|$$

$$
\begin{aligned}
&\le \frac{1}{|V|}\int_V \left|\varphi_i(h^{-1}g) - \varphi_i(g)\right| \mathrm{d}h \\
&\le \frac{1}{|V|}\int_V \sqrt{2 - 2\operatorname{Re}\varphi_i(h)}\, \mathrm{d}h \\
&\le \frac{1}{|V|}\left(\int_V \left(2 - 2\operatorname{Re}\varphi_i(h)\right) \mathrm{d}h\right)^{\frac{1}{2}} |V|^{\frac{1}{2}} \\
&\le \frac{1}{|V|}\left|2\int_V \left(1 - \varphi_i(h)\right) \mathrm{d}h\right|^{\frac{1}{2}} |V|^{\frac{1}{2}} \le \sqrt{2\delta}.
\end{aligned}
$$

Therefore, we get $\| f_V * \varphi_i - \varphi_i \|_\infty \le \sqrt{2\delta}$ for every $i \ge i_1$.

The correspondence $\varphi \in B(G) \longleftrightarrow \varphi^\vee \in B(G)$ is weak* bicontinuous. Hence the net $\{\varphi_i^\vee\}$ converges to $\varphi_0^\vee$ in the weak* topology. Since the subset $\lambda(K^{-1}) f_V \subset L^1(G)$ is compact in the norm topology, there exists $i_2 \in I$ such that

$$
\sup\left\{ \left|\langle \lambda(g^{-1}) f_V,\ \varphi_i^\vee - \varphi_0^\vee \rangle\right| : g \in K \right\} < \delta \quad \text{for every } i \ge i_2.
$$

On the other hand, we have, for any measurable function φ on G,

$$
\begin{aligned}
\langle \lambda(g^{-1}) f_V, \varphi^\vee \rangle &= \int_G \varphi^\vee(h) f_V(gh)\, \mathrm{d}h = \int_G \varphi^\vee(g^{-1}h) f_V(h)\, \mathrm{d}h \\
&= \int_G f_V(h)\varphi(h^{-1}g)\, \mathrm{d}h = f_V * \varphi(g).
\end{aligned}
$$

Hence we get for each $i \ge i_2$

$$
\sup_{g\in K} \left| f_V * \varphi_i(g) - f_V * \varphi_0(g) \right| \le \delta
$$

and therefore with $i_0 \in I$ dominating both i_1 and i_2 for each $i \ge i_0$

$$
\begin{aligned}
&\sup_{g\in K} \left|\varphi_i(g) - \varphi_0(g)\right| \\
&\quad \le \|\varphi_i - f_V * \varphi_i\|_\infty + \sup_{g\in K}\left\| f_V * \varphi_i(g) - f_V * \varphi_0(g)\right\| \| f_V * \varphi_0 - \varphi_0 \|_\infty \\
&\quad \le 2\sqrt{2\delta} + \delta.
\end{aligned}
$$

Therefore choosing $\delta > 0$ so small that $2\sqrt{2\delta} + \delta < \varepsilon$, we conclude that

$$
\sup_{g\in K} \left|\varphi_i(g) - \varphi_0(g)\right| \le \varepsilon, \quad i \ge i_0.
$$

Therefore the net $\{\varphi_i\}$ converges to φ_0 uniformly on every compact subset of G.

Q.E.D.

Exercise VII.3

1) Let G be a group equipped with a standard Borel structure such that the product map: $(g, h) \in G \times G \mapsto gh \in G$ and the inverse map: $g \in G \mapsto g^{-1} \in G$ are both Borel maps. Let μ be a σ-finite Borel measure on G which is quasi-invariant under the left translations in the sense that $\mu(gE) = 0 \iff \mu(E) = 0$ for every Borel set E in G and $g \in G$. For each $g \in G$, let $\chi(g, \cdot)$ be a Borel function on G such that

$$\mu(gE) = \int_E \chi(g, h)\, d\mu(h) .$$

(**a**) Show that for each pair $g, h \in G$

$$\chi(gh, k) = \chi(g, hk)\chi(h, k)$$

for μ-almost every $k \in G$, and for every $k \in G$

$$\chi(e, k) = 1,$$

where e is the unit element of G.

(**b**) Making use of (a), show that the function:

$$(g, k) \in G \times G \mapsto \chi(gk^{-1}, k)/\chi(k^{-1}, k) \in \mathbf{R}$$

is almost everywhere constant in $k \in G$. Let $f(g)$ be the common value of the above function for each $g \in G$, for example $f(g)$ can be given by

$$f(g) = \frac{1}{\mu(V)} \int_V \chi(gk^{-1}, k)/\chi(k^{-1}, k)\, d\mu(k)$$

with any Borel set V such that $0 < \mu(V) < +\infty$ and the integrand is bounded on V.

(**c**) Show that for every $g \in G$ and μ-almost every $h \in G$

$$\chi(g, h) = f(gh)/f(h).$$

(**d**) Define a new measure ν on G by

$$\nu(E) = \int_E f(h)^{-1}\, d\mu(h) .$$

Prove that the measure ν is invariant under the left translation on G.

2) Let G be as in Problem 1 and μ be a left invariant σ-finite measure on G. Show that if ν is another σ-finite left invariant measure on G, then μ and ν are proportional. (Hint: First observe that $\lambda = \mu + \nu$ is σ-finite and left invariant, then prove that the Radon-Nikodym derivatives $d\mu \,/\, d\lambda$ and $d\nu \,/\, d\lambda$ are both constant.)

3) Let G be as before and μ be a left invariant σ-finite measure on G. There exists a Borel map $\delta_G \colon g \in G \mapsto \delta_G(g) \in \mathbf{R}_+^*$ such that $\mu(Eg) = \delta_G(g)\mu(E)$ for every Borel set E in G.

4) With $\{G, \mu\}$ as above, let λ and ρ be the left and right regular representations of G on $\mathfrak{H} = L^2(G, \mu)$ defined as in the case of a locally compact group viewing μ as if the Haar measure. Let $\mathcal{R}_\ell(G)$ and $\mathcal{R}_r(G)$ be the von Neumann algebras generated by $\lambda(G)$ and $\rho(G)$ respectively. Set

$$(W\xi)(g, h) = \xi(g, gh), \quad \xi \in \mathfrak{H} \otimes \mathfrak{H}.$$

Following the line of the proof for Theorem 3.9, prove that the set $\mathcal{G}$ of all non-zero solutions $x \in \mathcal{L}(\mathfrak{H})$ of the equation:

$$W^*(x \otimes 1)W = x \otimes x$$

form a locally compact group which contains G.

5) Continuing the previous problem, show that $G = \mathcal{G}$. (Hint: The proof of Theorem 3.9 entails that the Haar measure of $\mathcal{G}$ is concentrated on G. Then use the fact that if E is a measurable subset of $\mathcal{G}$ with positive Haar measure, then $E^{-1}E$ contains the origin as an interior point, to prove that G is open in $\mathcal{G}$.)

6) From the above sequence of problems, conclude that if a group admits a quasi-invariant σ-finite standard Borel measure then it must be a locally compact group.

7) Prove that (i) an infinite dimensional Hilbert space can not admit a quasi-invariant σ-finite Borel measure; (ii) the unitary group of an infinite dimensional factor can not admit a quasi-invariant σ-finite Borel measure.

§ 4 Weights on a C^*-Algebra

A *weight* on a C^*-algebra A is defined to be a function φ on A_+ with values in $[0, \infty]$ satisfying the first two conditions in Definition 1.1. It is easy to see that φ gives rise to representation $\{\pi_\varphi, \mathfrak{H}_\varphi, \eta_\varphi\}$ of A via the construction in §1. But the non-degeneratedness of π_φ does not follow in this generality. At any rate, we have

$$\left.\begin{aligned} \left(\pi_\varphi(a)\eta_\varphi(x) \mid \eta_\varphi(y)\right) &= \dot{\varphi}(y^*ax), \quad a \in A, \quad x, y \in \mathfrak{n}_\varphi, \\ \varphi(x) &= \left\|\eta_\varphi(x^{\frac{1}{2}})\right\|^2, \quad x \in \mathfrak{m}_\varphi^+. \end{aligned}\right\} \tag{1}$$

We call $\{\pi_\varphi, \mathfrak{H}_\varphi, \eta_\varphi\}$ the *semi-cyclic representation* of A induced by φ.

Lemma 4.1. *If φ is a lower semi-continuous weight on a C^*-algebra A, then the semi-cyclic representation π_φ of A induced by φ is non-degenerate.*

PROOF: Let $\{u_i\}$ be an approximate identity of A. If $x \in \mathfrak{n}_\varphi$, then we have

$$\begin{aligned} \lim_i \left\|\eta_\varphi(x) - \pi_\varphi(u_i)\eta_\varphi(x)\right\|^2 &= \lim_i \varphi\left((x - u_i x)^*(x - u_i x)\right) \\ &= \lim_i \left(\varphi(x^*x) - 2\varphi(x^*u_i x) + \varphi(x^*u_i^2 x)\right). \end{aligned}$$

But we have

$$\varphi(x^*x) \leq \liminf \varphi(x^*u_ix) \leq \limsup \varphi(x^*u_ix) \leq \varphi(x^*x),$$
$$\varphi(x^*x) \leq \liminf \varphi(x^*u_i^2x) \leq \limsup \varphi(x^*u_i^2x) \leq \varphi(x^*x).$$

Hence we get

$$\lim_i \left\| \eta_\varphi(x) - \pi_\varphi(u_i)\eta_\varphi(x) \right\|^2 = 0.$$

Thus $\{\pi_\varphi(u_i)\}$ converges to 1 strongly on $\mathfrak{H}_\varphi$. Hence π_φ is non-degenerate. Q.E.D.

Proposition 4.2. *A lower semi-continuous weight φ on a C^*-algebra A is written in the form:*

$$\Phi_\varphi = \left\{ \omega \in A^* : \omega \leq \varphi \right\}, \tag{2}$$
$$\varphi(x) = \sup\left\{ \omega(x) : \omega \in \Phi_\varphi \right\}, \quad x \in A_+. \tag{3}$$

PROOF: We will show that the self-adjoint part A_h of A satisfies the condition (i) of Lemma 1.16 with respect to the norm topology. To this end, $\tilde{A}$ be the universal enveloping von Neumann algebra of A and $\tilde{E}$ be the σ-weak closure of a subset E of A.

Now, suppose F is a norm closed hereditary subset of A_+. By the Hahn-Banach separation theorem, we have $\tilde{F} \cap A = F$. Clearly $\tilde{F}$ is σ-weakly closed and convex. If $\tilde{F}$ is hereditary, then the proof of the implication (iii) $\Longrightarrow$ (iv) of Theorem 1.11 shows that $\tilde{F} = (\tilde{F} - \tilde{A}_+)^\sim \cap \tilde{A}_+$. Hence we have

$$\begin{aligned} F &\subset \left(F \cap A_+\right)^- \cap A_+ \subset \left(\tilde{F} - A_+\right)^\sim \\ &\subset \left(\tilde{F} - A_+\right)^\sim \cap \tilde{A}_+ \cap A = \tilde{F} \cap A = F, \end{aligned}$$

where the bar means the norm closure. Thus, we have only to prove the heredity of $\tilde{F}$. Suppose that $0 \leq y \leq x$ and $x \in \tilde{F}$. We then have $y^{1/2} = sx^{1/2}$, $\|s\| \leq 1$, for some $s \in \tilde{A}$. Choose a net $\{s_i\}$ (resp. $\{x_i\}$) from the unit ball of A (resp. F) such that $\{s_i\}$ (resp. $\{x_i\}$) converges to $x^{1/2}$. The boundedness of $\{s_i\}$ implies the σ-strong convergence of $\{s_ix_i^{1/2}\}$ to $sx^{1/2} = y^{1/2}$. Hence $\{x_i^{1/2}s_i^*s_ix_i^{1/2}\}$ converges to $x^{1/2}s^*sx^{1/2} = y$ σ-weakly (not necessarily σ-strongly though). The inequality $x_i^{1/2}s_i^*s_ix_i^{1/2} \leq x_i \in F$ yields that $x_i^{1/2}s_i^*s_ix_i^{1/2} \in F$ by the heredity of F. Thus y belongs to $\tilde{F}$. Q.E.D.

We fix a lower semi-continuous weight φ on a C^*-algebra A and set $\mathcal{M} = \pi_\varphi(A)''$. We set

$$\Phi_{\varphi,0} = \left\{ \omega \in A_+^* : (1+\varepsilon)\omega \leq \varphi \text{ for some } \varepsilon > 0 \right\}; \tag{4}$$

as in §2, we can show that $\Phi_{\varphi,0}$ is upward directed and $\varphi = \sup \Phi_{\varphi,0}$. Hence φ can be extended to a normal weight $\tilde{\varphi}$ on the universal enveloping von Neumann algebra $\tilde{A}$ by the formula:

$$\tilde{\varphi}(x) = \sup\left\{ \omega(x) : \omega \in \Phi_{\varphi,0} \right\}, \quad x \in \tilde{A}. \tag{5}$$

We can then view $\tilde{\varphi}$ as a normal weight on $\mathcal{M} = \pi_{\tilde{\varphi}}(\tilde{A})$ and construct the opposite weight φ' on $\mathcal{M}'$ which is faithful, semi-finite and normal.

Notes on Chapter VII

The theory of unbounded linear functionals is needed for many years as seen in the success of the theory of distributions. But the general theory in operator algebras was not yet available. In the early part of the history, the theory of semi-finite traces was developed along with the theory of Hilbert algebras. It was completed in the early fifties by the hand of Dixmier [85], Segal [326] and Ogasawara-Yoshinaga [269].

The theory of weights had to wait until the work of F. Combes, [65, 66, 69], and that of G. K. Pedersen, [276, 277, 280], started in the mid sixties. In these series, they established the framework for the general theory of weights by studying the relationship of a weights and the associated semi-cyclic representation. The final definitive form of weights as in Theorem 1.11 was proved by U. Haagerup, [538]. Before this work, the condition (iv) of Theorem 1.11 was taken as the definition of the normality for weights.

The correspondence of a faithful semi-finite normal weight and a left Hilbert algebra was independently established by F. Combes, [451], and M. Takesaki [34].

The Plancherel formula for non-abelian locally compact groups was a focal point of the theory of operator algebras and unitary group representation theory in the fifties and the sixties. In the early fifties, the two branches were not independent discipline of mathematics unlike today. As mentioned above, the theory of Hilbert algebras and the associated trace on the left von Neumann algebra were developed precisely aiming at the Plancherel formula. A good source of literatures of the fifties and the sixties which are directly related to the group representations is the two books of Dixmier [4, 5]. We treated the subject lightly in Section 3. The fact that the predual of the group von Neumann algebra of an unimodular locally compact group is a Banach algebra was first observed by W. F. Steinespring, [692]. He then proved Theorem 3.9 for unimodular locally compact groups extending the classical Tannaka duality theorem for compact groups. Theorem 3.9 combines results of many mathematicians and is called naturally Eymard-Steinespring-Tatsuuma duality theorem, [516, 692, 681, 718], The terminology "Fourier algebra $A(G)$" is due to Eymard. Further extension and comprehensive treatise of duality theory was given recently by Enock and Schwartz, [514].

Chapter VIII

Modular Automorphism Groups

§ 0 Introduction

As we have seen in the previous chapters, the modular operator gives rise to a one parameter automorphism group on the von Neumann algebra in question. First, we are going to identify this group. The first section is devoted to this task. This group is characterized by the *modular condition*, Definition 1.1. Unlike the classical integration theory, the non-commutative theory comes with the intrinsic dynamics of the modular automorphism group. This chapter is devoted to the elementary theory of the modular automorphism groups. In fact, the modular automorphism group is the key to the entire theory of this volume. Thus, the theory does not end in this chapter. It will continue to the rest of this book. Section 1 is devoted to the characterization of the modular automorphism group of a weight by the modular condition, Theorem 1.2. In Section 2, the analysis and the characterization of the fixed point algebra under the modular automorphism group, called the *centralizer*, of a weight are given. We also discuss the perturbation of a weight by an operator affiliated to the centralizer of the weight, which is needed in the following section. The classical Radon-Nikodym theorem in the integration theory is extended to the non-commutative setting in a surprisingly simple manner, Theorem 3.3. In the course of proving the theorem, we will see that the semi-cyclic representation of a von Neumann algebra induced by a faithful semi-finite normal weight is unique up to unitary equivalence, Theorem 3.2. One of the consequences of Theorem 3.3 is that the modular automorphism group is unique up to unitary one cocycles, hence their images in the quotient group $\mathrm{Out}(\mathcal{M}) = \mathrm{Aut}(\mathcal{M})/\,\mathrm{Int}(\mathcal{M})$ is an intrinsic quantity of the algebra. Many results follow from this theorem. Only immediate consequences are discussed in this section, and many others are spread to the rest of this book. The semi-finiteness of the algebra is then equivalent to the innerness of the modular automorphism group, Theorem 3.14.

The last section of the chapter is devoted to the discussion of tensor products of weights and direct integral. As it is unbounded, the tensor product of two weights has to be handled with care. It is done by considering the tensor product of the associated left Hilbert algebras.

§ 1 Modular Automorphism Group of a Weight

Throughout this section, we mean by a weight a faithful semi-finite normal weight unless otherwise is explicitly mentioned.

In the last chapter, we observed that a weight φ on a von Neumann algebra $\mathcal{M}$ and a full left Hilbert algebra $\mathfrak{A}_\varphi$ with $\mathcal{M} \cong \mathcal{R}_\ell(\mathfrak{A}_\varphi)$ correspond uniquely. In Chapter VI, it was shown that $\mathfrak{A}_\varphi$ gives rise to the modular operator Δ and the modular conjugation J. This section is devoted to the study of the relation between φ and Δ.

Definition 1.1. Let A be a C^*-algebra, equipped with a one parameter automorphism group $\{\sigma_t : t \in \mathbf{R}\}$. A lower semi continuous weight φ on A is said to satisfy the *modular condition* for $\{\sigma_t\}$ if the following two conditions hold:

(i) $\varphi = \varphi \circ \sigma_t,\ t \in \mathbf{R}$;

(ii) For every pair $x, y \in \mathfrak{n}_\varphi \cap \mathfrak{n}_\varphi^*$, there exists a bounded continuous function $F_{x,y}$ on the closed horizontal strip $\overline{\mathbf{D}}$ bounded by $\mathbf{R}$ and $\mathbf{R}+\mathrm{i}$ which is holomorphic on the open strip $\mathbf{D}$ such that

$$F_{x,y}(t) = \varphi\big(\sigma_t(x)y\big) \quad \text{and} \quad F_{x,y}(t+\mathrm{i}) = \varphi\big(y\sigma_t(x)\big), \quad t \in \mathbf{R}. \tag{1}$$

In the C^*-algebraic formalism of quantum statistical mechanics, the modular condition is known as the *Kubo-Martin-Schwinger boundary condition* which characterizes equilibrium states of the physical system at the inverse temperature $\beta = -1$. Namely, when the C^*-algebra A represents the algebra of observables and $\{\sigma_t\}$ describes the time evolution of the system, a state φ on A is said to satisfy the *KMS-condition* at the inverse temperature $\beta > 0$ if φ satisfies the modular condition for $\{\sigma_{-\beta t}\}$.

Theorem 1.2. *To each weight φ on a von Neumann algebra $\mathcal{M}$ there corresponds uniquely a one parameter automorphism group $\{\sigma_t\}$ of $\mathcal{M}$ for which φ satisfies the modular condition.*

PROOF: Let $\mathfrak{A}$ be the full left Hilbert algebra corresponding to φ. Considering the semi-cyclic representation $\{\pi_\varphi, \mathfrak{H}_\varphi, \eta_\varphi\}$, we identify $\mathcal{M}$ with $\pi_\varphi(\mathcal{M}) = \mathcal{R}_\ell(\mathfrak{A})$. By Theorem VI.1.19, the modular operator Δ of $\mathfrak{A}$ gives rise to a one parameter automorphism group $\{\sigma_t\}$ of $\mathcal{M}$ by the following:

$$\sigma_t(x) = \Delta^{\mathrm{i}t} x \Delta^{-\mathrm{i}t}, \quad x \in \mathcal{M}. \tag{2}$$

Since $\sigma_t(\mathfrak{n}_\varphi \cap \mathfrak{n}_\varphi^*) = \mathfrak{n}_\varphi \cap \mathfrak{n}_\varphi^*$, σ_t leaves $\mathfrak{m}_\varphi$ globally invariant. If $\xi, \eta \in \mathfrak{A}$, then we have

$$\begin{aligned}\varphi\Big(\sigma_t\big(\pi_\ell(\eta)^*\pi_\ell(\xi)\big)\Big) &= \varphi\big(\pi_\ell(\Delta^{\mathrm{i}t}\eta)^*\pi_\ell(\Delta^{\mathrm{i}t}\xi)\big)\\ &= \big(\Delta^{\mathrm{i}t}\xi \mid \Delta^{\mathrm{i}t}\eta\big) = (\xi \mid \eta).\end{aligned}$$

Therefore, φ is invariant under $\{\sigma_t\}$.

Let $\xi, \eta \in \mathfrak{A}$, and set $x = \pi_\ell(\xi)$ and $y = \pi_\ell(\eta)$. We consider a function F defined by the following

$$F(\alpha) = \left(\Delta^{-\frac{i\alpha}{2}}\xi \mid \Delta^{\frac{i\bar{\alpha}}{2}}\eta\right).$$

Since $\xi, \eta \in \mathfrak{D}^\sharp = \mathfrak{D}(\Delta^{1/2})$, the vector valued functions:

$$\xi(\alpha) = \Delta^{-\frac{i\alpha}{2}}\xi, \qquad \eta(\alpha) = \Delta^{-\frac{i\alpha}{2}}\eta$$

are both, by Lemma VI.2.3, bounded holomorphic on $\mathbf{D}$ and continuous on $\overline{\mathbf{D}}$. Hence F is holomorphic in $\mathbf{D}$ and bounded continuous on $\overline{\mathbf{D}}$. Now, we have

$$F(t) = \left(\Delta^{-\frac{it}{2}}\xi \mid \Delta^{\frac{it}{2}}\eta\right) = \left(\xi \mid \Delta^{it}\eta\right) = \varphi\left(\pi_\ell(\Delta^{it}\eta)^*\pi_\ell(\xi)\right) = \varphi\left(\sigma_t(y^*)x\right);$$

$$\begin{aligned} F(t+i) &= \left(\Delta^{-\frac{it}{2}}\Delta^{\frac{1}{2}}\xi \mid \Delta^{\frac{it}{2}}\Delta^{\frac{1}{2}}\eta\right) = \left(\Delta^{\frac{1}{2}}\xi \mid \Delta^{\frac{1}{2}}\Delta^{it}\eta\right) \\ &= \left(J\Delta^{\frac{1}{2}}\Delta^{it}\eta \mid J\Delta^{\frac{1}{2}}\xi\right) = \left((\Delta^{it}\eta)^\sharp \mid \xi^\sharp\right) = \varphi\left(\pi_\ell(\xi)\pi_\ell(\Delta^{it}\eta)^*\right) \\ &= \varphi\left(x\sigma_t(y^*)\right). \end{aligned}$$

We now prove the uniqueness of $\{\sigma_t\}$. Suppose that φ satisfies the modular condition for another one parameter automorphism group $\{\sigma'_t\}$ of $\mathcal{M}$. The invariance of φ under $\{\sigma'_t\}$ guarantees the existence of a one parameter unitary group $\{U(t)\}$ on $\mathfrak{A}_\varphi$ such that

$$U(t)\eta_\varphi(x) = \eta_\varphi\left(\sigma'_t(x)\right), \quad x \in \mathfrak{n}_\varphi, \quad t \in \mathbf{R}, \tag{3}$$

where the strong continuity of $\{U(t)\}$ follows from the continuity of functions F appearing in (ii) of Definition 1.1. By Stone's theorem, there exists a self-adjoint operator K such that $U(t) = \exp itK$. With $H = \exp K$ we want to prove $H = \Delta$ which yields $\sigma_t = \sigma'_t$.

Now, $\{\sigma'_t\}$ preserves the *-operation in $\mathcal{M}$ so that $U(t)\xi^\sharp = \left(U(t)\xi\right)^\sharp$ for each $\xi \in \mathfrak{A}$. Hence

$$U(t)\mathfrak{D}^\sharp = \mathfrak{D}^\sharp, \qquad \left(U(t)\xi\right)^\sharp = U(t)\xi^\sharp, \quad \xi \in \mathfrak{D}^\sharp. \tag{4}$$

Therefore, we have for each $\xi \in \mathfrak{D}^\sharp$,

$$\|\Delta^{\frac{1}{2}}\xi\| = \|S\xi\| = \|U(t)S\xi\| = \|SU(t)\xi\| = \|\Delta^{\frac{1}{2}}U(t)\xi\|,$$

so we get

$$\left(\Delta^{\frac{1}{2}}\xi \mid \Delta^{\frac{1}{2}}\eta\right) = \left(\Delta^{\frac{1}{2}}U(t)\xi \mid \Delta^{\frac{1}{2}}U(t)\eta\right), \quad \xi, \eta \in \mathfrak{D}^\sharp.$$

Hence $\Delta = U(t)\Delta U(-t)$ for every $t \in \mathbf{R}$, which means that the spectral projections of Δ and $\{U(t)\}$ commute; so we conclude

$$\Delta^{\frac{1}{2}}U(t) = U(t)\Delta^{\frac{1}{2}}, \quad t \in \mathbf{R}. \tag{5}$$

This implies, together with (4), that

$$JU(t) = U(t)J, \quad t \in \mathbf{R}. \tag{6}$$

If $\xi, \eta \in \mathfrak{D}^\sharp$, then we choose sequence $\{\xi_n\}$ and $\{\eta_n\}$ from $\mathfrak{A}$ so that

$$\lim_{n\to\infty} \|\xi - \xi_n\|_\sharp = 0 = \lim_{n\to\infty} \|\eta - \eta_n\|_\sharp . \tag{7}$$

Let $\{F_n\}$ be the sequence of bounded continuous functions on $\overline{\mathbf{D}}$, which are holomorphic in $\mathbf{D}$, satisfying the boundary conditions:

$$F_n(t) = \varphi\Big(\sigma'_t\big(\pi_\ell(\eta_n)^*\big)\pi_\ell(\xi_n)\Big) = \big(\xi_n \mid U(t)\eta_n\big);$$

$$F_n(t + \mathrm{i}) = \varphi\Big(\pi_\ell(\xi_n)\sigma'_t\big(\pi_\ell(\eta_n)^*\big)\Big) = \big(U(t)\eta_n^\sharp \mid \xi_n^\sharp\big).$$

The convergence in (7) implies that $\{F_n(t)\}$ and $\{F_n(t + \mathrm{i})\}$ both converge uniformly in t to the functions $\big(\xi \mid U(t)\eta\big)$ and $\big(U(t)\eta^\sharp \mid \xi^\sharp\big)$ respectively. The Phragmen-Lindelöf Theorem yields the uniform convergence of $\{F_n\}$ to $F_{\xi,\eta}$ on $\overline{\mathbf{D}}$. Hence $F_{\xi,\eta}$ is continuous, bounded on $\overline{\mathbf{D}}$, and holomorphic in $\mathbf{D}$, and has the boundary values:

$$F_{\xi,\eta}(t) = \big(\xi \mid U(t)\eta\big), \qquad F_{\xi,\eta}(t + \mathrm{i}) = \big(U(t)\eta^\sharp \mid \xi^\sharp\big). \tag{8}$$

By (5) and (6), we have

$$F_{\xi,\eta}(t + \mathrm{i}) = \big(\Delta^{\frac{1}{2}}\xi \mid \Delta^{\frac{1}{2}}U(t)\eta\big), \quad \xi \in \mathfrak{D}^\sharp. \tag{9}$$

We now consider the spectral decomposition of K, $K = \int_{\mathbf{R}} \lambda \, \mathrm{d}E(\lambda)$, so that

$$U(t) = \int_{\mathbf{R}} \mathrm{e}^{\mathrm{i}\lambda t} \, \mathrm{d}E(\lambda) \, .$$

We set $E_n = E(n) - E(-n)$. Since $\{E(\lambda)\}$ and Δ commute, we have $E_n\mathfrak{D}^\sharp \subset \mathfrak{D}^\sharp$, and $\mathfrak{D}_0 = \bigcup_{n=1}^\infty E_n\mathfrak{D}^\sharp$ is a core for $\Delta^{1/2}$ because $\lim_{n\to\infty}\big(1 + \Delta^{1/2}\big)E_n\xi = \lim_{n\to\infty} E_n\big(1 + \Delta^{1/2}\big)\xi$ for every $\xi \in \mathfrak{D}^\sharp$. Now, if $\xi \in \mathfrak{D}^\sharp$ and $\eta \in \mathfrak{D}_0$, then for a sufficiently large n we have

$$F_{\xi,\eta}(t) = \big(\xi \mid U(t)\eta\big) = \int_{-n}^{n} \mathrm{e}^{-\mathrm{i}\lambda t}\big(\xi \mid \mathrm{d}E(\lambda)\,\eta\big),$$

so $F_{\xi,\eta}$ is entire and

$$\begin{aligned} F_{\xi,\eta}(t + \mathrm{i}) &= \int_{-n}^{n} \mathrm{e}^{-\mathrm{i}\lambda(t+\mathrm{i})}\big(\xi \mid \mathrm{d}E(\lambda)\,\eta\big) \\ &= \int_{-n}^{n} \mathrm{e}^{\lambda}\mathrm{e}^{-\mathrm{i}\lambda t}\big(\xi \mid \mathrm{d}E(\lambda)\,\eta\big) = \big(\xi \mid HU(t)\eta\big), \end{aligned}$$

where we recall $H = \exp K$. Comparing this with (9), we conclude

$$\big(\Delta^{\frac{1}{2}}\xi \mid \Delta^{\frac{1}{2}}U(t)\eta\big) = \big(\xi \mid HU(t)\eta\big), \quad \xi \in \mathfrak{D}^\sharp, \quad \eta \in \mathfrak{D}_0.$$

In particular, setting $t = 0$, we get

$$\left(\Delta^{\frac{1}{2}}\xi \mid \Delta^{\frac{1}{2}}\eta\right) = \left(\xi \mid H\eta\right), \quad \xi \in \mathfrak{D}^{\sharp}, \quad \eta \in \mathfrak{D}_0.$$

Hence $\Delta^{1/2}\mathfrak{D}_0 \subset \mathfrak{D}^{\sharp}$ and $H\eta = \Delta\eta$ for each $\eta \in \mathfrak{D}_0$. On the other hand, $(1 + H)\mathfrak{D}_0 = \bigcup_{n=1}(1 + H)E_n\mathfrak{D}^{\sharp}$, and $(1 + H)E_n\mathfrak{D}^{\sharp}$ is dense in $E_n\mathfrak{H}_{\varphi}$, so that $\mathfrak{D}_0$ is dense in $\mathfrak{D}(H)$ with respect to the graph norm. Similarly, $\mathfrak{D}_0$ is dense in $\mathfrak{D}(\Delta)$ with respect to the graph norm. In other words, $\mathfrak{D}_0$ is a common core for H and Δ, on which H and Δ agree. Therefore, $H = \Delta$, which implies that $\Delta^{it} = H^{it} = U(t)$. Thus

$$\eta_{\varphi}\left(\sigma_t'(x)\right) = U(t)\eta_{\varphi}(x) = \Delta^{it}\eta_{\varphi}(x) = \eta_{\varphi}\left(\sigma_t(x)\right), \quad x \in \mathfrak{n}_{\varphi}.$$

Hence $\sigma_t'(x) = \sigma_t(x)$, $x \in \mathfrak{n}_{\varphi}$. Since $\mathfrak{n}_{\varphi}$ generates $\mathcal{M}$, we get $\sigma_t = \sigma_t'$ for every $t \in \mathbf{R}$. Q.E.D.

With the above theorem, the following definition is now in order:

Definition 1.3. The one parameter automorphism group $\{\sigma_t\}$ of $\mathcal{M}$ given by a weight φ is called the *modular automorphism group* associated with φ and denoted by $\{\sigma_t^{\varphi}\}$.

In order to avoid a repeated use of the lengthy phrase, let us fix a notation: we shall denote by $\mathcal{A}(\mathbf{D})$ the set of all bounded continuous functions on the closed horizontal strip $\overline{\mathbf{D}}$ which are holomorphic on the open strip $\mathbf{D}$. The function $F_{x,y} \in \mathcal{A}(\mathbf{D})$ of (1) is called the *two point function* of x and y.

The uniqueness of the modular automorphism group yields the following result at once.

Corollary 1.4. *Let π be an isomorphism of a von Neumann algebra $\mathcal{M}$ onto another $\mathcal{N}$. If ψ is a faithful semi-finite normal weight on $\mathcal{N}$, then*

$$\sigma_t^{\psi\circ\pi} = \pi^{-1} \circ \sigma_t^{\psi} \circ \pi, \quad t \in \mathbf{R}. \tag{10}$$

We leave the proof to the reader as an exercise.

Proposition 1.5. *Let A be a C^*-algebra equipped with one parameter automorphism group $\{\sigma_t\}$. If a faithful semi-finite lower semicontinuous weight φ on A satisfies the modular condition for $\{\sigma_t\}$, then the weight $\tilde{\varphi}$ on the von Neumann algebra $\mathcal{M} = \pi_{\varphi}(A)''$ given by (5) is a faithful semi-finite normal weight whose modular automorphism group satisfies the condition:*

$$\sigma_t^{\tilde{\varphi}} \circ \pi_{\varphi} = \pi_{\varphi} \circ \sigma_t.$$

PROOF: Let $\mathfrak{A}_{\varphi} = \eta_{\varphi}(\mathfrak{n}_{\varphi} \cap \mathfrak{n}_{\varphi}^*) \subset \mathfrak{H}_{\varphi}$. Suppose that a sequence $\{x_n\}$ in $\mathfrak{n}_{\varphi} \cap \mathfrak{n}_{\varphi}^*$ has the property:

$$\lim_{n\to\infty} \varphi(x_n^* x_n) = 0 \quad \text{and} \quad \lim_{n,m\to\infty} \varphi\left((x_n - x_m)(x_n - x_m)^*\right) = 0.$$

Let $\eta = \lim_{n\to\infty} \eta_\varphi(x_n^*) \in \mathfrak{H}_\varphi$. We want to show $\eta = 0$. For any $x \in \mathfrak{n}_\varphi \cap \mathfrak{n}_\varphi^*$, there exists a sequence F_n in $\mathcal{A}(\mathbf{D})$ such that

$$F_n(t) = \varphi\big(\sigma_t(y)x_n^*\big) \quad \text{and} \quad F_n(t+\mathrm{i}) = \varphi\big(x_n^*\sigma_t(y)\big).$$

As $n \to \infty$, $F_n(t)$ converges to $\big(\eta \mid \eta_\varphi(\sigma_t(y))\big)$ and $F_n(t+\mathrm{i})$ converges to zero uniformly in t. Hence $\{F_n\}$ converges to a function F in $\mathcal{A}(\mathbf{D})$ uniformly on $\mathbf{D}$ such that $F(t+\mathrm{i}) = 0,\ t \in \mathbf{R}$, so that F is the zero function. Thus $\big(\eta \mid \eta_\varphi(\sigma_t(y))\big) = 0$ for every $y \in \mathfrak{n}_\varphi \cap \mathfrak{n}_\varphi^*$. With $t = 0$, we have $\big(\eta \mid \eta_\varphi(y)\big) = 0$ for every $y \in \mathfrak{n}_\varphi \cap \mathfrak{n}_\varphi^*$. If $\{e_i\}$ is an approximate identity of $\mathfrak{n}_\varphi \cap \mathfrak{n}_\varphi^*$, then we have, for any $x \in \mathfrak{n}_\varphi$,

$$\big(\eta \mid \eta_\varphi(x)\big) = \lim_i \big(\pi_\varphi(e_i)\eta \mid \eta_\varphi(x)\big) = \lim_i \big(\eta \mid \eta_\varphi(e_i x)\big) = 0$$

since $e_i x \in \mathfrak{n}_\varphi \cap \mathfrak{n}_\varphi^*$. Hence $\eta = 0$. Therefore, the map: $\eta_\varphi(x) \mapsto \eta_\varphi(x^*),\ x \in \mathfrak{n}_\varphi \cap \mathfrak{n}_\varphi^*$, is preclosed. Hence $\mathfrak{A}_\varphi$ is a left Hilbert algebra. Clearly its left von Neumann algebra $\mathcal{R}_\ell(\mathfrak{A}_\varphi)$ is precisely $\mathcal{M}$ itself. The associated weight $\tilde{\varphi}$ on $\mathcal{M}$ is indeed the weight given by (VII.4.5). The uniqueness arguments in the proof of Theorem 1.2 also provides the proof for $\sigma_t^{\tilde{\varphi}} \circ \pi_\varphi = \pi_\varphi \circ \sigma_t$. Q.E.D.

Exercise VIII.1

1) Let A be a C^*-algebra equipped with a one parameter automorphism group $\{\sigma_t : t \in \mathbf{R}\}$. Show that in Definition 1.1, the condition (i) follows from (ii) if the weight φ in question is finite.

2) Let h be a non-singular positive nuclear operator on $\mathfrak{H}$. Define a faithful normal positive linear functional φ on $\mathcal{L}(\mathfrak{H})$ by $\varphi(x) = \mathrm{Tr}(xh),\ x \in \mathcal{L}(\mathfrak{H})$. Prove that the modular automorphism group $\{\sigma_t^\varphi\}$ on $\mathcal{L}(\mathfrak{H})$ is given by $\sigma_t^\varphi = \mathrm{Ad}(h^{\mathrm{i}t}),\ t \in \mathbf{R}$, where $\mathrm{Ad}(u)(x) = uxu^*,\ x \in \mathcal{L}(\mathfrak{H})$, for every unitary $u \in \mathcal{L}(\mathfrak{H})$.

3) Let A be a unital C^*-algebra equipped with a continuous one parameter automorphism group $\{\sigma_t : t \in \mathbf{R}\}$, where the continuity of $\{\sigma_t\}$ means $\lim_{t\to 0}\|\sigma_t(a) - a\| = 0$, $a \in A$. Prove that the set K of all states of A which satisfies the modular condition is a closed convex subset of the state space $\mathfrak{S}(A)$.

4) In the previous problem, show that a point ω of K is primary, i.e. $\pi_\omega(A)''$ is a factor, cf. Definition IV.6.30, if and only if ω is an extreme point of K.

5) Show that the σ-strong closure of the unitary group $\mathcal{U}(\mathcal{M})$ of a von Neumann algebra $\mathcal{M}$ is precisely the semi-group of all isometries in $\mathcal{M}$.

6) Show that the group $\mathrm{GL}(\mathcal{M})$ of invertible elements of $\mathcal{M}$ is σ-strongly dense, [504].

7) Show that for $\{\mathcal{M}, \mathfrak{H}\}$, the set $\mathrm{cyc}(\mathcal{M}, \mathfrak{H})$ of all cyclic vectors in $\mathfrak{H}$ is a G_δ-set. (Hint: If $\mathrm{Cyc}(\mathcal{M}, \mathfrak{H}) = \emptyset$, then $\mathrm{Cyc}(\mathcal{M}, \mathfrak{H})$ is a G_δ-set. If $\mathrm{Cyc}(\mathcal{M}, \mathfrak{H})$ contains $\xi_0 \in \mathfrak{H}$, then

$$\mathrm{Cyc}(\mathcal{M}, \mathfrak{H}) = \bigcap_{n \geq 1} \bigcup_{a \in \mathcal{M}} \{ \xi \in \mathfrak{H} : \|a\xi - \xi_0\| < 1/n \}. \quad) \qquad [504]$$

8) Show that in the last problem if $\mathrm{Cyc}(\mathcal{M}, \mathfrak{H}) \neq \emptyset$, then $\mathrm{Cyc}(\mathcal{M}, \mathfrak{H})$ is a dense G_δ-set in $\mathfrak{H}$. (Hint: Use (6), to prove $\mathrm{GL}(\mathcal{M})\xi_0$ is dense if $\xi_0 \in \mathrm{Cyc}(\mathcal{M})$.)

9) Prove that if $\mathcal{M}$ and $\mathcal{M}'$ are both σ-finite and properly infinite, then the set of cyclic and separating vectors of $\mathfrak{H}$ is a dense G_δ-set in $\mathfrak{H}$.

10) Show that if $\varphi, \psi \in \mathcal{M}_*$ are self-adjoint and $\varphi \leq \psi$, then the interval $[\varphi, \psi] = \{ \omega \in \mathcal{M}_*^h : \varphi \leq \omega \leq \psi \}$ is weakly compact.

11) Show that if φ is a faithful normal state on $\mathcal{M}$ with bounded modular operator Δ_φ, then the convex closure K of $\{ u^*\varphi u : u \in \mathcal{U}(\mathcal{M}) \}$ is weakly compact. (Hint: With $M = \|\Delta_\varphi\|$, show that $K \subset \left[\frac{1}{M}\varphi, M\varphi\right]$.)

12) Show that if $\mathcal{M}$ admits a faithful normal state φ with Δ_φ bounded, then $\mathcal{M}$ admits a faithful normal finite trace τ. (Hint: Use the Ryll-Nardzewski fixed point theorem in the last problem.)

13) Let A be a separable C^*-algebra acting on $\mathfrak{H}$ and $\mathcal{M}$ be the weak closure of A.

(a) Show that if a normal state on $\mathcal{M}$ is faithful whenever its restriction to A is faithful, then every non-zero $p \in \mathrm{Proj}(\mathcal{M})$ majorizes a non-zero element $h \in A_+$.

(b) Assume the property in (b) for $\mathcal{M}$ and A. Let $\tilde{A}$ be the universal enveloping von Neumann algebra of A, i.e. the second dual A^{**}. Let $\{\omega_n\}$ be a weak* dense sequence of pure states of A and set $\omega = \sum_{j=1}^{\infty} \omega_j / 2^j$. Let $\tilde{\omega}$ be a Hahn-Banach extension of ω to $\mathcal{M}$ and decompose it into the normal part and the singular part: $\tilde{\omega} = \tilde{\omega}_n + \tilde{\omega}_s$. Show that $\tilde{\omega}_n$ is faithful on $\mathcal{M}$.

(c) Let π be the normal homomorphism of $\tilde{A}$ onto $\mathcal{M}$ induced by the injection of A into $\mathcal{M}$, and let $\psi = {}^t\pi(\tilde{\omega}_n)$. Prove that $\tilde{A}_{s(\psi)}$ is atomic and π is an isomorphism of $\tilde{A}_{s(\psi)}$ onto $\mathcal{M}$, where $s(\psi)$ is the support of ψ in $\tilde{A}$. (Hint: $\psi \leq \omega$ and $\pi_\omega(A)''$ is atomic.)

(d) Show that A contains all minimal projections of $\mathcal{M}$.

§ 2 The Centralizer of a Weight

In this section, we mean by a weight a semi-finite normal one on a von Neumann algebra always.

We fix a faithful weight φ on a von Neumann algebra $\mathcal{M}$.

Definition 2.1. We first set

$$\mathcal{M}_\varphi = \left\{ x \in \mathcal{M} : \sigma_t^\varphi(x) = x, \ t \in \mathbf{R} \right\}. \tag{1}$$

It follows that $\mathcal{M}_\varphi$ is a von Neumann subalgebra of $\mathcal{M}$. We call this subalgebra the *centralizer* of φ. We say that x *and* φ *commute* if $x \in \mathcal{M}_\varphi$.

We want to give a criteria in terms of φ for an $x \in \mathcal{M}$ to commute with φ. We need some preparations.

Definition 2.2. We mean by a *one parameter automorphism group* $\{\sigma_t\}$ of $\mathcal{M}$ a homomorphism $\sigma : t \in \mathbf{R} \mapsto \sigma_t \in \mathrm{Aut}(\mathcal{M})$ from the additive group $\mathbf{R}$ into the group $\mathrm{Aut}(\mathcal{M})$ of automorphisms of $\mathcal{M}$ with the continuity requirement that for every $x \in \mathcal{M}$ and $\omega \in \mathcal{M}_*$ the function: $t \mapsto \omega(\sigma_t(x)) \in \mathbf{C}$ is continuous. An element $x \in \mathcal{M}$ is said to be *entire* if the function: $t \in t \in \mathbf{R} \mapsto \sigma_t(x) \in \mathcal{M}$ can be extended to an $\mathcal{M}$-valued entire function over $\mathbf{C}$. If this is the case, the value at $\alpha \in \mathbf{C}$ will be denoted by $\sigma_\alpha(x)$. We denote by $\mathcal{M}_a^\sigma$ the set of all entire elements of $\mathcal{M}$.

In the previous sections, we have used the terminology one parameter automorphism groups, without specifying the continuity requirement, which did not cause any problem so far. But we now need this requirement. In the C^*-algebra case, it is not clear what kind of continuity requirement should be imposed except for separable C^*-algebras. For a separable C^*-algebra A, the requirement

$$\lim_{t \to 0} \left\| \sigma_t(x) - x \right\| = 0, \qquad x \in A \tag{2}$$

is reasonable. But beyond this case, the condition (2) is too strong to cover interesting cases. For example, this condition for a von Neumann algebra implies the innerness as we will see later in Exercise XI.3.6. In the theory involving non-separable C^*-algebras, instead of imposing continuity of $\{\sigma_t\}$ on A, one works with positive linear functionals ω on A which generate continuous two point functions: $t \in \mathbf{R} \mapsto \omega(\sigma_t(x)y)$, $x, y \in A$, as we have done in the previous section.

Let us return to the original setting.

Lemma 2.3. *Let $\{\sigma_t\}$ be a one parameter automorphism group of $\mathcal{M}$.*

(i) *$\mathcal{M}_a^\sigma$ is a σ-weakly dense *-subalgebra of $\mathcal{M}$;*
(ii) *If $x, y \in \mathcal{M}_a^\sigma$ and $\alpha, \beta \in \mathbf{C}$, then*

$$\begin{aligned} \sigma_\alpha(xy) &= \sigma_\alpha(x)\sigma_\alpha(y), \\ \sigma_{\alpha+\beta}(x) &= \sigma_\alpha(\sigma_\beta(x)) = \sigma_\beta(\sigma_\alpha(x)), \\ \sigma_{\overline{\alpha}}(x) &= \sigma_\alpha(x^*)^* \end{aligned} \tag{3}$$

PROOF:

(i) For each $x \in \mathcal{M}_a^\sigma$, we set

$$\sigma'_\alpha(x) = \frac{\mathrm{d}}{\mathrm{d}\beta}\sigma_\beta(x)\Big|_{\beta=\alpha} = \lim_{\varepsilon\to 0}\frac{1}{\varepsilon}\big(\sigma_{\alpha+\varepsilon}(x) - \sigma_\alpha(x)\big).$$

For each $x, y \in \mathcal{M}_a^\sigma$, we have

$$\begin{aligned}
&\frac{1}{\varepsilon}\big(\sigma_{\alpha+\varepsilon}(x)\sigma_{\alpha+\varepsilon}(y) - \sigma_\alpha(x)\sigma_\alpha(y)\big)\\
&\quad = \frac{1}{\varepsilon}\Big(\big[\sigma_{\alpha+\varepsilon}(x) - \sigma_\alpha(x)\big]\sigma_{\alpha+\varepsilon}(y) + \sigma_\alpha(x)\big[\sigma_{\alpha+\varepsilon}(y) - \sigma_\alpha(y)\big]\Big)\\
&\quad \to \sigma'_\alpha(x)\sigma_\alpha(y) + \sigma_\alpha(x)\sigma'_\alpha(y)
\end{aligned}$$

as $\varepsilon \to 0$ in norm by the analyticity, A.1. Therefore, the function: $\alpha \in \mathbf{C} \to \sigma_\alpha(x)\sigma_\alpha(y) \in \mathcal{M}$ is entire; so xy is entire and $xy \in \mathcal{M}_a^\sigma$. The uniqueness of holomorphic extension implies $\sigma_\alpha(xy) = \sigma_\alpha(x)\sigma_\alpha(y)$. Hence $\mathcal{M}_a^\sigma$ is a subalgebra of $\mathcal{M}$. The self-adjointness of $\mathcal{M}_a^\sigma$ follows similarly.

Now, for each $x \in \mathcal{M}$, set

$$x_\gamma(\alpha) = \sqrt{\frac{\gamma}{\pi}}\int_{\mathbf{R}} \mathrm{e}^{-\gamma(t-\alpha)^2}\sigma_t(x)\,\mathrm{d}t, \quad \gamma > 0, \quad \alpha \in \mathbf{C}. \tag{4}$$

For each $\omega \in \mathcal{M}_*$, we have

$$\langle x_\gamma(\alpha), \omega\rangle = \sqrt{\frac{\gamma}{\pi}}\int_{\mathbf{R}} \mathrm{e}^{-\gamma(t-\alpha)^2}\langle\sigma_t(x), \omega\rangle\,\mathrm{d}t\,.$$

The boundedness of the function: $t \in \mathbf{R} \to \langle\sigma_t(x), \omega\rangle$ yields the analyticity of the right hand side as a function of α. Hence $x_\gamma(\,\cdot\,)$ is an $\mathcal{M}$-valued entire function which extends the function: $t \in \mathbf{R} \to \sigma_t\big(x_\gamma(0)\big) = x_\gamma(t)$. Thus, $x_\gamma = x_\gamma(0)$ is entire. By Lemma VI.2.4, $\{x_\gamma\}$ converges σ-weakly to x as $\gamma \to +\infty$. Hence $\mathcal{M}_a^\sigma$ is σ-weakly dense in $\mathcal{M}$.

We leave the rest to the reader. Q.E.D.

When we apply the above lemma to the modular automorphism group $\{\sigma_t^\varphi\}$ of a faithful weight φ, we write simply $\mathcal{M}_a^\varphi$ instead of $\mathcal{M}_a^{\sigma^\varphi}$. Fixing a faithful weight φ on $\mathcal{M}$, we use the previously established notations. So $\mathfrak{A}$ means the corresponding left Hilbert algebra and $\mathfrak{A}_0$ the maximal Tomita algebra, and so on. We set

$$\mathfrak{a}_0 = \pi_\ell(\mathfrak{A}_0), \qquad \mathfrak{a}_\varphi = \mathfrak{n}_\varphi \cap \mathfrak{n}_\varphi^* = \pi_\ell(\mathfrak{A}).$$

It then follows that $\mathfrak{a}_0 \subset \mathcal{M}_a^\varphi$ and

$$\sigma_\alpha^\varphi\big(\pi_\ell(\xi)\big) = \pi_\ell\big(\Delta^{\mathrm{i}\alpha}\xi\big), \quad \xi \in \mathfrak{A}_0. \tag{5}$$

Lemma 2.4.

(i) $\mathfrak{a}_\varphi$ *is an* $\mathcal{M}_a^\varphi$*-bimodule.*
(ii) $\mathfrak{m}_\varphi$ *is an* $\mathcal{M}_a^\varphi$*-bimodule.*
(iii) $\mathfrak{a}_0$ *is an ideal of* $\mathcal{M}_a^\varphi$.

PROOF:

(i) Since $\mathfrak{a}_\varphi$ and $\mathcal{M}_a^\varphi$ are both *-algebras, it suffices to prove that $\mathfrak{a}_\varphi$ is a left $\mathcal{M}_a^\varphi$-module. For this, it is also sufficient to prove $\mathcal{M}_a^\varphi \mathfrak{a}_\varphi \subset \mathfrak{n}_\varphi^*$, $\mathfrak{n}_\varphi$ being a left ideal. Let $a \in \mathcal{M}_a^\varphi$ and $x \in \mathfrak{a}_\varphi = \mathfrak{n}_\varphi \cap \mathfrak{n}_\varphi^*$. The assertion that $ax \in \mathfrak{n}_\varphi^*$ is equivalent to the statement that $\eta_\varphi(ax) \in \mathfrak{D}^\sharp = \mathfrak{D}(\Delta^{1/2})$. Let $\xi(t) = \Delta^{\mathrm{i}t}\eta_\varphi(ax)$, $t \in \mathbf{R}$. Then we have

$$\xi(t) = \eta_\varphi\big(\sigma_t^\varphi(ax)\big) = \sigma_t^\varphi(a)\eta_\varphi\big(\sigma_t^\varphi(x)\big) = \sigma_t^\varphi(a)\Delta^{\mathrm{i}t}\eta_\varphi(x).$$

Hence the $\mathfrak{H}$-valued function $\xi(\,\cdot\,)$ can be extended to an $\mathfrak{H}$-valued function $\{\xi(\alpha) : \alpha \in \overline{\mathbf{D}}_{1/2}\}$ of $\mathcal{A}_{\mathfrak{H}}(\mathbf{D}_{1/2})$ such that

$$\xi(\alpha) = \sigma_\alpha^\varphi(a)\Delta^{\mathrm{i}\alpha}\eta_\varphi(x), \qquad -\frac{1}{2} \le \operatorname{Im}\alpha \le 0,$$

where $\mathcal{A}_{\mathfrak{H}}(\mathbf{D}_{1/2})$ denotes the set of $\mathfrak{H}$-valued bounded continuous functions on the strip $\overline{\mathbf{D}}_{1/2} = \{\alpha \in \mathbf{C} : -\frac{1}{2} \le \operatorname{Im}\alpha \le 0\}$ which are holomorphic on the open strip $\mathbf{D}_{1/2}$. By Lemma VI.2.3, $\xi(0) = \eta_\varphi(ax)$ belongs to $\mathfrak{D}(\Delta^{1/2}) = \mathfrak{D}^\sharp$. Hence $ax \in \pi_\ell(\mathfrak{A}) = \mathfrak{a}_\varphi$.

(ii) Let $a \in \mathcal{M}_a^\varphi$ and $x \in \mathfrak{m}_\varphi^+$. Then $x^{1/2} \in \mathfrak{n}_\varphi \cap \mathfrak{n}_\varphi^* = \mathfrak{a}_\varphi$, so that (i) implies $ax = (ax^{1/2})x^{1/2} \in \mathfrak{m}_\varphi$. Since $\mathfrak{m}_\varphi^+$ spans linearly $\mathfrak{m}_\varphi$, $a\mathfrak{m}_\varphi \subset \mathfrak{m}_\varphi$, and so $\mathfrak{m}_\varphi a = (a^*\mathfrak{m}_\varphi)^* \subset \mathfrak{m}_\varphi$. Hence $\mathfrak{m}_\varphi$ is an $\mathcal{M}_a^\varphi$-bimodule.

(iii) This follows from the combination of (i) and (ii). Q.E.D.

Lemma 2.5.

(i) *If $a \in \mathcal{M}$ is a multiplier of $\mathfrak{m}_\varphi$ in the sense that $a\mathfrak{m}_\varphi \subset \mathfrak{m}_\varphi$ and $\mathfrak{m}_\varphi a \subset \mathfrak{m}_\varphi$, then for each $x, y \in \mathfrak{a}_0$ there exists an entire $F \in \mathcal{A}(\mathbf{D})$ such that*

$$F(t) = \varphi\big(\sigma_t^\varphi(a)xy^*\big), \qquad F(t+\mathrm{i}) = \varphi\big(xy^*\sigma_t^\varphi(a)\big). \tag{6}$$

(ii) *If $a \in \mathcal{M}_a^\varphi$ and $z \in \mathfrak{m}_\varphi$, then the function F_z defined by $\varphi\big(\sigma_\alpha^\varphi(a)z\big)$ is entire and bounded on $\mathbf{D}$, and further satisfies the condition:*

$$F_z(t) = \varphi\big(\sigma_t^\varphi(a)z\big), \qquad F_z(t+\mathrm{i}) = \varphi\big(z\sigma_t^\varphi(a)\big). \tag{7}$$

PROOF:

(i) Set

$$F(\alpha) = \big(a\Delta^{-\mathrm{i}\alpha}\eta_\varphi(x) \mid \Delta^{-\mathrm{i}\bar\alpha+1}\eta_\varphi(y)\big), \quad \alpha \in \mathbf{C}. \tag{8}$$

By the assumption on x and y, F is an entire function and belongs to $\mathcal{A}(\mathbf{D})$. We then simply compute:

$$\begin{aligned}F(t) &= \left(a\Delta^{-\mathrm{i}t}\eta_\varphi(x) \mid \Delta^{-\mathrm{i}t+1}\eta_\varphi(y)\right) \qquad (\text{by } ax \in \mathfrak{m}_\varphi)\\ &= \left(\sigma_t^\varphi(a)\eta_\varphi(x) \mid \Delta\eta_\varphi(y)\right) = \left(\Delta^{\frac{1}{2}}\eta_\varphi(\sigma_t^\varphi(a)x) \,\middle|\, \Delta^{\frac{1}{2}}\eta_\varphi(y)\right)\\ &= \left(S\eta_\varphi(y) \,\middle|\, S\eta_\varphi(\sigma_t^\varphi(a)x)\right) = \varphi(\sigma_t^\varphi(a)xy^*);\end{aligned}$$

$$\begin{aligned}F(t+\mathrm{i}) &= \left(a\Delta^{-\mathrm{i}t+1}\eta_\varphi(x) \mid \Delta_\varphi^{-\mathrm{i}t}\eta_\varphi(y)\right) = \left(\Delta^{\frac{1}{2}}\eta_\varphi(x) \mid \Delta^{\frac{1}{2}}\sigma_t^\varphi(a^*)\eta_\varphi(y)\right)\\ &= \left(S\eta_\varphi(\sigma_t^\varphi(a^*)y) \,\middle|\, S\eta_\varphi(x)\right) = \varphi(xy^*\sigma_t^\varphi(a)).\end{aligned}$$

(ii) We may assume that $z = xy^*$, $x, y \in \mathfrak{a}_\varphi$, because $\mathfrak{m}_\varphi^+$ spans $\mathfrak{m}_\varphi$ linearly and $z \mapsto F_z$ is linear in z. By the previous lemma $\eta_\varphi(ax) \in \mathfrak{D}(\Delta^{1/2})$ and

$$\sigma_\alpha^\varphi(a)\Delta^{\mathrm{i}\alpha}\eta_\varphi(x) = \Delta^{\mathrm{i}\alpha}\eta_\varphi(ax), \qquad \alpha \in \overline{\mathbf{D}}_{\frac{1}{2}}.$$

In particular, we have, applying the above to $\sigma_\beta^\varphi(a)$ with $\alpha = -\frac{\mathrm{i}}{2}$,

$$\Delta^{\frac{1}{2}}\sigma_\beta^\varphi(a)\eta_\varphi(x) = \sigma_{\beta-\frac{\mathrm{i}}{2}}^\varphi(a)\Delta^{\frac{1}{2}}\eta_\varphi(x), \qquad \beta \in \mathbf{C}.$$

We then compute:

$$\begin{aligned}F_z(\alpha) &= \varphi(\sigma_\alpha^\varphi(a)xy^*) = \left(S\eta_\varphi(y) \,\middle|\, S\eta_\varphi(\sigma_\alpha^\varphi(a)x)\right)\\ &= \left(\Delta^{\frac{1}{2}}\sigma_\alpha^\varphi(a)\eta_\varphi(x) \mid \Delta^{\frac{1}{2}}\eta_\varphi(y)\right)\\ &= \left(\sigma_{\alpha-\frac{\mathrm{i}}{2}}^\varphi(a)\Delta^{\frac{1}{2}}\eta_\varphi(x) \mid \Delta^{\frac{1}{2}}\eta_\varphi(y)\right).\end{aligned}$$

Hence F_z is entire and bounded on $\overline{\mathbf{D}}_{1/2}$. Now, we observe:

$$\begin{aligned}F_z(t+\mathrm{i}) &= \left(\sigma_{t+\frac{\mathrm{i}}{2}}^\varphi(a)\Delta^{\frac{1}{2}}\eta_\varphi(x) \mid \Delta^{\frac{1}{2}}\eta_\varphi(y)\right)\\ &= \left(\Delta^{\frac{1}{2}}\eta_\varphi(x) \mid \sigma_{t-\frac{\mathrm{i}}{2}}^\varphi(a^*)\Delta^{\frac{1}{2}}\eta_\varphi(y)\right)\\ &= \left(\Delta^{\frac{1}{2}}\eta_\varphi(x) \mid \Delta^{\frac{1}{2}}\sigma_t^\varphi(a^*)\eta_\varphi(y)\right) = \left(S\eta_\varphi(\sigma_t^\varphi(a^*)y) \,\middle|\, S\eta_\varphi(x)\right)\\ &= \varphi(xy^*\sigma_t^\varphi(a)).\end{aligned}$$

Thus (7) follows. Q.E.D.

We are now ready to characterize the centralizer of a weight.

Theorem 2.6. *Let φ be a faithful semi-finite normal weight on a von Neumann algebra $\mathcal{M}$. A necessary and sufficient condition for an element $a \in \mathcal{M}$ to belong to the centralizer $\mathcal{M}_\varphi$ of φ is that*

(i) *a is a multiplier of $\mathfrak{m}_\varphi$, i.e., $a\mathfrak{m}_\varphi \subset \mathfrak{m}_\varphi$ and $\mathfrak{m}_\varphi a \subset \mathfrak{m}_\varphi$;*
(ii) $\varphi(az) = \varphi(za)$, $z \in \mathfrak{m}_\varphi$. (9)

PROOF: Necessity: Suppose that a and φ commute. The constant function: $\alpha \in \mathbf{C} \mapsto \sigma_\alpha^\varphi(a) = a \in \mathcal{M}$ is entire, so a belongs to $\mathcal{M}_a^\varphi$; hence a is a multiplier of $\mathfrak{m}_\varphi$ by Lemma 2.4.(ii). By the last lemma, each $z \in \mathfrak{m}_\varphi$ gives rise to an entire $F_z \in \mathcal{A}(\mathbf{D})$ satisfying (7). But this F_z is the constant $\varphi(az)$ on $\mathbf{R}$, so that F_z is constant on the entire plane $\mathbf{C}$. Hence $F_z(t) = F_z(t + \mathrm{i})$ which yields (9).

Sufficiency: Suppose that the conditions (i) and (ii) hold. Each pair $x, y \in \mathfrak{a}_0$ gives rise to an entire $F \in \mathcal{A}(\mathbf{D})$ satisfying (6). The assumption means that

$$\begin{aligned} F(t) &= \varphi \circ \sigma_t^\varphi\big(a\sigma_{-t}^\varphi(xy^*)\big) = \varphi\big(a\sigma_{-t}^\varphi(xy^*)\big) \\ &= \varphi\big(\sigma_{-t}^\varphi(xy^*)a\big) = \varphi\big(xy^*\sigma_t^\varphi(a)\big) = F(t + \mathrm{i}). \end{aligned}$$

Hence F has the period i, so the boundedness of F on $\overline{\mathbf{D}}$ implies that of F on the entire plane $\mathbf{C}$. Therefore, the Liouville theorem shows that F is a constant. Thus

$$\varphi\big(\sigma_t^\varphi(a)xy^*\big) = \varphi(axy^*), \quad x, y \in \mathfrak{a}_0, \quad t \in \mathbf{R}.$$

Since the function F was given by (8), we have

$$\big(a\Delta^{-\mathrm{i}t}\eta_\varphi(x) \mid \Delta^{-\mathrm{i}t+1}\eta_\varphi(y)\big) = \big(a\eta_\varphi(x) \mid \Delta\eta_\varphi(y)\big), \quad x, y \in \mathfrak{a}_0,$$

so that

$$\Big(\big(\sigma_t^\varphi(a) - a\big)\eta_\varphi(x) \Bigm| \Delta\eta_\varphi(y)\Big) = 0, \quad x, y \in \mathfrak{a}_0.$$

Since $\eta_\varphi(\mathfrak{a}_0) = \mathfrak{A}_0$ and $\Delta\mathfrak{A}_0 = \mathfrak{A}_0$ are dense in $\mathfrak{H}_\varphi$, $\sigma_t^\varphi(a) = a$, $t \in \mathbf{R}$. Hence we conclude $a \in \mathcal{M}_\varphi$. Q.E.D.

We now consider to perturb φ by a positive self-adjoint operator affiliated with $\mathcal{M}_\varphi$. To this end, we introduce a new operator h_ε for any positive self-adjoint operator h on a Hilbert space $\mathfrak{H}$ by setting

$$h_\varepsilon = h\,(1 + \varepsilon h)^{-1}, \quad \varepsilon > 0. \tag{10}$$

It follows that h_ε is bounded and self-adjoint. Given two such operators h and k on $\mathfrak{H}$, we say that $h \leq k$ if there exists $\varepsilon > 0$ such that $h_\varepsilon \leq k_\varepsilon$, which is equivalent to the fact that $h_\varepsilon \leq k_\varepsilon$ for every $\varepsilon > 0$. This is then equivalent to the relation that

$$\mathfrak{D}\big(h^{\frac{1}{2}}\big) \supset \mathfrak{D}\big(k^{\frac{1}{2}}\big) \quad \text{and} \quad \big\|h^{\frac{1}{2}}\xi\big\| \leq \big\|k^{\frac{1}{2}}\xi\big\|, \quad \xi \in \mathfrak{D}\big(k^{\frac{1}{2}}\big). \tag{11}$$

Once again, we fix a faithful weight φ on $\mathcal{M}$.

Lemma 2.7. *For each $h \in \mathcal{M}_\varphi^+$, if we set*

$$\varphi_h(x) = \varphi\big(h^{\frac{1}{2}}xh^{\frac{1}{2}}\big), \quad x \in \mathcal{M}_+, \tag{12}$$

then φ_h is a weight on $\mathcal{M}$. The map: $h \in \mathcal{M}_\varphi^+ \mapsto \varphi_h$ is a monotone increasing affine map.

PROOF: By Lemma 2.4, φ_h takes finite values on $\mathfrak{m}_\varphi^+$, so that φ_h is semi-finite. The normality of φ_h follows from that of φ.

If $h, k \in \mathcal{M}_\varphi^+$, then there exist $u, v \in \mathcal{M}_\varphi$ such that $\|u\| \leq 1$, $\|v\| \leq 1$ and

$$h^{\frac{1}{2}} = u(h+k)^{\frac{1}{2}}, \qquad k^{\frac{1}{2}} = v(h+k)^{\frac{1}{2}}, \qquad u^*u + v^*v = s(h+k).$$

Suppose that $\varphi_{(h+k)}(x) < +\infty$, $x \in \mathcal{M}_+$. Namely, we have

$$y = (h+k)^{\frac{1}{2}} x (h+k)^{\frac{1}{2}} \in \mathfrak{m}_\varphi^+.$$

By Lemma 2.4, we get $uyu^* \in \mathfrak{m}_\varphi$ and $vyv^* \in \mathfrak{m}_\varphi$. By Theorem 2.6, we obtain

$$\begin{aligned}\varphi_h(x) + \varphi_k(x) &= \varphi\big(uyu^* + vyv^*\big) = \varphi\big(u^*uy + v^*vy\big)\\ &= \varphi\big((u^*u + v^*v)y\big) = \varphi(y) = \varphi_{(h+k)}(x).\end{aligned}$$

Conversely, suppose that $\varphi_h(x) < +\infty$ and $\varphi_k(x) < +\infty$. Then the inequality:

$$\begin{aligned}(h+k)^{\frac{1}{2}} x (h+k)^{\frac{1}{2}} &= \lim_{\varepsilon \to 0} \big(h+k+\varepsilon\big)^{-\frac{1}{2}} (h+k)\, x\, (h+k)\big(h+k+\varepsilon\big)^{-\frac{1}{2}}\\ &\leq 2 \lim_{\varepsilon \to 0} \big(h+k+\varepsilon\big)^{-\frac{1}{2}} \big(hxh + kxk\big)\big(h+k+\varepsilon\big)^{-\frac{1}{2}}\\ &= 2\Big(u^* h^{\frac{1}{2}} x h^{\frac{1}{2}} u + v^* k^{\frac{1}{2}} x k^{\frac{1}{2}} v\Big)\end{aligned}$$

yields, together with Lemma 2.4, that $\varphi_{(h+k)}(x) < +\infty$. Hence the first half of this paragraph implies that $\varphi_{(h+k)}(x) = \varphi_h(x) + \varphi_k(x)$. Thus we get $\varphi_h + \varphi_k = \varphi_{(h+k)}$.

If $h \leq k$, then $k = h + (k - h)$, so we have

$$\varphi_k = \varphi_h + \varphi_{(k-h)} \geq \varphi_h. \qquad \text{Q.E.D.}$$

Lemma 2.8. *Let h be a positive self-adjoint operator affiliated with the centralizer $\mathcal{M}_\varphi$ of φ. Then the right hand side of the following:*

$$\varphi_h(x) = \lim_{\varepsilon \searrow 0} \varphi\big(h_\varepsilon^{\frac{1}{2}} x h_\varepsilon^{\frac{1}{2}}\big), \qquad x \in \mathcal{M}_+, \tag{13}$$

converges in $[0, \infty]$ and gives a weight φ_h on $\mathcal{M}$. A necessary and sufficient condition for φ_h to be faithful is that h is non-singular, i.e. the range of h is dense in $\mathfrak{H}_\varphi$ or equivalently $h\xi \neq 0$ for every non-zero $\xi \in \mathfrak{D}(h)$.

PROOF: The inequality, $0 < \varepsilon < \delta$, implies $h_\delta \leq h_\varepsilon$, so that $\{\varphi_{h_\varepsilon} : \varepsilon > 0\}$ is monotone increasing by the previous lemma in the sense that $\varepsilon \searrow 0$ gives $\varphi_{h_\varepsilon} \nearrow \varphi_h$. Hence $\varphi_h = \sup \varphi_{h_\varepsilon} = \lim_{\varepsilon \searrow 0} \varphi_{h_\varepsilon}$ makes sense and is linear on $\mathcal{M}_+$ by the linearity of each φ_{h_ε}. The normality of φ_h follows from that of each φ_{h_ε}.

Now, we prove the semi-finiteness of φ_h. Let e_n be the spectral projection of h corresponding to the interval $[0, n]$. By Lemma 2.4, $e_n \mathfrak{m}_\varphi e_n$ is σ-weakly dense in $e_n \mathcal{M} e_n$, so that $\bigcup_{n=1}^\infty e_n \mathfrak{m}_\varphi e_n$ is σ-weakly dense in $\mathcal{M}$. But φ_h takes finite values on $e_n \mathfrak{m}_\varphi e_n$ because $he_n \in \mathcal{M}_\varphi^+$, which yields the semi-finiteness of φ_h. If $e = s(h)$ is the range projection of h, then $\varphi_h(1 - e) = 0$, so that the faithfulness of φ_h is equivalent to $e = 1$. Q.E.D.

Lemma 2.9. *If* $h \in \mathcal{M}_\varphi^+$ *is invertible, and if* $x \in \mathfrak{a}_0$ *and* $y \in \mathfrak{a}_\varphi$, *then the entire function* F:

$$F(\alpha) = \left(h^{\mathrm{i}\alpha+1}\Delta^{\mathrm{i}\alpha+1}\eta_\varphi(x) \mid Sh^{-\mathrm{i}\alpha}\eta_\varphi(y)\right) \tag{14}$$

belongs to $\mathcal{A}(\mathbf{D})$ *and satisfies the boundary condition:*

$$\left.\begin{aligned} F(t) &= \varphi\big(hh^{\mathrm{i}t}\sigma_t^\varphi(x)h^{-\mathrm{i}t}y\big),\\ F(t+\mathrm{i}) &= \varphi\big(hyh^{\mathrm{i}t}\sigma_t^\varphi(x)h^{-\mathrm{i}t}\big). \end{aligned}\right\} \tag{15}$$

This lemma follows from a routine direct computation, so we leave the proof to the reader.

Lemma 2.10. *If* $h \in \mathcal{M}_\varphi^+$ *is invertible, then the modular automorphism group* $\{\sigma_t^\psi\}$ *of* $\psi = \varphi_h$ *is given by the following:*

$$\sigma_t^\psi(x) = h^{\mathrm{i}t}\sigma_t^\varphi(x)h^{-\mathrm{i}t}, \quad x \in \mathcal{M}, \quad t \in \mathbf{R}.$$

PROOF: By Lemma 2.4, we have

$$\mathfrak{m}_\psi = h^{-\frac{1}{2}}\mathfrak{m}_\varphi h^{-\frac{1}{2}} \subset \mathfrak{m}_\varphi \qquad \text{and} \qquad \mathfrak{m}_\varphi = h^{-\frac{1}{2}}h^{\frac{1}{2}}\mathfrak{m}_\varphi h^{\frac{1}{2}}h^{-\frac{1}{2}} \subset \mathfrak{m}_\psi,$$

so that we get $\mathfrak{m}_\varphi = \mathfrak{m}_\psi$ and $\mathfrak{n}_\varphi = \mathfrak{n}_\psi$. Let $x, y \in \mathfrak{a}_\psi \cap \mathfrak{n}_\psi^*$.

We choose a sequence $\{x_n\}$ from $\mathfrak{a}_0$ so that $\lim\|\eta_\varphi(x) - \eta_\varphi(x_n)\|_\sharp = 0$. By the last lemma, we have a sequence $\{F_n\}$ of entire functions in $\mathcal{A}(\mathbf{D})$ such that

$$\begin{aligned} F_n(t) &= \varphi\big(hh^{\mathrm{i}t}\sigma_t^\varphi(x_n)h^{-\mathrm{i}t}y\big) = \big(\eta_\varphi(h^{-\mathrm{i}t}yh^{\mathrm{i}t}h) \mid \Delta^{\mathrm{i}t}\eta_\varphi(x_n^*)\big),\\ F_n(t+\mathrm{i}) &= \varphi\big(hyh^{\mathrm{i}t}\sigma_t^\varphi(x_n)h^{-\mathrm{i}t}\big) = \big(\Delta^{\mathrm{i}t}\eta_\varphi(x_n) \mid \eta_\varphi(h^{-\mathrm{i}t}y^*h^{\mathrm{i}t}h)\big). \end{aligned}$$

But the convergence:

$$\lim_{n\to\infty}\left\|\Delta^{\mathrm{i}t}\big(\eta_\varphi(x) - \eta_\varphi(x_n)\big)\right\|_\sharp = 0$$

is uniform in $t \in \mathbf{R}$, so the sequence $\{F_n\}$ converges uniformly on the boundaries $\mathbf{R}$ and $\mathbf{R} + \mathrm{i}$ of $\mathbf{D}$. Hence the Phragmen-Lindelöf theorem implies that $\{F_n\}$ converges to an $F \in \mathcal{A}(\mathbf{D})$ whose boundary values are given by:

$$F(t) = \lim_{n\to\infty} F_n(t) = \Big(\eta_\varphi(h^{-\mathrm{i}t}yh^{\mathrm{i}t}h) \;\Big|\; \eta_\varphi\big(\sigma_t^\varphi(x^*)\big)\Big) = \varphi\big(hh^{\mathrm{i}t}\sigma_t^\varphi(x)h^{-\mathrm{i}t}y\big);$$

$$F(t+\mathrm{i}) = \lim_{n\to\infty} F_n(t+\mathrm{i}) = \varphi\big(hyh^{\mathrm{i}t}\sigma_t^\varphi(x)h^{-\mathrm{i}t}\big).$$

This concludes the assertion of the lemma. Q.E.D.

We now have a result concerning perturbations of φ by positive self-adjoint operators affiliated with $\mathcal{M}_\varphi$:

Theorem 2.11. *Let φ be a faithful semi-finite normal weight on a von Neumann algebra $\mathcal{M}$. If h is a nonsingular positive self-adjoint operator affiliated with the centralizer $\mathcal{M}_\varphi$ of φ, then the modular automorphism group $\{\sigma_t^\psi\}$ of the faithful weight $\psi = \varphi_h$ given by* (13) *is of the form:*

$$\sigma_t^\psi(x) = h^{\mathrm{i}t}\sigma_t^\varphi(x)h^{-\mathrm{i}t}, \quad x \in \mathcal{M}, \quad t \in \mathbf{R}. \tag{16}$$

PROOF: By Lemma 2.8, $\psi = \varphi_h$ is a faithful weight. For each $n \in \mathbf{N}$ let e_n be the spectral projection of h corresponding to the interval $[1/n, n]$. The restriction of φ to $\mathcal{M}_{e_n}$ $(= e_n\mathcal{M}e_n)$ is a faithful weight on $\mathcal{M}_{e_n}$ whose modular automorphism group is nothing but the restriction of $\{\sigma_t^\varphi\}$ to $\mathcal{M}_{e_n}$. By Lemma 2.10, if $x \in \mathcal{M}_{e_n}$, then

$$\sigma_t^\psi(x) = (he_n)^{\mathrm{i}t}\sigma_t^\varphi(x)(he_n)^{-\mathrm{i}t} = h^{\mathrm{i}t}\sigma_t^\varphi(x)h^{-\mathrm{i}t}, \quad t \in \mathbf{R}.$$

Hence we get

$$\sigma_t^\psi(x) = h^{\mathrm{i}t}\sigma_t^\varphi(x)h^{-\mathrm{i}t}, \quad x \in \bigcup_{n=1}^{\infty} e_n\mathcal{M}e_n, \quad t \in \mathbf{R}.$$

Since $\bigcup e_n\mathcal{M}e_n$ is σ-weakly dense in $\mathcal{M}$, (16) follows. Q.E.D.

Exercise VIII.2

1) A semi-finite normal ψ on $\mathcal{M}$ is called *strictly* semi-finite if φ is the sum of normal positive linear functionals whose supports are mutually orthogonal. Show that the following conditions on a faithful semi-finite normal weight φ on $\mathcal{M}$ are equivalent:

(i) φ is strictly semi-finite;

(ii) The restriction of φ onto the centralizer $\mathcal{M}_\varphi$ is semi-finite;

(iii) The modular automorphism group $\{\sigma_t^\varphi\}$ admits sufficiently many invariant normal states;

(iv) For each $\omega \in \mathcal{M}_*$, the orbit $\left\{\omega \circ \sigma_t^\varphi : t \in \mathbf{R}\right\}$ is relatively weakly compact in $\mathcal{M}_*$. (Hint: Use the Ryll-Nardzewski fixed point theorem, and also Theorem III.5.4.(iii).) [452].

2) Let φ be a faithful semi-finite normal weight on $\mathcal{M}$, and $a \in \mathcal{M}$. Prove that the following conditions are equivalent:

(i) The $\mathcal{M}$-valued function: $t \in \mathbf{R} \mapsto \sigma_t^\varphi(a) \in \mathcal{M}$ can be extended to an $\mathcal{M}$-valued function: $z \in \bar{\mathbf{D}} = \left\{z \in \mathbf{C} : 0 \le \operatorname{Im} z \le 1\right\} \mapsto \sigma_z^\varphi(a) \in \mathcal{M}$ such that it is a bounded continuous function on $\bar{\mathbf{D}}$ and holomorphic in the interior $\mathbf{D}$.

(ii) $\mathfrak{n}_\varphi a \subset \mathfrak{n}_\varphi$ and there exists $b \in \mathcal{M}$ such that $\mathfrak{n}_\varphi b^* \subset \mathfrak{n}_\varphi$, and

$$\varphi(xa) = \varphi(bx), \quad x \in \mathcal{M}_\varphi.$$

If the above conditions hold, then $b = \sigma_{\mathrm{i}}^\varphi(a)$.

3) Let A be a unital C^*-algebra equipped with a continuous one parameter automorphism group $\{\alpha_t : t \in \mathbf{R}\}$ as in Exercise VIII.1.3 and let K_β the set of all states φ such that φ satisfies the modular condition relative to the group $\{\alpha_{\beta t} : t \in \mathbf{R}\}$, which is weakly* compact and convex by Exercise VIII.1.3.

(a) Show that two distinct extreme points of K_β give rise to disjoint primary representations of A.

(b) Show that K_β is a simplex in the sense that every point in K_β has a unique representing probability measure on the extreme boundary $\partial_e(K_\beta)$ of K_β.

(c) Show that if $\pi_\varphi(A)''$ is of type III with $\varphi \in K_{\beta_1}$, then for any $\psi \in K_{\beta_2}$ with $\beta_2 \neq \beta_1$ π_φ is disjoint from π_ψ, [710].

§ 3 The Connes Cocycle Derivative

The Radon-Nikodym Theorem concerning absolutely continuous measures plays important roles in many area of analysis. We will see that this important theorem can be generalized in the context of von Neumann algebras without change for its formulation. Before going into the details let us review the theorem quickly. Let (Γ, μ) be a σ-finite measure space and $\mathcal{L}^\infty(\Gamma, \mu)$ denote the $*$-algebra of all bounded measurable functions. With the norm: $\|f\|_\infty = \sup\{\,|f(\gamma)| : \gamma \in \Gamma\,\}$, $\mathcal{L}^\infty(\Gamma, \mu)$ is a C^*-algebra. Let $\mathcal{N}^\infty(\Gamma, \mu)$ be the algebra of all those functions f on Γ vanishing outside null sets. Then $\mathcal{N}^\infty(\Gamma, \mu)$ is a closed ideal of $\mathcal{L}^\infty(\Gamma, \mu)$, and the quotient C^*-algebra:

$$L^\infty(\Gamma, \mu) = \mathcal{L}^\infty(\Gamma, \mu) \big/ \mathcal{N}^\infty(\Gamma, \mu)$$

admits a faithful representation on the Hilbert space $L^2(\Gamma, \mu)$ as a von Neumann algebra by the natural multiplication. Identifying $f \in \mathcal{L}^\infty(\Gamma, \mu)$ and its image $f + \mathcal{N}^\infty(\Gamma, \mu)$ in $L^\infty(\Gamma, \mu)$, we define a weight φ on $L^\infty(\Gamma, \mu)$ by the following:

$$\varphi(f) = \int_\Gamma f(\gamma)\, \mathrm{d}\mu(\gamma), \qquad f \in \mathcal{L}^\infty(\Gamma, \mu)_+. \tag{1}$$

Now, suppose we have another measure ν in Γ. Then we define a weight ψ on $\mathcal{L}^\infty(\Gamma, \mu)$ by the integral:

$$\psi(f) = \int_\Gamma f(\gamma)\, \mathrm{d}\nu(\gamma)\,. \tag{2}$$

In order to view ψ as a weight on $L^\infty(\Gamma, \mu)$, we must have $\psi\big(\mathcal{N}^\infty(\Gamma, \mu)\big) = \{0\}$, which means the absolute continuity of ν with respect to μ. Furthermore, if this is the case, then the formula (2) gives a weight on $L^\infty(\Gamma, \mu)$, and the usual Radon-Nikodym theorem implies that there exists a measurable function h on $\{\Gamma, \mu\}$

$$\psi(f) = \int_\Gamma f(\gamma) h(\gamma)\, \mathrm{d}\mu(\gamma), \qquad f \in \mathcal{L}^\infty(\Gamma, \mu). \tag{3}$$

The faithfulness of ψ is equivalent to the claim that the zero set of h is of measure zero. Therefore, if we have two faithful semi-finite normal weights on an abelian von Neumann algebra, then one is given by an integral of a function with respect to the other. Furthermore, the function h in (3) is usually written $\mathrm{d}\nu/\mathrm{d}\mu$.

In order to have a natural extension of the Radon-Nikodym theorem (3) to the non-commutative setting, we need to reformulate (3) without losing its essential content. Considering only on the support of ψ, we may assume that ψ is faithful, which yields that $h(\gamma) > 0$. Now, let $u(t) = h^{\mathrm{i}t}$, $t \in \mathbf{R}$. It then follows that $\{u(t)\}$ is a one parameter unitary group in $L^\infty(\Gamma, \mu)$. If $x \in L^\infty(\Gamma, \mu)$ is integrable for both μ and ν, then there exists an $F \in \mathcal{A}(\mathbf{D})$ such that

$$F(t) = \psi\big(xu(t)\big), \qquad F(t+\mathrm{i}) = \varphi\big(xu(t)\big), \qquad t \in \mathbf{R} \tag{4}$$

Furthermore, the unitary group $\{u(t)\}$ is uniquely determined by (4). It is now easy to see that (4) is an equivalent reformulation of (3).

We now return to a general von Neumann algebra $\mathcal{M}$. We fix two faithful (semi-finite normal) weights φ and ψ on $\mathcal{M}$. (We will not repeat the terminology "semi-finite normal" except in the statements of theorems.) Let $M_2 = M(2; \mathbf{C})$, the 2×2 matrix algebra, and put $\mathcal{N} = \mathcal{M} \otimes M_2$. We fix the standard matrix units:

$$e_{11} = \begin{pmatrix} 1 & 0 \\ 0 & 0 \end{pmatrix}, \quad e_{12} = \begin{pmatrix} 0 & 1 \\ 0 & 0 \end{pmatrix}, \quad e_{21} = \begin{pmatrix} 0 & 0 \\ 1 & 0 \end{pmatrix}, \quad e_{22} = \begin{pmatrix} 0 & 0 \\ 0 & 1 \end{pmatrix}. \tag{5}$$

Each $x \in \mathcal{N}$ is represented by a matrix:

$$x = \begin{pmatrix} x_{11} & x_{12} \\ x_{21} & x_{22} \end{pmatrix} = \sum_{i,j} x_{ij} \otimes e_{ij} \tag{6}$$

The *balanced* weight $\varphi \oplus \psi$ of φ and ψ is defined to be the weight on $\mathcal{N}$ given by the following:

$$(\varphi \oplus \psi)\begin{pmatrix} x_{11} & x_{12} \\ x_{21} & x_{22} \end{pmatrix} = \varphi(x_{11}) + \psi(x_{22}) \tag{7}$$

We set $\rho = \varphi \oplus \psi$ for simplicity.

Lemma 3.1.

(i) *The weight* ρ *is faithful, semi-finite and normal.*

(ii) $\mathfrak{n}_\rho = \{ x \in \mathcal{N} : x_{11}, x_{21} \in \mathfrak{n}_\varphi$ *and* $x_{12}, x_{22} \in \mathfrak{n}_\psi \}$. (8)

PROOF: For each $x \in \mathcal{N}$, we have

$$(x^*x)_{11} = x_{11}^*x_{11} + x_{21}^*x_{21}, \qquad (x^*x)_{22} = x_{12}^*x_{12} + x_{22}^*x_{22}. \tag{9}$$

Hence ρ is faithful and (ii) holds. If $x \geq 0$, then $x_{11} \geq 0$ and $x_{22} \geq 0$, and we have

$$\rho(x) = \varphi(x_{11}) + \psi(x_{22}) = \sup\{ \omega(x_{11}) + \theta(x_{22}) : \omega \in \Phi_\varphi,\ \theta \in \Phi_\psi \}$$

Thus ρ is normal. Choosing approximate identities $\{a_i\}$ from $\mathfrak{m}_\varphi^+$ and $\{b_i\}$ from $\mathfrak{m}_\psi^+$ and with $e_i = a_i \otimes e_{11} + b_i \otimes e_{22}$, $\{e_i\}$ converges to 1 σ-strongly in $\mathcal{N}$. Thus ρ is semi-finite. Q.E.D.

We now study the semi-cyclic representation $\{\pi_\rho, \mathfrak{H}_\rho, \eta_\rho\}$ of $\mathcal{N}$ and the corresponding left Hilbert algebra $\mathfrak{A}_\rho$. First of all, the above lemma says:

$$\mathfrak{a}_\rho = \mathfrak{n}_\rho \cap \mathfrak{n}_\rho^* = \left\{ x \in \mathcal{N} : \begin{array}{ll} x_{11} \in \mathfrak{n}_\varphi \cap \mathfrak{n}_\varphi^* & x_{12} \in \mathfrak{n}_\varphi^* \cap \mathfrak{n}_\psi \\ x_{21} \in \mathfrak{n}_\varphi \cap \mathfrak{n}_\psi^*, & x_{22} \in \mathfrak{n}_\psi \cap \mathfrak{n}_\psi^* \end{array} \right\}. \tag{10}$$

Next, we consider the closed subspaces of $\mathfrak{H}_\rho$ defined by:

$$\mathfrak{H}_1 = \left[\eta_\rho\big((\mathfrak{n}_\varphi \cap \mathfrak{n}_\varphi^*) \otimes e_{11}\big)\right], \qquad \mathfrak{H}_3 = \left[\eta_\rho\big((\mathfrak{n}_\varphi^* \cap \mathfrak{n}_\psi) \otimes e_{12}\big)\right],$$

$$\mathfrak{H}_2 = \left[\eta_\rho\big((\mathfrak{n}_\varphi \cap \mathfrak{n}_\psi^*) \otimes e_{21}\big)\right], \qquad \mathfrak{H}_4 = \left[\eta_\rho\big((\mathfrak{n}_\psi \cap \mathfrak{n}_\psi^*) \otimes e_{22}\big)\right]$$

Then $\mathfrak{H}_1, \mathfrak{H}_2, \mathfrak{H}_3, \mathfrak{H}_4$ are mutually orthogonal and

$$\mathfrak{H}_\rho = \mathfrak{H}_1 \oplus \mathfrak{H}_2 \oplus \mathfrak{H}_3 \oplus \mathfrak{H}_4.$$

We define operators $U_1,\ U_2,\ U_3$ and U_4 as follows:

$$\left.\begin{aligned} U_1\eta_\varphi(x) &= \eta_\rho(x \otimes e_{11}), \quad x \in \mathfrak{n}_\varphi, \\ U_2\eta_\varphi(x) &= \eta_\rho(x \otimes e_{21}), \quad x \in \mathfrak{n}_\varphi, \\ U_3\eta_\psi(x) &= \eta_\rho(x \otimes e_{12}), \quad x \in \mathfrak{n}_\psi, \\ U_4\eta_\psi(x) &= \eta_\rho(x \otimes e_{22}), \quad x \in \mathfrak{n}_\psi \end{aligned}\right\} \tag{11}$$

Then U_1 and U_2 (resp. U_3 and U_4) can be extended to isometries of $\mathfrak{H}_\varphi$ (resp. $\mathfrak{H}_\psi$) onto $\mathfrak{H}_1$ and $\mathfrak{H}_2$ (resp. $\mathfrak{H}_3$ and $\mathfrak{H}_4$). We denote the extended isometries by $U_1, \dots, U_4$ again. For each x of the form (6), we have

$$x^* = x_{11}^* \otimes e_{11} + x_{12}^* \otimes e_{21} + x_{21}^* \otimes e_{12} + x_{22}^* \otimes e_{22}, \tag{6*}$$

so that the closure S of the $\sharp$-operation in $\mathfrak{A}_\rho$ and its domain $\mathfrak{D}^\sharp$ enjoy the properties:

$$\left.\begin{aligned} &\mathfrak{D}^\sharp = (\mathfrak{D}^\sharp \cap \mathfrak{H}_1) \oplus (\mathfrak{D}^\sharp \cap \mathfrak{H}_2) \oplus (\mathfrak{D}^\sharp \cap \mathfrak{H}_3) \oplus (\mathfrak{D}^\sharp \cap \mathfrak{H}_4), \\ &S(\mathfrak{D}^\sharp \cap \mathfrak{H}_1) = \mathfrak{D}^\sharp \cap \mathfrak{H}_1, \qquad S(\mathfrak{D}^\sharp \cap \mathfrak{H}_3) = \mathfrak{D}^\sharp \cap \mathfrak{H}_2, \\ &S(\mathfrak{D}^\sharp \cap \mathfrak{H}_2) = \mathfrak{D}^\sharp \cap \mathfrak{H}_3, \qquad S(\mathfrak{D}^\sharp \cap \mathfrak{H}_4) = \mathfrak{D}^\sharp \cap \mathfrak{H}_4. \end{aligned}\right\} \tag{12}$$

Therefore, S is given by the following matrix:

$$S = \begin{pmatrix} S_{11} & 0 & 0 & 0 \\ 0 & 0 & S_{23} & 0 \\ 0 & S_{32} & 0 & 0 \\ 0 & 0 & 0 & S_{44} \end{pmatrix}$$

where we represent a vector of $\mathfrak{H}_\rho$ by a column vector:

$$\eta_\rho(x) = \begin{pmatrix} \eta_\rho(x_{11} \otimes e_{11}) \\ \eta_\rho(x_{21} \otimes e_{21}) \\ \eta_\rho(x_{12} \otimes e_{12}) \\ \eta_\rho(x_{22} \otimes e_{22}) \end{pmatrix}.$$

It then follows immediately that the anti-linear operators:

$$\eta_\varphi(x) \mapsto \eta_\psi(x^*), \quad x \in \mathfrak{n}_\varphi \cap \mathfrak{n}_\psi^*, \qquad \text{and} \qquad \eta_\psi(x) \mapsto \eta_\varphi(x^*), \quad x \in \mathfrak{n}_\varphi^* \cap \mathfrak{n}_\psi,$$

are both preclosed. We denote their closures by $S_{\psi,\varphi}$ and $S_{\varphi,\psi}$ respectively. Then we get

$$\left.\begin{aligned} S_\varphi &= U_1^* S_{11} U_1, & S_{\psi,\varphi} &= U_3^* S_{3,2} U_2, \\ S_{\varphi,\psi} &= U_2^* S_{2,3} U_3, & S_\psi &= U_4^* S_{44} U_4, \end{aligned}\right\} \tag{13}$$

where S_φ and S_ψ should be understood naturally. Hence its adjoint operator F is given by the following:

$$F = \begin{pmatrix} F_{11} & 0 & 0 & 0 \\ 0 & 0 & F_{23} & 0 \\ 0 & F_{32} & 0 & 0 \\ 0 & 0 & 0 & F_{44} \end{pmatrix} \tag{14}$$

with

$$F_{11} = S_{11}^*, \qquad F_{23} = S_{32}^*, \qquad F_{32} = S_{23}^*, \qquad F_{44} = S_{44}^*. \tag{15}$$

Thus, the modular operator Δ and the modular conjugation J of $\mathfrak{A}_\rho$ are represented by the matrices:

$$\Delta = \begin{pmatrix} \Delta_{11} & 0 & 0 & 0 \\ 0 & \Delta_{22} & 0 & 0 \\ 0 & 0 & \Delta_{33} & 0 \\ 0 & 0 & 0 & \Delta_{44} \end{pmatrix}, \qquad J = \begin{pmatrix} J_{11} & 0 & 0 & 0 \\ 0 & 0 & J_{23} & 0 \\ 0 & J_{32} & 0 & 0 \\ 0 & 0 & 0 & J_{44} \end{pmatrix}. \tag{16}$$

The polar decomposition of S, $S = J\Delta^{1/2} = \Delta^{-1/2}J$, yields the following:

$$\left.\begin{aligned} S_{11} &= J_{11}\Delta_{11}^{\frac{1}{2}} = \Delta_{11}^{-\frac{1}{2}} J_{11}, & S_{32} &= J_{32}\Delta_{22}^{\frac{1}{2}} = \Delta_{33}^{-\frac{1}{2}} J_{32} \\ S_{23} &= J_{23}\Delta_{33}^{\frac{1}{2}} = \Delta_{22}^{-\frac{1}{2}} J_{23}, & S_{44} &= J_{44}\Delta_{44}^{\frac{1}{2}} = \Delta_{44}^{-\frac{1}{2}} J_{44} \end{aligned}\right\} \tag{17}$$

We now study the representation π_ρ of $\mathcal{N}$. First, consider

$$x = \sum x_{ij} \otimes e_{ij} \in \mathcal{N}, \qquad y = \sum y_{ij} \otimes e_{ij} \in \mathfrak{n}_\rho.$$

Then the matrix multiplication of x and y gives the matrix representation of $\pi_\rho(x)$ in the following form:

$$\pi_\rho(x) = \begin{pmatrix} U_1\pi_\varphi(x_{11})U_1^* & U_1\pi_\varphi(x_{12})U_2^* & 0 & 0 \\ U_2\pi_\varphi(x_{21})U_1^* & U_2\pi_\varphi(x_{22})U_2^* & 0 & 0 \\ 0 & 0 & U_3\pi_\psi(x_{11})U_3^* & U_3\pi_\psi(x_{12})U_4^* \\ 0 & 0 & U_4\pi_\psi(x_{21})U_3^* & U_4\pi_\psi(x_{22})U_4^* \end{pmatrix} \tag{18}$$

Hence

$$J\pi_\rho(x)J =$$

$$\begin{pmatrix} J_{11}U_1\pi_\varphi(x_{11})U_1^*J_{11} & 0 & J_{11}U_1\pi_\varphi(x_{12})U_2^*J_{23} & 0 \\ 0 & J_{23}U_3\pi_\psi(x_{11})U_3^*J_{32} & 0 & J_{23}U_3\pi_\psi(x_{12})U_4^*J_{44} \\ J_{32}U_2\pi_\varphi(x_{21})U_1^*J_{11} & 0 & J_{32}U_2\pi_\varphi(x_{22})U_2^*J_{23} & 0 \\ 0 & J_{44}U_4\pi_\psi(x_{21})U_3^*J_{32} & 0 & J_{44}U_4\pi_\psi(x_{22})U_4^*J_{44} \end{pmatrix} \tag{19}$$

Putting $x_{11} = x_{22} = 0$ and $x_{12} = x_{21} = 1$, we get an operator:

$$a = \begin{pmatrix} 0 & 0 & J_{11}U_1U_2^*J_{23} & 0 \\ 0 & 0 & 0 & J_{23}U_3U_4^*J_{44} \\ J_{32}U_2U_1^*J_{11} & 0 & 0 & 0 \\ 0 & J_{44}U_4U_3^*J_{32} & 0 & 0 \end{pmatrix}.$$

Since $J\pi_\rho(\mathcal{N})J = \pi_\rho(\mathcal{N})'$, a belongs to $\pi_\rho(\mathcal{N})'$. Putting

$$v_1 = J_{11}U_1U_2^*J_{23}, \qquad v_2 = J_{23}U_3U_4^*J_{44},$$

we obtain

$$U_1\pi_\varphi(x_{11})U_1^* = v_1U_3\pi_\psi(x_{11})U_3^*v_1^*,$$

$$U_2\pi_\varphi(x_{22})U_2^* = v_2U_4\pi_\psi(x_{22})U_4^*v_2^*$$

Hence with $W = U_1^*v_1U_3$, we conclude that W is an isometry of $\mathfrak{H}_\psi$ onto $\mathfrak{H}_\varphi$ such that

$$\pi_\varphi(x) = W\pi_\psi(x)W^*, \quad x \in \mathcal{M} \tag{20}$$

Thus we conclude the following result.

Theorem 3.2. *Let φ be a faithful semi-finite normal weight on $\mathcal{M}$. For any semi-finite normal weight ψ, there exists a linear map η from $\mathfrak{n}_\psi$ into $\mathfrak{H}_\varphi$ and an isometry u from $\mathfrak{H}_\psi$ onto $[\eta(\mathfrak{n}_\psi)]$ such that*

$$\left.\begin{aligned} \pi_\varphi(a)\eta(x) &= \eta(ax), && a \in \mathcal{M}, \quad x \in \mathfrak{n}_\psi, \\ u\eta_\psi(x) &= \eta(x), && x \in \mathfrak{n}_\psi, \\ \pi_\varphi(a)u &= u\pi_\psi(a), && a \in \mathcal{M}. \end{aligned}\right\} \tag{21}$$

If ψ is faithful, then η can be chosen so that $[\eta(\mathfrak{n}_\psi)] = \mathfrak{H}_\varphi$, hence π_φ and π_ψ are unitarily equivalent.

In particular, every $\omega \in \mathcal{M}_*^+$ is represented by a vector $\xi \in \mathfrak{H}_\varphi$ so that

$$\omega(x) = \big(\pi_\varphi(x)\xi \mid \xi\big), \quad x \in \mathcal{M}. \tag{22}$$

PROOF: If ψ is faithful, then the preceding arguments complete the proof simply by setting

$$u = W, \qquad \eta(x) = W\eta_\psi(x), \quad x \in \mathfrak{n}_\psi,$$

with the above W.

If ψ is not faithful, then we choose a weight ψ' on $\mathcal{M}$ with $s(\psi') = 1 - s(\psi)$ and consider $\overline{\psi} = \psi + \psi'$. Let $\overline{u}$ and $\overline{\eta}$ be the objects corresponding $\overline{\psi}$. Putting $e = s(\psi)$, we have $\psi(x) = \overline{\psi}(exe)$, $x \in \mathcal{M}$, and $e \in \mathcal{M}_{\overline{\psi}}$. Hence $\mathfrak{n}_\psi e = \mathfrak{n}_{\overline{\psi}} \subset \mathfrak{n}_{\overline{\psi}}$. So, setting $\eta(x) = \overline{\eta}(xe)$ and $u\eta_\psi(x)$ for each $x \in \mathfrak{n}_\psi$, we conclude the theorem. Q.E.D.

Theorem 3.3 (Cocycle Derivative Theorem). *If φ and ψ are faithful semifinite normal weights on a von Neumann algebra $\mathcal{M}$, then there exists a σ-strongly continuous one parameter family $\{u_t\}$ of unitaries in $\mathcal{M}$ with the following properties:*

a) $u_{s+t} = u_s\sigma_s^\psi(u_t), \quad s, t \in \mathbf{R};$ (23)

b) $u_s\sigma_s^\psi\big(\mathfrak{n}_\varphi^* \cap \mathfrak{n}_\psi\big) = \mathfrak{n}_\varphi^* \cap \mathfrak{n}_\psi;$

c) *For each $x \in \mathfrak{n}_\varphi \cap \mathfrak{n}_\psi^*$ and $y \in \mathfrak{n}_\varphi^* \cap \mathfrak{n}_\psi$, there exists an $F \in \mathcal{A}(\mathbf{D})$ such that*

$$F(t) = \varphi\big(u_t\sigma_t^\psi(y)x\big), \qquad F(t+\mathrm{i}) = \psi\big(xu_t\sigma_t^\psi(y)\big); \tag{24}$$

d) $\sigma_t^\varphi(x) = u_t\sigma_t^\psi(x)u_t^*, \quad x \in \mathcal{M}, \quad t \in \mathbf{R}.$ (25)

Furthermore, the family $\{u_t\}$ is uniquely determined by condition (c).

The uniqueness of $\{u_t\}$ makes the following definition possible.

Definition 3.4. The family $\{u_t\}$ of the theorem is called the *cocycle derivative* of φ with respect to ψ and denoted by the following:

$$\big(\mathrm{D}\varphi : \mathrm{D}\psi\big)_t = u_t, \quad t \in \mathbf{D}. \tag{26}$$

PROOF OF THEOREM 3.3: By Theorem 3.2, we may simplify the notations in the following way: $\mathcal{M}$ acts on a Hilbert space $\mathfrak{H}$ and there exist linear maps $\xi : x \in \mathfrak{n}_\varphi \to \xi(x) \in \mathfrak{H}$ and $\eta : x \in \mathfrak{n}_\psi \to \eta(x) \in \mathfrak{H}$ such that

$$\left.\begin{aligned} a\xi(x) &= \xi(ax), \qquad a\eta(x) = \eta(ax), && a \in \mathcal{M};\\ \varphi(y^*x) &= \big(\xi(x) \mid \xi(y)\big), && x, y \in \mathfrak{n}_\varphi;\\ \psi(y^*x) &= \big(\eta(x) \mid \eta(y)\big), && x, y \in \mathfrak{n}_\psi;\\ \mathfrak{H} &= \big[\xi(\mathfrak{n}_\varphi)\big] = \big[\eta(\mathfrak{n}_\psi)\big]. \end{aligned}\right\} \tag{27}$$

In this setting, $\mathcal{N} = \mathcal{M} \otimes M_2$ acts on $\mathfrak{H} \oplus \mathfrak{H} \oplus \mathfrak{H} \oplus \mathfrak{H} = \mathfrak{H}_\rho$ in the following way:

$$\pi_\rho \begin{pmatrix} x_{11} & x_{12} \\ x_{21} & x_{22} \end{pmatrix} = \begin{pmatrix} x_{11} & x_{12} & 0 & 0 \\ x_{21} & x_{22} & 0 & 0 \\ 0 & 0 & x_{11} & x_{12} \\ 0 & 0 & x_{21} & x_{22} \end{pmatrix}.$$

Hence Δ of (16) gives the following:

$$\left.\begin{aligned} \Delta_{11}^{it} x_{11} \Delta_{11}^{-it} &= \Delta_{33}^{it} x_{11} \Delta_{33}^{-it}, & \Delta_{11}^{it} x_{12} \Delta_{22}^{-it} &= \Delta_{33}^{it} x_{12} \Delta_{44}^{-it} \\ \Delta_{22}^{it} x_{21} \Delta_{11}^{-it} &= \Delta_{44}^{it} x_{21} \Delta_{33}^{-it}, & \Delta_{22}^{it} x_{22} \Delta_{22}^{-it} &= \Delta_{44}^{it} x_{22} \Delta_{44}^{-it} \end{aligned}\right\} \tag{28}$$

because $\Delta^{it} \pi_\rho(\mathcal{N}) \Delta^{-it} = \pi_\rho(\mathcal{N})$. Furthermore, $\mathfrak{H}_\varphi$ and $\mathfrak{H}_\psi$ are identified with the first and the fourth components of $\mathfrak{H}_\rho$ respectively, so that we have $\Delta_{11} = \Delta_\varphi$ and $\Delta_{44} = \Delta_\psi$. Thus, we obtain from (28) the following:

$$\left.\begin{aligned} \sigma_t^\varphi(x) &= \Delta_{11}^{it} x \Delta_{11}^{-it} \in \mathcal{M}, & \sigma_t^{\varphi,\psi}(x) &= \Delta_{11}^{it} x \Delta_{22}^{-it} \in \mathcal{M}, \\ \sigma_t^{\psi,\varphi}(x) &= \Delta_{22}^{it} x \Delta_{11}^{-it} \in \mathcal{M}, & \sigma_t^\psi(x) &= \Delta_{22}^{it} x \Delta_{22}^{-it} \in \mathcal{M}, \end{aligned}\right\} \tag{29}$$

for every $x \in \mathcal{M}$, so that

$$\sigma_t^\rho\left[\begin{pmatrix} x_{11} & x_{12} \\ x_{21} & x_{22} \end{pmatrix}\right] = \begin{pmatrix} \sigma_t^\varphi(x_{11}) & \sigma_t^{\varphi,\psi}(x_{12}) \\ \sigma_t^{\psi,\varphi}(x_{21}) & \sigma_t^\psi(x_{22}) \end{pmatrix}. \tag{30}$$

We now set

$$u_t = \sigma_t^{\varphi,\psi}(1), \quad t \in \mathbf{R}. \tag{31}$$

It then follows that $\{u_t\}$ is a σ-strongly continuous one parameter family of unitaries in $\mathcal{M}$. A simple 2×2 matrix calculation:

$$\begin{pmatrix} x & 0 \\ 0 & 0 \end{pmatrix}\begin{pmatrix} 0 & y \\ 0 & 0 \end{pmatrix} = \begin{pmatrix} 0 & xy \\ 0 & 0 \end{pmatrix} = \begin{pmatrix} 0 & x \\ 0 & 0 \end{pmatrix}\begin{pmatrix} 0 & 0 \\ 0 & y \end{pmatrix}$$

implies

$$\sigma_s^{\varphi,\psi}(xy) = \sigma_s^\varphi(x)\sigma_s^{\varphi,\psi}(y) = \sigma_s^{\varphi,\psi}(x)\sigma_s^\psi(y). \tag{32}$$

Hence the cocycle property (a) of $\{u_s\}$ follows from:

$$u_{s+t} = \sigma_{s+t}^{\varphi,\psi}(1) = \sigma_s^{\varphi,\psi}\left(\sigma_t^{\varphi,\psi}(1)\right) = \sigma_s^{\varphi,\psi}(1)\sigma_s^\psi\left(\sigma_t^{\varphi,\psi}(1)\right) = u_s \sigma_s^\psi(u_t).$$

The statement (b) is nothing but another expression of the fact $\sigma_t^\rho\left(\mathfrak{n} \cap \mathfrak{n}_\rho^*\right) = \mathfrak{n}_\rho \cap \mathfrak{n}_\rho^*$ in terms of the matrix components.

The statement (c) follows from the modular condition for $\{\sigma_t^\rho\}$ and the simple calculations:

$$\begin{pmatrix} xy & 0 \\ 0 & 0 \end{pmatrix} = \begin{pmatrix} 0 & x \\ 0 & 0 \end{pmatrix} \begin{pmatrix} 0 & 0 \\ y & 0 \end{pmatrix};$$

$$\begin{pmatrix} 0 & 0 \\ 0 & yx \end{pmatrix} = \begin{pmatrix} 0 & 0 \\ y & 0 \end{pmatrix} \begin{pmatrix} 0 & x \\ 0 & 0 \end{pmatrix}.$$

Now, the last statement (d) is checked by applying $\{\sigma_t^\rho\}$ of (30) to the following matrix product:

$$\begin{pmatrix} x & 0 \\ 0 & 0 \end{pmatrix} = \begin{pmatrix} 0 & 1 \\ 0 & 0 \end{pmatrix} \begin{pmatrix} 0 & 0 \\ 0 & x \end{pmatrix} \begin{pmatrix} 0 & 0 \\ 1 & 0 \end{pmatrix}.$$

The uniqueness of $\{u_t\}$ follows from the following lemma. Q.E.D.

Lemma 3.5.

(i) *Let φ and ψ be faithful weights on $\mathcal{M}$. Suppose that a σ-strong continuous function: $t \in \mathbf{R} \to y(t) \in \mathfrak{n}_\varphi^* \cap \mathfrak{n}_\psi$ satisfies the boundedness condition:*

$$\sup_t \varphi\big(y(t)y(t)^*\big) < +\infty, \qquad \sup_t \psi\big(y(t)^*y(t)\big) < +\infty$$

If for each $x \in \mathfrak{n}_\varphi \cap \mathfrak{n}_\psi^*$ there exists an $F \in \mathcal{A}(\mathbf{D})$ such that

$$F(t) = \varphi\big(y(t)x\big), \qquad F(t + \mathrm{i}) = \psi\big(xy(t)\big), \qquad t \in \mathbf{R},$$

then

$$y(t) = u_t\sigma_t^\psi\big(y(0)\big) = \sigma_t^{\varphi,\psi}\big(y(0)\big), \quad t \in \mathbf{R},$$

where $\{u_t\}$ and $\{\sigma_t^{\varphi,\psi}\}$ are given by (31) and (29) respectively.

(ii) *Let φ be a faithful weight on $\mathcal{M}$ and let $t \in \mathbf{R} \to y(t) \in \mathfrak{n}_\varphi^* \cap \mathfrak{n}_\varphi = \mathfrak{A}_\varphi$ be a σ-weakly continuous function such that*

$$\sup\Big[\varphi\big(y(t)^*y(t)\big) + \varphi\big(y(t)y(t)^*\big)\Big] < +\infty.$$

If for each $x \in \mathfrak{A}_\varphi$ there exists an $F \in \mathcal{A}(\mathbf{D})$ such that

$$F(t) = \varphi\big(y(t)x\big), \qquad F(t + \mathrm{i}) = \varphi\big(xy(t)\big), \qquad t \in \mathbf{R},$$

Then

$$y(t) = \sigma_t^\varphi\big(y(0)\big), \quad t \in \mathbf{R}$$

PROOF: The assertion (i) follows immediately from (ii) by considering the balanced weight. Thus we prove only (ii).

Now, let $\mathfrak{A}$ be the full left Hilbert algebra associated with $\{\mathcal{M}, \varphi\}$ and $\mathfrak{A}_0$ be the maximal Tomita algebra in $\mathfrak{A}$. We set $\eta(t) = \eta_\varphi\big(y(t)\big)$. By assumption, to each $\xi \in \mathfrak{A}$, there corresponds an $F \in \mathcal{A}(\mathbf{D})$ such that

$$F(t) = \big(\xi \mid J\Delta^{\frac{1}{2}}\eta(t)\big), \qquad F(t + \mathrm{i}) = \big(\eta(t) \mid J\Delta^{\frac{1}{2}}\xi\big).$$

Furthermore, the boundedness assumption on $\{y(t)\}$ means that $\sup\|\eta(t)\|_\sharp < +\infty$. Applying this to $J\xi$ for a fixed $\xi \in \mathfrak{A}_0$, we get a function $F_\xi \in \mathcal{A}(\mathbf{D})$ such that

$$F_\xi(t) = \left(\Delta^{\frac{1}{2}}\eta(t) \mid \xi\right), \qquad F_\xi(t+\mathrm{i}) = \left(\eta(t) \mid \Delta^{-\frac{1}{2}}\xi\right).$$

By the Phragmen-Lindelöf theorem, the map: $\xi \in \mathfrak{A}_0 \mapsto F_\xi(z) \in \mathbf{C}$, $z \in \mathbf{D}$, can be extended to a bounded anti-linear functional on $\mathfrak{D}^\flat$ with respect to the $\|\ \|_\flat$ norm, so that there corresponds an element $\zeta(z) \in \mathfrak{D}^\flat$ such that

$$F(z) = \left(\zeta(z) \mid \xi\right)_\flat = \left(\zeta(z) \mid \xi\right) + \left(\Delta^{-\frac{1}{2}}\zeta(z) \mid \Delta^{-\frac{1}{2}}\xi\right) = \left(\zeta(z) \mid (1+\Delta^{-1})\xi\right),$$

for every $z \in \mathbf{D}$ and $\xi \in \mathfrak{A}_0$. Now, with a fixed $\xi \in \mathfrak{A}_0$, we set

$$G(z, w) = \left(\zeta(z) \mid (1+\Delta^{-1})\Delta^{\mathrm{i}\overline{w}}\xi\right), \quad z \in \mathbf{D}, \quad w \in \mathbf{C}.$$

It then follows that G is holomorphic in $\mathbf{D} \times \mathbf{C}$. By the construction of $\zeta(z)$, we have

$$\left(\zeta(t) \mid (1+\Delta^{-1})\xi\right) = \left(\Delta^{\frac{1}{2}}\eta(t) \mid \xi\right),$$

so that $\zeta(t) \in \mathfrak{D}(\Delta^{-1})$ and $(1+\Delta^{-1})\zeta(t) = \Delta^{1/2}\eta(t)$. We further compute:

$$G(t, w) = \left(\Delta^{\frac{1}{2}}\eta(t) \mid \Delta^{\mathrm{i}\overline{w}}\xi\right) = \left(\eta(t) \mid \Delta^{\mathrm{i}\left(\overline{w}-\frac{\mathrm{i}}{2}\right)}\xi\right) = \left(\eta(t) \mid \Delta^{\mathrm{i}\left(w+\frac{\mathrm{i}}{2}\right)^-}\xi\right);$$

$$G(t+\mathrm{i}, w) = \left(\eta(t) \mid \Delta^{-\frac{1}{2}}\Delta^{\mathrm{i}\overline{w}}\xi\right) = \left(\eta(t) \mid \Delta^{\mathrm{i}\left(w-\frac{\mathrm{i}}{2}\right)^-}\xi\right) = G(t, w-\mathrm{i}).$$

We then put $f(z) = G(z, z)$ for $z \in \overline{\mathbf{D}}$. It follows that $f \in \mathcal{A}(\mathbf{D})$ and $f(t+\mathrm{i}) = f(t)$, $t \in \mathbf{R}$. Hence f can be extended to the whole $\mathbf{C}$ by the periodicity. The extended function is then bounded and holomorphic, so that Liouville's theorem entails that f is constant. Hence

$$\left(\Delta^{\frac{1}{2}}\eta(0) \mid \xi\right) = f(0) = f(t) = \left(\Delta^{\frac{1}{2}}\eta(t) \mid \Delta^{\mathrm{i}t}\xi\right) = \left(\Delta^{\frac{1}{2}}\Delta^{-\mathrm{i}t}\eta(t) \mid \xi\right).$$

This equality is true for all $\xi \in \mathfrak{A}_0$. Thus, we obtain $\Delta^{-\mathrm{i}t}\eta(t) = \eta(0)$, which means $\eta(t) = \Delta^{\mathrm{i}t}\eta(0)$, $t \in \mathbf{R}$. Therefore, we conclude that

$$y(t) = \pi_\ell\left(\eta(t)\right) = \pi_\ell\left(\Delta^{\mathrm{i}t}\eta(0)\right) = \sigma_t^\varphi\left(y(0)\right). \qquad \text{Q.E.D.}$$

Corollary 3.6. *For a pair φ, ψ of faithful semi-finite normal weights on $\mathcal{M}$, the following conditions are equivalent:*

(i) $\psi = \psi \circ \sigma_t^\varphi$, $t \in \mathbf{R}$;

(ii) $\psi = \varphi_h$ *for some non-singular positive self-adjoint operator h affiliated with $\mathcal{M}_\varphi$, where φ_h is given by* (13)*;*

(iii) $\varphi = \varphi \circ \sigma_t^\psi$, $t \in \mathbf{R}$.

PROOF: The implication (ii) $\Longrightarrow$ (i) was already shown in Theorem 2.11.

(i) $\Longrightarrow$ (ii): Suppose that $\psi = \psi \circ \sigma_t$, $t \in \mathbf{R}$. The uniqueness of the cocycle derivative yields that for any pair φ and ψ we have

$$\left(\mathrm{D}\psi \circ \alpha : \mathrm{D}\varphi \circ \alpha\right)_t = \alpha^{-1}\left((\mathrm{D}\psi : \mathrm{D}\varphi)_t\right), \quad \alpha \in \mathrm{Aut}(\mathcal{M}) \tag{33}$$

Therefore, we have

$$\sigma_s^{\varphi}\left((\mathrm{D}\psi : \mathrm{D}\varphi)_t\right) = \left(\mathrm{D}\psi \circ \sigma_{-s}^{\varphi} : \mathrm{D}\varphi \circ \sigma_{-s}^{\varphi}\right)_t = (\mathrm{D}\psi : \mathrm{D}\varphi)_t, \quad s, t \in \mathbf{R}.$$

Hence $\left\{(\mathrm{D}\psi : \mathrm{D}\varphi)_t\right\}$ is a one parameter unitary group in $\mathcal{M}_{\varphi}$. By Stone's theorem, there exists a non-singular positive self-adjoint operator h affiliated with $\mathcal{M}_{\varphi}$ such that $(\mathrm{D}\psi : \mathrm{D}\varphi)_t = h^{it}$, $t \in \mathbf{R}$. By the relative modular condition, ψ is uniquely determined by $(\mathrm{D}\psi : \mathrm{D}\varphi)_t = h^{it}$. But this is easily checked. Q.E.D.

Theorem 3.7 (Chain Rule). *If φ, ψ and ρ are faithful semi-finite normal weights on $\mathcal{M}$, then*

$$(\mathrm{D}\psi : \mathrm{D}\varphi)_t = (\mathrm{D}\psi : \mathrm{D}\rho)_t(\mathrm{D}\rho : \mathrm{D}\varphi)_t, \quad t \in \mathbf{R}. \tag{34}$$

PROOF: On the 3×3-matrix algebra $\mathcal{N} = \mathcal{M} \otimes M_3$, we consider the balanced weight $\chi = \psi \oplus \rho \oplus \varphi$, i.e.

$$\chi\left(\sum_{i,j=1}^{3} x_{ij} \otimes e_{ij}\right) = \psi(x_{11}) + \rho(x_{22}) + \varphi(x_{33}),$$

with $\{e_{i,j}\}$ the standard matrix unit in M_3. Then the formula (34) follows from the composition rule for matrix unit: $e_{1,3} = e_{1,2}e_{2,3}$ and the fact:

$$\begin{aligned}
(\mathrm{D}\psi : \mathrm{D}\varphi)_t \otimes e_{1,3} &= \sigma_t^{\chi}(1 \otimes e_{1,3}), \\
(\mathrm{D}\psi : \mathrm{D}\rho)_t \otimes e_{1,2} &= \sigma_t^{\chi}(1 \otimes e_{1,2}), \\
(\mathrm{D}\rho : \mathrm{D}\varphi)_t \otimes e_{2,3} &= \sigma_t^{\chi}(1 \otimes e_{2,3}).
\end{aligned}$$

Q.E.D.

We now state the converse of the cocycle derivative Theorem, Theorem 3.3, as follows:

Theorem 3.8. *Let φ be a faithful semi-finite normal weight on a von Neumann algebra $\mathcal{M}$. If $\{u_t : t \in \mathbf{R}\}$ is a σ-strongly continuous family of unitaries in $\mathcal{M}$ satisfying the cocycle identity:*

$$u_{s+t} = u_s \sigma_s^{\varphi}(u_t),$$

then there exists a faithful semi-finite normal weight ψ such that

$$(\mathrm{D}\psi : \mathrm{D}\varphi)_t = u_t, \quad t \in \mathbf{R}.$$

We prove the theorem in several steps. We shall use the notations established above such as $\{\pi_\varphi, \mathfrak{H}_\varphi, \eta_\varphi\}$, Δ, J, and so on. However, we identify $\mathcal{M}$ and $\pi_\varphi(\mathcal{M})$.

$$\left.\begin{aligned} \sigma_t(x) &= u_t\sigma_t^\varphi(x)u_t^*, \quad x \in \mathcal{M}, \\ \rho_t(x) &= u_t\sigma_t^\varphi(x), \quad t \in \mathbf{R}. \end{aligned}\right\} \tag{35}$$

It then follows that $\{\sigma_t\}$ (resp. $\{\rho_t\}$) is a one parameter group of automorphisms (isometric transformations) of $\mathcal{M}$. Furthermore, we have

$$\rho_t(xy) = \sigma_t(x)\rho_t(y) = \rho_t(x)\sigma_t^\varphi(y), \quad x, y \in \mathcal{M}. \tag{36}$$

For the notational simplicity, we denote $\mathcal{M}_a^\varphi$ and $\mathcal{M}_a$ by A and B respectively, where $\mathcal{M}_a^\varphi$ and $\mathcal{M}_a$ were the algebras of entire elements for σ^φ and σ respectively defined by Definition 2.2. Let C denote the set of all $x \in \mathcal{M}$ such that the $\mathcal{M}$-valued function: $t \in \mathbf{R} \to \rho_t(x) \in \mathcal{M}$ can be extended to an $\mathcal{M}$-valued entire function on $\mathbf{C}$. The value of the extended function at $\alpha \in \mathbf{C}$ is denoted by $\rho_\alpha(x)$ as before. It is easy to show that

$$\overline{\mathcal{M}}_a^{\bar\sigma} = B \otimes e_{11} + C^* \otimes e_{21} + C \otimes e_{12} + A \otimes e_{22},$$

where $\overline{\mathcal{M}} = \mathcal{M} \otimes M(2; \mathbf{C})$ and

$$\bar\sigma_t \begin{pmatrix} x_{11} & x_{12} \\ x_{21} & x_{22} \end{pmatrix} = \begin{pmatrix} u_t\sigma_t^\varphi(x_{11})u_t^* & u_t\sigma_t^\varphi(x_{12}) \\ \sigma_t^\varphi(x_{21})u_t^* & \sigma_t^\varphi(x_{22}) \end{pmatrix}. \tag{37}$$

Lemma 3.9. *The algebras A and B are both σ-weakly dense *-subalgebras of $\mathcal{M}$, and C is a σ-weakly dense subspace of $\mathcal{M}$. They enjoy the following properties:*

$$\begin{gathered} BCA \subset C, \qquad C^*C \subset A, \qquad CC^* \subset B; \\ \rho_\alpha(bca) = \sigma_\alpha(b)\rho_\alpha(c)\sigma_\alpha^\varphi(a), \quad a \in A, \quad b \in B, \quad c \in C. \end{gathered} \tag{38}$$

The proof is easy, so we leave it to the reader.

Next, we set

$$\begin{aligned} \mathfrak{n}_0 &= \{x \in C : \rho_\alpha(x) \in \mathfrak{n}_\varphi, \ \alpha \in \mathbf{C}\}, \\ \mathfrak{h}_0 &= \mathfrak{n}_0\mathfrak{n}_0^* = \left\{\sum_{k=1}^n x_k y_k^* : x_1, \dots, x_n, \ y_1, \dots, y_n \in \mathfrak{n}_0\right\}. \end{aligned} \tag{39}$$

Lemma 3.10.

(i) $\mathfrak{n}_\varphi A \subset \mathfrak{n}_\varphi$ *and*

$$\eta_\varphi(xa) = J\sigma^\varphi_{-\frac{i}{2}}(a^*)J\eta_\varphi(x), \quad x \in \mathfrak{n}_\varphi, \quad a \in A; \tag{40}$$

(ii) $B\mathfrak{n}_0 A \subset \mathfrak{n}_0$;

(iii) $\mathfrak{h}_0$ *is an ideal of* B *and* $\mathfrak{A}_0 = \mathfrak{n}_0^*\mathfrak{n}_0$ *is an ideal of* A.

PROOF: The statements (ii) and (iii) follow easily from the combination of (38) and the statement (i) together with the fact that $\mathfrak{n}_\varphi$ is a left ideal of $\mathcal{M}$. Thus we prove only (i).

By Lemma 2.4.(ii), $\mathfrak{m}_\varphi$ is an A-bimodule, so that $\mathfrak{n}_\varphi$ is a right A-module. The equation (40) for each $x \in \mathfrak{a}_{\varphi,0}$ follows from the formal calculation, where $\mathfrak{a}_\varphi = \eta_\varphi\big(\mathfrak{n}_\varphi \cap \mathfrak{n}_\varphi^*\big)$, i.e. $\mathfrak{a}_{\varphi,0} = \big\{ y \in A : \sigma_\alpha^\varphi(y) \in \mathfrak{n}_\varphi \cap \mathfrak{n}_\varphi^*,\ \alpha \in \mathbf{C} \big\}$. For a general $x \in \mathfrak{n}_\varphi$, we can prove (40) by making use of approximations of Theorem VI.1.26.(i).

Q.E.D.

We set

$$\eta\Big(\sum_{k=1}^n x_k y_k^*\Big) = \sum_{k=1}^n x_k J\eta_\varphi\big(\rho_{-\frac{i}{2}}(y_k)\big), \quad \sum_{k=1}^n x_k y_k^* \in \mathfrak{h}_0. \tag{41}$$

Lemma 3.11.

(i) *The formulas* (41) *give a well-defined linear map* η *from* $\mathfrak{h}_0$ *into* $\mathfrak{H}_\varphi$ *with dense range.*

(ii)

$$\left.\begin{aligned} &\eta(bz) = b\eta(z), \quad b \in B, \quad z \in \mathfrak{h}_0, \\ &\eta(zb) = J\sigma_{-\frac{i}{2}}(b^*)J\eta(z). \end{aligned}\right\} \tag{42}$$

PROOF:

(i) If we prove the well-definedness of η, then the linearity of η is automatic. Suppose $\sum_{k=1}^n x_k y_k^* = 0$ with $x_1, \dots, x_n,\ y_1, \dots, y_n \in \mathfrak{n}_0$. We then compute:

$$\begin{aligned}
\Big\| \sum_{k=1}^n x_k J\eta_\varphi\big(\rho_{-\frac{i}{2}}(y_i)\big)\Big\|^2 &= \sum_{k,\ell}\Big(x_k J\eta_\varphi\big(\rho_{-\frac{i}{2}}(y_k)\big) \,\Big|\, x_\ell J\eta_\varphi\big(\rho_{-\frac{i}{2}}(y_\ell)\big)\Big) \\
&= \sum_{k,\ell}\Big(x_\ell^* x_k J\eta_\varphi\big(\rho_{-\frac{i}{2}}(y_k)\big) \,\Big|\, J\eta_\varphi\big(\rho_{-\frac{i}{2}}(y_\ell)\big)\Big) \\
&= \sum_{k,\ell}\Big(\eta_\varphi\big(\rho_{-\frac{i}{2}}(y_\ell)\big) \,\Big|\, Jx_\ell^* x_k J\eta_\varphi\big(\rho_{-\frac{i}{2}}(y_k)\big)\Big) \\
&= \sum_{k,\ell}\Big(\eta_\varphi\big(\rho_{-\frac{i}{2}}(y_\ell)\big) \,\Big|\, \eta_\varphi\big(\rho_{-\frac{i}{2}}(y_k)\sigma^\varphi_{-\frac{i}{2}}(x_k^* x_\ell)\big)\Big) \qquad \text{by Lemma 3.9} \\
&= \sum_{\ell}\bigg(\eta_\varphi\big(\rho_{-\frac{i}{2}}(y_\ell)\big) \,\bigg|\, \eta_\varphi\Big(\sum_{k=1}^n \rho_{-\frac{i}{2}}\big(y_k x_k^* x_\ell\big)\Big)\bigg) \\
&= \sum_{\ell}\bigg(\eta_\varphi\big(\rho_{-\frac{i}{2}}(y_\ell)\big) \,\bigg|\, \eta_\varphi\Big(\rho_{-\frac{i}{2}}\Big(\Big(\sum_{k=1}^n x_k y_k^*\Big)^* x_\ell\Big)\Big)\bigg) = 0.
\end{aligned}$$

Hence η is well-defined. The density of the range of η follows from that of $\eta_\varphi(\mathfrak{n}_0)$ in $\mathfrak{A}_\varphi$ which is an easy consequence of the fact that $\eta_\varphi\big(\rho_t(x)\big) = u_t\Delta_\varphi^{it}\eta_\varphi(x),\ x \in \mathfrak{n}_\varphi$.

(ii) The first formula of (42) is trivial. Let $z = xy^*,\ x, y \in \mathfrak{n}_0$ and $b \in B$. We then have

$$\begin{aligned}\eta(zb) &= xJ\eta_\varphi\big(\rho_{-\frac{i}{2}}(b^*y)\big) = xJ\eta_\varphi\big(\sigma_{-\frac{i}{2}}(b^*)\rho_{-\frac{i}{2}}(y)\big) \qquad \text{by (38)}\\ &= xJ\sigma_{-\frac{i}{2}}(b^*)\eta_\varphi\big(\rho_{-\frac{i}{2}}(y)\big) = xJ\sigma_{-\frac{i}{2}}(b^*)JJ\eta_\varphi\big(\rho_{-\frac{i}{2}}(y)\big)\\ &= J\sigma_{-\frac{i}{2}}(b^*)JxJ\eta_\varphi\big(\rho_{-\frac{i}{2}}(y)\big) = J\sigma_{-\frac{i}{2}}(b^*)J\eta(z)\end{aligned}$$

Thus (42) follows by linearity. Q.E.D.

By the formula:

$$\sigma_t(xy^*) = \rho_t(x)\rho_t(y)^*, \quad x, y \in \mathcal{M}, \tag{43}$$

$\{\sigma_t\}$ leaves $\mathfrak{h}_0$ globally invariant. We set

$$\mathfrak{B}_0 = \eta(\mathfrak{h}_0), \qquad \eta(x)\eta(y) = \eta(xy), \quad x, y \in \mathfrak{h}_0,$$

$$\eta(x)^* = \eta(x^*), \qquad U(\alpha)\eta(z) = \eta\big(\sigma_\alpha(z)\big), \quad z \in \mathfrak{h}_0.$$

Lemma 3.12. *Equipped with the above structure, $\mathfrak{B}_0$ is a Tomita algebra whose left von Neumann algebra $\mathcal{R}_\ell(\mathfrak{B}_0)$ can be naturally identified with $\mathcal{M}$.*

PROOF: Since $\eta(xy) = x\eta(y)$ by (42), $\mathcal{R}_\ell(\mathfrak{B}_0) = \mathcal{M}$ if $\mathfrak{B}_0$ is a left Hilbert algebra. We prove the lemma in a few steps.

Step I. Analyticity of $U(\alpha)$: Let $x, y \in \mathfrak{n}_0$ and $z = xy^*$. We then have

$$U(\alpha)\eta(z) = \eta\big(\sigma_\alpha(xy^*)\big) = \eta\big(\rho_\alpha(x)\rho_{\overline{\alpha}}(y)^*\big) = \rho_\alpha(x)J\eta_\varphi\big(\rho_{\overline{\alpha}-\frac{i}{2}}(y)\big).$$

Since $\alpha \mapsto \rho_\alpha(x)$ is holomorphic, we have only to prove $\alpha \to \eta_\varphi\big(\rho_\alpha(x)\big),\ x \in \mathfrak{n}_0$, is holomorphic. If ξ_1 and ξ_2 are vectors in the maximal Tomita algebra $\mathfrak{A}_0$ associated with $\mathfrak{A}_\varphi = \xi_\varphi\big(\mathfrak{n}_\varphi \cap \mathfrak{n}_\varphi^*\big)$, then we have

$$\begin{aligned}\Big(\eta_\varphi\big(\rho_\alpha(x)\big) \,\Big|\, \xi_1\xi_2^\flat\Big) &= \Big(\eta_\varphi\big(\rho_\alpha(x)\big) \,\Big|\, \pi_r(\xi_2)^*\xi_1\Big) = \Big(\pi_r(\xi_2)\eta_\varphi\big(\rho_\alpha(x)\big) \,\Big|\, \xi_1\Big)\\ &= \big(\rho_\alpha(x)\xi_2 \,\big|\, \xi_1\big),\end{aligned}$$

so that the function $\alpha \in \mathbf{C} \mapsto \big(\eta_\varphi(\rho_\alpha(x)) \,\big|\, \xi_1\xi_2^\flat\big)$ is holomorphic. Since $\mathfrak{A}_0^2$ is dense in $\mathfrak{A}_\varphi$, we get the analyticity of the function $\alpha \in \mathbf{C} \mapsto \eta_\varphi\big(\rho_\alpha(x)\big) \in \mathfrak{H}_\varphi$.

Step II. The unitarity of $U(t),\ t \in \mathbf{R}$: Let x, y, u and $v \in \mathfrak{n}_0$. We then compute:

$$\begin{aligned}\big(U(t)\eta(xy^*) \,\big|\, U(t)\eta(uv^*)\big) &= \Big(\eta\big(\rho_t(x)\rho_t(y)^*\big) \,\Big|\, \eta\big(\rho_t(u)\rho_t(v)^*\big)\Big)\\ &= \Big(\rho_t(x)J\eta_\varphi\big(\rho_{t-\frac{i}{2}}(y)\big) \,\Big|\, \rho_t(u)J\eta_\varphi\big(\rho_{t-\frac{i}{2}}(v)\big)\Big)\\ &= \Big(\sigma_t^\varphi(x)J\eta_\varphi\big(\rho_{t-\frac{i}{2}}(y)\big) \,\Big|\, \sigma_t^\varphi(u)J\eta_\varphi\big(\rho_{t-\frac{i}{2}}(v)\big)\Big)\\ &= \Big(J\sigma_t^\varphi(x^*u)J\eta_\varphi\big(\rho_{t-\frac{i}{2}}(v)\big) \,\Big|\, \eta_\varphi\big(\rho_{t-\frac{i}{2}}(y)\big)\Big)\end{aligned}$$

$$= \left(\eta_\varphi\big(\rho_{t-\frac{i}{2}}(v)\sigma^\varphi_{t-\frac{i}{2}}(u^*x)\big)\,\Big|\, u_t\Delta_\varphi^{it}\eta_\varphi\big(\rho_{-\frac{i}{2}}(y)\big)\right)$$
$$= \left(\eta_\varphi\big(\rho_{t-\frac{i}{2}}(vu^*x)\big)\,\Big|\, u_t\Delta_\varphi^{it}\eta_\varphi\big(\rho_{-\frac{i}{2}}(y)\big)\right)$$
$$= \left(u_t\Delta_\varphi^{it}\eta_\varphi\big(\rho_{-\frac{i}{2}}(vu^*x)\big)\,\Big|\, u_t\Delta_\varphi^{it}\eta_\varphi\big(\rho_{-\frac{i}{2}}(y)\big)\right)$$
$$= \left(\eta_\varphi\big(\rho_{-\frac{i}{2}}(vu^*x)\big)\,\Big|\, \eta_\varphi\big(\rho_{-\frac{i}{2}}(y)\big)\right).$$

Hence the function: $t \mapsto \big(U(t)\eta(xy^*) \mid U(t)\eta(uv^*)\big)$ is constant, so that

$$\big(U(t)\eta(xy^*) \mid U(t)\eta(uv^*)\big) = \big(\eta(xy^*) \mid \eta(uv^*)\big)$$

By linearity, this means that $\{U(t)\}$ can be extended to a one parameter unitary group on $\mathfrak{H}_\varphi$, which will be denoted by $\{U(t)\}$ again.

Step III. Modular condition: We want to prove

$$\big(\xi_1^\sharp \mid \xi_2^\sharp\big) = \big(U(-i)\xi_2 \mid \xi_1\big), \qquad \xi_1, \xi_2 \in \mathfrak{B}_0.$$

To this end, we simply compute with x, y, u and v as above:

$$\big(U(-i)\eta(xy^*) \mid \eta(uv^*)\big) = \left(U\Big(-\frac{i}{2}\Big)\eta(xy^*)\,\Big|\, U\Big(-\frac{i}{2}\Big)^*\eta(uv^*)\right)$$
$$= \left(\eta\big(\sigma_{-\frac{i}{2}}(xy^*)\big)\,\Big|\,\eta\big(\sigma_{-\frac{i}{2}}(uv^*)\big)\right)$$
$$= \left(\eta\big(\rho_{-\frac{i}{2}}(x)\rho_{\frac{i}{2}}(y)^*\big)\,\Big|\,\eta\big(\rho_{-\frac{i}{2}}(u)\rho_{\frac{i}{2}}(v)^*\big)\right) \qquad (\text{by } \sigma_\alpha(xy^*) = \rho_\alpha(x)\rho_{\overline{\alpha}}(y)^*)$$
$$= \left(\rho_{-\frac{i}{2}}(x)J\eta_\varphi(y)\,\Big|\,\rho_{-\frac{i}{2}}(u)J\eta_\varphi(v)\right)$$
$$= \left(\rho_{-\frac{i}{2}}(u)^*\rho_{-\frac{i}{2}}(x)J\eta_\varphi(y)\,\Big|\,J\eta_\varphi(v)\right)$$
$$= \left(\sigma^\varphi_{\frac{i}{2}}\big(u^*\rho_{-i}(x)\big)J\eta_\varphi(y)\,\Big|\,J\eta_\varphi(v)\right) \qquad (\text{by } \sigma^\varphi_\alpha(x^*y) = \rho_{\overline{\alpha}}(x)^*\rho_\alpha(y))$$
$$= \left(J\sigma^\varphi_{\frac{i}{2}}\big(\rho_{-i}(x)^*u\big)J\eta_\varphi(v)\,\Big|\,\eta_\varphi(y)\right) = \big(\eta_\varphi(vu^*)\rho_{-i}(x) \mid \eta_\varphi(y)\big)$$
$$= \big(J\eta_\varphi(y) \mid J\eta_\varphi(vu^*)\rho_{-i}(x)\big) = \left(J\eta_\varphi(y)\,\Big|\,JvJS_\varphi\Delta_\varphi^{-\frac{1}{2}}\eta_\varphi\big(u^*\rho_{-i}(x)\big)\right)$$
$$= \left(J\eta_\varphi(y)\,\Big|\,JvJ\eta_\varphi\Big(\sigma^\varphi_{\frac{i}{2}}\big(u^*\rho_{-i}(x)\big)^*\Big)\right)$$
$$= \left(J\eta_\varphi(y)\,\Big|\,JvJ\eta_\varphi\Big(\sigma^\varphi_{-\frac{i}{2}}\big(\rho_{-i}(x)^*u\big)\Big)\right)$$
$$= \left(J\eta_\varphi(y)\,\Big|\,JvJ\eta_\varphi\big(\rho_{-\frac{i}{2}}(x)^*\rho_{-\frac{i}{2}}(u)\big)\right) \qquad (\text{by } \sigma^\varphi_\alpha(x^*y) = \rho_{\overline{\alpha}}(x)^*\rho_\alpha(y))$$
$$= \left(J\eta_\varphi(y)\,\Big|\,JvJ\rho_{-\frac{i}{2}}(x)^*\eta_\varphi\big(\rho_{-\frac{i}{2}}(u)\big)\right)$$
$$= \left(\rho_{-\frac{i}{2}}(x)J\eta_\varphi(y)\,\Big|\,JvJ\eta_\varphi\big(\rho_{-\frac{i}{2}}(u)\big)\right)$$

$$= \left(vJ\eta_\varphi\left(\rho_{-\frac{i}{2}}(u)\right) \;\middle|\; J\rho_{-\frac{i}{2}}(x)J\eta_\varphi(y)\right)$$

$$= \left(\eta(vu^*) \;\middle|\; yJ\eta_\varphi\left(\rho_{-\frac{i}{2}}(x)\right)\right) = \left(\eta(vu^*) \mid \eta(yx^*)\right),$$

where the last part follows from the general fact:

$$JxJ\eta_\varphi(y) = yJ\eta_\varphi(x), \quad x, y \in \mathfrak{n}_\varphi.$$

Therefore, the modular condition follows by linearity.

Now, it is easy to check the rest of the postulates for a Tomita algebra. Q.E.D.

COMPLETION OF THE PROOF OF THEOREM 3.8: With the notations established above, let ψ be the weight on $\mathcal{M} = \mathcal{R}_\ell(\mathfrak{B}_0)$ given by the fulfillment $\mathfrak{B}_0''$ of the Tomita algebra $\mathfrak{B}_0$. We complete the proof by showing that the identity $(D\psi : D\varphi)_t = u_t$, $t \in \mathbf{R}$, holds. By construction, we know that $\sigma_t^\psi = \sigma_t = \operatorname{Ad}(u_t) \circ \sigma_t^\varphi$, $t \in \mathbf{R}$.

We first observe that $\mathfrak{n}_0^* \subset \mathfrak{n}_\psi$ and

$$\eta_\psi(y^*) = J\eta_\varphi\left(\rho_{-\frac{i}{2}}(y)\right), \quad y \in \mathfrak{n}_0.$$

This can be seen by taking an approximate identity $\{x_j\} \subset \mathfrak{n}_0$ and

$$\lim_j \eta(x_j y^*) = \lim_j x_j J\eta_\varphi\left(\rho_{-\frac{i}{2}}(y)\right)$$

Hence we get

$$J\eta_\psi(y^*) = \eta_\varphi\left(\rho_{-\frac{i}{2}}(y)\right), \quad y \in \mathfrak{n}_0,$$

so that

$$\Delta_{\psi,\varphi}^{\frac{1}{2}}\eta_\varphi(y) = \eta_\varphi\left(\rho_{-\frac{i}{2}}(y)\right), \quad y \in \mathfrak{n}_0. \tag{44}$$

Consider the one parameter unitary group $V(t) = u_t\Delta_\varphi^{it}$, $t \in \mathbf{R}$. It follows that $\eta_\varphi(\mathfrak{n}_0)$ is a dense invariant subspace of entire vectors for $\{V(t)\}$. Writing $\mathfrak{B}_\varphi = \eta_\varphi(\mathfrak{n}_\varphi)$, we can characterize $\eta_\varphi(\mathfrak{n}_0)$ as follows:

$$\eta_\varphi(\mathfrak{n}_0) = \left\{\xi \in \bigcap_{\alpha \in \mathbf{C}}\left(\mathfrak{B}_\varphi \cap \mathfrak{D}\left(V(\alpha)\right)\right) : V(\alpha)\xi \in \mathfrak{B}_\varphi,\ \alpha \in \mathbf{C}\right\}.$$

By (44), it suffices to prove that $V(-\frac{i}{2})$ is essentially self-adjoint on $\eta_\varphi(\mathfrak{n}_0)$. We equip $\mathfrak{B}_\varphi$ with the norm:

$$|||\xi||| = \max\left\{\|\xi\|, \left\|\pi_\ell(\xi)\right\|\right\}.$$

Then $\mathfrak{B}_\varphi$ is a Banach space and $\{V(t)\}$ acts on $\mathfrak{B}_\varphi$ as a one parameter group of isometries. If $f \in L^1(\mathbf{R})$, then

$$V(f) = \int_{\mathbf{R}} f(s)V(s)\,ds$$

maps $\mathfrak{B}_\varphi$ into itself. Thus, the essential self-adjointness of $V\left(-\frac{i}{2}\right)$ on $\eta_\varphi(\mathfrak{n}_0)$ follows from the next general lemma. Q.E.D.

Lemma 3.13. *Let $\{V(t)\}$ be a one parameter unitary group on a Hilbert space $\mathfrak{H}$. Suppose that $\mathfrak{D}$ is a dense subspace of $\mathfrak{H}$ such that $V(t)\mathfrak{D} = \mathfrak{D}$ and $\mathfrak{D}$ is invariant under the operators:*

$$V(f) = \int_{\mathbf{R}} f(s)V(s)\,\mathrm{d}s, \qquad f \in L^1(\mathbf{R}).$$

Let $\mathfrak{D}_0$ be the set of all those $\xi \in \mathfrak{D}$ such that the function $t \in \mathbf{R} \mapsto V(t)\xi \in \mathfrak{D}$ is extended to a $\mathfrak{D}$-valued entire function $\alpha \in \mathbf{C} \mapsto V(\alpha)\xi \in \mathfrak{D}$. Then $\mathfrak{D}_0$ is a core of each closed operator $V(\alpha)$, $\alpha \in \mathbf{C}$, where

$$V(\alpha) = \exp(\mathrm{i}\alpha H)$$

with H the infinitesimal generator of $\{V(t)\}$.

PROOF: If the Fourier transform $\hat{f}$ of an $f \in \mathcal{S}(\mathbf{R})$, the Schwartz space of rapidly decreasing functions, is supported by a compact set in $\mathbf{R}$, then $V(f)$ maps $\mathfrak{D}$ in $\mathfrak{D}_0$ and

$$V(\alpha)V(f)\xi = V(f_\alpha)\xi, \quad \xi \in \mathfrak{H},$$

where f_α is given by the following:

$$f_\alpha(t) = f(t - \alpha)$$

which makes sense because f is entire. Hence we have

$$\mathfrak{D}_0 \supset \bigcup \{ V(f)\mathfrak{D} : f \in \mathcal{S}(\mathbf{R}),\ \hat{f} \in C_c^\infty(\mathbf{R}) \}$$

Now, fix a vector $\xi \in \mathfrak{D}(V(\alpha))$ and $\varepsilon > 0$. Choose an $f \in \mathcal{S}(\mathbf{R}) \cap \widehat{C_c^\infty}(\mathbf{R})$ such that

$$\|\xi - V(f)\xi\| < \varepsilon \qquad \text{and} \qquad \|V(\alpha)\xi - V(\alpha)V(f)\xi\| < \varepsilon,$$

which is possible because

$$V(\alpha)\xi = \lim V(f_n)V(\alpha)\xi = \lim V(\alpha)V(f_n)\xi,$$

when $\hat{f}_n \nearrow 1$ on $\mathbf{R}$ pointwise. Choose an $\eta \in \mathfrak{D}$ such that

$$\|\xi - \eta\| < \frac{\varepsilon}{\|V(f)\|} \qquad \text{and} \qquad \|V(\alpha)V(f)\xi - V(\alpha)V(f)\eta\| < \varepsilon.$$

This is possible because $V(\alpha)V(f)$ is bounded and $\mathfrak{D}$ is dense in $\mathfrak{H}$. Then we have

$$\|V(\alpha)\xi - V(\alpha)V(f)\eta\| < 2\varepsilon;$$

$$\|\xi - V(f)\eta\| \le \|\xi - V(f)\xi\| + \|V(f)(\xi - \eta)\| < \varepsilon + \varepsilon = 2\varepsilon.$$

Thus $\mathfrak{D}_0$ is a core of each closed operator $V(\alpha)$, $\alpha \in \mathbf{C}$. Q.E.D.

We now show a complete characterization of the semi-finiteness in terms of modular automorphisms groups.

Theorem 3.14. *For a von Neumann algebra $\mathcal{M}$, the following three conditions are equivalent:*

(i) *$\mathcal{M}$ is semi-finite;*
(ii) *The modular automorphism group $\{\sigma_t^\varphi\}$ of every faithful semi-finite normal weight φ on $\mathcal{M}$ is inner in the sense that there exists a continuous one parameter unitary group $\{u(t)\}$ in $\mathcal{M}$ such that $\sigma_t^\varphi = \mathrm{Ad}\big(u(t)\big)$, $t \in \mathbf{R}$;*
(iii) *There exists a faithful semi-finite normal weight φ on $\mathcal{M}$ such that $\{\sigma_t^\varphi\}$ is inner in the above sense.*

If this is the case, then with a fixed faithful semi-finite normal trace τ on $\mathcal{M}$, every semi-finite normal weight φ on $\mathcal{M}$ is written uniquely in the form:

$$\varphi = \tau_h. \tag{45}$$

with h a positive self-adjoint operator affiliated with $\mathcal{M}$.

PROOF:
(i) $\Longrightarrow$ (iii): Let τ be a faithful semi-finite normal trace on $\mathcal{M}$. Then $\sigma_t^\tau = \mathrm{id}$, $t \in \mathbf{R}$. Therefore, every weight φ is σ^τ-invariant. Hence $\varphi = \tau_h$ for some h by Corollary 3.6. If φ is faithful, then h is non-singular and $\sigma_t^\varphi = \mathrm{Ad}(h^{\mathrm{i}t})$ by Theorem 2.11. Thus (ii) follows.

(ii) $\Longrightarrow$ (iii): Trivial.

(iii) $\Longrightarrow$ (i): Suppose that $\sigma_t^\varphi = \mathrm{Ad}\big(u(t)\big)$, $t \in \mathbf{R}$. Since $\varphi = \varphi \circ \sigma_t^\varphi$, we have, $\mathfrak{n}_\varphi$ being a left ideal,

$$\mathfrak{n}_\varphi u(t)^* = u(t)\mathfrak{n}_\varphi u(t)^* = \sigma_t^\varphi(\mathfrak{n}_\varphi) = \mathfrak{n}_\varphi,$$

so that $u(t)$ is a multiplier of $\mathfrak{m}_\varphi$. Hence Theorem 2.6 yields that $u(t)$ belongs to $\mathcal{M}_\varphi$. Stone's Theorem then applies to conclude that there exists a positive non-singular self-adjoint operator h affiliated with $\mathcal{M}_\varphi$ such that $u(t) = h^{-\mathrm{i}t}$, $t \in \mathbf{R}$. We then put $\tau = \varphi_h$, so that

$$\sigma_t^\tau = \mathrm{Ad}(h^{\mathrm{i}t}) \circ \sigma_t^\varphi = \mathrm{Ad}(h^{\mathrm{i}t})\,\mathrm{Ad}(h^{-\mathrm{i}t}) = \mathrm{id},$$

which means precisely that τ is a trace on $\mathcal{M}$. Q.E.D.

Since weights are not continuous in general for the operator topologies, it is a delicate question how one identifies two weights from their coincidence on a dense subalgebra. The following result provides a useful sufficient condition.

Proposition 3.15. *Let φ be a faithful weight on $\mathcal{M}$. If ψ is a weight on $\mathcal{M}$ commuting with φ, i.e. $\psi = \psi \circ \sigma_t^\varphi$, $t \in \mathbf{R}$, and if there exists a σ-weakly dense $*$-subalgebra $\mathfrak{m}_0$ of $\mathfrak{m}_\varphi$ such that $\sigma_t^\varphi(\mathfrak{m}_0) = \mathfrak{m}_0$, $t \in \mathbf{R}$, and $\varphi(x) = \psi(x)$ for every $x \in \mathfrak{m}_0$, then $\varphi = \psi$.*

PROOF: By Corollary 3.6, ψ is of the form $\psi = \varphi_h$ with h a positive self-adjoint operator affiliated with $\mathcal{M}_\varphi$.

If a and b are in $\mathfrak{m}_0$, then the two functionals $x \in \mathcal{M} \to \varphi(axb)$ and $\psi(axb)$ are bounded and coincide on $\mathfrak{m}_0$, hence $\varphi(axb) = \psi(axb)$ for every $x \in \mathcal{M}$. Let $\{a_j\}$ be an increasing net in $\mathfrak{m}_0^+$ converging σ-strongly to 1. Put

$$e_j = \frac{1}{\sqrt{\pi}} \int_{-\infty}^{\infty} e^{-t^2} \sigma_t^\varphi(a_j)\, dt$$

Since $\mathfrak{m}_0$ is invariant under $\{\sigma_t^\varphi\}$ and the normality of φ and ψ allows us to change the order of integration and the evaluation of φ and ψ, we have

$$\varphi(e_j x e_j) = \psi(e_j x e_j), \quad x \in \mathcal{M}$$

Each e_j is now entire with respect to $\{\sigma_t^\varphi\}$ and

$$\sigma_\alpha^\varphi(e_j) = \frac{1}{\sqrt{\pi}} \int_{-\infty}^{\infty} e^{-(t-\alpha)^2} \sigma_t^\varphi(a_j)\, dt, \quad \alpha \in \mathbf{C}$$

For each $\xi \in \mathfrak{H}_\varphi$, the function $f_j(t) = \left\| \sigma_t^\varphi(1-a_i)\xi \right\|$ is continuous, and the net $\{f_j\}$ decreases to zero pointwise. Dini's Theorem then implies the uniform convergence of $\{f_j\}$ on compact sets. It follows therefore that

$$\begin{aligned}
\left\|(1 - \sigma_\alpha^\varphi(e_j))\xi\right\| &\le \frac{1}{\sqrt{\pi}} \int \left|e^{-(t-\alpha)^2}\right| \left\|\sigma_t^\varphi(1 - a_j)\xi\right\| dt \\
&= \frac{e^{(\operatorname{Im}\alpha)^2}}{\sqrt{\pi}} \int e^{-(t-\operatorname{Re}\alpha)^2} \left\|\sigma_t^\varphi(1 - a_j)\xi\right\| dt \searrow 0,
\end{aligned}$$

so that $\{\sigma_\alpha^\varphi(e_j)\}$, $\alpha \in \mathbf{C}$, converges strongly to 1 and is bounded. In particular, $e_j \nearrow 1$.

By the σ-weak lower semi-continuity of ψ and the analyticity of e_j, we have for each $x \in \mathfrak{m}_\varphi^+$

$$\begin{aligned}
\psi(x) &\le \lim \psi(e_j x e_j) = \lim \varphi(e_j x e_j) = \lim \left\| \eta_\varphi\left(x^{\frac{1}{2}} e_j\right) \right\|^2 \\
&= \lim \left\| S e_j S \eta_\varphi\left(x^{\frac{1}{2}}\right) \right\|^2 = \lim \left\| J \Delta_\varphi^{\frac{1}{2}} e_j \Delta_\varphi^{-\frac{1}{2}} J \eta_\varphi\left(x^{\frac{1}{2}}\right) \right\|^2 \\
&= \lim \left\| \sigma_{-\frac{i}{2}}^\varphi(e_j) J \eta_\varphi\left(x^{\frac{1}{2}}\right) \right\|^2 = \left\| J \eta_\varphi\left(x^{\frac{1}{2}}\right) \right\|^2 = \varphi(x).
\end{aligned}$$

Thus, we get $\psi \le \varphi$.

Let $q = 1 - s(\psi)$. By the general inequality:

$$axa \le 2\left[(1-a)x(1-a) + x\right], \quad x \in \mathcal{M}_+,$$

we have for each $x \in \mathfrak{m}_\varphi^+$ with $x \leq q$,

$$\begin{aligned}\varphi(x) &\leq \lim \varphi(e_j x e_j) = \lim \psi(e_j x e_j) \leq 2 \lim \psi((1-e_j)x(1-e_j)) \\ &\leq 2\lim \varphi((1-e_j)x(1-e_j)) = 2 \lim \left\| \eta_\varphi(x^{\frac{1}{2}}(1-e_j)) \right\|^2 \\ &= 2\lim \left\| S(1-e_j)S\eta_\varphi(x^{\frac{1}{2}}) \right\|^2 = 2 \lim \left\| (1-\sigma^\varphi_{-\frac{i}{2}}(e_j))\eta_\varphi(x^{\frac{1}{2}}) \right\|^2 = 0,\end{aligned}$$

so that $x = 0$. Since $q \in \mathcal{M}_\varphi$, $q\mathfrak{m}_\varphi^+ q \subset \mathfrak{m}_\varphi^+$, and so $q\mathfrak{m}_\varphi^+ q = \{0\}$. Hence $q = 0$. Thus ψ is faithful.

Now, we can change the role of φ and ψ, because φ is invariant under $\{\sigma_t^\psi\} = \{\mathrm{Ad}(h^{it})\sigma_t^\varphi\}$ to conclude $\varphi \leq \psi$ from the previous arguments. Hence $\varphi = \psi$.

Q.E.D.

In the above result, if we know that ψ is faithful and satisfies the modular condition for $\{\sigma_t^\varphi\}$, then we can drop the assumption of the invariance of $\mathfrak{m}_0$ under $\{\sigma_t^\varphi\}$ to conclude that $\varphi = \psi$, and the proof is simpler. For a later use, we state this remark:

Proposition 3.16. *Let φ and ψ be two faithful weights on $\mathcal{M}$ with the same modular automorphism group $\{\sigma_t\}$. If there exists a σ-weakly dense *-subalgebra $\mathfrak{m}_0$ of $\mathfrak{m}_\varphi$ on which φ and ψ agree, then $\varphi = \psi$.*

PROOF: In the proof of the previous proposition, we note that h is affiliated with the center $\mathcal{Z}$ of $\mathcal{M}$.

If $h \neq 1$, then there exist $\varepsilon > 0$ and $e \in \mathrm{Proj}(\mathcal{Z})$, $e \neq 0$, such that either $(1+\varepsilon)e \leq he$ or $(1-\varepsilon)e \geq he$. Suppose $(1+\varepsilon)e \leq he$. If $a \in \mathfrak{m}_0$, then we have, as before,

$$\varphi(a^* x a) = \psi(a^* x a), \qquad x \in \mathcal{M}$$

Putting $x = e$, we have

$$(1+\varepsilon)\varphi(a^* e a) \leq \varphi(a^* h e a) = \psi(a^* e a) = \varphi(a^* e a)$$

This is impossible. Similarly $(1-\varepsilon)e \geq he$ is impossible. Hence we must have $h = 1$.

Q.E.D.

Theorem 3.17. *For a pair φ, ψ of faithful semi-finite normal weights on $\mathcal{M}$, the following conditions are equivalent:*

(i) *There exists $M > 0$ such that*

$$\psi(x) \leq M\varphi(x), \qquad x \in \mathcal{M}_+; \tag{46}$$

(ii) *The cocycle derivative $(\mathrm{D}\psi : \mathrm{D}\varphi)_t = u_t$ can be extended to a member of $\mathcal{A}_\mathcal{M}(\mathbf{D}_{1/2})$ such that $\|u_{-i/2}\| \leq M^{1/2}$, where $\mathcal{A}_\mathcal{M}(\mathbf{D}_{1/2})$ means the set of all $\mathcal{M}$-valued σ-weakly continuous bounded functions on $\overline{\mathbf{D}}_{1/2}$ which are holomorphic in $\mathbf{D}_{1/2}$, and $\mathbf{D}_{1/2}$ means the horizontal strip bounded by $\mathbf{R}$ and $\mathbf{R} - \frac{i}{2}$.*

If these conditions hold, then

$$\psi(x) = \varphi\left(u^*_{-\frac{i}{2}} x u_{-\frac{i}{2}}\right), \quad x \in \mathfrak{m}_\psi. \tag{47}$$

We prove the theorem in several steps.

Lemma 3.18. *Let $a \in \mathcal{M}$ be fixed.*

(i) *The map: $t \in \mathbf{R} \mapsto \sigma_t^\varphi(a) \in \mathcal{M}$ can be extended to an $\mathcal{M}$-valued function of $\mathcal{A}_\mathcal{M}(\mathbf{D}_{1/2})$ with $\|\sigma^\varphi_{-i/2}(a)\| \le 1$ if and only if*

$$\varphi(axa^*) \le \varphi(x), \quad x \in \mathcal{M}_+. \tag{48}$$

(ii) *If a satisfies the conditions in* (i), *then*

$$\eta_\varphi(xa^*) = J\sigma^\varphi_{-\frac{i}{2}}(a)J\eta_\varphi(x), \quad x \in \mathfrak{n}_\varphi. \tag{49}$$

(iii) *If $a\,a^* \in \mathcal{M}_\varphi$ in addition to the above conditions, then*

$$\varphi(a\,a^*x) = \varphi\left(\sigma^\varphi_{-\frac{i}{2}}(a)^* x \sigma^\varphi_{-\frac{i}{2}}(a)\right), \quad x \in \mathfrak{m}_\varphi. \tag{50}$$

PROOF:

(i) and (ii): Suppose that the map: $t \in \mathbf{R} \mapsto \sigma_t^\varphi(a) \in \mathcal{M}$ is extended to a holomorphic function: $z \in \overline{\mathbf{D}}_{1/2} \mapsto \sigma_z^\varphi(a) \in \mathcal{M}$ such that $\|\sigma^\varphi_{-i/2}(a)\| \le 1$. Then for every $\xi \in \mathfrak{D}^\sharp = \mathfrak{D}(\Delta_\varphi^{1/2})$, $\sigma_t^\varphi(a)\xi$ belongs to $\mathfrak{D}^\sharp$ and $\sigma^\varphi_{-i/2}(a)\Delta_\varphi^{1/2}\xi = \Delta_\varphi^{1/2}a\xi$ for every $\xi \in \mathfrak{D}^\sharp$ by Lemma VI.2.3. Hence for every $x \in \mathfrak{n}_\varphi \cap \mathfrak{n}_\varphi^*$, we have

$$S_\varphi a\eta_\varphi(x^*) = J\sigma^\varphi_{-\frac{i}{2}}(a)J\eta_\varphi(x); \tag{49'}$$

hence for each $x \in \mathfrak{m}_\varphi^+$, we have

$$\varphi(axa^*) = \varphi\left(ax^{\frac{1}{2}}x^{\frac{1}{2}}a^*\right) = \left\|\eta_\varphi\left(x^{\frac{1}{2}}a^*\right)\right\|^2 \le \left\|\eta_\varphi\left(x^{\frac{1}{2}}\right)\right\|^2 = \varphi(x).$$

Thus we get the inequality (48). Furthermore, the last inequality also shows that $\mathfrak{n}_\varphi a^* \subset \mathfrak{n}_\varphi$, and that the operator: $\eta_\varphi(x) \mapsto \eta_\varphi(xa^*)$, $x \in \mathfrak{n}_\varphi \cap \mathfrak{n}_\varphi^*$, is bounded and

$$\eta_\varphi(xa^*) = J\sigma^\varphi_{-\frac{i}{2}}(a)J\eta_\varphi(x), \quad x \in \mathfrak{n}_\varphi.$$

This shows (ii) also.

Suppose that $\varphi(axa^*) \le \varphi(x)$, $x \in \mathcal{M}_+$. Then the map: $\eta_\varphi(x) \mapsto \eta_\varphi(xa^*)$, $x \in \mathfrak{n}_\varphi$, is bounded and its extension by continuity belongs to $\mathcal{M}'$ which we denote by b. Since $\mathfrak{n}_\varphi a^* \subset \mathfrak{n}_\varphi$, we have $(\mathfrak{n}_\varphi \cap \mathfrak{n}_\varphi^*)a \subset \mathfrak{n}_\varphi \cap \mathfrak{n}_\varphi^*$ and $J\Delta_\varphi^{1/2}\eta_\varphi(xa^*) = a\eta_\varphi(x)$ for every $x \in \mathfrak{n}_\varphi \cap \mathfrak{n}_\varphi^*$, so that

$$Jb\eta_\varphi(x) = J\eta_\varphi(xa^*) = \Delta_\varphi^{\frac{1}{2}}aJ\Delta_\varphi^{\frac{1}{2}}\eta_\varphi(x) = \Delta_\varphi^{\frac{1}{2}}a\Delta_\varphi^{-\frac{1}{2}}J\eta_\varphi(x), \quad x \in \mathfrak{n}_\varphi \cap \mathfrak{n}_\varphi^*.$$

Thus, we get, for each $\xi \in \mathfrak{D}^\sharp$, and $\eta \in \mathfrak{D}^\sharp$,

$$\left|\left(a\Delta_\varphi^{-\frac{1}{2}}\xi \mid \Delta_\varphi^{\frac{1}{2}}\eta\right)\right| \leq \|\xi\|\|\eta\|.$$

Hence we have, for any $z \in \overline{\mathbf{D}}_{1/2}$

$$\left|\left(a\Delta_\varphi^{-\mathrm{i}z}\xi \mid \Delta_\varphi^{-\mathrm{i}\bar{z}}\eta\right)\right| \leq \|\xi\|\|\eta\| \max(1, \|a\|).$$

Therefore, the function: $t \in \mathbf{R} \mapsto \sigma_t^\varphi(a) = \Delta_\varphi^{\mathrm{i}t}\, a\, \Delta_\varphi^{-\mathrm{i}t}$ can be extended to a homomorphic function: $z \in \overline{\mathbf{D}}_{1/2} \mapsto \sigma_z^\varphi(a) \in \mathcal{M}$.

(iii) Let $b = \sigma_{-\mathrm{i}/2}^\varphi(a) \in \mathcal{M}$. Then the function: $t \in \mathbf{R} \mapsto \sigma_t^\varphi(b^*)$ can be extended holomorphically to $\overline{\mathbf{D}}_{1/2}$ and $\sigma_{-\mathrm{i}/2}^\varphi(b^*) = a^*$. By (ii), we have

$$\eta_\varphi(xb) = Ja^*J\eta_\varphi(x), \quad x \in \mathfrak{n}_\varphi;$$

$$\begin{aligned}\|\eta_\varphi(xb)\|^2 &= \left(a^*J\eta_\varphi(x) \mid a^*J\eta_\varphi(x)\right) = \left(Jaa^*J\eta_\varphi(x) \mid \eta_\varphi(x)\right) \\ &= \left(\eta_\varphi(xaa^*) \mid \eta_\varphi(x)\right) \\ &= \varphi\left(x^*xaa^*\right) = \varphi\left(aa^*x^*x\right)\end{aligned}$$

by Theorem 2.6. Therefore, the assertion (iii) follows by the polarization identity.

Q.E.D.

Proof of Theorem 3.17: Replacing φ by $M^{-1}\varphi$, we may assume $M = 1$. On $\mathcal{N} = \mathcal{M} \otimes M_2(\mathbf{C})$, we consider the balanced weight $\rho = \psi \oplus \varphi$, i.e.

$$\rho\left(\sum_{i,j=1}^{2} x_{ij} \otimes e_{ij}\right) = \psi(x_{11}) + \varphi(x_{22}).$$

Let $a = 1 \otimes e_{12}$. Then we have

$$\sigma_t^\rho(a) = \left(\mathrm{D}\psi : \mathrm{D}\varphi\right)_t \otimes e_{12};$$

$$\rho\left(a\left(\sum x_{ij} \otimes e_{ij}\right)a^*\right) = \rho\left(x_{11} \otimes e_{11}\right) = \psi(x_{11}).$$

(i) $\Longrightarrow$ (ii): Lemma 3.18.(i) implies that the function: $t \in \mathbf{R} \mapsto u_t \in \mathcal{M}$ can be extended holomorphically to $\overline{\mathbf{D}}_{1/2}$ and $\|u_{-\mathrm{i}/2}\| \leq 1$. Since $aa^* = 1 \otimes e_{11} \in \mathcal{N}_\rho$, Lemma 3.18.(iii) applies to conclude that

$$\begin{aligned}\psi(x) &= \rho\left((1 \otimes e_{11})(x \otimes e_{11})(1 \otimes e_{11})\right) \\ &= \rho\left(\left(u_{-\frac{\mathrm{i}}{2}} \otimes e_{12}\right)^*(x \otimes e_{11})\left(u_{-\frac{\mathrm{i}}{2}} \otimes e_{12}\right)\right) \\ &= \varphi\left(u_{-\frac{\mathrm{i}}{2}}^* x u_{-\frac{\mathrm{i}}{2}}\right), \quad x \in m_\psi^+.\end{aligned}$$

(ii) $\Longrightarrow$ (i) : This implication also follows from Lemma 3.18.(iii). Q.E.D.

So far, we have been discussing only faithful weights and states. We now extend the theory to non-faithful ones. To this end, we first note that every normal weight φ is faithful and semi-finite on the reduced algebra $\mathcal{M}_{s(\varphi)}$ by the support $s(\varphi)$ of φ. Thus, the *modular automorphism group* $\{\sigma_t^\varphi\}$ of a normal weight φ always means that of φ on $\mathcal{M}_{s(\varphi)}$. The *centralizer* $\mathcal{M}_\varphi$ of φ means also the one in $\mathcal{M}_{s(\varphi)}$ for φ.

Theorem 3.19. *Let φ be a faithful semi-finite normal weight on $\mathcal{M}$. For any semi-finite normal weight ψ on $\mathcal{M}$, there exists uniquely a one parameter family $\{u_t\}$ of partial isometries such that*

(i) $t \in \mathbf{R} \mapsto u_t \in \mathcal{M}$ *is σ-strong $*$ continuous;*

(ii) $u_{s+t} = u_s \sigma_s^\varphi(u_t), \quad s, t \in \mathbf{R};$ (51)

(iii) $u_s u_s^* = s(\psi), \quad u_s^* u_s = \sigma_s^\varphi(s(\psi));$ (52)

(iv) $u_s \sigma_s^\varphi(\mathfrak{n}_\psi^* \cap \mathfrak{n}_\varphi) \subset \mathfrak{n}_\psi^* \cap \mathfrak{n}_\varphi;$ (53)

(v) *For each $x \in \mathfrak{n}_\psi \cap \mathfrak{n}_\varphi^*$ and $y \in \mathfrak{n}_\varphi \cap \mathfrak{n}_\psi^*$, there exists an $F \in \mathcal{A}(\mathbf{D})$ such that*

$$F(t) = \psi\big(u_t \sigma_t^\varphi(y) x\big), \qquad F(t + \mathrm{i}) = \varphi\big(x u_t \sigma_t^\varphi(y)\big); \tag{54}$$

(vi) $\sigma_t^\psi(x) = u_t \sigma_t^\varphi(x) u_t^*, \quad x \in \mathcal{M}_{s(\psi)}, t \in \mathbf{R}.$

Furthermore, the above property (v) *of $\{u_t\}$ determines the cocycle uniquely.*

PROOF: Let ψ' be a semi-finite normal weight with $s(\psi') = 1 - s(\psi)$, and set $\bar{\psi} = \psi + \psi'$. Let $e = s(\psi)$. Then $\bar{\psi}$ is a faithful semi-finite normal weight which commutes with e. Set $u_t = e(\mathrm{D}\bar{\psi} : \mathrm{D}\varphi)_t$. As $\bar{\psi}(ex) = \bar{\psi}(xe) = \psi(x)$ for $x \in \mathcal{M}_+$, the rest of our assertions follow from Theorem 3.3, except possibly the uniqueness of $\{u_t\}$. But the condition (iv) and Lemma 3.5.(i) applied to $\bar{\psi}$ and φ guarantees the uniqueness of $\{u_t\}$. Q.E.D.

Definition 3.20. The one parameter family $\{u_t\}$ of the above theorem will be called the *cocycle derivative* of ψ relative to φ and denoted by $(\mathrm{D}\psi : \mathrm{D}\varphi)_t, \ t \in \mathbf{R}$.

Theorem 3.21. *Let φ be a faithful semi-finite normal weight on $\mathcal{M}$. If $\{u_s\}$ is a σ-strong* continuous one parameter family of partial isometries such that*

$$\left.\begin{aligned} &u_{s+t} = u_s \sigma_s^\varphi(u_t); \\ &u_s u_s^* = e \in \operatorname{Proj}(\mathcal{M}), \quad u_s^* u_s = \sigma_s^\varphi(e), \quad s \in \mathbf{R}, \end{aligned}\right\} \tag{55}$$

then there exists a semi-finite normal weight ψ such that $s(\psi) = e$ and

$$\big(\mathrm{D}\psi : \mathrm{D}\varphi\big)_t = u_t .$$

PROOF: Choose a semi-finite normal weight ψ' such that $s(\psi') = 1 - e$, and set $v_t = (\mathrm{D}\psi' : \mathrm{D}\varphi)_t, \ t \in \mathbf{R}$. We then have $v_t v_t^* = (1 - e)$ and $v_t^* v_t = \sigma_t^\varphi(1 - e)$. Set $w_t = u_t + v_t, \ t \in \mathbf{R}$. Then $\{w_t\}$ is a unitary one cocycle relative to σ^φ, so that there exists a faithful semi-finite normal weight ψ'' such that $(\mathrm{D}\psi'' : \mathrm{D}\varphi)_t = w_t$.

As $\sigma_s^{\psi''}(e) = \mathrm{Ad}(w_t) \circ \sigma_t^{\varphi}(e) = e$, we have $e \in \mathcal{M}_{\psi''}$. Setting $\psi(x) = \psi''(exe)$, $x \in \mathcal{M}_+$, we obtain a semi-finite normal weight ψ on $\mathcal{M}$ with

$$(\mathrm{D}\psi : \mathrm{D}\varphi)_t = (\mathrm{D}\psi_e'' : \mathrm{D}\psi)_t = (\mathrm{D}\varphi'' : \mathrm{D}\varphi)_t = ew_t = u_t, \quad t \in \mathbf{R}. \qquad \text{Q.E.D.}$$

As we have observed at several occasions, the analytic continuation $\{\sigma_\alpha^\varphi(x)\}$ of the orbit $\{\sigma_t^\varphi(x)\}$ provides useful informations about the relationship of $x \in \mathcal{M}$ and $\varphi \in \mathfrak{W}(\mathcal{M})$, where $\mathfrak{W}(\mathcal{M})$ means the set of all semi-finite normal weights. We investigate this fact further. By restricting our attention to $\mathcal{M}_{s(\varphi)}$ for a non-faithful $\varphi \in \mathfrak{W}(\mathcal{M})$, we may assume φ to be faithful. Thus $\{\sigma_t^\varphi\}$ is indeed a one parameter automorphism group of $\mathcal{M}$. As we want to apply our theory to the relative modular group $\{\sigma_t^{\varphi,\psi}\}$, we consider a one parameter group $\{\sigma_t\}$ of isometries on $\mathcal{M}$ which is pointwise σ-weakly continuous in the sense that $t \in \mathbf{R} \mapsto \sigma_t(x) \in \mathcal{M}$ is σ-weakly continuous for each $x \in \mathcal{M}$. For each $\gamma = \alpha + \mathrm{i}\beta \in \mathbf{C}$, let $\mathcal{D}(\gamma)$ be the set of all those $x \in \mathcal{M}$ such that the function: $t \in \mathbf{R} \mapsto \sigma_t(x) \in \mathcal{M}$ admits an analytic extension to a member of $\mathcal{A}_{\mathcal{M}}(\mathcal{D}_\beta)$, where $\mathcal{A}_{\mathcal{M}}(\mathcal{D}_\beta)$ means the space of all $\mathcal{M}$-valued bounded functions which are continuous on and holomorphic in the horizontal domain $\mathbf{D}_\beta$ bounded by $\mathbf{R}$ and $\mathbf{R} - \mathrm{i}\beta$, the space similar to the one defined in Theorem 3.17 for $\beta = \frac{1}{2}$. We then define $\sigma_\gamma(x)$ naturally for $x \in \mathcal{D}(\sigma_\gamma)$. Among $\{\sigma_\gamma\}_{\gamma \in \mathbf{C}}$, we are specially interested in $\sigma_{-\mathrm{i}}$. We call $\sigma_{-\mathrm{i}}$ the *multiplicative* (sometimes *logarithmic* or *analytic*) *generator* of the one parameter group $\{\sigma_t\}$. In the case when $\{\sigma_t\}$ is norm continuous, i.e. $\lim_{t\to 0}\|\sigma_t - \mathrm{id}\| = 0$, $\sigma_{-\mathrm{i}}$ is precisely $e^{-\mathrm{i}\delta}$, where $\delta = \lim_{t\to 0}(\sigma_t - \mathrm{id})/t$; hence $\sigma_{-\mathrm{i}} = \mathrm{i}\log(\delta)$. The set $\mathcal{M}_a^\sigma = \bigcap_{n\in\mathbf{Z}} \mathcal{D}(\sigma_{-\mathrm{i}n})$ is then the set of all those $x \in \mathcal{M}$ such that the function: $t \in \mathbf{R} \mapsto \sigma_t(x) \in \mathcal{M}$ admits an extension to the entire $\mathbf{C}$ which is entire and bounded on every horizontal line $\mathbf{R} + \mathrm{i}\beta$. We have discussed this already in §2 for $\{\sigma_\alpha^\varphi\}$. We now introduce a more restricted class $\mathcal{M}_{\exp}^\sigma$ as the set of all these $x \in \mathcal{M}_a^\sigma$ such that for some constant $a > 0$

$$\sup_{\alpha\in\mathbf{C}} \|\sigma_\alpha(x)\| \mathrm{e}^{-a|\mathrm{Im}\,\alpha|} < +\infty. \tag{56}$$

We call each element x of $\mathcal{M}_{\exp}^\sigma$ of *exponential type* or more specifically exponential type $a > 0$. This means by the Paley-Wiener theorem that

$$\int_{\mathbf{R}}\left(\int_{\mathbf{R}} \mathrm{e}^{-\mathrm{i}st} f(t)\,\mathrm{d}t\right)\sigma_s(x)\,\mathrm{d}s = 0 \tag{56'}$$

for every $f \in C_c^\infty(\mathbf{R})$ with $\mathrm{supp}\, f \cap [-a, a] = \emptyset$. Equivalently, the Fourier transform of the $\mathcal{M}$-valued function: $t \in \mathbf{R} \mapsto \sigma_t(x) \in \mathcal{M}$ as a tempered distribution is supported by the closed interval $[-a, a]$.

Lemma 3.22. *In the above context, $\mathcal{M}_{\exp}^\sigma$ is a σ-weakly dense subspace of $\mathcal{M}$.*

PROOF: It is obvious that $\mathcal{M}_{\exp}^\sigma$ is a linear subspace of $\mathcal{M}$. Suppose that $\omega \in \mathcal{M}_*$ vanishes identically on $\mathcal{M}_{\exp}^\sigma$. We will prove $\omega = 0$. Let $x \in \mathcal{M}$ and $f \in C_c^\infty(\mathbf{R})$.

Set

$$\sigma_{\hat{f}}(x) = \int_{\mathbf{R}} \Big(\int_{\mathbf{R}} \mathrm{e}^{-\mathrm{i}st} f(t)\, \mathrm{d}t \Big) \sigma_s(x)\, \mathrm{d}s;$$

$$\hat{f}(s) = \int_{\mathbf{R}} \mathrm{e}^{-\mathrm{i}st} f(t)\, \mathrm{d}t \ .$$

As $\hat{f}$ belongs to the Schwartz space $\mathcal{S}(\mathbf{R})$ of rapidly decreasing functions, the above integral for $\sigma_{\hat{f}}(x)$ converges. As $\sigma_{\hat{f}}(x) \in \mathcal{M}^{\sigma}_{\mathrm{exp}}$, we have $\omega\big(\sigma_{\hat{f}}(x)\big) = 0$ for every $f \in C_c^{\infty}(\mathbf{R})$, i.e.

$$\int_{\mathbf{R}} \hat{f}(s)\omega\big(\sigma_s(x)\big)\, \mathrm{d}s = 0, \quad f \in C_c^{\infty}(\mathbf{R}).$$

As $\big\{ \hat{f} : f \in C_c^{\infty}(\mathbf{R}) \big\}$ is dense in $L^1(\mathbf{R})$, we conclude that $\omega\big(\sigma_s(x)\big) = 0$ for almost every $s \in \mathbf{R}$, which implies $\omega\big(\sigma_s(x)\big) = 0$ for every $s \in \mathbf{R}$ as $s \in \mathbf{R} \mapsto \omega\big(\sigma_s(x)\big) \in \mathbf{C}$ is continuous. Q.E.D.

Lemma 3.23. *If f is an entire function of exponential type and $f(\mathrm{i}n) = 0$, $n \in \mathbf{Z}_+$, then $f = 0$.*

PROOF: By assumption, there exist $M > 0$ and $r \geq 0$ such that $|f(x + \mathrm{i}y)| \leq M\mathrm{e}^{r|y|}$. By the Paley-Wiener Theorem for distributions, see [680, Chapter 7], there exists a distribution u on $\mathbf{R}$ with $\operatorname{supp} u \subset [-r, r]$ such that $f(z) = u(e_{-z})$, where $e_{-z}(x) = \mathrm{e}^{-\mathrm{i}zx}$. The assumption on f means that $f(\mathrm{i}n) = u(e_{-\mathrm{i}n}) = 0$ for $n \in \mathbf{Z}_+$. But $e_{-\mathrm{i}n}(x) = \mathrm{e}^{nx}$, $n \in \mathbf{Z}_+$, span a dense subspace of $C^{\infty}[-r - \varepsilon, r + \varepsilon]$, which means that $u = 0$ as a distribution; consequently $f = 0$. Q.E.D.

Proposition 3.24. *Let $\{\sigma_t^1\}$ and $\{\sigma_t^2\}$ be two pointwise σ-weakly continuous one parameter groups of isometries on a von Neumann algebra $\mathcal{M}$. If $\sigma_{\mathrm{i}}^1 \subset \sigma_{\mathrm{i}}^2$ in the sense that the graph of σ_{i}^2 contains that of σ_{i}^1, then $\sigma^1 = \sigma^2$.*

PROOF: We first claim that $\mathcal{M}^{\sigma^1}_{\mathrm{exp}} \subset \mathcal{M}^{\sigma^2}_{\mathrm{exp}}$. If $x \in \mathcal{M}^{\sigma^1}_{\mathrm{exp}}$, then there exist $M \geq 0$ and $r \geq 0$ such that $\|\sigma^1_{s+\mathrm{i}t}(x)\| \leq M\mathrm{e}^{r|t|}$. If $x \in \mathcal{M}^{\sigma^1}_{\mathrm{exp}}$, then $\sigma^1_{-\mathrm{i}n}(x) \in \mathcal{M}^{\sigma^1}_{\mathrm{exp}} \subset \mathcal{D}(\sigma^1_{\mathrm{i}n}) \subset \mathcal{D}(\sigma^2_{\mathrm{i}n})$, and

$$x = \sigma^1_{\mathrm{i}n}\big(\sigma^1_{-\mathrm{i}n}(x)\big) = \sigma^2_{-\mathrm{i}n}\big(\sigma^1_{-\mathrm{i}n}(x)\big),$$

so that $\sigma^1_{-\mathrm{i}n}(x) = \sigma^2_{-\mathrm{i}n}(x)$, $n \in \mathbf{Z}_+$. Hence we have $\mathcal{M}^{\sigma^1}_{\mathrm{exp}} \subset \bigcap_{n\in\mathbf{Z}} \mathcal{D}\big(\sigma^2_{\mathrm{i}n}\big) = \mathcal{M}^{\sigma^2}_{a}$. By the Phragmen-Lindelöf Theorem, we have, for $t = \lambda n$ with $0 \leq \lambda \leq 1$,

$$\|\sigma^2_{s+\mathrm{i}t}(x)\| \leq M(\mathrm{e}^{rn})^{\lambda} = M\mathrm{e}^{rt},$$

$$\|\sigma^2_{s-\mathrm{i}t}(x)\| \leq M(\mathrm{e}^{rn})^{\lambda} = M\mathrm{e}^{rt},$$

so that $x \in \mathcal{M}^{\sigma^2}_{\mathrm{exp}}$. Hence we proved the claim.

For each $x \in \mathcal{M}^{\sigma^1}_{\exp}$, and $\omega \in \mathcal{M}_*$, let

$$f(z) = \omega\big(\sigma^1_z(x) - \sigma^2_z(x)\big).$$

Then f is an entire function of exponential type and $f(\mathrm{i}n) = 0,\ n \geq 0$, by the assumption. By the last lemma, we have $f(z) = 0$ for every $z \in \mathbf{C}$. Hence $\omega\big(\sigma^1_t(x) - \sigma^2_t(x)\big) = 0$ for every $\omega \in \mathcal{M}_*$, i.e. $\sigma^1_t(x) = \sigma^2_t(x),\ x \in \mathbf{R}$, and $x \in \mathcal{M}^{\sigma^1}_{\exp}$. As $\mathcal{M}^{\sigma^1}_{\exp}$ is dense in $\mathcal{M}$, we conclude $\sigma^1 = \sigma^2$. Q.E.D.

We now apply the above result to the relative modular group $\{\sigma^{\varphi,\psi}_t\}$.

Theorem 3.25. *Let $\mathcal{M}$ be a von Neumann algebra, and let $\varphi, \psi \in \mathfrak{W}_0(\mathcal{M})$ where $\mathfrak{W}_0(\mathcal{M})$ means the set of all faithful semi-finite normal weights. For $a, b \in \mathcal{M}$, the following two conditions are equivalent:*

(i) *The pair (a, b) belongs to the graph $\mathcal{G}\big(\sigma^{\psi,\varphi}_{-\mathrm{i}}\big)$; i.e. $a \in \mathcal{D}\big(\sigma^{\psi,\varphi}_{-\mathrm{i}}\big)$ and $b = \sigma^{\psi,\varphi}_{-\mathrm{i}}(a)$;*

(ii) $a\mathfrak{n}^*_\varphi \subset \mathfrak{n}^*_\psi$, $\mathfrak{n}_\psi b \subset \mathfrak{n}_\varphi$ *and* $\psi(ax) = \varphi(xb),\ x \in \mathfrak{n}^*_\varphi \mathfrak{n}_\psi$. (57)

PROOF: By the standard 2×2-matrix technique and the balanced weight $\varphi \oplus \psi$, we may assume $\varphi = \psi$ without loss of generality.

(i) $\Longrightarrow$ (ii): Write $\sigma_t = \sigma^\varphi_t$ and assume that $(a, b) \in \mathcal{G}(\sigma_{-\mathrm{i}})$. It follows that $a \in \mathcal{D}\big(\sigma_{-\mathrm{i}/2}\big)$ and $b \in \mathcal{D}\big(\sigma_{\mathrm{i}/2}\big)$, and $\sigma_{-\mathrm{i}/2}(a) = \sigma_{\mathrm{i}/2}(b)$. As in Lemma 2.3, $b^* \in \mathcal{D}\big(\sigma_{-\mathrm{i}/2}\big)$ and $\sigma_{\mathrm{i}/2}(b)^* = \sigma_{-\mathrm{i}/2}(b^*)$. By Lemma 3.18, $\mathfrak{n}_\varphi a^* \subset \mathfrak{n}_\varphi$ and $\mathfrak{n}_\varphi b \subset \mathfrak{n}_\varphi$, which means that a (resp. b) is a left (resp. right) multiplier of $\mathfrak{m}_\varphi$, so that $\varphi(ax)$ and $\varphi(xb)$ are well-defined for $x \in \mathfrak{m}_\varphi$. For $x, y \in \mathfrak{n}_\varphi$, we have, as Lemma 3.18,

$$\begin{aligned}
\varphi(ay^*x) &= \big(\eta_\varphi(x) \mid \eta_\varphi(ya^*)\big) = \Big(\eta_\varphi(x) \,\Big|\, J\sigma_{-\frac{\mathrm{i}}{2}}(a)J\eta_\varphi(y)\Big) \\
&= \Big(\eta_\varphi(x) \,\Big|\, J\sigma_{\frac{\mathrm{i}}{2}}(b)J\eta_\varphi(y)\Big) = \Big(J\sigma_{\frac{\mathrm{i}}{2}}(b)^*J\eta_\varphi(x) \,\Big|\, \eta_\varphi(y)\Big) \\
&= \Big(J\sigma_{-\frac{\mathrm{i}}{2}}(b^*)J\eta_\varphi(x) \,\Big|\, \eta_\varphi(y)\Big) = \big(\eta_\varphi(xb) \mid \eta_\varphi(y)\big) = \varphi(y^*xb).
\end{aligned}$$

Thus we obtain (57).

(ii) $\Longrightarrow$ (i): Let $\mathfrak{A}$ be the full left Hilbert algebra associated with φ. We identify $\mathcal{M}$ with $\mathcal{R}_\ell(\mathfrak{A})$. Let $\mathfrak{A}_0$ be the Tomita algebra of $\mathfrak{A}$. For $\xi, \eta \in \mathfrak{A}_0$, set $x = \pi_\ell\big(\eta^\sharp\xi\big) = \pi_\ell(\eta)^*\pi_\ell(\xi) \in \mathfrak{m}_\varphi$. Assume that $a\mathfrak{n}^*_\varphi \subset \mathfrak{n}^*_\varphi$. Then

$$\begin{aligned}
\varphi(ax) &= \Big(\eta_\varphi\big(\pi_\ell(\xi)\big) \,\Big|\, \xi_\varphi\big(\pi_\ell(\eta)a^*\big)\Big) = \Big(\xi \,\Big|\, \eta_\varphi\big(\pi_\ell(\eta)a^*\big)\Big) \\
&= \Big(S\eta_\varphi\big(\pi_\ell(\eta)a^*\big) \,\Big|\, F\xi\Big) = \Big(\eta_\varphi\big(a\pi_\ell(\eta)^*\big) \,\Big|\, F\xi\Big) = \big(a\eta^\sharp \mid \xi^\flat\big).
\end{aligned}$$

Similarly, if $\mathfrak{n}_\varphi b \subset \mathfrak{n}_\varphi$, then

$$\varphi(xb) = \big(b\eta^\flat \mid \xi^\sharp\big).$$

Therefore, if (a, b) satisfies the condition (ii), then

$$\left(a\eta^{\sharp} \mid \xi^{\flat}\right) = \left(b\xi^{\flat} \mid \xi^{\sharp}\right), \quad \xi, \eta \in \mathfrak{A}_0. \tag{58}$$

Equivalently, with substitution $\xi_1 = \eta^{\flat}$ and $\xi_2 = \xi^{\sharp}$, we have

$$\left(a\Delta^{-1}\xi_1 \mid \Delta\xi_2\right) = \left(b\xi_1 \mid \xi_2\right), \quad \xi_1, \xi_2 \in \mathfrak{A}_0. \tag{58'}$$

For each $\alpha \in \mathbf{C}$, set

$$B_\alpha(\xi_1, \xi_2) = \left(a\Delta^{-\mathrm{i}\alpha}\xi_1 \mid \Delta^{-\mathrm{i}\bar{\alpha}}\xi_2\right), \quad \xi_1, \xi_2 \in \mathfrak{A}_0.$$

Then each B_α is a sesquilinear form on $\mathfrak{A}_0$ and $\alpha \in \mathbf{C} \mapsto B_\alpha(\xi_1, \xi_2) \in \mathbf{C}$ is entire for $\xi_1, \xi_2 \in \mathfrak{A}_0$ and bounded on each horizontal strip bounded by $\mathbf{R} + \mathrm{i}s$ and $\mathbf{R} + \mathrm{i}t$, $s < t$. For $t \in \mathbf{R}$, we have

$$B_t(\xi_1, \xi_2) = \left(\sigma_t^\varphi(a)\xi_1 \mid \xi_2\right), \quad \xi_1, \xi_2 \in \mathfrak{A}_0;$$

$$\begin{aligned} B_{t-\mathrm{i}}(\xi_1, \xi_2) &= \left(a\Delta^{-\mathrm{i}(t-\mathrm{i})}\xi_1 \mid \Delta^{-\mathrm{i}(t+\mathrm{i})}\xi_2\right) = \left(a\Delta^{-1}\Delta^{-\mathrm{i}t}\xi_1 \mid \Delta\Delta^{-\mathrm{i}t}\xi_2\right) \\ &= \left(b\Delta^{-\mathrm{i}t}\xi_1 \mid \Delta^{-\mathrm{i}t}\xi_2\right) = \left(\sigma_t^\varphi(b)\xi_1 \mid \xi_2\right). \end{aligned}$$

By the Phragmen-Lindelöf Theorem, we have

$$\left|B_\alpha(\xi_1, \xi_2)\right| \leq \max\left(\|a\|, \|b\|\right)\|\xi_1\|\|\xi_2\|, \quad \alpha \in \bar{\mathbf{D}}_1.$$

Hence there exists $x(\alpha) \in \mathcal{L}(\mathfrak{H}_\varphi)$ such that

$$\left(x(\alpha)\xi_1 \mid \xi_2\right) = \left(a\Delta^{-\mathrm{i}\alpha}\xi_2 \mid \Delta^{-\mathrm{i}\bar{\alpha}}\xi_2\right)$$

for $\alpha \in \bar{\mathbf{D}}_1$ and $\xi_1, \xi_2 \in \mathfrak{A}_0$, and $\|x(\alpha)\| \leq \max\left(\|a\|, \|b\|\right)$. It follows that $\alpha \in \mathbf{C} \mapsto x(\alpha) \in \mathcal{L}(\mathfrak{H}_\varphi)$ is continuous on $\bar{\mathbf{D}}_1$ and holomorphic in $\mathbf{D}_1$, and that $x(t) = \sigma_t^\varphi(a)$ and $x(t - \mathrm{i}) = \sigma_t^\varphi(b)$. This implies that $x(\alpha) \in M$, and $a \in \mathcal{D}(\sigma_{-\mathrm{i}}^\varphi)$, $b = \sigma_{-\mathrm{i}}^\varphi(a)$.

Q.E.D.

Applying the last theorem to bounded functionals, we obtain the following:

Corollary 3.26. *For faithful $\varphi, \psi \in \mathcal{M}_*^+$, and $a, b \in \mathcal{M}$, $(a, b) \in \mathcal{G}\left(\sigma_{-\mathrm{i}}^{\psi,\varphi}\right)$ if and only if $\psi(ax) = \varphi(xb)$ for every $x \in \mathcal{M}$.*

Exercise VIII.3

1) Show that if φ and ψ are faithful semi-finite normal weights on $\mathcal{M}$ whose modular automorphism groups commute, then $\varphi + \psi$ is semi-finite.

2) In the previous problem, show that if either φ or ψ is finite, then the following three conditions are equivalent:

(i) $\{\sigma_t^\varphi\}$ and $\{\sigma_s^\psi\}$ commute;

(ii) $\varphi = \varphi \circ \sigma_t^\psi, \quad t \in \mathbf{R}$;

(iii) $\psi = \psi \circ \sigma_t^\varphi, \quad t \in \mathbf{R}$.

3) Let $\varphi, \psi \in \mathcal{M}_*^+$ and φ be faithful. Show that the following conditions on ψ are equivalent;
(i) $\psi = \psi \circ \sigma_t^\varphi, \ t \in \mathbf{R}$;
(ii) $|\varphi + \mathrm{i}\psi| = |\varphi - \mathrm{i}\psi|$ in the polar decompositions.

4) Let $\varphi \in \mathcal{M}_*^+$ be faithful. Show that if $\mathcal{M}$ is a factor and $\mathrm{Aut}_\varphi(\mathcal{M}) = \{\alpha \in \mathrm{Aut}(\mathcal{M}) : \varphi \circ \alpha = \varphi\}$ acts ergodically in the sense that the fixed points under $\mathrm{Aut}_\varphi(\mathcal{M})$ reduces to the scalars, then either (i) φ is a trace or (ii) $\mathcal{M}$ is of type III.

5) Let φ be a faithful semi-finite weight on $\mathcal{M}$. Show that the following statements are equivalent:
(i) $\mathcal{M}$ admits a faithful semi-finite normal trace on $\mathcal{M}$;
(ii) For every non-zero $e \in \mathrm{Proj}(\mathcal{M})$, there exists a non-zero $f \in \mathrm{Proj}(\mathcal{M}_e)$ such that the modular operator Δ_{φ_f} for the restriction $\varphi|_{\mathcal{M}_f}$ is bounded.

6) Let $\mathfrak{H} = L^2(\mathbf{R})$ be the Hilbert space of all square integrable functions on $\mathbf{R}$ with respect to the Lebesgue measure. Define two self-adjoint operators H and K as follows:

$$(H\xi)(s) = s\xi(s), \qquad (K\xi)(s) = \frac{1}{\mathrm{i}}\xi'(s)$$

with the domains $\mathfrak{D}(K) = \{\xi \in \mathfrak{H} : \xi$ is differentiable and its derivative ξ' is square integrable$\}$. Set $h = \mathrm{e}^H$ and $k = \mathrm{e}^K$. Prove that the modular automorphism groups $\{\sigma_t^\varphi\}$ and $\{\sigma_s^\psi\}$ of $\varphi = \mathrm{Tr}(h\ \cdot\ )$ and $\psi = \mathrm{Tr}(k\ \cdot\ )$ on $\mathcal{L}(\mathfrak{H})$ commute, but φ and ψ do not commute.

7) Let $\mathcal{M}$ be a factor. Show that if the modular automorphism groups $\{\sigma_t^\varphi\}$ and $\{\sigma_s^\psi\}$ of two faithful semi-finite normal weights φ and ψ commute, then there exists $\theta \in \mathbf{R}$ such that $\psi \circ \sigma_t^\varphi = \mathrm{e}^{\theta t}\psi$ and $\varphi \circ \sigma_t^\psi = \mathrm{e}^{-\theta t}\varphi, \ t \in \mathbf{R}$.

8) Let $\mathcal{M}$ be a von Neumann algebra, and $\{\sigma_t : t \in \mathbf{R}\}$ be a pointwise σ-weakly continuous one parameter group of isometries of $\mathcal{M}$. For $\gamma = \alpha + \mathrm{i}\beta \in \mathbf{C}$, let $\mathcal{D}(\gamma)$ be the set of all $x \in \mathcal{M}$ such that the function: $t \in \mathbf{R} \mapsto \sigma_t(x) \in \mathcal{M}$ can be extended to a member of $\mathcal{A}_\mathcal{M}(\mathbf{D}_\beta)$, where $\mathcal{A}_\mathcal{M}(\mathbf{D}_\beta)$ means the space similar to the space defined in Theorem 3.17 for $\beta = 1/2$. We define $\sigma_\gamma(x)$ naturally. Let φ and ψ be two faithful semi-finite normal weights on $\mathcal{M}$, $a \in \mathcal{M}$ and $k \in \mathbf{R}_+$. Prove that the following conditions are equivalent:
(i) $\psi(axa^*) \le k^2\varphi(x), \ x \in \mathcal{M}_+$;
(ii) $\mathfrak{n}_\varphi a^* \subset \mathfrak{n}_\psi$ and $\|\eta_\psi(xa^*)\| \le k\|\eta_\varphi(x)\|, \ x \in \mathfrak{n}_\varphi$;
(iii) $a \in \mathcal{D}(\sigma_{-\mathrm{i}/2}^{\psi,\varphi})$ and $\|\sigma_{-\mathrm{i}/2}^{\psi,\varphi}(a)\| \le k$.
(Hint: Apply Lemma 3.18 to the balanced weight $\rho = k^2\varphi \oplus \psi$ and $\tilde{a} = a \otimes e_{21}$.)

9) Let φ be as before, and $a_1, a_2 \in \mathcal{D}(\sigma^{\varphi}_{-i/2})$ and set $\psi_1(x) = \varphi(a_1 x a_1^*)$ and $\psi_2(x) = \varphi(a_2 x a_2^*)$, $x \in \mathcal{M}_+$. Show that $\varphi = \psi_1 + \psi_2$ if and only if

$$\sigma^{\varphi}_{-i/2}(a_1)^* \sigma^{\varphi}_{-i/2}(a_1) + \sigma^{\varphi}_{-i/2}(a_2)^* \sigma^{\varphi}_{-i/2}(a_2) = 1.$$

§ 4 Tensor Product and Direct Integrals of Weights

Let $\mathcal{M}$ and $\mathcal{N}$ be von Neumann algebras equipped with faithful semi-finite normal weights φ and ψ respectively. We want to define and discuss the tensor product $\varphi \otimes \psi$ on $\mathcal{M} \overline{\otimes} \mathcal{N}$. To this end let $\mathfrak{A}$ and $\mathfrak{B}$ be the full left Hilbert algebras corresponding to φ and ψ respectively. It is easy to see that the algebraic tensor product $\mathfrak{A} \otimes \mathfrak{B}$ is a left Hilbert algebra such that

$$\mathcal{R}_\ell\big(\mathfrak{A} \otimes \mathfrak{B}\big) = \mathcal{R}_\ell(\mathfrak{A}) \overline{\otimes} \mathcal{R}_\ell(\mathfrak{B}).$$

To find the modular operator of $\mathfrak{A} \otimes \mathfrak{B}$, we shall prove the following:

Lemma 4.1. *Let T_1 and T_2 be two closed operators on $\mathfrak{H}_1$ and $\mathfrak{H}_2$. The algebraic tensor product $T_1 \otimes T_2$ on $\mathfrak{D}(T_1) \otimes \mathfrak{D}(T_2)$ is preclosed and its adjoint $(T_1 \otimes T_2)^*$ is the closure of $T_1^* \otimes T_2^*$ on $\mathfrak{D}(T_1^*) \otimes \mathfrak{D}(T_2^*)$.*

PROOF: Let $T_1 = u_1 H_1$ and $T_2 = u_2 H_2$ be the polar decomposition. If $\xi_i \in \mathfrak{D}(T_i)$ and $\eta_i \in \mathfrak{D}(T_i^*)$, $i = 1, 2$, then

$$\begin{aligned}\big((T_1 \otimes T_2)(\xi_1 \otimes \xi_2) \mid \eta_1 \otimes \eta_2\big) &= \big(T_1\xi_1 \mid \eta_1\big)\big(T_2\xi_2 \mid \eta_2\big)\\ &= \big(\xi_1 \mid T_1^*\eta_1\big)\big(\xi_2 \mid T_2^*\eta_2\big)\\ &= \big(\xi_1 \otimes \xi_2 \mid T_1^*\eta_1 \otimes T_2^*\eta_2\big),\end{aligned}$$

so that $\eta_1 \otimes \eta_2 \in \mathfrak{D}\big((T_1 \otimes T_2)^*\big)$. Hence $(T_1 \otimes T_2)^*$ is defined on $\mathfrak{D}(T_1^*) \otimes \mathfrak{D}(T_2^*)$ and extends $T_1^* \otimes T_2^*$. Thus, $T_1 \otimes T_2$ is preclosed.

Since $T_1^* = u_1^*(u_1 H_1 u_1^*)$ and $T_2^* = u_2^*(u_2 H_2 u_2^*)$ are the polar decompositions of T_1^* and T_2^* respectively, it suffices to prove that $H_1 \otimes H_2$ on $\mathfrak{D}(H_1) \otimes \mathfrak{D}(H_2)$ is essentially self-adjoint when H_1 and H_2 are positive self-adjoint.

Consider the operator:

$$A_\pm = \mathrm{i}\,(1 + H_1)^{-1} \otimes (1 + H_2)^{-1} \pm H_1(1 + H_1)^{-1} \otimes H_2(1 + H_2)^{-1}.$$

Then we have

$$A_\pm\big((1 + H_1)\mathfrak{D}(H_1) \otimes (1 + H_2)\mathfrak{D}(H_2)\big) = (\mathrm{i} \pm H_1 \otimes H_2)\big(\mathfrak{D}(H_1) \otimes \mathfrak{D}(H_2)\big).$$

Since $A_\pm$ is bounded and has a dense range, we conclude that $(\mathrm{i} \pm H_1 \otimes H_2)\big(\mathfrak{D}(H_1) \otimes \mathfrak{D}(H_2)\big)$ is dense in $\mathfrak{H}_1 \otimes \mathfrak{H}_2$. Hence $H_1 \otimes H_2$ is essentially self-adjoint. Q.E.D.

This lemma allows us to define the tensor product $T_1 \overline{\otimes} T_2$ of closed operators T_1 on $\mathfrak{H}_1$ and T_2 on $\mathfrak{H}_2$ as the closure of $T_1 \otimes T_2$, i.e.

$$T_1 \overline{\otimes} T_2 = (T_1 \otimes T_2)^{**}. \tag{1}$$

Then the lemma says that

$$(T_1 \overline{\otimes} T_2)^* = T_1^* \overline{\otimes} T_2^*. \tag{2}$$

Definition 4.2. Let $\mathcal{M}$ and $\mathcal{N}$ be von Neumann algebras. If φ and ψ are semi-finite normal weights on $\mathcal{M}$ and $\mathcal{N}$ respectively, then the *tensor product* $\varphi \otimes \psi$ of φ and ψ on $\mathcal{M}\overline{\otimes}\mathcal{N}$ is the weight ρ with support $s(\varphi)\otimes s(\psi)$ such that ρ on $(\mathcal{M}\overline{\otimes}\mathcal{N})_{s(\varphi)\overline{\otimes}s(\psi)}$ corresponds to the algebraic tensor product left Hilbert algebra $\mathfrak{A} \otimes \mathfrak{B}$ where $\mathfrak{A}$ and $\mathfrak{B}$ are the full left Hilbert algebras corresponding to $\{\mathcal{M}_{s(\varphi)}, \varphi\}$ and $\{\mathcal{N}_{s(\psi)}, \psi)\}$ respectively.

Proposition 4.3. *In the above definition, if φ and ψ are both faithful, then $\varphi \otimes \psi$ is also faithful and*

$$\sigma_t^{\varphi\otimes\psi} = \sigma_t^{\varphi} \otimes \sigma_t^{\psi}, \quad t \in \mathbf{R}. \tag{3}$$

This follows immediately from Lemma 4.1. We leave the detail to the reader.

We now discuss direct integrals of weights.

Definition 4.4. Let $\{\mathcal{M}(\gamma) : \gamma \in \Gamma\}$ be a measurable field of von Neumann algebras over a σ-finite standard measure space $\{\Gamma, \mu\}$. A field $\{\varphi_\gamma : \gamma \in \Gamma\}$ of faithful semi-finite normal weights on $\{\mathcal{M}(\gamma)\}$ is said to be *measurable* if there exist countably many measurable operator fields $\{x_j(\gamma) : \gamma \in \Gamma\}_{j\in\mathbf{N}} \subset \prod_{\gamma\in\Gamma} \mathcal{M}(\gamma)$ satisfying the following conditions:

a) each $x_j(\gamma)$ belongs to $\mathfrak{n}_{\varphi_\gamma} \cap \mathfrak{n}_{\varphi_\gamma}^*$;
b) the functions: $\gamma \in \Gamma \mapsto \varphi_\gamma(x_j(\gamma)^* x_k(\gamma))$, and $\gamma \in \Gamma \mapsto \varphi_\gamma(x_j(\gamma)x_k(\gamma)^*)$ are measurable for $j, k \in \mathbf{N}$;
c) for each $\gamma \in \Gamma$, $\{x_j(\gamma) : j \in \mathbf{N}\}$ is dense in $\mathfrak{n}_{\varphi_\gamma} \cap \mathfrak{n}_{\varphi_\gamma}^*$ with respect to the norm:

$$\|x\|_{\varphi_\gamma}^{\sharp} = \varphi_\gamma(x^*x + xx^*)^{\frac{1}{2}}, \quad x \in \mathfrak{n}_{\varphi_\gamma} \cap \mathfrak{n}_{\varphi_\gamma}^*. \tag{4}$$

One can talk about measurable fields of normal weights without assumption on the faithfulness or the semi-finiteness simply by restricting the algebras $\mathcal{M}(\gamma)$ to their reduced algebras by the support of φ_γ. However, we restrict ourselves only to the faithful semi-finite normal ones. Thus by weights we mean faithful semi-finite ones in this section.

We fix a measurable field $\{\varphi_\gamma : \gamma \in \Gamma\}$ of weights on $\{\mathcal{M}(\gamma) : \gamma \in \Gamma\}$. For each $\gamma \in \Gamma$, we set

$$\mathfrak{a}(\gamma) = \mathfrak{n}_{\varphi_\gamma} \cap \mathfrak{n}_{\varphi_\gamma}^*. \tag{5}$$

Each $\mathfrak{a}(\gamma)$ can be viewed as a full left Hilbert algebra with $\mathcal{R}_\ell(\mathfrak{a}(\gamma)) = \mathcal{M}(\gamma)$. Thus, we obtain a field $\{\mathfrak{a}(\gamma)\}$ of left Hilbert algebras.

Lemma 4.5. *Let $\{\mathcal{M}(\gamma):\gamma \in \Gamma\}$ be a measurable field of von Neumann algebras on $\{\Gamma, \mu\}$. For a field of weights $\{\varphi_\gamma\}$, the following conditions are equivalent:*

(i) *The field $\{\varphi_\gamma\}$ is measurable;*

(ii) *The field $\{\mathfrak{a}(\gamma)\}$ of left Hilbert algebras given by* (5) *is measurable in the sense of Definition VI.*3.1.

The proof is routine, so we leave it to the reader.

Definition 4.6. Given a measurable field $\{\varphi_\gamma\}$ of weights on $\{\mathcal{M}(\gamma):\gamma \in \Gamma\}$, we put

$$\mathcal{M} = \int_\Gamma^\oplus \mathcal{M}(\gamma)\, \mathrm{d}\mu(\gamma), \qquad \mathfrak{A} = \int_\Gamma^\oplus \mathfrak{A}(\gamma)\, \mathrm{d}\mu(\gamma), \qquad \text{see (VI.3.11).}$$

We call the weight φ on $\mathcal{M}$ corresponding to $\mathfrak{A}$ the *direct integral* of $\{\varphi_\gamma\}$ and write

$$\varphi = \int_\Gamma^\oplus \varphi_\gamma\, \mathrm{d}\mu(\gamma)\,. \tag{6}$$

Theorem 4.7. *Let $\{\mathcal{M}(\gamma), \varphi_\gamma : \gamma \in \Gamma\}$ be a measurable field of weights on a σ-finite standard measure space $\{\Gamma, \mu\}$ and let*

$$\{\mathcal{M}, \varphi\} = \int_\Gamma^\oplus \{\mathcal{M}(\gamma), \varphi_\gamma\}\, \mathrm{d}\mu(\gamma)\,. \tag{7}$$

(i) *If $\gamma \in \Gamma \mapsto x(\gamma) \in \mathcal{M}(\gamma)_+$ is a measurable field of operators, then the function: $\gamma \in \Gamma \mapsto \varphi_\gamma\big(x(\gamma)\big)$ is measurable.*

(ii) *If $x \in \mathcal{M}_+$ is represented by $\{x(\gamma)\}$, then*

$$\varphi(x) = \int_\Gamma^\oplus \varphi_\gamma\big(x(\gamma)\big)\, \mathrm{d}\mu(\gamma)\,. \tag{8}$$

PROOF:

(i) Let $\{\eta_j(\gamma)\}_{j\in\mathbf{N}}$ be a fundamental sequence of measurable fields of vectors relative to the measurable field $\{\mathfrak{A}(\gamma)'\}$ of right Hilbert algebras, where $\{\mathfrak{A}(\gamma)\}$ means the obvious one. Let $y_j(\gamma) = \pi_r\big(\eta_j(\gamma)\big) \in \mathcal{M}(\gamma)'$. For each $J \Subset \mathbf{N}$,[1] set

$$z_J(\gamma) = \Big(1 + \sum_{j\in J} y_j(\gamma)^* y_j(\gamma)\Big)^{-1} \sum_{j\in J} y_j(\gamma)^* y_j(\gamma) \in \mathcal{M}'^{+}_{\varphi'_\gamma},$$

1 $J \Subset \mathbf{N}$ means a finite subset as noted before.

where φ'_γ is the opposite weight of φ_γ on $\mathcal{M}(\gamma)'$. Then there exists $\eta_J(\gamma) \in \mathfrak{A}(\gamma)'$ such that $\pi_r(\eta_J(\gamma)) = z_J(\gamma)^{1/2}$. Since $\{z_J(\gamma)\}$ increases to the identity along the net $\mathcal{J} = \{J : J \subset\subset \mathbf{N}\}$, we have

$$\varphi_\gamma(x) = \sup_{J\in\mathcal{J}}(x\eta_J(\gamma) \mid \eta_J(\gamma)), \quad z \in \mathcal{M}(\gamma). \tag{9}$$

Since $\{z_J(\gamma)\}$ is a measurable field of operators, $\{\eta_J(\gamma)\}$ are measurable vector fields. Hence the countability of $\mathcal{J}$ and (9) imply the assertion (i).

(ii) Suppose that $x \in \mathcal{M}_+$ and $\varphi(x) < +\infty$. Then we have $x^{1/2} \in \mathfrak{n}_\varphi \cap \mathfrak{n}_\varphi^* = \mathfrak{a}$. Let $\{\xi(\gamma)\}$ be the vector field in $\{\mathfrak{a}(\gamma)\}$ such that $\eta_\varphi(x^{1/2}) \sim \{\xi(\gamma)\}$. We then have

$$\begin{aligned}\varphi(x) &= \left\|\eta_\varphi(x^{\frac{1}{2}})\right\| = \int_\Gamma \|\xi(\gamma)\|^2 \,\mathrm{d}\mu(\gamma)\\ &= \int_\Gamma \varphi\left(\pi_\ell(\xi(\gamma)^2)\right) \mathrm{d}\mu(\gamma) = \int_\Gamma \varphi(x(\gamma))\,\mathrm{d}\mu(\gamma),\end{aligned}$$

since $x(\gamma)^{1/2} = \pi_\ell(\xi(\gamma))$ if $x \sim \{x(\gamma)\}$.

Conversely, suppose that the right hand side of (8) is finite. Then $x(\gamma)^{1/2} \in \mathfrak{A}(\gamma)$ for almost every $\gamma \in \Gamma$. Let $\xi(\gamma) = \eta_{\varphi_\gamma}(x(\gamma)^{1/2}) \in \mathfrak{a}(\gamma)$. Then we have

$$\int_\Gamma \|\xi(\gamma)\|^2 \,\mathrm{d}\mu(\gamma) < +\infty,$$

so that $\{\xi(\gamma)\} \sim \xi \in \mathfrak{H} = \int_\Gamma^\oplus \mathfrak{H}(\gamma)\,\mathrm{d}\mu(\gamma)$, where $\mathfrak{H}(\gamma)$ is the completion of $\mathfrak{A}(\gamma)$. Since

$$\sup\left\|\pi_\ell(\xi(\gamma))\right\| = \sup\|x(\gamma)^{\frac{1}{2}}\| = \|x^{\frac{1}{2}}\| < +\infty, \quad \xi(\gamma)^\sharp = \xi(\gamma),$$

ξ belongs to $\mathfrak{A}$ and $\pi_\ell(\xi) = x^{1/2}$. Thus we get

$$\varphi(x) = \|\xi\|^2 = \int_\Gamma \|\xi(\gamma)\|^2 \,\mathrm{d}\mu(\gamma) = \int_\Gamma \varphi_\gamma(x(\gamma))\,\mathrm{d}\mu(\gamma). \qquad \text{Q.E.D.}$$

Combining the above results with Theorems VI.3.14 and VI.3.15, we conclude the following:

Theorem 4.8. *Let φ be a faithful semi-finite normal weight on a separable von Neumann algebra $\mathcal{M}$. To each von Neumann subalgebra $\mathcal{A}$ of the center $\mathcal{C}$ of $\mathcal{M}$, there corresponds uniquely a measurable field $\{\mathcal{M}(\gamma), \varphi(\gamma)\}$ of von Neumann algebras equipped with weights over a σ-finite standard measure space $\{\Gamma, \mu\}$ such that*

$$\{\mathcal{M}, \varphi\} = \int_\Gamma^\oplus \{\mathcal{M}(\gamma), \varphi_\gamma\}\,\mathrm{d}\mu(\gamma), \tag{10}$$

$\mathcal{A}$ is the diagonal algebra.

The modular automorphism group $\{\sigma_t^{\varphi}\}$ *of* φ *is given as follows:*

$$\sigma_t^{\varphi}\Big(\int_{\Gamma}^{\oplus} x(\gamma)\,\mathrm{d}\mu(\gamma)\Big) = \int_{\Gamma}^{\oplus} \sigma_t^{\varphi_\gamma}\big(x(\gamma)\big)\,\mathrm{d}\mu(\gamma)\,. \tag{11}$$

We now discuss disintegrations of a weight on a separable C^*-algebra A. We first prepare the following:

Lemma 4.9. *If* A *is a separable* C^* *-algebra, then the semi-cyclic representation* $\{\pi_\varphi, \mathfrak{H}_\varphi, \xi_\varphi\}$ *of* A *induced by a lower semi-continuous weight* φ *on* A *is separable.*

PROOF: Let $\mathcal{M} = \pi_\varphi(A)''$ and $\tilde{\varphi}$ be the normal weight on $\mathcal{M}$ given by (VII.4.5). Let φ' be the opposite weight on $\mathcal{M}'$ which is faithful. Let θ_φ be the complete positive map from $\mathfrak{m}_{\tilde{\varphi}}$ into $\mathcal{M}'_*$. Let $\{u_n\}$ be a sequential approximate identity of $\mathfrak{m}_\varphi$, where the separability of A guarantees the existence of the $\{u_n\}$. Set $\omega_n = \theta_\varphi\big(\pi_\varphi(u_n)\big)$. Then we have $\varphi' = \sup \omega_n$. Since φ' is faithful, $\{\omega_n\}$ is a separating family of normal positive functionals on $\mathcal{M}'$, so that $\mathcal{M}'$ is σ-finite. Hence the identity 1 is the sum of an orthogonal sequence of cyclic projections $\{e'_{\xi_n}\}$ in $\mathcal{M}'$. Since A is separable, $e'_{\xi_n}\mathfrak{H}_\varphi = \big[\pi_\varphi(A)\xi_n\big]$ is separable, so that the entire $\mathfrak{H}_\varphi = \sum^{\oplus}\big[\pi_\varphi(A)\xi_n\big]$ is also separable. Q.E.D.

We now fix a one parameter automorphism group $\big\{\sigma_t : t \in \mathbf{R}\big\}$ of a separable C^*-algebra A with norm continuity:

$$\lim_{t\to 0}\big\|\sigma_t(x) - x\big\| = 0, \quad x \in A. \tag{12}$$

An analogy of Proposition 3.15 also holds:

Lemma 4.10. *Suppose that* φ *and* ψ *are lower semi-continuous faithful semi-finite weights on* A *satisfying the modular condition relative to* $\{\sigma_t\}$. *If there exists a* $\{\sigma_t\}$*-invariant* **-subalgebra* B *of* $\mathfrak{m}_\varphi$, *which is dense and on which* φ *and* ψ *agree, then* $\varphi = \psi$.

PROOF: The hypothesis implies that $B \subset \mathfrak{m}_\psi$, so that the assumption is symmetric for φ and ψ, thus it suffices to prove that $\psi \le \varphi$. The proof goes very much parallel to that of Proposition 3.15 with only change; namely if $t \mapsto x(t) \in A_+$ is norm continuous and integrable, then $\varphi\big(\int_{\mathbf{R}} x(t)\,\mathrm{d}t\big) = \int_{\mathbf{R}} \varphi\big(x(t)\big)\,\mathrm{d}t$. Furthermore, if $x, y \in B$, then $\varphi(xzy) = \psi(xzy)$ for all $z \in A$, since $z \mapsto \varphi(xzy)$ and $z \mapsto \psi(xzy)$ are both continuous and agree on B. We leave the rest of the proof to the reader as an exercise. Q.E.D.

In the rest of this section, we shall consider only lower semicontinuous weights on A, so that we will mean by a weight always a lower semicontinuous one. Furthermore, if a weight φ on A satisfies the modular condition for $\{\sigma_t\}$, then the left kernel $N_\varphi = \{x \in A : \varphi(x^*x) = 0\}$ is indeed a closed two sided ideal, so that it factors through the quotient C^*-algebra A/N_φ. Namely, there exists a weight ψ on $B = A/N_\varphi$ such that $\varphi = \psi \circ \pi$, where π is the canonical map of A onto B. Thus, we shall consider only a faithful weight with the modular property for $\{\sigma_t\}$.

Now, the rest of this section is devoted to the proof of the next result:

Theorem 4.11. *Let A be a separable C^*-algebra equipped with a norm continuous one parameter automorphism group $\{\sigma_t\}$ in the sense of* (12). *Let φ be a faithful lower semi continuous semi-finite weight on A satisfying the modular condition for $\{\sigma_t\}$. Let $\mathcal{A}$ be a von Neumann subalgebra of the center $\mathcal{Z}$ of $\mathcal{M} = \pi_\varphi(A)''$. Relative to the disintegration:*

$$\pi_\varphi = \int_\Gamma^\oplus \pi_\gamma \, d\mu(\gamma) \tag{13}$$

over the measure theoretic spectrum $\{\Gamma, \mu\}$ of $\mathcal{A}$ with respect to the diagonal algebra $\mathcal{A}$, there corresponds a family $\{\varphi_\gamma : \gamma \in \Gamma\}$ of weights on A such that

(i) *for $x \in A_+$, the function: $\gamma \in \Gamma \mapsto \varphi_\gamma(x)$ is measurable and*

$$\varphi(x) = \int_\Gamma \varphi_\gamma(x) \, d\mu(\gamma); \tag{14}$$

(ii) *for almost every $\gamma \in \Gamma$, φ_γ is a lower semi-continuous semi-finite weight satisfying the modular condition for $\{\sigma_t\}$;*

(iii) *for almost every $\gamma \in \Gamma$, the semi-cyclic representation π_{φ_γ} induced by φ_γ is unitarily equivalent to π_γ.*

The proof of the theorem requires a few lemmas. But we first establish some notations.

The weight φ gives rise to a faithful semi-finite normal weight $\tilde{\varphi}$ on $\mathcal{M}$ such that $\sigma_t^{\tilde{\varphi}} \circ \pi_\varphi = \pi_\varphi \circ \sigma_t$. By Theorem 4.8 we have the disintegration relative to $\mathcal{A}$:

$$\{\mathcal{M}, \tilde{\varphi}, \sigma^{\tilde{\varphi}}\} = \int_\Gamma^\oplus \{\mathcal{M}(\gamma), \tilde{\varphi}_\gamma, \sigma^{\tilde{\varphi}_\gamma}\} \, d\mu(\gamma) \,. \tag{15}$$

At the same time, we also have the disintegration of π_φ of (13) relative to $\mathcal{A}$. The uniqueness of disintegration relative to $\mathcal{A}$ yields that for almost every $\gamma \in \Gamma$, $\pi_\gamma(A)$ is a σ-weakly dense C^*-subalgebra of $\mathcal{M}(\gamma)$, so that after deleting a null set from Γ, we can define a weight ψ_γ on A as follows:

$$\psi_\gamma(x) = \tilde{\varphi}_\gamma\big(\pi_\gamma(x)\big), \quad x \in A_+. \tag{16}$$

By its construction and Theorem 4.7, the function: $\gamma \in \Gamma \mapsto \psi_\gamma(x)$, $x \in A_+$, is measurable. Now define a weight ψ on A by the integral:

$$\psi(x) = \int_\Gamma \psi_\gamma(x) \, d\mu(\gamma), \quad x \in A_+. \tag{17}$$

Lemma 4.12. *We have* $\varphi = \psi$.

PROOF: For $x \in A_+$, let $T = \pi_\varphi(x) \in \mathcal{M}$. Relative to (15), T has the disintegration:

$$T = \int_\Gamma^\oplus T(\gamma)\, d\mu(\gamma),$$

and also $T = \int_\Gamma^\oplus \pi_\gamma(x)\, d\mu(\gamma)$. The uniqueness of disintegration entails that $T(\gamma) = \pi_\gamma(x)$ for almost every $\gamma \in \Gamma$. We now have

$$\begin{aligned}\varphi(x) = \tilde\varphi(T) &= \int_\Gamma \tilde\varphi_\gamma\big(T(\gamma)\big)\, d\mu(\gamma) = \int_\Gamma \tilde\varphi_\gamma\big(\pi_\gamma(x)\big)\, d\mu(\gamma)\\ &= \int_\Gamma \psi_\gamma(x)\, d\mu(x) = \psi(x).\end{aligned}$$

Since $\tilde\varphi_\gamma$ is a normal weight on $\mathcal{M}(\gamma)$, ψ_γ is lower semi continuous. Q.E.D.

Lemma 4.13. *Almost all the weights* ψ_γ *satisfy the modular condition for* $\{\sigma_t\}$.

PROOF: Since $\tilde\varphi_\gamma$ satisfies the modular condition for $\{\sigma_t^{\tilde\varphi_\gamma}\}$, it suffices to prove that

$$\sigma_t^{\tilde\varphi_\gamma} \circ \pi_\gamma = \pi_\gamma \circ \sigma_t \quad \text{for almost every } \gamma \in \Gamma.$$

But we have

$$\begin{aligned}\sigma_t^{\tilde\varphi}\big(\pi_\varphi(x)\big) &= \sigma_t^{\tilde\varphi}\left(\int_\Gamma^\oplus \pi_\gamma(x)\, d\mu(\gamma)\right) = \int_\Gamma^\oplus \sigma_t^{\tilde\varphi_\gamma}\big(\pi_\gamma(x)\big)\, d\mu(\gamma)\\ &= \pi_\varphi \circ \sigma_t(x) = \int_\Gamma^\oplus \pi_\gamma \circ \sigma_t(x)\, d\mu(\gamma).\end{aligned}$$

Therefore, the uniqueness of disintegration entails that

$$\sigma_t^{\tilde\varphi_\gamma} \circ \pi_\gamma(x) = \pi_\gamma \circ \sigma_t(x) \quad \text{for almost every } \gamma \in \Gamma.$$

Let B be a countable dense *-subalgebra of A over the rational complex field $\mathbf{Q}(\mathrm{i})$. For each $x \in B$ and $t \in \mathbf{Q}$, set

$$N(x,t) = \left\{\gamma \in \Gamma : \sigma_t^{\tilde\varphi_\gamma} \circ \pi_\gamma(x) \neq \pi_\gamma \circ \sigma_t(x)\right\},$$

and $N = \bigcup\{\, N(x,t) : x \in B,\ t \in \mathbf{Q}\}$. Then N is a null set in Γ and we have

$$\sigma_t^{\tilde\varphi_\gamma} \circ \pi_\gamma(x) = \pi_\gamma \circ \sigma_t(x), \quad x \in B, \quad t \in \mathbf{Q}, \quad \gamma \in \Gamma \setminus N.$$

The continuity in x implies that the above equality is valid for every $x \in A$ and $t \in \mathbf{Q}$ and $\gamma \in \Gamma \setminus N$. When t converges to $t_0 \in \mathbf{R}$, then the left hand side converges σ-strongly to $\sigma_{t_0}^{\tilde\varphi_\gamma}\big(\pi_\gamma(x)\big)$ and the right hand side converges in norm to $\pi_\gamma \circ \sigma_{t_0}(x)$. Hence the equality holds for every $x \in A$, $t \in \mathbf{R}$ and $\gamma \in \Gamma \setminus N$. Q.E.D.

PROOF OF THEOREM 4.11: The only thing remains to be proven is that almost all π_γ are equivalent to π_{φ_γ}. In the disintegration:

$$\{\mathcal{M}, \mathfrak{H}_{\tilde{\varphi}}\} = \int_\Gamma^\oplus \{\mathcal{M}(\gamma), \mathfrak{H}(\gamma)\}\, d\mu(\gamma),$$

almost all von Neumann algebras $\{\mathcal{M}(\gamma), \mathfrak{H}(\gamma)\}$ are given by the left Hilbert algebras $\mathfrak{A}(\gamma)$ given by $\tilde{\varphi}_\gamma$. This means that almost all representations π_γ are the semi-cyclic representations given by ψ_γ. Q.E.D.

Notes on Chapter VIII

The modular automorphism group of a faithful semi-finite normal weight on a von Neumann algebra was first introduced independently by F. Combes, [451] and M. Takesaki [709] as the next step following the theory of left Hilbert algebras. But the concept of the modular condition was introduced by mathematical physicists, Haag, Hugenholtz and Winnink, in their celebrated joint work [535] to describe an equilibrium state of a quantum statistical system under the term Kubo-Martin-Schwinger boundary condition, abbreviated to KMS-condition. A state φ on a C^*-algebra A equipped with a one parameter automorphism group $\{\alpha_t : t \in \mathbf{R}\}$ is said to satisfy the KMS-condition at the inverse temperature $\beta = 1/kT$ if φ satisfies the modular condition for $\{\alpha_{-\beta t}\}$. In statistical quantum physics, the time evolution of the system is described by a one parameter automorphism of the C^*-algebra of observables, which dictates the dynamics of the system and its equilibrium states. In mathematics on the other hand, we have observed that the state determines the modular automorphism group. There are extensive literatures in quantum statistical mechanics. We refer the interested reader to the two-volume book, [442], written by Bratteli and Robinson.

Chapter IX
Non-Commutative Integration

§ 0 Introduction

The theories of weights, traces and states are often referred as non commutative integration. If the von Neumann algebra in question is abelian, then our theory is precisely the theory of measures and integration. In fact, the weight value of a self-adjoint element is given precisely by the integration of the corresponding function on the spectrum relative to the measure corresponding to the weight. As there are many non-commuting self-adjoint elements in the algebra, we have to consider various spectral measures even if we fix one weight and we can not represent non-commuting self-adjoint elements as functions on the same space. The striking difference between the commutative case and the non-commutative case is the appearance of one parameter automorphism group which is determined by the weight. Namely, weights and/or states determine the dynamics of the system which does not have the commutative counter part. We have explored the relationship between weights and the modular automorphism groups so far. We now further investigate how the dynamics, i.e. the modular automorphism groups, of the algebra relate the different spaces associated with the algebras. First, we study the underlying Hilbert space of the algebra and find the intrinsic pointed convex cone there, which is called the natural cone, in the first section. The theory developed there allows us to view the standard Hilbert space as the square root of the predual of the algebra as well as to represent the automorphism group $\mathrm{Aut}(\mathcal{M})$, of a von Neumann algebra $\mathcal{M}$, as the group of unitaries which leaves the natural cone globally invariant. In §2, we consider the special case that the weight is a trace and see the very special character of a trace, which allows us to realize various spaces associated with the algebra $\mathcal{M}$ as the space of unbounded operators affiliated with $\mathcal{M}$ satisfying certain regulating condition which are called measurable operators. We will see that in the case of a finite von Neumann algebra $\mathcal{M}$ every closed operator affiliated with $\mathcal{M}$ is measurable. Measurable operators form an involutive algebra as we will see in this section. It is remarkable that the trace regulates the behavior of unbounded operator so strongly that closed symmetric measurable operators are automatically self-adjoint. Section 3 relates a von Neumann algebra $\mathcal{M}$ of operators on a Hilbert space $\mathfrak{H}$ to its commutant $\mathcal{M}'$ in very strong way. It will be shown that there uniquely corresponds a non-singular self-adjoint operator, called the spatial derivative, to any pair of a faithful semi-finite normal weights φ on $\mathcal{M}$ and ψ on $\mathcal{M}'$. The spatial derivative behaves very naturally even though they are unbounded operators. Also, we consider right actions, i.e. an anti-representation, of a von Neumann algebra on a Hilbert space and view it a right module over the von Neumann algebra. We then view a Hilbert space equipped with the usual action of a von Neumann algebra as a left module over the von Neumann algebra. Naturally, we consider bimodules over a pair of von Neumann algebras, as a Hilbert space equipped with commuting the left action of one algebra and the right action of the other. This view allows us to introduce the concept of relative tensor product of a pair of a right module and a left module over a fixed von Neumann algebra. It should be noticed that this corresponds to the relative tensor products in algebra but not straightforward way. There are subtle difference here and the situation in pure algebra. For example, the relative tensor product of an arbitrary pair of vectors does not make sense unless we are in the very special case of atomic von Neumann algebras. The last section, §4, discuss conditional expectations and unbounded operator valued weights as a generalization of conditional expectations which allows us to view our theory in a more balanced way. It allows us to factor the usual trace, Tr, on a Hilbert space $\mathfrak{H}$ through a faithful semi-finite normal trace τ on a semi-finite von Neumann algebra $\mathcal{M}$ and the corresponding operator valued weight to $\mathcal{M}$. This in a sense justifies to say that a factor is indeed a *factor* of $\mathcal{L}(\mathfrak{H})$. One should interpret an operator valued weight as a partial integral relative to the related non-commutative measure, which is not visible unless the algebra is abelian.

§ 1 Standard Form of a von Neumann Algebra

We begin our discussion with an example. Let $\mathcal{M} = \mathcal{L}(\mathfrak{H})$ with a Hilbert space $\mathfrak{H}$. We fix the usual trace Tr on $\mathcal{M}$ and denote it by τ, then consider the semi-cyclic representation $\{\pi_\tau, \mathfrak{H}_\tau, \eta_\tau\}$. The Hilbert space $\mathfrak{H}_\tau$ is then identified with the Hilbert space of operators of the Hilbert-Schmidt class, and the representation π_τ is given by the left multiplication. If φ is a faithful normal positive linear functional on $\mathcal{M}$, then φ is of the form:

$$\varphi(x) = \tau(xh) = \left(xh^{\frac{1}{2}} \mid h^{\frac{1}{2}}\right)_{\mathrm{HS}} \tag{1}$$

with h a non-singular nuclear positive operator on $\mathfrak{H}$, where $(\ \mid\ )_{\mathrm{HS}}$ denotes the inner product in $\mathfrak{H}_\tau$. Therefore, $h^{1/2}$ is the cyclic and separating vector associated with the functional φ. Hence we have

$$S_\varphi x = \left(xh^{-\frac{1}{2}}\right)^* h^{\frac{1}{2}} = h^{-\frac{1}{2}} x^* h^{\frac{1}{2}}, \tag{2}$$

for those $x \in \mathfrak{H}_\tau$ such that $xh^{-1/2}$ is bounded. We proceed our discussion formally without worrying about the domain questions and other analytical details, in order to figure out the picture of the entire mechanism. The adjoint involution F_φ of S_φ is then given by

$$F_\varphi x = h^{\frac{1}{2}}\left(h^{-\frac{1}{2}} x\right)^* = h^{\frac{1}{2}} x^* h^{-\frac{1}{2}}, \quad x \in \mathfrak{D}^\flat. \tag{3}$$

Hence we get

$$\Delta_\varphi x = hxh^{-1}, \qquad x \in \mathfrak{D}(\Delta_\varphi); \tag{4}$$

$$\Delta_\varphi^{it} x = h^{it} x h^{-it}, \quad x \in \mathfrak{H}_\tau. \tag{5}$$

We now observe that

$$\left[\Delta_\varphi^{\frac{1}{4}} \mathcal{M}_+ h^{\frac{1}{2}}\right] = \left[h^{\frac{1}{4}} \mathcal{M}_+ h^{\frac{1}{4}}\right] = (\mathfrak{H}_\tau)_+. \tag{6}$$

Thus, the positive cone $(\mathfrak{H}_\tau)_+$ of $\mathfrak{H}_\tau$ is recovered from $\mathcal{M}_+ h^{1/2}$ by multiplying $\Delta_\varphi^{1/4}$. The point here is that the positive cone $(\mathfrak{H}_\tau)_+$ does not depend on the choice of φ; it is intrinsic to the algebra $\mathcal{M} = \mathcal{L}(\mathfrak{H})$. We are going to see that this fact is not special to $\mathcal{L}(\mathfrak{H})$, but true for general von Neumann algebras.

Now, we fix a von Neumann algebra $\mathcal{M}$ and a faithful semi-finite normal weight φ on it; thus also the semi-cyclic representation $\{\pi_\varphi, \mathfrak{H}_\varphi, \eta_\varphi\}$. In the Hilbert space $\mathfrak{H}_\varphi$, we shall consider various pointed convex cones. Let $\mathfrak{A}$ be the full left Hilbert algebra associated with φ and $\mathfrak{A}_0$ the associated Tomita algebra. We identify $\mathcal{M}$ with $\mathcal{R}_\ell(\mathfrak{A})$. We put

$$\left.\begin{aligned} \mathfrak{P}_\varphi^\sharp &= \left\{\xi\xi^\sharp : \xi \in \mathfrak{A}\right\}^-, \\ \mathfrak{P}_\varphi^\flat &= \left\{\xi\xi^\flat : \xi \in \mathfrak{A}'\right\}^-, \\ \mathfrak{P}_\varphi &= \left\{\xi\xi^* : \xi \in \mathfrak{A}_0\right\}^-. \end{aligned}\right\} \tag{7}$$

Here $\xi^* = J\xi$, $\xi \in \mathfrak{H}_\varphi$ and the bar means the closure.

Definition 1.1. For a convex cone $\mathfrak{P}$ in a Hilbert space $\mathfrak{H}$, the *dual cone* $\mathfrak{P}^\circ$ is defined to be the set of all those vectors $\eta \in \mathfrak{H}$ such that $(\xi \mid \eta) \geq 0$ for every $\xi \in \mathfrak{P}$:

$$\mathfrak{P}^\circ = \left\{\eta \in \mathfrak{H} : (\xi \mid \eta) \geq 0, \ \xi \in \mathfrak{P}\right\}. \tag{8}$$

If $\mathfrak{P} = \mathfrak{P}^\circ$, then $\mathfrak{P}$ is called *self-dual.*

To motivate discussion, we state the next result concerning the convex cones defined in (7).

Theorem 1.2. *Under the above notations and assumptions, we have the following:*

(i) *The sets $\mathfrak{P}_\varphi^\sharp$ and $\mathfrak{P}_\varphi^\flat$ are mutually dual pointed convex cones;*
(ii) *$\mathfrak{P}_\varphi$ is a self-dual convex cone and every element of $\mathfrak{H}_\varphi$ is represented as a linear combination of four vectors of $\mathfrak{P}_\varphi$. Furthermore, each self-adjoint vector $\xi \in \mathfrak{H}_\varphi$, i.e. $\xi = \xi^*$, is uniquely represented as difference of two orthogonal vectors of $\mathfrak{P}_\varphi$;*
(iii) *$aJaJ\mathfrak{P}_\varphi \subset \mathfrak{P}_\varphi$ for every $a \in \mathcal{M}$;*
(iv) *To each $\omega \in \mathcal{M}_*^+$, there corresponds a unique $\xi \in \mathfrak{P}_\varphi$ with $\omega = \omega_\xi$. Furthermore, we have*

$$\|\xi - \eta\|^2 \leq \|\omega_\xi - \omega_\eta\| \leq \|\xi - \eta\| \, \|\xi + \eta\|, \quad \xi, \eta \in \mathfrak{P}_\varphi. \tag{9}$$

PROOF OF THEOREM 1.2.(i): Let $\xi \in \mathfrak{A}$ and $\eta \in \mathfrak{A}'$. We then have

$$\left(\xi^\sharp \xi \mid \eta\eta^\flat\right) = \left(\pi_\ell(\xi)^*\xi \mid \pi_r(\eta)^*\eta\right) = \left\|\pi_\ell(\xi)\eta\right\|^2 \geq 0.$$

Suppose that $\xi \in \mathfrak{H}_\varphi$ satisfies $\left(\xi \mid \eta\eta^\flat\right) \geq 0$ for every $\eta \in \mathfrak{A}'$. By the polarization identity, $\mathfrak{A}'^2$ is linearly spanned by $\left\{ \eta\eta^\flat : \eta \in \mathfrak{A}' \right\}$, so that Lemma VI.1.13 implies that ξ belongs to $\mathfrak{D}^\sharp$, and $\pi_\ell(\xi)$ is a symmetric operator affiliated with $\mathcal{M}$. The positivity assumption ξ entails that of $\pi_\ell(\xi)$. We denote the Friedricks extension of $\pi_\ell(\xi)$ by a. With its spectral decomposition $a = \int_0^\infty \lambda\, de(\lambda)$, set $\xi_n = e(n)\xi$. Then we get $\xi_n \in \mathfrak{A}$ and $\|\xi_n - \xi\| \to 0$ as $n \to \infty$. Furthermore, $\left(\xi_n \mid \eta\eta^\flat\right) = \left(ae(n)\eta \mid \eta\right) \geq 0,\ \eta \in \mathfrak{A}'$. Thus, to prove $\xi \in \mathfrak{P}_\varphi^\sharp$ it suffices to show that $\xi_n \in \mathfrak{P}_\varphi^\sharp$. This means that we are asked to prove $\xi \in \mathfrak{P}_\varphi^\sharp$ if $\left(\xi \mid \eta\eta^\flat\right) \geq 0,\ \eta \in \mathfrak{A}'$, and ξ is left bounded. As observed already, ξ belongs to $\mathfrak{A}$, so $a = \pi_\ell(\xi) \geq 0$ and $\varphi(a^2) < +\infty$. Since $\lambda\left(1 - e(\lambda)\right) \leq a,\ \lambda > 0$, we have $\varphi\left(1 - e(\lambda)\right) < +\infty$. Setting $b_n = \left(1 - e(1/n)\right)a^{1/2}$, we get $\varphi(b_n^2) < +\infty$ and $b_n^2 = \left(1 - e(1/n)\right)a$. Then with $\zeta_n = \eta_\varphi(b_n)$, we have $\zeta_n \in \mathfrak{A}$ and

$$\zeta_n^\sharp \zeta_n = \zeta_n^2 = \left(1 - e(1/n)\right)\xi.$$

Since $\xi \in \left[a\mathfrak{H}_\varphi\right]$, we get $\left\|\xi - \left(1 - e(1/n)\right)\xi\right\| \to 0$ as $n \to \infty$, concluding that $\xi \in \mathfrak{P}_\varphi^\sharp$. Thus $\mathfrak{P}_\varphi^\sharp = \left(\mathfrak{P}_\varphi^\flat\right)^0$. By symmetry, we have $\mathfrak{P}_\varphi^\flat = \left(\mathfrak{P}_\varphi^\sharp\right)^0$. Q.E.D.

Lemma 1.3. $\mathfrak{P}_\varphi = \left(\Delta^{1/4}\mathfrak{P}_\varphi^\sharp\right)^- = \left(\Delta^{-1/4}\mathfrak{P}_\varphi^\flat\right)^-$. *In particular,* $\mathfrak{P}_\varphi$ *is a closed convex cone.*

PROOF: First, we observe that $\mathfrak{P}_\varphi^\sharp \subset \mathfrak{D}^\sharp = \mathfrak{D}\left(\Delta^{1/2}\right)$ and $\mathfrak{P}_\varphi^\flat \subset \mathfrak{D}^\flat = \mathfrak{D}\left(\Delta^{-1/2}\right)$. Hence $\Delta^{1/4}\mathfrak{P}_\varphi^\sharp$ and $\Delta^{-1/4}\mathfrak{P}_\varphi^\flat$ make sense. Furthermore, we have

$$\Delta^{\frac{1}{2}}\mathfrak{P}_\varphi^\sharp = JS\mathfrak{P}_\varphi^\sharp = J\mathfrak{P}_\varphi^\sharp = \mathfrak{P}_\varphi^\flat,$$

so that $\Delta^{1/4}\mathfrak{P}_\varphi^\sharp = \Delta^{-1/4}\mathfrak{P}_\varphi^\flat$. Thus, it only remains to be proven that $\mathfrak{P}_\varphi = \left(\Delta^{1/4}\mathfrak{P}_\varphi^\sharp\right)^-$. Now, if $\xi \in \mathfrak{A}_0$, then

$$\Delta^{\frac{1}{4}}\left(\xi\xi^\sharp\right) = \left(\Delta^{\frac{1}{4}}\xi\right)\left(\Delta^{\frac{1}{4}}\xi^\sharp\right) = \left(\Delta^{\frac{1}{4}}\xi\right)\left(\Delta^{\frac{1}{4}}\xi\right)^*,$$

so that $\Delta^{1/4}\left\{ \xi\xi^\sharp : \xi \in \mathfrak{A}_0 \right\} = \left\{ \xi\xi^* : \xi \in \mathfrak{A}_0 \right\}$. Hence

$$\mathfrak{P}_0 = \left\{ \xi\xi^* : \xi \in \mathfrak{A}_0 \right\} \subset \Delta^{\frac{1}{4}}\mathfrak{P}_\varphi^\sharp.$$

If $\xi \in \mathfrak{A}$, set

$$\xi_r = \sqrt{\frac{r}{\pi}} \int_{-\infty}^{\infty} e^{-rt^2}\Delta^{it}\xi\, dt, \quad r > 0. \tag{10}$$

We then have, as $r \to \infty$,

$$\begin{cases} \|\xi - \xi_r\| \to 0, & \|\xi^\sharp - \xi_r^\sharp\| \to 0 \\ \|\pi_\ell(\xi_r)\| \le \|\pi_\ell(\xi)\|, & \pi_\ell(\xi_r) \to \pi_\ell(\xi)\ \sigma\text{-strongly}^*, \end{cases}$$

and that $\xi_r \in \mathfrak{A}_0$. Therefore we conclude

$$\mathfrak{P}_\varphi^\sharp = \{\xi\xi^\sharp : \xi \in \mathfrak{A}_0\}^-, \qquad \mathfrak{P}_\varphi = \{\xi\xi^* : \xi \in \mathfrak{A}_0\}^-, \qquad \mathfrak{P}_\varphi^\flat = \{\xi\xi^\flat : \xi \in \mathfrak{A}_0\}^-.$$

Thus, we obtain the inclusion: $\mathfrak{P}_\varphi \subset (\Delta^{1/4}\mathfrak{P}_\varphi^\sharp)^-$. To show the reversed inclusion, let $\xi \in \mathfrak{P}_\varphi^\sharp$ and choose $\xi_n \in \mathfrak{P}_0^\sharp = \{\eta\eta^\sharp : \eta \in \mathfrak{A}_0\}$ so that $\xi_n \to \xi$. Then we have $S\xi = \xi$, $S\xi_n = \xi_n$, so that $\Delta^{1/2}\xi_n = J\xi_n \to J\xi = \Delta^{1/2}\xi$, which yields the convergence:

$$\left\|\Delta^{\frac{1}{4}}\xi - \Delta^{\frac{1}{4}}\xi_n\right\|^2 = \left(\Delta^{\frac{1}{2}}(\xi - \xi_n) \mid \xi - \xi_n\right) \to 0.$$

Since $\Delta^{1/4}\xi_n \in \mathfrak{P}_0$, $\Delta^{1/4}\xi \in \mathfrak{P}_\varphi$. Thus $\Delta^{1/4}\mathfrak{P}_\varphi^\sharp \subset \mathfrak{P}_\varphi$. Q.E.D.

Lemma 1.4. *$\mathfrak{P}_\varphi$ is a self-dual cone and $\Delta^{it}\mathfrak{P}_\varphi = \mathfrak{P}_\varphi$.*

PROOF: For any $\xi \in \mathfrak{P}_\varphi^\sharp$ and $\eta \in \mathfrak{P}_\varphi^\flat$,

$$\left(\Delta^{\frac{1}{4}}\xi \mid \Delta^{-\frac{1}{4}}\eta\right) = (\xi \mid \eta) \ge 0$$

by Theorem 1.2.(i). Hence $(\xi \mid \eta) \ge 0$ for every pair $\xi, \eta \in \mathfrak{P}_\varphi$.

Now suppose that $\xi \in \mathfrak{H}_\varphi$ satisfies the inequality

$$(\xi \mid \eta) \ge 0 \quad \text{for every } \eta \in \mathfrak{P}_\varphi.$$

For each $r > 0$, define ξ_r by (10). Then for each $\eta \in \mathfrak{A}_0$, we have

$$\begin{aligned}(\xi_r \mid \eta\eta^*) &= \sqrt{\frac{r}{\pi}} \int_{-\infty}^{\infty} e^{-rt^2}(\Delta^{it}\xi \mid \eta\eta^*)\, dt \\ &= \sqrt{\frac{r}{\pi}} \int_{-\infty}^{\infty} e^{-rt^2}\left(\xi \,\middle|\, (\Delta^{-it}\eta)(\Delta^{-it}\eta)^*\right) dt \ge 0.\end{aligned}$$

Now, we get

$$\left(\Delta^{-\frac{1}{4}}\xi_r \mid \eta\eta^\flat\right) = \left(\xi_r \mid \Delta^{-\frac{1}{4}}(\eta\eta^\flat)\right) = \left(\xi_r \,\middle|\, (\Delta^{-\frac{1}{4}}\eta)(\Delta^{-\frac{1}{4}}\eta)^*\right) \ge 0.$$

Hence $\zeta_r = \Delta^{-1/4}\xi_r \in \mathfrak{P}_\varphi^\sharp$ because $\mathfrak{P}_0^\flat = \{\eta\eta^\flat : \eta \in \mathfrak{A}_0\}$ is dense in $\mathfrak{P}_\varphi^\flat$, so that $\xi_r = \Delta^{1/4}\zeta_r \in \mathfrak{P}_\varphi$. Therefore ξ belongs to $\mathfrak{P}_\varphi$ because $\|\xi - \xi_r\| \to 0$. Thus $\mathfrak{P}_\varphi$ is self-dual.

The invariance: $\Delta^{it}\mathfrak{P}_\varphi = \mathfrak{P}_\varphi$ follows from the fact that $\Delta^{it}(\xi\xi^*) = (\Delta^{it}\xi)(\Delta^{it}\xi)^*$, $\xi \in \mathfrak{A}_0$. Q.E.D.

Lemma 1.5. *If ψ is another faithful semi-finite normal weight on $\mathcal{M}$, then there exists uniquely a unitary $U_{\varphi,\psi}$ from $\mathfrak{H}_\psi$ onto $\mathfrak{H}_\varphi$ such that*

(i) $U_{\varphi,\psi}\pi_\psi(x)U_{\varphi,\psi}^* = \pi_\varphi(x), \quad x \in \mathcal{M}$;

(ii) $U_{\varphi,\psi}\mathfrak{P}_\psi = \mathfrak{P}_\varphi$.

PROOF: As in the previous chapter, we consider $\mathcal{N} = \mathcal{M} \otimes M(2;\mathbf{C})$ and the balanced weight $\rho = \varphi \oplus \psi$ given by (VIII.3.7). Then define S_φ, $S_{\varphi,\psi}$, $S_{\psi,\varphi}$ and S_ψ by (VIII.3.13) and consider their polar decompositions:

$$\begin{aligned} S_\varphi &= J_\varphi \Delta_\varphi^{\frac{1}{2}}, & S_{\varphi,\psi} &= J_{\varphi,\psi}\Delta_{\varphi,\psi}^{\frac{1}{2}}, \\ S_{\psi,\varphi} &= J_{\psi,\varphi}\Delta_{\psi,\varphi}^{\frac{1}{2}}, & S_\psi &= J_\psi \Delta_\psi^{\frac{1}{2}}. \end{aligned} \tag{11}$$

As in Theorem VIII.3.2, the operator

$$U_{\varphi,\psi} = J_\varphi J_{\varphi,\psi} \tag{12}$$

implements the unitary equivalence of π_φ and π_ψ. We recall the fact that $U_{\varphi,\psi}$ is also given by:

$$U_{\varphi,\psi} = J_{\varphi,\psi} J_\psi. \tag{12'}$$

Since $\mathfrak{P}_\varphi$ is the closure of $\mathfrak{P}_0$, we get

$$\mathfrak{P}_\varphi = \left\{\pi_\varphi(x)J_\varphi\eta_\varphi(x) : x \in \mathfrak{n}_\varphi \cap \mathfrak{n}_\varphi^*\right\}^-,$$

$$\mathfrak{P}_\psi = \left\{\pi_\psi(y)J_\psi\eta_\psi(y) : y \in \mathfrak{n}_\psi \cap \mathfrak{n}_\psi^*\right\}^-.$$

If $x \in \mathfrak{n}_\varphi \cap \mathfrak{n}_\varphi^*$ and $y \in \mathfrak{n}_\psi \cap \mathfrak{n}_\psi^*$, then we get

$$\begin{aligned}
&\left(\pi_\varphi(x)J_\varphi\eta_\varphi(x) \mid U_{\varphi,\psi}\pi_\psi(y)J_\psi\eta_\psi(y)\right) \\
&= \left(J_\varphi\eta_\varphi(x) \mid U_{\varphi,\psi}\pi_\psi(x^*)\pi_\psi(y)J_\psi\eta_\psi(y)\right) \\
&= \left(J_\varphi\eta_\varphi(x) \mid U_{\varphi,\psi}\pi_\psi(x^*)J_\psi\pi_\psi(y)J_\psi\eta_\psi(y)\right) \quad [1] \\
&= \left(J_\varphi\eta_\varphi(x) \mid U_{\varphi,\psi}J_\psi\pi_\psi(y)J_\psi\pi_\psi(x^*)\eta_\psi(y)\right) \\
&= \left(J_\varphi\eta_\varphi(x) \mid J_\varphi U_{\varphi,\psi}\pi_\psi(y)J_\psi\eta_\psi(x^*y)\right) \\
&= \left(\pi_\varphi(y)U_{\varphi,\psi}J_\psi\eta_\psi(x^*y) \mid \eta_\varphi(x)\right) \\
&= \left(J_{\varphi,\psi}\eta_\psi(x^*y) \mid \eta_\varphi(y^*x)\right) = \left(\Delta_{\varphi,\psi}^{\frac{1}{2}}\eta_\varphi(y^*x) \,\middle|\, \eta_\varphi(y^*x)\right) \geq 0,
\end{aligned}$$

where we used the fact that $y^*x \in \mathfrak{n}_\varphi \cap \mathfrak{n}_\psi$. Hence, the self-duality of $\mathfrak{P}_\varphi$ implies that $U_{\varphi,\psi}\mathfrak{P}_\psi \subset \mathfrak{P}_\varphi$. The converse inclusion relation follows from the fact that $U_{\varphi,\psi}^* = U_{\psi,\varphi}$.

[1] $\pi_\psi(y)J_\psi\eta_\psi(y) = J_\psi\pi_\psi(y)J_\psi\eta_\psi(y)$.

Now, suppose that U is a unitary from $\mathfrak{H}_\psi$ onto $\mathfrak{H}_\varphi$ satisfying (i) and (ii). To prove $U = U_{\varphi,\psi}$, set $V = U_{\varphi,\psi}U^*$. Then $V \in \pi_\varphi(\mathcal{M})'$ and $V\mathfrak{P}_\varphi = \mathfrak{P}_\varphi$. Hence for any $\xi \in \mathfrak{P}_\varphi$, we have $J_\varphi V\xi = V\xi = J_\varphi V J_\varphi \xi$. By the polarization identity, $\mathfrak{A}_0^2$ is spanned by $\{\xi\xi^* : \xi \in \mathfrak{A}_0\}$ and $\mathfrak{A}_0^2$ is dense in $\mathfrak{H}_\varphi$. Thus, we obtain $V = J_\varphi V J_\varphi$. Therefore, $V = V^* \in \pi_\varphi(\mathcal{M}) \cap \pi_\varphi(\mathcal{M})'$. So we write $V = E - F$ as a difference of orthogonal projections E, F (in the center of $\pi_\varphi(\mathcal{M})$ with $E + F = 1$). Since $F = FJ_\varphi FJ_\varphi$, for each $\xi \in \mathfrak{A}_0$ we have

$$F(\xi\xi^*) = (F\xi)(F\xi)^* \in \mathfrak{P}_\varphi,$$

so that $F\mathfrak{P}_\varphi \subset \mathfrak{P}_\varphi$. If $0 \neq F\xi = \xi$, $\xi \in \mathfrak{P}_\varphi$, then

$$\big(V\xi \mid \xi\big) = \big((E - F)\xi \mid \xi\big) = -\|F\xi\|^2 < 0,$$

which is a contradiction. Hence $F\mathfrak{P}_\varphi = \{0\}$. Since $[\mathfrak{P}_\varphi] = \mathfrak{H}_\varphi$ as we just observed, $F = 0$. Thus $V = 1$, which means that $U = U_{\varphi,\psi}$. Q.E.D.

Lemma 1.6. *For each $\omega \in \mathcal{M}_*^+$, there exists $\xi \in \mathfrak{P}_\varphi$ such that*

$$\omega(x) = \big(\pi_\varphi(x)\xi \mid \xi\big).$$

PROOF: We first prove the case where ω and φ commute and $\omega \leq \varphi$. As in §VII.2, there exists $h_\omega \in \pi_\varphi(\mathcal{M})'_+$ and $\eta_\omega \in \mathfrak{H}_\varphi$ such that

$$h_\omega^{\frac{1}{2}}\eta_\varphi(x) = \pi_\varphi(x)\eta_\omega, \qquad \omega(x) = \big(\pi_\varphi(x)\eta_\omega \mid \eta_\omega\big), \qquad x \in \mathcal{M}.$$

It follows easily that $\eta_\omega \in \mathfrak{P}_\varphi^\flat$. By assumption, ω and φ commute, so that $\Delta_\varphi^{it}\eta_\omega = \eta_\omega$. Hence $\eta_\omega \in \Delta_\varphi^{-1/4}\mathfrak{P}_\varphi^\flat \subset \mathfrak{P}_\varphi$.

We next prove the case that ω is faithful. We then consider the cyclic representation $\{\pi_\omega, \mathfrak{H}_\omega, \xi_\omega\}$ and set $\eta_\omega = U_{\varphi,\omega}\xi_\omega$. Since $\xi_\omega \in \mathfrak{P}_\omega$, Lemma 1.5 implies that $\eta_\omega \in \mathfrak{P}_\varphi$ and $\omega(x) = \big(\pi_\varphi(x)\eta_\omega \mid \eta_\omega\big)$.

Finally we prove the general case. If ω is not faithful, then we choose a semifinite normal weight ψ' on $\mathcal{M}$ such that $s(\psi') = 1 - s(\omega)$ and set $\psi = \psi' + \omega$. Then ω and ψ commute and $\omega \leq \psi$, so that there exists $\eta_\omega \in \mathfrak{P}_\psi$ such that $\omega(x) = \big(\pi_\psi(x)\eta_\omega \mid \eta_\omega\big)$. Now, we put $\xi = U_{\varphi,\psi}\eta_\omega$, which is the required vector by the previous lemma. Q.E.D.

In the above discussion, we used the notations π_φ and π_ψ in order to avoid confusion. But we now identify $\pi_\varphi(x)$ and x, so we write x alone instead of $\pi_\varphi(x)$. Thus $\mathfrak{H}$ means $\mathfrak{H}_\varphi$.

Lemma 1.7. *If $J\xi = \xi$, $\xi \in \mathfrak{H}$, then ξ is uniquely written as a difference $\xi = \xi_+ - \xi_-$ of two orthogonal vectors $\xi_+, \xi_- \in \mathfrak{P}_\varphi$.*

PROOF: Since $\mathfrak{P}_\varphi$ is a closed convex cone, the distance d from ξ to $\mathfrak{P}_\varphi$ is given by a vector $\xi_+ \in \mathfrak{P}_\varphi$, so that $d = \|\xi - \xi_+\|$ and ξ_+ is orthogonal to $\xi - \xi_+$. If $\eta \in \mathfrak{P}_\varphi$, $\eta \neq 0$, then

$$\|\xi - \xi_+\|^2 = d^2 < \|\xi - (\xi_+ + \lambda\eta)\|^2, \quad \lambda > 0,$$

so

$$\lambda^2\|\eta\|^2 - 2\lambda(\xi - \xi_+ \mid \eta) > 0, \quad \lambda > 0,$$

which is possible only if $(\xi - \xi_+ \mid \eta) \leq 0$. Thus $\xi_- = \xi_+ - \xi$ belongs to $\mathfrak{P}_\varphi$ and $\xi = \xi_+ - \xi_-$.

Suppose that $\xi = \xi'_+ - \xi'_-$, $\xi'_+ \perp \xi'_-$ and $\xi'_+, \xi'_- \in \mathfrak{P}_\varphi$. For each $\eta \in \mathfrak{P}_\varphi$, we have

$$\begin{aligned}\|\xi - \eta\|^2 &= \|\xi'_+ - \eta\|^2 + \|\xi'_-\|^2 - 2(\xi'_+ - \eta \mid \xi'_-) \\ &= \|\xi'_+ - \eta\|^2 + \|\xi'_-\|^2 + 2(\eta \mid \xi'_-) \geq \|\xi'_-\|^2.\end{aligned}$$

The last inequality becomes an equality only when $\eta = \xi'_+$. Hence ξ'_+ is the vector in $\mathfrak{P}_\varphi$ which gives the distance from ξ to $\mathfrak{P}_\varphi$. Thus $\xi'_+ = \xi_+$ and so $\xi'_- = \xi_-$. Q.E.D.

For each $\xi \in \mathfrak{P}_\varphi$, we consider $\omega = \omega_\xi \in \mathcal{M}_*^+$ and set $e = s(\omega)$. Then ω can be viewed as a faithful positive normal functional on the reduced algebra $\mathcal{M}_e$. Viewing $\{\pi_\omega, \mathfrak{H}_\omega, \xi_\omega\}$ as the cyclic representation of $\mathcal{M}_e$, we consider naturally $\{J_\omega, \Delta_\omega, \mathfrak{P}_\omega, \mathfrak{P}_\omega^\sharp, \mathfrak{P}_\omega^\flat\}$. Now put

$$U_{\varphi,\omega}\pi_\omega(x)\xi_\omega = x\xi, \quad x \in \mathcal{M}_e. \tag{13}$$

Lemma 1.8. *$U_{\varphi,\omega}$ is an isometry from $\mathfrak{H}_\omega$ onto $[\mathcal{M}\xi] \cap [\mathcal{M}'\xi]$ which enjoys the following properties:*

(i) $U_{\varphi,\omega}\pi_\omega(x) = xU_{\varphi,\omega}$, $x \in \mathcal{M}_e$;

(ii) $U_{\varphi,\omega}\mathfrak{P}_\omega = \mathfrak{P}_\varphi \cap [\mathcal{M}\xi] \cap [\mathcal{M}'\xi]$;

(iii) *$U_{\varphi,\omega}$ is unique, subject to conditions* (i) *and* (ii).

PROOF: We note that $e = [\mathcal{M}'\xi]$. If e' is the projection to $[\mathcal{M}\xi]$, then $e' = JeJ$ and $f = ee'$ commutes with J. It is clear that $U_{\varphi,\omega}$ is an isometry of $\mathfrak{H}_\omega$ onto $f\mathfrak{H}$ and satisfies (i).

We put $J_0 = U_{\varphi,\omega}^* J U_{\varphi,\omega}$. Then it follows that:

$$\left.\begin{aligned}&J_0\xi_\omega = \xi_\omega;\\ &J_0\pi_\omega(\mathcal{M}_e)J_0 = \pi_\omega(\mathcal{M}_e)';\\ &(J_0\pi_\omega(x)J_0\pi_\omega(x)\xi_\omega \mid \xi_\omega) \geq 0, \quad x \in \mathcal{M}_e.\end{aligned}\right\} \tag{14}$$

From this, we want to conclude that $J_0 = J_\omega$. To this end, put $H = J_0S_\omega$ on $\mathfrak{H}_\omega$, where S_ω means of course the $\sharp$-operation in $\mathfrak{H}_\omega$ determined by ω. For each $x \in \mathcal{M}_e$, we have

$$\big(H\pi_\omega(x)\xi_\omega \mid \pi_\omega(x)\xi_\omega\big) = \big(J_0\pi_\omega(x^*)\xi_\omega \mid \pi_\omega(x)\xi_\omega\big) = \big(\pi_\omega(x^*)J_0\pi_\omega(x^*)\xi_\omega \mid \xi_\omega\big)$$
$$= \big(J_0\pi_\omega(x^*)J_0\pi_\omega(x^*)\xi_\omega \mid \xi_\omega\big) \geq 0.$$

Since $\pi_\omega(\mathcal{M}_e)\xi_\omega$ is a core for both H and $\Delta_\omega^{1/2}$, we get $H = \Delta_\omega^{1/2}$ and $J_0 = J_\omega$ from the uniqueness of polar decomposition.

We have seen already

$$\mathfrak{P}_\omega = \overline{\Big\{ J_\omega\pi_\omega(x)J_\omega\pi_\omega(x)\xi_\omega : x \in \mathcal{M}_e \Big\}},$$

and also

$$U_{\varphi,\omega}J_\omega\pi_\omega(x)J_\omega\pi_\omega(x)\xi_\omega = JxJx\xi \in \mathfrak{P}_\varphi, \quad x \in \mathcal{M}_e,$$

so that $U_{\varphi,\omega}\mathfrak{P}_\omega \subset \mathfrak{P}_\varphi$. By the self-duality of $\mathfrak{P}_\omega$, we have

$$U_{\varphi,\omega}^*\big(\mathfrak{P}_\varphi \cap [\mathcal{M}\xi] \cap [\mathcal{M}'\xi]\big) \subset \mathfrak{P}_\omega.$$

Thus we obtain assertion (ii).

The uniqueness of $U_{\varphi,\omega}$ follows from the arguments similar to that for $U_{\varphi,\psi}$ in Lemma 1.5. We leave the detail to the reader. Q.E.D.

Remark 1.9. We note here that the property (14) characterizes the modular conjugation J_ω for a faithful $\omega \in \mathcal{M}_*^+$.

We shall use the following notations:

$$\Delta_\xi = U_{\varphi,\omega}\Delta_\omega U_{\varphi,\omega}^*, \qquad S_\xi = U_{\varphi,\omega}S_\omega U_{\varphi,\omega}^*, \qquad \mathfrak{P}_\xi = U_{\varphi,\omega}\mathfrak{P}_\omega. \tag{15}$$

Corollary 1.10. *For each $\omega \in \mathcal{M}_*^+$, the vector $\xi \in \mathfrak{P}_\varphi$ with $\omega = \omega_\xi$ is unique.*

PROOF: Suppose that $\omega = \omega_\xi = \omega_\eta$ for some $\xi, \eta \in \mathfrak{P}_\varphi$. Then there exists a partial isometry u of $\mathcal{M}'$ such that $u\xi = \eta$ and $u^*\eta = \xi$. Hence $[\mathcal{M}'\xi] = [\mathcal{M}'\eta]$. It then follows that

$$[\mathcal{M}\xi] = J[\mathcal{M}'\xi] = J[\mathcal{M}'\eta] = [\mathcal{M}\eta].$$

Therefore, if we define $V_{\varphi,\omega}$ by (13) replacing ξ with η, then $V_{\varphi,\omega}$ enjoys the exactly same property as $U_{\varphi,\omega}$ and $V_{\varphi,\omega}$ has the same range as $U_{\varphi,\omega}$. Now $V_{\varphi,\omega}^*U_{\varphi,\omega}$ leaves $\mathfrak{P}_\omega$ invariant. Hence, we have $V_{\varphi,\omega}^*U_{\varphi,\omega} = 1$ by Lemma 1.5. Thus $\xi = \eta$. Q.E.D.

This result allows us to identify $\{\mathfrak{H}_\omega, \xi_\omega\}$ with $\{[\mathcal{M}\xi] \cap [\mathcal{M}'\xi], \xi\}$. Hence we shall view ξ_ω as a vector in $\mathfrak{P}_\varphi$ for any $\omega \in \mathcal{M}_*^+$, so that $\mathfrak{P}_\omega = \mathfrak{P}_\xi$.

Lemma 1.11. *If $\xi \in \mathfrak{P}_\varphi$, then*

$$\mathfrak{P}_\xi = \Big\{(\mathbf{R}_+\xi - \mathfrak{P}_\varphi) \cap \mathfrak{P}_\varphi\Big\}^-.$$

PROOF: Putting $e = s(\omega_\xi)$, we have

$$\mathfrak{P}_\xi = \left\{ \Delta_\xi^{\frac{1}{4}} x\xi : x \in \mathcal{M}_e^+ \right\}^-.$$

Now, from the identity:

$$\|x\|\xi = \Delta_\xi^{\frac{1}{4}} x\xi + \Delta_\xi^{\frac{1}{4}}(\|x\| - x)\xi, \quad x \in \mathcal{M}_e^+,$$

it follows that $\Delta_\xi^{1/4} x\xi \in (\mathbf{R}_+\xi - \mathfrak{P}_\varphi) \cap \mathfrak{P}_\varphi$. Hence we get

$$\mathfrak{P}_\xi \subset \left\{ (\mathbf{R}_+\xi - \mathfrak{P}_\varphi) \cap \mathfrak{P}_\varphi \right\}^-.$$

To prove the converse inclusion, suppose that $\xi, \eta \in \mathfrak{P}_\varphi$ and $\zeta = \xi - \eta \in \mathfrak{P}_\varphi$. We choose a semi-finite normal weight ψ' with $s(\psi') = 1 - s(\omega_\xi)$ and put $\psi = \omega_\xi + \psi'$. Transferring all the structure in $\mathfrak{H}_\psi$ into $\mathfrak{H}$ by $U_{\varphi,\psi}$, we can replace φ by ψ, so that we may assume that ω and φ commute, which means that Δ leaves ξ invariant. For each $r > 0$, we consider ξ_r, η_r and ζ_r given by (10). But $\xi = \xi_r$ since $\xi = \Delta^{it}\xi$, so that

$$\xi = \eta_r + \zeta_r, \quad \eta_r, \zeta_r \in \mathfrak{P}_\varphi, \quad r > 0.$$

Now, we have

$$\xi = \Delta^{-\frac{1}{4}}\eta_r + \Delta^{-\frac{1}{4}}\zeta_r \in \mathfrak{P}_\varphi^\sharp; \quad \Delta^{-\frac{1}{4}}\eta_r \in \mathfrak{P}_\varphi^\sharp, \quad \Delta^{-\frac{1}{4}}\zeta_r \in \mathfrak{P}_\varphi^\sharp.$$

Since ξ is left and right bounded, $\Delta^{1/4}\eta_r$ and $\Delta^{-1/4}\zeta_r$ are both left bounded, and satisfy the inequality:

$$\pi_\ell(\Delta^{-\frac{1}{4}}\eta_r) \le \pi_\ell(\xi) \le s(\omega_\xi), \qquad \pi_\ell(\Delta^{-\frac{1}{4}}\zeta_r) \le \pi_\ell(\xi).$$

Setting $e = s(\omega_\xi)$, we have $e\Delta^{-1/4}\eta_r = \Delta^{-1/4}\eta_r$ and $e\Delta^{-1/4}\zeta_r = \zeta_r$. Since e and Δ commute, $\eta_r = e\eta_r$ and $\zeta_r = e\zeta_r$. As $r \to \infty$, we get $\eta_r \to \eta$ and $\zeta_r \to \zeta$, so that $\eta = e\eta = JeJ\eta$ and $\zeta = e\zeta = JeJ\zeta$. Thus, η and ζ both belong to $\mathfrak{P}_\xi$. Therefore, we get $(\mathbf{R}_+\xi - \mathfrak{P}_\varphi) \cap \mathfrak{P}_\varphi \subset \mathfrak{P}_\xi$. Q.E.D.

Lemma 1.12. *For each pair $\xi_1, \xi_2 \in \mathfrak{P}_\varphi$, the following three conditions are equivalent:*

(i) $\xi_1 \perp \xi_2$;

(ii) $\mathfrak{P}_{\xi_1} \perp \mathfrak{P}_{\xi_2}$;

(iii) $s(\omega_{\xi_1}) \perp s(\omega_{\xi_2})$.

PROOF: The implications (iii) $\Longrightarrow$ (ii) $\Longrightarrow$ (i) are obvious. In general $[\mathfrak{P}_\xi] = [\mathcal{M}\xi] \cap [\mathcal{M}'\xi]$ for $\xi \in \mathfrak{P}_\varphi$, so that (ii) $\Longrightarrow$ (iii) follows.

(i) $\Longrightarrow$ (ii): By the previous lemma, it is enough to show that

$$(\mathbf{R}_+\xi_1 - \mathfrak{P}_\varphi) \cap \mathfrak{P}_\varphi \perp (\mathbf{R}_+\xi_2 - \mathfrak{P}_\varphi) \cap \mathfrak{P}_\varphi.$$

If $\xi_1, \eta_2 \in \mathfrak{P}_\varphi$ and $\lambda\xi_1 - \eta_1 \in \mathfrak{P}_\varphi,\ \lambda > 0$, and $\mu\xi_2 - \eta_2 \in \mathfrak{P}_\varphi,\ \mu > 0$, then we have

$$0 \le (\eta_1 \mid \eta_2) \le \lambda\,(\xi_1 \mid \eta_2) \le \lambda\mu\,(\xi_1 \mid \xi_2) = 0,$$

so that $\eta_1 \perp \eta_2$. Q.E.D.

End of Proof of Theorem 1.2: We have already proved almost all of Theorem 1.2 except inequality (9). The first half of the inequality follows from the general fact:

$$\omega_\xi - \omega_\eta = \frac{1}{2}\big(\omega_{\xi+\eta,\,\xi-\eta} + \omega_{\xi-\eta,\,\xi+\eta}\big).$$

For a given pair $\xi, \eta \in \mathfrak{P}_\varphi$, choose orthogonal $\zeta_+, \zeta_- \in \mathfrak{P}_\varphi$ such that $\xi - \eta = \zeta_+ - \zeta_-$ by Lemma 1.7. By Lemma 1.12, $e = s(\omega_{\zeta_+}) \perp f = s(\omega_{\zeta_-})$. Putting $a = e - f$, we have

$$\begin{aligned}
\big\|\omega_\zeta - \omega_\eta\big\| &\ge \big|\omega_\xi(a) - \omega_\eta(a)\big| \qquad (\text{since } \|a\| = 1)\\
&= \frac{1}{2}\Big|\big(a\,(\xi+\eta) \mid \xi - \eta\big) + \big(a\,(\xi-\eta) \mid \xi+\eta\big)\Big|\\
&= \frac{1}{2}\Big|\big(a\,(\xi+\eta) \mid \zeta_+ - \zeta_-\big) + \big(a\,(\zeta_+ - \zeta_-) \mid \xi+\eta\big)\Big|\\
&= \frac{1}{2}\Big|\big(\xi+\eta \mid \zeta_+ + \zeta_-\big) + \big(\zeta_+ + \zeta_- \mid \xi+\eta\big)\Big| \quad ^{2}\\
&= \big(\xi+\eta \mid \zeta_+ + \zeta_-\big)\\
&\ge (\xi \mid \zeta_+) - (\xi \mid \zeta_-) - (\xi \mid \zeta_-) - (\eta \mid \zeta_+) + (\eta \mid \zeta_-) = \|\xi - \eta\|^2.
\end{aligned}$$

This completes the proof for the first part of inequality (9). The second part of (9) is a general fact. Q.E.D.

Based on the theorem just proven, we make the following definition:

Definition 1.13. Given a von Neumann algebra $\{\mathcal{M}, \mathfrak{H}\}$, a quadruple $\{\mathcal{M}, \mathfrak{H}, J, \mathfrak{P}\}$ of a unitary involution[3] J, called *modular conjugation*, and a self-dual cone $\mathfrak{P}$ in $\mathfrak{H}$ is said to be a *standard form* of $\mathcal{M}$ if the following requirements are satisfied:

(i) $J\mathcal{M}J = \mathcal{M}'$;

(ii) $JaJ = a^*,\ \ a \in \mathcal{C}_\mathcal{M}\ (= \mathcal{M} \cap \mathcal{M}')$;

(iii) $J\xi = \xi,\ \ \xi \in \mathfrak{P}$;

(iv) $aJaJ\mathfrak{P} \subset \mathfrak{P},\ \ a \in \mathcal{M}$.

---

2 $a(\zeta_+ - \zeta_-) = \zeta_+ + \zeta_-$.

3 Anti linear isometry J with $J^2 = 1$.

Theorem 1.14. *Suppose that* $\{\mathcal{M}_1, \mathfrak{H}_1, J_1, \mathfrak{P}_1\}$ *and* $\{\mathcal{M}_2, \mathfrak{H}_2, J_2, \mathfrak{P}_2\}$ *are standard forms. If* π *is an isomorphism of* $\mathcal{M}_1$ *onto* $\mathcal{M}_2$*, then there exists uniquely a unitary* u *from* $\mathfrak{H}_1$ *onto* $\mathfrak{H}_2$ *such that*

(i) $\pi(x) = uxu^*, \quad x \in \mathcal{M}_1;$

(ii) $J_2 = uJ_1u^*;$

(iii) $\mathfrak{P}_2 = u\mathfrak{P}_1.$

PROOF: Let $\{\mathcal{M}, \mathfrak{H}, J, \mathfrak{P}\}$ be a standard form. We shall prove that

$$\{\mathcal{M}, \mathfrak{H}, J, \mathfrak{P}\} \cong \{\pi_\varphi(\mathcal{M}), \mathfrak{H}_\varphi, J_\varphi, \mathfrak{P}_\varphi\}$$

with φ a faithful weight on $\mathcal{M}$, which in turn shows the existence of u. The uniqueness of u then follows from that of $U_{\varphi,\psi}$ in Lemma 1.5.

For each $\xi \in \mathfrak{P}$, set $\mathfrak{H}(\xi) = [\mathcal{M}\xi] \cap [\mathcal{M}'\xi]$ and $e = s(\omega_\xi)$. We define S_ξ naturally and set $H = JS_\xi$. The arguments that follow (14) show $\Delta_\xi^{1/2} = H$ and $J_\xi = J$. Since $\mathfrak{P}_\xi = \{xJx\xi : x \in \mathcal{M}_e\}^-$, we have $\mathfrak{P}_\xi \subset \mathfrak{P}$. The self-duality of $\mathfrak{P}_\xi$ in $\mathfrak{H}(\xi)$ then implies $\mathfrak{P}_\xi = \mathfrak{P} \cap \mathfrak{H}(\xi)$.

The σ-finite case: Assume the σ-finiteness for $\mathcal{M}$. Take a maximal orthogonal family $\{\xi_i : i \in I\}$ in $\mathfrak{P}$. Since $s(\omega_{\xi_i}) \perp s(\omega_{\xi_j}),\ i \neq j,\ I$ is countable. Adjusting norms, we may assume that $\sum \|\xi_i\|^2 < +\infty$. Put $\xi = \sum_{i\in I} \xi_i$. The maximality of $\{\xi_i\}$ implies, by Lemma 1.12, that ω_ξ is faithful. Hence we have $\mathfrak{H} = \mathfrak{H}(\xi)$, so that $\mathfrak{P} = \mathfrak{P}_\xi$.

The general case: For each $e \in \mathrm{Proj}(\mathcal{M})$, we set $\mathfrak{H}(e) = e\mathfrak{H} \cap Je\mathfrak{H}$. If e is σ-finite, then $\mathfrak{P} \cap \mathfrak{H}(e) = \mathfrak{P}_\xi$ for some $\xi \in \mathfrak{P}$ and there exists uniquely a unitary U_e from $\mathfrak{H}(e)$ onto $\pi_\varphi(e)J_\varphi\pi_\varphi(e)\mathfrak{H}_\varphi$ such that $U_e x = \pi_\varphi(x)U_e,\ x \in \mathcal{M}_e$, and $U_e\mathfrak{P}_\xi = \mathfrak{P}_\varphi \cap U_e\mathfrak{H}(e)$. The uniqueness of U_e means that U_e extends U_f if $f \leq e$. Since the family of σ-finite projections is upward directed and has supremum 1, there exists a common extension U of all possible U_e's. Since $\mathfrak{P} = \bigcup \mathfrak{P}_\xi$, this U enjoys the required properties. Q.E.D.

We shall close this section with an application of the above theorem to the automorphism group. Given a von Neumann algebra $\mathcal{M}$, we denote by $\mathrm{Aut}(\mathcal{M})$ the group of all automorphisms of $\mathcal{M}$. The group of all inner automorphism groups will be denoted by $\mathrm{Int}(\mathcal{M})$. It is easy to see that $\mathrm{Int}(\mathcal{M})$ is a normal subgroup, so we can form the quotient group $\mathrm{Out}(\mathcal{M})$ of $\mathrm{Aut}(\mathcal{M})$ by $\mathrm{Int}(\mathcal{M})$. To each finite subset $\{\omega_1, \omega_2, \dots, \omega_n\}$ of $\mathcal{M}_*$, and each $\alpha \in \mathrm{Aut}(\mathcal{M})$, we set

$$U(\alpha; \omega_1, \dots, \omega_n) = \left\{ \beta \in \mathrm{Aut}(\mathcal{M}) : \begin{array}{l} \|\omega_i \circ \alpha - \omega_i \circ \beta\| < 1; \\ \|\omega_i \circ \alpha^{-1} - \omega_i \circ \beta^{-1}\| < 1, \end{array} \ i = 1, \dots, n \right\}. \tag{16}$$

The family $\{U(\alpha; \omega_1, \dots, \omega_n) : \omega_1, \dots, \omega_n \in \mathcal{M}_*\}$ gives rise to a topology in $\mathrm{Aut}(\mathcal{M})$ which makes $\mathrm{Aut}(\mathcal{M})$ a topological group. It is not difficult to show that $\mathrm{Aut}(\mathcal{M})$ is complete with respect to this uniform structure.

Theorem 1.15. *If* $\{\mathcal{M}, \mathfrak{H}, J, \mathfrak{P}\}$ *is a standard form, then the group* $\mathcal{U}$ *of all unitaries* u *satisfying*

$$u\mathcal{M}u^* = \mathcal{M}, \qquad uJu^* = J, \qquad u\mathfrak{P} = \mathfrak{P}$$

is isomorphic to $\mathrm{Aut}(\mathcal{M})$ *under the map:* $u \in \mathcal{U} \mapsto \alpha_u \in \mathrm{Aut}(\mathcal{M})$, *where*

$$\alpha_u(x) = uxu^*, \quad x \in \mathcal{M}.$$

Furthermore, this map is a homeomorphism of $\mathcal{U}$ *equipped with the strong operator topology onto* $\mathrm{Aut}(\mathcal{M})$.

PROOF: The map: $u \in \mathcal{U} \mapsto \alpha_u \in \mathrm{Aut}(\mathcal{M})$ is clearly a homomorphism. By Theorem 1.14, it is surjective. The uniqueness in the theorem yields the injectivity. By inequality (9), the map: $\xi \in \mathfrak{P} \mapsto \omega_\xi \in \mathcal{M}_*^+$ is a homeomorphism, which in turn means that the map: $u \mapsto \alpha_u$ is a homeomorphism. Q.E.D.

Definition 1.16. The inverse map of the above map: $u \mapsto \alpha_u \in \mathrm{Aut}(\mathcal{M})$ is called the *standard implementation* and denoted by $U(\theta)$ for each $\theta \in \mathrm{Aut}(\mathcal{M})$.

The standard implementation $U(\theta)$ of $\theta \in \mathrm{Aut}(\mathcal{M})$ is characterized by the following:

$$\begin{aligned} \theta(x) &= U(\theta)xU(\theta)^*, \quad x \in \mathcal{M}, \\ U(\theta)\mathfrak{P} &= \mathfrak{P}. \end{aligned} \tag{17}$$

If we fix a standard form $\{\mathcal{M}, \mathfrak{H}, J, \mathfrak{P}\}$, then for any faithful weight φ, there exists uniquely a unitary U_φ of $\mathfrak{H}_\varphi$ onto $\mathfrak{H}$ such that

$$U_\varphi \pi_\varphi(x) U_\varphi^* = x, \quad x \in \mathcal{M}, \qquad U_\varphi \mathfrak{P} = \mathfrak{P}.$$

With this U_φ, we identify $\mathfrak{H}_\varphi$ and $\mathfrak{H}$. We then realize $\{\mathfrak{H}_\varphi, \eta_\varphi, J_\varphi, \mathfrak{P}_\varphi\}$ in $\mathfrak{H}$ so that

$$\mathfrak{H}_\varphi = \mathfrak{H}, \qquad J_\varphi = J, \qquad \mathfrak{P}_\varphi = \mathfrak{P}$$

and η_φ is a map from $\mathfrak{n}_\varphi$ into $\mathfrak{H}$ such that

$$\Delta_\varphi^{\frac{1}{4}} \eta_\varphi(x^*x) \in \mathfrak{P}, \quad x \in \mathfrak{n}_\varphi.$$

Proposition 1.17. *In the above setting, the standard implementation* $U(\theta)$ *of* $\theta \in \mathrm{Aut}(\mathcal{M})$ *has the property:*

$$U(\theta)\eta_\varphi(x) = \eta_{\varphi\circ\theta^{-1}}\big(\theta(x)\big), \quad x \in \mathfrak{n}_\varphi. \tag{18}$$

In particular, we have

$$U(\theta)\xi_\omega = \xi_{\omega\circ\theta^{-1}}, \quad \omega \in \mathcal{M}_*^+, \tag{19}$$

where ξ_ω *is the vector in* $\mathfrak{P}$ *such that* $\omega = \omega_{\xi_\omega}$.

PROOF: Since $\mathfrak{n}_{\varphi\circ\theta^{-1}} = \theta(\mathfrak{n}_\varphi)$, the right hand side of (18) makes sense. We define V to be the operator given by the right hand side of (18). It follows that V can be extended to a unitary on $\mathfrak{H}$ denoted by V again, which implements θ on $\mathcal{M}$ because for each $a \in \mathcal{M}$ and $x \in \mathfrak{n}_\varphi$

$$\begin{aligned}VaV^*\eta_\varphi(x) &= Va\eta_{\varphi\circ\theta}\big(\theta^{-1}(x)\big) = V\eta_{\varphi\circ\theta}\big(a\theta^{-1}(x)\big)\\ &= \eta_\varphi\big(\theta(a)x\big) = \theta(a)\eta_\varphi(x).\end{aligned}$$

To show $V = U(\theta)$, we have only to prove $V\mathfrak{P} = \mathfrak{P}$. For each $x \in \mathfrak{n}_\varphi \cap \mathfrak{n}_\varphi^*$, we have

$$\begin{aligned}VS_\varphi\eta_\varphi(x) &= V\eta_\varphi(x^*) = \eta_{\varphi\circ\theta^{-1}}\big(\theta(x^*)\big)\\ &= S_{\varphi\circ\theta^{-1}}\eta_{\varphi\circ\theta^{-1}}\big(\theta(x)\big) = S_{\varphi\circ\theta^{-1}}V\eta_\varphi(x).\end{aligned}$$

Since $\eta_\varphi(\mathfrak{n}_\varphi \cap \mathfrak{n}_\varphi^*)$ is a core for S_φ and $\eta_{\varphi\circ\theta^{-1}}\big(\theta(\mathfrak{n}_\varphi \cap \mathfrak{n}_\varphi^*)\big)$ is also a core for $S_{\varphi\circ\theta^{-1}}$, we have $VS_\varphi = S_{\varphi\circ\theta^{-1}}V$. Hence the uniqueness of the polar decomposition implies that

$$VJ = JV \qquad \text{and} \qquad V\Delta_\varphi V^* = \Delta_{\varphi\circ\theta^{-1}}.$$

Recalling that $\mathfrak{P} = \big\{xJ\eta_\varphi(x) : x \in \mathfrak{n}_\varphi\big\}^-$, we compute:

$$\begin{aligned}V\mathfrak{P} &= \big\{VxJ\eta_\varphi(x) : x \in \mathfrak{n}_\varphi\big\}^- = \big\{\theta(x)JV\eta_\varphi(x) : x \in \mathfrak{n}_\varphi\big\}^-\\ &= \Big\{\theta(x)J\eta_{\varphi\circ\theta^{-1}}\big(\theta(x)\big) : x \in \mathfrak{n}_\varphi\Big\}^-\\ &= \big\{yJ\eta_{\varphi\circ\theta^{-1}}(y) : y \in \mathfrak{n}_{\varphi\circ\theta^{-1}}\big\}^- = \mathfrak{P}_{\varphi\circ\theta^{-1}} = \mathfrak{P}.\end{aligned}$$

Q.E.D.

Let $\mathfrak{W}(\mathcal{M})$ be the set of semi-finite normal weights on $\mathcal{M}$ and $\mathfrak{W}_0(\mathcal{M})$ be the set of faithful semi-finite normal weights on $\mathcal{M}$. To each pair $(\psi, \varphi) \in \mathfrak{W}_0(\mathcal{M}) \times \mathfrak{W}_0(\mathcal{M})$, there corresponds canonically a unitary $U_{\psi,\varphi} \colon \mathfrak{H}_\varphi \mapsto \mathfrak{H}_\psi$ which carries the natural cone $\mathfrak{P}_\varphi$ of $\mathfrak{H}_\varphi$ onto $\mathfrak{P}_\psi$ of $\mathfrak{H}_\psi$ and J_φ onto J_ψ, and intertwines π_φ and π_ψ. We now define the canonical Hilbert space $L^2(\mathcal{M})$ associated with a given von Neumann algebra $\mathcal{M}$ as the Hilbert space:

$$\Big\{\zeta = \{\zeta_\varphi\} \in \prod_{\varphi\in\mathfrak{W}_0(\mathcal{M})} \mathfrak{H}_\varphi : \zeta_\psi = U_{\psi,\varphi}\zeta_\varphi \ \text{ for every } (\psi,\varphi) \in \mathfrak{W}_0(\mathcal{M}) \times \mathfrak{W}_0(\mathcal{M})\Big\} \tag{20}$$

equipped with the inner product defined by:

$$\big(\{\zeta_{1,\varphi}\} \bigm| \{\zeta_{2,\varphi}\}\big) = \big(\zeta_{1,\varphi_0} \bigm| \zeta_{2,\varphi_0}\big)$$

with any fixed $\varphi_0 \in \mathfrak{W}_0(\mathcal{M})$. The positive cone $L^2(\mathcal{M})_+$ of $L^2(\mathcal{M})$ is then defined as the set of those $\zeta = \{\zeta_\varphi\} \in L^2(\mathcal{M})$ such that $\zeta_\varphi \in \mathfrak{P}_\varphi$, $\varphi \in \mathfrak{W}_0(\mathcal{M})$. The action of $\mathcal{M}$ on $L^2(\mathcal{M})$ is then given by

$$x\{\zeta_\varphi\} = \big\{\pi_\varphi(x)\zeta_\varphi\big\}, \quad x \in \mathcal{M}.$$

By virtue of Theorem 1.14, for any fixed $\varphi_0 \in \mathfrak{W}_0(\mathcal{M})$ the map $U \colon \xi \in \mathfrak{H}_{\varphi_0} \mapsto \big\{U_{\psi,\varphi_0}\xi\big\} \in L^2(\mathcal{M})$ gives rise to the unitary equivalence of

$$\big\{\pi_{\varphi_0}(\mathcal{M}),\ \mathfrak{H}_{\varphi_0},\ \mathfrak{P}_{\varphi_0},\ J_{\varphi_0}\big\}$$

and the canonical $\big\{\mathcal{M},\ L^2(\mathcal{M}),\ L^2(\mathcal{M})_+,\ J\big\}$.

Definition 1.18. We call $\{\mathcal{M},\ L^2(\mathcal{M}),\ L^2(\mathcal{M})_+,\ J\}$ *the* standard form of $\mathcal{M}$ or the *canonical* standard form.

For any $\varphi \in \mathfrak{W}_0(\mathcal{M})$, the projection $U_\varphi \colon \{\xi_\psi\} \mapsto \xi_\varphi \in \mathfrak{H}_\varphi$ gives unitary equivalence between the canonical standard form $\{\mathcal{M},\ L^2(\mathcal{M}),\ L^2(\mathcal{M})_+,\ J\}$ and the one $\{\pi_{\varphi_0}(\mathcal{M}),\ \mathfrak{H}_{\varphi_0},\ \mathfrak{P}_{\varphi_0},\ J_{\varphi_0}\}$ associated with φ_0. The map: $x \in \mathfrak{n}_\varphi \mapsto U_\varphi^* \eta_\varphi(x) \in L^2(\mathcal{M})$ gives the realization of the semi-cyclic representation π_φ on $L^2(\mathcal{M})$. We will identify this map with η_φ and therefore the semi-cyclic representation π_φ with $\{L^2(\mathcal{M}),\ \eta_\varphi\}$ equipped with the canonical action of $\mathcal{M}$. In the case that we want to emphasize the dependence on φ, we will write $L^2(\mathcal{M}, \varphi)$. To cover a non-faithful weight, we use the following trick. Let φ be a faithful semi-finite normal weight on $\mathcal{M}$ and $e \in \mathrm{Proj}(\mathcal{M}_\varphi)$. For the semi-finite normal weight ψ defined by $\psi(x) = \varphi(exe),\ x \in \mathcal{M}_+$, we set:

$$\eta_\psi(x) = \eta_\varphi(xe), \quad x \in \mathfrak{n}_\psi. \tag{21}$$

It follows immediately that the action of $\mathcal{M}$ on $JeJL^2(\mathcal{M})$ together with the map η_ψ is the realization of the semi-cyclic representation $\{\pi_\psi,\ \mathfrak{H}_\psi,\ \eta_\psi\}$ inside the standard form.

Theorem 1.19. *Let $\mathcal{M}$ be a σ-finite von Neumann algebra in the standard form $\{\mathcal{M},\ \mathfrak{H},\ \mathfrak{H}_+,\ J\}$. Suppose that a sequence $\{\varphi_n\}$ of faithful normal positive functionals on $\mathcal{M}$ converges to a faithful φ in norm. Then for any fixed faithful semi-finite normal weight ψ we have the following convergence:*

(i) *the sequence $\{\Delta_{\varphi_n,\psi}\}$ of relative modular operators converges to $\Delta_{\varphi,\psi}$ in the strong resolvent sense.*

(ii) *The sequence $\{(\mathrm{D}\varphi_n:\mathrm{D}\psi)_t\}$ of the cocycle derivatives converges to $(\mathrm{D}\varphi:\mathrm{D}\psi)_t$ σ-strongly and uniformly in t on any bounded interval.*

(iii) *The sequence $\{\sigma^{\varphi_n}\}$ of modular automorphism groups converges to the modular automorphism group σ^φ of φ in the following sense:*

$$\lim_{n\to\infty} \left\| \sigma_t^{\varphi_n}(x)\eta - \sigma_t^{\varphi}(x)\eta \right\| = 0, \quad \eta \in \mathfrak{H}, \quad x \in \mathcal{M},$$

and the convergence is uniform on any bounded interval of t.

PROOF:

(i) Let $\mathfrak{a}_\psi = \mathfrak{n}_\psi \cap \mathfrak{n}_\psi^*$ and $\mathfrak{A}_\psi = \eta_\psi(\mathfrak{a}_\psi) \subset \mathfrak{H}$, a full left Hilbert algebra. Then with ξ_φ the representing vector of $\varphi \in \mathcal{M}_*$ in the cone $\mathfrak{H}_+$, we have $J\Delta_{\varphi,\psi}^{1/2}\eta_\psi(x) = x^*\xi_\varphi$. By inequality (9), we have

$$\lim_{n\to\infty} \left\| \xi_{\varphi_n} - \xi_\varphi \right\| = 0.$$

Hence we get, for any $x \in \mathfrak{a}_\psi$,

$$\begin{aligned}\lim_{n\to\infty} \left\| \left(\Delta_{\varphi_n,\psi} - \Delta_{\varphi,\psi}\right)\eta_\psi(x) \right\| &= \lim_{n\to\infty} \left\| \left(J\Delta_{\varphi_n,\psi} - J\Delta_{\varphi,\psi}\right)\eta_\psi(x) \right\| \\ &= \lim_{n\to\infty} \left\| x^*\xi_{\varphi_n} - x^*\xi_\varphi \right\| = 0.\end{aligned}$$

Since $\mathfrak{A}_\psi = \eta_\psi(\mathfrak{a}_\psi)$ is a core for $\Delta_{\varphi,\psi}^{1/2}$ and contained in $\bigcap_{n=1}^\infty \mathfrak{D}(\Delta_{\varphi_n,\psi}^{1/2})$, the sequence $\{\Delta_{\varphi_n,\psi}\}$ of relative modular operators converges to $\Delta_{\varphi,\psi}$ in the strong resolvent sense by Theorem A.6.(iii).

(ii) and (iii): We have $(\mathrm{D}\varphi_n : \mathrm{D}\psi)_t = \Delta_{\varphi_n,\psi}^{it}\Delta_\psi^{-it}$, $t \in \mathbf{R}$. The strong resolvent convergence of $\{\Delta_{\varphi_n,\psi}^{it}\}$ to $\Delta_{\varphi,\psi}$ implies the strong convergence of the sequence $\{\Delta_{\varphi_n,\psi}^{it}\}$ of one parameter unitary groups to the one parameter unitary group $\{\Delta_{\varphi,\psi}^{it}\}$ uniformly on any bounded interval of t by Theorem A.6. Thus our assertion follows.

Q.E.D.

Given a standard form $\{\mathcal{M}, \mathfrak{H}, J, \mathfrak{P}\}$ of a von Neumann algebra $\mathcal{M}$, we define an action of $\mathcal{M}$ on $\mathfrak{H}$ from the right as follows:

$$\xi x = Jx^* J\xi, \quad x \in \mathcal{M}, \quad \xi \in \mathfrak{H}. \tag{22}$$

Since $J\mathcal{M}J = \mathcal{M}'$, $\mathfrak{H}$ becomes a two sided $\mathcal{M}$-module and the commutant of $\mathcal{M}$ is precisely given by the right action of $\mathcal{M}$. Writing $J\xi = \xi^*$ as before, we have

$$(x\xi)^* = \xi x^*, \quad x \in \mathcal{M}, \quad \xi \in \mathfrak{H}. \tag{23}$$

We must note however that when $\{\mathcal{M}, \mathfrak{H}, J, \mathfrak{P}\}$ is given by a Tomita algebra $\mathfrak{A}$, for $\xi, \eta \in \mathfrak{A}$, we have

$$\xi\pi_\ell(\eta) = J\pi_\ell(\eta)^* J\xi = J(\eta^\sharp J\xi) = \xi\Delta^{\frac{1}{2}}\eta = \pi_r(\Delta^{\frac{1}{2}}\eta)\xi.$$

Thus, the right action of $x = \pi_\ell(\eta)$ is not $\pi_r(\eta)$ but $\pi_r(\Delta^{1/2}\eta)$.

For each $\varphi \in \mathcal{M}_*^+$, we denote the representing vector of φ in $\mathfrak{P}$ by ξ_φ, i.e. $\varphi(x) = (x\xi_\varphi \mid \xi_\varphi)$, $x \in \mathcal{M}$. It is easy to see that $x\xi_\varphi = \xi_\varphi x \iff x\varphi = \varphi x$. For each pair $x \in \mathcal{M}$ and $\xi \in \mathfrak{H}$ we write

$$[x, \xi] = x\xi - \xi x.$$

Proposition 1.20. *If h and k are self-adjoint elements of $\mathcal{M}$, then for each $\varphi \in \mathcal{M}_*^+$, there exists a Radon measure μ on $\mathbf{R} \times \mathbf{R}$ such that*

$$\|f(h)\xi_\varphi - \xi_\varphi g(k)\|^2 = \iint_{\mathbf{R}\times\mathbf{R}} |f(x) - g(y)|\, \mathrm{d}\mu(x, y) \tag{24}$$

for every bounded Borel functions f and g on $\mathbf{R}$.

PROOF: Let π (resp. π') be the left (resp. right) representation of $\mathcal{M}$ on $\mathfrak{H}$, i.e. $\pi(x)\xi = x\xi$ and $\pi'(x)\xi = \xi x$. Let A be the C^*-algebra generated by $\pi(h)$ and $\pi'(k)$ and 1. Since $\pi(h)$ and $\pi'(k)$ commute, A is abelian. Hence $A \cong C(\mathrm{Sp}(A))$. For each $\omega \in \mathrm{Sp}(A)$ put

$$x(\omega) = \omega(\pi(h)), \qquad y(\omega) = \omega(\pi'(k)).$$

Then we have $\big(x(\omega), y(\omega)\big) \in \mathbf{R}^2$. Since A is generated by $\pi(h)$ and $\pi'(k)$ and 1, the map: $\omega \in \mathrm{Sp}(A) \mapsto \big(x(\omega), y(\omega)\big) \in \mathbf{R}^2$ is injective and continuous. Hence under this map, $\mathrm{Sp}(A)$ is identified with a compact subset of $\mathbf{R}^2$, called the joint spectrum of $\pi(h)$ and $\pi'(k)$. The operators $\pi(h)$ and $\pi'(k)$ are respectively identified with the projection function from $\mathrm{Sp}(A)$ to the first (resp. second) coordinate of $\omega \in \mathrm{Sp}(A)$. Let μ be the spectral measure corresponding to the vector ξ_φ, i.e.

$$\|a\xi_\varphi\|^2 = \int_{\mathrm{Sp}(A)} |a(\omega)| \, \mathrm{d}\mu(\omega), \qquad a \in A.$$

Viewing the measure μ as a measure on $\mathbf{R}^2$ supported by $\mathrm{Sp}(A)$, we have

$$\left\| f\big(\pi(h), \pi'(k)\big)\xi_\varphi \right\|^2 = \iint_{\mathbf{R}^2} |f(x, y)|^2 \, \mathrm{d}\mu(x, y)$$

for every bounded Borel function f on $\mathbf{R}^2$. In particular, we have (24). Q.E.D.

Proposition 1.21. *Writing*

$$[x, \varphi] = x\varphi - \varphi x, \qquad x \in \mathcal{M}, \qquad \varphi \in \mathcal{M}_*,$$

we have

$$\big\|[x, \varphi]\big\| \leq 2\|\varphi\|^{\frac{1}{2}} \big\|[x, \xi_\varphi]\big\|, \qquad \varphi \in \mathcal{M}_*^+, \qquad x \in \mathcal{M}. \tag{25}$$

PROOF: For each $y \in \mathcal{M}$, we have

$$\begin{aligned}
\big|\langle y, [x, \varphi]\rangle\big| &= |\langle yx - xy, \varphi\rangle| = \big|(yx\xi_\varphi - xy\xi_\varphi \mid \xi_\varphi)\big| \\
&\leq \big|(y(x\xi_\varphi - \xi_\varphi x) \mid \xi_\varphi)\big| + \big|(y\xi_\varphi x - xy\xi_\varphi \mid \xi_\varphi)\big| \\
&\leq \|y\| \big\|[x, \xi_\varphi]\big\| \|\xi_\varphi\| + \big|(y\xi_\varphi \mid \xi_\varphi x^* - x^*\xi_\varphi)\big| \\
&\leq \|y\| \|\xi_\varphi\| \big\|[x, \xi_\varphi]\big\| + \|y\| \|\xi_\varphi\| \big\|[x^*, \xi_\varphi]\big\|.
\end{aligned}$$

But we have

$$\big\|[x^*, \xi_\varphi]\big\| = \big\|J[x^*, \xi_\varphi]\big\| = \big\|[J\xi_\varphi, x]\big\| = \big\|[\xi_\varphi, x]\big\|.$$ Q.E.D.

For each $a \in \mathbf{R}_+$ let E_a denote the characteristic function of the half line $[a, \infty[$. For each $x \in \mathcal{M}$ let $x = u(x)|x|$ be the polar decomposition, and define

$$u_a(x) = u(x)E_a\big(|x|\big), \qquad x \in \mathcal{M}. \tag{26}$$

Proposition 1.22. *For any $x \in \mathcal{M}$ and $\varphi \in \mathcal{M}_*^+$ we have*

(i)
$$\int_0^\infty \varphi\big(E_{\sqrt{a}}(|x|)\big)\, \mathrm{d}a = \varphi(x^*x), \tag{27}$$

(ii)
$$\int_0^\infty \big\|\big[u_{\sqrt{a}}(x), \xi_\varphi\big]\big\|^2\, \mathrm{d}a \le 4\big\|[x, \xi_\varphi]\big\|\varphi(x^*x + xx^*)^{\frac{1}{2}}. \tag{28}$$

PROOF:

(i) We have $E_{\sqrt{a}}\big(|x|\big) = E_a(x^*x)$ and

$$x^*x = \int_0^\infty E_a(x^*x)\, \mathrm{d}a\,.$$

This implies (i).

(ii) First, we assume $x = x^*$. Put $F_a(t) = \operatorname{sign} t\, E_a\big(|t|\big)$ for $a \in \mathbf{R}_+$ and $t \in \mathbf{R}$. Let μ be the measure on $\mathbf{R}^2$ determined by $h = k = x$ in Proposition 1.20. Then we have

$$\int_0^\infty \big\|\big[u_{\sqrt{a}}(x), \xi_\varphi\big]\big\|^2\, \mathrm{d}a = \int_0^\infty \iint_{\mathbf{R}^2} \big|F_{\sqrt{a}}(\alpha) - F_{\sqrt{a}}(\beta)\big|^2\, \mathrm{d}\mu(\alpha,\beta)\, \mathrm{d}a$$
$$= \iint_{\mathbf{R}^2} \left(\int_0^\infty \big|F_{\sqrt{a}}(\alpha) - F_{\sqrt{a}}(\beta)\big|^2\, \mathrm{d}a\right) \mathrm{d}\mu(\alpha,\beta)\,.$$

If $\operatorname{sign}(\alpha) = \operatorname{sign}(\beta)$, then

$$\int_0^\infty \big|F_{\sqrt{a}}(\alpha) - F_{\sqrt{a}}(\beta)\big|^2\, \mathrm{d}a = \int_0^\infty \big|E_{\sqrt{a}}(\alpha) - E_{\sqrt{a}}(\beta)\big|^2\, \mathrm{d}a$$
$$= \int_0^\infty \big|E_a(\alpha^2) - E_a(\beta^2)\big|^2\, \mathrm{d}a = \big|\alpha^2 - \beta^2\big| = |\alpha - \beta|\big(|\alpha| + |\beta|\big).$$

If $\operatorname{sign}(\alpha) = -\operatorname{sign}(\beta)$, then we have

$$\big|F_{\sqrt{a}}(\alpha) - F_{\sqrt{a}}(\beta)\big|^2 \le 2\big(E_a(\alpha^2) + E_a(\beta^2)\big),$$

so that

$$\int_0^\infty \big|F_{\sqrt{a}}(\alpha) - F_{\sqrt{a}}(\beta)\big|^2\, \mathrm{d}a \le 2\big(\alpha^2 + \beta^2\big)$$
$$\le 4(\alpha - \beta)^2 = 4|\alpha - \beta|\big(|\alpha| + |\beta|\big).$$

Therefore, we obtain

$$\begin{aligned}
\int_0^\infty \left\| [u_{\sqrt{a}}(x), \xi_\varphi] \right\|^2 \mathrm{d}a &\le 4 \iint_{\mathbf{R}^2} |\alpha - \beta| \big(|\alpha| + |\beta|\big) \,\mathrm{d}\mu(\alpha, \beta) \\
&\le 4 \Big(\iint_{\mathbf{R}^2} |\alpha - \beta|^2 \,\mathrm{d}\mu(\alpha, \beta) \Big)^{\frac{1}{2}} \\
&\quad \times \bigg(\Big(\iint_{\mathbf{R}^2} |\alpha|^2 \,\mathrm{d}\mu(\alpha, \beta) \Big)^{\frac{1}{2}} + \Big(\iint_{\mathbf{R}^2} |\beta|^2 \,\mathrm{d}\mu(\alpha, \beta) \Big)^{\frac{1}{2}} \bigg)^{\frac{1}{2}} \\
&= 4 \big\| [x, \xi_\varphi] \big\| \big(\|x\xi_\varphi\| + \|\xi_\varphi x\| \big) = 4 \big\| [x, \xi_\varphi] \big\| \big(\|x\xi_\varphi\| + \|x\xi_\varphi\| \big)^{\frac{1}{2}} \\
&= 4\sqrt{2} \big\| [x, \xi_\varphi] \big\| \varphi(x^2)^{\frac{1}{2}}.
\end{aligned}$$

This proves (28) in this self-adjoint case.

For the case $x \neq x^*$, we consider the 2×2-matrix algebra $\mathcal{M}_2 = M_2(\mathbf{C}) \otimes \mathcal{M}$. The standard form for $\mathcal{M}_2$ is given by 2×2-matrices over $\mathfrak{H}$. So let $\widetilde{\mathfrak{H}} = M_2(\mathbf{C}) \otimes \mathfrak{H}$ be the vector space of all 2×2-matrices with entries from $\mathfrak{H}$. Thus each element $\xi \in \widetilde{\mathfrak{H}}$ is of the form:

$$\xi = \begin{pmatrix} \xi_{11} & \xi_{12} \\ \xi_{21} & \xi_{22} \end{pmatrix}, \quad \xi_{i,j} \in \mathfrak{H}, \quad i, j = 1, 2;$$

$$\xi^* = \widetilde{J}\xi = \begin{pmatrix} J\xi_{11} & J\xi_{21} \\ J\xi_{12} & J\xi_{22} \end{pmatrix}.$$

Each element of $\mathcal{M}_2$ acts on $\widetilde{\mathfrak{H}}$ from the left by the matrix multiplication. The commutant $\mathcal{M}_2'$ of $\mathcal{M}_2$ is given by the right multiplication of $\mathcal{M}_2$. The natural positive cone $\widetilde{\mathfrak{P}}$ is then the closure of the set of matrices:

$$\begin{pmatrix} x_{11}\xi x_{11}^* + x_{12}\eta x_{12}^* & x_{11}\xi x_{21}^* + x_{12}\eta x_{22}^* \\ x_{21}\xi x_{11}^* + x_{22}\eta x_{12}^* & x_{21}\xi x_{21}^* + x_{22}\eta x_{22}^* \end{pmatrix}, \quad \xi, \eta \in \mathfrak{P}, \quad x = [x_{i,j}] \in \mathcal{M}_2.$$

Now with this setting, we consider

$$\widetilde{x} = \begin{pmatrix} 0 & x^* \\ x & 0 \end{pmatrix}, \qquad \xi_{\widetilde{\varphi}} = \begin{pmatrix} \xi_\varphi & 0 \\ 0 & \xi_\varphi \end{pmatrix}.$$

Then we get

$$\widetilde{x}^2 = \begin{pmatrix} x^*x & 0 \\ 0 & xx^* \end{pmatrix}, \qquad |\widetilde{x}| = \begin{pmatrix} |x| & 0 \\ 0 & |x^*| \end{pmatrix}, \qquad u(\widetilde{x}) = \begin{pmatrix} 0 & u(x)^* \\ u(x) & 0 \end{pmatrix},$$

$$E_a\big(|\widetilde{x}|\big) = \begin{pmatrix} E_a\big(|x|\big) & 0 \\ 0 & E_a\big(|x^*|\big) \end{pmatrix}, \qquad u_a(\widetilde{x}) = \begin{pmatrix} 0 & u_a(x)^* \\ u_a(x) & 0 \end{pmatrix}.$$

Furthermore, we have

$$u_a(\widetilde{x})\xi_{\widetilde{\varphi}} = \begin{pmatrix} 0 & u_a(x)^*\xi_\varphi \\ u_a(x)\xi_\varphi & 0 \end{pmatrix}, \qquad \xi_{\widetilde{\varphi}} u_a(\widetilde{x}) = \begin{pmatrix} 0 & \xi_\varphi u_a(x)^* \\ \xi_\varphi u_a(x) & 0 \end{pmatrix},$$

so that

$$\left\| [u_a(\widetilde{x}), \xi_{\widetilde{\varphi}}] \right\|^2 = \left\| [u_a(x), \xi_\varphi] \right\|^2 + \left\| [u_a(x)^*, \xi_\varphi] \right\|^2 = 2\left\| [u_a(x), \xi_\varphi] \right\|^2.$$

Hence we get

$$\begin{aligned} 2\int_0^\infty \left\| [u_{\sqrt{a}}(x), \xi_\varphi] \right\|^2 da &= \int_0^\infty \left\| [u_{\sqrt{a}}(\widetilde{x}), \xi_{\widetilde{\varphi}}] \right\|^2 da \\ &\leq 4\sqrt{2}\|[\widetilde{x}, \xi_{\widetilde{\varphi}}]\|\widetilde{\varphi}(\widetilde{x}^2)^{\frac{1}{2}} = 8\|[x, \xi_\varphi]\|\varphi(x^*x + xx^*)^{\frac{1}{2}}. \end{aligned}$$

Q.E.D.

Corollary 1.23. *If $x \in \mathcal{M}$, $x \neq 0$, and $\varphi \in \mathcal{M}_*^+$ satisfy the inequality for $\varepsilon > 0$:*

$$\|[x, \xi_\varphi]\| \leq \varepsilon\varphi(x^*x + xx^*)^{\frac{1}{2}},$$

then there exists $a > 0$ such that

$$\left\| [u_{\sqrt{a}}(x), \xi_\varphi] \right\| \leq 2\varepsilon\varphi\Big(u_{\sqrt{a}}(x)^* u_{\sqrt{a}}(x) + u_{\sqrt{a}}(x) u_{\sqrt{a}}(x)^*\Big)^{\frac{1}{2}}, \qquad u_a(x) \neq 0.$$

PROOF: By (27) and (28), we have

$$\begin{aligned} \int_0^\infty \left\| [u_{\sqrt{a}}(x), \xi_\varphi] \right\|^2 da &\leq 4\|[x, \xi_\varphi]\|\varphi(x^*x + xx^*)^{\frac{1}{2}} \\ &= \|[x, \xi_\varphi]\|\left(\int_0^\infty \varphi\Big(E_{\sqrt{a}}(|x|) + E_{\sqrt{a}}(|x^*|)\Big)\, da \right)^{\frac{1}{2}} \\ &= 4\|[x, \xi_\varphi]\|\left(\int_0^\infty \varphi\Big(u_{\sqrt{a}}(x)^* u_{\sqrt{a}}(x) + u_{\sqrt{a}}(x) u_{\sqrt{a}}(x)^*\Big)\right)^{\frac{1}{2}} \\ &\leq 4\varepsilon \int_0^\infty \varphi\Big(u_{\sqrt{a}}(x)^* u_{\sqrt{a}}(x) + u_{\sqrt{a}}(x) u_{\sqrt{a}}(x)^*\Big)\, da\,. \end{aligned}$$

Thus, we have for some $a > 0$

$$\left\| \left[u_{\sqrt{a}}(x), \xi_\varphi \right] \right\|^2 \leq 4\varepsilon\varphi\left(u_{\sqrt{a}}(x)^* u_{\sqrt{a}}(x) + u_{\sqrt{a}}(x) u_{\sqrt{a}}(x)^* \right) \neq 0.$$

Hence with this $a > 0$ we obtain:

$$\left\| \left[u_{\sqrt{a}}(x), \xi_\varphi \right] \right\| \leq 2\sqrt{\varepsilon}\varphi\left(u_{\sqrt{a}}(x)^* u_{\sqrt{a}}(x) + u_{\sqrt{a}}(x) u_{\sqrt{a}}(x)^* \right)^{\frac{1}{2}}.$$

Q.E.D.

The following lemma will be used later, in Chapter XVIII.

Lemma 1.24. *There exists a universal constant* $C > 0$ *such that*

$$\left\| \left(\Delta_\varphi^{it} - 1 \right) x \xi_\varphi \right\| \leq C\left(1 + |t| \right) \left\| [x, \xi_\varphi] \right\|, \quad x \in \mathcal{M}, \tag{29}$$

for any faithful $\varphi \in \mathcal{M}_*^+$.

PROOF: Since $\left\| [x, \xi_\varphi] \right\| = \left\| \left(1 - \Delta_\varphi^{1/2} \right) x \xi \right\|$, we simply estimate a bound of the function: $(\lambda, t) \in \mathbf{R}_+^* \times \mathbf{R} \mapsto \left(\lambda^{it} - 1 \right) \left(\lambda^{1/2} - 1 \right)^{-1} \in \mathbf{C}$. Thus we consider the function:

$$f(\lambda, t) = \frac{\mathrm{e}^{it} - 1}{\left(1 + |t| \right) \left(\mathrm{e}^{\frac{1}{2}\lambda} - 1 \right)}, \quad (\lambda, t) \in \mathbf{R}^2,$$

and observe that this function is indeed bounded on $\mathbf{R}^2$. Q.E.D.

Exercise IX.1

1) Let $\{\mathcal{M}, \mathfrak{H}, J, \mathfrak{P}\}$ be a von Neumann algebra in a standard form. In $\mathfrak{H}$, define an order relation $\xi \geq \eta$ by $\xi - \eta \in \mathfrak{P}$. For each $\omega \in \mathcal{M}_*^+$, let $\xi(\omega)$ be the vector in $\mathfrak{P}$ such that $\omega = \omega_{\xi(\omega)}$. Show that the map: $\omega \in \mathcal{M}_*^+ \mapsto \xi(\omega) \in \mathfrak{P}$ enjoys the following properties:

$$\xi(\lambda\omega) = \sqrt{\lambda}\xi(\omega), \quad \lambda \geq 0, \quad \omega \in \mathcal{M}_*^+;$$

$$\omega_1 \geq \omega_2 \Longrightarrow \xi(\omega_1) \geq \xi(\omega_2), \quad \omega_1, \omega_2 \in \mathcal{M}_*^+.$$

(Hint: Use Exercise VIII.3.8.)

$$\xi\left(\lambda\omega_1 + (1 - \lambda)\omega_2 \right) \geq \lambda\xi(\omega_1) + (1 - \lambda)\xi(w_2), \quad 0 \leq \lambda \leq 1,$$

(Hint: $\lambda\omega_{\xi_1} + (1 - \lambda)\omega_{\xi_2} \geq \omega_{\lambda\xi_1 + (1-\lambda)\xi_2}$ in general.)

2) With $\{\mathcal{M}, \mathfrak{H}, J, \mathfrak{P}\}$ as above, show that for every $\xi \in \mathfrak{H}$ there exist uniquely a partial isometry $u \in \mathcal{M}$ and $|\xi| \in \mathfrak{P}$ such that $\xi = u|\xi|$, u^*u is the cyclic projection to $\left[\mathcal{M}'|\xi| \right]$ and uu^* is the cyclic projection to $[\mathcal{M}'\xi]$. The vector $|\xi|$ is called the *absolute value* of ξ and u the *phase* of ξ. (Hint: Consider $\omega = \omega_\xi \in (\mathcal{M}'_*)^+$ and $|\xi| = \xi(\omega_\xi)$ relative to $\mathcal{M}'$.)

3) Show that if $\xi = u|\xi|$ is the polar decomposition, then $|\xi^*| = u|\xi|u^*$ and $\xi^* = u^*|\xi^*|$ is the polar decomposition of ξ^*, and $u|\xi| = |\xi^*|u$.

4) When we consider non faithful elements of $\mathcal{M}_*^+$, their modular automorphism groups and related objects mean the ones in the reduced algebra by their support. Let φ and ψ be elements in $\mathcal{M}_*^+$. Show that if $\varphi \le M\psi$ for some M, then

$$\xi(\varphi) = (\mathrm{D}\varphi : \mathrm{D}\psi)_{-\frac{i}{2}}\xi(\psi) = (\mathrm{D}\varphi : \mathrm{D}\psi)_{-\frac{i}{4}}\xi(\psi)(\mathrm{D}\varphi : \mathrm{D}\psi)^*_{-\frac{i}{4}}.$$

(Hint: For the first identity, compare $(\mathrm{D}\varphi : \mathrm{D}\psi)_z\xi(\psi)$ and $\Delta_{\varphi,\psi}^{iz}\xi(\psi)$. Then observe that they agree on $\mathbf{R}$ and conclude that

$$\xi(\varphi) = \Delta_{\varphi,\psi}^{\frac{1}{2}}\xi(\psi) = (\mathrm{D}\varphi : \mathrm{D}\psi)_{-\frac{i}{2}}\xi(\psi).$$

For the second identity, show that

$$a\xi(\psi) = \xi(\psi)\sigma^{\psi}_{\frac{i}{2}}(a), \quad a \in \mathfrak{D}\big(\sigma^{\psi}_{\frac{i}{2}}\big).$$

5) Assume that $\mathcal{M}$ is σ-finite and ψ is faithful. Prove that $\Theta_\psi : x \in \mathcal{M} \mapsto \Delta_\psi^{1/4}x\xi(\psi)$ is an order isomorphism of $\mathcal{M}$ onto $\Delta_\psi^{1/4}\mathcal{M}\xi(\psi)$, which is the linear span of the face $\bigcup_{\lambda>0}\big[0, \lambda\xi(\psi)\big]$ of $\mathfrak{P}$ generated by $\xi(\psi)$, where $[0,\xi] = \big\{\eta \in \mathfrak{H} : 0 \le \eta \le \xi\big\}$ in the ordered Hilbert space $\mathfrak{H}$ given by $\mathfrak{P}$.

6) Let $\big\{\mathcal{M}_i, \mathfrak{H}_i, J_i, \mathfrak{P}_i\big\}$, $i = 1, 2$, be the standard form of two σ-finite von Neumann algebras. Following the steps described below, prove that U is an isometry of $\mathfrak{H}_1$ onto $\mathfrak{H}_2$ with $U(\mathfrak{P}_1) = \mathfrak{P}_2$, then U gives rise to an order isomorphism θ, i.e. a Jordan $*$-isomorphism, of $\mathcal{M}_1$ onto $\mathcal{M}_2$ such that

$$\big(\theta(x)U\xi \mid U\xi\big) = \big(x\xi \mid \xi\big), \quad x \in \mathcal{M}_1, \quad \xi \in \mathfrak{H}_1.$$

In other words, the order structure of the standard form determines the order (Jordan) structure of a von Neumann algebra. [459]

Before going to the next step, recall that an order isomorphism between C^*-algebras is precisely a Jordan isomorphism by Exercise IV.1.2.

(a) Prove that U maps the order interval $\big[0, \xi(\psi_1)\big]$ given by a faithful $\psi_1 \in \mathcal{M}_*^+$ onto the order interval $\big[0, U\xi(\psi_1)\big]$ which generates a dense face of $\mathfrak{P}_2$, so that $\psi_2 = \omega_{U\xi(\psi_1)}$ is faithful.

(b) Prove that $\theta = \Theta_{\psi_2}^{-1} \circ U \circ \Theta_{\psi_1}$ is the required order isomorphism of $\mathcal{M}_1$ onto $\mathcal{M}_2$.

7) Show that if θ is a Jordan isomorphism of a von Neumann algebra $\mathcal{M}_1$ onto another $\mathcal{M}_2$, then there exists a unitary U_θ which maps the standard form $\big\{\mathcal{M}_1, \mathfrak{H}_1, J_1, \mathfrak{P}_1\big\}$ of $\mathcal{M}_1$ onto that $\big\{\mathcal{M}_2, \mathfrak{H}_2, J_2, \mathfrak{P}_2\big\}$ of $\mathcal{M}_2$ in such a way that

$$\big(\theta(x)\xi_2 \mid \xi_2\big) = \big(xU_\theta^{-1}\xi_2 \mid U_\theta^{-1}\xi_2\big), \quad \xi_2 \in \mathfrak{P}_2, \quad x \in \mathcal{M}_1.$$

(Hint: By Exercise IV.1.1 and 2, there exists a central projection $z \in \mathcal{M}_1$ such that θ is an isomorphism of $\mathcal{M}_1 z_1$ and an anti-isomorphism of $\mathcal{M}_1(1 - z_1)$. Use the isomorphism $\bar{\theta}(x) = \theta(xz_1) + J\theta\big(x(1-z_1)\big)^* J$, $x \in \mathcal{M}_1$, of $\mathcal{M}_1$, onto $\mathcal{M}_2\theta(z_1) \oplus \mathcal{M}_2'\big(1 - \theta(z_1)\big)$ to construct the required unitary U_θ.)

8) Keep the notations and the σ-finiteness assumption of Problem 5. Following the steps described below, prove that if $T \in \mathcal{L}(\mathfrak{H})$ is positive and invertible such that $T(\mathfrak{P}_1) = \mathfrak{P}_2$, then there exists a unique invertible $h \in \mathcal{M}_+$ such that $T = hJhJ$. [459]

(a) Show that if Θ is a linear map of $\mathcal{M}$ into $\mathfrak{H}$ such that $\Theta(\mathcal{M}_+)$ is a dense face of $\mathfrak{P}$, then there exist a unitary $U \in \mathcal{L}(\mathfrak{H})$ with $U(\mathfrak{P}) = \mathfrak{P}$ and a faithful $\psi \in \mathcal{M}_*^+$ such that

$$\Theta = U \circ \Theta_\psi .$$

(Hint: As $\Theta(\mathcal{M}_+)$ is dense in $\mathfrak{P}$, $\big[0, \Theta(1)\big]$ is total. So $\varphi = \omega_{\Theta(1)}$ is faithful. As Θ_φ is an order isomorphism from $\mathcal{M}$ onto the face of $\mathfrak{P}$ generated by $\xi(\varphi) = \Theta(1)$, $\theta = \Theta_\varphi^{-1} \circ \Theta$ is an order isomorphism of $\mathcal{M}$ onto itself such that $\theta(1) = 1$. Let $U = U_\theta$ be the unitary given by Problem 7, and $\psi = \omega_{U^{-1}\xi(\varphi)} = \varphi \circ \theta$. Then $U\Theta_\psi(x) = \Theta_\varphi \circ \Theta(x) = \Theta(x)$.)

(b) Let $\varphi \in \mathcal{M}_*^+$ be faithful. Show that there exists a faithful $\psi \in \mathcal{M}_*^+$ such that the absolute value of the bounded invertible operator: $\Delta_\varphi^{1/4} x\xi(\varphi) \mapsto \Delta_\psi^{1/4} x\xi(\psi)$ is exactly T. (Hint: Let $\Theta = T \circ \Theta_\varphi$ and apply (a) to Θ to find a unitary U and ψ such that

$$T\Delta_\varphi^{\frac{1}{4}} x\xi(\varphi) = U\Delta_\psi^{\frac{1}{4}} x\xi(\psi), \quad x \in \mathcal{M}.$$

Hence $U^* T\Delta_\varphi^{1/4} x\xi(\varphi) = \Delta_\psi^{1/4} x\xi(\psi)$, $x \in \mathcal{M}$.)

(c) Show the equivalence of the following statements:

(i) $\big\|\Delta_\varphi^{1/4} x\xi(\varphi)\big\| \geq \big\|\Delta_\psi^{1/4} x\xi(\psi)\big\|$, $x \in \mathcal{M}$;

(ii) The function: $t \in \mathbf{R} \mapsto u_t = (\mathrm{D}\psi : \mathrm{D}\varphi)_t \in \mathcal{M}$ has an analytic extension to a member $\{u_z\}$ of $\mathcal{A}_\mathcal{M}(\mathbf{D}_{1/4})$ such that $\|u_{-\mathrm{i}/4}\| \leq 1$.

Then show that if the above conditions hold, then

$$\Delta_\psi^{\frac{1}{4}} x\xi(\psi) = u_{-\frac{\mathrm{i}}{4}}\Big(\Delta_\varphi^{\frac{1}{4}} x\xi(\varphi)\Big)u_{-\frac{\mathrm{i}}{4}}^* .$$

(Hint: Making use of the 2×2-matrix technique, justify the following formal computations:

$$J\Delta_{\psi,\varphi}^{\frac{1}{2}}\Delta_\varphi^{-\frac{1}{2}} J = \Delta_\psi^{-\frac{1}{2}}\Delta_{\psi,\varphi}^{\frac{1}{2}}$$

as seen below:

$$\begin{aligned} J\Delta_{\psi,\varphi}^{\frac{1}{2}}\Delta_\varphi^{-\frac{1}{2}} Jx\xi(\varphi) &= J\Delta_{\psi,\varphi}^{\frac{1}{2}} x^*\xi(\varphi) = x\xi(\psi) = \Delta_\psi^{-\frac{1}{2}} Jx^*\xi(\varphi) \\ &= \Delta_\psi^{-\frac{1}{2}}\Delta_{\psi,\varphi}^{\frac{1}{2}} x\xi(\varphi); \end{aligned}$$

$$\begin{aligned}\Delta_\psi^{\frac14}x\xi(\psi) &= \Delta_\psi^{\frac14}J\Delta_{\psi,\varphi}^{\frac12}x^*\xi(\varphi) = \Delta_\psi^{\frac14}J\Delta_{\psi,\varphi}^{\frac12}\Delta_\varphi^{-\frac12}Jx\xi(\varphi)\\ &= \Delta_\psi^{\frac14}\Delta_\psi^{-\frac12}\Delta_{\psi,\varphi}^{\frac12}x\xi(\varphi) = \Delta_\psi^{-\frac14}\Delta_{\psi,\varphi}^{\frac12}\Delta_\varphi^{-\frac14}\Delta_\varphi^{\frac14}x\xi(\varphi)\\ &= \Delta_\psi^{-\frac14}\Delta_{\psi,\varphi}^{\frac14}\Delta_{\psi,\varphi}^{\frac14}\Delta_\varphi^{-\frac14}\Delta_\varphi^{\frac14}x\xi(\varphi)\\ &= \left(J\Delta_{\psi,\varphi}^{\frac14}\Delta_\varphi^{-\frac14}J\right)\left(\Delta_{\psi,\varphi}^{\frac14}\Delta_\varphi^{-\frac14}\right)\Delta_\varphi^{\frac14}x\xi(\varphi)\\ &= u_{-\frac{i}{4}}\left(\Delta_\varphi^{\frac14}x\xi(\varphi)\right)u_{-\frac{i}{4}}^*.\end{aligned}$$

(d) In view of (c), T is the absolute value of the operator $aJaJ$ for some $a \in \mathcal{M}$. But the absolute value $|aJaJ|$ is given by $T = |aJaJ| = |a|J|a|J$.

(e) If $aJaJ = 1$, $a \in \mathcal{M}$, then a is invertible and $a^{-1} = JaJ \in \mathcal{M} \cap \mathcal{M}' = \mathfrak{Z}$, so that a is central. If $h, k \in \mathcal{M}_+$ are invertible and if $hJhJ = kJkJ$, then $h = ak$ with some $a \in \mathfrak{Z}_+$ such that $1 = aJaJ = a^2$, so that $a = 1$. Thus the uniqueness of h with $T = hJhJ$ follows.

9) Keep the notations and the assumptions in the previous problem. Assume the innerness of a derivation on a von Neumann algebra which will be proven later, Theorem XI.3.5. Following the arguments presented below, prove that if $\delta \in \mathcal{L}(\mathfrak{H})$ generates a one parameter group $\exp(t\delta)$ of invertible operators such that $\exp(t\delta)(\mathfrak{P}) = \mathfrak{P}$, $t \in \mathbf{R}$, then $\delta = x + JxJ$ for some $x \in \mathcal{M}$.

(a) Based on the formula, called the *Lie-Trotter formula*:

$$\exp\big(t(\delta+\delta^*)\big) = \lim_{n\to\infty}\big(\exp(t\delta/n)\exp(t\delta/n)^*\big)^n;$$

$$\exp\big(t(\delta-\delta^*)\big) = \lim_{n\to\infty}\big(\exp(t\delta/n)\exp(-t\delta/n)^*\big)^n,$$

observe that it suffices to prove the claim for a self-adjoint δ and a skew-adjoint δ separately.

(b) If $\delta = \delta^*$, then $T_t = \exp(t\delta)$ is of the form $T_t = H_tJH_tJ$, $H_t \in \mathcal{M}_+$, by Problem 8. By the uniqueness of H_t, we have $H_{s+t} = H_sH_t$. Hence with $h = \lim_{t\to 0}(H_t - 1)/t$, we obtain $\delta = h + JhJ$.

(c) Assume $\delta = -\delta^*$. Then $U(t) = \exp(t\delta)$ is a one parameter unitary group such that $U(t)(\mathfrak{P}) = \mathfrak{P}$. By Problem 6, $U(t)$ gives rise to a one parameter group $\{\theta_t\}$ of Jordan *-automorphisms of $\mathcal{M}$ such that $U(t) = U_{\theta_t}$. As $\theta_t = (\theta_{t/2})^2$, each θ_t is the square of a Jordan *-automorphism which is an automorphism. Hence $\{\theta_t\}$ is a one parameter automorphism group of $\mathcal{M}$ which is given by $\theta_t(x) = U(t)xU(t)^*$, $t \in \mathbf{R}$, $x \in \mathcal{M}$.

(d) Differentiating the last equation, we get

$$\left.\frac{\mathrm{d}}{\mathrm{d}t}\theta_t(x)\right|_{t=0} = \delta x - x\delta = [\delta, x] \in \mathcal{M}, \quad x \in \mathcal{M},$$

so that $\mathrm{ad}(\delta) = [\delta, \cdot\,]$ is a derivation of $\mathcal{M}$.

(e) By Theorem XI.3.5, there exists $h \in \mathcal{M}_h$ such that $\delta x - x\delta = \mathrm{i}\,[h, x],\ x \in \mathcal{M}$. Hence $\delta - \mathrm{i}h = \mathrm{i}k \in \mathcal{M}'$. As $U(t)$ preserves $\mathfrak{P}$, $JU(t)J = U(t)$, so that $J\delta J = \delta$. Hence

$$\delta = \mathrm{i}h + \mathrm{i}k = J\big(\mathrm{i}h + \mathrm{i}h + \mathrm{i}k\big)J = -\mathrm{i}J(h+k)J.$$

Thus $a = h + JkJ = -(JhJ + k) \in \mathcal{M} \cap \mathcal{M}' = \mathfrak{Z}$, so that

$$a = a^* = JaJ = JhJ + k = -a.$$

Therefore $a = 0$, which means that

$$\delta = \mathrm{i}h - \mathrm{i}JhJ = \mathrm{i}h + J(\mathrm{i}h)J.$$

10) Keep the above notations and the assumptions. Observe that the set $\mathfrak{g} = \mathfrak{g}(\mathfrak{P})$ of all $\delta \in \mathcal{L}(\mathfrak{H})$ with $\exp(t\delta)(\mathfrak{P}) = \mathfrak{P},\ t \in \mathbf{R}$, is a Lie algebra under the Lie bracket operation: $[\delta_1, \delta_2] = \delta_1\delta_2 - \delta_2\delta_1$. By the previous problem, the map $x \in \mathcal{M} \mapsto x + JxJ \in \mathfrak{g}$ is a real linear surjective Lie algebra homomorphism from the Lie algebra $\mathcal{M}$ onto $\mathfrak{g}$ when $\mathcal{M}$ is viewed as a Lie algebra under the Lie product $[x, y] = xy - yx,\ x, y \in \mathcal{M}$.

(a) Prove that the kernel of the above Lie homomorphism is precisely $\{a \in \mathfrak{Z} : a^* = -a\}$ and that the center $\mathfrak{c}$ of $\mathfrak{g}$ is $\{a + JaJ : a \in \mathfrak{Z},\ a = a^*\}$.

(b) Consider the quotient Lie algebra $\hat{\mathfrak{g}} = \mathfrak{g}/\mathfrak{c}$, and denote the coset $\delta + \mathfrak{c} \in \hat{\mathfrak{g}}$ of δ by $\hat{\delta}$. Prove that the homomorphism: $x \in \mathcal{M} \mapsto x + JxJ \in \mathfrak{g}$ gives rise to an isomorphism j of the quotient Lie algebra $\mathcal{M}/\mathfrak{Z} = \hat{\mathcal{M}}$, i.e. the Lie algebra of derivations of $\mathcal{M}$, onto $\hat{\mathfrak{g}}$.

(c) Prove that if $a = [x, y],\ x, y \in \mathcal{L}(\mathfrak{H})$, commutes with both x and y, then $\mathrm{Sp}(a) = \{0\}$. Hence if $x, y \in \mathcal{M}$ and $a \in \mathfrak{Z}$, then $a = 0$. (Hint: As a and x commute, $[\mathrm{e}^{tx}, y] = ta\mathrm{e}^{tx}$, so that $\mathrm{e}^{tx}y\mathrm{e}^{-tx} = y + ta$. If $\lambda \in \mathrm{Sp}(a)$, then $t\lambda + \mathrm{Sp}(y) \subset \mathrm{Sp}(\mathrm{e}^{tx}y\mathrm{e}^{-tx}) = \mathrm{Sp}(y)$. Hence the boundedness of $\mathrm{Sp}(y)$ implies $\lambda = 0$.) Therefore, $\hat{\mathcal{M}}$ has no center other than $\{0\}$.

(d) Define $I_{\mathcal{M}} : \hat{\mathfrak{g}} \to \hat{\mathfrak{g}}$ by

$$I_{\mathcal{M}}(\hat{\delta}) = j\big(\mathrm{i}j^{-1}(\hat{\delta})\big).$$

Prove that $I = I_{\mathcal{M}}$ enjoys the following property:

$$\left.\begin{aligned} &I^2 = -\,\mathrm{id}, \qquad I(\hat{\delta}^*) = -I(\hat{\delta})^*; \\ &\big[\pm\delta_1, \delta_2\big] = \big[\delta_1, I\delta_2\big] = I\big[\delta_1, \delta_2\big]. \end{aligned}\right\} \tag{$*$}$$

(e) Call a real linear map I of $\hat{\mathfrak{g}}$ onto itself satisfying the above condition $(*)$ an *orientation* of $\mathfrak{P}$. Prove that if I_1 and I_2 are two orientations of $\mathfrak{P}$, then there exists a central projection $e \in \mathcal{M}$ such that $I_1 = I_2$ on $\hat{\mathfrak{g}}(e\mathfrak{P})$ and $I_1 = -I_2$ on $\hat{\mathfrak{g}}\big((1-e)\mathfrak{P}\big)$.

(Hint: Set $\mathcal{E} = I_2 \circ I_1^{-1}$. Prove that $\mathcal{E}^2(\hat{\delta}) - \hat{\delta}$ belongs to the center of $\hat{\mathfrak{g}}$ which is $\{0\}$ by (c). Thus $\hat{\mathfrak{g}}$ is the direct sum of two ideals

$$\hat{\mathfrak{g}}_1 = \big\{\hat{\delta} \in \hat{\mathfrak{g}} : \mathcal{E}(\hat{\delta}) = \hat{\delta}\big\} \quad \text{and} \quad \hat{\mathfrak{g}}_{-1} = \big\{\hat{\delta} \in \hat{\mathfrak{g}} : \mathcal{E}(\hat{\delta}) = -\hat{\delta}\big\}.$$

Set

$$\mathcal{M}_1 = \{ x \in \mathcal{M} : j \circ \mathrm{ad}(x) \in \hat{\mathfrak{g}}_1 \} \quad \text{and} \quad \mathcal{M}_{-1} = \{ x \in \mathcal{M} : j \circ \mathrm{ad}(x) \in \hat{\mathfrak{g}}_{-1} \}.$$

If $x \in \mathcal{M}_1$ and $y \in \mathcal{M}_{-1}$, then $j \circ \mathrm{ad}[x, y] = 0$, so $[x, y] \in \mathfrak{Z}$; hence $[x, y] = 0$ by (c). As $\mathcal{E}(\hat{\delta}^*) = \mathcal{E}(\hat{\delta})^*$, $\hat{\mathfrak{g}}_1$, $\hat{\mathfrak{g}}_{-1}$, $\mathcal{M}_1$ and $\mathcal{M}_1$ are all closed under the *-operation. If $x \in \mathcal{M}_1' \cap \mathcal{M}$, then $j \circ \mathrm{ad}(x)$ commutes with $\hat{\mathfrak{g}}_1$, hence it belongs to $\hat{\mathfrak{g}}_{-1}$. Hence $\mathcal{M}_{-1} = \mathcal{M}_1' \cap \mathcal{M}$. Similarly, $\mathcal{M}_1 = \mathcal{M}_{-1}' \cap \mathcal{M}$. Thus $\mathcal{M}_1$ and $\mathcal{M}_{-1}$ are both von Neumann subalgebras of $\mathcal{M}$, which contain $\mathfrak{Z}$. As $j \circ \mathrm{ad}(x) \in \hat{\mathfrak{g}}_1 + \hat{\mathfrak{g}}_{-1}$ for every $x \in \mathcal{M}$, we have $\mathcal{M} = \mathcal{M}_1 + \mathcal{M}_{-1}$. If $x = x_1 + x_{-1}$ and $y = y_1 + y_{-1}$ with $x_1, y_1 \in \mathcal{M}_1$ and $x_{-1}, y_{-1} \in \mathcal{M}_{-1}$, then $[x, y] = [x_1, y_1] + [x_{-1}, y_{-1}]$; hence $[x_1, y] = [x_1, y_1] \in \mathcal{M}_1$. If $z = z_1 + z_{-1}$ with $z_1 \in \mathcal{M}_1$ and $z_{-1} \in \mathcal{M}_{-1}$, then

$$[x_1, y_1]z = [x_1, y_1]z_1 + x_1 y_1 z_{-1} - y_1 z_{-1} x_1 = [x_1, y_1]z_1 + [x_1, y_1 z_{-1}] \in \mathcal{M}_1.$$

Hence if $e = \bigvee \{ u s_\ell([x, y]) u^* : x, y \in \mathcal{M}_1, \; u \in \mathcal{U}(\mathcal{M}_1) \}$, then e is a central projection of $\mathcal{M}_1$ such that $\mathcal{M}_{1,e}$ is an ideal of $\mathcal{M}$. As $\mathcal{M}_1 \cap \mathcal{M}_1' = \mathfrak{Z}$, $e \in \mathfrak{Z}$ and $\mathcal{M}e = \mathcal{M}_1 e$. Since $\mathcal{M}_1(1-e) = \mathfrak{Z}(1-e) = \mathfrak{Z}(1-e)$, we have $\mathcal{M}_1 = \mathcal{M}(1-e) + \mathfrak{Z}$. Now we have seen that $\mathcal{E} = \mathrm{id}$ on $\hat{\mathfrak{g}}(e\mathfrak{P})$ and $\mathcal{E} = -\mathrm{id}$ on $\hat{\mathfrak{g}}((1-e)\mathfrak{P})$.)

(f) Prove that if with $e \in \mathrm{Proj}(\mathfrak{Z})$ $I_1 = I_{\mathcal{M}}$ on $\hat{\mathfrak{g}}(e\mathfrak{P})$ and $I_1 = -I_{\mathcal{M}}$ on $\hat{\mathfrak{g}}((1-e)(\mathfrak{P}))$, then $I = I_{\mathcal{N}}$ where $\mathcal{N} = \mathcal{M}e + \mathcal{M}'(1-e)$.

(g) Prove, with $I = I_{\mathcal{M}}$, that

$$\mathcal{M} = \{ \delta_1 - I\delta_2 : \hat{\delta}_2 = I(\hat{\delta}_1), \; \delta_1, \delta_2 \in \mathfrak{g}(\mathfrak{P}) \};$$

$$\mathcal{M}' = \{ \delta_1 + I\delta_2 : \hat{\delta}_2 = I(\hat{\delta}_1), \; \delta_2 \in \mathfrak{g}(\mathfrak{P}) \}.$$

Therefore, the natural positive cone $\mathfrak{P}$ together with the orientation I determines the von Neumann algebra. [459].

11) Drop the σ-finiteness assumption from Problems (6), (8), (9) and (10).

12) Consider the standard form $\{ \tilde{\mathcal{M}}, \tilde{\mathfrak{H}}, \tilde{J}, \tilde{\mathfrak{P}} \}$ of $\tilde{\mathcal{M}} = \mathcal{M} \otimes M_2(\mathbf{C})$, viewing $\tilde{\mathfrak{H}}$ as the Hilbert space of 2×2-matrices with entries from $\mathfrak{H}$ and letting $\tilde{\mathcal{M}}$ acts by the left multiplication, where the inner product in $\tilde{\mathfrak{H}}$ is defined as follows:

$$([\xi_{i,j}] \mid [\eta_{k,\ell}]) = \sum_{i,j=1}^{2} (\xi_{ij} \mid \eta_{i,j}).$$

(a) Observe that

$$(\xi^*)_{i,j} = (\xi_{j,i})^*, \qquad \xi = [\xi_{i,j}] \in \tilde{\mathfrak{H}}.$$

(b) Show that $\begin{pmatrix} a & b^* \\ b & c \end{pmatrix} \in \tilde{\mathcal{M}}$ is positive if and only if $a, b \geq 0$ and

$$|(b\xi \mid \eta)| \leq (a\xi \mid \xi)^{\frac{1}{2}} (c\eta \mid \eta)^{\frac{1}{2}}, \quad \xi, \eta \in \mathfrak{H};$$

if and only if there exists a unique $s \in \mathcal{M}$ with $\|s\| \leq 1$ such that $b = c^{1/2} s a^{1/2}$.

(c) Show that $\begin{pmatrix} \varphi & \rho^* \\ \rho & \psi \end{pmatrix} \in \widetilde{\mathcal{M}}_*$ is positive when the pairing with $\widetilde{\mathcal{M}}$ is given by

$$\left\langle \begin{pmatrix} x_{11} & x_{12} \\ x_{21} & x_{22} \end{pmatrix}, \begin{pmatrix} \omega_{11} & \omega_{12} \\ \omega_{21} & \omega_{22} \end{pmatrix} \right\rangle = \sum_{i,j=1}^{2} \omega_{i,j}(x_{i,j}),$$

if and only if $\varphi, \psi \geq 0$ and

$$\left|\rho(y^*x)\right| \leq \varphi(x^*x)^{\frac{1}{2}}\psi(y^*y)^{\frac{1}{2}}, \quad x, y \in \mathcal{M}.$$

(d) Suppose that $\varphi \in \mathcal{M}_*^+$ is faithful. Show that $\begin{pmatrix} \xi & \zeta^* \\ \zeta & \eta \end{pmatrix} \in \widetilde{\mathfrak{H}}$ belongs to $\widetilde{\mathfrak{P}}$ if and only if $\xi, \eta \in \mathfrak{P}$ and

$$\left|\left(x^*\zeta y \mid \xi(\varphi)\right)\right| \leq \left(x^*\xi x \mid \xi(\varphi)\right)^{\frac{1}{2}}\left(y^*\eta y \mid \xi(\varphi)\right)^{\frac{1}{2}}, \quad x, y \in \mathcal{M}.$$

13†) Let $\{\mathcal{M}, \mathfrak{H}, J, \mathfrak{P}\}$ be a standard von Neumann algebra. If $\xi_0 \in \mathfrak{P}$ is cyclic, hence separating, then $\mathfrak{P}$ is precisely the set of all vectors of the form

$$\xi = TJTJ\xi_0$$

where T is a densely defined closed operator affiliated with $\mathcal{M}$ such that $\xi_0 \in \mathfrak{D}(T)$ and $JTJ\xi_0 = JT\xi_0 = JT\xi_0 \in \mathfrak{D}(T)$. [601, 602]

§ 2 Measurable Operators and Integral for a Trace

We fix a faithful semi-finite normal trace τ on a semi-finite von Neumann algebra $\mathcal{M}$.

Definition 2.1. The *measure topology* of $\mathcal{M}$ with respect to τ (or simply τ*-measure topology*) is the uniform topology given by a neighborhood system $\{x + N(\varepsilon, \delta) : \varepsilon, \delta > 0\}$, $x \in \mathcal{M}$, where $N(\varepsilon, \delta)$ is the set of all operators $a \in \mathcal{M}$ such that

$$\|ap\| < \varepsilon \quad \text{and} \quad \tau(p^\perp) < \delta$$

for some $p \in \text{Proj}(\mathcal{M})$. The convergence with respect to this topology is called τ-*measure* convergence. The completion of $\mathcal{M}$ with respect to this topology is denoted by $\mathfrak{M}(\mathcal{M})$.

When $\mathcal{M}$ acts on $\mathfrak{H}$, we define the $(\mathcal{M}, \tau)$*-measure* (or simply *measure*) *topology* of $\mathfrak{H}$ as the uniform topology given by a neighborhood system $\{\xi + O(\varepsilon, \delta) : \varepsilon > 0, \delta > 0\}$, $\xi \in \mathfrak{H}$, where $O(\varepsilon, \delta)$ is the set of all $\eta \in \mathfrak{H}$ such that

$$\|p\eta\| < \varepsilon \quad \text{and} \quad \tau(p^\perp) < \delta$$

for some $p \in \text{Proj}(\mathcal{M})$. The completion of $\mathfrak{H}$ is denoted by $\mathfrak{M}(\mathfrak{H})$. We naturally define the boundedness for subsets of $\mathcal{M}$ and $\mathfrak{H}$ with respect to the measure topology.

Theorem 2.2. *Consider the following*

$$a \in \mathcal{M} \mapsto a^* \in \mathcal{M}; \tag{1}$$

$$(a, b) \in \mathcal{M} \times \mathcal{M} \mapsto a + b \in \mathcal{M}; \tag{2}$$

$$(a, b) \in \mathcal{M} \times \mathcal{M} \mapsto ab \in \mathcal{M}; \tag{3}$$

$$(\xi, \eta) \in \mathfrak{H} \times \mathfrak{H} \mapsto \xi + \eta \in \mathfrak{H}; \tag{4}$$

$$(a, \xi) \in \mathcal{M} \times \mathfrak{H} \mapsto a\xi \in \mathfrak{H}. \tag{5}$$

These maps can respectively be extended to maps: $\mathfrak{M}(\mathcal{M}) \to \mathfrak{M}(\mathcal{M})$, $\mathfrak{M}(\mathcal{M}) \times \mathfrak{M}(\mathcal{M}) \to \mathfrak{M}(\mathcal{M})$, $\mathfrak{M}(\mathcal{M}) \times \mathfrak{M}(\mathcal{M}) \to \mathfrak{M}(\mathcal{M})$, $\mathfrak{M}(\mathfrak{H}) \times \mathfrak{M}(\mathfrak{H}) \to \mathfrak{M}(\mathfrak{H})$ *and* $\mathfrak{M}(\mathcal{M}) \times \mathfrak{M}(\mathfrak{H}) \to \mathfrak{M}(\mathfrak{H})$. *These extensions are unique, and maps* (1), (2) *and* (4) *together with their extensions are uniformly continuous. Maps* (3) *and* (5) *together with their extensions are uniformly continuous on the product set of bounded subsets. Thus,* $\mathfrak{M}(\mathcal{M})$ *is a topological involutive algebra with a continuous representation on a topological vector space* $\mathfrak{M}(\mathfrak{H})$.

PROOF: To prove the theorem, we shall show the following inclusions:

$$N(\varepsilon, \delta)^* \subset N(\varepsilon, 2\delta) \tag{1'}$$

$$N(\varepsilon_1, \delta_1) + N(\varepsilon_2, \delta_2) \subset N(\varepsilon_1 + \varepsilon_2, \delta_1 + \delta_2) \tag{2'}$$

$$N(\varepsilon_1, \delta_1)N(\varepsilon_2, \delta_2) \subset N(\varepsilon_1\varepsilon_2, \delta_1 + \delta_2) \tag{3'}$$

$$O(\varepsilon_1, \delta_1) + O(\varepsilon_2, \delta_2) \subset O(\varepsilon_1 + \varepsilon_2, \delta_1 + \delta_2) \tag{4'}$$

$$N(\varepsilon_1, \delta_1)O(\varepsilon_2, \delta_2) \subset O(\varepsilon_1\varepsilon_2, \delta_1 + \delta_2) \tag{5'}$$

Assume these inclusions for a moment. It then follows that the maps (1), (2), (4) are uniformly continuous. Suppose S_1 and S_2 are bounded subsets of $\mathcal{M}$. We want to show that the map (3) is uniformly continuous on $S_1 \times S_2$. To this end, for any $\varepsilon, \delta > 0$ we want to find $\varepsilon_1, \delta_1, \varepsilon_2, \delta_2 > 0$ so that

$$\big(a + N(\varepsilon_1, \delta_1)\big)\big(b + N(\varepsilon_2, \delta_2)\big) \subset ab + N(\varepsilon, \delta) \tag{6}$$

for every $a \in S_1$ and $b \in S_2$.

The boundedness of S_1 and S_2 means that for any $\alpha, \beta > 0$ there exists $\gamma > 0$ such that $\gamma S_1 \subset N(\alpha, \beta)$ and $\gamma S_2 \subset N(\alpha, \beta)$, which is equivalent to $S_1 \subset N(\alpha/\gamma, \beta)$ and $S_2 \subset N(\alpha/\gamma, \beta)$. Thus, for each $\alpha_1, \alpha_2 > 0$, we can find $\eta_1, \eta_2 > 0$ such that $S_1 \subset N(\eta_1, \alpha_1)$ and $S_2 \subset N(\eta_2, \alpha_2)$. By (2′) and (3′), we have, if $a \in S_1$ and $b \in S_2$,

$$\begin{aligned}
&\big(a + N(\varepsilon_1, \delta_1)\big)\big(b + N(\varepsilon_2, \delta_2)\big) - ab \\
&\quad\subset N(\eta_1, \alpha_1)N(\varepsilon_2, \delta_2) + N(\varepsilon_1, \delta_1)N(\eta_2, \alpha_2) + N(\varepsilon_1, \delta_1)N(\varepsilon_2, \delta_2) \\
&\quad\subset N\big(\eta_1\varepsilon_2, \alpha_1 + \delta_2\big) + N\big(\varepsilon_1\eta_2, \delta_1 + \alpha_2\big) + N\big(\varepsilon_1\varepsilon_2, \delta_1 + \delta_2\big) \\
&\quad\subset N\big(\eta_1\varepsilon_2 + \varepsilon_1\eta_2 + \varepsilon_1\varepsilon_2, \alpha_1 + \alpha_2 + 2\delta_1 + 2\delta_2\big).
\end{aligned}$$

Thus, we choose $\alpha_1, \alpha_2, \delta_1, \delta_2 > 0$ in such a way that $\alpha_1 + \alpha_2 + 2\delta_1 + 2\delta_2 \leq \delta$, and then choose $\varepsilon_1, \varepsilon_2 > 0$ further so that $\eta_1\varepsilon_2 + \eta_2\varepsilon_1 + \varepsilon_1\varepsilon_2 \leq \varepsilon$, to conclude (6). In a similar way, the uniform continuity of map (5) on a bounded set can be shown. Now, we shall show (1′), (2′) and (3′). The proof for (4′) and (5′) is similar, so we leave it to the reader.

(1′) If $a \in N(\varepsilon, \delta)$, then $\|ap\| < \varepsilon$ and $\tau(p^\perp) < \delta$ for some $p \in \mathrm{Proj}(\mathcal{M})$. We put

$$p_a = \bigvee \{ q \in \mathrm{Proj}(\mathcal{M}) : aq = paq \}. \tag{7}$$

It then follows that

$$p_a^\perp = \bigwedge \{ q \in \mathrm{Proj}(\mathcal{M}) : qa^* p^\perp = a^* p^\perp \},$$

so that $p_a^\perp$ is the range projection of $a^* p^\perp$. Hence

$$p_a^\perp \precsim p^\perp. \tag{8}$$

Applying (8) to p_{a^*}, we get

$$\tau((p \wedge p_{a^*})^\perp) = \tau(p^\perp \vee p_{a^*}^\perp) \leq \tau(p^\perp) + \tau(p_{a^*}^\perp) \leq 2\tau(p^\perp) < 2\delta,$$

and

$$a^*(p \wedge p_{a^*}) = pa^*(p \wedge p_{a^*}); \qquad \|a^*(p \wedge p_{a^*})\| \leq \|pa^*\| = \|ap\| < \varepsilon.$$

Thus, a^* belongs to $N(\varepsilon, 2\delta)$.

(2′) Suppose $a \in N(\varepsilon_1, \delta_1)$ and $b \in N(\varepsilon_2, \delta_2)$. Choose $p, q \in \mathrm{Proj}(\mathcal{M})$ so that

$$\|ap\| < \varepsilon_1, \qquad \|bq\| < \varepsilon_2, \qquad \tau(p^\perp) < \delta_1, \qquad \tau(q^\perp) < \delta_2.$$

Since $(a+b)(p \wedge q) = (ap + bq)(p \wedge q)$, we have $\|(a+b)(p \wedge q)\| < \varepsilon_1 + \varepsilon_2$ and

$$\tau((p \wedge q)^\perp) = \tau(p^\perp \vee q^\perp) \leq \tau(p^\perp) + \tau(q^\perp) \leq \delta_1 + \delta_2.$$

(3′) We use p_b given by (7) with b in the place of a. Now we have

$$ab(q \wedge p_b) = apb(q \wedge p_b),$$

so

$$\|ab(q \wedge p_b)\| < \varepsilon_1\varepsilon_2.$$

By (8), we get

$$\tau((q \wedge p_b)^\perp) = \tau(q^\perp \vee p_b^\perp) \leq \tau(q^\perp) + (p_b^\perp) \leq \tau(q^\perp) + \tau(p^\perp) < \delta_1 + \delta_2,$$

thus $ab \in N(\varepsilon_1\varepsilon_2, \delta_1 + \delta_2)$. Q.E.D.

Lemma 2.3.

(i) *The measure topologies of* $\mathcal{M}$ *and* $\mathfrak{H}$ *both satisfy the Hausdorff separation axiom. Thus* $\mathcal{M}$ *and* $\mathfrak{H}$ *can be viewed as subsets of their completion* $\mathfrak{M}(\mathcal{M})$ *and* $\mathfrak{M}(\mathfrak{H})$ *respectively.*

(ii) *For each* $a \in \mathfrak{M}(\mathcal{M})$ *and* $\varepsilon > 0$, *there exists* $p \in \mathrm{Proj}(\mathcal{M})$ *such that* $ap \in \mathcal{M}$ *and* $\tau(p^{\perp}) < \varepsilon$.

PROOF:

(i) Since they are both uniform topologies, it suffices to prove the T_1-separation axiom. Suppose $\xi \in \bigcap\{ O(\varepsilon, \delta) : \varepsilon > 0,\ \delta > 0\}$. For each $n = 1, 2, \ldots$, choose $q_n \in \mathrm{Proj}(\mathcal{M})$ so that $\|q_n\xi\| < 2^{-n}$ and $\tau(q_n^{\perp}) < 2^{-n}$. We set $p_n = \bigwedge_{k=n}^{\infty} q_k$. Then $\tau(p_n^{\perp}) < 2^{-n+1}$ and $p_1 \leq p_2 \leq \cdots$, so that $p_n^{\perp} \searrow 0$ which means $p_n \nearrow 1$. But $\|p_n\xi\| \leq \|q_k\xi\| < 2^{-k}$ for every $k \geq n$, so that $p_n\xi = 0$. Thus $\xi = 0$.

If $a \neq 0$, then $a\xi \neq 0$ for some $\xi \in \mathfrak{H}$. If $a \in \bigcap N(\varepsilon, \delta)$, then $a\xi \in \bigcap O(\varepsilon, \delta)$ by Theorem 2.2, which is impossible as just shown. Thus $\mathcal{M}$ satisfies the T_1-axiom.

(ii) If $a \in \mathfrak{M}(\mathcal{M})$, then a is the measure convergence limit of a sequence $\{a_n\}$ of $\mathcal{M}$. By selecting a subsequence, we may assume that

$$a = a_0 + \sum_{k=1}^{\infty}(a_{k+1} - a_k), \qquad a_{k+1} - a_k \in N(2^{-k}, 2^{-k}).$$

Put $b_k = a_{k+1} - a_k$, $k = 1, 2, \ldots$, and choose $q_k \in \mathrm{Proj}(\mathcal{M})$ so that $\|b_k q_k\| < 2^{-k}$ and $\tau(q_k^{\perp}) < 2^{-k}$. We then set $p_n = \bigwedge_{k \geq n} q_k$. It follows that $\{p_n\}$ is increasing and $\tau(p_n^{\perp}) < 2^{-n+1} \to 0$, so that $p_n \nearrow 1$. By Theorem 2.2, we get

$$ap_n = a_0 p_n + \sum_{k=1}^{\infty} b_k p_n = a_0 p_n + \sum_{k=1}^{n-1} b_k p_n + \sum_{k=n}^{\infty} b_k q_k p_n.$$

The last summation converges in norm, so that $ap_n \in \mathcal{M}$. Q.E.D.

Definition 2.4. For each $a \in \mathfrak{M}(\mathcal{M})$, we set

$$\mathfrak{D}\big(M(a)\big) = \big\{ \xi \in \mathfrak{H} : a\xi \in \mathfrak{H} \big\}, \qquad M(a)\xi = a\xi, \quad \xi \in \mathfrak{D}\big(M(a)\big). \tag{9}$$

Theorem 2.5.

(i) *For each* $a \in \mathfrak{M}(\mathcal{M})$, $M(a)$ *is a densely defined closed operator affiliated with* $\mathcal{M}$ *and maximal in the sense that* $M(a)$ *has no proper closed extension affiliated with* $\mathcal{M}$.

(ii) *For each pair* $a, b \in \mathfrak{M}(\mathcal{M})$, *we have*

$$M(a^*) = M(a)^*,$$

$$M(a + b) = \overline{M(a) + M(b)},$$

$$M(ab) = \overline{M(a)M(b)},$$

where the bar on the right hand side means the closure.

(iii) *If a sequence* $\{p_n\}$ *in* $\mathrm{Proj}(\mathcal{M})$ *is increasing and* $\tau(p_n^{\perp}) \searrow 0$ *and if a linear operator* A *defined on* $\mathfrak{D} = \bigcup p_n \mathfrak{H}$ *has the property that* $Ap_n \in \mathcal{M}$, *then* A *is preclosed and there exists uniquely an element* $a \in \mathfrak{M}(\mathcal{M})$ *with* $M(a) = \bar{A}$.

We need a few lemmas.

Lemma 2.6.

(i) *If* $p \wedge q = 0$, $p, q \in \mathrm{Proj}(\mathcal{M})$, *then* $p \precsim q^{\perp}$ *and* $q \precsim p^{\perp}$.

(ii) *Suppose that* e *and* f *are projections in* $\mathcal{M}$. *If for any* $\varepsilon > 0$ *there exists* $p \in \mathrm{Proj}(\mathcal{M})$ *with*

$$e \wedge p = f \wedge p \quad \textit{and} \quad \tau(p^{\perp}) < \varepsilon,$$

then $e = f$.

PROOF:

(i) If $p \wedge q = 0$, then $p^{\perp} \vee q^{\perp} = 1$, so

$$\begin{aligned} p &= p^{\perp} \vee q^{\perp} - p^{\perp} \sim q^{\perp} - q^{\perp} \wedge p^{\perp} \leq q^{\perp}, \\ q &= p^{\perp} \vee q^{\perp} - q^{\perp} \sim p^{\perp} - q^{\perp} \wedge p^{\perp} \leq p^{\perp}. \end{aligned}$$

(ii) We have

$$(e - e \wedge f) \wedge p \leq e \wedge p - (e \wedge f) \wedge p = f \wedge p - f \wedge p = 0,$$

so (i) implies that $e - e \wedge f \precsim p^{\perp}$. Hence we get $\tau(e - e \wedge f) < \varepsilon$ for any $\varepsilon > 0$, so $e = e \wedge f$. Similarly, we get $f = e \wedge f$. Q.E.D.

Lemma 2.7. *Given two densely defined closed operators A and B affiliated with* $\mathcal{M}$, *if for any* $\varepsilon > 0$, *there exists* $p \in \mathrm{Proj}(\mathcal{M})$ *with* $\tau(p^{\perp}) < \varepsilon$ *such that*

$$\begin{aligned} p\mathfrak{H} \cap \mathfrak{D}(A) \cap A^{-1}(p\mathfrak{H}) &= p\mathfrak{H} \cap \mathfrak{D}(B) \cap B^{-1}(p\mathfrak{H}) = \mathfrak{D}, \\ A\xi &= B\xi \quad \textit{for every } \xi \in \mathfrak{D}, \end{aligned}$$

then $A = B$. *In particular, if* $p\mathfrak{H} \subset \mathfrak{D}(A) \cap \mathfrak{D}(B)$ *and* $Ap = Bp$ *for some* $p \in \mathrm{Proj}(\mathcal{M})$ *with* $\tau(p^{\perp}) < \varepsilon$, *then* $A = B$.

PROOF: Consider $\tilde{\mathcal{M}} = \mathcal{M} \otimes M_2(\mathbf{C})$ on $\tilde{\mathfrak{H}} = \mathfrak{H} \oplus \mathfrak{H}$ and the trace $\tilde{\tau}$ defined by

$$\tilde{\tau}\begin{pmatrix} a & b \\ c & d \end{pmatrix} = \tau(a) + \tau(b).$$

It follows that $\tilde{\tau}$ is a faithful semi-finite normal trace on $\tilde{\mathcal{M}}$. Let $g(A)$ and $g(B)$ be the projections of $\tilde{\mathfrak{H}}$ onto the graphs of A and B respectively. The direct computation shows that

$$g(A) = \begin{pmatrix} (1 + A^*A)^{-1} & A^*(1 + AA^*)^{-1} \\ A(1 + A^*A)^{-1} & AA^*(1 + AA^*)^{-1} \end{pmatrix}. \tag{10}$$

Hence $g(A)$ and $g(B)$ both belong to $\tilde{M}$. Now, set

$$\tilde{p} = p \otimes 1 = \begin{pmatrix} p & 0 \\ 0 & p \end{pmatrix} \in \tilde{\mathcal{M}}.$$

Then we have $\tilde{\tau}(\tilde{p}^{\perp}) = 2\tau(p^{\perp}) < 2\varepsilon$. The assumption on A and B means here that $g(A) \wedge \tilde{p} = g(B) \wedge \tilde{p}$. The previous lemma then shows that $g(A) = g(B)$, hence the conclusion. Q.E.D.

We are now ready to complete the proof of Theorem 2.5.

PROOF OF THEOREM 2.5:

(i) If $a \in \mathfrak{M}(\mathcal{M})$, then we can find a sequence $\{a_n\}$ in $\mathcal{M}$ which converges to a in the measure convergence. By Theorem 2.2, $\{a_n\xi\}$ converges to $a\xi \in \mathfrak{M}(\mathfrak{H})$ in measure. If $\xi \in \mathfrak{D}(M(a))$, then $\{a_n\xi\}$ converges to $a\xi = M(a)\xi \in \mathfrak{H}$ in measure. If $u \in \mathcal{U}(\mathcal{M}')$, then $a_n u\xi \to au\xi$ in measure and $a_n u\xi = ua_n\xi$. Since operators in $\mathcal{M}'$ are continuous in the measure topology, we get the measure convergence: $ua_n\xi \to uM(a)\xi$. Thus we have $u\xi \in \mathfrak{D}(M(a))$ and $M(a)u\xi = uM(a)\xi$, so that $M(a)$ is affiliated with $\mathcal{M}$. Suppose now that $\{\xi_n\} \subset \mathfrak{D}(M(a))$, $\xi = \lim \xi_n$ and $\eta = \lim M(a)\xi_n$. Since $\xi_n \to \xi$ in measure, $a\xi_n \to a\xi \in \mathfrak{M}(\mathfrak{H})$ in measure. The convergence of $M(a)\xi_n$ to η implies the measure convergence. The separation property of the measure topology in $\mathfrak{M}(\mathfrak{H})$ yields that $\eta = a\xi$. Thus $\xi \in \mathfrak{D}(M(a))$ and $\eta = M(a)\xi$, which means that $M(a)$ is closed. Finally, if $ap \in \mathcal{M}$, $p \in \mathrm{Proj}(\mathcal{M})$, then $p\mathfrak{H} \subset \mathfrak{D}(M(a))$, so that Lemma 2.3.(ii) guarantees the density of $\mathfrak{D}(M(a))$ in $\mathfrak{H}$.

Suppose that A is a closed operator affiliated with $\mathcal{M}$ and $A \supset M(a)$. If $ap \in \mathcal{M}$, $p \in \mathrm{Proj}(\mathcal{M})$, then $Ap = ap = M(a)p \in \mathcal{M}$. By Lemma 2.7, we have $A = M(a)$. Thus $M(a)$ is maximal.

(ii) Let $\varepsilon > 0$. Choose $p \in \mathrm{Proj}(\mathcal{M})$ by Lemma 2.3.(ii) so that $a^*p \in \mathcal{M}$ and $\tau(p^{\perp}) < \varepsilon$. Then $p\mathfrak{H} \subset \mathfrak{D}(M(a^*))$ and $a^*p = M(a^*)p$. From the spectral analysis of the absolute value $|M(a)|$ of $M(a)$, it follows that there exists an increasing sequence $\{q_n\}$ in $\mathrm{Proj}(\mathcal{M})$ with $1 = \lim q_n$ such that $M(a)q_n = aq_n \in \mathcal{M}$ and $\xi \in \mathfrak{H}$ belongs to $\mathfrak{D}(M(a))$ if and only if $M(a)q_n\xi$ converges, and $M(a)\xi = \lim M(a)q_n\xi$. Now if $\xi \in \mathfrak{D}(M(a))$ and $\eta \in p\mathfrak{H}$, then we have

$$\begin{aligned}\big(M(a)\xi \mid \eta\big) &= \lim\big(aq_n\xi \mid \eta\big) = \lim\big(\xi \mid (q_n a)^* p\eta\big) = \lim\big(\xi \mid q_n(a^*p)\eta\big)\\ &= \lim\big(q_n\xi \mid a^*p\eta\big) = \big(\xi \mid a^*\eta\big) = \big(\xi \mid M(a^*)\eta\big).\end{aligned}$$

Hence $p\mathfrak{H} \subset \mathfrak{D}(M(a)^*)$ and $M(a)^*p = M(a^*)p$. Lemma 2.7 now implies that $M(a)^* = M(a^*)$.

Trivially, we have

$$M(a) + M(b) \subset M(a+b) \qquad \text{and} \qquad M(a)M(b) \subset M(ab).$$

Thus, $M(a) + M(b)$ and $M(a)M(b)$ are both preclosed. Choose $p, q \in \mathrm{Proj}(\mathcal{M})$ with $\tau(p^{\perp}) < \varepsilon$ and $\tau(q^{\perp}) < \varepsilon$ such that $ap \in \mathcal{M}$ and $bq \in \mathcal{M}$. Put $r = p \wedge q$. Then we get $\tau(r^{\perp}) < 2\varepsilon$ and

$$r\mathfrak{H} \subset \mathfrak{D}\big(M(a)\big) \cap \mathfrak{D}\big(M(b)\big),$$

$$\big[M(a) + M(b)\big]r = ar + br = M(a+b)r.$$

Thus, Lemma 2.7 implies that $M(a+b) = \overline{M(a) + M(b)}$. With p_b given by applying (7) to b, we put $s = q \wedge p_b$. By (8), we have $\tau(s^{\perp}) < 2\varepsilon$, and furthermore

$$s\mathfrak{H} \subset \mathfrak{D}\big(M(a)M(b)\big), \qquad M(a)M(b)s = apbs = M(ab)s.$$

Thus, we obtain $\overline{M(a)M(b)} = M(ab)$ by Lemma 2.7.

(iii) Put $a_n = Ap_n \in \mathcal{M}$. Then $\{a_n\}$ is a Cauchy sequence in measure, so that it converges to $a \in \mathfrak{M}(\mathcal{M})$ in measure. If $\xi \in \mathfrak{D}$, then for a sufficiently large n we have $a_n\xi = a_{n+1}\xi = \cdots = A\xi$, so that $\{a_n\xi\}$ converges to $A\xi$ strongly. On the other hand, $\{a_n\xi\}$ also converges to $a\xi$ in measure. Hence $a\xi = A\xi \in \mathfrak{H}$ and $\xi \in \mathfrak{D}\big(M(a)\big)$, so $A\xi = M(a)\xi$. Therefore, we get $M(a) \supset A$. By Lemma 2.7, we conclude $M(a) = \bar{A}$. Q.E.D.

Definition 2.8. When $\{\mathcal{M}, \mathfrak{H}\}$ and τ are given as before, a closed operator A of the form $A = M(a)$, $a \in \mathfrak{M}(\mathcal{M})$, is said to be *$\tau$-measurable*.

By Theorems 2.2 and 2.5, the measurable operators form a *-algebra.

Corollary 2.9. *Let A be a densely defined closed operator affiliated with $\mathcal{M}$ and*

$$A = uH, \qquad H = \int_0^{\infty} \lambda \, de(\lambda)$$

be its polar decomposition and the spectral decomposition of the absolute value. A necessary and sufficient condition for A to be τ-measurable is that

$$\lim_{\lambda \to \infty} \tau\big(e(\lambda)^{\perp}\big) = 0. \tag{11}$$

This condition is equivalent to saying that

$$\tau\big(e(\lambda)^{\perp}\big) < +\infty \quad \textit{for large } \lambda > 0. \tag{12}$$

The proof is straightforward. We leave it to the reader.

Corollary 2.10. *In the algebra $\mathfrak{M}(\mathcal{M})$, let*

$$\mathfrak{M}(\mathcal{M})_+ = \big\{a^*a : a \in \mathfrak{M}(\mathcal{M})\big\}. \tag{13}$$

Then $\mathfrak{M}(\mathcal{M})_+$ is a pointed convex cone and each element a of $\mathfrak{M}(\mathcal{M})_+$ has a unique square root in $\mathfrak{M}(\mathcal{M})_+$ denoted by $a^{1/2}$. Every $a \in \mathfrak{M}(\mathcal{M})$ has the polar decomposition

$$a = u|a|, \quad \textit{with} \quad |a| = (a^*a)^{\frac{1}{2}}.$$

The proof may be carried out by considering $\{ M(a) : a \in \mathfrak{M}(\mathcal{M}) \}$. We call $|a| = (a^*a)^{1/2}$ the *absolute value* of $a \in \mathfrak{M}(\mathcal{M})$.

Definition 2.11. The algebra $\mathfrak{M}(\mathcal{M})$ is called the *τ-measurable operator algebra*. If $\mathcal{M}$ acts on $\mathfrak{H}$, then $\{ M(a) : a \in \mathfrak{M}(\mathcal{M}) \}$ is also called the *τ-measurable operator algebra*. When we need to distinguish them, we call the former the *abstract* τ-measurable operator algebra and the latter the *concrete* τ-measurable operator algebra.

For the sum and the product of two measurable operators A and B, we use the closures of the algebraic sum and product. When we emphasize this fact, we write them as

$$A \dotplus B, \qquad A \cdot B$$

and call them the *strong* sum and the *strong* product respectively.

We now extend the trace τ on $\mathcal{M}_+$ to $\mathfrak{M}(\mathcal{M}_+)$ as follows:

$$\tau(h) = \lim_{\varepsilon \to 0} \tau\big(h\,(1 + \varepsilon h)^{-1}\big), \quad h \in \mathfrak{M}(\mathcal{M})_+. \tag{14}$$

For each $x \in \mathfrak{M}(\mathcal{M})$, we define

$$x_\varepsilon = x\big(1 + \varepsilon|x|\big)^{-1} \in \mathcal{M}, \quad \varepsilon > 0. \tag{15}$$

As before, we consider $\mathfrak{m}_\tau$ and $\mathfrak{n}_\tau$. The trace property of τ, $\tau(x^*x) = \tau(xx^*)$, implies that $\mathfrak{m}_\tau$ and $\mathfrak{n}_\tau$ are both ideals of $\mathcal{M}$. By the polarization identity, we have

$$\tau(xy) = \tau(yx), \quad x, y \in \mathfrak{n}_\tau. \tag{16}$$

By the polar decomposition, we know that every $a \in \mathfrak{m}_\tau$ is of the form $a = xy$, $x, y \in \mathfrak{n}_\tau$, so that for every $b \in \mathcal{M}$, we have

$$\tau(ab) = \tau(xyb) = \tau\big(x(yb)\big) = \tau\big((yb)x\big) = \tau\big(y(bx)\big) = \tau(bxy) = \tau(ba).$$

so we get

$$\tau(ab) = \tau(ba), \quad a \in \mathfrak{m}_\tau, \quad b \in \mathcal{M}. \tag{16'}$$

For $x \in \mathfrak{m}_\tau$ and $y \in \mathcal{M}$, let $x = u|x|$, $y = v|y|$ and $xy = w|xy|$ be the polar decompositions. By the Cauchy-Schwarz inequality, we have

$$\begin{aligned}
\big|\tau(xy)\big| &= \Big|\tau\big(u\,|x|^{\frac12}\,|x|^{\frac12}\,v\,|y|^{\frac12}\,|y|^{\frac12}\big)\Big| = \Big|\tau\big(|y|^{\frac12}\,u\,|x|^{\frac12}\,|x|^{\frac12}\,v\,|y|^{\frac12}\big)\Big| \\
&\le \tau\big(|y|^{\frac12}\,u\,|x|\,u^*|y|^{\frac12}\big)^{\frac12}\,\tau\big(|y|^{\frac12}\,v^*|x|\,v\,|y|^{\frac12}\big)^{\frac12} \\
&= \tau\big(|x^*||y|\big)^{\frac12}\,\tau\big(|x||y^*|\big)^{\frac12}.
\end{aligned}$$

Namely, we have the inequality:

$$\big|\tau(xy)\big|^2 \le \tau\big(|x^*||y|\big)\tau\big(|x||y^*|\big), \quad x \in \mathfrak{m}_\tau, \quad y \in \mathcal{M}. \tag{17}$$

The right hand side is further bounded by the following:

$$\leq \|y\|\tau(|x^*|)\|y\|\tau(|x|) = \|y\|^2\tau(|x|)^2;$$

thus we get

$$|\tau(xy)| \leq \|y\|\tau(|x|), \quad x \in \mathfrak{m}_\tau, \quad y \in \mathcal{M}. \tag{18}$$

Hence each $x \in \mathfrak{m}_\tau$ gives rise to an element τ_x of $\mathcal{M}_*$: $y \in \mathcal{M} \mapsto \tau_x(y) = \tau(xy) \in \mathbf{C}$.

Lemma 2.12.

(i) *Each* $x \in \mathfrak{m}_\tau$ *gives rise to* $\tau_x \in \mathcal{M}_*$ *by the formula:*

$$\tau_x(y) = \tau(xy), \quad x \in \mathfrak{m}_\tau, \quad y \in \mathcal{M}, \tag{19}$$

whose norm is given by:

$$\|\tau_x\| = \tau(|x|), \quad x \in \mathfrak{m}_\tau. \tag{20}$$

(ii) *The set*

$$L^1(\mathcal{M}, \tau) = \left\{x \in \mathfrak{M}(\mathcal{M}) : \tau(|x|) < +\infty\right\}$$

is a two sided $\mathcal{M}$*-submodule of* $\mathfrak{M}(\mathcal{M})$ *and the function:*

$$\|x\|_1 = \tau(|x|), \quad x \in L^1(\mathcal{M}, \tau)$$

is a complete norm of $L^1(\mathcal{M}, \tau)$ *with respect to which* $\mathfrak{m}_\tau$ *is a dense subspace.*

(iii) *We can extend* τ *on* $\mathfrak{m}_\tau$ *to* $L^1(\mathcal{M}, \tau)$ *continuously to a linear functional and the bilinear form:*

$$(x, y) \in \mathcal{M} \times L^1(\mathcal{M}, \tau) \mapsto \tau(xy) \in \mathbf{C}$$

identifies $L^1(\mathcal{M}, \tau)$ *with the predual* $\mathcal{M}_*$.

PROOF:

(i) We have proved $\|\tau_x\| \leq \tau(|x|)$ for $x \in \mathfrak{m}_\tau$. By the polar decomposition $x = u|x|$, we have

$$\tau(|x|) = \tau(u^*x) = \tau(xu^*) = \tau_x(u^*),$$

so that we have $\|\tau_x\| \geq \tau(|x|)$. Hence $\|x\|_1 = \tau(|x|)$ gives the norm of $\tau_x \in \mathcal{M}_*$ for $x \in \mathfrak{m}_\tau$.

(ii) and (iii): Fix an $x \in \mathfrak{M}(\mathcal{M})$ and let $|x| = \int_0^\infty \lambda \, de(\lambda)$ be the spectral decomposition of $|x|$. Consider the abelian von Neumann subalgebra $\mathcal{A}$ generated by $\{e(\lambda)\}$. If $x \in L^1(\mathcal{M}, \tau)$, then $\tau(|x|_\varepsilon) \nearrow \tau(|x|) < +\infty$ as $\varepsilon \searrow 0$. This means that the increasing function: $\lambda \in \mathbf{R}_+ \mapsto \tau(e(\lambda)) \in [0, +\infty]$ gives rise to a measure μ on the open half line $\mathbf{R}_+^* =]0, +\infty[$. Hence as $\varepsilon \searrow 0$ we have

$$\tau(|x|_\varepsilon) = \int_0^\infty \frac{\lambda}{1 + \varepsilon\lambda} \, d\mu(\lambda) \nearrow \tau(|x|) = \int_0^\infty \lambda \, d\mu(\lambda) < +\infty.$$

Therefore, we get

$$\big\| |x| - |x|_\varepsilon \big\|_1 \to 0 \quad \text{as} \quad \varepsilon \to 0$$

and

$$\big\|x - x_\varepsilon\big\|_1 = \tau(|x - x_\varepsilon|) = \tau(|x| - |x|_\varepsilon) = \| \, |x| - |x|_\varepsilon \|_1 \to 0.$$

Thus we have

$$\lim_{\varepsilon \to 0} \big\|x - x_\varepsilon\big\|_1 = 0 \quad \text{and} \quad \lim_{\varepsilon \to 0} \|x_\varepsilon\|_1 = \|x\|_1.$$

Therefore, x is approximated by $x_\varepsilon \in \mathfrak{m}_\tau$. This means that $L^1(\mathcal{M}, \tau)$ can be viewed as a subspace of the completion of the normed space $\{\mathfrak{m}_\tau, \|\cdot\|_1\}$. Hence $\|\cdot\|_1$ is a norm on $L^1(\mathcal{M}, \tau)$ and τ can be extended continuously to $L^1(\mathcal{M}, \tau)$.

By Corollary VIII.3.6, every semi-finite normal weight φ on $\mathcal{M}$ is of the form $\varphi = \tau_h$ with a closed self-adjoint positive operator affiliated with $\mathcal{M}$. If $\varphi \in \mathcal{M}_*^+$, then $\tau(h) = \lim_{\varepsilon\to 0} \tau(h_\varepsilon) = \varphi(1) < +\infty$. With the spectral decomposition $h = \int_0^\infty \lambda \, de(\lambda)$, we have $\lambda e(\lambda)^\perp \le h$, so that $\tau\big(e(\lambda)^\perp\big)$ is finite. Thus h is τ-measurable and belongs to $L^1(\mathcal{M}, \tau)$. So $\mathcal{M}_*^+$ can be identified with $L^1(\mathcal{M}, \tau)_+$ under the correspondence $h \in L^1(\mathcal{M}, \tau)_+ \longleftrightarrow \tau_h \in \mathcal{M}_*^+$.

Consider the polar decomposition $x = u|x| = |x^*|u$ of $x \in \mathfrak{M}(\mathcal{M})$. Then we have $|x^*|_\varepsilon = u|x|_\varepsilon u^*$, so that

$$\tau\big(|x^*|\big) = \lim \tau\big(|x^*|_\varepsilon\big) = \lim \tau\big(|x|_\varepsilon\big) = \tau\big(|x|\big).$$

Hence $L^1(\mathcal{M}, \tau)$ is self-adjoint. For any unitary $a \in \mathcal{M}$ and $x \in L^1(\mathcal{M}, \tau)$, we have $|ax| = |x|$, so that $L^1(\mathcal{M}, \tau)$ is a left $\mathcal{M}$-module. The self-adjointness of $L^1(\mathcal{M}, \tau)$ then yields the two sided module property over $\mathcal{M}$.

Now, the polar decompositions in $\mathcal{M}_*$ and in $L^1(\mathcal{M}, \tau)$ allow us to identify them. Q.E.D.

Theorem 2.13.

(i) *The function τ on $\mathfrak{M}(\mathcal{M})_+$ enjoys the following properties:*

$$\left.\begin{aligned} \tau(a + b) &= \tau(a) + \tau(b), & a, b &\in \mathfrak{M}(\mathcal{M})_+, \\ \tau(\lambda a) &= \lambda\tau(a), & \lambda &\ge 0, \\ \tau(x^*x) &= \tau(xx^*), & x &\in \mathfrak{M}(\mathcal{M}). \end{aligned}\right\} \tag{21}$$

(ii) *For $1 \le p < +\infty$, set*

$$\begin{aligned} \|x\|_p &= \tau\big(|x|^p\big)^{1/p}, \quad x \in \mathfrak{M}(\mathcal{M}), \\ L^p(\mathcal{M}, \tau) &= \big\{x \in \mathfrak{M}(\mathcal{M}) : \|x\|_p < +\infty\big\}. \end{aligned} \tag{22}$$

Then $L^p(\mathcal{M}, \tau)$ is a Banach space in which $\mathcal{M} \cap L^p(\mathcal{M}, \tau)$ is dense. Furthermore, $L^p(\mathcal{M}, \tau)$ is invariant under the multiplications of $\mathcal{M}$ from both sides, and

$$\|ax\|_p \le \|a\| \|x\|_p, \qquad \|xa\|_p \le \|a\| \|x\|_p \tag{23}$$

for each $a \in \mathcal{M}$, $x \in L^p(\mathcal{M}, \tau)$.

(iii) *The extended trace* τ *on* $L^1(\mathcal{M}, \tau)$ *identifies* $\mathcal{M}_*$ *with* $L^1(\mathcal{M}, \tau)$ *by the bilinear form:*

$$(x, y) \in \mathcal{M} \times L^1(\mathcal{M}, \tau) \mapsto \tau(xy) \in \mathbf{C}.$$

(iv) *If* $1/p + 1/q = 1$ *and* $p > 1$, *then the product of* $L^p(\mathcal{M}, \tau)$ *and* $L^q(\mathcal{M}, \tau)$ *agrees with* $L^1(\mathcal{M}, \tau)$ *and we have the Hölder's inequality:*

$$\left|\tau(xy)\right| \leq \|x\|_p \|y\|_q, \quad x \in L^p(\mathcal{M}, \tau), \quad y \in L^q(\mathcal{M}, \tau). \tag{24}$$

Furthermore, $L^p(\mathcal{M}, \tau)$ *and* $L^q(\mathcal{M}, \tau)$ *are the conjugate space of each other.*

PROOF:

(i) Take a and b from $\mathfrak{M}(\mathcal{M})_+$. If $\tau(a) < +\infty$ and $\tau(b) < +\infty$ then a and b are both in $L^1(\mathcal{M}, \tau)$. The linearity of τ on $L^1(\mathcal{M}, \tau)$ implies that $\tau(a) + \tau(b) = \tau(a + b)$. If $\tau(a + b) < +\infty$, then $a \leq a + b$ and $b \leq a + b$, so $a_\varepsilon \leq (a + b)_\varepsilon$ and $b_\varepsilon \leq (a + b)_\varepsilon$. Thus we get

$$\tau(a) = \lim_{\varepsilon \to 0} \tau(a_\varepsilon) \leq \lim_{\varepsilon \to 0} \tau\big((a + b)_\varepsilon\big) = \tau(a + b) < +\infty,$$

similarly $\tau(b) < +\infty$. Therefore, $\tau(a) + \tau(b) = \tau(a + b)$.

The homogeneity of τ on $\mathfrak{M}(\mathcal{M})_+$ is trivial.

For any $x \in \mathfrak{M}(\mathcal{M})$, let $x = u|x|$ be the polar decomposition. Then we have $xx^* = ux^*xu^*$, so $u(x^*x)_\varepsilon u^* = (xx^*)_\varepsilon$. Hence

$$\tau(xx^*) = \lim_{\varepsilon \to 0} \tau\big((xx^*)_\varepsilon\big) = \lim_{\varepsilon \to 0} \tau\big(u(x^*x)_\varepsilon u^*\big) = \lim_{\varepsilon \to 0} \tau\big((x^*x)_\varepsilon\big) = \tau(x^*x).$$

(ii) Assume $p > 1$, since we have proven the case $p = 1$. Choose q so that $1/p + 1/q = 1$. Take $a, b \in \mathfrak{m}_\tau$ and $c, d \in \mathfrak{m}_\tau^+$. Assume that $c \geq \varepsilon s(c)$ and $d \geq \varepsilon s(d)$ for some $\varepsilon > 0$, where $s(c)$ and $s(d)$ are the support of c and d respectively. In this case, we can define $\mathfrak{m}_\tau$-valued entire functions: $\lambda \in \mathbf{C} \to c^\lambda$ and $\lambda \in \mathbf{C} \to d^\lambda$. For each x and y in $\mathcal{M}$ with $\|x\| \leq 1$ and $\|y\| \leq 1$, the function f:

$$f(\lambda) = \tau\big(xc^\lambda yd^{1-\lambda}\big), \quad \lambda \in \mathbf{C},$$

is entire, to which the Phragmen-Lindelöf theorem applies. Thus, we get

$$\left|\tau\big(xc^\sigma yd^{1-\sigma}\big)\right| \leq \sup_{\mathrm{Re}\, s=1} \left|\tau\big(xc^s yd^{1-s}\big)\right|^\sigma \sup_{\mathrm{Re}\, s=0} \left|\tau\big(xc^s yd^{1-c}\big)\right|^{1-\sigma}$$

for $0 \leq \sigma \leq 1$. By (18), the right hand side is bounded by $\|c\|_1^\sigma \|d\|_1^{1-\sigma}$. Now, consider the polar decomposition $a = u|a|$ and $b = v|b|$ and put $c = |a|^p$ and $d = |b|^q$, and further $\sigma = 1/q$. Then under the hypothesis for c and d, we have the Hölder type inequality:

$$\left|\tau(ab)\right| \leq \tau\big(|a|^p\big)^{1/p} \tau\big(|b|^q\big)^{1/q} = \|a\|_p \|b\|_q. \tag{25}$$

By passing to the limit, the above inequality holds for general pair $a, b \in \mathfrak{m}_\tau$.

Let $\mathfrak{m}_0 = \{x \in \mathcal{M} : s_\ell(x) \in \mathfrak{m}_\tau\} \subset \mathfrak{m}_\tau$. It follows that $\mathfrak{m}_0$ is an ideal of $\mathcal{M}$ and dense in $L^1(\mathcal{M}, \tau)$. If $a \in \mathfrak{m}_0$, then we put

$$b = \|a\|_p^{-p/q} |a|^{p-1} u^* \in \mathfrak{m}_0,$$

and obtain $\tau(ab) = \|a\|_p$ and $\|b\|_q = 1$. Thus, we obtain

$$\|a\|_p = \sup\Big\{ \big|\tau(ab)\big| : b \in \mathfrak{m}_\tau,\ \|b\|_q \leq 1 \Big\}, \quad a \in \mathfrak{m}_0. \tag{26}$$

For a general $a \in \mathfrak{m}_\tau$, we have

$$\begin{aligned} \|a\|_p &= \sup\{ \|ea\|_p : e \in \operatorname{Proj}(\mathcal{M}) \cap \mathfrak{m}_\tau \} \\ &= \sup\Big\{ \big|\tau(eab)\big| : b \in \mathfrak{m}_\tau,\ \|b\|_q \leq 1,\ e \in \operatorname{Proj}(\mathcal{M}) \cap \mathfrak{m}_\tau \Big\} \\ &= \sup\Big\{ \big|\tau(ab)\big| : b \in \mathfrak{m}_\tau,\ \|b\|_q \leq 1 \Big\}. \end{aligned}$$

Therefore, we have the Minkowski type inequality:

$$\|a + b\|_p \leq \|a\|_p + \|b\|_p, \quad a, b \in \mathfrak{m}_\tau.$$

We now want to show that the embedding of $\{\mathfrak{m}_\tau, \|\cdot\|_p\}$ into $\mathfrak{M}(\mathcal{M})$ is continuous, so that the identity map of $\mathfrak{m}_\tau$ is extendable to a continuous map of the completion E_p of $\{\mathfrak{m}_\tau, \|\cdot\|_p\}$ into $\mathfrak{M}(\mathcal{M})$. Let $\varepsilon, \delta > 0$. Suppose $\|x\|_p < \varepsilon\delta^{1/p}$, $x \in \mathfrak{m}_\tau$, and let $x = uh$ be the polar decomposition and $h = \int_0^\infty \lambda\, de(\lambda)$ the spectral decomposition. Then we have, since $\lambda e(\lambda)^\perp \leq h$,

$$\varepsilon^p \tau\big(e(\varepsilon)^\perp\big) \leq \tau(h^p) < \varepsilon^p \delta,$$

so that $\tau\big(e(\varepsilon)^\perp\big) < \delta$; hence $x \in N(2\varepsilon, \delta)$ because $\big\|xe(\varepsilon)\big\| \leq \varepsilon$.

We prove next the injectivity of the extended map of E_p into $\mathfrak{M}(\mathcal{M})$. Suppose that a Cauchy sequence $\{x_n\}$ in $\{\mathfrak{m}_\tau, \|\cdot\|_p\}$ converges to zero in measure. For a fixed $\varepsilon > 0$, choose n_0 so that $\|x_n - x_m\|_p < \varepsilon$ for $n, m \geq n_0$. By (26) extended to $\mathfrak{m}_\tau$, there exists $y \in \mathfrak{m}_\tau$ with $\|y\|_q \leq 1$ such that $\big|\tau(x_{n_0} y)\big| \geq \|x_{n_0}\|_p - \varepsilon$. Then we have for $n \geq n_0$

$$\big|\tau(x_n y)\big| \geq \big|\tau(x_{n_0} y)\big| - \Big|\tau\big((x_n - x_{n_0})y\big)\Big| \geq \|x_{n_0}\|_p - 2\varepsilon.$$

Therefore, if $x = \lim x_n \in E_p$ is not zero, then there exists $y \in \mathfrak{m}_\tau$, $\|y\|_q \leq 1$, such that $\{\tau(x_n y)\}$ is bounded away from zero, i.e. $\big|\tau(x_n y)\big| \geq \delta > 0$ for $n \geq n_0$ and some $\delta > 0$. But the measure convergence of $\{x_n\}$ to zero implies that for any $\varepsilon > 0$ there exists $e \in \operatorname{Proj}(\mathcal{M})$ with $\|x_n e\| < \varepsilon$ and $\tau(e^\perp) < \varepsilon^q$. But

$$\begin{aligned} \delta &\leq \big|\tau(x_n y)\big| \leq \big|\tau(x_n e y)\big| + \big|\tau(x_n e^\perp y)\big| \\ &\leq \|x_n e\|_\infty \|y\|_1 + \|y\|_\infty \|x_n\|_p \|e^\perp\|_q \leq \varepsilon\big(\|y_1\| + \|x_n\|_p \|y\|_\infty\big). \end{aligned}$$

Since $\{\|x_n\|_p\}$ is bounded and y is fixed, we can choose $\varepsilon > 0$ small enough so that the above inequality does not hold. Therefore, $\|x_n\|_p$ must converge to zero, that is, the map of E_p into $\mathfrak{M}(\mathcal{M})$ is injective. So we embed E_p into $\mathfrak{M}(\mathcal{M})$.

Before going further, we observe the following:

$$\|axb\|_p \leq \|a\|\|b\|\|x\|_p, \quad x \in \mathfrak{m}_\tau, \quad a, b \in \mathcal{M}. \tag{27}$$

By the inequalities (26) and (25) extended to $\mathfrak{m}_\tau$, we have

$$\begin{aligned}\|ax\|_p &= \sup\Big\{ \big|\tau(axy)\big| : y \in \mathfrak{m}_\tau,\ \|y\|_q \leq 1 \Big\} \\ &\leq \|a\| \sup\{ \|xy\|_1 : y \in \mathfrak{m}_\tau,\ \|y\|_q \leq 1 \} = \|a\|\|x\|_p;\end{aligned}$$

$$\begin{aligned}\|xb\|_p &= \sup\Big\{ \big|\tau(xby)\big| : y \in \mathfrak{m}_\tau,\ \|y\|_q \leq 1 \Big\} \\ &= \sup\Big\{ \big|\tau(yxb)\big| : y \in \mathfrak{m}_\tau,\ \|y\|_q \leq 1 \Big\} \\ &\leq \|b\| \sup\{ \|yx\|_1 : y \in \mathfrak{m}_\tau,\ \|y\|_q \leq 1 \} = \|b\|\|x\|_p.\end{aligned}$$

Therefore, E_p is a two sided $\mathcal{M}$-module. We define the actions of $\mathcal{M}$ on $(E_q)^*$ by the following:

$$\big\langle x, afb \big\rangle = \big\langle bxa, f \big\rangle, \quad x \in E_q, \quad a, b \in \mathcal{M}, \quad f \in E_q^*.$$

Suppose that a bounded net $\{a_i\}$ in $\mathcal{M}$ converges to zero $*$-strongly. If $x = exe \in \mathfrak{m}_0$, $e \in \mathrm{Proj}(\mathcal{M}) \cap \mathfrak{m}_\tau$, then we have

$$\|a_i x\|_p = \tau\big((x^* a_i^* a_i x)^{p/2}\big)^{1/p} = \tau_e\big((x^* a_i^* a_i x)^{p/2}\big)^{1/p} \to 0;$$

$$\|x a_i\|_p = \tau\big((a_i^* x^* x a_i)^{p/2}\big)^{1/p} = \tau_e\big((a_i^* x^* x a_i)^{p/2}\big)^{1/p} \to 0.$$

Since $\mathfrak{m}_0$ is dense in E_p, the actions of $\mathcal{M}$ on E_p are *-strongly continuous on bounded sets.

Fix an $x \in L^p(\mathcal{M}, \tau)$. Let $x = uh$ and $h = \int_0^\infty \lambda\, de(\lambda)$ be the polar and spectral decompositions. Set $e_n = e(n) - e(1/n)$, $n = 2, 3, \ldots$. We then have $\lim_{n\to\infty} \|x - xe_n\|_p = 0$ and $xe_n \in \mathfrak{m}_\tau$. Hence x belongs to E_p, that is $L^p(\mathcal{M}, \tau) \subset E_p$.

Fix an $f \in (E_q)^*$. We define f^* by

$$\langle x, f^* \rangle = \overline{\langle x^*, f \rangle}, \quad x \in \mathfrak{m}_\tau.$$

It follows that $f^* \in (E_q)^*$ and $\|f\| = \|f^*\|$. Now, if $e \in \mathrm{Proj}(\mathcal{M}) \cap \mathfrak{m}_\tau$, then the map: $x \in \mathcal{M} \mapsto \langle x, fe \rangle = \langle ex, f \rangle$ is σ-strongly* continuous, so that fe belongs to $L^1(\mathcal{M}, \tau) \subset \mathfrak{M}(\mathcal{M})$. Furthermore, the polar decomposition of fe and the spectral decomposition of $|fe|$, together with the Hölder type inequality yield the estimate:

$$\begin{aligned}\|fe\|_p &= \sup\Big\{ \big|\tau(xfe)\big| : x \in \mathfrak{m}_\tau,\ \|x\|_q \leq 1 \Big\} \\ &= \sup\Big\{ \big|\langle x, fe \rangle\big| : x \in \mathfrak{m}_\tau,\ \|x\|_q \leq 1 \Big\} = \|fe\|_{(E_q)^*} < +\infty.\end{aligned}$$

Thus fe belongs to $L^p(\mathcal{M},\tau)$. Hence f gives rise to a densely defined closed operator $M(fe)$ for each $e \in \mathrm{Proj}(\mathcal{M}) \cap \mathfrak{m}_\tau$. Let $\mathfrak{D}_A = \bigcup\{\mathfrak{D}(M(fe)) \cap e\mathfrak{H} : e \in \mathrm{Proj}(\mathcal{M}) \cap \mathfrak{m}_\tau\}$. Since $f(e_1 \vee e_2)e_1 = fe_1$ and $f(e_1 \vee e_2)e_2 = fe_2$ for e_1 and e_2, there exists a linear operator A on $\mathfrak{D}$ such that $A\xi = M(fe)\xi$ if $\xi \in \mathfrak{D}\big(M(fe)\big) \cap e\mathfrak{H}$. It is clear that A commutes with every unitary in $\mathcal{M}'$, so that A is affiliated with $\mathcal{M}$. Based on f^*, we then define B by

$$B\eta = M(f^*g)\eta \quad \text{for} \quad \eta \in \mathfrak{D}_B = \bigcup\Big\{\mathfrak{D}\big(M(f^*g)\big) \cap g\mathfrak{H} : g \in \mathrm{Proj}(\mathcal{M}) \cap \mathfrak{m}_\tau\Big\}.$$

We observe here that $(fe)^* = ef^*$ since $L^p(\mathcal{M},\tau) \subset E_p \subset (E_q)^*$. For each $\xi \in \mathfrak{D}\big(M(fe)\big) \cap e\mathfrak{H}$ and $\eta \in \mathfrak{D}\big(M(f^*g)\big) \cap g\mathfrak{H}$ with $e, g \in \mathrm{Proj}(\mathcal{M}) \cap \mathfrak{m}_\tau$, we have

$$\big(A\xi \,|\, \eta\big) = \big(fe\xi \,|\, g\eta\big) = \big(gfe\xi \,|\, \eta\big) = \big((f^*g)^*e\xi \,|\, \eta\big) = \big(e\xi \,|\, f^*g\eta\big) = \big(\xi \,|\, B\eta\big).$$

Thus, $A^* \supset B$ and $B^* \supset A$. In particular, A is preclosed. Identifying A^{**} with f, we view f as a densely defined closed operator affiliated with $\mathcal{M}$. We now consider the polar and spectral decompositions $f = u|f|$ and $|f| = \int_0^\infty \lambda\, de(\lambda)$. If $\tau\big(e(\lambda)^\perp\big) = +\infty$, $\lambda > 0$, then $\lambda e(\lambda)^\perp \le |f|$ implies $\tau\big(|f|^p\big) = +\infty$. So we have $\sup\big\{\big|\tau(x|f|)\big| : x \in \mathfrak{m}_\tau,\ \|x\|_q \le 1\big\} = +\infty$. But $|f| = u^*f \in (E_q)^*$, which is a contradiction. Thus, we conclude that $\tau\big(e(\lambda)^\perp\big) < +\infty$ for $\lambda > 0$. Therefore, f is measurable and

$$\|f\|_p = \tau\big(|f|^p\big)^{1/p} = \sup\Big\{\big|\tau(xf)\big| : x \in \mathfrak{m}_\tau,\ \|x\|_q \le 1\Big\} = \|f\|_{(E_q)^*}.$$

Thus, f belongs to $L^p(\mathcal{M},\tau)$. We now conclude that $(E_q)^* = L^p(\mathcal{M},\tau)$; hence $L^p(\mathcal{M},\tau) = \big(L^q(\mathcal{M},\tau)\big)^*$. In particular, $L^p(\mathcal{M},\tau)$ is a Banach space. Q.E.D.

In the case $p = 2$, $L^2(\mathcal{M},\tau)$ is a Hilbert space which can canonically be identified with the Hilbert space $\mathfrak{H}_\tau$.

We will use the following result later:

Theorem 2.14. *Let $\{\mathcal{M},\tau\}$ be as before, and let E_a be the characteristic function of $[a, +\infty[$. Then we have*

$$\int_0^\infty \big\|E_{\sqrt{a}}(h)\big\|_2^2\, da = \|h\|_2^2, \quad h \in L^2(\mathcal{M},\tau)_+; \tag{28}$$

$$\int_0^\infty \big\|E_{\sqrt{a}}(h) - E_{\sqrt{a}}(k)\big\|_2^2\, da \le \|h-k\|_2\|h+k\|_2, \quad h, k \in L^2(\mathcal{M},\tau). \tag{29}$$

PROOF: Let $X = (\mathbf{R}_+ \setminus \{0\}) \times (\mathbf{R}_+ \setminus \{0\})$. Let H and K be the projections of X to the first and second components respectively, i.e. $H(x, y) = x$ and $K(x, y) = y$. Let h and k be positive elements of $L^2(\mathcal{M}, \tau)$, and consider the von Neumann algebra $\mathcal{A}$ generated by h and JkJ, or more precisely by their spectral projections. Then $\mathcal{A}$ is an abelian von Neumann algebra. Let $\mathcal{K}$ be the algebra of all continuous functions on $]0, +\infty[$ with compact support, and consider the algebraic tensor product $\mathcal{K} \otimes \mathcal{K}$ identified with a subalgebra of $C_\infty(X)$. Let $\mathfrak{A}$ be the subset of $L^2(\mathcal{M}, \tau)$ consisting of all operators of the form:

$$\sum_{i=1}^{n} f_i(h)g_i(k) = \sum_{i=1}^{n} (f_i \otimes g_i)(h, k).$$

In $\mathfrak{A}$, we define a new product:

$$\Big(\sum_{i=1}^{n} f_i(h)g_i(k)\Big) \cdot \Big(\sum_{j=1}^{m} f'_j(h)g'_j(k)\Big) = \sum_{i,j} f_i(h)f'_j(h)g_i(k)g'_j(k),$$

and a new involution:

$$\Big(\sum_{i=1}^{n} f_i(h)g_i(k)\Big)^{\circ} = \sum_{i=1}^{n} \bar{f}_i(h)\bar{g}_i(k).$$

It then follows that $\mathfrak{A}$ is a commutative Hilbert algebra and its left von Neumann algebra $\mathcal{R}_\ell(\mathfrak{A})$ is isomorphic to $\mathcal{A}$. In fact, the operator corresponding to $\sum_{i=1}^n f_i(h)g_i(k)$ is exactly the restriction of $\sum_{i=1}^n f_i(h)J\bar{g}_i(k)J$ on the closure $\mathfrak{K}$ of $\mathfrak{A}$. Therefore the faithful semifinite normal trace on $\mathcal{A}$ corresponding to the Hilbert algebra $\mathfrak{A}$ gives rise to a measure μ on X such that

$$\Big\| \sum_{i=1}^{n} f_i(h)g_i(k) \Big\|_2^2 = \int_X \Big| \sum_{i=1}^{n} f_i(x)g_i(y) \Big|^2 \, \mathrm{d}\mu(x, y)\,.$$

Furthermore, the subspace $\mathfrak{K}$ is identified with $L^2(X, \mu)$ under the identification: $\sum_{i=1}^n f_i \otimes g_i \in L^2(X, \mu) \longleftrightarrow \sum_{i=1}^n f_i(h)g_i(k)$. Therefore we get for any Borel functions f and g:

$$\|f(h) - g(k)\|_2^2 = \int_X |f(x) - g(y)|^2 \, \mathrm{d}\mu(x, y) = \|f \circ H - g \circ K\|_{L^2(X,\mu)}^2.$$

Now, we compute as in the last part of the last section:

$$\|h\|_2^2 = \tau\Big(\int_0^\infty E_a(h^2)\, \mathrm{d}a\Big) = \int_0^\infty \tau(E_{\sqrt{a}}(h))\, \mathrm{d}a = \int_0^\infty \|E_{\sqrt{a}}(h)\|_2^2 \, \mathrm{d}a;$$

$$\begin{aligned}
\int_0^\infty \|E_{\sqrt{a}}(h) - E_{\sqrt{a}}(k)\|_2^2 \, da &= \int_0^\infty \int_X |E_{\sqrt{a}}(x) - E_{\sqrt{a}}(y)|^2 \, d\mu(x, y) \, da \\
&= \int_X \int_0^\infty \left|E_{\sqrt{a}}(x) - E_{\sqrt{a}}(y)\right| \, da \, d\mu(x, y) \\
&= \int_X \left|x^2 - y^2\right| \, d\mu(x, y) \\
&\leq \left(\int_X |x - y|^2 \, d\mu(x, y)\right)^{\frac{1}{2}} \left(\int_X |x + y|^2 \, d\mu(x, y)\right)^{\frac{1}{2}} \\
&= \|h - k\|_2 \, \|h + k\|_2.
\end{aligned}$$

Q.E.D.

Exercise IX.2

Assume that $\{\mathcal{M}, \tau\}$ is a von Neumann algebra equipped with a faithful semi-finite normal trace τ.

1) Show that if $\mathcal{M}$ is a factor of type I, then the τ-measure convergence in $\mathcal{M}$ is precisely the norm convergence, so that $\mathfrak{M}(\mathcal{M}) = \mathcal{M}$.

2) (**a**) Show that the identity map of $\mathcal{M}$ is continuous as a map from $\mathcal{M}$ with the norm topology to $\mathcal{M}$ with the τ-topology, i.e. the norm topology is stronger than the τ-topology.

(**b**) Show that if $\mathcal{M}$ has no non-zero minimal projection and τ is finite, then there is no non-zero linear functional on $\mathfrak{M}(\mathcal{M})$ which is continuous relative to the τ-topology.

(**c**) Show that if $\mathcal{M}$ has no non-zero minimal projection and τ is infinite, then a necessary and sufficient condition for a linear functional φ on $\mathcal{M}$ to be τ-continuous is that $\varphi(\mathfrak{m}_\tau) = 0$.

3) Let $\mathcal{M} = \ell^\infty(\mathbf{N})$ be the von Neumann algebra of bounded sequences and τ be a faithful semi-finite normal trace. Show that τ is finite if and only if every sequence is τ-measurable.

4) Let $\mathcal{M} = L^\infty(X, \mu)$ with $\{X, \mu\}$ a σ-finite measure space and τ be the trace on $\mathcal{M}$ corresponding to the integration relative to the measure μ. Show that τ-measurable operators are precisely those functions f on X such that (i) f is μ-measurable as a function on X and (ii) $\mu(\{x \in X : |f(x)| \geq n\}) \to 0$ as $n \to \infty$.

5) Let $\mathfrak{H}$ be the underlying Hilbert space of $\mathcal{M}$. A subspace $\mathfrak{D}$ of $\mathfrak{H}$, not necessarily closed, is said to be τ-*dense* if there exists an increasing sequence $\{p_n\}$ in $\mathrm{Proj}(\mathcal{M})$ such that (i) $\lim_{n\to\infty} \tau(p_n^\perp) = 0$ and (ii) $p_n\mathfrak{H} \subset \mathfrak{D}$.

(a) Show that a subspace $\mathfrak{D}$ of $\mathfrak{H}$ is τ-dense if and only if for any $\varepsilon > 0$ there exists $p \in \mathrm{Proj}(\mathcal{M})$ such that $\tau(p^\perp) < \varepsilon$ and $p\mathfrak{H} \subset \mathfrak{D}$.

(b) Show that if $\{\mathfrak{D}_n\}$ is a sequence of τ-dense subspaces then $\mathfrak{D} = \bigcap_{n=1}^\infty \mathfrak{D}_n$ is also τ-dense. (Hint: Use (a). For each $\varepsilon > 0$, let $p_n \in \mathrm{Proj}(\mathcal{M})$ be such that $\tau(p_n^\perp) < \varepsilon/2^n$ and $p_n\mathfrak{H} \subset \mathfrak{D}_n$. Set $p = \bigwedge_{n=1}^\infty p_n$ and observe $\tau(p^\perp) \le \sum_{n=1}^\infty \tau(p_n^\perp) < \varepsilon$.)

6) A sequence $\{A_n\}$ in $\mathfrak{M}(\mathcal{M})$ is said to converge τ-*nearly everywhere* if there exists a τ-dense subspace $\mathfrak{D} \subset \mathfrak{H}$ such that $\mathfrak{D} \subset \bigcap_{n=1}^\infty \mathfrak{D}(A_n)$ and $A_n\xi$ converges in norm for every $\xi \in \mathfrak{D}$. Define $A\xi = \lim_{n\to\infty} A_n\xi$ for those $\xi \in \bigcap_{n=1}^\infty \mathfrak{D}(A_n)$ such that $\lim_{n\to\infty} A_n\xi$ exists.

(a) Show that A is affiliated with $\mathcal{M}$.

(b) Show that if $p\mathfrak{H} \subset \mathfrak{D}$, $p \in \mathrm{Proj}(\mathcal{M})$, then $\|A_n p\| < +\infty$ and $\|Ap\| < +\infty$. (Hint: Use the closed graph theorem for the boundedness of $A_n p$ and the uniform boundedness theorem for $\{A_n p\}$).

(c) Prove that A is τ-measurable.

(d) Show that if $\tau(e) < +\infty$, $e \in \mathrm{Proj}(\mathcal{M})$, then for any $\varepsilon > 0$ and $\delta > 0$ there exists N and $p \in \mathrm{Proj}(\mathcal{M}_e)$ such that $\tau(e - p) < \delta$ and $\|(A_n - A)p\| < \varepsilon$, $n \ge N$. (Hint: $\tau(e - e \wedge q) = \tau(e \vee q - q) \le \tau(q^\perp)$ for those q with $q\mathfrak{H} \subset \mathfrak{D}$. Set $p_1 = e \wedge q$ to get p_1 with $\tau(e - p_1) < \delta/2$. Observe that $A_n p_1$ converges strongly to Ap_1. Apply Lemma II.4.12 to $A_n p_1$.)

(e) Show that if $\tau(e) < +\infty$, $e \in \mathrm{Proj}(\mathcal{M})$, then for any $\varepsilon > 0$, there exists $p \in \mathrm{Proj}(\mathcal{M}_e)$ such that

$$\tau(e - p) < \varepsilon \qquad \text{and} \qquad \lim_{n\to\infty} \|(A_n - A)p\| = 0\,.$$

(f) Let $\mathcal{M} = L^\infty(X, \mu)$ with $\{X, \mu\}$ a σ-finite measure space and τ the integration relative to μ. Let $\{A_n\}$ be a sequence of measurable operators and $\{f_n\}$ be the corresponding sequence of functions on X. Show that if $\{f_n\}$ converges μ-almost everywhere, then $\{A_n\}$ converges τ-nearly everywhere; conversely if $\{A_n\}$ converges τ-nearly everywhere, then some subsequence $\{f_{n_k}\}$ of $\{f_n\}$ converges almost everywhere, and the limit function does not depend on the choice of a convergent subsequence.

(g) Show that if $\mathcal{M}$ is non-atomic, i.e. does not admit a non-zero minimal projection then the τ-measure convergence does not imply the τ-nearly everywhere convergence. (Hint: If $\mathcal{M}$ is non-atomic, then $L^\infty(0, 1)$, the algebra of essentially bounded functions on $[0, 1]$ relative to the Lebesgue measure, can be embedded as a von Neumann subalgebra of a reduced subalgebra $\mathcal{M}_e$ along with the trace on $L^\infty(0, 1)$ given by the Lebesgue measure. Consider the sequence $\{I_n\}$ of intervals defined by

$$I_{j+k(k+1)/2} = \left[\frac{j-1}{k+1}, \frac{j}{k+1}\right], \quad 1 \le j \le n+1,$$

and set

$$f_n = |I_n|^{-2}\chi_{I_n}\,.$$

Then $\{f_n\}$ converges to zero in measure, but not τ-nearly everywhere.)

(**h**) Show that if $\mathcal{M} = L^\infty(0,1)$ and τ is the integration relative to the Lebesgue measure, then the τ-nearly everywhere convergence does not imply the almost everywhere convergence. (Hint: In the hint for (g), set $f_n = \chi_{I_n}$. Then $\{f_n\}$ converges strongly to zero, so that it converges to zero τ-nearly everywhere. But it does not converge to zero at any point of $[0,1]$.)

(**i**) Show that the τ-nearly everywhere convergence of self-adjoint measurable operators implies the resolvent convergence. (cf. A.6.(iii).)

7) For a τ-measurable operator T, set

$$\lambda_t(T) = \tau\Big(E_{]t,+\infty[}\big(|T|\big)\Big), \quad t \geq 0,$$

where $E_{]t,+\infty[}\big(|T|\big)$ is the spectral projection of $|T|$ corresponding to the open half line $]t,+\infty[$;

$$\mu_t(T) = \inf\big\{\, \|Te\| : e \in \mathrm{Proj}(\mathcal{M}),\ \tau(e^\perp) \leq t \,\big\}.$$

Following the steps described below, prove that

$$\mu_t(T) = \inf\big\{\, s \geq 0 : \lambda_s(T) \leq t \,\big\}; \quad \lambda_{\mu_t(T)}(T) \leq t, \quad t > 0. \qquad [517]$$

(**a**) First, observe that $s \mapsto \lambda_s(T)$ is continuous from the right. With $a = \inf\big\{\, s \geq 0 : \lambda_s(T) \leq t \,\big\}$ and $E = E_{[0,a]}(|T|)$, prove

$$\|TE\| \leq a \quad \text{and} \quad \tau(1-E) \leq t.$$

(**b**) Given $\varepsilon > 0$, let $E \in \mathrm{Proj}(\mathcal{M})$ such that $\tau(1-E) \leq t$ and

$$\|TE\| < \mu_t(T) + \varepsilon = \alpha\,.$$

Show that $E \wedge E_{]\alpha,\infty[}\big(|T|\big) = 0$. (Hint: If $\xi \in E\mathfrak{H} \cap E_{]\alpha,\infty[}(|T|)\mathfrak{H}$, $\|\xi\| = 1$, then $(T^*T\xi \mid \xi) \geq \alpha^2 > \|T\xi\|^2 = (T^*T\xi \mid \xi)$.)

(**c**) Show that $\tau(E_{]\alpha,\infty[}\big(|T|\big)) \leq t$. (Hint: $\lambda_\alpha(T) = \tau(E_{]\alpha,\infty[}(|T|)) \leq t$, so $a \leq \alpha$.)

(**d**) Show that if T is a positive τ-measurable operator, then

$$\tau(T) = \int_0^\infty \mu_t(T)\,\mathrm{d}t; \qquad \mu_t\big(f(T)\big) = f\big(\mu_t(T)\big)$$

for any continuous increasing function f on $[0,\infty[$ with $f(0) \geq 0$;

$$\mu_t(T) \leq \mu_t(S); \qquad \tau\big(f(T)\big) \leq \tau\big(f(S)\big) \quad \text{if } 0 \leq T \leq S \text{ in addition.}$$

8) Let $\{\mathcal{M}, \mathfrak{H}\}$ be a semi-finite von Neumann algebra equipped with a faithful semi-finite normal trace τ. Let $\mathcal{N}$ be the commutant $\pi_\tau(\mathcal{M})'$ of $\mathcal{M}$ represented on $L^2(\mathcal{M}, \tau)$ and let $\tau_{\mathcal{N}}$ be the trace on $\mathcal{N}$ opposite to τ.

(a) Show that there exists a unique faithful semi-finite normal trace $\tau_{\mathfrak{H}}$ on the commutant $\pi_{\mathfrak{H}}(\mathcal{M})'$ of $\mathcal{M}$ on $\mathfrak{H}$ such that

$$\tau_{\mathfrak{H}}(yy^*) = \tau_{\mathcal{N}}(y^*y)$$

for every $y \in \mathcal{L}\big(L^2(\mathcal{M}, \tau), \mathfrak{H}\big)$ such that $xy = y\pi_\tau(x),\ x \in \mathcal{M}$.

(b) Observe that for every $\xi \in \mathfrak{H}$ there exists a unique $h_\xi \in L^2(\mathcal{M}, \tau)_+$ such that $(x\xi \mid \xi) = \big(xh_\xi \mid h_\xi\big)_\tau,\ x \in \mathcal{M}$. Show that each $\xi \in \mathfrak{H}$ gives rise to an operator: $\eta_\tau(x) \mapsto x\xi,\ x \in \mathfrak{n}_\tau$, which is preclosed.

(c) Let $\mathfrak{K}$ be another Hilbert space on which $\mathcal{M}$ acts faithfully and normally. Normalize the faithful semi-finite normal trace $\tau_{\mathfrak{K}}$ on the commutant $\{\mathcal{M}'_{\mathfrak{K}}, \mathfrak{K}\}$ of $\pi_{\mathfrak{K}}(\mathcal{M})$ according to (a). Show that if $y \in \mathcal{L}(\mathfrak{H}, \mathfrak{K})$ intertwines the actions of $\mathcal{M}$, i.e. $y\pi_{\mathfrak{H}}(x) = \pi_{\mathfrak{K}}(x)y,\ x \in \mathcal{M}$, then $\tau_{\mathfrak{H}}(y^*y) = \tau_{\mathfrak{K}}(yy^*)$. Set $\mathcal{I}_{\mathcal{M}}(\mathfrak{H}, \mathfrak{K}) = \big\{\, y \in \mathcal{L}(\mathfrak{H}, \mathfrak{K}) : y\pi_{\mathfrak{H}}(x) = \pi_{\mathfrak{K}}(x)y,\ x \in \mathcal{M} \,\big\}$.

(d) Define the measure topology on $\mathcal{I}_{\mathcal{M}}(\mathfrak{H}, \mathfrak{K})$ by the family of a neighborhood system $\big\{\, N(\varepsilon, \delta) : \varepsilon > 0,\ \delta > 0 \,\big\}$ of 0 given by:

$$N(\varepsilon, \delta) = \big\{\, y \in \mathcal{I}_{\mathcal{M}}(\mathfrak{H}, \mathfrak{K}) : \|yp\| < \varepsilon \text{ and } \tau_{\mathfrak{H}}(p^\perp) < \delta \text{ for some } p \in \mathrm{Proj}(\mathcal{M}'_{\mathfrak{H}}) \,\big\}.$$

Prove the statements on $\mathcal{I}_{\mathcal{M}}(\mathfrak{H}, \mathfrak{K})$ corresponding to Theorem 2.2 with respect to the measure topology defined above. But the measure topology on both $\mathfrak{H}$ and $\mathfrak{K}$ should be referred to $\big\{\mathcal{M}'_{\mathfrak{H}}, \tau_{\mathfrak{H}}\big\}$ and $\big\{\mathcal{M}'_{\mathfrak{K}}, \tau_{\mathfrak{K}}\big\}$ respectively instead of $\mathcal{M}$. (Hint: Consider the direct sum $\tilde{\mathfrak{H}} = L^2(\mathcal{M}, \tau) \oplus \mathfrak{H} \oplus \mathfrak{K}$, $\pi_{\tilde{\mathfrak{H}}} = \pi_\tau \oplus \pi_{\mathfrak{H}} \oplus \pi_{\mathfrak{K}}$ and $\mathcal{M}'_{\tilde{\mathfrak{H}}} = \pi_{\tilde{\mathfrak{H}}}(\mathcal{M})'$ and identify $\mathcal{I}_{\mathcal{M}}(\mathfrak{H}, \mathfrak{K}) = p_{\mathfrak{K}}\mathcal{M}'_{\tilde{\mathfrak{H}}}p_{\mathfrak{H}}$ where $p_{\mathfrak{H}}$ and $p_{\mathfrak{K}}$ are the projections from $\tilde{\mathfrak{H}}$ onto $\mathfrak{H}$ and $\mathfrak{K}$ respectively.)

(e) Let $\mathfrak{M}_{\mathcal{M}}(\mathfrak{H}, \mathfrak{K})$ be the completion of $\mathcal{I}_{\mathcal{M}}(\mathfrak{H}, \mathfrak{K})$ relative to the measure topology. Prove the statements on $\mathfrak{M}_{\mathcal{M}}(\mathfrak{H}, \mathfrak{K})$ corresponding to Theorem 2.5.

(f) Prove that each $\xi \in \mathfrak{H}$ gives rise to an element $R(\xi) \in \mathfrak{M}_{\mathcal{M}}\big(L^2(\mathcal{M}, \tau), \mathfrak{H}\big)$ such that $R(\xi)\eta_\tau(x) = x\xi,\ x \in \mathfrak{n}_\tau$.

(g) Prove that $R(\xi)^*R(\xi) \in L^1(\mathcal{N}, \tau_{\mathcal{N}})$.

(h) Prove that every $R \in \mathfrak{M}_{\mathcal{M}}\big(L^2(\mathcal{M}, \tau), \mathfrak{H}\big)$ with $\tau_{\mathcal{N}}(R^*R) < \infty$, called *square integrable*, corresponds uniquely to a vector $\xi \in \mathfrak{H}$ such that $R = R(\xi)$. Let $L^2_{\mathcal{M}}(\mathfrak{H}) = \big\{\, R \in \mathfrak{M}_{\mathcal{M}}\big(L^2(\mathcal{M}, \tau), \mathfrak{H}\big) : \tau_{\mathcal{N}}(R^*R)^{1/2} = \|R\|_2 < \infty \,\big\}$.

(i) Prove that $\xi \in \mathfrak{H} \mapsto R(\xi) \in L^2_{\mathcal{M}}(\mathfrak{H})$ is a surjective isometry of $\mathfrak{H}$ onto $L^2_{\mathcal{M}}(\mathfrak{H})$.

(j) Observe that on the Hilbert space $L^2_{\mathcal{M}}(\mathfrak{H})$, $\mathcal{M}'_{\mathfrak{H}}$ acts from the left and $\mathcal{N}$ acts from the right by operator multiplication.

(k) Observe that it is possible to do the same for $\mathcal{M}'_{\mathfrak{H}}$ so that $\mathcal{M}$ acts on $L^2_{\mathcal{M}'_{\mathfrak{H}}}(\mathcal{M}'_{\mathfrak{H}}, \tau_{\mathfrak{H}})$ from the left. This will be studied more in detail in the next section.

§ 3 Bimodules, Spatial Derivatives and Relative Tensor Products

We are going to consider a right action of a von Neumann algebra on a Hilbert space. To this end, we first set:

Definition 3.1.

(i) Given a von Neumann algebra $\mathcal{N}$, the *opposite* von Neumann algebra $\mathcal{N}^\circ$ means the von Neumann algebra obtained by reversing the product in $\mathcal{N}$, i.e. as a linear space equipped with the *-operation we take $\mathcal{N}$ to be $\mathcal{N}^\circ$, denote by x° the element in $\mathcal{N}^\circ$ corresponding to $x \in \mathcal{N}$ and then define the product in $\mathcal{N}^\circ$ by:
$$x^\circ y^\circ = (yx)^\circ, \quad x, y \in \mathcal{N}. \tag{1}$$

(ii) A right $\mathcal{N}$*-module* means a Hilbert space $\mathfrak{H}$ equipped with a normal anti-representation, $\pi'_{\mathfrak{H}}$, of $\mathcal{N}$ on $\mathfrak{H}$, equivalently a Hilbert space equipped with a normal representation of $\mathcal{N}^\circ$. To avoid uninteresting notational complexity, we consider only *faithful* right $\mathcal{N}$-modules $\mathfrak{H}$ in the sense that $\pi'_{\mathfrak{H}}(x) \neq 0$ for every non-zero $x \in \mathcal{N}$. We denote the right $\mathcal{N}$-module $\mathfrak{H}$ by $\mathfrak{H}_{\mathcal{N}}$ to emphasize that $\mathfrak{H}$ is being viewed as a right $\mathcal{N}$-module.

(iii) For a pair $\mathcal{M}$, $\mathcal{N}$ of von Neumann algebras, an $\mathcal{M}$-$\mathcal{N}$*-bimodule* means a Hilbert space $\mathfrak{H}$, denoted by ${}_{\mathcal{M}}\mathfrak{H}_{\mathcal{N}}$, equipped with a normal representation of π of $\mathcal{M}$ on $\mathfrak{H}$ and a normal anti-representation π' of $\mathcal{N}$ on $\mathfrak{H}$ such that $\pi(\mathcal{M})$ and $\pi'(\mathcal{N})$ commute. We write:
$$x\xi y = \pi(x)\pi'(y)\xi, \quad x \in \mathcal{M}, \quad y \in \mathcal{N}. \tag{2}$$

The commutativity of $\pi(\mathcal{M})$ and $\pi'(\mathcal{N})$ is equivalent to the associativity: $x(\xi y) = (x\xi)y$, $x \in \mathcal{M}$ and $y \in \mathcal{N}$. Once again, we are going to consider only faithful modules. In the case that $\pi(\mathcal{M})' = \pi'(\mathcal{N})$, the bimodule ${}_{\mathcal{M}}\mathfrak{H}_{\mathcal{N}}$ is said to be *full*.

Let us fix von Neumann algebras $\mathcal{M}$ and $\mathcal{N}$. If $\mathfrak{H}$ is an $\mathcal{M}$-$\mathcal{N}$-bimodule, then its Banach space dual $\bar{\mathfrak{H}}$ is canonically an $\mathcal{N}$-$\mathcal{M}$-bimodule by the action:
$$x\bar{\xi}y = \overline{y^*\xi x^*}, \quad x \in \mathcal{M}, \quad y \in \mathcal{N}, \tag{3}$$

where $\bar{\xi}$ means the vector in $\bar{\mathfrak{H}}$ corresponding to $\xi \in \mathfrak{H}$ by the pairing: $\langle \eta, \bar{\xi} \rangle = (\eta \mid \xi)$, $\eta \in \mathfrak{H}$ and $\bar{\xi} \in \bar{\mathfrak{H}}$. This $\mathcal{M}$-$\mathcal{N}$-module $\bar{\mathfrak{H}}$ will be called the *conjugate* bimodule or the bimodule *dual* to the original bimodule $\mathfrak{H}$.

Of special interest is a von Neumann algebra in a standard form. Let us fix a faithful semi-finite normal weight ψ on $\mathcal{N}$ and consider the standard form, which we will denote by $\{L^2(\mathcal{N}), L^2(\mathcal{N})_+, J\}$. The right action of $\mathcal{N}$ is given by:
$$\xi x = Jx^*J\xi, \quad x \in \mathcal{N}. \tag{4}$$

Thus we obtain an $\mathcal{N}$-$\mathcal{N}$-bimodule $L^2(\mathcal{N})$, which will be called the *standard bimodule*. Sometimes, we write ξ^* for $J\xi$, $\xi \in L^2(\mathcal{N})$. We state here the following easy but important proposition:

Proposition 3.2. *For a von Neumann algebra* $\mathcal{N}$, *the standard bimodule* $L^2(\mathcal{N})$ *is self-dual under the correspondence:* $\xi^* \longleftrightarrow \bar{\xi}$, $\xi \in L^2(\mathcal{N})$.

The proof is straightforward, so it will be left to the reader.

With ψ a faithful semi-finite normal weight on $\mathcal{N}$, the left action on $L^2(\mathcal{N})$ is nothing but the semi-cyclic representation π_ψ on $\mathfrak{H}_\psi$. The right action π'_ψ of $\mathcal{N}$ is then given by:

$$\pi'_\psi(x) = J\pi_\psi(x^*)J, \quad x \in \mathcal{N}. \tag{5}$$

It then follows from Lemma VIII.3.18 that the right action of $\mathcal{N}$ is given by the following:

$$\eta_\psi(x)b = \eta_\psi\big(x\sigma^\psi_{-\frac{i}{2}}(b)\big), \quad x \in \mathfrak{n}_\psi, \quad b \in \mathcal{D}\big(\sigma^\psi_{-\frac{i}{2}}\big). \tag{6}$$

This twist on the right action suggests that we write $x\psi^{1/2}$ for $\eta_\psi(x)$, $x \in \mathfrak{n}_\psi$, viewing $\psi^{1/2}$ as an infinitely long vector "in" $L^2(\mathcal{N})$. Then the formula (6) is simply:

$$x\psi^{\frac{1}{2}}b = \big(x\psi^{\frac{1}{2}}b\psi^{-\frac{1}{2}}\big)\psi^{\frac{1}{2}} = \big(x\sigma^\psi_{-\frac{i}{2}}(b)\big)\psi^{\frac{1}{2}}, \quad x \in \mathfrak{n}_\psi, \quad b \in \mathcal{D}\big(\sigma^\psi_{-\frac{i}{2}}\big). \tag{6$'$}$$

We then introduce a new notation:

$$\eta'_\psi(x) = J\eta_\psi(x^*), \quad x \in \mathfrak{n}^*_\psi, \tag{7}$$

which can be written as $\psi^{1/2}x$, $x \in \mathfrak{n}^*_\psi$. This new map $\eta'_\psi : x \in \mathfrak{n}^*_\psi \mapsto \eta'_\psi(x) \in L^2(\mathcal{N})$ allows us to write (5) in a simple form:

$$\pi'_\psi(b)\eta'_\psi(x) = \eta'_\psi(xb) = \eta'_\psi(x)b, \quad x \in \mathfrak{n}^*_\psi, \quad b \in \mathcal{N}. \tag{8}$$

We now consider a general right $\mathcal{N}$-module $\mathfrak{H}$. First, we set for a pair $\{\mathfrak{H}_1, \mathfrak{H}_2\}$ of right $\mathcal{N}$-modules:

$$\mathcal{L}\big(\mathfrak{H}_{1,\mathcal{N}}, \mathfrak{H}_{2,\mathcal{N}}\big) = \big\{T \in \mathcal{L}(\mathfrak{H}_1, \mathfrak{H}_2) : T(\xi y) = (T\xi)y, \ y \in \mathcal{N}\big\} \tag{9}$$

and for $\mathcal{L}\big(\mathfrak{H}_\mathcal{N}, \mathfrak{H}_\mathcal{N}\big)$, we write simply $\mathcal{L}(\mathfrak{H}_\mathcal{N})$. With this notation, the right $\mathcal{N}$-module $\mathfrak{H}$ becomes canonically an $\mathcal{L}(\mathfrak{H}_\mathcal{N})$-$\mathcal{N}$-bimodule. Also, we note that $\mathcal{L}\big(L^2(\mathcal{N})_\mathcal{N}\big) = \mathcal{N}$, a fact that will be used heavily later. For the pair $\{\mathfrak{H}_1, \mathfrak{H}_2\}$, we consider the direct sum right $\mathcal{N}$-module: $\mathfrak{H}_\mathcal{N} = \mathfrak{H}_1 \oplus \mathfrak{H}_2$ and let e_1 and e_2 be the projection of $\mathfrak{H}$ to $\mathfrak{H}_1$ and $\mathfrak{H}_2$ respectively. Then we have $\mathcal{L}\big(\mathfrak{H}_{1,\mathcal{N}}, \mathfrak{H}_{2,\mathcal{N}}\big) = e_2\mathcal{L}(\mathfrak{H}_\mathcal{N})e_1$.

Now let $\{\mathcal{M}, \mathfrak{H}\}$ be a von Neumann algebra. We want to study the relation between a semi-finite normal weight φ on $\mathcal{M}$ and a faithful semi-finite normal weight ψ' on $\mathcal{M}'$. Set $\mathcal{N} = (\mathcal{M}')^\circ$, which allows us to view $\mathfrak{H}$ as an $\mathcal{M}$-$\mathcal{N}$-bimodule. Let ψ be the weight on $\mathcal{N}$ defined by $\psi(y) = \psi'(y^\circ)$, $y \in \mathcal{N}$. We first relate the von Neumann algebra $\{\mathcal{M}, \mathfrak{H}\}$ to the one in a standard form. Let $\tilde{\mathfrak{H}} = L^2(\mathcal{N}) \oplus \mathfrak{H}$ as a right $\mathcal{N}$-module. Let $\mathcal{R} = \mathcal{L}(\tilde{\mathfrak{H}}_\mathcal{N})$. It is then straightforward to observe that

$\mathcal{L}(L^2(\mathcal{N})_{\mathcal{N}}, \mathfrak{H}_{\mathcal{N}}) = f\mathcal{R}e$, where e and f are the projections of $\tilde{\mathfrak{H}}$ onto $L^2(\mathcal{N})$ and $\mathfrak{H}$ respectively. The semi-finite normal weights ψ on $\mathcal{N}$ and φ on $\mathcal{M}$ give rise to a semi-finite normal weight ρ on $\mathcal{R}$ by: $\rho(x) = \psi(exe) + \varphi(fxf)$, $x \in \mathcal{R}$. We set

$$\left.\begin{aligned} \mathfrak{n}_\psi(\mathfrak{H}) &= f(\mathcal{R} \cap \mathfrak{n}_\rho)e = \left\{x \in \mathcal{L}(L^2(\mathcal{N})_{\mathcal{N}}, \mathfrak{H}_{\mathcal{N}}) : \psi(x^*x) < \infty\right\}; \\ \mathfrak{D}(\mathfrak{H}, \psi) &= \left\{\xi \in \mathfrak{H} : \|\xi x\| \le C_\xi \left\|\eta'_\psi(x)\right\|,\ x \in \mathfrak{n}^*_\psi \text{ for some } C_\xi \ge 0\right\}. \end{aligned}\right\} \tag{10}$$

Each $\xi \in \mathfrak{D}(\mathfrak{H}, \psi)$ gives rise to an operator, denoted by $L_\psi(\xi)$, which belongs to $\mathcal{L}(L^2(\mathcal{N})_{\mathcal{N}}, \mathfrak{H}_{\mathcal{N}})$, defined by the equation:

$$L_\psi(\xi)\eta'_\psi(x) = \xi x, \quad x \in \mathfrak{n}^*_\psi, \quad \xi \in \mathfrak{D}(\mathfrak{H}, \psi). \tag{11}$$

Lemma 3.3.

(i) *We have*

$$\mathfrak{n}_\psi(\mathfrak{H}) = \mathcal{L}(L^2(\mathcal{N})_{\mathcal{N}}, \mathfrak{H}_{\mathcal{N}})\mathfrak{n}_\psi \quad \text{and} \quad \mathfrak{D}(\mathfrak{H}, \psi) = \mathcal{L}(L^2(\mathcal{N})_{\mathcal{N}}, \mathfrak{H}_{\mathcal{N}})\mathfrak{B}_\psi,$$

where $\mathfrak{B}_\psi = \eta_\psi(\mathfrak{n}_\psi)$ *in* $L^2(\mathcal{N})$.

(ii) *The map* $\eta_\psi \colon (x, y) \in \mathcal{L}(L^2(\mathcal{N})_{\mathcal{N}}, \mathfrak{H}_{\mathcal{N}}) \times \mathfrak{n}_\psi \mapsto x\eta_\psi(y) \in \mathfrak{D}(\mathfrak{H}, \psi)$ *gives rise to a map, denoted by* η_ψ *again, from* $\mathfrak{n}_\psi(\mathfrak{H})$ *onto* $\mathfrak{D}(\mathfrak{H}, \psi)$ *such that*

$$\left.\begin{aligned} \eta_\psi(ax) &= a\eta_\psi(x), \quad a \in \mathcal{M}, \quad x \in \mathfrak{n}_\psi(\mathfrak{H}); \\ \eta_\psi\left(x\sigma^\psi_{-\frac{i}{2}}(b)\right) &= \eta_\psi(x)b, \quad x \in \mathfrak{n}_\psi(\mathfrak{H}), \quad b \in \mathcal{D}\left(\sigma^\psi_{-\frac{i}{2}}\right). \end{aligned}\right\} \tag{12}$$

(iii) $\mathfrak{D}(\mathfrak{H}, \psi)$ *is dense in* $\mathfrak{H}$.

(iv) *The maps* $L_\psi \colon \xi \in \mathfrak{D}(\mathfrak{H}, \psi) \mapsto L_\psi(\xi) \in \mathfrak{n}_\psi(\mathfrak{H})$ *and* $\eta_\psi \colon x \in \mathfrak{n}_\psi(\mathfrak{H}) \mapsto \eta_\psi(x) \in \mathfrak{D}(\mathfrak{H}, \psi)$ *are the inverse of each other.*

(v)

$$L_\psi\left(\xi\sigma^\psi_{\frac{i}{2}}(b)\right) = L_\psi(\xi)b, \quad \xi \in \mathfrak{D}(\mathfrak{H}, \psi), \quad b \in \mathcal{D}\left(\sigma^\psi_{\frac{i}{2}}\right). \tag{12'}$$

(vi) *With the semi-finite normal weight* $\bar{\psi}$ *on* $\mathcal{R}$ *defined by*

$$\bar{\psi}(x) = \psi(exe) = \rho(exe), \quad x \in \mathcal{R},$$

we have $\mathfrak{n}_{\bar{\psi}} = \mathfrak{n}_\psi \oplus \mathfrak{n}_\psi(\mathfrak{H}) \oplus \mathcal{R}f$, $\mathcal{R}f \subset N_{\bar{\psi}}$ *where* $N_{\bar{\psi}}$ *means the left kernel* $\left\{y \in \mathcal{N} : \bar{\psi}(y^*y) = 0\right\}$ *of* $\bar{\psi}$ *and the action of* $\mathcal{R}$ *on* $\tilde{\mathfrak{H}}$ *is semi-cyclic relative to the semi-finite normal weight* $\bar{\psi}$ *under the identification:*

$$\eta_{\bar{\psi}}(x+y) \in \mathfrak{H}_{\bar{\psi}} \longleftrightarrow \eta_\psi(x) \oplus \eta_\psi(y) \in L^2(\mathcal{N}) \oplus \mathfrak{H}, \quad x \in \mathfrak{n}_\psi, \quad y \in \mathfrak{n}_\psi(\mathfrak{H}).$$

PROOF:

(i) If x is in $\mathfrak{n}_\psi(\mathfrak{H})$, then the absolute value $|x|$ belongs to $\mathfrak{n}_\psi$ by definition, so that the polar decomposition of x shows that x belongs to $\mathcal{L}(L^2(\mathcal{N})_{\mathcal{N}}, \mathfrak{H})\mathfrak{n}_\psi$. Conversely, if $a \in \mathcal{L}(L^2(\mathcal{N})_{\mathcal{N}}, \mathfrak{H}_{\mathcal{N}})$ and $x \in \mathfrak{n}_\psi$, then the inequality: $x^*a^*ax \le \|a\|^2x^*x$ implies that $ax \in \mathfrak{n}_\psi(\mathfrak{H})$. If $\xi \in \mathfrak{n}_\psi(\mathfrak{H})$, then $a = L_\psi(\xi)$ belongs to

$\mathcal{L}(L^2(\mathcal{N})_{\mathcal{N}}, \mathfrak{H})$ and with the polar decomposition $a = uh$ we conclude first that h belongs to $\mathfrak{n}_\psi$ and also that $\xi = u\eta_\psi(h) \in \mathcal{L}(L^2(\mathcal{N})_{\mathcal{N}}, \mathfrak{H})\mathfrak{B}_\psi$.

(ii) If $xy = 0$ with $x \in \mathcal{L}(L^2(\mathcal{N})_{\mathcal{N}}, \mathfrak{H}_{\mathcal{N}})$ and $y \in \mathfrak{n}_\psi$, then with $x = uh$ the polar decomposition we have $hy = 0$ and

$$x\eta_\psi(y) = uh\eta_\psi(y) = u\eta_\psi(hy) = 0.$$

This means that if $x_1y_1 = x_2y_2$ with $x_1, x_2 \in \mathcal{L}(L^2(\mathcal{N})_{\mathcal{N}}, \mathfrak{H}_{\mathcal{N}})$ and $y_1, y_2 \in \mathfrak{n}_\psi$, then we have $x_1\eta_\psi(y_1) = x_2\eta_\psi(y_2)$, so that the map η_ψ is well-defined. The rest follows easily.

(iii) From (i) it follows that

$$[\mathfrak{D}(\mathfrak{H}, \psi)] = [\mathfrak{D}(\mathfrak{H}, \psi)\mathfrak{B}_\psi] = [\mathfrak{D}(\mathfrak{H}, \psi)L^2(\mathcal{N})].$$

Let $\xi \in \mathfrak{H}$. Consider $\omega = \omega_\xi$ as a functional over $\mathcal{N}$. Let $\xi(\omega)$ be the representing vector in $L^2(\mathcal{N})_+$ of ω for the right action of $\mathcal{N}$ on $L^2(\mathcal{N})$. Then we have a partial isometry u in $\mathcal{L}(L^2(\mathcal{N})_{\mathcal{N}}, \mathfrak{H}_{\mathcal{N}})$ such that $u\xi(\omega) = \xi$. This means that $\mathcal{L}(L^2(\mathcal{N})_{\mathcal{N}}, \mathfrak{H}_{\mathcal{N}})L^2(\mathcal{N}) = \mathfrak{H}$; hence $\mathfrak{D}(\mathfrak{H}, \psi)$ is dense in $\mathfrak{H}$. The first formula of (9) follows from the construction of the map η_ψ.

(iv) Let $\xi = \eta_\psi(x)$ with $x \in \mathfrak{n}_\psi(\mathfrak{H})$ and $x = uh$ be the polar decomposition. Then $h \in \mathfrak{n}_\psi$ and $u \in \mathcal{L}(L^2(\mathcal{N})_{\mathcal{N}}, \mathfrak{H}_{\mathcal{N}})$ and $\xi = u\eta_\psi(h)$ by (ii). Now for each $y \in \mathfrak{n}_\psi \cap \mathfrak{n}_\psi^* \cap \mathcal{D}(\sigma^\psi_{-i/2})$ such that $\sigma^\psi_{-i/2}(y) \in \mathfrak{n}_\psi$, we have

$$\begin{aligned} L_\psi(\xi)J\eta_\psi(y^*) &= \xi y = u\eta_\psi(h)y = u\eta_\psi\big(h\sigma^\psi_{-\frac{i}{2}}(y)\big) \qquad \text{by (5)} \\ &= uh\eta_\psi\big(\sigma^\psi_{-\frac{i}{2}}(y)\big) = x\Delta_\psi^{\frac{1}{2}}\eta_\psi(y) = xJ\eta_\psi(y^*). \end{aligned}$$

Therefore, we have $x = L_\psi(\xi)$. Conversely, suppose $x = L_\psi(\xi)$ with $\xi \in \mathfrak{D}(\mathfrak{H}, \psi)$. With $x = uh$ polar decomposition, we have, for each $y \in \mathfrak{n}_\psi \cap \mathfrak{n}_\psi^* \cap \mathcal{D}(\sigma^\psi_{-i/2})$ such that $\sigma^\psi_{-i/2}(y) \in \mathfrak{n}_\psi$,

$$hJ\eta_\psi(y^*) = u^*xJ\eta_\psi(y^*) = u^*(\xi y) = (u^*\xi)y,$$

so that the vector $u^*\xi \in L^2(\mathcal{N})$ is left bounded relative to the left Hilbert algebra $\mathfrak{A}_\psi$ and $h = \pi_\ell(u^*\xi)$. This means that $h \in \mathfrak{n}_\psi$, and so $x \in \mathfrak{n}_\psi(\mathfrak{H})$. It is easy to see now that $\xi = \eta_\psi(x)$.

(v) This follows from (ii), (iv) and Lemma VIII.3.18.

(vi) By now, the assertion follows from a routine calculation of actions of $\mathcal{R}$ on $\tilde{\mathfrak{H}}$ and $\mathfrak{H}_{\bar{\psi}}$. Q.E.D.

We continue to study the action of $\mathcal{R}$ on $\tilde{\mathfrak{H}}$. The direct sum decomposition, $\tilde{\mathfrak{H}} = L^2(\mathcal{N}) \oplus \mathfrak{H}$, entails the matrix representation of $\mathcal{R}$:

$$x = \begin{pmatrix} x_{11} & x_{12} \\ x_{21} & x_{22} \end{pmatrix}, \quad \begin{array}{ll} x_{11} \in \mathcal{N}, & x_{12} \in \mathcal{L}(L^2(\mathcal{N})_{\mathcal{N}}, \mathfrak{H}_{\mathcal{N}}), \\ x_{21} \in \mathcal{L}(\mathfrak{H}_{\mathcal{N}}, L^2(\mathcal{N})_{\mathcal{N}}), & x_{22} \in \mathcal{M} = \mathcal{L}(\mathfrak{H}_{\mathcal{N}}) \end{array}$$

for each $x \in \mathcal{R}$.

We have seen that $\mathfrak{a}_\psi = \mathfrak{n}_\psi \cap \mathfrak{n}_\psi^*$ or more precisely its image $\eta_\psi(\mathfrak{a}_\psi)$ form a left Hilbert algebra. Likewise, $\mathcal{A} = \mathfrak{D}(\sigma^\psi_{-i/2}) \cap \mathfrak{D}(\sigma^\psi_{i/2}) = \mathfrak{D}(\sigma^\psi_{i/2}) \cap \mathfrak{D}(\sigma^\psi_{i/2})^*$ is a self-adjoint subalgebra of $\mathcal{N}$ which multiplies $\mathfrak{a}_\psi$ and $\mathfrak{n}_\psi^*$ from both sides. We then have the following tautological statement:

Lemma 3.4. *The anti-representation π'_ψ of $\mathcal{A}$ defined by:*

$$\pi'_\psi(b)\eta_\psi(x) = \eta_\psi\left(x\sigma^\psi_{-\frac{i}{2}}(b)\right), \quad x \in \mathfrak{n}_\psi(\mathfrak{H}), \quad b \in \mathcal{A}, \tag{12''}$$

extends to the original right action of $\mathcal{N}$ on $\mathfrak{H}$.

PROOF: Our assertion follows directly from (6), (12) and (12′). We leave the detail to the reader. Q.E.D.

Observe that we have used only the semi-finite normal weight ψ on $\mathcal{N}$ and not at all the semi-finite normal weight φ on $\mathcal{M}$. The "balanced" weight $\rho = \psi \oplus \varphi$ on $\mathcal{R}$ then gives a semi-cyclic representation $\{\pi_\rho, \mathfrak{H}_\rho\}$ of $\mathcal{R}$. We want to identify the representation π_ρ in terms of $\mathfrak{H}$ and π_φ. First we consider the weights $\bar{\psi}$ and $\bar{\varphi}$ on $\mathcal{R}$ given by $\bar{\psi}(x) = \psi(exe)$ and $\bar{\varphi}(x) = \varphi(fxf)$, $x \in \mathcal{R}$. We have now:

$$\begin{aligned}\mathfrak{H}_\rho &= \begin{pmatrix} \eta_\rho(e\mathcal{R}e \cap \mathfrak{n}_\rho) + \eta_\rho(e\mathcal{R}f \cap \mathfrak{n}_\rho) \\ + \\ \eta_\rho(f\mathcal{R}e \cap \mathfrak{n}_\rho) + \eta_\rho(f\mathcal{R}f \cap \mathfrak{n}_\rho) \end{pmatrix} \\ &= \begin{pmatrix} L^2(\mathcal{N}) \oplus [\eta_\rho(e\mathcal{R}f \cap \mathfrak{n}_\rho)] \\ \oplus \\ \mathfrak{H} \quad \oplus \quad \mathfrak{H}_\varphi \end{pmatrix},\end{aligned} \tag{13}$$

where $[\,\cdot\,]$ stands for the closure in the Hilbert space as usual. We have seen already that $\eta_\rho(f\mathcal{R}e \cap \mathfrak{n}_\rho) = \mathfrak{D}(\mathfrak{H}, \psi)$ and $f\mathcal{R}e \cap \mathfrak{n}_\rho = \mathfrak{n}_\psi(\mathfrak{H})$. Also we knew that $f\mathcal{R}e = \mathcal{L}(L^2(\mathcal{N})_\mathcal{N}, \mathfrak{H}_\mathcal{N})$ and $e\mathcal{R}f = \mathcal{L}(\mathfrak{H}_\mathcal{N}, L^2(\mathcal{N})_\mathcal{N})$. We now want to know about $\mathfrak{n}_\rho \cap e\mathcal{R}f$ and its image under the map η_ρ. As we did not assume the faithfulness for φ, we don't have symmetry between φ and ψ. At any rate, we have

$$\begin{aligned} e\mathcal{R}f \cap \mathfrak{n}_\rho &= \left\{ y \in \mathcal{L}(\mathfrak{H}_\mathcal{N}, L^2(\mathcal{N})_\mathcal{N}) : \varphi(y^*y) < \infty \right\} \\ &= \left\{ x^* : x \in \mathcal{L}(L^2(\mathcal{N})_\mathcal{N}, \mathfrak{H}_\mathcal{N}) \text{ with } \varphi(xx^*) < \infty \right\}. \end{aligned} \tag{10'}$$

Following the decomposition (13), we naturally set:

$$\begin{aligned} \mathfrak{H}_{1,1} &= L^2(\mathcal{N}), & \mathfrak{H}_{1,2} &= [\eta_\rho(e\mathcal{R}f \cap \mathfrak{n}_\rho)] \\ \mathfrak{H}_{2,1} &= \mathfrak{H}, & \mathfrak{H}_{2,2} &= \mathfrak{H}_\varphi. \end{aligned}$$

Lemma 3.5.

(i) *The restriction of π_ρ to the second column space of* (13), $\mathfrak{H}_{1,2} \oplus \mathfrak{H}_{2,2}$, *is semi-cyclic relative to the weight $\bar{\varphi}$.*

(ii) *The Hilbert space* $\mathfrak{H}_{1,2}$ *is isomorphic to* $\overline{s(\varphi)\mathfrak{H}_{2,1}}$ *dual to* $s(\varphi)\mathfrak{H}_{2,1}$ *as an* $\mathcal{N}$*-*$\mathcal{M}_{s(\varphi)}$*-bimodule under the natural map.*

PROOF: First consider the case that φ is faithful. Then with $\mathfrak{a}_\rho = \mathfrak{n}_\rho \cap \mathfrak{n}_\rho^*$ and $\mathfrak{A}_\rho = \eta_\rho(\mathfrak{a}_\rho)$, we have a left Hilbert algebra $\mathfrak{A}_\rho$. Furthermore, the $\mathcal{R}$-$\mathcal{R}$-bimodule $L^2(\mathcal{R})$ is naturally identified with $\mathfrak{H}_\rho$. Then the components of $\mathfrak{H}_\rho$ of (13) are identified with

$$\mathfrak{H}_{1,1} = eL^2(\mathcal{R})e, \qquad \mathfrak{H}_{1,2} = eL^2(\mathcal{R})f,$$

$$\mathfrak{H}_{2,1} = fL^2(\mathcal{R})e, \qquad \mathfrak{H}_{2,2} = fL^2(\mathcal{R})f.$$

The modular conjugation J implements the desired isomorphism between $\bar{\mathfrak{H}}_{2,1}$ and $\mathfrak{H}_{1,2}$. This proves the assertion (ii) for the faithful case. The assertion (i) follows from the symmetry between ψ on $\mathcal{N}$ and φ° on $\mathcal{M}^\circ$.

If φ is not faithful, then with $q = s(\varphi)$ we consider an auxiliary semi-finite normal weight φ_1 on $\mathcal{M}$ with $e_1 = s(\varphi_1) = 1_{\mathfrak{H}} - q \leq e$ to have a faithful semi-finite normal weight $\rho_1 = \rho + \bar{\varphi}_1$ on $\mathcal{R}$ such that $s(\bar{\varphi}_1)$ belongs to $\mathcal{R}_{\rho_1}$ where $\bar{\varphi}_1$ is given by $\bar{\varphi}_1(x) = \varphi_1(exe)$, $x \in \mathcal{R}_+$. Observe that η_ρ and η_{ρ_1} agree on $e\mathcal{R}e$ and $f\mathcal{R}e$ and that we have $\eta_\rho(x) = \eta_{\rho_1}(xq)$, $x \in (e\mathcal{R}f + f\mathcal{R}f) \cap \mathfrak{n}_\rho$. Hence we get

$$\mathfrak{H}_{1,1} = eL^2(\mathcal{R})e, \qquad \mathfrak{H}_{2,1} = fL^2(\mathcal{R})e,$$

$$\mathfrak{H}_{1,2} = eL^2(\mathcal{R})q, \qquad \mathfrak{H}_{2,2} = fL^2(\mathcal{R})q.$$

Therefore, we have $Jq\mathfrak{H}_{2,1} = JqL^2(\mathcal{R})e = eL(\mathcal{R})q = \mathfrak{H}_{1,2}$. This completes the proof for (i). For (ii), we have $\mathfrak{H}_{1,2} \oplus \mathfrak{H}_{2,2} = L^2(\mathcal{R})q$ as an $\mathcal{N}$-$\mathcal{M}_q$-bimodule. Therefore, the representation $\{\pi_\rho, \mathfrak{H}_{1,2} \oplus \mathfrak{H}_{2,2}\}$ is nothing but the semi-cyclic representation $\{\pi_{\bar{\varphi}}, \mathfrak{H}_{\bar{\varphi}}, \eta_{\bar{\varphi}}\}$. Q.E.D.

If we look at the origin of the above proof, we recognize that the conjugation operator J comes from the conjugate linear operator

$$S_{\varphi,\psi}\colon \eta_\psi(x) \in \eta_\psi\big(\mathcal{L}(L^2(\mathcal{N})_\mathcal{N}, \mathfrak{H}_\mathcal{N}) \cap \mathfrak{n}_\rho \cap \mathfrak{n}_\rho^*\big) \mapsto \eta_\varphi(x^*) \in \mathfrak{H}_{1,2},$$

which is the restriction of the $S_{\bar{\varphi}_1,\bar{\psi}}$-map for ρ_1 to the smaller domain $qL^2(\mathcal{R})e$, i.e. $S_{\varphi,\psi}$ can be directly defined as the closure of the operator given by $S_{\varphi,\psi}\eta_\psi(x) = \eta_\varphi(x^*)$ for $x \in \mathfrak{n}_\psi(\mathfrak{H}) \cap \mathfrak{n}_\varphi(\mathfrak{H})^*$, where

$$\mathfrak{n}_\varphi(\mathfrak{H}) = \Big\{x \in \mathcal{L}\big(\mathfrak{H}_\mathcal{N}, L^2(\mathcal{N})_\mathcal{N}\big) : \varphi(x^*x) < \infty\Big\}.$$

Thus we come to the following:

Definition 3.6. The absolute value $\Delta_{\varphi,\psi}$ of $S_{\varphi,\psi}$ is called *spatial derivative* of the semi-finite normal weight φ on $\mathcal{M}$ relative to the faithful semi-finite normal weight ψ' on the commutant $\mathcal{M}'$ and denoted by $\frac{d\varphi}{d\psi'}$ as it is determined by ψ' on $\mathcal{M}'$ and φ on $\mathcal{M}$.

Dualizing (10), we set:

$$\mathfrak{D}'(\mathfrak{H}, \varphi) = \left\{ \xi \in \mathfrak{H} : \begin{array}{l} \|x\xi\|^2 \leq C_\xi \varphi(x^* x), \ x \in \mathcal{M}, \\ \text{for some constant } C_\xi \geq 0 \end{array} \right\}. \tag{10''}$$

To each $\xi \in \mathfrak{D}'(\mathfrak{H}, \varphi)$ there corresponds an operator $R_\varphi(\xi)$ defined by

$$R_\varphi(\xi)\eta_\varphi(x) = x\xi, \quad x \in \mathfrak{n}_\varphi,$$

which belongs to $\mathcal{L}\big({}_{\mathcal{M}}L^2(\mathcal{M}),\ {}_{\mathcal{M}}\mathfrak{H}\big)$. As φ is not assumed to be faithful, ψ and φ are not symmetric. In fact, we have the following:

Lemma 3.7. *The closure of* $\mathfrak{D}'(\mathfrak{H}, \varphi)$ *is the range of the projection* $s(\varphi)$, *i.e* $\big[\mathfrak{D}'(\mathfrak{H}, \varphi)\big] = s(\varphi)\mathfrak{H}$.

PROOF: If $\xi \in \mathfrak{D}'(\mathfrak{H}, \varphi)$, then we have

$$\big\|\big(1 - s(\varphi)\big)\xi\big\|^2 \leq C_\xi \varphi\big((1 - s(\varphi)\big) = 0.$$

Hence $\mathfrak{D}'(\mathfrak{H}, \varphi) \subset s(\varphi)\mathfrak{H}$. Conversely, suppose $\xi \perp \mathfrak{D}'(\mathfrak{H}, \varphi)$. With $\Phi = \big\{\omega : \omega \leq \varphi\big\}$, we know $\varphi = \sup \Phi$, and that

$$\eta \in \mathfrak{D}'(\mathfrak{H}, \varphi) \Longleftrightarrow \omega_\eta \in \bigcup_{C>0} C\Phi.$$

Also every $\omega \in \Phi$ is a countable sum of ω_{η_n} with $\eta_n \in \mathfrak{D}'(\mathfrak{H}, \varphi)$, so that $s(\varphi) = \sup\big\{ s(\omega_\eta) : \eta \in \mathfrak{D}'(\mathfrak{H}, \varphi)\big\}$; thus we have $s(\varphi)\xi = 0$. Q.E.D.

We now state one of the main results of the section:

Theorem 3.8. *Let* $\{\mathcal{M}, \mathfrak{H}\}$ *be a von Neumann algebra,* φ *a semi-finite normal weight on* $\mathcal{M}$ *and* ψ' *a faithful semi-finite normal weight on the commutant* $\mathcal{M}'$. *Then the spatial derivative* $\mathrm{d}\varphi/\mathrm{d}\psi'$ *has the following property:*

(i) *The support* $s\big(\mathrm{d}\varphi/\mathrm{d}\psi'\big)$ *of the spatial derivative* $\mathrm{d}\varphi/\mathrm{d}\psi'$ [4] *is equal to* $s(\varphi)$.

(ii) *On the reduced von Neumann algebra* $\big\{\mathcal{M}_{s(\varphi)}, s(\varphi)\mathfrak{H}\big\}$ *and its commutant* $\mathcal{M}'_{s(\varphi)}$, *we have:*

$$\left.\begin{aligned} \left(\frac{\mathrm{d}\varphi}{\mathrm{d}\psi'}\right)^{it} x \left(\frac{\mathrm{d}\varphi}{\mathrm{d}\psi'}\right)^{-it} &= \sigma_t^\varphi(x), \quad x \in \mathcal{M}_{s(\varphi)}, \\ \left(\frac{\mathrm{d}\varphi}{\mathrm{d}\psi'}\right)^{it} y \left(\frac{\mathrm{d}\varphi}{\mathrm{d}\psi'}\right)^{-it} &= \sigma_{-t}^{\psi'}(y), \quad y \in \mathcal{M}'_{s(\varphi)}. \end{aligned}\right\} \tag{14}$$

(iii) *If* φ_1 *and* φ_2 *are faithful semi-finite normal weights on* $\mathcal{M}$, *then*

$$\left(\frac{\mathrm{d}\varphi_2}{\mathrm{d}\psi'}\right)^{it} = \big(\mathrm{D}\varphi_2 : \mathrm{D}\varphi_1\big)_t \left(\frac{\mathrm{d}\varphi_1}{\mathrm{d}\psi'}\right)^{it}. \tag{15}$$

---

[4] The support of a self-adjoint operator means the projection to the closure of the range.

(iv) *If φ is also faithful, then*

$$\frac{d\psi'}{d\varphi} = \left(\frac{d\varphi}{d\psi'}\right)^{-1}. \tag{16}$$

(v) *With $\mathcal{N} = (\mathcal{M}')^\circ$ and $\psi = (\psi')^\circ$, the square root of the spatial derivative, $\left(d\varphi / d\psi'\right)^{1/2}$, is essentially self-adjoint on*

$$\mathfrak{D}_{\varphi,\psi}(\mathfrak{H}) = \left\{ \xi \in \mathfrak{D}(\mathfrak{H}, \psi) : L_\psi(\xi) \in \mathfrak{n}_\varphi^* \right\}$$

and determined by:

$$\left(\left(\frac{d\varphi}{d\psi'}\right)^{\frac{1}{2}} \xi \,\middle|\, \left(\frac{d\varphi}{d\psi'}\right)^{\frac{1}{2}} \eta \right) = \varphi(L_\psi(\xi) L_\psi(\eta)^*), \quad \xi, \eta \in \mathfrak{D}_{\varphi,\psi}(\mathfrak{H}). \tag{17}$$

Therefore, the spatial derivative $d\varphi / d\psi'$ of φ relative to ψ' is directly computable from φ and ψ'.

PROOF: From the previous arguments to $\mathcal{M}$, $\mathcal{N}$, φ and ψ, we know that the spatial derivative $d\varphi / d\psi'$ is defined to be the relative modular operator $\Delta_{\varphi,\psi}$ on the subspace $s(\varphi)\mathfrak{H}$. Now we replace $\mathfrak{H}$ by $s(\varphi)\mathfrak{H}$ and assume that φ is faithful. Then the assertions (i), (ii), (iii) and (iv) are nothing but the statements about the relative modular operator which was already proven in the last chapter. Statement (v) also follows from the definition of the relative modular operator $\Delta_{\varphi,\psi}$ in the last chapter, i.e. $\Delta_{\psi,\varphi}^{1/2}$ is the absolute value of $S_{\varphi,\psi}$. Q.E.D.

From the statement (v) in the above theorem it follows that the spatial derivative $d\varphi / d\psi'$ is directly determined by φ and ψ' without making use of the auxiliary von Neumann algebra $\mathcal{R}$ in the preceding discussion.

Lemma 3.9. *The linear span $\mathcal{J}_\psi$ of $L_\psi(\xi) L_\psi(\eta)^*$, $\xi, \eta \in \mathfrak{D}(\mathfrak{H}, \psi)$, is a σ-weakly dense ideal of $\mathcal{M}$ and we have:*

$$\mathcal{J}_\psi^+ = \left\{ \sum_{i=1}^{n} L_\psi(\xi_i) L_\psi(\xi_i)^* : \xi_i \in \mathfrak{D}(\mathfrak{H}, \psi),\ i = 1, \dots, n \right\}.$$

PROOF: It is easy to see that $L_\psi(a\xi) = aL_\psi(\xi)$ for any $a \in \mathcal{M}$ and $\xi \in \mathfrak{D}(\mathfrak{H}, \psi)$. Hence $\mathcal{J}_\psi$ is an ideal of $\mathcal{M}$. The rest follows from the polarization techniques which we have been using repeatedly. See for example, Chapter VII, §1. As $L_\psi\left(\mathfrak{D}(\mathfrak{H}, \psi)\right) = \mathfrak{n}_\psi(\mathfrak{H})$, for the σ-weak density of $\mathcal{J}_\psi$ it suffices to prove that $a\mathfrak{n}_\psi(\mathfrak{H}) = \{0\}$, $a \in \mathcal{M}$, implies that $a = 0$. Suppose $a\mathfrak{n}_\psi(\mathfrak{H}) = \{0\}$, $a \in \mathcal{M}$. This means that $aL_\psi(\xi) = 0$ for every $\xi \in \mathfrak{D}(\mathfrak{H}, \psi)$. Thus for every $x \in \mathfrak{n}_\psi^*$ we have

$$0 = aL_\psi(\xi)\eta_\psi'(x) = a(\xi x) = (a\xi)x.$$

As $\pi_{\mathfrak{H}}'(\mathfrak{n}_\psi^*)$ is σ-weakly dense in $\pi_{\mathfrak{H}}'(\mathcal{N})$, we have $a\xi = 0$. The density of $\mathfrak{D}(\mathfrak{H}, \psi)$ in $\mathfrak{H}$ implies $a = 0$. Q.E.D.

Proposition 3.10. *The spatial derivative* $(\mathrm{d}\varphi/\mathrm{d}\psi')$ *has the following properties further:*

(i)
$$\varphi_1 \le \varphi_2 \iff \frac{\mathrm{d}\varphi_1}{\mathrm{d}\psi'} \le \frac{\mathrm{d}\varphi_2}{\mathrm{d}\psi'}.$$

(ii) *If* φ_1 *and* φ_2 *are both finite, then*

$$\frac{\mathrm{d}(\varphi_1+\varphi_2)}{\mathrm{d}\psi'} = \frac{\mathrm{d}\varphi_1}{\mathrm{d}\psi'} + \frac{\mathrm{d}\varphi_2}{\mathrm{d}\psi'}, \tag{18}$$

where the above sum is a form sum, see A.9.

(iii) *If* $a \in \mathcal{M}$ *is invertible, then*

$$\frac{\mathrm{d}(a\varphi a^*)}{\mathrm{d}\psi'} = a\left(\frac{\mathrm{d}\varphi}{\mathrm{d}\psi'}\right)a^*. \tag{19}$$

(iv) *The support of* $\frac{\mathrm{d}\varphi}{\mathrm{d}\psi'}$ *is equal to the support* $s(\varphi)$ *of* φ, *where the support of* $\frac{\mathrm{d}\varphi}{\mathrm{d}\psi'}$ *is the projection to the closure of the range of the operator* $\frac{\mathrm{d}\varphi}{\mathrm{d}\psi'}$ *and denoted by* $s\left(\frac{\mathrm{d}\varphi}{\mathrm{d}\psi'}\right)$.

PROOF:

(i) Suppose $\varphi_1 \le \varphi_2$. Then we have

$$\left\|\left(\frac{\mathrm{d}\varphi_1}{\mathrm{d}\psi'}\right)^{\frac{1}{2}}\xi\right\|^2 = \varphi_1\big(L_\psi(\xi)L_\psi(\xi)^*\big) \le \varphi_2\big(L_\psi(\xi)L_\psi(\xi)^*\big) = \left\|\left(\frac{\mathrm{d}\varphi_2}{\mathrm{d}\psi'}\right)^{\frac{1}{2}}\xi\right\|^2$$

for each $\xi \in \mathfrak{D}(\mathfrak{H},\psi)$. Hence $\mathrm{d}\varphi_1/\mathrm{d}\psi \le \mathrm{d}\varphi_2/\mathrm{d}\psi$ follows.

Conversely, suppose $\mathrm{d}\varphi_1/\mathrm{d}\psi' \le \mathrm{d}\varphi_2/\mathrm{d}\psi'$. It then follows that $\varphi_1(a) \le \varphi_2(a)$ for every $a \in \mathcal{M}_+$ of the form $a = \sum_{i=1}^n L_\psi(\xi_i)L_\psi(\xi_i)^*$. Our assertion then follows from the last lemma.

(ii) Suppose that $\varphi_1, \varphi_2 \in \mathcal{M}_*^+$ and $\varphi = \varphi_1 + \varphi_2$. The boundedness of φ_1 and φ_2 imply that φ is bounded and the square roots of all the spatial derivatives $\frac{\mathrm{d}\varphi_1}{\mathrm{d}\psi'}$, $\frac{\mathrm{d}\varphi_2}{\mathrm{d}\psi'}$ and $\frac{\mathrm{d}\varphi}{\mathrm{d}\psi'}$ are essentially self-adjoint on $\mathfrak{D}(\mathfrak{H},\psi)$. Let $H_1 = \mathrm{d}\varphi_1/\mathrm{d}\psi'$, $H_2 = \mathrm{d}\varphi_2/\mathrm{d}\psi'$ and $H = \mathrm{d}\varphi/\mathrm{d}\psi'$. Then we have $\|H^{1/2}\xi\|^2 = \|H_1^{1/2}\xi\|^2 + \|H_2^{1/2}\xi\|^2$, $\xi \in \mathfrak{D}(\mathfrak{H},\psi)$. Hence our assertion follows.

(iii) Let $H = \mathrm{d}\varphi/\mathrm{d}\psi'$. It follows that aHa^* is a positive self-adjoint operator with domain $a^{*^{-1}}\mathfrak{D}(H)$, and that for each $\xi \in \mathfrak{D}(\mathfrak{H},\psi)$

$$\begin{aligned}\|H^{\frac{1}{2}}a^*\xi\|^2 &= \varphi\big(L_\psi(a^*\xi)L_\psi(a^*\xi)^*\big) = \varphi\big(a^*L_\psi(\xi)L_\psi(\xi)^*a\big)\\ &= (a\varphi a^*)\big(L_\psi(\xi)L_\psi(\xi)^*\big).\end{aligned}$$ [5]

Since a is invertible, $\mathfrak{D}\big((aHa^*)^{1/2}\big) = \mathfrak{D}\big(H^{1/2}a^*\big)$ and the absolute value of $H^{1/2}a^*$ is precisely $(aHa^*)^{1/2}$. Hence we conclude (19).

5 $\|H^{1/2}a^*\xi\|^2$ can be $+\infty$ if φ is not finite. In fact, $\|H^{1/2}a^*\xi\|^2 < +\infty$ if and only if $a^*\xi \in \mathfrak{D}(H^{1/2})$.

(iv) Let p be the support of $H = \mathrm{d}\varphi/\mathrm{d}\psi'$ and $q = s(\varphi)$. Then p is characterized by the fact that $1 - p$ is the projection of $\mathfrak{H}$ onto the null space of H, i.e. onto the subspace: $\mathfrak{N} = \{\xi \in \mathfrak{H} : H\xi = 0\}$. Let $\mathfrak{A}_1$ be the maximal Tomita algebra associated with the left Hilbert algebra $\mathfrak{A}_{\rho_1} = \eta_{\rho_1}(\mathfrak{n}_{\rho_1} \cap \mathfrak{n}_{\rho_1}^*)$. As $f \in \mathcal{R}_{\rho_1}$, we have, with $f' = JfJ$, $f'\mathfrak{A}_1 = J\mathfrak{A}_1 \subset \mathfrak{A}_1$. If $\xi \in \mathfrak{N}$, then there exists a sequence $\{\xi_n\}$ in $\mathfrak{D}(\mathfrak{H}, \psi)$ such that $\xi = \lim \xi_n$ and $\lim H\xi_n = 0$ as $\mathfrak{D}(\mathfrak{H}, \psi)$ is a core for H. For each $\eta \in f'\mathfrak{A}_1$, we have $\pi_r(\eta)\xi = \lim_n \pi_r(\eta)\xi_n$ and

$$\begin{aligned}\left\| H^{\frac{1}{2}}\pi_r(\eta)\xi \right\|^2 &= \lim \left\| H^{\frac{1}{2}}\pi_r(\eta)\xi_n \right\|^2 = \lim \left\| H^{\frac{1}{2}}\pi_\ell(\xi_n)\eta \right\|^2 \\ &= \lim \varphi\big(\pi_\ell(\xi_n\eta)\pi_\ell(\xi_n\eta)^*\big) \\ &\le \left\|\pi_\ell(\eta)\right\|^2 \lim \varphi\big(\pi_\ell(\xi_n)\pi_\ell(\xi_n)^*\big) = 0,\end{aligned}$$

so that $\pi_r(\eta)\xi \in \mathfrak{N}$. Since $\{\pi_r(\eta)_e : \eta \in f'\mathfrak{A}_1\}$ is σ-weakly dense in $\mathcal{N}$, the projection p belongs to $\mathcal{M} = \mathcal{N}'$.

If $\xi \in (1 - q)\mathfrak{D}(\mathfrak{H}, \psi)$, then

$$\varphi\big(L_\psi(\xi)L_\psi(\xi)^*\big) = \varphi\big(qL_\psi(\xi)L_\psi(\xi)^*q\big) = \varphi\big(L_\psi(q\xi)L_\psi(q\xi)^*\big) = 0,$$

so $\left(\frac{\mathrm{d}\varphi}{\mathrm{d}\psi'}\right)^{1/2}\xi = 0$. If $\xi \in (1 - q)\mathfrak{H}$, then we choose a sequence $\xi_n \in \mathfrak{D}(\mathfrak{H}, \psi)$ with $\xi = \lim \xi_n$. It follows that $\xi = \lim(1 - q)\xi_n$ and $\left(\mathrm{d}\varphi/\mathrm{d}\psi'\right)^{1/2}(1 - q)\xi_n = 0$, so that $\xi \in \mathfrak{D}\big((\mathrm{d}\varphi/\mathrm{d}\psi')^{1/2}\big)$ and $\left(\mathrm{d}\varphi/\mathrm{d}\psi'\right)^{1/2}\xi = 0$. Thus $1 - q \le 1 - p$, i.e. $p \le q$. On the other hand, φ is a faithful weight on $\mathcal{M}_q$. It is then easy to check that we can view the weight ψ as one on $\mathcal{N}_{s(\varphi)}$ without changing $\mathrm{d}\varphi/\mathrm{d}\psi'$ other than changing the space from $\mathfrak{H}$ to $s(\varphi)\mathfrak{H}$. By Theorem 3.8, $\mathrm{d}\varphi/\mathrm{d}\psi'$ is non-singular on $s(\varphi)\mathfrak{H}$, which means $p = s(\varphi)$. Q.E.D.

Theorem 3.11. *Let $\{\mathcal{M}, \mathfrak{H}\}$ and $\mathcal{N} = \mathcal{M}'$ be as before and fix a faithful weight ψ on $\mathcal{N}$. For a positive self-adjoint operator H on $\mathfrak{H}$, the following three conditions are equivalent:*

(i) *There exists a weight φ on $\mathcal{M}$ with $H = \mathrm{d}\varphi/\mathrm{d}\psi'$;*

(ii) *For every $y \in \mathcal{N}$, $H^{\mathrm{i}t}\sigma_t^\psi(y) = yH^{\mathrm{i}t}$, $t \in \mathbf{R}$, where $H^{\mathrm{i}t}$ is considered only on the closure of the range of H;*

(iii) *$\mathfrak{D}(\mathfrak{H}, \psi) \cap \mathfrak{D}(H^{1/2})$ is a core for $H^{1/2}$ and the scalar $\sum_{i=1}^n \left\| H^{1/2}\xi_i \right\|^2$ depends only on the operator:*

$$\sum_{i=1}^n L_\psi(\xi_i)L_\psi(\xi_i)^* \quad \textit{for } \xi_1, \dots, \xi_n \in \mathfrak{D}(\mathfrak{H}, \psi) \cap \mathfrak{D}\big(H^{\frac{1}{2}}\big).$$

PROOF:

(i) $\Longleftrightarrow$ (ii): Let p be the support of H. Each of conditions (i) and (ii) implies $p \in \mathcal{M}$. Hence we may and do assume the non-singularity of H. The implication (i) $\Longrightarrow$ (ii) follows from Theorem 3.8. Now assume (ii), and take any faithful weight φ_1 on $\mathcal{M}$. Put $H_1 = \mathrm{d}\varphi_1/\mathrm{d}\psi'$, and $u_t = H^{\mathrm{i}t}H_1^{-\mathrm{i}t}$, $t \in \mathbf{R}$. It then follows that u_t is a $\{\sigma_t^{\varphi_1}\}$-cocycle in $\mathcal{M}$. By Theorem VIII.3.7 there exists a faithful weight φ on $\mathcal{M}$ with $(\mathrm{D}\varphi : \mathrm{D}\varphi_1)_t = u_t,\ t \in \mathbf{R}$, and hence $H = \mathrm{d}\varphi/\mathrm{d}\psi'$.

(i) $\Longrightarrow$ (iii): By construction, we have

$$\sum_{i=1}^{n} \left\| H^{\frac{1}{2}} \xi_i \right\|^2 = \varphi\Big(\sum_{i=1}^{n} L_\psi(\xi_i) L_\psi(\xi_i)^* \Big), \quad \xi_1, \dots, \xi_n \in \mathfrak{D}(\mathfrak{H}, \psi),$$

so that the assertion follows.

(iii) $\Longrightarrow$ (i): We need only to construct a semi-finite normal weight φ on $\mathcal{M}$ such that

$$\left\| H^{\frac{1}{2}} \xi \right\|^2 = \varphi\big(L_\psi(\xi) L_\psi(\xi)^* \big), \quad \xi \in \mathfrak{D}(\mathfrak{H}, \psi) \cap \mathfrak{D}\big(H^{\frac{1}{2}}\big).$$

The weight with this property is automatically semi-finite because $\mathfrak{D}\big(H^{1/2}\big) \cap \mathfrak{D}(\mathfrak{H}, \psi)$ is dense in $\mathfrak{H}$ which means that $\big\{ L_\psi(\xi) L_\psi(\xi)^* : \xi \in \mathfrak{D}\big(H^{1/2}\big) \cap \mathfrak{D}(\mathfrak{H}, \psi) \big\}$ is non-degenerate. The rest of the proof then follows from the next result.

Q.E.D.

Lemma 3.12.

(i) *Let H be as in Theorem 3.11.(iii). Then there exists a weight*[6] φ_1 *on* $\mathfrak{J}_\psi$ *such that*

$$\varphi_1\big(L_\psi(\xi) L_\psi(\xi)^* \big) = \left\| H^{\frac{1}{2}} \xi \right\|^2, \quad \xi \in \mathfrak{D}(\mathfrak{H}, \psi), \tag{20}$$

where $\left\| H^{1/2} \xi \right\|^2 = +\infty$ *if* $\xi \notin \mathfrak{D}\big(H^{1/2}\big)$. *The weight* φ_1 *has the property that for any net* $\{x_i\}$ *of* $\mathcal{M}$ *converging strongly to* 1

$$\liminf \varphi_1\big(x_i y x_i^* \big) \geq \varphi_1(y), \quad y \in \mathfrak{J}_\psi^+. \tag{21}$$

(ii) *Any weight* φ_1 *on* $\mathfrak{J}_\psi$ *with the above property extends to a normal weight* φ *on* $\mathcal{M}$.

PROOF:

(i) By assumption, φ_1 defined by (20) on $L_\psi(\xi) L_\psi(\xi)^*$ extends to a weight on $\mathfrak{J}_\psi^+$ by Lemma 3.5.(ii), which is denoted again by φ_1. Suppose $y = \sum_{k=1}^{n} L_\psi(\xi_k) L_\psi(\xi_k)^* \in \mathfrak{J}_\psi^+$. We then have

$$x_i y x_i^* = \sum_{k=1}^{n} L_\psi(x_i \xi_k) L_\psi(x_i \xi_k)^*$$

so that

$$\varphi(x_i y x_i^*) = \sum_{k=1}^{n} \left\| H^{\frac{1}{2}} x_i \xi_k \right\|^2.$$

Hence the inequality (21) follows from the lower semi-continuity of the positive quadratic form associated with H, see A.8.

---

6 We do not consider the normality nor the semi-finiteness for φ_1 here.

(ii) Since $\mathcal{J}_\psi$ is a σ-weakly dense ideal of $\mathcal{M}$, every element of $\mathcal{M}_+$ is approximated by $\mathcal{J}_\psi^+$ from below. So we put

$$\varphi(x) = \sup\{\, \varphi_1(y) : y \in \mathcal{J}_\psi^+,\ y \leq x \,\}, \quad x \in \mathcal{M}_+.$$

It follows that φ agrees with φ_1 on $\mathcal{J}_\varphi^+$. Since $x_i \nearrow x$ and $y_i \nearrow y$ imply $(x_i + y_i) \nearrow (x + y)$, the additivity of φ follows from that of φ_1. We only have to check the normality of φ. Suppose that $x_i \nearrow x$ in $\mathcal{M}_+$. Then $x_i^{1/2} = a_i x^{1/2}$ for a unique $a_i \in \mathcal{M}$ with $s_r(a_i) \leq s(x)$. Put $b_i = a_i + \big(1 - s(x)\big)$. Then $x_i = b_i x b_i^*$ and $\{b_i\}$ converges strongly to 1. For any $y \in \mathcal{J}_\psi^+$ with $y \leq x$ we must show that $\sup \varphi(x_i) \geq \varphi_1(y)$. But we have, by (21),

$$\varphi_1(y) \leq \liminf \varphi_1(b_i y b_i^*) \leq \liminf \varphi(b_i x b_i^*),$$

and $b_i x b_i^* = x_i$. Q.E.D.

Corollary 3.13. *Let $\mathcal{M}$, $\mathcal{N}$ and $\mathfrak{H}$ be as before, and ψ' a fixed faithful weight on $\mathcal{M}'$. If $\{\varphi_n\}$ is an increasing sequence of faithful weights on $\mathcal{M}$ and if $\varphi = \sup \varphi_n$ is semi-finite, then $\{\mathrm{d}\varphi_n/\mathrm{d}\psi'\}$ is increasing and converges to $\mathrm{d}\varphi/\mathrm{d}\psi'$ in the strong resolvent sense, and therefore $\{\sigma_t^{\varphi_n}\}$ converges to $\{\sigma_t^{\varphi}\}$ in $\mathrm{Aut}(\mathcal{M})$ uniformly on any finite interval of t.*

PROOF: Let $H_n = \mathrm{d}\varphi_n/\mathrm{d}\psi'$. By Proposition 3.10, $\{H_n\}$ is increasing and bounded by $H = \mathrm{d}\varphi/\mathrm{d}\psi'$ from above. Hence $\{H_n\}$ converges to a positive self-adjoint operator K in the strong resolvent sense by A.11. Since $H_n^{\mathrm{i}t} y H_n^{-\mathrm{i}t} = \sigma_{-t}^{\psi}(y)$ for every $y \in \mathcal{N}$, $K^{\mathrm{i}t} y K^{\mathrm{i}t} = \sigma_{-t}^{\psi}(y)$, $y \in \mathcal{N}$. By Theorem 3.11, there exists uniquely a weight ρ on $\mathcal{M}$ with $K = \mathrm{d}\rho/\mathrm{d}\psi'$. The inequalities, $H_n \leq K \leq H$, show that $\varphi_n \leq \rho \leq \varphi$. Hence $\rho = \varphi$ and $K = H$. The rest follows from the general fact about monotone convergence, see A.11. Q.E.D.

We are now going to study a very special case that $\mathcal{M}$ and $\mathcal{N}$ are both finite.

Theorem 3.14. *Suppose that a finite von Neumann algebra $\{\mathcal{M}, \mathfrak{H}\}$ has finite commutant $\mathcal{M}' = \mathcal{N}^\circ$. If τ and τ' are both faithful normal tracial states on $\mathcal{M}$ and on $\mathcal{M}'$ respectively which agree on the center $\mathcal{Z} = \mathcal{M} \cap \mathcal{M}'$, then the spatial derivative $\frac{\mathrm{d}\tau}{\mathrm{d}\tau'}$ is equal to the coupling operator $c_{\mathcal{M}}$ in the sense of §3 of Chapter V.*

PROOF: To avoid possible confusions, we use the notations φ for τ and ψ' for τ' and set $\psi = \psi'^\circ$ on $\mathcal{N}$. First we note that as the modular automorphism groups of φ and ψ' are both trivial, the spatial derivative $\left(\frac{\mathrm{d}\varphi}{\mathrm{d}\psi'}\right)^{\mathrm{i}t}$ must belong to the center $\mathcal{Z}$ of $\mathcal{M}$. Hence $\frac{\mathrm{d}\varphi}{\mathrm{d}\psi'}$ is affiliated with $\mathcal{Z}$.

As in the previous arguments, we consider the von Neumann algebra $\mathcal{R} = \mathcal{L}\big((L^2(\mathcal{N}) \oplus \mathfrak{H})_{\mathcal{N}}\big)$. The projections e and f of $\tilde{\mathfrak{H}} = L^2(\mathcal{N}) \oplus \mathfrak{H}$ onto $L^2(\mathcal{N})$ and $\mathfrak{H}$ respectively are both finite in $\mathcal{R}$. As $1 = e + f$, we conclude by Theorem V.1.37 that $\mathcal{R}$ is finite and σ-finite. Let $\tilde{\rho}$ be a normalized faithful normal trace on $\mathcal{R}$. Then there exist non-singular positive self-adjoint operators a and b affiliated with $\mathcal{Z}$ such that $\bar{\varphi} = \tilde{\rho}(ae\,\cdot\,)$ and $\bar{\psi} = \tilde{\rho}(bf\,\cdot\,)$, where $\bar{\varphi}$ and $\bar{\psi}$ mean the ones we have defined above. Recall also the decomposition (13) of $\mathfrak{H}_\rho$ with $\rho = \bar{\varphi} + \bar{\psi}$.

Choose $\xi \in \mathfrak{H}$ and consider its polar decomposition: $\xi = u|\xi|$ in $L^2(\mathcal{R})$. As $\xi \in \mathfrak{H} = \mathfrak{H}_{2,1}$, its absolute value $|\xi|$ belongs to $L^2(\mathcal{N})_+$ and $|\xi^*| \in L^2(\mathcal{M})_+$. The initial projection $p = u^*u$ of u is equal to $\pi_\psi\big(s(\omega_\xi|_{\mathcal{M}'})^\circ\big)$ and the final projection $q = uu^*$ of u is equal to $\pi_\varphi\big(s(\omega_\xi|_{\mathcal{M}})\big)$. Hence, with $T_{\mathcal{R}}$ the center valued trace of $\mathcal{R}$, we have $T_{\mathcal{R}}(p) = T_{\mathcal{R}}(q)$, so that $\varphi\big(s(\omega_\xi|_{\mathcal{M}})a^{-1}x\big) = \psi\big(s(\omega_\xi|_{\mathcal{M}'})^\circ b^{-1}x\big)$, $x \in \mathcal{Z}$. Thus we have

$$T_{\mathcal{M}}\big(s(\omega_\xi|_{\mathcal{M}})a^{-1}\big) = T_{(\mathcal{M}')^\circ}\big(s(\omega_\xi|_{\mathcal{M}'})^\circ b^{-1}\big) = T_{\mathcal{M}'}\big(s(\omega_\xi|_{\mathcal{M}'})b^{-1}\big).$$

Therefore, we conclude that $c_{\mathcal{M}} = ab^{-1}$.

Suppose now that $\xi, \eta \in \mathfrak{D}_{\psi,\varphi}(\mathfrak{H})$. Let $\{z_n\}$ be an increasing sequence of projections in $\mathcal{Z}$ converging σ-strongly to $1_{\mathcal{R}}$ such that $az_n \in \mathcal{Z}$, $bz_n \in \mathcal{Z}$ and $\big(\frac{d\varphi}{d\psi'}\big)z_n \in \mathcal{Z}$ for each $n \in \mathbf{N}$. Then we have

$$\begin{aligned}
\bar{\rho}\left(b\left(\frac{d\varphi}{d\psi'}\right)z_nL_\psi(\eta)^*L_\psi(\xi)\right) &= \bar{\psi}\left(\left(\frac{d\varphi}{d\psi'}\right)z_nL_\psi(\eta)^*L_\psi(\xi)\right)\\
&= \left(\left(\frac{d\varphi}{d\psi'}\right)\eta_{\bar{\psi}}(L_\psi(\xi))\,\Big|\,\eta_{\bar{\psi}}(L_\psi(\xi))\right)\\
&= \left(\left(\frac{d\varphi}{d\psi'}\right)z_n\xi\,\Big|\,\eta\right) = \varphi\big(z_nL_\psi(\xi)L_\psi(\eta)^*\big)\\
&= \bar{\rho}\big(az_nL_\psi(\xi)L_\psi(\eta)^*\big).
\end{aligned}$$

As $\big(\mathfrak{n}_\rho \cap \mathfrak{n}_\rho^* \cap f\mathcal{R}e\big)^*\big(\mathfrak{n}_\rho \cap \mathfrak{n}_\rho^* \cap f\mathcal{R}e\big)$ is σ-weakly dense in $\mathcal{R}_e$ and also the same is true for $\big(\mathfrak{n}_\rho \cap \mathfrak{n}_\rho^* \cap f\mathcal{R}e\big)\big(\mathfrak{n}_\rho \cap \mathfrak{n}_\rho^* \cap f\mathcal{R}e\big)^*$ in $\mathcal{R}_f$, the above calculation shows that $b\big(\frac{d\varphi}{d\psi'}\big)z_n = az_n$. Hence we have $\frac{d\varphi}{d\psi'} = ab^{-1} = c_{\mathcal{M}}$. Q.E.D.

We now move on to relative tensor products of a right module and a left module over the same von Neumann algebra $\mathcal{N}$. Unlike the ordinary tensor product case, the construction of the relative tensor product depends on the choice of a faithful semi-finite normal weight on $\mathcal{N}$ as mentioned above. Furthermore, the tensor product of an arbitrary pair of vectors of $\mathfrak{H}$ and $\mathfrak{K}$ cannot be defined. It is restricted to certain pairs of vectors of $\mathfrak{H}$ and $\mathfrak{K}$ depending on the faithful semi-finite normal weight. It behaves like the product of closed unbounded operators. Let us prepare some notations first.

As in the case of right modules, for two left $\mathcal{N}$-modules ${}_{\mathcal{N}}\mathfrak{K}_1$ and ${}_{\mathcal{N}}\mathfrak{K}_2$ we consider $\mathcal{L}({}_{\mathcal{N}}\mathfrak{K}_1, {}_{\mathcal{N}}\mathfrak{K}_2) = \big\{T \in \mathcal{L}(\mathfrak{K}_1, \mathfrak{K}_2) : Ta\eta = aT\eta,\ \eta \in \mathfrak{K}_1,\ a \in \mathcal{N}\big\}$. For $\mathcal{L}({}_{\mathcal{N}}\mathfrak{K}, {}_{\mathcal{N}}\mathfrak{K})$ we write simply $\mathcal{L}({}_{\mathcal{N}}\mathfrak{K})$. Throughout the rest of this section, we denote by $\mathfrak{H}$ a right $\mathcal{N}$-module and by $\mathfrak{K}$ a left $\mathcal{N}$-module. Observe that a right $\mathcal{N}$-module $\mathfrak{H}$ means canonically an $\mathcal{L}(\mathfrak{H}_{\mathcal{N}})$-$\mathcal{N}$-bimodule and a left $\mathcal{N}$-module $\mathfrak{K}$ means canonically an $\mathcal{N}$-$\mathcal{L}({}_{\mathcal{N}}\mathfrak{K})^\circ$-bimodule. We are now going to construct the relative tensor product $\mathfrak{H} \otimes_\psi \mathfrak{K}$ of a right $\mathcal{N}$-module $\mathfrak{H}$ and a left $\mathcal{N}$-module $\mathfrak{K}$ which will depend on the choice of a faithful semi-finite normal weight ψ on $\mathcal{N}$. To put things in perspective, let us consider a standard von Neumann algebra $\{\mathcal{R}, L^2(\mathcal{R})\}$ and a pair of projections $e, f \in \mathcal{R}$ such that $\{\mathcal{N}, \mathfrak{K}\} \cong \{\mathcal{R}_{eJfJ}, eL^2(\mathcal{R})f\}$ and $\mathcal{N} \cong \mathcal{R}_e$. Here we are considering the case that the both $\mathfrak{H}$ and $\mathfrak{K}$ are faithful modules, so that the central support $c(e)$ and $c(f)$ are both the identity. Thus after identifying $\mathcal{N}$ with $\mathcal{R}_e$ as an abstract von Neumann algebra, the left $\mathcal{N}$-module $\mathfrak{K}$ is identified with $eL^2(\mathcal{R})f$. This identification identifies also $\mathcal{L}({}_{\mathcal{N}}\mathfrak{K})$ with $\{\mathcal{R}'_{eJfJ}, eL^2(\mathcal{R})f\}$. Also the right $\mathcal{N}$-module $\mathfrak{H}$ is identified with $gL^2(\mathcal{R})e$ for another projection $g \in \mathcal{R}$ which yields the identification of $\mathcal{L}(\mathfrak{H}_{\mathcal{N}})$ with $\{\mathcal{R}_g, gL^2(\mathcal{R})e\}$. The relative tensor product of $\mathfrak{H}$ and $\mathfrak{K}$ means roughly the Hilbert space $g[L^2(\mathcal{R})eL^2(\mathcal{R})]f = gL^2(\mathcal{R})f$ with the $\mathcal{R}_g$-$\mathcal{R}_f$-bimodule structure. Here we observe that we cannot multiply an arbitrary pair $\xi \in gL^2(\mathcal{R})e$ and $\eta \in gL^2(\mathcal{R})f$ to have a vector "$\xi\eta$" $\in gL(\mathcal{R})f$. To associate the third vector "$\xi \otimes_{\mathcal{N}} \eta$" in $gL^2(\mathcal{R})f$, one needs to convert ξ into an operator $\pi_\ell(\xi)$, which belongs to $\mathcal{N}$, after fixing a faithful semi-finite normal weight ψ on $\mathcal{N}$, and then set $\xi \otimes_{\mathcal{N}} \eta$ to be $\pi_\ell(\xi)\eta$. In doing this, we see that only certain vectors ξ can be converted to $\pi_\ell(\xi)$, i.e. ξ must be left bounded relative to a certain left Hilbert algebra. However, in this way we will have too many choices. First, the projections e, f, g are determined only up to equivalence. The modules $\mathfrak{H}$ and $\mathfrak{K}$ alone can not determine them. So we have to take an approach which relates the construction more directly to the modules $\mathfrak{H}$ and $\mathfrak{K}$.

We fix a von Neumann algebra $\mathcal{N}$, a right $\mathcal{N}$-module $\mathfrak{H}$ and a left $\mathcal{N}$-module $\mathfrak{K}$. We also need a faithful semi-finite normal weight ψ on $\mathcal{N}$ to be fixed. We know from Lemma 3.3 that the right module $\mathfrak{H}$ can be recovered from $\mathfrak{D}(\mathfrak{H}, \psi)$ and that the left module $\mathfrak{K}$ is also recoverable from $\mathfrak{D}'(\mathfrak{K}, \psi)$ (observe that ψ and ψ° are symmetric as they are both faithful, unlike the previous case of φ). We collect here a few facts about $\mathfrak{D}(\mathfrak{H}, \psi)$ (resp. $\mathfrak{D}'(\mathfrak{H}, \psi)$) and L_ψ (resp. R_ψ) which have been implicit in the previous arguments:

$$\left.\begin{aligned}
(\xi_1 \mid \xi_2) &= \psi\big(L_\psi(\xi_2)^* L_\psi(\xi_1)\big), && \xi_1, \xi_2 \in \mathfrak{D}(\mathfrak{H}, \psi) \\
\big(\text{resp. } (\eta_1 \mid \eta_2) &= \psi\big(J R_\psi(\eta_1)^* R_\psi(\eta_2) J\big), && \eta_1, \eta_2 \in \mathfrak{D}'(\mathfrak{K}, \psi)\big); \\
\eta_\psi\big(L_\psi(\xi_2)^* L_\psi(\xi_1)\big) &= L_\psi(\xi_2)^* \xi_1, && \xi_1, \xi_2 \in \mathfrak{D}(\mathfrak{H}, \psi) \\
\big(\text{resp. } \eta'_\psi\big(J R_\psi(\eta_1)^* R_\psi(\eta_2) J\big) &= R_\psi(\eta_2)^* \eta_1, && \eta_1, \eta_2 \in \mathfrak{D}'(\mathfrak{K}, \psi)\big).
\end{aligned}\right\} \quad (22)$$

It is easy to see that if $\mathfrak{H} = L^2(\mathcal{N}, \psi)$ with the right action of $\mathcal{N}$ (resp. $\mathfrak{K} = L^2(\mathcal{N}, \psi)$ with the left action of $\mathcal{N}$) as a right (resp. left) module, then

$$\begin{aligned}
&\mathfrak{D}(\mathfrak{H}, \psi) = \eta_\psi(\mathfrak{n}_\psi) = \mathfrak{B}_\psi, \quad && L_\psi(\xi) = \pi_\ell(\xi), \quad && \xi \in \mathfrak{B}_\psi \\
\big(\text{resp. } &\mathfrak{D}'(\mathfrak{K}, \psi) = \mathfrak{B}'_\psi, && R_\psi(\eta) = \pi_r(\eta), && \eta \in \mathfrak{B}'_\psi\big),
\end{aligned}$$

where $\mathfrak{B}_\psi$ (resp. $\mathfrak{B}'_\psi$) means the algebra of all left (resp. right) bounded vectors in $L^2(\mathcal{N}, \psi)$.

Proposition 3.15.

(i) *The sesquilinear form, say B, on the algebraic tensor product $\mathfrak{D}(\mathfrak{H}, \psi)\otimes_{\mathrm{alg}}\mathfrak{K}$ determined by:*

$$B(\xi_1 \otimes \eta_1, \xi_2 \otimes \eta_2) = \left(\pi_{\mathfrak{K}}(L_\psi(\xi_2)^* L_\psi(\xi_1))\eta_1 \,\middle|\, \eta_2\right) \in \mathbf{C} \qquad (23)$$

is positive, so that it defines an inner product on $\mathfrak{D}(\mathfrak{H}, \psi) \otimes_{\mathrm{alg}} \mathfrak{K}$, which is in many cases degenerated.

(ii) *If $\xi_1, \xi_2 \in \mathfrak{D}(\mathfrak{H}, \psi)$ and $\eta_1, \eta_2 \in \mathfrak{D}'(\mathfrak{K}, \psi)$, then*

$$\left(\pi_{\mathfrak{K}}(L_\psi(\xi_2)^* L_\psi(\xi_1))\eta_1 \,\middle|\, \eta_2\right) = \left(\pi'_{\mathfrak{H}}(J R_\psi(\eta_1)^* R_\psi(\eta_2) J)\xi_1 \,\middle|\, \xi_2\right). \qquad (24)$$

(i′) *Dual to* (i), *the sesquilinear form B' on $\mathfrak{H} \otimes_{\mathrm{alg}} \mathfrak{D}'(\mathfrak{K}, \psi)$ determined by:*

$$B'(\xi_1 \otimes \eta_1, \xi_2 \otimes \eta_2) = \left(\pi_{\mathfrak{K}}(J R_\psi(\eta_1)^* R_\psi(\eta_2) J)\xi_1 \,\middle|\, \xi_2\right) \in \mathbf{C} \qquad (23')$$

is positive and agrees with B on $\mathfrak{D}(H, \psi) \otimes_{\mathrm{alg}} \mathfrak{D}'(\mathfrak{K}, \psi)$.

PROOF:

(i) Suppose that $\xi_1, \dots, \xi_n \in \mathfrak{D}(\mathfrak{H}, \psi)$. Let $a_{k,j} = L_\psi(\xi_k)^* L_\psi(\xi_j)$, $k, j = 1, \dots, n$, and $a = [a_{k,j}]$ be an $n \times n$ matrix over $\mathcal{N}$. If $x_1, \dots, x_n \in \mathcal{A} = \mathcal{D}(\sigma^\psi_{i/2}) \cap \mathcal{D}(\sigma^\psi_{-i/2})$, then we have, by (12′),

$$\sum_{k,j=1}^n x_j^* L_\psi(\xi_j)^* L_\psi(\xi_k) x_k = \sum_{k,j=1}^n L_\psi\left(\xi_j \sigma^\psi_{\frac{i}{2}}(x_j)\right)^* L_\psi\left(\xi_k \sigma^\psi_{\frac{i}{2}}(x_k)\right)$$
$$= \left(\sum_{k=1}^n L_\psi\left(\xi_k \sigma^\psi_{\frac{i}{2}}(x_k)\right)\right)^* \left(\sum_{k=1}^n L_\psi\left(\xi_k \sigma^\psi_{\frac{i}{2}}(x_k)\right)\right) \geq 0.$$

As $\mathcal{A}$ is σ-weakly dense in $\mathcal{N}$, the matrix a is positive in $M_n(\mathcal{N})$, so that there exists $b = [b_{k,j}] \in M_n(\mathcal{N})$ such that $a = b^* b$, i.e.,

$$a_{k,j} = \sum_{\ell=1}^n b^*_{\ell,k} b_{\ell,j}, \quad k, j = 1, \dots, n.$$

We then have, for $\eta_1, \dots, \eta_n \in \mathfrak{K}$,

$$B\left(\sum_{k=1}^n \xi_k \otimes \eta_k, \sum_{k=1}^n \xi_k \otimes \eta_k\right) = \sum_{k,j=1}^n (a_{k,j}\eta_j \mid \eta_k) = \sum_{k=1}^n \left\| \sum_{j=1}^n b_{k,j}\eta_j \right\|^2 \geq 0.$$

Hence the sesquilinear form B is positive.

(ii) Suppose that ξ's and η's are as in the Proposition. We then have, as $L_\psi(\xi_2)^* L_\psi(\xi_1) \in \mathfrak{m}_\psi \subset \mathfrak{n}_\psi \cap \mathfrak{n}_\psi^*$ and also $J R_\psi(\eta_2)^* R_\psi(\eta_1) J \in \mathfrak{m}_\psi$,

$$\begin{aligned}
\Big(\pi_{\mathfrak{K}}(L_\psi(\xi_2)^* L_\psi(\xi_1))\eta_1 \,\Big|\, \eta_2\Big) &= \Big(R_\psi(\eta_1)\eta_\psi(L_\psi(\xi_2)^* L_\psi(\xi_1)) \,\Big|\, \eta_2\Big) \\
&= \big(L_\psi(\xi_2)^*\xi_1 \mid R_\psi(\eta_1)^*\eta_2\big) = \Big(L_\psi(\xi_2)^*\xi_1 \,\Big|\, \eta'_\psi\big(J R_\psi(\eta_2)^* R_\psi(\eta_1) J\big)\Big) \\
&= \Big(\xi_1 \,\Big|\, L_\psi(\xi_2)\eta'_\psi\big(J R_\psi(\eta_2)^* R_\psi(\eta_1) J\big)\Big) \\
&= \Big(\xi_1 \,\Big|\, \pi'_{\mathfrak{H}}\big(J R_\psi(\eta_2)^* R_\psi(\eta_1) J\big)\xi_2\Big) = \Big(\pi'_{\mathfrak{H}}\big(J R_\psi(\eta_1)^* R_\psi(\eta_2) J\big)\xi_1 \,\Big|\, \xi_2\Big).
\end{aligned}$$

(i′) The positivity follows from (i) by symmetry. The second assertion follows from (ii). Q.E.D.

Definition 3.16. Let $\mathfrak{N}$ be the subspace of $\mathfrak{D}(\mathfrak{H}, \psi) \otimes_{\mathrm{alg}} \mathfrak{K}$ consisting of those vectors ζ with $B(\zeta, \zeta) = 0$. The Hilbert space obtained as the completion of the quotient space $\mathfrak{D}(\mathfrak{H}, \psi) \otimes_{\mathrm{alg}} \mathfrak{K}/\mathfrak{N}$ relative to the inner product induced by the positive sequilinear form B will be called the *relative tensor product* of the right $\mathcal{N}$-module $\mathfrak{H}$ and the left $\mathcal{N}$-module $\mathfrak{K}$ with respect to the faithful semi-finite normal weight ψ and will be written as $\mathfrak{H} \otimes_\psi \mathfrak{K}$ and the image of $\xi \otimes \eta$ as $\xi \otimes_\psi \eta$ for $\xi \in \mathfrak{D}(\mathfrak{H}, \psi)$, $\eta \in \mathfrak{K}$. By Proposition 3.15, the relative tensor product $\mathfrak{H} \otimes_\psi \mathfrak{K}$ can also be obtained as the completion of the quotient space of the algebraic tensor product $\mathfrak{H} \otimes_{\mathrm{alg}} \mathfrak{D}'(\mathfrak{K}, \psi)$ by the subspace $\mathfrak{N}'$ consisting of null vectors with respect to the positive sequilinear form B'. In this way, we can consider the tensor product $\xi \otimes_\psi \eta$ for a pair $\xi \in \mathfrak{H}$, $\eta \in \mathfrak{D}'(\mathfrak{K}, \psi)$.

Theorem 3.17. *Let $\mathcal{N}$ be a von Neumann algebra equipped with a faithful semi-finite normal weight ψ, $\mathfrak{H}$ a right $\mathcal{N}$-module and $\mathfrak{K}$ a left $\mathcal{N}$-module. Set $\mathcal{P} = \mathcal{L}(\mathfrak{H}_{\mathcal{N}})$ and $\mathcal{Q} = \mathcal{L}({}_{\mathcal{N}}\mathfrak{K})$. Consider the direct sum: $\tilde{\mathfrak{H}} = L^2(\mathcal{N}, \psi) \oplus \mathfrak{H} \oplus \bar{\mathfrak{K}}$ as a right $\mathcal{N}$-module and also $\mathcal{R} = \mathcal{L}(\tilde{\mathfrak{H}}_{\mathcal{N}})$ together with the "balanced" faithful semi-finite normal weight $\rho = \psi \oplus \varphi \oplus \nu$ where φ is a faithful semi-finite normal weight on $\mathcal{P}$ and ν is a faithful semi-finite normal weight on $\mathcal{Q}$. Let e, f and g be respectively the projections of $\tilde{\mathfrak{H}}$ onto $L^2(\mathcal{N}, \psi)$, $\mathfrak{H}$ and $\bar{\mathfrak{K}}$, which belong to $\mathcal{R}$. Represent the standard Hilbert space $\mathfrak{H}_\rho$ as the space of 3×3 matrices:*

$$\mathfrak{H}_\rho = \begin{pmatrix} L^2(\mathcal{N}, \psi) & \bar{\mathfrak{H}} & \mathfrak{K} \\ \mathfrak{H} & L^2(\mathcal{P}, \varphi) & \mathfrak{H}_{2,3} \\ \bar{\mathfrak{K}} & \mathfrak{H}_{3,2} & L^2(\mathcal{Q}, \nu) \end{pmatrix}. \tag{13′}$$

Then there exists a natural isomorphism between $\mathfrak{H} \otimes_\psi \mathfrak{K}$ and $\mathfrak{H}_{2,3}$.

PROOF: Let $\mathfrak{A}(= \mathfrak{A}_\rho)$ be the left Hilbert algebra associated with ρ, $\mathfrak{B}\ (= \mathfrak{B}_\rho)$ the algebra of left bounded elements in $L^2(\mathcal{R}, \rho)$ and $\mathfrak{n}_\rho = \{x \in \mathcal{R} : \varphi(x^*x) < \infty\}$. As e and f are both in $\mathcal{R}_\rho$, $\mathfrak{A}$ and $\mathfrak{B}$ are both decomposed into the matrix direct sum relative to (13′):

$$\mathfrak{A} = \begin{pmatrix} \mathfrak{A}_{11} & \mathfrak{A}_{12} & \mathfrak{A}_{13} \\ \mathfrak{A}_{21} & \mathfrak{A}_{22} & \mathfrak{A}_{23} \\ \mathfrak{A}_{31} & \mathfrak{A}_{32} & \mathfrak{A}_{33} \end{pmatrix}; \qquad \mathfrak{B} = \begin{pmatrix} \mathfrak{B}_{11} & \mathfrak{B}_{12} & \mathfrak{B}_{13} \\ \mathfrak{B}_{21} & \mathfrak{B}_{22} & \mathfrak{B}_{23} \\ \mathfrak{B}_{31} & \mathfrak{B}_{32} & \mathfrak{B}_{33} \end{pmatrix}. \tag{13″}$$

It follows from Lemma 3.3 that $\mathfrak{B}_{21} = \mathfrak{D}(\mathfrak{H}, \psi)$ and $\mathfrak{B}_{31} = \mathfrak{D}(\bar{\mathfrak{K}}, \psi)$. Also we have $L_\psi(\xi) = \pi_\ell(\xi)|_{\mathfrak{H}_{11}}$, $\xi \in \mathfrak{D}(\mathfrak{H}, \psi) = \mathfrak{B}_{21}$ and $L_\psi(\bar{\eta}) = \pi_\ell(\bar{\eta})|_{\mathfrak{H}_{12}}$, $\bar{\eta} \in \mathfrak{D}(\bar{\mathfrak{K}}, \psi) = \mathfrak{B}_{31}$, where π_ℓ means the left multiplication representation of $\mathfrak{B}$ on $\mathfrak{H}_\rho$. At this point, one should note that the right Hilbert algebra $\mathfrak{A}'$ and the algebra $\mathfrak{B}'$ of right bounded vectors admit also the similar matrix decompositions and that $\mathfrak{B}'_{21} = \mathfrak{D}'(\bar{\mathfrak{H}}, \psi)$, $\mathfrak{B}'_{31} = \mathfrak{D}'(\mathfrak{K}, \psi)$ and $R_\psi(\eta) = \pi_r(\eta)|_{\mathfrak{H}_{11}}$, $\eta \in \mathfrak{D}'(\mathfrak{K}, \psi)$.

We claim that $\xi \otimes_\psi \eta$, with $\xi \in \mathfrak{D}(\mathfrak{H}, \psi)$ and $\eta \in \mathfrak{K}$, is identified with $\pi_\ell(\xi)\eta \in \mathfrak{H}_{23}$. Let U_0 be the map from $\mathfrak{D}(\mathfrak{H}, \psi) \otimes_{\mathrm{alg}} \mathfrak{K}$ into $\mathfrak{H}_\rho$ determined by $U_0(\xi \otimes \eta) = \pi_\ell(\xi)\eta$ for $\xi \in \mathfrak{D}(\mathfrak{H}, \psi)$, $\eta \in \mathfrak{K}$. As $\xi \in \mathfrak{B}_{21}$ and $\eta \in \mathfrak{K} = \mathfrak{H}_{13}$, $\pi_\ell(\xi)\eta$ belongs to $\mathfrak{H}_{23}$. Now we have, for $\xi_1, \xi_2 \in \mathfrak{D}(\mathfrak{H}, \psi)$ and $\eta_1, \eta_2 \in \mathfrak{K}$,

$$\begin{aligned}\big(U_0(\xi_1 \otimes \eta_1) \mid U_0(\xi_2 \otimes \eta_2)\big) &= \big(\pi_\ell(\xi_1)\eta_1 \mid \pi_\ell(\xi_2)\eta_2\big) = \big(\pi_\ell(\xi_2)^*\pi_\ell(\xi_1)\eta_1 \mid \eta_2\big) \\ &= \Big(\pi_{\mathfrak{K}}\big(L_\psi(\xi_2)^* L_\psi(\xi_1)\big)\eta_1 \Bigm| \eta_2\Big) = \big(\xi_1 \otimes_\psi \eta_1 \mid \xi_2 \otimes_\psi \eta_2\big).\end{aligned}$$

Therefore, the map U_0 gives rise to an isometry U of $\mathfrak{H} \otimes_\psi \mathfrak{K}$ into $\mathfrak{H}_{23}$. Let $\mathfrak{M} = U(\mathfrak{H} \otimes_\psi \mathfrak{K}) = \big[\pi_\ell(\mathfrak{B}_{21})\mathfrak{K}\big]$. First, we observe $\mathfrak{H}_{23} = eL^2(\mathcal{R}, \psi)g$, $\mathcal{P} = \mathcal{R}_e$ and $\mathcal{Q} = \mathcal{R}_g$. Hence $\pi_{\mathfrak{H}_{23}}(\mathcal{P})' = \pi'_{\mathfrak{H}_{23}}(\mathcal{Q})$. We know that $\mathfrak{M}$ is invariant under the right action of $\mathcal{Q}$. Hence the projection p of $\mathfrak{H}_{23}$ onto $\mathfrak{M}$ belongs to $\pi_{\mathfrak{H}_{23}}(\mathcal{P})$, i.e. p can be identified with the left multiplication by a projection in $\mathcal{P}$, which will be denoted by p again. This means that $\mathfrak{M} = p\mathfrak{H}_{23}$ with $p \in (\mathrm{Proj}\,\mathcal{P})$. But as $\pi_\ell(a\xi) = a\pi_\ell(\xi)$ for $a \in \mathcal{P}$ and $\xi \in \mathfrak{D}(\mathfrak{H}, \psi) = \mathfrak{B}_{21}$, $\mathfrak{B}_{23}$ is invariant under the left multiplication by $\mathcal{P}$, which in turn implies the invariance of $\mathfrak{M}$ under the left multiplication by $\mathcal{P}$. Hence, the projection p belongs to the center $\mathcal{C}_\mathcal{P}$ of $\mathcal{P}$, which is of the form: $\mathcal{C}_\mathcal{P} = (\mathcal{C}_\mathcal{R})_f$. Thus p can be viewed as a projection in $\mathcal{C}_\mathcal{R}$. Now we have $(1-p)\mathfrak{M} = \{0\}$, so that $0 = (1-p)\pi_\ell(\xi)\eta = \pi_\ell\big((1-p)\xi\big)\eta$ for every $\xi \in \mathfrak{B}_{21}$ and $\eta \in \mathfrak{K}$. Thus, $\pi_{\mathfrak{K}}\big(\pi_\ell((1-p)\xi)^*\pi_\ell((1-p)\xi)\big) = 0$. As $\mathfrak{K}$ is a faithful left $\mathcal{N}$-module, we have $\pi_\ell\big((1-p)\xi\big) = 0$, $\xi \in \mathfrak{D}(\mathfrak{H}, \psi)$, which means that $1 - p = 0$. Therefore, we have $\mathfrak{M} = \mathfrak{H}_{23}$.

Thus under the isometry U, $\mathfrak{H} \otimes_\psi \mathfrak{K}$ is identified with $\mathfrak{H}_{23}$. Q.E.D.

Corollary 3.18.

(i) *If $\mathfrak{H}$ and $\mathfrak{K}$ are respectively a right $\mathcal{N}$-module and a left $\mathcal{N}$-module for a von Neumann algebra $\mathcal{N}$ equipped with a faithful semi-finite normal weight ψ, then the relative tensor product $\mathfrak{H} \otimes_\psi \mathfrak{K}$ is naturally an $\mathcal{L}(\mathfrak{H}_\mathcal{N})$-$\mathcal{L}({}_\mathcal{N}\mathfrak{K})^\circ$-bimodule which is determined by:*

$$\begin{gathered} a(\xi \otimes_\psi \eta)b = (a\xi) \otimes_\psi (\eta b), \\ a \in \mathcal{L}(\mathfrak{H}_\mathcal{N}), \quad b \in \mathcal{L}({}_\mathcal{N}\mathfrak{K})^\circ, \quad \xi \in \mathfrak{D}(\mathfrak{H}, \psi), \quad \eta \in \mathfrak{K}. \end{gathered} \tag{25}$$

(ii) *In terms of operators acting from the left as usual, if $x \in \mathcal{L}(\mathfrak{H}_\mathcal{N})$ and $y \in \mathcal{L}({}_\mathcal{N}\mathfrak{K})$, then there exists a unique operator $x \otimes_\psi y \in \mathcal{L}(\mathfrak{H} \otimes_\psi \mathfrak{K})$ such that:*

$$(x\xi) \otimes_\psi (y\eta) = (x \otimes_\psi y)(\xi \otimes_\psi \eta), \quad \xi \in \mathfrak{D}(\mathfrak{H}, \psi), \quad \eta \in \mathfrak{K}. \tag{26}$$

The map: $(x, y) \in \mathcal{L}(\mathfrak{H}_\mathcal{N}) \times \mathcal{L}({}_\mathcal{N}\mathfrak{K}) \mapsto x \otimes_\psi y \in \mathcal{L}(\mathfrak{H} \otimes_\psi \mathfrak{K})$ extends canonically to an injective $*$-homomorphism from the algebraic tensor product, $\mathcal{L}(\mathfrak{H}_\mathcal{N}) \otimes_{\mathrm{alg}} \mathcal{L}({}_\mathcal{N}\mathfrak{K})$, into $\mathcal{L}(\mathfrak{H} \otimes_\psi \mathfrak{K})$.

(iii) *Although* $\mathcal{N}$ *does not act on the relative tensor product* $\mathfrak{H} \otimes_\psi \mathfrak{K}$, *we have:*

$$(\xi b) \otimes_\psi \eta = \xi \otimes_\psi \left(\sigma^\psi_{\frac{i}{2}}(b)\eta\right), \quad b \in \mathcal{D}\left(\sigma^\psi_{\frac{i}{2}}\right), \quad \xi \in \mathfrak{D}(\mathfrak{H}, \psi), \quad \eta \in \mathfrak{K}. \tag{27}$$

We leave the proof to the reader as an exercise.

Summarizing the above arguments, we restate the matrix decomposition of $\mathfrak{H}_\rho$ in the following form:

$$L^2(\mathcal{R}, \rho) = \begin{pmatrix} L^2(\mathcal{N}, \psi) & \bar{\mathfrak{H}} & \mathfrak{K} \\ \mathfrak{H} & L^2(\mathcal{P}, \varphi) & \mathfrak{H} \otimes_\psi \mathfrak{K} \\ \bar{\mathfrak{K}} & \bar{\mathfrak{K}} \otimes_\psi \bar{\mathfrak{H}} & L^2(\mathcal{Q}, \nu) \end{pmatrix}. \tag{28}$$

Proposition 3.19.

(i) *Viewing* $L^2(\mathcal{N}, \psi)$ *as a right* $\mathcal{N}$*-module, the map*

$$V^\psi_{\mathfrak{K}} : \eta_\psi(y) \otimes_\psi \eta \in L^2(\mathcal{N}, \psi) \otimes_\psi \mathfrak{K} \mapsto y\eta \in \mathfrak{K}, \quad y \in \mathfrak{n}_\psi, \quad \eta \in \mathfrak{K},$$

gives rise to an isomorphism of $L^2(\mathcal{N}, \psi) \otimes_\psi \mathfrak{K}$ *onto* $\mathfrak{K}$ *as* $\mathcal{N}$-$\mathcal{L}({}_\mathcal{N}\mathfrak{K})^\circ$-*bimodules.*

(ii) *If we look at* $L^2(\mathcal{N}, \psi)$ *as a left* $\mathcal{N}$*-module, then the map* $U^\psi_{\mathfrak{H}}$:

$$\xi \otimes_\psi \eta'_\psi(y) \in \mathfrak{H} \otimes_\psi L^2(\mathcal{N}, \psi) \mapsto \xi y \in \mathfrak{H}, \quad \xi \in \mathfrak{H}, \quad y \in \mathfrak{n}^*_\psi$$

extends to an isomorphism of $\mathfrak{H} \otimes_\psi L^2(\mathcal{N}, \psi)$ *onto* $\mathfrak{H}$ *as* $\mathcal{L}(\mathfrak{H}_\mathcal{N})$-$\mathcal{N}$-*bimodules.*

The proof is now routine, so we leave it to the reader.

Also routine arguments show the following identities with canonical identifications:

$$\left.\begin{aligned} (\mathfrak{H}_1 \oplus \mathfrak{H}_2) \otimes_\psi \mathfrak{K} &= (\mathfrak{H}_1 \otimes_\psi \mathfrak{K}) \oplus (\mathfrak{H}_2 \otimes_\psi \mathfrak{K}); \\ \mathfrak{H} \otimes_\psi (\mathfrak{K}_1 \oplus \mathfrak{K}_2) &= (\mathfrak{H} \otimes_\psi \mathfrak{K}_1) \oplus (\mathfrak{H} \otimes_\psi \mathfrak{K}_2), \end{aligned}\right\} \tag{29}$$

where $\mathfrak{H}$, $\mathfrak{H}_1$ and $\mathfrak{H}_2$ are right $\mathcal{N}$-modules whilst $\mathfrak{K}$, $\mathfrak{K}_1$ and $\mathfrak{K}_2$ are left $\mathcal{N}$-modules.

Theorem 3.20. *Let* $\mathcal{M}$ *and* $\mathcal{N}$ *be two von Neumann algebras equipped with faithful semi-finite normal weights* φ *and* ψ *respectively. If* $\mathfrak{H}$, $\mathfrak{K}$ *and* $\mathfrak{M}$ *are respectively a right* $\mathcal{M}$*-module, an* $\mathcal{M}$-$\mathcal{N}$*-bimodule and a left* $\mathcal{N}$*-module. Then under a natural identification we have*

$$(\mathfrak{H} \otimes_\varphi \mathfrak{K}) \otimes_\psi \mathfrak{M} = \mathfrak{H} \otimes_\varphi (\mathfrak{K} \otimes_\psi \mathfrak{M}) \tag{30}$$

as $\mathcal{L}(\mathfrak{H}_\mathcal{M})$-$\mathcal{L}({}_\mathcal{N}\mathfrak{M})^\circ$-*bimodules.*

PROOF: For each $\xi \in \mathfrak{D}(\mathfrak{H}, \varphi)$, $\eta \in \mathfrak{K}$ and $\zeta \in \mathfrak{D}'(\mathfrak{M}, \psi)$, set

$$U\big((\xi \otimes_\varphi \eta) \otimes_\psi \zeta\big) = \xi \otimes_\varphi (\eta \otimes_\psi \zeta).$$

Let ξ's, η's and ζ's be as above. We want to show that:

$$\Big(U\big((\xi_1 \otimes_\varphi \eta_1) \otimes_\psi \zeta_1\big) \,\Big|\, U\big((\xi_2 \otimes_\varphi \eta_2) \otimes_\psi \zeta_2\big)\Big) = \Big((\xi_1 \otimes_\varphi \eta_1) \otimes_\psi \zeta_1 \,\Big|\, (\xi_2 \otimes_\varphi \eta_2) \otimes_\psi \zeta_2\Big),$$

as this shows that U is well-defined and a unitary. Let us compute:

$$\begin{aligned}
&\Big(U\big((\xi_1 \otimes_\varphi \eta_1) \otimes_\psi \zeta_1\big) \,\Big|\, U\big((\xi_2 \otimes_\varphi \eta_2) \otimes_\psi \zeta_2\big)\Big) \\
&= \big(\xi_1 \otimes_\varphi (\eta_1 \otimes_\psi \zeta_1) \mid \xi_2 \otimes_\varphi (\eta_2 \otimes_\psi \zeta_2)\big) \\
&= \Big(\pi_{\mathfrak{K}\otimes_\psi \mathfrak{M}}\big(L_\varphi(\xi_2)^* L_\varphi(\xi_1)\big)(\eta_1 \otimes_\psi \zeta_1) \,\Big|\, \eta_2 \otimes_\psi \zeta_2\Big) \\
&= \Big(\Big(\pi_{\mathfrak{K}}\big(L_\varphi(\xi_2)^* L_\varphi(\xi_1)\big)\eta_1\Big) \otimes_\psi \zeta_1 \,\Big|\, \eta_2 \otimes_\psi \zeta_2\Big) \\
&= \Big(\pi'_{\mathfrak{K}}\big(J R_\psi(\zeta_1)^* R_\psi(\zeta_2) J\big)\pi_{\mathfrak{K}}\big(L_\varphi(\xi_2)^* L_\varphi(\xi_1)\big)\eta_1 \,\Big|\, \eta_2\Big) \qquad \text{by (24)} \\
&= \Big(\pi_{\mathfrak{K}}\big(L_\varphi(\xi_2)^* L_\varphi(\xi_1)\big)\pi'_{\mathfrak{K}}\big(J R_\psi(\zeta_1)^* R_\psi(\zeta_2) J\big)\eta_1 \,\Big|\, \eta_2\Big) \\
&= \Big(\xi_1 \otimes_\varphi \Big(\pi'_{\mathfrak{K}}\big(J R_\psi(\zeta_1)^* R_\psi(\zeta_2) J\big)\eta_1 \,\Big|\, \xi_2 \otimes_\varphi \eta_2\Big) \\
&= \Big(\pi'_{\mathfrak{H}\otimes_\varphi \mathfrak{K}}\big(J R_\psi(\zeta_1)^* R_\psi(\zeta_2) J\big)\big(\xi_1 \otimes_\varphi \eta_1\big) \,\Big|\, \xi_2 \otimes_\varphi \eta_2\Big) \\
&= \Big(\big(\xi_1 \otimes_\varphi \eta_1\big) \otimes_\psi \zeta_1 \,\Big|\, \big(\xi_2 \otimes_\varphi \eta_2\big) \otimes_\psi \zeta_2\Big).
\end{aligned}$$

Now we have, for each $a \in \mathcal{L}(\mathfrak{H}_{\mathcal{M}})$ and $b \in \mathcal{L}({}_{\mathcal{N}}\mathfrak{M})^\circ$,

$$\begin{aligned}
U\Big(a\big((\xi \otimes_\varphi \eta) \otimes_\psi \zeta\big)b\Big) &= U\Big(\big((a\xi) \otimes_\varphi \eta\big) \otimes_\psi (\zeta b)\Big) = (a\xi) \otimes_\varphi \big(\eta \otimes_\psi (\zeta b)\big) \\
&= a\big(\xi \otimes_\varphi (\eta \otimes_\psi \zeta)\big)b = a\Big(U\big((\xi \otimes_\varphi \eta) \otimes_\psi \zeta\big)\Big)b.
\end{aligned}$$

Therefore, U is indeed an isomorphism of $(\mathfrak{H} \otimes_\varphi \mathfrak{K}) \otimes_\psi \mathfrak{M}$ onto $\mathfrak{H} \otimes_\varphi (\mathfrak{K} \otimes_\psi \mathfrak{M})$ as $\mathcal{L}(\mathfrak{H}_{\mathcal{M}})$-$\mathcal{L}({}_{\mathcal{N}}\mathfrak{M})^\circ$-bimodules. Q.E.D.

We now want to study what happens on relative tensor products when we change the reference faithful semi-finite normal weight ψ on a von Neumann algebra $\mathcal{N}$. To this end, we first fix a couple of notations. Let $\mathfrak{W}(\mathcal{M})$ denote the set of all semi-finite normal weights on a von Neumann algebra $\mathcal{M}$, and $\mathfrak{W}_0(\mathcal{M})$ the set of all faithful semi-finite normal weights on $\mathcal{M}$. With a fixed von Neumann algebra $\mathcal{N}$, we continue to study the relative tensor product of a right $\mathcal{N}$-module $\mathfrak{H}$ and a left $\mathcal{N}$-module $\mathfrak{K}$ relative to a faithful semi-finite normal weight ψ.

Theorem 3.21. *Let $\mathcal{N}$ be a von Neumann algebra, $\mathfrak{H}$ and $\mathfrak{K}$ be respectively right and left $\mathcal{N}$-modules. To each pair $(\psi_1, \psi_2) \in \mathfrak{W}_0(\mathcal{N}) \times \mathfrak{W}(\mathcal{N})$, there corresponds uniquely an $\mathcal{L}(\mathfrak{H}_{\mathcal{N}})$-$\mathcal{L}({}_{\mathcal{N}}\mathfrak{K})^\circ$-bimodule isomorphism $U^{\psi_2,\psi_1}_{\mathfrak{H},\mathfrak{K}}$ of $\mathfrak{H} \otimes_{\psi_1} \mathfrak{K}$ onto $\mathfrak{H} \otimes_{\psi_2} \mathfrak{K}$ which makes the following diagram commutative for every pair $(a_i, b_i) \in \mathcal{L}\big(\mathfrak{H}_{\mathcal{N}}, L^2(\mathcal{N}, \psi_i)_{\mathcal{N}}\big) \times \mathcal{L}\big({}_{\mathcal{N}}L^2(\mathcal{N}, \psi_i), {}_{\mathcal{N}}\mathfrak{K}\big)$, $i = 1, 2$, such that $a_2 = U_{\psi_2,\psi_1} a_1$, $b_2 = b_1 U_{\psi_2,\psi_1}$ with U_{ψ_2,ψ_1} the canonical unitary implementing the equivalence of the standard forms $\{\mathcal{N}, L^2(\mathcal{N}, \psi_1), \mathfrak{P}_{\psi_1}, J_{\psi_1}\}$ and $\{\mathcal{N}, L^2(\mathcal{N}, \psi_2), \mathfrak{P}_{\psi_2}, J_{\psi_2}\}$:*

$$\left.\begin{array}{ccc}
\mathfrak{H} \otimes_{\psi_1} \mathfrak{K} & \xrightarrow{U^{\psi_2,\psi_1}_{\mathfrak{H},\mathfrak{K}}} & \mathfrak{H} \otimes_{\psi_2} \mathfrak{K} \\
{\scriptstyle (a_1 \otimes_{\psi_1} b_1^*)}\downarrow & & {\scriptstyle (a_2 \otimes_{\psi_2} b_2^*)}\downarrow \\
L^2(\mathcal{N}, \psi_1) \otimes_{\psi_1} L^2(\mathcal{N}, \psi_1) & & L^2(\mathcal{N}, \psi_2) \otimes_{\psi_2} L^2(\mathcal{N}, \psi_2) \\
{\scriptstyle U^{\psi_1}_{L^2(\mathcal{N},\psi_1)}}\downarrow & & {\scriptstyle U^{\psi_2}_{L^2(\mathcal{N},\psi_2)}}\downarrow \\
L^2(\mathcal{N}, \psi_1) & \xrightarrow{U_{\psi_2,\psi_1}} & L^2(\mathcal{N}, \psi_2)
\end{array}\right\} \tag{31}$$

The correspondence: $(\psi_1, \psi_2) \in \mathfrak{W}_0(\mathcal{N}) \times \mathfrak{W}_0(\mathcal{N}) \mapsto U^{\psi_2,\psi_1}_{\mathfrak{H},\mathfrak{K}}$ satisfies the chain rule:

$$U^{\psi_3,\psi_2}_{\mathfrak{H},\mathfrak{K}} U^{\psi_2,\psi_1}_{\mathfrak{H},\mathfrak{K}} = U^{\psi_3,\psi_1}_{\mathfrak{H},\mathfrak{K}}, \quad \psi_1, \psi_2, \psi_3 \in \mathfrak{W}_0(\mathcal{N}). \tag{32}$$

PROOF: Existence: We use the notations of Theorem 3.17. Choose faithful semi-finite normal weights $\varphi \in \mathfrak{W}_0(\mathcal{P})$ and $\nu \in \mathfrak{W}(\mathcal{Q})$ for $i = 1, 2$ and set $\rho_i = \psi_i \oplus \varphi \oplus \nu$, $i = 1, 2$ on $\mathcal{R}$, where one should observe that the construction of $\mathcal{R}$ does not depend on the choice of the faithful semi-finite normal weights ψ's. We then have the canonical isometry U_{ρ_2,ρ_1} from $L^2(\mathcal{R}, \rho_1)$ onto $L^2(\mathcal{R}, \rho_2)$ which implements an $\mathcal{R}$-$\mathcal{R}$-bimodule isomorphism. As the projections e, f and g commute with the faithful semi-finite normal weights ρ_1 and ρ_2, it is easy to check that the unitary U_{ρ_2,ρ_1} preserve the matrix decompositions (13′) of $L^2(\mathcal{R}, \rho_i)$, $i = 1, 2$. With J the conjugation operator: $\bar\eta \in \bar{\mathfrak{K}} \mapsto \eta \in \mathfrak{K}$, set $b_i^\circ = J b_i^* J \in \mathcal{L}\big(\bar{\mathfrak{K}}_{\mathcal{N}}, L^2(\mathcal{N}, \psi_i)\big)$, $i = 1, 2$. Then we have $A_i = \begin{pmatrix} 0 & a_i & 0 \\ 0 & 0 & 0 \\ 0 & 0 & 0 \end{pmatrix}$, $B_i = \begin{pmatrix} 0 & 0 & b_i^\circ \\ 0 & 0 & 0 \\ 0 & 0 & 0 \end{pmatrix} \in \mathcal{R}$, $i = 1, 2$. We then see that the restriction of the operator $\pi_{\rho_i}(A_i)\pi'_{\rho_i}(B_i)^*$ to the (2,3)-component of $L^2(\mathcal{R}, \rho_i)$ is equal to $U^{\psi_i}_{L^2(\mathcal{N},\psi_i)}\big(a \otimes_{\psi_i} b_i^*\big)$, with π'_{ρ_i} the semi-cyclic anti-representation of $\mathcal{R}$ defined by $\pi'_{\rho_i}(x) = J\pi_{\rho_i}(x)^* J$, $x \in \mathcal{R}$. As U_{ρ_2,ρ_1} is an $\mathcal{R}$-$\mathcal{R}$-bimodule isomorphism of $L^2(\mathcal{R}, \rho_1)$ onto $L^2(\mathcal{R}, \rho_2)$ and carries the matrix decomposition (13′) of $L^2(\mathcal{R}, \rho_1)$ onto that of $L^2(\mathcal{R}, \rho_2)$, its restrictions to the (1,1)-component and (2,3)-component of (13′) gives U_{ψ_2,ψ_1} and the one for $U^{\psi_2,\psi_1}_{\mathfrak{H},\mathfrak{K}}$ which satisfy (31).

Uniqueness: Let $\mathfrak{A}_i$ $(= \mathfrak{A}_{\psi_i})$, $i = 1, 2$, be the left Hilbert algebra associated wtih $\{\mathcal{N}, \psi_i\}$ and $\mathfrak{A}_i^0$ be the associated Tomita algebra. Set $\mathfrak{a}_i = \mathfrak{n}_{\psi_i} \cap \mathfrak{n}_{\psi_i}^* = \pi_\ell(\mathfrak{A}_i)$

and $\mathfrak{a}_i^0 = \pi_\ell(\mathfrak{A}_i^0)$, $i = 1, 2$. For each $\xi \in \mathfrak{D}(\mathfrak{H}, \psi_1)$, $\eta \in \mathfrak{D}'(\mathfrak{K}, \psi_1)$ and each $y_1, y_2 \in \mathfrak{a}_i^0$, we have with $a_1 = L_{\psi_1}(\xi)^*$ and $b_1 = R_{\psi_1}(\eta)$ in (31):

$$\begin{aligned}
U_{\mathfrak{H},\mathfrak{K}}^{\psi_2,\psi_1}(\xi y_1 \otimes_{\psi_1} y_2\eta) &= U_{\mathfrak{H},\mathfrak{K}}^{\psi_2,\psi_1}\big(L_{\psi_1}(\xi)\eta'_{\psi_1}(y_1) \otimes_{\psi_1} R_{\psi_1}(\eta)\eta_{\psi_1}(y_2)\big)\\
&= U_{\mathfrak{H},\mathfrak{K}}^{\psi_2,\psi_1}\big(L_{\psi_1}(\xi) \otimes_{\psi_1} R_{\psi_1}(\eta)\big)\big(\eta'_{\psi_1}(y_1) \otimes_{\psi_1} \eta_{\psi_1}(y_2)\big)\\
&= U_{\mathfrak{H},\mathfrak{K}}^{\psi_2,\psi_1}\big(a_1^* \otimes_{\psi_1} b_1\big)\big(U_{L^2(\mathcal{N},\psi_1)}^{\psi_1}\big)^*\big(\sigma_{-\frac{i}{2}}^{\psi_1}(y_1)\eta_{\psi_1}(y_2)\big)\\
&= \big(a_2^* \otimes_{\psi_2} b_2\big)\big(U_{L^2(\mathcal{N},\psi_2)}^{\psi_2}\big)^* U_{\psi_2,\psi_1}\big(\sigma_{-\frac{i}{2}}^{\psi_1}(y_1)\eta_{\psi_1}(y_2)\big).
\end{aligned}$$

This means that $U_{\mathfrak{H},\mathfrak{K}}^{\psi_2,\psi_1}$ is uniquely determined on the vectors of the form: $\xi y_1 \otimes_{\psi_1} y_2\eta$ with ξ, η, y_1 and y_2 as above, which are dense in $\mathfrak{H} \otimes_{\psi_1} \mathfrak{K}$. Hence, $U_{\mathfrak{H},\mathfrak{K}}^{\psi_2,\psi_1}$ is uniquely determined by the commutative diagram of (31).

The chain rule (32) follows from the uniqueness of $U_{\mathfrak{H},\mathfrak{K}}^{\psi_2,\psi_1}$. Q.E.D.

Remark 3.22. The bimodule isomorphism $U_{\mathfrak{H},\mathfrak{K}}^{\psi_2,\psi_1}$ does not send $\xi \otimes_{\psi_1} \eta$ into $\xi \otimes_{\psi_2} \eta$ for $\xi \in \mathfrak{H}$ and $\eta \in \mathfrak{K}$.

At this point, one might puzzle if one can construct the relative tensor products $\mathfrak{H} \otimes_{\mathcal{N}} \mathfrak{K}$ directly from the right $\mathcal{N}$-module $\mathfrak{H}$ and the left $\mathcal{N}$-module $\mathfrak{K}$. It is in fact possible to do so if one gives up the tensor product $\xi \otimes_{\mathcal{N}} \eta$ of vectors. We will discuss a weight free construction of the relative tensor product $\mathfrak{H} \otimes_{\mathcal{N}} \mathfrak{K}$ in the exercise.

Before closing the section, we discuss an example. Let $\mathcal{A}$ be an abelian separable von Neumann algebra and let it act on a separable Hilbert space $\mathfrak{H}$. As $\mathcal{A}$ is abelian, $\mathfrak{H}$ can be viewed as two sided module over $\mathcal{A}$. Let $\mathfrak{K}$ be another separable Hilbert space on which $\mathcal{A}$ acts. Now we view $\mathfrak{H}$ as a right $\mathcal{A}$-module and $\mathfrak{K}$ as a left $\mathcal{A}$-module. Let ψ be a faithful semi-finite normal weight on $\mathcal{A}$ and consider $\mathfrak{H} \otimes_\psi \mathfrak{K}$. We want to identify $\mathfrak{H} \otimes_\psi \mathfrak{K}$ more explicitly. First, consider the direct disintegration of $\mathfrak{H}$ and $\mathfrak{K}$ relative to $\mathcal{A}$:

$$\mathfrak{H} = \int_\Gamma^\oplus \mathfrak{H}(\gamma)\, d\mu(\gamma); \qquad \mathfrak{K} = \int_\Gamma^\oplus \mathfrak{K}(\gamma)\, d\mu(\gamma),$$

where $\mathcal{A} = L^\infty(\Gamma, \mu)$ is the measure theoretic spectral representation of $\mathcal{A}$ and μ is the measure corresponding to the weight ψ.

Proposition 3.23. *Under the above situation:*

$$\mathfrak{H} \otimes_\psi \mathfrak{K} = \int_\Gamma^\oplus \mathfrak{H}(\gamma) \otimes \mathfrak{K}(\gamma)\, d\mu(\gamma)$$

and

$$\xi \otimes_\psi \eta = \int_\Gamma^\oplus \xi(\gamma) \otimes \eta(\gamma)\, d\mu(\gamma)$$

if

$$\xi = \int_\Gamma^\oplus \xi(\gamma)\,\mathrm{d}\mu(\gamma) \in \mathfrak{H}; \qquad \eta = \int_\Gamma^\oplus \eta(\gamma)\,\mathrm{d}\mu(\gamma) \in \mathfrak{K}$$

are such that $\int_\Gamma \|\xi(\gamma)\|^2\, \|\eta(\gamma)\|^2\, \mathrm{d}\mu(\gamma) < \infty$.

We leave the proof to the reader as it is an easy good exercise. Here the reader should note why the relative tensor product of an arbitrary pair of vectors does not make sense and also why one need to fix a faithful semi-finite normal weight on $\mathcal{A}$ before considering the relative tensor product of vectors.

Exercise IX.3

Throughout this exercise, let $\mathcal{M}$ be a fixed von Neumann algebra.

1) In the positive cone $\mathcal{M}_*^+$, denote each element of $\mathcal{M}_*^+$ by $\varphi^{1/2}$, $\varphi \in \mathcal{M}_*^+$. Define for each pair $\varphi, \psi \in \mathcal{M}_*^+$:

$$\varphi^{\frac12} + \psi^{\frac12} = \chi,$$

where $\chi \in \mathcal{M}_*^+$ is given by

$$\chi(x) = (\varphi+\psi)(a^* x a), \qquad x \in \mathcal{M},$$

$$a = \big(\mathrm{D}\varphi : \mathrm{D}(\varphi+\psi)\big)_{-\frac{i}{2}} + \big(\mathrm{D}\psi : \mathrm{D}(\varphi+\psi)\big)_{-\frac{i}{2}};$$

$$\lambda \cdot \varphi^{\frac12} = \big(\lambda^2\varphi\big)^{\frac12}, \qquad \lambda \ge 0;$$

$$\big(\varphi^{\frac12} \mid \psi^{\frac12}\big) = (\varphi+\psi)(a^* a).$$

(a) Let $\{\mathcal{M}, \mathfrak{H}, J, \mathcal{P}\}$ be a standard form and, for each $\varphi \in \mathcal{M}_*^+$, $\xi(\varphi)$ be the representing vector of φ in $\mathcal{P}$. By making use of the map: $\varphi \in \mathcal{M}_*^+ \to \xi(\varphi) \in \mathcal{P}$, show that

$$\varphi^{\frac12} + \psi^{\frac12} = \psi^{\frac12} + \varphi^{\frac12};$$

$$\varphi^{\frac12} + 0^{\frac12} = \varphi^{\frac12};$$

$$\big(\varphi^{\frac12} + \psi^{\frac12}\big) + \chi^{\frac12} = \varphi^{\frac12} + \big(\psi^{\frac12} + \chi^{\frac12}\big);$$

$$\varphi^{\frac12} + \chi^{\frac12} = \psi^{\frac12} + \chi^{\frac12} \iff \varphi = \psi;$$

$$0 \cdot \varphi^{\frac12} = 0^{\frac12};$$

$$(\lambda\mu) \cdot \varphi^{\frac12} = \lambda \cdot \big(\mu \cdot \varphi^{\frac12}\big);$$

$$(\lambda+\mu) \cdot \varphi^{\frac12} = \lambda \cdot \varphi^{\frac12} + \mu \cdot \varphi^{\frac12};$$

$$\lambda \cdot \big(\varphi^{\frac12} + \psi^{\frac12}\big) = \lambda \cdot \varphi^{\frac12} + \lambda \cdot \varphi^{\frac12}.$$

(b) Show that the relation: $\left(\varphi_1^{1/2}, \psi_1^{1/2}\right) \sim \left(\varphi_2^{1/2}, \psi_2^{1/2}\right)$ in $\mathcal{M}_*^+ \times \mathcal{M}_*^+$ defined by $\varphi_1^{1/2} + \psi_2^{1/2} = \varphi_2^{1/2} + \psi_1^{1/2}$ is an equivalence relation. Denote the equivalence class of $\left(\varphi^{1/2}, \psi^{1/2}\right)$ by $\varphi^{1/2} - \psi^{1/2}$.

(c) Show that the quotient space $\mathcal{M}_*^+ \times \mathcal{M}_*^+ /$"$\sim$" is a real Hilbert space under the inner product:

$$\left(\varphi_1^{\frac{1}{2}} - \psi_1^{\frac{1}{2}} \mid \varphi_2^{\frac{1}{2}} - \psi_2^{\frac{1}{2}}\right) = \left(\varphi_1^{\frac{1}{2}} \mid \varphi_2^{\frac{1}{2}}\right) - 2\left(\varphi_1^{\frac{1}{2}} \mid \psi_2^{\frac{1}{2}}\right) + \left(\psi_1^{\frac{1}{2}} \mid \psi_2^{\frac{1}{2}}\right),$$

and the multiplication by scalars:

$$(\lambda - \mu) \cdot \left(\varphi^{\frac{1}{2}} - \psi^{\frac{1}{2}}\right) = \lambda \cdot \varphi^{\frac{1}{2}} + \mu \cdot \psi^{\frac{1}{2}} - \lambda \cdot \psi^{\frac{1}{2}} - \mu \cdot \varphi^{\frac{1}{2}}$$

for $\lambda, \mu \geq 0$. Denote this real Hilbert space by $L^2_{\mathbf{R}}(\mathcal{M})$, and the complexified Hilbert space: $L^2_{\mathbf{R}}(\mathcal{M}) \otimes_{\mathbf{R}} \mathbf{C}$ simply by $L^2(\mathcal{M})$.

(d) Observe that the canonical Hilbert space $L^2(\mathcal{M})$ constructed above is indeed canonical to the von Neumann algebra $\mathcal{M}$ which depends solely on $\mathcal{M}$ alone but nothing else.

(e) Let $L^2(\mathcal{M})_+$ denote the convex cone $\left\{\varphi^{1/2} : \varphi \in \mathcal{M}_*^+\right\}$ in $L^2(\mathcal{M})$, and define:

$$J_{\mathcal{M}}(\xi + \mathrm{i}\eta) = \xi - \mathrm{i}\eta, \quad \xi, \eta \in L^2_{\mathbf{R}}(\mathcal{M}).$$

2) Considering a standard form $\{\mathcal{M}, \mathfrak{H}, J, \mathcal{P}\}$, proceed the following.

(a) For a fixed $\varphi \in \mathcal{M}_*^+$ and $x, y \in \mathcal{M}$, show that the function:

$$f_{x,y}^{\varphi}(t) = \varphi\left(\sigma_{-t}^{\varphi}(x)\, x^*\, y\, x\, \sigma_t^{\varphi}(x)\right), \quad t \in \mathbf{R},$$

can be extended to a function $f_{x,y}^{\varphi} \in \mathcal{A}\left(\mathbf{D}_{1/2}\right)$, where $\mathcal{D}_{1/2}$ is the horizontal strip bounded by $\mathbf{R}$ and $\mathbf{R} - \frac{\mathrm{i}}{2}$ and $\mathcal{A}\left(\mathbf{D}_{1/2}\right)$ is the space of functions homomorphic on $\mathbf{D}_{1/2}$, bounded and continuous on the closure $\overline{\mathbf{D}_{1/2}}$.

(b) Show that the map: $y \in \mathcal{M} \to f_{x,y}^{\varphi}\left(-\frac{\mathrm{i}}{2}\right) = \left(\alpha(x)\varphi\right)(y)$ is a positive linear functional of y, so that $\alpha(x)\varphi \in \mathcal{M}_*^+$. (Hint: $\xi(\alpha(x)\varphi) = x\xi(\varphi)x^*$.)

(c) Show that the map $\rho(x)\colon \varphi^{1/2} \in L^2(\mathcal{M})_+ \to \left(\alpha(x)\varphi\right)^{1/2} \in L^2(\mathcal{M})_+$ can be uniquely extended to a bounded linear operator, denoted again by $\rho(x)$, on $L^2(\mathcal{M})$.

(d) Show that $\rho(x_1x_2) = \rho(x_1)\rho(x_2)$ for $x_1, x_2 \in \mathcal{M}$ and $\rho(1) = 1$.

(e) Show that for each $x \in \mathcal{M}$, $\{\rho(\exp tx)\}_{t\in\mathbf{R}}$ is a norm continuous one parameter group in $\mathcal{L}\left(L^2(\mathcal{M})\right)$. (Hint: $\rho(\exp tx)$ corresponds to $\exp(tx)J\exp(tx)J$.)

(f) Set $\delta(x) = \frac{\mathrm{d}}{\mathrm{d}t}\left(\rho(\exp tx)\right)\big|_{t=0}$, $x \in \mathcal{M}$. Show that δ is a real linear Lie algebra homomorphism of $\mathcal{M}$ into $\mathcal{L}(L^2)$). (Hint: Recall Exercise IX.1.)

(g) Set

$$\pi(x) = \delta(x) - \mathrm{i}\delta(\mathrm{i}x), \quad x \in \mathcal{M};$$
$$\pi'(x) = \delta(x^*) + \mathrm{i}\delta(\mathrm{i}x^*).$$

Prove that π (resp. π') is a faithful normal representation (resp. anti-representation) of $\mathcal{M}$ on $L^2(\mathcal{M})$ such that $\{\pi(\mathcal{M}),\ L^2(\mathcal{M}),\ J_{\mathcal{M}},\ L^2(\mathcal{M})_+\}$ is a standard from of $\mathcal{M}$ and

$$J_{\mathcal{M}}\pi(x)^*J_{\mathcal{M}} = \pi'(x), \quad x \in \mathcal{M}.$$

We call $\{\pi(\mathcal{M}),\ L^2(\mathcal{M}),\ J_{\mathcal{M}},\ L^2(\mathcal{M})_+\}$ *the* standard form of $\mathcal{M}$.

3) For a semi-finite normal weight φ on $\mathcal{M}$, the map $\eta_\varphi : x \in \mathfrak{n}_\varphi \to \eta_\varphi(x) \in \mathfrak{H}_\varphi$ can be interpreted as $x\varphi^{1/2}$, viewing $\varphi^{1/2}$ as an infinitely long vector of $L^2(\mathcal{M})$ unless φ is finite, i.e.

$$\|\varphi^{\frac{1}{2}}\|_{L^2(\mathcal{M})} = \varphi(1)^{\frac{1}{2}} = +\infty.$$

Observe that this new interpretation of $\eta_\varphi(x)$ as $x\varphi^{1/2}$ is consistent with the previously established concepts and notations.

4†) It is also possible to construct two sided Banach $\mathcal{M}$-modules $L^p(\mathcal{M}),\ p \geq 1$, from $\{\varphi : \varphi \in \mathcal{M}_*^+\}$, [596].

5) Let $\{\mathcal{M}, \mathfrak{H}\}$ be a von Neumann algebra and set $\mathcal{N} = (\mathcal{M}')^\circ$. View $\mathfrak{H}$ as an $\mathcal{M}$-$\mathcal{N}$-bimodule. Let ψ' be a faithful semi-finite normal weight on $\mathcal{M}'$ and $\psi = (\psi')^\circ$ on $\mathcal{N}$. With φ a faithful semi-finite normal weight on $\mathcal{M}$, let $\tilde{\mathfrak{H}},\ \mathcal{R},\ \rho$ and others as in the earlier part of the section. Prove that if $\xi \in \mathfrak{D}(\mathfrak{H}, \psi) = \mathfrak{H} \cap \mathfrak{B}$, then $\pi_\ell(\xi) = f\pi_\ell(\xi)e$ and $L_\psi(\xi) = \pi_\ell(\xi)|_{L^2(\mathcal{N},\psi)}$.

6) In the previous problem, consider the relative tensor product: $\mathfrak{H} \otimes_\psi \bar{\mathfrak{H}}$. Let J be the conjugate linear involution of $\mathfrak{H} \otimes_\psi \bar{\mathfrak{H}}$ defined by $J(\xi \otimes_\psi \bar{\eta}) = (\eta \otimes_\psi \bar{\xi})$, $\xi, \eta \in \mathfrak{D}(\mathfrak{H}, \psi)$ and $(\mathfrak{H} \otimes_\psi \bar{\mathfrak{H}})_+$ be the closed convex cone in $\mathfrak{H} \otimes_\psi \bar{\mathfrak{H}}$ generated by $\{\xi \otimes_\psi \bar{\xi} : \xi \in \mathfrak{D}(\mathfrak{H}, \psi)\}$. Prove that $\{\mathfrak{H} \otimes_\psi \bar{\mathfrak{H}},\ (\mathfrak{H} \otimes_\psi \bar{\mathfrak{H}})_+,\ J\}$ is isormorphic to the standard form $\{L^2(\mathcal{M}),\ L^2(\mathcal{M})_+,\ J\}$ of $\mathcal{M}$ as an $\mathcal{M}$-$\mathcal{M}$-bimodule.

7) Keep the set-up of the last two problems. Let $L^2(\mathcal{N})$ be the standard form of $\mathcal{N}$. Let $\tilde{\mathfrak{H}} = L^2(\mathcal{N}) \oplus \mathfrak{H}$ and $\mathcal{R} = \mathcal{L}(\tilde{\mathfrak{H}}_\mathcal{N})$. Notice that in doing this we do not pick up a semi-finite normal weight ψ on $\mathcal{N}$ unlike the last problem or in the main part of the section. Following the steps described below, show that $\tilde{\mathfrak{H}}$ can be canonically embedded into the standard form $L^2(\mathcal{R})$ such a way that $J\tilde{\mathfrak{H}} = L^2(\mathcal{R}) \oplus \bar{\mathfrak{H}}$: $J(\xi \oplus \eta) = \xi^* \oplus \bar{\eta},\ \xi \in L^2(\mathcal{N}),\ \eta \in \mathfrak{H}$, where ξ^* means the modular conjugation in $L^2(\mathcal{N})$ and $\bar{\eta}$ means the element in $\bar{\mathfrak{H}}$ corresponding to $\eta \in \mathfrak{H}$ under the canonical conjugation.

(a) To each $\xi \in \tilde{\mathfrak{H}}$, there corresponds a unique element $|\xi| \in L^2(\mathcal{N})_+$ and a partial isometry $u \in \mathcal{R}$ such that $(\xi x \mid \xi) = (|\xi|x \mid |\xi|),\ x \in \mathcal{N},\ \xi = u|\xi|$ and u^*u is the projection from $\tilde{\mathfrak{H}}$ onto $[|\xi|\mathcal{N}]$, the smallest closed linear subspace containing $|\xi|\mathcal{N}$.

(b) As $\mathcal{N} = \mathcal{R}_e$, where e is the projection of $\tilde{\mathfrak{H}}$ onto $L^2(\mathcal{N})$ as before, we have $L^2(\mathcal{N}) = eL^2(\mathcal{R})e$.

(c) Set $U\xi = u|\xi|$, $\xi \in \tilde{\mathfrak{H}}$ and show that U is the required isometry embedding $\tilde{\mathfrak{H}}$ into $L^2(\mathcal{R})$ and $U\mathfrak{H} = fL^2(\mathcal{R})e$.

8) Let $\mathcal{N}$ be a von Neumann algebra, $\mathfrak{H}$ a right $\mathcal{N}$-module and $\mathfrak{K}$ a left $\mathcal{N}$-module. Let $\tilde{\mathfrak{H}} = L^2(\mathcal{N}) \oplus \mathfrak{H} \oplus \bar{\mathfrak{K}}$ and $\mathcal{R} = \mathcal{L}(\tilde{\mathfrak{H}}_{\mathcal{N}})$. By (7), we view $L^2(\mathcal{N})$, $\mathfrak{H}$ and $\bar{\mathfrak{K}}$ as closed subspaces of $L^2(\mathcal{R})$ and let e, f and g be respectively the projection from $L^2(\mathcal{R})$ onto these subspaces. With this identification, we have $\mathfrak{H} = fL^2(\mathcal{R})e$ and $\bar{\mathfrak{K}} = gL^2(\mathcal{R})e$, which means that $\mathfrak{K} = eL^2(\mathcal{R})g$. Define the relative tensor product of $\mathfrak{H}$ and $\mathfrak{K}$ with respect to $\mathcal{N}$ in the following way:

$$\mathfrak{H} \otimes_{\mathcal{N}} \mathfrak{K} = fL^2(\mathcal{R})g,$$

which is an $\mathcal{L}(\mathfrak{H}_{\mathcal{N}})$-$\mathcal{L}({}_{\mathcal{N}}\mathfrak{K})^{\circ}$-bimodule in the natural fashion. In this way, we eliminated the dependence of the relative tensor product on faithful semi-finite normal weights on $\mathcal{N}$. Prove that there exists a natural bimodule isomorphism of $\mathfrak{H} \otimes_{\psi} \mathfrak{K}$ onto $\mathfrak{H} \otimes_{\mathcal{N}} \mathfrak{K}$.

9) Assume that $\mathcal{N}$ is equipped with faithful semi-finite normal trace $\tau_{\mathcal{N}}$. Let $\mathfrak{H}$ be a right $\mathcal{N}$-module and $\mathcal{M} = \mathcal{L}(\mathfrak{H}_{\mathcal{N}})$. Consider $\tilde{\mathfrak{H}}$ and $\mathcal{R}$ as before. Normalize the trace $\tau_{\mathcal{M}}$ on $\mathcal{M}$ so that

$$\tau_{\mathcal{M}}(xx^*) = \tau'_{\mathcal{N}}(x^*x), \quad x \in \mathcal{L}\big(L^2(\mathcal{N}, \tau_{\mathcal{N}})_{\mathcal{N}}, \mathfrak{H}_{\mathcal{N}}\big), \qquad \tau_{\mathcal{R}} = \tau_{\mathcal{N}} \oplus \tau_{\mathcal{M}},$$

and

$$\tau_{\mathcal{R}}\begin{pmatrix} x_{11} & x_{12} \\ x_{21} & x_{22} \end{pmatrix} = \tau_{\mathcal{N}}(x_{11}) + \tau_{\mathcal{M}}(x_{22}).$$

Identify $\mathfrak{H}$ with the (2, 1)-component of $L^2(\mathcal{R}, \tau_{\mathcal{R}})$ and $L^2(\mathcal{M}, \tau_{\mathcal{M}})$ with the (2, 2)-component of $L^2(\mathcal{R}, \tau_{\mathcal{R}})$ so that $\mathfrak{H}$ can be viewed as the L^2-space of measurable operators from $L^2(\mathcal{N}, \tau_{\mathcal{N}})$ into $L^2(\mathcal{M}, \tau_{\mathcal{M}})$ relative to the trace $\tau_{\mathcal{R}}$.

10) Consider a standard von Neumann algebra $\{\mathcal{N},\ L^2(\mathcal{N}),\ J,\ L^2(\mathcal{N})_+\}$ and the standard bimodule ${}_{\mathcal{N}}L^2(\mathcal{N})_{\mathcal{N}}$. For each $\alpha \in \mathrm{Aut}(\mathcal{N})$, let $\mathfrak{H}(\alpha)$ be the $\mathcal{N}$-$\mathcal{N}$ full bimodule obtained by:

$$x\xi y = xJ\alpha(y^*)J\xi, \quad x, y \in \mathcal{N}, \quad \xi \in L^2(\mathcal{N}).$$

Prove that if $\alpha, \beta \in \mathrm{Aut}(\mathcal{N})$, then $\mathfrak{H}(\alpha) \otimes_{\mathcal{N}} \mathfrak{H}(\beta) = \mathfrak{H}(\alpha\beta)$.

11) Show that if $\mathfrak{H}$ is a full $\mathcal{N}$-bimodule, then there exists $\alpha \in \mathrm{Aut}(\mathcal{N})$ such that $\mathfrak{H} = \mathfrak{H}(\alpha)$.

§ 4 Conditional Expectations and Operator Valued Weights

In probability theory, conditional expectations play a fundamental role. We discuss a non-commutative analogue of conditional expectations. In fact, it will play a key role in the structure theory of factors.

Definition 4.1. Let φ be a faithful (semi-finite normal) weight on a von Neumann algebra $\mathcal{M}$ and $\mathcal{N}$ be a von Neumann subalgebra of $\mathcal{M}$ such that the restriction $\varphi|_{\mathcal{N}}$ of φ to $\mathcal{N}$ is semi-finite. A linear map $\mathcal{E}$ of $\mathcal{M}$ onto $\mathcal{N}$ is called the *conditional expectation of $\mathcal{M}$ onto $\mathcal{N}$ with respect to φ* if the following conditions are satisfied:

$$\|\mathcal{E}(x)\| \leq \|x\|, \qquad x \in \mathcal{M}; \tag{1}$$

$$\mathcal{E}(x) = x, \qquad x \in \mathcal{N}; \tag{2}$$

$$\varphi = \varphi \circ \mathcal{E}. \tag{3}$$

By Theorem III.3.4, a conditional expectation $\mathcal{E}$ of $\mathcal{M}$ onto $\mathcal{N}$ enjoys the following properties:

$$\left.\begin{aligned} &\mathcal{E}(x^*x) \geq 0, && x \in \mathcal{M}; \\ &\mathcal{E}(axb) = a\mathcal{E}(x)b, && a, b \in \mathcal{N}, \quad x \in \mathcal{M}; \\ &\mathcal{E}(x)^*\mathcal{E}(x) \leq \mathcal{E}(x^*x), && x \in \mathcal{M}. \end{aligned}\right\} \tag{4}$$

In fact, a conditional expectation is completely positive.

Theorem 4.2. *Let $\mathcal{M}$, $\mathcal{N}$ and φ be as in Definition* 4.1. *The existence of a conditional expectation $\mathcal{E}$ of $\mathcal{M}$ onto $\mathcal{N}$ with respect to φ is equivalent to the global invariance, $\sigma_t^\varphi(\mathcal{N}) = \mathcal{N}$, $t \in \mathbf{R}$, of $\mathcal{N}$ under the modular automorphism group. If this is the case, then the conditional expectation $\mathcal{E}$ is normal and uniquely determined by φ.*

PROOF: Existence $\Longrightarrow$ Invariance: Assume that $\mathcal{E}$ exists and $\varphi = \varphi \circ \mathcal{E}$. Let $\mathfrak{A}$ and $\mathfrak{B}$ be the full left Hilbert algebras corresponding to $\{\mathcal{M}, \varphi\}$, and $\{\mathcal{N}, \varphi|_{\mathcal{N}}\}$. It follows that $\mathfrak{B}$ is a self-adjoint subalgebra of $\mathfrak{A}$. Let $\mathfrak{H}$ and $\mathfrak{K}$ be the completions of $\mathfrak{A}$ and $\mathfrak{B}$ respectively, and E be the projection of $\mathfrak{H}$ onto $\mathfrak{K}$. For each $x \in \mathfrak{n}_\varphi$ and $y \in \mathcal{N} \cap \mathfrak{n}_\varphi$, we have

$$\begin{aligned} \big(\eta_\varphi(x) \mid \eta_\varphi(y)\big) &= \varphi(y^*x) = \varphi\big(\mathcal{E}(y^*x)\big) = \varphi\big(y^*\mathcal{E}(x)\big) \\ &= \Big(\eta_\varphi\big(\mathcal{E}(x)\big) \Bigm| \eta_\varphi(y)\Big) \qquad \text{by (4),} \end{aligned}$$

so that

$$E\eta_\varphi(x) = \eta_\varphi\big(\mathcal{E}(x)\big), \quad x \in \mathfrak{n}_\varphi.$$

Since $\mathcal{E}$ preserves the $*$-operation, we have

$$E\xi^\sharp = (E\xi)^\sharp, \quad \xi \in \mathfrak{A}.$$

Employing the notations $\mathfrak{D}^\sharp(\mathfrak{A})$, $\mathfrak{D}^\sharp(\mathfrak{B}), \ldots$ to indicate the natural objects corresponding to $\mathfrak{A}$ and $\mathfrak{B}$, we have

$$ES\xi = SE\xi, \quad \xi \in \mathfrak{D}^\sharp(\mathfrak{A}).$$

Namely, E leaves $\mathfrak{D}^\sharp(\mathfrak{A})$ invariant globally and $ES \subset SE$. Hence we get $(1-2E)S \subset S(1-2E)$. Applying $(1-2E)$ from the right, we get $(1-2E)S(1-2E) \subset S$, so that

$$S = (1-2E)^2 S(1-2E)^2 \subset (1-E)S(1-E) \subset S,$$

hence we conclude

$$S = (1-2E)S(1-2E), \tag{5}$$

consequently

$$F = (1-2E)F(1-2E), \qquad \Delta = (1-2E)\Delta(1-2E). \tag{6}$$

Hence Δ and E commute, so Δ^{it} leaves $\mathfrak{K}$ invariant. Since $E\eta_\varphi(\mathfrak{n}_\varphi) = \eta_\varphi(\mathfrak{n}_\varphi \cap \mathcal{N})$ by (3), $E\mathfrak{A} = \mathfrak{B}$ and

$$\Delta^{it}\mathfrak{B} = \Delta^{it}E\mathfrak{A} = E\Delta^{it}\mathfrak{A} = E\mathfrak{A} = \mathfrak{B}.$$

Hence Δ^{it} leaves $\mathfrak{B}$ invariant. Hence with the equation:

$$\eta_\varphi\big(\sigma_t^\varphi(x)\big) = \Delta^{it}\eta_\varphi(x), \quad x \in \mathfrak{n}_\varphi \cap \mathfrak{n}_\varphi^* \cap \mathcal{N},$$

we conclude that $\{\sigma_t^\varphi\}$ leaves $\mathfrak{n}_\varphi \cap \mathfrak{n}_\varphi^* \cap \mathcal{N}$ invariant. Thus the semi-finiteness of φ on $\mathcal{N}$ yields the invariance of $\mathcal{N}$ under $\{\sigma_t^\varphi\}$ by the σ-weak density of $\mathfrak{n}_\varphi \cap \mathfrak{n}_\varphi^* \cap \mathcal{N}$ in $\mathcal{N}$.

Invariance $\Longrightarrow$ Existence: Suppose $\sigma_t^\varphi(\mathcal{N}) = \mathcal{N}$, $t \in \mathbf{R}$. We use the notations: $\mathfrak{A}$, $\mathfrak{B}$, $\mathfrak{H}$, $\mathfrak{K}$, $\Delta_\mathfrak{A}$, $\Delta_\mathfrak{B}$ and so on as before. For each $x \in \mathfrak{n}_\varphi \cap \mathfrak{n}_\varphi^*$, we have $\Delta_\mathfrak{A}^{it}\eta_\varphi(x) = \eta_\varphi\big(\sigma_t^\varphi(x)\big)$, so that $\Delta_\mathfrak{A}^{it}\mathfrak{B} = \mathfrak{B}$. Hence $\Delta_\mathfrak{A}^{it}\mathfrak{K} = \mathfrak{K}$, so the projection E of $\mathfrak{H}$ onto $\mathfrak{K}$ and $\Delta_\mathfrak{A}^{it}$ commute. Since the modular automorphism group of $\varphi|_\mathcal{N}$ is the restriction of $\{\sigma_t^\varphi\}$ to $\mathcal{N}$, $\Delta_\mathfrak{B}^{it}$ is nothing but the restriction of $\Delta_\mathfrak{A}^{it}$ to $\mathfrak{K}$. Hence

$$\mathfrak{D}^\sharp(\mathfrak{B}) = \mathfrak{D}\big(\Delta_\mathfrak{B}^{\frac{1}{2}}\big) = \mathfrak{D}\big(\Delta_\mathfrak{A}^{\frac{1}{2}}\big) \cap \mathfrak{K} = \mathfrak{D}^\sharp(\mathfrak{A}) \cap \mathfrak{K}, \tag{7}$$

which means that $\Delta_\mathfrak{B}^{1/2}$ is precisely the restriction of $\Delta_\mathfrak{A}^{1/2}$ to $\mathfrak{K}$ and $S_\mathfrak{B} = S_\mathfrak{A}\big|_\mathfrak{K}$. Therefore, we get $J_\mathfrak{B} = J_\mathfrak{A}$. Thus, we denote them simply by S and J. Furthermore, we have

$$\mathfrak{D}^\flat(\mathfrak{B}) = \mathfrak{D}^\flat(\mathfrak{A}) \cap \mathfrak{K} = \mathfrak{D}\big(\Delta_\mathfrak{A}^{-\frac{1}{2}}\big) \cap \mathfrak{K}. \tag{8}$$

Therefore, we can safely omit the subscripts $\mathfrak{A}$ and $\mathfrak{B}$ from Δ, $\mathfrak{K}$ and others as well.

We now prove

$$\mathfrak{B} = \mathfrak{A} \cap \mathfrak{K}, \qquad \mathfrak{B}' = \mathfrak{A}' \cap \mathfrak{K}. \tag{9}$$

Clearly $\mathfrak{B} \subset \mathfrak{A} \cap \mathfrak{K}$. Suppose $\xi \in \mathfrak{A}' \cap \mathfrak{K}$. There exists a constant $\gamma > 0$ such that $\big\|\pi_\ell^\mathfrak{A}(\eta)\xi\big\| \le \gamma\|\eta\|$ for every $\eta \in \mathfrak{A}$. In particular, $\big\|\pi_\ell^\mathfrak{B}(\eta)\xi\big\| \le \gamma\|\eta\|$ for every $\eta \in \mathfrak{B}$. Hence ξ is right bounded with respect to $\mathfrak{B}$. Since $\xi \in \mathfrak{D}^\flat(\mathfrak{A}) \cap \mathfrak{K} = \mathfrak{D}^\flat(\mathfrak{B})$, we have $\xi \in \mathfrak{B}'$. Thus $\mathfrak{A}' \cap \mathfrak{K} \subset \mathfrak{B}'$. Applying J, we get

$$\mathfrak{A} \cap \mathfrak{K} = J(\mathfrak{A}' \cap \mathfrak{K}) \subset J\mathfrak{B}' = \mathfrak{B}.$$

Thus $\mathfrak{A} \cap \mathfrak{K} = \mathfrak{B}$. Applying J once again, we conclude $\mathfrak{B}' = \mathfrak{A}' \cap \mathfrak{K}$.

Now, we consider the associated Tomita algebras $\mathfrak{A}_0$ and $\mathfrak{B}_0$ respectively. By definition, we have

$$\mathfrak{A}_0 = \left\{ \xi \in \bigcap_{n \in \mathbf{Z}} \mathfrak{D}(\Delta^n) : \Delta^n \xi \in \mathfrak{A} \right\}.$$

Hence $\mathfrak{B}_0 = \mathfrak{A}_0 \cap \mathfrak{K}$ follows from the previous discussion. We now prove

$$\left.\begin{aligned} &\mathfrak{B} = E\mathfrak{A}, \qquad \mathfrak{B}' = E\mathfrak{A}', \qquad \mathfrak{B}_0 = E\mathfrak{A}_0; \\ &E(\xi\eta) = \xi E\eta, \qquad \xi \in \mathfrak{B}, \quad \eta \in \mathfrak{A}; \\ &E(\xi\eta) = (E\xi)\eta, \quad \xi \in \mathfrak{A}, \quad \eta \in \mathfrak{B}'. \end{aligned}\right\} \tag{10}$$

First, we note that $\mathfrak{K}$ is invariant under $\pi_\ell^{\mathfrak{A}}(\mathfrak{B})$ and $\pi_\ell^{\mathfrak{A}}(\mathfrak{B})$ and $\pi_r^{\mathfrak{A}}(\mathfrak{B}')$ by (9). Hence E and $\pi_\ell^{\mathfrak{A}}(\mathfrak{B})$ (resp. $\pi_r^{\mathfrak{A}}(\mathfrak{B}')$) commute, which means precisely the last two identities in (10). If $\xi \in \mathfrak{A}$ and $\eta \in \mathfrak{B}'$, then

$$\|\pi_r^{\mathfrak{B}}(\eta)E\xi\| = \|E\pi_r^{\mathfrak{A}}(\eta)\xi\| = \|E\pi_\ell^{\mathfrak{A}}(\xi)\eta\| \le \|\pi_\ell^{\mathfrak{A}}(\xi)\|\|\eta\|,$$

so that $E\xi$ is left bounded with respect to $\mathfrak{B}$. Since $\mathfrak{B} = E\mathfrak{B} \subset E\mathfrak{A}$, we conclude $E\mathfrak{A} = \mathfrak{B}$. Applying J, we get $E\mathfrak{A}' = \mathfrak{B}'$. Now, the last claim $E\mathfrak{A}_0 = \mathfrak{B}_0$ in (10) follows from the construction of Tomita algebras and the above established facts. Thus (10) now follows.

The second formula in (10) means

$$\pi_\ell^{\mathfrak{B}}(\xi)E = E\pi_\ell^{\mathfrak{A}}(\xi)E, \quad \xi \in \mathfrak{B}. \tag{11}$$

Hence we have $E\mathcal{R}_\ell(\mathfrak{A})E\big|_{\mathfrak{K}} = \mathcal{R}_\ell(\mathfrak{B})$. With $\pi = \pi_\varphi\big|_{\mathcal{N}}$, we set

$$\mathcal{E}(x) = \pi^{-1}\big(E\pi_\varphi(x)E\big), \quad x \in \mathcal{M}. \tag{12}$$

It then follows that $\mathcal{E}$ is a normal projection of norm one from $\mathcal{M}$ onto $\mathcal{N}$.

If $x \in \mathcal{M}$ and $y, z \in \mathfrak{n}_\varphi \cap \mathcal{N}$, then

$$\begin{aligned} \varphi(z^*xy) &= \big(\pi_\varphi(x)\eta_\varphi(y) \mid \eta_\varphi(z)\big) = \big(E\pi_\varphi(x)E\eta_\varphi(y) \mid \eta_\varphi(z)\big) \\ &= \Big(\pi_\varphi\big(\mathcal{E}(x)\big)\eta_\varphi(y) \,\Big|\, \eta_\varphi(z)\Big) = \varphi\big(z^*\mathcal{E}(x)y\big), \end{aligned}$$

so we get

$$\varphi(z^*xy) = \varphi\big(z^*\mathcal{E}(x)y\big), \quad x \in \mathcal{M}, \quad y, z \in \mathfrak{n}_\varphi \cap \mathcal{N}.$$

We now prove

$$\varphi(x) = \sup\Big\{ \big(\pi_\varphi(x)\eta \mid \eta\big) : \eta \in \mathfrak{B}', \ \|\pi_r^{\mathfrak{B}}(\eta)\| < 1 \Big\}, \quad x \in \mathfrak{m}_\varphi^+. \tag{13}$$

Since $\mathfrak{B}' = \mathfrak{A}' \cap \mathfrak{K}$ and $\|\pi_r^{\mathfrak{B}}(\eta)\| = \|\pi_r^{\mathfrak{A}}(\eta)\|$ for $\eta \in \mathfrak{B}'$, it follows from the proof of Theorem VII.1.17 that the left hand side of (13) majorizes the right hand side of (13). If $\xi \in \mathfrak{m}_\varphi^+$, then the non-degeneracy of $\pi_r^{\mathfrak{A}}(\mathfrak{B}')$ yields

$$\begin{aligned}\varphi(x) = \left\|\eta_\varphi(x^{\frac{1}{2}})\right\|^2 &= \sup\left\{ \left\|\pi_r^{\mathfrak{A}}(\eta)\eta_\varphi(x^{\frac{1}{2}})\right\|^2 : \eta \in \mathfrak{B}', \ \left\|\pi_r^{\mathfrak{A}}(\eta)\right\| < 1 \right\} \\ &= \sup\left\{ \left\|\pi_\varphi(x^{\frac{1}{2}})\eta\right\|^2 : \eta \in \mathfrak{B}', \ \left\|\pi_r^{\mathfrak{B}}(\eta)\right\| < 1 \right\} \\ &= \sup\left\{ \left(\pi_\varphi(x)\eta \mid \eta\right) : \eta \in \mathfrak{B}', \ \left\|\pi_r^{\mathfrak{B}}(\eta)\right\| < 1 \right\}.\end{aligned}$$

Hence (13) follows.

Now, if $x \in \mathfrak{m}_\varphi^+$, then we have

$$\begin{aligned}\varphi(x) &= \sup\left\{ \left(\pi_\varphi(x)\eta \mid \eta\right) : \eta \in \mathfrak{B}', \ \left\|\pi_r^{\mathfrak{B}}(\eta)\right\| < 1 \right\} \\ &= \sup\left\{ \left(E\pi_\varphi(x)E\eta \mid \eta\right) : \eta \in \mathfrak{B}', \ \left\|\pi_r^{\mathfrak{B}}(\eta)\right\| < 1 \right\} \\ &= \sup\left\{ \left(\pi_\varphi\big(\mathcal{E}(x)\big)\eta \,\Big|\, \eta\right) : \eta \in \mathfrak{B}', \ \left\|\pi_r^{\mathfrak{B}}(\eta)\right\| < 1 \right\} = \varphi \circ \mathcal{E}(x).\end{aligned}$$

If $\mathcal{E}(x) = 0$ for an $x \in \mathcal{M}_+$, then $E\pi_\varphi(x)E = 0$. Since $\pi_\varphi(x) \geq 0$, $\pi_\varphi(x)\mathfrak{K} = \{0\}$. Namely, $xy = 0$ for every $y \in \mathfrak{n}_\varphi \cap \mathcal{N}$. But $\mathfrak{n}_\varphi \cap \mathcal{N}$ is σ-weakly dense in $\mathcal{N}$, so that 1 is in the σ-weak closure of $\mathfrak{n}_\varphi \cap \mathcal{N}$; thus $x = 0$. Therefore, $\mathcal{E}$ is faithful.

Finally, we complete the proof of the theorem by showing $\varphi = \varphi \circ \mathcal{E}$. To this end, set $\psi = \varphi \circ \mathcal{E}$. The commutativity of Δ and $\mathcal{E}$ implies the invariance of ψ under $\{\sigma_t^\varphi\}$. Hence Corollary VIII.3.6 implies the existence of a non-singular positive self-adjoint operator h affiliated with $\mathcal{M}_\varphi$ such that $\psi = \varphi_h$. As seen above, $\varphi(x) = \psi(x)$ for every $x \in \mathfrak{m}_\varphi$, so that for each $x \in \mathfrak{n}_\varphi$ we have, with $h_\varepsilon = h\,(1 + \varepsilon h)^{-1}$,

$$\lim_{\varepsilon \to 0} \left(\pi_\varphi(h_\varepsilon)\eta_\varphi(x) \mid \eta_\varphi(x)\right) = \varphi_h(x^*x) = \psi(x^*x) = \varphi(x^*x) = \left\|\eta_\varphi(x)\right\|^2.$$

Hence $\eta_\varphi(\mathfrak{n}_\varphi) \subset \mathfrak{D}\left(\pi_\varphi(h)^{1/2}\right)$ and $\left\|\pi_\varphi(h)^{1/2}\eta_\varphi(x)\right\|^2 = \left\|\eta_\varphi(x)\right\|^2$, $x \in \mathfrak{n}_\varphi$. Hence $h = 1$, so $\varphi = \psi$. Q.E.D.

Proposition 4.3. *Let $\mathcal{M}$ be a von Neumann algebra, and $\mathcal{N}$ be a von Neumann subalgebra such that $\mathcal{N}' \cap \mathcal{M} = \mathfrak{C}_\mathcal{N} \subset \mathcal{N}$. If $\mathcal{E}$ is a normal projection of norm one from $\mathcal{M}$ onto $\mathcal{N}$, then $\mathcal{E}$ is necessarily faithful and unique. Furthermore, if ψ is a faithful semi-finite normal weight on $\mathcal{N}$, then $\mathcal{E}$ is the conditional expectation of $\mathcal{M}$ onto $\mathcal{N}$ relative to $\varphi = \psi \circ \mathcal{E}$.*

PROOF: Let ψ be a faithful semi-finite normal weight on $\mathcal{N}$ and put $\varphi = \psi \circ \mathcal{E}$. Then φ is semi-finite and normal. Let N_φ be the left kernel of φ, i.e. $N_\varphi = \left\{x \in \mathcal{M} : \varphi(x^*x) = 0\right\}$. Then with $e = s(\varphi)$, the support of φ, we have $N_\varphi = \mathcal{M}(1 - e)$. As ψ is faithful, $x \in \mathcal{M}$ belongs to N_φ if and only if $\mathcal{E}(x^*x) = 0$. If $b \in \mathcal{N}$, then for any $x \in N_\varphi$ we have $\mathcal{E}\left((xb)^*(xb)\right) = b^*\mathcal{E}(x^*x)b = 0$, so that N_φ is invariant under the multiplication of $\mathcal{N}$ from the right. This means that e commutes with $\mathcal{N}$, so that by the assumption on the relative commutant of $\mathcal{N}$, e belongs to the center $\mathfrak{C}_\mathcal{N}$ of $\mathcal{N}$. But ψ and φ agree on $\mathcal{N}$, so that $e = 1$. Thus φ is faithful.

Suppose $\mathcal{E}_1$ is another normal projection of norm one from $\mathcal{M}$ onto $\mathcal{N}$. Set $\varphi_1 = \psi \circ \mathcal{E}_1$. As $\sigma_t^\varphi\big|_{\mathcal{N}} = \sigma_t^\psi = \sigma_t^{\varphi_1}\big|_{\mathcal{N}}$, we have $\big(\mathrm{D}\varphi_1 : \mathrm{D}\varphi\big)_t \in \mathcal{N}' \cap \mathcal{M} = \mathcal{C}_{\mathcal{N}}$. As σ_t^φ acts trivially on $\mathcal{C}_{\mathcal{N}}$, being the same as σ_t^ψ on $\mathcal{C}_{\mathcal{N}}$, $\big\{(\mathrm{D}\varphi_1 : \mathrm{D}\varphi)_t\big\}$ is a one parameter unitary group in $\mathcal{C}_{\mathcal{N}}$, so that there exists a self-adjoint non-singular positive operator h affiliated with $\mathcal{C}_{\mathcal{N}}$ such that $\big(\mathrm{D}\varphi_1 : \mathrm{D}\varphi\big)_t = h^{it}$, which means that $\varphi_1 = \varphi(h\ \cdot\ )$. But φ_1 and φ are precisely ψ on $\mathcal{N}$, so that $h = 1$. Hence $\varphi_1 = \varphi$. Thus we have $\psi \circ \mathcal{E}_1 = \psi \circ \mathcal{E}$. Theorem 4.2 implies that $\mathcal{E}_1 = \mathcal{E}$. Q.E.D.

In this second volume, we have learned that the study of unbounded operators and unbounded functionals – weights – provides powerful tools as well as good insight of the subject. In this section, we just established a criteria for the existence of conditional expectations. In this new domain, it is also true that the extension of our study to "unbounded conditional expectations" will give us useful tools.

To study weights, we considered the extended positive real numbers $\mathbf{R}_+ \cup \{\infty\}$. To study "unbounded conditional expectations", we need to consider the "extended positive part" $\widehat{\mathcal{N}}_+$ of the von Neumann subalgebra $\mathcal{N}$ of $\mathcal{M}$. We begin by the following:

Definition 4.4. For a von Neumann algebra $\mathcal{M}$, the *extended positive cone* $\widehat{\mathcal{M}}_+$ of $\mathcal{M}$ is the set of maps $m\colon \mathcal{M}_*^+ \mapsto [0, \infty]$ with the following properties:

(i) $m(\lambda\varphi) = \lambda m(\varphi),\quad \varphi \in \mathcal{M}_*^+,\quad \lambda \geq 0,$
(ii) $m(\varphi + \psi) = m(\varphi) + m(\psi),\quad \varphi, \psi \in \mathcal{M}_*^+,$
(iii) m is lower semi-continuous.

Clearly, the positive part $\mathcal{M}_+$ of $\mathcal{M}$ is a subset of $\widehat{\mathcal{M}}_+$. It is easy to see that $\widehat{\mathcal{M}}_+$ is closed under addition, multiplication by non-negative scalars and increasing limits.

Example 4.5. Let $\{\mathcal{M}, \mathfrak{H}\}$ be a von Neumann algebra and A a positive self-adjoint operator on $\mathfrak{H}$ affiliated with $\mathcal{M}$. Suppose that

$$A = \int_0^\infty \lambda\, \mathrm{d}e(\lambda) \tag{14}$$

is the spectral decomposition of A. For each $\varphi \in \mathcal{M}_*^+$, put

$$m_A(\varphi) = \int_0^\infty \lambda\, \mathrm{d}\varphi\big(e(\lambda)\big)\,. \tag{15}$$

Then m_A satisfies the conditions (i), (ii) and (iii) of Definition 4.4. The last condition, the lower semi-continuity, follows from

$$m_A(\varphi) = \sup_n \varphi(A_n) \quad \text{with} \quad A_n = \int_0^n \lambda\, \mathrm{d}e(\lambda) \in \mathcal{M}_+.$$

It now follows that

$$m_A(\omega_\xi) = \int_0^\infty \lambda \,\mathrm{d}\big(e(\lambda)\xi \mid \xi\big) = \begin{cases} \|A^{\frac12}\xi\|^2, & \xi \in \mathfrak{D}(A^{\frac12}), \\ +\infty, & \xi \notin \mathfrak{D}(A^{\frac12}). \end{cases} \tag{16}$$

Hence if B is another positive self-adjoint operator on $\mathfrak{H}$ affiliated with $\mathcal{M}$, then the equality $m_A = m_B$ means precisely $A = B$. Hence the map: $A \mapsto m_A \in \widehat{\mathcal{M}}_+$ is injective. Thus, the set of positive self-adjoint operators affiliated with $\mathcal{M}$ can be identified with a subset of the extended positive cone $\widehat{\mathcal{M}}_+$.

Definition 4.6. For $m, n \in \widehat{\mathcal{M}}_+$, $\lambda \geq 0$ and $a \in \mathcal{M}$, we define the following operations:

$$\left.\begin{aligned} (\lambda m)(\varphi) &= \lambda m(\varphi), && \varphi \in \mathcal{M}_*^+, \\ (m+n)(\varphi) &= m(\varphi) + n(\varphi), && \varphi \in \mathcal{M}_*^+, \\ (a^* m a)(\varphi) &= m(a\varphi a^*), && \varphi \in \mathcal{M}_*^+. \end{aligned}\right\} \tag{17}$$

We also note here that $\sup m_i$ of an increasing net in $\widehat{\mathcal{M}}_+$ can be naturally defined.

Lemma 4.7. *Let $\{\mathcal{M}, \mathfrak{H}\}$ be a von Neumann algebra. To each $m \in \widehat{\mathcal{M}}_+$, there corresponds uniquely a pair $\{A, \mathfrak{K}\}$ of a closed subspace $\mathfrak{K}$ of $\mathfrak{H}$ and a positive self-adjoint operator on $\mathfrak{K}$ such that:*

(i) *$\mathfrak{K}$ is affiliated with $\mathcal{M}$, in the sense that the projection to $\mathfrak{K}$ belongs to $\mathcal{M}$, and A is affiliated with $\mathcal{M}$;*

(ii)

$$m(\omega_\xi) = \begin{cases} \big\|A^{\frac12}\xi\big\|^2, & \xi \in \mathfrak{D}(A^{\frac12}), \\ +\infty, & \text{otherwise,} \end{cases} \tag{18}$$

where ω_ξ means, of course, the functional $x \in \mathcal{M} \mapsto (x\xi \mid \xi)$.

PROOF: Consider the extended positive real valued function $q : \xi \in \mathfrak{H} \mapsto m(\omega_\xi)$. Then it follows easily that q is a lower semi continuous quadratic form invariant under the unitary group $\mathcal{U}(\mathcal{M}')$ of $\mathcal{M}'$, cf. A.7–A.8. Let $\mathfrak{K}$ be the closure of $\mathfrak{D}(q)$. Then $\mathfrak{K}$ is affiliated with $\mathcal{M}$ by the invariance of q under $\mathcal{U}(\mathcal{M}')$. Then by A.8, there exists a densely defined self-adjoint positive operator A on $\mathfrak{K}$ such that $\mathfrak{D}\big(A^{1/2}\big) = \mathfrak{D}(q)$ and $\big\|A^{1/2}\xi\big\|^2 = q(\xi)$, $\xi \in \mathfrak{K}$. Once again, the invariance of q implies that A is affiliated with $\mathcal{M}$. Q.E.D.

We say that an element $m \in \widehat{\mathcal{M}}_+$ is *semi-finite* if $\big\{\varphi \in \mathcal{M}_*^+ : m(\varphi) < +\infty\big\}$ is dense in $\mathcal{M}_*^+$; *faithful* if $m(\varphi) > 0$ for every non-zero $\varphi \in \mathcal{M}_*^+$.

Theorem 4.8. *Let $\mathcal{M}$ be a von Neumann algebra. Each $m \in \widehat{\mathcal{M}}_+$ has a unique spectral decomposition of the form:*

$$m(\varphi) = \int_0^\infty \lambda \, \mathrm{d}\varphi\big(e(\lambda)\big) + \infty\, \varphi(p), \qquad \varphi \in \mathcal{M}_*^+, \tag{19}$$

where $\{e(\lambda) : \lambda \in \mathbf{R}_+\}$ *is an increasing family of projections in* $\mathcal{M}$ *which is* σ*-strongly continuous from the right, and* $p = 1 - \lim_{\lambda\to\infty} e(\lambda)$. *Furthermore,* $e(0) = 0$ *if and only if* m *is faithful, and* $p = 0$ *if and only if* m *is semi-finite.*

PROOF: Representing $\mathcal{M}$ on a Hilbert space $\mathfrak{H}$, we apply Lemma 4.7 to m, so that we obtain the pair $\{A, \mathfrak{K}\}$ of the lemma. Let $A = \int_0^{+\infty} \lambda \, de(\lambda)$ be the spectral decomposition of A and let p be the projection of $\mathfrak{H}$ onto $\mathfrak{K}^\perp$. Since $e(\lambda) \in \mathcal{M}_{\mathfrak{K}}$, $e(\lambda)$ may be regarded as a projection of $\mathcal{M}$. Now we have

$$A^{\frac{1}{2}} = \int_0^\infty \sqrt{\lambda}\, de(\lambda),$$

$$\mathfrak{D}\big(A^{\frac{1}{2}}\big) = \left\{\xi \in \mathfrak{K} : \int_0^\infty \lambda \, \mathrm{d}\big(e(\lambda)\xi \mid \xi\big) < \infty\right\},$$

$$\|A^{\frac{1}{2}}\xi\|^2 = \int_0^\infty \lambda \, \mathrm{d}\big(e(\lambda)\xi \mid \xi\big), \qquad \xi \in \mathfrak{D}\big(A^{\frac{1}{2}}\big).$$

Hence, we get

$$m(\omega_\xi) = \int_0^\infty \lambda \, \mathrm{d}\big(e(\lambda)\xi \mid \xi\big), \qquad \xi \in \mathfrak{K},$$

and if $p\xi \neq 0$, then $m(\omega_\xi) = +\infty$, so that we obtain the formula:

$$m(\omega_\xi) = \int_0^\infty \lambda \, \mathrm{d}\big(e(\lambda)\xi \mid \xi\big) + \infty\, (p\xi \mid \xi), \qquad \xi \in \mathfrak{H}.$$

If we choose the semi-cyclic representation induced by a faithful semi-finite normal weight for $\{\mathcal{M}, \mathfrak{H}\}$, then every $\varphi \in \mathcal{M}_*^+$ is a vectorial functional by Theorem VIII.3.2. Thus the formula (19) follows.

The uniqueness of $\{e(\lambda)\}$ follows from the uniqueness of $\{A, \mathfrak{K}\}$ and the uniqueness of the spectral decomposition of A.

Now, we have the implications:

$$e(0) = 0 \iff A \text{ is nonsingular} \iff m(\omega_\xi) > 0, \quad \xi \neq 0.$$

If $p = 0$, then $\mathfrak{K} = \mathfrak{H}$; so $\mathfrak{D}\big(A^{1/2}\big)$ is dense in $\mathfrak{H}$; hence $\{\varphi : m(\varphi) < +\infty\}$ is dense in $\mathcal{M}_*^+$, which means the semi-finiteness of m. If $p \neq 0$, then $m(\varphi) = \infty$ for every non-zero $\varphi \in p\mathcal{M}_*^+ p$. But we have

$$\{\psi \in \mathcal{M}_*^+ : m(\psi) < +\infty\} \subset (1-p)\mathcal{M}_*^+(1-p),$$

so that m is not semi-finite. Q.E.D.

To simplify the notation, we write

$$m = h + \infty p, \qquad h = \int_0^\infty \lambda \, de(\lambda), \tag{19'}$$

when m has the form of (19). We keep the convention: $0 \cdot (+\infty) = 0$. Although we consider h as an operator affiliated with $\mathcal{M}$, we use the following abuse notation:

$$\mathfrak{D}(h^{\frac{1}{2}}) = \{\xi \in \mathfrak{H} : m(\omega_\xi) < +\infty\},$$

as long as the circumstance allows us to do this.

Corollary 4.9. *Any normal weight φ on $\mathcal{M}$ has a unique extension, denoted by φ again, to $\widehat{\mathcal{M}}_+$ such that*

$$\varphi(\lambda m) = \lambda \varphi(m), \qquad \lambda \geq 0, \quad m \in \widehat{\mathcal{M}}_+;$$

$$\varphi(m + n) = \varphi(m) + \varphi(n), \quad m, n \in \widehat{\mathcal{M}}_+;$$

$$\varphi(\sup m_i) = \sup \varphi(m_i)$$

for any increasing net $\{m_i\}$ in $\widehat{\mathcal{M}}_+$ in the sense that $\{m_i(\omega)\}$ is increasing for every $\omega \in \mathcal{M}_*^+$.

We leave the proof to the reader as an exercise.

We observed that the extended positive cone $\widehat{\mathcal{M}}_+$ of $\mathcal{M}$ is nothing but the set of pairs $(A, \mathfrak{K})$ of closed subspaces $\mathfrak{K}$ and positive self-adjoint operators A, both of which are affiliated with $\mathcal{M}$. One of the main advantages of $\widehat{\mathcal{M}}_+$ over the collection of pairs $\{A, \mathfrak{K}\}$ is that one can freely add elements in $\widehat{\mathcal{M}}_+$ without worrying about domains. Of course, the sum of two semi-finite elements of $\widehat{\mathcal{M}}_+$ need not be semi-finite. In the theory of unbounded operators, there are several ways to define the "sum" of two positive self-adjoint operators A and B. When we define $A \dotplus B$ as the operator corresponding to $m_A + m_B$, the operator $A \dotplus B$ is called the *form sum* of A and B, which is not, in general, the closure of the algebraic sum $A + B$. It is completely possible that $A \dotplus B$ has no definition domain other than just $\{0\}$.

Now, for each $m = h + \infty p$, we put

$$m_0 = (1 + h)^{-1}(1 - p), \tag{20}$$

$$m_\varepsilon = h\,(1 + \varepsilon h)^{-1}(1 - p) + \frac{1}{\varepsilon} p, \quad \varepsilon > 0. \tag{21}$$

We notice that both m_0 and m_ε are bounded.

Lemma 4.10.

(i) *For each $m, n \in \widehat{\mathcal{M}}_+$, we have the following equivalence:*

$$m \leq n \iff m_0 \geq n_0 \iff m_\varepsilon \leq n_\varepsilon, \quad \varepsilon > 0. \tag{22}$$

(ii) *Let* $\{m_i\} = \{h_i + \infty p_i\}$ *be an increasing net in* $\widehat{\mathcal{M}}_+$ *and* $m = h + \infty p$. *Then we have*

$$m_i \nearrow m \iff (m_i)_0 \searrow m_0 \iff (m_i)_\varepsilon \nearrow m_\varepsilon, \quad \varepsilon > 0. \tag{23}$$

PROOF:

(i) Suppose $m \leq n$. Let $m = h + \infty p$ and $n = k + \infty q$. By assumption, we have $p \leq q$. Let $\mathfrak{K} = (1-p)\mathfrak{H}$ and $\mathfrak{M} = (1-q)\mathfrak{H}$, where $\mathfrak{H}$ means of course the underlying Hilbert space of $\mathcal{M}$. For any $\xi \in \mathfrak{M}$, set $\eta = (\varepsilon + h)^{-1}\xi \in \mathfrak{D}(h)$ and $\zeta = (\varepsilon + k)^{-1}\xi \in \mathfrak{D}(k)$. We then have

$$\begin{aligned}
\big((\varepsilon + k)^{-1}\xi \mid \xi\big)^2 &= \big(\xi \mid (\varepsilon + k)^{-1}\xi\big)^2 = \big((\varepsilon + h)\eta \mid \zeta\big)^2 \\
&= \big((\varepsilon + h)^{\frac{1}{2}}\eta \mid (\varepsilon + h)^{\frac{1}{2}}\zeta\big)^2 \\
&\qquad \Big(\zeta \in \mathfrak{D}(k) \subset \mathfrak{D}\big(k^{\frac{1}{2}}\big) \subset \mathfrak{D}\big(h^{\frac{1}{2}}\big) = \mathfrak{D}\big((\varepsilon + h)^{\frac{1}{2}}\big)\Big) \\
&\leq \big\|(\varepsilon + h)^{\frac{1}{2}}\eta\big\|^2 \big\|(\varepsilon + h)^{\frac{1}{2}}\zeta\big\|^2 \\
&= \big((\varepsilon + h)\eta \mid \eta\big)\big((\varepsilon + h)\zeta \mid \zeta\big) \\
&\leq \big((\varepsilon + h)\eta \mid \eta\big)\big((\varepsilon + k)\zeta \mid \zeta\big) \\
&= \big(\xi \mid (\varepsilon + h)^{-1}\xi\big)\big(\xi \mid (\varepsilon + k)^{-1}\xi\big),
\end{aligned}$$

so that

$$\big((\varepsilon + k)^{-1}\xi \mid \xi\big) \leq \big((\varepsilon + h)^{-1}\xi \mid \xi\big), \quad \xi \in \mathfrak{M}.$$

Hence we get $n_0(\omega_\xi) \leq m_0(\omega_\xi)$ for any $\xi \in \mathfrak{M}$ by setting $\varepsilon = 1$. If $\xi \in \mathfrak{M}^\perp$, then $n_0(\omega_\xi) = 0$. Thus we conclude $n_0 \leq m_0$.

Suppose $n_0 \leq m_0$. It follows that $p \leq q$. Let $\mathfrak{K}$ and $\mathfrak{M}$ be as before. The assumption means that

$$\big((1 + h)^{-1}\xi \mid \xi\big) \geq \big((1 + k)^{-1}\xi \mid \xi\big), \quad \xi \in \mathfrak{H}.$$

Setting $(1 + k)^{-1}\xi = 0$ for $\xi \in \mathfrak{M}^\perp \cap \mathfrak{K}$, we view $(1 + k)^{-1}$ as an operator on $\mathfrak{K}$ and have $(1 + h)^{-1} \geq (1 + k)^{-1}$. Then we have $(1 + h)^{-1/2} \geq (1 + k)^{-1/2}$ (see Proposition I.6.3), and

$$\mathfrak{D}\big(h^{\frac{1}{2}}\big) = (1 + h)^{-\frac{1}{2}}\mathfrak{K} \supset (1 + k)^{\frac{1}{2}}\mathfrak{K} = \mathfrak{D}\big(k^{\frac{1}{2}}\big).$$

The argument in the first paragraph shows

$$\begin{aligned}
(1 + h)\big(1 + \varepsilon(1 + h)^{-1}\big)^{-1} &= \big(\varepsilon + (1 + h)^{-1}\big)^{-1} \\
&\leq \big(\varepsilon + (1 + k)^{-1}\big)^{-1} = (1 + k)\big(1 + \varepsilon(1 + k)^{-1}\big).
\end{aligned}$$

If $\xi \in \mathfrak{D}(k^{1/2})$, we have

$$\left\|(1+k)^{\frac{1}{2}}\xi\right\|^2 = \lim_{\varepsilon\to 0}\left\|(1+k)^{\frac{1}{2}}\left(1+\varepsilon(1+k)^{-1}\right)^{-\frac{1}{2}}\xi\right\|^2$$

$$\geq \lim_{\varepsilon\to 0}\left\|(1+h)^{\frac{1}{2}}\left(1+\varepsilon(1+h)^{-1}\right)^{-\frac{1}{2}}\xi\right\|^2 = \left\|(1+h)^{\frac{1}{2}}\xi\right\|^2,$$

so that we conclude $1+n \geq 1+m$, equivalently $n \geq m$.

Now we have seen the equivalence: $m \leq n \iff m_0 \geq n_0$. For a fixed $\varepsilon > 0$, we have then

$$m \leq n \iff \varepsilon m \leq \varepsilon n \iff (\varepsilon m)_0 \geq (\varepsilon n)_0$$

$$\iff 1-(\varepsilon m)_0 \leq 1-(\varepsilon n)_0$$

$$\iff m_\varepsilon = \frac{1}{\varepsilon}\left(1-(\varepsilon m)_0\right) \leq \frac{1}{\varepsilon}\left(1-(\varepsilon n)_0\right) = n_\varepsilon.$$

(ii) By (i), the net $\{(m_i)_0\}$ is decreasing. If $\ell = \inf(m_i)_0$, then there exists $n \in \widehat{\mathcal{M}}_+$ such that $n_0 = \ell$ because $(m_i)_0 \leq 1$ implies $\ell \leq 1$. If $m = \sup m_i$, then we have $m_0 \leq (m_i)_0$, so $m_0 \leq n_0$, which implies $n \leq m$ by (i). Hence $m_0 = \inf(m_i)_0 = \lim(m_i)_0$. Thus we proved the equivalence: $m_i \nearrow m \iff (m_i)_0 \searrow m_0$. Finally, the equality:

$$m_\varepsilon = \frac{1}{\varepsilon}\left(1-(\varepsilon m)_0\right) \tag{24}$$

shows the remaining equivalence. Q.E.D.

Proposition 4.11. *Let φ be a faithful semi-finite normal weight on $\mathcal{M}$, and set $\mathcal{N} = \mathcal{M}_\varphi$. For each $m \in \widehat{\mathcal{N}}_+$, set*

$$\varphi_m(x) = \lim_{\varepsilon\to 0}\varphi_{m_\varepsilon}(x), \quad x \in \mathcal{M}_+. \tag{25}$$

Then the map: $m \in \widehat{\mathcal{N}}_+ \mapsto \varphi_m$ is an order preserving bijection from $\widehat{\mathcal{N}}_+$ onto the set of all $\{\sigma_t^\varphi\}$-invariant, not necessarily faithful nor semi-finite, normal weights on $\mathcal{M}$. Furthermore, we have

$$m_i \nearrow m \ \text{ in } \widehat{\mathcal{N}}_+ \iff \varphi_{m_i} \nearrow \varphi_m \ \text{ pointwise on } \mathcal{M}_+.$$

PROOF: For a fixed $x \in \mathcal{M}_+$, we define a normal weight φ_x on $\mathcal{N}$ by:

$$\varphi_x(a) = \varphi\left(a^{\frac{1}{2}}xa^{\frac{1}{2}}\right), \quad a \in \mathcal{N}_+. \tag{26}$$

If we prove the additivity of φ_x, then the normality follows from that of φ. Let $a, b \in \mathcal{N}_+$ and $c = a+b$. Choose $s, t \in \mathcal{N}$ as usual so that $a^{1/2} = sc^{1/2}$, $b^{1/2} = tc^{1/2}$ and $s^*s + t^*t$ is the range projection of c. If $\varphi_x(c) < +\infty$, then $y = c^{1/2}xc^{1/2} \in \mathfrak{m}_\varphi$; so sys^* and tyt^* both belong to $\mathfrak{m}_\varphi$ by Lemma VIII.2.4.(ii), and we get

$$\varphi(sys^*) + \varphi(tyt^*) = \varphi(ys^*s) + \varphi(yt^*t) = \varphi\left(y(s^*s+t^*t)\right) = \varphi(y) = \varphi_x(c);$$

$$\varphi(sys^*) = \varphi\big(a^{\frac{1}{2}}xa^{\frac{1}{2}}\big) = \varphi_x(a); \qquad \varphi(tyt^*) = \varphi\big(b^{\frac{1}{2}}xb^{\frac{1}{2}}\big) = \varphi_x(b).$$

Thus, $\varphi_x(a) + \varphi_x(b) = \varphi_x(c)$. Now, we have

$$s = \lim_{\varepsilon \to 0} a^{\frac{1}{2}}(c+\varepsilon)^{-\frac{1}{2}}, \qquad t = \lim_{\varepsilon \to 0} b^{\frac{1}{2}}(c+\varepsilon)^{-\frac{1}{2}},$$

so that

$$\begin{aligned} c^{\frac{1}{2}}xc^{\frac{1}{2}} &= \lim_{\varepsilon \to 0}(c+\varepsilon)^{-\frac{1}{2}}cxc\,(c+\varepsilon)^{-\frac{1}{2}} \\ &= \lim_{\varepsilon \to 0}(c+\varepsilon)^{-\frac{1}{2}}(a+b)x\,(a+b)(c+\varepsilon)^{-\frac{1}{2}} \\ &\leq 2\lim_{\varepsilon \to 0}(c+\varepsilon)^{-\frac{1}{2}}\big[axa+bxb\big](c+\varepsilon)^{-\frac{1}{2}} \\ &= 2\Big(s^*a^{\frac{1}{2}}xa^{\frac{1}{2}}s + t^*b^{\frac{1}{2}}xb^{\frac{1}{2}}t\Big). \end{aligned}$$

Therefore, if $\varphi_x(a) < +\infty$ and $\varphi_x(b) < +\infty$, then $\varphi_x(c) < +\infty$; hence $\varphi_x(a+b) = \varphi_x(a) + \varphi_x(b)$ for $a, b \in \mathcal{N}_+$.

Now, for each $m \in \widehat{\mathcal{N}}_+$, we set

$$\varphi_m(x) = \varphi_x(m), \qquad x \in \mathcal{M}_+. \tag{27}$$

This makes sense by Corollary 4.9. First, by Corollary 4.9, we have

$$\varphi_x\big(\sup m_i\big) = \sup \varphi_x(m_i) \quad \text{if} \quad m_i \nearrow m.$$

Hence we have

$$\varphi_m(x) = \varphi_x(m) = \lim_{\varepsilon \to 0} \varphi_x(m_\varepsilon) = \lim_{\varepsilon \to 0} \varphi_{m_\varepsilon}(x).$$

If $\{x_i\} \nearrow x$ in $\mathcal{M}_+$, then

$$\begin{aligned} \varphi_x(m) &= \sup_{\varepsilon > 0} \varphi_{m_\varepsilon}(x) = \sup_{\varepsilon > 0} \sup_i \varphi_{m_\varepsilon}(x_i) \\ &= \sup_i \sup_{\varepsilon > 0} \varphi_{m_\varepsilon}(x_i) = \sup_i \varphi_m(x_i), \end{aligned}$$

so that φ_m is normal. The additivity of φ_m follows from the convergence (25).

The invariance of φ_m under $\{\sigma_t^\varphi\}$ follows from that of φ_{m_ε}. The rest is now easy.

Q.E.D.

Definition 4.12. Let $\mathcal{M}$ be a von Neumann algebra and $\mathcal{N}$ a von Neumann subalgebra of $\mathcal{M}$. An *operator valued weight* from $\mathcal{M}$ to $\mathcal{N}$ is a map $T\colon \mathcal{M}_+ \mapsto \widehat{\mathcal{N}}_+$ which satisfies the following conditions:

$$\left.\begin{aligned} &\text{(a)} \quad T(\lambda x) = \lambda T(x), && \lambda \geq 0, \quad x \in \mathcal{M}_+ \\ &\text{(b)} \quad T(x+y) = T(x) + T(y), && x, y \in \mathcal{M}_+ \\ &\text{(c)} \quad T(a^*xa) = a^*T(x)a, && x \in \mathcal{M}_+, \quad a \in \mathcal{N}. \end{aligned}\right\} \tag{28}$$

We say that T is *normal* if

$$\text{(d)}\quad T(x_i) \nearrow T(x) \quad \text{whenever} \quad x_i \nearrow x, \quad x_i, x \in \mathcal{M}_+.$$

As in the case of ordinary weights, we set

$$\left.\begin{aligned} \mathfrak{n}_T &= \left\{ x \in \mathcal{M} : \left\| T(x^*x) \right\| < +\infty \right\} \\ \mathfrak{m}_T &= \mathfrak{n}_T^* \mathfrak{n}_T = \left\{ \sum_{i=1}^{n} y_i^* x_i : x_1, \dots, x_n,\ y_1, \dots, y_n \in \mathfrak{n}_T \right\}. \end{aligned}\right\} \tag{29}$$

By now, the next lemma is an easy exercise:

Lemma 4.13.
(i) $\mathfrak{m}_T$ *is spanned by its positive part:*

$$\mathfrak{m}_T^+ = \left\{ x \in \mathcal{M}_+ : \left\| T(x) \right\| < +\infty \right\}.$$

(ii) $\mathfrak{m}_T$ *and* $\mathfrak{n}_T$ *are two sided modules over* $\mathcal{N}$.
(iii) T *has a unique linear extension* $\dot{T} : \mathfrak{m}_T \mapsto \mathcal{N}$, *which enjoys the module map property:*

$$\dot{T}(axb) = a\dot{T}(x)b, \quad a, b \in \mathcal{N}, \quad x \in \mathfrak{m}_T. \tag{30}$$

In particular, if $T(1) = 1$, *then* T *is a projection of norm one from* $\mathcal{M}$ *onto* $\mathcal{N}$.

In the sequel, we shall not distinguish T and $\dot{T}$ unless we need to.

Definition 4.14. We say that T is *semi-finite* if $\mathfrak{n}_T$ is σ-weakly dense in $\mathcal{M}$; *faithful* if $T(x^*x) \neq 0$ for $x \neq 0$. We denote by $\mathfrak{W}(\mathcal{M}, \mathcal{N})$ (resp. $\mathfrak{W}_0(\mathcal{M}, \mathcal{N})$) the set of (resp. faithful semi-finite) normal operator valued weights from $\mathcal{M}$ to $\mathcal{N}$. In the case that $\mathcal{N} = \mathbf{C}$, we write $\mathfrak{W}(\mathcal{M})$ (resp. $\mathfrak{W}_0(\mathcal{M})$) for ($\mathfrak{W}(\mathcal{M}, \mathbf{C})$ (resp. $\mathfrak{W}_0(\mathcal{M}, \mathbf{C})$).

Remark 4.15. If $T : \mathcal{M}_+ \mapsto \widehat{\mathcal{N}}_+$ is a normal operator valued weight, it can be naturally extended to a normal "linear" map from $\widehat{\mathcal{M}}_+ \mapsto \widehat{\mathcal{N}}_+$. Therefore, if $\mathcal{P} \subset \mathcal{N} \subset \mathcal{M}$ and if $T \in \mathfrak{W}(\mathcal{M}, \mathcal{N})$ and $S \in \mathfrak{W}(\mathcal{N}, \mathcal{P})$ then we have $S \circ T \in \mathfrak{W}(\mathcal{M}, \mathcal{P})$.

Proposition 4.16. *If* $\mathcal{P} \subset \mathcal{N} \subset \mathcal{M}$ *are von Neumann subalgebras and if* $T \in \mathfrak{W}_0(\mathcal{M}, \mathcal{N})$ *and* $S \in \mathfrak{W}_0(\mathcal{N}, \mathcal{P})$, *then* $S \circ T \in \mathfrak{W}_0(\mathcal{M}, \mathcal{P})$.

PROOF: The only non-trivial part is the semi-finiteness of $S \circ T$. If $x \in \mathfrak{n}_T$, then $T(x^*x) \in \mathcal{N}_+$. Choose a net $\{a_i\}$ in $\mathfrak{n}_S$ such that $a_i \to 1$ σ-strongly. Then we have

$$S \circ T\big(a_i^* x^* x a_i\big) = S\big(a_i^* T(x^*x) a_i\big) \le \left\| T(x^*x) \right\| S(a_i^* a_i),$$

so that $xa_i \in \mathfrak{n}_{S \circ T}$. Hence $\mathfrak{n}_{S \circ T}$ is σ-strongly dense in $\mathcal{M}$ because $\mathfrak{n}_T$ is. Q.E.D.

Proposition 4.17. *If $\mathcal{N} \subset \mathcal{M}$ are von Neumann algebras and if $T \in \mathfrak{W}_0(\mathcal{M}, \mathcal{N})$, then*

(i) $\dot{T}(\mathfrak{m}_T)$ *is a* σ*-weakly dense ideal of* $\mathcal{N}$;
(ii) *After extending* T *to the map of* $\hat{\mathcal{M}}_+$ *to* $\hat{\mathcal{N}}_+$, *as in Remark* 4.15,

$$T(\hat{\mathcal{M}}_+) = \hat{\mathcal{N}}_+ .$$

PROOF:

(i) From the module map property of T:

$$\dot{T}(axb) = a\dot{T}(x)b, \quad a, b \in \mathcal{N}, \quad x \in \mathfrak{m}_T,$$

it follows that $\dot{T}(\mathfrak{m}_T)$ is an ideal of $\mathcal{N}$. Let z denote the greatest projection of the σ-weak closure of $\dot{T}(\mathfrak{m}_T)$, which is central in $\mathcal{N}$. Assume $z \neq 1$. As $\mathfrak{n}_T$ is σ-weakly dense in $\mathcal{M}$, there exists $x \in \mathfrak{n}_T$ with $x(1-z) \neq 0$, so that $(1-z)x^*x(1-z) \in \mathfrak{m}_T \setminus \{0\}$ and as T is faithful

$$0 \neq T\big((1-z)x^*x(1-z)\big) = (1-z)T(x^*x)(1-z) = 0,$$

which is a contradiction. Hence $z = 1$, which means that $\dot{T}(\mathfrak{m}_T)$ is σ-weakly dense in $\mathcal{N}$.

(ii) Let $b \in \dot{T}(\mathfrak{m}_T)_+$. Then b is of the form $b = \dot{T}(h),\ h \in \mathfrak{m}_T$. Replacing h by $\frac{1}{2}(h+h^*)$, h can be chosen to be self-adjoint. As $b \leq T\big(|h|\big)$, we can find $s \in \mathcal{N}$ such that $b = sT\big(|h|\big)s^*$. With $a = s|h|s^*$, we have $b = T(a),\ a \in \mathfrak{m}_T^+$. Hence $\dot{T}(\mathfrak{m}_T)_+ = \dot{T}(\mathfrak{m}_T^+)$.

Let $\{b_i\}_{i\in I}$ be a maximal family in the ideal $\dot{T}(\mathfrak{m}_T)_+$ such that $\sum_{i\in I} b_i \leq 1$. The maximality and Proposition II.3.13 entail $1 = \sum_{i\in I} b_i$ in the σ-strong topology. Every $y \in \mathcal{N}_+$ is then of the form: $y = \sum_{i\in I} y^{1/2} b_i y^{1/2}$, so that it is of the form: $y = \sum_{i\in I} Tx_i$ with $\{x_i\} \subset \mathfrak{m}_T^+$.

Let $z \in \hat{\mathcal{N}}_+$. From Theorem 4.8, it follows that there exists a sequence $y_n \in \mathcal{N}_+$ such that $y_n \nearrow z$. Set $z_1 = y_1$ and $z_n = y_n - y_{n-1},\ n \geq 2$. Then we have $z = \sum_{n=1}^{\infty} y_n$. Each y_n is of the form: $y_n = \sum Tx_{n,i}$ with $\{x_{n,i}\} \subset \mathfrak{m}_T^+$. Hence we have

$$z = T\left(\sum_{n=1}^{\infty}\sum_i x_{n,i}\right),$$

where T means the extended one to $\hat{\mathcal{M}}_+$. Thus T maps $\hat{\mathcal{M}}_+$ onto $\hat{\mathcal{N}}_+$. Q.E.D.

The next theorem generalizes Theorem 4.2.

Theorem 4.18. *Let $\mathcal{M} \supset \mathcal{N}$ be von Neumann algebras. There exists a faithful semi-finite normal operator valued weight $T: \mathcal{M}_+ \mapsto \hat{\mathcal{N}}_+$ if and only if there exist faithful semi-finite normal weights $\tilde{\varphi}$ on $\mathcal{M}$ and φ on $\mathcal{N}$ such that*

$$\sigma_t^{\varphi}(x) = \sigma_t^{\tilde{\varphi}}(x), \quad x \in \mathcal{N}. \tag{31}$$

If this condition is satisfied, then T can be chosen such a way that $\tilde{\varphi} = \varphi \circ T$, and T is uniquely determined by this identity.

Lemma 4.19. *Let* $m\colon \mathcal{M}_*^+ \mapsto [0, +\infty]$ *be an extended positive real valued function such that*

$$m(\lambda\varphi) = \lambda m(\varphi), \qquad \lambda \geq 0, \quad \varphi \in \mathcal{M}_*^+;$$

$$m(\varphi + \psi) = m(\varphi) + m(\psi), \qquad \varphi, \psi \in \mathcal{M}_*^+$$

is lower semi-continuous; hence a member of the extended positive cone $\hat{\mathcal{M}}_+$ *if and only if* m *is countably additive in the sense that*

$$m\left(\sum_{n=1}^{\infty} \varphi_n\right) = \sum_{n=1}^{\infty} m(\varphi_n)$$

for every $\{\varphi_n\} \subset \mathcal{M}_*^+$ *with* $\sum_{n=1}^{\infty} \|\varphi_n\| < +\infty$.

PROOF: The "only part" is trivial, so we prove the only "if" part. Suppose m is countably additive. Represent $\mathcal{M}$ on a Hilbert space $\mathfrak{H}$ and consider $\bar{m}\colon \omega \in \mathcal{L}(\mathfrak{H})_*^+ \mapsto m(\omega|_{\mathcal{M}}) \in [0, +\infty]$. We claim first that m is lower semi continuous if $\bar{m}$ is. Suppose $\bar{m}$ is lower semi-continuous. Then by Lemma 4.7, there exists a unique pair $\{A, \mathfrak{K}\}$ of a closed subspace $\mathfrak{K}$ of $\mathfrak{H}$ and a positive self-adjoint operator A on $\mathfrak{K}$ such that (18) holds. As $u\omega u^*\big|_{\mathcal{M}} = \omega|_{\mathcal{M}}$ for every $u \in \mathcal{U}(\mathcal{M}')$, $u^*\bar{m}u = \bar{m}$, so that $\{A, \mathfrak{K}\}$ is affiliated to $\mathcal{M}$, and $m = m_A$. Thus, m is a member of $\hat{\mathcal{M}}_+$. Therefore, it suffices to prove the lemma for $\mathcal{M} = \mathcal{L}(\mathfrak{H})$. Replacing m by m' defined by $m'(\varphi) = m(\varphi) + \varphi(1)$, we may assume $m(\varphi) \geq \|\varphi\|$, $\varphi \in \mathcal{M}_*^+$. As $\mathcal{M}_* = L^1(\mathcal{M}, \mathrm{Tr})$ is an ideal of $\mathcal{M}$, we can define a map $\varphi\colon x \in \mathcal{M}_+ \mapsto [0, +\infty]$ by

$$\varphi(x) = \begin{cases} m(\omega_x), & x \in L^1(\mathcal{M}, \mathrm{Tr})_+; \\ +\infty & \text{otherwise,} \end{cases}$$

where $\omega_x(a) = \mathrm{Tr}(ax)$, $a \in \mathcal{M}$, $x \in L^1(\mathcal{M}, \mathrm{Tr})$. Then φ is a weight on $\mathcal{M}$, and $\omega \geq \mathrm{Tr}$. Let $\{x_i\}_{i \in I}$ be a family of positive operators with $x = \sum x_i \in \mathcal{M}_+$. If $\mathrm{Tr}(x) = +\infty$, then $\sum_i \mathrm{Tr}(x_i) = \infty$ and

$$\infty = \sum \mathrm{Tr}(x_i) \leq \sum \omega(x_i) \leq \omega(x),$$

so that $\omega(x) = \infty = \sum \omega(x_i)$. If $\mathrm{Tr}(x) < +\infty$, then $\sum \mathrm{Tr}(x_i) < +\infty$, so that $x_i \neq 0$ for at most countably infinite i's. Hence we have

$$\sum_{i \in I} \varphi(x_i) = \sum_i m(\omega_{x_i}) = m\left(\sum \omega_{x_i}\right) = m(\omega_x) = \varphi(x).$$

Thus, φ is a completely additive weight on $\mathcal{M}$. By Theorem VII.1.11, φ is normal. If $\{x_n\}$ is a sequence in $L^1(\mathcal{M}, \mathrm{Tr})_+$ with $\lim_{n\to\infty} \|x - x_n\|_1 = 0$, then $\{x_n\}$ converges to x σ-weakly, so that

$$m(\omega_x) = \varphi(x) \leq \liminf_{n\to\infty} \varphi(x_n) = \liminf_{n\to\infty} m(\omega_{x_n}).$$

Hence m is lower semi-continuous. Q.E.D.

Lemma 4.20. *Let* $\varphi \in \mathfrak{W}(\mathcal{M})$ *and* $\{a_n\}$ *be a sequence in* $\mathcal{M}_{s(\varphi)}$. *The following two conditions are equivalent:*

(i) $\varphi(x) = \sum_{n=1}^{\infty} \varphi(a_n x a_n^*), \quad x \in \mathcal{M}_+ ;$ (32)

(ii) $\{a_n\} \subset \mathfrak{D}\big(\sigma^{\varphi}_{-i/2}\big)$ *and* $\sum_{n=1}^{\infty} \sigma^{\varphi}_{-i/2}(a_n)^* \, \sigma^{\varphi}_{-i/2}(a_n) = s(\varphi)$. (33)

PROOF: As we can view the problem in the reduced algebra $\mathcal{M}_{s(\varphi)}$, we may assume that φ is faithful.

(i) $\Longrightarrow$ (ii): By Theorem VIII.3.17, each a_n belongs to $\mathfrak{D}\big(\sigma^{\varphi}_{-i/2}\big)$ and satisfies the inequality $\big\|\sigma^{\varphi}_{-i/2}(a_n)\big\| \leq 1$. We then have, for each $x \in \mathfrak{n}_{\varphi}$,

$$\begin{aligned}\|\eta_{\varphi}(x)\|^2 &= \varphi(x^*x) = \sum_{n=1}^{\infty} \varphi\big(a_n x^* x a_n^*\big) = \sum_{n=1}^{\infty} \|\eta_{\varphi}(x a_n^*)\|^2 \\ &= \sum_{n=1}^{\infty} \big\|J\sigma^{\varphi}_{-i/2}(a_n) J\eta_{\varphi}(x)\big\|^2 \qquad \text{by (VIII.3.49)} \\ &= \Big(J\sum_{n=1}^{\infty} \sigma^{\varphi}_{-i/2}(a_n)^* \sigma^{\varphi}_{-i/2}(a_n) J\eta_{\varphi}(x) \,\Big|\, \eta_{\varphi}(x)\Big).\end{aligned}$$

Hence we have

$$1 = \sum_{n=1}^{\infty} \sigma^{\varphi}_{-i/2}(a_n)^* \sigma^{\varphi}_{-i/2}(a_n).$$

(ii) $\Longrightarrow$ (i): This implication follows from the above computation traced backward. Q.E.D.

PROOF OF THEOREM 4.18: Suppose that $\mathcal{M} \supset \mathcal{N}$ are von Neumann algebras, and

$$\sigma_t^{\tilde{\varphi}}(x) = \sigma_t^{\varphi}(x), \qquad x \in \mathcal{N},$$

for some $\tilde{\varphi} \in \mathfrak{W}_0(\mathcal{M})$ and $\varphi \in \mathfrak{W}_0(\mathcal{N})$. We fix $\tilde{\varphi}$ and φ. For each $\psi \in \mathfrak{W}(\mathcal{N})$, we have the cocycle derivative $u_t = (\mathrm{D}\psi : \mathrm{D}\varphi)_t \in \mathcal{N}$, $t \in \mathbf{R}$. By Theorem VIII.3.21, there corresponds $\tilde{\psi} \in \mathfrak{W}(\mathcal{M})$ such that $s(\tilde{\psi}) = s(\psi)$ and $(\mathrm{D}\tilde{\psi} : \mathrm{D}\tilde{\varphi})_t = u_t = (\mathrm{D}\psi : \mathrm{D}\varphi)_t$. As we have

$$\big(\mathrm{D}(\lambda\psi) : \mathrm{D}\varphi\big)_t = \lambda^{it}\big(\mathrm{D}\psi : \mathrm{D}\varphi\big)_t = \lambda^{it}\big(\mathrm{D}\tilde{\psi} : \mathrm{D}\tilde{\varphi}\big)_t = \big(\mathrm{D}(\lambda\tilde{\psi}) : \mathrm{D}\tilde{\varphi}\big)_t,$$

we obtain $\widetilde{\lambda\psi} = \lambda\tilde{\psi}$, $\lambda > 0$.

By the chain rule for cocycle derivatives, we have $(\mathrm{D}\tilde{\psi}_1 : \mathrm{D}\tilde{\psi}_2)_t = (\mathrm{D}\psi_1 : \mathrm{D}\psi_2)_t$ for any $\psi_1 \in \mathfrak{W}(\mathcal{N})$ and $\psi_2 \in \mathfrak{W}_0(\mathcal{N})$. Let $\{\psi_n\}$ be a sequence in $\mathcal{N}_*^+$ such that $\psi = \sum_{n=1}^{\infty} \psi_n \in \mathcal{N}_*^+$. Let $u_t^n = (\mathrm{D}\psi_n : \mathrm{D}\psi)_t$ by restricting our consideration to $\mathcal{N}_{s(\psi)}$ and $\mathcal{M}_{s(\tilde{\psi})}$. Then by Theorem VIII.3.17, $u^n_{-i/2}$ is defined and

$$\psi_n(x) = \psi\Big(\big(u^n_{-i/2}\big)^* x \,\big(u^n_{-i/2}\big)\Big), \qquad x \in \mathcal{N}.$$

Therefore, Lemma 4.20 entails that

$$\begin{aligned}
s(\psi) &= \sum_{n=1}^{\infty} \sigma^{\psi}_{-i/2}\Big(\big(u^n_{-i/2}\big)^*\Big)^* \sigma^{\psi}_{-i/2}\Big(\big(u^n_{-i/2}\big)^*\Big) \\
&= \sum_{n=1}^{\infty} \sigma^{\psi}_{i/2}(u^n_{-i/2})\sigma^{\psi}_{i/2}(u^n_{-i/2})^* \\
&= \sum_{n=1}^{\infty} \sigma^{\tilde\psi}_{i/2}((\mathrm{D}\tilde\psi_n : \mathrm{D}\tilde\psi)_{-i/2})\, \sigma^{\tilde\psi}_{i/2}\big((\mathrm{D}\tilde\psi_n : \mathrm{D}\tilde\psi)_{-i/2}\big)^*.
\end{aligned}$$

As $s(\psi) = s(\tilde\psi)$, the above calculation shows that

$$\tilde\psi(x) = \sum_{n=1}^{\infty} \tilde\psi_n(x), \qquad x \in \mathcal{M}_+.$$

Hence the map: $\psi \in \mathcal{N}_*^+ \mapsto \tilde\psi \in \mathfrak{W}(\mathcal{M})$ is homogeneous and countably additive. Hence, the map: $\psi \in \mathcal{N}_*^+ \mapsto \tilde\psi(x) \in [0, +\infty]$, $x \in \mathcal{M}_+$, gives rise to a map $T: \mathcal{M}_+ \mapsto \hat{\mathcal{N}}_+$ by Lemma 4.19.

For every $u \in \mathcal{U}(\mathcal{N})$, we have

$$\begin{aligned}
\big(\mathrm{D}(u\psi u^*)^{\sim} : \mathrm{D}\tilde\varphi\big)_t &= \big(\mathrm{D}(u\psi u^*) : \mathrm{D}\varphi\big)_t = u\,(\mathrm{D}\psi : \mathrm{D}\varphi)_t\, \sigma^{\varphi}_t(u^*) \\
&= u\big(\mathrm{D}\tilde\psi : \mathrm{D}\tilde\varphi\big)_t \sigma^{\tilde\varphi}_t(u^*),
\end{aligned}$$

so that $(u\psi u^*)^{\sim} = u\tilde\psi u^*$, $u \in \mathcal{U}(\mathcal{N})$. Thus, we have $T(uxu^*) = uT(x)u^*$, $x \in \mathcal{M}_+$, $u \in \mathcal{U}(\mathcal{N})$. Hence $\mathfrak{n}_T$ is invariant by the right multiplication of $\mathcal{U}(\mathcal{N})$, and $T(ux) = uT(x)$, $T(xu) = T(x)u$ for every $x \in \mathfrak{m}_T = \mathfrak{n}_T^*\mathfrak{n}_T$ and $u \in \mathcal{U}(\mathcal{N})$. As $\mathcal{U}(\mathcal{N})$ spans $\mathcal{N}$ linearly, $\mathfrak{m}_T$ is a two sided module over $\mathcal{N}$ and $T(axb) = aT(x)b$ for $a, b \in \mathcal{N}$ and $x \in \mathfrak{m}_T$.

We now want to prove

$$T(axa^*) = aT(x)a^*, \quad a \in \mathcal{N}, \quad x \in \mathcal{M}_+. \tag{34}$$

First, we observe that if $a \in \mathcal{N}_\psi$ with ψ faithful, then $\big(\mathrm{D}a^*\psi a : \mathrm{D}\psi\big)_t = \big(a^*a\big)^{it}$, where $\big(a^*a\big)^{it}$ should be considered in the reduced algebra $\mathcal{N}_{s_r(a)}$ by the right support $s_r(a)$ of a, so that

$$\big(\mathrm{D}(a^*\psi a)^{\sim} : \mathrm{D}\tilde\psi\big)_t = \big(\mathrm{D}a^*\psi a : \mathrm{D}\psi\big)_t = \big(a^*a\big)^{it};$$

hence $(a^*\psi a)^{\sim} = a^*\tilde\psi a$. If ψ is not faithful, then we consider $\psi'' = \psi + \psi'$ with $\psi' \in \mathfrak{W}(\mathcal{N})$ such that $s(\psi') = 1 - s(\psi)$; and apply the above argument to ψ to conclude that $(a^*\psi a)^{\sim} = a^*\tilde\psi a$, $a \in \mathcal{N}_\psi$. Now, we apply the above whole argument to $\mathcal{N} \otimes M_2(\mathbf{C})$ and $\mathcal{M} \otimes M_2(\mathbf{C})$ with $\psi \otimes \mathrm{Tr}$ and $\tilde\psi \otimes \mathrm{Tr}$ and observe that

$\sigma_t^{\psi\otimes\mathrm{Tr}}(x) = \sigma_t^{\tilde{\psi}\otimes\mathrm{Tr}}(x)$ for every $x \in \mathcal{N} \otimes M_2(\mathbf{C})$. This implies that there exists a map $S\colon (\mathcal{M} \otimes M_2)_+ \mapsto (\mathcal{N} \otimes M_2)\widehat{_+}$ such that

$$\big(\psi \otimes \mathrm{Tr}\big) \circ S(x) = \big(\tilde{\psi} \otimes \mathrm{Tr}\big)(x), \quad x \in (\mathcal{M} \otimes M_2)_+.$$

As $\mathbf{C} \otimes M_2 \subset \mathcal{N}_\psi$ for any faithful ψ, we have $(\psi \otimes \mathrm{Tr})^\sim = \tilde{\psi} \otimes \mathrm{Tr}$. If $a \in \mathcal{N}$ has $\|a\| \le 1$, then set

$$u = \begin{pmatrix} a & \big(1 - aa^*\big)^{\frac{1}{2}} \\ -\big(1 - a^*a\big)^{\frac{1}{2}} & a^* \end{pmatrix} \in \mathcal{N} \otimes M_2.$$

Then u is unitary. Therefore, we have

$$\big(u^*(\psi \otimes \mathrm{Tr})u\big)^\sim = u^*\big(\tilde{\psi} \otimes \mathrm{Tr}\big)u.$$

With $\{e_{i,j}\}$ the standard matrix unit of M_2, we have

$$\Big[\big(1 \otimes e_{ij}\big)^*\big(\psi \otimes \mathrm{Tr}\big)\big(1 \otimes e_{ij}\big)\Big]^\sim = \big(1 \otimes e_{ij}\big)^*\big(\tilde{\psi} \otimes \mathrm{Tr}\big)\big(1 \otimes e_{ij}\big)$$

as $(1 \otimes e_{ij}) \in (\mathcal{N} \otimes M_2)_{\psi\otimes\mathrm{Tr}}$. Hence we conclude that

$$\Big[\big(1 \otimes e_{11}\big)u^*\big(1 \otimes e_{11}\big)\big(\psi \otimes \mathrm{Tr}\big)\big(1 \otimes e_{11}\big)u\,\big(1 \otimes e_{11}\big)\Big]^\sim$$
$$= \big(a \otimes e_{11}\big)^*\big(\tilde{\psi} \otimes \mathrm{Tr}\big)\big(a \otimes e_{11}\big),$$

i.e.

$$\Big[\big(a^* \otimes e_{11}\big)\big(\psi \otimes \mathrm{Tr}\big)\big(a \otimes e_{11}\big)\Big]^\sim = \big(a^* \otimes e_{11}\big)\big(\tilde{\psi} \otimes \mathrm{Tr}\big)\big(a \otimes e_{11}\big).$$

This means that

$$(a^*\psi a)^\sim = a^*\tilde{\psi}a$$

for every $a \in \mathcal{N}$ with $\|a\| \le 1$, which entails (34). Therefore T is an operator valued weight of $\mathcal{M}_+$ onto $\hat{\mathcal{N}}_+$ such that $\tilde{\psi} = \psi \circ T$, $\psi \in \mathfrak{W}_0(\mathcal{N})$. As $\tilde{\psi}$ is faithful for any $\psi \in \mathfrak{W}_0(\mathcal{N})$, T is faithful.

The semi-finiteness and the uniqueness of T together with the converse follow from the following lemma. Q.E.D.

Lemma 4.21. *Let $\mathcal{N} \subset \mathcal{M}$ be von Neumann algebras and $T\colon \mathcal{M}_+ \to \hat{\mathcal{N}}_+$ be a normal operator valued weight.*

(i) *If $\psi \circ T$ is semi-finite for some $\psi \in \mathfrak{W}_0(\mathcal{N})$, then T is semi-finite;*

(ii) *If $\psi \in \mathfrak{W}_0(\mathcal{N})$ and $S \in \mathfrak{W}(\mathcal{M}, \mathcal{N})$ satisfy $\psi \circ T = \psi \circ S$, then $T = S$;*

(iii) *If $T \in \mathfrak{W}_0(\mathcal{M}, \mathcal{N})$, then*

$$\sigma_t^{\psi\circ T}(x) = \sigma_t^{\psi}(x), \quad x \in \mathcal{N}. \tag{35}$$

PROOF:

(i) Set $\tilde{\psi} = \psi \circ T$. Let $h \in \mathfrak{m}_{\tilde{\psi}}^+$ and

$$T(h) = \int_0^\infty \lambda \, de(\lambda) + \infty p$$

be the spectral decomposition. As $\psi \circ T(h) < +\infty$, and ψ is faithful, we have $p = 0$. Hence $e(\lambda) \nearrow 1$ as $\lambda \nearrow \infty$, so that $e(\lambda)he(\lambda) \to h$ strongly as $\lambda \nearrow \infty$, and $T\big(e(\lambda)he(\lambda)\big) = e(\lambda)T(h)e(\lambda) \in \mathcal{N}_+$; hence $e(\lambda)he(\lambda) \in \mathfrak{m}_T^+$. Therefore $\mathfrak{m}_T^+$ is σ-strongly dense in $\mathcal{M}_+$.

(ii) For any $x \in \mathcal{M}_+$ and $a \in \mathcal{N}$, we have

$$\psi\big(a^* T(x) a\big) = \psi\big(T(a^* x a)\big) = \psi\big(S(a^* x a)\big) = \psi\big(a^* S(x) a\big),$$

so that $a\psi a^* \circ T = a\psi a^* \circ S$ for any $a \in \mathcal{N}$. Therefore if $x \in \mathfrak{m}_T^+ \cup \mathfrak{m}_S^+$, then $T(x) = S(x)$ since $\{a\psi a^* : a \in \mathfrak{n}_\psi\}$ is a dense subset of $\mathcal{N}_*^+$. For a general $x \in \mathcal{M}_+$, consider the spectral decomposition:

$$m = S(x) = \int_0^\infty \lambda \, de(\lambda) + \infty p \in \hat{\mathcal{N}}_+ ;$$

$$n = T(x) = \int_0^\infty \lambda \, df(\lambda) + \infty q \in \hat{\mathcal{N}}_+ .$$

Then we have $e(\lambda)xe(\lambda) \in \mathfrak{m}_S \cap \mathfrak{m}_T$ and

$$\begin{aligned} me(\lambda) &= e(\lambda)S(x)e(\lambda) = S\big(e(\lambda)xe(\lambda)\big) = T\big(e(\lambda)xe(\lambda)\big) \\ &= e(\lambda)T(x)e(\lambda) = e(\lambda)ne(\lambda); \end{aligned}$$

similarly

$$nf(\lambda) = f(\lambda)mf(\lambda), \quad \lambda \geq 0.$$

Hence we have, for every $\xi \in \bigcup_{\lambda \geq 0} e(\lambda)\mathfrak{H}_\psi$,

$$\|m^{\frac{1}{2}}\xi\|^2 = m(\omega_\xi) = n(\omega_\xi) = \|n^{\frac{1}{2}}\xi\|^2;$$

similarly

$$\|m^{\frac{1}{2}}\xi\|^2 = \|n^{\frac{1}{2}}\xi\|^2, \quad \xi \in \bigcup_{\lambda \geq 0} f(\lambda)\mathfrak{H}_\psi .$$

But $m^{1/2}$ (resp. $n^{1/2}$) is essentially self-adjoint on $\bigcup e(\lambda)\mathfrak{H}_\psi$ (resp. $\bigcup f(\lambda)\mathfrak{H}_\psi$), so we get

$$\|m^{\frac{1}{2}}\xi\|^2 = \|n^{\frac{1}{2}}\xi\|^2, \quad \xi \in \mathcal{D}(m^{\frac{1}{2}}) \cup \mathcal{D}(n^{\frac{1}{2}}).$$

Hence there exists a partial isometry $u \in \mathcal{N}$ such that $m^{1/2} = un^{1/2}$ and $n^{1/2} = u^* m^{1/2}$, so that the uniqueness of the polar decomposition implies $m^{1/2} = n^{1/2}$ and $(1-p) = (1-q)$. Thus $m = n$ as elements of $\hat{\mathcal{N}}_+$. Hence $S = T$.

(iii) We prove $\mathcal{G}(\sigma_{-\mathrm{i}}^{\psi}) \subset \mathcal{G}(\sigma_{-\mathrm{i}}^{\psi\circ T})$. Let $\tilde{\psi} = \psi \circ T$, and $(a, b) \in \mathcal{G}(\sigma_{-\mathrm{i}}^{\psi})$. As $a \in \mathcal{D}(\sigma_{-\mathrm{i}/2}^{\psi})$ and $b^* \in \mathcal{D}(\sigma_{-\mathrm{i}/2}^{\psi}) = \mathcal{D}(\sigma_{\mathrm{i}/2}^{\psi})^*$, by Lemma VIII.3.18, there exists $M \geq 0$ such that for every $x \in \mathcal{N}_+$

$$\psi(axa^*) \leq M^2\psi(x), \qquad \psi(b^*xb) \leq M^2\psi(x).$$

Taking increasing limit, we see that the above inequalities are valid for every $x \in \hat{\mathcal{N}}_+$, so that

$$\tilde{\psi}(axa^*) \leq M^2\tilde{\psi}(x), \qquad \tilde{\psi}(b^*xb) \leq M^2\tilde{\psi}(x) \tag{36}$$

for every $x \in \mathcal{M}_+$; therefore $\mathfrak{n}_{\tilde{\psi}}a^* \subset \mathfrak{n}_{\tilde{\psi}}$, $\mathfrak{n}_{\tilde{\psi}}b \subset \mathfrak{n}_{\tilde{\psi}}$ and

$$\left.\begin{aligned} &\|\eta_{\tilde{\psi}}(xa^*)\| \leq M\|\eta_{\tilde{\psi}}(x)\|;\\ &\|\eta_{\tilde{\psi}}(xb)\| \leq M\|\eta_{\tilde{\psi}}(x)\|, \quad x \in \mathfrak{n}_{\tilde{\psi}}. \end{aligned}\right\} \tag{36'}$$

We will prove that $(a, b) \in \mathcal{G}(\sigma_{-\mathrm{i}}^{\tilde{\psi}})$. By Theorem VIII.3.25, it suffices to show

$$\tilde{\psi}(ax) = \tilde{\psi}(xb), \quad x \in \mathfrak{m}_{\tilde{\psi}}. \tag{37}$$

Fix $x_0 = y_0^*z_0$ with $y_0, z_0 \in \mathfrak{n}_{\tilde{\psi}} \cap \mathfrak{n}_T$. As $\mathfrak{n}_{\tilde{\psi}}a^* \subset \mathfrak{n}_{\tilde{\psi}}$ and $\mathfrak{n}_T$ is a right $\mathcal{N}$-module, we have

$$ax_0 = (y_0a^*)^*z_0 \in (\mathfrak{n}_{\tilde{\psi}} \cap \mathfrak{n}_T)^*(\mathfrak{n}_{\tilde{\psi}} \cap \mathfrak{n}_T) \subset \mathfrak{m}_{\tilde{\psi}} \cap \mathfrak{m}_T.$$

Similarly, we get

$$x_0b = y_0^*(z_0b) \in \mathfrak{m}_{\tilde{\psi}} \cap \mathfrak{m}_T.$$

Since $(a, b) \in \mathcal{G}(\sigma_{-\mathrm{i}}^{\psi})$, we have

$$\psi \circ \dot{T}(ax_0) = \psi(a\dot{T}(x_0)) = \psi(\dot{T}(x_0)b) = \psi \circ \dot{T}(x_0b),$$

so that

$$\tilde{\psi}(ax_0) = \tilde{\psi}(x_0b). \tag{38}$$

Now suppose $x = y^*z$ with $y, z \in \mathfrak{n}_{\tilde{\psi}}$. Since $\psi \circ T(y^*y) < +\infty$, we have the spectral decomposition of $T(y^*y)$: $T(y^*y) = \int_0^\infty \lambda\, \mathrm{d}e(\lambda)$. For any $\lambda > 0$, we have $ye(\lambda) \in \mathfrak{n}_T$ and

$$\psi \circ T(e(\lambda)y^*ye(\lambda)) = \psi(e(\lambda)T(y^*y)e(\lambda)) \leq \psi \circ T(y^*y) < +\infty,$$

so that $ye(\lambda) \in \mathfrak{n}_{\tilde{\psi}}$. Furthermore,

$$\begin{aligned} \left\|\eta_{\tilde{\psi}}(ye(\lambda) - y)\right\|^2 &= \psi \circ T\Big((ye(\lambda) - y)^*(ye(\lambda) - y)\Big)\\ &= \psi\Big((1 - e(\lambda))T(y^*y)\Big)(1 - e(\lambda))\Big)\\ &= \psi\Big(\int_\lambda^\infty \mu\, \mathrm{d}e(\mu)\Big) \to 0 \quad \text{as} \quad \lambda \to \infty. \end{aligned}$$

Similarly, with the spectral decomposition of $T(z^*z)$:

$$T(z^*z) = \int_0^\infty \lambda \, \mathrm{d}f(\lambda),$$

we have $zf(\lambda) \in \mathfrak{n}_{\tilde{\psi}} \cap \mathfrak{n}_T$ and

$$\lim_{\lambda\to\infty} \left\| \eta_{\tilde{\psi}}(zf(\lambda) - z) \right\| = 0.$$

By (36′), we get

$$\lim_{\lambda\to\infty} \left\| \eta_{\tilde{\psi}}(ye(\lambda)a^*) - \eta_{\tilde{\psi}}(ya^*) \right\| = 0; \qquad \lim_{\lambda\to\infty} \left\| \eta_{\tilde{\psi}}(zf(\lambda)b) - \eta_{\tilde{\psi}}(zb) \right\| = 0.$$

Therefore, we have, by the previous arguments for x_0,

$$\begin{aligned}
\tilde{\psi}(ax) &= (\eta_{\tilde{\psi}}(z) \mid \eta_{\tilde{\psi}}(ya^*)) = \lim_{\lambda\to\infty} \left(\eta_{\tilde{\psi}}(zf(\lambda)) \,\middle|\, \eta_{\tilde{\psi}}(ye(\lambda)a^*) \right) \\
&= \lim_{\lambda\to\infty} \tilde{\psi}\left(a(ye(\lambda))^*(zf(\lambda)) \right) = \lim_{\lambda\to\infty} \tilde{\psi}\left((ye(\lambda))^*(zf(\lambda))b \right) \\
&= \lim_{\lambda\to\infty} \left(\eta_{\tilde{\psi}}(zf(\lambda)b) \,\middle|\, \eta_{\tilde{\psi}}(ye(\lambda)) \right) = (\eta_\psi(zb) \mid \eta_\psi(y)) = \tilde{\psi}(xb).
\end{aligned}$$

Therefore, we conclude that $(a, b) \in \mathcal{G}(\sigma_{-\mathrm{i}}^{\tilde{\psi}})$, i.e. $\mathcal{G}(\sigma_{-\mathrm{i}}^{\psi}) \subset \mathcal{G}(\sigma_{-\mathrm{i}}^{\tilde{\psi}})$.

As seen in the proof of Proposition VIII.3.24, $x \in \mathcal{M}$ is of exponential type relative to $\{\sigma_t^{\tilde{\psi}}\}$ if and only if

$$x \in \bigcap_{n\in\mathbf{Z}} \mathcal{D}(\sigma_{-ni}^{\tilde{\psi}}), \qquad \sup_{n\in\mathbf{Z}} \|\sigma_{-ni}^{\tilde{\psi}}(x)\| \mathrm{e}^{-c|n|} < +\infty \tag{39}$$

for some $c > 0$. Hence $\mathcal{N}_{\exp}^{\psi} \subset \mathcal{M}_{\exp}^{\tilde{\psi}}$. For each $x \in \mathcal{N}_{\exp}^{\psi}$, we consider

$$y(\alpha) = \sigma_\alpha^\psi(x) - \sigma_\alpha^{\tilde{\psi}}(x), \quad \alpha \in \mathbf{C}.$$

From the above arguments, it follows that the function $f\colon \alpha \in \mathbf{C} \mapsto \omega(y(\alpha)) \in \mathbf{C}$ for any $\omega \in \mathcal{M}_*$ is an entire function of exponential type and that $f(-in) = 0$, $n \in \mathbf{Z}$. Hence $f(\alpha) = 0$ for every $\alpha \in \mathbf{C}$, thus $y(\alpha) = 0$. This means that

$$\sigma_t^\psi(x) = \sigma_t^{\tilde{\psi}}(x), \quad x \in \mathcal{N}_{\exp}^{\psi}.$$

As $\mathcal{N}_{\exp}^{\psi}$ is σ-weakly dense in $\mathcal{N}$, we conclude (35). Q.E.D.

Corollary 4.22. *Suppose $\mathcal{N} \subset \mathcal{M}$ is a pair of von Neumann algebras, and $T \in \mathfrak{W}_0(\mathcal{M}, \mathcal{N})$.*

(i) *For every $\varphi \in \mathfrak{W}_0(\mathcal{N})$,*

$$\sigma_t^{\varphi\circ T}(x) = \sigma_t^\varphi(x), \quad x \in \mathcal{N}; \tag{35}$$

(ii) *For every pair* $\varphi, \psi \in \mathfrak{W}_0(\mathcal{N})$,

$$\big(\mathrm{D}\psi \circ T : \mathrm{D}\varphi \circ T\big)_t = \big(\mathrm{D}\psi : \mathrm{D}\varphi\big)_t, \quad t \in \mathbf{R}. \tag{40}$$

(iii) *If* S *is another element of* $\mathfrak{W}_0(\mathcal{M}, \mathcal{N})$, *then for any* $\varphi \in \mathfrak{W}_0(\mathcal{N})$, $\big(\mathrm{D}\varphi \circ T : \mathrm{D}\varphi \circ S\big)_t$, $t \in \mathbf{R}$, *belongs to* $\mathcal{N}^c = \mathcal{N}' \cap \mathcal{M}$ *and does not depends on the choice of* $\varphi \in \mathfrak{W}_0(\mathcal{N})$, *i.e.*

$$\big(\mathrm{D}\varphi \circ T : \mathrm{D}\varphi \circ S\big)_t = \big(\mathrm{D}\psi \circ T : \mathrm{D}\psi \circ S\big)_t, \quad t \in \mathbf{R}$$

for any other $\psi \in \mathfrak{W}_0(\mathcal{N})$.

PROOF: We have proved (i) in Lemma 4.21.

(ii) Consider the 2×2-matrix algebras $\mathcal{N} \otimes M_2 \subset \mathcal{M} \otimes M_2$. As $\sigma_t^{\varphi \otimes \mathrm{Tr}} = \sigma_t^{\varphi} \otimes \mathrm{id}$ and $\sigma_t^{(\varphi \circ T) \otimes \mathrm{Tr}} = \sigma_t^{\varphi \circ T} \otimes \mathrm{id}$ agree on $\mathcal{N} \otimes M_2$, there exists $T' \in \mathfrak{W}_0\big(\mathcal{M} \otimes M_2, \mathcal{N} \otimes M_2\big)$ by Theorem 4.18 such that

$$(\varphi \circ T) \otimes \mathrm{Tr} = (\varphi \otimes \mathrm{Tr}) \circ T'.$$

We denote this T' by $T \otimes \mathrm{id}$. Now consider the balanced weight $\rho = \varphi \oplus \psi$ on $\mathcal{N} \otimes M_2$ and $\tilde{\rho} = \rho \circ (T \otimes \mathrm{id})$, and observe that $\tilde{\rho} = \tilde{\varphi} \oplus \tilde{\psi}$, where $\tilde{\varphi} = \varphi \circ T$ and $\tilde{\psi} = \psi \circ T$. As $(1 \otimes e_{21}) \in \mathcal{N} \otimes M_2$, we have

$$\big(\mathrm{D}\psi \circ T : \mathrm{D}\varphi \circ T\big)_t = \sigma_t^{\tilde{\rho}}(1 \otimes e_{21}) = \sigma_t^{\rho}(1 \otimes e_{21}) = (\mathrm{D}\psi : \mathrm{D}\varphi)_t$$

by (i).

(iii) For $x = \sum x_{ij} \otimes e_{ij} \in \mathcal{M} \otimes M_2$, set

$$R\Big(\sum_{i,j=1}^{2} x_{ij} \otimes e_{ij}\Big) = \big(T(x_{11}) + S(x_{22})\big) \otimes 1.$$

Then it follows that $R \in \mathfrak{W}_0\big(\mathcal{M} \otimes M_2, \mathcal{N} \otimes \mathbf{C}\big)$. Identifying $\mathcal{N}$ with $\mathcal{N} \otimes \mathbf{C}$, we consider $\varphi \circ R$ and observe that $\varphi \circ R = \varphi \circ T \oplus \varphi \circ S$. Hence

$$\big(\mathrm{D}\varphi \circ T : \mathrm{D}\varphi \circ S\big)_t \otimes e_{12} = \sigma_t^{\varphi \circ R}(1 \otimes e_{12}).$$

As $1 \otimes e_{12} \in \big(\mathcal{N} \otimes \mathbf{C}\big)' \cap \mathcal{M} \otimes M_2$ and $\sigma_t^{\varphi \circ R}$ leaves $\mathcal{N} \otimes \mathbf{C}$ invariant, $\sigma_t^{\varphi \circ R}(1 \otimes e_{12})$ belongs to $\big(\mathcal{N} \otimes \mathbf{C}\big)' \cap \mathcal{M} \otimes M_2 = \mathcal{N}^c \otimes M_2$, which means that

$$\big(\mathrm{D}\varphi \circ T : \mathrm{D}\varphi \circ S\big)_t \in \mathcal{N}^c.$$

For another $\psi \in \mathfrak{W}_0(\mathcal{N})$, we have

$$\begin{aligned}
&\big(\mathrm{D}\psi \circ T : \mathrm{D}\psi \circ S\big)_t \\
&= \big(\mathrm{D}\psi \circ T : \mathrm{D}\varphi \circ T\big)_t \big(\mathrm{D}\varphi \circ T : \mathrm{D}\varphi \circ S\big)_t \big(\mathrm{D}\varphi \circ S : \mathrm{D}\psi \circ S\big)_t \\
&= (\mathrm{D}\psi : \mathrm{D}\varphi)_t \big(\mathrm{D}\varphi \circ T : \mathrm{D}\psi \circ S\big)_t (\mathrm{D}\varphi : \mathrm{D}\psi)_t \\
&= (\mathrm{D}\psi : \mathrm{D}\varphi)_t (\mathrm{D}\varphi : \mathrm{D}\psi)_t \big(\mathrm{D}\varphi \circ T : \mathrm{D}\varphi \circ S\big)_t \\
&= \big(\mathrm{D}\varphi \circ T : \mathrm{D}\varphi \circ S\big)_t
\end{aligned}$$

because $\big(\mathrm{D}\varphi \circ T : \mathrm{D}\varphi \circ S\big)_t \in \mathcal{N}^c$. Q.E.D.

Definition 4.23. Let $\mathcal{N} \subset \mathcal{M}$ be von Neumann algebras.

(i) For $T \in \mathfrak{W}_0(\mathcal{M}, \mathcal{N})$, let σ_t^T be the restriction of $\sigma_t^{\varphi\circ T}$, $\varphi \in \mathfrak{W}_0(\mathcal{M})$, to the relative commutant $\mathcal{N}^c = \mathcal{N}' \cap \mathcal{M}$, which is independent of the choice of φ. We call it the *modular automorphism group* of T.

(ii) For $T_1, T_2 \in \mathfrak{W}_0(\mathcal{M}, \mathcal{N})$, we set

$$\left(\mathrm{D}T_1 : \mathrm{D}T_2\right)_t = \left(\mathrm{D}\varphi \circ T_1 : \mathrm{D}\varphi \circ T_2\right)_t, \quad t \in \mathbf{R},$$

with $\varphi \in \mathfrak{W}_0(\mathcal{N})$ which does not depend on the choice of φ, and call it the *cocycle derivative* of T_1 relative to T_2.

Theorem 4.24. *Let $\mathcal{N} \subset \mathcal{M}$ be von Neumann algebras. There exists a bijection: $T \in \mathfrak{W}_0(\mathcal{M}, \mathcal{N}) \longleftrightarrow T' \in \mathfrak{W}_0(\mathcal{N}', \mathcal{M}')$ such that*

$$\sigma_t^T = \sigma_{-t}^{T'}, \quad T \in \mathfrak{W}_0(\mathcal{M}, \mathcal{N});$$

$$\left(\mathrm{D}T_1 : \mathrm{D}T_2\right)_t = \left(\mathrm{D}T_2' : \mathrm{D}T_1'\right)_{-t}, \quad T_1, T_2 \in \mathfrak{W}_0(\mathcal{M}, \mathcal{N}).$$

In particular, $\mathfrak{W}_0(\mathcal{M}, \mathcal{N}) \neq \emptyset$ if and only if $\mathfrak{W}_0(\mathcal{N}', \mathcal{M}') \neq \emptyset$.

PROOF: Suppose $\mathfrak{W}_0(\mathcal{M}, \mathcal{N}) \neq \emptyset$. Let $T \in \mathfrak{W}_0(\mathcal{M}, \mathcal{N})$. Fix $\varphi \in \mathfrak{W}_0(\mathcal{N})$, and $\psi \in \mathfrak{W}_0(\mathcal{M}')$. We then have, by Theorem 3.8, with $H = \mathrm{d}\varphi \circ T / \mathrm{d}\psi$, that

$$H^{\mathrm{i}t} x H^{-\mathrm{i}t} = \sigma_t^{\varphi\circ T}(x), \quad x \in \mathcal{M}, \qquad H^{\mathrm{i}t} y H^{-\mathrm{i}t} = \sigma_{-t}^{\psi}(y), \quad y \in \mathcal{M}'.$$

Choose $\omega \in \mathfrak{W}_0(\mathcal{N}')$ and let $K = \mathrm{d}\varphi/\mathrm{d}\omega$. Then

$$K^{\mathrm{i}t} x K^{-\mathrm{i}t} = \sigma_t^{\varphi}(x) = H^{\mathrm{i}t} x H^{-\mathrm{i}t}, \quad x \in \mathcal{N}; \qquad K^{\mathrm{i}t} y K^{-\mathrm{i}t} = \sigma_{-t}^{\omega}(y), \quad y \in \mathcal{N}'.$$

Hence $u_t = H^{-\mathrm{i}t} K^{\mathrm{i}t}$ belongs $\mathcal{N}'$ and

$$\begin{aligned} u_{s+t} &= H^{-\mathrm{i}s} H^{-\mathrm{i}t} K^{\mathrm{i}s} K^{\mathrm{i}t} = H^{-\mathrm{i}s} K^{\mathrm{i}s} K^{-\mathrm{i}s} H^{-\mathrm{i}t} K^{\mathrm{i}t} K^{\mathrm{i}s} \\ &= u_s \sigma_s^{\omega}(u_t), \quad s, t \in \mathbf{R}. \end{aligned}$$

Hence $\{u_s\}$ is a one cocycle for σ^{ω}, so that there exists $\tilde{\psi} \in \mathfrak{W}_0(\mathcal{N}')$ such that $\left(\mathrm{D}\tilde{\psi} : \mathrm{D}\omega\right)_t = u_t$. Then we have for every $y \in \mathcal{N}'$

$$\begin{aligned} \sigma_t^{\tilde{\psi}}(y) &= \left(\mathrm{D}\tilde{\psi} : \mathrm{D}\omega\right)_t^* \sigma_t^{\omega}(y) \left(\mathrm{D}\tilde{\psi} : \mathrm{D}\omega\right)_t^* \\ &= H^{-\mathrm{i}t} K^{\mathrm{i}t} K^{-\mathrm{i}t} y K^{\mathrm{i}t} K^{-\mathrm{i}t} H^{\mathrm{i}t} = H^{-\mathrm{i}t} y H^{\mathrm{i}t}. \end{aligned}$$

Therefore, we obtain $\sigma_t^{\tilde{\psi}}(y) = \sigma_t^{\psi}(y)$ for every $y \in \mathcal{M}'$. Theorem 4.18 entails the existence of $T' \in \mathfrak{W}_0(\mathcal{N}', \mathcal{M}')$ such that $\tilde{\psi} = \psi \circ T'$.

Now for $x \in \mathcal{N}^c = \mathcal{N}' \cap \mathcal{M} = (\mathcal{M}')' \cap \mathcal{N}'$,

$$\sigma_t^T(x) = \sigma_t^{\varphi\circ T}(x) = H^{\mathrm{i}t} x H^{-\mathrm{i}t} = \sigma_{-t}^{\psi\circ T'}(x) = \sigma_{-t}^{T'}(x).$$

Choose another $S \in \mathfrak{W}_0(\mathcal{M}, \mathcal{N})$ and observe that $u_t = (\mathrm{D}S : \mathrm{D}T)_t = \big(\mathrm{D}\varphi \circ S : \mathrm{D}\varphi \circ T\big)_t$ belongs to $\mathcal{N}^c$, and it is a $\sigma^{\varphi\circ T}$-cocycle. Namely,

$$u_{s+t} = u_s \sigma_s^{\varphi\circ T}(u_t) = u_s H^{\mathrm{i}s} u_t H^{-\mathrm{i}s}.$$

Set $v_t = u_{-t}^* = (\mathrm{D}T : \mathrm{D}S)_{-t}$. Then we have

$$\begin{aligned} v_{s+t} &= u_{-(s+t)}^* = u_{-t}^* \sigma_{-t}^{\varphi\circ T}(u_{-s}^*) = v_t H^{-\mathrm{i}t} u_{-s}^* H^{\mathrm{i}t} \\ &= v_t \sigma_t^{\psi\circ T}(u_{-s}^*) = v_t \sigma_t^{\psi\circ T}(v_t). \end{aligned}$$

Hence there exists $\omega \in \mathfrak{W}_0(\mathcal{N}')$ with $v_t = (\mathrm{D}\omega : \mathrm{D}\psi \circ T)_t$. Then we have, for every $y \in \mathcal{M}'$,

$$\sigma_t^{\omega}(y) = v_t \sigma_t^{\psi\circ T'}(y) v_t^* = v_t \sigma_t^{\psi}(y) v_t^* = \sigma_t^{\psi}(y)$$

because $v_t \in \mathcal{M}$. Hence there exists $S' \in \mathfrak{W}_0(\mathcal{N}', \mathcal{M}')$ by Theorem 4.18 such that $\omega = \psi \circ S'$. We therefore obtain

$$(\mathrm{D}S' : \mathrm{D}T')_t = \big(\mathrm{D}\psi \circ S' : \mathrm{D}\psi\circ!T'\big)_t = \big(\mathrm{D}\omega : \mathrm{D}(\psi\circ T)\big)_t = v_t = (\mathrm{D}T : \mathrm{D}S)_{-t}.$$

As S' is uniquely determined by the formula $\omega = \psi \circ S'$, S' is determined by $(\mathrm{D}S' : \mathrm{D}T')_t$. Therefore, the map: $S \in \mathfrak{W}_0(\mathcal{M}, \mathcal{N}) \mapsto S' \in \mathfrak{W}_0(\mathcal{N}', \mathcal{M}')$ is an injection. By symmetry, this correspondence is bijective. Q.E.D.

Corollary 4.25. *Let $\{\mathcal{M}, \mathfrak{H}\}$ be a von Neumann algebra. Then there exists a bijection: $\varphi \in \mathfrak{W}_0(\mathcal{M}) \mapsto T_\varphi \in \mathfrak{W}_0\big(\mathcal{L}(\mathfrak{H}), \mathcal{M}'\big)$ such that*

$$\sigma_t^{\varphi} = \sigma_{-t}^{T_\varphi};$$

$$(\mathrm{D}\psi : \mathrm{D}\varphi)_t = (\mathrm{D}T_\varphi : \mathrm{D}T_\psi)_{-t}, \quad t \in \mathbf{R};$$

$$\psi \circ T_\varphi = \mathrm{Tr}\left(\left(\frac{\mathrm{d}\psi}{\mathrm{d}\varphi}\right) \cdot \right), \quad \psi \in \mathfrak{W}_0(\mathcal{M}'),$$

where the weight $\mathrm{Tr}\big((\frac{\mathrm{d}\psi}{\mathrm{d}\varphi} \cdot)\big)$ means the one given in Lemma VIII.2.8 *relative to the standard trace* Tr *on* $\mathcal{L}(\mathfrak{H})$.

This is an immediate consequence of Theorem 4.24 and its proof.

Exercise IX.4

1) Let $\mathcal{N}$ be a von Neumann subalgebra of a von Neumann algebra $\mathcal{M}$. Suppose that there exists a normal projection $\mathcal{E}$ of norm one from $\mathcal{M}$ onto $\mathcal{N}$. Set $N_{\mathcal{E}} = \{x \in \mathcal{M} : \mathcal{E}(x^*x) = 0\}$.

(a) Show that there exists a projection $e \in \mathcal{M} \cap \mathcal{N}'$ such that the central support of e in $\mathcal{N}'$ is the identity and $N_{\mathcal{E}} = (1 - e)\mathcal{M}$.

(b) Prove that the map $\mathcal{E}'$ given by $\mathcal{E}'(x) = \mathcal{E}(x)e$, $x \in \mathcal{M}_e$, is a faithful projection of norm one from $\mathcal{M}_e$ onto $\mathcal{N}_e$.

(c) Prove that if $\mathcal{M}$ is semi-finite, then $\mathcal{N}$ must be also semi-finite. (Hint: Recall the proof of Theorem V.2.30.)

(d) Prove that if $\mathcal{M}$ is of type I, then $\mathcal{N}$ must be of type I. (Hint: By (a) and (b) we may assume that $\mathcal{E}$ is faithful. By (c), there exists a finite projection e in $\mathcal{N}$ with the central support in $\mathcal{N}$ equal to 1. Restricting $\mathcal{E}$ to $\mathcal{M}_e$ and $\mathcal{N}_e$, we may assume that $\mathcal{N}$ is finite. If τ is a normal trace on $\mathcal{N}$, then $\varphi = \tau \circ \mathcal{E}$ is a normal positive functional on $\mathcal{M}$ such that $s(\varphi) \in \mathcal{N}' \cap \mathcal{M}$ and $\mathcal{M}_\varphi \supset \mathcal{N}_{s(\varphi)}$. Restricting our attention to $\mathcal{M}_{s(\varphi)}$ and $\mathcal{N}_{s(\varphi)}$, we may assume that φ is faithful. Thus the situation is that τ is a faithful norm trace on $\mathcal{N}$ and $\mathcal{E}$ is a faithful conditional expectation relative to φ. Let $\tilde{\tau}$ be a faithful semi-finite normal trace on $\mathcal{M}$ and h be the element in $L^1(\mathcal{M}, \tilde{\tau})_+$ corresponding to φ. Then $\mathcal{N} \subset \{h\}' \cap \mathcal{M}$. We complete the proof by showing that $\mathcal{M}_\varphi = \{h\}' \cap \mathcal{M}$ contains only von Neumann algebras of type I based on the fact that h is integrable relative to $\tilde{\tau}$ in the von Neumann algebra of type I, which entails that $\mathcal{M}_\varphi$ is finite and of type I.)

(e) Prove that if $\mathcal{M}$ is atomic, then $\mathcal{N}$ is also atomic.

2) Let $\mathcal{M}$ be a von Neumann algebra and G a group of automorphisms of $\mathcal{M}$, i.e. a subgroup of $\mathrm{Aut}(\mathcal{M})$. Following the line of arguments suggested below, show that if $\mathcal{M}$ admits sufficiently many G-invariant normal states in the sense that for any non-zero $x \in \mathcal{M}$ there exists a normal state φ such that $\varphi(x^*x) > 0$ and $\varphi \circ \alpha = \varphi, \alpha \in G$, then there exists a unique faithful normal projection $\mathcal{E}$ of norm one from $\mathcal{M}$ onto the fixed point subalgebra $\mathcal{M}^G = \left\{ x \in \mathcal{M} : \alpha(x) = x,\ \alpha \in G \right\}$ such that $\mathcal{E} \circ \alpha = \mathcal{E},\ \alpha \in G$.

(a) Let $\mathfrak{S}_*^G$ be the set of all G-invariant normal states. Show that the support, $s(\varphi)$, of each $\varphi \in \mathfrak{S}_*^G$ is a projection in $\mathcal{M}^G$.

(b) Let $\mathcal{F}$ be a maximal family of elements of $\mathfrak{S}_*^G$ with orthogonal support and set $\psi = \sum_{\varphi \in \mathcal{F}} \varphi$. Show that ψ is a faithful semi-finite normal weight on $\mathcal{M}$ such that ψ is semi-finite on $\mathcal{M}^G$ and the modular automorphism group $\left\{ \sigma_t^\psi : t \in \mathbf{R} \right\}$ leaves $\mathcal{M}^G$ globally invariant.

(c) Apply Theorem 4.2 to $\mathcal{M}$, $\mathcal{M}^G$ and ψ.

3) Under the same hypothesis and the notations, for each $x \in \mathcal{M}$ let $K(x)$ be the σ-weak convex closure of the orbit $\mathrm{Orb}(x) = \left\{ \alpha(x) : \alpha \in G \right\}$ of x. Following the arguments suggested below, i.e. without making use of the G-invariant faithful projection $\mathcal{E}$ of norm one, shown in the last problem, prove that $K(x) \cap \mathcal{M}^G \neq \emptyset$, in fact $K(x) \cap \mathcal{M}^G = \left\{ \mathcal{E}(x) \right\}$ as this will give alternative proof, which does make use of the modular theory, for the existence and the uniqueness of a faithful G-invariant normal projection of norm one.

(a) Observe first that for each $\varphi \in \mathfrak{S}_*^G$ the function: $x \in \mathcal{M} \mapsto \|x\|_\varphi = \varphi(x^*x)^{1/2}$ is lower semi-continuous.

(b) Observe that the set $K_F(x) = \left\{ y \in K(x) : \|y\|_\varphi = \inf\{\|z\|_\varphi : z \in K(x)\},\ \varphi \in F \right\}$ is a non-empty closed face of the σ-weakly compact convex set $K(x)$ for any finite subset F of $\mathfrak{S}_*^G$ and that $\alpha\big(K_F(x)\big) = K_F(x),\ \alpha \in G$.

(**c**) Show that $K_{\mathfrak{S}_*^G}(x) = \left\{ y \in K(x) : \|y\|_\varphi = \inf\{ \|z\|_\varphi : z \in K(x) \},\ \varphi \in \mathfrak{S}_*^G \right\}$ $\neq \emptyset$ and in fact that $K_{\mathfrak{S}_*^G}(x)$ is the singleton set $\{\mathcal{E}(x)\}$.

4) Let $\mathfrak{H} = L^2(\mathbf{R})$ and $\mathcal{A} = L^\infty(\mathbf{R})$ act on $\mathfrak{H}$ by multiplication. Let $u(t)$ be the element of $\mathcal{A}$ corresponding to the function: $s \in \mathbf{R} \mapsto e^{ist} \in \mathbf{T}$. Let φ be the faithful semi-finite normal weight on $\mathcal{A}$ obtained by the integration relative to the Lebesgue measure on $\mathbf{R}$. Show that the operator valued weight T from $\mathcal{L}(\mathfrak{H})$ to $\mathcal{A}$ such that $\mathrm{Tr} = \varphi \circ T$ is given by $T(x) = \int_{\mathbf{R}} u(t)xu(t)^*\, dt,\ t \in \mathbf{R},\ x \in \mathcal{L}(\mathfrak{H})_+$.

5) Let G be an abelian locally compact group with $\hat{G}$ the dual group. Let $\mathfrak{H} = L^2(G)$ be the Hilbert space of square integrable functions relative to the Haar measure and set $\{u(p)\xi\}(s) = \langle s, p\rangle\xi(s),\ \xi \in \mathfrak{H},\ s \in G$ and $p \in \hat{G}$. Let $\mathcal{A} = L^\infty(G)$ act on $\mathfrak{H}$ by multiplication as in the last problem. Show that if φ is the faithful semi-finite normal weight on $\mathcal{A}$ given by the integration relative to the Haar measure on G, then the operator valued weight T from $\mathcal{L}(\mathfrak{H})$ to $\mathcal{A}$ such that $\mathrm{Tr} = \varphi \circ T$ is given by: $T(x) = \int_{\hat{G}} u(p)xu(p)^*\, dp,\ x \in \mathcal{L}(\mathfrak{H})_+$.

6) Let G be a discrete countable group, $\mathfrak{H} = \ell^2(G)$ and $\mathcal{M} = \mathcal{R}_\ell(G)$. Let τ be the normalized trace given by $\tau(x) = \left(x\xi_0 \mid \xi_0\right),\ x \in \mathcal{M}$ where $\xi_0(s) = 1$ for $s = e$ and 0 for $s \neq e$. Let λ and ρ be the left and right regular representations of G on $\mathfrak{H}$ respectively.

(**a**) Show that the operator valued weight T of $\mathcal{L}(\mathfrak{H})$ to $\mathcal{M}$ such that $\mathrm{Tr} = \tau \circ T$ is given by: $T(x) = \sum_{s \in G} \rho(s)x\rho(s)^*,\ x \in \mathcal{L}(\mathfrak{H})_+$.

(**b**) Show that the operator valued weight T in (a) is also given by:
$x = \sum_{s \in G} \mathrm{Tr}\big(\lambda(s)^*x\big)\lambda(s),\ x \in \mathcal{L}(\mathfrak{H})_+$.

Notes on Chapter IX

The cones $\mathfrak{P}^\sharp$ and $\mathfrak{P}^\flat$ were introduced by Takesaki in [708] in the case of a cyclic and separating vector, to connect vectors in the Hilbert space and normal positive functional on the von Neumann algebra in question. The systematic study of one parameter family of convex cones in the Hilbert space $\mathfrak{H}$, on which $\mathcal{M}$ acts, with a distinguished cyclic and separating vector ξ_0 was done by Araki, [424]. Independently, Connes, [459], and Haagerup, [537], considered the natural cone $\mathfrak{P}^\natural$. Theorems 1.2 and 1.14 were proven by them. The inequality (11) is referred as Powers-Størmer inequality as it was proved first by them for $\mathcal{L}(\mathfrak{H})$, [671].

Before the Tomita-Takesaki theory was established in the late sixties and the early seventies, non-commutative integration meant the theory presented in Section 2. One can trace the history of the subject to the work of Murray and von Neumann. They proved for example that every symmetric closed operator affiliated with a factor of type II_1 is automatically self-adjoint and that the collection of densely defined operators affiliated with a factor of type II_1 forms an involutive algebra under the natural operations. It was this fact that von Neumann thought the theory of operator algebras would play a significant role in quantum physics by accommodating algebraic operations for unbounded physical quantities, a dream which was not realized in its original form. Araki and Woods in a joint work, [430], showed in 1963 that the most of factors appearing in quantum physics is either of type I or type III, a fact that shattered the long held dream of many specialists. Nevertheless, the theory presented in Section 3 really tells that the theory related to traces is indeed non-commutative analogue of the usual integration-measure theory. The theory was completed by the hands of Dixmier, [502], Dye, [99], Segal, [326], and many others. Theory presented here followed the approach given by Nelson, [647].

The theory of spatial derivatives is due to Connes, [472], inspired by the work of Woronowicz on phase system, [741], and the work of Haagerup on operator valued weights, [541]. The theory presented here follows the line set by Falcone in his thesis, [518, 519], which gives more streamlined approach to the theory of relative tensor product of Sauvageot, [688]. The theory of bimodules goes further far beyond the one presented here. It was originally developed by Connes although its full account has never been published. He called it a *correspondence.* The motivation and its applications are presented in his book, [480].

The concept of conditional expectations has played significant roles in the development of the theory of operator algebras. It was first discovered by Dixmier, [85], in the course of his theory of non-commutative integration for traces. But it was Umegaki, [392, 726], who viewed the projection of norm one determined by a normal tracial state as a non-commutative counter part of the conditional expectation in probability theory. Subsequently, he developed a theory of conditional expectations closely following the analogy with probability theory. It was, however, the remarkable bimodule property which yielded a wide range of applications as we will see in the later chapters in the structure theory of factors and subfactors. The theory of operator valued weights was developed by Haagerup, [541]. But it had appeared long in disguise before his formal treatise. For example, the theory of integrable groups actions on a von Neumann algebra which will be treated later, Chapter X, made use of operator valued weights without the full theory. Haagerup's original approach however used the structure theory of von Neumann algebras of type III, i.e. the duality theorem, Theorem XII.1.1, applied to the modular automorphism groups. We took an approach given by Falcone and Takesaki, [518, 519], which develops the theory within the framework of non-commutative integration.

Chapter X
Crossed Products and Duality

§ 0 Introduction

We begin now systematic study of automorphism actions of a locally compact group on a von Neumann algebra. For those readers who are unfamiliar with a locally compact group or unwilling to spend extra minute on the general theory of locally compact groups should assume that your group is simply the additive group **R** of real numbers or **Z** of integers. We will take these two groups for applications in the later chapters.

At the dawn of the subject, i.e. when Murray and von Neumann started their fundamental work on operator algebras, abstract algebra saw its peak, more specifically the theory of centrally simple algebras flourished by the hand of Artin, Burnside, Brauer, Jacobson, Nakayama, Noether, Wedderburn and many others. The work of Murray and von Neumann followed closely the development of abstract algebra. One of the central items in the theory of centrally simple algebras was crossed product. It was no surprise to see that Murray and von Neumann gave their thought on the so-called group measure space construction of factors, which is nothing but the crossed product of an abelian von Neumann algebra by a countable discrete group relative to an ergodic action, although it was their deep insight to foresee that the ergodicity of the action gives rise to a factor.

We will begin the chapter with discussion on the construction of crossed product $\mathcal{M} \rtimes_\alpha G$ of a von Neumann algebra $\mathcal{M}$ by a locally compact group G acting via $\alpha\colon s \in G \mapsto \alpha_s \in \mathrm{Aut}(\mathcal{M})$ and the weight $\tilde{\varphi}$ on $\mathcal{M} \rtimes_\alpha G$ dual to a semi-finite normal weight φ on $\mathcal{M}$ and analyze their relationship. We will give a precise description of the modular automorphism group of the dual weight. In the second section, we discuss first the Heisenberg-Weyl commutation relation and prove the uniqueness of the representation of the Heisenberg-Weyl commutation relation up to multiplicity. Then we will move on to the discussion of the duality theorem for crossed products of von Neumann algebra by an abelian locally compact group G. First, we will see that there is a natural action $\hat{\alpha}$ of the dual group $\hat{G}$, called the dual action or dual to the original action α of G. When we form the second crossed product $\left(\mathcal{M} \rtimes_\alpha G\right) \rtimes_{\hat{\alpha}} \hat{G}$, then we have the duality: Theorem 2.3. This theorem will play a fundamental role in the structure analysis of von Neumann algebra of type III in later chapters, for example Chapter XII.

Section 3 is devoted to the discussion of equivariant disintegrations of covariant systems. It is important to have a concrete picture of the system when a group G acts on a Hilbert space $\mathfrak{H}$ by unitaries leaving an abelian von Neumann algebra $\mathcal{A}$ globally invariant so that the Hilbert space $\mathfrak{H}$ is disintegrated over the measure theoretic spectrum of $\mathcal{A}$. We will discuss how the action of G looks like relative to the disintegration of $\mathfrak{H}$. We then specialize the study in §4 to the case that the action of G on $\mathcal{A}$ is transitive. In this case, it will be shown that the covariant system $\{\mathcal{M}, G, \alpha\}$ can be constructed from a smaller covariant system $\{\mathcal{N}, H, \beta\}$ as the induced system, Definition 4.10. As a preparation, the theory of induced representation of Mackey is presented. We then analyze the crossed product of the induced covariant system and relate to the one on the original smaller covariant system. This will be applied to the case that $G = \mathbf{R}$ and $H = T\mathbf{Z}$, $T \in \mathbf{R}$, in the structure analysis of a factor of type III.

§ 1 Crossed Products and Dual Weights

We recall first that given a von Neumann algebra $\mathcal{M}$, the automorphism group $\mathrm{Aut}(\mathcal{M})$ is a topological group with respect to the neighborhood system

$$\left\{ U(\alpha; \omega_1, \ldots, \omega_n) : \omega_1, \ldots, \omega_n \in \mathcal{M}_* \right\},$$

where $U(\alpha; \omega_1, \ldots, \omega_n)$ is given by:

$$U(\alpha; \omega_1, \ldots, \omega_n) = \left\{ \beta \in \mathrm{Aut}(\mathcal{M}) : \begin{array}{l} \|\omega_i \circ \alpha - \omega_i \circ \beta\| < 1, \\ \|\omega_i \circ \alpha^{-1} - \omega_i \circ \beta^{-1}\| < 1, \end{array} \quad i = 1, \ldots, n \right\}. \tag{1}$$

Definition 1.1. An *action* of a locally compact group G means a continuous homomorphism α of G into $\mathrm{Aut}(\mathcal{M})$. The triplet $\{\mathcal{M}, G, \alpha\}$ is called a *covariant system* or more precisely a *von Neumann covariant system* over G.

Given $\{\mathcal{M}, \mathfrak{H}\}$ and a unitary representation U of G [1] on $\mathfrak{H}$ such that

$$U(s)\mathcal{M}U(s)^* = \mathcal{M}, \quad s \in G, \tag{2}$$

if we set

$$\alpha_s(x) = U(s)xU(s)^*, \quad x \in \mathcal{M}, \quad s \in G, \tag{3}$$

then $\alpha : s \in G \mapsto \alpha_s \in \mathrm{Aut}(\mathcal{M})$ is an action of G.

Proposition 1.2. *Let G be a locally compact group and $\alpha : s \in G \mapsto \alpha_s \in \mathrm{Aut}(\mathcal{M})$ a homomorphism. The following conditions are equivalent for α:*

(i) *α is continuous;*
(ii) *For every $x \in \mathcal{M}$ and $\omega \in \mathcal{M}_*$, the function: $s \in G \mapsto \langle \alpha_s(x), \omega \rangle$ is continuous on G.*

[1] The strong continuity of a unitary representation is always assumed.

This follows from the next general result:

Proposition 1.2′. *Let G be a locally compact group and E be a Banach space. Suppose that the map: $s \in G \mapsto U(s) \in \mathcal{L}(E)$ is a homomorphism of G into the group of invertible bounded operators on E. Then the following two conditions on the continuity of U are equivalent:*

(i) *For every $x \in E$, the map: $s \in G \mapsto U(s)x \in E$ is continuous in norm;*
(ii) *For every $x \in E$ and $x^* \in E^*$, the map: $s \in G \mapsto \langle U(s)x, x^*\rangle$ is continuous.*

PROOF: The implication (i) $\Longrightarrow$ (ii) is just a triviality. We have only to prove (ii) $\Longrightarrow$ (i). So we assume (ii). For each $f \in \mathcal{K}(G)$,[2] we consider the integral;

$$\int_G f(s)\langle U(s)x, x^*\rangle \, ds = \langle U(f)x, x^*\rangle.$$

We claim that the integral $U(f)x$ belongs to E rather than E^{**}. To see this, let $\{x_i^*\}$ be a net in E^* converging to zero in $\tau(E^*, E)$-topology[3]. Let $K = \operatorname{supp} f$. Then $\langle U(s)x, x_i^*\rangle$ converges to zero uniformly for $s \in K$ because $\{U(s)x : s \in K\}$ is $\sigma(E, E^*)$-compact and the convex closure of a weakly compact set is weakly compact by Krein-Šmulian's Theorem. Hence $\langle U(f)x, x_i^*\rangle$ converges to zero. Thus $U(f)x$ is a $\tau(E^*, E)$-continuous functional on E^*, so that it belongs to E. Furthermore, we know that

$$\|U(f)\| \le \|f\|_1 \sup\{\|U(s)\| : s \in K\}.$$

Hence we have

$$\lim_{s\to e}\|U(s)U(f) - U(f)\| = \lim_{s\to e}\|U(\lambda_s f) - U(f)\| = 0,$$

where $(\lambda_s f)(t) = f(s^{-1}t)$. Thus U is strongly continuous on the range of $U(f)$. Let $E_0 = \{x \in E : \lim_{s\to e}\|U(s)x - x\| = 0\}$. Clearly E_0 is a norm closed subspace of E because for any compact neighborhood V of e, $\{U(s) : s \in V\}$ is bounded. We observed that $U(f)E \subset E_0$ for any $f \in \mathcal{K}(G)$. If $E_0 \ne E$, then there exists an $x^* \in E^*$, $x^* \ne 0$, such that $\langle x, x^*\rangle = 0$ for every $x \in E_0$. In particular, we have

$$\int_G f(s)\langle U(s)x, x^*\rangle \, ds = \langle U(f)x, x^*\rangle = 0$$

for every $f \in \mathcal{K}(G)$. Hence $\langle U(s)x, x^*\rangle = 0$ for every $x \in E$, which contradicts to $x^* \ne 0$. Thus $E = E_0$. Q.E.D.

---

2 $\mathcal{K}(G)$ means the space of continuous functions with compact support.

3 Arens-Mackey topology.

Definition 1.3. Given a covariant system $\{\mathcal{M}, G, \alpha\}$, a normal representation $\{\pi, \mathfrak{H}\}$ of $\mathcal{M}$ and a unitary representation U of G on $\mathfrak{H}$ are said to be *covariant* if

$$\pi \circ \alpha_s(x) = U(s)\pi(x)U(s)^*, \quad x \in \mathcal{M}, \quad s \in G. \tag{4}$$

If $\{\mathcal{M}, \mathfrak{H}, \mathfrak{P}, J\}$ is a standard form of $\mathcal{M}$, then there exists uniquely a unitary representation U of G on $\mathfrak{H}$ such that

$$\begin{aligned} \alpha_s(x) &= U(s)xU(s)^*, \quad x \in \mathcal{M}, \quad s \in G, \\ U(s)\mathfrak{P} &= \mathfrak{P}, \qquad JU(s) = U(s)J. \end{aligned} \tag{5}$$

In this case, we call $\{U, \mathfrak{H}\}$ the *standard form* of $\{\mathcal{M}, G, \alpha\}$, or the *standard* representation.

Definition 1.4. For a covariant system $\{\mathcal{M}, G, \alpha\}$, a strongly continuous $\mathcal{U}(\mathcal{M})$-valued function $u\colon s \in G \mapsto u_s \in \mathcal{U}(\mathcal{M})$ is called an *α-cocycle* or more precisely *α-one cocycle* if we have

$$u_{st} = u_s\alpha_s(u_t), \quad s, t \in G. \tag{6}$$

We denote by $\mathrm{Z}^1(\mathcal{M}, G, \alpha)$ or simply by $\mathrm{Z}^1(G, \alpha)$ the set of all α-cocycles. Two cocycles u and v are said to be *equivalent* and written $u \simeq v$ if there exists $w \in \mathcal{U}(\mathcal{M})$ such that

$$v_s = w^*u_s\alpha_s(w), \quad s \in G. \tag{7}$$

It follows that the relation "$\simeq$" is an equivalence relation. The quotient space of $Z^1(G,\alpha)$ under "$\simeq$" is called the *one cohomology space* and denoted by $\mathrm{H}^1(\mathcal{M},G,\alpha)$ or $\mathrm{H}^1(G, \alpha)$. For each $u \in \mathrm{Z}^1(G, \alpha)$, if we set

$${}_u\alpha_s(x) = u_s\alpha_s(x)u_s^*, \quad x \in \mathcal{M}, \quad s \in G, \tag{8}$$

then ${}_u\alpha$ is a new action of G on $\mathcal{M}$. We call it the *perturbation* of α by u.

Definition 1.5. For two covariant systems $\{\mathcal{M}, G, \alpha\}$ and $\{\mathcal{N}, G, \beta\}$ over G, if there exists an isomorphism π of $\mathcal{M}$ onto $\mathcal{N}$ such that

$$\pi \circ \alpha_s = \beta_s \circ \pi, \quad s \in G, \tag{9}$$

then we say that α and β are *conjugate* and write

$$\{\mathcal{M}, G, \alpha\} \cong \{\mathcal{N}, G, \beta\}$$

or $\alpha \cong \beta$. If ${}_u\alpha \cong \beta$ for some $u \in \mathrm{Z}^1(G, \alpha)$, then we say that α and β are *cocycle conjugate* and write $\alpha \sim \beta$.

We now fix a covariant system $\{\mathcal{M}, G, \alpha\}$ and consider a left invariant Haar measure as before. Let $\{\mathcal{M}, \mathfrak{H}, J, \mathfrak{P}, U\}$ be the standard form. On the Hilbert space

$L^2(G, \mathfrak{H})$ of all square integrable $\mathfrak{H}$-valued functions, we define representations π_α of $\mathcal{M}$ and λ of G as follows:

$$\begin{aligned} \big(\pi_\alpha(x)\xi\big)(s) &= \alpha_s^{-1}(x)\xi(s), && x \in \mathcal{M}, \quad s \in G, \\ \big(\lambda(t)\xi\big)(s) &= \xi(t^{-1}s), && s, t \in G, \quad \xi \in L^2(G, \mathfrak{H}). \end{aligned} \tag{10}$$

It follows that $\{\pi_\alpha, \lambda\}$ is covariant.

Definition 1.6. The von Neumann algebra generated by $\pi_\alpha(\mathcal{M})$ and $\lambda(G)$ on $L^2(G, \mathfrak{H})$ is called the *crossed product* of $\mathcal{M}$ by α and denoted by $\mathcal{M} \rtimes_\alpha G$.

The next result tells us that the algebraic structure of the crossed product is independent of the representing space of the original algebra (i.e. one can use any representing Hilbert space), and of the cocycle perturbations.

Theorem 1.7. *For a covariant system $\{\mathcal{M}, G, \alpha\}$ over a locally compact group G, we have the following conclusions:*

(i) *For any faithful normal representation $\{\rho, \mathfrak{K}\}$ of $\mathcal{M}$, if we define a covariant representation $\{\rho_\alpha, \lambda_\mathfrak{K}\}$ on $L^2(G, \mathfrak{K})$ by the following:*

$$\begin{aligned} \big(\rho_\alpha(x)\xi\big)(s) &= \rho\big(\alpha_s^{-1}(x)\big)\xi(s), && x \in \mathcal{M}, \quad s \in G, \\ \big(\lambda_\mathfrak{K}(t)\xi\big)(s) &= \xi(t^{-1}s), && t \in G, \quad \xi \in L^2(G, \mathfrak{K}), \end{aligned} \tag{11}$$

then there exists uniquely an isomorphism Φ of $\mathcal{M} \rtimes_\alpha G$ onto $\{\rho_\alpha(\mathcal{M}) \cup \lambda_\mathfrak{K}(G)\}''$ such that

$$\Phi \circ \pi_\alpha(x) = \rho_\alpha(x), \quad x \in \mathcal{M}; \qquad \Phi\big(\lambda(s)\big) = \lambda_\mathfrak{K}(s), \quad s \in G. \tag{12}$$

(ii) *For each $u \in Z^1(\mathcal{M}, G, \alpha)$, if we put*

$$(U\xi)(s) = u_{s^{-1}}\xi(s), \quad \xi \in L^2(G, \mathfrak{K}), \tag{13}$$

then we have

$$U\big(\mathcal{M} \rtimes_\alpha G\big)U^* = \mathcal{M} \rtimes_{u_\alpha} G.$$

Therefore, the algebraic structure of the crossed product is stable under cocycle perturbations.

PROOF:

(i) If $\mathfrak{H}_0$ is a Hilbert space with dimension large enough, then the representations $\tilde{\rho} = \mathrm{id} \otimes \rho$ of $\mathcal{M}$ on $\mathfrak{H}_0 \otimes \mathfrak{K}$ and $x \in \mathcal{M} \mapsto 1_{\mathfrak{H}_0} \otimes x \in \mathcal{L}(\mathfrak{H}_0 \otimes \mathfrak{H})$ are unitarily equivalent, where $1_{\mathfrak{H}_0}$ means the identity operator on $\mathfrak{H}_0$. Hence there exists a unitary U of $\mathfrak{H}_0 \otimes \mathfrak{K}$ onto $\mathfrak{H}_0 \otimes \mathfrak{H}$ such that

$$U\tilde{\rho}(x)U^* = U\big(1_{\mathfrak{H}_0} \otimes \rho(x)\big)U^* = 1_{\mathfrak{H}_0} \otimes x, \quad x \in \mathcal{M}.$$

We now put $\widetilde{U} = U \otimes 1_{L^2(G)}$. Then $\widetilde{U}$ is a unitary of $L^2(G, \mathfrak{H}_0 \otimes \mathfrak{K}) = \mathfrak{H}_0 \otimes \mathfrak{K} \otimes L^2(G)$ onto $L^2(G, \mathfrak{H}_0 \otimes \mathfrak{H})$ such that

$$\widetilde{U}\tilde{\rho}_\alpha(x)\widetilde{U}^* = 1_{\mathfrak{H}_0} \otimes \pi_\alpha(x), \quad x \in \mathcal{M},$$

$$\widetilde{U}\lambda_{(\mathfrak{H}_0\otimes\mathfrak{K})}(s)\widetilde{U}^* = 1_{\mathfrak{H}_0} \otimes \lambda(s), \quad s \in G.$$

Hence we get

$$\widetilde{U}\big(\tilde{\rho}_\alpha(\mathcal{M}) \cup \lambda_{(\mathfrak{H}_0\otimes\mathfrak{K})}(G)\big)''\widetilde{U}^* = \mathbf{C}_{\mathfrak{H}_0} \otimes \big(\mathcal{M} \rtimes_\alpha G\big),$$

where $\mathbf{C}_{\mathfrak{H}_0} = \mathbf{C}1_{\mathfrak{H}_0}$. But

$$\tilde{\rho}_\alpha(\mathcal{M}) = \mathbf{C}_{\mathfrak{H}_0} \otimes \rho_\alpha(\mathcal{M}), \qquad \lambda_{(\mathfrak{H}_0\otimes\mathfrak{K})}(s) = 1_{\mathfrak{H}_0} \otimes \lambda_{\mathfrak{K}}(s), \quad s \in G,$$

so that

$$\big\{\tilde{\rho}_\alpha(\mathcal{M}) \cup \lambda_{(\mathfrak{H}_0\otimes\mathfrak{K})}(G)\big\}'' = \mathbf{C}_{\mathfrak{H}_0} \otimes \big\{\rho_\alpha(\mathcal{M}) \cup \lambda_{\mathfrak{K}}(G)\big\}''.$$

Thus assertion (i) follows.

(ii) For a fixed $u \in Z^1(\mathcal{M}, G, \alpha)$, put $\beta =_u \alpha$ and define U by (13). We then compute:

$$\big(U\pi_\alpha(x)U^*\xi\big)(s) = u_{s^{-1}}\alpha_s^{-1}(x)u_{s^{-1}}^*\xi(s) = \beta_s^{-1}(x)\xi(s) = \big(\pi_\beta(x)\xi\big)(s);$$

$$\begin{aligned}\big(U\lambda(t)U^*\xi\big)(s) &= u_{s^{-1}}u_{s^{-1}t}^*\xi(t^{-1}s) = u_{s^{-1}}\big(u_{s^{-1}}\alpha_{s^{-1}}(u_t)\big)^*\xi(t^{-1}s)\\ &= u_{s^{-1}}\alpha_{s^{-1}}(u_t^*)u_{s^{-1}}^*\xi(t^{-1}s) = \big(\pi_\beta(u_t^*)\lambda(t)\xi\big)(s).\end{aligned}$$

Hence we get

$$\left.\begin{aligned} U\pi_\alpha(x)U^* &= \pi_\beta(x), && x \in \mathcal{M},\\ U\lambda(t)U^* &= \pi_\beta(u_t^*)\lambda(t), && t \in G,\end{aligned}\right\} \tag{14}$$

so that $U\big(\mathcal{M} \rtimes_\alpha G\big)U^* \subset \mathcal{M} \rtimes_\beta G$. Now, we observe that $u_{st}^* = u_s^*\beta_s(u_t^*)$, and so $u^* \in Z^1(G, \beta)$ and $\alpha = {}_{u^*}\beta$. Thus $U^*\big(\mathcal{M} \rtimes_\beta G\big)U \subset \mathcal{M} \rtimes_\alpha G$, which completes the proof. Q.E.D.

We want to study the commutant of $\mathcal{M} \rtimes_\alpha G$ or more generally the commutant of $\big(\rho_\alpha(\mathcal{M}) \cup \lambda_{\mathfrak{K}}(G)\big)'$ when ρ is the part of a covariant representation $\{\rho, V\}$ of $\big\{\mathcal{M}, G, \alpha\big\}$. To this end, we need to study the weights on $\mathcal{M}$ and $\mathcal{M} \rtimes_\alpha G$.

We fix a covariant system $\big\{\mathcal{M}, G, \alpha\big\}$. Let $\mathcal{K}(G, \mathcal{M})$ be the vector space of all σ-strongly* continuous $\mathcal{M}$-valued functions on G with compact support. We define the algebra structure in $\mathcal{K}(G, \mathcal{M})$ as follows:

$$\left.\begin{aligned} x * y(t) &= \int_G \alpha_s\big(x(ts)\big)y(s^{-1})\,ds, \quad x, y \in \mathcal{K}(G, \mathcal{M})\\ x^\sharp(t) &= \delta_G(t)^{-1}\alpha_t^{-1}\big(x(t^{-1})^*\big),\end{aligned}\right\} \tag{15}$$

where δ_G means of course the modular function of G. It is easy to check that $\mathcal{K}(G, \mathcal{M})$, is an involutive algebra with respect to (15). Furthermore, if we define

$$\left.\begin{aligned} &(x \cdot a)(s) = x(s)a, && x \in \mathcal{K}(G, \mathcal{M}), \quad a \in \mathcal{M}, \\ &(a \cdot x)(s) = \alpha_s^{-1}(a)x(s), && s \in G, \end{aligned}\right\} \tag{16}$$

then $\mathcal{K}(G, \mathcal{M})$ is a two sided module over $\mathcal{M}$ such that

$$\begin{aligned} &a \cdot (x * y) = (a \cdot x) * y, && (x * y) \cdot a = x * (y \cdot a), && a \in \mathcal{M}, \\ &(x \cdot a)^\sharp = a^* \cdot x^\sharp, && (a \cdot x)^\sharp = x^\sharp \cdot a^*, && x, y \in \mathcal{K}(G, \mathcal{M}). \end{aligned} \tag{17}$$

Lemma 1.8. *For a covariant representation $\{\rho, V, \mathfrak{K}\}$, if we put*

$$\tilde{\rho}(x) = \int_G V(s)\rho(x(s))\, ds, \quad x \in \mathcal{K}(G, \mathcal{M}), \tag{18}$$

then $\tilde{\rho}$ is a *-representation of $\mathcal{K}(G, \mathcal{M})$ and

$$\{\rho(\mathcal{M}) \cup V(G) :'\}' = \tilde{\rho}(\mathcal{K}(G, \mathcal{M}))''. \tag{19}$$

Furthermore, we have

$$\tilde{\rho}(a \cdot x) = \rho(a)\tilde{\rho}(x), \quad \tilde{\rho}(x \cdot a) = \tilde{\rho}(x)\rho(a), \quad a \in \mathcal{M}, \quad x \in \mathcal{K}(G, \mathcal{M}). \tag{20}$$

PROOF: We first observe that $\mathcal{K}(G)$ is a *-subalgebra of $\mathcal{K}(G, \mathcal{M})$ and

$$\tilde{\rho}(f) = \int_G f(s)V(s)\, ds = V(f), \quad f \in \mathcal{K}(G).$$

Since $\mathcal{K}(G)$ has an approximate identity $\{f_i\}$ we know $V(\mathcal{K}(G))'' = V(G)''$. A direct calculation shows that $\tilde{\rho}$ defined by (18) is indeed a *-representation of $\mathcal{K}(G, \mathcal{M})$, and that $\tilde{\rho}$ satisfies (20). For each $a \in \mathcal{M}$, we have $\tilde{\rho}(a \cdot f_i) = \rho(a)\tilde{\rho}(f_i)$ and $\{\rho(f_i)\}$ converges to 1 σ-strongly*. Hence $\rho(a) \in \tilde{\rho}(\mathcal{K}(G, \mathcal{M}))''$, so that $\rho(\mathcal{M}) \subset \tilde{\rho}(\mathcal{K}(G, M))''$. Therefore, $(\rho(\mathcal{M}) \cup V(G))''$ is contained in $\tilde{\rho}(\mathcal{K}(G, \mathcal{M}))''$. The converse inclusion follows from the fact that the integral (18) can be approximated by Riemann sum in the σ-strong* topology. Q.E.D.

From this point on, we fix a covariant representation $\{\rho, V, \mathfrak{K}\}$. We then consider the space $\mathcal{K}(G, \mathfrak{K})$ of all continuous $\mathfrak{K}$-valued functions on G with compact support, and view it as a dense subspace of $L^2(G, \mathfrak{K})$. We define $\tilde{\rho}_\alpha$ by (18) based on $\{\rho_\alpha, \lambda_{\mathfrak{K}}, L^2(G, \mathfrak{K})\}$.

Lemma 1.9. *If $x \in \mathcal{K}(G, \mathcal{M})$ and $\xi \in \mathcal{K}(G, \mathfrak{K})$, then $\tilde{\rho}_\alpha(x)\xi$ belongs to $\mathcal{K}(G, \mathfrak{K})$ and is given by the integral:*

$$(\tilde{\rho}_\alpha(x)\xi)(s) = \int_G \rho(\alpha_r(x(sr))\xi(r^{-1})\, dr\,. \tag{21}$$

PROOF: It is clear that the right hand side of (21) gives an element of $\mathcal{K}(G, \mathfrak{K})$. If $\eta \in \mathcal{K}(G, \mathfrak{K})$, then we have

$$\begin{aligned}(\tilde{\rho}_\alpha(x)\xi \mid \eta) &= \int_G \left(\int_G \left(\left\{ \lambda_{\mathfrak{K}}(r)\rho_\alpha\big(x(r)\big)\xi \right\}(s) \,\Big|\, \eta(s) \right) ds \right) dr \\ &= \int_G \left(\int_G \left(\rho\Big(\alpha_{s^{-1}r}\big(x(r)\big)\Big)\xi(r^{-1}s) \,\Big|\, \eta(s) \right) dr \right) ds \\ &= \int_G \left(\int_G \left(\rho\Big(\alpha_r\big(x(sr)\big)\Big)\xi(r^{-1}) \,\Big|\, \eta(s) \right) dr \right) ds,\end{aligned}$$

which implies (21). Q.E.D.

We now fix a faithful weight φ on $\mathcal{M}$ and consider the standard representation $\{\pi_\varphi, U_\varphi, \mathfrak{H}_\varphi, J_\varphi, \mathfrak{P}_\varphi\}$ of $\{\mathcal{M}, G, \alpha\}$. Since we look at only $\pi_\varphi(\mathcal{M})$ for the moment, we identify x and $\pi_\varphi(x)$ for each $x \in \mathcal{M}$. Set

$$\mathfrak{b}_\varphi = \mathcal{K}(G, \mathcal{M}) \cdot \mathfrak{n}_\varphi = \operatorname{span}\{ x \cdot a : x \in \mathcal{K}(G, \mathcal{M}),\ a \in \mathfrak{n}_\varphi \}, \tag{22}$$

where span means the algebraic linear span. For each $y \in \mathcal{K}(G, \mathcal{M})$ and $a \in \mathfrak{n}_\varphi$, $y(s)a$ belongs to $\mathfrak{n}_\varphi$, so that $\eta_\varphi\big(y(s)a\big) = y(s)\eta_\varphi(a)$ makes sense. Hence each $x \in \mathfrak{b}_\varphi$ gives rise to a continuous function: $s \in G \mapsto \eta_\varphi\big(x(s)\big) \in \mathfrak{H}_\varphi$, which belongs to $\mathcal{K}(G, \mathfrak{H}_\varphi)$. Thus we define the map $\tilde{\eta}_\varphi : x \in \mathfrak{b}_\varphi \mapsto \tilde{\eta}_\varphi(x) \in L^2(G, \mathfrak{H}_\varphi)$ by

$$\tilde{\eta}_\varphi(x)(s) = \eta_\varphi\big(x(s)\big), \quad x \in \mathfrak{b}_\varphi, \quad s \in G. \tag{23}$$

With $V = \lambda$ and $\rho = \pi_\alpha$, we define $\tilde{\pi}_\alpha$ by (18).

Lemma 1.10.

(i) $\tilde{\eta}_\varphi(x * y) = \tilde{\pi}_\alpha(x)\tilde{\eta}_\varphi(y), \quad x \in \mathcal{K}(G, \mathcal{M}), \quad y \in \mathfrak{b}_\varphi.$ (24)

(ii) *If* $x, y \in \mathfrak{b}_\varphi$, *then* $(y * x)(e) \in \mathfrak{m}_\varphi$ *and*

$$\big(\tilde{\eta}_\varphi(x) \mid \tilde{\eta}_\varphi(y)\big) = \varphi\big((y^\sharp * x)(e)\big) \tag{25}$$

(iii) $\tilde{\pi}_\alpha(\mathfrak{b}_\varphi \cap \mathfrak{b}_\varphi^\sharp)$ *generates* $\mathcal{M} \rtimes_\alpha G$.

(iv) $\tilde{\eta}_\varphi(\mathfrak{b}_\varphi \cap \mathfrak{b}_\varphi^\sharp)$ *is dense in* $L^2(G, \mathfrak{H}_\varphi)$.

PROOF:

(i) We may and do assume $y = z \cdot a$, $z \in \mathcal{K}(G, \mathcal{M})$ and $a \in \mathfrak{n}_\varphi$. By Lemma 1.9, we have

$$\begin{aligned}\tilde{\eta}_\varphi(x * y)(s) &= \eta_\varphi\big(x * y(s)\big) = \eta_\varphi\big((x * z)(s)a\big) = (x * z)(s)\eta_\varphi(a) \\ &= \int_G \alpha_r\big(x(sr)\big)z(r^{-1})\eta_\varphi(a)\, dr = \big(\tilde{\pi}_\alpha(x)\xi\big)(s),\end{aligned}$$

where

$$\xi(s) = z(s)\eta_\varphi(a) = \eta_\varphi\big((z\cdot a)(s)\big) = \tilde\eta_\varphi(z\cdot a)(s) = \tilde\eta_\varphi(y)(s).$$

Hence (24) holds.

(ii) Suppose $x = x_1 \cdot a$ and $y = y_1 \cdot b$ with $x_1, y_1 \in \mathcal{K}(G, \mathcal{M})$ and $a, b \in \mathfrak{n}_\varphi$. Then we have

$$(y^\sharp * x)(s) = \alpha_s^{-1}(b^*)(y_1^\sharp * x_1)(s)a,$$

so we get

$$(y^\sharp * x)(e) = b^*(y_1^\sharp * x_1)(e)a \in \mathfrak{n}_\varphi^*\mathfrak{n}_\varphi = \mathfrak{m}_\varphi.$$

Now we have

$$\begin{aligned}\big(\tilde\eta_\varphi(x) \mid \tilde\eta_\varphi(y)\big) &= \int_G \big(x_1(s)\eta_\varphi(a) \mid y_1(s)\eta_\varphi(b)\big)\,ds\\ &= \int_G \varphi\big(b^* y_1(s)^* x_1(s) a\big)\,ds = \varphi\left(b^*\left(\int_G y_1(s)^* x_1(s)\,ds\right)a\right)\\ &= \varphi\big(b^*(y_1^\sharp * x_1)(e)a\big) = \varphi\big((y^\sharp * x)(e)\big).\end{aligned}$$

(iii) First we observe $\mathfrak{b}_\varphi \cap \mathfrak{b}_\varphi^\sharp \supset \mathfrak{n}_\varphi^* \cdot \mathcal{K}(G, \mathcal{M}) \cdot \mathfrak{n}_\varphi$ and

$$\tilde\pi_\alpha(b^* x a) = \pi_\alpha(b^*)\tilde\pi_\alpha(x)\pi_\alpha(a), \quad a, b \in \mathfrak{n}_\varphi, \quad x \in \mathcal{K}(G, \mathcal{M}).$$

Since $\tilde\pi_\alpha\big(\mathcal{K}(G, \mathcal{M})\big)$ generates $\mathcal{M} \rtimes_\alpha G$ by Lemma 1.8 and $\mathfrak{n}_\varphi$ is dense in $\mathcal{M}$, $\tilde\pi_\alpha\big(\mathfrak{b}_\varphi \cap \mathfrak{b}_\varphi^\sharp\big)$ generates $\mathcal{M} \rtimes_\alpha G$.

(iv) If $a, b \in \mathfrak{n}_\varphi$ and $f \in \mathcal{K}(G)$, then $b^* \cdot f \cdot a \in \mathfrak{b}_\varphi \cap \mathfrak{b}_\varphi^\sharp$ and

$$\begin{aligned}\tilde\eta_\varphi\big(b^* \cdot f \cdot a\big)(s) &= \eta_\varphi\big(\alpha_s^{-1}(b^*)f(s)a\big) = \alpha_s^{-1}(b^*)f(s)\eta_\varphi(a)\\ &= \Big[\pi_\alpha(b^*)\big(\eta_\varphi(a) \otimes f\big)\Big](s)\end{aligned}$$

by identifying $L^2(G, \mathfrak{H}_\varphi)$ and $\mathfrak{H}_\varphi \otimes L^2(G)$. Since 1 is in the σ-weak closure of $\pi_\alpha(\mathfrak{n}_\varphi^*)$ and the algebraic tensor product $\eta_\varphi(\mathfrak{n}_\varphi) \otimes \mathcal{K}(G)$ is dense in $L^2(G, \mathfrak{H}_\varphi)$, we conclude that $\tilde\eta_\varphi(\mathfrak{b}_\varphi \cap \mathfrak{b}_\varphi^\sharp)$ is dense in $L^2(G, \mathfrak{H}_\varphi)$. Q.E.D.

Lemma 1.11. *The map:* $(s, t) \in G \times \mathbf{R} \mapsto (D\varphi \circ \alpha_s : D\varphi)_t$ *is* σ*-strongly continuous.*

PROOF: In general, for any faithful weigh ψ on a von Neumann algebra $\mathcal{N}$ and $\theta \in \mathrm{Aut}(\mathcal{N})$, the uniqueness of the modular automorphism group implies the equality:

$$\sigma_t^{\psi\circ\theta} = \theta^{-1}\sigma_t^{\psi}\theta, \quad t \in \mathbf{R}. \tag{26}$$

Applying this to the 2×2-matrix algebra over $\mathcal{N}$, we have

$$\big(D\psi_1 \circ \theta : D\psi_2 \circ \theta\big)_t = \theta^{-1}\big((D\psi_1 : D\psi_2)_t\big), \quad t \in \mathbf{R}, \tag{27}$$

for any faithful weights ψ_1 and ψ_2 on $\mathcal{N}$.

Now, we consider $\mathcal{N} = \mathcal{M} \overline{\otimes} L^\infty(G)$. We define the action λ of G on $L^\infty(G)$ by:

$$(\lambda_s f)(t) = f(s^{-1}t), \quad f \in L^\infty(G), \quad s, t \in G. \tag{28}$$

We set

$$\beta_s = \alpha_s \otimes \lambda_s, \quad s \in G. \tag{29}$$

With the standard representation U of G on $\mathfrak{H}_\varphi$, we put

$$(W\xi)(s) = U(s)\xi(s), \quad \xi \in L^2(G, \mathfrak{H}_\varphi). \tag{30}$$

Representing $\mathcal{N}$ on $L^2(G, \mathfrak{H}_\varphi)$, we have $W^*\mathcal{N}W = \mathcal{N}$, so that we can define $\gamma \in \mathrm{Aut}(\mathcal{N})$ by

$$\gamma(x) = W^* x W, \quad x \in \mathcal{N}. \tag{31}$$

It is easy to see that $\gamma(a)(s) = \alpha_s^{-1}\big(a(s)\big)$ for any $a \in C_b(G, \mathcal{M}) \subset \mathcal{N}$, where $C_b(G, \mathcal{M})$ means the algebra of $\mathcal{M}$-valued bounded σ-strong* continuous functions on G. Hence for any $r, s \in G$, we have

$$\begin{aligned}
\big[\gamma\beta_r\gamma^{-1}(a)\big](s) &= \alpha_s^{-1}\Big(\big(\beta_r \circ \gamma^{-1}(a)\big)(s)\Big) = \alpha_s^{-1}\alpha_r\Big(\big(\gamma^{-1}(a)\big)(r^{-1}s)\Big)\\
&= \alpha_s^{-1} \circ \alpha_r \circ \alpha_{r^{-1}s}\big(a(r^{-1}s)\big) = \big[(\mathrm{id} \otimes \lambda_r)(a)\big](s),
\end{aligned}$$

thus

$$\gamma\beta_r\gamma^{-1} = \mathrm{id} \otimes \lambda_r, \quad r \in G. \tag{32}$$

Let μ be the faithful weight on $L^\infty(G)$ obtained by the left Haar measure integration:

$$\mu(f) = \int_G f(s)\,\mathrm{d}s, \quad f \in L^\infty(G)_+. \tag{33}$$

Then the weight $\psi = \varphi \otimes \mu$ on $\mathcal{N}$ is invariant with respect to $\mathrm{id} \otimes \lambda$, so that $\chi = \psi \circ \gamma$ is invariant under β. Now, by the chain rule for the cocycle derivatives, Theorem VIII.3.7, we have

$$\begin{aligned}
\big(\mathrm{D}(\psi \circ \beta_s) : \mathrm{D}\psi\big)_t &= \big(\mathrm{D}(\psi \circ \beta_s) : \mathrm{D}(\chi \circ \beta_s)\big)_t \big(\mathrm{D}(\chi \circ \beta_s) : \mathrm{D}\chi\big)_t (\mathrm{D}\chi : \mathrm{D}\psi)_t\\
&= \beta_s^{-1}\big((\mathrm{D}\psi : \mathrm{D}\chi)_t\big)(\mathrm{D}\chi : \mathrm{D}\psi)_t.
\end{aligned}$$

On the other hand, we have

$$\begin{aligned}
\big(\mathrm{D}(\psi \circ \beta_s) : \mathrm{D}\psi\big)_t &= \Big(\mathrm{D}(\varphi \circ \alpha_s \otimes \mu \circ \lambda_s) : \mathrm{D}(\varphi \otimes \mu)\Big)_t\\
&= \big(\mathrm{D}(\varphi \circ \alpha_s) : \mathrm{D}\varphi\big)_t \otimes \big(\mathrm{D}(\mu \circ \alpha_s) : \mathrm{D}\mu\big)_t\\
&= \big(\mathrm{D}(\varphi \circ \alpha_s) : \mathrm{D}\varphi\big)_t \otimes 1.
\end{aligned}$$

Therefore, we have

$$\big(\mathrm{D}(\varphi \circ \alpha_s) : \mathrm{D}\varphi\big)_t \otimes 1 = \beta_s^{-1}\big((\mathrm{D}\psi : \mathrm{D}\chi)_t\big)(\mathrm{D}\chi : \mathrm{D}\psi)_t.$$

The right hand side of the above equality is a continuous function on $G \times \mathbf{R}$. Hence we complete the proof. Q.E.D.

Recalling the arguments in §3, Chapter VIII, we observe that

$$\sigma_t^{\varphi\circ\alpha_s,\varphi}(x) = \big(\mathrm{D}(\varphi\circ\alpha_s) : \mathrm{D}\varphi\big)_t \sigma_t^{\varphi}(x), \quad x \in \mathcal{M}, \quad t \in \mathbf{R} \tag{34}$$

is a one parameter group in $\mathcal{L}(\mathcal{M})$ for each $s \in G$ and that the map: $t \in \mathbf{R} \mapsto \sigma_t^{\varphi\circ\alpha_s,\varphi}(x)$, $x \in \mathcal{M}$, is σ-strongly* continuous. We then define the one parameter group $\{\rho_t^{\varphi}\}$ of transformations on $\mathcal{K}(G, \mathcal{M})$ by the following:

$$\big[\rho_t^{\varphi}(x)\big](s) = \delta_G(s)^{it}\sigma_t^{\varphi\circ\alpha_s,\varphi}\big(x(s)\big), \quad x \in \mathcal{K}(G, \mathcal{M}), \quad (s,t) \in G \times \mathbf{R}. \tag{35}$$

Lemma 1.12.

(i) *The family* $\{\rho_t^{\varphi}\}$ *is a one parameter automorphism group of the* **-algebra* $\mathcal{K}(G, \mathcal{M})$.

(ii) *For each* $x \in \mathcal{K}(G, \mathcal{M})$ *and* $a \in \mathcal{M}$, *we have*

$$\rho_t^{\varphi}(a \cdot x) = \sigma_t^{\varphi}(a) \cdot \rho_t^{\varphi}(x), \qquad \rho_t^{\varphi}(x \cdot a) = \rho_t^{\varphi}(x)\sigma_t^{\varphi}(a). \tag{36}$$

(iii) $\rho_t^{\varphi}\big(\mathfrak{b}_{\varphi} \cap \mathfrak{b}_{\varphi}^{\sharp}\big) = \mathfrak{b}_{\varphi} \cap \mathfrak{b}_{\varphi}^{\sharp}$.

The proof is a straightforward calculation. We leave it to the reader.

We shall show in several steps that $\tilde{\mathfrak{A}}_{\varphi} = \tilde{\eta}_{\varphi}\big(\mathfrak{b}_{\varphi} \cap \mathfrak{b}_{\varphi}^{\sharp}\big)$ is a left Hilbert algebra with left von Neumann algebra $\mathcal{R}_{\ell}(\tilde{\mathfrak{A}}_{\varphi}) = \mathcal{M} \rtimes_{\alpha} G$, and construct its modular operator $\tilde{\Delta}_{\varphi}$ and the modular conjugation $\tilde{J}$ explicitly.

Lemma 1.13. *There exists a non-singular positive self-adjoint operator* $\tilde{\Delta}_{\varphi}$ *and a conjugation* $\tilde{J}$ [4] *on* $L^2(G, \mathfrak{H}_{\varphi})$ *such that*

$$\left.\begin{aligned} \big(\tilde{\Delta}_{\varphi}^{it}\xi\big)(s) &= \delta_G(s)^{it}\Delta_{\varphi\circ\alpha_s,\varphi}^{it}\xi(s), && \xi \in L^2(G, \mathfrak{H}_{\varphi}), \quad t \in \mathbf{R}, \\ \big(\tilde{J}\xi\big)(s) &= \delta_G(s)^{-\frac{1}{2}}U_{\varphi}(s)^*J\xi(s^{-1}), && s \in G. \end{aligned}\right\} \tag{37}$$

PROOF: We know $\Delta_{\varphi\circ\alpha_s,\varphi}^{it} = \big(\mathrm{D}(\varphi\circ\alpha_s) : \mathrm{D}\varphi\big)_t \Delta_{\varphi}^{it}$. So we define a one parameter unitary group $u_{\varphi}(t)$ on $L^2(G, \mathfrak{H}_{\varphi})$ by:

$$\big(u_{\varphi}(t)\xi\big)(s) = \delta_G(s)^{it}\Delta_{\varphi\circ\alpha_s,\varphi}^{it}\xi(s), \quad \xi \in L^2(G, \mathfrak{H}_{\varphi}).$$

It then follows that $\{u_{\varphi}(t)\}$ is strongly continuous. By Stone's theorem, $\{u_{\varphi}(t)\}$ is of the form: $u_{\varphi}(t) = \tilde{\Delta}_{\varphi}^{it}$.

If we define $\tilde{J}$ by the second formula of (37), then $\tilde{J}$ is a conjugate linear isometry. By the commutativity of J and $U_{\varphi}(s)$, we have

$$\tilde{J}^2\xi(s) = \delta_G(s)^{-\frac{1}{2}}U_{\varphi}(s)^*J\big(\delta_G(s)^{\frac{1}{2}}U_{\varphi}(s)J\xi(s)\big) = \xi(s). \qquad \text{Q.E.D.}$$

[4] A conjugation means a conjugate linear involutive isometry.

Lemma 1.14. *For any* $\tau \in \mathbf{R}$, $\tilde{\Delta}_\varphi^\tau$ *is characterized by the following:*

(i) $\mathfrak{D}(\tilde{\Delta}_\varphi^\tau)$ *consists of those* $\xi \in L^2(G, \mathfrak{H}_\varphi)$ *such that* $\xi(s) \in \mathfrak{D}(\Delta_{\varphi\circ\alpha_s,\varphi}^\tau)$ *for almost every* $s \in G$, *and* $\eta(s) = \delta_G(s)^\tau \Delta_{\varphi\circ\alpha_s,\varphi}^\tau \xi(s)$ *belongs to* $L^2(G, \mathfrak{H}_\varphi)$;

(ii) $$\big(\tilde{\Delta}_\varphi^\tau \xi\big)(s) = \delta_G(s)^\tau \Delta_{\varphi\circ\alpha_s,\varphi}^\tau \xi(s), \quad \xi \in \mathfrak{D}(\tilde{\Delta}_\varphi^\tau) \tag{38}$$

The proof is a straightforward calculation based on (37), and Lemma VI.2.3. We leave the detail to the reader.

Lemma 1.15.

(i) $\tilde{\mathfrak{A}}_\varphi = \tilde{\eta}_\varphi(\mathfrak{b}_\varphi \cap \mathfrak{b}_\varphi^\sharp)$ *is a left Hilbert algebra with respect to the following operations:*

$$\begin{aligned} \tilde{\eta}_\varphi(x)\tilde{\eta}_\varphi(y) &= \tilde{\eta}_\varphi(x * y), \quad x, y \in \mathfrak{b}_\varphi \cap \mathfrak{b}_\varphi^\sharp, \\ \tilde{\eta}_\varphi(x)^\sharp &= \tilde{\eta}_\varphi(x^\sharp). \end{aligned} \tag{39}$$

(ii) *The closure* $\tilde{S}_\varphi$ *of the* $\sharp$*-operation of* $\tilde{\mathfrak{A}}_\varphi$ *has the polar decomposition:*

$$\tilde{S}_\varphi = \tilde{J}\tilde{\Delta}_\varphi^{\frac{1}{2}}. \tag{40}$$

(iii) $$\tilde{\Delta}_\varphi^{it}\tilde{\eta}_\varphi(x) = \tilde{\eta}_\varphi(\rho_t^\varphi(x)), \quad x \in \mathfrak{b}_\varphi \cap \mathfrak{b}_\varphi^\sharp, \quad t \in \mathbf{R}. \tag{41}$$

(iv) $\mathcal{R}_\ell(\tilde{\mathfrak{A}}_\varphi) = \mathcal{M} \rtimes_\alpha G$.

PROOF:

(iii) If $x \in \mathfrak{b}_\varphi \cap \mathfrak{b}_\varphi^\sharp$ and $t \in \mathbf{R}$, then

$$\begin{aligned} \tilde{\eta}_\varphi(\rho_t^\varphi(x))(s) &= \eta_\varphi\Big(\delta_G(s)^{it}\sigma_t^{\varphi\circ\alpha_s,\varphi}\big(x(s)\big)\Big) \\ &= \delta_G(s)^{it}\big(\mathrm{D}(\varphi\circ\alpha_s) : \mathrm{D}\varphi\big)_t \eta_\varphi\Big(\sigma_t^\varphi\big(x(s)\big)\Big) \\ &= \delta_G(s)^{it}\big(\mathrm{D}(\varphi\circ\alpha_s) : \mathrm{D}\varphi\big)_t \Delta_\varphi^{it}\eta_\varphi\big(x(s)\big) \\ &= \delta_G(s)^{it}\Delta_{\varphi\circ\alpha_s,\varphi}^{it}\eta_\varphi\big(x(s)\big) \\ &= \big[\tilde{\Delta}_\varphi^{it}\eta_\varphi(x)\big](s) \qquad \text{by (37).} \end{aligned}$$

(ii) Based on Proposition IX1.17, we compute for $x \in \mathfrak{b}_\varphi \cap \mathfrak{b}_\varphi^\sharp$:

$$\begin{aligned} \big\{\tilde{J}\tilde{\eta}_\varphi(x^\sharp)\big\}(s) &= \delta_G(s)^{-\frac{1}{2}}U_\varphi(s^{-1})J\eta_\varphi\big(x^\sharp(s^{-1})\big) \\ &= \delta_G(s)^{\frac{1}{2}}JU_\varphi(s^{-1})\eta_\varphi\big(\alpha_s(x(s)^*)\big) \\ &= \delta_G(s)^{\frac{1}{2}}J\eta_{\varphi\circ\alpha_s}\big(x(s)^*\big) = \delta_G(s)^{\frac{1}{2}}JS_{\varphi\circ\alpha_s,\varphi}\eta_\varphi\big(x(s)\big) \\ &= \delta_G(s)^{\frac{1}{2}}\Delta_{\varphi\circ\alpha_s,\varphi}^{\frac{1}{2}}\eta_\varphi\big(x(s)\big) = \delta_G(s)^{\frac{1}{2}}\Delta_{\varphi\circ\alpha_s,\varphi}^{\frac{1}{2}}\big\{\tilde{\eta}_\varphi(x)\big\}(s). \end{aligned}$$

By Lemma 1.14, $\tilde{\eta}_\varphi(x) \in \mathfrak{D}(\tilde{\Delta}_\varphi^{1/2})$ and

$$\tilde{J}\tilde{\eta}_\varphi(x^\sharp) = \tilde{\Delta}_\varphi^{\frac{1}{2}}\tilde{\eta}_\varphi(x),$$

so

$$\tilde{\eta}_\varphi(x^\sharp) = \tilde{J}\tilde{\Delta}_\varphi^{\frac{1}{2}}\tilde{\eta}_\varphi(x), \qquad x \in \mathfrak{b}_\varphi \cap \mathfrak{b}_\varphi^\sharp.$$

Therefore, the $\sharp$-operation in $\tilde{\mathfrak{A}}_\varphi$ is preclosed and its closure $\tilde{S}_\varphi$ is a restriction of $\tilde{J}\tilde{\Delta}_\varphi^{1/2}$. Hence, if $\tilde{\mathfrak{A}}_\varphi$ is a core of $\tilde{\Delta}_\varphi^{1/2}$, then we get (40). But this follows by A.4 from the invariance of $\tilde{\mathfrak{A}}_\varphi$ under $\tilde{\Delta}_\varphi^{it}$ as seen in (iii) above.

(i) By (24), we have $\tilde{\eta}_\varphi(x*y) = \tilde{\pi}_\alpha(x)\tilde{\eta}_\varphi(y)$ for $x, y \in \mathfrak{b}_\varphi \cap \mathfrak{b}_\varphi^\sharp$. Hence the left multiplication operator $\pi_\ell(\xi)\colon \eta \in \tilde{\mathfrak{A}}_\varphi \to \xi\eta \in \tilde{\mathfrak{A}}_\varphi$ is bounded and $\pi_\ell(\tilde{\eta}_\varphi(x)) = \tilde{\pi}_\alpha(x)$. Furthermore, $\tilde{\pi}_\alpha(x^\sharp) = \tilde{\pi}_\alpha(x)^*$ by Lemma 1.8, equivalently $(\xi\eta \mid \zeta) = (\eta \mid \xi^\sharp\zeta)$, $\xi, \eta, \zeta \in \tilde{\mathfrak{A}}_\varphi$. Since $\pi_\ell(\tilde{\mathfrak{A}}_\varphi) = \tilde{\pi}_\alpha(\mathfrak{b}_\varphi \cap \mathfrak{b}_\varphi^\sharp)$ is a σ-weakly dense *-subalgebra of $\mathcal{M} \rtimes_\alpha G$, $\tilde{\mathfrak{A}}_\varphi^2$ is dense in $L^2(G, \mathfrak{H}_\varphi)$. Therefore, we conclude that $\tilde{\mathfrak{A}}_\varphi$ is a left Hilbert algebra.

(iv) This is now a triviality. Q.E.D.

Definition 1.16. Given a covariant system $\{\mathcal{M}, G, \alpha\}$, for a faithful weight φ on $\mathcal{M}$. The weight $\tilde{\varphi}$ on $\mathcal{M} \rtimes_\alpha G$ associated with the left Hilbert algebra $\tilde{\mathfrak{A}}_\varphi$ constructed in Lemma 1.15 is called the *dual weight* of φ or the weight *dual* to φ.

Theorem 1.17.

(i) *For each* $x \in \mathfrak{b}_\varphi$, *we have*

$$\tilde{\varphi}\big(\tilde{\pi}_\alpha(x)^*\tilde{\pi}_\alpha(x)\big) = \varphi\big((x^\sharp * x)(e)\big). \tag{42}$$

(ii)

$$\begin{aligned} \sigma_t^{\tilde{\varphi}}\big(\pi_\alpha(x)\big) &= \pi_\alpha\big(\sigma_t^\varphi(x)\big), \qquad x \in \mathcal{M}, \quad t \in \mathbf{R} \\ \sigma_t^{\tilde{\varphi}}\big(\lambda(s)\big) &= \delta_G(s)^{it}\lambda(s)\pi_\alpha\big((\mathrm{D}(\varphi\circ\alpha_s) : \mathrm{D}\varphi)_t\big), \qquad s \in G. \end{aligned} \tag{43}$$

(iii) *If* ψ *is another faithful weight on* $\mathcal{M}$, *then*

$$\big(\mathrm{D}\tilde{\psi} : \mathrm{D}\tilde{\varphi}\big)_t = \pi_\alpha\big((\mathrm{D}\psi : \mathrm{D}\varphi)_t\big), \qquad t \in \mathbf{R}. \tag{44}$$

(iv) *There exists uniquely an operator valued weight* T *from* $\mathcal{M} \rtimes_\alpha G$ *onto* $\pi_\alpha(\mathcal{M})$ *such that*

$$\tilde{\varphi}(x) = \varphi\circ\pi_\alpha^{-1}\big(T(x)\big), \qquad x \in (\mathcal{M} \rtimes_\alpha G)_+.$$

for any faithful semi-finite normal weight φ *on* $\mathcal{M}$.

PROOF:

(i) Let $\mathfrak{B}$ be the algebra of all left bounded vectors in $L^2(G, \mathfrak{H}_\varphi)$ with respect to $\tilde{\mathfrak{A}}_\varphi$. For each $x \in \mathcal{R}_\ell(\tilde{\mathfrak{A}}_\varphi)_+ = (\mathcal{M} \rtimes_\alpha G)_+$, we have

$$\tilde{\varphi}(x) = \begin{cases} \|\xi\|^2 & \text{if } x = \pi_\ell(\xi)^*\pi_\ell(\xi), \quad \xi \in \mathfrak{B}, \\ +\infty & \text{otherwise.} \end{cases}$$

Fix an $x \in \mathfrak{b}_\varphi$. Let $\{a_i\}_{i\in I}$ be a net in $\mathfrak{n}_\varphi^*$ converging to 1 strongly. We then have

$$a_i \cdot x \in \mathfrak{n}_\varphi^* \cdot \mathcal{K}(G, \mathcal{M}) \cdot \mathfrak{n}_\varphi \subset \mathfrak{b}_\varphi \cap \mathfrak{b}_\varphi^\sharp,$$

and $\tilde{\pi}_\alpha(a_i \cdot x) = \pi_\alpha(a_i)\tilde{\pi}_\alpha(x)$ converges to $\tilde{\pi}_\alpha(x)$ strongly, and

$$\left\|\tilde{\eta}_\varphi(x) - \tilde{\eta}_\varphi(a_i \cdot x)\right\| = \left\|\tilde{\eta}_\varphi(x) - \pi_\alpha(a_i)\tilde{\eta}_\varphi(x)\right\| \to 0.$$

Hence, if $\zeta \in \tilde{\mathfrak{A}}'_\varphi$, then we have

$$\begin{aligned} \pi_r(\zeta)\tilde{\eta}_\varphi(x) &= \lim_i \pi_r(\zeta)\pi_\alpha(a_i)\tilde{\eta}_\varphi(x) = \lim_i \pi_r(\zeta)\tilde{\eta}_\varphi(a_i \cdot x) \\ &= \lim_i \tilde{\pi}_\alpha(a_i \cdot x)\zeta = \tilde{\pi}_\alpha(x)\zeta, \end{aligned}$$

so that $\tilde{\eta}_\varphi(x) \in \mathfrak{B}$ and $\pi_\ell\big(\tilde{\eta}_\varphi(x)\big) = \tilde{\pi}_\alpha(x)$. Hence by (25) we get (42).

(ii) This follows directly from $\sigma_t^{\tilde{\varphi}}(x) = \tilde{\Delta}_\varphi^{it} x \tilde{\Delta}_\varphi^{-it}$ for $x \in \mathcal{M} \rtimes_\alpha G$. [STOP]

To prove (iii) and (iv), we need some preparations.

Lemma 1.18. *On $\mathcal{M} \rtimes_\alpha G$, the dual weight $\tilde{\varphi}$ of a faithful weight φ on $\mathcal{M}$ is uniquely determined by* (42) *and* (43).

PROOF: Let ψ be a faithful weight on $\mathcal{M} \rtimes_\alpha G$ satisfying (42) and (43) replacing $\tilde{\varphi}$ by ψ. We want to prove $\psi = \tilde{\varphi}$. The modular automorphism groups $\{\sigma_t^{\tilde{\varphi}}\}$ and $\{\sigma_t^\psi\}$ agree on the generators of $\mathcal{M} \rtimes_\alpha G$ by (43), so that we have $\sigma_t^{\tilde{\varphi}} = \sigma_t^\psi$, $t \in \mathbf{R}$. By Corollary VIII.3.6, there exists a positive self-adjoint operator h affiliated with the center $\mathcal{C}$ of $\mathcal{M} \rtimes_\alpha G$ such that $\tilde{\varphi} = \psi_h$. If $h \neq 1$, then there exists a non-zero projection $p \in \mathcal{C}$ and $\varepsilon > 0$ such that either $(1+\varepsilon)p \leq hp$ or $(1-\varepsilon)p \geq hp$. We assume the former case. The latter case can be disproved similarly. By (42), we know that ψ and $\tilde{\varphi}$ agree on $\tilde{\pi}_\alpha\big(\mathfrak{b}_\varphi^\sharp * \mathfrak{b}_\varphi\big)$. Hence if $a, x \in \tilde{\pi}_\alpha\big(\mathfrak{b}_\varphi^\sharp * \mathfrak{b}_\varphi\big)$, then we have

$$\psi(a^*xa) = \tilde{\varphi}(a^*xa) = \psi_h(a^*xa).$$

Then both sides of the above equality are continuous functions of x, so that it holds for every $x \in \mathcal{M} \rtimes_\alpha G$. Setting $x = p$, we have

$$(1+\varepsilon)\psi(a^*pa) \leq \psi(a^*hpa) = \tilde{\varphi}(a^*pa) = \psi(a^*pa).$$

Hence $\psi(a^*pa) = 0$, i.e. $pa = 0$ for every $a \in \tilde{\pi}_\alpha\big(\mathfrak{b}_\varphi^\sharp * \mathfrak{b}_\varphi\big)$. Therefore, we get $p = 0$, a contradiction. Thus $h = 1$. Q.E.D.

Now with $M_2 = M(2, \mathbf{C})$, we construct a new covariant system

$$\{\mathcal{N}, G, \beta\} = \{\mathcal{M} \otimes M_2, G, \alpha \otimes \mathrm{id}\}.$$

Naturally we identify the crossed products:

$$(\mathcal{M} \otimes M_2) \rtimes_\beta G = (\mathcal{M} \rtimes_\alpha G) \otimes M_2.$$

Lemma 1.19. *For faithful weights φ and ψ on $\mathcal{M}$, we consider the balanced weight $\chi = \varphi \oplus \psi$ on $\mathcal{N} = \mathcal{M} \otimes M_2$. Then we have*

$$\left(\mathrm{D}(\chi \circ \beta_s) : \mathrm{D}\chi\right)_t = \begin{pmatrix} \left(\mathrm{D}(\varphi \circ \alpha_s) : \mathrm{D}\varphi\right)_t & 0 \\ 0 & \left(\mathrm{D}(\psi \circ \alpha_s) : \mathrm{D}\psi\right)_t \end{pmatrix}. \tag{45}$$

PROOF: With matrix unit $\{e_{i,j}\}$ of M_2, for each $x = \sum x_{ij} \otimes e_{ij} \in \mathcal{N}_+$ we have

$$\chi \circ \beta_s(x) = \varphi \circ \alpha_s(x_{11}) + \chi \circ \alpha_s(x_{22}),$$

so that $1 \otimes e_{11}$ and $1 \otimes e_{22}$ both belong to the centralizer $\mathcal{N}_{\chi \circ \beta_s}$. By the formula

$$\sigma_t^{\varphi,\psi}(xy) = \sigma_t^{\varphi,\omega}(x)\,\sigma_t^{\omega,\psi}(y), \quad x, y \in \mathcal{M}, \tag{46}$$

for the third faithful weight ω on $\mathcal{M}$ (cf. Chapter VIII, §3), we have

$$\begin{aligned}\sigma_t^{\chi\circ\beta_s,\chi}(1 \otimes e_{ii}) &= \sigma_t^{\chi\circ\beta_s}(1 \otimes e_{ii})\,\sigma_t^{\chi\circ\beta_s,\chi}(1 \otimes e_{ii})\,\sigma_t^{\chi}(1 \otimes e_{ii}) \\ &= (1 \otimes e_{ii})\,\sigma_t^{\chi\circ\beta_s,\chi}(1 \otimes e_{ii})(1 \otimes e_{ii}),\end{aligned}$$

so that we get, with a one parameter continuous family $\{u_t^i\}$ in $\mathcal{U}(\mathcal{M})$,

$$\sigma_t^{\chi\circ\beta_s,\chi}(1 \otimes e_{ii}) = u_t^i \otimes e_{ii}.$$

By the relative modular condition, we can conclude that

$$u_t^1 = \left(\mathrm{D}(\varphi \circ \alpha_s) : \mathrm{D}\varphi\right)_t, \qquad u_t^2 = \left(\mathrm{D}(\psi \circ \alpha_s) : \mathrm{D}\psi\right)_t.$$

Hence we have

$$\begin{aligned}\left(\mathrm{D}(\chi \circ \beta_s) : \mathrm{D}\chi\right)_t &= \sigma_t^{\chi\circ\beta_s,\chi}(1) = \sigma_t^{\chi\circ\beta_s,\chi}\left(1 \otimes e_{11} + 1 \otimes e_{22}\right) \\ &= u_t^1 \otimes e_{11} + u_t^2 \otimes e_{22}.\end{aligned}$$

Q.E.D.

Lemma 1.20. *With φ, ψ and χ as before, let $\tilde{\varphi}$, $\tilde{\psi}$ and $\tilde{\chi}$ be their dual weights. Then*

$$\tilde{\chi}\left(\sum y_{ij} \otimes e_{ij}\right) = \tilde{\varphi}(y_{11}) + \tilde{\psi}(y_{22}), \qquad y = \sum y_{ij} \otimes e_{ij} \in \mathcal{N} \rtimes_\beta G,$$

i.e. $\tilde{\chi}$ is the balanced weight of $\tilde{\varphi}$ and $\tilde{\psi}$.

PROOF: Let $\{\pi_\beta, \bar\lambda\}$ be the covariant representation of $\{\mathcal{N}, G, \beta\}$ associated with the crossed product $\mathcal{N} \rtimes_\beta G$. It then follows that

$$\pi_\beta\Big(\sum x_{ij} \otimes e_{ij}\Big) = \sum \pi_\alpha(x_{ij}) \otimes e_{ij}, \quad x_{ij} \in \mathcal{M},$$

$$\bar\lambda(s) = \lambda(s) \otimes 1, \quad s \in G.$$

Consequently, the representation $\tilde\pi_\beta$ of $\mathcal{K}(G, \mathcal{N}) = \mathcal{K}(G, \mathcal{M}) \otimes M_2$ is given by

$$\tilde\pi_\beta\Big(\sum y_{ij} \otimes e_{ij}\Big) = \sum \tilde\pi_\alpha(y_{ij}) \otimes e_{ij}, \quad y_{ij} \in \mathcal{K}(G, \mathcal{M}).$$

Hence, by Theorem 1.17.(ii) we have

$$\sigma_t^{\tilde\chi}(1 \otimes e_{ii}) = \pi_\beta\big(\sigma_t^{\chi}(1 \otimes e_{ii})\big) = \pi_\beta(1 \otimes e_{ii}) = 1 \otimes e_{ii}, \quad i = 1, 2.$$

Thus, $1 \otimes e_{11}$ and $1 \otimes e_{22}$ both belong to $(\mathcal{N} \rtimes_\beta G)$. By Theorem VIII.2.6, we have for each $x \in (\mathcal{M} \rtimes_\alpha G) \otimes M_2$

$$\tilde\chi(x) = \tilde\chi\big((1 \otimes e_{11})x(1 \otimes e_{11})\big) + \tilde\chi\big((1 \otimes e_{22})x(1 \otimes e_{22})\big).$$

This means that $\tilde\chi$ is the balanced weight of ω_1 and ω_2 where ω_1 and ω_2 are the weights on $\mathcal{M} \rtimes_\alpha G$ given by:

$$\omega_1(x) = \tilde\chi(x \otimes e_{11}) \quad \text{and} \quad \omega_2(x) = \tilde\chi(x \otimes e_{22}).$$

We now want to prove $\omega_1 = \tilde\varphi$ and $\omega_2 = \tilde\psi$. For each $x \in \mathcal{M}$, we have

$$\begin{aligned}\sigma_t^{\omega_1}\big(\pi_\alpha(x)\big) \otimes e_{11} &= \sigma_t^{\tilde\chi}\big(\pi_\alpha(x) \otimes e_{11}\big) = \sigma_t^{\tilde\chi}\big(\pi_\beta(x \otimes e_{11})\big) = \pi_\beta\big(\sigma_t^{\chi}(x \otimes e_{11})\big)\\ &= \pi_\beta\big(\sigma_t^{\varphi}(x) \otimes e_{11}\big) = \pi_\alpha\big(\sigma_t^{\varphi}(x)\big) \otimes e_{11},\end{aligned}$$

so that $\sigma_t^{\omega_1}\big(\pi_\alpha(x)\big) = \sigma_t^{\tilde\varphi}\big(\pi_\alpha(x)\big)$, $x \in \mathcal{M}$. By (45), we get

$$\begin{aligned}\sigma_t^{\omega_1}\big(\lambda(s)\big) \otimes e_{11} &= \sigma_t^{\tilde\chi}\big(\lambda(s) \otimes e_{11}\big) = \sigma_t^{\tilde\chi}\big(\bar\lambda(s)\big)(1 \otimes e_{11})\\ &= \delta_G(s)^{it}\bar\lambda(s)\pi_\beta\Big(\big(\mathrm{D}(\chi \circ \beta_s) : \mathrm{D}\chi\big)_t\Big)(1 \otimes e_{11})\\ &= \delta_G(s)^{it}\Big(\lambda(s)\pi_\alpha\big(\big(\mathrm{D}(\varphi \circ \alpha_s) : \mathrm{D}\varphi\big)_t\big) \otimes e_{11}\Big)\\ &= \sigma_t^{\tilde\varphi}\big(\lambda(s)\big) \otimes e_{11}.\end{aligned}$$

If $x \in \mathfrak{b}_\varphi = \mathcal{K}(G, \mathcal{M}) \cdot \mathfrak{n}_\varphi$, then

$$x \otimes e_{11} \in \big(\mathcal{K}(G, \mathcal{M})(1 \otimes e_{11}) \cdot (\mathfrak{n}_\varphi \otimes e_{11})\big) \subseteq \mathcal{K}\big(G, \mathcal{M} \otimes M_2\big) \cdot \mathfrak{n}_\chi = \mathfrak{b}_\chi.$$

With $y = x^\sharp * x$, $y \otimes e_{11} = (x \otimes e_{11})^\sharp * (x \otimes e_{11})$. By Theorem 1.17.(i), we have

$$\begin{aligned}\omega_1\big(\tilde\pi_\alpha(x^\sharp * x)\big) &= \tilde\chi\big(\tilde\pi_\alpha(y) \otimes e_{11}\big) = \tilde\chi\big(\tilde\pi_\beta(y \otimes e_{11})\big) = \chi\big((y \otimes e_{11})(e)\big)\\ &= \chi\big(y(e) \otimes e_{11}\big) = \varphi\big((x^\sharp * x)(e)\big) = \tilde\varphi\big(\tilde\pi_\alpha(x^\sharp * x)\big).\end{aligned}$$

By Lemma 1.18, we conclude $\omega_1 = \tilde\varphi$. Similarly $\omega_2 = \tilde\psi$. Q.E.D.

PROOF OF THEOREM 1.17.(iii) AND (iv):

(iii) Let $\tilde{\varphi}$, $\tilde{\psi}$ and $\tilde{\chi}$ be as before. We then have

$$\begin{aligned}(D\tilde{\psi} : D\tilde{\varphi})_t \otimes e_{21} &= \sigma_t^{\tilde{\chi}}(1 \otimes e_{21}) = \sigma_t^{\tilde{\chi}}\big(\pi_\beta(1 \otimes e_{21})\big) = \pi_\beta\big(\sigma_t^{\chi}(1 \otimes e_{21})\big)\\ &= \pi_\beta\big((D\psi : D\varphi)_t \otimes e_{21}\big) = \pi_\alpha\big((D\psi : D\varphi)_t\big) \otimes e_{21}.\end{aligned}$$

Hence (44) follows.

(iv) This follows immediately from Theorem IX.4.18. Q.E.D.

We conclude this section with the following description of the commutant of $\mathcal{M} \rtimes_\alpha G$.

Theorem 1.21. *Suppose that a covariant system $\{\mathcal{M}, G, \alpha\}$ is represented by a covariant representation $\{\rho, V, \mathfrak{K}\}$. Define*

$$(W\xi)(s) = V(s)\xi(s), \quad \xi \in L^2(G, \mathfrak{K}). \tag{47}$$

Then with ρ_α and $\lambda_\mathfrak{K}$ defined by (11), *we have*

$$\{\rho_\alpha(\mathcal{M}) \cup \lambda_\mathfrak{K}(G)\}'' = \Big\{W^*\big(\rho(\mathcal{M}) \otimes \mathbf{C}\big)W\Big\} \vee \big(\mathbf{C} \otimes \mathcal{R}_\ell(G)\big), \tag{48}$$

$$\{\rho_\alpha(\mathcal{M}) \cup \lambda_\mathfrak{K}(G)\}' = \big(\rho(\mathcal{M})' \otimes \mathbf{C}\big) \vee W^*\big(\mathbf{C} \otimes \mathcal{R}_r(G)\big)W, \tag{49}$$

where $\mathcal{R}_\ell(G)$ and $\mathcal{R}_r(G)$ are respectively the left and right von Neumann algebras of G.

PROOF: Let λ_ℓ and λ_r be the left and right regular representations of G, i.e.,

$$\big(\lambda_\ell(s)\xi\big)(r) = \xi(s^{-1}r), \quad \xi \in L^2(G), \qquad \big(\lambda_r(s)\xi\big)(r) = \delta_G(r)^{\frac{1}{2}}\xi(rs), \quad r, s \in G.$$

For each $x \in \mathcal{M}$ and $\xi \in L^2(G, \mathfrak{K})$, we have

$$\begin{aligned}\big(\rho_\alpha(x)\xi\big)(s) &= \rho\big(\alpha_s^{-1}(x)\big)\xi(s) = V(s)^*\rho(x)V(s)\xi(s)\\ &= \Big(W^*\big(\rho(x) \otimes 1\big)W\xi\Big)(s).\end{aligned}$$

so that

$$\rho_\alpha(x) = W^*\big(\rho(x) \otimes 1\big)W, \quad x \in \mathcal{M}. \tag{50}$$

Clearly, we have $\lambda_\mathfrak{K}(s) = \lambda_\ell(s) \otimes 1,\ s \in G$. Hence

$$\big[\rho_\alpha(\mathcal{M}) \cup \lambda_\mathfrak{K}(G)\big]'' = W^*\big(\rho(\mathcal{M}) \otimes \mathbf{C}\big)W \vee \big(\mathbf{C} \otimes \mathcal{R}_\ell(G)\big).$$

We first show (49) when $\{\rho, V, \mathfrak{K}\}$ is the standard representation $\{\pi_\varphi, U_\varphi, \mathfrak{H}_\varphi\}$. We omit π_φ as before, and assume $\mathcal{M}$ acts on $\mathfrak{H}_\varphi$. So we use the previous notations. Then we know $\mathcal{M} \rtimes_\alpha G = \mathcal{R}_\ell(\tilde{\mathfrak{A}}_\varphi)$, so that

$$(\mathcal{M} \rtimes_\alpha G)' = \tilde{J}(\mathcal{M} \rtimes_\alpha G)\tilde{J}.$$

Hence $(\mathcal{M} \rtimes_\alpha G)'$ is generated by $\tilde{J}\pi_\alpha(\mathcal{M})\tilde{J}$ and $\tilde{J}\big(\mathbf{C} \otimes \mathcal{R}_\ell(G)\big)\tilde{J}$. Now, we have, for each $x \in \mathcal{M}$,

$$\begin{aligned}
\big(\tilde{J}\pi_\alpha(x)\tilde{J}\xi\big)(s) &= \delta_G(s)^{-\frac{1}{2}} U_\varphi(s)^* J\big(\pi_\alpha(x)\tilde{J}\xi\big)(s^{-1}) \\
&= \delta_G(s)^{-\frac{1}{2}} J U_\varphi(s)^* \alpha_s(x)(\tilde{J}\xi)(s^{-1}) \\
&= J U_\varphi(s)^* \alpha_s(x) U_\varphi(s) J\xi(s) = JxJ\xi(s) = \big\{(JxJ \otimes 1)\xi\big\}(s);
\end{aligned}$$

$$\begin{aligned}
\big(\tilde{J}\lambda(r)\tilde{J}\xi\big)(s) &= \delta_G(s)^{-\frac{1}{2}} U_\varphi(s)^* J\big\{\lambda(r)\tilde{J}\xi\big\}(s^{-1}) \\
&= \delta_G(s)^{-\frac{1}{2}} U_\varphi(s)^* J\big\{\tilde{J}\xi\big\}\big(r^{-1}s^{-1}\big) \\
&= \delta_G(r)^{\frac{1}{2}} U_\varphi(s)^* J U_\varphi(sr) J\xi(sr) \\
&= \delta_G(r)^{\frac{1}{2}} U_\varphi(r)\xi(sr) = \big\{U_\varphi(r) \otimes \lambda_r(r)\xi\big\}(s).
\end{aligned}$$

Hence we obtain

$$\tilde{J}\big(\mathcal{M} \rtimes_\alpha G\big)\tilde{J} = \Big\{\big(\mathcal{M}' \otimes \mathbf{C}\big) \cup \big(U_\varphi \otimes \lambda_r\big)(G)\Big\}''.$$

A direct computation shows that $W^*\big(1 \otimes \lambda_r(s)\big)W = U_\varphi(s) \otimes \lambda_r(s)$, so that (49) follows in the case of the standard representation.

The case $\rho \cong \pi_\varphi$: In this case, we identify ρ and π_φ and omit π_φ as before. We want to show that $\big(\mathcal{M} \rtimes_\alpha G\big)'$ is generated by $\mathcal{M}' \otimes \mathbf{C}$ and $(V \otimes \lambda_r)(G)$. Since V and U_φ give rise to the same action α of G on $\mathcal{M}$, $v(s) = V(s)U_\varphi(s)^*$ belongs to $\mathcal{M}'$. Hence we get

$$\begin{aligned}
V(s) \otimes \lambda_r(s) &= \big(v(s) \otimes 1\big)\big(U_\varphi(s) \otimes \lambda_r(s)\big), \\
U_\varphi(s) \otimes \lambda_r(s) &= \big(v(s)^* \otimes 1\big)\big(V(s) \otimes \lambda_r(s)\big),
\end{aligned}$$

so that

$$\begin{aligned}
\Big\{\big(\mathcal{M}' \otimes \mathbf{C}\big) \cup \big((U_\varphi \otimes \lambda_r)(G)\big)\Big\}'' &= \Big\{\big(\mathcal{M}' \otimes \mathbf{C}\big) \cup \big(V \otimes \lambda_r\big)(G)\Big\}'' \\
&= \big(\mathcal{M}' \otimes \mathbf{C}\big) \vee W^*\big(\mathbf{C} \otimes \mathcal{R}_r(G)\big)W.
\end{aligned}$$

The case $\rho \cong \mathrm{id} \otimes \pi_\varphi$: Let Φ be the isomorphism of (12). We identify ρ and $\mathrm{id} \otimes \pi_\varphi$. Then we get

$$v(s) = V(s)\big(1 \otimes U_\varphi(s)\big)^* \in \rho(\mathcal{M})' = \big(\mathbf{C} \otimes \mathcal{M}\big)'.$$

With this, it is easy to determine $\Phi(\mathcal{M} \rtimes_\alpha G)'$.

General case: Let $\mathfrak{H}_0$ be a Hilbert space with sufficiently large dimension so that $\big\{\mathrm{id} \otimes \rho, \mathfrak{H}_0 \otimes \mathfrak{K}\big\} \cong \big\{\mathrm{id} \otimes \pi_\varphi, \mathfrak{H}_0 \otimes \mathfrak{K}_\varphi\big\}$. We set $\bar{\rho} = \mathrm{id} \otimes \rho$ and $\bar{V} = \mathrm{id} \otimes V$. We then have by the previous arguments

$$\big\{\mathbf{C} \otimes \Phi(\mathcal{M} \rtimes_\alpha G)\big\}' = \Big\{\big(\bar{\rho}(\mathcal{M})' \otimes \mathbf{C}\big) \cup \big((\bar{V} \otimes \lambda_r)(G)\big)\Big\}''.$$

If p is a projection of $\mathfrak{H}_0$ to a one-dimensional subspace, then $p \otimes 1_{\mathfrak{K}} \in \bar{\rho}(\mathcal{M})' \cap \bar{V}(G)'$ and we have, in the sense of spatial isomorphism,

$$\begin{aligned}\Phi\big(\mathcal{M} \rtimes_\alpha G\big)' &\cong \Big[\big(\bar{\rho}(\mathcal{M})' \otimes \mathbf{C}\big) \cup \big(\bar{V} \otimes \lambda_r\big)(G)\Big]''_{p\otimes 1} \qquad \text{on } L^2(G, \mathfrak{H}_0 \otimes \mathfrak{K}) \\ &= \Big[\big(\bar{\rho}(\mathcal{M})'_e \otimes \mathbf{C}\big) \cup \big(\bar{V}_e \otimes \lambda_r\big)(G)\Big]'' \\ &\cong \Big[\big(\rho(\mathcal{M})' \otimes \mathbf{C}\big) \cup \big(V \otimes \lambda_r\big)(G)\Big]'' \\ &= \big(\rho(\mathcal{M})' \otimes \mathbf{C}\big) \vee W^*\big(\mathbf{C} \otimes \mathcal{R}_r(G)\big)W. \end{aligned}$$

Q.E.D.

The formula (49) is equivalent to

$$W\big(\rho_\alpha(\mathcal{M}) \cup \lambda_{\mathfrak{K}}(G)\big)' W^* = W\big(\rho(\mathcal{M})' \otimes \mathbf{C}\big) W^* \vee \mathbf{C} \otimes \mathcal{R}_r(G).$$

Now, we define

$$C\xi(s) = \delta_G(s)^{-\frac{1}{2}} \xi(s^{-1}), \quad s \in G, \quad \xi \in L^2(G, \mathfrak{K}); \qquad \tilde{W} = CW. \tag{51}$$

Then for each $y \in \rho(\mathcal{M})'$, we have

$$\big(\tilde{W}(y \otimes 1)\tilde{W}^*\xi\big)(s) = V(s)^* y V(s)\xi(s),$$

$$\Big(\tilde{W}\big(1 \otimes \lambda_r(t)\big)\tilde{W}^*\xi\Big)(s) = \big(1 \otimes \lambda_\ell(t)\big)\xi(s).$$

Therefore, with $\alpha'_s(y) = V(s) y V(s)^*$, $y \in \rho(\mathcal{M})'$, $s \in G$, we obtain

$$\tilde{W}\big(\rho_\alpha(\mathcal{M}) \cup \lambda_{\mathfrak{K}}(G)\big)' \tilde{W}^* \cong \rho(\mathcal{M})' \rtimes_{\alpha'} G.$$

Thus, we conclude the following:

Corollary 1.22.

(i) *If a covariant system* $\{\mathcal{M}, G, \alpha\}$ *is represented on* $\mathfrak{K}$ *via a covariant representation* $\{\rho, V\}$, *then*

$$\big[\rho_\alpha(\mathcal{M}) \cup \lambda_{\mathfrak{K}}(G)\big]'' \cong \mathcal{M} \rtimes_\alpha G, \qquad \big[\rho_\alpha(\mathcal{M}) \cup \lambda_{\mathfrak{K}}(G)\big]' \cong \rho(\mathcal{M})' \rtimes_{\alpha'} G,$$

where α' *is the action of* G *on* $\rho(\mathcal{M})'$ *induced by the unitary representation* V.

(ii) *Let* $\tilde{\mathcal{M}} = \mathcal{M} \mathbin{\overline{\otimes}} \mathcal{L}\big(L^2(G)\big)$ *and*

$$\tilde{\alpha}_s = \alpha_s \otimes \mathrm{Ad}\big(\lambda_r(s)\big), \quad s \in G. \tag{52}$$

Then the fixed point algebra $\tilde{\mathcal{M}}^{\tilde{\alpha}} = \big\{x \in \tilde{\mathcal{M}} : \tilde{\alpha}_s(x) = x, \ s \in G\big\}$ *is isomorphic to* $\mathcal{M} \rtimes_\alpha G$.

Exercise X.1

1) Let $\{\mathcal{M}, G, \alpha\}$ be a covariant system with G a finite group. Assume that the action α is free in the sense that α_s in not inner, $s \in G$, unless $s = e$. Show that if $\{\pi, U, \mathfrak{H}\}$ is a covariant representation, then the von Neumann algebra $\mathcal{N}$ generated by $\pi(\mathcal{M}) \cup U(G)$ is isomorphic to the crossed product $\mathcal{M} \rtimes_\alpha G$ under the natural correspondence.

2) Show by a counter example that the above isomorphism is false unless G is finite. (Hint: In fact, for any infinite locally compact group, the above claim is false. But, at this point we should simply construct an example of an infinite group which violates the above claim. For example, we can take $G = \mathbf{Z}$ and $\mathcal{M} = L^\infty(\mathbf{R}/2\pi\mathbf{Z})$ and $\{\alpha_n(f)\}(t) = f(t + n\theta)$, $f \in \mathcal{M}$, $n \in \mathbf{Z}$, where θ is a fixed irrational number, and $\mathfrak{H} = L^2(\mathbf{R}/2\pi\mathbf{Z})$ acted upon by $\mathcal{M}$ via multiplication and $\{U(n)\xi\}(t) = \xi(t + n\theta)$, $\xi \in \mathfrak{H}$, $n \in \mathbf{Z}$.)

3) Show that if the group G is discrete in the covariant system $\{\mathcal{M}, G, \alpha\}$, the operator valued weight T in Theorem 1.17 is indeed a conditional expectation.

4) Show that if G is a finite group, $\mathcal{M} = L^\infty(G)$ and the action α of G on $\mathcal{M}$ is by translation, then $\mathcal{M} \rtimes_\alpha G$ is isomorphic to the $n \times n$ matrix algebra $M_n(\mathbf{C})$ with $n =$ the order of G.

5) Let G be a locally compact group and $\mathcal{M}$ a von Neumann algebra. Let Act$(G, \mathcal{M})$ be the set of all actions α of G on $\mathcal{M}$ and $Z(G, \mathcal{U}(\mathcal{M})) = \{(u, \alpha) : u \in Z^1(\mathcal{M}, G, \alpha),\ \alpha \in \text{Act}(G, \mathcal{M})\}$. In the notation of Theorem 1.7, for each $(u, \alpha) \in$ Act$(G, \mathcal{M})$, set:

$$U(u, \alpha)\xi(s) = u_{s^{-1}}\xi(s), \quad \xi \in L^2(G, \mathfrak{K}), \quad s \in G$$

as in (13). Observe that the proof for Theorem 1.7.(ii) shows that

$$\left.\begin{aligned} U(u, \alpha)\pi_\alpha(x)U(u, \alpha)^* &= \pi_{{}_u\alpha}(x), & x &\in \mathcal{M} \\ U(u, \alpha)\pi_\alpha(u_t)\lambda(t)U(u, \alpha)^* &= \lambda(t), & t &\in G. \end{aligned}\right\} \tag{E.1}$$

Show that if $(v, {}_u\alpha) \in Z^1(\mathcal{M}, G, {}_u\alpha)$ then $vu \in Z^1(\mathcal{M}, G, \alpha)$ and $U(v, {}_u\alpha)U(u, \alpha) = U(vu, \alpha)$.

6) In the previous problem, let $\mathcal{C}$ be the center of $\mathcal{M}$ and observe that $Z^1(\mathcal{C}, G, \alpha)$ is a group as ${}_u\alpha = \alpha$ for any $u \in Z^1(\mathcal{C}, G, \alpha)$. Let $B^1(\mathcal{C}, G, \alpha) = \{u \in Z^1(\mathcal{C}, G, \alpha) :$ there exists $v \in \mathcal{U}(\mathcal{C})$ with $u_s = (\partial v)_s = v^*\alpha_s(v)\}$.

(a) Show that each $u \in Z^1(\mathcal{C}, G, \alpha)$ gives rise to an automorphism $\widehat{\alpha_u}(x) = U(u, \alpha)xU(u, \alpha)^*$, $x \in \mathcal{M} \rtimes_\alpha G$, of $\mathcal{M} \rtimes_\alpha G$ and that the map: $u \in Z^1(\mathcal{C}, G, \alpha) \mapsto \widehat{\alpha_u} \in \text{Aut}(\mathcal{M} \rtimes_\alpha G)$ is a homomorphism.

(b) Show that if $u \in B^1(\mathcal{C}, G, \alpha)$ then $\widehat{\alpha_u} \in \text{Int}(\mathcal{M} \rtimes_\alpha G)$.

(c) Show under the assumption that $\pi_\alpha(\mathcal{M})' \cap \mathcal{M} \rtimes_\alpha G = \pi_\alpha(\mathcal{C})$ that if $\widehat{\alpha_u} \in \text{Int}(\mathcal{M} \rtimes_\alpha G)$ then $u \in B^1(\mathcal{C}, G, \alpha)$.

§ 2 Duality for Crossed Products by Abelian Groups

In this section, we consider crossed products by abelian groups. We fix a locally compact abelian group G written additively. The dual group $\widehat{G}$ of G is also viewed as an additive group. We recall the Fourier transform $\mathcal{F}$ and the Fourier inverse transform $\widehat{\mathcal{F}}$:

$$(\mathcal{F} f)(p) = \int_G \overline{\langle s, p\rangle} f(s)\, \mathrm{d}s, \quad f \in L^1(G), \tag{1}$$

$$(\widehat{\mathcal{F}} f)(s) = \int_{\widehat{G}} \langle s, p\rangle f(p)\, \mathrm{d}p, \quad f \in L^1(\widehat{G}). \tag{2}$$

The measures $\mathrm{d}s$ and $\mathrm{d}p$ are so adjusted that the Plancherel formula holds:

$$\|f\|_2 = \|\mathcal{F} f\|_2, \quad f \in L^1(G) \cap L^2(G). \tag{3}$$

We also view $\mathcal{F}$ as a unitary from $L^2(G)$ onto $L^2(\widehat{G})$, by extending $\mathcal{F}$ of (1) by L^2-continuity.

On the Hilbert space $L^2(G)$, we define

$$\left.\begin{aligned} \big(\lambda_G(s)\xi\big)(r) &= \xi(r - s), \quad \xi \in L^2(G), \quad r, s \in G, \\ \big(\mu_G(p)\xi\big)(r) &= \overline{\langle r, p\rangle}\xi(r), \quad p \in \widehat{G}. \end{aligned}\right\} \tag{4}$$

It then follows that with $U(s) = \lambda_G(s)$, $s \in G$, and $V(p) = \mu_G(p)$, $p \in \widehat{G}$, U and V satisfy the commutation relation:

$$U(s)V(p)U(s)^* V(p)^* = \langle s, p\rangle, \quad s \in G, \quad p \in \widehat{G}. \tag{5}$$

Definition 2.1. In general, a pair of unitary representations U of G and V of $\widehat{G}$ on the same Hilbert space $\mathfrak{H}$ is said to be *covariant* if the commutation relation (5) is satisfied. The commutation relation (5) is called the *Heisenberg-Weyl commutation relation*.

In other words, λ_G and μ_G are covariant. The next result then asserts that covariant representations are unique up to multiplicity.

Proposition 2.2. *The covariant representation* $\{\lambda_G, \mu_G\}$ *generates the factor* $\mathcal{L}(L^2(G))$ *of all bounded operators. If* $\{U, V, \mathfrak{H}\}$ *is a covariant representation, then there exists a Hilbert space* $\mathfrak{H}_0$ *such that*

$$\{\lambda_G \otimes \mathrm{id},\ \mu_G \otimes \mathrm{id},\ L^2(G) \otimes \mathfrak{H}_0\} \cong \{U, V, \mathfrak{H}\}.$$

The dimension of $\mathfrak{H}_0$ *is called the multiplicity of* $\mathfrak{H}$.

PROOF: Let $A(G)$ be the Fourier algebra of G and recall $A(G) = \widehat{\mathcal{F}}L^1(\widehat{G})$ (cf. Chapter VII, §3). When we identify $A(G)$ and $L^1(\widehat{G})$ however, we should note that the definition of $\mathcal{F}$ and $\widehat{\mathcal{F}}$ is not symmetric. In fact, $f^\vee \in L^1(\widehat{G})$ and $\widehat{\mathcal{F}} f \in A(G)$ must be identified, instead of f and $\widehat{\mathcal{F}} f$. For each $f \in L^1(\widehat{G})$, we define

$$V(f) = \int_{\widehat{G}} f(p)V(p)\,\mathrm{d}p\,.$$

Then V is a *-representation of $L^1(\widehat{G})$, so that it can be extended to the enveloping C^*-algebra $C_\infty(G)$. We shall denote the extended representation of $C_\infty(G)$ by V again. In the case that $V = \mu_G$, $\mu_G(f)$, $f \in C_\infty(G)$, is the multiplication of f on $L^2(G)$. Hence the von Neumann algebra $\mathcal{A}$ generated by $\{\mu_G(f) : f \in C_\infty(G)\}$ is the multiplication algebra $L^\infty(G)$ on $L^2(G)$. So it is maximal abelian. Now, we have

$$\lambda_G(s)\,\mu_G(f)\,\lambda_G(s)^* = \mu_G(\lambda_s f), \quad s \in G, \quad f \in L^\infty(G) \tag{6}$$

where $(\lambda_s f)(r) = f(r - s)$. Hence the operators of $\mathcal{A}$ commuting with $\lambda_G(G)$ are only scalars. Therefore, $\{\lambda_G(G) \cup \mu_G(\widehat{G})\}' = \mathbf{C}$, so that $\{\lambda_G, \mu_G\}$ is irreducible. Thus $\{\lambda_G, \mu_G\}$ generates $\mathcal{L}(L^2(G))$.

In $\mathcal{K}(G \times G)$, the vector space of all continuous functions on $G \times G$ with compact support, we define the involutive algebra structure:

$$\begin{aligned}(f * g)(r, s) &= \int_G f(r, t)g(t, s)\,\mathrm{d}t\,, \quad f, g \in \mathcal{K}(G \times G),\\ f^*(r, s) &= \overline{f(s, r)}, \quad r, s \in G.\end{aligned} \tag{7}$$

For each $f \in \mathcal{K}(G \times G)$, define $f_r \in \mathcal{K}(G)$ by:

$$f_r(s) = f(r + s, s), \quad r, s \in G, \tag{8}$$

and further put

$$\pi_{U,V}(f) = \int_G U(r)V(f_r)\,\mathrm{d}r\,, \quad f \in \mathcal{K}(G \times G). \tag{9}$$

It then follows that $\pi_{U,V}$ is a *-representation of the *-algebra $\mathcal{K}(G \times G)$ of (7). We leave the detail to the reader as an exercise.

In the case that $U = \lambda_G$ and $V = \mu_G$, $\pi_{U,V}(f)$, $f \in \mathcal{K}(G \times G)$, is given by the integral operator $\pi_G(f)$:

$$\pi_G(f)\xi(r) = \int_G f(r, s)\xi(s)\,\mathrm{d}s\,.$$

Hence if $f(r, s) = f_1(r)f_2(s)$, i.e. $f = f_1 \otimes f_2$, then $\pi_G(f)$ is an operator of rank one. With $g = g_1 \otimes g_2$ and $f = f_1 \otimes f_2$, $g_1, g_2, f_1, f_2 \in \mathcal{K}(G)$, $\pi_G(g^* * \mathcal{K}(G \times G) * f)$ are rank one operators mapping $\mathbf{C}\bar{f_2}$ onto $\mathbf{C}g_1$.

Now, let $\mathcal{B}$ be the von Neumann algebra generated by $U(G)$ and $V(\widehat{G})$, which is also generated by $\pi_{U,V}(\mathcal{K}(G \times G))$. If $f = f_1 \otimes f_2$ and $g = g_1 \otimes g_2$ as before, then $\pi_{U,V}(g)\, \mathcal{B}\, \pi_{U,V}(f)^*$ is of dimension one and consists of scalar multiples of a partial isometry from the range of $\pi_{U,V}(f)$ onto the range of $\pi_{U,V}(g)$. Hence $\mathcal{B}$ is a factor of type I, and isomorphic to $\mathcal{L}(L^2(G))$.

For a minimal projection e of $\mathcal{B}'$, putting $\mathfrak{H}_0 = e\mathfrak{H}$, we get

$$\{\pi_{U,V}, \mathfrak{H}\} \cong \{\pi_G \otimes \mathrm{id},\ L^2(G) \otimes \mathfrak{H}_0\}.$$

Q.E.D.

We now consider a covariant system $\{\mathcal{M}, G, \alpha\}$ over a locally compact abelian group G. Let $\mathcal{M}$ act on a Hilbert space $\mathfrak{H}$. On $L^2(G, \mathfrak{H})$, we define

$$\begin{aligned} (\lambda(r)\xi)(s) &= \xi(s - r), \qquad s, r \in G, \quad \xi \in L^2(G, \mathfrak{H}), \\ (\mu(p)\xi)(s) &= \overline{\langle s, p\rangle}\xi(s), \qquad p \in \widehat{G}. \end{aligned} \tag{10}$$

It follows that λ and μ are covariant. The representation π_α of $\mathcal{M}$ is given by the following:

$$(\pi_\alpha(x)\xi)(s) = \alpha_s^{-1}(x)\xi(s), \qquad x \in \mathcal{M}. \tag{11}$$

The crossed product $\mathcal{M} \rtimes_\alpha G$ is then generated by $\pi_\alpha(\mathcal{M})$ and $\lambda(G)$. Here, we use Theorem 1.7.(i) to make the construction of the crossed product free from the standard representation of $\mathcal{M}$.

Theorem 2.3. *The representation μ of $\widehat{G}$ on $L^2(G, \mathfrak{H})$ given by* (10) *induces an action $\widehat{\alpha}$ of $\widehat{G}$ on $\mathcal{M} \rtimes_\alpha G$ as follows:*

$$\widehat{\alpha}_p(x) = \mu(p)x\mu(p)^*, \qquad x \in \mathcal{M} \rtimes_\alpha G, \quad p \in \widehat{G}. \tag{12}$$

The action $\widehat{\alpha}$ of $\widehat{G}$ has the following properties:

(i) *The fixed point algebra of $\mathcal{M} \rtimes_\alpha G$ under $\widehat{\alpha}$ is precisely $\pi_\alpha(\mathcal{M})$;*
(ii) *A faithful weight ω on $\mathcal{M} \rtimes_\alpha G$ is dual to some faithful weight φ on $\mathcal{M}$ if and only if ω is invariant under $\widehat{\alpha}$;*
(iii) *Considering the second crossed product $(\mathcal{M} \rtimes_\alpha G) \rtimes_{\widehat{\alpha}} \widehat{G}$, there exists a unique isomorphism Φ of $(\mathcal{M} \rtimes_\alpha G) \rtimes_{\widehat{\alpha}} \widehat{G}$ onto $\mathcal{M} \overline{\otimes} \mathcal{L}(L^2(G))$ such that*

$$\left.\begin{aligned} \{\Phi(\pi_{\widehat{\alpha}} \circ \pi_\alpha(x))\xi\}(s) &= \alpha_s^{-1}(x)\xi(s), \qquad \xi \in L^2(G, \mathfrak{H}), \quad x \in \mathcal{M}, \\ \{\Phi(\pi_{\widehat{\alpha}} \circ \lambda(r))\xi\}(s) &= \xi(s - r), \qquad s, r \in G, \\ \{\Phi(\lambda(p))\xi\}(s) &= \overline{\langle s, p\rangle}\xi(s), \qquad p \in \widehat{G}. \end{aligned}\right\} \tag{13}$$

Here $\lambda(p)$, $p \in \widehat{G}$, means the unitary representation of $\widehat{G}$ in the second crossed product.

(iv) *The action* $\widehat{\widehat{\alpha}}$ *of* $G = \widehat{\widehat{G}}$ *on* $(\mathcal{M} \rtimes_\alpha G) \rtimes_{\widehat{\alpha}} \widehat{G}$ *given by the above procedure is transformed into* $\alpha \otimes \rho$ *by the isomorphism* Φ :

$$\Phi \circ \widehat{\widehat{\alpha}}_s \circ \Phi^{-1} = \alpha_s \otimes \rho_s, \quad s \in G. \tag{14}$$

where the action ρ *of* G *on* $\mathcal{L}(L^2(G))$ *is given by*

$$\begin{aligned} \rho_s &= \operatorname{Ad} \lambda_G(s)^{-1}, && s \in G, \\ \big(\lambda_G(s)\xi\big)(r) &= \xi(r - s), && \xi \in L^2(G). \end{aligned} \tag{15}$$

(v) *For a faithful weight* φ *on* $\mathcal{M}$, *the second dual weight* $\tilde{\tilde{\varphi}}$ *on* $(\mathcal{M} \rtimes_\alpha G) \rtimes_{\widehat{\alpha}} \widehat{G}$ *enjoys the property:*

$$\Big(\mathrm{D}(\tilde{\tilde{\varphi}} \circ \Phi^{-1}) : \mathrm{D}(\varphi \otimes \mathrm{Tr})\Big)_t \xi(s) = \big(\mathrm{D}\varphi \circ \alpha_s : \mathrm{D}\varphi\big)_t \xi(s), \tag{16}$$

where Tr *is the usual trace on* $\mathcal{L}(L^2(G))$.

Before going into the proof, we make the following:

Definition 2.4. The new action $\widehat{\alpha}$ is called the *dual* action and $\{\mathcal{M} \rtimes_\alpha G, \widehat{G}, \widehat{\alpha}\}$ is called the *dual* covariant system.

PROOF OF THEOREM 2.3: We look at $\widehat{\alpha}_p$, $p \in \widehat{G}$, on the generators of $\mathcal{M} \rtimes_\alpha G$: for each $x \in \mathcal{M}$ and $s \in G$,

$$\begin{aligned} \Big\{\widehat{\alpha}_p\big(\pi_\alpha(x)\big)\xi\Big\}(r) &= \big(\mu(p)\,\pi_\alpha(x)\,\mu(p)^*\,\xi\big)(r) = \overline{\langle r, p\rangle}\alpha_r^{-1}(x)\langle r, p\rangle\xi(r) \\ &= \alpha_r^{-1}(x)\xi(r) = \big\{\pi_\alpha(x)\xi\big\}(r), \\ \Big\{\widehat{\alpha}_p\big(\lambda(s)\big)\xi\Big\}(r) &= \overline{\langle s, p\rangle}\xi(r - s) = \overline{\langle s, p\rangle}\big\{\lambda(s)\xi\big\}(r), \end{aligned}$$

so that

$$\widehat{\alpha}_p\{\pi_\alpha(x)\} = \pi_\alpha(x), \quad x \in \mathcal{M}, \qquad \widehat{\alpha}_p\big(\lambda(s)\big) = \overline{\langle s, p\rangle}\lambda(s), \quad s \in G, \quad p \in \widehat{G}.$$

Hence $\widehat{\alpha}_p$ maps the generators of $\mathcal{M} \rtimes_\alpha G$ into $\mathcal{M} \rtimes_\alpha G$. Therefore $\widehat{\alpha}_p$ maps $\mathcal{M} \rtimes_\alpha G$ into $\mathcal{M} \rtimes_\alpha G$. Similarly, $\widehat{\alpha}_{-p}$ also sends $\mathcal{M} \rtimes_\alpha G$ into itself. Thus $\widehat{\alpha}_p$ is an automorphism of $\mathcal{M} \rtimes_\alpha G$. Since μ is a unitary representation of $\widehat{G}$, $\widehat{\alpha}$ is a continuous homomorphism of $\widehat{G}$ into $\mathrm{Aut}(\mathcal{M} \rtimes_\alpha G)$. Thus, $\widehat{\alpha}$ is an action of $\widehat{G}$.

(i) We fix the standard form $\{\mathcal{M}, U_\varphi, \mathfrak{H}_\varphi, \eta_\varphi, J, \mathcal{P}_\varphi\}$ of $\{\mathcal{M}, G, \alpha\}$ by a faithful weight φ. The crossed product $\mathcal{M} \rtimes_\alpha G$ acts on $L^2(G, \mathfrak{H}_\varphi)$ and its commutant is generated by $\mathcal{M}' \otimes \mathbf{C}$ and $\{U_\varphi(s) \otimes \lambda_G\}$ by Theorem 1.21. The dual action $\widehat{\alpha}$ of $\widehat{G}$ is induced by the unitary representation $\mu = \mathrm{id} \otimes \mu_G$ on $\mathfrak{H}_\varphi \otimes L^2(G) = L^2(G, \mathfrak{H}_\varphi)$. Hence the fixed point algebra $(\mathcal{M} \rtimes_\alpha G)^{\widehat{\alpha}}$ is given by the following:

$$\begin{aligned}
\big(\mathcal{M} \rtimes_\alpha G\big)^{\widehat{\alpha}} &= \big(\mathcal{M} \rtimes_\alpha G\big) \cap \big\{1 \otimes \mu_G(p) : p \in \widehat{G}\big\}' \\
&= \big(\mathcal{M} \rtimes_\alpha G\big) \cap \big(\mathbf{C} \otimes L^\infty(G)\big)' \\
&= \big(\mathcal{M} \rtimes_\alpha G\big) \cap \big\{\mathcal{L}(\mathfrak{H}_\varphi) \overline{\otimes} L^\infty(G)\big\} \\
&= \big(\mathcal{M}' \otimes \mathbf{C}\big)' \cap \big\{U_\varphi(s) \otimes \lambda_G(-s) : s \in G\big\}' \cap \big\{\mathcal{L}(\mathfrak{H}_\varphi) \overline{\otimes} L^\infty(G)\big\} \\
&= \big\{\mathcal{M} \overline{\otimes} L^\infty(G)\big\} \cap \big\{U_\varphi(s) \otimes \lambda_G(-s) : s \in G\big\}'.
\end{aligned}$$

Thus, $\big(\mathcal{M} \rtimes_\alpha G\big)^{\widehat{\alpha}}$ is precisely the fixed point algebra in $\mathcal{M} \overline{\otimes} L^\infty(G)$ under the action $\widetilde{\alpha}$ of G given by:

$$\widetilde{\alpha}_s = \alpha_s \otimes \operatorname{Ad}\big(\lambda_G(-s)\big), \quad s \in G. \tag{17}$$

We now define a unitary W on $L^2(G, \mathfrak{H}_\varphi)$ by

$$(W\xi)(s) = U_\varphi(s)\xi(s), \quad \xi \in L^2(G, \mathfrak{H}_\varphi), \quad s \in G. \tag{18}$$

It then follows that

$$\begin{aligned}
W\big(U_\varphi(s) \otimes \lambda_G(-s)\big)W^* &= 1 \otimes \lambda_G(-s), \quad s \in G, \\
W\big(\mathcal{M} \overline{\otimes} L^\infty(G)\big)W^* &= \mathcal{M} \overline{\otimes} L^\infty(G).
\end{aligned}$$

This then means that the fixed point algebra $\big(\mathcal{M} \overline{\otimes} L^\infty(G)\big)^{\widetilde{\alpha}}$ is given by

$$W\big(\mathcal{M} \overline{\otimes} L^\infty(G)\big)^{\widetilde{\alpha}} W^* = \big(\mathcal{M} \overline{\otimes} L^\infty(G)\big) \cap \big(\mathbf{C} \otimes \mathcal{R}(G)\big)' = \mathcal{M} \overline{\otimes} \mathbf{C},$$

where of course $\mathcal{R}(G) = \lambda_G(G)''$. Therefore, we conclude

$$\big(\mathcal{M} \rtimes_\alpha G\big)^{\widehat{\alpha}} = \big(\mathcal{M} \overline{\otimes} L^\infty(G)\big)^{\widetilde{\alpha}} = W^*\big(\mathcal{M} \otimes \mathbf{C}\big)W = \pi_\alpha(\mathcal{M}).$$

(ii) Let $\widetilde{\varphi}$ be the weight on $\mathcal{M} \rtimes_\alpha G$ dual to φ. By Theorem 1.17.(ii), the dual action $\widehat{\alpha}$ and the modular automorphism group $\sigma^{\widetilde{\varphi}}$ commute, so that we have

$$\sigma_t^{\widetilde{\varphi} \circ \widehat{\alpha}_p} = \widehat{\alpha}_p^{-1} \circ \sigma_t^{\widetilde{\varphi}} \circ \widehat{\alpha}_p = \sigma_t^{\widetilde{\varphi}}, \quad t \in \mathbf{R}, \quad p \in \widehat{G}.$$

With $\mathfrak{H}_\varphi$, $\widetilde{\pi}_\alpha$ and $\widetilde{\eta}_\varphi$ as in the last section, (1.22), (1.18) and (1.23),

$$\widehat{\alpha}_p\big(\widetilde{\pi}_\alpha(x)\big) = \pi_\ell\big(\mu(p)\widetilde{\eta}_\varphi(x)\big), \quad x \in \mathfrak{b}_\varphi \cap \mathfrak{b}_\varphi^\sharp,$$

so that

$$\widetilde{\varphi}\big(\widetilde{\pi}_\alpha(y)^* \widetilde{\pi}_\alpha(x)\big) = \widetilde{\varphi} \circ \widehat{\alpha}_p\big(\widetilde{\pi}_\alpha(y)^* \widetilde{\pi}_\alpha(x)\big), \quad x, y \in \mathfrak{b}_\varphi^\sharp \cap \mathfrak{b}_\varphi.$$

Hence $\widetilde{\varphi} = \widetilde{\varphi} \circ \widehat{\alpha}_\varphi$ by Proposition VIII.3.15.

Conversely, suppose that ω is a faithful weight on $\mathcal{M} \rtimes_\alpha G$ invariant under $\widehat{\alpha}$. For a fixed faithful weight φ on $\mathcal{M}$, we consider the dual weight $\widetilde{\varphi}$. Then, we have

$$\widehat{\alpha}_{-p}\big((\mathrm{D}\omega : \mathrm{D}\widetilde{\varphi})_t\big) = \big(\mathrm{D}(\omega \circ \widehat{\alpha}_p) : \mathrm{D}(\widetilde{\varphi} \circ \widehat{\alpha}_p)\big)_t = (\mathrm{D}\omega : \mathrm{D}\widetilde{\varphi})_t, \quad p \in \widehat{G}.$$

By (i), $(\mathrm{D}\omega : \mathrm{D}\widetilde{\varphi})_t \in \pi_\alpha(\mathcal{M})$. Put $u_t = \pi_\alpha^{-1}\big((\mathrm{D}\omega : \mathrm{D}\widetilde{\varphi})_t\big)$. It then follows that $\{u_t\}$ is a cocycle with respect to $\{\sigma_t^\varphi\}$. By Theorem VIII.3.7, there exists a faithful weight ψ on $\mathcal{M}$ such that $u_t = (\mathrm{D}\psi : \mathrm{D}\varphi)_t$. We then have

$$\big(\mathrm{D}\widetilde{\psi} : \mathrm{D}\widetilde{\varphi}\big)_t = \pi_\alpha\big((\mathrm{D}\psi : \mathrm{D}\varphi)_t\big) = (\mathrm{D}\omega : \mathrm{D}\widetilde{\varphi})_t,$$

so that $\widetilde{\psi} = \omega$.

(iii) Put $U(s) = \pi_{\widehat{\alpha}}\big(\lambda(s)\big)$, $s \in G$, and $V(p) = \lambda(p)$, $p \in \widehat{G}$. It follows that U and V are covariant, so that $U(G)$ and $V(\widehat{G})$ together generate a subfactor of $\big(\mathcal{M} \rtimes_\alpha G\big) \rtimes_{\widehat{\alpha}} \widehat{G}$ isomorphic to $\mathcal{L}\big(L^2(G)\big)$. To simplify the notation, we set $\mathcal{N} = \big(\mathcal{M} \rtimes_\alpha G\big) \rtimes_{\widehat{\alpha}} \widehat{G}$. Here $\mathcal{N}$ acts on $L^2\big(\widehat{G}, L^2(G, \mathfrak{H}_\varphi)\big) = \mathfrak{H}_\varphi \otimes L^2(G) \otimes L^2(\widehat{G})$. We map this Hilbert space onto $\mathfrak{H}_\varphi \otimes L^2(G) \otimes L^2(G) = L^2(G \times G, \mathfrak{H}_\varphi) = \mathfrak{K}$ by the partial Fourier inverse transform $1 \otimes \widehat{\mathcal{F}}$. Then U, V and $\pi_{\widehat{\alpha}} \circ \pi_\alpha(x)$, $x \in \mathcal{M}$, are identified with the following operators:

$$\left.\begin{aligned} U(s) &= 1 \otimes \lambda_G(s) \otimes \lambda_G(s), && s \in G,\\ V(p) &= 1 \otimes 1 \otimes \mu_G(p), && p \in G,\\ \pi_{\widehat{\alpha}} \circ \pi_\alpha(x) &= \pi_\alpha(x) \otimes 1, && x \in \mathcal{M}. \end{aligned}\right\} \tag{19}$$

We define a unitary W on $\mathfrak{K}$ as follows:

$$(W\xi)(r, s) = \xi(r + s, s), \quad \xi \in \mathfrak{K}, \quad r, s \in G. \tag{20}$$

We then have

$$\left.\begin{aligned} W\big(1 \otimes 1 \otimes \mu_G(p)\big)W^* &= 1 \otimes 1 \otimes \mu_G(p), && p \in \widehat{G},\\ W\big(1 \otimes \lambda_G(s) \otimes \lambda_G(s)\big)W^* &= 1 \otimes 1 \otimes \lambda_G(s), && s \in G,\\ W\big(\pi_\alpha(x) \otimes 1\big)W^* &= \pi(x), && x \in \mathcal{M}, \end{aligned}\right\} \tag{21}$$

where

$$\big(\pi(x)\xi\big)(r, s) = \alpha_{r+s}^{-1}(x)\xi(r, s), \quad x \in \mathcal{M}, \quad r, s \in G. \tag{22}$$

If we would have proved

$$\pi(\mathcal{M}) \vee \big(\mathbf{C} \otimes L^\infty(G)\big) = \pi_\alpha(\mathcal{M}) \mathbin{\overline{\otimes}} L^\infty(G), \tag{23}$$

we should get

$$\begin{aligned} W\mathcal{N}W^* &= \big(\mathbf{C} \otimes \mathbf{C} \otimes L^\infty(G)\big) \vee \big(\mathbf{C} \otimes \mathbf{C} \otimes \mathcal{R}(G)\big) \vee \pi(\mathcal{M})\\ &= \big(\pi_\alpha(\mathcal{M}) \mathbin{\overline{\otimes}} L^\infty(G)\big) \vee \big(\mathbf{C} \otimes \mathbf{C} \otimes \mathcal{R}(G)\big) = \pi_\alpha(\mathcal{M}) \mathbin{\overline{\otimes}} \mathcal{L}\big(L^2(G)\big). \end{aligned}$$

Thus, it suffices to prove (23).

We observe first that $\pi(x)$, $x \in \mathcal{M}$, defined by (22) is precisely $\pi_\alpha\big(\alpha_s^{-1}(x)\big)$ as a function of r when $s \in G$ is fixed. Namely, $\pi(x)$ can be identified with the function: $s \in G \mapsto \pi_\alpha\big(\alpha_s^{-1}(x)\big) \in \pi_\alpha(\mathcal{M})$, so that $\pi(x) \in \pi_\alpha(\mathcal{M}) \mathbin{\overline{\otimes}} L^\infty(G)$. Hence we have

$$W\big(\pi_\alpha(\mathcal{M}) \otimes \mathbf{C}\big)W^* = \pi(\mathcal{M}) \subset \pi_\alpha(\mathcal{M}) \mathbin{\overline{\otimes}} L^\infty(G).$$

Since W commutes with $\mathbf{C} \otimes \mathbf{C} \otimes L^\infty(G)$, we have

$$W\big(\pi_\alpha(\mathcal{M}) \mathbin{\overline{\otimes}} L^\infty(G)\big)W^* \subset \pi_\alpha(\mathcal{M}) \mathbin{\overline{\otimes}} L^\infty(G).$$

Similarly,

$$W^*\big(\pi_\alpha(\mathcal{M})\,\overline{\otimes}\,L^\infty(G)\big)W \subset \pi_\alpha(\mathcal{M})\,\overline{\otimes}\,L^\infty(G).$$

Thus, we obtain

$$W\big(\pi_\alpha(\mathcal{M})\,\overline{\otimes}\,L^\infty(G)\big)W^* = \pi_\alpha(\mathcal{M})\,\overline{\otimes}\,L^\infty(G).$$

Now, we have

$$\begin{aligned}\pi_\alpha(\mathcal{M})\,\overline{\otimes}\,L^\infty(G) &= \big(\pi_\alpha(\mathcal{M})\otimes\mathbf{C}\big)\vee\big(\mathbf{C}\otimes L^\infty(G)\big)\\ &= W\big(\pi_\alpha(\mathcal{M})\otimes\mathbf{C}\big)W^*\vee\big(\mathbf{C}\otimes L^\infty(G)\big)\\ &= \pi(\mathcal{M})\vee\big(\mathbf{C}\otimes L^\infty(G)\big).\end{aligned}$$

(iv) We continue the above discussion. The second dual action $\hat{\hat{\alpha}}$ of G acts on the generators of $\big(\mathcal{M}\rtimes_\alpha G\big)\rtimes_{\hat{\alpha}}\widehat{G}$ as follows:

$$\begin{aligned}\widehat{\widehat{\alpha}}_s\big(\pi_\alpha(x)\otimes 1\big) &= \hat{\hat{\alpha}}_s\big(\pi_{\hat{\alpha}}\circ\pi_\alpha(x)\big) = \pi_{\hat{\alpha}}\circ\pi_\alpha(x) \quad = \pi_\alpha(x)\otimes 1, \quad x\in\mathcal{M},\\ \widehat{\widehat{\alpha}}_s\big(U(t)\big) &= U(t), \quad t\in G,\\ \widehat{\widehat{\alpha}}_s\big(V(p)\big) &= \overline{\langle s,p\rangle}V(p), \quad p\in\widehat{G}.\end{aligned}$$

Hence $\hat{\hat{\alpha}}_s$ is given by the restriction of $\mathrm{Ad}\big(1\otimes 1\otimes\lambda_G(-s)\big)$ to $\mathcal{N} = (\mathcal{M}\rtimes_\alpha G)\rtimes_{\hat{\alpha}}\widehat{G}$. Since

$$W\big(1\otimes 1\otimes\lambda_G(-s)\big)W^* = 1\otimes\lambda_G(s)\otimes\lambda_G(-s), \quad s\in G,$$

the isomorphism $\Phi : x\in\mathcal{N}\mapsto WxW^*\in\pi_\alpha(\mathcal{M})\,\overline{\otimes}\,\mathcal{L}\big(L^2(G)\big)$ transform $\hat{\hat{\alpha}}_s$ into

$$\mathrm{Ad}\big(1\otimes\lambda_G(s)\otimes\lambda_G(-s)\big),$$

which means

$$\Phi\widehat{\widehat{\alpha}}_s\Phi^{-1} = \alpha_s\otimes\rho_s, \quad s\in G,$$

after identifying $\mathcal{M}$ with $\pi_\alpha(\mathcal{M})$.

(v) It is routine to construct the second dual weight based on an appropriate left Hilbert algebra based on $\mathcal{K}(G\times G,\mathfrak{m}_\varphi)$. We leave the proof of assertion (v) to the reader. We remark here that one uses Proposition VIII.3.15 at an appropriate stage. Q.E.D.

Proposition 2.5. *The dual covariant systems of cocycle conjugate covariant systems are conjugate.*

We leave the proof to the reader as an exercise.

Proposition 2.6. *A covariant system* $\{\mathcal{M}, G, \alpha\}$ *is dual to some covariant system* $\{\mathcal{N}, \widehat{G}, \beta\}$ *over the dual group* $\widehat{G}$ *if and only if there exists a continuous unitary representation* $\{u(p) : p\in\widehat{G}\}$ *with values in* $\mathcal{M}$ *such that*

$$\alpha_s\big(u(p)\big) = \overline{\langle s,p\rangle}u(p), \quad p\in\widehat{G}, \quad s\in G. \tag{24}$$

PROOF: The "only if" part is a part of the definition of a dual action. So we assume the existence of $\{u(p)\}$ satisfying (24). Put $\mathcal{N} = \mathcal{M}^\alpha$ and $\theta_p = \mathrm{Ad}(u(p))$, $p \in \widehat{G}$. Since θ_p and α_s commute, $\theta_p(\mathcal{N}) = \mathcal{N}$, so that $\{\mathcal{N}, \theta\}$ is a covariant system over $\widehat{G}$. We shall prove

$$\{\mathcal{N} \rtimes_\theta \widehat{G}, \widehat{\theta}\} \cong \{\mathcal{M}, \alpha\}.$$

Let $\mathcal{M}$ act on $\mathfrak{H}$. Of course, $\mathcal{N}$ acts on $\mathfrak{H}$ too. The crossed product $\mathcal{N} \rtimes_\theta \widehat{G}$ is then generated by the operators on $L^2(\widehat{G}, \mathfrak{H})$:

$$\begin{aligned} \{\pi_\theta(x)\xi\}(p) &= \theta_{-p}(x)\xi(p), \quad x \in \mathcal{N}, \quad \xi \in L^2(\widehat{G}, \mathfrak{H}), \\ \{\lambda(q)\xi\}(p) &= \xi(p-q), \qquad p, q \in \widehat{G}. \end{aligned} \tag{25}$$

We define a unitary operator U from $L^2(\widehat{G}, \mathfrak{H})$ onto $L^2(G, \mathfrak{H})$ by the following:

$$(U\xi)(s) = \int_{\widehat{G}} \langle s, p\rangle u(p)\xi(p)\, dp, \quad \xi \in \mathcal{K}(\widehat{G}, \mathfrak{H}). \tag{26}$$

We then have

$$(U^*\eta)(p) = u(p)^* \int_G \overline{\langle t, p\rangle}\eta(t)\, dt, \quad \xi \in \mathcal{K}(G, \mathfrak{H}). \tag{27}$$

Now, representing $\mathcal{M}$ on $L^2(G, \mathfrak{H})$ by π_α, we have for each $x \in \mathcal{N}$, $q \in \widehat{G}$ and $\eta \in \mathcal{K}(G, \mathfrak{H})$,

$$\begin{aligned} \{U^*\pi_\alpha(x)\eta\}(p) &= u(p)^* \int_G \overline{\langle t, p\rangle}\alpha_{-t}(x)\eta\, dt = u(p)^* \int_G \overline{\langle t, p\rangle} x\eta(t)\, dt \\ &= u(p)^* x \int_G \overline{\langle t, p\rangle}\eta(t)\, dt \\ &= \theta_{-p}(x)u(p)^* \int_G \overline{\langle t, p\rangle}\eta(t)\, dt = \{\pi_\theta(x)U^*\eta\}(p), \end{aligned}$$

$$\begin{aligned} \Big\{U^*\pi_\alpha\big(u(q)\big)\eta\Big\}(p) &= u(p)^* \int_G \overline{\langle t, p\rangle}\alpha_{-t}\big(u(q)\big)\eta(t)\, dt \\ &= u(p)^* \int_G \overline{\langle t, p-q\rangle} u(q)\eta(t)\, dt \\ &= u(p-q)^* \int_G \overline{\langle t, p-q\rangle}\eta(t)\, dt = (U^*\eta)(p-q). \end{aligned}$$

Hence we obtain

$$U^*\pi_\alpha(x)U = \pi_\theta(x), \quad x \in \mathcal{N}; \qquad U^*\pi_\alpha\big(u(p)\big)U = \lambda(p), \quad p \in \widehat{G}, \tag{28}$$

so that

$$U^*\left(\pi_\alpha(\mathcal{N}) \cup \pi_\alpha\big(u(\widehat{G})\big)\right)'' U = \mathcal{N} \rtimes_\alpha \widehat{G}.$$

Furthermore, the isomorphism of $\left[\pi_\alpha(\mathcal{N}) \cup \pi_\alpha\big(u(\widehat{G})\big)\right]''$ onto $\mathcal{N} \rtimes_\theta \widehat{G}$ induced by U^* transforms α into $\widehat{\theta}$. Thus all we have to do is to show that $\mathcal{N}$ and $u(\widehat{G})$ generate $\mathcal{M}$.

We say that an $x \in \mathcal{M}_+$ is *α-integrable* if

$$\mathcal{E}_K(x) = \int_K \alpha_t(x)\, \mathrm{d}t \in \mathcal{M}_+$$

is bounded when K runs all compact subsets of G. Since $\{\mathcal{E}_K(x)\}$ is upward directed, we define

$$\mathcal{E}_\alpha(x) = \sup_K \mathcal{E}_K(x) = \int_G \alpha_t(x)\, \mathrm{d}t \in \mathcal{M}_+ \tag{29}$$

for each α-integrable $x \in \mathcal{M}_+$. It is easy to see that $\mathcal{E}_\alpha$ is an operator valued normal weight on $\mathcal{M}_+$ with values in $\widehat{\mathcal{N}_+}$ in the sense of §4, Chapter VIII. We say that α is *integrable* if $\mathcal{E}_\alpha$ is semi-finite. We shall prove that α is integrable, that is the set $\mathfrak{p}_+$ of all α-integrable elements in $\mathcal{M}_+$ generates $\mathcal{M}$. By the uniqueness of Heisenberg-Weyl commutation relation, the map: $p \in \widehat{G} \mapsto u(p) \in \mathcal{M}$ is uniquely extended to a normal isomorphism: $f \in L^\infty(G) \mapsto u(f) \in \mathcal{A} = u(G)''$, and we have

$$\alpha_t\big(u(f)\big) = u(\lambda_{-t} f), \quad f \in L^\infty(G), \quad t \in G,$$

where $(\lambda_t f)(s) = f(s-t),\ \ s, t \in G$. Hence, if $f \in L^1(G) \cap L^\infty(G)$, then $u(f)$ belongs to $\mathfrak{p}_\alpha = \mathfrak{p}_+ - \mathfrak{p}_+ + \mathrm{i}\,(\mathfrak{p}_+ - \mathfrak{p}_+)$, and

$$\mathcal{E}_\alpha\big(u(f)\big) = \left(\int_G f(s)\, \mathrm{d}s\right), \quad f \in L^1(G) \cap L^\infty(G).$$

Hence $\mathfrak{p}_+$ contains an increasing net converging to $1 \in \mathcal{M}$. Therefore, we get $\mathcal{M} = (\mathfrak{p}_+)''$, i.e. α is integrable.

If $x \in \mathfrak{p}_\alpha$, then the function: $s \in G \mapsto \langle \alpha_s(x), \omega \rangle,\ \ \omega \in \mathcal{M}_*$, is integrable, so that the integral:

$$\widehat{x}(p) = \int_G \overline{\langle s, p \rangle} \alpha_s(s)\, \mathrm{d}s, \quad x \in \mathfrak{p}_\alpha, \quad p \in \widehat{G},$$

makes sense, and

$$\alpha_t\big(\widehat{x}(p)\big) = \langle t, p \rangle \widehat{x}(p).$$

If we put $\widetilde{x}(p) = \widehat{x}(p) u(p)$, then we have

$$\widehat{x}(p) = \widetilde{x}(p) u(p)^* \in \mathcal{N} u(p)^*.$$

The uniqueness of the Fourier transform implies that if $\omega \in \mathcal{M}_*$ vanishes on all $\mathcal{N} u(p)$'s, then $\omega(x) = 0$. Hence $\mathfrak{p}_\alpha$ is contained in the σ-weak closure of $\left[\mathcal{N} u(p) : p \in \widehat{G}\right]$. Thus $\mathcal{M}$ is generated by $\mathcal{N}$ and $u(\widehat{G})$. Q.E.D.

We study integrable actions a little further. Note that the integrability of $\{\mathcal{M}, G, \alpha\}$ is equivalent to the existence of a net of integrable elements converging σ-weakly to 1. With this remark, we have the following:

Lemma 2.7. *For a covariant system $\{\mathcal{M}, G, \alpha\}$ to be integrable, any one of the following conditions is sufficient:*

(i) *There exists a globally invariant von Neumann subalgebra $\mathcal{N}$ of $\mathcal{M}$ such that the restriction of α to $\mathcal{N}$ is integrable.*
(ii) *$\{\mathcal{M}, G, \alpha\} \cong \{\mathcal{M}_1, G, \alpha_1\} \overline{\otimes} \{\mathcal{M}_2, G, \alpha_2\}$ and either α_1 or α_2 is integrable.*
(iii) *There exists an integrable covariant system $\{\mathcal{N}, G, \beta\}$ and a projection $e \in \mathcal{N}^\beta$ such that α is conjugate to the restriction of β to the reduced algebra $\mathcal{N}_e$ of $\mathcal{N}$.*

We drop the commutativity assumption from G for the rest of this section. Given a covariant system $\{\mathcal{M}, G, \alpha\}$ and $u \in Z^1_\alpha(G, \mathcal{M})$, we recall the perturbation ${}_u\alpha$ of α by u defined by (8):

$$ {}_u\alpha_s(x) = u_s \alpha_s(x) u_s^*, \quad x \in \mathcal{M}, \quad s \in G. $$

For simplicity, we denote by $\mathcal{M}^u$ the fixed point algebra of the perturbed action ${}_u\alpha$. For $u, v \in Z^1_\alpha(G, \mathcal{M})$, we write $u \precsim v$ if there exists $w \in \mathcal{M}$ such that

$$ w^* w = 1, \quad w w^* = e \in \mathcal{M}^v; \qquad u_s = w^* v_s \alpha_s(w), \quad s \in G. \tag{30} $$

This means that the cocycle u is equivalent to the reduced cocycle v^e of v. We also say that a cocycle $u \in Z^1_\alpha(G, \mathcal{M})$ is *square integrable* if the perturbed action ${}_u\alpha$ is integrable.

Theorem 2.8. *For a separable covariant system $\{\mathcal{M}, G, \alpha\}$, consider a new system $\{\overline{\mathcal{M}}, G, \bar{\alpha}\}$ as follows:*

$$ \overline{\mathcal{M}} = \mathcal{M} \overline{\otimes} \mathcal{L}(L^2(G)), \qquad \bar{\alpha}_s = \alpha_s \otimes \mathrm{id}, \quad s \in G. \tag{31} $$

Let $u \in Z^1_\alpha(G, \mathcal{M})$. If $\mathcal{M}^u$ is properly infinite, then the following two conditions for α are mutually equivalent:

(i) *u is square integrable;*
(ii) *$u \otimes 1 \precsim 1 \otimes \lambda_G$, where $1 \otimes \lambda_G$ is viewed as an element of $Z^1_{\bar{\alpha}}(G, \overline{\mathcal{M}})$.*

The implication (ii) $\Longrightarrow$ (i) follows from Lemma 2.7.(iii) because $1 \otimes \lambda_G$ is square integrable, which is a consequence of the fact that any element of $L^\infty(G) \cap L^1(G)$ is integrable for the action $\mathrm{Ad}\,\lambda_G$. For the proof of the reversed implication, we extend slightly the notion of cocycles.

Definition 2.9. With $\{\mathcal{M}, G, \alpha\}$ fixed, a σ-strong* continuous function $a\colon s \in G \mapsto a(s) \in \mathcal{M}$ is called an *α-twisted *-representation* of G in $\mathcal{M}$ if

$$ a(st) = a(s)\alpha_s(a_s(t)), \quad s, t \in G; \qquad a(s^{-1}) = \alpha_s^{-1}(a(s)^*). \tag{32} $$

If all $a(s)$'s are unitaries, then a is an α-unitary cocycle. We denote by $Z_\alpha(G, \mathcal{M})$ (resp. $Z_\alpha(G, \mathcal{U}(\mathcal{M}))$) the set of all α-twisted *-representations (resp. α-unitary cocycles) of G in $\mathcal{M}$.

It follows that if $a \in Z_\alpha(G, \mathcal{M})$, then $a(s)$ are partial isometries such that

$$a(s)a(s)^* = a(1), \qquad a(s)^*a(s) = \alpha_s\big(a(1)\big), \qquad s \in G,$$

where 1 denotes the identity of G. We denote $a(1)$ by e_a. Each $a \in Z_\alpha(G, \mathcal{M})$ gives rise to an action ${}_a\alpha$ of G on $\mathcal{M}_{e_a}$ by

$${}_a\alpha_s(x) = a(s)\alpha_s(x)a(s)^*, \quad x \in \mathcal{M}_{e_a}, \quad s \in G.$$

We denote by $\mathcal{M}^a$ the fixed point subalgebra of $\mathcal{M}_{e_a}$ under the new action ${}_a\alpha$. For each $p \in \operatorname{Proj}(\mathcal{M}^a)$, the map $a^p : s \in G \mapsto pa(s) \in \mathcal{M}$ is a new α-twisted *-representation of G in $\mathcal{M}$, which we denote by a^p.

Definition 2.10. For $a, b \in Z_\alpha(G, \mathcal{M})$, we say that a and b are *equivalent* and write $a \simeq b$ if there exists $c \in \mathcal{M}$ such that

$$a(s) = c^*b(s)\alpha_s(c); \qquad b(s) = ca(s)\alpha_s(c^*), \qquad s \in G. \tag{33}$$

We write $a \precsim b$ if $a \simeq b^q$ for some $q \in \operatorname{Proj}(\mathcal{M}^b)$.

Lemma 2.11. *Let $\mathcal{P} = \mathcal{M} \otimes M_2$ be the 2×2-matrix algebra over $\mathcal{M}$ and $\bar{\alpha} = \alpha \otimes \mathrm{i}$. Given $a, b \in Z_a(G, \mathcal{M})$, we define $c \in Z_{\bar{\alpha}}(G, \mathcal{P})$ by*

$$c(s) = a(s) \otimes e_{11} + b(s) \otimes e_{22}, \tag{34}$$

and write $c = a \oplus b$, where $\{e_{i,j}\}$ is of course the standard matrix unit of M_2. Then the following two statements are equivalent:

(i) *$a \precsim b$ (resp. $a \sim b$);*
(ii) *$e_a \otimes e_{11} \precsim e_b \otimes e_{22}$ (resp. $e_a \otimes e_{11} \sim e_b \otimes e_{22}$) in $\mathcal{P}^c$.*

Since the proof is routine, we leave it to the reader.

Definition 2.12. We say that $a, b \in Z_\alpha(G, \mathcal{M})$ are *disjoint* and write $a \;\delta\; b$ if $e_a \otimes e_{11}$ and $e_b \otimes e_{22}$ are centrally orthogonal in $\mathcal{P}^c$, i.e. if there exists a central projection z in $\mathcal{P}^c$ such that $z \geq e_a \otimes e_{11}$ and $1 - z \geq e_b \otimes e_{22}$. We say that a and b are *quasi equivalent* if the central supports of $e_a \otimes e_{11}$ and $e_b \otimes e_{22}$ agree (i.e. $e_a \otimes e_{11} + e_b \otimes e_{22}$).

We introduce another notation $I(b, a)$ for $a, b \in Z_\alpha(G, \mathcal{M})$:

$$I(b, a) = \big\{x \in e_b\mathcal{M}e_a : xa(s) = b(s)\alpha_s(x),\ s \in G\big\}. \tag{35}$$

The following properties of $I(b, a)$ follow immediately:

Lemma 2.13.

(i) *For $a, b, c \in Z_\alpha(G, \mathcal{M})$, we have the following:*

$$I(b,a)^* = I(a,b); \qquad I(a,a) = \mathcal{M}^a;$$

$$x = \begin{pmatrix} x_{11} & x_{12} \\ x_{21} & x_{22} \end{pmatrix} \in \mathcal{P}^{a\oplus b} \iff \begin{cases} x_{11} \in I(a,a), & x_{12} \in I(a,b), \\ x_{21} \in I(b,a), & x_{22} \in I(b,b); \end{cases}$$

$$a \perp b \iff I(a,b) = \{0\};$$

$$I(a,b)I(b,c) \subset I(a,c).$$

(ii) *If $x = uh$ is the polar decomposition of $x \in I(a,b)$, then $u \in I(a,b)$ and $h \in I(b,b)$.*

Definition 2.14. We say that $a \in Z_\alpha(G, \mathcal{M})$ are of *infinite multiplicity*, if $\mathcal{M}^a$ is properly infinite.

Lemma 2.15. *If $a, b \in Z_\alpha(G, \mathcal{M})$ are of infinite multiplicity, then the two conditions are equivalent:*

(i) $a \simeq b$;
(ii) $a \sim b$.

PROOF: The implication (i) $\Longrightarrow$ (ii) is trivial.

(ii) $\Longrightarrow$ (i): Suppose $a \sim b$. In $\mathcal{P} = \mathcal{M} \otimes M_2$, let $c = a \oplus b$. By the infinite multiplicity of a and b, $e_a \otimes e_{11}$ and $e_b \otimes e_{22}$ are both properly infinite in $\mathcal{P}^c$. Since $\mathcal{P}^c$ is σ-finite by the standing assumption, $e_a \otimes e_{11}$ and $e_b \otimes e_{22}$ are equivalent to their central support. Therefore, $e_a \otimes e_{11} \sim e_b \otimes e_{22}$. Q.E.D.

Lemma 2.16. *Let ρ be the right regular representation of G on $L^2(G)$. If $b \in Z_\alpha(G, \mathcal{M})$ is square integrable, then*

$$\bigvee \{ s_r(x) : x \in I(b \otimes \rho, b \otimes 1) \} = e_b \otimes 1 \in \mathcal{M} \,\overline{\otimes}\, \mathcal{L}(L^2(G)).$$

PROOF: Let e be the projection of the left hand side. By Lemma 2.13.(ii), e belongs to $\overline{\mathcal{M}}^{(b\otimes 1)}$, where $\overline{\mathcal{M}} = \mathcal{M} \,\overline{\otimes}\, \mathcal{L}(L^2(G))$. For any unitary $u \in \widetilde{\mathcal{M}}^{(b\otimes 1)}$, we have

$$I(b \otimes \rho,\ b \otimes 1)u = I(b \otimes \lambda_r,\ b \otimes 1),$$

so that $u^* e u = e$; hence e is a central projection of $\widetilde{\mathcal{M}}^{(b\otimes 1)}$. Since $I(b\otimes\rho,\ b\otimes 1)e = I(b \otimes \rho,\ b \otimes 1)$, we only have to prove

$$I(b \otimes \rho,\ b \otimes 1) f \neq \{0\}$$

for every non-zero central projection f of $\overline{\mathcal{M}}^{(b\otimes 1)}$. Since $\overline{\mathcal{M}}^{(b\otimes 1)} = \mathcal{M}^b \overline{\otimes} \mathcal{L}(L^2(G))$, f is of the form $f = p \otimes 1$ with a central projection p of $\mathcal{M}^b$. We now represent $\mathcal{M}$ on a Hilbert space $\mathfrak{H}$, and consider $L^2(G)$ with respect to the right Haar measure $d_r s = d_\ell s^{-1}$. Choose an $x \in \mathfrak{p}_{b\alpha}$ such that $x = xp \neq 0$, and $f \in \mathcal{K}(G)$. Put

$$(y\xi)(s) = {}_b\alpha_s^{-1}(x) \int_G f(t)\xi(t)\, \mathrm{d}_r t, \quad \xi \in L^2(G, \mathfrak{H}). \tag{36}$$

We then have, with $\mathcal{E}_{b\alpha}$ of (29) for ${}_b\alpha$,

$$\begin{aligned}
\|y\xi\|^2 &= \int_G \left\| {}_b\alpha_s^{-1}(x)\left(\int_G f(t)\xi(t)\, \mathrm{d}_r t \right) \right\|^2 \mathrm{d}_r s \\
&\leq \int_G \int_G \| f(t){}_b\alpha_s^{-1}(x)\xi(t) \|^2\, \mathrm{d}_r t\, \mathrm{d}_r s \\
&= \int_G |f(t)|^2 \left(\int_G \| {}_b\alpha_s^{-1}(x)\xi(t) \|^2\, \mathrm{d}_r s \right) \mathrm{d}_r t \\
&= \int_G |f(t)|^2 \left(\int_G \| {}_b\alpha_s(x)\xi(t) \|^2\, \mathrm{d}_\ell s \right) \mathrm{d}_r t \\
&= \int_G |f(t)|^2 \left(\mathcal{E}_{b\alpha}(x^*x)\xi(t) \mid \xi(t) \right) \mathrm{d}_r t \\
&= \int_G |f(t)|^2 \| \mathcal{E}_{b\alpha}(x^*x)^{\frac{1}{2}}\xi(t) \|^2\, \mathrm{d}_r t \\
&\leq \|f\|_\infty^2 \|\mathcal{E}_{b\alpha}(x^*x)\| \|\xi\|^2.
\end{aligned}$$

Hence y is bounded and belongs to $\overline{\mathcal{M}}$. Furthermore, we have, for $\xi \in L^2(G, \mathfrak{H})$ and $r, s \in G$,

$$\begin{aligned}
\left[y\left(pb(r) \otimes 1\right)\xi \right](s) &= {}_b\alpha_s^{-1}(x)pb(r) \int_G f(t)\xi(t)\, \mathrm{d}_r t \\
&= {}_b\alpha_s^{-1}(x)b(r) \int_G f(t)\xi(t)\, \mathrm{d}_r t;
\end{aligned}$$

$$\begin{aligned}
\left\{ \left[b(r) \otimes \rho(r) \right](\alpha_r \otimes \mathrm{id})(y)\xi \right\}(s) &= b(r)\left((\alpha_r \otimes \mathrm{id}\right)(y)(\xi)(sr) \\
&= b(r)\alpha_r\left({}_b\alpha_{sr}^{-1}(x) \right) \int_G f(t)\xi(t)\, \mathrm{d}_r t \\
&= {}_b\alpha_r \circ {}_b\alpha_{sr}^{-1}(x) \int_G f(t)\xi(t)\, \mathrm{d}_r t \\
&= {}_b\alpha_s^{-1}(x)b(r) \int_G f(t)\xi(t)\, \mathrm{d}_r t\, .
\end{aligned}$$

Hence y belongs to $I\left(b \otimes \rho, b \otimes 1\right)$ and $y(p \otimes 1) = y$. Clearly $y \neq 0$ as long as $f \neq 0$. Q.E.D.

PROOF OF THEOREM 2.8 (i) $\Longrightarrow$ (ii): Let $a = u \otimes 1$ and $b = u \otimes \rho$ with ρ the right regular representation of G on $L^2(G)$. Consider $c = a \oplus b$ in $\overline{\mathcal{M}} \otimes M_2 = \mathcal{P}$. By Lemma 2.16, the initial support of $1 \otimes 1 \otimes e_{11}$ in $\mathcal{P}^c$ is majorized by the central support of $1 \otimes 1 \otimes e_{22}$ in $\mathcal{P}^c$, where $\mathcal{P}^c$ is of course the fixed point subalgebra of $\mathcal{P}$ under ${}_c(\bar{\alpha} \otimes \mathrm{id})$. By the proper infiniteness of $\mathcal{M}^u$, $1 \otimes 1 \otimes e_{11}$ and $1 \otimes 1 \otimes e_{22}$ are both properly infinite projections of $\mathcal{P}^c$. Hence $1 \otimes 1 \otimes e_{11} \precsim 1 \otimes 1 \otimes e_{22}$ in $\mathcal{P}^c$. This means that $a \prec b$.

We now prove that for any unitary α-cocycle u

$$u \otimes \rho \simeq 1 \otimes \rho. \tag{37}$$

Representing $\mathcal{M}$ on $\mathfrak{H}$, we define $v \in \overline{\mathcal{M}} = \mathcal{M} \overline{\otimes} \mathcal{L}(L^2(G))$ as follows:

$$(v\xi)(s) = u_{s^{-1}}\xi(s), \quad s \in G, \quad \xi \in L^2(G, \mathfrak{H}).$$

We then compute:

$$\begin{aligned}\left\{\left[u_s \otimes \rho(s)\right]\bar{\alpha}_s(v)\xi\right\}(t) &= u_s\left(\bar{\alpha}_s(v)\xi\right)(ts) = v_s a_s\left(u_{(ts)^{-1}}\right)\xi(ts)\\ &= u_s\alpha_s(u_{s^{-1}})u_{t^{-1}}\xi(ts) = u_{t^{-1}}\xi(ts) = \left\{v\left(1 \otimes \rho(s)\right)\xi\right\}(t).\end{aligned}$$

Hence we get

$$u_s \otimes \rho(s) = v\left(1 \otimes \rho(s)\right)\bar{\alpha}_s(v^*), \quad s \in G,$$

which means (37). Q.E.D.

Definition 2.17. A unitary cocycle $a \in Z^1_\alpha(G, \mathcal{M})$ is called *dominant* if $\mathcal{M}^a$ is property infinite and $a \sim 1 \otimes \lambda_G$. The action α itself is called *dominant* if α is conjugate to $\alpha \otimes \rho$, where ρ is given by the right regular representation ρ_G of G on $L^2(G)$.

Theorem 2.18. *Let G be a separable locally compact abelian group. For a separable covariant system $\{\mathcal{M}, G, \alpha\}$, the following conditions are equivalent:*

(i) *α is dominant;*
(ii) *$\mathcal{M}^\alpha$ is properly infinite and for every $p \in \widehat{G}$, there exists a unitary u such that $\alpha_s(u) = \langle s, p\rangle u$, $s \in G$.*

PROOF: The implication (i) $\Longrightarrow$ (ii) is trivial; so we prove the reverse implication, (ii) $\Longrightarrow$ (i).

Let $\mathfrak{H}$ be the Hilbert space on which $\mathcal{M}$ acts. Since $\mathcal{M}^\alpha$ is properly infinite, α is conjugate to $\bar{\alpha} = \alpha \otimes \mathrm{id}$, where id is the trivial action of G on $\mathcal{L}(L^2(G))$. By the separability, we can choose a measurable function: $p \in \widehat{G} \mapsto u(p) \in \mathcal{U}(\mathcal{M})$ such that $\alpha_s(u(p)) = \langle s, p\rangle u(p)$. Consider the direct integral:

$$w = \int_{\widehat{G}}^{\oplus} u(p)\, dp \in \mathcal{M} \overline{\otimes} \mathcal{L}(L^2(G)).$$

Then we have

$$\begin{aligned}\bar{\alpha}_s(w) &= \int_{\widehat{G}}^{\oplus} \alpha_s\big(u(p)\big)\,\mathrm{d}p = \int_{\widehat{G}}^{\oplus} \langle s, p\rangle u(p)\,\mathrm{d}p \\ &= w\int_{\widehat{G}}^{\oplus} \langle s, p\rangle\,\mathrm{d}p = w\big(1\otimes\rho_G(s)\big)\end{aligned}$$

by identifying $L^2(G)$ with $L^2(\widehat{G})$ under the Fourier transform. Therefore, we have $w^*\bar{\alpha}_s(w) = 1\otimes\rho_G(s),\ s\in G$. Hence we have

$$\alpha_s\otimes\rho_s = \mathrm{Ad}\big(1\otimes\rho_G(s)\big)\circ\bar{\alpha}_s = \mathrm{Ad}(w)^{-1}\circ\bar{\alpha}_s\circ\mathrm{Ad}(w).$$

Hence $\alpha\otimes\rho\simeq\bar{\alpha}\simeq\alpha$. Hence α is dominant. Q.E.D.

Definition 2.19.

(i) A covariant system $\{\mathcal{M}, G, \alpha\}$ is called *stable* if every α-cocycle is a coboundary, i.e. for every α-cocycle $\{a_s\}$ there exists $b\in\mathcal{U}(\mathcal{M})$ such that $a_s = b^*\alpha_s(b)$.

(ii) We say that $\{\mathcal{M}, G, \alpha\}$ is *free* if for any non-zero $a\in\mathcal{M}$ and $s\in G,\ s\neq e$, there exists $x\in\mathcal{M}$ such that $\alpha_s(x)a\neq ax$.

Proposition 2.20. *A free covariant system $\{\mathcal{M}, G, \alpha\}$ over a finite group G is always stable.*

PROOF: The freeness of α is equivalent to the relative commutant property: $\mathcal{M}'\cap\big(\mathcal{M}\rtimes_\alpha G\big) = \mathcal{C}(\mathcal{M})$, where $\mathcal{C}(\mathcal{N})$ means the center of a von Neumann algebra $\mathcal{N}$. Let e be the projection of $\ell^2(G)$ onto the subspace of constant functions on G, which is also given by $e = 1/|G|\sum_{s\in G}\rho(s)$. Then we have

$$\{\mathcal{M}, G, \alpha\}\cong\Big\{\mathcal{M}\otimes\mathcal{L}(\ell^2(G)),\ G,\ \alpha\otimes\rho\Big\}_{1\otimes e}.$$

Hence we have, under the above conjugation,

$$\mathcal{M}^\alpha\Big[\mathcal{M}\otimes\mathcal{L}\big(\ell^2(G)\big)\Big]^{\alpha\otimes\rho}_{(1\otimes e)} = (\mathcal{M}\rtimes_\alpha G)_{(1\otimes e)}.$$

Hence we have

$$\mathcal{C}(\mathcal{M}^\alpha)\cong\mathcal{C}\big(\mathcal{M}\rtimes_\alpha G\big)_{(1\otimes e)}\subseteq\mathcal{C}(\mathcal{M})_{(1\otimes e)}.$$

Since $\mathcal{M}^\alpha$ goes to $\mathcal{M}^\alpha\otimes\mathbf{C}$ in the identification of $\mathcal{M}$ as the subalgebra of $\mathcal{M}\rtimes_\alpha G$, we conclude $\mathcal{C}(\mathcal{M}^\alpha)\subset\mathcal{C}(\mathcal{M})$, so that $\mathcal{C}(\mathcal{M}^\alpha) = \mathcal{C}(\mathcal{M})^\alpha$.

We separate the proof into the properly infinite case and the finite case by decomposing $\mathcal{M}$ into the direct sum of these two components.

a) Properly infinite case: Assume that $\mathcal{M}$ is properly infinite. First we claim that $\mathcal{M}^\alpha$ is also properly infinite. Since $\mathcal{M}^\alpha$ is isomorphic to a reduced algebra of $\mathcal{M} \rtimes_\alpha G$, if $\mathcal{M}$ is of type III, then $\mathcal{M}^\alpha$ is also of type III. Suppose $\mathcal{M}$ is semi-finite, and $\mathfrak{T}$ is a faithful semi-finite normal trace. Set $\mathcal{E} = 1/|G| \sum_{s\in G} \alpha_s$. Then $\mathcal{E}$ is an α-invariant conditional expectation of $\mathcal{M}$ onto $\mathcal{M}^\alpha$. Let $\mathfrak{m}$ be the definition ideal of $\mathfrak{T}$, which is σ-weakly dense in $\mathcal{M}$. It then follows that $\bigcap_{s\in G} \alpha_s(\mathfrak{m})$ is also σ-weakly dense in $\mathcal{M}$. Hence $\mathfrak{T} \circ \mathcal{E}$ is a faithful semi-finite normal trace on $\mathcal{M}$. Thus, we may assume that $\mathfrak{T}$ is invariant under α, and semi-finite on $\mathcal{M}^\alpha$. If τ is a tracial normal state on $\mathcal{M}^\alpha$, then we set $\varphi = \tau \circ \mathcal{E}$. Then there exists $h \in L^1(\mathcal{M}, \mathfrak{T})_+$ such that $\varphi = \mathfrak{T}(h\ \cdot\ )$. Since φ and $\mathfrak{T}$ are both invariant, h is also α-invariant, so that $h \in L^1(\mathcal{M}^\alpha, \mathfrak{T})_+$. Since φ agrees with τ on $\mathcal{M}^\alpha$, h must be central in $\mathcal{M}^\alpha$, so that $h \in L^1\big(\mathcal{C}(\mathcal{M}), \mathfrak{T}\big)_+$ by the first part of our arguments. Therefore φ is a tracial normal state on $\mathcal{M}$, which contradicts the proper infiniteness of $\mathcal{M}$. Therefore, $\mathcal{M}^\alpha$ is also properly infinite.

Given unitary α-cocycles a and b, consider $\bar{\alpha} = \alpha \otimes \mathrm{id}$ on $\mathcal{P} = \mathcal{M} \otimes M_2$, and $c = a \oplus b$ as before. It then follows that $\bar{\alpha}$ is also a free action of G on $\mathcal{P}$. We want to prove $1 \otimes e_{11} \sim 1 \otimes e_{22}$ in $\mathcal{P}^c$. Since $\mathcal{C}(\mathcal{P}^c) \subset \mathcal{C}(\mathcal{P})$, the central support of $1 \otimes e_{ii}$, $i = 1, 2$, in $\mathcal{P}^c$ majorizes their central supports in $\mathcal{P}$ which are 1. Hence the proper infiniteness of $1 \otimes e_{11}$ and $1 \otimes e_{22}$ implies $1 \otimes e_{11} \sim 1 \otimes e_{22}$, which is equivalent to $a \sim b$.

b) Finite case: Suppose $\mathcal{M}$ is finite. Let $\mathfrak{T}$ be the canonical center valued trace of $\mathcal{M}$. By the uniqueness of $\mathfrak{T}$, it is equivariant under α, i.e. $\mathfrak{T} \circ \alpha_s = \alpha_s \circ \mathfrak{T}$, $s \in G$. Hence $\mathfrak{T}$ maps $\mathcal{M}^\alpha$ onto $\mathcal{C}(\mathcal{M})^\alpha = \mathcal{C}(\mathcal{M}^\alpha)$. Therefore, if $e, f \in \mathrm{Proj}(\mathcal{M}^\alpha)$ are equivalent in $\mathcal{M}$, then they are equivalent in $\mathcal{M}^\alpha$. Applying this fact to $\mathcal{P} = \mathcal{M} \otimes M_2$, $\bar{\alpha} = \alpha \otimes 2$ and $c = a \oplus b$ for unitary α-cocycles a and b, we conclude $1 \otimes e_{11} \sim e_{22}$ in $\mathcal{P}^c$, so that $a \simeq b$. Q.E.D.

Exercise X.2

1) Let E be an even dimensional real vector space viewed as a locally compact abelian group relative to the addition. Let σ be a symplectic bilinear form on E in the sense that $\sigma : (r, s) \in E \mapsto \sigma(s, t) \in \mathbf{R}$ is a continuous bilinear map such that $\sigma(r, s) = -\sigma(s, r)$, $r, s \in E$ and the map: $r \in E \mapsto \sigma(r,\ \cdot\ ) \in E^*$ is injective, where E^* means the vector space dual of E, not the Pontrjagin dual which will be denoted by $\widehat{E}$ instead. Let W be a strongly continuous map from E into the group $\mathcal{U}(\mathfrak{H})$ of unitaries on a Hilbert space $\mathfrak{H}$ such that

$$W(r)W(s) = \mathrm{e}^{-\mathrm{i}\sigma(r,s)} W(r+s), \quad r, s \in E,$$

i.e. with $m(r, s) = \mathrm{e}^{-\mathrm{i}\sigma(r,s)}$, $r, s \in E$, m is a multiplier of the additive group E. We call such a map W a *Weyl representation* of $\{E, \sigma\}$. Following the argument below, show that a Weyl representation of $\{E, \sigma\}$ is unique up to multiplicity, which gives an alternative proof for Proposition 2.2 applied to the additive group of a real vector space.

(a) In the Banach space $L^1(E)$ of all Lebesque integrable functions on E, define two operations:

$$f *_\sigma g(r) = \int_E e^{i\sigma(r,p)} f(p) g(r-p)\, dp; \qquad f, g \in L^1(E), \quad r \in E.$$
$$f^*(r) = \overline{f(-r)},$$

Show that $L^1(E)$ is an involutive Banach algebra relative to the above twisted convolution product and the $*$-operation. Denote this algebra by $L^1(E, \sigma)$.

(b) Show that the integral:

$$W(f) = \int_E f(r) W(r)\, dr, \quad f \in L^1(E, \sigma)$$

gives a $*$-representation of the involutive Banach algebra $L^1(E, \sigma)$.

(c) Show that every non-degenerate $*$-representation of $L^1(E, \sigma)$ is of the above form.

(d) Find a basis $\{e_1, \ldots, e_{2n}\}$ of E such that

$$\sigma(e_{2k-1}, e_{2k}) = 1 = -\sigma(e_{2k}, e_{2k-1}), \quad 1 \leq k \leq n;$$
$$\sigma(e_k, e_\ell) = 0, \quad |k - \ell| \geq 2, \qquad \sigma(e_{2k}, e_{2k+1}) = 0, \quad 1 \leq k \leq n-1,$$

and define an automorphism $I \in \mathrm{Aut}(E)$ by setting:

$$I e_{2k-1} = e_{2k}, \qquad I e_{2k} = -e_{2k-1}, \qquad 1 \leq k \leq n,$$

to make E into a complex vector space by setting $(\lambda + i\mu)r = (\lambda + \mu I)r$, $\lambda, \mu \in \mathbf{R}$, $r \in E$. Also define an involutive automorphism $J \in \mathrm{Aut}(E)$ by

$$J e_{2k-1} = e_{2k-1}, \qquad J e_{2k} = -e_{2k}, \qquad 1 \leq k \leq n.$$

(e) Set $\left\langle \sum_k \lambda_k e_k \,\middle|\, \sum_\ell \mu_\ell e_\ell \right\rangle = \sum_k \lambda_k \bar{\mu}_k$, $(\lambda_k), (\mu_\ell) \in \mathbf{C}^n$, so that the symplectic form σ becomes the imaginary part of the inner product of an n-dimensional complex vector space E.

(f) Set $\varphi_0(r) = C \exp\left(-\frac{1}{2}\|r\|^2\right)$ with $C = \left(\int_E \exp(-\|r\|^2)\, dr\right)^{-1}$.

(g) Show that φ_0 is a self-adjoint idempotent of $L^1(E, \sigma)$ such that

$$\varphi_0 *_\sigma f *_\sigma \varphi_0 = \left(\int_E \exp\left(-\frac{1}{2}\|r\|^2\right) f(r)\, dr \right) \varphi_0$$
$$= \langle C\varphi_0, f \rangle \varphi_0, \quad f \in L^1(E, \sigma),$$

where $\langle f, g \rangle$ is the pairing of $g \in L^1(E)$ and $f \in L^\infty(E)$.

(h) Show that the left and right translations $\left\{ f *_\sigma \varphi_0 *_\sigma g : f, g \in L^1(E, \sigma) \right\}$ of φ_0 span a dense subspace of $L^1(E, \sigma)$, which can be seen by pairing against $L^\infty(E)$.

(i) Show that $W(\varphi_0)$ is non-zero and for any non-zero $\xi_0 \in W(\varphi_0)\mathfrak{H}$ the representation:

$$f \in L^1(E,\sigma) \mapsto W(f)\big|_{[W(L^1(E,\sigma)\xi_0)]}$$

is unitarily equivalent to the cyclic representation $\{\pi_{\varphi_0}, \mathfrak{H}_{\varphi_0}\}$. Thus W and $\pi_{\varphi_0} \otimes \mathrm{id}_{W(\varphi_0)\mathfrak{H}}$ are unitarily equivalent.

(j) The representation $\{\pi_{\varphi_0}, \mathfrak{H}_{\varphi_0}\}$ is called the *Schrödinger* representation of the Weyl relation.

2) Keep the notation of the last problem. Show that $L^2(E,\sigma)$ with the structure defined by (a) in the last problem 1 is a unimodular Hilbert algebra, following the suggestion below. First, call the representation W of $L^1(E,\sigma)$ on $L^2(E,\sigma)$ the *regular* Weyl representation.

(a) Let F be the fixed point subspace of E under J and F^* be the anti-fixed point subspace $\{x \in E : Jx = -x\}$ of E under J. It is easy to see that $\{e_{2k-1} : 1 \le k \le n\}$ is a basis for F and $\{e_{2k} : 1 \le k \le n\}$ is the corresponding basis for F^*. The symplectic form σ restricted to $F \times F^*$ gives the paring of the vector space duality between F and F^*. We then identify E with $F \times F^*$. Set $\mathfrak{H} = L^2(F)$ be the Hilbert space of square integrable functions relative to the Lebesques measure with respect to the coordinate system $\{e_1, e_3, \dots, e_{2n-1}\}$. For simplicity, we write $r \cdot p = \sigma(r,p)$, $r \in F$, $p \in F^*$, so that $\sigma(r+p, s+q) = r \cdot q - s \cdot p$, $r, s \in F$, $p, q \in F^*$. Check that if W is a Weyl representation of $\{E,\sigma\}$, then

$$W(r)W(p) = e^{2ir\cdot p} W(p)W(r), \quad r \in F, \quad p \in F^*,$$

so that $U = W|_F$ and $V = W|_{F^*}$ satisfy the Heisenberg-Weyl commutation relation relative to the pairing: $\langle s, p\rangle = e^{2is\cdot p}$, $s \in F$, $p \in F^*$.

(b) Show the converse of (a) in the sense that if $\{U, V\}$ is a covariant representation on a Hilbert space $\mathfrak{H}$ of F and F^* in the sense of Definition 2.1 relative to the pairing in (a), then the map $W : s+p \in E \mapsto W(s+p) = e^{is\cdot p}U(s)V(p)$, $s \in F$, $p \in F^*$, is a Weyl representation of $\{E,\sigma\}$. We denote this W by $W_{U,V}$.

(c) In (b), let $W_0 = W_{\lambda_F,\mu_F}$ in the notation of (4) with $G = F$ and set

$$\psi_0(s) = \left(\frac{1}{\sqrt{2\pi}}\right)^n \exp\left(-\frac{1}{2}\|s\|^2\right), \quad s \in F,$$

so that $\|\psi_0\|_2 = 1$ in $L^2(F)$. Show that

$$\langle f, \varphi_0\rangle = \left(W_0(f)\psi_0 \mid \psi_0\right), \quad f \in L^1(E,\sigma),$$

and that the representation W_0 is unitarily equivalent to the irreducible representation $\{\pi_{\varphi_0}, \mathfrak{H}_{\varphi_0}\}$.

(d) Set $\mathfrak{A}_0 = L^1(E,\sigma) \cap L^2(E,\sigma)$ and observe that $\mathfrak{A}_0$ is a unimodular Hilbert algebra such that $\mathcal{R}_\ell(\mathfrak{A}_0) \cong \mathcal{L}(L^2(F))$ by the unitary equivalence of W_0 and π_{φ_0}, so that it is a factor of type I.

(e) Based on the fact that the trace τ on $\mathcal{B} = \mathcal{R}_\ell(\mathfrak{A}_0)$ associated with the Hilbert algebra $\mathfrak{A}_0$ is proportional to the trace Tr transfered from the usual trace on $\mathcal{L}(L^2(F))$, which is characterized by the fact that it takes value one on a minimal projection of $\mathcal{B}$, show that the full Hilbert algebra $\mathfrak{A} = \mathfrak{A}_0''$ associated with $\mathfrak{A}_0$ is indeed $L^2(E, \sigma)$, which means that $f *_\sigma g \in L^2(E, \sigma)$ if $f, g \in L^2(E, \sigma)$. Furthermore, observing that $\|\varphi_0\|_2 = \sqrt{C}$, prove that $\tau = C\,\mathrm{Tr}$.

3) Keep the above notation. Set $A(E, \sigma) = \{ f *_\sigma g : f, g \in L^2(E, \sigma) \}$.

(a) Show that $A(E, \sigma)$ is a dense subspace of both $L^2(E, \sigma)$ and $C_\infty(E)$, the Banach space of all continuous functions on E vanishing at infinity. Define a norm on $A(E, \sigma)$ by:

$$\|f\|_{A(E,\sigma)}$$
$$= \inf\left\{ \sum_{j=1}^{N} \|g_j\|_2 \|h_j\|_2 : f = \sum_{j=1}^{N} g_j *_\sigma h_j,\ \ g_j, h_j \in L^2(E, \sigma),\ 1 \le j \le N \right\}.$$

(b) Show that $A(E, \sigma)$ with the norm just defined is an involutive Banach algebra under the twisted convolution and the involution * which is isometrically isomorphic to the involutive Banach algebra $\mathcal{LT}(\mathfrak{H})$ of all nuclear operators on $\mathfrak{H}$ under the Weyl representation W_0 of Problem 2(a).

(c) Show that the isomorphism W_0 transforms the Tr on $\mathcal{LT}(\mathfrak{H})$ in the following fashion:

$$\mathrm{Tr}\big(W_0(f)\big) = Cf(0), \quad f \in A(E, \sigma).$$

(d) Show that if a finite Borel measure μ on E is σ-positive definite in the sense that $\mu(f^* *_\sigma f) \ge 0,\ f \in L^2(E, \sigma)$, then μ is absolutely continuous relative to the Lebesgue measure and its Radon Nikodym derivaltive is an element of $A(E, \sigma)$ of the form: $g *_\sigma g^*,\ g \in L^2(E, \sigma)$.

4) On the factor $\mathcal{L}(\mathfrak{H})$, define an action α of E by:

$$\alpha_s(x) = W_0(s) W_0(x) W_0(s)^*, \quad x \in \mathcal{L}(\mathfrak{H}), \quad s \in E$$

(a) Show that the action α of E is integrable.

(b) Show

$$\mathrm{Tr}(x) = \int_E \alpha_s(x)\, ds, \quad x \in \mathcal{L}(\mathfrak{H})_+.$$

Hence the trace on $\mathcal{L}(\mathfrak{H})$ is indeed an integration.

5) Let us consider now more general antisymmetric forms σ on E, i.e. we remove the non-degeneracy assumption from the antisymmetric form σ. The Weyl relation of Problem 1 still makes sense although the uniqueness and other nice features of the representation of Weyl relation are lost. In $L^1(E, \sigma) \cap L^2(E, \sigma)$ we consider the unimodular Hilbert algebra structure of Problem 1(a) and denote it by $\mathfrak{A}_0(E, \sigma)$.

Denote the associated left von Neumann algebra by $\mathcal{R}(E,\sigma)$ and the fullfilment of $\mathfrak{A}_0(E,\sigma)$ by $\mathfrak{A}(E,\sigma)$. We also have the associated trace τ_σ on $\mathcal{R}(E,\sigma)$. Set

$$A(E,\sigma) = \{\, f *_\sigma g : f, g \in L^2(E,\sigma)\,\};$$

$$\|f\|_{A(E,\sigma)}$$
$$= \inf\left\{ \sum_{j=1}^{N} \|g_j\|_2 \|h_j\|_2 : f = \sum_{j=1}^{N} g_j *_\sigma h_j,\ \ g_j, h_j \in L^2(E,\sigma),\ 1 \le j \le N \right\}.$$

(a) Show that $A(E,\sigma)$ is a Banach space and isometrically isomorphic to the predual of $\mathcal{R}(E,\sigma)$.

(b) Show that $A(E,\sigma) \subset C_\infty(E)$ and $\overline{A(E,\sigma)} = A(E,-\sigma)$.

(c) Show that if σ_1 and σ_2 are both antisymmetric bilinear forms on E, then we have:

$$A(E,\sigma_1)A(E,\sigma_2) \subset A(E,\sigma_1+\sigma_2);$$

$$\|fg\|_{A(E,\sigma_1+\sigma_2)} \le \|f\|_{A(E,\sigma_1)} \|g\|_{A(E,\sigma_2)}, \quad f \in A(E,\sigma_1), \quad g \in A(E,\sigma_2),$$

where the product is taken place as elements of $C_\infty(E)$.

(d) Let Σ be the real vector space of all antisymmetric bilinear forms on E considered as a locally compact abelian group relative to the addition. Let $A(E,\Sigma)$ be the set of all measurable functions f on $E \times \Sigma$ such that:

(i) for each fixed $\sigma \in \Sigma$, $f(\,\cdot\,,\sigma)$ belongs to $A(E,\sigma)$ as a function of the first variable;

(ii)
$$\|f\|_{A(E,\Sigma)} = \int_\Sigma \left\| f(\,\cdot\,,\sigma) \right\|_{A(E,\sigma)} \mathrm{d}\sigma < \infty,$$

where $\mathrm{d}\sigma$ means a Haar measure on Σ, i.e. the Lebesgue measure relative to any but fixed coordinate system in Σ.

Show that $A(E,\Sigma)$ is a commutative involutive Banach algebra relative to the following convolution and involution:

$$(f * g)(s,\sigma) = \int_\Sigma f(s,\rho) g(s,\sigma-\rho)\, \mathrm{d}\rho, \quad f, g \in A(E,\Sigma);$$

$$f^*(s,\sigma) = \overline{f(s,-\sigma)}.$$

(e) Show that $A(E,\Sigma)$ is semi-simple and identify its spectrum with $E \times \widehat{\Sigma}$ where $\widehat{\Sigma}$ is of course the Pontrjagin dual of Σ.

(f) Show that $A(E,\Sigma)$ is the predual of a von Neumann algebra, say $\mathcal{R}(E,\Sigma)$. Identify the von Neumann algebra $\mathcal{R}(E,\Sigma)$.

(g) Show that the von Neumann algebra $\mathcal{R}(E,\Sigma)$ is of type I.

6) Find what parts of the arguments in Problems 1 through 5 will be valid for a locally compact abelian groups G if we replace antisymmetric biliniear forms by continuous antisymmetric bicharacters χ in the sense that χ is a continuous function from $G \times G$ into $\mathbf{T}$ such that (i) $\chi(r, s_1 + s_2) = \chi(r, s_1)\chi(r, s_2)$ and (ii) $\chi(r, s) = \overline{\chi(s, r)}$, $r, s, s_1, s_2 \in G$. (Obviously, the Weyl relation make sense, but the function φ_0 is lost. Does this mean that all other claims lost?)

7) Let G be an abelian separable locally compact group and m be a multiplier of G, i.e. an element of $Z^2(G, \mathbf{T})$. Set $\chi_m(r, s) = m(r, s)\overline{m(s, r)}$, $r, s \in G$.

(a) Show that χ_m is a continuous antisymmetric bicharacter of G. (Hint: With $\mathcal{M} = \mathcal{L}\big(L^2(G)\big)$, set $\alpha_s^m = \mathrm{Ad}\big(\lambda_m(s)\big)$, $s \in G$ for the left regular m-representation of G on $L^2(G)$ and observe that (i) α^m is an action of G on $\mathcal{M}$ and (ii) $\chi_m(s)\lambda(r) = \alpha_s^m\big(\lambda(r)\big)$, $r, s \in G$.)

(b) Show that if $\chi_m = 1$, then m is a coboundary. (Hint: Observe that $\chi_m = 1 \Longleftrightarrow$ "$G \times_m \mathbf{T}$ is abelian". If $G \times_m \mathbf{T}$ is abelian, then the exact sequence:

$$1 \to \mathbf{T} \to G \times_m \mathbf{T} \to G \to 0$$

splits becauses the identity map: $\mathbf{T} \xrightarrow{\mathrm{id}} \mathbf{T}$ extends to $G \times_m \mathbf{T}$ as a character of the abelian group $G \times_m \mathbf{T}$.)

(c) Show that for a pair $m_1, m_2 \in Z^2(G, \mathbf{T})$ $[m_1] = [m_2]$ if and only if $\chi_{m_1} = \chi_{m_2}$. Thus the map: $m \in Z^2(G, \mathbf{T}) \mapsto \chi_m \in X^2(G, \mathbf{T})$ is an isomorphism of the second cohomology group $H^2(G, \mathbf{T})$ into the group $X^2(G, \mathbf{T})$ of all continuous antisymmetric bicharacters of G.

(d) Show that $X^2(G, \mathbf{T})$ is an abelian locally compact group under the topology of uniform convergence on compact subsets of G.

8) Consider the problem of whether or not the map: $[m] \in H^2(G, \mathbf{T}) \mapsto \chi_m \in X^2(G, \mathbf{T})$ is an isomorphism.

(a) Show that if G is discrete then the map: $[m] \mapsto \chi_m$ is an isomorphism.

(b) Show that if there exists a continuous endomorphism: $g \in G \mapsto g' \in G$ such that $g = g' + g'$ then the map: $[m] \mapsto \chi_m$ is an isomorphism. Such a group is said to be *divisible by* 2.

9) Each $m \in Z^2(G, \mathbf{T})$ determines uniquely an exact sequence:

$$1 \longrightarrow \mathbf{T} \xrightarrow{i} E_m \xrightarrow{\pi} G \longrightarrow 1.$$

Show that if

$$1 \longrightarrow \mathbf{T} \xrightarrow{i_1} E_m \xrightarrow{\pi_1} G \longrightarrow 1;$$

$$1 \longrightarrow \mathbf{T} \xrightarrow{i_2} E_m \xrightarrow{\pi_2} G \longrightarrow 1$$

are the exact sequences corresponding to $m_1, m_2 \in Z^2(G, \mathbf{R})$, then the exact sequence:

$$1 \longrightarrow \mathbf{T} \xrightarrow{i} E_m \xrightarrow{\pi} G \longrightarrow 1$$

with $m = m_1 m_2 \in Z^2(G, \mathbf{T})$ is identified with the one obtained by the following:

$$E = \left\{ (g_1, g_2) \in E_{m_1} \times E_{m_2} : \pi_1(g_1) = \pi_2(g_2) \right\};$$
$$i(\lambda) = \left(i_1(\lambda), i_2(\lambda)\right) \in E, \quad \lambda \in \mathbf{T};$$
$$1 \longrightarrow \mathbf{T} \xrightarrow{i} E \xrightarrow{\pi} G \longrightarrow 1.$$

10) With the same setting of the last problem, let $\{\mathcal{M}, G, \alpha\}$ be a separable covariant system.

(a) Show that if $m \in Z^2(G, \mathbf{T})$ is fixed, then the twisted crossed product $\mathcal{M} \rtimes_{\alpha,m} G$ admits an action $\widehat{\alpha}$ of the Pontrjagin dual $\widehat{G}$ of G, called *dual action*, such that $\hat{\alpha}_p\left(\pi_\alpha(x)\right) = \pi_\alpha(x)$, $x \in \mathcal{M}$ and $\hat{\alpha}_p\left(u_m(s)\right) = \overline{\langle s, p\rangle} u_m(s)$, $s \in G$, $p \in \widehat{G}$.

(b) Show that the duality

$$\left\{\left(\mathcal{M} \rtimes_{\alpha,m} G\right) \rtimes_{\widehat{\alpha}} \widehat{G},\ G,\ \widehat{\widehat{\alpha}}\right\} \cong \left\{\mathcal{M} \,\overline{\otimes}\, \mathcal{L}\left(L^2(G)\right),\ G,\ \alpha \otimes \rho\right\}$$

holds.

11) Let $\{\mathcal{M}, G, \alpha\}$ be a covariant system with G a general locally compact group. Consider the crossed product $\mathcal{N} = \mathcal{M} \rtimes_\alpha G$. Use the notations from §1. Consider the unitary W on $L^2(G, \mathfrak{H}) \otimes L^2(G) = L^2\left(G \times G, \mathfrak{H}\right)$ defined by:

$$(W\xi)(s, t) = \xi(s, st), \quad \xi \in L^2\left(G \times G, \mathfrak{H}\right), \quad s, t \in G,$$

where $\mathfrak{H}$ is the original Hilbert space on which $\mathcal{M}$ is represented.

(a) Show that $W^*\left(\mathcal{N} \otimes \mathbf{C}_{L^2(G)}\right)W \subset \mathcal{N} \,\overline{\otimes}\, \mathcal{R}_\ell(G)$. Set $\pi(x) = W^*(x \otimes 1)W$, $x \in \mathcal{N}$.

(b) Show that π is an isomorphism of $\mathcal{N}$ into $\mathcal{N} \,\overline{\otimes}\, \mathcal{R}_\ell(G)$ such that $(\mathrm{id} \otimes \pi_G) \circ \pi = (\pi \otimes \mathrm{id}) \circ \pi$, where π_G is the isomorphism of $\mathcal{R}_\ell(G)$ into $\mathcal{R}_\ell(G) \,\overline{\otimes}\, \mathcal{R}_\ell(G)$ defined similarly by viewing $\mathcal{R}_\ell(G)$ as the crossed product: $\mathcal{R}_\ell(G) = \mathbf{C} \rtimes G$.

(c) Let $A(G) = \mathcal{R}_\ell(G)_*$ the Fourier algebra of G. Observe that each $\varphi \in A(G)$ acts on $\mathcal{N}$ denoted by $\widehat{\alpha_\varphi}$:

$$\left\langle \widehat{\alpha_\varphi}(x), \omega \right\rangle = \left\langle \pi(x), \omega \otimes \varphi \right\rangle, \quad x \in \mathcal{N}, \quad \omega \in \mathcal{N}_*.$$

(d) Show that $\hat{\alpha} : \varphi \in A(G) \mapsto \widehat{\alpha_\varphi} \in \mathcal{L}_w(\mathcal{N})$ is an action of the Banach algebra $A(G)$ on $\mathcal{N}$.

(e) Show that $\widehat{\alpha_\varphi}\left(u(s)\right) = \varphi(s)u(s)$, $s \in G$, where $\left\{ u(s) : s \in G \right\}$ means the unitary representation of G in $\mathcal{N}$ associated with the crossed product.

§ 3 Equivariant Disintegration

Let G be a separable locally compact group and A a separable C^*-algebra. Suppose that we have a continuous homomorphism $\alpha\colon s \in G \mapsto \alpha_s \in \mathrm{Aut}(A)$ in the sense that

$$\lim_{s\to e} \|\alpha_s(x) - x\| = 0, \quad x \in A. \tag{1}$$

We call $\{A, G, \alpha\}$ a *C^*-covariant system.* Of course, for the definition of a C^*-covariant system, we don't need the separability for A. But, since we are going to discuss the disintegration of representations of $\{A, G, \alpha\}$, we assume the countability condition to avoid unnecessary complications.

Proposition 3.1. *For a separable representation $\{\pi, \mathfrak{H}\}$ of A, the following two conditions are equivalent:*

(i) *$\pi \circ \alpha_s$ and π are quasi-equivalent for every $s \in G$;*
(ii) *There exists an action $\tilde{\alpha}$ of G on the von Neumann algebra $\mathcal{M} = \pi(A)''$ generated by $\pi(A)$ such that $\tilde{\alpha}_s \circ \pi = \pi \circ \alpha_s$, $s \in G$.*

PROOF: Except for the σ-weak continuity: $s \in G \mapsto \alpha_s(x) \in \mathcal{M}$, the equivalence of (i) and (ii) is an easy consequence of Chapter III. By the separability of $\mathfrak{H}$, $\mathcal{M}$ admits a faithful normal state φ. By considering the quotient C^*-algebra $A/\pi^{-1}(0)$, we view A as a σ-weakly dense C^*-subalgebra of $\mathcal{M}$ on which G acts continuously in norm. The unit ball of $\mathcal{M}$ is metrizable in the σ-strong operator topology, so that for every $x \in \mathcal{M}$ there exists a sequence $\{x_n\}$ in A such that $\{x_n\}$ converges to x σ-strongly. Hence for every $\omega \in \mathcal{M}_*$, we have $\omega \circ \tilde{\alpha}_s(x) = \lim_{n\to\infty} \omega \circ \alpha_s(x_n)$, so that the function $s \in G \mapsto \omega \circ \alpha_s(x) \in \mathbf{C}$ is measurable. Therefore, for every $f \in L^1(G)$ the integral

$$\int_G f(s)\omega\big(\alpha_s(x)\big)\, ds, \quad x \in \mathcal{M},$$

makes sense and gives an element $\omega \circ \alpha_f$ of $\mathcal{M}^*$. If $\omega \in \mathcal{M}_*^+$ and $f \geq 0$, then $\omega \circ \alpha_f$ is also positive. By the monotone convergence theorem, $\omega \circ \alpha_f$ is normal, so that $\omega \circ \alpha_f \in \mathcal{M}_*$. With δ_G the modular function of G, we have

$$\omega \circ \alpha_f \circ \alpha_s = \delta_G(s)\omega \circ \alpha_{f_s},$$

where $f_s(t) = f(ts^{-1})$, so that $s \in G \mapsto \omega \circ \alpha_f \circ \alpha_s \in \mathcal{M}_*$ is norm continuous if $f \in \mathcal{K}(G)$, the space of all continuous functions with compact support. Since $\{\omega \circ \alpha_f : f \in \mathcal{K}(G)\}$ is total in $\mathcal{M}_*$, we conclude that $s \in G \mapsto \tilde{\alpha}_s \in \mathrm{Aut}(\mathcal{M})$ is continuous. Q.E.D.

Remark 3.2. In the above proposition, the separability of π is essential. Namely, if π is a universal representation of A, then the canonical extension $\tilde{\alpha}_s$ of α_s to $\pi(A)'' = A^{**}$ does not give a continuous action of G unless G is discrete.

Definition 3.3. We say that a representation $\{\pi, \mathfrak{H}\}$ of A is *quasi-invariant* if the conditions of Proposition 3.1 are satisfied. We also say that $\varphi \in A_+^*$ is *quasi-invariant* if the associated cyclic representation π_φ is quasi-invariant.

Associated with the action α of G on A is the action T of G on the state space $\mathfrak{S}$ of A defined by

$$T_s\omega = \omega \circ \alpha_s^{-1}, \quad \omega \in \mathfrak{S}, \quad s \in G. \tag{2}$$

To each $\omega \in \mathfrak{S}$, we set

$$U_{s,\omega}\pi_\omega(x)\xi_\omega = \pi_{T_s\omega}\big(\alpha_s(x)\big)\xi_{T_s\omega}, \quad x \in A. \tag{3}$$

It then follows that $U_{s,\omega}$ can be extended uniquely to a unitary of $\mathfrak{H}_\omega$ onto $\mathfrak{H}_{T_s\omega}$, denoted by $U_{s,\omega}$ again, such that

$$U_{s,\omega}\pi_\omega(x)U_{s,\omega}^* = \pi_{T_s\omega}\big(\alpha_s(x)\big), \quad x \in A. \tag{4}$$

Hence it gives rise to an isomorphism $\alpha_{s,\omega}$ of $\mathcal{M}(\omega) = \pi_\omega(A)''$ onto $\mathcal{M}(T_s\omega) = \pi_{T_s(\omega)}(A)''$:

$$\alpha_{s,\omega}(x) = U_{s,\omega}xU_{s,\omega}^*, \quad x \in \mathcal{M}(\omega), \tag{5}$$

such that

$$\alpha_{s,\omega}\big(\pi_\omega(x)\big) = \pi_{T_s\omega}\big(\alpha_s(x)\big), \quad x \in A. \tag{6}$$

On the canonical measurable field $\{\mathcal{M}(\omega), \mathfrak{H}_\omega\}$ over the state space $\mathfrak{S}$ which is a Polish space, we define

$$\{\tilde{\alpha}_s(x)\}(\omega) = \alpha_{s,T_s^{-1}\omega}\big(x(T_s^{-1}\omega)\big), \qquad \{\widetilde{U}(s)\xi\}(\omega) = U_{s,T_s^{-1}\omega}\xi(T_s^{-1}\omega) \tag{7}$$

for measurable fields $\omega \in \mathfrak{S} \mapsto x(\omega) \in \mathcal{M}(\omega)$ (resp. $\omega \in \mathfrak{S} \mapsto \xi(\omega) \in \mathfrak{H}_\omega$).

Lemma 3.4. *In the above context, for every $s \in G$, the new operator (resp. vector) field $\tilde{\alpha}_s(x)$ (resp. $\widetilde{U}(s)\xi$) is measurable.*

PROOF: Suppose that $\xi(\omega) = \pi_\omega(x)\xi_\omega$ for some $x \in A$. Then

$$\big(\widetilde{U}(s)\xi\big)(\omega) = \pi_\omega\big(\alpha_s(x)\big)\xi_\omega,$$

so that $\widetilde{U}(x)\xi$ is a measurable vector field. Suppose $\omega \in \mathfrak{S} \mapsto \eta(\omega) \in \mathfrak{H}_\omega$ is measurable. Then we have

$$\begin{aligned}\Big(\big(\widetilde{U}(s)\eta\big)(\omega) \;\Big|\; \xi(\omega)\Big) &= \Big(U_{s,T_s^{-1}\omega}\eta\big(T_s^{-1}\omega\big) \;\Big|\; \xi(\omega)\Big)\\ &= \Big(\eta\big(T_s^{-1}\omega\big) \;\Big|\; U_{s,T_s^{-1}\omega}^*\pi_\omega(x)\xi_\omega\Big)\\ &= \Big(\eta\big(T_s^{-1}\omega\big) \;\Big|\; \pi_{T_s^{-1}\omega}\big(\alpha_s^{-1}(x)\big)\xi_{T_s^{-1}\omega}\Big),\end{aligned}$$

so that the function: $\omega \mapsto \big((\widetilde{U}(s)\eta)(\omega) \mid \xi(\omega)\big) \in \mathbf{C}$ is measurable. Therefore $\widetilde{U}(s)\eta$ is a measurable field in view of Lemma IV.8.10.

Since $\tilde{\alpha}_s(x) = \widetilde{U}(s)x\widetilde{U}(s^{-1})$ for any measurable field $x : \omega \in \mathfrak{S} \mapsto x(\omega) \in \mathcal{M}(\omega)$, the measurability of $\tilde{\alpha}_s(x)$ follows from the measurability of $\widetilde{U}(s)$. Q.E.D.

Lemma 3.5. *In the above context, we conclude the following cocycle identity:*

$$\left.\begin{aligned} \alpha_{st,\omega} &= \alpha_{s,T_t\omega} \circ \alpha_{t,\omega}\,, \\ U_{st,\omega} &= U_{s,T_t\omega} U_{t,\omega}\,, \end{aligned} \qquad s, t \in G, \quad \omega \in \mathfrak{S}. \right\} \tag{8}$$

This follows from a simple calculation, so we leave the proof to the reader.

Proposition 3.6. *Let μ be an orthogonal probability measure on $\mathfrak{S}$ with barycenter $\varphi \in \mathfrak{S}$. If μ is quasi-invariant under the action of G, then we have the following conclusion:*

(i) *The cyclic representation $\{\pi_\varphi, \mathfrak{H}_\varphi\}$ of A admits a covariant unitary representation $\overline{U}$ of G given by*

$$\big(\overline{U}(s)\xi\big)(\omega) = \sqrt{\frac{\mathrm{d}\mu \circ T_s^{-1}}{\mathrm{d}\mu}(\omega)}\; U_{s,T_s^{-1}\omega}\xi\big(T_s^{-1}\omega\big), \tag{9}$$

where

$$\{\pi_\varphi, \mathfrak{H}_\varphi, \xi_\varphi\} = \int_{\mathfrak{S}}^{\oplus} \{\pi_\omega, \mathfrak{H}_\omega, \xi_\omega\}\, \mathrm{d}\mu(\omega). \tag{10}$$

(ii) *The von Neumann algebra $\mathcal{M} = \pi_\varphi(A)''$ admits a continuous action $\overline{\alpha}$ of G such that*

$$\overline{\alpha}_s \circ \pi_\varphi = \pi_\varphi \circ \alpha_s\,. \tag{11}$$

(iii) *Relative to the disintegration:*

$$x = \int_{\mathfrak{S}}^{\oplus} x(\omega)\, \mathrm{d}\mu(\omega) \tag{12}$$

of each $x \in \mathcal{M}$, $\overline{\alpha}_s(x)$ is given by:

$$\overline{\alpha}_s(x) = \int_{\mathfrak{S}}^{\oplus} \alpha_{s,T_s^{-1}\omega}\big(x(T_s^{-1}\omega)\big)\, \mathrm{d}\mu(\omega)\,. \tag{13}$$

Note that since we do not assume μ to be central, $\mathcal{M}$ is not necessarily the direct integral $\int_{\mathfrak{S}}^{\oplus} \mathcal{M}(\omega)\, \mathrm{d}\mu(\omega)$ in (iii).

PROOF:

(i) By Theorem IV.8.31, we have

$$\{\pi_\varphi, \mathfrak{H}_\varphi, \xi_\varphi\} = \int_{\mathfrak{S}}^{\oplus} \{\pi_\omega, \mathfrak{H}_\omega, \xi_\omega\}\, \mathrm{d}\mu(\omega)\,.$$

We define

$$\big(\overline{U}(s)\eta\big)(\omega) = \sqrt{\frac{\mathrm{d}\mu \circ T_s^{-1}}{\mathrm{d}\mu}(\omega)}\; U_{s,T_s^{-1}\omega}\eta\big(T_s^{-1}\omega\big)$$

for each $\eta \in \mathfrak{H}_\varphi$. It then follows that $\{\overline{U}(s)\}$ is a unitary representation such that

$$\overline{U}(s)\pi_\varphi(a)\overline{U}(s)^* = \pi_\varphi(\alpha_s(a)), \quad a \in A.$$

Hence $\overline{U}$ gives rise to an action $\overline{\alpha}$ of G on $\mathcal{M}$. From its construction, $\overline{\alpha}$ has the form of (9).

Statements (ii) and (iii) follow directly from (i). Q.E.D.

Definition 3.7. Let $\{\pi, U, \mathfrak{H}\}$ be a covariant representation of $\{A, G, \alpha\}$. An *equivariant disintegration* of $\{\pi, U, \mathfrak{H}\}$ means a disintegration of $\{\pi, \mathfrak{H}\}$

$$\{\pi, \mathfrak{H}\} = \int_\Omega^\oplus \{\pi_\omega, \mathfrak{H}(\omega)\}\, d\mu(\omega) \tag{14}$$

together with

(i) an action $\{T_s : s \in G\}$ on $\{\Omega, \mu\}$ leaving the measure μ quasi-invariant;

(ii) a system $\{U_{s,\omega} : s \in G,\ \omega \in \Omega\}$ such that

a) $U_{s,\omega}$ is a unitary of $\mathfrak{H}(\omega)$ onto $\mathfrak{H}(T_s\omega)$;

b) $U_{st,\omega} = U_{s,T_t\omega}U_{t,\omega},\ \ s, t \in G$, for almost every $\omega \in \Omega$;

c) For each measurable vector field $\{\xi(\omega)\}$, the new field:

$$\omega \mapsto U_{s,T_s^{-1}\omega}\xi(T_s^{-1}\omega) \in \mathfrak{H}(\omega)$$

is measurable;

d) The unitary representation $U(s)$ is given by

$$(U(s)\xi)(\omega) = \sqrt{\frac{d\mu \circ T_s^{-1}}{d\mu}(\omega)}\, U_{s,T_s^{-1}\omega}\xi(T_s^{-1}\omega) \tag{15}$$

for every $\xi = \int_\Omega^\oplus \xi(\omega)\, d\mu(\omega) \in \mathfrak{H}$.

We write

$$U(s) = \int_\Omega^\oplus U_{s,\omega}\, d\mu(\omega), \quad s \in G, \tag{16}$$

even though it is a somewhat abusive notation.

Theorem 3.8. *Let $\{\pi, U, \mathfrak{H}\}$ be a covariant representation of a C^*-covariant system $\{A, G, \alpha\}$. To each abelian von Neumann subalgebra $\mathcal{A}$ of $\pi(A)'$ such that $U(s)\mathcal{A}U(s)^* = \mathcal{A}$, there corresponds an equivariant disintegration of $\{\pi, U, \mathfrak{H}\}$:*

$$\{\pi,\ U,\ \mathfrak{H}\} = \int_\Omega^\oplus \{\pi_\omega,\ U_{\cdot,\omega},\ \mathfrak{H}(\omega)\}\, d\mu(\omega) \tag{17}$$

such that the diagonal algebra of the disintegration is $\mathcal{A}$.

PROOF: Let $\mathcal{B} = \mathcal{A}'$, and $\mathfrak{T}$ be the center valued trace on $\mathcal{B}$ normalized in such a way that $\mathfrak{T}(e) = 1$ for any abelian projection of $\mathcal{B}$ with central support 1. Since the normalization of $\mathfrak{T}$ uniquely determines $\mathfrak{T}$, $\mathfrak{T}$ is equivariant under $\mathrm{Aut}(\mathcal{B})$, i.e. for any $\sigma \in \mathrm{Aut}(\mathcal{B})$, we have $\sigma \circ \mathfrak{T} = \mathfrak{T} \circ \sigma$. Since the unitary representation U normalizes $\mathcal{A}$, it induces an action $\tilde{\alpha}$ of G on $\mathcal{B}$ by $\tilde{\alpha}_s(x) = U(s)xU(s)^*,\ s \in G$.

Let B be a separable σ-weakly dense C^*-subalgebra of $\mathcal{B}$ such that

a) $B \supset \pi(A)$;

b) $\tilde{\alpha}_s(B) = B$ and $\{B, G, \tilde{\alpha}\}$ is a C^*-covariant system;

c) $\{x \in B : \mathfrak{T}(x^*x) \in \mathcal{A}\}$ is σ-weakly dense in $\mathcal{B}$.

The existence of such B follows from the smoothing operation, i.e. applications of $\tilde{\alpha}_f,\ f \in \mathcal{K}(G)$, as seen before. The only nontrivial part would be to see the condition (c). But this can be seen by the fact

$$\tilde{\alpha}_f \circ \mathfrak{T} = \mathfrak{T} \circ \tilde{\alpha}_f, \quad f \in \mathcal{K}(G)_+.$$

Let ξ_0 be a cyclic vector of $\mathfrak{H}$ for $\mathcal{B}$, i.e. a separating vector for $\mathcal{A}$. We normalize it, so that $\|\xi_0\| = 1$.

Let e be the cyclic projection of $\mathcal{B}$ with range $[\mathcal{A}\xi_0]$. Let $\{\Gamma, \mu\}$ be the measure theoretic spectrum of $\mathcal{A}$, i.e. $\mathcal{A} = L^\infty(\Gamma, \mu)$ where Γ is a Polish space, say compact and metrizable, and μ is given by the state

$$\varphi(a) = \left(a\xi_0 \mid \xi_0\right), \quad a \in \mathcal{B},$$

restricted to $\mathcal{A}$. We then have for each $x \in \mathcal{B}$

$$\varphi(x) = \varphi\big(\mathfrak{T}(xe)\big) = \int_\Gamma \mathfrak{T}(xe)(\gamma)\, d\mu(\gamma),$$

so that we have

$$\varphi = \int_\Gamma^\oplus \varphi_\gamma\, d\mu(\gamma),$$

with

$$\varphi_\gamma(x) = \mathfrak{T}(xe)(\gamma), \quad x \in B.$$

By applying the Gelfand-Naimark-Segal construction, Definition I.9.15, to $\{B, \varphi_\gamma\}$, we obtain the γ-component, i.e.

$$\{\pi_\gamma, \mathfrak{H}(\gamma)\} = \{\pi_{\varphi_\gamma}, \mathfrak{H}_{\varphi_\gamma}\}, \quad \gamma \in \Gamma.$$

The vector ξ_0 is disintegrated into:

$$\xi_0(\gamma) = \xi_{\varphi_\gamma}.$$

Fix $s \in G$ and get the disintegration:

$$U(s)\xi_0 = \int_\Gamma^\oplus \big(U(s)\xi_0\big)(\gamma)\, d\mu(\gamma).$$

Let T_s be the nonsingular transformation on $\{\Gamma, \mu\}$ corresponding to $\tilde{\alpha}_s$ on $\mathcal{A}$, i.e.

$$\tilde{\alpha}_s(f)(\gamma) = f\left(T_s^{-1}\gamma\right), \quad f \in \mathcal{A}.$$

We then have, for each $f \in \mathcal{A}$ and $x, y \in B$,

$$\begin{aligned}
\int_\Gamma f(\gamma)\big(\pi_\gamma(x)\xi_0(\gamma) \mid \pi_\gamma(y)\xi_0(\gamma)\big)\, \mathrm{d}\mu(\gamma) &= \big(\tilde{\alpha}_s(f)U(s)x\xi_0 \mid U(s)y\xi_0\big) \\
&= \int_\Gamma f(T_s^{-1}\gamma)\Big(\big(U(s)x\xi_0\big)(\gamma) \,\Big|\, \big(U(s)y\xi_0\big)(\gamma)\Big)\, \mathrm{d}\mu(\gamma) \\
&= \int_\Gamma f(\gamma)\Big(\big(U(s)x\xi_0\big)(T_s\gamma) \,\Big|\, \big(U(s)y\xi_0\big)(T_s\gamma)\Big)\frac{\mathrm{d}\mu\circ T_s}{\mathrm{d}\mu}(\gamma)\, \mathrm{d}\mu(\gamma),
\end{aligned}$$

so that with $\rho(s,\gamma) = \frac{\mathrm{d}\mu\circ T_s}{\mathrm{d}\mu}(\gamma)$

$$\big(\pi_\gamma(x)\xi_0(\gamma) \mid \pi_\gamma(y)\xi_0(\gamma)\big) = \Big(\big(U(s)x\xi_0\big)(T_s\gamma) \,\Big|\, \big(U(s)y\xi_0\big)(T_s\gamma)\Big)\rho(s,\gamma)$$

for each $x, y \in B$, $s \in G$ and almost every $\gamma \in \Gamma$. The separability of B allows us to find a null set $N(s)$ in Γ such that the above identity holds for every $\gamma \in \Gamma\backslash N(s)$. We then set

$$U_{s,\gamma}\, \pi_\gamma(x)\, \xi_0(\gamma) = \rho(s,\gamma)^{\frac{1}{2}}\, \big(U(s)x\xi_0\big)(T_s\gamma), \quad \gamma \in \Gamma \setminus N(s),$$

and see that $U_{s,\gamma}$ can be viewed as isometry of $\mathfrak{H}(\gamma)$ onto $\mathfrak{H}(T_s\gamma)$. It is now straightforward to check that

$$U_{st,\gamma} = U_{s,T_t\gamma}\, U_{t,\gamma}, \quad \gamma \in \Gamma \setminus \big(N(s) \cup N(t)\big); \tag{18}$$

$$\big(U(s)x\xi_0\big)(\Gamma_s\gamma) = \rho(s,\gamma)^{-\frac{1}{2}}\, U_{s,\gamma}\, \pi_\gamma(x)\, \xi_0(\gamma). \tag{19}$$

Restricting π_γ to $\pi(A)$, we obtain equivariant disintegration:

$$\{\pi, U, \mathfrak{H}\} = \int_\Gamma^\oplus \{\pi_\gamma, U_{\cdot,\gamma}, \mathfrak{H}(\gamma)\}\, \mathrm{d}\mu(\gamma)\,. \tag{20}$$

Q.E.D.

Remark 3.9. Note that the map: $\gamma \in \Gamma \mapsto \varphi_\gamma \in \mathfrak{S}$ appeared in the proof is not necessarily equivariant, i.e. the equation $\varphi_{T_s\gamma} = \varphi_\gamma \circ \alpha_s^{-1}$ need not hold, so that the induced measure on $\mathfrak{S}$ by the map: $\gamma \in \Gamma \mapsto \varphi_\gamma \in \mathfrak{S}$ is not necessarily quasi-invariant.

Proposition 3.10. *Let $\{A, G, \alpha\}$ be as before and $\varphi \in \mathfrak{S}$ be invariant, i.e. $\varphi \circ \alpha_s = \varphi$, $s \in G$. Then any G-invariant abelian von Neumann subalgebra $\mathcal{A}$ of $\pi_\varphi(A)'$ gives rise to an equivariant disintegration of $\{\pi_\varphi, U_\varphi, \mathfrak{H}_\varphi, \xi_\varphi\}$ over its measure theoretic spectrum $\{\Gamma, \mu\}$ such that*

(i) *the cocycle identity:*

$$U_{st,\gamma} = U_{s,T_t\gamma} U_{t,\gamma} \tag{21}$$

holds for every $s, t \in G$ *and* $\gamma \in \Gamma$;

(ii) $U_{s,\gamma}\xi_\varphi(\gamma) = \xi_\varphi(T_s\gamma)$ *for every* $s \in G$ *and* $\gamma \in \Gamma$, *where of course*

$$\xi_\varphi = \int_\Gamma^\oplus \xi_\varphi(\gamma)\, d\mu(\gamma)\,.$$

PROOF: As in the proof of the last theorem, we choose ξ_φ for ξ_0. In the choice of the C^*-subalgebra B of $\mathcal{B}$ we require one more condition:

d) $B \cap \mathcal{A}$ is σ-weakly dense in $\mathcal{A}$.

Let $\Gamma = \mathrm{Sp}(B \cap \mathcal{A})$ and μ be the Radon measure given by $\varphi(x) = \left(x\xi_\varphi \mid \xi_\varphi\right)$, $x \in B \cap \mathcal{A}$. Since $\mathfrak{T} \circ \tilde{\alpha}_s = \tilde{\alpha}_s \circ \mathfrak{T}$ and $\tilde{\alpha}_s(e) = e$, we have $(\varphi \circ \tilde{\alpha}_s)_\gamma = \varphi_\gamma \circ \tilde{\alpha}_s$, $\gamma \in \Gamma$. The measure μ is invariant under G. Therefore, the map: $\gamma \mapsto \varphi_\gamma \in \mathfrak{S}(B)$ is equivariant, and we have $\varphi = \int_\Gamma \varphi_\gamma\, d\mu(\gamma)$ in $\mathfrak{S}(B)$. The rest is now routine.

Q.E.D.

Theorem 3.11. *Let* $\{\mathcal{M}, G, \alpha\}$ *be a separable covariant system, and also let* $\{\mathcal{M}, U, \mathfrak{P}^\natural, \mathfrak{H}\}$ *be a standard form. If* $\mathcal{A}$ *is an invariant von Neumann subalgebra of the center* $\mathcal{Z}$ *of* $\mathcal{M}$, *then the system gets the following equivariant disintegration:*

$$\{\mathcal{M}, U, \mathfrak{P}^\natural, \mathfrak{H}\} = \int_\Gamma^\oplus \{\mathcal{M}(\gamma), U_{\cdot,\gamma}, \mathfrak{P}^\natural(\gamma), \mathfrak{H}(\gamma)\}\, d\mu(\gamma)$$

over the measure theoretic spectrum $\{\Gamma, \mu\}$ *of* $\mathcal{A}$ *with respect to the diagonal algebra* $\mathcal{A}$ *such that*

$$\left.\begin{aligned}
\mathcal{M}(T_s\gamma) &= U_{s,\gamma}\mathcal{M}(\gamma)U_{s,\gamma}^*, && \gamma \in \Gamma,\\
\mathfrak{P}^\natural(T_s\gamma) &= U_{s,\gamma}\mathfrak{P}^\natural(\gamma), &&\\
\alpha_s(f)(T_s\gamma) &= f(\gamma), && f \in \mathcal{A},\\
U_{st,\gamma} &= U_{s,T_t\gamma}U_{t,\gamma}, && s, t \in G \text{ for almost every } \gamma \in \Gamma.
\end{aligned}\right\} \tag{22}$$

PROOF: Choose a faithful normal state φ on $\mathcal{M}$ and consider the disintegration relative to $\mathcal{A}$:

$$\{\mathcal{M}, \varphi\} = \int_\Gamma^\oplus \{\mathcal{M}(\gamma), \varphi_\gamma\}\, d\mu(\gamma)\,.$$

By Theorem VIII.4.8, almost all φ_γ are faithful on $\mathcal{M}(\gamma)$. Let ξ_φ be the representing vector of φ in $\mathfrak{P}^\natural$. Then we have

$$\xi_\varphi = \int_\Gamma^\oplus \xi_\varphi(\gamma)\, d\mu(\gamma)\,.$$

For each $\gamma \in \Gamma$, we define

$$\mathfrak{P}^\natural(\gamma) = \overline{\Delta_{\varphi_\gamma}^{\frac{1}{4}} \mathcal{M}(\gamma)_+ \xi_\varphi(\gamma)}.$$

Then we have the disintegration of standard form:

$$\{\mathcal{M}, \mathfrak{H}, \mathfrak{P}^\natural\} = \int_\Gamma^\oplus \{\mathcal{M}(\gamma), \mathfrak{H}(\gamma), \mathfrak{P}^\natural(\gamma)\} \, d\mu(\gamma) \tag{23}$$

in the natural sense. In order to decompose the unitary representation U of G, we consider a σ-weakly dense separable C^*-algebra A in $\mathcal{M}$ such that

a) $\alpha_s(A) = A, \ s \in G,$
b) $\{A, G, \alpha\}$ is a covariant C^*-system:

$$\lim_{s \to e} \|\alpha_s(x) - x\| = 0, \quad x \in A;$$

c) $C = A \cap \mathcal{A}$ is σ-weakly dense in $\mathcal{A}$.

We replace Γ by the spectrum of C, so that G acts on Γ as a topological transformation group:

$$\alpha_s(f)(\gamma) = f(T_s^{-1}\gamma), \quad f \in C, \quad s \in G, \quad \gamma \in \Gamma.$$

To each normal $\psi \in \mathcal{M}_*^+$, there corresponds uniquely a Radon measure μ_ψ on Γ which is absolutely continuous with respect to μ_φ. It then follows that

$$\mu_{\psi \circ \alpha_s} = \mu_\psi \circ T_s. \tag{24}$$

The disintegration (25) of $\mathfrak{H}$ gives rise to a disintegration of the identity representation π of A:

$$\pi = \int_\Gamma^\oplus \pi_\gamma \, d\mu_\varphi(\gamma) \, . \tag{25}$$

Observe that for any $\psi \in \mathcal{M}_*^+$ and $x \in A$, we have

$$\begin{aligned}\psi(x) = (\pi(x)\xi_\psi \mid \xi_\psi) &= \int_\Gamma (\pi_\gamma(x)\xi_\psi(\gamma) \mid \xi_\psi(\gamma)) \, d\mu_\varphi(\gamma) \\ &= \int_\Gamma \psi_\gamma(x) \, d\mu_\psi(\gamma) = \int_\Gamma \psi_\gamma(x) \frac{d\mu_\psi}{d\mu_\varphi}(\gamma) \, d\mu_\varphi(\gamma),\end{aligned}$$

so that the representing vector of ψ_γ in $\mathfrak{P}^\natural(\gamma)$ is given by

$$\xi_{\psi_\gamma} = \left(\frac{d\mu_\psi}{d\mu_\varphi}(\gamma)\right)^{-\frac{1}{2}} \xi_\psi(\gamma) \tag{26}$$

not by $\xi_\psi(\gamma)$. We now compute $\psi\big(\alpha_s(xf)\big)$, $x \in A$, $f \in C$, in two ways:

$$\begin{aligned}\psi\big(\alpha_s(fx)\big) &= \int_\Gamma f(\gamma)(\psi \circ \alpha_s)_\gamma(x)\, \mathrm{d}\mu_{\psi\circ\alpha_s}(\gamma)\\ &= \int_\Gamma f(\gamma)(\psi \circ \alpha_s)_\gamma(x)\, \mathrm{d}\mu_\psi \circ T_s(\gamma)\\ &= \int_\Gamma f(T_s^{-1}\gamma)(\psi \circ \alpha_s)_{T_s^{-1}\gamma}(x)\, \mathrm{d}\mu_\psi(\gamma);\end{aligned}$$

$$\begin{aligned}\psi\big(\alpha_s(fx)\big) &= \int_\Gamma \alpha_s(f)(\gamma)\psi_\gamma\big(\alpha_s(x)\big)\, \mathrm{d}\mu_\psi(\gamma)\\ &= \int_\Gamma f(T_s^{-1}\gamma)\psi_\gamma \circ \alpha_s(x)\, \mathrm{d}\mu_\psi(\gamma);\end{aligned}$$

hence we have

$$(\psi_\gamma \circ \alpha_s) = (\psi \circ \alpha_s)_{T_s^{-1}\gamma}, \quad s \in G, \quad \text{for almost every } \gamma \in \Gamma. \tag{27}$$

This means that the two representations of A:

$$\begin{aligned}&x \in A \mapsto \pi_{(\psi\circ\alpha_s)_\gamma}(x) \in \mathcal{L}\big(\mathfrak{H}(\gamma)\big),\\ &x \in A \mapsto \pi_{\psi_{T_s\gamma}}\big(\alpha_s(x)\big) \in \mathcal{L}\big(\mathfrak{H}(T_s\gamma)\big)\end{aligned}$$

are unitarily equivalent for any faithful normal state ψ on $\mathcal{M}$. But $\pi_{(\psi\circ\alpha_s)_\gamma}$ and π_γ (resp. $\pi_{\psi_{T_s\gamma}}$ and $\pi_{T_s\gamma}$) can be canonical identified by looking at the representing vector $\xi_{(\psi\circ\alpha_s)_\gamma}$ (resp. $\xi_{\psi_{T_s\gamma}}$). Since $\mathcal{M}(\gamma) = \pi_\gamma(A)''$ and $\mathcal{M}(T_s\gamma) = \pi_{T_s\gamma}\big(\alpha_s(A)\big)''$, there exists an isomorphism

$$\alpha_{s,\gamma}\colon \mathcal{M}(\gamma) \mapsto \mathcal{M}(T_s\gamma), \tag{28}$$

which sends $\pi_\gamma(x)$ into $\pi_{T_s\gamma}\big(\alpha_s(x)\big)$, $x \in A$, $s \in G$. Furthermore, this isomorphism $\alpha_{s,\gamma}$ is implemented by a unitary $U_{s,\gamma}$ given by the formula:

$$U_{s,\gamma}\pi_\gamma(x)\xi_{(\psi\circ\alpha_s)_\gamma} = \pi_{T_s\gamma}\big(\alpha_s(x)\big)\xi_{\psi_{T_s\gamma}}. \tag{29}$$

Since $U_{s,\gamma}$ maps $\xi_{(\psi\circ\alpha_s)_\gamma}$ of $\mathfrak{P}^\natural(\gamma)$ into $\xi_{\psi_{T_s\gamma}}$ of $\mathfrak{P}^\natural(T_s\gamma)$ and sends $\mathcal{M}(\gamma)$ onto $\mathcal{M}(T_s\gamma)$, it maps $\mathfrak{P}^\natural(\gamma)$ onto $\mathfrak{P}^\natural(T_s\gamma)$. Therefore we have

$$U_{s,\gamma}\xi_{(\psi\circ\alpha_s)_\gamma} = \xi_{\psi_{T_s\gamma}} \tag{29'}$$

for any $\psi \in \mathcal{M}_*^+$, which determines $U_{s,\gamma}$. From this, the cocycle identity for $\{U_{s,\gamma}\}$ follows. Making use of the separability of A, one can easily take care of the almost everywhere question. Q.E.D.

Corollary 3.12. *Let $\{\mathcal{M}, G, \alpha\}$ be a separable covariant system in the sense that $\mathcal{M}$ and G are both separable. If $\mathcal{A}$ is an α-invariant von Neumann subalgebra of the center $\mathcal{Z}$ of $\mathcal{M}$, then we have the following equivariant disintegration:*

$$\{\mathcal{M}, \alpha\} = \int_{\Gamma}^{\oplus} \{\mathcal{M}(\gamma), \alpha_{\cdot,\gamma}\}\, \mathrm{d}\mu(\gamma) \tag{30}$$

over the measure theoretic spectrum $\{\Gamma, \mu\}$ of $\mathcal{A}$, i.e.

$$\mathcal{M} = \int_{\Gamma}^{\oplus} \mathcal{M}(\gamma)\, \mathrm{d}\mu(\gamma);$$

for each $s \in G$, $\alpha_{s,\gamma}$ is an isomorphism of $\mathcal{M}(\gamma)$ onto $\mathcal{M}(T_s\gamma)$ for almost every $\gamma \in \Gamma$ such that

$$\alpha_s(x)(T_s\gamma) = \alpha_{s,\gamma}\big(x(\gamma)\big)$$

for each $x = \int_{\Gamma}^{\oplus} x(\gamma)\, \mathrm{d}\mu(\gamma) \in \mathcal{M}$ and for each $s, t \in G$

$$\alpha_{st,\gamma} = \alpha_{s,T_t\gamma} \circ \alpha_{t,\gamma} \quad \text{for almost every } \gamma \in \Gamma. \tag{31}$$

PROOF: Represent $\mathcal{M}$ in a standard form and apply Theorem 3.11. Then set $\alpha_{s,\gamma}(x) = U_{s,\gamma} x U_{s,\gamma}^*$ for each $x \in \mathcal{M}(\gamma)$. Q.E.D.

Definition 3.13. A covariant system $\{\mathcal{M}, G, \alpha\}$ is called *ergodic* if $\mathcal{M}^\alpha = \mathbf{C}$; *centrally ergodic* if α is ergodic on the center $\mathcal{Z}$, i.e. $\mathcal{Z}^\alpha = \mathbf{C}$.

Corollary 3.14. *In Corollary* 3.12*, if $\mathcal{A} = \mathcal{Z}^\alpha$, the fixed point subalgebra of the center $\mathcal{Z}$ of $\mathcal{M}$, then almost all component $\{\mathcal{M}(\gamma), \alpha_{\cdot,\gamma}\}$ of the disintegration* (30) *are centrally ergodic covariant systems.*

PROOF: Since α on $\mathcal{A}$ is trivial, G acts on Γ trivially, so that for each $s, t \in G$,

$$U_{st,\gamma} = U_{s,\gamma} U_{t,\gamma} \quad \text{for almost every } \gamma \in \Gamma.$$

By Fubini's theorem, for almost every fixed $\gamma \in \Gamma$, we have $U_{st,\gamma} = U_{s,\gamma} U_{t,\gamma}$ for almost every pair $s, t \in G$ relative to the Haar measure of G. Then the next lemma takes care of the rest of the proof. Q.E.D.

Lemma 3.15. *Let G be a locally compact group. If $U : s \in G \mapsto U(s) \in \mathcal{U}(\mathfrak{H})$, the unitary group of a Hilbert space $\mathfrak{H}$, satisfies the following conditions:*

(i) $s \in G \mapsto \big(U(s)\xi \mid \eta\big)$ *is measurable for every $\xi, \eta \in \mathfrak{H}$;*

(ii) $U(st) = U(s)U(t)$ *for almost every pair $s, t \in G$ relative to the Haar measure of G, then there exists a unitary representation V of G on $\mathfrak{H}$ such that $U(s) = V(s)$ for almost every $s \in G$.*

PROOF: The measurability condition on U enables us to define $U(f)=\int_G f(s)U(s)\,ds$ for each $f \in L^1(G)$. Since the convolution involves integration, the almost multiplicativity of U entails $U(f * g) = U(f)U(g)$ and $U(f^*) = U(f)^*$. Hence U is a *-representation of the group algebra $L^1(G)$. Therefore, there exists a unitary representation V of G on $\mathfrak{H}$ such that $V(f) = \int_G f(s)V(s)\,ds = U(f)$. This means that $U(s) = V(s)$ for almost every $s \in G$. Q.E.D.

Definition 3.16. The disintegration in Corollary 3.12 is called the *centrally ergodic disintegration.* In the case that $\mathcal{M}$ is abelian, i.e. $\mathcal{M} = \mathcal{Z}$, it is called simply the *ergodic disintegration* or *decomposition.*

Exercise X.3

1) Let $G = \left\{ g = \begin{pmatrix} a & b \\ 0 & 1 \end{pmatrix} : a \in \mathbf{Q},\ a \neq 0,\ b \in \mathbf{R} \right\}$ be the locally compact group equipped with the discrete topology on $\mathbf{Q}$ and the usual topology on $\mathbf{R}$. Set $N = \left\{ g = \begin{pmatrix} 1 & b \\ 0 & 1 \end{pmatrix} : b \in \mathbf{R} \right\}$. Identify the Pontrjagin dual $\widehat{N}$ of N with $\mathbf{R}$ by the pairing: $\left\langle \begin{pmatrix} 1 & b \\ 0 & 1 \end{pmatrix}, p \right\rangle = e^{ibp},\ b, p \in \mathbf{R}$. Identify the multiplicative group $\mathbf{Q}^*$ with the subgroup $\left\{ \begin{pmatrix} a & 0 \\ 0 & 1 \end{pmatrix} : a \in \mathbf{Q},\ a \neq 0 \right\}$ naturally, so that $G = \mathbf{Q}^* \ltimes N$.

Consider a separable unitary representation $\{U, \mathfrak{H}\}$ of G and set $\mathcal{A} = U(N)''$.

(**a**) Show that there exists a Borel measure μ on $\mathbf{R}$ such that $L^\infty(\mathbf{R}, \mu) \cong \mathcal{A}$.

(**b**) Observe that $U(g)\mathcal{A}U(g)^* = \mathcal{A},\ g \in G$, so that G acts on $\mathcal{A}$ as an automorphism group. Denote the action by α.

(**c**) Determine the action α on $\mathcal{A}$ in terms of $L^\infty(\mathbf{R}, \mu)$ and prove that the measure μ on $\mathbf{R}$ is quasi-invariant under the action of G.

(**d**) Construct the equivariant disintegration of U relative to $\mathcal{A}$.

2) Fix a positive irrational number θ. Let u and v be a pair of unitaries on a separable Hilbert space $\mathfrak{H}$ such that

$$uv = e^{2\pi i\theta} vu.$$

Consider the von Neumann algebras $\mathcal{R}$, $\mathcal{A}$ and $\mathcal{B}$ generated respectively by $\{u, v\}$, v and u.

(**a**) Show that $\mathfrak{H}$ cannot be finite dimensional. (Hint: As $uvu^* = e^{2\pi i\theta}v$, the spectrum of v is invariant under the multiplication of $e^{2\pi i\theta}$ and there is no finite set of complex numbers other than the singleton set $\{0\}$ which is invariant under the multiplication of $e^{2\pi i\theta}$.)

(**b**) Denote by α the automorphism of $\mathcal{A}$ given by $\alpha(x) = uxu^*,\ x \in \mathcal{A}$. Show that there exists a finite Borel measure μ on $\mathbf{T} = \mathbf{R}/\mathbf{Z}$ and an isomorphism π from $\mathcal{A}$ onto $L^\infty(\mathbf{T}, \mu)$ such that (i) the measure μ is quasi invariant under the translation by θ and (ii) $\{\pi(\alpha(a))\}(t) = \{\pi(a)\}(t - \theta),\ t \in \mathbf{T},\ a \in \mathcal{A}$.

(**c**) Suppose that $\mathcal{R}$ admits a faithful normal trace τ with $\tau(1) = 1$. Show that $\tau(u^m v^n) = 0$ for any $(m, n) \in \mathbf{Z}^2$ with $(m, n) \neq (0, 0)$.

(d) Keep the assumption of (c) for the rest of this problem. Show that the measure on $\mathbf{T}$ corresponding to the restriction of τ to $\mathcal{A}$ is the Lebesgue measure.

(e) Show that $\mathcal{R}$ is a factor of type II_1 and isomorphic to $\mathcal{A} \rtimes_\alpha \mathbf{Z}$ under the natural isomorphism.

§ 4 Induced Covariant System and Crossed Product

This section is devoted to a construction of a new covariant system $\{M, G, \alpha\}$ from a covariant system $\{\mathcal{N}, H, \beta\}$ over a closed subgroup H of G, and to a characterization as well as to analysis of such covariant systems. It is closely related to the theory of induced representations. We thus review quickly induced unitary representations.

Let G be a locally compact group and H a closed subgroup. We denote the left quotient space $G/H = \{sH : s \in G\}$ by Γ and the coset sH by $\gamma(s)$. Let δ_G and δ_H be respectively the modular functions of G and H. We consider the left Haar measures $\mathrm{d}_G s$ and $\mathrm{d}_H s$ on G and H respectively. When we need to emphasize the left (or right) Haar measure, we write $\mathrm{d}_G^\ell s$ (or $\mathrm{d}_H^r s$).

On the abelian von Neumann algebra $L^\infty(G)$, G acts in two ways: the one λ by left translation and the one ρ by right translation. Restricting ρ to the subgroup H, the fixed point subalgebra $\mathcal{A}_H = L^\infty(G)^{\rho(H)}$ is identified with the abelian von Neumann algebra $L^\infty(G/H)$ of all bounded measurable functions over $\Gamma = G/H$. Since ρ is integrable as an action of G, it is integrable over H, so that we have an $\mathcal{A}_H$-valued weight $\mathcal{E}_H$ of $L^\infty(G)$ onto $L^\infty(\Gamma)$, i.e.

$$\mathcal{E}_H(f)\big(\gamma(s)\big) = \int_H f(sh)\,\mathrm{d}_H h\,. \tag{1}$$

We will use the following result in the general theory of locally compact groups.

Theorem 4.1. *Let G be a locally compact group and H a closed subgroup of G.*

(i) *There exists a unique class, up to equivalence of measures, of quasi-invariant measures on the homogeneous space $\Gamma = G/H$ under the left translation by G. A subset E of Γ is negligible under the quasi-invariant measures if and only if its inverse image $\gamma^{-1}(E)$ is negligible in the Haar measure of G.*

(ii) *Let φ be a non-negative locally integrable function on G. The measure μ_φ defined by*

$$\mu_\varphi(f) = \int_G f(s)\varphi(s)\,\mathrm{d}s\,, \qquad f \in \mathcal{K}(G) \tag{2}$$

is of the form $\mu \circ \mathcal{E}_H$ for some measure μ on Γ if and only if

$$\varphi(sr) = \delta_G(r)^{-1}\,\delta_H(r)\,\varphi(s), \quad r \in H, \quad s \in G. \tag{3}$$

If condition (3) *is satisfied, then the measure μ on Γ is uniquely determined by the equation:*

$$\int_{\Gamma}\left(\int_{H} f(sr)\,\mathrm{d}_H r\right)\mathrm{d}\mu(\dot{s}) = \int_{G} f(s)\varphi(s)\,\mathrm{d}s, \quad f \in \mathcal{K}(G). \tag{4}$$

(iii) *There exists an everywhere positive function* φ *on* G *satisfying condition* (3).

(iv) *If* $\varphi > 0$ *satisfies* (3), *then the function* $\varphi(st)/\varphi(t)$ *depends only on* $s \in G$ *and* $\gamma(t) \in \Gamma$, *so that*

$$\chi\big(s, \gamma(t)\big) = \frac{\varphi(st)}{\varphi(t)}, \tag{5}$$

defines a function χ *on* $G \times \Gamma$. *Furthermore, if* μ *is the measure on* Γ *defined by* (4), *then*

$$\frac{\mathrm{d}\mu \circ T_s}{\mathrm{d}\mu}(\gamma) = \chi(s, \gamma), \quad s \in G, \quad \gamma \in \Gamma, \tag{6}$$

where T_s *means the left translation by* $s \in G$.

We refer the reader to [441, Chapter VII, § 2, no. 5].

We now consider a unitary representation $\{V, \mathfrak{K}\}$ of a closed subgroup H of a locally compact group G. Let $\mathfrak{F}^*$ be the set of all $\mathfrak{K}$-valued measurable, in the sense of Definition IV.7.1, functions ξ on G such that

a)
$$\xi(sr) = \delta_G(r)^{-\frac{1}{2}}\,\delta_H(r)^{\frac{1}{2}}\,V(r)^{-1}\,\xi(s), \tag{7}$$
for every $r \in H$ and locally almost every $s \in G$;

b) the function: $s \in G \mapsto \|\xi(s)\|^2$ is locally integrable. By the above theorem, there exists uniquely a positive Radon measure μ_ξ on Γ such that

$$\int_{\Gamma} \mathcal{E}_H(f)(\gamma)\,\mathrm{d}\mu_\xi(\gamma) = \int_{G} f(s)\|\xi(s)\|^2\,\mathrm{d}s, \quad f \in \mathcal{K}(G).$$

We set

$$\|\xi\| = \mu_\xi(\Gamma)^{\frac{1}{2}}, \tag{8}$$

and $\mathfrak{F} = \big\{\xi \in \mathfrak{F}^* : \|\xi\| < +\infty\big\}$. It is easy to see that $\mathfrak{F}$ is a linear space under the natural pointwise vector operations. It then follows that $\|\xi\| = 0$ if and only if $\xi(s) = 0$ locally almost everywhere. Let $\mathfrak{H}$ be the quotient vector space $\mathfrak{F} \big/ \big\{ f \in \mathfrak{F} : \|f\| = 0 \big\}$ of $\mathfrak{F}$ by the null vectors.

Lemma 4.2.

(i) *For each compact subset* K *of* G, *there exists a constant* $C_K > 0$ *such that*

$$\int_{K} \|\xi(s)\|\,\mathrm{d}s \leq C_K\|\xi\|, \quad \xi \in \mathfrak{F}. \tag{9}$$

(ii) $\mathfrak{H}$ *is a Hilbert space, i.e. it is complete.*

PROOF:

(i) Choose $g \in \mathcal{K}(G)_+$ such that $g = 1$ on K, and compute for $\xi \in \mathfrak{F}$:

$$\begin{aligned}\int_K \|\xi(s)\|\, \mathrm{d}s &\le |K|^{\frac{1}{2}}\Big(\int_K \|\xi(s)\|^2\, \mathrm{d}s\Big)^{\frac{1}{2}} \le |K|^{\frac{1}{2}}\Big(\int_G g(s)\|\xi(s)\|^2\, \mathrm{d}s\Big)^{\frac{1}{2}} \\ &\le |K|^{\frac{1}{2}}\Big(\int_\Gamma \mathcal{E}_H(g)(\gamma)\, \mathrm{d}\mu_\xi(\gamma)\Big)^{\frac{1}{2}} \\ &\le |K|^{\frac{1}{2}}\|\mathcal{E}_H(g)\|_\infty^{\frac{1}{2}}\, \mu_\xi(\Gamma)^{\frac{1}{2}} = \Big(|K|\,\|\mathcal{E}_H(g)\|_\infty\Big)^{\frac{1}{2}}\|\xi\|,\end{aligned}$$

where $|K| = \int_K \mathrm{d}s$.

(ii) The polarization identity, (II.1.3), allows us to find a measure $\mu_{\xi,\eta}$ on Γ for each $\xi, \eta \in \mathfrak{F}$ such that

$$\int_G g(s)\big(\xi(s) \mid \eta(s)\big)\, \mathrm{d}s = \int_G \mathcal{E}_H(g)(\gamma)\, \mathrm{d}\mu_{\xi,\eta}(\gamma), \quad g \in \mathcal{K}(G), \tag{10}$$

so that

$$(\xi \mid \eta) = \mu_{\xi,\eta}(\Gamma) \tag{8'}$$

gives an inner product in $\mathfrak{F}$ which is naturally passed onto $\mathfrak{H}$ to make it a pre-Hilbert space. Fixing a continuous function $f_0 \ge 0$ such that $\mathcal{E}_H(f_0) = 1$, i.e.

$$\int_H f_0(sr)\, \mathrm{d}_H r = 1, \quad s \in G, \tag{11}$$

we have, for each pair $\xi, \eta \in \mathfrak{F}$,

$$(\xi \mid \eta) = \int_G f_0(s)\big(\xi(s) \mid \eta(s)\big)\, \mathrm{d}s\,. \tag{8''}$$

Let $\{\xi_n\}$ be a Cauchy sequence in $\mathfrak{F}$. Passing to a subsequence, we may assume that $\|\xi_n - \xi_{n+1}\| \le 2^{-n}$. Let K be any compact subset of G. Then by (i), we have

$$\int_K \big\|\xi_n(s) - \xi_{n+1}(s)\big\|\, \mathrm{d}s \le 2^{-n} C_K,$$

so that

$$\int_K \Big(\sum_{n=1}^{\infty} \big\|\xi_n(s) - \xi_{n+1}(s)\big\|\Big)\, \mathrm{d}s \le C_K.$$

Therefore, $\{\xi_n(s)\}$ is a Cauchy sequence in $\mathfrak{K}$ for almost every $s \in K$. Let $\xi(s) = \lim_{n\to\infty} \xi_n(s)$ or 0 according to the existence of the limit.

It is clear that ξ satisfies condition (a) for $\mathfrak{F}^*$. Iterating use of the parallelogram law for $\mathfrak{K}$, we have, for $g \in L^\infty(G)_+$,

$$\int_G \left\|\xi_n(s) - \xi_{n+k}(s)\right\|^2 g(s)\,\mathrm{d}s \le \sum_{j=1}^{\infty} 2^j \int_G \left\|\xi_{n+j-1}(s) - \xi_{n+j}(s)\right\|^2 g(s)\,\mathrm{d}s$$

$$\le \sum_{j=1}^{\infty} 2^j \left\|\xi_{n+j-1} - \xi_{n+j}\right\|^2 \left\|\mathcal{E}_H(g)\right\|_\infty \le 2^{-2n+2} \left\|\mathcal{E}_H(g)\right\|_\infty.$$

Letting $k \to \infty$, we have

$$\int_G \left\|\xi_n(s) - \xi(s)\right\|^2 g(s)\,\mathrm{d}s \le 2^{-2(n-1)} \left\|\mathcal{E}_H(g)\right\|_\infty.$$

For a compact subset K of G, choose $g \in \mathcal{K}(G)_+$ with $g = 1$ on K, we have

$$\int_K \left\|\xi_n(s) - \xi(s)\right\|^2 \mathrm{d}s \le 2^{-2(n-1)} \left\|\mathcal{E}_H(g)\right\|_\infty,$$

so that $\left\|\xi_n(\,\cdot\,) - \xi(\,\cdot\,)\right\|^2$ is locally integrable, and hence ξ is also. With $g = f_0$, we have

$$\|\xi_n - \xi\|^2 \le 2^{-2(n-1)} \to 0 \quad \text{as} \quad n \to \infty.$$

Thus, ξ belongs to $\mathfrak{F}$ and $\{\xi_n\}$ converges to ξ. Q.E.D.

For the later use, we introduce a new notation χ or χ_H, when we need to specify the subgroup H:

$$\chi(r) = \delta_G(r)^{-1}\delta_H(r), \quad r \in H. \tag{12}$$

Thus the formula (7) becomes

$$\xi(sr) = \chi(r)^{\frac{1}{2}} V(r)^* \xi(s), \quad r \in H, \quad s \in G. \tag{7'}$$

Define a bilinear map Λ on $\mathcal{K}(G) \times \mathfrak{K}$ by the integral:

$$\Lambda(f, \eta)(s) = \int_H \chi(r)^{-\frac{1}{2}} f(sr) V(r)\eta\,\mathrm{d}r, \quad f \in \mathcal{K}(G), \quad \eta \in \mathfrak{K}. \tag{13}$$

It then follows that $\Lambda(f, \eta)$ is an element of $\mathfrak{H}$.

Lemma 4.3.

(i) *If $f \in \mathcal{K}(G)$ is supported by a compact set K in G, then $\Lambda(f, \eta)$ is supported by KH and*

$$\left\|\Lambda(f, \eta)\right\| \le C_K \|f\|_\infty \|\eta\|, \quad \eta \in \mathfrak{K}. \tag{14}$$

(ii) *$\Lambda\big(\mathcal{K}(G) \times \mathfrak{K}\big)$ is total in $\mathfrak{H}$.*

PROOF:

(i) Choose $g \in \mathcal{K}(G)_+$ such that $\mathcal{E}_H(g) = 1$ on $\gamma(K)$ and compute for $\xi \in \mathfrak{H}$ and $\zeta = \Lambda(f, \eta)$:

$$\begin{aligned}
\big(\Lambda(f,\eta) \mid \xi\big) &= \mu_{\zeta,\xi}\big(\gamma(K)\big) \qquad \text{since} \quad \operatorname{supp}\mu_{\zeta,\xi} \subset \gamma(K)\\
&= \mu_{\zeta,\xi}\big(\mathcal{E}_H(g)\big) = \int_G g(s)\big(\Lambda(f,\eta)(s) \mid \xi(s)\big)\,\mathrm{d}s\\
&= \int_G\int_H g(s)\chi(r)^{-\frac{1}{2}} f(sr)\big(V(r)\eta \mid \xi(s)\big)\,\mathrm{d}_H r\,\mathrm{d}_G s\\
&= \int_H\int_G g(s)\chi(r)^{-1} f(sr)\big(\eta \mid \xi(sr)\big)\,\mathrm{d}_G s\,\mathrm{d}_H r\\
&= \int_H\int_G g(sr^{-1})\Delta_H(r)^{-1} f(s)\big(\eta \mid \xi(s)\big)\,\mathrm{d}_G s\,\mathrm{d}_H r\\
&= \int_G\int_H g(sr) f(s)\big(\eta \mid \xi(s)\big)\,\mathrm{d}_H r\,\mathrm{d}_G s\\
&= \int_G \mathcal{E}_H(g)\big(\gamma(s)\big) f(s)\big(\eta \mid \xi(s)\big)\,\mathrm{d}s\\
&= \int_G f(s)\big(\eta \mid \xi(s)\big)\,\mathrm{d}s \qquad \text{by the choice of } g.
\end{aligned}$$

Thus we have

$$\big(\Lambda(f,\eta) \mid \xi\big) = \int_G f(s)\big(\eta \mid \xi(s)\big)\,\mathrm{d}s, \tag{15}$$

for every $f \in \mathcal{K}(G)$ with $\operatorname{supp} f \subset K$, $\eta \in \mathfrak{K}$ and $\xi \in \mathfrak{H}$. Therefore, we obtain

$$\begin{aligned}
\Big|\big(\Lambda(f,\eta) \mid \xi\big)\Big| &\le \int_G |f(s)|\,\|\eta\|\,\big\|\xi(s)\big\|\,\mathrm{d}s \le \|f\|_\infty\|\eta\| \int_K \big\|\xi(s)\big\|\,\mathrm{d}s\\
&\le C_K\|f\|_\infty\|\eta\|\,\|\xi\|
\end{aligned}$$

by inequality (9).

(ii) Suppose that $\big(\Lambda(f,\eta) \mid \xi\big) = 0$ for every $f \in \mathcal{K}(G)$ and $\eta \in \mathfrak{K}$. Then the above calculation in (i) shows that

$$\int_G f(s)\big(\eta \mid \xi(s)\big)\,\mathrm{d}s = 0, \qquad f \in \mathcal{K}(G), \qquad \eta \in \mathfrak{K},$$

so that $\big(\eta \mid \xi(s)\big) = 0$ for every $\eta \in \mathfrak{K}$ and locally almost every $s \in G$. Hence, by Proposition IV.7.2, we have $\xi(s) = 0$ for locally almost every $s \in G$. Q.E.D.

Finally, we come to the following:

Theorem 4.4. *The left translation U on $\mathfrak{H}$:*

$$\big(U(t)\xi\big)(s) = \xi\big(t^{-1}s\big), \quad s, t \in G, \quad \xi \in \mathfrak{H}, \tag{16}$$

is a unitary representation of G.

PROOF: Clearly $U(t)$ maps $\mathfrak{H}$ into itself. Since $\mathcal{E}_H$ commutes with the left translations by G, $\mathcal{E}_H(\lambda_t f_0) = 1$ if $\mathcal{E}_H(f_0) = 1$, so that

$$\begin{aligned}\|U(t)\xi\|^2 &= \mu_{U(t)\xi}(\Gamma) = \int_G f_0(s)\|\xi(t^{-1}s)\|^2 \,\mathrm{d}s = \int_G f_0(ts)\|\xi(s)\|^2 \,\mathrm{d}s \\ &= \int_\Gamma \mathcal{E}_H\big(\lambda_t^{-1} f_0\big)\big(\gamma(s)\big)\,\mathrm{d}\mu_{\xi(\gamma)} = \mu_\xi(\Gamma) = \|\xi\|^2.\end{aligned}$$

Hence $U(t)$ is an isometry. Since $U(t^{-1})U(t) = 1$, $U(t)$ is unitary.

For any $f \in \mathcal{K}(G)$ and $\eta \in \mathfrak{K}$, we have $U(t)\Lambda(f, \eta) = \Lambda(\lambda_t f, \eta)$, so that for any compact subset L of G we get, with $K = \operatorname{supp} f$,

$$\begin{aligned}\|U(t)\Lambda(f, \eta) - \Lambda(f, \eta)\| &= \|\Lambda(\lambda_t f - f, \eta)\| \\ &\le C_{LK}\|\lambda_t f - f\|_\infty \|\eta\|, \quad t \in LH, \quad \text{by (14)}.\end{aligned}$$

Choosing L to be a compact neighborhood of $e \in G$, we have

$$\lim_{t\to e}\|U(t)\Lambda(f, \eta) - \Lambda(f, \eta)\| \le C_{LK} \lim_{t\to e}\|\lambda_t f - f\|_\infty \|\eta\| = 0.$$

Hence U is strongly continuous by Lemma 4.3.(ii). Q.E.D.

Definition 4.5. We call the unitary representation $\{U, \mathfrak{H}\}$ of G the *induced* representation of G from the unitary representation $\{V, \mathfrak{K}\}$ of H and write

$$U = \operatorname{Ind}_H^G V. \tag{17}$$

We continue to study the induced representation $\{U, \mathfrak{H}\}$. For each $f \in C_\infty(\Gamma)$ and $\xi \in \mathfrak{F}$, we set

$$\big(\pi(f)\xi\big)(s) = f\big(\gamma(s)\big)\xi(s), \quad s \in G. \tag{18}$$

It follows easily that $\pi(f)\xi$ belongs to $\mathfrak{F}$ and

$$\mathrm{d}\mu_{\pi(f)\xi} = |f|^2 \,\mathrm{d}\mu_\xi, \tag{19}$$

so that

$$\|\pi(f)\xi\| \le \|f\|_\infty \|\xi\|, \quad f \in C_\infty(\Gamma), \quad \xi \in \mathfrak{F}. \tag{20}$$

Therefore, π becomes a representation of $C_\infty(\Gamma)$ on the Hilbert space $\mathfrak{H}$. Let α denote the translation action of G on $C_\infty(\Gamma)$, i.e.

$$\left(\alpha_t f\right)\left(\gamma(s)\right) = f\left(\gamma(t^{-1}s)\right), \quad s, t \in G, \quad f \in C_\infty(\Gamma). \tag{21}$$

Then the pair $\{\pi, U\}$ is a covariant representation of $\left\{C_\infty(\Gamma), G, \alpha\right\}$.

Definition 4.6. We call the covariant representation $\{\pi, U\}$ the *imprimitivity* system associated with the induced representation U. We also say that π is the *imprimitivity* system of U.

The next result then characterizes the induced representation:

Theorem 4.7. *Let G be a locally compact group and H a closed subgroup. Let $\Gamma = G/H$ be the left coset space $\left\{sH : s \in G\right\}$. Let $\{U, \mathfrak{H}\}$ be a unitary representation of G. If π is a representation of the abelian C^*-algebra $C_\infty(\Gamma)$ on the same space $\mathfrak{H}$ such that $\{\pi, U\}$ is a covariant representation of $\left\{C_\infty(\Gamma), G, \alpha\right\}$, then there exists a unitary representation $\{V, \mathfrak{K}\}$ of the subgroup H such that $\{\pi, U\}$ is unitarily equivalent to the imprimitivity system associated with the induced representation $\mathrm{Ind}_H^G V$ of G from V. Furthermore, the representation V of H is unique up to unitary equivalence.*

PROOF: Since the kernel $\pi^{-1}(0)$ is an α-invariant ideal of $C_\infty(\Gamma)$, the hull of $\pi^{-1}(0)$ is a left translation invariant closed subset of Γ, which is either Γ itself or empty by the transitivity of the action of G. Since π is not a zero representation, $\pi^{-1}(0)$ must be trivial, so that $\left\|\pi(f)\right\| = \|f\|_\infty$ for every $f \in C_\infty(\Gamma)$. Consider

$$\mathfrak{D} = \operatorname{span}\left\{ U(f)\xi : \xi \in \mathfrak{H},\ f \in \mathcal{K}(G) \right\},$$

where $U(f) = \int_G f(s)U(s)\, ds$. To each $\xi, \eta \in \mathfrak{H}$, we associate a Radon measure $\mu_{\xi,\eta}$ on G by

$$\int_G f(s)\, d\mu_{\xi,\eta}(s) = \left(\pi\left(\mathcal{E}_H(f)\right)\xi \,\middle|\, \eta\right), \quad f \in \mathcal{K}(G). \tag{22}$$

We claim that if $\xi, \eta \in \mathfrak{D}$, then $\mu_{\xi,\eta}$ is given by a continuous function. To see this, for any $f \in \mathcal{K}(G \times G)$ set

$$f_t\left(\gamma(s)\right) = \int_H f(sh, t)\, dh, \tag{23}$$

and define a linear functional $\nu_{\xi,\eta}$ of $\mathcal{K}(G \times G)$ by

$$\nu_{\xi,\eta}(f) = \int_G \left(\pi(f_t)U(t)\xi \mid \eta\right) dt, \tag{24}$$

where the right hand side integration makes sense because $t \in G \mapsto \pi(f_t)$ is continuous and has a compact support. Furthermore, if f is supported by a compact set $K \times L$, then

$$\left|\nu_{\xi,\eta}(f)\right| \leq \sup_{t \in L} \|f_t\|_\infty |L|,$$

where $|L|$ means the volume of L relative to the Haar measure. Therefore, $\nu_{\xi,\eta}$ is a Radon measure on $G \times G$. We now compute with $f, g, h \in \mathcal{K}(G)$:

$$\begin{aligned}
\int_G f(s)\, \mathrm{d}\mu_{U(g)\xi, U(h)\eta}(s) &= \left(\pi\left(\mathcal{E}_H(f)\right)U(g)\xi \;\middle|\; U(h)\eta\right) \\
&= \int_G \overline{h(t)}\left(\pi\left(\mathcal{E}_H(f)\right)U(g)\xi \;\middle|\; U(t)\eta\right) \mathrm{d}t \\
&= \int_G \overline{h(t)}\left(\pi\left(\alpha_t^{-1}\left(\mathcal{E}_H(f)\right)\right)U(t)^* U(g)\xi \;\middle|\; \eta\right) \mathrm{d}t \\
&= \int_G \overline{h(t)}\left(\pi\left(\mathcal{E}_H(\lambda_t^{-1} f)\right)U(\lambda_t^{-1} g)\xi \;\middle|\; \eta\right) \mathrm{d}t \\
&= \int_G \overline{h(t)}\left(\int_{G\times G} f(tr)g(ts)\, \mathrm{d}\nu_{\xi,\eta}(r,s)\right) \mathrm{d}t \\
&= \int_G f(t) \int_{G\times G} \overline{h(tr^{-1})}\, g\left(tr^{-1}s\right)\delta_G(r)^{-1}\, \mathrm{d}\nu_{\xi,\eta}(r,s)\, \mathrm{d}t,
\end{aligned}$$

where we have used Fubini's theorem together with a change of variable in the t-integration. Hence with

$$k_{U(g)\xi, U(h)\eta}(t) = \int_{G\times G} \overline{h(tr^{-1})}\, g\left(tr^{-1}s\right)\delta_G(r)^{-1}\, \mathrm{d}\nu_{\xi,\eta}(r,s), \tag{25}$$

we have

$$\mathrm{d}\mu_{U(g)\xi, U(h)\eta}(t) = k_{U(g)\xi, U(h)\eta}(t)\, \mathrm{d}t\,. \tag{26}$$

Since $k_{U(g)\xi, U(h)\eta}$ is a continuous function, we have proved the claim. It is now easy to see that the function:

$$(r, s, t) \in G \times G \times G \mapsto k_{U(r)U(g)\xi,\, U(s)U(h)\eta}(t) \in \mathbf{C}$$

is continuous. Therefore for any $\xi, \eta \in \mathfrak{D}$ we have a continuous function $k_{V(r)\xi, U(s)\eta}(t)$ in the three variables r, s, t such that

$$\mathrm{d}\mu_{\xi,\eta}(t) = k_{\xi,\eta}(t)\, \mathrm{d}t\,. \tag{27}$$

We now define a sesquilinear form B on $\mathfrak{D} \times \mathcal{D}$ by

$$B(\xi, \eta) = k_{\xi,\eta}(e), \quad \xi, \eta \in \mathfrak{D}. \tag{28}$$

Since $\mu_{\xi,\xi}$ is positive, B is positive. Since $U(s)U(g) = U(\lambda_s g)$, the formula (25) implies

$$k_{\xi,\eta}(s) = B\big(U(s)^*\xi,\ U(s)^*\eta\big), \quad s \in G, \quad \xi, \eta \in \mathfrak{D}. \tag{29}$$

Hence we get, for each $f \in \mathcal{K}(G)$,

$$\Big(\pi(\mathcal{E}_H f)\xi \;\Big|\; \eta\Big) = \int_G f(s) B\big(U(s)^*\xi, V(s)^*\eta\big)\, \mathrm{d}s, \quad \xi, \eta \in \mathfrak{D}. \tag{30}$$

On the other hand, the left hand side is given by the integration over the homogeneous space Γ relative to the spectral measure $\lambda_{\xi,\eta}$:

$$\Big(\pi(\mathcal{E}_H f)\xi \;\Big|\; \eta\Big) = \int_\Gamma (\mathcal{E}_H f)(\gamma)\, \mathrm{d}\lambda_{\xi,\eta}(\gamma)\,.$$

Thus, Theorem 4.1 entails that for $\xi, \eta \in \mathfrak{D}$

$$B\big(U(r)\xi, U(r)\eta\big) = \delta_G(r)\, \delta_H(r)^{-1}\, B(\xi, \eta), \quad r \in H. \tag{31}$$

Now, let $\mathfrak{K}$ be the Hilbert space completion of the quotient space $\mathfrak{D}/\operatorname{Ker} B$ of $\mathfrak{D}$ by the subspace of B-null vectors. Let $[\xi]$ be the class of $\xi \in \mathfrak{D}$ in $\mathfrak{K}$. Set

$$V(r)[\xi] = \delta_G(r)^{\frac{1}{2}}\, \delta_H(r)^{\frac{1}{2}}\, \big[U(r)\xi\big] \tag{32}$$

for each $r \in H$ and $\xi \in \mathfrak{D}$. Then it follows that the function: $r \in H \mapsto \big(V(r)[\xi] \mid [\eta]\big) = \chi(r)^{1/2} B\big(U(r)\xi, \eta\big)$ is continuous for $\xi, \eta \in \mathcal{D}$. Hence by (31), V is a unitary representation of H on $\mathfrak{K}$.

For each $\xi \in \mathfrak{D}$, define a $\mathfrak{K}$-valued function $W\xi$ by

$$(W\xi)(s) = \big[U(s)^*\xi\big], \quad s \in G, \quad \xi \in \mathfrak{D}. \tag{33}$$

By (32), we have, for each $r \in H$ and $s \in G$,

$$(W\xi)(sr) = \chi(r)^{\frac{1}{2}}\, V(r)^*\big[U(s)^*\xi\big] = \chi(r)^{\frac{1}{2}}\, V(r)^*(W\xi)(s),$$

so that $W\xi$ satisfies (7). Since $W\xi$ is a $\mathfrak{K}$-valued continuous function, it belongs to $\mathfrak{F}$. With f_0 as in the proof of Lemma 4.2.(ii), i.e. $\mathcal{E}_H(f_0) = 1$, we have

$$\begin{aligned}\|W\xi\|^2 &= \int_G f_0(s)\|W\xi(s)\|^2\, \mathrm{d}s = \int_G f_0(s)\Big\|\big[U(s)^*\xi\big]\Big\|^2\, \mathrm{d}s \\ &= \int_G f_0(s) B\big(U(s)^*\xi, U(s)^*\xi\big)\, \mathrm{d}s = \Big(\pi\big(\mathcal{E}_H(f_0)\big)\xi \;\Big|\; \xi\Big) = \|\xi\|^2.\end{aligned}$$

Therefore, W is an isometry of $\mathfrak{D}$ into $\mathfrak{F}$, so that it can be extended to an isometry of $\mathfrak{H}$ into the Hilbert space $\tilde{\mathfrak{H}}$ of the induced representation $\tilde{U} = \operatorname{Ind}_H^G V$. It then follows from (33) that W intertwines U and the induced representation $\tilde{U}$. Let $\tilde{\pi}$ be the imprimitivity system of $\tilde{U}$. We then have, for $\xi, \eta \in \mathfrak{D}$ and $f \in \mathcal{K}(G)$,

$$\left(\tilde{\pi}(\mathcal{E}_H f)W\xi \;\middle|\; W\eta\right)_{\tilde{\mathfrak{H}}} = \int_\Gamma (\mathcal{E}_H f)(\gamma)\, d\mu_{W\xi, W\eta}(\gamma)$$
$$= \int_G f(s)\big(W\xi(s) \mid W\eta(s)\big)_{\mathfrak{K}}\, ds$$
$$= \int_G f(s)B\big(U(s)^*\xi, U(s)^*\eta\big)\, ds = \big(\pi(f)\xi \mid \eta\big)_{\mathfrak{H}}.$$

Therefore, W intertwines π and $\tilde{\pi}$. Suppose $\eta \in \mathfrak{K}$ is orthogonal to $W\mathfrak{D}$. Since $W\mathfrak{D}$ is invariant under $\tilde{\pi}(f)$, $f \in \mathcal{K}(\Gamma)$, we have, for any $\xi \in \mathfrak{D}$ and $f \in \mathcal{K}(G)$,

$$0 = \left(\tilde{\pi}(\mathcal{E}_H f)W\xi \;\middle|\; \eta\right) = \int_G f(s)\Big(\big[U(s)^*\xi\big] \;\Big|\; \eta(s)\Big)_{\mathfrak{K}}\, ds,$$

so that $\big(\big[U(s)^*\xi\big] \mid \eta(s)\big)_{\mathfrak{K}} = 0$ for locally almost every $s \in G$, so that $\eta(s) \perp [\mathfrak{D}]$ locally almost everywhere. Hence $\eta(s) = 0$ locally almost everywhere. Thus η is a zero vector in $\tilde{\mathfrak{H}}$. Therefore W is a unitary of $\mathfrak{H}$ onto $\tilde{\mathfrak{H}}$ which intertwines (π, U) and $(\tilde{\pi}, \tilde{U})$.

The uniqueness of V follows from the next finer result. Q.E.D.

Theorem 4.8. *Let G, H and Γ be as in the previous theorem. Let $\{V_i, \mathfrak{K}_i\}$, $i = 1, 2$, be two unitary representations of H and let $\{\pi_i, U_i, \mathfrak{H}_i\}$, $i = 1, 2$, be the corresponding induced representations and the associated covariant systems. Consider*

$$\mathcal{I}(V_1; V_2) = \left\{x \in \mathcal{L}(\mathfrak{K}_1, \mathfrak{K}_2) : xV_1(r) = V_2(r)x,\ r \in H\right\};$$
$$\mathcal{I}(\pi_1, U_1; \pi_2, U_2) = \left\{x \in \mathcal{L}(\mathfrak{H}_1, \mathfrak{H}_2) : \begin{matrix} x\pi_1(f) = \pi_2(f)x, & f \in C_\infty(\Gamma), \\ xU_1(s) = U_2(s)x, & s \in G. \end{matrix}\right\}. \tag{34}$$

Then there exists a natural σ-weakly continuous isometric bijection Φ from $\mathcal{I}(V_1;V_2)$ onto $\mathcal{I}\big(\pi_1, U_1;\ \pi_2, U_2\big)$ such that

$$\left.\begin{aligned} &\Phi(x^*) = \Phi(x)^*, && x \in \mathcal{I}(V_1, V_2); \\ &\Phi(yx) = \Phi(y)\Phi(x), && x \in \mathcal{I}(V_1, V_2), \quad y \in \mathcal{I}(V_2, V_3). \end{aligned}\right\} \tag{35}$$

if $\{V_3, \mathfrak{H}_3\}$ is another unitary representation of H. In other words, the association of $\{\pi, \operatorname{Ind} V\}$ to a unitary representation V of H gives rise to an equivalence of two categories: the category of unitary representations of H and the category of covariant representations of $\{C_\infty(\Gamma),\ G,\ \alpha\}$.

Considering the direct sums $V_1 \oplus V_2 = V$, $\pi_1 \oplus \pi_2 = \pi$ and $U_1 \oplus U_2 = U$ since the inducing process respects the direct sum operation, it suffices to prove the statement that $V(G)' \cong \pi\big(C_\infty(\Gamma)\big)' \cap U(G)'$.

Lemma 4.9. *With* $\{\pi, V, \mathfrak{H}\} = \mathrm{Ind}_H^G\{V, \mathfrak{K}\}$, *set*

$$\mathfrak{D} = \mathrm{span}\{ U(f)\xi : \xi \in \mathfrak{H},\ f \in \mathcal{K}(G) \}. \tag{36}$$

Then $\mathfrak{D}$ *is a dense subspace of* $\mathfrak{H}$ *and each member of* $\mathfrak{D}$ *is a continuous* $\mathfrak{K}$-*valued function. Every operator of* $U(G)'$ *maps* $\mathfrak{D}$ *into* $\mathfrak{D}$.

The proof follows the standard argument that the convolution of any measurable locally integrable function against a continuous function with compact support is continuous. So we leave it to the reader as an exercise.

PROOF OF THEOREM 4.8: Let $\{\pi, U, \mathfrak{H}\} = \mathrm{Ind}_H^G\{V, \mathfrak{K}\}$, and set $\mathcal{N} = V(H)'$. For each $x \in \mathcal{N}$, we define an operator $\Phi(x)$ on $\mathfrak{H}$ by the formula:

$$\big(\Phi(x)\xi\big)(s) = x\xi(s), \quad \xi \in \mathfrak{H}, \quad s \in G. \tag{37}$$

Since x commutes with $V(H)$, $\Phi(x)\xi$ belongs to $\mathfrak{F}$. With the function f_0 of (11), we have

$$\|\Phi(x)\xi\|^2 = \int_G f_0(s)\|x\xi(s)\|^2\, ds \le \|x\|^2 \int_G f_0(s)\|\xi(s)\|^2\, ds = \|x\|^2\|\xi\|^2,$$

so that $\Phi(x)\xi \in \mathfrak{H}$. It is easy to check that Φ is an isomorphism of $\mathcal{N}$ into $\mathcal{M} = U(G)' \cap \pi\big(C_\infty(\Gamma)\big)'$. We now show the normality of Φ. Let $\mathfrak{M}$ be the set of all $\mathfrak{K}$-valued continuous functions ξ of $\mathfrak{F}$ such that $\gamma(\mathrm{supp}\,\xi)$ is relatively compact in Γ. Then $\Lambda\big(\mathcal{K}(G) \times \mathfrak{K}\big)$ is contained in $\mathfrak{M}$, so that $\mathfrak{M}$ is dense in $\mathfrak{H}$. For any pair $\xi_1, \xi_2 \in \mathfrak{M}$, we consider the integral:

$$\omega = \int_G f_0(s)\omega_{\xi_1(s),\xi_2(s)}\, ds \in \mathcal{N}_*.$$

Then we have

$$\big(\Phi(x)\xi_1 \mid \xi_2\big) = \int_G f_0(s)\big(x\xi_1(s) \mid \xi_2(s)\big)\, ds = \omega(x), \quad x \in \mathcal{N}.$$

Therefore, Φ is σ-weakly continuous.

We now want to show that Φ maps $\mathcal{N}$ onto $\mathcal{M}$. Let $x \in \mathcal{M}$. Then x maps $\mathfrak{D}$ into $\mathfrak{D}$ by Lemma 4.9. We first show that if $\xi \in \mathfrak{D}$ takes value zero at e, i.e. $\xi(e) = 0$, then $(x\xi)(e) = 0$. For any $\eta \in \mathfrak{D}$, we have, for each $f \in \mathcal{K}(G)$,

$$\begin{aligned}\int_G f(s)\big((x\xi)(s) \mid \eta(s)\big)\, ds &= \big(\pi(\mathcal{E}_H f)x\xi \mid \eta\big) = \big(\pi(\mathcal{E}_H f)\xi \mid x^*\eta\big)\\ &= \int_G f(s)\big(\xi(s) \mid (x^*\eta)(s)\big)\, ds .\end{aligned}$$

For each compact neighborhood K of e in G, choose $f_K \in \mathcal{K}(G)_+$ such that $\operatorname{supp} f_K \subset K$ and $\int_G f_K(s)\,ds = 1$. Since the functions: $s \in G \mapsto \big((x\xi)(s) \mid \eta(s)\big)$ and $s \in G \mapsto \big(\xi(s) \mid (x^*\eta)(s)\big)$ are both continuous, we have

$$\big((x\xi)(e) \mid \eta(e)\big) = \lim_{K\to e} \int_G f_K(s)\big((x\xi)(s) \mid \eta(s)\big)\,ds = \big(\xi(e) \mid (x^*\eta)(e)\big) = 0.$$

Since $\{\eta(e) : \eta \in \mathfrak{D}\}$ is total in $\mathfrak{K}$, we conclude that $(x\xi)(e) = 0$. Therefore, we have a linear map $y\colon \xi(e) \in \mathfrak{K} \mapsto (x\xi)(e) \in \mathfrak{K}$. We then have

$$(x\xi)(s) = \big(U(s)^*x\xi\big)(e) = \big(xU(s)^*\xi\big)(e) = y\big(U(s)^*\xi\big)(e) = y\xi(s),$$

so that

$$\begin{aligned}
\|y\xi(e)\|^2 &= \lim_{K\to e} \int_G f_K(s)\|(x\xi)(s)\|^2\,ds = \lim_{K\to e} \|\pi(\mathcal{E}_H f)x\xi\|^2 \\
&\le \|x\|^2 \lim_{K\to e} \|\pi(\mathcal{E}_H f)\xi\|^2 \\
&= \|x\|^2 \lim_{K\to e} \int_G f_K(s)\|\xi(s)\|^2\,ds = \|x\|^2\|\xi(e)\|^2.
\end{aligned}$$

Therefore, we have $\|y\| \le \|x\|$, and y is extended to the entire space $\mathfrak{K}$ by continuity. If $r \in H$, then

$$\begin{aligned}
V(r)y\xi(e) &= V(r)(x\xi)(e) = \chi(r)^{-\frac{1}{2}}(x\xi)(r^{-1}) = \chi(r)^{-\frac{1}{2}}\big(U(r)x\xi\big)(e) \\
&= \chi(r)^{-\frac{1}{2}}\big(xU(r)\xi\big)(e) \\
&= \chi(r)^{\frac{1}{2}}y\big(U(r)\xi\big)(e) = y\chi(r)^{-\frac{1}{2}}\xi(r^{-1}) = yV(r)\xi(e),
\end{aligned}$$

so that y belongs to $\mathcal{N} = V(H)'$. Thus $x = \Phi(y)$. Q.E.D.

We now move on to the discussion of induced covariant systems. Let G be a locally compact group and H a closed subgroup. Given a covariant system $\{\mathcal{N}, H, \beta\}$ over H, we want to construct a new covariant system $\{\mathcal{M}, G, \alpha\}$ over G. Set

$$\bar{\mathcal{N}} = \mathcal{N} \,\bar{\otimes}\, L^\infty(G), \qquad \bar{\beta}_r = \beta_r \otimes \rho_r, \quad r \in H, \tag{38}$$

where ρ is the right translation action of G on $L^\infty(G)$ as usual. Let $\mathcal{M}$ be the fixed point algebra $\bar{\mathcal{N}}^H$ under $\rho(H)$. Since the left translation action λ of G and ρ commute, $\iota \otimes \lambda$, say α, is an action of G.

Definition 4.10. We call the new covariant system $\{\mathcal{M}, G, \alpha\}$ the *induced* covariant system from $\{\mathcal{N}, H, \beta\}$ and write

$$\{\mathcal{M}, G, \alpha\} = \operatorname{Ind}_H^G\{\mathcal{N}, H, \beta\}. \tag{39}$$

Proposition 4.11. *Let G be a locally compact group and H a closed subgroup. For a covariant system $\{\mathcal{M}, G, \alpha\}$ over G, the following two statements are equivalent:*

(i) $\{\mathcal{M}, G, \alpha\} \cong \mathrm{Ind}_H^G\{\mathcal{N}, H, \beta\}$ *for some covariant system* $\{\mathcal{N}, H, \beta\}$ *over* H;
(ii) *There exists a normal isomorphism π of $L^\infty(G/H)$ into the center $\mathcal{Z}$ of $\mathcal{M}$ which intertwines the actions of G, the left translation action of G on $L^\infty(G/H)$, and the restriction of α to $\mathcal{Z}$.*

If either condition of the above (i) *or* (ii) *holds, then the covariant system $\{\mathcal{N}, H, \beta\}$ is unique up to conjugacy.*

PROOF:
(i) $\Longrightarrow$ (ii): Suppose (i) holds. Then $\mathcal{M}$ is the fixed point subalgebra of $\mathcal{N} \overline{\otimes} L^\infty(G)$ under the action $\{\beta_r \otimes \rho_r : r \in H\}$. Since $L^\infty(G/H)$ can be viewed as the subalgebra of $L^\infty(G)$ consisting of those elements fixed by $\{\rho_r : r \in H\}$, $\mathbf{C} \otimes L^\infty(G/H)$ is a von Neumann subalgebra of the center $\mathcal{Z}$ of $\mathcal{M}$. Then our assertion follows from the definition of the induced action α.

(ii) $\Longrightarrow$ (i): We present here a proof for the separable case only. The proof for non-separable case requires either a different approach which is more C^*-algebraic, or a more powerful lifting theorem in measure theory. In any case, we are mainly interested in the separable case, at least for groups.

Thus, we assume that G is separable and (ii) holds. We represent $\mathcal{M}$ on a Hilbert space $\mathfrak{H}$ in such a way that the action α is implemented by a unitary representation U on $\mathfrak{H}$. By Theorem 4.7, there exists a unitary representation $\{V, \mathfrak{K}\}$ of H such that $\{U, \mathfrak{H}\} \cong \mathrm{Ind}_H^G\{V, \mathfrak{K}\}$. Thus we identify $\{V, \mathfrak{H}\}$ with $\mathrm{Ind}_H^G\{V, \mathfrak{K}\}$, which yields the identification of the α-invariant subalgebra of the center $\mathcal{Z}$ with $L^\infty(G/H)$ under the map π.

By the separability of G, there exists a Borel subset E of G which meets with every left H-coset exactly once. Hence $G = EH$. Let φ be the function of Theorem 4.1.(iii) and μ be the corresponding quasi invariant measure on Γ given by (4). Thus the measure $\varphi(s)\,\mathrm{d}s$ is identified with the product measure $\mathrm{d}h\,\mathrm{d}\mu(s)$. We now want to identify $\mathfrak{H}$ with $L^2_{\mathfrak{K}}(\Gamma, \mu)$. Identifying Γ with the cross-section E, for each $\xi \in L^2_{\mathfrak{K}}(\Gamma, \mu)$ we set

$$\bar{\xi}(\gamma r) = \varphi(\gamma r)^{\frac{1}{2}} V(r)^* \xi(\gamma). \tag{40}$$

Clearly $\bar{\xi}$ satisfies (7) or rather (7′), and also

$$\begin{aligned}\|\bar{\xi}\|^2 &= \int_G f_0(s)\|\bar{\xi}(s)\|^2\,\mathrm{d}s = \int_G f_0(s)\varphi(s)\|\xi(\gamma(s))\|^2\,\mathrm{d}s \\ &= \int_\Gamma \|\xi(\gamma)\|^2\,\mathrm{d}\mu(\gamma),\end{aligned}$$

where f_0 is the function of (11), i.e. $\mathcal{E}_H(f) = 1$, and $\gamma(s) = sH$. Therefore, $W : \xi \in L^2_{\mathfrak{K}}(\Gamma) \mapsto \bar{\xi} \in \mathfrak{H}$ is a unitary. We consider $W^*\mathcal{M}W$ and $\{W^*U(s)W : s \in G\}$.

Then the operator $W^*\pi(f)W$, $f \in L^\infty(G/H)$, is nothing but the multiplication by f on $L^2_{\mathfrak{K}}(\Gamma, \mu)$. Since $L^\infty(G/H)$ is a subalgebra of the center of $W^*\mathcal{M}W$, each $x \in W^*\mathcal{M}W$ is decomposable on $L^2_{\mathfrak{K}}(\Gamma, \mu)$ by Theorem IV.7.10, which means that there exists an $\mathcal{L}(\mathfrak{K})$-valued bounded measurable function $x(\,\cdot\,)$ on Γ such that $(x\xi)(\gamma) = x(\gamma)\xi(\gamma)$, $\xi \in L^2_{\mathfrak{K}}(\Gamma, \mu)$, for almost every $\gamma \in \Gamma$. This means that each $x \in \mathcal{M}$ is decomposable in the sense that there exists an $\mathcal{L}(\mathfrak{K})$-valued bounded measurable function $x(\,\cdot\,)$ on G such that

$$\left.\begin{aligned} x(sr) &= V(r)^*x(s)V(r), \quad && s \in G, \quad && r \in H \\ (x\xi)(s) &= x(s)\xi(s), && x \in \mathcal{M}, && \xi \in \mathfrak{H}. \end{aligned}\right\} \tag{41}$$

It then follows that

$$\big(\alpha_s(x)\xi\big)(t) = \big(U(s)xU(s)^*\xi\big)(t) = x(s^{-1}t)\xi(t).$$

Therefore, we have

$$\big\{\alpha_s(x)\big\}(t) = x(s^{-1}t), \quad s, t \in G, \quad x \in \mathcal{M}. \tag{42}$$

Let A be the C^*-algebra of all those elements x of $\mathcal{M}$ such that

$$\lim_{s \to e} \big\|\alpha_s(x) - x\big\| = 0.$$

It then follows that $x \in \mathcal{M}$ belongs to A if and only if the function: $s \in G \mapsto x(s) \in \mathcal{L}(\mathfrak{K})$ is uniformly continuous, i.e. $\lim_{s \to e} \sup_t \big\|x(s^{-1}t) - x(t)\big\| = 0$. Let $\mathcal{N}$ be the von Neumann algebra generated by $\big\{x(e) : x \in A\big\}$. The first equality of (41) implies that $V(r)\mathcal{N}V(r)^* = \mathcal{N}$, so that we can define an action β of H on $\mathcal{N}$ by

$$\beta_h(x) = V(h)xV(h)^*, \quad x \in \mathcal{N}, \quad h \in H. \tag{43}$$

Since A is σ-weakly dense in $\mathcal{M}$ and globally invariant under α, we can easily conclude that

$$\big\{\mathcal{M}, G, \alpha\big\} = \operatorname{Ind}_H^G\big\{\mathcal{N}, H, \beta\big\}.$$

In fact, A is the C^*-algebra of all uniformly continuous $\mathcal{N}$-valued functions on G satisfying (41).

The uniqueness follows from two facts: the uniqueness of representation V and the construction of $\mathcal{N}$. We leave the detail to the reader. Q.E.D.

Theorem 4.12. *Let G be a separable locally compact group and H a closed subgroup of G. Given a covariant system $\big\{\mathcal{N}, H, \beta\big\}$ over H, set*

$$\big\{\mathcal{M}, G, \alpha\big\} = \operatorname{Ind}_H^G\big\{\mathcal{N}, H, \beta\big\}.$$

Then we have

$$\mathcal{M} \rtimes_\alpha G \cong \big(\mathcal{N} \rtimes_\beta H\big) \overline{\otimes} \mathcal{L}\big(L^2(G/H, \mu)\big)$$

where μ is a unique quasi-invariant measure on the homogeneous space $\Gamma = G/H$.

PROOF: We continue to use the notation in the previous Proposition. It follows that

$$\begin{aligned}&\mathcal{M} \rtimes_\alpha G\\ &\quad= \left[\mathcal{N} \overline{\otimes} L^\infty(G) \overline{\otimes} \mathcal{L}\big(L^2(G)\big)\right] \cap \big(V \otimes \rho_G \otimes \mathrm{id}\big)(H)' \cap \big(1 \otimes \lambda_G \otimes \rho_G(G)\big).\end{aligned}$$

We first look at $L^\infty(G) \overline{\otimes} \mathcal{L}\big(L^2(G)\big) \cap (\lambda_G \otimes \rho_G)(G)' = L^\infty(G) \rtimes_\lambda G$. On $L^2(G \times G)$, $L^\infty(G) \rtimes_\lambda G$ is generated by the following operators:

$$\left.\begin{aligned}\big(\pi(f)\xi\big)(s,t) &= f(ts)\xi(s,t), \quad f \in L^\infty(G), \quad \xi \in L^2(G \times G)\\ \big(1 \otimes \lambda_G(r)\xi\big)(s,t) &= \xi\big(s, r^{-1}t\big), \qquad r, s, t \in G.\end{aligned}\right\} \tag{44}$$

Set

$$(W\xi)(s,t) = \delta_G(t)^{-\frac{1}{2}}\xi(t, st^{-1}). \tag{45}$$

Then we have

$$(W^*\xi)(s,t) = \delta_G(s)^{\frac{1}{2}}\xi(ts, s). \tag{45$'$}$$

It follows that

$$\left.\begin{aligned}\big(W\pi(f)W^*\xi\big)(s,t) &= f(s)\xi(s,t), \qquad f \in L^\infty(G),\\ \Big(W\big(1 \otimes \lambda_G(r)\big)W^*\xi\Big)(s,t) &= \xi(r^{-1}s, t), \qquad r, s, t \in G,\\ \Big(W\big(\rho_G(r) \otimes 1\big)W^*\xi\Big)(s,t) &= \delta_G(r)\xi(sr, tr).\end{aligned}\right\} \tag{46}$$

Therefore, we get

$$\left.\begin{aligned}W\pi(f)W^* &= f \otimes 1, \qquad f \in L^\infty(G),\\ W\big(1 \otimes \lambda_G(s)\big)W^* &= \lambda_G(s) \otimes 1, \qquad s \in G,\\ W\big(\rho_G(s) \otimes 1\big)W^* &= \rho_G(s) \otimes \rho_G(s).\end{aligned}\right\} \tag{46$'$}$$

Hence, we have $W\big(L^\infty(G) \rtimes_\lambda G\big)W^* = \mathcal{L}\big(L^2(G)\big) \otimes \mathbf{C}$ on $L^2(G \times G)$, and also with $\bar{W} = 1 \otimes W$ on $\mathfrak{K} \otimes L^2(G \times G)$ we get

$$\begin{aligned}\bar{W}\big(\mathcal{M} \rtimes_\alpha G\big)\bar{W}^* &= \Big(\mathcal{N} \overline{\otimes} \mathcal{L}\big(L^2(G)\big) \otimes \mathbf{C}\Big) \cap \big(V \otimes \rho_G \otimes \rho_G\big)(H)'\\ &\cong \Big(\mathcal{N} \overline{\otimes} \mathcal{L}\big(L^2(G)\big)\Big) \cap \big(V \otimes \rho_G\big)(H)'.\end{aligned}$$

We now look at $\rho_G(H)$ on $L^2(G)$ more in detail. Decomposing $G = EH$ as in Proposition 4.11, we have $L^2(G) \cong L^2(\Gamma, \mu) \otimes L^2(H)$, which yields the decomposition of $\rho_G(H)$: $\rho_G(r) \cong 1 \otimes \rho_H(r)$, $r \in H$.

Therefore we have

$$\begin{aligned}
&\mathcal{N} \overline{\otimes} \mathcal{L}\big(L^2(G)\big) \cap (V \otimes \rho_G)(H)' \\
&\cong \Big[\mathcal{N} \overline{\otimes} \mathcal{L}\big(L^2(\Gamma, \mu)\big) \overline{\otimes} \mathcal{L}\big(L^2(H)\big)\Big] \cap \Big[\big(V \otimes 1 \otimes \rho_H\big)(H)\Big]' \\
&\cong \Big[\mathcal{N} \overline{\otimes} \mathcal{L}\big(L^2(H)\big) \overline{\otimes} \mathcal{L}\big(L^2(\Gamma, \mu)\big)\Big] \cap \Big[\big(V \otimes \rho_H \otimes 1\big)(H)\Big]' \\
&= \big(\mathcal{N} \rtimes_\beta H\big) \overline{\otimes} \mathcal{L}\big(L^2(\Gamma, \mu)\big).
\end{aligned}$$

Q.E.D.

We now consider the ergodic disintegration of an integrable covariant system $\{\mathcal{A}, G, \alpha\}$ with an abelian carrier algebra $\mathcal{A}$.

Lemma 4.13. *If $\{\mathcal{A}, G, \alpha\}$ is an ergodic integrable covariant system with $\mathcal{A}$ abelian, then $\{\mathcal{A}, G, \alpha\}$ is conjugate to the covariant system induced from the trivial action of some compact subgroup H of G, i.e. there exists a compact subgroup H of G such that $\{\mathcal{A}, G, \alpha\}$ is conjugate to the translation action of G on $L^\infty(G/H)$.*

PROOF: Let $\mathcal{E}$ be the map: $\mathcal{A}_+ \mapsto \widehat{\mathcal{A}}_+^\alpha = [0, \infty]$ defined by $\mathcal{E}(a) = \int_G \alpha_g(a)\, dg$, $a \in \mathcal{A}_+$. By the integrability of α, $\mathcal{E}$ is a faithful semi-finite normal weight on $\mathcal{A}$, so it is more appropriate to denote $\mathcal{E}$ by φ. Let B be the C^*-subalgebra of all norm continuous elements in $\mathcal{A}$, i.e. $B = \big\{a \in \mathcal{A} : \lim_{g\to e}\|\alpha_g(a) - a\| = 0\big\}$, and let A be the norm closure of $B \cap m_\varphi$. Set $\Gamma = \mathrm{Sp}(A)$. Then the algebra $\mathcal{K}(\Gamma)$ of all continuous functions on Γ with compact support is the smallest norm dense ideal of $A = C_\infty(\Gamma)$, so that $\mathcal{K}(\Gamma) \subset m_\varphi$. Hence the weight φ restricted to A gives rise a Radon measure μ on Γ such that $\mathcal{A}$ can be canonically identified with $L^\infty(\Gamma, \mu)$. Since $\{A, G, \alpha\}$ is a C^*-covariant system, α is given by an action of G on Γ, i.e. $\alpha_g(a)(\gamma) = a(T_y^{-1}\gamma)$, $a \in A$ and $g \in G$. It then follows that for every $f \in \mathcal{K}(\Gamma)$,

$$\int_\Gamma f(\gamma)\, d\mu(\gamma) = \int_G \alpha_g(f)\, dg = \varepsilon(f)(\gamma_0) = \int_G f\big(T_g^{-1}\gamma_0\big)\, dg$$

for every fixed $\gamma_0 \in T$. This means that μ is concentrated on a single orbit $G\gamma_0$, and given by the Haar measure on G. Fix $\gamma_0 \in \Gamma$ and let $H = \big\{g \in G : T_g\gamma_0 = \gamma_0\big\}$. We want to show that H is compact. Set $\pi(a)(g) = a\big(T_g^{-1}\gamma_0\big)$. Then $\pi\big(\mathcal{K}(\Gamma)\big) \subset L^p(G) \cap C_b(G)$, $p \ge 1$. Since $\rho_g \circ \pi = \pi \circ \alpha_g$, $g \in G$, we have $\pi \circ \alpha_f(a) = \pi(a) * f^\upsilon$ for any $f \in L^1(G)$. Hence if $f \in \mathcal{K}(G)$, then $\pi \circ \alpha_f(a)$ belongs to the Fourier algebra $A(G)$, so that it vanishes at infinity, which is possible only when H is compact. Q.E.D.

Let $\{\mathcal{A}, G, \alpha\}$ be a separable integrable covariant system with $\mathcal{A}$ abelian. Let $\mathcal{B} = \mathcal{A}^\alpha$ be the fixed point algebra. Let $\mathcal{E}$ be the $\mathcal{B}$-valued weight given by the integral:

$$\mathcal{E}(a) = \int_G \alpha_g(a)\, dg, \quad a \in \mathcal{A}.$$

Let $\mathfrak{p}$ be the definition domain of $\mathcal{E}$, which is a σ-weakly dense ideal of $\mathcal{A}$. Let A_1 be the C^*-algebra of all norm continuous elements of $\mathcal{A}$ under α. Let A be a σ-weakly dense C^*-subalgebra of $\mathcal{A}$ such that

(i) A is a separable,
(ii) $A \subset A_1$,
(iii) $\alpha_g(A) = A, \ g \in G$, and
(iv) $A \cap \mathfrak{p}$ is norm dense in A.

Set $\Gamma = \mathrm{Sp}(A)$. As before, G acts on Γ topologically and $\mathcal{K}(\Gamma) \subset \mathfrak{p}$ because $\mathcal{K}(\Gamma)$ is the smallest dense ideal of $A = C_\infty(\Gamma)$ as seen in the proof of Theorem VII.3.14. Let B be the C^*-subalgebra of $\mathcal{B}$ generated by $\mathcal{E}\big(\mathcal{K}(\Gamma)\big)$.

Lemma 4.14. *The C^*-algebra B is contained in the multiplier algebra $M(A)$ of A.*

PROOF: Let $f \in \mathcal{K}(\Gamma)$ and $K = \operatorname{supp} f$. Let U be relatively compact open subset of Γ. Since $\mathcal{E}(\chi_U)$ exists, the subset $E = \big\{g \in G : gU \cap U \neq \emptyset\big\}$ has finite volume relative to the left Haar measure. Let $\{\gamma_n\}$ be a sequence converging to $\gamma \in U$. We may assume that $\{\gamma_n\}$ is contained in U, and also that $U \supset K$. Then we have

$$\int_G f(g^{-1}\gamma_n)\, dg = \int_E f(g^{-1}\gamma_n)\, dg .$$

As functions on E, $f(g^{-1}\gamma_n)$ is bounded and converges to $f(g^{-1}\gamma)$ pointwise, so that by the Lebesgue dominated convergence theorem

$$\int_G f(g^{-1}\gamma_n)\, dg = \int_E f(g^{-1}\gamma_n)\, dg \to \int_E f(g^{-1}\gamma)\, dg = \int_G f(g^{-1}\gamma)\, dg .$$

Hence $\mathcal{E}(f)$ is a bounded continuous function on Γ. Thus $\mathcal{E}(f) \in M(A)$ for every $f \in \mathcal{K}(\Gamma)$. Q.E.D.

Set $\Omega = \mathrm{Sp}(B)$. Then Ω is a separable locally compact space. Since $B \subset M(A) = C_b(\Gamma)$, each point $\gamma \in \Gamma$ gives a point $\omega = \pi(\gamma) \in \Omega$ as the evaluation of each element of B at γ. Of course, the map $\pi : \gamma \in \Gamma \mapsto \pi(\gamma) = \omega \in \Omega$ is continuous. Each element of B is constant on the fibre $\Gamma(\omega) = \pi^{-1}(\omega)$ over each $\omega \in \Omega$. If $\omega = \pi(\gamma)$, then $g\gamma$ and γ both give the same character ω on B, so that $\pi(g\gamma) = \omega$; hence each $\Gamma(\omega)$ is stable under the action of G.

Let ψ be a faithful normal state on $\mathcal{B}$. Since B is σ-weakly dense in $\mathcal{B}$ by the σ-weak density of $\mathcal{K}(\Gamma)$ in $\mathcal{A}$, $\mathcal{B}$ is then identified with $L^\infty(\Omega, \nu)$, where ν is the Radon measure corresponding to the state ψ restricted to B. Let $\varphi = \psi \circ \mathcal{E}$ and μ be the corresponding Radon measure on Γ. Then we have

$$\int_\Gamma f(\gamma)\, d\mu(\gamma) = \int_\Omega \int_G f(g^{-1}\gamma)\, dg\, d\mu(\pi(\gamma)), \quad f \in \mathcal{K}(\Gamma).$$

Lemma 4.15. *For each* $\omega \in \Omega$, *and* $\gamma \in \Gamma(\omega)$, *the map:* $f \in \mathcal{K}(\Gamma(\omega)) \mapsto \int_G f(g^{-1}\gamma)\,\mathrm{d}g \in \mathbf{C}$ *is well-defined and independent of the choice of* γ *in* $\Gamma(\omega)$; *hence it gives rise to a Radon measure* μ_ω *on* $\Gamma(\omega)$ *such that*

$$\int_\Gamma f(\gamma)\,\mathrm{d}\mu(\gamma) = \int_\Omega \Bigg(\int_{\Gamma(\omega)} f(\gamma)\,\mathrm{d}\mu_\omega(\gamma) \Bigg)\,\mathrm{d}\nu(\omega), \qquad f \in \mathcal{K}(\Gamma).$$

PROOF: Each $f \in \mathcal{K}(\Gamma(\omega))$ can be extended to an element $f' \in \mathcal{K}(\Gamma)$ by the Tietz extension theorem. For each γ_1 and $\gamma_2 \in \Gamma(\omega)$, we have

$$\begin{aligned}\int_G f(g^{-1}\gamma_1)\,\mathrm{d}g &= \int_G f'(g^{-1}\gamma_1)\,\mathrm{d}g = \mathcal{E}(f')(\pi(\gamma_1)) = \mathcal{E}(f')(\pi(\gamma_2)) \\ &= \int_G f'(g^{-1}\gamma_2)\,\mathrm{d}g = \int_G f(g^{-1}\gamma_2)\,\mathrm{d}g\,.\end{aligned}$$

Hence our assertion follows. Q.E.D.

Lemma 4.16. *For each* $\gamma \in \Gamma$, *let* $H(\gamma)$ *be the stabilizer subgroup* $\{g \in G : g\gamma = \gamma\}$ *of* G *at* γ. *Then* $\{H(\gamma) : \gamma \in \Gamma\}$ *is a measurable field of subgroups of* G *in the sense that there exists a sequence* $\{h_n(\gamma)\}$ *of* G*-valued measurable functions over* Γ *such that for almost every* $\gamma \in \Gamma$, $\{h_n(\gamma) : \gamma \in \Gamma\}$ *is a dense subset of* $H(\gamma)$.

PROOF: In the cartesian product $G \times \Gamma$, consider the subset $\mathcal{H} = \{(h,\gamma) \in G \times \Gamma : h\gamma = \gamma\}$. Clearly $\mathcal{H}$ is a closed subset of $G \times \Gamma$ and $H(\gamma) = \mathrm{pr_I}(\mathcal{H} \cap (G \times \{\gamma\}))$, where $\mathrm{pr_I}$ means the projection to the first variable. Then the measurable cross section theorem, Theorem A.16, takes care of the rest. Q.E.D.

Theorem 4.17. *If* $\{\mathcal{A}, G, \alpha\}$ *is an integrable separable covariant system with an abelian von Neumann algebra* $\mathcal{A}$, *then*

(i) *the ergodic disintegration:*

$$\{\mathcal{A}, G, \alpha\} = \int_\Omega^\oplus \{\mathcal{A}(\omega), G, \alpha^\omega\}\,\mathrm{d}\nu(\omega)$$

has the following properties:

a) *to each* $\omega \in \Omega$ *there corresponds a compact subgroup* $H(\omega)$ *of* G *in such a way that the map:* $\omega \in \Omega \mapsto H(\omega)$ *is a measurable field of closed subgroups of* G *in the sense that there exists a sequence* $\{h_n(\omega)\}$ *of measurable* G*-valued functions such that* $H(\omega)$ *is the closure of* $\{h_n(\omega) : n \in \mathbf{N}\}$;

b) *the covariant system* $\{\mathcal{A}(\omega), G, \alpha^\omega\}$ *is conjugate to the translation action of* G *on the homogeneous space* $L^\infty(G/H(\omega))$ *for almost every* $\omega \in \Omega$;

(ii) *the crossed product is then given by:*

$$\mathcal{A} \rtimes_\alpha G \cong \int_\Omega^\oplus \mathcal{R}\big(H(\omega)\big) \overline{\otimes} \mathcal{L}\Big(L^2\big(G/H(\omega)\big)\Big)\, d\nu(\omega),$$

where $\mathcal{R}\big(H(\omega)\big)$ *is the von Neumann algebra generated by the left regular representation of* $H(\omega)$.

PROOF:
(i) We keep the notations established above. By the last lemma, we have

$$\{\mathcal{A}, G, \alpha\} = \int_\Omega^\oplus \Big\{L^\infty\big(\Gamma(\omega), \mu_\omega\big),\ G,\ \alpha^\omega\Big\}\, d\nu(\omega)\,.$$

This disintegration is precisely the ergodic disintegration of the system. By construction, $\{L^\infty\big(\Gamma(\omega), \mu_\omega\big),\ G,\ \alpha^\omega\}$ is integrable and in fact the measure μ_ω is given by the integral along the orbit relative to the Haar measure of G. By Lemma 4.13, $\{G, \Gamma(\omega)\}$ is conjugate to the translation action of G on the left coset space $G/H(\omega)$ by a compact subgroup $H(\omega)$. To choose $\{H(\omega)\}$ in the measurable fashion, let $r\colon \omega \in \Omega \mapsto r(\omega) \in \Gamma(\omega)$ be a measurable right inverse of π, i.e. $\pi \circ \gamma(\omega) = \omega$, and set $H(\omega)$ to be the stabilizer subgroup at $r(\omega)$. Then Lemma 4.16 guarantees the measurability of the field: $w \in \Omega \mapsto H(\omega)$.

The statement (b) follows from Lemmas 4.13 and 4.15.

(ii) The crossed product operation and the direct integral operation commute, so that Theorem 4.12 yields the assertion because for almost every $\omega \in \Omega$, $\{\mathcal{A}(\omega), G, \alpha^\omega\} = \mathrm{Ind}_{H(\omega)}^G\{\mathbf{C}, H(\omega), \mathrm{id}\}$ and $\mathbf{C} \rtimes H(\omega) \cong \mathcal{R}\big(H(\omega)\big)$. Q.E.D.

Exercise X.4

1) Let $\{A, G, \alpha\}$ be a separable C^*-covariant system in the sense that A and G are both separable. Let $L^1(G, A)$ be the Banach space of A-valued integrable functions on G relative to the left Haar measure ds. In $L^1(G, A)$, define:

$$\left.\begin{aligned} f * g(s) &= \int_G f(t)\alpha_t\big(g(t^{-1}s)\big)\, dt;\\ f^*(s) &= \delta_G(s)^{-1}\alpha_s^{-1}\big(f(s^{-1})\big)^*, \quad f, g \in L^1(G, A). \end{aligned}\right\} \tag{47}$$

(**a**) Prove that $L^1(G, A)$ is an involutive Banach algebra such that for every covariant representation $\{\pi, U, \mathfrak{H}\}$ of $\{A, G, \alpha\}$ the integral:

$$\rho_{\pi,U}(f) = \int_G \pi\big(f(s)\big)U(s)\, ds, \quad f \in L^1(G, A)$$

gives a representation of $L^1(G, A)$ and that every representation of $L^1(G, A)$ is of this form.

(**b**) Define the *reduced C^*-crossed product* of $\{A, G, \alpha\}$ as the C^*-algebra generated by $\rho_0(L^1(G, A))$ where $\rho_0 = \rho_{\pi_0, U_0}$ is the representation of the involutive Banach algebra $L^1(G, A)$ such that $\{\pi_0, U_0\} = \mathrm{Ind}_{\{e\}}^G\{\pi, \mathrm{id}\}$ with π a faithful representation and show that this C^*-algebra does not depend on the choice of a faithful representation π of A. Denote the reduced crossed product of $\{A, G, \alpha\}$ by $A \rtimes_{\alpha,r} G$.

(**c**) Observe that the representation U_0 of G is naturally identified with $\mathrm{id} \otimes \lambda$ where λ is the left regular representation of G on $L^2(G)$ and that this identification gives the representation of the Hilbert space: $\mathfrak{H}_{\rho_0} = L^2(G, \mathfrak{H}_\pi)$.

2) In the last problem, the *full* crossed product $A \rtimes_\alpha G$ is defined to be the enveloping C^*-algebra of $L^1(G, A)$. Prove that if G is amenable in the sense that the regular representation of G gives a faithful representation of the group C^*-algebra $C^*(G)$ of G, the enveloping C^*-algebra of the group algebra $L^1(G)$. Note that abelian groups are amenable.

3) In Problem 1, assume that G is abelian and write the group operation additively. Let $\widehat{G}$ be the Kampen-Pontrjagin dual of G.

(**a**) Show that the unitary representation μ of $\widehat{G}$ in §2 gives rise to a continuous action $\hat{\alpha}$ of $\widehat{G}$ on $A \rtimes_\alpha G$. Call $\hat{\alpha}$ the *dual* action of $\widehat{G}$ on $A \rtimes_\alpha G$ or the action of $\widehat{G}$ on $A \rtimes_\alpha G$ *dual* to α.

(**b**) Prove that

$$\left\{(A \rtimes_\alpha G) \rtimes_{\hat{\alpha}} \widehat{G},\ G,\ \hat{\hat{\alpha}}\right\} \cong \left\{A \otimes_{\min} \mathcal{LC}\big(L^2(G)\big),\ G,\ \alpha \otimes \rho\right\},$$

where $\rho_s = \mathrm{Ad}\big(\rho_G(s)\big),\ s \in G$.

Notes on Chapter X

As mentioned in the introduction, the theory of crossed product in operator algebras goes back to the fundamental work of Murray and von Neumann. But the systematic study of crossed products was originated by the Japanese school of operator algebras lead by M. Nakamura, following the their tradition in abstract algebras established by T. Nakayama, G. Azumaya and others. First, T. Turumaru gave a general construction of the crossed product of a C^*-algebra by a discrete group of automorphisms, [725], which was soon followed by N. Suzuki, [702], and Nakamura-Takeda, [644, 645], in which they considered the crossed product of a von Neumann algebra by a discrete group. Prior to their work, I. M. Singer, [332], gave a construction of an isomorphism from the first cohomology group into the automorphism group $\mathrm{Aut}(\mathcal{R})$ of the factor $\mathcal{R}$ associated with a measure preserving ergodic system via the group measure space construction, i.e. $\mathcal{R} = L^\infty(X) \rtimes G$ and gave the criteria of the innerness of the automorphism in terms of the coboundary property. As they were mainly interested in factors of type II_1, they restricted their study to discrete groups and focused their attention to the analysis of intermediate algebras trying to establish the Galois type correspondence, which they succeeded for finite groups acting on a finite factor, [645].

In the mid sixties, mathematical physicists lead by H. Borchers, R. Haag, D. Kastler and others started to look at group actions on an operator algebra to treat the symmetries and/or the time evolution of a physical system. It has a rich history. The interested reader is referred to the monograph of O. Bratteli-D. W. Robinson, [442]. Influenced by this new trend, Zeller-Meier developed the theory of twisted crossed products of a C^*-algebra by a discrete group actions along with $\mathbf{T}$-valued 2-cocycles, [749]. It was Takesaki, [706], who combined the theory of covariant representations and Mackey's theory, which was developed by G. W. Mackey in the fifties, of induced unitary representations of a locally compact group, [630]. The materials covered in §3 and §4 are taken from Takesaki [362] and modified according to the further development.

The structure analysis of a von Neumann algebra of type III in the early seventies brought about the duality theory of crossed products. We presented here only the theory for abelian locally compact group crossed product duality in §2 which was first discovered by Takesaki [362]. The duality theorem for non abelian group crossed products and/or more general duality theorems have been proved. A book by Enock-Schwartz, [514], provides a good references. The duality in Exercise 4.3 was proven by Takai, [703]. It is a vast research area, which is still active, beyond the scope of the present book. The theory of dual weights on crossed products was developed by T. Digernes, [500, 501], for the separable case and U. Haagerup for the general case, [540].

Chapter XI

Abelian Automorphism Group

§ 0 Introduction

This chapter is devoted to the spectral analysis of actions of abelian groups on a von Neumann algebra which will be used in the subsequent chapters. As a Banach space, an operator algebra presents a great challenge to functional analysis. Although an operator algebra lives on a Hilbert space, it behaves very pathologically from the point of view of Banach spaces. For example, the spectral decomposition of a one parameter group of automorphisms on a C^*-algebra is out of question. Unless the group is periodic, any attempt of decomposition of the algebra relative to spectrum has been defeated. Thus, we take a moderate approach by examining the concept of spectrum of an element of the algebra relative to a given action of a locally compact abelian group. To this end, first we will look at an abelian group action on a Banach space and then move to analysis of automorphism actions of such a group on a von Neumann algebra $\mathcal{M}$. The formal definition of the spectrum $\mathrm{Sp}_\alpha(x)$ of an element of $x \in \mathcal{M}$ is defined in somewhat convoluted way. But it is, roughly speaking, nothing but the simple oscillation component of the function: $s \in G \mapsto \alpha_s(x) \in \mathcal{M}$. The reader should make the techniques in this chapter as reliable tools. For those readers who are not comfortable with general locally compact groups, it is advised to assume that all the groups involved are the additive group $\mathbf{R}$ of real numbers.

The multiplication in the algebra behaves like tensor products relative to the action. This means that a locally compact abelian group action on von Neumann algebra $\mathcal{M}$ gives rise to a kind of "grading" as seen in Corollary 1.7.

Section 1 is devoted to a formal definition of the spectrum of an element $x \in \mathcal{M}$ relative to an action α of a locally compact abelian group G on a von Neumann algebra $\mathcal{M}$ and its easy consequences. We then further sharpen the techniques in §2 based on the special feature of a von Neumann algebra. The Connes spectrum $\Gamma(\alpha)$ of α will be introduced there and seen to be invariant under cocycle perturbations. It is then proved in §3 that every derivation of a von Neumann algebra is inner, Theorem 3.5, and that the group of inner automorphisms, $\mathrm{Int}(\mathcal{M})$, is the principal component (i.e. connected component of the identity) of the group $\mathrm{Aut}(\mathcal{M})$ of automorphisms of $\mathcal{M}$ relative to the norm topology, Corollary 3.8. A characterization of inner automorphisms and the cohomology vanishing theorem, Theorem 3.11, are also given in §3.

§ 1 Spectral Analysis

To study further a covariant system $\{\mathcal{M}, G, \alpha\}$ over a locally compact abelian group G, we make a small detour.

Let X be a Banach space, which is the dual space of another Banach space X_*. By $x, y, \dots$, we denote the elements of X, and by $\varphi, \psi, \dots$ the elements of X_*. The canonical bilinear form $X \times X_*$ is denoted by $\langle x, \varphi \rangle$. The Banach algebra of all $\sigma(X, X_*)$-continuous linear operators on X is denoted by $\mathcal{L}_w(X)$. We write $\mathrm{GL}_w(X)$ for the group of all invertible elements of $\mathcal{L}_w(X)$.

Definition 1.1. A *representation* of a locally compact group G on X means a homomorphism α of G into $\mathrm{GL}_w(X)$ such that

$$\sup_{s \in G} \|\alpha_s\| < +\infty, \qquad \lim_{s \to e} \|\varphi \circ \alpha_s - \varphi\| = 0, \quad \varphi \in X_*. \tag{1}$$

In the case that we do not consider X_*, i.e. the case that X is not a dual space, then we replace the continuity requirement by a stronger one:

$$\lim_{s \to e} \|\alpha_s(x) - x\| = 0, \quad x \in X. \tag{1'}$$

Throughout this section, we consider only the case that $X = X_*^*$. But all the results are valid in the non-dual case if we replace the continuity condition by (1′). In fact, it is easier to handle this case. We shall leave the non-dual case to the reader as exercises.

In the sequel, we assume that G is abelian and write $\widehat{G}$ for the dual group of G. The Fourier algebra $A(\widehat{G})$ of $\widehat{G}$ is the image of $L^1(G)$ under the Fourier transform:

$$(\mathcal{F} f)(p) = \int_G \overline{\langle s, p \rangle} f(s)\, ds, \quad f \in L^1(G).$$

The norm $\|\mathcal{F} f\|_{A(\widehat{G})}$ of $\mathcal{F} f$, $f \in L^1(G)$, is defined to be $\|f\|_1$. For each $f \in A(\widehat{G})$, we denote its inverse image $\mathcal{F}^{-1} f$ by $\tilde{f}$, which can be computed by the Fourier inverse transform $\widehat{\mathcal{F}}$ if f is integrable:

$$\tilde{f}(s) = \int_{\widehat{G}} \langle s, p \rangle f(p)\, dp, \quad f \in A(\widehat{G}) \cap L^1(\widehat{G}).$$

Now, let α be a representation of G on X. For each $x \in X$ and $f \in A(\widehat{G})$, $\alpha_f(x)$ is defined to be:

$$\alpha_f(x) = \int_G \tilde{f}(-s) \alpha_s(x)\, ds\,. \tag{2}$$

The map $\alpha \colon f \in A(\widehat{G}) \mapsto \alpha_f \in \mathcal{L}_w(X)$ is a homomorphism of $A(\widehat{G})$ into $\mathcal{L}_w(X)$.

For each $x \in X$, putting

$$I(x) = \left\{ f \in A(\widehat{G}) : \alpha_f(x) = 0 \right\}, \tag{3}$$

we obtain a closed ideal of $A(\widehat{G})$.

Definition 1.2.

(i) The *α-spectrum* $\mathrm{Sp}_\alpha(x)$ of each $x \in X$ is defined to be the zero point set $\left\{ p \in \widehat{G} : f(p) = 0,\ f \in I(x) \right\}$ of $I(x)$. The *spectrum* $\mathrm{Sp}(\alpha)$, called *Arveson spectrum*, of α is defined to be the zero point set $\left\{ p \in \widehat{G} : f(p) = 0,\ f \in I(\alpha) \right\}$ of the kernel $I(\alpha) = \bigcap \left\{ I(x) : x \in X \right\}$ of α.

(ii) For a subset E of $\widehat{G}$, put

$$X^\alpha(E) = \left\{ x \in X : \mathrm{Sp}_\alpha(x) \subset E \right\}.$$

For an open subset U of $\widehat{G}$, $X_0^\alpha(U)$ is defined to be the weak* closed subspace spanned by the range of $\left\{ \alpha_f : f \in A(\widehat{G}) \text{ with } \operatorname{supp} f \subset U \right\}$.

We shall fix a representation $\left\{X, X_*, \alpha\right\}$ of G.

Lemma 1.3.

(i) *For each neighborhood U of $0 \in G$, we choose $\tilde{f}_U \in L^1(G)$ in such a way that*

$$\operatorname{supp} \tilde{f}_U \subset U, \qquad \tilde{f}_U(s) \geq 0 \qquad \text{and} \qquad \int_G \tilde{f}_U(s)\, ds = 1.$$

Then for each $x \in X$, $\left\{\alpha_{f_U}(x)\right\}$ converges to x weakly along the directed set of neighborhoods of $0 \in G$.*

(ii) $\mathrm{Sp}_\alpha\big(\alpha_f(x)\big) \subset \operatorname{supp} f \cap \mathrm{Sp}_\alpha(x),\ \ f \in A(\widehat{G}),\ \ x \in X.$

(iii) *When $\mathrm{Sp}_\alpha(x)$ is compact, if $f \in A(\widehat{G})$ takes value 1 on a neighborhood of $\mathrm{Sp}_\alpha(x)$, then $\alpha_f(x) = x$.*

(iv) *If $U \subset E \subset V \subset \widehat{G}$ with U and V open and E closed, then $X_0^\alpha(U) \subset X^\alpha(E) \subset X_0^\alpha(V)$.*

(v) *A point $p \in \widehat{G}$ belongs to $\mathrm{Sp}(\alpha)$ if and only if $X_0^\alpha(U) \neq \{0\}$ for every neighborhood U of p.*

PROOF:

(i) Given $\varphi \in X_*$ and $\varepsilon > 0$, choose a neighborhood U of $0 \in G$ such that $\|\varphi \circ \alpha_t - \varphi\| < \varepsilon,\ t \in U$, and $U = -U$. Then we have

$$\begin{aligned}\|\varphi \circ \alpha_{f_U} - \varphi\| &= \left\| \int_G \tilde{f}_U(-s)\big(\varphi \circ \alpha_s - \varphi\big)\, ds \right\| \\ &\leq \int_U \tilde{f}_U(-s) \|\varphi \circ \alpha_s - \varphi\|\, ds \leq \varepsilon \int_U \tilde{f}_U(-s)\, ds = \varepsilon,\end{aligned}$$

so that the net $\{{}^{t}\alpha_{f_U}\}$ of transposed maps converges to the identity operator on X_* strongly; hence $\{\alpha_{f_U}\}$ converges to 1 on X in the pointwise weak* topology.

(ii) Since $\alpha_{fg} = \alpha_f \circ \alpha_g = \alpha_g \circ \alpha_f$ for any $f, g \in A(\widehat{G})$, $I\big(\alpha_f(x)\big) \supset I(x)$ and $g \in I\big(\alpha_f(x)\big)$ if $fg = 0$. Thus our assertion follows.

(iii) For a compact subset K of $\widehat{G}$, let $I_0(K)$ be the ideal of $A(\widehat{G})$ consisting of all functions in $A(\widehat{G})$ vanishing on a neighborhood of K. It then follows from the general theory of commutative Banach algebras that $I_0(K)$ is smallest among the ideals whose zero point set is K. Hence $I(x) \supset I_0\big(\mathrm{Sp}_\alpha(x)\big)$. If $f \in A(\widehat{G})$ takes value 1 on a neighborhood U of $\mathrm{Sp}_\alpha(x)$, then $fg - g$, $g \in A(\widehat{G})$, takes value zero on U, so that $fg - g \in I_0\big(\mathrm{Sp}_\alpha(x)\big) \subset I(x)$. Hence $\alpha_g\big(\alpha_f(x) - x\big) = 0$. By (i) we have $\alpha_f(x) - x = 0$.

(iv) It follows from (ii) that $X_0^\alpha(U) \subset X^\alpha(\overline{U})$. Suppose that $\varphi \in X_*$ vanishes on $X_0^\alpha(V)$. We want to prove that $\varphi \in X^\alpha(E)^\perp$. Let

$$J = \big\{ f \in A(\widehat{G}) \cap \mathcal{K}(\widehat{G}) : \mathrm{supp}\, f \subset V \big\}.$$

Then J is a closed ideal of $A(\widehat{G})$ with zero point set $\complement V$. Suppose that $x \in X^\alpha(E)$. For each $f \in J$, $\alpha_f(x)$ belongs to $X_0^\alpha(V)$, so that

$$0 = \langle \alpha_f(x), \varphi \rangle = \int_G \tilde{f}(-s) \langle \alpha_s(x), \varphi \rangle \, \mathrm{d}s\,.$$

Since $I_0(E) \subset I(x)$, $\alpha_g(x) = 0$ for every $g \in I_0(E)$. Hence $\langle \alpha_f(x), \varphi \rangle = 0$ for every $f \in I_0(E) + J$. But the zero point set of $I_0(E) + J$ is $E \cap \complement V = \emptyset$, so that $I_0(E) + J$ is dense in $A(\widehat{G})$. Thus $\langle \alpha_f(x), \varphi \rangle = 0$ for every $f \in A(\widehat{G})$, which means that $\langle x, \varphi \rangle = 0$ by (i). Therefore, x belongs to $X_0^\alpha(V)^{\perp\perp} = X_0^\alpha(V)$ as we wanted.

(v) Suppose $X_0^\alpha(U) \neq \{0\}$ for every neighborhood U of $p \in \widehat{G}$. If $0 \neq x \in X_0^\alpha(U)$, then $\mathrm{Sp}_\alpha(x) \subset U$, so that $U \cap \mathrm{hull}\big(I(\alpha)\big) \neq \emptyset$ where $\mathrm{hull}\big(I(\alpha)\big)$ denotes the zero point set of $I(\alpha)$. Hence $p \in \mathrm{hull}\big(I(\alpha)\big) = \mathrm{Sp}(\alpha)$.

Conversely, suppose $p \in \mathrm{Sp}(\alpha)$. If U is an open neighborhood of p, then we can find $f \in A(\widehat{G})$ such that $f(p) = 1$ and $\mathrm{supp}\, f \subset U$. By the assumption on p, $f \notin I(\alpha)$, there exists a nonzero $x \in X$ with $\alpha_f(x) \neq 0$. But $\alpha_f(x) \in X_0^\alpha(U)$; hence $X_0^\alpha(U) \neq \{0\}$. Q.E.D.

From this lemma, it follows immediately that

$$X^\alpha(E) = \cap \big\{ X_0^\alpha(U) : E \subset U \big\}. \tag{4}$$

Now, we consider another pair $\{Y, Y_*\}$ of Banach spaces similar to $\{X, X_*\}$ and the space $\mathcal{L}_w(X, Y)$ of all weakly* continuous operators from X into Y. For an operator $T \in \mathcal{L}(X, Y)$ to be weakly* continuous, it is necessary and sufficient that the transpose ${}^{t}T$ of T maps Y_* into X_*. So the correspondence: $T \in \mathcal{L}_w(X, Y) \mapsto T_* = {}^{t}T|_{Y_*} \in \mathcal{L}(Y_*, X_*)$ is a Banach space isomorphism. Each $T \in \mathcal{L}(X, Y)$ corresponds to a bounded linear functional $\widetilde{T}$ on the projective tensor product $X \otimes_\gamma Y_*$ of X and Y_* by the formula:

$$\langle \widetilde{T}, x \otimes \psi \rangle = \langle Tx, \psi \rangle, \quad T \in \mathcal{L}(X, Y), \quad x \in X, \quad \psi \in Y_*. \tag{5}$$

Identifying T with $\widetilde{T}$, we equip $\mathcal{L}_w(X, Y)$ with the relative topology of $\sigma\big(\mathcal{L}(X, Y), X \otimes_\gamma Y_*\big)$. We call this the *$\sigma$-topology*, and denote $\mathcal{L}_w(X, Y)$ simply by $\mathcal{L}_w$ for short.

We now consider representations $\{X, X_*, \alpha\}$ and $\{Y, Y_*, \beta\}$ of G and define a new representation γ of G on $\mathcal{L}(X, Y)$ by

$$\gamma_t(T) = \beta_t \circ T \circ \alpha_t^{-1}, \quad t \in G, \quad T \in \mathcal{L}(X, Y). \tag{6}$$

In general, the representation[1] γ of G on $\mathcal{L}(X, Y)$ is not continuous in the point norm topology. But, γ leaves the subspace $\mathcal{L}_w$ of $\mathcal{L}(X, Y)$ invariant, so that we may restrict γ to $\mathcal{L}_w$. Then, since for each $T \in \mathcal{L}_w(X, Y)$

$$\langle \gamma_t(T), x \otimes \psi\rangle = \langle \beta_t \circ T \circ \alpha_t^{-1}(x), \psi\rangle = \langle \alpha_t^{-1}(x), T_*(\psi \circ \beta_t)\rangle,$$

the map: $t \in G \mapsto \langle \gamma_t(T), \rho\rangle$, $\rho \in X \otimes_\gamma Y_*$, is continuous. Hence, the representation $\{\mathcal{L}_w, \gamma\}$ is continuous with respect to the σ-topology. But in general $\mathcal{L}_w \subsetneqq (X \otimes_\gamma Y_*)^*$, so that we cannot apply the previous discussion to $\{\mathcal{L}_w, \gamma\}$ directly. The next result enables us to get around most of the difficulties:

Lemma 1.4.

(i) *In the above setting, every σ-compact subset K of $\mathcal{L}_w$ is contained in a σ-compact convex set.*

(ii) *If $f \in A(\widehat{G})$, then the integral*

$$\int_G \tilde{f}(-s)\gamma_s(T)\,\mathrm{d}s = \gamma_f(T), \quad T \in \mathcal{L}_w, \tag{7}$$

converges in $\mathcal{L}_w$, and the map: $T \in \mathcal{L}_w \mapsto \gamma_f(T) \in \mathcal{L}_w$ is σ-continuous.

PROOF:

(i) We view $\mathcal{L}_w$ as a subspace of $(X \otimes_\gamma Y_*)^* = \mathcal{L}(X, Y)$. Then for $T \in \mathcal{L}(X, Y)$ to belong to $\mathcal{L}_w$ it is necessary and sufficient that with the bilinear form B_T:

$$B_T(x, \psi) = \langle Tx, \psi\rangle = \langle T, x \otimes \psi\rangle, \quad x \in X, \quad \psi \in Y_*,$$

for each $\psi \in Y_*$, the function: $x \in X \mapsto B_T(x, \psi)$ is weakly* continuous.

Suppose $K \subset \mathcal{L}_w$ is σ-compact. Let $M_1^+(K)$ be the set of all regular positive Borel measures on K with total mass one. It is a weak* compact convex set. For each $\mu \in M_1^+(K)$, put

$$T_\mu = \int_K A\,\mathrm{d}\mu(A) \in \mathcal{L}(X, Y).$$

By definition,

$$\langle T_\mu x, \psi\rangle = \int_K \langle Ax, \psi\rangle\,\mathrm{d}\mu(A) = \int_K \langle x, A_*\psi\rangle\,\mathrm{d}\mu(A)\,.$$

[1] This representation γ and the symbol γ for the projective tensor product of Banach spaces are mere coincidence, i.e. not related.

With fixed $x \in X$ and $\psi \in Y_*$, the map: $A \in K \mapsto \langle x, A_*\psi\rangle$ is a continuous function. Hence the map: $A \in K \mapsto A_*\psi \in X_*$ is continuous with respect to the σ-topology on K and the weak topology on X_*. Hence $\{ A_*\psi : A \in K \}$ is weakly compact. Thus, the Krein-Šmulian Theorem implies that $K_*\psi$ is contained in a weakly compact convex set in X_*. Hence, the integral:

$$\int_K A_*\psi \, d\mu(A) = (T_\mu)_*\psi \in X_*$$

converges within X_*. Thus, $T_\mu \in \mathcal{L}_w(X, Y)$. Furthermore, if $x \in X$ and $\psi \in Y_*$, then the function:

$$\mu \in M_1^+(K) \mapsto \langle T_\mu x, \psi \rangle = \int_K \langle Ax, \psi\rangle \, d\mu(A)$$

is continuous on $M_1^+(K)$. Therefore, the map: $\mu \mapsto T_\mu \in \mathcal{L}_w$ is continuous with respect to the weak* topology on $M_1^+(K)$ and the σ-topology on $\mathcal{L}_w$. The weak* compactness of $M_1^+(K)$ then implies the compactness of the range $\{ T_\mu : \mu \in M_1^+(K) \}$, which contains K.

(ii) If $\tilde{f} \in \mathcal{K}(G)$, then the continuity: $t \in G \mapsto \gamma_t(T) \in \mathcal{L}_w$ relative to the σ-topology implies the σ-compactness of $\{ \tilde{f}(-t)\gamma_t(T) : t \in G \} = K$, so that (i) implies the convergence of (7). On the other hand, we have

$$\|\gamma_f(T\| \le \int_G |\tilde{f}(-s)| \|\gamma_s(T)\| \, ds \le \sup\|\gamma_s\| \|\tilde{f}\|_1 \|T\|,$$

and $\mathcal{K}(G)$ is dense in $L^1(G)$. Since $\mathcal{L}_w$ is closed in norm, we conclude the convergence of (7) for every $f \in A(\widehat{G})$. Q.E.D.

Theorem 1.5. *Let G, $\{X, X_*, \alpha\}$ and $\{Y, Y_*, \beta\}$ be as before, and construct $\{\mathcal{L}_w(X, Y), \gamma\}$ by the above procedure. If a closed subset $E \subset \widehat{G}$ has the properties: $E = \overline{E^{\rm int}}$, i.e., the closure of the interior $E^{\rm int}$ of E, and $0 \in E$, then the following two conditions on $A \in \mathcal{L}_w$ are equivalent:*

(i) $A \in \mathcal{L}_w^\gamma(\bigcap_{p\in E}(E - p))$;

(ii) $AX^\alpha(E + p) \subseteqq Y^\beta(E + p)$, $p \in \widehat{G}$.

Before the proof, we state an application.

Theorem 1.6. *Given $\{X, X_*\}$, consider a pair of representations $\{X, \alpha\}$ and $\{X, \beta\}$ of G. A necessary and sufficient condition for $\alpha = \beta$ is that $X^\alpha(E) \subset X^\beta(E)$ for every compact subset E of $\widehat{G}$.*

PROOF: The necessity is clear, so we shall show only the sufficiency. On $\mathcal{L}_w(X) = \mathcal{L}_w(X, X)$, consider $\gamma_t(T) = \beta_t \circ T \circ \alpha_{-t}$, $t \in G$. Let $\{E_j\}$ be a system of compact neighborhoods of 0 in $\widehat{G}$ with $E_j = \overline{E_j^{\rm int}}$. By assumption, $X^\alpha(E_j + p) \subset X^\beta(E_j + p)$, $p \in \widehat{G}$. By the Theorem, with

$$S_E = \bigcap \{ E - p : p \in E \}, \quad E \subset \widehat{G}, \tag{8}$$

we have $1 \in \mathcal{L}_w^\gamma(S_{E_j}) \subset \mathcal{L}_w^\gamma(E_j)$, so that

$$1 \in \bigcap \mathcal{L}_w^\gamma(E_j) = \mathcal{L}_w^\gamma\Big(\bigcap E_j\Big) = \mathcal{L}_w^\gamma(\{0\}).$$

Hence $\mathrm{Sp}_\gamma(1) = \{0\}$, i.e. $\gamma_t(1) = 1$. This means precisely $\alpha_t = \beta_t$, $t \in G$. Q.E.D.

PROOF OF THEOREM 1.5:

(i) $\Longrightarrow$ (ii) Define S_E by (8) and note $S_E = \{ p \in \widehat{G} : p + E \subset E \}$. Since $0 \in E$, $S_E \subset E$ and S_E is a semigroup. Suppose $A \in \mathcal{L}_w^\gamma(S_E)$ and $x \in X^\alpha(E + p)$ for a fixed $p \in \widehat{G}$.

Since $X^\alpha(\bigcap E_i) = \bigcap X^\alpha(E_i)$ in general, to prove $Ax \in Y^\beta(E + p)$ it suffices to show that $Ax \in Y^\beta(E + p + V)$ for every compact neighborhood V of 0 in $\widehat{G}$. So, fixing V, choose a compact neighborhood W of 0 with $W + W \subset V$ and $W = -W$. By Lemma 1.3.(iv), x is expressed as a weak* convergence limit of linear combinations of $\{ \alpha_f(y) : f \in A(\widehat{G}),\ \mathrm{supp}\, f \subset E + p + W,\ y \in X \}$. Since A is weakly* continuous and $Y^\beta(E + p + V)$ is weakly* closed, it is sufficient to prove that

$$A\alpha_f(y) \in Y^\beta(E + p + V), \qquad f \in A(\widehat{G}), \qquad \mathrm{supp}\, f \subset E + p + W. \tag{$*$}$$

Now, we fix f and y. The map: $T \in \mathcal{L}_w(X, Y) \mapsto T\alpha_f(y) \in Y$ is continuous with respect to the σ-topology of $\mathcal{L}_w$ and the weak* topology in Y. We have $\mathcal{L}_w^\gamma(S_E) = \bigcap \{ (\mathcal{L}_w)_0^\gamma(S_E + U) : U \subset \widehat{G}$ runs open neighborhoods of $0 \in \widehat{G} \}$. Thus to show $(*)$ it suffices to prove that for each $g \in A(\widehat{G})$ with $\mathrm{supp}\, g \subset S_E + W$,

$$\gamma_g(B)\alpha_f(y) \in Y^\beta(E + p + V), \quad B \in \mathcal{L}_w.$$

Now, putting $z = \gamma_g(B)\alpha_f(y)$, we prove $\beta_h(z) = 0$ for any $h \in A(\widehat{G})$ with $\mathrm{supp}\, h \cap (E + p + V) = \emptyset$. For each $\psi \in Y_*$, we compute

$$\begin{aligned}\langle \beta_h(z), \psi \rangle &= \iiint \tilde{h}(-t)\, \tilde{g}(-s)\, \tilde{f}(-r) \langle \beta_{s+t} \circ B \circ \alpha_{r-s}(y), \psi \rangle\, dr\ ds\ dt \\ &= \iiint \tilde{h}(s - t)\, \tilde{g}(-s)\, \tilde{f}(-r - s) \langle \beta_t \circ B \circ \alpha_r(y), \psi \rangle\, dr\ ds\ dt\ .\end{aligned}$$

With $\tilde{f}_r(s) = \tilde{f}(r + s)$, put

$$\tilde{k}_r(t) = \int \tilde{h}(s + t)\, \tilde{g}(-s)\, \tilde{f}(-r - s)\, ds = \Big\{ \tilde{h} * \big(\tilde{g}\tilde{f}_{-r}\big) \Big\}(t).$$

We then have

$$\langle \beta_h(z), \psi \rangle = \iint \tilde{k}_r(-t) \langle \beta_t \circ B \circ \alpha_r(y), \psi \rangle\, dt\ dr\ .$$

Note that the function $k_r \in A(\widehat{G})$ corresponding to $\tilde{k}_r$ is given by $k_r = h(g * f_{-r})$ provided $g, f \in L^1(\widehat{G})$. Furthermore, $f_{-r}(q) = \langle r, q\rangle f(r),\ q \in \widehat{G}$. The restriction for the choice of f and g was that $\operatorname{supp} f \subset E + p + W$ and $\operatorname{supp} g \subset S_E + W$. Since $A(\widehat{G}) \cap L^1(\widehat{G})$ is a dense ideal of $A(\widehat{G})$, the above f and g can be approximated by functions in $L^1(\widehat{G}) \cap A(\widehat{G})$ with the same support conditions for f and g respectively. Thus to prove $\langle \beta_h(z), \psi\rangle = 0$ we may assume that $k_r = h(g * f_{-r})$. Now we have

$$\operatorname{supp}(g * f_{-r}) \subseteq E + p + W + S_E + W \subset E + p + V,$$

so that $k_r = h(g * f_{-r}) = 0$. Hence the above integral vanishes for every $\psi \in Y_*$. Thus $\beta_h(z) = 0$ whenever $\operatorname{supp} h \cap (E + p + V) = \emptyset$, i.e. $\operatorname{Sp}_\beta(z) = \operatorname{hull}(I(z)) \subset E + p + V$. This completes the proof of (i) $\Longrightarrow$ (ii).

(ii) $\Longrightarrow$ (i) Suppose that an $A \in \mathcal{L}_w(X, Y)$ satisfies $AX^\alpha(E+p) \subseteq Y^\beta(E+p)$, $p \in \widehat{G}$. Consider an arbitrary $f \in A(\widehat{G})$ with $\operatorname{supp} f \subset E$ and let

$$f_p(q) = f(p+q) = (\rho_p f)(q), \quad p, q \in \widehat{G}.$$

Note that $\operatorname{supp} f_p = \operatorname{supp} f - p \subset E - p$. If $x \in X$, then $\alpha_f(x) \in X^\alpha(E)$. Put $y_p = \alpha_{f_p}(x)$ for each $p \in \widehat{G}$. Then $y_p \in X^\alpha(E - p)$. Consider another $g \in A(\widehat{G})$ with $\operatorname{supp} g \cap E = \emptyset$, and let $\varphi_p = \psi \circ \beta_{g_p} \in Y_*$ with $\psi \in Y_*$. Then $\varphi_p \in Y^\beta(E - p)^\perp,\ p \in \widehat{G}$. Now, by the assumption on A, we have

$$\begin{aligned} 0 = \langle Ay_p, \varphi_p\rangle &= \iint_{G\times G} \langle s + t, p\rangle \tilde{f}(-s)\, \tilde{g}(-t) \langle \beta_t A\alpha_s x, \psi\rangle \, \mathrm{d}s\, \mathrm{d}t \\ &= \int_G \langle s, p\rangle \left(\int_G \tilde{f}(t - s)\, \tilde{g}(-t)\langle \beta_t A\alpha_{s-t} x, \psi\rangle \, \mathrm{d}t \right) \mathrm{d}s \end{aligned}$$

for every $p \in \widehat{G}$. If we choose f from $A(\widehat{G}) \cap L^1(\widehat{G})$, then the integral inside the parenthesis is an integrable function of s, whose Fourier inverse transform vanishes identically. Hence the uniqueness of the Fourier inverse transform shows that

$$\int_G \tilde{f}(t - s)\, \tilde{g}(-t)\langle \beta_t A\alpha_{s-t} x, \psi\rangle \, \mathrm{d}t = 0$$

for every $s \in G$. With $s = 0$, we have

$$\int_G \tilde{f}(t)\, \tilde{g}(-t)\langle \beta_t A\alpha_{-t} x, \psi\rangle \, \mathrm{d}t = 0.$$

Replacing f by $(\rho_s f)(p) = \langle s, p\rangle f(p)$ and applying the above conclusion to $\rho_s f$, we get

$$\int_G \tilde{f}(s + t)\, \tilde{g}(-t)\langle \gamma_t(A) x, \psi\rangle \, \mathrm{d}t = 0, \quad s \in G.$$

Now, put

$$\tilde{h}(t) = \tilde{g}(t)\langle \gamma_{-t}(A) x, \psi\rangle, \quad t \in G.$$

Then $\tilde{h} \in L^1(G)$ and $\tilde{f} * \tilde{h} = 0$, so that $fh = 0$. The only condition for f was that $f \in L^1(\widehat{G}) \cap A(\widehat{G})$ and $\operatorname{supp} f \subset E$. Hence h vanishes identically on the interior of E. By the assumption, $E = \overline{E^{\text{int}}}$, we conclude that $h(p) = 0, \; p \in E$. This precisely means

$$\int_G \overline{\langle s, p\rangle}\, \tilde{g}(s) \langle \gamma_{-s}(A)x, \psi\rangle \, ds = 0, \qquad p \in E.$$

Since $x \in X$ and $\psi \in Y_*$ are arbitrary, we have

$$\gamma_{g_p}(A) = \int \langle s, p\rangle\, \tilde{g}(-s)\, \gamma_s(A)\, ds = 0, \qquad p \in E.$$

The only requirement for g was that $g \in A(\widehat{G})$ and $\operatorname{supp} g \cap E = \emptyset$. Hence we have $A \in \mathcal{L}_w^\gamma(E - p)$ for every $p \in E$, so that

$$A \in \bigcap_{p \in E} \mathcal{L}_w^\gamma(E - p) = \mathcal{L}_w^\gamma\Big(\bigcap_{p \in E} E - p\Big) = \mathcal{L}_w^\gamma(S_E). \qquad \text{Q.E.D.}$$

Corollary 1.7. *Let $\{X, X_*, \alpha\}$ and $\{Y, Y_*, \beta\}$ be as in Theorem 1.5, and form $\{\mathcal{L}_w(X, Y), \gamma\}$ as before. If E and F are closed subsets of $\widehat{G}$, then*

$$\mathcal{L}_w^\gamma(E) X^\alpha(F) \subset Y^\beta\big((E + F)^-\big).$$

PROOF: Let $A \in \mathcal{L}_w^\gamma(E)$ and $x \in X^\alpha(F)$. As in the proof of Theorem 1.5, it suffices to show that $Ax \in Y^\beta(E + F + V)$ for every compact neighborhood V of $0 \in \widehat{G}$. Choose a neighborhood W of $0 \in \widehat{G}$ such that $W + W \subset V$ and $W = -W$. Then x is weak* limit of linear combinations of $\{\alpha_f(y) : y \in X,\; f \in A(\widehat{G}),\; \operatorname{supp} f \subset F + W\}$ and A is a σ-convergence limit of linear combinations of $\{\gamma_g(B) : B \in \mathcal{L}_w(X, Y),\; g \in A(\widehat{G}),\; \operatorname{supp} g \subset E + W\}$. Hence it suffices to show that $\gamma_g(B)\alpha_f(y) \in Y^\beta(E + F + V)$ if $\operatorname{supp} g \subset E + W$ and $\operatorname{supp} f \subset F + W$. Let $z = \gamma_g(B)\alpha_f(y)$. We shall show that $\beta_h(z) = 0$ whenever $h \in A(\widehat{G})$ and $\operatorname{supp} h \cap (E + F + V) = \emptyset$. For each $\psi \in Y_*$, we compute

$$\langle \beta_h(z), \psi\rangle = \iiint \tilde{h}(-t)\, \tilde{g}(-s)\, \tilde{f}(-r) \langle \beta_{s+t} \circ B \circ \alpha_{r-s}(y), \psi\rangle \, dr\, ds\, dt\,.$$

Following exactly the same line of the arguments of the proof of Theorem 1.5, we conclude that the above integral vanishes. We leave the detail to the reader. Q.E.D.

Instead of continuing the general discussion, we state a few applications to a covariant system $\{\mathcal{M}, G, \alpha\}$.

Corollary 1.8. *For a covariant system $\{\mathcal{M}, G, \alpha\}$ over a locally compact abelian group G, the following statements hold:*

(i) $\operatorname{Sp}_\alpha(x^*) = -\operatorname{Sp}_\alpha(x), \; x \in \mathcal{M}$, (i)′ $\mathcal{M}^\alpha(E)^* = \mathcal{M}^\alpha(-E)$,

(ii) $\operatorname{Sp}_\alpha(xy) \subset [\operatorname{Sp}_\alpha(x) + \operatorname{Sp}_\alpha(y)]^-$, (ii)′ $\mathcal{M}^\alpha(E)\mathcal{M}^\alpha(F) \subset \mathcal{M}^\alpha\big((E + F)^-\big)$.

PROOF:

(i) For each $f \in A(\widehat{G})$, put

$$f^{\vee}(p) = f(-p), \qquad \overline{f}(p) = \overline{f(p)}, \qquad f^* = \overline{f}^{\vee}. \tag{9}$$

It then follows that $\alpha_f(x)^* = \alpha_{f^*}(x^*)$, $x \in \mathcal{M}$. This implies our assertions immediately.

(ii) Putting $L_x y = xy$, $x, y \in \mathcal{M}$, we obtain $L_x \in \mathcal{L}_w(\mathcal{M})$. Setting $\gamma_t(T) = \alpha_t \circ T \circ \alpha_t^{-1}$, $t \in G$, $T \in \mathcal{L}_w$, we have $\gamma_t(L_x) = L_{\alpha_t(x)}$. Hence $\gamma_f(L_x) = L_{\alpha_f(x)}$, $f \in A(\widehat{G})$. Thus we have $\mathrm{Sp}_\gamma(L_x) = \mathrm{Sp}_\alpha(x)$. Then Corollary 1.7 completes the proof. Q.E.D.

Proposition 1.9. *With the notations above, set*

$$\left.\begin{aligned} p^\alpha(E) &= \sup\{\, s_r(x) : x \in \mathcal{M}^\alpha(E)\,\}, \\ q^\alpha(E) &= \sup\{\, s_\ell(x) : x \in \mathcal{M}^\alpha(E)\,\} \end{aligned}\right\} \tag{10}$$

for each closed subset E of $\widehat{G}$. Then $p^\alpha(E)$ and $q^\alpha(E)$ both belong to the center $\mathcal{C}_\alpha$ of the fixed point algebra $\mathcal{M}^\alpha$.

PROOF: By Corollary 1.8, $\mathcal{M}^\alpha(E)$ is a two sided module over $\mathcal{M}^\alpha$. Hence for every $u \in \mathcal{U}(\mathcal{M}^\alpha)$, we have

$$\begin{aligned} up^\alpha(E)u^* &= \sup\{\, us_r(x)u^* : x \in \mathcal{M}^\alpha(E)\,\} = \sup\{\, s_r(uxu^*) : x \in \mathcal{M}^\alpha(E)\,\} \\ &= \sup\{\, s_r(x) : x \in \mathcal{M}^\alpha(E)\,\} = p^\alpha(E), \end{aligned}$$

so that $p^\alpha(E) \in (\mathcal{M}^\alpha)' \cap \mathcal{M}$. On the other hand, $\alpha_s\alpha_f = \alpha_f\alpha_s$ implies that $\alpha_s\big(\mathcal{M}^\alpha(E)\big) = \mathcal{M}^\alpha(E)$, so that $\alpha_s\big(p^\alpha(E)\big) = p^\alpha(E)$, $s \in G$. Thus $p^\alpha(E) \in \mathcal{M}^\alpha$. Together with the above, we have $p^\alpha(E) \in \mathcal{C}_\alpha$. By Corollary 1.8.(i), we have $q^\alpha(E) = p^\alpha(-E) \in \mathcal{C}_\alpha$. Q.E.D.

With the above preparation, we are ready to characterize inner one parameter automorphism groups of a von Neumann algebra.

Theorem 1.10. *For a one parameter automorphism group $\{\mathcal{M}, \mathbf{R}, \alpha\}$ of $\mathcal{M}$, the following two conditions are equivalent:*

(i) *There exists a one parameter unitary group $\{u(t)\}$ in $\mathcal{M}$ such that*

$$\alpha_t(x) = u(t)xu(t)^*, \quad x \in \mathcal{M}, \quad t \in \mathbf{R},$$

and $\{u(t)\}$ has a positive spectrum in the sense that

$$u(t) = \int_{[0,\infty[} e^{i\lambda t}\, de(\lambda) \tag{11}$$

gives the spectral decomposition of $u(t)$ over $[0, \infty[$;

(ii) $$\inf_{t \in \mathbf{R}} q^\alpha\big([t, \infty[\big) = 0.$$

If condition (ii) *holds, then the spectral measure* e *of* $\{u(t)\}$ *is given by*

$$e\big([t,\infty[\big) = \inf\Big\{ q^\alpha\big([s,\infty[\big) : s < t \Big\}. \tag{12}$$

PROOF:

(i) $\Longrightarrow$ (ii): Assume that $\mathcal{M}$ acts on $\mathfrak{H}$, and let $\{u(t)\}$ be a one parameter unitary group on $\mathfrak{H}$ with spectral decomposition:

$$u(t) = \int_0^\infty e^{i\lambda t}\, de(\lambda)\,.$$

Suppose that $u(t)\mathcal{M}u(t)^* = \mathcal{M},\ t \in \mathbf{R}$, and set $\alpha_t(x) = u(t)xu(t)^*,\ x \in \mathcal{M}$. We shall prove that $\alpha_t \in \operatorname{Int}(\mathcal{M}),\ t \in \mathbf{R}$.

If $t \le 0$, then $e\big([t,\infty[\big) = 1$, so that we have $\inf e\big([t,\infty[\big) = 0$. In Corollary 1.7, we set $X = \mathfrak{H}$ and $F = [0,\infty[$. If $x \in \mathcal{M}^\alpha\big([t,\infty[\big)$, then $\operatorname{Sp}_\alpha(x) \subset [t,\infty[= t + F$, and so

$$x\mathfrak{H}^u(F) \subset \mathfrak{H}^u(t+F) = e\big([t,\infty[\big)\mathfrak{H}.$$

Hence $s_\ell(x) \le e\big([t,\infty[\big)$ for every $x \in \mathcal{M}^\alpha\big([t,\infty[\big)$, so that $q^\alpha\big([t,\infty[\big) \le e\big([t,\infty[\big)$. Thus $\inf q_\alpha\big([t,\infty[\big) = 0$.

(ii) $\Longrightarrow$ (i): Assume (ii) and define $e\big([t,\infty[\big)$ by (12). The map: $t \in \mathbf{R} \mapsto e\big([t,\infty[\big) = e'(t)$ is strongly continuous from the left. Since $1 \in \mathcal{M}^\alpha\big([t,\infty[\big)$ for $t \le 0$, $e'(t) = 1$ for $t \le 0$. By assumption, $\lim_{t\to\infty} e'(t) = 0$. Hence $e(t) = 1 - e'(t)$ is a spectral measure on $\mathbf{R}$. Now, define $\{u(t)\}$ by (11), and set $\beta_t(x) = u(t)xu(t)^*,\ x \in \mathcal{M}$. By Corollary 1.8, we have $\mathcal{M}^\alpha\big([s,\infty[\big)\mathcal{M}^\alpha\big([t,\infty[\big) \subseteq \mathcal{M}^\alpha\big([s+t,\infty[\big)$, which yields that

$$\mathcal{M}^\alpha\big([s,\infty[\big)e\big([t,\infty[\big)\mathfrak{H} \subset e\big([s+t,\infty[\big)\mathfrak{H}$$

because $q^\alpha\big([t,\infty[\big)\mathfrak{H} = \big[\mathcal{M}^\alpha\big([t,\infty[\big)\mathfrak{H}\big]$. Theorem 1.5 implies $\mathcal{M}^\alpha\big([s,\infty[\big) \subset \mathcal{M}^\beta\big([s,\infty[\big)$. By Theorem 1.6, we have $\alpha = \beta$. Q.E.D.

We now return to the general discussion of $\{X, X_*, G, \alpha\}$.

Lemma 1.11. *For each* $\varepsilon > 0$, $p \in \widehat{G}$ *and a compact subset* K *of* G, *there exists a neighborhood* U *of* p *such that*

$$\big\|\alpha_t(x) - \langle t, p\rangle x\big\| \le \varepsilon\|x\|, \quad x \in X^\alpha(U), \quad t \in K.$$

PROOF: Let V be a compact neighborhood of p and choose $f \in A(\widehat{G})$ such that $f(q) = 1$ for every $q \in V$. If $U \subset V$, then $\alpha_f(x) = x,\ x \in X^\alpha(U)$. Now, put $f_s(q) = \langle s, q\rangle f(q)$ for each $s \in G$, and note that $\alpha_s\big(\alpha_f(x)\big) = \alpha_{f_s}(x),\ x \in X$. Define

$$g_s(q) = f_s(q) - \langle s, p\rangle f(q).$$

It follows that

$$\alpha_s\big(\alpha_f(x)\big) - \langle s, p\rangle \alpha_f(x) = \alpha_{g_s}(x), \quad x \in X.$$

Since $g_s(p) = 0$ for each $s \in G$, there exists an $h_s \in A(\widehat{G})$ and a neighborhood U_s of p such that

$$\big\|h_s g_s\big\|_{A(\widehat{G})} < \varepsilon \quad \text{and} \quad h_s(q) = 1, \quad q \in U_s, \quad \|h_s\|_{A(\widehat{G})} < 2.$$

Since the map: $s \in G \mapsto g_s \in A(\widehat{G})$ is continuous, for each $s \in K$, we can choose a neighborhood W_s of s so that

$$\big\|h_s g_t\big\|_{A(\widehat{G})} < \varepsilon, \quad t \in W_s.$$

By the compactness of K, we can choose $s_1, s_2, \dots, s_n \in K$ so that $K \subset \bigcup_{i=1}^n W_{s_i}$. Let $U = \bigcap_{j=1}^n U_{s_j}$. For each $x \in X^\alpha(U)$, we have $\alpha_{h_{s_j}}(x) = x, \ 1 \le j \le n$, and for any $t \in W_{s_j}$, we have

$$\begin{aligned}\big\|\langle t, p\rangle x - \alpha_t(x)\big\| &= \big\|\langle t, p\rangle \alpha_{h_{s_j}}(x) - \alpha_t \circ \alpha_{h_{s_j}}(x)\big\| \\ &= \big\|\alpha_{h_{s_j}} \circ \alpha_{g_t}(x)\big\| \le \big\|h_{s_j} g_t\big\|_{A(\widehat{G})} \|x\| < \varepsilon \|x\|.\end{aligned}$$

Thus, we obtain

$$\big\|\langle t, p\rangle x - \alpha_t(x)\big\| < \varepsilon \|x\|, \quad t \in K, \quad x \in X^\alpha(U). \qquad \text{Q.E.D.}$$

Lemma 1.12. *Under the same setting as before, the next four conditions on* $p \in \widehat{G}$ *are equivalent:*

(i) $p \in \mathrm{Sp}(\alpha)$;

(ii) *There exists a net* $\{x_i\}$ *in* X *with* $\|x_i\| = 1$ *such that the limit:*

$$\lim_i \big\|\alpha_t(x_i) - \langle t, p\rangle x_i\big\| = 0$$

converges uniformly on every compact subset of G;

(iii) *For every finite measure* μ *on* G,

$$\big|\hat{\mu}(p)\big| \le \|\alpha_\mu\|, \tag{13}$$

where $\hat{\mu}$ *and* α_μ *are defined by the integrals:*

$$\hat{\mu}(q) = \int_G \overline{\langle s, q\rangle}\, d\mu(s), \qquad \alpha_\mu(x) = \int_G \alpha_{-s}(x)\, d\mu(s);$$

(iv) *For each* $f \in A(\widehat{G})$,

$$\big|f(p)\big| \le \|\alpha_f\|. \tag{13'}$$

PROOF:

(i) $\Longrightarrow$ (ii): This follows directly from the previous lemma.

(ii) $\Longrightarrow$ (iii): For a compact $K \subset G$, we have

$$\left\|\alpha_\mu(x_i) - \hat{\mu}(p)x_i\right\| = \left\| \int_G \left(\alpha_{-t}(x_i) - \overline{\langle t, p\rangle} x_i\right) d\mu(t) \right\|$$

$$\leq \sup_{t\in K} \left\|\alpha_t(x_i) - \langle t, p\rangle x_i\right\| + 2|\mu|(K^c).$$

By assumption (ii) and the regularity of the measure μ, we have

$$\left|\hat{\mu}(p)\right| - \varepsilon \leq \left\|\alpha_\mu(x_i)\right\| \leq \|\alpha_\mu\|$$

for any preassigned $\varepsilon > 0$ and a large enough i. Hence we conclude (13).

(iii) $\Longrightarrow$ (iv): For each $f \in A(\widehat{G})$, with $d\mu(t) = \tilde{f}(t)\,dt$ the inequality (13) contains (13′) as a special case.

(iv) $\Longrightarrow$ (i): With $I(\alpha) = \left\{ f \in A(\widehat{G}) : \alpha_f = 0 \right\}$, $\mathrm{Sp}(\alpha)$ is the hull of $I(\alpha)$. If p satisfies (13′), then p is in the hull of $I(\alpha)$, so that $p \in \mathrm{Sp}(\alpha)$. Q.E.D.

Lemma 1.13. *For a fixed $s \in G$, the spectrum $\mathrm{Sp}_{\mathcal{L}(X)}(\alpha_s)$ of α_s in the Banach algebra $\mathcal{L}(X)$ is precisely the closure of $\left\{ \langle s, p\rangle : p \in \mathrm{Sp}(\alpha) \right\}$.*

PROOF: Set $S = \mathrm{Sp}_{\mathcal{L}(X)}(\alpha_s) \subset \mathbf{C}$. Since α_s is an isometry of X, $S \subset \mathbf{T} = \left\{ z \in \mathbf{C} : |z| = 1 \right\}$. If $p \in \mathrm{Sp}(\alpha)$, then by (ii) of the previous lemma, we have

$$\lim \left\|\alpha_s(x_i) - \langle s, p\rangle x_i\right\| = 0, \qquad \|x_i\| = 1.$$

Hence $\langle s, p\rangle \in S$.

Let $K = \left\{ \langle s, p\rangle : p \in \mathrm{Sp}(\alpha) \right\}^{-} \subset \mathbf{T}$ and $\lambda \notin K$. Note that $\mathbf{T} = \widehat{\mathbf{Z}}$ and $C^\infty(\mathbf{T}) \subset A(\mathbf{T})$. We extend the function: $z \in K \mapsto (z - \lambda)^{-1} \in \mathbf{C}$ to a function $f \in C^\infty(\mathbf{T})$. Then f is expanded to an absolutely convergent Fourier series: $f(z) = \sum_{n\in\mathbf{Z}} a_n z^n$, $\sum_{n\in\mathbf{Z}} |a_n| = \|f\|_{A(\mathbf{T})} < +\infty$. Put

$$T = \sum_{n\in\mathbf{Z}} a_n \alpha_{ns} \in \mathcal{L}_w(X), \qquad \mu = \sum_{n\in\mathbf{Z}} a_n \delta_{ns} \in M^1(G),$$

where δ_t, $t \in G$, means the Dirac measure at t. Then we have $T = \alpha_\mu$. With $\nu = \left(\delta_s - \lambda\delta_0\right) * \mu$ we have

$$T(\alpha_s - \lambda) = (\alpha_s - \lambda)T = \alpha_\nu$$

and

$$\hat{\nu}(p) = \left(\langle s, p\rangle - \lambda\right) f\left(\langle s, p\rangle\right) = 1, \quad p \in \mathrm{Sp}(\alpha).$$

Hence $\alpha_\nu = 1$. Thus $\lambda \notin S$. Q.E.D.

Corollary 1.14. *If an isometry σ of X has spectrum $\mathrm{Sp}_{\mathcal{L}(X)}(\sigma) = \{1\}$, then $\sigma = 1$.*

PROOF: Consider the action α of $\mathbf{Z}$: $n \in \mathbf{Z} \mapsto \sigma^n = \alpha_n$, and apply the last lemma to α. Q.E.D.

Corollary 1.15. *Given $\{X, G, \alpha\}$ as before, consider the Banach algebra $\mathfrak{A}$ generated by $\{\alpha_f \in \mathcal{L}(X) : f \in A(\widehat{G})\}$. Then $\mathrm{Sp}(\alpha)$ can be naturally viewed as the spectrum $\mathrm{Sp}(\mathfrak{A})$ of the commutative Banach algebra $\mathfrak{A}$.*

PROOF: If $p \in \mathrm{Sp}(\alpha)$, then the map: $f \in A(\widehat{G}) \mapsto f(p) \in \mathbf{C}$ can be viewed as a character $\omega_p\colon \alpha_f \in \mathfrak{A} \mapsto f(p) \in \mathbf{C}$ by (13′). Conversely, if $\omega \in \mathrm{Sp}(\mathfrak{A})$, then the map: $f \in A(\widehat{G}) \mapsto \omega(\alpha_f)$ is a character of $A(\widehat{G})$. Since $\widehat{G} = \mathrm{Sp}(A(\widehat{G}))$, ω is of the form $\omega = \omega_p$ for some $p \in \widehat{G}$. Then by Lemma 1.12, p belongs to $\mathrm{Sp}(\alpha)$. Since $\{\alpha_f : f \in A(\widehat{G})\}$ is dense in $\mathfrak{A}$, the map: $p \in \mathrm{Sp}(\alpha) \mapsto \omega_p \in \mathrm{Sp}(\mathfrak{A})$ is a homeomorphism. Q.E.D.

Corollary 1.16. *For $\{X, G, \alpha\}$, the following two conditions are equivalent:*

(i) *The map: $s \in G \mapsto \alpha_s \in \mathcal{L}(X)$ is continuous in norm;*
(ii) *$\mathrm{Sp}(\alpha)$ is compact.*

PROOF:
(i) $\Longrightarrow$ (ii): Let $\{f_U\} \subset A(\widehat{G})$ be the system chosen in Lemma 1.3.(i). If α is continuous in norm, then the proof of the same lemma shows that α_{f_U} converges to the identity operator 1 in norm. Hence 1 belongs to $\mathfrak{A}$, the Banach algebra generated by $\{\alpha_f : f \in A(\widehat{G})\}$. Hence $\mathrm{Sp}(\mathfrak{A})$ is compact. The last corollary then implies the compactness of $\mathrm{Sp}(\alpha)$.

(ii) $\Longrightarrow$ (i): Let $K = \mathrm{Sp}(\alpha)$, and choose $f \in A(\widehat{G})$ such that $f(p) = 1$ on a neighborhood U of K. Then $\alpha_f = 1$ by Lemma 1.3.(iii). Let $f_s(q) = \langle s, q\rangle f(q)$ for $s \in G$ and $q \in \widehat{G}$. We then have $\alpha_s = \alpha_s\alpha_f = \alpha_{f_s}$. Since $s \in G \mapsto f_s \in A(\widehat{G})$ is continuous in norm, the map: $s \in G \mapsto \alpha_s = \alpha_{f_s} \in \mathcal{L}(X)$ is continuous in norm. Q.E.D.

We now consider the case $G = \mathbf{R}$, the additive group of real numbers. To treat the norm continuous case, we consider here a one parameter group of invertible elements of a Banach algebra.

Lemma 1.17. *Let A be a unital Banach algebra. If $h \in A$, then*

$$g(t) = \exp(th) = \sum_{n=0}^{\infty} \frac{(th)^n}{n!}, \quad t \in \mathbf{R}, \tag{14}$$

is a norm continuous one parameter group of invertible elements of A. Furthermore, it is differentiable in norm in the sense that

$$\lim_{\varepsilon\to 0}\left\|\frac{1}{\varepsilon}\big(g(t+\varepsilon)-g(t)\big)-hg(t)\right\|=0. \tag{15}$$

Conversely, if $\{g(t)\}$ *is a norm continuous one parameter subgroup of* $\mathrm{GL}(A)$*, then* $\{g(t)\}$ *is of the above form, i.e.* $g(t)=\exp(th)$ *for some* $h\in A$.

If $\|g(t)\|=1$, $t\in\mathbf{R}$*, then the spectrum of* h *is pure imaginary.*

PROOF: A direct computation as in the elementary calculus shows that (15) follows from (14).

Suppose $\{g(t)\}$ is a norm continuous one parameter subgroup of $\mathrm{GL}(A)$. Choose $\delta>0$ so that $\|g(t)-1\|<1$ for $|t|<\delta$. Then $\mathrm{Sp}_A\big(g(t)\big)$ is contained in the unit disk around the center 1, in particular, contained in the right half plane. Let Log be the principal logarithm there and set $h(t)=\mathrm{Log}\,g(t)$, $|t|<\delta$. Since the $g(t)$'s are commutative, we have $h(s+t)=h(s)+h(t)$ if s,t and $s+t$ are all in $]-\delta,\delta[$. Furthermore, $h(\,\cdot\,)$ is continuous on $]-\delta,\delta[$. Thus, we can extend $h(\,\cdot\,)$ uniquely to the entire $\mathbf{R}$ by

$$h(t)=nh(t/n)$$

for sufficiently large n. Note that $h(\,\cdot\,)$ is well-defined. Put $h=h(1)$. Since $\exp(\mathrm{Log}\,z)=z$ for $z\in\mathrm{Sp}_A\big(g(t)\big)$, $|t|<\delta$, we have $\exp(th)=g(t)$ for $|t|<\delta$ by the spectral mapping theorem. Then the group property of g implies that $g(t)=\exp(th)$ for every $t\in\mathbf{R}$.

Suppose now $\|g(t)\|=1$ for every $t\in\mathbf{R}$. Then we have $\mathrm{Sp}_A\big(g(t)\big)\subset\{z\in\mathbf{C}:|z|\le 1\}=\mathbf{D}$. Since $g(-t)=g(t)^{-1}$, the spectral mapping theorem yields that $\mathrm{Sp}_A\big(g(t)\big)\subset\mathbf{T}$. Hence $\mathrm{Sp}_A\big(h(t)\big)\subset i\mathbf{R}$, $|t|<\delta$, so that $\mathrm{Sp}_A(h)=\frac{1}{t}\,\mathrm{Sp}_A\big(h(t)\big)\subset \mathrm{i}\mathbf{R}$, $|t|<\delta$, $t\neq 0$. Q.E.D.

We now return to the case $\{X,X_*,\mathbf{R},\alpha\}$. To proceed, we fix the pairing of $\mathbf{R}$ and $\widehat{\mathbf{R}}$ by

$$\langle s,p\rangle=\mathrm{e}^{\mathrm{i}sp},\quad s,p\in\mathbf{R}. \tag{16}$$

Although this pairing identifies $\mathbf{R}$ and its dual $\widehat{\mathbf{R}}$, we often write $\widehat{\mathbf{R}}$ in order to distinguish the role of $\mathbf{R}$ and $\widehat{\mathbf{R}}$. We now consider the differential operator δ defined as follows: The domain $\mathcal{D}(\delta)=\mathcal{D}$ consists of all those $x\in X$ such that the function: $t\in\mathbf{R}\mapsto\langle\alpha_t(x),\varphi\rangle$ is differentiable for every $\varphi\in X_*$ and $\frac{\mathrm{d}}{\mathrm{d}t}\langle\alpha_t(x),\varphi\rangle\big|_{t=0}$ is a bounded functional of φ, so that there exists $\delta(x)\in X$ such that

$$\langle\delta(x),\varphi\rangle=\frac{\mathrm{d}}{\mathrm{d}t}\langle\alpha_t(x),\varphi\rangle\bigg|_{t=0},\quad\varphi\in X_*,\quad x\in\mathcal{D}. \tag{17}$$

It is easy to see that $\mathcal{D}$ is weakly* dense in X and δ is a weakly* closed operator on X.

Definition 1.18. We call δ the *infinitesimal generator* of $\{\alpha_t\}$.

Lemma 1.19. *The spectrum of* δ *is pure imaginary.*

PROOF: If we replace $\{\alpha_t\}$ by $\{e^{i\beta t}\alpha_t\}$ with $\beta \in \mathbf{R}$, then δ is replaced by $\delta + i\beta$. Hence to show the claim, it suffices to prove that non-zero real numbers are resolvents of δ. Let $\lambda \in \mathbf{R}$. Suppose $\lambda > 0$. Consider the Laplace transform:

$$R_\lambda = \int_0^\infty e^{-\lambda t}\alpha_t \, dt, \quad \lambda > 0. \tag{18}$$

For every $x \in X$, we have for each $s \in \mathbf{R}$,

$$\begin{aligned} \alpha_s R_\lambda(x) &= \int_s^\infty e^{-\lambda(t-s)}\alpha_t(x)\, dt \\ &= e^{\lambda s}\int_0^\infty e^{-\lambda t}\alpha_t(x)\, dt - e^{\lambda s}\int_0^s e^{-\lambda t}\alpha_t(x)\, dt \\ &= e^{\lambda s}R_\lambda(x) - e^{\lambda s}\int_0^s e^{-\lambda t}\alpha_t(x)\, dt\, . \end{aligned}$$

Hence $R_\lambda x \in \mathcal{D}$ and we have

$$\delta R_\lambda(x) = \lambda R_\lambda(x) - x, \quad x \in X.$$

Thus, R_λ is the right inverse of $\lambda - \delta$. Since R_λ and α_t commute, R_λ is also the left inverse in the sense that if $x \in \mathcal{D}$, then

$$R_\lambda(\lambda - \delta)(x) = (\lambda - \delta)R_\lambda(x) = x.$$

Hence $R_\lambda = (\lambda - \delta)^{-1}$. If $\lambda < 0$, then we consider

$$R_\lambda = \int_0^\infty e^{\lambda t}\alpha_{-t}\, dt,$$

and find $\delta R_\lambda(x) = \lambda R_\lambda(x) - x$, $x \in X$. Thus we conclude that $R_\lambda = (\lambda - \delta)^{-1}$.

Q.E.D.

In many applications, it so happens that a candidate for the infinitesimal generator of an unknown one parameter group is given and we are asked if the candidate is indeed a generator or not. For this reason, the following theorem, called Hille-Yosida Theorem, is important. But it is treated in many standard textbooks in functional analysis, so we omit the proof.

Theorem 1.20. *Let $\{X, X_*\}$ be a dual pair of Banach spaces, and δ a weakly* closed operator on X with weakly* dense domain $\mathcal{D}$. Then the following two conditions are equivalent:*

(i) *The operator δ is the infinitesimal generator of a weakly* continuous one parameter group $\{\alpha_t\}$ of isometries in $\mathcal{L}_w(X)$;*

(ii) *For* $s \in \mathbf{R}$, *we have*

$$\|(1 - s\delta)x\| \geq \|x\|, \quad x \in \mathcal{D}$$

and for some $r_1, r_2 > 0$

$$(1 + r_1\delta)\mathcal{D} = X \quad \text{and} \quad (1 - r_2\delta)\mathcal{D} = X.$$

If condition (ii) *holds, then* $(1 - s\delta)^{-1}$ *exists for every* $s \in \mathbf{R}$ *and the group* $\{\alpha_t\}$ *is given by either of the limits:*

$$\alpha_t(x) = \lim_{\varepsilon \to 0} \exp\{t\delta(1 - \varepsilon\delta)^{-1}\}(x) = \lim_{n \to \infty} \left(1 - \frac{t}{n}\delta\right)^{-n}(x), \quad x \in X, \tag{19}$$

where the limit converges in the weak topology uniformly for* t *in compact sets. If* x *belongs to the norm closure of* $\mathcal{D}$, *then the limit converges in norm.*

We now want to show $\mathrm{Sp}(\delta) = \mathrm{i}\,\mathrm{Sp}(\alpha)$. To this end, we concentrate ourselves on the canonical action of $\mathbf{R}$ on $A(\widehat{\mathbf{R}})$. Namely, we define the action ρ of $\mathbf{R}$ on $A(\widehat{\mathbf{R}})$ by:

$$(\rho_s f)(p) = \mathrm{e}^{\mathrm{i}sp} f(p), \quad f \in A(\widehat{\mathbf{R}}), \quad s, p \in \mathbf{R}. \tag{20}$$

Namely, ρ is the Fourier transform of the translation of $L^1(\mathbf{R})$: $f \mapsto \rho_s f$ with $(\rho_s f)(t) = f(s + t)$. Furthermore, if $\{X, X^*, \mathbf{R}, \alpha\}$ is given, then we have

$$\alpha_s \alpha_f = \alpha_{\rho_s f}, \quad f \in A(\widehat{\mathbf{R}}). \tag{21}$$

In order to study further, we prepare some lemmas:

Lemma 1.21. *For* $\varepsilon > 0$, *let* ψ *be a* C^∞*-function on* $\widehat{\mathbf{R}}$ *such that the restriction of* ψ *on* $]-\infty, -\varepsilon] \cup [\varepsilon, +\infty[$ *is* $1/p$, *i.e.* $\psi(p) = 1/p$ *if* $|p| \geq \varepsilon$. *Then* $\psi \in A(\widehat{\mathbf{R}})$.

Proof: We know that ψ and all its derivatives $\psi^{(n)}$ are square integrable with respect to the Lebesgue measure. Let $\widehat{\mathcal{F}}$ be the extended Fourier inverse transform on $L^2(\widehat{\mathbf{R}})$ and put $f = \widehat{\mathcal{F}}\psi$ and $f_n = \widehat{\mathcal{F}}\psi^{(n)}$. Then we have $f_n(s) = (-\mathrm{i}s)^n f(s)$, $s \in \mathbf{R}$. Let $g = f - f_2$. Then we have $f(s) = g(s)/(1 + s^2)$. Since $g \in L^2(\mathbf{R})$, we have $f \in L^2(\mathbf{R}) \cap L^1(\mathbf{R})$. Thus $\psi = \mathcal{F} f \in A(\widehat{\mathbf{R}})$. Q.E.D.

Corollary 1.22. *For a closed subset* S *of* $\widehat{\mathbf{R}}$, *let* $A(S)$ *denote the quotient algebra* $A(\widehat{\mathbf{R}})/I(S)$, *where* $I(S)$ *denotes the ideal of* $A(\widehat{\mathbf{R}})$ *consisting of all those* $f \in A(\widehat{\mathbf{R}})$ *vanishing on* S. *The canonical action* ρ *of* $\mathbf{R}$ *on* $A(\widehat{\mathbf{R}})$ *induces naturally an action* ρ^S *of* $\mathbf{R}$ *on* $A(S)$. *Let* δ_S *be the generator of* ρ^S. *Then we have* $\mathrm{Sp}(\delta_S) = \mathrm{i}S$.

Proof: We know that $\mathrm{Sp}(\rho^S) = S$ and $\mathrm{Sp}(A(S)) = S$. We consider $A(S)$ as a function algebra on S. Then we have $(\rho_t^S f)(p) = \mathrm{e}^{\mathrm{i}pt} f(p)$, $f \in A(S)$. Hence the generator δ_S is the multiplication operator on $A(S)$ by the function: $p \in S \mapsto \mathrm{i}p \in \mathbf{C}$. Hence we have $\mathrm{i}S \subset \mathrm{Sp}(\delta_S)$.

Suppose $p \notin S$. Then there exists $\varepsilon > 0$ such that $S \subset]-\infty, p-\varepsilon] \cup [p+\varepsilon, +\infty[$. For this ε, we consider a C^∞-function ψ on $\mathbf{R}$ such that $\psi(q) = 1/q$ if $|q| \geq \varepsilon/2$. By the last lemma, $\psi \in A(\widehat{\mathbf{R}})$. Let $f(q) = \psi(q-p)$. Then $f \in A(\widehat{\mathbf{R}})$ and $f(q) = 1/(q-p)$ if $|q-p| \geq \varepsilon/2$. Let g be an element of $A(\widehat{\mathbf{R}})$ such that $g(q) = 1$ if $|q-p| \leq \varepsilon/2$ and $g(q) = 0$ if $|q-p| \geq \varepsilon$. Then $f(1-g)A(\widehat{\mathbf{R}}) \subset \mathcal{D}(\delta)$ and

$$\left\{(\delta - \mathrm{i}p)\big(f(1-g)\big)h\right\}(q) = \mathrm{i}\,(1-g)(q)h(q), \quad h \in A(\widehat{\mathbf{R}}).$$

Hence on $A(S)$, the multiplication by f/i is the inverse of $\delta_S - \mathrm{i}p$. Hence $\mathrm{i}p$ does not belong to $\mathrm{Sp}(\delta_S)$. Q.E.D.

Now, we return to the study of $\{X, X^*, \mathbf{R}, \alpha\}$. Let δ_α be the infinitesimal generator of α and δ_ρ the same for ρ on $A(\widehat{\mathbf{R}})$. It then follows from (21) that

$$\delta_\alpha \alpha_f = \alpha_{\delta_\rho f}, \quad f \in \mathcal{D}(\delta_\rho). \tag{21$'$}$$

Theorem 1.23. *In the above setting, we have*

$$\mathrm{Sp}(\delta_\alpha) = \mathrm{i}\,\mathrm{Sp}(\alpha).$$

Proof: Let $S = \mathrm{Sp}(\alpha) \subset \widehat{\mathbf{R}}$. Let $p \in S$ be fixed. For any $\varepsilon > 0$, choose $x_\varepsilon \in X^\alpha(p-\varepsilon, p+\varepsilon)$ with $\|x_\varepsilon\| = 1$. Let f_ε be a C^∞-function supported by $[p-2\varepsilon, p+2\varepsilon]$ such that $f_\varepsilon(q) = 1$ for $|q-p| \leq 3\varepsilon/2$. Then we know $\alpha_{f_\varepsilon}(x_\varepsilon) = x_\varepsilon$ and $f_\varepsilon \in \mathcal{D}(\delta_\rho)$. We then have

$$\lim_{\varepsilon \to 0} \big(f_{\varepsilon/2}\delta_\rho f_\varepsilon - \mathrm{i}pf_\varepsilon \cdot f_{\varepsilon/2}\big) = 0$$

in the Schwartz space $\mathcal{S}(\widehat{\mathbf{R}})$ of rapidly decreasing smooth functions. The embedding of $\mathcal{S}(\widehat{\mathbf{R}})$ in $A(\widehat{\mathbf{R}})$ is continuous, so that we get

$$\lim_{\varepsilon \to 0} \|\delta_\alpha x_\varepsilon - \mathrm{i}px_\varepsilon\| = \lim_{\varepsilon \to 0} \|\alpha_{(f_{\varepsilon/2}\cdot\delta_\rho f_\varepsilon - \mathrm{i}pf_\varepsilon \cdot f_{\varepsilon/2})}(x)\| = 0.$$

Hence $\mathrm{i}p \in \mathrm{Sp}(\delta_\alpha)$.

By Lemma 1.19, we know that $\mathrm{Sp}(\delta_\alpha)$ is pure imaginary. Let $p \in \widehat{\mathbf{R}} \setminus S$. By Corollary 1.22, we know $\mathrm{Sp}(\delta_S) = \mathrm{i}S$. Let $f \in A(\widehat{\mathbf{R}})$ such that

$$\big(\delta_\rho - \mathrm{i}p\big)fg = g, \quad g \in A(\widehat{\mathbf{R}}).$$

By (21$'$), we have

$$\big(\delta_\alpha - \mathrm{i}p\big)\alpha_f\alpha_g = \alpha_g, \quad g \in A(\widehat{\mathbf{R}}).$$

Since $\delta_\alpha - \mathrm{i}p$ is weakly* closed and $\{\alpha_g\}$ contains a weak* approximate identity in $\mathcal{L}_w(X)$, the above equality means that $\alpha_f(X) \subset \mathcal{D}(\delta_\alpha)$ and α_f is the right inverse of δ_α. Since α_f and δ_α commute, we conclude $\alpha_f = (\delta_\alpha - \mathrm{i}p)^{-1}$. Q.E.D.

Returning to the covariant system $\{\mathcal{M}, G, \alpha\}$, we prove the following result to conclude this section:

Proposition 1.24. *If a covariant system* $\{\mathcal{M}, G, \alpha\}$ *admits an invariant faithful weight, then the standard form* $\{\mathcal{M}, \mathfrak{H}, J, \mathcal{P}, U\}$ *realizes the spectrum of* α *in the sense that* $\mathrm{Sp}(U) = \mathrm{Sp}(\alpha)$.

PROOF: Let φ be a faithful invariant weight, and $\mathfrak{A}$ be the associated left Hilbert algebra. We represent $\mathcal{M}$ by $\mathcal{R}_\ell(\mathfrak{A})$. If $f \in A(\widehat{G})$ does not belong to $I(\alpha)$, then $\alpha_f \neq 0$. By the density of $\mathfrak{n}_\varphi$, we can find an $x \in \mathfrak{n}_\varphi$ such that $\alpha_f(x) \neq 0$. For each $\xi, \zeta \in \mathfrak{A}'$, we have

$$\begin{aligned}\big(\alpha_f(x)\xi \mid \zeta\big) &= \int_G \tilde{f}(-s)\big(\alpha_s(x)\xi \mid \zeta\big)\,ds = \int_G \tilde{f}(-s)\Big(\eta_\varphi\big(\alpha_s(x)\big) \Bigm| \zeta\xi^b\Big)\,ds \\ &= \int_G \tilde{f}(-s)\big(U(s)\eta_\varphi(x) \mid \zeta\xi^b\big)\,ds = \big(U(f)\eta_\varphi(x) \mid \zeta\xi^b\big),\end{aligned}$$

so that $\alpha_f(x) \in \mathfrak{n}_\varphi$ and

$$U(f)\eta_\varphi(x) = \eta_\varphi\big(\alpha_f(x)\big).$$

Since $\alpha_f(x) \neq 0$, $U(f)\eta_\varphi(x) \neq 0$. Hence $U(f) \neq 0$. Thus $f \notin I(U)$. Conversely, if $U(f) \neq 0$, then the density of $\eta_\varphi(\mathfrak{n}_\varphi)$ in $\mathfrak{H}_\varphi$ implies the existence of $x \in \mathfrak{n}_\varphi$ with $U(f)\eta_\varphi(x) \neq 0$, which in turn means $\alpha_f(x) \neq 0$. Thus $f \notin I(\alpha)$. Hence, we conclude that $I(U) = I(\alpha)$. Namely, $\mathrm{Sp}(U) = \mathrm{Sp}(\alpha)$. Q.E.D.

Exercise XI.1

1) Given a dual pair $\{X, X_*\}$, let $\{X, \mathbf{R}, \alpha\}$ be a one parameter group of isometries in $\mathcal{L}_w(X)$. Following the steps described below, show that if $\lim_{t\to 0}\|\alpha_t(x) - x\| = 0$ for every $x \in X$, then $\lim_{t\to 0}\|\alpha_t - \mathrm{id}\| = 0$.

(**a**) First let S be the closed unit ball of X. For an arbitrary $\varepsilon > 0$ and $n \in \mathbf{N}$, set $A_n(\varepsilon) = \Big\{x \in S : \sup\big\{\|\alpha_t(x) - x\| : |t| < \frac{1}{n}\big\} \le \varepsilon\Big\}$. By the weak* lower semi-continuity of the norm in X, $A_n(\varepsilon)$ is w*-closed.

(**b**) Observe that $\bigcup_{n\in\mathbf{N}} A_n(\varepsilon) = S$.

(**c**) Use the Baire category theorem to show that there exists some $m \in \mathbf{N}$ such that $A_m(\varepsilon)$ contains an interior point relative to the w*-topology, i.e. $A_m(\varepsilon) \supset (x_0 + V)$ for some $x_0 \in S$ and a w*-neighborhood V of 0. This means that $\|\alpha_t(x_0 + v) - (x_0 + v)\| \le \varepsilon$ for $|t| < \frac{1}{m}$ and $v \in V$; hence for every $v \in V$ and $t \in \left]-1/m, 1/m\right[$ we have :

$$\|\alpha_t(v) - v\| \le \|\alpha_t(x_0 + v) - (x_0 + v)\| + \|\alpha_t(x_0) - x_0\| \le 2\varepsilon.$$

(**d**) Use the weak* compactness of S to conclude the existence of finitely many $x_1, x_2, \ldots, x_k \in S$ such that $\bigcup_{i=1}^k (x_i + V) \supset S$.

(**e**) Show that there exists $\delta > 0$ such that $\|\alpha_t(x) - x\| \le 3\varepsilon$ for every $x \in S$ whenever $|t| < \delta$. This means that $\|\alpha_t - \mathrm{id}\| \le 3\varepsilon$ if $|t| < \delta$.

2) Let $\mathcal{A} = L^\infty(\mathbf{T})$ and $\alpha_t(f)(s) = f(s+t)$, $f \in \mathcal{A}$, $s, t \in \mathbf{T}$ where $\mathbf{T} = \mathbf{R}/\mathbf{Z}$ is written additively. Identify $\widehat{\mathbf{T}}$ with $\mathbf{Z}$ under the pairing: $\langle s, n\rangle = \exp(2\pi i n s)$, $n \in \mathbf{Z}$, $s \in \widehat{\mathbf{T}}$. Let u_0 be the element of $\mathcal{A}$ given by $u_0(s) = \exp(2\pi i s)$, $s \in \widehat{\mathbf{T}}$. Show that $\mathcal{A}^\alpha(n) = \mathbf{C}u_0^n$, $n \in \mathbf{Z}$.

3) Let $\{\mathcal{M}, \mathbf{T}, \alpha\}$ be a covariant system such that $\mathrm{Ker}(\alpha) = \{0\}$ and $\mathcal{M}^\alpha = \mathbf{C}$, i.e. an ergodic covariant system over $\mathbf{T}$.

(a) Fix an integer $n \in \mathbf{Z}$. Show that if $x \in \mathcal{M}^\alpha(n)$, then $x^*x = xx^* = \|x\|^2 \in \mathbf{R}_+$, so that if $\|x\| = 1$ in addition, then $x \in \mathcal{U}(\mathcal{M})$.

(b) Show that if $\mathcal{M}^\alpha(n) \neq \{0\}$, then there exists $u_n \in \mathcal{M}^\alpha(n) \cap \mathcal{U}(\mathcal{M})$ such that $\mathcal{M}^\alpha(n) = \mathbf{C}u_n$.

(c) Show that $J = \{n \in \mathbf{Z} : \mathcal{M}^\alpha(n) \neq \{0\}\}$ is a subgroup of $\mathbf{Z}$; hence of the form $J = N\mathbf{Z}$ for some $N \in \mathbf{Z}_+$.

(d) Show that $J = Z$. (Hint: Use the faithfulness of α.)

(e) Show that $u_1 \in \mathcal{M}^\alpha(1) \cap \mathcal{U}(\mathcal{M})$ generates $\mathcal{M}$ and $\mathcal{M}^\alpha(n) = \mathbf{C}u_1^n$.

(f) Show that the system $\{\mathcal{M}, \mathbf{T}, \alpha\}$ is conjugate to the system $\{\mathcal{A}, \mathbf{T}, \alpha\}$ in Problem 1.

4) Try to generalize Problem 2 to a general compact abelian group or to $\mathbf{T}^2$ and see what happens.

5) A covariant system $\{\mathcal{M}, G, \alpha\}$ is called *homogeneous* if $\mathrm{Aut}_\alpha(\mathcal{M}) = \{\sigma \in \mathrm{Aut}(\mathcal{M}) : \sigma \circ \alpha_s = \alpha_s \circ \sigma,\ s \in G\}$ is ergodic on $\mathcal{M}$. An ergodic action of an abelian group is always homogeneous. Suppose that the covariant system $\{\mathcal{M}, G, \alpha\}$ is homogeneous and G is compact and abelian. Following the steps indicated below, show that if $\sigma \in \mathrm{Aut}(\mathcal{M})$ commutes with $\mathrm{Aut}_\alpha(\mathcal{M})$ and leaves the fixed point algebra $\mathcal{M}^\alpha$ pointwise invariant, then σ belongs to the group $\alpha(G)$. Assume that $\mathcal{M}$ is σ-finite to avoid unnecessary complications.

(a) By considering the quotient group $G/\mathrm{Ker}(\alpha)$, we may assume that the action α of G is faithful.

(b) Show that $\mathrm{Sp}(\alpha)$ is the entire dual group $\widehat{G}$.

(c) Fix $p \in \widehat{G}$. Show that $\sup\{s_\ell(a) : a \in \mathcal{M}^\alpha(p)\} = 1$ and $\sup\{s_r(a) : a \in \mathcal{M}^\alpha(p)\} = 1$.

(d) Keep $p \in \widehat{G}$ fixed. Consider the 2×2 matrix algebra $\mathcal{M}_2 = \mathcal{M} \overline{\otimes} M_2(\mathbf{C})$ and the action α^p defined by:

$$\alpha_s^p \begin{pmatrix} x_{11} & x_{12} \\ x_{21} & x_{22} \end{pmatrix} = \begin{pmatrix} \alpha_s(x_{11}) & \langle s, p\rangle \alpha_s(x_{12}) \\ \overline{\langle s, p\rangle} \alpha_s(x_{21}) & \alpha_s(x_{22}) \end{pmatrix}, \quad s \in G, \quad (x_{ij}) \in \mathcal{M}_2.$$

With $\{e_{ij}\}$ the standard matrix unit in $M_2(\mathbf{C})$, prove that $x \in \mathcal{M}$ belongs to $\mathcal{M}^\alpha(p)$ if and only if $x \otimes e_{21}$ belongs to the fixed point algebra $\mathcal{M}_2^{\alpha^p}$ and that the existence of a unitary $u \in \mathcal{M}^\alpha(p)$ is equivalent to $1 \otimes e_{11} \sim 1 \otimes e_{22}$ in $\mathcal{M}_2^{\alpha^p}$. Observe that $\mathcal{M}_2^{\alpha^p}$ has the matrix decomposition:

$$\mathcal{M}_2^{\alpha^p} = \begin{pmatrix} \mathcal{M}^\alpha & \mathcal{M}^\alpha(-p) \\ \mathcal{M}^\alpha(p) & \mathcal{M}^\alpha \end{pmatrix}.$$

(e) Show that $c_{\mathcal{M}_2^{\alpha^p}}(1 \otimes e_{11}) = c_{\mathcal{M}_2^{\alpha^p}}(1 \otimes e_{22})$ where $c_{\mathcal{M}_2^\alpha}(e),\ e \in \operatorname{Proj}(\mathcal{M}_2^{\alpha^p})$, means the central support of $e \in \operatorname{Proj}(\mathcal{M}_2^{\alpha^p})$ in $\mathcal{M}_2^{\alpha^p}$. (Hint: Use the ergodicity of $\operatorname{Aut}_\alpha(\mathcal{M})$ on $\mathcal{M}$.)

(f) Assume the σ-finiteness of $\mathcal{M}$ and the proper infiniteness of $\mathcal{M}^\alpha$. Show that $\mathcal{M}^\alpha(p) \cap \mathcal{U}(\mathcal{M}) \neq \emptyset$ and that if $u \in \mathcal{M}^\alpha(p) \cap \mathcal{U}(\mathcal{M})$ then $\mathcal{M}^\alpha(p) = u\mathcal{M}^\alpha$.

(g) Keep the assumption in (f). Show that there exists a scalar $\lambda(p) \in \mathbf{T}$ such that $\sigma(x) = \lambda(p)x,\ x \in \mathcal{M}^\alpha(p)$. (Hint: Choose $u \in \mathcal{M}^\alpha(p) \cap \mathcal{U}(\mathcal{M})$ and set $v = u^*\sigma(u)$. For $\beta \in \operatorname{Aut}_\alpha(\mathcal{M})$, compute:

$$\begin{aligned} v^*\beta(v) &= \sigma(u^*)\,u\,\beta(u^*)\,\beta \circ \sigma(u) = \sigma(u^*)\,u\,\beta(u^*)\sigma \circ \beta(u) \\ &= \sigma\Big(u^*\big(u\beta(u^*)\big)\Big)\sigma \circ \beta(u) \qquad \big(\text{as } u\beta(u^*) \in \mathcal{M}^\alpha\big) \\ &= 1. \end{aligned}$$

Hence the ergodicity of $\operatorname{Aut}_\alpha(\mathcal{M})$ implies that $v \in \mathbf{T}$.

(h) Continue (g). Show that $\lambda\colon p \in \widehat{G} \mapsto \mathbf{T}$ is a character of $\widehat{G}$; hence of the form that $\lambda(p) = \langle s_0, p\rangle$ for some $s_0 \in G$. Prove that $\sigma = \alpha_{s_0}$.

(i) Prove the claim that $\sigma \in \alpha(G)$ by considering the system $\big\{\mathcal{M}\overline{\otimes}\mathcal{L}(\ell^2), G, \alpha\otimes\mathrm{id}\big\}$ and $\sigma \otimes \mathrm{id}$.

6) Let G be a compact abelian group. Show that if $\big\{\mathcal{M}, G, \alpha\big\}$ is an ergodic covariant system, then the integral:

$$\tau(x) = \int_G \alpha_s(x)\,\mathrm{d}s, \qquad x \in \mathcal{M}$$

gives a normalized faithful normal trace τ on $\mathcal{M}$.

7) Let $\big\{\mathcal{M}, G, \alpha\big\}$ be a covariant system with G a general locally compact group. Let $\mathcal{N} = \mathcal{M} \rtimes_\alpha G$. Consider the action $\hat{\alpha}$ of $A(G)$ on $\mathcal{N}$ defined in Exercise X.2.10. For each $x \in \mathcal{N}$, let $\operatorname{Sp}_{\hat{\alpha}}(x) =$ the hull of $I_{\hat{\alpha}}(x)$ with $I_{\hat{\alpha}}(x) = \big\{\varphi : \widehat{\alpha_\varphi}(x) = 0\big\}$, i.e. $\operatorname{Sp}_{\hat{\alpha}}(x) = \big\{s \in G : \varphi(s) = 0 \text{ whenever } \widehat{\alpha_\varphi}(x) = 0\big\}$. For a closed subset $E \subset G$, set $\mathcal{N}^{\hat{\alpha}}(S) = \big\{x \in \mathcal{N} : \operatorname{Sp}_{\hat{\alpha}}(x) \subset E\big\}$.

(a) Show that if $\big\{u(s) : s \in G\big\}$ is the associated unitary representation of G in $\mathcal{N}$, then $\operatorname{Sp}_{\hat{\alpha}}\big(u(s)\big) = \{s\}$ and that $\mathcal{N}^{\hat{\alpha}}\big(\{s\}\big) = \mathcal{M}u(s),\ s \in G$.

(b) Show that $\operatorname{Sp}_{\hat{\alpha}}\big(\widehat{\alpha_\varphi}(x)\big) = \operatorname{supp}(\varphi) \cap \operatorname{Sp}_{\hat{\alpha}}(x),\ x \in \mathcal{N},\ \varphi \in A(G)$.

§ 2 Connes Spectrum $\Gamma(\alpha)$

We now come back to the study of a covariant system $\{\mathcal{M}, G, \alpha\}$ over a locally compact abelian group G. We continue to use the notations established above. If the reader is not familiar with locally compact abelian groups, one may assume that G is the additive group $\mathbf{R}$ of real numbers.

For each $e \in \text{Proj}(\mathcal{M}^\alpha)$, the restriction of α to the reduced algebra $\mathcal{M}_e$ gives rise to a new covariant system $\{\mathcal{M}_e, G, \alpha^e\}$, which we shall call the *reduced* covariant system or the *reduced* action of α, and denote it by α^e.

Definition 2.1. The intersection:

$$\Gamma(\alpha) = \bigcap \{\, \text{Sp}(\alpha^e) : e \in \text{Proj}(\mathcal{M}^\alpha),\ e \neq 0 \,\}$$

is called the *Connes spectrum* of α or the *essential spectrum* of α.

The importance of the Connes spectrum can be seen from the fact that $\Gamma(\alpha)$ is precisely the kernel of the restriction of the dual action $\hat{\alpha}$ of $\widehat{G}$ to the center $\mathcal{C}_{\mathcal{M} \rtimes_\alpha G}$ of $\mathcal{M} \rtimes_\alpha G$, Theorem 2.7.

Lemma 2.2.

(i) *For each* $e \in \text{Proj}(\mathcal{M}^\alpha)$, *let* $\overline{e}$ *denote the central support of* e *in* $\mathcal{M}^\alpha$. *Then for each closed* $E \subset \widehat{G}$,

$$\mathcal{M}^\alpha(E) \cap \mathcal{M}_e \neq \{0\} \iff \mathcal{M}^\alpha(E) \cap \mathcal{M}_{\overline{e}} \neq \{0\}.$$

(ii) $\Gamma(\alpha) = \bigcap \{\, \text{Sp}(\alpha^e) : e \in \text{Proj}(\mathcal{C}_{\mathcal{M}^\alpha}) \,\}$.

(iii) *If* $\mathcal{M}^\alpha$ *is a factor, then* $\Gamma(\alpha) = \text{Sp}(\alpha)$.

(iv) *If* $e, f \in \text{Proj}(\mathcal{M}^\alpha)$ *are equivalent in* $\mathcal{M}$, *then*

$$\Gamma(\alpha^e) = \Gamma(\alpha^f).$$

PROOF:

(i) We have $\overline{e} = \bigvee \{\, ueu^* : u \in \mathcal{U}(\mathcal{M}^\alpha) \,\}$. If $x \in \mathcal{M}^\alpha(E) \cap \mathcal{M}_{\overline{e}},\ x \neq 0$, then there exist $u, v \in \mathcal{U}(\mathcal{M}^\alpha)$ with $ueu^*xvev^* \neq 0$. With $y = eu^*xve \neq 0$, we have $y \in \mathcal{M}_e \cap \mathcal{M}^\alpha(E) = \mathcal{M}_e^\alpha(E)$.

The statements (ii) and (iii) follow immediately from (i).

(iv) By assumption, there exists $u \in \mathcal{M}$ such that $e = u^*u$ and $f = uu^*$. Suppose $p \in \Gamma(\alpha^e)$. For any closed neighborhood U of p and $f_1 \in \text{Proj}(\mathcal{M}_f^\alpha)$, we shall show that $\mathcal{M}^\alpha(U) \cap \mathcal{M}_{f_1} \neq \{0\}$. Choose neighborhoods W of $0 \in \widehat{G}$ and V of p such that $V + W \subset U$. Since $f_1 u \neq 0$, choose $q \in \text{Sp}_\alpha(f_1 u)$, and let U_1 be a neighborhood of q such that $U_1 - U_1 \subset W$. We then choose $g \in A(\widehat{G})$ with $\operatorname{supp} g \subset U_1$ and $\alpha_g(f_1 u) \neq 0$. Then $x = f_1 \alpha_g(u)$ has the property that $\text{Sp}_\alpha(x) \subset U_1$, $s_\ell(x) \leq f_1$ and $s_r(x) \leq e$. Put

$$e_1 = \bigvee \Big\{\, s_r\big(\alpha_t(x)\big) : t \in G \,\Big\} \in \text{Proj}(\mathcal{M}^\alpha).$$

Then $e \geq e_1 \neq 0$. By the assumption on p, there exists a nonzero $y \in \mathcal{M}^\alpha(V) \cap \mathcal{M}_{e_1}$. By the construction of e_1, we can find $t_1 \in G$ with $\alpha_{t_1}(x)y \neq 0$ and also $t_2 \in G$ with $z = \alpha_{t_1}(x)y\alpha_{t_2}(x)^* \neq 0$. Now, $z \in \mathcal{M}_{f_1}$ and

$$\mathrm{Sp}_\alpha(z) \subset \mathrm{Sp}_\alpha(x) + \mathrm{Sp}_\alpha(y) - \mathrm{Sp}_\alpha(x) \subset U_1 + V - U_1 \subset V + W \subset U.$$

Hence $\mathcal{M}^\alpha(U) \cap \mathcal{M}_{f_1} \neq \{0\}$, so that $p \in \mathrm{Sp}(\alpha^{f_1})$. Since $f_1 \in \mathrm{Proj}(\mathcal{M}_f^\alpha)$ was arbitrary, $p \in \Gamma(\alpha^f)$. Hence $\Gamma(\alpha^e) \subset \Gamma(\alpha^f)$. By symmetry, $\Gamma(\alpha^f) \subset \Gamma(\alpha^e)$, and so $\Gamma(\alpha^e) = \Gamma(\alpha^f)$. Q.E.D.

Lemma 2.3. *For each $u \in Z^1_\alpha(G, \mathcal{M})$, we put*

$$\gamma_t \begin{pmatrix} x_{11} & x_{12} \\ x_{21} & x_{22} \end{pmatrix} = \begin{pmatrix} \alpha_t(x_{11}) & \alpha_t(x_{12})u_t^* \\ u_t\alpha_t(x_{21}) & u_t\alpha_t(x_{22})u_t^* \end{pmatrix}, \quad t \in G, \tag{1}$$

for each $x = \begin{pmatrix} x_{11} & x_{12} \\ x_{21} & x_{22} \end{pmatrix} \in \mathcal{M}_2 = \mathcal{M} \otimes M(2; \mathbf{C})$. Then γ is an action of G on $\mathcal{M}_2$ and

$$e_{11} = \begin{pmatrix} 1 & 0 \\ 0 & 0 \end{pmatrix} \in \mathcal{M}_2^\gamma, \qquad e_{22} = \begin{pmatrix} 0 & 0 \\ 0 & 1 \end{pmatrix} \in \mathcal{M}_2^\gamma.$$

The proof is routine. So we leave it to the reader.

Definition 2.4.

(i) A *C^*-covariant system* means a triplet $\{A, G, \alpha\}$ of a C^*-algebra A, a locally compact group G and a homomorphism $\alpha : s \in G \mapsto \alpha_s \in \mathrm{Aut}(A)$ such that $\lim_{s\to t}\|\alpha_s(a) - \alpha_t(a)\| = 0, \ a \in A, \ t \in G$.

(ii) For a C^*-covariant system $\{A, G, \alpha\}$, a representation $\{\pi, \mathfrak{H}\}$ of A is said to be an *α-representation* if there exists an action $\tilde{\alpha}$ of G on the von Neumann algebra $\mathcal{M} = \pi(A)''$ such that $\tilde{\alpha}_s \circ \pi(a) = \pi \circ \alpha_s(a), \ a \in A, \ s \in G$. Here we demand the continuity of $\tilde{\alpha}$.

Note that the universal representation $\tilde{\pi}$ of A is not in general an α-representation because of the continuity requirement for $\tilde{\alpha}$.

For the next lemma, we do not assume the commutativity for G.

Lemma 2.5.

(i) *Let $\{A, G, \alpha\}$ be a C^*-covariant system based on an abelian C^*-algebra A, and $\{\pi, \mathfrak{H}\}$ an α-representation of A. If $\tilde{\alpha}_s \neq \mathrm{id}$, then there exists a neighborhood U of s and a nonzero projection $p \in \mathcal{A} = \pi(A)''$ such that $\alpha_t(p)p = 0, \ t \in U$.*

(ii) *Let $\{\mathcal{M}, G, \tilde{\alpha}\}$ be a von Neumann covariant system and put*

$$A = \left\{x \in \mathcal{M} : \lim_{s\to e}\|\alpha_s(x) - x\| = 0\right\},$$

where $e \in G$ is the unit of G. Then A is a σ-weakly dense C^*-subalgebra of $\mathcal{M}$ which is globally invariant under $\tilde{\alpha}$. Furthermore, the restriction α of $\tilde{\alpha}$ to A makes $\{A, G, \alpha\}$ a C^*-covariant system, and the inclusion map of A into $\mathcal{M}$ is an α-representation if $\mathcal{M}$ acts on a Hilbert space $\mathfrak{H}$.

PROOF:

(i) Putting $\Omega = \mathrm{Sp}(A)$, we identify A with $C_\infty(\Omega)$. The action α gives rise to an action T of G on Ω such that $\alpha_s(a)(\omega) = a(T_s^{-1}\omega)$, $a \in A$, $s \in G$. The continuity of α implies the continuity of the map: $(s, \omega) \in G \times \Omega \mapsto T_s\omega \in \Omega$, and with a fixed $s \in G$, the map: $\omega \in \Omega \mapsto T_s\omega \in \Omega$ is a homeomorphism of Ω. Now, if $\tilde{\alpha}_s \neq \mathrm{id}$, then $\alpha_s \neq \mathrm{id}$ by the σ-weak density of $\pi(A)$ in $\mathcal{A}$. Hence there exists a point $\omega_0 \in \Omega$ outside of the hull of $\pi^{-1}(0)$ such that $T_s^{-1}\omega_0 \neq \omega_0$. By the continuity of $(s, \omega) \mapsto T_s\omega$, there exists an open neighborhood V of ω_0 and U of s such that $T_t^{-1}(V) \cap V = \emptyset$ for every $t \in U$. Choose an $f \in C_\infty^+(\Omega)$ such that $f(\omega_0) = 1$ and $\mathrm{supp}\, f \subset V$, and put $p = s\big(\pi(f)\big)$. It is clear that p has the required property.

(ii) It suffices to prove the σ-weak density of A in $\mathcal{M}$. For each $f \in \mathcal{K}(G)$, we consider $\alpha_f(x) = \int_G f(s)\alpha_s(x)\, ds$. Then we have

$$\alpha_t\big(\alpha_f(x)\big) = \int_G f(s)\alpha_{ts}(x)\, ds = \int_G f(t^{-1}s)\alpha_s(x)\, ds = \alpha_{\lambda(t)f}(x);$$

$$\Big\|\alpha_t\big(\alpha_f(x)\big) - \alpha_f(x)\Big\| \leq \|\lambda(t)f - f\|_1 \|x\| \to 0 \quad \text{as } t \to e.$$

Hence $\alpha_f(x) \in A$. For each neighborhood U of e, we choose $f_U \in \mathcal{K}(G)^+$ so that $\mathrm{supp}\, f_U \subset U$ and $\int_G f_U(s)\, ds = 1$. Then the net $\{\alpha_{f_U}(x)\}$ converges to x σ-strongly for every $x \in \mathcal{M}$. Q.E.D.

We now return to the case that G is abelian.

Lemma 2.6. *Suppose that G is a locally compact abelian group. If $\{\mathcal{M}, G, \alpha\}$ is an integrable covariant system over G, then the following two statements hold:*

(i) *For each non-empty open subset V of $\widehat{G}$, $\mathcal{M}^\alpha(V) \neq \{0\}$ if and only if V contains a point p with $\mathcal{M}^\alpha(p) \neq \{0\}$.*

(ii) $\mathrm{Sp}(\alpha) = \big\{ p \in \widehat{G} : \mathcal{M}^\alpha(p) \neq \{0\} \big\}^-$.

PROOF: If $x \in \mathfrak{p}_\alpha$, then the function: $s \in G \mapsto \langle \alpha_s(x), \omega\rangle$, $\omega \in \mathcal{M}_*$, is integrable, so that we can define an element $\hat{x}(p) \in \mathcal{M}$ by the integral:

$$\langle \hat{x}(p), \omega\rangle = \int_G \overline{\langle s, p\rangle}\langle \alpha_s(x), \omega\rangle\, ds, \quad x \in \mathfrak{p}_\alpha. \tag{2}$$

It then follows that

$$\alpha_s\big(\hat{x}(p)\big) = \langle s, p\rangle\hat{x}(p), \quad p \in \widehat{G}, \quad s \in G. \tag{3}$$

Hence $\hat{x}(p)$ belongs to $\mathcal{M}^\alpha(p)$. Furthermore, for each $\omega \in \mathcal{M}_*$, the function: $p \in \widehat{G} \mapsto \langle \hat{x}(p), \omega \rangle$ belongs to $A(\widehat{G})$. Therefore, the map: $p \in \widehat{G} \mapsto \hat{x}(p) \in \mathcal{M}$ is σ-weakly continuous and vanishes at infinity, and

$$\mathrm{Sp}_\alpha(x) = \mathrm{supp}(\hat{x}), \quad x \in \mathfrak{p}_\alpha, \tag{4}$$

when we view $\hat{x}$ as an $\mathcal{M}$-valued function over $\widehat{G}$. It is then straightforward to check that if $f \in A(\widehat{G})$, then $\alpha_f(\mathfrak{p}_\alpha) \subset \mathfrak{p}_\alpha$ and

$$\alpha_f(x)\widehat{\ }(p) = f(p)\hat{x}(p), \quad f \in A(\widehat{G}), \quad x \in \mathfrak{p}_\alpha, \quad p \in \widehat{G}. \tag{5}$$

Therefore, we get

$$\alpha_f(x) = 0 \iff f(\mathrm{supp}\,\hat{x}) = 0, \quad f \in A(\widehat{G}), \quad x \in \mathfrak{p}_\alpha. \tag{6}$$

Now, suppose that $\hat{x}(p) = 0$ for every $x \in \mathfrak{p}_\alpha$ and $p \in V$. Then we have $\mathfrak{p}_\alpha \subset \mathcal{M}^\alpha(\complement V)$. Since $\mathcal{M}^\alpha(\complement V)$ is σ-weakly closed, $\mathcal{M} \subset \mathcal{M}^\alpha(\complement V)$. Hence $\mathcal{M}^\alpha(V) = \{0\}$. This proves (i).

Let $E = \{ p \in \widehat{G} : \mathcal{M}^\alpha(p) \neq \{0\} \}^-$. Then $\hat{x}(p) = 0$ for every $p \notin E$ and $x \in \mathfrak{p}_\alpha$. Hence $\mathfrak{p}_\alpha \subset \mathcal{M}^\alpha(E)$, and so $\mathcal{M} \subset \mathcal{M}^\alpha(E)$. Hence $\mathrm{Sp}(\alpha) \subset E$. Conversely, if $\mathcal{M}^\alpha(p) \neq \{0\}$, then $p \in \mathrm{Sp}(\alpha)$. Hence $E \subset \mathrm{Sp}(\alpha)$ by the closedness of $\mathrm{Sp}(\alpha)$. Thus (ii) follows. Q.E.D.

Theorem 2.7. *Let $\{\mathcal{M}, G, \alpha\}$ be a covariant system over a locally compact abelian group G.*

(i) *The Connes spectrum $\Gamma(\alpha)$ of α is a closed subgroup of $\widehat{G}$.*
(ii) $\Gamma(\alpha) + \mathrm{Sp}(\alpha) = \mathrm{Sp}(\alpha)$.
(iii) *If β is another action of G on $\mathcal{M}$ which is cocycle conjugate to α, then $\Gamma(\alpha) = \Gamma(\beta)$.*
(iv) *$\Gamma(\alpha)$ is precisely the kernel of the restriction $\{\mathcal{C}_\mathcal{N}, \widehat{G}, \hat{\alpha}\}$ of the dual action $\hat{\alpha}$ to the center $\mathcal{C}_\mathcal{N}$ of the crossed product $\mathcal{N} = \mathcal{M} \rtimes_\alpha G$.*

PROOF:
(i) The statement (i) follows from either (ii) or (iv).

(ii) Since $\mathrm{Sp}_\alpha(1) = \{0\}$, $0 \in \Gamma(\alpha)$. Suppose $p \in \Gamma(\alpha)$ and $q \in \mathrm{Sp}(\alpha)$. We want to prove $p + q \in \mathrm{Sp}(\alpha)$. To this end, we prove $\mathcal{M}^\alpha(U) \neq \{0\}$ for any neighborhood U of $p + q$. Choose neighborhoods V of p and W of q such that $V + W \subset U$. Take a nonzero $x \in \mathcal{M}^\alpha(W)$, and set $e = \bigvee \{ s_r(\alpha_t(x)) : t \in G \}$. Then $e \in \mathcal{M}^\alpha$ and $e \neq 0$. Since $p \in \mathrm{Sp}(\alpha^e)$, there exists a nonzero $y \in \mathcal{M}^\alpha(V) \cap \mathcal{M}_e$. By the definition of e, we have $z = \alpha_t(x)y \neq 0$ for some $t \in G$. Then z belongs to $\mathcal{M}^\alpha(V+W)$ by Corollary 1.8. Hence $\mathcal{M}^\alpha(U) \neq \{0\}$ as desired. Thus, $p+q \in \mathrm{Sp}(\alpha)$.

(iii) This follows from Lemmas 2.2 and 2.3.

(iv) Let ρ be the action of G on $\mathcal{L}(L^2(G))$ implemented by the right regular representation ρ_G of G on $L^2(G)$, i.e.

$$(\rho_G(s)\xi)(r) = \xi(s + r), \quad s, r \in G, \quad \xi \in L^2(G). \tag{7}$$

We call $\{\mathcal{L}(L^2(G)), G, \rho\}$ the *regular* action of G. Put

$$\widetilde{\mathcal{M}} = \mathcal{M} \otimes \mathcal{L}(L^2(G)), \qquad \overline{\alpha} = \alpha \otimes \mathrm{id} \qquad \text{and} \qquad \tilde{\alpha} = \alpha \otimes \rho.$$

If e is a minimal projection of $\mathcal{L}(L^2(G))$, then $\{\mathcal{M}, G, \alpha\} \cong \{\widetilde{\mathcal{M}}_{1\otimes e}, G, \overline{\alpha}^{1\otimes e}\}$, so that $\Gamma(\alpha) = \Gamma(\overline{\alpha})$. Since $\tilde{\alpha}$ is the cocycle perturbation of $\overline{\alpha}$ by the cocycle $1 \otimes \rho_G$, we have $\Gamma(\overline{\alpha}) = \Gamma(\tilde{\alpha})$, hence $\Gamma(\alpha) = \Gamma(\tilde{\alpha})$. Since the regular action is integrable; in fact $\mathfrak{p}_\rho \cap L^\infty(G) = L^1(G) \cap L^\infty(G)$, $\tilde{\alpha}$ is integrable. By Corollary X.1.22, $\mathcal{M} \rtimes_\alpha G = \widetilde{\mathcal{M}}^{\tilde{\alpha}}$. The dual action $\hat{\alpha}$ on $\mathcal{M} \rtimes_\alpha G$ is implemented by the unitary representation μ of $\widehat{G}$ into $\widetilde{\mathcal{M}}$ given by (X.2.10) and μ satisfies the equation:

$$\tilde{\alpha}_s(\mu(p)) = \langle s, p\rangle \mu(p), \quad s \in G, \quad p \in \widehat{G}. \tag{8}$$

Hence $\mu(p) \in \widetilde{\mathcal{M}}^{\tilde{\alpha}}(p)$. Since $\widetilde{\mathcal{M}}^{\tilde{\alpha}}(p)\mu(p)^* \subset \widetilde{\mathcal{M}}^{\tilde{\alpha}}$, we have

$$\widetilde{\mathcal{M}}^{\tilde{\alpha}}(p) = \widetilde{\mathcal{M}}^{\tilde{\alpha}}\mu(p), \quad p \in \widehat{G}. \tag{9}$$

If $e \in \widetilde{\mathcal{M}}^{\tilde{\alpha}}$ is a central projection, then

$$e\widetilde{\mathcal{M}}^{\tilde{\alpha}}(p)e = \widetilde{\mathcal{M}}^{\tilde{\alpha}} e\hat{\alpha}_p(e)\mu(p),$$

so that

$$\widetilde{\mathcal{M}}^{\tilde{\alpha}}(p) \cap \widetilde{\mathcal{M}}_e \neq \{0\} \iff \hat{\alpha}_p(e)e \neq 0.$$

Hence if $\hat{\alpha}_p = \mathrm{id}$ on the center $\mathcal{C}$ of $\widetilde{\mathcal{M}}^{\tilde{\alpha}}$, then $\hat{\alpha}_p(e)e \neq 0$ for every nonzero $e \in \mathrm{Proj}(\mathcal{C})$, so that $p \in \mathrm{Sp}(\tilde{\alpha}^e)$; hence $p \in \Gamma(\tilde{\alpha}) = \Gamma(\alpha)$. If $\hat{\alpha}_p \neq \mathrm{id}$ on $\mathcal{C}$, then there exists a nonzero $e \in \mathrm{Proj}(\mathcal{C})$ and a neighborhood U of p such that $\hat{\alpha}_q(e)e = 0$ for every $q \in U$. Hence $\widetilde{\mathcal{M}}^{\tilde{\alpha}}(q) \cap \widetilde{\mathcal{M}}_e = \{0\}$ for every $q \in U$. By Lemma 2.6.(i), we have $\widetilde{\mathcal{M}}^{\tilde{\alpha}}(U) \cap \widetilde{\mathcal{M}}_e = \{0\}$ because $\tilde{\alpha}^e$ is integrable. Hence $p \notin \mathrm{Sp}(\tilde{\alpha}^e)$, i.e. $p \notin \Gamma(\tilde{\alpha})$. Q.E.D.

Corollary 2.8. *If the action α of G on $\mathcal{M}$ is centrally ergodic in the sense that the restriction of α to the center $\mathcal{C}$ of $\mathcal{M}$ is ergodic, then the following two conditions for α are equivalent:*

(i) $\Gamma(\alpha) = \widehat{G}$;
(ii) $\mathcal{M} \rtimes_\alpha G$ *is a factor.*

PROOF:
(i) $\Longrightarrow$ (ii): Let $\mathcal{N} = \mathcal{M} \rtimes_\alpha G = \widetilde{\mathcal{M}}^{\tilde{\alpha}}$, and $\mathcal{C}_\mathcal{N}$ be the center of $\mathcal{N}$. If $\Gamma(\alpha) = \widehat{G}$, then $\mathcal{C}_\mathcal{N} \subset \mathcal{N}^{\hat{\alpha}}$ by Theorem 2.7.(iv). Hence $\mathcal{C}_\mathcal{N}$ is contained in the center $\mathcal{C} \otimes \mathbf{C}$ of $\mathcal{N} \rtimes_{\hat{\alpha}} \widehat{G} = \mathcal{M} \overline{\otimes} \mathcal{L}(L^2(G))$. But $\mathcal{N} = \widetilde{\mathcal{M}}^{\tilde{\alpha}}$, so that $\mathcal{C}_\mathcal{N} \subset (\mathcal{C} \otimes \mathbf{C})^{\tilde{\alpha}} = \mathcal{C}^\alpha \otimes \mathbf{C} = \mathbf{C}$. Hence $\mathcal{C}_\mathcal{N} = \mathbf{C}$.

(ii) $\Longrightarrow$ (i): If $\mathcal{M} \rtimes_\alpha G = \mathcal{N}$ is a factor, then the kernel of $\{\mathcal{C}_\mathcal{N}, \widehat{G}, \hat{\alpha}\}$ must be the entire $\widehat{G}$. By Theorem 2.7.(iv), we get $\widehat{G} = \Gamma(\alpha)$. Q.E.D.

Theorem 2.9. *Suppose that $\{\mathcal{M}, G, \alpha\}$ is centrally ergodic. Then we have the following conclusions:*

(i) $\Gamma(\alpha) = \cap\{\mathrm{Sp}({}_u\alpha) : u \in Z^1_\alpha(G, \mathcal{M})\}$.

(ii) *If $\mathcal{W}$ denotes the neighborhood system of* 0 *in* $\widehat{G}$, *then the family* $\{\mathrm{Sp}(\alpha^e)+U : U \in \mathcal{W},\ e \in \mathrm{Proj}(\mathcal{C}_{\mathcal{M}^\alpha}),\ e \neq 0\}$ *is a downward directed net.*

(iii) *If the canonical image* $\mathrm{Sp}(\alpha)/\Gamma(\alpha)$ *of* $\mathrm{Sp}(\alpha)$ *in the quotient group* $\widehat{G}/\Gamma(\alpha)$ *is compact, then*

$$\Gamma(\alpha)^\perp = \left\{ s \in G : \alpha_s \in \mathrm{Ad}\big(\mathcal{U}(\mathcal{C}_{\mathcal{M}^\alpha})\big) \right\},$$

where $\Gamma(\alpha)^\perp = \{ s \in G : \langle s, p\rangle = 1,\ \ p \in \Gamma(\alpha) \}$.

(iv) *If* $\widehat{G}/\Gamma(\alpha)$ *is compact, then there exists a cocycle* $u \in Z^1_\alpha(G, \mathcal{M})$ *such that* $\mathrm{Ker}({}_u\alpha) = \Gamma(\alpha)^\perp$.

We shall prove the theorem in several steps.

Lemma 2.10. *Let* $\mathcal{N}$ *be a von Neumann algebra and* $\sigma \in \mathrm{Aut}(\mathcal{N})$. *If* $e \in \mathrm{Proj}(\mathcal{N}^\sigma)$ *has the central support* 1 *and* $\sigma^e = \mathrm{Ad}(v)$ *for some* $v \in \mathcal{U}(\mathcal{N}_e)$, *then there exists a unique* $u \in \mathcal{U}(\mathcal{N})$ *such that* $\sigma = \mathrm{Ad}(u)$ *and* $v = ue = eu$.

PROOF: Uniqueness: Suppose that $w \in \mathcal{U}(\mathcal{N})$ satisfies the requirement for u. Then $c = w^*u \in \mathcal{C}_\mathcal{N}$ and $ce = e$. Hence $(c-1)e = 0$. Since the central support of e is 1, we have $c = 1$.

Existence: Represent $\mathcal{N}$ on $\mathfrak{H}$ in such a way that σ is implemented by a unitary U on $\mathfrak{H}$, so that

$$\sigma(x) = UxU^*, \qquad x \in \mathcal{N}.$$

By assumption, $Uv^* = w$ is a unitary in $\mathcal{N}_e'$. Since the induction: $y \in \mathcal{N}' \mapsto ye \in \mathcal{N}_e'$ is an isomorphism, there exists $W \in \mathcal{U}(\mathcal{N}')$ such that $Uv^* = We$. Then we have $W^*Ue = v \in \mathcal{N}_e$. Let $u = W^*U$. Now, we have $U\mathcal{N}U^* = \mathcal{N}$, hence $U\mathcal{N}'U^* = \mathcal{N}'$; so $U\mathcal{N}'u^* = \mathcal{N}'$. For each $y \in \mathcal{N}'$, we have

$$uyu^*e = W^*UyU^*We = W^*UyeU^*W = W^*Uyv^* = W^*Uv^*ye = ye.$$

Hence $uyu^* = y,\ \ y \in \mathcal{N}'$. Thus $u \in \mathcal{N}$. Q.E.D.

Lemma 2.11. *Let* α *and* β *be two actions of* G *on* $\mathcal{M}$. *If* β^e *is a cocycle perturbation of* α^e *for some nonzero* $e \in \mathrm{Proj}(\mathcal{M}^\alpha \cap \mathcal{M}^\beta)$ *with central support* $z(e) = 1$, *then* β *is a cocycle perturbation of* α *on the entire algebra* $\mathcal{M}$.

PROOF: Let v be an α^e-cocycle in $\mathcal{U}(\mathcal{M}_e)$ such that $\mathrm{Ad}(v_t)\alpha_t^e = \beta_t^e$. Applying Lemma 2.10 to $\beta_t\alpha_t^{-1}$, we get a unitary $u_t \in \mathcal{M}$ such that $\mathrm{Ad}(u_t) = \beta_t\alpha_t^{-1}$ for each $t \in G$. Now, we want to prove that $u \in Z^1_\alpha(G, \mathcal{M})$.

To this end, we represent $\mathcal{M}$ on $\mathfrak{H}$ which carries unitary representations U and V of G implementing α and β respectively. As in the proof of the previous lemma, $w_t = V(t)U(t)^*v_t^*,\ t \in G$, belongs to $\mathcal{M}_e'$. Since $z(e) = 1$, w_t is of the form: $w_t = W_te$ for a unique $W_t \in \mathcal{M}'$, so that we have $V(t)U(t)^*e = W_tv_t$. Put $\tilde{u}_t = W_t^*V(t)U(t)^*,\ t \in G$. As in the previous case, $\tilde{u}_t$ belongs to $\mathcal{M}$, and by its construction, the map: $t \in G \mapsto \tilde{u}_t \in \mathcal{M}$ is σ-strongly continuous. Furthermore, we have $\mathrm{Ad}(\tilde{u}_t) = \beta_t \alpha_t^{-1}$, and $\tilde{u}_t e = e\tilde{u}_t = v_t$. Hence the uniqueness in Lemma 2.10 implies that $\tilde{u}_t$ and u_t agree. Thus, $\{u_t\}$ is σ-strongly continuous. Now, we have

$$u_s \alpha_s(u_t) e = e u_s \alpha_s(u_t) = v_s \alpha_s^e(v_t) = v_{s+t} = u_{s+t} e = e u_{s+t}.$$

Hence, the uniqueness again implies that $u_s \alpha_s(u_t) = u_{s+t}$. Thus we get $u \in Z_\alpha^1(G, \mathcal{M})$. Q.E.D.

Remark 2.12. In the above lemma, we did not use the commutativity of G. Thus, it holds for non abelian groups as well.

Lemma 2.13. *Let A and B be abelian C^*-algebras with $\Omega_A = \mathrm{Sp}(A)$ and $\Omega_B = \mathrm{Sp}(B)$. For an open continuous map Φ of Ω_A onto Ω_B, B may be regarded as a C^*-subalgebra of the multiplier algebra $M(A)$ by the transpose map of Φ. Any representation $\{\pi, \mathfrak{H}\}$ of B can be extended to a representation $\overline{\pi}$ of $M(A)$ such that $\overline{\pi}(A)'' = \pi(B)''$. In particular, for a closed subgroup H of a locally compact abelian group G, any unitary representation π of H is extended to that $\overline{\pi}$ of G with $\pi(H)'' = \overline{\pi}(G)''$.*

PROOF: Let $\mathcal{B} = \pi(B)''$. By decomposing 1 to a sum of cyclic projections, we may assume that $\mathcal{B}$ admits a cyclic vector. So π can be viewed as the multiplication representation of $B = C_\infty(\Omega_B)$ on $L^2(\Omega_B, \nu)$ with $\nu(\Omega_B) = 1$. Further decomposing ν into disjoint sum, we may assume that $\mathrm{supp}\,\nu = K$ is compact.

Since Φ is open and continuous by assumption, there exists a compact subset L of Ω_A such that $\Phi(L) = K$. Thus, we may assume that $\Omega_B = K$ and $\Omega_A = L$, so that $B = C(\Omega_B) \subset C(\Omega_A) = A$. Let φ be the state on B corresponding to ν. The set K of all state extensions of φ to A is a compact convex subset of the state space $\mathfrak{S}(A)$ of A. Let $\overline{\varphi}$ be an extreme point of K and μ be the corresponding Radon measure on Ω_A. We shall prove that $\pi_{\overline{\varphi}}(A)'' = L^\infty(\Omega_A, \mu) = \mathcal{A}$ and $\pi_\varphi(B)'' = L^\infty(\Omega_B, \nu)$ are isomorphic under the natural correspondence.

Let $\{\pi_{\overline{\varphi}}, \mathfrak{H}_{\overline{\varphi}}, \xi_{\overline{\varphi}}\}$ be the cyclic representation of A induced by $\overline{\varphi}$. Since $\xi_{\overline{\varphi}}$ is cyclic for $\pi_{\overline{\varphi}}(A) \subset \pi_{\overline{\varphi}}(B)'$, $\xi_{\overline{\varphi}}$ is a separating vector for $\pi_{\overline{\varphi}}(B)''$. Hence, if e is the projection of $\mathfrak{H}_{\overline{\varphi}}$ onto $[\pi_{\overline{\varphi}}(B)\xi_{\overline{\varphi}}] = \mathfrak{K}$, then the induction: $x \in \pi_{\overline{\varphi}}(B)'' \mapsto xe \in \pi_{\overline{\varphi}}(B)''e$ is an isomorphism. Since π_φ and $\{\pi_{\overline{\varphi}}, \mathfrak{K}\}$ are unitarily equivalent as representations of B, we may identify π_φ with $\{\pi_{\overline{\varphi}}, \mathfrak{K}\}$. Hence we have $\mathcal{B} = \pi_{\overline{\varphi}}(B)''_e$. We want to prove $\mathcal{A} = \pi_{\overline{\varphi}}(B)''$. Let $\mathcal{E}$ be the conditional expectation of $\mathcal{A}$ onto $\pi_{\overline{\varphi}}(B)'' = \overline{\mathcal{B}}$ relative to φ. If $\mathcal{A} \neq \overline{\mathcal{B}}$, then there exists $p \in \mathrm{Proj}(\mathcal{A})$ with $\mathrm{Sp}(\mathcal{E}(p)) \neq \{0, 1\}$, so that we can find $\lambda > 0$ and $q \in \mathrm{Proj}(\overline{\mathcal{B}})$ such that

$$0 < \lambda q \leq \mathcal{E}(p), \qquad \lambda q \leq \mathcal{E}(1-p) = 1 - \mathcal{E}(p).$$

Choose $a, b \in \overline{\mathcal{B}}$ such that $0 \leq a \leq 1$, $0 \leq b \leq 1$ and $\lambda q = a\mathcal{E}(p)$ and $\lambda q = b\mathcal{E}(1-p)$ and put

$$c = (1-a)p + (1+b)(1-p), \qquad d = (1+a)p + (1-b)(1-p).$$

Then we have $\mathcal{E}(c) = 1 = \mathcal{E}(d)$. Put

$$\varphi_1(x) = \varphi(xc), \qquad \varphi_2(x) = \varphi(xd), \qquad x \in \mathcal{A}.$$

It then follows that φ_1 and φ_2 are both extensions of φ from $\overline{\mathcal{B}}$ to $\mathcal{A}$ and $\varphi = \frac{1}{2}(\varphi_1 + \varphi_2)$. Since $c \neq 1 \neq d$, $\varphi_1 \neq \varphi_2$. This contradicts to the choice of φ in K. Hence $\mathcal{E}(p) \in \mathrm{Proj}(\overline{\mathcal{B}})$ for every $p \in \mathrm{Proj}(\mathcal{A})$. Thus $\mathcal{A} = \overline{\mathcal{B}}$, which in turn implies that $e = 1$ and $\mathcal{A} = \mathcal{B}$.

For the second assertion, we set $A = C^*(G)$ and $B = C^*(H)$. Then $\mathrm{Sp}(A) = \widehat{G}$ and $\mathrm{Sp}(B) = \widehat{H} \cong G/H^\perp$. Therefore, the first assertion entails the second one. Q.E.D.

Lemma 2.14. *Let $\{\mathcal{M}, G, \alpha\}$ be as before and centrally ergodic. If $\mathcal{M}^\alpha$ contains a minimal projection, then there exists $u \in Z^1_\alpha(G, \mathcal{M})$ such that $\mathrm{Sp}({}_u\alpha) = \Gamma(\alpha) = \mathrm{Ker}({}_u\alpha)^\perp$.*

PROOF: Let e be a minimal projection of $\mathcal{M}^\alpha$. Then α^e is ergodic, so that $\Gamma(\alpha) = \Gamma(\alpha^e) = \mathrm{Sp}(\alpha^e)$. By Lemma 1.13, $\mathrm{Sp}(\alpha^e_t) = \{1\}$ if and only if $t \in \mathrm{Sp}(\alpha^e)^\perp$. Hence $\alpha^e_t = \mathrm{id}$ if and only if $t \in \mathrm{Sp}(\alpha^e)^\perp$ by Corollary 1.14. Let $H = \Gamma(\alpha)^\perp$. Then the trivial action of H and the restriction of α to H have the same reduced action by α. Hence by Lemma 2.11, there exists a unitary representation $v: t \in H \mapsto v_t \in \mathcal{M}$ such that $\alpha_t = \mathrm{Ad}\, v_t$, $t \in H$. By the uniqueness of each v_t in Lemma 2.10, we have $\alpha_s(v_t) = v_t$, $s \in G$ and $t \in H$. Thus v is a representation of H into $\mathcal{M}^\alpha$. Applying Lemma 2.13 to v, we extend v to a unitary representation u of G into $\mathcal{M}^\alpha$, and obtain $u \in Z^1_\alpha(G, \mathcal{M})$ such that ${}_u\alpha$ enjoys the required property. Q.E.D.

PROOF OF THEOREM 2.9.(i): To avoid tedious arguments involving cardinal numbers, we present here a proof only for the σ-finite case. For the general case, we leave it to the interested reader. Thus, we assume that $\mathcal{M}$ is σ-finite.

It suffices to prove that for any nonzero $e \in \mathrm{Proj}(\mathcal{M}^\alpha)$, there exists $u \in Z^1_\alpha(G, \mathcal{M})$ such that $\mathrm{Sp}({}_u\alpha) \subset \mathrm{Sp}(\alpha^e)$.

Case 1, $e \sim 1$: Choose $v \in \mathcal{M}$ with $v^*v = 1$ and $vv^* = e$. Put $u_t = v^*\alpha_t(v)$, $t \in G$. We then have, for any $t \in G$,

$$u_t^*u_t = \alpha_t(v^*)vv^*\alpha_t(v) = \alpha_t(v^*)e\alpha_t(v) = \alpha_t(v^*ev) = 1;$$

$$u_tu_t^* = v^*\alpha_t(v)\alpha_t(v^*)v = v^*\alpha_t(vv^*)v = v^*ev = 1.$$

Hence $u_t \in \mathcal{U}(\mathcal{M})$. Now, we compute:

$$\begin{aligned} u_{s+t} &= v^*\alpha_{s+t}(v) = v^*\alpha_s\alpha_t(v) = v^*\alpha_s\big(e\alpha_t(v)\big) = v^*\alpha_s\big(vv^*\alpha_t(v)\big) \\ &= v^*\alpha_s(v)\alpha_s\big(v^*\alpha_t(v)\big) = u_s\alpha_s(u_t). \end{aligned}$$

Thus, we get $u \in Z^1_\alpha(G, \mathcal{M})$. For any $x \in \mathcal{M}_e$, we have

$$\begin{aligned} v{}_u\alpha_t(v^*xv)v^* &= vu_t\alpha_t(v^*xv)u_t^*v^* = vv^*\alpha_t(v)\alpha_t(v^*xv)\alpha_t(v^*)vv^* \\ &= e\alpha_t(exe)e = \alpha_t(x), \end{aligned}$$

hence ${}_u\alpha \cong \alpha^e$. Thus $\mathrm{Sp}(\alpha^e) = \mathrm{Sp}({}_u\alpha)$.

Case 2, $e \not\sim 1$ and $\mathcal{M}_e^\alpha$ has a minimal projection: Lemma 2.14 implies our assertion.

Case 3, $e \not\sim 1$, $\mathcal{M}^\alpha$ is non-atomic and $\mathcal{M}$ is finite: Let T denote the center valued trace on $\mathcal{M}$. By the uniqueness of T, T is invariant under any automorphism of $\mathcal{M}$, so that we have $\alpha_s\bigl(T(e)\bigr) = T\bigl(\alpha_s(e)\bigr) = T(e)$, $s \in G$. Hence $T(e) = \lambda$ for some $\lambda > 0$. Choose $n \in \mathbf{Z}_+$ with $1/n \leq \lambda$. Let $\{f_i\}$ be a maximal orthogonal family in $\mathrm{Proj}(\mathcal{M}_e^\alpha)$ such that $T\bigl(\sum f_i\bigr) \leq 1/n$, and put $e_1 = \sum f_i$. By the maximality of $\{f_i\}$ and the assumption on $\mathcal{M}^\alpha$, we have $T(e_1) = 1/n$. This means that there exist mutually equivalent orthogonal projections $e_1, e_2, \ldots, e_n$ in $\mathcal{M}$ such that $\sum_{i=1}^n e_i = 1$. Let $\{u_i\}_{1\leq i\leq n}$ be a family of partial isometries in $\mathcal{M}$ such that $u_i^* u_i = e_1$ and $u_i u_i^* = e_i$. Set $e_{ij} = u_i^* u_j$. Then $\{e_{ij}\}$ is an $n \times n$ matrix unit in $\mathcal{M}$, so that it generates a subalgebra $\mathcal{B}$ isomorphic to $M_n(\mathbf{C})$, and $\mathcal{M} \cong \mathcal{M}_{e_1} \otimes M_n(\mathbf{C})$. Thus, we have an action β on $\mathcal{M}$ such that $\mathcal{B} \subset \mathcal{M}^\beta$ and $\beta^{e_1} = \alpha^{e_1}$, namely β is conjugate to $\alpha^{e_1} \otimes \mathrm{id}$ on $\mathcal{M}_{e_1} \otimes M_n(\mathbf{C})$. We then have $\mathrm{Sp}(\beta) = \mathrm{Sp}(\alpha^{e_1}) \subset \mathrm{Sp}(\alpha^e)$. By Lemma 2.11, we have $\beta = {}_u\alpha$ for some $u \in Z_\alpha^1(G, \mathcal{M})$.

Case 4, $e \not\sim 1$, $\mathcal{M}^\alpha$ is non-atomic and $\mathcal{M}$ is properly infinite: From the argument in Case 3, we have only to prove that e majorizes a nonzero $e_1 \in \mathrm{Proj}(\mathcal{M}^\alpha)$ such that e_1 is a member of mutually equivalent orthogonal family of projections with sum up to 1. If e is not properly infinite, then e has a decomposition $e = e_1 + e_2$ with e_1 finite and e_2 properly infinite and $e_1 \neq 0$. Since the decomposition is unique, e_1 and e_2 fall in $\mathcal{M}^\alpha$. Since $z(e_1) = 1$, e_1 has the required property. Suppose that e is properly infinite. Then $e \sim z(e)$ by the σ-finiteness of $\mathcal{M}$. But $z(e) = 1$. Hence $e \sim 1$. This contradicts the assumption.

The central ergodicity of α implies that $\mathcal{M}$ is either finite or properly infinite. Hence the above four cases cover all possibilities. Q.E.D.

Lemma 2.15. *Let $\mathcal{D}$ be the center $\mathcal{C}_{\mathcal{M}^\alpha}$ of $\mathcal{M}^\alpha$. If α is centrally ergodic, then for any nonzero $e_1, e_2 \in \mathrm{Proj}(\mathcal{D})$ and a neighborhood U of $0 \in \widehat{G}$, there exist nonzero $f_1, f_2 \in \mathrm{Proj}(\mathcal{D})$ with $f_1 \leq e_1$ and $f_2 \leq e_2$ such that*

$$\mathrm{Sp}(\alpha^{f_1}) \subset \mathrm{Sp}(\alpha^{f_2}) + U, \qquad \mathrm{Sp}(\alpha^{f_2}) \subset \mathrm{Sp}(\alpha^{f_1}) + U.$$

PROOF: Since $z(e_1) = 1 = z(e_2)$, $e_1 \mathcal{M} e_2 \neq \{0\}$. Choose $x = e_1 x e_2 \neq 0$, and $f \in A(\widehat{G})$ such that $y = \alpha_f(x) \neq 0$ and $\mathrm{supp}\, f - \mathrm{supp}\, f \subset U$. Since $e_1, e_2 \in \mathcal{M}^\alpha$, we have $y = e_1 y e_2$. Put

$$f_1 = \bigvee\Bigl\{ s_\ell\bigl(z_1 \alpha_t(y) z_2\bigr) : t \in G,\ z_1, z_2 \in \mathcal{M}^\alpha \Bigr\},$$

$$f_2 = \bigvee\Bigl\{ s_r\bigl(z_1 \alpha_t(y) z_2\bigr) : t \in G,\ z_1, z_2 \in \mathcal{M}^\alpha \Bigr\}.$$

Clearly, $0 \neq f_1 \leq e_1$ and $0 \neq f_2 \leq e_2$ because e_1 and e_2 are central in $\mathcal{M}^\alpha$. We shall show that f_1 and f_2 enjoy the required property.

Without loss of generality, we assume $U = -U$. By symmetry, it suffices to show $\mathrm{Sp}(\alpha^{f_1}) \subset \mathrm{Sp}(\alpha^{f_2}) + U$. Take $p \in \mathrm{Sp}(\alpha^{f_1})$ and a neighborhood V of p.

By considering a compact neighborhood U of 0 from the beginning, to prove $p \in \mathrm{Sp}(\alpha^{f_2})+U$, it suffices to show that $V\cap(\mathrm{Sp}(\alpha^{f_2})+U) \neq \emptyset$, which is equivalent to $(V+U)\cap \mathrm{Sp}(\alpha^{f_2}) \neq \emptyset$. Choose a nonzero $x_1 \in \mathcal{M}^\alpha(V)\cap \mathcal{M}_{f_1}$. By the definition of f_1, there exist $s, t \in G$ and $z_i, z_i' \in \mathcal{M}^\alpha$, $i = 1, 2$, such that

$$x_2 = z_1'\alpha_s(y^*)z_2'x_1z_1\alpha_t(y)z_2 \neq 0.$$

It then follows that $x_2 \in \mathcal{M}_{f_2}$ and

$$\begin{aligned}\mathrm{Sp}_\alpha(x_2) &\subset \mathrm{Sp}_\alpha\big(z_1\alpha_t(y)z_2\big) + \mathrm{Sp}_\alpha\big(z_1'\alpha_s(y^*)z_2'\big) + \mathrm{Sp}_\alpha(x_1)\\ &\subset \mathrm{supp}\, f + V - \mathrm{supp}\, f \subset V+U.\end{aligned}$$

Hence $x_2 \in \mathcal{M}^\alpha(V+U)\cap \mathcal{M}_{f_2}$. Thus $(V+U)\cap \mathrm{Sp}(\alpha^{f_2}) \neq \emptyset$. Q.E.D.

PROOF OF THEOREM 2.9.(ii): Let $\mathcal{D}$ be the center $\mathcal{C}_{\mathcal{M}^\alpha}$ of $\mathcal{M}^\alpha$. For each pair U_1, U_2 of neighborhoods of $0 \in \widehat{G}$ and nonzero $e_1, e_2 \in \mathrm{Proj}(\mathcal{D})$, choose a neighborhood U of 0 with $U+U \subset U_1\cap U_2$, and nonzero $f_1, f_2 \in \mathrm{Proj}(\mathcal{D})$ with $f_1 \leq e_1$ and $f_2 \leq e_2$ such that $\mathrm{Sp}(\alpha^{f_1}) \subset \mathrm{Sp}(\alpha^{f_2}) + U$ and $\mathrm{Sp}(\alpha^{f_2}) \subset \mathrm{Sp}(\alpha^{f_1}) + U$. We then have

$$\begin{aligned}\mathrm{Sp}(\alpha^{f_1}) + U &\subset \mathrm{Sp}(\alpha^{f_2}) + U + U \subset \mathrm{Sp}(\alpha^{f_2}) + U_2,\\ \mathrm{Sp}(\alpha^{f_2}) + U &\subset \mathrm{Sp}(\alpha^{f_1}) + U + U \subset \mathrm{Sp}(\alpha^{f_1}) + U_1.\end{aligned}$$

Q.E.D.

Lemma 2.16.

(i) *For any $u \in \mathcal{U}(\mathcal{M})$, the spectrum $\mathrm{Sp}_{\mathcal{L}(\mathcal{M})}(\mathrm{Ad}(u))$ of $\mathrm{Ad}(u)$ in $\mathcal{L}(\mathcal{M})$ is contained in $\{\lambda\bar{\mu} : \lambda, \mu \in \mathrm{Sp}(u)\}$. If $\mathcal{M}$ is a factor, then*

$$\mathrm{Sp}_{\mathcal{L}(\mathcal{M})}(\mathrm{Ad}(u)) = \mathrm{Sp}(u)\,\mathrm{Sp}(u)^{-1}.$$

(ii) *Let $\{\mathcal{M}, G, \alpha\}$ be a covariant system and fix an element $t \in G$. If α_t is of the form $\alpha_t = \mathrm{Ad}(u)$, $u \in \mathcal{U}(\mathcal{M}^\alpha)$, then $t \in \Gamma(\alpha)^\perp$.*

PROOF:

(i) Put $L_ux = ux$ and $R_{u^*}x = xu^*$, $x \in \mathcal{M}$. Then $\mathrm{Sp}(L_u) \subset \mathrm{Sp}(u)$ and $\mathrm{Sp}(R_{u^*}) \subset \mathrm{Sp}(u)^{-1}$. Since L_u and R_{u^*} commute, we have

$$\mathrm{Sp}(\mathrm{Ad}(u)) = \mathrm{Sp}(L_uR_{u^*}) \subset \mathrm{Sp}(L_u)\,\mathrm{Sp}(R_{u^*}) \subset \mathrm{Sp}(u)\,\mathrm{Sp}(u)^{-1}.$$

Suppose that $\mathcal{M}$ is a factor. For $\lambda, \mu \in \mathrm{Sp}(u)$, choose spectral projections e_1 and e_2 of u so that $\|(u-\lambda)e_1\| < \varepsilon$ and $\|(u-\mu)e_2\| < \varepsilon$ for a preassigned $\varepsilon > 0$. Let v be a partial isometry of $\mathcal{M}$ with $vv^* \leq e_1$ and $v^*v \leq e_2$. Then we have

$$\|uvu^* - \lambda\bar{\mu}v\| = \|uv - \lambda\bar{\mu}vu\| \leq \|(u-\lambda)e_1v\| + \|\lambda ve_2(\mu-u)\| < 2\varepsilon,$$

so that $\lambda\bar{\mu} \in \mathrm{Sp}(\mathrm{Ad}(u))$.

(ii) Choose $\lambda \in \mathrm{Sp}(u)$ and a spectral projection e for $\varepsilon > 0$ such that $\|(u-\lambda)e\| < \varepsilon$. Since $\alpha_t^e = \mathrm{Ad}(eue)$, we have, by (i),

$$\mathrm{Sp}(\alpha_t^e) \subset \{\mu_1\overline{\mu}_2 : |\lambda - \mu_i| < \varepsilon,\ i = 1,2\} \subset \{z \in \mathbf{C} : |z-1| < 2\varepsilon\}.$$

By Lemma 1.13, we have

$$\{\langle t, p\rangle : p \in \mathrm{Sp}(\alpha^e)\}^- = \mathrm{Sp}(\alpha^e) \subset \{z : |z-1| < 2\varepsilon\}.$$

Thus, we get $\langle t, p\rangle = 1,\ p \in \Gamma(\alpha)$. Q.E.D.

Lemma 2.17. *Let $\mathcal{M} = \mathcal{M}_1 \overline{\otimes} \mathcal{M}_2$ and $\sigma = \sigma_1 \otimes \sigma_2$ with $\sigma_i \in \mathrm{Aut}(\mathcal{M}_i)$, $i = 1, 2$. Then σ is inner if and only σ_1 and σ_2 are both inner.*

PROOF: Suppose $\sigma_1 = \mathrm{Ad}(u_1)$ and $\sigma_2 = \mathrm{Ad}(u_2)$ with $u_1 \in \mathcal{U}(\mathcal{M}_1)$ and $u_2 \in \mathcal{U}(\mathcal{M}_2)$, put $u = u_1 \otimes u_2 \in \mathcal{U}(\mathcal{M})$. Then we have $\sigma = \mathrm{Ad}(u)$.

Conversely, suppose that σ is inner. So $\sigma = \mathrm{Ad}(u)$ with $u \in \mathcal{U}(\mathcal{M})$. Let $\mathcal{E}$ be a normal conditional expectation of $\mathcal{M}$ onto $\mathcal{M}_1$, and put $a = \mathcal{E}(u)$. Let $a = vh$ be the polar decomposition. We then have, for any $x \in \mathcal{M}_1$,

$$ax = \mathcal{E}(u)x = \mathcal{E}(ux) = \mathcal{E}(\sigma(x)u) = \sigma_1(x)\mathcal{E}(u) = \sigma_1(x)a.$$

If x is unitary, then $vxx^*hx = \sigma_1(x)vh$, so that the uniqueness of the polar decomposition implies that $vx = \sigma_1(x)v$ and $x^*hx = h$ for every $x \in \mathcal{U}(\mathcal{M}_1)$. Thus $vx = \sigma_1(x)v$ and h falls in the center $\mathcal{C}_1$ of $\mathcal{M}_1$. Hence $e = v^*v = s(h)$ and $vv^* = s(vhv^*) = s(h) = e$, so that $e \in \mathcal{C}_1$, and $v = ve = \sigma_1(e)v$. Thus $e \leq \sigma_1(e)$. Since $a^* = \mathcal{E}(u^*)$ and $a^*x = \sigma_1^{-1}(x)a^*$ for every $x \in \mathcal{M}_1$, we have $e \leq \sigma_1^{-1}(e)$ from the above discussion. Hence $e = \sigma_1(e)$ and $\sigma_1^e = \mathrm{Ad}(v)$.

Let $\{e_i\}$ be a maximal orthogonal family of central projections such that $\sigma_1(e_i) = e_i$ and $\sigma_1^{e_i} \in \mathrm{Int}(\mathcal{M}_{1,e_i})$. Let $e = \sum e_i$. If $e \neq 1$, then we consider $\mathcal{M}_{1,(1-e)} \overline{\otimes} \mathcal{M}_2$ and $\sigma_1^{(1-e)} \otimes \sigma_2 = \sigma^{(1-e)} \in \mathrm{Int}(\mathcal{M}_{(1-e)})$. Choosing a conditional expectation $\mathcal{E}$ so that $\mathcal{E}(u(1-e)) \neq 0$, we get a nonzero projection $e_0 \leq 1 - e$ such that $\sigma_1^{(1-e_0)}$ is inner. This contradicts to the maximality of e. Thus $e = 1$. Hence $\sigma_1 \in \mathrm{Int}(\mathcal{M}_1)$. Q.E.D.

Lemma 2.18. *Consider $\sigma \in \mathrm{Aut}(\mathcal{M})$ as an action of $\mathbf{Z}$: $n \in \mathbf{Z} \mapsto \sigma^n \in \mathrm{Aut}(\mathcal{M})$. Let $\mathrm{Aut}_\sigma(\mathcal{M}) = \{\alpha \in \mathrm{Aut}(\mathcal{M}) : \alpha\sigma = \sigma\alpha\}$. If $\Gamma(\sigma) = \{0\}$ and $\mathrm{Aut}_\sigma(\mathcal{M})$ is centrally ergodic, then σ is inner.*

PROOF: Put $\widetilde{\mathcal{M}} = \mathcal{M} \overline{\otimes} \mathcal{L}(\ell^2(\mathbf{Z}))$ and

$$(U\xi)(n) = \xi(n+1), \quad \xi \in \ell^2(\mathbf{Z}), \quad n \in \mathbf{Z},$$

$$(V(\theta)\xi)(n) = e^{in\theta}\xi(n), \quad 0 \leq \theta < 2\pi.$$

Set $\tilde{\sigma} = \sigma \otimes \mathrm{Ad}(U)$ and $\hat{\sigma}_\theta = \mathrm{Ad}\big(1 \otimes V(\theta)\big)$. By the last lemma it suffices to prove that $\tilde{\sigma}$ is inner. Note that $\mathcal{G} = \big\{\alpha \in \mathrm{Aut}(\widetilde{\mathcal{M}}) : \tilde{\alpha}\sigma = \tilde{\sigma}\alpha \text{ and } \alpha\hat{\sigma}_\theta = \hat{\sigma}_\theta\alpha\big\}$ acts ergodically on the center $\mathcal{C}$ of $\mathcal{M}$ because $\mathcal{G}$ contains $\mathrm{Aut}_\sigma(\mathcal{M}) \otimes \mathrm{id}$. Let $\mathcal{N} = \widetilde{\mathcal{M}}^{\tilde{\sigma}}$. Then $\mathcal{N}$ can be identified with $\mathcal{M} \rtimes_\sigma \mathbf{Z}$ and $\hat{\sigma}|_{\mathcal{N}}$ with the dual action of σ. Let $\mathcal{D}$ be the center of $\mathcal{N}$. It then follows that $\mathcal{G}$ acts on $\mathcal{N}$, hence on $\mathcal{D}$. The fixed point algebra $\mathcal{D}^{\mathcal{G}}$ is a subalgebra of $\mathcal{C}$ because $\hat{\sigma}_\theta \in \mathcal{G}$, $0 \le \theta < 2\pi$, so that the central ergodicity implies that $\mathcal{D}^{\mathcal{G}} = \mathbf{C}$. Hence $\mathcal{G}$ is centrally ergodic on $\mathcal{N}$, and commutes with $\hat{\sigma}$ by definition. By Theorem 2.7.(iv), $\mathrm{Ker}\big(\hat{\sigma}|_{\mathcal{D}}\big) = \Gamma(\sigma) = \{0\}$, so that $\hat{\sigma}|_{\mathcal{D}}$ is a faithful action of $\mathbf{T} = \mathbf{R}/2\pi\mathbf{Z}$. Let $\mathcal{D}_n = \mathcal{D}^{\hat{\sigma}}(n) = \big\{x \in \mathcal{D} : \hat{\sigma}_\theta(x) = \mathrm{e}^{in\theta}x\big\}$. For the moment, assume that $\mathcal{D}_1$ contains a unitary. Then we have

$$uxu^* = x, \quad x \in \mathcal{N}, \qquad u\big(1 \otimes V(\theta)\big)u^* = \mathrm{e}^{-\mathrm{i}\theta}V(\theta).$$

Hence $\tilde{\sigma} = \mathrm{Ad}(u^*)$ on the generators of $\widetilde{\mathcal{M}}$, so $\tilde{\sigma} = \mathrm{Ad}(u^*)$ on $\widetilde{\mathcal{M}}$, which in turn implies that σ is inner.

We now want to prove the last assumption on $\mathcal{D}_1$. Put $e_n = \bigvee\big\{s(x) : x \in \mathcal{D}_n\big\}$. Since $\alpha(\mathcal{D}_n) = \mathcal{D}_n$ for every $\alpha \in \mathcal{G}$, we have $\alpha(e_n) = e_n$, so that either $e_n = 1$ or 0, by ergodicity. Hence $\mathrm{Sp}\big(\hat{\sigma}|_{\mathcal{D}}\big) = \big\{n \in \mathbf{Z} : e_n = 1\big\}$. Thus $\mathcal{D}_n \neq \{0\}$ and $\mathcal{D}_m \neq \{0\}$ imply $\mathcal{D}_{n+m} \neq \{0\}$ and $\mathcal{D}_{-n} \neq \{0\}$. Hence $\mathrm{Sp}(\hat{\sigma}|_{\mathcal{D}})$ is a subgroup of $\mathbf{Z}$, which must be of the form $k\mathbf{Z}$ with $k \in \mathbf{Z}_+$. Since $\{\mathcal{D}_n\}$ span a σ-weakly dense subspace of $\mathcal{D}$, we have $\{0\} = \mathrm{Ker}\big(\hat{\sigma}|_{\mathcal{D}}\big) = \mathrm{Sp}\big(\hat{\sigma}|_{\mathcal{D}}\big)^\perp$. Hence we have $k = 1$. Thus, we get $e_1 = 1$. Let $\{x_i\}$ be a maximal family in $\mathcal{D}_1$ such that $s(x_i) \perp s(x_j)$, $i \neq j$. Put $x = \sum x_i/\|x_i\| \in \mathcal{D}_1$, and let $x = uh$ be the polar decomposition of x. Then u is the required unitary in $\mathcal{D}_1$. Q.E.D.

Lemma 2.19. *Let $u \in \mathcal{U}(\mathcal{M})$ and $\sigma = \mathrm{Ad}(u)$. If $\mathrm{e}^{\mathrm{i}(\theta_1-\theta_2)} \notin \mathrm{Sp}(\sigma)$, $-\pi \le \theta_1, \theta_2 < \pi$, then there exists $\varepsilon > 0$ such that the spectral projections p_1 and p_2 of u corresponding to the intervals $[\theta_1 - \varepsilon, \theta_1 + \varepsilon]$ and $[\theta_2 - \varepsilon, \theta_2 + \varepsilon]$ are centrally orthogonal, i.e. $z(p_1) \perp z(p_2)$.*

PROOF: Let $u = \int_{-\pi}^{\pi} \mathrm{e}^{\mathrm{i}\theta}\,dp(\theta)$ be the spectral decomposition of u. For each $\varepsilon > 0$, let $p_i(\varepsilon) = p\big([\theta_i - \varepsilon, \theta_i + \varepsilon]\big)$, $i = 1, 2$. If $p_1(\varepsilon)\mathcal{M}p_2(\varepsilon) \neq \{0\}$ for every $\varepsilon > 0$, then we choose an $x_\varepsilon \in p_1(\varepsilon)\mathcal{M}p_2(\varepsilon)$, $\|x_\varepsilon\| = 1$, so that

$$\big\|ux_\varepsilon - \mathrm{e}^{\mathrm{i}\theta_1}x_\varepsilon\big\| \le \varepsilon, \qquad \big\|x_\varepsilon u - \mathrm{e}^{\mathrm{i}\theta_2}x_\varepsilon\big\| \le \varepsilon$$

Then we get

$$\begin{aligned}\big\|ux_\varepsilon u^* - \mathrm{e}^{\mathrm{i}(\theta_1-\theta_2)}x_\varepsilon\big\| &= \big\|ux_\varepsilon - \mathrm{e}^{\mathrm{i}(\theta_1-\theta_2)}x_\varepsilon u\big\| \\ &\le \big\|ux_\varepsilon - \mathrm{e}^{\mathrm{i}\theta_1}x_\varepsilon\big\| + \big\|\mathrm{e}^{\mathrm{i}\theta_1}x_\varepsilon - \mathrm{e}^{\mathrm{i}(\theta_1-\theta_2)}xu_\varepsilon\big\| \le 2\varepsilon.\end{aligned}$$

Thus, $\mathrm{e}^{\mathrm{i}(\theta_1-\theta_2)} \in \mathrm{Sp}(\sigma)$. Hence $p_1(\varepsilon)\mathcal{M}p_2(\varepsilon) = \{0\}$ for a sufficiently small $\varepsilon > 0$. Q.E.D.

Lemma 2.20. *In the previous lemma, if $\mathrm{Sp}(\sigma) \subset \big\{\mathrm{e}^{\mathrm{i}\theta} : |\theta| < \varepsilon\big\} = V(\varepsilon)$, $0 < \varepsilon < \pi/2$, then we can choose $v \in \mathcal{U}(\mathcal{M})$ so that $\sigma = \mathrm{Ad}(v)$ and $\mathrm{Sp}(v) \subset V(2\varepsilon)$.*

PROOF: We use the notations of the previous lemma. For each $\delta > 0$ and $-\pi \le \theta < \pi$, let $p(\theta, \delta) = p\big([\theta - \delta, \theta + \delta]\big)$. By the previous lemma, $p(\theta, \delta)$ and $1 - p(\theta, \delta + \varepsilon)$ are centrally orthogonal. Hence $\mathrm{Sp}\big(uz(p(\theta, \delta))\big) \subset \mathrm{e}^{\mathrm{i}\theta} V(\varepsilon)$. Viewing $uz\big(p(\theta, \delta)\big)$ as a unitary in $\mathcal{M}_{z(p(\theta,\delta))}$, we have $\mathrm{Sp}\big(\mathrm{e}^{-\mathrm{i}\theta} uz(p(\theta, \delta))\big) \subset V(\varepsilon)$. Let $-\pi = \theta_0 < \theta_1 < \cdots < \theta_n = \pi$ and $\max\big(\theta_{i+1} - \theta_i\big) < \delta$, and put, with $0 < \delta < \min\big\{\varepsilon, \pi/2 - \varepsilon\big\}$,

$$q_0 = 0, \qquad q_j = z\big(p(\theta_j, \delta)\big)(1 - q_{j-1}), \quad 1 \le j \le n.$$

$$w = \sum_{j=1}^{n} \mathrm{e}^{-\mathrm{i}\theta_j} q_j \in \mathcal{U}(\mathcal{C}_{\mathcal{M}}).$$

Then $v = uw$ enjoys the required property. Q.E.D.

Definition 2.21. Given a covariant system $\big\{\mathcal{M}, G, \alpha\big\}$ an *additive one cocycle* means a σ-weakly continuous function $h\colon s \in G \mapsto h(s) \in \mathcal{M}$ such that

$$h(s + t) = h(s) + \alpha_s\big(h(t)\big), \quad s, t \in G. \tag{10}$$

An additive one cocycle of the form:

$$h(s) = \alpha_s(k) - k, \quad s \in G,$$

for some $k \in \mathcal{M}$ is called an *additive coboundary* or simply a *coboundary*.

Note that the definition of additive cocycles and coboundary is free from the commutativity of G, although we do not pick it up here.

Lemma 2.22.

(i) *Any bounded additive one cocycle is a coboundary.*
(ii) *If* $\sigma \in \mathrm{Int}(\mathcal{M})$ *commutes with* α *and* $\mathrm{Sp}(\sigma) \subset V(\varepsilon)$, $0 < \varepsilon < \pi/4$, *then there exists* $u \in \mathcal{U}(\mathcal{M}^{\alpha})$ *with* $\sigma = \mathrm{Ad}(u)$.

PROOF:
(i) Given a bounded additive cocycle h, we set

$$T_s(x) = \alpha_s(x) - h(s), \quad s \in G.$$

Then the map: $s \in G \mapsto T_s$ is a representation of G by affine transformations on $\mathcal{M}$. The equality (10) implies immediately $T_s \circ T_t = T_{s+t}$, $s, t \in G$. For each $x \in \mathcal{M}$, the orbit $\big\{ T_s(x) : s \in G \big\}$ is bounded. Let $K(x)$ denote the σ-weakly closed convex closure of $\big\{ T_s(x) : s \in G \big\}$. It follows that $K(x)$ is σ-weakly compact and convex. By the Kakutani fixed point theorem, there exists $k \in K(x)$ such that $T_s(k) = k$, $s \in G$, which means that $\alpha_s(k) - h(s) = k$. Hence $h(s) = \alpha_s(k) - k$ is a coboundary.

(ii) By Lemma 2.20, choose $v \in \mathcal{U}(\mathcal{M})$ so that $\sigma = \mathrm{Ad}(v)$ and $\mathrm{Sp}(v) \subset V(2\varepsilon)$. Since α and σ commute, we have

$$\mathrm{Ad}(v) = \alpha_s \,\mathrm{Ad}(v) \alpha_s^{-1} = \mathrm{Ad}\big(\alpha_s(v)\big), \quad s \in G,$$

so that $a_s = v^*\alpha_s(v)$ belongs to the center $\mathcal{C}$ of $\mathcal{M}$. We further have $\mathrm{Sp}(a_s) \subset V(4\varepsilon)$. Let Log be the principal value of logarithm. We set $h(s) = \frac{1}{\mathrm{i}}\operatorname{Log} a_s,\ s \in G$. Since Log is continuous on $V(4\varepsilon)$, h is well-defined and σ-weakly continuous, and

$$-\pi < h(s) < \pi, \qquad h(s) \in \mathcal{C}.$$

Since Log maps multiplications into additions, the cocycle identity for a is mapped into the additive cocycle identity (10) for h. Thus, we can choose $k \in \mathcal{C}$ such that $h(s) = \alpha_s(k) - k$ by (i). Since $h(s) = h(s)^*$, we can choose k to be self adjoint. Let $w = \exp(\mathrm{i}k)$ and $u = vw^* \in \mathcal{U}(\mathcal{M})$. We then have

$$\alpha_s(u) = va_s\alpha_s(w^*) = v\exp\Big[\mathrm{i}\,\big(h(s) - \alpha_s(k)\big)\Big]$$

$$= v\exp(-\mathrm{i}k) = vw^* = u, \quad s \in G,$$

$$\mathrm{Ad}(u) = \mathrm{Ad}(v)\,\mathrm{Ad}(w^*) = \mathrm{Ad}(v) = \sigma. \qquad \text{Q.E.D.}$$

PROOF OF THEOREM 2.9.(iii): Let π denote the canonical quotient map of $\widehat{G}$ onto $\widehat{G}/\Gamma(\alpha)$. By assumption, $\pi\big(\mathrm{Sp}(\alpha)\big)$ is compact.

Fix $t \in \Gamma(\alpha)^\perp$ arbitrarily, and put $\sigma = \alpha_t$. For each $\varepsilon > 0$, set $U_\varepsilon = \big\{ p \in \widehat{G} : \mathrm{Re}\langle t, q\rangle > 1 - \varepsilon \big\}$. Then $\pi^{-1}\pi(U_\varepsilon) = U_\varepsilon$. With the system $\mathcal{W}$ of compact neighborhoods of $0 \in \widehat{G}$, $\big\{\pi\big(\mathrm{Sp}(\alpha^e) + U\big) : e \in \mathrm{Proj}(\mathcal{C}_{\mathcal{M}^\alpha}) \text{ and } U \in \mathcal{W}\big\}$ is a decreasing directed family of compact subsets of $\widehat{G}/\Gamma(\alpha)$ with intersection $\{0\}$. Hence for some $u \in \mathcal{W}$ and $e \in \mathrm{Proj}(\mathcal{C}_{\mathcal{M}^\alpha})$, we have $\pi\big(\mathrm{Sp}(\alpha^e) + U\big) \subset \pi(U_\varepsilon)$, so that $\mathrm{Sp}(\alpha^e) \subset U_\varepsilon$. By Lemma 1.13, we get

$$\mathrm{Sp}_{\mathcal{L}(\mathcal{M}_e)}(\sigma^e) \subset \big\{ \langle t, p\rangle : p \in \Gamma(\alpha^e) \big\}^- \subset \big\{ z \in \mathbf{T} : \mathrm{Re}\, z \geq 1 - \varepsilon \big\}.$$

Viewing σ^e as an action of $\mathbf{Z}$ on $\mathcal{M}_e$: $n \in \mathbf{Z} \mapsto (\sigma^e)^n$, and identifying $\widehat{\mathbf{Z}} = \mathbf{T}$ with $\big\{ z \in \mathbf{C} : |z| = 1 \big\}$, we have

$$\mathrm{Sp}(\sigma^e) = \mathrm{Sp}_{\mathcal{L}(\mathcal{M}_e)}(\sigma^e).$$

Since $\mathbf{T}$ contains no subgroup in the right half plane other than $\{1\}$, we have $\Gamma(\sigma^e) = \{1\}$; hence $\Gamma(\sigma) = \{1\}$. Since $\mathrm{Aut}_\sigma(\mathcal{M})$ contains $\alpha(G)$, $\mathrm{Aut}_\sigma(\mathcal{M})$ is centrally ergodic. Thus Lemma 2.17 applies to σ, implying that σ is inner. If we choose ε so small that $0 < \varepsilon < 1/\sqrt{2}$, then $\mathrm{Sp}(\sigma^e) \subset V(\delta)$ in the notation of Lemma 2.22 with $\delta = \cos^{-1}(1 - \varepsilon) < \pi/4$. By Lemma 2.22.(ii), we have $\sigma^e = \mathrm{Ad}(v)$ for some $v \in \mathcal{U}(\mathcal{M}_e^\alpha)$.

Now, the central ergodicity of α implies $z(e) = 1$. Hence Lemma 2.10 yields that there exists uniquely $u \in \mathcal{U}(\mathcal{M})$ such that $\sigma = \mathrm{Ad}(u)$ and $eu = ue = v$. Since $\sigma = \alpha_s\sigma\alpha_s^{-1} = \mathrm{Ad}\big(\alpha_s(u)\big)$ and $e\alpha_s(u) = \alpha_s(eu) = \alpha_s(ue) = \alpha_s(v) = v$, the uniqueness of u implies that $u = \alpha_s(u),\ s \in G$. Hence $u \in \mathcal{M}^\alpha$. Since $\sigma = \alpha_t$, σ leaves $\mathcal{M}^\alpha$ pointwise fixed, so that u must be in $\mathcal{C}_{\mathcal{M}^\alpha}$. Q.E.D.

Lemma 2.23. *Any subgroup H of G isomorphic to $\mathbf{T}$ is a direct summand of G.*

PROOF: Let f be an isomorphism of H onto $\mathbf{T}$. Then $f \in \widehat{H} = \widehat{G}/H^{\perp}$. Thus f is extended to a continuous homomorphism g of G onto $\mathbf{T}$. Put $\pi(s) = f^{-1} \circ g(s)$, $s \in G$. Then π is a projection of G onto H and $G \simeq H \times \pi^{-1}(0)$. Q.E.D.

PROOF OF THEOREM 2.9.(iv): By assumption, $\Gamma(\alpha)^{\perp} \simeq \big(\widehat{G}/\Gamma(\alpha)\big)^{\widehat{}}$ is discrete. By (iii), for each $t \in \Gamma(\alpha)^{\perp}$ we choose $v(t) \in \mathcal{U}(\mathcal{C}_{\mathcal{M}^\alpha})$ so that $\alpha_t = \mathrm{Ad}\big(v(t)\big)$. Since

$$\mathrm{Ad}\big(v(s)v(t)\big) = \alpha_{s+t} = \mathrm{Ad}\big(v(s+t)\big), \quad s, t \in \Gamma(\alpha)^{\perp},$$

we get $\lambda(s,t) \in \mathcal{U}\big(\mathcal{C}_{\mathcal{M}} \cap \mathcal{C}_{\mathcal{M}^\alpha}\big) = \mathbf{T}$ such that

$$v(s)v(t) = \lambda(s,t)v(s+t), \quad s, t \in \Gamma(\alpha)^{\perp}.$$

Since $v(s)v(t) = v(t)v(s)$, we have $\lambda(s,t) = \lambda(t,s)$. In the cartesian product $\mathbf{T} \times \Gamma(\alpha)^{\perp} = H$, we define a product by the formula:

$$(a,s)(b,t) = \big(ab\lambda(s,t),\ s+t\big).$$

Then H is an abelian locally compact group. Clearly $K = \big\{(a,0) : a \in \mathbf{T}\big\}$ is isomorphic to $\mathbf{T}$. Hence by the last lemma, we have $H = K \oplus K' \cong K \oplus H/K = \mathbf{T} \oplus \Gamma(\alpha)^{\perp}$. Let f be the isomorphism of $\Gamma(\alpha)^{\perp}$ onto K'. Then f is of the form:

$$f(s) = \big(g(s), s\big) \in \mathbf{T} \times \Gamma(\alpha)^{\perp}.$$

Hence we get

$$\begin{aligned}\big(g(s)g(t)\lambda(s,t), s+t\big) &= \big(g(s), s\big)\big(g(t), t\big) = f(s)f(t)\\ &= f(s+t) = \big(g(s+t), s+t\big).\end{aligned}$$

so that $\lambda(s,t) = g(s+t)/g(s)g(t)$. Now, put

$$w(s) = g(s)v(s), \quad s \in \Gamma(\alpha)^{\perp}.$$

Then we have, for each $s, t \in \Gamma(\alpha)^{\perp}$,

$$\begin{aligned}w(s)w(t) &= g(s)g(t)v(s)v(t) = \lambda(s,t)g(s)g(t)v(s+t)\\ &= g(s+t)v(s+t) = w(s+t).\end{aligned}$$

Thus, w is a unitary representation of $\Gamma(\alpha)^{\perp}$ into $\mathcal{C}_{\mathcal{M}^\alpha}$. We then extend w to a unitary representation u of G into $\mathcal{C}_{\mathcal{M}^\alpha}$ by Lemma 2.13. Now, we perturb α by the cocycle $\big\{u(s)^*\big\}$ to get $\beta_s = \mathrm{Ad}\big(u(s)^*\big)\alpha_s$, $s \in G$, which has the kernel $\Gamma(\alpha)^{\perp}$. Q.E.D.

Definition 2.24.

(i) An automorphism α of a von Neumann algebra $\mathcal{M}$ is called *free* if for every non-zero $a \in \mathcal{M}$ there exists $x \in \mathcal{M}$ such that $ax \neq \alpha(x)a$.

(ii) An action α of a locally compact group G on $\mathcal{M}$ is called *free* if for every $s \in G$, $s \neq e$, α_s is a free automorphism of G.

Proposition 2.25. *For a covariant system $\{\mathcal{M}, G, \alpha\}$ on a discrete group G, the following two conditions are equivalent:*

(i) *α is free;*
(ii) *The relative commutant $\mathcal{M}' \cap (\mathcal{M} \rtimes_\alpha G)$ of $\mathcal{M}$ in the crossed product $\mathcal{M} \rtimes_\alpha G$ is precisely the center $C_{\mathcal{M}}$ of $\mathcal{M}$.*

PROOF: Since G is discrete, there exists a normal faithful conditional expectation $\mathcal{E}$ from $\mathcal{N} = \mathcal{M} \rtimes_\alpha G$ onto $\mathcal{M}$ such that $\mathcal{E}(u(s)xu(s)^*) = \alpha_s(\mathcal{E}(x))$, where $\{u(s)\}$ is the unitary representation associated with the crossed product. With $x(s) = \mathcal{E}(xu(s)^*)$, x is uniquely determined by $\{x(s)\}$. In fact, x is written in the form $x = \sum_{s \in G} x(s)u(s)$ with respect to the locally convex topology given by the L^2-semi-norms $\|x\|_{\hat{\varphi}} = \hat{\varphi}(x^*x)^{1/2}$, where $\hat{\varphi} = \varphi \circ \mathcal{E}$, $\varphi \in \mathfrak{S}_*(\mathcal{M})$. For each $a \in \mathcal{M}$, we have

$$(ax)(s) = \mathcal{E}(axu(s)^*) = a\mathcal{E}(xu(s)^*) = ax(s),$$

$$(xa)(s) = \mathcal{E}(xau(s)^*) = \mathcal{E}(xu(s)^*\alpha_s(a)) = \mathcal{E}(xu(s)^*)\alpha_s(a) = x(s)\alpha_s(a).$$

Therefore, x belongs to $\mathcal{M}' \cap \mathcal{N}$ if and only if $ax(s) = x(s)\alpha_s(a)$ for every $a \in \mathcal{M}$. Thus if α is free, then $x(s) = 0$, $s \neq e$, for $x \in \mathcal{M}' \cap \mathcal{N}$, so that $x \in \mathcal{M}$. Conversely, if α is not free, then there exists a pair $(a, s) \in \mathcal{M} \times G$, $s \neq e$, such that $ay = \alpha_s(y)a$ for every $y \in \mathcal{M}$. Then $x = au(s)^*$ belongs to the relative commutant $\mathcal{M}' \cap \mathcal{N}$ but not to $\mathcal{M}$. Q.E.D.

We will use the next result later in the non-separable setting. Thus, we *do not* assume the separability here.

Proposition 2.26. *If $\{\mathcal{M}, G, \alpha\}$ is a free covariant system on a finite group G, then α is stable in the sense that every unitary α-one cocycle is a coboundary. Therefore, every perturbation of α by one cocycle is conjugate to α itself under* $\mathrm{Int}(\mathcal{M})$.

PROOF: Although we do not assume the separability for $\mathcal{M}$, we do assume the σ-finiteness for $\mathcal{M}$ to avoid arguments involving cardinal numbers. But the conclusion remains true for the general case, which we will leave to the interested reader. Decomposing $\mathcal{M}$ into a direct sum, we may assume that $\mathcal{M}$ is either finite or properly infinite.

Let $\{u_s\}$ be a unitary α-cocycle in $\mathcal{M}$. Consider the action $\tilde{\alpha}$ of G on $\mathcal{N} = \mathcal{M} \otimes M_2(\mathbf{C})$ given by the following:

$$\tilde{\alpha}_s \begin{pmatrix} x_{11} & x_{12} \\ x_{22} & x_{22} \end{pmatrix} = \begin{pmatrix} \alpha_s(x_{11}) & \alpha_s(x_{12})u_s^* \\ u_s\alpha_s(x_{21}) & u_s\alpha_s(x_{22})u_s^* \end{pmatrix}.$$

We denote by $\{e_{ij}\}$ the standard matrix unit of $M_2(\mathbf{C})$. It then follows that $\{u_s\}$ is a coboundary if and only if $e_1 = 1 \otimes e_{11} \sim e_2 = 1 \otimes e_{22}$ in $\mathcal{N}^{\tilde{\alpha}}$. Since α is free, $\tilde{\alpha}$ is also free.

Let $\{\mathcal{N}, \mathfrak{H}, U\}$ be a standard representation of $\{\mathcal{N}, G, \tilde{\alpha}\}$. Since G is finite, we have $(\mathcal{N} \cup U(G))'' \cong \mathcal{N} \rtimes_{\tilde{\alpha}} G$. Let $\tilde{\alpha}'$ be the action of G on $\mathcal{N}'$ induced by $U(G)$. Then we have also $(\mathcal{N}' \cup U(G))'' \cong \mathcal{N}' \rtimes_{\tilde{\alpha}'} G$. Now, we have

$$\mathcal{N}^{\tilde{\alpha}} = \mathcal{N} \cap U(G)' = [\mathcal{N}' \cup U(G)]',$$

so that $[\mathcal{N}' \cup U(G)]'' = (\mathcal{N}^{\tilde{\alpha}})'$. Hence $(\mathcal{N}^{\tilde{\alpha}})' \cap \mathcal{N}$ is the relative commutant of $\mathcal{N}'$ in $[\mathcal{N}' \cup U(G)]''$, which is the center $\mathcal{C}$ of $\mathcal{N}'$ by the freeness of $\tilde{\alpha}'$. Hence $(\mathcal{N}^{\tilde{\alpha}})' \cap \mathcal{N} = \mathcal{C}$. Therefore, the center of $\mathcal{N}^{\tilde{\alpha}}$ is contained in $\mathcal{C}$, so that $\mathcal{C}^{\tilde{\alpha}}$ is precisely the center of $\mathcal{N}^{\tilde{\alpha}}$. We now split the cases.

Properly Infinite Case. Assume that $\mathcal{M}$ is properly infinite. Then $\mathcal{N}$ is properly infinite too. Since G is finite, $\mathcal{N}^{\tilde{\alpha}}$ is properly infinite. Since $e_1 \sim e_2$ in $\mathcal{N}$ and $e_1 + e_2 = 1$, the central support of e_1 and e_2 in $\mathcal{N}$ are both 1, so that the central support of e_1 and e_2 in $\mathcal{N}^{\tilde{\alpha}}$ are also 1. Since e_1 and e_2 are both properly infinite, (because M^{α} is), the σ-finiteness of e_1 and e_2 implies $e_1 \sim e_2$ in $\mathcal{N}^{\tilde{\alpha}}$.

Finite Case. Suppose that $\mathcal{M}$ is finite. Then $\mathcal{N}$ is also finite. Let T be the center valued trace of $\mathcal{N}$. Then we have $T(e_1) = 1/2 = T(e_2)$. Since $\mathcal{C}_{\mathcal{N}^{\tilde{\alpha}}} = \mathcal{C}^{\tilde{\alpha}}$, the center valued trace $\tilde{T}$ of $\mathcal{N}^{\tilde{\alpha}}$ is given by

$$\tilde{T}(x) = \frac{1}{|G|} \sum_{s \in G} \tilde{\alpha}_s(T(x)), \quad x \in \mathcal{N}^{\tilde{\alpha}}.$$

Hence we have $\tilde{T}(e_1) = 1/2 = \tilde{T}(e_2)$. Thus $e_1 \sim e_2$. Q.E.D.

Exercise XI.2

1) A Hilbert space *in* a von Neumann algebra $\mathcal{M}$ means a linear subspace $\mathfrak{H}$ of $\mathcal{M}$ such that

(i) $y^*x = (x \mid y)$, $x, y \in \mathfrak{H}$, belongs to the scalar field $\mathbf{C}$;
(ii) $ax = 0$ for every $x \in \mathfrak{H}$ implies $a = 0$ for any $a \in \mathcal{M}$;
(iii) $\mathfrak{H}$ is complete relative to the norm given by the inner product of (i).

Observe that the space $\mathfrak{H}^*$ of adjoint of a Hilbert space $\mathfrak{H}$ in $\mathcal{M}$ need not be a Hilbert space in $\mathcal{M}$.

(**a**) Show that (i) if $u \in \mathcal{M}$ is unitary, then $\mathbf{C}u$ is a one dimensional Hilbert space in $\mathcal{M}$ and that every one dimensional subspace of $\mathcal{M}$ is of this form. (ii) Show that if $u, v \in \mathcal{M}$ satisfy the relation: $u^*u = v^*v = 1$, $uu^* + vv^* = 1$ then the subspace $\mathfrak{H} = \mathbf{C}u + \mathbf{C}v$ is a two dimensional Hilbert space in $\mathcal{M}$ and that every two dimensional Hilbert space in $\mathcal{M}$ is of this form.

(**b**) Show that a Hilbert space in $\mathcal{M}$ is closed under any of the operator topologies of $\mathcal{M}$.

(**c**) Let $\mathfrak{H}$ be a Hilbert space in $\mathcal{M}$. Show that if $\{u_i : i \in I\}$ is an orthonormal basis of $\mathfrak{H}$, then first each u_i is an isometry of $\mathcal{M}$ and second the range projections $\{u_i u_i^* : i \in I\}$ are orthogonal and $\sum_{i \in I} uu_i^* = 1$.

(**d**) For two Hilbert spaces $\mathfrak{H}$ and $\mathfrak{K}$ in $\mathcal{M}$, choose two isometries v and w in $\mathcal{M}$ with $vv^* + ww^* = 1$. Show that $\mathfrak{M} = v\mathfrak{H} + w\mathfrak{K}$ is a Hilbert space in $\mathcal{M}$ which is naturally isometric to the direct sum $\mathfrak{H} \oplus \mathfrak{K}$. Generalize this formulation of direct sum of Hilbert spaces in $\mathcal{M}$ to an arbitrary family of Hilbert spaces in $\mathcal{M}$.

(**e**) Show that if $\mathfrak{H}$ and $\mathfrak{K}$ are two Hilbert spaces in $\mathcal{M}$, then the closure $[\mathfrak{H}\mathfrak{K}]$ of the linear span of $\left\{ xy : x \in \mathfrak{H},\ y \in \mathfrak{K} \right\}$ is a Hilbert space in $\mathcal{M}$ naturally isometric to the Hilbert space tensor product $\mathfrak{H} \otimes \mathfrak{K}$ under the map: $xy \mapsto x \otimes y,\ x \in \mathfrak{H},\ y \in \mathfrak{K}$.

2) Let $\mathfrak{H}$ and $\mathfrak{K}$ be two Hilbert spaces in a von Neumann algebra $\mathcal{M}$.

(**a**) Show that the linear span $\mathfrak{H}\mathfrak{K}^*$ of $\left\{ xy^* : x \in \mathfrak{H},\ y \in \mathfrak{K} \right\}$ is canonically identified with the space $\mathcal{LF}(\mathfrak{K}, \mathfrak{H})$ of finite rank operators from $\mathfrak{K}$ into $\mathfrak{H}$ under the correspondence: $(xy^*)z = xy^*z = (z \mid y)x,\ x \in \mathfrak{H},\ y, z \in \mathfrak{K}$; thus $\mathcal{LF}(\mathfrak{K}, \mathfrak{H}) \subset \mathcal{M}$.

(**b**) Show further that the norm closure $\overline{\mathfrak{H}\mathfrak{K}^*}$ is then identified with $\mathcal{LC}(\mathfrak{K}, \mathfrak{H})$ the space of compact operators from $\mathfrak{K}$ into $\mathfrak{H}$ and that the σ-weak closure $\widetilde{\mathfrak{H}\mathfrak{K}^*}$ is identified with $\mathcal{L}(\mathfrak{K}, \mathfrak{H})$ the space of all bounded operators from $\mathfrak{K}$ into $\mathfrak{H}$; thus $\mathcal{L}(\mathfrak{K}, \mathfrak{H}) \subset \mathcal{M}$.

3) Let $\mathfrak{H}$ be a Hilbert space in a von Neumann algebra $\mathcal{M}$ and $\left\{ u_i : i \in I \right\}$ an orthonormal basis of $\mathfrak{H}$.

(**a**) Show that the map ρ:

$$\rho(x) = \sum_{i \in I} u_i x u_i^*, \quad x \in \mathcal{M},$$

is an injective endomorphism of $\mathcal{M}$ with range $\rho(\mathcal{M}) = \mathcal{L}(\mathfrak{H})' \cap \mathcal{M}$, the relative commutant of $\mathcal{L}(\mathfrak{H})$ in $\mathcal{M}$, so that $\mathcal{M} = \mathcal{L}(\mathfrak{H}) \overline{\otimes} \rho(\mathcal{M})$.

(**b**) Show that the endomorphism ρ of (a) does not depend on the choice of a basis $\left\{ u_i : i \in I \right\}$ of $\mathfrak{H}$. Thus denote it by $\rho_{\mathfrak{H}}$.

(**c**) Show that if $\alpha \in \mathrm{Aut}(\mathcal{M})$ leaves $\mathfrak{H}$ globally invariant, i.e. $\alpha(\mathfrak{H}) = \mathfrak{H}$, then the restriction $\alpha|_{\mathfrak{H}}$ is a unitary on $\mathfrak{H}$ and that $\rho_{\mathfrak{H}}$ maps the fixed point algebra $\mathcal{M}^\alpha$ of α into itself, i.e. $\rho_{\mathfrak{H}}(\mathcal{M}^\alpha) \subset \mathcal{M}^\alpha$.

4) Let G be a compact group and $\left\{\mathcal{M}, G, \alpha\right\}$ a homogeneous, (cf. Exercise XI.1.5), covariant system over G. To avoid triviality, assume that the action α of G is faithful. Following the steps indicated below, prove that if $\sigma \in \mathrm{Aut}(\mathcal{M})$ commutes with $\mathrm{Aut}_\alpha(\mathcal{M})$ and leaves $\mathcal{M}^\alpha$ pointwise invariant, then σ belongs to $\alpha(G)$. To avoid unnecessary complication, assume that the system is separable in the sense that G and $\mathcal{M}$ are both separable.

(**a**) Replace the system with $\left\{\mathcal{M} \overline{\otimes} \mathcal{L}(\ell^2),\ G,\ \alpha \otimes \mathrm{id}\right\}$ and σ by $\sigma \otimes \mathrm{id}$, so that $\mathcal{M}^\alpha$ is properly infinite. Show that the original claim follows if one proves the claim in the new setting. So assume that $\mathcal{M}^\alpha$ is properly infinite.

(**b**) Let $\mathcal{H}_\alpha$ be the set of Hilbert spaces $\mathfrak{H}$ in $\mathcal{M}$ such that $\alpha_s(\mathfrak{H}) = \mathfrak{H},\ s \in G$ and $\mathcal{H}(\mathcal{M}^\alpha)$ be the set of Hilbert spaces in $\mathcal{M}^\alpha$. With $\mathfrak{H} \in \mathcal{H}_\alpha$ with $\dim \mathfrak{H} < \infty$, let $\{U_\alpha, \mathfrak{H}\}$ be the unitary representation of G on $\mathfrak{H}$ given by the restriction of α to $\mathfrak{H}$, i.e. $U_\alpha(s)x = \alpha_s(x),\ x \in \mathfrak{H}$. Show that there exists a Hilbert space $\mathfrak{K} \in \mathcal{H}(\mathcal{M}^\alpha)$ and a unitary representation U of G on $\mathfrak{K}$ such that $\{U_\alpha, \mathfrak{H}\} \cong \{U, \mathfrak{K}\}$ and that $w \in \mathcal{U}(\mathfrak{H}, \mathfrak{K})$ implements the unitary equivalence if and only if $w \in \mathcal{U}(\mathcal{M})$ and

$\alpha_s(w) = wU(s),\ s \in G$. (Hint: First choose a Hilbert space $\mathfrak{K} \in \mathcal{H}(\mathcal{M}^\alpha)$ such that $\dim \mathfrak{H} = \dim \mathfrak{K}$. Let w be an isometry which maps $\mathfrak{H}$ onto $\mathfrak{K}$. By Exercise 3 above, w belongs to $\mathcal{M}$. Set $U(s) = wU_\alpha(s)w^*,\ s \in G$.)

(**c**) Let $\mathcal{H}(\mathcal{M}^\alpha, G)$ be the collection of unitary representations $\{U, \mathfrak{K}\}$ of G with $\mathfrak{K} \in \mathcal{H}(\mathcal{M}^\alpha)$ and $\dim \mathfrak{K} < \infty$. Set

$$\mathcal{M}_\alpha(U) = \{x \in \mathcal{M} : \alpha_s(x) = xU(s),\ s \in G\}, \quad \{U, \mathfrak{K}\} \in \mathcal{H}(\mathcal{M}^\alpha, G).$$

Show that if $\mathcal{M}_\alpha(U) \neq \{0\}$ with $\{U, \mathfrak{K}\} \in \mathcal{H}(\mathcal{M}^\alpha, G)$ irreducible, then there exists $\mathfrak{H} \in \mathcal{H}_\alpha$ such that $\{U_\alpha, \mathfrak{H}\} \cong \{U, \mathfrak{K}\}$. (Hint: Recalling the tensor product split: $\{\mathcal{M}, G, \alpha\} = \{\mathcal{L}(\mathfrak{K}) \overline{\otimes} \rho_{\mathfrak{K}}(\mathcal{M}),\ G,\ \mathrm{id} \otimes \alpha \mid \rho(\mathcal{M})\}$, consider a new covariant system $\{\mathcal{M}_2 = M_2(\mathbf{C}) \otimes \mathcal{M},\ G,\ \bar{\alpha}\}$:

$$\bar{\alpha}_s \begin{pmatrix} x_{11} & x_{12} \\ x_{21} & x_{22} \end{pmatrix} = \begin{pmatrix} \alpha_s(x_{11}) & U(s)\alpha_s(x_{12}) \\ \alpha_s(x_{21})U(s)^* & \mathrm{Ad}\big(U(s)\big) \circ \alpha_s(x_{22}) \end{pmatrix}$$

for every $s \in G,\ (x_{ij}) \in \mathcal{M}_2$. Observe that $e_{12} \otimes x \in \mathcal{M}_2^{\bar{\alpha}} \iff x \in \mathcal{M}_\alpha(U)$, $x \in \mathcal{M}$, with $\{e_{ij}\}$ the standard matrix unit in $M_2(\mathbf{C})$. As in Exercise XI.1.5, show that $e_{11} \otimes 1 \sim e_{22} \otimes 1$ in $\mathcal{M}_2^{\bar{\alpha}}$ which shows the existence of a unitary $w \in \mathcal{M}_\alpha(U)$. Then set $\mathfrak{H} = w\mathfrak{K}$.)

(**d**) For each $x \in \mathcal{M}$, set $\mathrm{Orb}_\alpha(x) = \{\alpha_s(x) : s \in G\}$, the orbit of x. Let $\mathcal{M}_\alpha^f = \{x \in \mathcal{M} : \mathrm{Orb}_\alpha(x) \text{ lies in a finite dimensional subspace}\}$. Show that $\mathcal{M}_\alpha^f$ is a σ-weakly dense $*$-subalgebra of $\mathcal{M}$. (Hint: If $f \in A(G)$ is of the form: $f(s) = \big(U(s)\xi \mid \eta\big),\ \xi, \eta \in \mathfrak{H}$ with $\{U, \mathfrak{H}\}$ a finite dimensional unitary representation of G, then $\alpha_f(x) \in \mathcal{M}_\alpha^f$ for any $x \in \mathcal{M}$.)

(**e**) Recall that if $\mathcal{V} \subset \mathcal{M}$ is an α-invariant finite dimensional subspace then there exists a basis $\{a_i : 1 \leq i \leq n\}$ of $\mathcal{V}$ such that $\alpha_s(a_i) = \sum_{j=1}^n a_j u_{ji}(s),\ s \in G$ with $\{(u_{ij}(s)) : s \in G\}$ a unitary matrix representation of G. Show that if the above representation $U = \{(u_{ij}(s)) : s \in G\}$ is irreducible, then $\mathcal{M}_\alpha(U)$ contains a unitary. (Hint: In view of (c), we need only to show that $\mathcal{M}_\alpha(U) \neq \{0\}$. Let $\mathfrak{K} \in \mathcal{H}(\mathcal{M}^\alpha, G)$ which supports the representation U of G, i.e. $\mathfrak{K}$ is a Hilbert space in $\mathcal{M}^\alpha$ and $\dim \mathfrak{K} = n$. With $\{e_i : 1 \leq i \leq n\}$ an orthonormal basis of $\mathfrak{K}$, set:

$$a = \sum_{i=1}^n e_1 e_i^* \rho_{\mathfrak{K}}(a_i).$$

and prove that $a \in \mathcal{M}_\alpha(U)$).

(**f**) Show that if $\mathcal{M}_\alpha(U) \neq \{0\}$ then $\mathcal{M}_\alpha(\bar{U}) \neq \{0\}$ where $\bar{U}$ means the representation conjugate to U, i.e. if U is given by a matrix representation $\{(u_{ij}(s)) : s \in G\}$ then $\bar{U}$ is given by $\{\overline{(u_{ij}(s))} : s \in G\}$. (Hint: Suppose that $\mathfrak{H} \in \mathcal{H}_\alpha(\mathcal{M})$ carries the representation U, i.e. $\{U_\alpha, \mathfrak{H}\}$ is unitarily equivalent to U. Then $\mathcal{V} = \mathfrak{H}^*$ is an α-invariant subspace of $\mathcal{M}$ such that the restriction of α to $\mathcal{V}$ gives the representation $\bar{U}$.)

(**g**) Show that the collection of representations of G given by a element of $\mathcal{H}_\alpha$ is closed under direct sum operations, tensor product operations and conjugation operations, i.e. it is a self adjoint subring of the ring of unitary representations of G.

(Hint: If $\{\mathfrak{H}_i : 1 \le i \le n\}$ is a finite family of members of $\mathcal{H}_\alpha$, then choose a family $\{w_i : 1 \le i \le n\}$ of isometries in $\mathcal{M}^\alpha$ with $\sum_{i=1}^n w_i w_i^* = 1$ and set $\mathfrak{H} = \sum_{i=1}^n w_i \mathfrak{H}_i \in \mathcal{H}_\alpha$ which shows that $\{U_\alpha, \mathfrak{H}\} = \sum_{1\le i\le n}^{\oplus} \{U_\alpha, \mathfrak{H}_i\}$. For tensor product, put $\mathfrak{H} = [\mathfrak{H}_1\mathfrak{H}_2]$ and show that $\{U_\alpha, \mathfrak{H}\} \cong \{U_\alpha, \mathfrak{H}_1\} \otimes \{U_\alpha, \mathfrak{H}_2\}$. The stability under the conjugation is somewhat tricky. Making use of the homogeneity of α, show that any subrepresentation of a representation carried by a member of $\mathcal{H}_\alpha$ is carried by a member of $\mathcal{H}_\alpha$. If a finite dimensional representation U is carried by a member of $\mathcal{H}_\alpha$, then the n-th power $U^{\otimes n}$, with $n = \dim(U)$, contains the determinant representation $\det(U(\,\cdot\,))$ which is carried by the one dimensional Hilbert space in $\mathcal{H}_\alpha$ and therefore by the space spanned by a single unitary operator, say $u \in \mathcal{U}(\mathcal{M})$. Hence its conjugate is carried by the one dimensional Hilbert space generated by the adjoint u^*, so that it is carried by a member of $\mathcal{H}_\alpha$. Therefore the product representation $U^{\otimes n} \otimes \overline{\det}(U)$ contains the trivial representaion; hence $U^{\otimes(n-1)} \otimes \overline{\det}(U)$ contains the conjugate representation $\bar{U}$ of the original representation.)

(**h**) Show that every irreducible unitary representation of G is carried by some element of $\mathcal{H}_\alpha$, up to unitary equivalence. (Hint: Use the faithfulness of α and the fact that the matrix coefficients of representations carried by members of $\mathcal{H}_\alpha$ form a self-adjoint subalgebra of $A(G)$ which separates the point by (g).)

(**i**) Show that there exists $\mathfrak{H} \in \mathcal{H}_\alpha$ such that $\{U_\alpha, \mathfrak{H}\} \cong \{\lambda, L^2(G)\}$ with λ the regular representation of G.

(**j**) Show that the automorphism σ leaves $\mathfrak{H} \in \mathcal{H}_\alpha$ globally invariant. (Hint: Choose $\mathfrak{K} \in \mathcal{H}(\mathcal{M}^\alpha)$ with $\dim \mathfrak{H} = \dim \mathfrak{K}$ and a unitary $w \in \mathcal{L}(\mathfrak{K}, \mathfrak{H}) \subset \mathcal{M}$ and set $U(s) = w^* U_\alpha(s) w$, $s \in G$ so that $\alpha_s(w) = wU(s)$, $s \in G$. Applying the techniques in Exercise XI.1.4, show that $w^*\sigma(w) \in \mathcal{L}(\mathfrak{K})$ so that $\sigma(\mathfrak{H}) = \sigma(w)\mathfrak{K} = w\mathfrak{K} = \mathfrak{H}$.)

(**k**) Complete the proof by applying the Tannaka duality theorem.

5) Let $\{U, V, L^2(\mathbf{R})\}$ be the pair of one parameter unitary groups defined in §X.2 satisfying the Heisenberg-Weyl commutation relation, i.e.

$$\begin{cases} U(s)\xi(r) = \xi(r+s), & \xi \in L^2(\mathbf{R}); \\ V(t)\xi(r) = \mathrm{e}^{irt}\xi(r), & r, s, t \in \mathbf{R}. \end{cases}$$

For each $(s,t) \in \mathbf{R}^2$, set $\alpha_{(s,t)} = \mathrm{Ad}\big(U(s)V(t)\big)$ on $\mathcal{B} = \mathcal{L}\big(L^2(\mathbf{R})\big)$.

(**a**) Show that α is an integrable action of $\mathbf{R}^2$ on $\mathcal{B}$.

(**b**) Show that $\Gamma(\alpha) = \mathbf{R}^2$.

(**c**) Show that $\{\beta \in \mathrm{Aut}(\mathcal{B}) : \beta \circ \alpha_{(s,t)} = \alpha_{(s,t)} \circ \beta,\ (s,t) \in \mathbf{R}^2\} = \alpha(\mathbf{R}^2)$.

(**d**) Show that the correspondence: $H \longleftrightarrow \mathcal{B}^H = \{x \in \mathcal{B} : x = \alpha_{(s,t)}(x),\ (s,t) \in H\}$ is a bijection between closed subgroups H of $\mathbf{R}^2$ and α-invariant von Neumann subalgebras of $\mathcal{B}$.

(**e**) Show that $\{x \in \mathcal{B} : \lim_{g\to 0}\|x - \alpha_g(x)\| = 0\} = \mathcal{LC}(L^2(\mathbf{R}))$.

§ 3 Derivations and Inner Automorphisms

Definition 3.1. A *derivation* of a C^*-algebra A is a linear map δ of A into itself such that

$$\delta(xy) = \delta(x)y + x\delta(y), \quad x, y \in A. \tag{1}$$

If $\delta(x^*) = \delta(x)^*$, then it is called a **-derivation*. For each $h \in A$, we define a derivation $\mathrm{ad}(h)$ by:

$$\mathrm{ad}(h)(x) = [h, x] = hx - xh, \quad x \in A. \tag{2}$$

Such a derivation is said to be *inner*.

By setting $\delta^*(x) = \delta(x^*)^*$ for any derivation δ, we get a new derivation δ^*, and δ is uniquely decomposed into *-derivations:

$$\delta = \frac{1}{2}(\delta + \delta^*) + \frac{\mathrm{i}}{2\mathrm{i}}(\delta - \delta^*).$$

Unless otherwise stated explicitly, we shall consider only *-derivations in the sequel.

Proposition 3.2. *A derivation of a C^*-algebra is bounded.*

PROOF: Let δ be a *-derivation of A. If A is not unital, we extend δ to $A_I = A + \mathbf{C}1$ by setting $\delta(1) = 0$. Note that any derivation of a unital algebra kills the identity because $\delta(1) = \delta(1)1 + 1\delta(1) = 2\delta(1)$.

We first show that $\varphi(\delta(x)) = 0$ if $x \in A_+$ and $\varphi \in \mathfrak{S}(A)$ satisfies $\varphi(x) = \|x\|$. Assuming $\|x\| = 1$ which we may, we put $y = (1 - x)^{1/2}$, and compute:

$$|\varphi(\delta(x))| = \left|\varphi(\delta(y^2))\right| = \left|\varphi(\delta(y)y + y\delta(y))\right|,$$

$$\left|\varphi(\delta(y)y)\right|^2 \leq \varphi(\delta(y)^2)\varphi(y^2) = \varphi(\delta(y)^2)(1 - \varphi(x)) = 0,$$

$$\left|\varphi(y\delta(y))\right|^2 \leq \varphi(y^2)\varphi(\delta(y)^2) = 0.$$

Thus we get $\varphi(\delta(x)) = 0$.

Let $\{x_n\}$ be a sequence in A_h such that $x_n \to 0$ and $y_n = \delta(x_n) \to y$. If we have proved $y = 0$ for all such $\{x_n\}$, then the closed graph theorem implies the boundedness of δ on A_h, hence on A. Suppose $y \neq 0$. Replacing x_n by $x_n/\|y\|$, we may assume $\|y\| = 1$. If necessary, replacing y by $-y$, we may assume that $1 \in \mathrm{Sp}_A(y)$. Since $\delta(x_n + \lambda) = \delta(x_n)$, we may replace x_n by $x_n + \|x_n\|$, so that we assume $x_n \in A_+$. Choose $\varphi_n \in \mathfrak{S}(A)$ so that $\varphi_n(y_+ + x_n) = \left\|y_+ + x_n\right\|$ where y_+ means of course the positive part of y, and let φ be an accumulation point of $\{\varphi_n\}$ in $\mathfrak{S}(A)$. By the above arguments, we have

$$\varphi_n(\delta(y_+) + \delta(x_n)) = 0.$$

Since $\|x_n\| \to 0$ and $\left\|y - \delta(x_n)\right\| \to 0$, we have

$$\varphi(y_+) = \|y_+\| \qquad \text{and} \qquad \varphi(\delta(y_+) + y) = 0.$$

Hence again by the above arguments, $\varphi\big(\delta(y_+)\big) = 0$, so that $\varphi(y) = 0$. But $\varphi(y_+) = \|y_+\| = 1$. Hence $\varphi(y_-) = 1$. Then $\varphi(y_+ + y_-) = 2$ whilst $\big\||y|\big\| = 1$. This is impossible because φ is a state. Q.E.D.

Proposition 3.3. *If δ is a derivation of A, then*

$$\alpha_t = \exp(t\delta) = \sum_{n=0}^{\infty} \frac{(t\delta)^n}{n!}, \qquad t \in \mathbf{R}, \tag{3}$$

defines a uniformly continuous one parameter automorphism group of A. Conversely, every uniformly continuous one parameter automorphism group of A is of this form. Here the uniform continuity refers to the following convergence:

$$\lim_{t \to 0} \big\|\alpha_t - \mathrm{id}\big\| = 0.$$

PROOF: Since δ is bounded, the power series (3) converges and defines in fact an entire $\mathcal{L}(A)$-valued function over $\mathbf{C}$. Hence in particular, $t \mapsto \alpha_t$ is differentiable. Furthermore, we know that $\alpha_{s+t} = \alpha_s \circ \alpha_t$ for $s, t \in \mathbf{R}$. Hence $\{\alpha_t\}$ is a uniformly continuous one parameter group of invertible $*$-preserving operators on A.

By induction, the derivation formula (1) is extended to the Leibniz formula:

$$\delta^n(xy) = \sum_{k=0}^{n} \binom{n}{k} \delta^k(x)\delta^{n-k}(y). \tag{4}$$

We then have

$$\begin{aligned}
(\exp\delta)(x)\,(\exp\delta)(y) &= \sum_{k=0}^{\infty}\sum_{j=0}^{\infty} \frac{1}{k!}\frac{1}{j!}\,\delta^k(x)\,\delta^j(y) \\
&= \sum_{n=0}^{\infty}\sum_{k=0}^{n} \frac{1}{k!}\frac{1}{(n-k)!}\,\delta^k(x)\,\delta^{n-k}(y) \\
&= \sum_{n=0}^{\infty} \frac{1}{n!}\,\delta^n(xy) = (\exp\delta)(xy).
\end{aligned}$$

Thus each α_t is an automorphism of A.

Conversely, suppose that $\{\alpha_t\}$ is a uniformly continuous one parameter automorphism group of A. Then $\{\alpha_t\}$ is of the form: $\alpha_t = \exp(t\delta)$ for some $\delta \in \mathcal{L}(A)$, and δ is given by the differentiation in norm:

$$\delta = \lim_{t \to 0} \frac{1}{t}\big(\alpha_t - \mathrm{id}\big).$$

Then the derivation identity (1) can be seen by differentiating both sides of the identity $\alpha_t(xy) = \alpha_t(x)\alpha_t(y)$ at $t = 0$. The fact that δ preserves the $*$-operation follows from that of α_t. Q.E.D.

Lemma 3.4. *If δ is a derivation of a C^*-algebra A, then δ leaves every closed ideal $\mathcal{I}$ of A globally invariant. Hence it induces naturally a derivation $\bar{\delta}$ of the quotient C^*-algebra $A/\mathcal{I}$. Furthermore, if π is a representation of A, then there exists a derivation $\tilde{\delta}$ on the von Neumann algebra $\pi(A)''$ such that $\pi \circ \delta = \tilde{\delta} \circ \pi$. Furthermore, any derivation on a von Neumann algebra is σ-weakly continuous.*

PROOF: Let $\mathcal{I}$ be a closed ideal of A. Let $\{e_i\}$ be an approximate identity of $\mathcal{I}$. We then have, for every $x \in \mathcal{I}$,

$$\delta(x) = \lim \delta(e_i x) = \lim\bigl(\delta(e_i)x + e_i\delta(x)\bigr) \in \mathcal{I}.$$

Thus $\delta(\mathcal{I}) \subset \mathcal{I}$.

Let $\tilde{\delta}$ be the second transpose of δ on the universal enveloping von Neumann algebra $\widetilde{A} = A^{**}$. Then $\tilde{\delta}$ is also a derivation of $\widetilde{A}$. If V is a norm closed invariant subspace of A^*, then the polar $\mathcal{I} = V^0$ is a closed ideal of A, so that $\tilde{\delta}(\mathcal{I}) \subset \mathcal{I}$. Hence the bipolar theorem for norm closed convex sets in A^* implies that ${}^t\delta(V) \subset V$. The rest follows from §2, Chapter III. Q.E.D.

Theorem 3.5. *Every derivation δ of a von Neumann algebra $\mathcal{M}$ is inner. More precisely, $\delta = \mathrm{i}\,\mathrm{ad}(h)$ for some $h \in \mathcal{M}_h$.*

PROOF: Let δ be a derivation of a von Neumann algebra $\mathcal{M}$. As the innerness of all *-derivations on $\mathcal{M}$ implies the innerness of every derivation on $\mathcal{M}$, we may assume that δ is a *-derivation. Put $\alpha_t = \exp(t\delta)$, $t \in \mathbf{R}$. By Theorem 1.23, we have $\mathrm{Sp}(\alpha) = -\mathrm{i}\,\mathrm{Sp}(\delta) \subset \bigl[-\|\delta\|, \|\delta\|\bigr]$. With $q^\alpha\bigl([t, +\infty[\bigr)$ defined by 1.10), $q^\alpha\bigl([t, \infty[\bigr) = 0$ for $t > \|\delta\|$. Hence by Theorem 1.10, there exists a one parameter unitary group $\bigl\{u(t)\bigr\}$ in the center $\mathcal{C}_\alpha$ of $\mathcal{M}^\alpha$ such that $\alpha_t = \mathrm{Ad}\,u(t)$. Furthermore, $u(t)$ is constructed by the spectral integral:

$$u(t) = \int \mathrm{e}^{\mathrm{i}t\lambda}\, de(\lambda), \qquad 1 - e(\lambda) = \inf\Bigl\{ q^\alpha\bigl([s, \infty[\bigr) : s < \lambda \Bigr\}.$$

Hence $h = \int \lambda\, de(\lambda)$ is bounded and $u(t) = \exp(ith)$. Now we have for any $x \in \mathcal{M}$,

$$\begin{aligned}\delta(x) &= \frac{\mathrm{d}}{\mathrm{d}t}\bigl[\exp(ith)x\exp(-ith)\bigr]\Big|_{t=0}\\ &= \frac{\mathrm{d}}{\mathrm{d}t}\exp(ith)x\Big|_{t=0} + \frac{\mathrm{d}}{\mathrm{d}t}x\exp(-ith)\Big|_{t=0}\\ &= ihx - ixh = \bigl[ih, x\bigr].\end{aligned}$$

Thus $\delta = \mathrm{ad}(ih)$, so that δ is inner. Q.E.D.

Let Log be the principal logarithm defined on the domain $\mathbf{C}'$ obtained by deleting the negative real axis from $\mathbf{C}$.

Theorem 3.6. *Let A be a C^*-algebra and $\alpha \in \mathrm{Aut}(A)$. If $\|\alpha - \mathrm{id}\|_{\mathrm{sp}} < \sqrt{3}$ or $\|\alpha - \mathrm{id}\| < 2$, then $\mathrm{Log}(\alpha) = \delta$ is a derivation of A.*

To prove the theorem, we prepare the next lemma.

Lemma 3.7.

(i) *Every automorphism α of $\mathcal{L}(\mathfrak{H})$ is inner.*

(ii) *If $\alpha \in \mathrm{Aut}(\mathcal{L}(\mathfrak{H}))$ satisfies the condition of Theorem 3.6, then there exists $u \in \mathcal{U}(\mathfrak{H})$ such that $\alpha = \mathrm{Ad}(u)$ and $\mathrm{Sp}(u)$ is contained in the right half plane.*

PROOF:

(i) Let $e \in \mathcal{L}(\mathfrak{H})$ be a projection with $\dim e = 1$. Then $\alpha(e)$ is also minimal in $\mathcal{L}(\mathfrak{H})$ so that there exists a unitary $u \in \mathcal{L}(\mathfrak{H})$ such that $u\alpha(e)u^* = e$. Let $\beta = \mathrm{Ad}(u) \circ \alpha$. Then $e \in \mathcal{L}(\mathfrak{H})^\beta$ and $\mathcal{L}(\mathfrak{H})_e = \mathbf{C}$. Hence $\beta^e = \mathrm{id}$. Thus, Lemma 2.11 implies that β is inner, hence α itself is inner.

(ii) Choose $u \in \mathcal{U}(\mathfrak{H})$ so that $\alpha = \mathrm{Ad}(u)$. By Lemma 2.16, we have $\mathrm{Sp}(\alpha) = \mathrm{Sp}(u)\,\mathrm{Sp}(u)^{-1}$. Let

$$u = \int_{-\pi}^{\pi} e^{i\lambda}\, dp(\lambda)$$

be the spectral decomposition and set $h = \int_{-\pi}^{\pi} \lambda\, dp(\lambda)$. Let $a = \inf \mathrm{Sp}(h)$ and $b = \sup \mathrm{Sp}(h)$. Since any rotation u by $\mathbf{T}$ does not have an effect on α, we may assume $a = -b$, so that $\mathrm{Sp}(h) \subset [-b, b]$. Then we have $\pm b \in \mathrm{Sp}(h)$. Hence $e^{2ib} \in \mathrm{Sp}(\alpha)$. Since $-1 \notin \mathrm{Sp}(\alpha)$, we have $2b < \pi$, i.e. $b < \pi/2$. Therefore, $\mathrm{Sp}(u) \subset \{e^{i\lambda} : -\pi/2 < \lambda < \pi/2\}$. Q.E.D.

PROOF OF THEOREM 3.6: Let $\mathcal{M}$ denote the universal enveloping von Neumann algebra A^{**} of A, and $\tilde{\alpha}$ the second transpose of α. We first note that $\mathrm{Sp}(\tilde{\alpha}) = \mathrm{Sp}(\alpha)$ and $\|\tilde{\alpha} - \mathrm{id}\| = \|\alpha - \mathrm{id}\|$.

We prove that $\tilde{\alpha}$ leaves the center $\mathcal{C}$ pointwise fixed under the assumption of the theorem. First, assume $\|\tilde{\alpha} - \mathrm{id}\| < 2$. For any $e \in \mathrm{Proj}(\mathcal{C})$, $1 - 2e \in \mathcal{U}(\mathcal{C})$ and

$$2 > \|1 - 2e - 1 - 2\alpha(e)\| = 2\|e - \alpha(e)\|,$$

so that $\|e - \alpha(e)\| < 1$. Since e and $\alpha(e)$ commute, we have $e = \alpha(e)$. Thus α is the identity on $\mathcal{C}$. Next, assume $\|\alpha - \mathrm{id}\|_{\mathrm{sp}} < \sqrt{3}$. If $\lambda \in \mathrm{Sp}(\alpha|_{\mathcal{C}})$, then for any $\varepsilon > 0$ there exists an $x \in \mathcal{C}$ with $\|x\| = 1$ such that $\|\alpha(x) - \lambda x\| < \varepsilon$. Since x is normal, we have $\|x^n\| = \|x\|^n = 1$ and for $n \in \mathbf{Z}_+$

$$\begin{aligned}\|\alpha(x^n) - \lambda^n x^n\| &= \|\alpha(x)^n - (\lambda x)^n\| \\ &= \left\|\big(\alpha(x) - \lambda x\big)\big(\alpha(x)^{n-1} + \lambda x \alpha(x)^{n-2} + \cdots + (\lambda x)^{n-1}\big)\right\| \le n\varepsilon.\end{aligned}$$

Thus $\lambda^n \in \mathrm{Sp}(\alpha|_{\mathcal{C}})$. Therefore we have $|\lambda^n - 1| < \sqrt{3}$ for all $n \in \mathbf{Z}$ and $\lambda \in \mathrm{Sp}(\alpha|_{\mathcal{C}})$, since $\mathrm{Sp}(\alpha|_{\mathcal{C}}) = \mathrm{Sp}(\alpha|_{\mathcal{C}})^{-1}$. This means that $\mathrm{Sp}(\alpha|_{\mathcal{C}}) = \{1\}$. Hence α is the identity on $\mathcal{C}$ by Corollary 1.14.

Let $z_0 \in \mathcal{M}$ be the support projection of the universal atomic representation of A and put $\mathcal{M}_0 = \mathcal{M}_{z_0}$. Then we have $\mathcal{M}_0 \cong \sum_{j\in I}^{\oplus} \mathcal{L}(\mathfrak{H}_i)$ for some family of Hilbert spaces. Let π_j be the irreducible representation of A on $\mathfrak{H}_j$ for each $j \in I$. By the above arguments, $\tilde{\alpha}$ leaves each $\mathcal{L}(\mathfrak{H}_j)$ globally invariant and $\tilde{\alpha} \circ \pi_j = \pi_j \circ \alpha$. Let u_j be a unitary on $\mathfrak{H}_j$ such that $\tilde{\alpha}|_{\mathcal{L}(\mathfrak{H}_j)} = \mathrm{Ad}(u_j)$. By Lemma 3.7, we choose each u_j in such a way that $\mathrm{Sp}(u_j)$ lies in the right half plane. Let $u = \sum_{j\in I}^{\oplus} u_j \in \mathcal{M}_0$. Then $\tilde{\alpha} = \mathrm{Ad}(u)$ on $\mathcal{M}_0$. Let $h = \frac{1}{\mathrm{i}} \operatorname{Log} u$. Now put $L_a x = ax$ and $Rx = xa$ for each $a, x \in \mathcal{M}_0$. Then we have

$$\operatorname{Log} \tilde{\alpha} = \operatorname{Log} L_u + \operatorname{Log} R_u^{-1} = \mathrm{i}\,(L_h - R_h) = \mathrm{i}\,\mathrm{ad}(h).$$

Since Log is approximated uniformly by polynomials on $\mathrm{Sp}(\tilde{\alpha})$, $\tilde{\delta}$ leaves $\pi_0(A)$ globally invariant where $\pi_0 = \sum^{\oplus} \pi_j$. Since $\pi_0 \circ \alpha = \tilde{\alpha} \circ \pi_0$, we have $\pi_0 \circ \delta = \delta \circ \pi_0$. Since $\tilde{\delta} = \mathrm{i}\,\mathrm{ad}(h)$ is a derivation of $\mathcal{M}_0$, δ is also a derivation of A. Q.E.D.

Corollary 3.8.

(i) *If* $\alpha \in \mathrm{Aut}(\mathcal{M})$ *satisfies either* $\|\alpha - \mathrm{id}\| < 2$ *or* $\|\alpha - \mathrm{id}\|_{\mathrm{sp}} < \sqrt{3}$, *then* α *is inner.*

(ii) *In the norm topology, the principal connected component of* $\mathrm{Aut}(\mathcal{M})$ *is precisely the group* $\mathrm{Int}(\mathcal{M})$ *of all inner automorphisms.*

PROOF:

(i) Let $\delta = \operatorname{Log} \alpha$. Then δ is a derivation of $\mathcal{M}$, hence $\delta = \mathrm{i}\,\mathrm{ad}(h)$ with $h \in \mathcal{M}_h$ by Theorem 3.5. Let $u = \exp(\mathrm{i}h)$. Then we have

$$\alpha = \exp \delta = \exp(\mathrm{i}\,\mathrm{ad}(h)) = \mathrm{ad}(\exp(\mathrm{i}h)) = \mathrm{ad}(u).$$

(ii) By (i), $\{\alpha \in \mathrm{Aut}(\mathcal{M}) : \|\alpha - \mathrm{id}\| < 2\}$ is contained in $\mathrm{Int}(\mathcal{M})$. Hence $\mathrm{Int}(\mathcal{M})$ is an open subgroup of $\mathrm{Aut}(\mathcal{M})$, hence it is closed. On the other hand, $\mathcal{U}(\mathcal{M})$ is connected because every $u \in \mathcal{U}(\mathcal{M})$ is of the form $u = \exp(\mathrm{i}h)$ with $h \in \mathcal{M}_h$, so that $\mathrm{Int}(\mathcal{M})$ is connected, being a homomorphism image of $\mathcal{U}(\mathcal{M})$. Q.E.D.

Remark 3.9. On $L^\infty(\mathbf{T})$, consider an automorphism α defined by

$$(\alpha f)(\lambda) = f(e^{2\pi \mathrm{i}/3}\lambda), \quad f \in L^\infty(\mathbf{T}), \quad \lambda \in \mathbf{T}.$$

Clearly $\alpha^3 = \mathrm{id}$ and $\mathrm{Sp}(\alpha) = \{1, e^{2\pi\mathrm{i}/3}, e^{4\pi\mathrm{i}/3}\}$, but $\alpha \notin \mathrm{Int}(L^\infty(\mathbf{T}))$. Hence $\|\alpha - \mathrm{id}\| = 2$ and $\|\alpha - \mathrm{id}\|_{\mathrm{sp}} = \sqrt{3}$. This shows a remarkably delicate nature of the spectral analysis of automorphisms.

Proposition 3.10. *If* $\alpha \in \mathrm{Aut}(\mathcal{M})$, *then the following conditions are equivalent:*

(i) α *is inner;*

(ii) *For every* $\varepsilon > 0$, *there exists* $e \in \mathrm{Proj}(\mathcal{M}^\alpha)$ *with* $z(e) = 1$, *where* $z(e)$ *means the central support of* e, *such that* $\left\|(\mathrm{id} - \alpha)|_{\mathcal{M}_e}\right\| < \varepsilon$.

PROOF:

(i) $\Longrightarrow$ (ii): Suppose $\alpha = \mathrm{Ad}(u)$ with $u \in \mathcal{U}(\mathcal{M})$. Let e_1 be a spectral projection of u such that $\left\|ue_1 - \lambda e_1\right\| < \varepsilon/2$ for some $\lambda \in \mathbf{C}$, $|\lambda| = 1$. Then we have that $e_1 \in \mathcal{M}^\alpha$ and for any $x \in \mathcal{M}_{e_1}$,

$$\begin{aligned}\left\|uxu^* - x\right\| &\le \left\|(u - \lambda e_1)xu^*\right\| + \left\|\lambda e_1 x(u^* - \bar{\lambda} e_1)\right\| \\ &\le \frac{\varepsilon}{2}\|x\| + \frac{\varepsilon}{2}\|x\| = \varepsilon\|x\|.\end{aligned}$$

Let $\{e_i\}$ be a maximal family of spectral projections of u such that $\left\|ue_i - \lambda_i\right\| < \varepsilon/2$ for some $\lambda_i \in \mathbf{C}$ with $|\lambda_i| = 1$ and $\{e_i\}$ are centrally orthogonal. Put $e = \sum_{i \in I} e_i$. It then follows that e satisfies the requirement.

(ii) $\Longrightarrow$ (i): By Corollary 3.8, α^e is inner if (ii) holds with $\varepsilon = 2$. It then follows from Lemma 2.11 that α is inner. Q.E.D.

Theorem 3.11. *If* $\{\mathcal{M}, \mathbf{R}, \alpha\}$ *is a separable covariant system over the additive group* $\mathbf{R}$ *of real numbers such that each individual* α_t *is inner, then there exists a* σ*-strongly continuous one parameter unitary group* $\{u(s)\}$ *such that* $\alpha_t = \mathrm{Ad}\big(u(t)\big)$, $t \in \mathbf{R}$.

In the theorem, the separability of $\mathcal{M}$ is essential.

By the separability of $\mathcal{M}$, there exists a Borel map: $s \in \mathbf{R} \mapsto u(s) \in \mathcal{U}(\mathcal{M})$ with $\alpha_s = \mathrm{Ad}\big(u(s)\big)$.

Lemma 3.12. *The unitary* $u(s)$ *belongs to* $\mathcal{M}^\alpha$ *for each* $s \in \mathbf{R}$. *In particular, we have*

$$u(s)u(t) = u(t)u(s), \qquad s, t \in \mathbf{R}.$$

PROOF: Since $\mathrm{Ad}\big(\alpha_t(u(s)\big) = \alpha_t \circ \alpha_s \circ \alpha_t^{-1} = \alpha_s = \mathrm{Ad}\big(u(s)\big)$, there exists a Borel map $\chi : (s, t) \in \mathbf{R} \times \mathbf{R} \mapsto \chi(s, t) \in \mathcal{U}(\mathcal{C})$, such that $\alpha_t\big(u(s)\big) = \chi(s, t)u(s)$, where $\mathcal{C}$ is the center of $\mathcal{M}$. It then follows that

$$\begin{cases} \chi(s, t) = \chi(t, s)^*, \\ \chi(r, s + t) = \chi(r, s)\chi(r, t), \qquad r, s, t \in \mathbf{R}. \end{cases}$$

Since α_t is continuous in t, $\chi(r, s)$ is σ-strongly separately continuous in r and s. For each $m \in \mathbf{Z}$, we have $\chi(mr, s) = \chi(r, ms)$, so that for any $m, n \in \mathbf{Z}$, $n \neq 0$,

$$\chi\left(\frac{m}{n}, s\right) = \chi\left(\frac{1}{n}, n\left(\frac{m}{n}\right)s\right) = \chi\left(1, \frac{m}{n}s\right);$$

thus $\chi(p, q) = \chi(1, pq)$ for any $p, q \in \mathbf{Q}$. By continuity, we have $\chi(s, t) = \chi(1, st)$ for every $s, t \in \mathbf{R}$. The anti-symmetry of χ then implies that $\chi(s, t) = 1$ for every $s, t \in \mathbf{R}$. Hence $\alpha_t\big(u(s)\big) = u(s)$. Q.E.D.

PROOF OF THEOREM 3.11: Consider $\mathcal{N} = \mathcal{M} \rtimes_\alpha \mathbf{R}$, and let $\{v(s) : s \in \mathbf{R}\}$ be the one parameter unitary group in $\mathcal{N}$ corresponding to the crossed product. Let $\hat{\alpha}$ be the dual action of $\mathbf{R} = \widehat{\mathbf{R}}$ on $\mathcal{N}$, so that $\hat{\alpha}_p(v(s)) = \mathrm{e}^{-\mathrm{i}sp}v(s),\ s \in \mathbf{R},\ p \in \widehat{\mathbf{R}}$. By the previous lemma $v(s)$ and $u(t)$ commute. Set $w(s) = u(s)v(s)^*,\ s \in \mathbf{R}$. Then $\{w(s)\}$ belong to the relative commutant $\mathcal{M}' \cap \mathcal{N}$ and generate an abelian von Neumann algebra $\mathcal{A}$. Since $\hat{\alpha}_p(w(s)) = \mathrm{e}^{\mathrm{i}sp}w(s)$, $\hat{\alpha}$ leaves $\mathcal{A}$ globally invariant.

We claim that

$$\left\{\mathcal{A} \,\overline{\otimes}\, \mathcal{L}(L^2(\mathbf{R})),\ \mathbf{R},\ \hat{\alpha} \otimes \mathrm{id}\right\} \cong \left\{\mathcal{A} \,\overline{\otimes}\, \mathcal{L}(L^2(\mathbf{R})),\ \mathbf{R},\ \hat{\alpha} \otimes \rho\right\},$$

where ρ is given by the one parameter unitary group $\{\rho(p)\}$:

$$\rho(p)\xi(s) = \mathrm{e}^{\mathrm{i}sp}\xi(s), \quad \xi \in L^2(\mathbf{R}).$$

Set $W = \int_{\mathbf{R}}^{\oplus} w(s)\,\mathrm{d}s \in \mathcal{A} \,\overline{\otimes}\, L^\infty(\mathbf{R}) \subset \mathcal{A} \,\overline{\otimes}\, \mathcal{L}(L^2(\mathbf{R}))$. Then we have

$$(\hat{\alpha}_p \otimes \mathrm{id})(W) = \int_{\mathbf{R}}^{\oplus} \mathrm{e}^{\mathrm{i}sp} w(s)\,\mathrm{d}s = \big(1 \otimes \rho(p)\big)W,$$

so that we obtain

$$\hat{\alpha}_p \otimes \mathrm{Ad}\,\rho(p) = \mathrm{Ad}(W^*) \circ (\hat{\alpha}_p \otimes \mathrm{id}) \circ \mathrm{Ad}(W),$$

which establishes the claim.

Let e be a minimal projection of $\mathcal{L}(L^2(\mathbf{R}))$. Then $\{\mathcal{A}, \mathbf{R}, \hat{\alpha}\} \cong \{\mathcal{A} \overline{\otimes} \mathcal{L}(L^2(\mathbf{R})), \mathbf{R},\ \hat{\alpha} \otimes \mathrm{id}\}_{1 \otimes e}$. If $\mathfrak{p}$ is the algebra of $(\hat{\alpha} \otimes \mathrm{id})$-integrable elements, then $\mathfrak{p}$ is σ-weakly dense in $\mathcal{A} \,\overline{\otimes}\, \mathcal{L}(L^2(\mathbf{R}))$ and $(1 \otimes e)\mathfrak{p}(1 \otimes e) \subset \mathfrak{p}$. Hence $\{\mathcal{A},\ \mathbf{R},\ \hat{\alpha}\}$ is integrable. Thus, Theorem X.4.17 applies to $\{\mathcal{A}, \mathbf{R}, \hat{\alpha}\}$ so that the measure theoretic spectrum $\{\Gamma, \mu\}$ and the corresponding flow $\{T_p\}$ are disintegrated into transitive pieces: $\{\Gamma(\omega),\ \mathbf{R},\ T_p^\omega,\ \mu_\omega\}$. Since we have $\hat{\alpha}_p(w(s)) = \mathrm{e}^{-\mathrm{i}sp}w(s)$, almost every T^ω is aperiodic, so that we have $\{\Gamma, \mathbf{R}, T\} \cong \{\Omega \times \mathbf{R},\ \mathbf{R},\ \mathrm{id} \times \text{translation}\}$, where $\mathcal{A}^\alpha = L^\infty(\Omega, \nu)$. Therefore, there exists a one parameter unitary group $\bar{w}(s)$ in $\mathcal{A}$ such that $\hat{\alpha}_p(\bar{w}(s)) = \mathrm{e}^{-\mathrm{i}sp}\bar{w}(s)$. Let $\bar{u}(s) = \bar{w}(s)v(s)$. Then $\{\bar{u}(s)\}$ is a one parameter unitary group because $v(s)$ and $\mathcal{A}$ commute, and furthermore, $\hat{\alpha}_p(\bar{u}(s)) = \bar{u}(s)$, so that $\bar{u}(s) \in \mathcal{M}$. Since $\bar{w}(s) \in \mathcal{A} \subset \mathcal{M}' \cap \mathcal{N}$, we have

$$\mathrm{Ad}(\bar{u}(s)) = \mathrm{Ad}(\bar{w}(s)) \circ \mathrm{Ad}(v(s)) = \alpha_s, \quad s \in \mathbf{R}.$$

Thus, $\{\bar{u}(s)\}$ is the one parameter unitary group implementing α. Q.E.D.

Exercise XI.3

1†) A derivation δ of a C^*-algebra A is said to be *inner* if there exists an element $h \in M(A)$ such that $\delta(x) = [h, x],\ x \in A$, where $M(A)$ is the multiplier algebra of A. Every derivation of a simple C^*-algebra A is inner. [685]

2) Let $\{u(t) : t \in \mathbf{R}\}$ be a σ-strongly differentiable curve in $\mathcal{U}(\mathcal{M})$ in the sense that $u'(t) = \lim_{s\to 0} \frac{1}{s}\big(u(t+s) - u(t)\big)$ converges in the σ-strong topology. Show that $h(t) = \frac{1}{i}u(t)^*u'(t)$ is a self-adjoint for every $t \in \mathbf{R}$. (Hint: $\|u(t)\xi\|^2 = \|\xi\|^2$, $t \in \mathbf{R}$.)

3) Let $\{a(t) : t \in \mathbf{R}\}$ is a σ-strongly continuous curve of skew adjoint elements in $\mathcal{M}$. Show that the differential equation:

$$\frac{\mathrm{d}}{\mathrm{d}t}u(t) = u(t)a(t), \quad t \in \mathbf{R}$$

with the initial condition: $u(0) = 1$ has a unique solution $\{u(t) : t \in \mathbf{R}\}$ in $\mathcal{U}(\mathcal{M})$ written in the series expansion:

$$u(t) = 1 + \sum_{n=1}^{\infty} \int_0^t \int_0^{t_1} \cdots \int_0^{t_{n-1}} a(t_n)a(t_{n-1})\cdots a(t_1)\,\mathrm{d}t_n\,\mathrm{d}t_{n-1}\cdots \mathrm{d}t_1\,.$$

(Hint: As the volume of the simplex,

$$\Delta_n\big(|t|\big) = \big\{(t_1, t_2, \ldots, t_n) : 0 \le t_n \le t_{n-1} \le \cdots \le t_1 \le |t|\big\}$$

is $\frac{|t|^n}{n!}$ and $\{a(t) : |t| \le N\}$ is bounded for any $N \ge 0$, the integrals in the series converge in the σ^*-strong topology and the summation converges in norm. It is easy to check that the series gives indeed a solution. If $\{v(t) : t \in \mathbf{R}\}$ is a solution, then recurrent use of the identity:

$$v(t) = 1 + \int_0^t v(t_1)a(t_1)\,\mathrm{d}t_1,$$

and an easy estimate of the remainder gives the series expansion of $\{v(t)\}$ and hence $v(t) = u(t)$, $t \in \mathbf{R}$. The differential equation gives $\frac{\mathrm{d}}{\mathrm{d}t}u(t)^* = -a(t)u(t)^*$, $t \in \mathbf{R}$, hence $\frac{\mathrm{d}}{\mathrm{d}t}\|u(t)^*\xi\|^2 = 0$, $t \in \mathbf{R}$ which means that $u(t)^*$ is an isometry because $u(0)^* = u(0) = 1$. Now with $0 < s < t \le T$, we have $\|u(t) - u(s)\| \le \sum_{n=1}^{\infty} K(T)^n\big(t^n - s^n\big)/n! \le \exp\big(K(T)(t-s)\big) \to 0$ as $t - s \to 0$, where $K(T) = \sup\{\|a(r)\| : 0 \le r \le T\}$. This means that the curve $\{u(t)\}$ is continuous in norm. Thus the set $J = \{t \in \mathbf{R} : u(t) \text{ is invertible}\} = \{t \in \mathbf{R} : u(t) \in \mathcal{U}(\mathcal{M})\}$ is open in $\mathbf{R}$ and contains 0. Hence if $t \in \mathbf{R}$ is small, then $u(t)$ is a unitary. As $\mathcal{U}(\mathcal{M})$ is norm closed in $\mathcal{M}$, the set J is also closed. Hence $J = \mathbf{R}$, i.e. $\{u(t) : t \in \mathbf{R}\} \subset \mathcal{U}(\mathcal{M})$.)

4) Let $\{\mathcal{M}, \mathbf{R}, \alpha\}$ be a one parameter automorphism group of a von Neumann algebra $\mathcal{M}$. Let $\mathcal{M}_\infty^\alpha$ be the set of all those $x \in \mathcal{M}$ such that the map: $t \in \mathbf{R} \mapsto \alpha_t(x) \in \mathcal{M}$ is indefinitely differentiable relative to the σ^*-strong topology of $\mathcal{M}$. Recall that $\mathcal{M}_\infty^\alpha$ is a σ^*-strongly dense *-subalgebra of $\mathcal{M}$. For each $x \in \mathcal{M}_\infty^\alpha$, set $\delta(x)\ \big(= \delta_\alpha(x)\big) = \lim_{t\to 0} \frac{1}{t}\big(\alpha_t(x) - x\big)$.

(a) Show that δ is a *-derivation of $\mathcal{M}_\infty^\alpha$.

(b) Show that if $h \in \mathcal{M}_\infty^\alpha \cap \mathcal{M}_h$ then $u = \exp(ih) \in \mathcal{M}_\infty^\alpha \cap \mathcal{U}(\mathcal{M})$ and

$$\delta(u) = \sum_{n=0}^{\infty} \frac{1}{n!}\Big(\sum_{k=0}^{n-1} (\mathrm{i}h)^k \delta(\mathrm{i}h)(\mathrm{i}h)^{n-k-1}\Big),$$

which converges in norm. Set $k = \frac{1}{\mathrm{i}} u^* \delta(u) \in \mathcal{M}_h \cap \mathcal{M}_\infty^\alpha$.

(c) Show that if $u \in \mathcal{M}_\infty^\alpha \cap \mathcal{U}(\mathcal{M})$ then with $k = \frac{1}{\mathrm{i}} u^*\delta(u)$ the series expansion:

$$u^*\alpha_t(u) = + \sum_{n=1}^{\infty} \mathrm{i}^n \int_0^t \int_0^{t_1} \cdots \int_0^{t_{n-1}} \alpha_{t_n}(k)\,\alpha_{t_{n-1}}(k) \cdots \alpha_{t_1}(k)\, \mathrm{d}t_n\, \mathrm{d}t_{n-1} \cdots \mathrm{d}t_1$$

converges in norm for the summation whilst the integral of each term converges in the σ-strong topology. (Hint: As in the previous problem use the recurrent relation:

$$u^*\alpha_t(u) = 1 + \int_0^t u^*\delta\big(\alpha_{t_1}(u)\big)\, \mathrm{d}t_1 = 1 + \mathrm{i} \int_0^t u^*\alpha_{t_1}(u)\,\alpha_{t_1}(k)\, \mathrm{d}t_1$$

to prove the series expansion.)

(d) Show that if $e \in \mathrm{Proj}(\mathcal{M}) \cap \mathcal{M}_\infty^\alpha$, then with $a = \big[\delta(e), e\big] \in \mathcal{M}_\infty^\alpha$ we have $\delta(e) = \big[a, e\big]$.

5) Let $\{\mathcal{M}, \mathbf{R}, \alpha\}$ be a one parameter automorphism group of a von Neumann algebra $\mathcal{M}$.

(a) Making use of Problem 3, show that if $a \in \mathcal{M}$ is skew adjoint, i.e. $a = -a^*$ then the series:

$$u_t = 1 + \sum_{n=1}^{\infty} \int_0^t \int_0^{t_1} \cdots \int_0^{t_{n-1}} \alpha_{t_n}(a)\,\alpha_{t_{n-1}}(a) \cdots \alpha_{t_1}(a)\, \mathrm{d}t_n\, \mathrm{d}t_{n-1} \cdots \mathrm{d}t_1$$

for $t \in \mathbf{R}$ gives a norm continuous one parameter family of unitaries which satisfies:

$$\begin{cases} \dfrac{\mathrm{d}}{\mathrm{d}t} u_t = u_t \alpha_t(a); \\[2ex] u_{s+t} = u_s \alpha_s(u_t), \quad s, t \in \mathbf{R}. \end{cases}$$

(Hint: Set $f(t) = u_s\alpha_s(u_t)$ and $g(t) = u_{s+t}$ for $t \in \mathbf{R}$. Observe that $f'(t) = f(t)\alpha_t(a)$ and $g'(t) = g(t)\alpha_t(a)$, where the differentiations are taken in the σ^*-strong topology, and that $u_s = f(0) = g(0)$. Hence the uniqueness of solutions of the $\mathcal{M}$-valued first order linear differential equation implies that f and g agree, i.e. we have $u_{s+t} = u_s\alpha_s(u_t)$, $s, t \in \mathbf{R}$.)

(**b**) Set $\beta_t = \mathrm{Ad}(u_t) \circ \alpha_t,\ t \in \mathbf{R}$. Show that $\mathcal{M}^\alpha_\infty = \mathcal{M}^\beta_\infty$ and that

$$\delta_\beta(x) = \delta_\alpha(x) + [a, x], \quad x \in \mathcal{M}^\alpha_\infty.$$

(Hint: Show first that $\mathcal{M}^\alpha_\infty \subset \mathcal{M}^\beta_\infty$ and observe that $\alpha_t = \mathrm{Ad}(u_t)^{-1} \circ \beta_t,\ t \in \mathbf{R}$ and $\delta_\alpha(x) = \delta_\beta(x) - [a, x],\ x \in \mathcal{M}^\beta_\infty$.)

6†) Let $\{\mathcal{M}, \mathbf{R}, \alpha\}$ be a one parameter automorphism group of a von Neumann algebra $\mathcal{M}$ such that $\lim_{t\to 0}\|\alpha_t(x) - x\| = 0$ for every $x \in \mathcal{M}$. Show that $\lim_{t\to 0}\sup_{\|x\|\leq 1}\|\alpha_t(x) - x\| = 0$. Consequently $\{\alpha_t\}$ is given by a one parameter unitary group in $\mathcal{M}$. [585, 587].

Notes on Chapter XI

As mentioned in the introduction, an operator algebra, C^*-algebra and/or von Neumann algebra behaves mysteriously as a Banach space. One needs extreme care to deal with its analysis. It was W. Arveson who provided techniques for analysis of a locally compact abelian group action on a von Neumann algebra. The materials in Section 1 were mostly due to him, [433, 434]. However, the origin of these techniques goes further back to the theorem of F. and M. Riesz which asserts that a finite Borel measure on $\mathbf{T}$ whose negative Fourier-Stieltjes coefficients vanish is absolutely continuous relative to the Lebesgue measure. Arveson was motivated to give a constructive proof for a theorem of H. Borchers, [440], which states that if a one parameter unitary group $\{U(t) : t \in \mathbf{R}\}$ on a Hilbert space $\mathfrak{H}$ is generated by a positive self-adjoint operator H, i.e. $U(t) = \exp(itH)$ and $H \geq 0$, and if it normalizes a von Neumann algebra $\mathcal{M}$, i.e. $U(t)\mathcal{M}U(t)^* = \mathcal{M}$, then there exists a positive self-adjoint operator K affiliated with $\mathcal{M}$ such that the one parameter automorphism group $\{\alpha_t : t \in \mathbf{R}\}$ of $\mathcal{M}$ given by $\alpha_t(x) = U(t)xU(t)^*,\ x \in \mathcal{M}$, is given by the one parameter unitary group $\{\exp(itK)\} \subset \mathcal{U}(\mathcal{M})$. This theorem has a consequence in the quantum field theory that the energy is a limit of local observables. The original proof of Borchers relied on the derivation theorem, Theorem 3.5, which had only non-constructive proof at that time. Thus, Arveson tried to give a constructive proof for Theorem 3.5 which we presented in §3. The history of this section is fascinating. The derivation theorem, Theorem 3.5, was one of the major results in the mid sixties proven by R. V. Kadison, [576], and S. Sakai, [684]. However, the original proof was based on a skillful use of the compactness and fixed point arguments. Hence it was not constructive. As bounded derivations appear rarely in applications, it was originally thought that a beautiful theorem without applications, but as mentioned above Borchers showed an immediate application to the algebraic quantum field theory.

It was Arveson's theory which motivated Takesaki to take a fresh look at the modular automorphism group of a factor of type III and eventually to discover the duality theorem, Theorem X.2.3, which was then applied to the structure analysis of a von Neumann algebra of type III.

Arveson's spectral analysis was then further sharpened by A. Connes in his thesis, [458], to the theory of Connes spectrum, $\Gamma(\alpha)$, which was covered in §2.

Theory of derivations of an operator algebra is an interesting subject. We touched lightly in §3. Interested readers are referred to Books of Pedersen, [652], and Sakai, [687]. For example, there are a large family of C^*-algebras whose derivations are inner, including simple C^*-algebras, and it is possible to give a complete structural characterization for those C^*-algebras whose derivations are always inner, [418].

Chapter XII

Structure of a von Neumann Algebra of Type III

§ 0 Introduction

We are now going to apply the results of Chapters X and XI to the structure analysis of a von Neumann algebra $\mathcal{M}$ of type III. It turns out that the crossed product $\mathcal{N} = \mathcal{M} \rtimes_{\sigma^\varphi} \mathbf{R}$ relative to the modular automorphism group of a faithful semi-finite normal weight φ is a von Neumann algebra of type II_∞ equipped with a faithful semi-finite normal trace τ such that $\tau \circ \theta_s = \mathrm{e}^{-s}\tau$, $s \in \mathbf{R}$, where θ is the action of $\mathbf{R}$ on $\mathcal{N}$ dual to σ^φ, and that $\mathcal{M} \cong \mathcal{N} \rtimes_\theta \mathbf{R}$, Theorem 1.1. Furthermore, this decomposition of $\mathcal{M}$ is unique in the strongest sense. This chapter is devoted to the analysis of consequences of Theorem 1.1. In §1 we discuss general facts on this decomposition. Besides Theorem 1.1, the stability of a trace scaling one parameter automorphism group of a von Neumann algebra $\mathcal{N}$ of II_∞, Lemma 1.2, together with the relative commutant theorem, Theorem 1.7, will have strong implications on the structure of the automorphism group $\mathrm{Aut}(\mathcal{M})$ of a factor $\mathcal{M}$ of type III. The uniqueness of $\{\mathcal{N}, \mathbf{R}, \theta\}$ gives an immediate algebraic invariant to $\mathcal{M}$, i.e. the flow $\{\mathcal{C}_\mathcal{N}, \mathbf{R}, \theta\}$ on the center $\mathcal{C}_\mathcal{N}$ of $\mathcal{N}$ is an invariant of $\mathcal{M}$. If it is periodic with period $T > 0$, then the factor $\mathcal{M}$ is said to be of type III_λ with $\lambda = \mathrm{e}^{-2\pi/T}$. In this case, $\mathcal{M}$ admits a discrete decomposition: $\mathcal{M} \cong \mathcal{N}_0 \rtimes_{\theta_0} \mathbf{Z}$ as $\{\mathcal{N}, \mathbf{R}, \theta\}$ is induced from a covariant system $\{\mathcal{N}_0, \mathbf{Z}, \theta_0\}$, and $\mathcal{N}_0$ is a factor. Section 2 is devoted to the study of such factors. If $\mathcal{N}$ is a factor, then $\mathcal{M}$ is called of type III_1. In this case, we don't have a discrete decomposition of $\mathcal{M}$ in general. If the ergodic flow $\{\mathcal{C}_\mathcal{N}, \mathbf{R}, \theta\}$ is properly ergodic, then the factor $\mathcal{M}$ is called of type III_0. We will handle such factors in §3. Thanks to the Ambrose-Kakutani-Kubo-Krengel Theorem, Theorem 3.2, concerning the representation of an ergodic flow, we still have a discrete decomposition of a factor of type III_0, Theorem 3.7.

In §4, we will discuss the canonical construction of the flow $\{\mathcal{C}_\mathcal{N}, \mathbf{R}, \theta\}$, called the *flow of weights*, directly from $\mathcal{M}$ without going through the crossed product decomposition of $\mathcal{M}$. What we need to do turns out to develop a comparison theory of weights on a factor $\mathcal{M}$ like the comparison theory of projections which is a familiar subject for us. Section 5, is devoted to the analysis of the action of $\mathrm{Int}(\mathcal{M})$ on the normal state space $\mathfrak{S}_*$ and its quotient space $\widetilde{\mathfrak{S}}_* = \mathfrak{S}_* / \mathrm{Int}(\mathcal{M})$. It turns out that the diameter of the quotient space $\widetilde{\mathfrak{S}}_*$ for a factor $\mathcal{M}$ of type III_λ is $2\left(\frac{1-\sqrt{\lambda}}{1+\sqrt{\lambda}}\right)$ and zero for a factor $\mathcal{M}$ of type III_1, Theorem 5.12.

The last section, §6, is devoted to analysis of the functoriality of the association of the core to a factor. It turns out that to each factor $\mathcal{M}$ one can associate canonically the core $\{\widetilde{\mathcal{M}}, \mathbf{R}, \theta, \tau\}$ of semi-finite von Neumann algebra equipped with one parameter automorphism group $\{\theta_t\}$ and trace τ such that $\mathcal{M} = \widetilde{\mathcal{M}}^{\theta}$ and $\tau \circ \theta_t = e^{-t}\tau$ without fixing a faithful semi-finite normal weight. The system $\{\widetilde{\mathcal{M}}, \mathbf{R}, \theta, \tau\}$ will be called the *non-commutative flow of weights* on $\mathcal{M}$.

§ 1 Structure of a von Neumann Algebra of Type III, Part I

We are now going to describe the general structure of a von Neumann algebra of type III.

Theorem 1.1.

(i) *Let* $\{\mathcal{N}, \mathbf{R}, \theta\}$ *be a covariant system such that*

a) $\mathcal{N}$ *admits a faithful semi-finite normal trace* τ ;

b) θ *transforms* τ *in such a way that*

$$\tau \circ \theta_s = e^{-s}\tau, \quad s \in \mathbf{R}. \tag{1}$$

Then the crossed product $\mathcal{M} = \mathcal{N} \rtimes_{\theta} \mathbf{R}$ *is properly infinite and the center* $\mathcal{C}_{\mathcal{M}}$ *is precisely the fixed point subalgebra* $\mathcal{C}_{\mathcal{N}}^{\theta}$ *of the center* $\mathcal{C}_{\mathcal{N}}$ *of* $\mathcal{N}$ *under the canonical embedding of* $\mathcal{N}$ *into* $\mathcal{M}$. *Furthermore,* $\mathcal{M}$ *is of type* III *if and only if the central covariant system* $\{\mathcal{C}_{\mathcal{N}}, \mathbf{R}, \theta\}$ *does not contain an invariant subalgebra* $\mathcal{A}$ *such that the subsystem* $\{\mathcal{A}, \mathbf{R}, \theta\}$ *is isomorphic to* $L^{\infty}(\mathbf{R})$ *equipped with the translation action of* $\mathbf{R}$. *In the case that* $\mathcal{M}$ *is of type* III, $\mathcal{N}$ *is necessarily of type* II_{∞}.

(ii) *If* $\mathcal{M}$ *is a von Neumann algebra of type* III, *then there exists a unique, up to conjugacy, covariant system* $\{\mathcal{N}, \mathbf{R}, \theta\}$ *satisfying the conditions of* (i).

PROOF OF (i): $\mathcal{C}_{\mathcal{M}} = \mathcal{C}_{\mathcal{N}}^{\theta}$: We identify $\mathcal{N}$ with $\pi_{\theta}(\mathcal{N})$, so that $\mathcal{N}$ is viewed as the fixed point subalgebra $\mathcal{M}^{\hat{\theta}}$ of $\mathcal{M}$ under the dual action $\hat{\theta}$ of $\mathbf{R}$. Let φ be the weight on $\mathcal{M}$ dual to τ. It follows from Theorem X.1.17 that

$$\sigma_t^{\varphi}(x) = \sigma_t^{\tau}(x) = x, \quad x \in \mathcal{N}, \quad t \in \mathbf{R}$$

$$\sigma_t^{\varphi}\big(u(s)\big) = \big(D\tau \circ \theta_s : D\tau\big)_t u(s) = e^{-ist}u(s), \quad s \in \mathbf{R},$$

where $\{u(s)\}$ is the one parameter unitary group in $\mathcal{M}$ associated with the crossed product. This means that the modular automorphism group $\{\sigma_t^{\varphi}\}$ is precisely the dual action $\hat{\theta}$ of θ. Hence $\mathcal{N}$ coincides with the centralizer $\mathcal{M}_{\varphi}$ of φ, which contain the center $\mathcal{C}_{\mathcal{M}}$ of $\mathcal{M}$. Thus $\mathcal{C}_{\mathcal{M}} \subset \mathcal{C}_{\mathcal{N}}$. Since $\mathcal{C}_{\mathcal{M}}$ commutes with $\{u(s)\}$, we have $\mathcal{C}_{\mathcal{M}} \subset \mathcal{C}_{\mathcal{N}}^{\theta}$. By construction, $\mathcal{C}_{\mathcal{N}}^{\theta} \subset \mathcal{C}_{\mathcal{M}}$. Hence $\mathcal{C}_{\mathcal{M}} = \mathcal{C}_{\mathcal{N}}^{\theta}$.

Proper Infiniteness: We shall show that $\mathcal{M}$ admits no non-zero finite normal trace. Let ψ be a finite normal trace on $\mathcal{M}$. Let $e = s(\psi)$ be the support of ψ. Then

$e \in \mathcal{C}_{\mathcal{M}} = \mathcal{C}_{\mathcal{N}}^{\theta}$. The restriction of ψ to $\mathcal{N}$ is also a finite normal trace and e is the support of $\psi|_{\mathcal{N}}$. Since τ is faithful, there exists a self-adjoint positive operator h affiliated with $\mathcal{C}_{\mathcal{N}}$ such that $\psi|_{\mathcal{N}} = \tau(h\,\cdot\,)$. Since $\mathrm{Ad}(u(s))$ leaves ψ invariant on $\mathcal{M}$, $\psi|_{\mathcal{N}}$ is invariant under $\{\theta_s\}$. This implies that for every $x \in \mathcal{N}_+$,

$$\tau(hx) = \psi(x) = \psi \circ \theta_s(x) = \tau\big(h\theta_s(x)\big) = \tau \circ \theta_s\big(\theta_{-s}(h)x\big) = \mathrm{e}^{-s}\tau\big(\theta_{-s}(h)x\big).$$

Hence we get $\theta_s(h) = \mathrm{e}^{-s}h,\ s \in \mathbf{R}$. Considering h in $\mathcal{M}_e$ and $\mathcal{N}_e$, we get $\theta_s(h^{it}) = \mathrm{e}^{-ist}h^{it}$. Thus $\{h^{it}\}$ and $\{u(s)\}$ satisfies the Heisenberg-Weyl commutation relation, so that they generate a von Neumann algebra $\mathcal{B}$ isomorphic to $\mathcal{L}\big(L^2(\mathbf{R})\big)$. But $\mathcal{B}$ is a subalgebra of $\mathcal{M}_e$ on which ψ is a faithful finite normal trace. This is impossible. Thus, $\mathcal{M}$ does not admit a non-zero finite trace. This is equivalent to the proper infiniteness of $\mathcal{M}$.

Type Question: Suppose that there exists an isomorphism π of $L^\infty(\mathbf{R})$ into $\mathcal{C}_{\mathcal{N}}$ which intertwines the translation α on $L^\infty(\mathbf{R})$ and θ. Here we do not assume that $\pi(1) = 1$. Thus $e = \pi(1)$ is a projection of $\mathcal{C}_{\mathcal{N}}^{\theta}$. Let $\{w(t)\}$ be the one parameter unitary group in $L^\infty(\mathbf{R})$ defined by

$$w(t)(\lambda) = \mathrm{e}^{it\lambda}, \qquad \lambda \in \mathbf{R}.$$

We then have $\alpha_s\big(w(t)\big) = \mathrm{e}^{ist}w(t),\ s \in \mathbf{R}$. Putting $v(t) = \pi\big(w(t)\big)$, we get $\theta_s\big(v(t)\big) = \mathrm{e}^{ist}v(t)$. In $\mathcal{M}_e = \mathcal{N}_e \rtimes_\theta \mathbf{R}$, we have

$$v(t)xv(t)^* = x, \quad x \in \mathcal{N}_e, \qquad v(t)u(s)v(t)^* = \mathrm{e}^{-ist}u(s), \quad s, t \in \mathbf{R}.$$

This implies that $\mathrm{Ad}\big(v(t)\big)$ and σ_t^φ agree on the generators of $\mathcal{M}_e$. Hence $\{\sigma_t^\varphi\}$ is implemented by $\{v(t)\}$. Thus $\mathcal{M}_e$ is semi-finite by Theorem VIII.3.14.

Conversely, suppose that $\mathcal{M}$ admits a non-zero semi-finite central projection e. Since $e \in \mathcal{C}_{\mathcal{M}} = \mathcal{C}_{\mathcal{N}}^{\theta}$, we have $\mathcal{M}_e = \mathcal{N}_e \rtimes_\theta \mathbf{R}$. Hence we may assume that $e = 1$. The semi-finiteness of $\mathcal{M}$ then implies that $\{\sigma_t^\varphi\}$ is implemented by a one parameter unitary group $\{v(t)\}$ in $\mathcal{M}_\varphi = \mathcal{N}$. Since $\mathrm{Ad}\big(v(t)\big)$ leaves $\mathcal{N}$ pointwise invariant, each $v(t)$ belongs to the center $\mathcal{C}_{\mathcal{N}}$ of $\mathcal{N}$. We then have

$$\mathrm{e}^{-ist}u(s) = \sigma_t^\varphi\big(u(s)\big) = v(t)u(s)v(t)^*, \quad s, t \in \mathbf{R}.$$

Thus the pair $\{u(s), v(t)\}$ of one parameter unitary groups satisfies the Heisenberg-Weyl commutation relation (X.2.5). The uniqueness of a representation of the Heisenberg-Weyl relation implies that the von Neumann algebra $\mathcal{A}$ generated by $\{v(t)\}$ is isomorphic to $L^\infty(\mathbf{R})$ and the isomorphism of $\mathcal{A}$ onto $L^\infty(\mathbf{R})$ intertwines $\{\mathrm{Ad}\big(u(s)\big)\} = \{\theta_s\}$ on $\mathcal{A}$ and the translation $\{\alpha_s\}$.

Suppose that $e \in \mathrm{Proj}(\mathcal{C}_{\mathcal{N}})$ be the greatest projection such that $\mathcal{N}_e$ is of type II_∞. It then follows that $e \in \mathcal{C}_{\mathcal{N}}^{\theta}$. We want to show that $\mathcal{M}_{(1-e)}$ is semi-finite. Looking at $\{\mathcal{N}_{(1-e)}, \mathbf{R}, \theta\}$, we assume that $e = 0$. Thus $\mathcal{N}$ is a direct sum of the type I part and the type II_1 part. It then follows that $\mathcal{N}$ admits a center valued trace T which is normalized in such a way that T commutes with $\mathrm{Aut}(\mathcal{N})$, in particular with $\{\theta_s\}$.

We now decompose the trace τ by a faithful semi-finite normal weight ψ on $\mathcal{C}_{\mathcal{N}}$ and T so that $\tau = \psi \circ T$, i.e. $\tau(x) = \psi\big(T(x)\big)$, $x \in \mathcal{N}_+$. We then have

$$e^{-s}\tau(x) = \tau \circ \theta_s(x) = \psi \circ T\big(\theta_s(x)\big) = \psi\Big(\theta_s\big(T(x)\big)\Big), \quad x \in \mathcal{N}_+.$$

Since the range of T exhausts $\mathcal{C}_{\mathcal{N}}$, we have

$$\psi \circ \theta_s = e^{-s}\psi, \quad s \in \mathbf{R}.$$

This implies, by the next lemma, that $\mathcal{C}_{\mathcal{N}}$ contains a one parameter unitary group $\{v(t)\}$ such that $\theta_s\big(v(t)\big) = e^{ist}v(t)$. The above arguments then show that $\mathcal{M}$ is semi-finite. This completes the proof of part (i) provided the next lemma. [STOP]

Lemma 1.2. *Let $\{\mathcal{Q}, \mathbf{R}, \theta\}$ be a covariant system over $\mathbf{R}$. If $\mathcal{Q}$ admits a faithful semi-finite normal trace ψ such that*

$$\psi \circ \theta_s = e^{-s}\psi, \quad s \in \mathbf{R},$$

then there exists a one parameter unitary group $\{v(t)\}$ in $\mathcal{Q}$ such that

$$\theta_s\big(v(t)\big) = e^{ist}v(t), \quad s, t \in \mathbf{R}. \tag{2}$$

PROOF: Let $T > 0$, and $\lambda = e^{-T}$. We first prove that any non-zero $e \in \text{Proj}(\mathcal{Q}^{\theta_T})$ dominates a non-zero $f \in \text{Proj}(\mathcal{Q})$ such that $\theta_T(f) < f$ and $\psi(f) < +\infty$. Clearly e dominates a non-zero $p \in \text{Proj}(\mathcal{Q})$ such that $\psi(p) < +\infty$. Let $f = \bigvee_{n\geq 0}\theta_{nT}(p)$. Since we have

$$\psi(f) \leq \sum_{n=0}^{\infty} \psi \circ \theta_{nT}(p) = \sum_{n=0}^{\infty} \lambda^n \psi(p) = \frac{1}{1-\lambda}\psi(p) < +\infty,$$

we get $\theta_T(f) \leq f$ and

$$\psi\big(\theta_T(f)\big) = \lambda\psi(f) < \psi(f).$$

Hence we have $\theta_T(f) < f$. Put $q = f - \theta_T(f)$. Then $\{\theta_{nT}(q) : n \in \mathbf{Z}\}$ is orthogonal and $\sum_{n\in Z}\theta_{nT}(q) = e_0$ is a projection in $\mathcal{Q}^{\theta_T}$ and $q \leq e$. Hence we proved that any non-zero $e \in \text{Proj}(\mathcal{Q}^{\theta_T})$ dominates a projection $q \in \text{Proj}(\mathcal{Q})$, $q \neq 0$, such that $\{\theta_{nT}(q) : n \in \mathbf{Z}\}$ is orthogonal. Such a projection q is called *wondering* under θ_T. Let $s(q) = \sum_{n\in\mathbf{Z}}\theta_{nT}(q)$ for a wondering projection q. Clearly $s_T(q) \in \mathcal{Q}^{\theta_T}$. Let $\{q_i\}_{i\in I}$ be a maximal family of wondering projections such that $\{s_T(q_i)\}_{i\in I}$ is orthogonal. Let $p = \sum_{i\in I} q_i$. By the maximality and the orthogonality of $\{s_T(q_i)\}$, we have $\sum_{n\in\mathbf{Z}}\theta_{nT}(p) = 1$. Set $v(t) = \sum_{n\in\mathbf{Z}}\lambda^{int}\theta_{nT}(p)$. Then $\{v(t) : t \in \mathbf{R}\}$ is a one parameter unitary group in $\mathcal{Q}$ with period $2\pi/T$, and satisfies the condition:

$$\theta_T\big(v(t)\big) = \lambda^{-it}v(t).$$

This means by Proposition X.2.6 that θ_T is a dual action and that $\mathcal{Q}^{\theta_T}$ together with $\{ v(t) : 0 \leq t < 2\pi/T \}$ generates $\mathcal{Q}$. Hence θ_T gives an integrable action of $\mathbf{Z}$. Let

$$\mathcal{E}(x) = \int_{\mathbf{R}} \theta_t(x)\, \mathrm{d}t\,, \quad x \in \mathcal{Q}^+.$$

Then we have

$$\mathcal{E}(x) = \int_0^T \theta_t\Big(\sum_{n\in Z} \theta_{nT}(x)\Big)\, \mathrm{d}t\,.$$

If $x \in \mathcal{Q}_{p_n}$ where $p_n = \sum_{|k|\leq n} \theta_{kT}(p)$, then $\mathcal{E}(x)$ converges and falls in $\mathcal{Q}^\theta$. Since $p_n \nearrow 1$, θ itself is integrable. Let ω be a faithful semi-finite normal weight on $\mathcal{Q}^\theta$ and set $\varphi = \omega \circ \mathcal{E}$. Then φ is a faithful semi-finite normal weight on $\mathcal{Q}$. Since ψ is a faithful semi-finite normal trace on $\mathcal{Q}$, there exists a non-singular self-adjoint positive operator h affiliated with $\mathcal{Q}$ such that $\varphi = \psi(h\,\cdot\,)$. We then have, for each $x \in \mathcal{Q}^+$,

$$\begin{aligned}\psi(hx) = \varphi(x) = \varphi\big(\theta_s(x)\big) = \psi\big(h\theta_s(x)\big) = \psi \circ \theta_s\big(\theta_{-s}(h)x)\big)\\ = \mathrm{e}^{-s}\psi\big(\theta_{-s}(h)x\big),\end{aligned}$$

so that $\theta_s(h) = \mathrm{e}^{-s}h,\ s \in \mathbf{R}$. Thus, putting $v(t) = h^{-\mathrm{i}t}$, we obtain a one parameter unitary group $\{v(t)\}$ satisfying the condition (2). Q.E.D.

PROOF OF THEOREM 1.1.(ii): Suppose that $\mathcal{M}$ is a von Neumann algebra of type III. Being properly infinite, $\mathcal{M}$ admits a tensor product split: $\mathcal{M} = \mathcal{M}_1 \overline{\otimes} \mathcal{L}(L^2(\mathbf{R}))$. Let ω be a faithful weight on $\mathcal{M}_1$ and $\sigma_t = \{\sigma_t^\omega\}$ be the modular automorphism group of ω. Put $\mathcal{N} = \mathcal{M}_1 \rtimes_\sigma \mathbf{R}$. Then the duality theorem, Theorem X.2.3, implies that $\mathcal{M} \cong \mathcal{N} \rtimes_{\hat\sigma} \mathbf{R}$. Set $\theta = \hat\sigma$, so that $\mathcal{M} \cong \mathcal{N} \rtimes_\theta \mathbf{R}$.

We want to prove that $\{\mathcal{N}, \mathbf{R}, \theta\}$ satisfies the conditions of (i). Let $\hat\omega$ be the weight on $\mathcal{N}$ dual to ω. By Theorem X.1.17, we have

$$\begin{aligned}\sigma_t^{\hat\omega}(x) &= \sigma_t(x), \quad x \in \mathcal{M}_1,\\ \sigma_t^{\hat\omega}\big(u(s)\big) &= \big(\mathrm{D}\omega \circ \sigma_s : \mathrm{D}\omega\big)_t u(s) = u(s), \quad s, t \in \mathbf{R},\end{aligned}$$

where $\{u(s)\}$ means the unitary group associated with the crossed product $\mathcal{M}_1 \rtimes_\sigma \mathbf{R}$. Therefore, $\sigma_t^{\hat\omega} = \mathrm{Ad}\big(u(t)\big),\ t \in \mathbf{R}$. Thus, $\mathcal{N}$ is semi-finite by Theorem VIII.3.14. Let h be the non-singular positive self-adjoint operator affiliated with $\mathcal{N}$ such that $u(t) = h^{\mathrm{i}t},\ t \in \mathbf{R}$. Put $\tau = \hat\omega(h^{-1}\,\cdot\,)$. Since $\sigma_t^\tau = \mathrm{Ad}(h^{-\mathrm{i}t})\sigma_t^{\hat\omega} = \mathrm{id},\ t \in \mathbf{R},\ \tau$ is a faithful semi-finite normal trace on $\mathcal{N}$. Since θ is dual to σ, we have $\theta_s\big(u(t)\big) = \mathrm{e}^{-\mathrm{i}st}u(t)$, so that $\theta_s(h) = \mathrm{e}^{-s}h,\ s \in \mathbf{R}$. Hence we get, for $x \in \mathcal{N}_+$,

$$\tau \circ \theta_s(x) = \hat\omega\big(h^{-1}\theta_s(x)\big) = \hat\omega\Big(\theta_s\big(\theta_{-s}(h^{-1})x\big)\Big) = \hat\omega\big(\mathrm{e}^{-s}h^{-1}x\big) = \mathrm{e}^{-s}\tau(x).$$

Therefore, $\{\mathcal{N}, \mathbf{R}, \theta\}$ and τ satisfy the condition in (i). Since $\mathcal{M}$ is of type III, $\mathcal{N}$ must be of type II_∞ by (i).

We shall prove the uniqueness. Let $\{\mathcal{N}, \mathbf{R}, \theta\}$ be a covariant system satisfying the condition in (i). We then proved in (i) that the dual action $\hat{\theta}$ on $\mathcal{M} = \mathcal{N} \rtimes_\theta \mathbf{R}$ is precisely the modular automorphism group $\{\sigma_t^\varphi\}$ of the weight φ dual to τ. Hence we have, by Theorem X.2.3,

$$\left\{\mathcal{N} \overline{\otimes} \mathcal{L}(L^2(\mathbf{R})),\ \mathbf{R},\ \theta \otimes \rho\right\} \cong \{\mathcal{M} \rtimes_{\sigma^\varphi} \mathbf{R},\ \mathbf{R},\ \hat{\sigma}^\varphi\}.$$

On the other hand, it follows from Lemma 1.2 that $\{\mathcal{N}, \mathbf{R}, \theta\}$ is a dual covariant system by Proposition X.2.6. Let $\{\mathcal{M}_1, \mathbf{R}, \alpha\}$ be a covariant system such that

$$\{\mathcal{N}, \mathbf{R}, \theta\} \cong \{\mathcal{M}_1 \rtimes_\alpha \mathbf{R},\ \mathbf{R},\ \hat{\alpha}\}.$$

We then have that

$$\mathcal{M} = \mathcal{N} \rtimes_\theta \mathbf{R} \cong (\mathcal{M}_1 \rtimes_\alpha \mathbf{R}) \rtimes_{\hat{\alpha}} \mathbf{R} \cong \mathcal{M}_1 \overline{\otimes} \mathcal{L}(L^2(\mathbf{R})),$$

so that $\mathcal{M}_1$ is of type III and isomorphic to $\mathcal{M}$. At the same time, we also have $\hat{\theta} = \alpha \otimes \rho$. Writing $\mathcal{M}_1 = \mathcal{M}_2 \overline{\otimes} \mathcal{B}$ with $\mathcal{B}$ a factor of type I_∞, we decompose $\{\mathcal{N}, \mathbf{R}, \theta\}$ into a tensor product:

$$\{\mathcal{N}, \mathbf{R}, \theta\} \cong \{\mathcal{N}_1 \overline{\otimes} \mathcal{B},\ \mathbf{R}, \theta^1 \otimes \mathrm{id}\},$$

where $\mathcal{N}_1 = \mathcal{B}' \cap \mathcal{N}$ and $\theta^1 = \theta|_{\mathcal{N}_1}$. Therefore, we have,

$$\theta \simeq \theta \otimes \mathrm{id} \simeq \theta \otimes \rho \simeq \hat{\hat{\theta}}.$$

This completes the proof of the uniqueness. Q.E.D.

Definition 1.3. For a von Neumann algebra $\mathcal{M}$ of type III, the covariant system $\{\mathcal{N}, \mathbf{R}, \theta\}$ in Theorem 1.1 is called the *associated covariant system* of $\mathcal{M}$, or the *non-commutative flow of weights of* $\mathcal{M}$ as explained later in detail, see §6.

We call the crossed product $\mathcal{M} = \mathcal{N} \rtimes_\theta \mathbf{R}$ the *continuous decomposition* of $\mathcal{M}$.

A one parameter automorphism group $\{\theta_s\}$ of $\mathcal{N}$ is said to *scale* a trace if (1) holds. Similarly, we say that an automorphism θ of $\mathcal{N}$ *scales* a trace τ if $\tau \circ \theta = \lambda\tau$ for some $\lambda \neq 1$. We call the number λ^{-1} the *modulus* of θ and denote it by $\mathrm{mod}(\theta)$.

Since $\{\mathcal{N}, \mathbf{R}, \theta\}$ is a complete invariant for the algebraic type of $\mathcal{M}$, the flow $\{\mathcal{C}_\mathcal{N}, \mathbf{R}, \theta\}$ is also an algebraic invariant of $\mathcal{M}$. We call it the *flow of weights* of $\mathcal{M}$.

Corollary 1.4. *Let $\mathcal{M}$ be a von Neumann algebra with associated covariant system $\{\mathcal{N}, \mathbf{R}, \theta\}$. Then $\mathcal{M}$ is a factor if and only if θ is centrally ergodic.*

Therefore, the flow of weights of a factor of type III is ergodic. Every conjugacy invariant for ergodic flows provides an algebraic invariant for factors of type III. Among those invariant, the simplest of all is the period and the point spectrum. Thus we make the following definition:

Definition 1.5. Let $\mathcal{M}$ be a factor of type III.

(i) If $T > 0$ is the period of the flow of weights, then with $\lambda = \mathrm{e}^{-T}$, $\mathcal{M}$ is said to be of type III_λ.

(ii) If the flow of weights has no period, then $\mathcal{M}$ is called of type III_0.
(iii) If the flow of weights is trivial, i.e. the carrier algebra $\mathcal{N}$ of the associated covariant system is a factor, then $\mathcal{M}$ is said to be of type III_1.
(iv) Let $\mathcal{W}_0$ be the set of all faithful semi-finite weights on $\mathcal{M}$. Then the set

$$S(\mathcal{M}) = \bigcap \{ \operatorname{Sp}(\Delta_\varphi) : \varphi \in \mathcal{W}_0 \} \tag{3}$$

is called the *modular spectrum* of $\mathcal{M}$.
(v) The *modular period group* of $\mathcal{M}$ is defined to be the subset of $\mathbf{R}$:

$$T(\mathcal{M}) = \{ t \in \mathbf{R} : \sigma_t^\varphi \in \operatorname{Int}(\mathcal{M}) \}, \tag{4}$$

where φ is a faithful weight on $\mathcal{M}$.

The results of the last chapter readily yield the next:

Theorem 1.6. *Let $\mathcal{M}$ be a factor of type* III.

(i) *$\mathcal{M}$ is of type III_λ, $0 < \lambda < 1$, if and only if $S(\mathcal{M}) = \{0\} \cup \{ \lambda^n : n \in \mathbf{Z} \}$.*
(ii) *$\mathcal{M}$ is of type III_0 if and only if $S(\mathcal{M}) = \{0, 1\}$.*
(iii) *$\mathcal{M}$ is of type III_1 if and only if $S(\mathcal{M}) = \mathbf{R}_+$.*
(iv) *$p \in T(\mathcal{M})$ if and only if there exists $u \in \mathcal{U}(\mathcal{C})$ with $\theta_s(u) = \mathrm{e}^{\mathrm{i}ps}u$, where $\{\mathcal{C}, \theta\}$ is the flow of weights.*

PROOF: Let $\mathcal{M} = \mathcal{N} \rtimes_\theta \mathbf{R}$ be the continuous decomposition of $\mathcal{M}$. By Proposition XI.1.24, we have

$$\exp\big(\operatorname{Sp}(\sigma^\varphi)\big) = \operatorname{Sp}\big(\Delta_\varphi\big) \setminus \{0\}, \quad \varphi \in \mathcal{W}_0.$$

Hence with $\varphi = \hat{\tau}$, we have

$$\exp\big(\Gamma(\sigma^\varphi)\big) = \bigcap \{ \operatorname{Sp}(\Delta\varphi_e) : e \in \operatorname{Proj}(\mathcal{C}) \} \setminus \{0\} \supset S(\mathcal{M}) \setminus \{0\}$$

because $\mathcal{M}_e \cong \mathcal{M}$ if $e \neq 0$. Conversely, we have

$$\log\big(S(\mathcal{M}) \setminus \{0\}\big) = \bigcap \{ \operatorname{Sp}(\sigma^\psi) : \psi \in \mathcal{W}_0 \} \supset \Gamma(\sigma^\varphi).$$

Hence $S(\mathcal{M}) \setminus \{0\} = \exp\big(\Gamma(\sigma^\varphi)\big) = \exp\big(\operatorname{Ker} \theta|_{\mathcal{C}}\big)$.

Since $\mathcal{M}$ is of type III, Δ_ψ is unbounded for any $\psi \in \mathcal{W}_0$, and $J\Delta_\psi J = \Delta_\psi^{-1}$ means then that $\operatorname{Sp}(\Delta_\psi)$ always contains 0. Thus $S(\mathcal{M})$ must contain 0. Therefore, we get

$$S(\mathcal{M}) = \exp\big(\Gamma(\sigma^\varphi)\big) \cup \{0\} = \exp\big(\operatorname{Ker} \theta|_{\mathcal{C}}\big) \cup \{0\}.$$

Thus, we conclude (i), (ii) and (iii).

(iv) The image δ_t of σ_t^ψ in $\operatorname{Out}(\mathcal{M}) = \operatorname{Aut}(\mathcal{M})/\operatorname{Int}(\mathcal{M})$ does not depend on $\psi \in \mathcal{W}_0$. So we choose $\varphi = \hat{\tau}$ for ψ.

Suppose $p \in T(\mathcal{M})$. Then there exists $a \in \mathcal{U}(\mathcal{M})$ such that $\sigma_p^\varphi = \mathrm{Ad}(a)$. Since $\varphi \circ \mathrm{Ad}(a) = \varphi \circ \sigma_p^\varphi = \varphi$, a belongs to $\mathcal{M}_\varphi = \mathcal{N}$. Furthermore, $\mathrm{Ad}(a) = \sigma_p^\varphi$ leaves $\mathcal{M}_\varphi$ pointwise invariant, so that a belongs to the center $\mathcal{C}$ of $\mathcal{N} = \mathcal{M}_\varphi$. Let $\{u(s)\}$ be the one parameter unitary group in $\mathcal{M}$ associated with the crossed product. Then we have

$$au(s)a^* = \sigma_p^\varphi\big(u(s)\big) = \mathrm{e}^{-\mathrm{i}sp}u(s),$$

so that

$$\theta_s(a) = u(s)\,a\,u(s)^* = \mathrm{e}^{\mathrm{i}sp}a.$$

Conversely, if $\theta_s(a) = \mathrm{e}^{\mathrm{i}sp}a$ with $a \in \mathcal{U}(\mathcal{C})$, then we have

$$axa^* = x, \quad x \in \mathcal{N}, \qquad au(s)a^* = \mathrm{e}^{-\mathrm{i}sp}u(s), \quad s \in \mathbf{R},$$

which means that $\sigma_p^\varphi = \mathrm{Ad}(a)$. Q.E.D.

Theorem 1.7. *If $\mathcal{M} = \mathcal{N} \rtimes_\theta \mathbf{R}$ is the continuous decomposition of a factor $\mathcal{M}$ of type* III, *then*

$$\mathcal{N}' \cap \mathcal{M} = \mathcal{C}_\mathcal{N}. \tag{5}$$

In particular, if $\mathcal{M}$ is of type III_1, *then*

$$\mathcal{N}' \cap \mathcal{M} = \mathbf{C}.$$

PROOF: Every factor is a tensor product of a σ-finite factor and a factor of type I, so that we may assume that $\mathcal{M}$ is σ-finite.

For the real additive group $\mathbf{R}$, we fix a few notations:

$$\left.\begin{aligned}
&\langle s, p\rangle = \mathrm{e}^{\mathrm{i}sp}, && s, p \in \mathbf{R};\\
&\big(U(s)f\big)(t) = f(s+t), && f \in L^2(\mathbf{R}), \quad t \in \mathbf{R};\\
&\big(V(s)f\big)(t) = \langle s, t\rangle f(t);\\
&\big(Ff\big)(p) = \frac{1}{\sqrt{2\pi}}\int_\mathbf{R} \overline{\langle s, p\rangle} f(s)\,\mathrm{d}s, && f \in L^2(\mathbf{R}) \cap L^1(\mathbf{R});\\
&\big(F^*f\big)(s) = \frac{1}{\sqrt{2\pi}}\int_\mathbf{R} \langle s, p\rangle f(p)\,\mathrm{d}p, && f \in L^2(\mathbf{R}) \cap L^1(\mathbf{R}).
\end{aligned}\right\} \tag{6}$$

The the one parameter unitary groups $\{U, V\}$ satisfy the Heisenberg-Weyl commutation relation with respect to the pairing $\langle\ ,\ \rangle$, and generates the factor $\mathcal{L}\big(L^2(\mathbf{R})\big)$. We extend U and V to $L^\infty(\mathbf{R})$, identifying

$$U(s) = U\big(\langle s, \cdot\rangle\big);$$
$$V(s) = V\big(\langle s, \cdot\rangle\big).$$

Namely, we view U and V as normal isomorphisms of $L^\infty(\mathbf{R})$ into $\mathcal{L}\big(L^2(\mathbf{R})\big)$ such that $U\big(L^\infty(\mathbf{R})\big) = \mathcal{A}$ and $V\big(L^\infty(\mathbf{R})\big) = \mathcal{B}$ are both maximal abelian.

We know that $\{\mathcal{N}, \mathbf{R}, \theta\}$ is a dual covariant system. Let $\{\mathcal{P}, \mathbf{R}, \alpha\}$ be a covariant system such that

$$\{\mathcal{N}, \mathbf{R}, \theta\} \cong \{\mathcal{P} \rtimes_\alpha \mathbf{R}, \mathbf{R}, \hat{\alpha}\}.$$

By the duality theorem, we have, with $\varphi = \hat{\tau}$,

$$\{\mathcal{M}, \sigma^\varphi\} \cong \{\mathcal{N} \rtimes_\theta \mathbf{R}, \hat{\theta}\} \cong \Big\{\mathcal{P} \overline{\otimes} \mathcal{L}\big(L^2(\mathbf{R})\big), \alpha \otimes \mathrm{Ad}(U)\Big\}.$$

Let $U(t) = h^{it}$, $t \in \mathbf{R}$, and $\psi = \varphi\big((1 \otimes h)^{-1} \cdot\big)$. We then have $\sigma^\psi = \alpha \otimes \mathrm{id}$, so that α may be identified with the modular automorphism group σ^ω of a weight ω on $\mathcal{P}$. The dual action is invariant up to conjugacy under cocycle perturbations of the original action α. Hence we may take a faithful normal state as ω because $\mathcal{P}$ is σ-finite by the first reduction. Then we have $\varphi = \omega \otimes \mathrm{Tr}(h\ \cdot\ )$. We represent $\mathcal{P}$ on a Hilbert space $\mathfrak{H}$ in the standard form and define a representation π of $\mathcal{P}$ on $L^2(\mathbf{R}, \mathfrak{H})$ as follows:

$$\big(\pi(x)\xi\big)(s) = \sigma^\omega_{-s}(x)\xi(s), \quad x \in \mathcal{P}, \quad \xi \in L^2(\mathbf{R}, \mathfrak{H}).$$

Then $\mathcal{N}$ is generated by $\pi(\mathcal{P})$ and $\mathbf{C} \otimes U\big(L^\infty(\mathbf{R})\big) = \mathbf{C} \otimes \mathcal{A}$. Hence we get

$$\begin{aligned}\mathcal{N}' \cap \mathcal{M} &= \Big[\pi(\mathcal{P})' \cap \Big(\mathbf{C} \otimes U\big(L^\infty(\mathbf{R})\big)\Big)'\Big] \cap \Big[\mathcal{P}\overline{\otimes}\mathcal{L}\big(L^2(\mathbf{R})\big)\Big]\\ &= \Big[\pi(\mathcal{P})' \cap \big(\mathcal{L}(\mathfrak{H}) \overline{\otimes} \mathcal{A}\big)\Big] \cap \Big[\mathcal{P} \overline{\otimes} \mathcal{L}\big(L^2(\mathbf{R})\big)\Big] = \pi(\mathcal{P})' \cap \big(\mathcal{P} \overline{\otimes} \mathcal{A}\big).\end{aligned}$$

Therefore, (5) follows from the next inclusion:

$$\pi(\mathcal{P})' \cap \big(\mathcal{P} \overline{\otimes} \mathcal{A}\big) \subset \mathcal{P}_\omega \overline{\otimes} \mathcal{A}. \tag{7}$$

Since $\theta_s = \mathrm{Ad}\big(1 \otimes V(s)\big)$ on $\mathcal{N}$, we extend it to $\mathcal{M}$ as an inner automorphism for each $s \in \mathbf{R}$. Now, we consider the isomorphism $\tilde{U} = \mathrm{id} \otimes U : \mathcal{P}\overline{\otimes}L^\infty(\mathbf{R}) \mapsto \mathcal{P}\overline{\otimes}\mathcal{A}$ and the action $\tilde{\lambda}_t = \mathrm{id} \otimes \lambda_t \in \mathrm{Aut}\big(\mathcal{P} \overline{\otimes} L^\infty(\mathbf{R})\big)$, where $(\lambda_t f)(s) = f(s - t)$, $f \in L^\infty(\mathbf{R})$. We then have $U(\lambda_t f) = V(t)U(f)V(t)^*$, so that we obtain:

$$\theta_t\big(\tilde{U}(a)\big) = \tilde{U}\big(\tilde{\lambda}_t(a)\big), \quad a \in L^\infty(\mathbf{R}, \mathcal{P}) = \mathcal{P} \overline{\otimes} L^\infty(\mathbf{R}). \tag{8}$$

Since θ_t leaves $\pi(\mathcal{P})$ pointwise fixed, $\pi(\mathcal{P})' \cap \mathcal{M}$ is globally invariant under θ. Hence

$$\theta_t\Big(\tilde{U}\big(L^\infty(\mathbf{R}, \mathcal{P})\big) \cap \pi(\mathcal{P})'\Big) = \theta_t\Big(\big(\mathcal{P} \overline{\otimes} \mathcal{A}\big) \cap \pi(\mathcal{P})'\Big) = \big(\mathcal{P} \overline{\otimes} \mathcal{A}\big) \cap \pi(\mathcal{P})'.$$

Thus, for each $f \in \mathcal{K}(\mathbf{R})$, we have

$$\theta_f\Big(\tilde{U}\big(L^\infty(\mathbf{R}, \mathcal{P})\big) \cap \pi(\mathcal{P})'\Big) \subset \tilde{U}\big(L^\infty(\mathbf{R}, \mathcal{P})\big) \cap \pi(\mathcal{P})'.$$

But we have $\tilde{\lambda}_f\big(L^\infty(\mathbf{R}, \mathcal{P})\big) \subset C(\mathbf{R}, \mathcal{P})$, where $C(\mathbf{R}, \mathcal{P})$ means the C^*-algebra of all σ-strong* continuous $\mathcal{P}$-valued functions over $\mathbf{R}$. Since $\mathcal{K}(\mathbf{R})$ contains an approximate identity, $\tilde{U}\big(C(\mathbf{R}, \mathcal{P})\big) \cap \pi(\mathcal{P})'$ is σ-strongly dense in $\big(\mathcal{P}\overline{\otimes}\mathcal{A}\big) \cap \pi(\mathcal{P})'$.

Thus to show (7) it is sufficient to prove the following:

$$\tilde{U}\big(C(\mathbf{R},\ \mathcal{P})\big) \cap \pi(\mathcal{P})' \subset \tilde{U}\big(C(\mathbf{R},\ \mathcal{P}_\omega)\big). \tag{9}$$

With $\tilde{F} = 1 \otimes F$, (9) is equivalent to the following:

$$C(\mathbf{R},\ \mathcal{P}) \cap \big(\tilde{F}\pi(\mathcal{P})\tilde{F}^*\big)' \subset C(\mathbf{R},\ \mathcal{P}_\omega), \tag{10}$$

where $C(\mathbf{R},\ \mathcal{P})$ is acting on $L^2(\mathbf{R},\ \mathfrak{H})$ by:

$$\big(a\xi\big)(s) = a(s)\xi(s), \quad \xi \in L^2(\mathbf{R},\ \mathfrak{H}), \quad a \in C(\mathbf{R},\ \mathcal{P}). \qquad \text{[STOP]}$$

We now need a couple of lemmas.

Lemma 1.8. *Let $a \in C(\mathbf{R},\ \mathcal{P})$ and $x \in \mathcal{P}$. For each $f, g \in C_c^\infty(\mathbf{R})$, the algebra of C^∞-functions on $\mathbf{R}$ with compact support, we have*

(i)

$$\begin{aligned}\big(\tilde{F}\pi(x)\tilde{F}^*(\xi_\omega \otimes f) \mid a^*(\xi_\omega \otimes g)\big)\\ = (2\pi)^{-3/2} \iiint e^{it(p-q)} f(p)\overline{g(q)}\,\omega\big(a(q)\sigma_{-t}^\omega(x)\big)\,dp\,dq\,dt,\end{aligned} \tag{11}$$

where $\xi_\omega \in \mathfrak{H}$ is the representing vector of ω.

(ii) *If a and $\tilde{F}\pi(x)\tilde{F}^*$ commute, then*

$$\begin{aligned}\iiint e^{i(s-i)(p-q)} f(p)\,g(q)\,\omega\big(a(p)\sigma_{-s}^\omega(x)\big)\,dp\,dq\,ds\\ = \iiint e^{is(p-q)} f(p)\,g(q)\,\omega\big(a(q)\sigma_{-s}^\omega(x)\big)\,dp\,dq\,ds.\end{aligned} \tag{12}$$

In the above integrals, we can change the order of integrations of p and q.

PROOF:

(i) For short, we write σ_t for σ_t^ω. We first observe

$$\Big[\pi(x)\tilde{F}^*\big(\xi_\omega \otimes f\big)\Big](t) = \sigma_{-t}(x)\big(F^*f\big)(t)\xi_\omega;$$

$$\begin{aligned}\Big[\tilde{F}\pi(x)\tilde{F}^*\big(\xi_\omega \otimes f\big)\Big](q) &= \frac{1}{\sqrt{2\pi}} \int e^{-itq}\sigma_{-t}(x)\big(F^*f\big)(t)\xi_\omega\,dt\\ &= \frac{1}{2\pi} \iint e^{-itq}e^{itp} f(p)\sigma_{-t}(x)\xi_\omega\,dp\,dt.\end{aligned}$$

Hence (11) follows from the above and the equality:

$$\big(\sigma_{-t}(x)\xi_\omega \mid a^*(q)\xi_\omega\big) = \omega\big(a(q)\sigma_{-t}(x)\big), \quad q, t \in \mathbf{R}.$$

(ii) By the modular condition, for each $p \in \mathbf{R}$, there exists $G(\,\cdot\,, p) \in \mathcal{A}(\mathbf{D})$, $\mathbf{D} = \{ z \in \mathbf{C} : -1 < \operatorname{Im} z < 0 \}$, such that

$$G(s, p) = \omega\big(\sigma_{-s}(x)a(p)\big), \qquad G(s - \mathrm{i}, p) = \omega\big(a(p)\sigma_{-s}(x)\big).$$

We then compute

$$\begin{aligned}
&(2\pi)^{-\frac{3}{2}} \iiint \mathrm{e}^{\mathrm{i}s(p-q)}\, f(p)\, g(q)\, \omega\big(\sigma_{-s}(x)a(p)\big)\, \mathrm{d}p\, \mathrm{d}q\, \mathrm{d}s \\
&\quad = (2\pi)^{-\frac{3}{2}} \iiint \mathrm{e}^{\mathrm{i}s(p-q)}\, f(p)\, g(q)\, G(s, p)\, \mathrm{d}p\, \mathrm{d}q\, \mathrm{d}s \\
&\quad = \frac{1}{2\pi} \iint \mathrm{e}^{\mathrm{i}sp}\, f(p)\, (Fg)(s)\, G(s, p)\, \mathrm{d}p\, \mathrm{d}s \\
&\quad = \frac{1}{2\pi} \iint \mathrm{e}^{\mathrm{i}sp}\, (Fg)(s)\, f(p)\, G(s, p)\, \mathrm{d}s\, \mathrm{d}p = I.
\end{aligned}$$

The function: $z \in \mathbf{D} \mapsto \mathrm{e}^{\mathrm{i}zp}\, (Fg)(z)\, G(z, p) \in \mathbf{C}$ is holomorphic on $\mathbf{D}$ and vanishes at infinity exponentially. Hence the above integral can be computed as follows:

$$\begin{aligned}
I &= \frac{1}{2\pi} \iint \mathrm{e}^{\mathrm{i}\,(s-\mathrm{i})p}\, (Fg)(s - \mathrm{i})\, f(p)\, G(s - \mathrm{i}, p)\, \mathrm{d}s\, \mathrm{d}p \\
&= \frac{1}{2\pi} \iint \mathrm{e}^{\mathrm{i}\,(s-\mathrm{i})p}\, (Fg)(s - \mathrm{i})\, f(p)\, G(s - \mathrm{i}, p)\, \mathrm{d}p\, \mathrm{d}s \\
&= (2\pi)^{-\frac{3}{2}} \iiint \mathrm{e}^{\mathrm{i}\,(s-\mathrm{i})(p-q)}\, g(q)\, f(p)\, \omega\big(a(p)\sigma_{-s}(x)\big)\, \mathrm{d}q\, \mathrm{d}p\, \mathrm{d}s\,.
\end{aligned}$$

On the other hand, we have, by the commutativity of a and $\tilde{F}\pi(x)\tilde{F}^*$,

$$\begin{aligned}
&(2\pi)^{-\frac{3}{2}} \iiint \mathrm{e}^{\mathrm{i}s(p-q)}\, f(p)\, g(q)\, \omega\big(\sigma_{-s}(x)a(p)\big)\, \mathrm{d}p\, \mathrm{d}q\, \mathrm{d}s \\
&\quad = (2\pi)^{-\frac{3}{2}} \iint \mathrm{e}^{\mathrm{i}s(p-q)}\, f(p)\, g(q)\, \overline{\omega\big(a(p)^*\sigma_{-s}(x^*)\big)}\, \mathrm{d}p\, \mathrm{d}q\, \mathrm{d}s \\
&\quad = (2\pi)^{-\frac{3}{2}} \overline{\left\{ \iiint \mathrm{e}^{\mathrm{i}s(p-q)}\, \overline{f(p)\, g(q)}\, \big(\sigma_{-s}(x^*)\xi_\omega \mid a(p)\xi_\omega\big)\, \mathrm{d}p\, \mathrm{d}q\, \mathrm{d}s \right\}} \\
&\quad = \overline{\Big(\big(\tilde{F}\pi(x^*)\tilde{F}^*\big)\big(\xi_\omega \otimes \overline{g}\big) \,\Big|\, a\big(\xi_\omega \otimes f\big)\Big)} \\
&\quad = \Big(a\big(\xi_\omega \otimes f\big) \mid \tilde{F}\pi(x)^*\tilde{F}^*\big(\xi_\omega \otimes \overline{g}\big)\Big) \\
&\quad = (2\pi)^{-\frac{3}{2}} \iiint \mathrm{e}^{\mathrm{i}s(p-q)}\, f(p)\, g(q)\, \omega\big(a(q)\sigma_{-s}(x)\big)\, \mathrm{d}p\, \mathrm{d}q\, \mathrm{d}s\,.
\end{aligned}$$

Q.E.D.

Lemma 1.9. *Let H be a bounded continuous function on $\mathbf{R} \times \mathbf{R}$. If H satisfies the integral equation for each pair $f, g \in C_c^\infty(\mathbf{R})$:*

$$\begin{aligned} &\iiint e^{is(p-q)} f(p) g(q) H(q,s)\, dp\, dq\, ds \\ &\quad = \iiint e^{i(s-i)(p-q)} f(p) g(q) H(p,s)\, dp\, dq\, ds, \end{aligned} \tag{13}$$

then H is constant in the second variable, i.e. $H(q,s) = H(q,0)$, $(q,s) \in \mathbf{R} \times \mathbf{R}$.

PROOF: Define two distributions $\tilde{H}_1$ and $\tilde{H}_2$ on $\mathbf{R}^2$ by the integrals:

$$\langle f, \tilde{H}_1 \rangle = \iiint e^{is(p-q)} f(p,q) H(q,s)\, dp\, dq\, ds;$$

$$\langle f, \tilde{H}_2 \rangle = \iiint e^{i(s-i)(p-q)} f(p,q) H(p,s)\, dp\, dq\, ds.$$

By assumption, we have

$$\langle f \otimes g, \tilde{H}_1 \rangle = \langle f \otimes g, \tilde{H}_2 \rangle, \quad f, g \in C_c^\infty(\mathbf{R}).$$

But the algebraic tensor product $C_c^\infty(\mathbf{R}) \otimes C_c^\infty(\mathbf{R})$ is dense in $C_c^\infty(\mathbf{R}^2)$, so that $\tilde{H}_1 = \tilde{H}_2$. Hence we get, for $f \in C_c^\infty(\mathbf{R}^2)$,

$$\begin{aligned} &\iiint e^{isr} f(q+r, q) H(q,s)\, dr\, dq\, ds \\ &\quad = \iiint e^{isr} e^r f(q+r, q) H(q+r, s)\, dr\, dq\, ds. \end{aligned}$$

Set $g(r,q) = f(q+r, q)$. (We note that when f runs over all of $C_c^\infty(\mathbf{R}^2)$, g exhausts $C_c^\infty(\mathbf{R}^2)$.) We then compute

$$\begin{aligned} \iiint e^{isr} g(r,q) H(q,s)\, dr\, dq\, ds &= \iiint e^{isr} e^r g(r,q) H(q+r, s)\, dr\, dq\, ds \\ &= \iiint e^{isr} e^r g(r, q-r) H(q,s)\, dr\, dq\, ds. \end{aligned}$$

Put

$$\langle g, \tilde{H} \rangle = \iiint e^{isp} g(p,q) H(q,s)\, dp\, dq\, ds.$$

Then $\tilde{H}$ is the Fourier transform of a bounded continuous function H with respect to the second variable, so that $\tilde{H}$ is a tempered distribution. We then define a linear transformation T on $C_c^\infty(\mathbf{R}^2)$ as follows:

$$(Tg)(p,q) = e^p g(p, q-p), \quad g \in C_c^\infty(\mathbf{R}^2).$$

Then the above equation for H implies that $\langle (1-T)g, \tilde{H}\rangle = 0$ for every $g \in C_c^\infty(\mathbf{R}^2)$. We now show that $(1-T)C_c^\infty(\mathbf{R}^2)$ approximates well any function $f \in C_c^\infty(\mathbf{R}^2)$ with $\operatorname{supp} f \cap (\{0\} \times \mathbf{R}) = \emptyset$ in the topology of the Schwartz space $\mathcal{S}(\mathbf{R}^2)$. Suppose $f \in C_c^\infty(\mathbf{R}^2)$ has the support off $\{0\} \times \mathbf{R}$. Put

$$g_n(p,q) = \sum_{k=0}^{n} \mathrm{e}^{kp} f(p, q - kp).$$

Then $g_n \in C_c^\infty(\mathbf{R}^2)$ and

$$\big[(1-T)g_n\big](p,q) = f(p,q) - \mathrm{e}^{(n+1)p} f\big(p, q-(n+1)p\big).$$

Hence for each $(p,q) \in \mathbf{R}^2$, $p \neq 0$, $(1-T)g_n(p,q) = f(p,q)$ for a large n. Since $\operatorname{supp} f$ is compact and disjoint from $\{0\} \times \mathbf{R}$, we can choose n so large that $(1-T)g_n = f$. Hence $\langle f, \tilde{H}\rangle = 0$ for any $f \in C_c^\infty(\mathbf{R}^2)$ with $\operatorname{supp} f \cap (\{0\} \times \mathbf{R}) = \emptyset$. Thus $\operatorname{supp} \tilde{H} \subset \{0\} \times \mathbf{R}$, equivalently

$$\iint \mathrm{e}^{\mathrm{i}sp} f(p) H(q,s)\, \mathrm{d}p\, \mathrm{d}s = 0$$

for any $f \in C_c^\infty(\mathbf{R})$ with $\operatorname{supp} f \not\ni 0$. Thus, for each q, $H(q,s)$ is a polynomial of s. But the boundedness of H implies that $H(q,s)$ must be constant in s. Q.E.D.

End of the Proof of Theorem 1.7: Let $a \in C(\mathbf{R}, \mathcal{P}) \cap \tilde{F}\pi(\mathcal{P})'\tilde{F}^*$. For each $x \in \mathcal{P}$, we set $H(p,s) = \omega\big(a(p)\sigma_{-s}(x)\big)$, $(p,s) \in \mathbf{R}^2$. We then apply the last two lemmas to H to conclude that

$$\omega\big(a(p)\sigma_{-s}(x)\big) = \omega\big(a(p)x\big), \quad (p,s) \in \mathbf{R}^2.$$

This means that

$$\omega\Big(\sigma_s\big(a(p)\big)x\Big) = \omega\big(a(p)x\big), \quad (p,s) \in \mathbf{R}^2.$$

Hence we get $\sigma_s\big(a(p)\big) = a(p)$, i.e. $a(p)$ belongs to $\mathcal{P}_\omega$. Namely, the inclusion (10) follows. Q.E.D.

Let $\mathcal{M} = \mathcal{N} \rtimes_\theta \mathbf{R}$ be the continuous decomposition of a factor $\mathcal{M}$ of type III, and $\{u(s)\}$ be the associated one parameter unitary group. We view $\mathcal{N}$ as the fixed point subalgebra of $\mathcal{M}$ under the dual action $\hat{\theta}$ which is the modular automorphism group $\{\sigma_t^\varphi\}$ of the weight $\varphi = \hat{\tau}$ dual to the trace τ of $\mathcal{N}$ scaled by $\{\theta_s\}$. Hence $\mathcal{N} = \mathcal{M}_\varphi$. Let $\mathcal{C}$ be the center $\mathcal{C}_\mathcal{N}$ of $\mathcal{N}$. We recall that $\mathrm{Z}_\theta^1\big(\mathbf{R}, \mathcal{U}(\mathcal{C})\big)$ is the group of all σ-strongly continuous $\mathcal{U}(\mathcal{C})$-valued functions $c\colon t \in \mathbf{R} \mapsto c(t) \in \mathcal{U}(\mathcal{C})$ such that

$$c(s+t) = c(s)\theta_s\big(c(t)\big), \quad s,t \in \mathbf{R}. \tag{14}$$

For each $a \in \mathcal{U}(\mathcal{C})$, we set

$$\big(\partial a\big)(s) = a^*\theta_s(a). \tag{15}$$

Clearly ∂a belongs to $Z^1_\theta(\mathbf{R}, \mathcal{U}(\mathcal{C}))$. The group of all such ∂a's is denoted by $B^1_\theta(\mathbf{R}, \mathcal{U}(\mathcal{C}))$. The first cohomology group is defined to be the quotient group:

$$H^1_\theta(\mathbf{R}, \mathcal{U}(\mathcal{C})) = Z^1_\theta(\mathbf{R}, \mathcal{U}(\mathcal{C})) / B^1_\theta(\mathbf{R}, \mathcal{U}(\mathcal{C})). \tag{16}$$

For each $c \in Z^1_\theta(\mathbf{R}, \mathcal{U}(\mathcal{C}))$, we define

$$\left.\begin{aligned} \sigma^\varphi_c(x) &= x, & x &\in \mathcal{N}; \\ \sigma^\varphi_c(u(s)) &= c(s)u(s), & s &\in \mathbf{R}. \end{aligned}\right\} \tag{17}$$

Theorem 1.10. *With the notations as above, we have the following conclusions:*

(i) *The formula* (17) *gives indeed an automorphism* σ^φ_c *of* $\mathcal{M}$, *which fixes* $\mathcal{N}$ *pointwise.*
(ii) *Every automorphism of* $\mathcal{M}$ *leaving* $\mathcal{N}$ *pointwise fixed is of this form* σ^φ_c *for some* $c \in Z^1_\theta(\mathbf{R}, \mathcal{U}(\mathcal{C}))$.
(iii) *The automorphism* σ^φ_c *is inner if and only if the cocycle* c *is a coboundary, i.e.* $c \in B^1_\theta(\mathbf{R}, \mathcal{U}(\mathcal{C}))$.
(iv) *If we denote by* $\delta_{[c]}$ *the coset of* σ^φ_c *in* $\mathrm{Out}(\mathcal{M}) = \mathrm{Aut}(\mathcal{M})/\mathrm{Int}(\mathcal{M})$, *then we have a natural short exact sequence:*

$$1 \longrightarrow H^1_\theta(\mathbf{R}, \mathcal{U}(\mathcal{C})) \xrightarrow{\ \delta\ } \mathrm{Out}(\mathcal{M}) \xrightarrow{\ \gamma\ } \mathrm{Out}_{\theta,\tau}(\mathcal{N}) \longrightarrow 1, \tag{18}$$

where $\mathrm{Out}_{\theta,\tau}(\mathcal{N})$ *is the canonical image of*

$$\mathrm{Aut}_{\theta,\tau}(\mathcal{N}) = \left\{\alpha \in \mathrm{Aut}(\mathcal{N}) : \alpha \circ \theta_s = \theta_s \circ \alpha,\ s \in \mathbf{R},\ \tau \circ \alpha = \tau\right\}, \tag{19}$$

in the quotient group $\mathrm{Out}(\mathcal{N})$.

The automorphism σ^φ_c will be called the *extended modular automorphism* of φ corresponding to the cocycle $c \in Z^1_\theta(\mathbf{R}, \mathcal{U}(\mathcal{C}))$.

PROOF:

(i) We consider a standard form $\{\mathcal{N}, \mathfrak{K}, J, \mathcal{P}\}$ of $\mathcal{N}$ and let $\mathcal{M}$ act on $L^2(\mathbf{R}, \mathfrak{K})$ as usual. The action of $\mathcal{N}$ on $\mathfrak{H} = L^2(\mathbf{R}, \mathfrak{K})$ is given by

$$(x\xi)(s) = \theta_{-s}(x)\xi(s), \quad x \in \mathcal{N}, \quad \xi \in \mathfrak{H}.$$

The one parameter unitary group $\{u(s)\}$ is of the form:

$$(u(t)\xi)(s) = \xi(s-t).$$

Put

$$(C\xi)(s) = c(-s)^*\xi(s), \quad \xi \in L^2(\mathbf{R}, \mathfrak{K}), \quad s \in \mathbf{R}. \tag{20}$$

Then a simple calculation shows that the unitary C on $\mathfrak{H}$ gives rise to σ_c^φ, i.e.

$$CxC^* = \sigma_c^\varphi(x), \quad x \in \mathcal{M}. \tag{21}$$

Hence $\sigma_c^\varphi \in \mathrm{Aut}(\mathcal{M})$. Furthermore, since $c(s)$ belongs to the center of $\mathcal{N}$, σ_c^φ fixed $\mathcal{N}$ pointwise.

(ii) Suppose that $\alpha \in \mathrm{Aut}(\mathcal{M})$ leaves $\mathcal{N}$ pointwise invariant. Set $c(s) = \alpha\big(u(s)\big)u(s)^*$, $s \in \mathbf{R}$. Then for each $x \in \mathcal{N}$, we have

$$\begin{aligned} c(s)x &= \alpha\big(u(s)\big)u(s)^*x = \alpha\big(u(s)\big)\theta_{-s}(x)u(s)^* = \alpha\big(u(s)\theta_{-s}(x)\big)u(s)^* \\ &= \alpha\big(xu(s)\big)u(s)^* = x\alpha\big(u(s)\big)u(s)^* = xc(s), \end{aligned}$$

so that $c(s) \in \mathcal{N}' \cap \mathcal{M}$. By Theorem 1.7, $c(s)$ belongs to $\mathcal{U}(\mathcal{C})$. The cocycle identity (14) follows directly from the one parameter group property of $\{u(s)\}$. Hence $c \in Z_\theta^1\big(\mathbf{R}, \mathcal{U}(\mathcal{C})\big)$ and $\alpha = \sigma_c^\varphi$.

(iii) Suppose that $\sigma_c^\varphi = \mathrm{Ad}(v)$ for some $v \in \mathcal{U}(\mathcal{M})$. Then we have, for any $x \in \mathcal{N}$,

$$vxv^* = \sigma_c^\varphi(x) = x,$$

so that $v \in \mathcal{N}' \cap \mathcal{M} = \mathcal{C}$. We further compute:

$$c(s)u(s) = \sigma_c^\varphi\big(u(s)\big) = vu(s)v^* = v\theta_s(v^*)u(s),$$

so that $c(s) = v\theta_s(v^*)$; hence $c \in B_\theta^1\big(\mathbf{R}, \mathcal{U}(\mathcal{C})\big)$.

Conversely, if $c(s) = v\theta_s(v^*)$ with $v \in \mathcal{U}(\mathcal{C})$, then a simple calculation shows that $\sigma_c^\varphi = \mathrm{Ad}(v)$. Hence σ_c^φ is inner.

(iv) The proof of this assertion requires a stronger uniqueness theorem, which will be proved in §4, Theorem 4.18.[1] If we assume the uniqueness of φ modulo $\mathrm{Int}(\mathcal{M})$, where φ is the weight on $\mathcal{M}$ dual to τ, then the rest of the proof is easy. Namely, if $\alpha \in \mathrm{Aut}(\mathcal{M})$, then Theorem 4.18, to be proven, implies that there exists $w \in \mathcal{U}(\mathcal{M})$ such that $\varphi \circ \alpha = \varphi \circ \mathrm{Ad}(w)$. Hence $\alpha \circ \mathrm{Ad}(w)^{-1}$ leaves φ invariant, so that it induces an automorphism α' of $\mathcal{N}$. If $\varphi \circ \alpha = \varphi \circ \mathrm{Ad}(w')$ with another $w' \in \mathcal{U}(\mathcal{M})$, then $\mathrm{Ad}(w'w^*)$ leaves φ invariant, so that $w'w^* \in \mathcal{N}$. Hence $\alpha \circ \mathrm{Ad}(w)^{-1}$ and $\alpha \circ \mathrm{Ad}(w')^{-1}$ are congruent modulo $\mathrm{Int}(\mathcal{N})$. Hence the class $\dot\alpha'$ of α' in $\mathrm{Out}(\mathcal{N})$ does not depend on the choice of w, so that we get a map $\gamma_0\colon \alpha \in \mathrm{Aut}(\mathcal{M}) \mapsto \dot\alpha' \in \mathrm{Out}(\mathcal{N})$. Once again, a perturbation of α by $\mathrm{Int}(\mathcal{M})$ does not affect $\dot\alpha'$ at all. Hence the map γ_0 gives rise to a map $\gamma\colon \mathrm{Out}(\mathcal{M}) \mapsto \mathrm{Out}(\mathcal{N})$. It is straightforward to see that γ is actually a homomorphism. For $\alpha \in \mathrm{Aut}(\mathcal{M})$, $\gamma(\dot\alpha) = \mathrm{id}$ if and only if there exists $w \in \mathcal{U}(\mathcal{M})$ such that $\alpha \circ \mathrm{Ad}(w)|_{\mathcal{N}} \in \mathrm{Int}(\mathcal{N})$; if and only if there exists $v \in \mathcal{U}(\mathcal{N})$ such that $\alpha \circ \mathrm{Ad}(w) \circ \mathrm{Ad}(v) = \alpha \circ \mathrm{Ad}(wv)$ leaves $\mathcal{N}$ pointwise invariant; if and only if $\dot\alpha \in \delta\big(H_\theta^1(\mathbf{R}, \mathcal{U}(\mathcal{C})\big)$ by (ii) and (iii).

[1] Theorem 4.18 assumes the separability on the factor $\mathcal{M}$. But what we need here is the uniqueness of a dominant weight up to the perturbation by $\mathrm{Int}(\mathcal{M})$, which can be guaranteed without the separability assumption on the factor $\mathcal{M}$ by the existence of a one parameter unitary group $\{w(s)\}$ such that $\sigma_t^\varphi\big(w(s)\big) = e^{ist}w(s)$, $s, t \in \mathbf{R}$.

Let $G = \{\sigma \in \mathrm{Aut}(\mathcal{N}) : \dot{\sigma} \in \mathrm{Image}(\gamma)\}$. Then we have

$$G = \{\alpha|_{\mathcal{N}} : \alpha \in \mathrm{Aut}(\mathcal{M}),\ \varphi \circ \alpha = \varphi\}.$$

But if $\varphi \circ \alpha = \varphi$, then α commutes with σ^φ, hence with the integral:

$$\mathcal{E}_\varphi(x) = \int_{\mathbf{R}} \sigma_t^\varphi(x)\, dt, \quad x \in \mathcal{M}_+,$$

the operator valued weight associated with the crossed product decomposition of $\mathcal{M}$. We know that

$$\varphi(x) = \tau \circ \mathcal{E}_\varphi(x), \quad x \in \mathcal{M}_+.$$

Since the range of $\mathcal{E}_\varphi$ exhausts $\mathcal{N}_+$, $\tau \circ \alpha = \tau$ follows from $\varphi \circ \alpha = \varphi$. Thus, every $\alpha \in G$ preserves τ.

Finally, if $\varphi \circ \alpha = \varphi$, then set $c(s) = \alpha(u(s))u(s)^*$, $s \in \mathbf{R}$. It is a straightforward computation to check $c \in \mathrm{Z}_\theta^1(\mathbf{R}, \mathcal{U}(\mathcal{N}))$, and $\alpha\theta_s\alpha^{-1} = \mathrm{Ad}(c(s)) \circ \theta_s$. But the next result, Theorem 1.11, shows that c is a coboundary, i.e. $c(s) = v^*\theta_s(v)$ for some $v \in \mathcal{U}(\mathcal{N})$. Hence $\mathrm{Ad}(v) \circ \alpha$ commutes with $\{\theta_s\}$, so that $\mathrm{Ad}(v) \circ \alpha \in \mathrm{Aut}_{\theta,\tau}(\mathcal{N})$.

Q.E.D.

Theorem 1.11. *If a one parameter automorphism group $\{\theta_s\}$ of $\mathcal{N}$ scales a faithful semi-finite normal trace τ, then θ is stable in the sense that*

$$\mathrm{Z}_\theta^1(\mathbf{R}, \mathcal{U}(\mathcal{N})) = \mathrm{B}_\theta^1(\mathbf{R}, \mathcal{U}(\mathcal{N})).$$

In particular, every cocycle perturbation of θ is conjugate to θ under $\mathrm{Int}(\mathcal{N})$.

PROOF: By Lemma 1.2, $\{\mathcal{N}, \theta\}$ is a dual covariant system. Furthermore, Theorem 1.1 and Lemma 1.2 together imply that $\mathcal{N}^\theta$ is semi-finite if and only if $\{\mathcal{N}, \theta\} \cong \{\mathcal{N}^\theta \overline{\otimes} L^\infty(\mathbf{R}), \mathrm{id} \otimes \text{translation}\}$. In this case, let $c \in \mathrm{Z}_\theta^1(\mathbf{R}, \mathcal{U}(\mathcal{N}))$. Considering the von Neumann subalgebra generated by $\{\theta_t(c(s))\}$ and $\mathbf{C} \otimes L^\infty(\mathbf{R})$, we may assume that $\mathcal{N}$ is separable. (If θ is given by $\mathrm{id} \otimes$ translation, we don't need a trace.) Let $\mathcal{P} = \mathcal{N}^\theta$ and $\mathcal{N} = \mathcal{P} \overline{\otimes} L^\infty(\mathbf{R})$. We then view $\mathcal{N}$ as $L^\infty(\mathbf{R}, \mathcal{P})$. Let $c(\,\cdot\,, s)$ be the $\mathcal{U}(\mathcal{P})$-valued function on $\mathbf{R}$. The cocycle identity means that for every pair $s, t \in \mathbf{R}$

$$c(r, s+t) = c(r, s)c(r+s, t)$$

for almost every r. If this equation holds for every $r, s, t \in \mathbf{R}$, then with $b(r) = c(0, r)$, we get

$$c(s, t) = b(s)^*b(s+t),$$

i.e. $c(t) = b^*\theta_t(b)$. But, in general, the above cocycle identity holds only for every $s, t \in \mathbf{R}$ and almost every $r \in \mathbf{R}$. We thus proceed in the following way. By the translation invariance of the Lebesgue measure and by Fubini's theorem, we have, for every $s \in \mathbf{R}$, and almost every $p, r \in \mathbf{R}$,

$$c(p, r-p+s) = c(p, r-p)c(r, s),$$

so that for every $s \in \mathbf{R}$ and almost every $r, p \in \mathbf{R}$,

$$c(r, s) = c\big(p, r - p\big)^* c\big(p, r - p + s\big).$$

Hence there exists $p \in \mathbf{R}$ such that the above identity is true for almost every $r, s \in \mathbf{R}$. With this particular p, we set $b(r) = c(p, r - p)$, and we have

$$c(r, s) = b(r)^* b(r + s)$$

for almost every $r, s \in \mathbf{R}$. Hence $c(s) = b^* \theta_s(b)$ for almost every $s \in \mathbf{R}$. But the σ-weak continuity of both sides implies that $c(s) = b^* \theta_s(b)$ for every $s \in \mathbf{R}$.

If $\mathcal{N}^\theta$ is of type III, then θ is conjugate to $\hat{\hat{\theta}} = \theta \otimes \rho$. But every cocycle perturbation yields the conjugate dual action. Q.E.D.

Exercise XII.1

Let $\mathcal{M}$ be a factor and $\{\mathcal{M}, \mathfrak{H}, J, \mathfrak{P}\}$ be a standard form which gives rise to the standard implementation $\alpha \in \mathrm{Aut}(\mathcal{M}) \mapsto U(\alpha) \in \mathcal{U}(\mathfrak{H})$. With $\varphi \in \mathfrak{W}_0$ set $\mathcal{N} = \mathcal{M} \rtimes_{\sigma^\varphi} \mathbf{R}$ and $\mathfrak{K} = \mathfrak{H} \otimes L^2(\mathbf{R}) = L^2(\mathbf{R}, \mathfrak{H})$ on which $\mathcal{N}$ acts. Let $\{u(t) : t \in \mathbf{R}\}$ be the one parameter unitary group in $\mathcal{N}$ corresponding to the crossed product implementing the modular automorphism $\sigma_t^{\hat{\varphi}}$ of the dual weight $\hat{\varphi}$ on $\mathcal{N}$.

1) Following the steps suggested below, prove that to each $\alpha \in \mathrm{Aut}(\mathcal{M})$ there corresponds an element $\widetilde{\alpha} \in \mathrm{Aut}(\mathcal{N})$ and that the map: $\alpha \in \mathrm{Aut}(\mathcal{M}) \mapsto \widetilde{\alpha} \in \mathrm{Aut}(\mathcal{N})$ is an injective homomorphism of $\mathrm{Aut}(\mathcal{M})$:

(a) To each $\alpha \in \mathrm{Aut}(\mathcal{M})$ associate a unitary $V(\alpha)$ defined by:

$$\{V(\alpha)\xi\}(r) = \big(\mathrm{D}\varphi \circ \alpha^{-1} : \mathrm{D}\varphi\big)^*_{-r} U(\alpha)\xi(r), \quad \xi \in \mathfrak{K}.$$

(b) Show that

$$V(\alpha\beta) = V(\alpha)V(\beta), \quad \alpha, \beta \in \mathrm{Aut}(\mathcal{M}).$$

(c) Show that

$$V(\alpha)xV(\alpha)^* = \alpha(x), \quad x \in \mathcal{M}, \quad \alpha \in \mathrm{Aut}(\mathcal{M}).$$

(d) Show that

$$V(\alpha)u(t)V(\alpha)^* = \big(\mathrm{D}\varphi \circ \alpha^{-1} : \mathrm{D}\varphi\big)_t u(t), \quad t \in \mathbf{R}, \quad \alpha \in \mathrm{Aut}(\mathcal{M}).$$

(e) Show that the map $\widetilde{\alpha} : x \in \mathcal{N} \mapsto V(\alpha)xV(\alpha)^* \in \mathcal{N}$ is an automorphism of $\mathcal{N}$, and that the map: $\alpha \in \mathrm{Aut}(\mathcal{M}) \mapsto \widetilde{\alpha} \in \mathrm{Aut}(\mathcal{N})$ is an injective homomorphism.

(f) Show that the map: $\alpha \in \mathrm{Aut}(\mathcal{M}) \mapsto V(\alpha) \in \mathcal{U}(\mathfrak{K})$ is continuous relative to the topology in $\mathrm{Aut}(\mathcal{M})$ defined in §1, Chapter X, and the strong topology in $\mathcal{U}(\mathfrak{K})$. Hence, the map: $\alpha \in \mathrm{Aut}(\mathcal{M}) \mapsto \widetilde{\alpha} \in \mathrm{Aut}(\mathcal{N})$ is a continuous injective homomorphism.

2) Show that

$$\left\{ \widetilde{\alpha} \in \mathrm{Aut}(\mathcal{N}) : \alpha \in \mathrm{Aut}(\mathcal{M}) \right\} = \mathrm{Aut}_{\theta,\tau}(\mathcal{N})$$

where

$$\mathrm{Aut}_{\theta,\tau}(\mathcal{N}) = \left\{ \alpha \in \mathrm{Aut}(\mathcal{N}) : \theta_s \circ \alpha = \alpha \circ \theta_s,\ s \in \mathbf{R},\ \tau \circ \alpha = \tau \right\}.$$

§ 2 Structure of Factors of Type III$_\lambda$, $0 < \lambda < 1$

In this section, we consider a factor of Type III$_\lambda$, $0 < \lambda < 1$. First, we recall that a factor $\mathcal{M}$ is, by definition, of type III$_\lambda$, $0 < \lambda < 1$, if

$$S(\mathcal{M}) = \bigcap \{ \mathrm{Sp}(\Delta_\varphi) : \varphi \in \mathcal{W}_0 \} = \{ \lambda^n : n \in \mathbf{Z} \} \cup \{0\}, \tag{1}$$

where $\mathcal{W}_0$ denotes the set of all faithful semi-finite normal weights on $\mathcal{M}$. This is then equivalent to the fact:

$$\Gamma(\sigma^\varphi) = (\log \lambda)\mathbf{Z}. \tag{2}$$

Theorem 2.1.

(i) *If $\mathcal{N}_0$ is a factor of type* II$_\infty$, *and if $\theta \in \mathrm{Aut}(\mathcal{N}_0)$ scales a faithful semi-finite normal trace τ at the rate λ, i.e.*

$$\tau \circ \theta = \lambda \tau, \quad 0 < \lambda < 1, \tag{3}$$

or equivalently $\mathrm{mod}(\theta) = \lambda^{-1}$, *then $\mathcal{N}_0 \rtimes_\theta \mathbf{Z}$ is a factor of type* III$_\lambda$.

(ii) *If $\mathcal{M}$ is a factor of type* III$_\lambda$, $0 < \lambda < 1$, *then there exists a unique covariant system $\{\mathcal{N}_0, \theta\}$ of a factor $\mathcal{N}_0$ of type* II$_\infty$ *and $\theta \in \mathrm{Aut}(\mathcal{N}_0)$ with* $\mathrm{mod}\,\theta = \lambda^{-1}$ *such that*

$$\mathcal{M} \cong \mathcal{N}_0 \rtimes_\theta \mathbf{Z}. \tag{4}$$

PROOF:

(i) Let φ be the weight on $\mathcal{M} = \mathcal{N}_0 \rtimes_\theta \mathbf{Z}$ dual to τ. Let u be the unitary associated with the crossed product. Since $\left(\mathrm{D}\tau \circ \theta : \mathrm{D}\tau\right)_t = \lambda^{\mathrm{i}t}$, we have $\sigma_t^\varphi(u) = \lambda^{\mathrm{i}t} u$, and $\sigma_t^\varphi(x) = x$, $x \in \mathcal{N}_0$. Hence σ^φ is periodic with period $T = -2\pi/\log\lambda$. Since $\mathcal{M}_\varphi = \mathcal{N}_0$ is a factor, we have $\Gamma(\sigma^\varphi) = \mathrm{Sp}(\sigma^\varphi) = (\log\lambda)\mathbf{Z}$. Hence $\mathcal{M}$ is a factor of type III$_\lambda$.

(ii) Suppose that $\mathcal{M}$ is a factor of type III$_\lambda$, $0 < \lambda < 1$. Let $\{\mathcal{N}, \mathbf{R}, \theta\}$ be the associated covariant system. Then the flow of weights $\{\mathcal{C}, \mathbf{R}, \theta\}$ has period $T = -2\pi/\log\lambda$. Hence $\mathrm{Sp}(\theta|_{\mathcal{C}}) = (\log\lambda)\mathbf{Z}$. Let $a \in \mathcal{C}$, $a \neq 0$, be such that $\theta_s(a) = \mathrm{e}^{-2\pi \mathrm{i}s/T} a = \lambda^{\mathrm{i}s} a$. Then $a^*a = aa^* \in \mathcal{C}^\theta = \mathbf{C}$. Hence a is a scalar multiple of a unitary. We replace a by this unitary, so that we may assume that $a^*a = aa^* = 1$ and $\theta_s(a) = \lambda^{\mathrm{i}s} a$, $s \in \mathbf{R}$. Therefore, if $\{u(s)\}$ is the one parameter unitary group in $\mathcal{M}$ associated with the continuous decomposition, then

$$a u(s) a^* = \lambda^{-\mathrm{i}s} u(s), \quad s \in \mathbf{R}, \tag{5}$$

so that $\mathrm{Ad}(a) = \sigma_T^\varphi$, where φ is the weight on $\mathcal{M}$ dual to τ. Thus we have $T \in T(\mathcal{M})$, the modular period group. Suppose now ψ is an arbitrary faithful semi-finite normal weight on $\mathcal{M}$ with $\psi(1) = +\infty$. We then have $\sigma_T^\psi \in \mathrm{Int}(\mathcal{M})$. Choose $b \in \mathcal{U}(\mathcal{M})$ with $\sigma_T^\psi = \mathrm{Ad}(b)$. It then follows that b belongs to the center $\mathcal{C}_\psi$ of $\mathcal{M}_\psi$. Choose an invertible $h \in \mathcal{C}_\psi^+$ such that $b = h^{\mathrm{i}T}$. Put $\varphi = \psi(h^{-1}\cdot)$. This φ might be different from the above one. By the invertibility of h, we have $\varphi(1) = \psi(h^{-1}) = +\infty$, and also $\sigma_T^\varphi = \mathrm{Ad}(h^{-\mathrm{i}T}) \circ \sigma_T^\psi = \mathrm{Ad}(h^{-\mathrm{i}T}) \circ \mathrm{Ad}(b) = \mathrm{id}$. Hence $\{\sigma_t^\varphi\}$ has period T. Therefore, the integral

$$\mathcal{E} = \frac{1}{T}\int_0^T \sigma_t^\varphi \, \mathrm{d}t \tag{6}$$

gives a faithful conditional expectation from $\mathcal{M}$ onto $\mathcal{M}_\varphi$ such that $\varphi = \varphi \circ \mathcal{E}$. This implies that $\mathcal{E}(\mathfrak{m}_\varphi) = \mathcal{M}_\varphi \cap \mathfrak{m}_\varphi$, so that φ is semi-finite on $\mathcal{M}_\varphi$. Therefore, the restriction of φ to $\mathcal{M}_\varphi$ is a faithful semi-finite normal trace, say τ.

Since $\left\{\sigma_t^\varphi\right\}$ has period T, we have

$$\mathrm{Sp}(\sigma^\varphi) \subset \left(\frac{2\pi}{T}\right)\mathbf{Z} = (\log\lambda)\mathbf{Z};$$

on the other hand, we have, by definition,

$$\mathrm{Sp}(\sigma^\varphi) \supset \Gamma(\sigma^\varphi) = (\log\lambda)\mathbf{Z}.$$

Hence we get $\mathrm{Sp}(\sigma^\varphi) = \Gamma(\sigma^\varphi) = (\log\lambda)\mathbf{Z}$.

Viewing σ^φ as an action of the torus group $\mathbf{T} = \mathbf{R}/T\mathbf{Z}$, we consider the crossed product $\mathcal{M} \rtimes_{\sigma^\varphi} \mathbf{T}$. Since $\Gamma(\sigma^\varphi) = \hat{\mathbf{T}} = (\log\lambda)\mathbf{Z}$, $\mathcal{M} \rtimes_{\sigma^\varphi} \mathbf{T}$ must be a factor, Corollary XI.2.8. Since $\mathbf{T}$ is compact, $\left\{\mathcal{M}, \mathbf{T}, \sigma^\varphi\right\}$ is conjugate to a reduced system of the second dual action $\sigma^\varphi \otimes \rho$ of $\mathbf{T}$, see §2, Chapter X. Hence the fixed point algebra $\mathcal{M}_\varphi$ of σ^φ is a factor, being isomorphic to a reduced algebra of the factor $\mathcal{M} \rtimes_{\sigma^\varphi} \mathbf{T}$. Since $\tau(1) = +\infty$, $\mathcal{M}_\varphi$ must be infinite.

From this point on, we can proceed in two ways: one way is through the duality, $(\mathcal{M} \rtimes_{\sigma^\varphi} \mathbf{T}) \rtimes_{\hat{\sigma}^\varphi} \hat{\mathbf{T}}$, and the other is through the Fourier analysis of $\left\{\mathcal{M}, \sigma^\varphi\right\}$. Since we are now familiar with the first approach, we shall take the second one to expand our scope, even though this path is slightly longer.

We recall first that every factor is a tensor product of a σ-finite factor and a factor of type I. So we split the centralizer $\mathcal{M}_\varphi$ into the tensor product of a σ-finite factor and a factor $\mathcal{B}$ of type I. We choose $\mathcal{B}$ in such a way that the restriction $\tau|_{\mathcal{B}}$ of τ to $\mathcal{B}$ is semi-finite, so that the restriction $\varphi|_{\mathcal{B}'\cap\mathcal{M}}$ is finite and a scalar multiple normalization of φ gives the usual trace Tr of $\mathcal{B}$ when it is restricted to $\mathcal{B}$. This means that $\varphi = \overline{\varphi} \otimes \mathrm{Tr}$ where $\overline{\varphi}$ is a faithful normal state on $\mathcal{B}' \cap \mathcal{M} = \overline{\mathcal{M}}$.

Since $\sigma_t^\varphi = \sigma_t^{\overline{\varphi}} \otimes \mathrm{id}$, $\left\{\sigma_t^{\overline{\varphi}}\right\}$ has the same period T, and $\mathrm{Sp}(\sigma^{\overline{\varphi}}) = \Gamma(\sigma^\varphi) = (\log\lambda)\mathbf{Z}$. Choose $a \in \overline{\mathcal{M}}_{\overline{\varphi}}(1)$, $a \neq 0$, i.e.

$$\sigma_t^{\overline{\varphi}}(a) = \lambda^{\mathrm{i}t}a, \quad t \in \mathbf{R}.$$

Let $a = uh$ be the polar decomposition of a. Then we have $\sigma_t^{\overline{\varphi}}(u) = \lambda^{\mathrm{i}t}u$ and $h \in \overline{\mathcal{M}}_{\overline{\varphi}}$. Set $\overline{e} = u^*u$ and $\overline{f} = uu^*$. Then we have

$$F(t) = \lambda^{\mathrm{i}t}\overline{\varphi}(ux) = \overline{\varphi}\big(\sigma_t^{\overline{\varphi}}(u)x\big), \quad x \in \overline{\mathcal{M}},$$

$$\lambda^{\mathrm{i}\,(t+\mathrm{i})}\overline{\varphi}(ux) = F(t+\mathrm{i}) = \overline{\varphi}\big(x\sigma_t^{\overline{\varphi}}(u)\big),$$

so that

$$\lambda\overline{\varphi}(xu) = \overline{\varphi}(ux), \quad x \in \overline{\mathcal{M}}.$$

Thus, we get

$$\overline{\varphi}(uxu^*) = \lambda\overline{\varphi}(x), \quad x \in \overline{\mathcal{M}}_e. \tag{7}$$

In $\mathcal{M}_\varphi = \overline{\mathcal{M}}_{\overline{\varphi}}\overline{\otimes}\mathcal{B},\ 1 \sim \overline{e} \sim \overline{f}$. So we choose partial isometries v and w in $\mathcal{M}_\varphi$ such that $v^*v = 1,\ vv^* = \overline{e},\ w^*w = \overline{f}$ and $ww^* = 1$, and set $U = wuv \in \mathcal{U}(\mathcal{M})$. Since $\sigma_t^\varphi = \sigma_t^{\overline{\varphi}} \otimes \mathrm{id}$, we have $\sigma_t^\varphi(U) = \lambda^{\mathrm{i}t}U,\ t \in \mathbf{R}$. The automorphism $\theta = \mathrm{Ad}(U)$ leaves $\mathcal{M}_\varphi$ globally invariant, and

$$\varphi(UxU^*) = \lambda\varphi(x), \quad x \in \mathcal{M}.$$

Hence the restriction of θ to $\mathcal{M}_\varphi$ scales the trace $\tau = \varphi|_{\mathcal{M}_\varphi}$ at the rate $\lambda = \mathrm{mod}(\theta)^{-1}$. By Proposition X.2.6, we have

$$\mathcal{M} = \mathcal{M}_\varphi \rtimes_\theta \mathbf{Z}.$$

Since $\mathrm{mod}(\theta)^{-1} = \lambda < 1$, $\mathcal{M}_\varphi$ must be of type II_∞. With $\mathcal{N}_0 = \mathcal{M}_\varphi$, we get the desired decomposition.

The uniqueness of the decomposition (4) follows from the next result immediately. Q.E.D.

Theorem 2.2. *Let $\mathcal{M}$ be a factor of type III_λ, $0 < \lambda < 1$, and set $T = -2\pi/\log\lambda$. If φ and ψ are faithful semi-finite normal weights on $\mathcal{M}$ such that $\sigma_T^\varphi = \sigma_T^\psi = \mathrm{id}$ and $\varphi(1) = \psi(1) = +\infty$, then there exists a unique α, $\lambda < \alpha \leq 1$, and a unitary $u \in \mathcal{M}$ such that*

$$\alpha\psi(x) = \varphi\big(uxu^*\big), \quad x \in \mathcal{M}_+. \tag{8}$$

PROOF: As in the proof of the previous theorem, the proof is reduced to the σ-finite case. Thus, we assume the σ-finiteness for $\mathcal{M}$.

Let $v_t = (\mathrm{D}\psi : \mathrm{D}\varphi)_t,\ t \in \mathbf{R}$. Since $\mathrm{Ad}(v_T) = \mathrm{Ad}(v_T) \circ \sigma_T^\varphi = \sigma_T^\psi = \mathrm{id}$, v_T must be a scalar of absolute value one. Choose $\alpha \in]\lambda, 1]$ so that $v_T = \alpha^{\mathrm{i}T}$. Note that such an α is unique. Now, we have

$$\big(\mathrm{D}\psi : \mathrm{D}\alpha\varphi\big)_t = (\mathrm{D}\psi : \mathrm{D}\varphi)_t\big(\mathrm{D}\varphi : \mathrm{D}\alpha\varphi\big)_t = \alpha^{-\mathrm{i}t}v_t.$$

Thus, we get $\big(\mathrm{D}\psi : \mathrm{D}\alpha\varphi\big)_t = 1$. Replace φ by $\alpha\varphi$, so that we have $(\mathrm{D}\psi : \mathrm{D}\varphi)_T = 1$.

We now consider the balanced weight $\rho = \varphi \oplus \psi$ on $\overline{\mathcal{M}} = \mathcal{M} \otimes \mathcal{M}_2(\mathbf{C})$. The assumption $(\mathrm{D}\psi : \mathrm{D}\varphi)_T = 1$, together with $\sigma_T^\varphi = \sigma_T^\psi = \mathrm{id}$, entails that $\sigma_T^\rho = \mathrm{id}$. Hence $\mathrm{Sp}(\sigma^\rho) \subset (\log \lambda)\mathbf{Z}$. But the inclusion $(\log \lambda)\mathbf{Z} = \Gamma(\sigma^\rho) \subset \mathrm{Sp}(\sigma^\rho)$ implies that $\mathrm{Sp}(\sigma^\rho) = (\log \lambda)\mathbf{Z}$. Hence the proof of the previous theorem shows that $\overline{\mathcal{M}}_\rho$ is a factor and the restriction of ρ to $\overline{\mathcal{M}}_\rho$ is a trace. Since $\varphi(1) = \psi(1) = +\infty$, e_{11} and e_{22} are both infinite projections in $\overline{\mathcal{M}}_\rho$. Hence the σ-finiteness of $\overline{\mathcal{M}}_\rho$ implies that $e_{11} \sim e_{22}$. Let $v \in \overline{\mathcal{M}}_\rho$ with $vv^* = e_{11}$ and $v^*v = e_{22}$. Then there exists a unitary $u \in \mathcal{M}$ such that $v = u \otimes e_{12}$. Since $v \in \overline{\mathcal{M}}_\rho$, we have

$$\begin{aligned} \varphi\big(uxu^*\big) &= \rho\big(uxu^* \otimes e_{11}\big) = \rho\big((u \otimes e_{12})(x \otimes e_{22})(u^* \otimes e_{21})\big) \\ &= \rho(x \otimes e_{22}) = \psi(x). \end{aligned}$$

This proves the existence of u.

To prove the uniqueness of α, suppose $\alpha\varphi(x) = \varphi(uxu^*)$ for some $u \in \mathcal{U}(\mathcal{M})$, and $\alpha > 0$. This means that

$$\alpha^{it} = \big(\mathrm{D}\alpha\varphi : \mathrm{D}\varphi\big)_t = \big(\mathrm{D}\varphi \circ \mathrm{Ad}(u) : \mathrm{D}\varphi\big)_t.$$

Consider the balance weight $\rho = \varphi \oplus \varphi \circ \mathrm{Ad}(u)$ on $\overline{\mathcal{M}} = \mathcal{M} \otimes M_2(\mathbf{C})$. We then have $u \otimes e_{21} \in \overline{\mathcal{M}}_\rho$, so that

$$\begin{aligned} u \otimes e_{21} &= \sigma_t^\rho\big(u \otimes e_{21}\big) = \sigma_t^\rho\big((1 \otimes e_{21})(u \otimes e_{11})\big) \\ &= \Big[\big(\mathrm{D}\varphi \circ \mathrm{Ad}(u) : \mathrm{D}\varphi\big)_t \otimes e_{21}\Big]\big[\sigma_t^\varphi(u) \otimes e_{11}\big] \\ &= \big(\mathrm{D}\varphi \circ \mathrm{Ad}(u) : \mathrm{D}\varphi\big)_t \sigma_t^\varphi(u) \otimes e_{21}. \end{aligned}$$

Thus, we have

$$\big(\mathrm{D}\varphi \circ \mathrm{Ad}(u) : \mathrm{D}\varphi\big)_t = u\sigma_t^\varphi(u^*), \quad t \in \mathbf{R}. \tag{9}$$

Therefore, we have

$$\sigma_t^\varphi(u) = \alpha^{-it}u.$$

Hence $\log \alpha \in \mathrm{Sp}(\sigma^\varphi) = (\log \lambda)\mathbf{Z}$, so that $\alpha = \lambda^n$ for some $n \in \mathbf{Z}$. Therefore, if $\lambda < \alpha \leq 1$, then $\alpha = 1$. Q.E.D.

Definition 2.3.

(i) The crossed product decomposition $\mathcal{M} = \mathcal{N}_0 \rtimes_\theta \mathbf{Z}$ of Theorem 2.1 for a factor $\mathcal{M}$ of type III_λ, $0 < \lambda < 1$, is called the *discrete* decomposition of $\mathcal{M}$. The covariant system $\{\mathcal{N}, \theta_0\}$ is called the *discrete* associated system, or *discrete core*.

(ii) The weight φ of Theorem 2.2 is called a *generalized trace* of a factor $\mathcal{M}$ of type III_λ, $0 < \lambda < 1$.

(iii) A faithful semi-finite normal weight φ is called *periodic* if $\sigma_T^\varphi = \mathrm{id}$ for some $T > 0$, and the smallest such T is naturally called the *period* of φ.

Exercise XII.2

Throughout this exercise, let $\mathcal{M}$ be a factor of type III_λ, $T = -\frac{2\pi}{\log\lambda}$, so that $\lambda^{\mathrm{i}T} = 1$, and $\mathcal{M} = \mathcal{N} \rtimes_\theta \mathbf{Z}$ be a discrete decomposition. Let $\mathfrak{S}_*$ be the set of normal states on $\mathcal{M}$ and $\mathfrak{S}_*^0$ be the set of faithful normal states on $\mathcal{M}$.

1) Show that if $\psi \in \mathfrak{S}_*^0$, then there exists an invertible $h \in \mathcal{M}_\psi$ such that $\varphi = \psi(h\,\cdot\,)$ is periodic with period T.

2) Show that if $\varphi \in \mathfrak{S}_*^0$ has period T, then $\mathcal{M}_\varphi$ is a subfactor of type II_1 and $\mathcal{M}_\varphi' \cap \mathcal{M} = \mathbf{C}$.

3) Show that if $\varphi \in \mathfrak{S}_*^0$ has period T, then there exists an isometry $u \in \mathcal{M}$ such that $\sigma_t^\varphi(u) = \lambda^{\mathrm{i}t} u$.

4) In the above situation, show that every maximal abelian self-adjoint subalgebra $\mathcal{A}$ of $\mathcal{M}_\varphi$ is maximal abelian in $\mathcal{M}$.

5) Keep the notations and assumptions. For a pair of projections $e, f \in \mathrm{Proj}(\mathcal{M}_\varphi)$ there exists a partial isometry $v \in \mathcal{M}$ such that $v^*v = e$, $vv^* = f$ and $v(\mathcal{M}_\varphi)_e v^* = (\mathcal{M}_\varphi)_f$ if and only if $\varphi(e) = \lambda^n \varphi(f)$ for some $n \in \mathbf{Z}$.

§ 3 Structure of Factors of Type III_0

This section is devoted to the description of the structure of a factor of type III_0 in terms of a discrete cross product of a von Neumann algebra of type II_∞ by a single automorphism. To this end, we need to discuss some part of ergodic flows. In order to avoid unnecessary complications, we consider only separable von Neumann algebras.

Let $\{\Omega, \mu\}$ be a standard σ-finite measure space. Let T be a non-singular transformation of $\{\Omega, \mu\}$, i.e. T is a bijective Borel map of Ω onto itself which preserves the class of all μ-null sets. Suppose T is ergodic. Let r be a positive Borel function on Ω such that $\inf\{ r(\omega) : \omega \in \Omega \} > 0$. We fix r and define inductively:

$$\left.\begin{aligned} r_0(\omega) &= 0, \\ r_n(\omega) &= r(\omega) + r_{n-1}(T\omega), \qquad n \geq 0, \\ r_{-n}(\omega) &= -r_n(T^{-n}\omega), \qquad n = 1, 2, \dots . \end{aligned}\right\} \tag{1}$$

It is easy to check:

$$r_{n+m}(\omega) = r_n(\omega) + r_m(T^n\omega). \tag{2}$$

Namely, $\{r_n\}$ is the one cocycle generated by r. We consider the new space $\tilde{\Gamma} = \mathbf{R} \times \Omega$, equipped with the product measure $m \otimes \mu$ of the Lebesgue measure m on $\mathbf{R}$ and μ. On $\tilde{\Gamma}$, we define

$$\begin{aligned} \tilde{T}(s, \omega) &= \big(s - r(\omega), T\omega\big), \qquad (s, \omega) \in \tilde{\Gamma}, \\ \tilde{S}_t(s, \omega) &= (s + t, \omega), \qquad t \in \mathbf{R}. \end{aligned} \tag{3}$$

Clearly, $\tilde{T}$ and $\tilde{S}_t$ commute and

$$\tilde{T}^n(s, \omega) = \big(s - r_n(\omega), T^n\omega\big).$$

We then form the quotient space $\Gamma = \tilde{\Gamma}/\tilde{T}$, i.e. the space of all $\tilde{T}$-orbits. In terms of von Neumann algebras, we consider $\tilde{\mathcal{A}} = L^\infty\big(\tilde{\Gamma}, m \otimes \mu\big)$ and

$$\tilde{\alpha}(f)(\omega) = f\big(\tilde{T}^{-1}\omega\big), \quad f \in \tilde{\mathcal{A}},$$

$$\tilde{\beta}_t(f)(\omega) = f\big(\tilde{S}_t^{-1}\omega\big).$$

We let $\mathcal{C} = \tilde{\mathcal{A}}^{\tilde{\alpha}}$, the fixed point subalgebra of $\tilde{\mathcal{A}}$ under $\tilde{\alpha}$. Since $\tilde{\alpha}$ and $\{\tilde{\beta}_t\}$ commute, $\{\tilde{\beta}_t\}$ leaves $\mathcal{C}$ globally invariant, so that the restriction $\{\beta_t\}$ of $\{\tilde{\beta}_t\}$ to $\mathcal{C}$ makes sense. The measure theoretic spectrum of $\mathcal{C}$ can be visualized in the following figure:

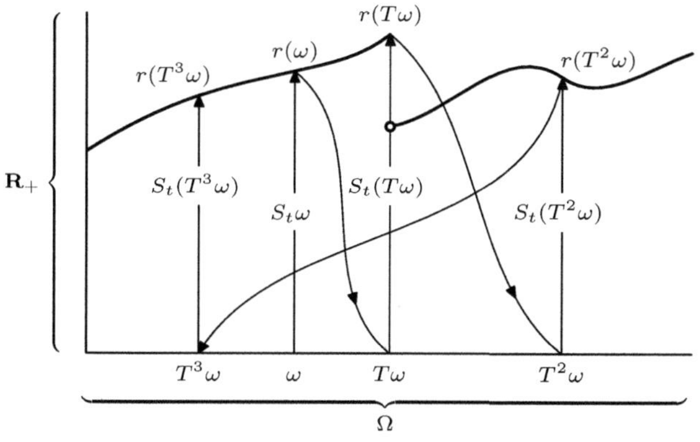

Namely, consider the subgraph $\Gamma = \big\{(s, \omega) : 0 \leq s \leq r(\omega)\big\}$ of the function r, and identify each top end $\big(r(\omega), \omega\big)$ with the bottom end $(0, T\omega)$. Then the algebra $L^\infty(\Gamma, m \otimes \mu)$ is naturally identified with $\mathcal{C}$ and the flow S_t corresponding to $\{\beta_t\}$ is the upward vertical flow on Γ with unit speed.

Definition 3.1. The flow $\{\Gamma, m \otimes \mu, S_t\}$ constructed above is called the *flow built under the ceiling function r with base transformation* $\{\Omega, \mu, T\}$.

If $\{\Omega, \mu, T\}$ is ergodic, then $\{\tilde{\Gamma}, m \otimes \mu, \tilde{T}, \{\tilde{S}_t\}\}$ is ergodic as an action of $\mathbf{Z} \times \mathbf{R}$, so that $\{\Gamma, m \otimes \mu, S_t\}$ is ergodic.

Theorem 3.2. *Every properly*[2] *ergodic flow is conjugate to the flow built under a ceiling function r on an ergodic single transformation* $\{\Omega, \mu, T\}$.

[2] If an ergodic flow is not conjugate to a transitive flow, then it is called *properly* ergodic. A flow is transitive by definition, if the measure is concentrated on a single orbit.

PROOF: Let $\{\mathcal{C}, \mathbf{R}, \theta\}$ be an ergodic flow, i.e. a triple consisting of a separable abelian von Neumann algebra $\mathcal{C}$, $\mathbf{R}$ and an ergodic action θ of $\mathbf{R}$ on $\mathcal{C}$. Let $\mathcal{A}$ be a separable σ-weekly dense C^*-subalgebra of $\mathcal{C}$ such that $\lim_{t\to 0}\|\theta_t(a) - a\| = 0$, $a \in \mathcal{A}$. Let $\Gamma = \mathrm{Sp}(\mathcal{A})$, and ν be the Radon measure on Γ corresponding to a faithful normal state on $\mathcal{C}$ restricted to $\mathcal{A}$. Then there exists a continuous action $S\colon (t, \gamma) \in \mathbf{R} \times \Gamma \mapsto S_t\gamma \in \Gamma$ such that $(\theta_t f)(\gamma) = f(S_t^{-1}\gamma)$. Since ν and $\nu \circ \theta_t$ give unitarily equivalent representations of $\mathcal{A}$ via the cyclic representations, the measures ν and $\nu \circ S_t$ are absolutely continuous to each other. The ergodicity of $\{\theta_t\}$ implies that of $\{S_t\}$ with respect to μ. Since ν is faithful, $\nu(U) > 0$ for every open set U. Choose $\gamma_0 \in \Gamma$ and $t_0 > 0$ such that $S_{t_0}\gamma_0 \neq \gamma_0$. There exists an open set $U \ni \gamma_0$ and $\delta > 0$ such that $S_t U \cap U = \emptyset$ for $|t - t_0| < \delta$. Choose an open set V such that $\gamma_0 \in V$ and $\overline{V} \subset U$. Then there exists $a > 0$ such that $S_t\overline{V} \subset U$ for $|t| < a$ because $\overline{V}$ is compact. (We assumed that Ω is compact, i.e. $\mathcal{A}$ is unital.) By ergodicity, $\bigcup_{t\in\mathbf{R}} S_t V$ is conull in Γ. Since $\{S_t\}$ is properly ergodic, $\bigcup_{t\geq 0} S_t V$ is conull in Γ. We define

$$\psi(\gamma) = \frac{1}{a}\int_0^a \chi_U(S_t\gamma)\,\mathrm{d}t\,.$$

We then have

$$\left.\begin{aligned} &\psi(\gamma) = 1, \qquad \psi(S_{t_0}\gamma) = 0, \qquad \gamma \in V, \\ &\left|\psi(S_t\gamma) - \psi(S_s\gamma)\right| \leq 2|s - t|/a. \end{aligned}\right\} \tag{4}$$

Put

$$\Omega = \left\{\gamma \in \Gamma : \psi(\gamma) = \frac{1}{2},\ \psi(S_t\gamma) < \frac{1}{2},\ 0 < t < \frac{a}{4}\right\}. \tag{5}$$

We shall show first that Ω is a Borel subset of Γ.

Since U is open, ψ is lower semi-continuous, so that the level set $\Omega_1 = \left\{\gamma \in \Gamma : \psi(\gamma) = \frac{1}{2}\right\}$ is a Borel set, and

$$\Omega_2 = \bigcup_{n=1}^{\infty}\left\{\gamma \in \Gamma : \psi(S_t\gamma) \leq \frac{1}{2} - \frac{1}{n},\ 0 < t < \frac{a}{4}\right\}$$

is an F_σ-set. Thus $\Omega = \Omega_1 \cap \Omega_2$ is a Borel subset of Γ.

We next show that $\bigcup_{t\geq 0} S_t\Omega$ is conull in Γ. Let $\gamma \in V$. Then $\psi(\gamma) = 1$ and $\psi(S_{t_0}\gamma) = 0$. Put $t_1 = \inf\{t > 0 : \psi(S_t\gamma) = 0\}$. We know that $0 < t_1 \leq t_0$ and $\psi(S_{t_1}\gamma) = 0$. Set

$$t_2 = \sup\left\{t \geq 0 : \psi(S_t\gamma) = \frac{1}{2},\ t \leq t_1\right\}.$$

By inequality (4), $t_1 - t_2 \geq \frac{a}{4}$ and $\psi(S_t\gamma) < \frac{1}{2}$ for every $t \in [t_2, t_1]$, so that $S_{t_2}\gamma \in \Omega$. Thus $\gamma \in S_{-t_2}\Omega$. Therefore, we get $V \subset \bigcup_{0\leq t\leq t_0} S_{-t}\Omega$, so that $\bigcup_{t\geq -t_0} S_t\Omega \supset \bigcup_{t\geq 0} S_t V$; hence $\bigcup_{t\geq 0} S_t\Omega = S_{t_0}\left(\bigcup_{t\geq -t_0} S_t\Omega\right)$ is conull.

We then define the *hitting time* $r_\Omega(\gamma)$ of γ of Ω by

$$r_\Omega(\gamma) = \inf\{ t > 0 : S_t\gamma \in \Omega \}, \quad \gamma \in \Gamma. \tag{6}$$

For each $\omega \in \Omega$, $r_\Omega(\omega)$ is also called the *return time* of ω to Ω. The second condition in (5) says that $\{ t \in \mathbf{R} : S_t\gamma \in \Omega \} = R(\gamma)$ is discrete and $r_\Omega(\omega) \geq \frac{a}{4}$. For simplicity, we write $r(\gamma)$ for $r_\Omega(\gamma)$. Then $S_{r(\gamma)}\gamma \in \Omega$. Since $R(\gamma)$ is discrete, in fact $|s-t| \geq \frac{a}{4}$ whenever $s, t \in R(\gamma)$, $s \neq t$, we arrange $R(\gamma)$ in the increasing order:

$$\begin{aligned} \cdots < r_\Omega(-1,\gamma) < r_\Omega(0,\gamma) \leq 0 < r_\Omega(\gamma) = r_\Omega(1,\gamma) \\ < r_\Omega(2,\gamma) < \cdots < r_\Omega(n,\gamma) < \cdots . \end{aligned} \tag{7}$$

We call $r_\Omega(n,\gamma)$ the *n-th hitting time* of γ and write $r(n,\gamma)$ or $r_n(\gamma)$ whichever more convenient. If $\omega \in \Omega$, then $r_0(\omega) = 0$.

We now define a Borel automorphism T of Ω by

$$T\omega = S_{r(\omega)}\omega, \quad \omega \in \Omega. \tag{8}$$

Discarding the complement of $\bigcup_{t\in\mathbf{R}} S_t\Omega$, which is null, we replace $\{\Gamma, \nu\}$ by $\{\bigcup_{t\in\mathbf{R}} S_t\Omega,\ \nu\}$, i.e. we may and do assume $\Gamma = \bigcup_{t\in\mathbf{R}} S_t\Omega$. Thus, the flow $\{\Gamma, \nu, S_t\}$ is formally conjugate to the flow built under the ceiling function r based on $\{\Omega, T\}$. It now remains to show that we can introduce a σ-finite measure μ on Ω quasi-invariant under T such that the map S: $\{ (t,\omega) \in \mathbf{R}_+ \times X : 0 \leq t < r(\omega) \} \mapsto S_t\omega \in \Gamma$ carries the measure class of $m \otimes \mu$ into that of ν.

To this end, we define the *hitting number* of γ of Ω in time t:

$$n_\Omega(t,\gamma) = \sup\{ n \in \mathbf{Z} : r(n,\gamma) \leq t \}, \quad t \in \mathbf{R}. \tag{9}$$

We also write $n(t,\gamma)$ for $n_\Omega(t,\gamma)$ when Ω is fixed. We then have

$$n(s+t,\gamma) = n(s, S_t\gamma) + n(t,\gamma), \quad s,t \in \mathbf{R}, \quad \gamma \in \Gamma. \tag{9$'$}$$

This allows us to define a flow $\{\tilde{S}_t\}$ on $\tilde{\Gamma} = \mathbf{Z} \times \Gamma$ by:

$$\tilde{S}_t(m,\gamma) = \big(m + n(t,\gamma), S_t\gamma\big). \tag{10}$$

On $\tilde{\Gamma}$, we consider the product measure $\delta \otimes \nu$ of the counting measure δ on $\mathbf{Z}$ and the original measure ν on Γ. Put

$$\varphi(t,\omega) = \big(n(t,\omega), S_t\omega\big) \in \tilde{\Gamma}, \quad (t,\omega) \in \mathbf{R} \times \Omega. \tag{11}$$

It then follows that φ is a Borel isomorphism of $\tilde{\Omega} = \mathbf{R} \times \Omega$ onto $\tilde{\Gamma} = \mathbf{Z} \times \Gamma$ and its inverse φ^{-1} is given by:

$$\varphi^{-1}(m,\gamma) = \big(-r(-m,\gamma), S_{r(-m,\gamma)}\gamma\big) \in \tilde{\Omega}, \quad (m,\gamma) \in \tilde{\Gamma}. \tag{12}$$

On $\tilde{\Gamma}$ and $\tilde{\Omega}$, we define actions of $\mathbf{Z}$ and $\mathbf{R}$ as follows:

$$\begin{aligned} L(m, \gamma) &= (m-1, \gamma), \qquad (m, \gamma) \in \tilde{\Gamma}; \\ R_t(s, \omega) &= (s+t, \omega), \qquad (s, \omega) \in \tilde{\Omega}. \end{aligned} \tag{13}$$

We then have

$$\begin{aligned} \varphi R_t(s, \omega) = \varphi(s+t, \omega) &= \big(n(s+t, \omega), \\ S_{s+t}\omega) = \big(n(t, S_s\omega) + n(s, \omega), S_t S_s \omega\big) &= \tilde{S}_t \varphi(s, \omega). \end{aligned}$$

With $\tilde{T}(s, \omega) = \big(s - r(\omega), T\omega\big) = \big(s - r(\omega), S_{r(\omega)}\omega\big)$, we have

$$\begin{aligned} \varphi\tilde{T}(s, \omega) &= \varphi\big(s - r(\omega),\ S_{r(\omega)}\omega\big) = \Big(n\big(s - r(\omega),\ S_{r(\omega)}\omega\big),\ S_{s-r(\omega)}S_{r(\omega)}\omega\Big) \\ &= \big(n(s, \omega) - 1,\ S_s\omega\big) = L\varphi(s, \omega). \end{aligned}$$

Thus we get

$$\varphi R_t \varphi^{-1} = \tilde{S}_t \qquad \text{and} \qquad \varphi\tilde{T}\varphi^{-1} = L. \tag{14}$$

Therefore, the measure $(\delta \otimes \nu) \circ \varphi$ on $\tilde{\Omega}$ is quasi-invariant under $\{R_t\}$. Since the Lebesgue measure class on $\mathbf{R}$ is the unique measure class of translation quasi-invariant measures, the measure $(\delta \otimes \nu) \circ \varphi$ is equivalent to the product measure $m \otimes \mu$ for some σ-finite measure μ on Ω. This means that $\varphi(m \otimes \mu) = (m \otimes \mu) \circ \varphi^{-1}$ is equivalent to $\delta \otimes \nu$, or equivalently the map $S\colon \big\{ (t, \omega) \in \tilde{\Omega} : 0 \le t < r(\omega) \big\} \mapsto S_t\omega \in \Gamma$ carries $m \otimes \mu$ into a measure equivalent to ν. Thus, replacing ν by $S(m \otimes \mu) = (m \otimes \mu) \circ S^{-1}$, we obtain a representation of $\big\{\Gamma, \nu, S_t\big\}$ by the flow built under the ceiling function r based on $\big\{\Omega, \mu, T\big\}$. The ergodicity of the measure μ with respect to T follows from that of ν by construction. Q.E.D.

The base transformation $\big\{\Omega, \mu, T\big\}$ is not unique. As we shall see later, it is possible to characterize when two ergodic transformations $\big\{\Omega_1, \mu_1, T_1\big\}$ and $\big\{\Omega_2, \mu_2, T_2\big\}$ give rise to conjugate flows under suitable ceiling functions.

In the above proof, we have proved more than the assertion of the theorem. We summarize the consequence of the proof.

Corollary 3.3. *If $\big\{\mathcal{C}, \mathbf{R}, \theta\big\}$ is a properly ergodic flow, then there exists a covariant system $\big\{\mathcal{D}, \mathbf{Z} \times \mathbf{R}, \tilde{\theta}\big\}$ carried by a separable abelian von Neumann algebra $\mathcal{D}$ such that*

(i) *$\tilde{\theta}$ is ergodic;*

(ii) *The restriction of $\tilde{\theta}$ to $\mathbf{Z} \times \{0\}$, say $\tilde{\alpha}$, is free and integrable;*

(iii) *The restriction of $\tilde{\theta}$ to $\{0\} \times \mathbf{R}$, say $\overline{\theta}$, is free and integrable;*

(iv) *$\big\{\mathcal{C}, \mathbf{R}, \theta\big\}$ is conjugate to $\big\{\mathcal{D}^{\tilde{\alpha}}, \mathbf{R}, \overline{\theta}\big\}$.*

By restricting $\tilde{\alpha}$ to $\mathcal{D}^{\overline{\theta}} = \mathcal{A}$, we obtain an ergodic system $\big\{\mathcal{A}, \alpha\big\}$ which gives rise to $\big\{\mathcal{C}, \mathbf{R}, \theta\big\}$ as the flow built under a ceiling function based on $\{\mathcal{A}, \alpha\}$.

The above study motivates us to consider the following situation. Let $\{\Gamma, \nu\}$ be a standard measure space on which a separable locally compact group G acts by non singular transformations.[3] Suppose ρ is a Borel map from $G \times \Gamma$ into a separable locally compact group H such that for every $g_1, g_2 \in G$

$$\rho(g_1 g_2, \gamma) = \rho(g_1, g_2\gamma)\rho(g_2, \gamma) \tag{15}$$

for ν-almost every $\gamma \in \Gamma$. Such a ρ is called an *H-valued one cocycle* on $\{\Gamma, \nu, G\}$. Consider $\tilde{\Gamma} = H \times \Gamma$ and the product measure $dh \otimes d\nu$ of the left Haar measure dh on H and the original measure $d\nu$ on Γ. We define an action of G on $\tilde{\Gamma}$ by

$$\tilde{g}(h, \gamma) = \left(h\rho(g, \gamma)^{-1}, g\gamma\right), \quad g \in G, \quad (h, \gamma) \in \tilde{\Gamma}. \tag{16}$$

It is easy to see that $\tilde{g}$ is a Borel automorphism of $\tilde{\Gamma}$. For each pair $g_1, g_2 \in G$, we have

$$(g_1 g_2)^{\sim}(h, \gamma) = \tilde{g}_1 \cdot \tilde{g}_2(h, \gamma)$$

for almost every (h, γ). Hence G acts on the von Neumann algebra $L^\infty\big(H \times \Gamma, dh \otimes d\nu\big) = L^\infty(H) \mathbin{\overline{\otimes}} L^\infty(\Gamma, \nu)$ by

$$\tilde{\alpha}_g(f)(h, \gamma) = f\left(\tilde{g}^{-1}(h, \gamma)\right), \quad f \in L^\infty\left(H \times \Gamma, dh \otimes d\nu\right).$$

Let $\mathcal{A} = L^\infty(\Gamma, \nu)$ and $\tilde{\mathcal{A}} = L^\infty(H) \mathbin{\overline{\otimes}} \mathcal{A}$. On $\tilde{\mathcal{A}}$, we have the translation action $\tilde{\lambda}$ of H, i.e.

$$\left(\tilde{\lambda}_k f\right)(h, \gamma) = f\left(k^{-1}h, \gamma\right), \quad k \in H, \quad (h, \gamma) \in \tilde{\Gamma}. \tag{17}$$

Clearly, $\tilde{\lambda}_h$, $h \in H$, and $\tilde{\alpha}_g$, $g \in G$, commute. Hence $\tilde{\lambda}$ induces an action β of H on the fixed point algebra $\mathcal{C} = \tilde{\mathcal{A}}^{\tilde{\alpha}}$ of $\tilde{\alpha}$. The covariant system $\{\mathcal{C}, H, \beta\}$ is called the *stable range* of the cocycle ρ.

Notice that the flow $\{\mathcal{C}, \mathbf{R}, \theta\}$ built under a ceiling function r on a base transformation $\{\Omega, \mu, T\}$ is nothing else but the stable range of the return time cocycle $\{r_n\}$, and also that $\{\Omega, \mu, T\}$ is the stable range of the hitting number cocycle $n(t, \gamma)$ of the flow $\{\Gamma, \nu, S_t\}$.

Lemma 3.4. *Let $\{\mathcal{P}, G, \alpha\}$ be a covariant system and $\mathcal{A} = L^\infty(\Gamma, \nu)$ be a globally α-invariant subalgebra of the center $\mathcal{C}$ of $\mathcal{P}$. Let ρ be an H-valued one cocycle on $\{\Gamma, \nu, G\}$, where H is another locally compact group. Then on the tensor product von Neumann algebra $L^\infty(H) \mathbin{\overline{\otimes}} \mathcal{P}$, the action α of G on $\mathcal{C} \otimes \mathcal{P}$ can be extended to an action $\tilde{\alpha}$ of G on $L^\infty(H) \mathbin{\overline{\otimes}} \mathcal{P}$ in such a way that*

(i) *$\tilde{\alpha}$ commutes with the left translation action $\tilde{\lambda} = \lambda \otimes \mathrm{id}$ of H on $L^\infty(H) \mathbin{\overline{\otimes}} \mathcal{P}$, and*

(ii) *the restriction of $\tilde{\lambda}$ to $\left(L^\infty(H) \mathbin{\overline{\otimes}} \mathcal{A}\right)^{\tilde{\alpha}}$ is precisely the stable range action of ρ.*

---

[3] A bijective null set preserving Borel map is called a non-singular transformation of a measure space.

PROOF: Consider the equivariant direct disintegration of $\{\mathcal{P}, G, \alpha\}$ on $\{\Gamma, \nu\}$:

$$\mathcal{P} = \int_{\Gamma}^{\oplus} \mathcal{P}(\gamma)\, d\nu(\gamma),$$

$$\alpha_g \sim \{\alpha_{g,\gamma}\}, \qquad \alpha_{g,\gamma} : \mathcal{P}(\gamma) \mapsto \mathcal{P}(g\gamma),$$

as in Corollary X.3.12. We then have

$$L^\infty(H) \overline{\otimes} \mathcal{P} = \int_{H\times\Gamma}^{\oplus} \mathcal{P}(h, \gamma)\, dh \otimes d\nu(\gamma),$$

where $\mathcal{P}(h, \gamma) = \mathcal{P}(\gamma),\ h \in H$. We put

$$\tilde{\alpha}_{g,(h,\gamma)} = \alpha_{g,\gamma} : \mathcal{P}(h, \gamma) \mapsto \mathcal{P}\big(h\rho(g, \gamma)^{-1}, g\gamma\big),$$

and define

$$\{\tilde{\alpha}_g(a)\}\big(\tilde{g}(h, \gamma)\big) = \tilde{\alpha}_{g,(h,\gamma)}\big(a(h, \gamma)\big) = \alpha_{g,\gamma}\big(a(h, \gamma)\big),$$

for $a \sim \{a(h, \gamma)\} \in L^\infty(H)\overline{\otimes}\mathcal{P}$. It follows that $\tilde{\alpha}$ satisfies the required conditions.

Q.E.D.

We now apply this lemma to the flow built under ceiling function on an ergodic transformation. Suppose $\{\mathcal{P}, \mathbf{R}, \theta\}$ be a centrally ergodic covariant system over $\mathbf{R}$. Let $\mathcal{C}$ be the center of $\mathcal{P}$. We represent $\{\mathcal{C}, \mathbf{R}, \theta\}$ by $\{\Gamma, \nu, S_t\}$.[4] By Theorem 3.2, the flow $\{\Gamma, \nu, S_t\}$ can be built under a ceiling function r on an ergodic transformation $\{\Omega, \nu, T\}$. Let $\mathcal{A} = L^\infty(\Omega, \mu)$, and $\mathcal{Q} = \ell^\infty(\mathbf{Z}) \overline{\otimes} \mathcal{P}$. By the last lemma, the action θ on $\mathcal{P}$ can be extended to an action $\tilde{\theta}$ of $\mathbf{R}$ on $\mathcal{Q}$ based on the hitting number cocycle $n_\Omega(t, \gamma)$ of $\mathbf{R} \times \Gamma$, and also $\mathbf{Z}$ acts on $\mathcal{Q}$ by translation $\tilde{\lambda}$ whose restriction to $\big(\ell^\infty(\mathbf{Z}) \overline{\otimes} \mathcal{C}\big)^{\theta}$ is conjugate to the action α of $\mathbf{Z}$ on $\mathcal{A}$ corresponding to T. We identify $\mathcal{A}$ with $\big(\ell^\infty(\mathbf{Z}) \overline{\otimes} \mathcal{C}\big)^{\tilde{\theta}}$ and put $\mathcal{D} = \ell^\infty(\mathbf{Z}) \overline{\otimes} \mathcal{C}$.

Lemma 3.5. *In the above setting, the action $\tilde{\lambda}$ of $\mathbf{Z}$ on $\mathcal{Q}$ can be extended to $\mathcal{Q} \rtimes_{\tilde{\theta}} \mathbf{R} = \mathcal{R}$ in such a way that*

(i) $\mathcal{R} \rtimes_{\tilde{\lambda}} \mathbf{Z} \cong \big(\mathcal{P} \rtimes_\theta \mathbf{R}\big) \overline{\otimes} \mathcal{L}\big(\ell^2(\mathbf{Z})\big)$;
(ii) *The restriction of the extended action $\tilde{\lambda}$ of $\mathbf{Z}$ to the center of $\mathcal{R}$ is conjugate to $\{\mathcal{A}, \mathbf{Z}, \alpha\}$;*
(iii) *If $\mathcal{P}$ is of type i, $i =$ I, II, III, then $\mathcal{R}$ is also of type i.*

PROOF:

(i) We use the notations in the proof of Theorem 3.2, and the previous lemma.

---

4 We keep the proper ergodicity assumption for $\{S_t\}$ as before.

We then have

$$\mathcal{Q} = \ell^\infty(\mathbf{Z}) \overline{\otimes} \mathcal{P} = \sum_{k\in\mathbf{Z}}^{\oplus} \int_\Gamma^{\oplus} \mathcal{P}(k,\gamma)\,\mathrm{d}\nu(\gamma) = \int_{\mathbf{R}\times\Omega}^{\oplus} \mathcal{P}\big(\varphi(t,\omega)\big)\,\mathrm{d}t\,\mathrm{d}\nu(\omega)\,.$$

The action $\tilde{\theta}$ of $\mathbf{R}$ on $\mathcal{Q}$ is then disintegrated to the system:

$$\tilde{\theta}_{t,(k,\gamma)}\colon \mathcal{P}(k,\gamma) \mapsto \mathcal{P}\big(k+n(t,\gamma), S_t\gamma\big),$$

which further gives rise to the system:

$$\tilde{\theta}_{t,(s,\omega)}\colon \mathcal{Q}(s,\omega) \mapsto \mathcal{Q}(s+t,\omega),$$

where $\mathcal{Q}(s,\omega) = \mathcal{P}\big(\varphi(s,\omega)\big)$. Hence with

$$\mathcal{Q}(s) = \int_\Omega^{\oplus} Q(s,\omega)\,\mathrm{d}\mu(\omega),$$

we have

$$\tilde{\theta}_{t,s} = \int_\Omega^{\oplus} \tilde{\theta}_{t,(s,\omega)}\,\mathrm{d}\mu(\omega)\colon \mathcal{Q}(s) \mapsto \mathcal{Q}(s+t).$$

Namely, almost $\mathcal{Q}(s)$'s are mutually isomorphic, so that we have, by Corollary IV.8.30,

$$\mathcal{Q} \cong L^\infty(\mathbf{R}) \overline{\otimes} Q_0, \qquad \tilde{\theta}_t \cong \rho_t \otimes \mathrm{id}$$

where $\mathcal{Q}_0$ is isomorphic to $\int_\Omega^\oplus \mathcal{Q}(s,\omega)\,\mathrm{d}\nu(\omega)$ for almost every s, and ρ_t corresponds to the translation: $s \mapsto s+t$. Therefore, we have

$$\mathcal{R} = \mathcal{Q} \rtimes_{\tilde{\theta}} \mathbf{R} = \big(L^\infty(\mathbf{R}) \rtimes_\rho \mathbf{R}\big) \overline{\otimes} \mathcal{Q}_0 \cong \mathcal{L}\big(L^2(\mathbf{R})\big) \overline{\otimes} \mathcal{Q}_0.$$

Since the actions $\tilde{\lambda}$ of $\mathbf{Z}$ and $\tilde{\theta}$ of $\mathbf{R}$ on $\mathcal{Q}$ commute, $\tilde{\lambda}$ can be lifted canonically to $\mathcal{Q} \rtimes_{\tilde{\theta}} \mathbf{R} = \mathcal{R}$. The commutativity of $\tilde{\lambda}$ and $\tilde{\theta}$ also entails that

$$\mathcal{R} \rtimes_{\tilde{\lambda}} \mathbf{Z} = \big(\mathcal{Q} \rtimes_{\tilde{\theta}} \mathbf{R}\big) \rtimes_{\tilde{\lambda}} \mathbf{Z} \cong \big(\mathcal{Q} \rtimes_{\tilde{\lambda}} \mathbf{Z}\big) \rtimes_{\tilde{\theta}} \mathbf{R},$$

where we extend $\tilde{\theta}$ to $\mathcal{Q} \rtimes_{\tilde{\lambda}} \mathbf{Z}$ naturally. But we have, since $\tilde{\lambda} = \lambda \otimes \mathrm{id}$ on $\ell^\infty(\mathbf{Z}) \overline{\otimes} \mathcal{P}$,

$$\mathcal{Q} \rtimes_{\tilde{\lambda}} \mathbf{Z} = \big(\ell^\infty(\mathbf{Z}) \rtimes_{\tilde{\lambda}} \mathbf{Z}\big) \overline{\otimes} \mathcal{P} = \mathcal{L}\big(\ell^2(\mathbf{Z})\big) \overline{\otimes} \mathcal{P} = \mathcal{B} \overline{\otimes} \mathcal{P}.$$

Furthermore, we observe that $\tilde{\theta}$ on $\mathcal{Q}$ is given by the natural extension of the system:

$$\tilde{\theta}_{t,\gamma}(a)\big(k+n(t,\gamma), S_t\gamma\big) = \theta_{t,\gamma}\big(a(k,\gamma)\big), \quad a \in \mathcal{Q},$$

to $\mathcal{B} \overline{\otimes} \mathcal{P}$. With the disintegration:

$$\tilde{\mathcal{P}} = \mathcal{B} \overline{\otimes} \mathcal{P} = \int_\Gamma^{\oplus} \mathcal{B} \overline{\otimes} \mathcal{P}(\gamma)\,\mathrm{d}\nu(\gamma),$$

the extended $\tilde{\theta}_{t,\gamma}\colon \tilde{\mathcal{P}}(\gamma) = \mathcal{B} \overline{\otimes} \mathcal{P}(\gamma) \mapsto \tilde{\mathcal{P}}(S_t\gamma) = \mathcal{B} \overline{\otimes} \mathcal{P}(S_t\gamma)$ is given by $\mathrm{Ad}\big(u(t,\gamma)\big) \otimes \theta_{t,\gamma}$, where $u(t,\gamma)$ is the unitary in $\mathcal{B} = \mathcal{L}(\ell^2)$ corresponding to the translation: $k \mapsto k + n(t, \gamma)$ on $\mathbf{Z}$. The cocycle property of $n(t, \gamma)$ entails that $u(s + t, \gamma) = u(s, S_t\gamma)u(t, \gamma)$. With

$$u_t = \int_\Gamma^\oplus u(t, S_t\gamma)\, \mathrm{d}\nu(\gamma) \in \mathcal{B} \overline{\otimes} \mathcal{C},$$

we have $u_{s+t} = u_s(\mathrm{id} \otimes \theta_s)(u_t)$ and $\tilde{\theta}_s = \mathrm{Ad}(u_s) \circ (\mathrm{id} \otimes \theta_s)$ on $\mathcal{B} \overline{\otimes} \mathcal{P}$. Thus, $\tilde{\theta}$ is a cocycle perturbation of $\mathrm{id} \otimes \theta$, so that we obtain

$$\left(\mathcal{B} \overline{\otimes} \mathcal{P}\right) \rtimes_{\tilde{\theta}} \mathbf{R} \cong \mathcal{B} \overline{\otimes} \left(\mathcal{P} \rtimes_\theta \mathbf{R}\right),$$

which in turn implies that $\mathcal{R} \rtimes_{\tilde{\lambda}} \mathbf{Z} \cong \mathcal{B} \overline{\otimes} (\mathcal{P} \rtimes_\theta \mathbf{R})$. Thus we complete the proof of (i).

(ii) This follows from the decomposition of $\mathcal{Q}$:

$$\mathcal{Q} = L^\infty(\mathbf{R}) \overline{\otimes} \mathcal{Q}_0 \qquad \text{and} \qquad \mathcal{R} \cong \mathcal{L}\left(L^2(\mathbf{R})\right) \overline{\otimes} \mathcal{Q}_0.$$

We leave the detail to the reader.

(iii) If $\mathcal{P}$ is of type i, then $\mathcal{Q}$ is also of type i and so is $\mathcal{Q}_0$. Thus, $\mathcal{R}$ is of type i.

Q.E.D.

Corollary 3.6. *If $\mathcal{M}$ is a separable factor of type* III_0, *then there exists a covariance system* $\{\mathcal{N}_0, \theta_0\}$ *of a von Neumann algebra* $\mathcal{N}_0$ *of type* II_∞ *and* $\theta_0 \in \mathrm{Aut}(\mathcal{N}_0)$ *such that*

(i) $\mathcal{M} \cong \mathcal{N}_0 \rtimes_{\theta_0} \mathbf{Z}$;

(ii) *The flow of weights of $\mathcal{M}$, $\{\mathcal{C}, \mathbf{R}, \theta\}$, is built under a ceiling function on the restriction of θ_0 to the center of $\mathcal{N}_0$;*

(iii) *There exists a faithful semi-finite normal trace τ_0 on $\mathcal{N}_0$ and $0 < \lambda < 1$ such that*

$$\tau_0 \circ \theta_0 \le \lambda \tau_0. \tag{18}$$

PROOF: Assertions (i) and (ii) have been already established. Thus we have only to prove (iii).

Let τ be a faithful semi-finite normal trace on $\mathcal{N}$ such that $\tau \circ \theta_s = \mathrm{e}^{-s}\tau$. We use the above established notations. We further identify Γ with $\{(t, \omega) \in \mathbf{R} \times \Omega : 0 \le t \le r(\omega),\ (r(\omega), \omega) = (0, T\omega)\}$. We then disintegrate τ on Γ:

$$\tau = \int_\Gamma^\oplus \tau_{s,\omega}\, \mathrm{d}s\, \mathrm{d}\mu(\omega)\,.$$

We then compute, with $\nu = m \otimes \mu$,

$$\begin{aligned} \mathrm{e}^{-t}\tau(x) = \tau\left(\theta_t(x)\right) &= \int_\Gamma \tau_{S_t\gamma}\left(\theta_t(x)(S_t\gamma)\right) \mathrm{d}\nu \circ S_t(\gamma) \\ &= \int_\Gamma \tau_{S_t\gamma}\left(\theta_{t,\gamma}\left(x(\gamma)\right)\right) \frac{\mathrm{d}\nu \circ S_t}{\mathrm{d}\nu}(\gamma)\, \mathrm{d}\nu(\gamma), \end{aligned}$$

so that

$$\int_\Gamma^\oplus \mathrm{e}^{-t}\tau_\gamma \, \mathrm{d}\nu(\gamma) = \int_\Gamma^\oplus \left(\tau_{S_t\gamma} \circ \theta_{t,\gamma}\right)\frac{\mathrm{d}\nu \circ S_t}{\mathrm{d}\nu}(\gamma)\, \mathrm{d}\nu(\gamma)\,.$$

Hence we get, for each $t \in \mathbf{R}$,

$$\mathrm{e}^{-t}\tau_\gamma = \frac{\mathrm{d}\nu \circ S_t}{\mathrm{d}\nu}(\gamma)\tau_{S_t\gamma} \circ \theta_{t,\gamma} \quad \text{for } \nu\text{-almost every } \gamma. \tag{19}$$

On the other hand, the center $\mathcal{C}_0$ of $\mathcal{N}_0$ is identified with $L^\infty(\Omega, \mu)$; in fact $\mathcal{N}_0$ is identified with $\mathcal{L}\big(L^2(\mathbf{R})\big) \mathbin{\overline{\otimes}} \int_\Omega^\oplus \mathcal{N}(s, \omega)\, \mathrm{d}\nu(\omega)$ for almost every $s \in \mathbf{R}$. Furthermore, we have

$$\frac{\mathrm{d}\nu \circ S_t}{\mathrm{d}\nu}(\gamma) = \frac{\mathrm{d}\mu \circ T^{n(t,\gamma)}}{\mathrm{d}\mu}(\gamma);$$

$$S_t(s, \omega) = \Big(s + t - r\big(n(s + t, \omega), \omega\big),\, T^{n(s+t,\omega)}\omega\Big),$$

so that (19) becomes the following

$$\mathrm{e}^{-t}\tau_{(s,\omega)} = \frac{\mathrm{d}\mu \circ T^{n(s+t,\omega)}}{\mathrm{d}\mu}(\omega)\tau_{(s+t-r(n(s+t,\omega),\omega),\, T^{n(s+t,\omega)}\omega)} \circ \theta_{t,(s,\omega)}$$

for almost s and ω. With $t = r(\omega)$, we have

$$\mathrm{e}^{-r(\omega)}\tau_{(s,\omega)} = \frac{\mathrm{d}\mu \circ T}{\mathrm{d}\mu}(\omega)\tau_{(s,T\omega)} \circ \theta_{r(\omega),(s,\omega)}$$

for almost every s, $0 \le s < \frac{a}{4} \le \inf r(\omega)$, and $\omega \in \Omega$. Setting $\tau_\omega = \frac{1}{\delta}\int_0^\delta \tau_{s,\omega}\, \mathrm{d}s$ on $\mathcal{N}(\omega)$ with small $\delta > 0$ (recall $\mathcal{N}(s, \omega) = \mathcal{N}(\omega)$), we get

$$\mathrm{e}^{-r(\omega)}\tau_\omega = \frac{\mathrm{d}\mu \circ T}{\mathrm{d}\mu}(\omega)\tau_{T\omega} \circ \theta_{T,\omega},$$

where $\theta_{T,\omega} = \theta_{r(\omega),(0,\omega)}$. We set

$$\tau_0 = \int_\Omega^\oplus \tau_\omega \, \mathrm{d}\mu(\omega),$$

and obtain, for $x \in \mathcal{N}_0^+$,

$$\begin{aligned}
\tau_0 \circ \theta_0(x) &= \int_\Omega \tau_{T\omega}\big(\theta_0(x)(T\omega)\big)\, \mathrm{d}\mu \circ T(\omega) \\
&= \int_\Omega \tau_{T\omega}\Big(\theta_{T,\omega}\big(x(\omega)\big)\Big)\frac{\mathrm{d}\mu \circ T}{\mathrm{d}\mu}(\omega)\, \mathrm{d}\mu(\omega) \\
&= \int_\Omega \mathrm{e}^{-r(\omega)}\tau_\omega\big(x(\omega)\big)\, \mathrm{d}\mu(\omega) \le \mathrm{e}^{-a/4}\int_\Omega \tau_\omega\big(x(\omega)\big)\, \mathrm{d}\mu(\omega) \\
&= \lambda\tau_0(x).
\end{aligned}$$

This completes the proof. Q.E.D.

Theorem 3.7.

(i) *If* $\mathcal{M}$ *is a separable factor of type* III_0, *then there exists a covariant system* $\{\mathcal{N}, \mathbf{Z}, \theta\}$ *such that*

a) $\mathcal{N}$ *is of type* II_∞;

b) θ *is centrally ergodic;*

c) *the center* $\mathcal{C}$ *of* $\mathcal{N}$ *is non atomic;*

d) $\mathcal{N}$ *admits a faithful semi-finite normal trace* τ *which is transformed under* θ *in such a way that* $\tau \circ \theta \leq \lambda\tau$ *for some* $0 < \lambda < 1$, *and*

e) $\mathcal{M} \cong \mathcal{N} \rtimes_\theta \mathbf{Z}$.

(ii) *If* $\{\mathcal{N}, \mathbf{Z}, \theta\}$ *is a covariant system satisfying* (a), (b), (c) *and* (d), *then* $\mathcal{M} = \mathcal{N} \rtimes_\theta \mathbf{Z}$ *is a factor of type* III_0.

(iii) *Let* $\{\mathcal{N}_1, \mathbf{Z}, \theta_1\}$ *and* $\{\mathcal{N}_2, \mathbf{Z}, \theta_2\}$ *be covariant systems satisfying* (a), (b) *and* (c). *Then*

$$\mathcal{N}_1 \rtimes_{\theta_1} \mathbf{Z} \cong \mathcal{N}_2 \rtimes_{\theta_2} \mathbf{Z}$$

if and only if there exist $e_1 \in \mathrm{Proj}(\mathcal{C}_{\mathcal{N}_1})$ *and* $e_2 \in \mathrm{Proj}(\mathcal{C}_{\mathcal{N}_2})$ *such that* $e_1 \neq 0$, $e_2 \neq 0$ *and*

$$\{\mathcal{N}_{e_1}, \theta_{1,e_1}\} \cong \{\mathcal{N}_{e_2}, \theta_{2,e_2}\},$$

where for a centrally ergodic system $\{\mathcal{N}, \mathbf{Z}, \theta\}$ *and a non-zero* $e \in \mathrm{Proj}(\mathcal{C}_{\mathcal{N}})$, θ_e *is defined as follows:*

$$\theta_e(x) = \sum_{n=1}^{\infty} \theta^n(xe_n), \qquad x \in \mathcal{N}_e; \tag{20}$$

where

$$\begin{aligned} e_1 &= e\,\theta^{-1}(e), \\ e_2 &= (e - e_1)\,\theta^{-2}(e), \ldots \\ e_n &= (e - e_1 - \cdots - e_{n-1})\,\theta^{-n}(e), \ldots . \end{aligned} \tag{21}$$

PROOF: We have proved (i) in the previous Corollary.

(ii) Let ρ be the positive self-adjoint operator affiliated with the center $\mathcal{C}$ of $\mathcal{N}$ such that $(\mathrm{D}\tau \circ \theta : \mathrm{D}\tau)_t = \rho^{it}$. Namely, we have $\tau \circ \theta = \tau(\rho\,\cdot\,)$. By assumption, we have $0 < \rho \leq \lambda$. Let $r(\omega) = -\log \rho(\omega)$, where $\mathcal{C} = L^\infty(\Omega, \mu)$. With the central decomposition:

$$\{\mathcal{N}, \tau\} = \int_\Omega^\oplus \{\mathcal{N}(\omega), \tau_\omega\}\, \mathrm{d}\mu(\omega),$$

and the ergodic transformation T of $\{\Omega, \omega\}$ corresponding to $\{\mathcal{C}, \mathbf{Z}, \theta\}$, we have, for $x \in \mathcal{N}_+$,

$$\begin{aligned}\int_\Omega e^{-r(\omega)}\tau_\omega\big(x(\omega)\big)\,d\mu(\omega) &= \tau(\rho x) = \tau\big(\theta(x)\big)\\ &= \int_\Omega \tau_{T\omega}\Big(\big(\theta(x)\big)(T\omega)\Big)\,d\mu\circ T(\omega)\\ &= \int_\Omega \tau_{T\omega}\Big(\theta_{T,\omega}\big(x(\omega)\big)\Big)\frac{d\mu\circ T}{d\mu}(\omega)\,d\mu(\omega)\,.\end{aligned}$$

Hence we get

$$e^{-r(\omega)}\tau_\omega = \frac{d\mu\circ T}{d\mu}(\omega)\tau_{T\omega}\circ\theta_{T,\omega}.$$

This means that if we use r as a ceiling function to build a flow $\{\Gamma, \mathbf{R}, S_t\}$ from $\{\Omega, \mu, T\}$ and the construction of the covariant system $\{\tilde{\mathcal{N}}, \mathbf{R}, \tilde{\theta}\}$ from this data in Lemma 3.4, then the trace $\tilde{\tau}$:

$$\tilde{\tau} = \int_\Omega^\oplus \Big(\int_{[0,r(\omega)[}^\oplus \tau_\omega\, ds\Big)\,d\mu(\omega) \quad \text{on} \quad \tilde{\mathcal{N}} = \int_\Omega^\oplus L^\infty\big[0, r(\omega)\big[\,\overline{\otimes}\,\mathcal{N}(\omega)\,d\mu(\omega)$$

is scaled by $\{\tilde{\theta}_t\}$. Then Lemma 3.5 entails that $\tilde{\mathcal{N}} \rtimes_{\tilde{\theta}} \mathbf{R} \cong \mathcal{N} \rtimes_\theta \mathbf{Z} = \mathcal{M}$. Hence $\mathcal{M}$ is of type III_0.

(iii) Let $\{\mathcal{N}, \mathbf{Z}, \theta\}$ be as above, and $e \in \mathrm{Proj}(\mathcal{C})$. We represent $\mathcal{C} = L^\infty(\Omega, \mu)$ and θ by an ergodic transformation T of $\{\Omega, \mu\}$ as before. Let E be the Borel set in Ω corresponding to e. Let $\mathbf{Z}(\omega) = \{n \in \mathbf{Z} : T^n\omega \in E\}$ and arrange $\mathbf{Z}(\omega)$ in the increasing order:

$$\cdots < r_E(-1,\omega) < r_E(0,\omega) \le 0 < r_E(1,\omega) = r_E(\omega) < r_E(2,\omega) < \cdots.$$

We call $r_E(n,\omega)$ the n-th hitting time of ω of E. The number of hitting E of ω in time k is denoted by $n_E(k,\omega)$ and defined by:

$$n_E(k,\omega) = \sup\{n : r(n,\omega) \le k\}.$$

We now let $E_n = \{\omega \in E : r_E(\omega) = n\}$, $n \in \mathbf{N}$. Since T is properly ergodic, we have $\bigcup_{n\ge 0} T^n E = \Omega$, and so $\{T^n E_n\}$ is a partition of E. The set E_n is the Borel set corresponding to the projection e_n of (21). Since $\{e_n\}$ and $\{\theta^n(e_n)\}$ are both partitions of e, (20) gives an automorphism θ_e of $\mathcal{N}_e$. The automorphism θ_e is said to be induced on $\mathcal{N}_e$ by θ. It should not be confused with the reduced action. We observe that the original covariant system $\{\mathcal{N}, \mathbf{Z}, \theta\}$ can be recovered completely from $\{\mathcal{N}_e, \mathbf{Z}, \theta_e\}$ and the return time r_E. This reconstruction of $\{\mathcal{N}, \mathbf{Z}, \theta\}$ is precisely the discrete version of the flow built under a ceiling function. In fact, our case is easier than the continuous time case, because we don't have to worry about the null set problem. Thus we leave to the reader the details of the proof of the assertion:

$$\mathcal{N} \rtimes_\theta \mathbf{Z} \cong \mathcal{N}_e \rtimes_{\theta_e} \mathbf{Z}.$$

This completes the proof of the "if" part.

Suppose that $\mathcal{N}_1 \rtimes_{\theta_1} \mathbf{Z} \cong \mathcal{N}_2 \rtimes_{\theta_2} \mathbf{Z}$. Since the continuous covariant system $\{\tilde{\mathcal{N}}_i, \mathbf{R}, \tilde{\theta}_i\}$ associated with $\mathcal{N}_i \rtimes_{\theta_i} \mathbf{Z}$ are obtained as the "stable range" of ceiling function cocycles on $\mathfrak{C}_i$, $i = 1, 2$, the uniqueness of the associated continuous covariant system allows us to view $\{\mathcal{N}_1, \mathbf{Z}, \theta_1\}$ and $\{\mathcal{N}_2, \mathbf{Z}, \theta_2\}$ as the systems obtained from the same continuous covariant system $\{\tilde{\mathcal{N}}, \mathbf{R}, \tilde{\theta}\}$. Let $\tilde{\mathfrak{C}} = L^\infty(\Gamma, \nu)$ be the center of $\tilde{\mathcal{N}}$, and $\mathfrak{C}_i = L^\infty(\Omega_i, \mu_i)$, $i = 1, 2$. Then, Ω_1 and Ω_2 are both identified with subsets of Γ. Let $r_i = r_{\Omega_i}$ be the respective hitting time functions. If the trace τ_i of $\mathcal{N}_i$ is transformed under θ_i in such a way that $\tau_i \circ \theta_i \leq \lambda_i \tau_i$ with $0 < \lambda_i < 1$, then the return time r_i satisfies the inequality, $e^{-r_i(\omega)} \leq \lambda_i$, $\omega \in \Omega_i$, as seen above. Hence we have $r_i(\omega) \geq -\log \lambda_i = \delta_i > 0$, $\omega \in \Omega_i$. This means that the gap between the n-th hitting time $r_i(n, \gamma)$ and the $(n+1)$-th hitting time $r_i(n+1, \gamma)$, $\gamma \in \Gamma$, is greater than or equal to δ_i for every $\gamma \in \Gamma$, i.e. $r_i(n+1, \gamma) - r_i(n, \gamma) \geq \delta_i > 0$, $\gamma \in \Gamma$. Let $E_1 = \{\omega \in \Omega_1 : r_2(\omega) < r_1(\omega)\}$. We claim that $\bigcup_{n \in \mathbf{Z}} T_1^n E_1$ is conull[5] in Ω_1, so that we have $\mu_1(E_1) > 0$ in particular. By the above arguments, for each $\omega \in \Omega_1$, there exists uniquely $n \in \mathbf{Z}$ such that $r_1(n, \omega) \leq r_2(\omega) < r_1(n+1, \omega)$. We then have

$$\begin{aligned} r_2(T_1^n \omega) &= r_2(S_{r_1(n,\omega)}\omega) = r_2(\omega) - r_1(n, \omega) \\ &< r_1(n+1, \omega) - r_1(n, \omega) = r_1(T_1^n \omega). \end{aligned}$$

Hence $T_1^n \omega \in E_1$. Thus, $\bigcup_{n \in \mathbf{Z}} T_1^n E_1$ covers all of Ω_1. The following picture illustrates the relation of Ω_1 and Ω_2:

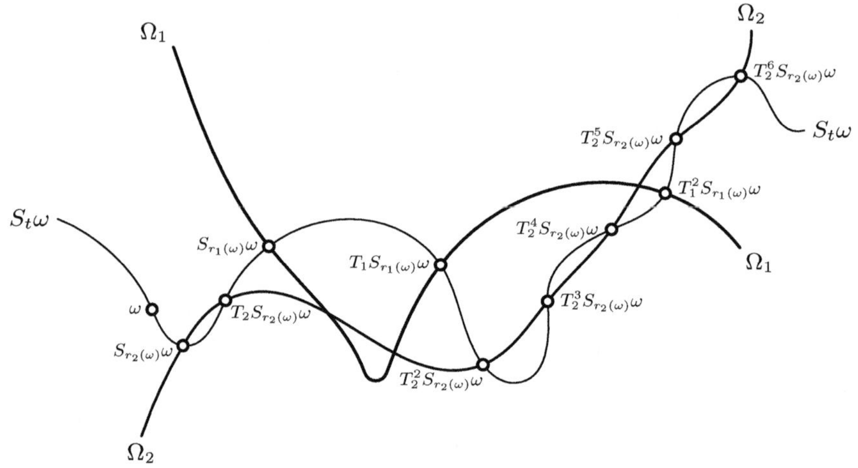

For each $\omega \in E_1$, we define

$$\begin{aligned} W\omega &= S_{r_2(\omega)}\omega \in \Omega_2, \quad \omega \in E_1, \\ E_2 &= W(E_1) \subset \Omega_2. \end{aligned} \tag{22}$$

5 T_1 and T_2 are the corresponding ergodic transformations on $\{\Omega_1, \mu_1\}$ and $\{\Omega_2, \mu_2\}$, respectively.

The image E_2 of W is characterized as follows:

$$E_2 = \{\omega \in \Omega_2 : r_2(-1, \omega) < r_1(0, \omega)\}.$$

It follows that W is a bijection from E_1 onto E_2. If $T_i^{E_i}$, $i = 1, 2$, is the induced transformation on E_i from T_i, then $T_i^{E_i}$ is directly obtained from S_t replacing Ω_i by E_i, so that

$$T_i^{E_i}\omega = S_{r_{E_i}(\omega)}\omega, \quad \omega \in E_i, \quad i = 1, 2.$$

From (22), it follows that

$$r_{E_2}(W\omega) = r_{E_1}(\omega) - r_2(\omega) + r_2(S_{r_{E_1}(\omega)}\omega), \quad \omega \in E_1;$$

and also

$$T_2^{E_2}W\omega = S_{r_{E_2}(W\omega)}W\omega = S_{r_{E_1}(\omega) - r_2(\omega) + r_2(S_{r_{E_1}(\omega)}\omega)}S_{r_2(\omega)}\omega$$

$$= S_{r_2(T_1^{E_1}\omega)}T_1^{E_1}\omega = WT_1^{E_1}\omega.$$

Hence W conjugates $T_1^{E_1}$ and $T_1^{E_2}$. Let e_1 and e_2 be the projections of $\mathcal{C}_1$ and $\mathcal{C}_2$ corresponding to E_1 and E_2 respectively. Then $\{\mathcal{C}_{1,e_1}, \theta_{1,e_1}\}$ and $\{\mathcal{C}_{2,e_2}, \theta_{2,e_2}\}$ correspond to the induced ergodic systems $\{E_1, \mu_{1,E_1}, T_1^{E_1}\}$ and $\{E_2, \mu_{2,E_2}, T_2^{E_2}\}$, which in turn entails that $\{\mathcal{N}_{1,e_1}, \theta_{1,e_1}\} \cong \{\mathcal{N}_{2,e_2}, \theta_{2,e_2}\}$. We leave the measure theoretic details, such as the absolute continuity of $W(\mu_{1,E_1})$ with respect to μ_{2,E_2}, to the reader. Q.E.D.

Definition 3.8. For a factor $\mathcal{M}$ of type III_0, a covariant system $\{\mathcal{N}, \mathbf{Z}, \theta\}$ of Theorem 3.7.(i) is called a *discrete associated covariant system.*

With $\mathcal{M} = \mathcal{N} \rtimes_\theta \mathbf{Z}$, the weight $\varphi = \hat{\tau}$ dual to the trace τ on $\mathcal{N}$ with $\tau \circ \theta \leq \lambda\tau$ has the property that 1 is an isolated point spectrum of Δ_φ.

Definition 3.9. A faithful semi-finite normal weight φ on $\mathcal{M}$ is said to be *lacunary* if 1 is an isolated point spectrum of Δ_φ.

Theorem 3.10. *If φ is a lacunary weight on a separable factor $\mathcal{M}$ of type* III_0 *with infinite multiplicity, then*

(i) *the centralizer $\mathcal{M}_\varphi$ is of type* II_∞ ;

(ii) *there exists a unitary $U \in \mathcal{M}$ such that*

a) $U\mathcal{M}_\varphi U^* = \mathcal{M}_\varphi, \quad \varphi \circ \mathrm{Ad}(U) \leq \lambda\varphi, \quad 0 < \lambda < 1,$

b) $\mathcal{M} \cong \mathcal{M}_\varphi \rtimes_\theta \mathbf{Z}, \quad \theta = \mathrm{Ad}(U)|_{\mathcal{M}_\varphi},$

c) $\varphi = \widehat{\varphi|_{\mathcal{M}_\varphi}}.$

We shall prove the theorem in several steps. First, we fix

$$\lambda_0 = \lambda(\varphi) = \sup\{e^{-\mu} : \mathrm{Sp}(\sigma^\varphi) \cap]0, \mu] \neq \emptyset\} > 0.$$

We recall that the infinite multiplicity of φ means the proper infiniteness of $\mathcal{M}_\varphi$. By the lacunary assumption on φ, we have $\lambda_0 > 0$. We set

$$\mu_0 = -\log \lambda_0 > 0.$$

Let $f \in A(\mathbf{R})$ such that $f(\lambda) = 1$ for $|\lambda| < \mu_0/3$ and $f(\lambda) = 0$ for $|\lambda| > 2\mu_0/3$. Then $\sigma_f^\varphi = \mathcal{E}$ is a faithful normal conditional expectation of $\mathcal{M}$ onto $\mathcal{M}_\varphi$ such that $\varphi = \varphi \circ \mathcal{E}$. Hence φ is semi-finite on $\mathcal{M}_\varphi$. We shall write $\mathcal{M}_\varphi(V)$, $V \subset \mathbf{R}$, for $\mathcal{M}^{\sigma^\varphi}(V)$ and $\mathrm{Sp}_\varphi(x)$ for $\mathrm{Sp}_{\sigma^\varphi}(x)$.

Lemma 3.11. *If $x \in \mathcal{M}_\varphi\big]t - \mu_0/3,\ t + \mu_0/3\big[$, then $h = |x|$ belongs to $\mathcal{M}_\varphi$ and there exists a partial isometry $v \in \mathcal{M}$ and $k \in \mathcal{M}_\varphi$ such that*

(i) $x = vk$,

(ii) *v^*v is the central support of $s_r(x)$ in $\mathcal{M}_\varphi$,*

(iii) *vv^* is the central support of $s_\ell(x)$ in $\mathcal{M}_\varphi$,*

(iv) $\mathrm{Sp}_\varphi(v) \subset \mathrm{Sp}_\varphi(x) \subset \big[t - \mu_0/3,\ t + \mu_0/3\big]$,

(v) $v\mathcal{M}_\varphi v^* \subset \mathcal{M}_\varphi, \quad v^*\mathcal{M}_\varphi v \subset \mathcal{M}_\varphi$.

PROOF: Let $x = uh$ be the polar decomposition. Since $\mathrm{Sp}_\varphi(x^*x) \subset \big[-2\mu_0/3,\ 2\mu_0/3\big]$, x^*x belongs to $\mathcal{M}_\varphi\big[-2\mu_0/3,\ 2\mu_0/3\big] = \mathcal{M}_\varphi$, so that $h = (x^*x)^{1/2} \in \mathcal{M}_\varphi$. Hence $u = \lim_{\varepsilon \to 0} x\,(\varepsilon + h)^{-1}$ entails $\mathrm{Sp}_\varphi(u) = \mathrm{Sp}_\varphi(x)$, so that $e = u^*u \in \mathcal{M}_\varphi$ and $e' = uu^* \in \mathcal{M}_\varphi$. Let c and c' be the central supports of e and e' in $\mathcal{M}_\varphi$ respectively. By the infinite multiplicity of φ, c and c' are both properly infinite, so that there exist mutually orthogonal projections $\{e_i\}$ and $\{e_i'\}$ in $\mathcal{M}_\varphi$ such that

$$e \sim e_i \quad \text{and} \quad e' \sim e_i' \quad \text{in } \mathcal{M}_\varphi;$$

$$\sum_{i=1}^\infty e_i = c, \qquad \sum_{i=1}^\infty e_i' = c'.$$

Let $\{v_i\}$ (resp. $\{v_i'\}$) be a family of partial isometries in $\mathcal{M}_\varphi$ such that $v_i v_i^* = e$ and $v_i^* v_i = e_i$ (resp. $v_i'^* v_i' = e'$ and $v_i' v_i'^* = e_i'$). We then set

$$v = \sum_{i=1}^\infty v_i' u v_i \in \mathcal{M}_\varphi\big(\mathrm{Sp}_\varphi(u)\big);$$

$$k = v^* x \in \mathcal{M}_\varphi.$$

Then we have $v^*v = c$ and $vv^* = c'$, $v\mathcal{M}_\varphi v^* \subset \mathcal{M}_\varphi$ and $v^*\mathcal{M}_\varphi v \subset \mathcal{M}_\varphi$. Q.E.D.

We shall write $\mathcal{C}_\varphi$ for the center of $\mathcal{M}_\varphi$. Let $\mathcal{J}$ be the set of all those partial isometries of $\mathcal{M}$ such that

a) $v^*v,\ vv^* \in \mathrm{Proj}(\mathcal{C}_\varphi)$;

b) $v\mathcal{M}_\varphi v^* \subset \mathcal{M}_\varphi,\ v^*\mathcal{M}_\varphi v \subset \mathcal{M}_\varphi$;

c) $v \in \mathcal{M}_\varphi\big([\mu_0, +\infty[\big)$.

By the above lemma, $\mathcal{J}$ is non-empty. For a pair $v_1, v_2 \in \mathcal{J}$, we define $v_1 < v_2$ by the fact that $s_r(v_1) \leq s_r(v_2)$ and $v_1 = v_2 s_r(v_1)$. Clearly the relation "<" is an order relation in $\mathcal{J}$. If $s_r(v_1) \perp s_r(v_2)$ and $s_\ell(v_1) \perp s_\ell(v_2)$ for $v_1, v_2 \in \mathcal{J}$, then $v_1 + v_2 \in \mathcal{J}$ and $v_1, v_2 < v_1 + v_2$.

We next define two subsets $\mathcal{H}$ and $\mathcal{G}$ of $\mathcal{J}$ as follows:

$$\mathcal{H} = \left\{ v \in \mathcal{J} : v \neq 0,\ \mathrm{Sp}(v) - \mathrm{Sp}(v) \subset \left]-\mu_0/2,\ \mu_0/2)\right[\right\};$$

$$\mathcal{G} = \left\{ v \in \mathcal{H} : v_1, v_2 \in \mathcal{H} \text{ and } v_1 v_2 < v \implies v_1 v_2 = 0 \right\}.$$

Before going further, let us explain our strategy. To prove the theorem, we want to construct a unitary $U \in \mathcal{M}_\varphi([\mu_0, +\infty[)$, which normalizes $\mathcal{M}_\varphi$. We observed already that each $\mathcal{M}_\varphi\left[t - \mu_0/3,\ t + \mu_0/3\right]$ contains a non-zero partial isometry v. It is not difficult to see that if $v, w \in \mathcal{M}\left[t - \mu_0/3,\ t + \mu_0/3\right]$ are partial isometries with $s_r(v) = s_r(w) \in \mathcal{C}_\varphi$, then $s_\ell(v) = s_\ell(w)$ and $vxv^* = wxw^*$ for every $x \in \mathcal{C}_\varphi$. This means that with

$$p_\varphi(E) = \bigvee \left\{ s_\ell(x) : x \in \mathcal{M}_\varphi(E) \right\} \in \mathcal{C}_\varphi;$$

$$q_\varphi(E) = \bigvee \left\{ s_r(x) : x \in \mathcal{M}_\varphi(E) \right\} \in \mathcal{C}_\varphi.$$

There exists an isomorphism from $\mathcal{C}_{\varphi, p_\varphi]t-\mu_0/3,\, t+\mu_0/3)[}$ to $\mathcal{C}_{\varphi, q_\varphi]t-\mu_0/3,\, t+\mu_0/3)[}$ "induced" by the subspace $\mathcal{M}_\varphi\left[t - \mu_0/3,\ t + \mu_0/3\right]$, which would be induced by the action of the unitary U. But this transformation is given by patching the iterates of automorphism $\theta = \mathrm{Ad}(U)$. Thus, we want to find out primitive transformations which are a part of θ. Namely, $\mathcal{G}$ is the set of "primitive" elements of $\mathcal{H}$. We now resume our step by step arguments.

Lemma 3.12. *For every $v \in \mathcal{H}$ there exist $v_1, v_2, \ldots, v_k \in \mathcal{G}$ such that $0 \neq v_1 v_2 \cdots v_k < v$.*

PROOF: By the definition of $\mathcal{H}$, $\mathrm{Sp}_\varphi(v)$ is compact, so that $\mathrm{Sp}_\varphi(v) \subset]0, n\mu_0]$ for some $n \in \mathbf{N}$. We shall prove the lemma by induction.

Suppose $n = 1$. Then $\mathcal{M}_\varphi(]0, \mu_0[) \cap \mathcal{J} = \emptyset$, so that we have nothing to prove. Suppose that the lemma is true for every $v \in \mathcal{H} \cap \mathcal{M}_\varphi(]0, n\mu_0])$, and let $v \in \mathcal{H} \cap \mathcal{M}_\varphi(]0, (n+1)\mu_0])$. If $v \in \mathcal{G}$, then we take $v_1 = v$ and $k = 1$, so that the conclusion of the lemma holds. Suppose $v \notin \mathcal{G}$, and choose $w_1, w_2 \in \mathcal{H}$ with $0 \neq w_1 w_2 < v$. Set $e = s_r(w_1)$, and $w_3 = e w_2 \in \mathcal{H}$. Then $f = s_\ell(w_1 w_3) \in \mathcal{C}_\varphi$ and

$$w_3 = w_1^* w_1 w_3 = w_1^* f v, \qquad \mathrm{Sp}_\varphi(w_3) \subset \mathrm{Sp}_\varphi(v) - \mathrm{Sp}_\varphi(w_1),$$

so that $\mathrm{Sp}_\varphi(w_3) \subset]0, n\mu_0]$. By the induction hypothesis, there exist elements $v_1, v_2, \ldots, v_k \in \mathcal{H}$ with $0 \neq v_1 v_2 \cdots v_k < w_3$. Hence we have $0 \neq w_1 (v_1 v_2 \cdots v_k) < w_3$. Putting $f' = s_\ell(v_1 \cdots v_k)$, we have $w_1 f' = v (v_1 v_2 \cdots v_k)^*$, so that

$$\mathrm{Sp}_\varphi(w_1 f') \subset \mathrm{Sp}_\varphi(v) - \mathrm{Sp}_\varphi(v_1 v_2 \cdots v_k) \subset \mathrm{Sp}_\varphi(v) - \left[k\mu_0, +\infty\right[;$$

hence $w_1 f' \in \mathcal{M}_\varphi(]0, n\mu_0])$. By the induction hypothesis, we have elements $v'_1, \dots, v'_\ell \in \mathcal{G}$ with $0 \neq v'_1 \cdots v'_\ell < w_1 f'$. Hence we obtain

$$0 \neq v'_1 v'_1 \cdots v'_\ell v_1 \cdots v_k < v. \qquad \text{Q.E.D.}$$

Lemma 3.13.

(i) *If* $u \in \mathcal{J}, u \neq 0$, *and* $\mathrm{Sp}_\varphi(u) - \mathrm{Sp}_\varphi(u) \subset]-\mu_0, \mu_0[$, *then there exists* $w \in \mathcal{H}$ *such that* $w < u$.

(ii) *If* $v_1, v_2 \in \mathcal{G}$, *then* $v_1^* v_2 \in \mathcal{M}_\varphi$ *and* $v_1 v_2^* \in \mathcal{M}_\varphi$.

(iii) $\mathcal{M}$ *is generated by* $\mathcal{M}_\varphi$ *and* $\mathcal{G}$.

PROOF:

(i) Choose $g \in A(\mathbf{R})$ so that $\operatorname{supp} g - \operatorname{supp} g \subset [-\mu_0/3, \mu_0/3]$ and $\sigma_g^\varphi(u)^* u \neq 0$. We apply Lemma 3.11 to $x = \sigma_g^\varphi(u)$ to obtain $v \in \mathcal{J}$ and $k \in \mathcal{M}_\varphi$ such that $x = vk$, and

$$\mathrm{Sp}_\varphi(v) \subset \mathrm{Sp}_\varphi(x) \subset \mathrm{Sp}_\varphi(u) \cap \operatorname{supp} g.$$

Hence $v \in \mathcal{H}$. Furthermore, we have

$$\mathrm{Sp}_\varphi(v^* u) \subset \mathrm{Sp}_\varphi(u) - \mathrm{Sp}_\varphi(u) \subset]-\mu_0, \mu_0[,$$

and so $v^* u \in \mathcal{M}_\varphi$. With $w = vv^* u$, we get $w \in \mathcal{J}$ and $\mathrm{Sp}_\varphi(v) \supset \mathrm{Sp}_\varphi(w)$, so that $w \in \mathcal{H}$. By construction, we have $w < u$.

(ii) Suppose $v_1, v_2 \in \mathcal{G}$. We then have

$$\begin{aligned}\mathrm{Sp}_\varphi(v_1^* v_2) - \mathrm{Sp}_\varphi(v_1^* v_2) &\subset \mathrm{Sp}_\varphi(v_2) - \mathrm{Sp}_\varphi(v_1) - \mathrm{Sp}_\varphi(v_2) - \mathrm{Sp}_\varphi(v_1)\\ &\subset \mathrm{Sp}_\varphi(v_2) - \mathrm{Sp}_\varphi(v_2) - \mathrm{Sp}_\varphi(v_1) - \mathrm{Sp}_\varphi(v_1)\\ &\subset]-\mu_0/2, \mu_0/2[+]-\mu_0/2, \mu_0/2[\subset]-\mu_0, \mu_0[.\end{aligned}$$

Hence we have one of the following three cases:

$$v_1^* v_2 \in \mathcal{M}_\varphi, \qquad v_1^* v_2 \in \mathcal{M}_\varphi([\mu_0, +\infty[), \qquad v_2^* v_1 \in \mathcal{M}_\varphi([\mu_0, +\infty[).$$

We shall disprove the case, $0 \neq v_1^* v_2 \in \mathcal{M}_\varphi([\mu_0, +\infty[)$. Set $u = v_1^* v_2$. By (i), there exists $w \in \mathcal{H}$ with $w < u$. Then we get $v_1 w < v_1 u = v_1 v_1^* v_2 < v_2$ and $v, w \neq 0$, which contradicts the fact that $v_2 \in \mathcal{G}$. Hence $0 \neq v_1^* v_2 \in \mathcal{M}_\varphi([\mu_0, +\infty[)$ is impossible. Similarly $0 \neq v_2^* v_1 \in \mathcal{M}_\varphi([\mu_0, +\infty[)$ is impossible, so that $v_1^* v_2 \in \mathcal{M}_\varphi$. By symmetry, we get also $v_2^* v_1 \in \mathcal{M}_\varphi$.

(iii) Let $\mathcal{P} = (\mathcal{M}_\varphi \cup \mathcal{G})''$. From a partition of unity in $A(\mathbf{R})$, it follows that $\mathcal{M}$ is generated by the subset $\{x \in \mathcal{M} : \mathrm{Sp}_\varphi(x) - \mathrm{Sp}_\varphi(x) \subset]-\mu_0/3, \mu_0/3[\}$. Hence it suffices to prove that if $x \in \mathcal{M}_\varphi([\mu_0, +\infty[)$ satisfies the spectrum condition, $\mathrm{Sp}_\varphi(x) - \mathrm{Sp}_\varphi(x) \subset]-\mu_0/3, \mu_0/3[$, then x belongs to $\mathcal{P}$. Let x be such an element. By Lemma 3.11, $x = vk$ for some $v \in \mathcal{H}$ and $k \in \mathcal{M}_\varphi$. Thus, we want to prove $\mathcal{H} \subset \mathcal{P}$. Let $v \in \mathcal{H}$ and $e = s_r(v)$. Put

$$P = \left\{ p \in \mathrm{Proj}(\mathcal{C}_\varphi) : vp \in \mathcal{P} \text{ and } p \leq e \right\}.$$

Clearly P is an inductive subset of $\mathrm{Proj}(\mathcal{C}_\varphi)$, so that P admits a maximal element p. If $e = p$, then we have $v \in \mathcal{P}$. If $e - p \neq 0$, then $v(e - p) \in \mathcal{H}$. By the previous lemma, there exist $v_1, \dots, v_k \in \mathcal{G}$ with $0 \neq v_1 \cdots v_k < v(e-p)$, so that $v_1 \cdots v_k = v(e - p)s_r(v_1 \cdots v_k) \in \mathcal{P}$ which contradicts the maximality of p. Hence $p = e$.

Q.E.D.

Lemma 3.14. *The factor $\mathcal{M}$ is generated by $\mathcal{M}_\varphi$ and some unitary $U \in \mathcal{J}$.*

PROOF: Let $\{v_j\}$ be a maximal family of elements of $\mathcal{G}$ such that $v_i^* v_j = v_i v_j^* = 0$, $i \neq j$. Put $U = \sum v_j \in \mathcal{M}_\varphi\big([\mu_0, +\infty[\big)$. We shall prove first that U is a unitary. Put

$$e_j = s_r(v_j), \qquad e_j' = s_\ell(v_j), \qquad e = \sum e_j, \qquad e' = \sum e_j'.$$

Assume that $1 - e \neq 0$. Choose $0 \neq x \in \mathcal{M}_{(1-e)} \cap \mathcal{M}_\varphi\big([\mu_0, +\infty[\big)$ with $\mathrm{Sp}_\varphi(x) - \mathrm{Sp}_\varphi(x) \subset \big]-\mu_0/3, \mu_0/3\big[$. Lemma 3.11 entails the existence of $v \in \mathcal{H}$ with $v^*v \leq 1 - e$ and $vv^* \leq 1 - e$. By Lemma 3.12, there exist $w_1, \dots, w_k \in \mathcal{G}$ with $0 \neq w_1 \cdots w_k < v$. With $w = w_k \in \mathcal{G}$, we have $w^*w \leq 1 - e$, so that $we_j = 0$; hence $v_j w^* = 0$ for every j. By Lemma 3.13.(ii), we have $wv_j^* \in \mathcal{M}_\varphi$, and $s_r(w^*v_j) \leq e_j$ and $s_\ell(w^*v_j) \leq 1 - e$. Thus $w^*v_j \in (1 - e)\mathcal{M}_\varphi e = \{0\}$, so that $w^*v_j = 0$ for every j. This contradicts the maximality of $\{v_j\}$. Thus $e = 1$. Similarly, we have $e' = 1$. Hence U is a unitary. Since $\mathcal{J}$ is closed under orthogonal sum, U belongs to $\mathcal{J}$, which means that $U\mathcal{M}_\varphi U^* = \mathcal{M}_\varphi$. For every $v \in \mathcal{G}$, we have $U^*v = \sum v_j^* v \in \mathcal{M}_\varphi$, so that $\mathcal{G} \subset U\mathcal{M}_\varphi$. Hence $\mathcal{M}$ is generated by U and $\mathcal{M}_\varphi$ as seen in Lemma 3.13.(iii).

Q.E.D.

PROOF OF THEOREM 3.10: We shall use the notations established above. Let $\theta = \mathrm{Ad}(U)$. Then θ leaves $\mathcal{M}_\varphi$ globally invariant, so that we may consider θ as an automorphism of $\mathcal{M}_\varphi$. Note that θ is in $\mathrm{Int}(\mathcal{M})$, whilst θ is not in $\mathrm{Int}(\mathcal{M}_\varphi)$. Since $U \in \mathcal{M}_\varphi\big([\mu_0, +\infty[\big)$, we have $U^n \in \mathcal{M}_\varphi\big([n\mu_0, +\infty[\big)$, $n = 1, 2, \dots$, so that $\mathcal{E}(U^n) = 0$ for $n \geq 1$, and hence $\mathcal{E}(U^{-n}) = \mathcal{E}(U^n)^* = 0$ for every $n \in \mathbf{Z}$, $n \neq 0$. Thus we have

$$U^n \eta_\varphi\big(\mathfrak{n}_\varphi \cap \mathfrak{n}_\varphi^* \cap \mathcal{M}_\varphi)\big) \perp U^m \eta_\varphi\big(\mathfrak{n}_\varphi \cap \mathfrak{n}_\varphi^* \cap \mathcal{M}_\varphi\big), \quad n \neq m.$$

Denoting the unitary associated with the crossed product $\mathcal{M}_\varphi \rtimes_\theta \mathbf{Z}$ by u, the correspondence:

$$\sum_{n \in \mathbf{Z}} \pi_\theta(x_n)u^n \longleftrightarrow \sum_{n \in \mathbf{Z}} x_n U^n, \quad x_n \in \mathcal{M}_\varphi,$$

gives an isomorphism between $\mathcal{M}_\varphi \rtimes_\theta \mathbf{Z}$ and $\mathcal{M}$. Furthermore, under this isomorphism, the dual weight $\hat{\tau}$ of $\tau = \varphi|_{\mathcal{M}_\varphi}$ corresponds exactly to φ itself because $\varphi = \tau \circ \mathcal{E}$.

Since $\mathcal{M}$ is a factor, θ is centrally ergodic on $\mathcal{M}_\varphi$. Since $\tau \circ \theta$ is a faithful semi-finite normal trace on $\mathcal{M}_\varphi$, there exists a positive self-adjoint operator ρ affiliated with $\mathcal{C}_\varphi$ such that $\tau \circ \theta = \tau(\rho\,\cdot\,)$, i.e. $\left(D\tau\circ\theta : D\tau\right)_t = \rho^{it}$. Since $\varphi = \hat{\tau}$, Theorem X.1.17 entails $\sigma_t^\varphi(U) = \rho^{it}U$. Since $U \in \mathcal{M}_\varphi\left([\mu_0, +\infty[\right)$, we have $\rho \leq e^{-\mu_0} = \lambda < 1$. Thus we have $\tau\circ\theta \leq \lambda\tau$.

We now prove that $\mathcal{M}_\varphi$ is of type II_∞. Since θ is centrally ergodic, $\mathcal{M}_\varphi$ is either of type I_∞ or of type II_∞, because we assumed that $\mathcal{M}_\varphi$ is properly infinite. If $\mathcal{M}_\varphi$ is of type I_∞, then $\mathcal{M}_\varphi \cong \mathcal{C}_\varphi \overline{\otimes} \mathcal{B}$ where $\mathcal{B}$ is a factor of type I_∞. Let $\theta_0 = \theta|_{\mathcal{C}_0} \otimes \mathrm{id}$. Then $\theta\theta_0^{-1}$ leaves the center $\mathcal{C}_\varphi$ pointwise invariant, so that $\theta\circ\theta_0^{-1} = \mathrm{Ad}(v)$ for some $v \in \mathcal{U}(\mathcal{M}_\varphi)$. Thus we have $\theta = \mathrm{Ad}(v)\circ\theta_0$. Hence we have $\lambda\tau \geq \tau\circ\theta = \tau\circ\theta_0$. The uniqueness of a trace on $\mathcal{B}$ implies that τ is of the form, $\tau = \psi\otimes\mathrm{Tr}$, with the standard trace Tr on $\mathcal{B}$. Hence we have $\psi\circ\theta \leq \lambda\psi$ on $\mathcal{C}_\varphi$. This means that $\left\{\mathcal{C}_\varphi, \theta\right\}$ is conjugate to $\ell^\infty(\mathbf{Z})$ with translation. Therefore, we have

$$\begin{aligned}\mathcal{M} &\cong \left(\ell^\infty(\mathbf{Z})\overline{\otimes}\mathcal{B}\rtimes_\theta \mathbf{Z}\right) \cong \left(\ell^\infty(\mathbf{Z})\overline{\otimes}\mathcal{B}\right)\rtimes_{\theta_0}\mathbf{Z}\\ &= \left(\ell^\infty(\mathbf{Z})\rtimes_\theta\mathbf{Z}\right)\overline{\otimes}\mathcal{B} = \mathcal{L}\left(\ell^2(\mathbf{Z})\right)\overline{\otimes}\mathcal{B},\end{aligned}$$

which means that $\mathcal{M}$ is of type I_∞. This is a contradiction. Therefore, $\mathcal{M}_\varphi$ must be of type II_∞. Q.E.D.

Proposition 3.15. *If $\mathcal{M}$ is a factor of type* III_0, *then the center $\mathcal{C}_\varphi$ of the centralizer of any faithful semi-finite normal weight φ is non-atomic.*

PROOF: If $e \in \mathrm{Proj}(\mathcal{C}_\varphi)$ is minimal, then $\mathcal{M}_{\varphi,e}$ is a factor, which means that $\mathrm{Sp}(\Delta_{\varphi_e})$ is the modular spectrum $S(\mathcal{M})$ of $\mathcal{M}$. By assumption, $S(\mathcal{M}) = \{0, 1\}$. But $\mathrm{Sp}(\Delta_{\varphi_e})$ cannot be reduced to $\{0, 1\}$ because $\mathcal{M}$ is of type III and Δ_{φ_e} is not a projection. Thus $\mathcal{C}_\varphi$ cannot admit a minimal projection. Q.E.D.

Exercise XII.3

1†) Let $\left\{\mathcal{A}, \mathbf{R}, \alpha\right\}$ be an ergodic covariant system with $\mathcal{A}$ an abelian separable von Neumann algebra. Show that any cocycle $c \in Z\left(\mathbf{R}, \mathcal{U}(\mathcal{A}), \alpha\right)$ is cohomologous to an element $\bar{c} \in Z\left(\mathbf{R}, \mathcal{U}(\mathcal{A}), \alpha\right)$ such that the $\mathcal{A}$-valued function: $t \in \mathbf{R} \mapsto \bar{c}(t) \in \mathcal{A}$ is indefinitely differentiable relative to norm. (Hint: Represent the flow by the flow built under ceiling function over a single ergodic transformation.) [700]

2†) Let $\mathcal{M}$ be a separable factor of type III_0 and $\mathcal{N} = \mathcal{M}\rtimes_{\sigma^\varphi}\mathbf{R}$ with φ a faithful semi-finite normal weight on $\mathcal{M}$. Consider the ergodic flow $\left\{\mathcal{C}, \mathbf{R}, \theta\right\}$ where $\mathcal{C}$ is the center of $\mathcal{N}$. With $\mathcal{M}_0 = \mathcal{M}$, $\varphi_0 = \varphi$, $\mathcal{N}_0 = \mathcal{N}$, $\mathcal{C}_0 = \mathcal{C}$ and $\theta^0 = \theta$, define the covariant systems $\left\{\mathcal{C}_n, \mathbf{R}, \theta^n\right\}$ as follows, provided that $\mathcal{M}_{n-1}$ is still of type III_0:

$$\left\{\begin{aligned}\mathcal{M}_n &= \mathcal{C}_{n-1}\rtimes_{\theta^{n-1}}\mathbf{R};\\ \mathcal{N}_n &= \mathcal{M}_n\rtimes_{\sigma^{\varphi_n}}\mathbf{R};\\ \mathcal{C}_n &= \text{The Center of } \mathcal{N}_n; \qquad \varphi_n \in \mathfrak{W}_0(\mathcal{M}_n);\\ \theta^n &= \widehat{\sigma^{\varphi_n}} = \text{The action of } \mathbf{R} \text{ on } \mathcal{N}_n \text{ dual to } \sigma^{\varphi_n}.\end{aligned}\right.$$

(**a**) Show that $\mathcal{M}_1$ is an infinite factor of type II_∞ or III. (Hint: The covariant system $\{\mathcal{C}, \mathbf{R}, \theta\}$ is a non transitive ergodic flow.)

(**b**) Observe that the $N \in \mathbf{N}$ such that $\mathcal{M}_N$ is of type III_0 but $\mathcal{M}_{N+1}$ is not is an algebraic invariant of $\mathcal{M}$ ($N = \infty$ if the induction process does not stop at a finite stage). Call the sequence of ergodic flows, $\{\mathcal{C}_0, \mathbf{R}, \theta^0\}, \dots, \{\mathcal{C}_N, \mathbf{R}, \theta^N\}$, *Krieger's hierarchy tower* of $\mathcal{M}$. [621]

3) Let $\mathcal{M}_1$ be a factor of type III_0 and $\mathcal{M}_2$ a factor of type III_λ, $0 < \lambda \leq 1$. Show that $\mathcal{M} = \mathcal{M}_1 \bar{\otimes} \mathcal{M}_2$ is a factor of type III_λ.

§ 4 The Flow of Weights

We consider only separable von Neumann algebras in this section. Let $\mathcal{M}$ be a von Neumann algebra, and $\mathfrak{W}$ be the set of all semi-finite normal weights on $\mathcal{M}$ and $\mathfrak{W}_0$ be the set of all faithful ones in $\mathfrak{W}$. When we want to specify the dependence on $\mathcal{M}$, we write $\mathfrak{W}(\mathcal{M})$ or $\mathfrak{W}_0(\mathcal{M})$.

For each $\varphi \in \mathfrak{W}(\mathcal{M})$, we consider the reduced algebra $\mathcal{M}_e$ with $e = s(\varphi)$ to define the modular automorphism group $\{\sigma_t^\varphi\}$ of φ; the centralizer of φ and others. Namely, $\{\sigma_t^\varphi\}$, $\mathcal{M}_\varphi$ and the other related objects are defined on $\mathcal{M}_e$. For a projection $p \in \mathcal{M}_\varphi$, φ_p means the weight defined by $\varphi_p(x) = \varphi(pxp)$, $x \in \mathcal{M}$.

Definition 4.1. For $\varphi, \psi \in \mathfrak{W}$, we say that φ and ψ are *equivalent,* and write $\varphi \sim \psi$ if there exists a partial isometry $u \in \mathcal{M}$ such that $s(\varphi) = u^*u$ and $s(\psi) = uu^*$, and furthermore $\varphi(x) = \psi(uxu^*)$, $x \in \mathcal{M}$. In this case, we write $\psi_u = \varphi$. If there exists $e \in \mathrm{Proj}(\mathcal{M}_\psi)$ such that $\varphi \sim \psi_e$, then we write $\varphi \precsim \psi$ or $\psi \succsim \varphi$.

As in the comparison of two projections, we have the following Cantor-Bernstein type result:

Proposition 4.2. *If $\varphi \precsim \psi$ and $\varphi \succsim \psi$, then $\varphi \sim \psi$.*

We leave the proof to the reader.

To relate the equivalence relation $\varphi \sim \psi$ to the algebraic apparatus, we extend the notion of the cocycle derivative $(\mathrm{D}\varphi : \mathrm{D}\psi)$ to the case where ψ is faithful and φ is not necessarily faithful.

With $M_2 = M_2(\mathbf{C})$, the 2×2-matrix algebra, let $\mathcal{P} = \mathcal{M} \otimes M_2$ and set the balanced weight $\rho = \varphi \oplus \psi$ to be:

$$\rho\left(\sum_{i,j=1}^{2} x_{ij} \otimes e_{ij}\right) = \varphi(x_{11}) + \psi(x_{22}).$$

We then have

$$s(\rho) = s(\varphi) \otimes e_{11} + 1 \otimes e_{22}$$

so that $s(\varphi) \otimes e_{11} \in \mathcal{P}_{s(\rho)}$. There exists a strongly continuous family of partial isometries $\{u_t\}$ in $\mathcal{M}$ such that

$$\begin{cases} u_t u_t^* = s(\varphi), & u_t^* u_t = \sigma_t^{\psi}\big(s(\varphi)\big); \\ \sigma_t^{\rho}\big(s(\varphi) \otimes e_{12}\big) = u_t \otimes e_{12}, & t \in \mathbf{R}. \end{cases}$$

We denote u_t by $(\mathrm{D}\varphi : \mathrm{D}\psi)_t$. It is a straightforward exercise to check that $\{(\mathrm{D}\varphi : \mathrm{D}\psi)_t\}$ satisfies the cocycle identity (VIII.3.23) and is characterized by the relative modular condition in Theorem VIII.3.3.(c).

Lemma 4.3. *Let φ_1, φ_2 and $\psi \in \mathfrak{W}$ and assume ψ is faithful. On $\mathcal{P} = \mathcal{M} \otimes M_2$, put $\varphi = \varphi_1 \oplus \varphi_2$, the balanced weight of φ_1 and φ_2.*

(i) *For a partial isometry $w \in \mathcal{M}$ with $w^* w = s(\varphi_2)$ and $ww^* \in \mathcal{M}_{\varphi_1}$, the following conditions are equivalent:*
  a) $\varphi_2 = \varphi_{1,w}$;
  b) $w \otimes e_{12} \in \mathcal{P}_\varphi$;
  c) $\big(\mathrm{D}\varphi_2 : \mathrm{D}\psi\big)_t = w^* \big(\mathrm{D}\varphi_1 : \mathrm{D}\psi\big)_t \sigma_t^{\psi}(w), \quad t \in \mathbf{R}$.

(ii) $\varphi_2 \precsim \varphi_1$ *if and only if* $s(\varphi_2) \otimes e_{22} \precsim s(\varphi_1) \otimes e_{11}$ *in* $\mathcal{P}_\varphi$.

(iii) $\varphi_1 \sim \varphi_2$ *if and only if* $s(\varphi_1) \otimes e_{11} \sim s(\varphi_2) \otimes e_{22}$ *in* $\mathcal{P}_\varphi$.

(iv) *For partial isometries $u, v \in \mathcal{M}$ with $p = u^* u \in \mathcal{M}_{\varphi_1}$ and $q = vv^* \in \mathcal{M}_{\varphi_1}$,*

$$\varphi_{1,u} \precsim \varphi_{1,v} \quad \textit{if and only if} \quad p \precsim q \ \textit{in}\ \mathcal{M}_{\varphi_1}.$$

PROOF: We give a proof only for (i), and leave the rest to the reader.

We have $s(\varphi) = s(\varphi_1) \otimes e_{11} + s(\varphi_2) \otimes e_{22}$, so that $w \otimes e_{12} \in \mathcal{P}_{s(\varphi)}$. Hence Theorem VIII.2.6, applied to $\mathcal{P}_{s(\varphi)}$, implies that $w \otimes e_{12} \in \mathcal{P}_\varphi$ if and only if for every $x \in \mathcal{M}$

$$\varphi\big((w \otimes e_{12})(x \otimes e_{22})(w \otimes e_{12})^*\big) = \varphi\Big(\big(s(\varphi_2) \otimes e_{22}\big)(x \otimes e_{22})\big(s(\varphi_2) \otimes e_{22}\big)\Big),$$

which is equivalent to $\varphi_{1,w} = \varphi_2$. Put now $\mathcal{Q} = \mathcal{M} \otimes M_3$ and $\chi = \varphi_1 \oplus \varphi_2 \oplus \psi$:

$$\chi\Big(\sum_{i,j=1}^{3} x_{ij} \otimes e_{ij}\Big) = \varphi_1(x_{11}) + \varphi_2(x_{22}) + \psi(x_{33}).$$

Since we have

$$\sigma_t^{\chi}\big(s(\varphi_j) \otimes e_{j,3}\big) = \big(\mathrm{D}\varphi_j : \mathrm{D}\psi\big)_t \otimes e_{j,3}, \quad j = 1, 2,$$

we get

$$\begin{aligned} \sigma_t^{\chi}(w \otimes e_{12}) &= \sigma_t^{\chi}\Big(\big(s(\varphi_1) \otimes e_{13}\big)(w \otimes e_{33})\big(s(\varphi_2) \otimes e_{32}\big)\Big) \\ &= \big(\mathrm{D}\varphi_1 : \mathrm{D}\psi\big)_t \sigma_t^{\psi}(w)\big(\mathrm{D}\varphi_2 : \mathrm{D}\psi\big)_t^* \otimes e_{12}, \end{aligned}$$

so that $w \otimes e_{12} \in \mathcal{P}_\varphi$ if and only if $w \otimes e_{12} \in \mathcal{Q}_\chi$, if and only if

$$\big(\mathrm{D}\varphi_2 : \mathrm{D}\psi\big)_t = w^* \big(\mathrm{D}\varphi_1 : \mathrm{D}\psi\big)_t \sigma_t^{\psi}(w), \quad t \in \mathbf{R}.$$

Q.E.D.

We recall that $\varphi \in \mathfrak{W}(\mathcal{M})$ is of *infinite multiplicity* if $\mathcal{M}_\varphi$ is properly infinite. We denote by $\mathfrak{W}_{\text{inf}}$ the set of all semi-finite normal weights of infinite multiplicity.

Theorem 4.4. *Let $\mathcal{M}$ be a separable factor of type* III.

(i) *There exist uniquely a non separable abelian von Neumann algebra $\mathcal{A}$ and a map $p_\mathcal{M}$ from $\mathfrak{W}_{\text{inf}}$ onto the lattice $\text{Proj}_\sigma(\mathcal{A})$ of σ-finite projections of $\mathcal{A}$ such that*

$$p_\mathcal{M}(\varphi_1) \leq p_\mathcal{M}(\varphi_2) \iff \varphi_1 \precsim \varphi_2.$$

(ii) *There exist a non-separable semi-finite von Neumann algebra $\mathcal{F} = \mathcal{F}(\mathcal{M})$ and a map $q_\mathcal{M}$ from $\mathcal{M}_*^+ \setminus \{0\}$ into the lattice $\text{Proj}_f(\mathcal{F})$ of finite projections of $\mathcal{F}$ such that $\varphi_1 \precsim \varphi_2$ if and only if*

$$q_\mathcal{M}(\varphi_1) \precsim q_\mathcal{M}(\varphi_2) \quad \textit{in } \mathcal{F}.$$

If $\mathfrak{C}$ is the center $\mathfrak{C}_\mathcal{F}$ of $\mathcal{F}$, then $\mathfrak{C}$ can be identified with a direct summand of $\mathcal{A}$ in such a way that for each $\varphi \in M_*^+$, $\varphi \neq 0$, $z\big(q_\mathcal{M}(\varphi)\big) = p_\mathcal{M}(\check{\varphi})$, where $z(e)$, $e \in \text{Proj}(\mathcal{F})$, denotes the central support of e in $\mathcal{F}$ and $\check{\varphi} = \varphi \otimes \text{Tr}$ under an identification of $\mathcal{M}$ and $\mathcal{M} \otimes \mathcal{L}\big(\ell^2(\mathbf{N})\big)$.

(iii) *To each $\varphi \in \mathcal{M}_*^+$, $\varphi \neq 0$, there corresponds a normal positive linear functional μ_φ on $\mathfrak{C}$ in such a way that*

a) $\varphi_1 \precsim \varphi_2 \iff \mu_{\varphi_1} \leq \mu_{\varphi_2}$;

b) $\mu_{\varphi_1+\varphi_2} = \mu_{\varphi_1} + \mu_{\varphi_2}$ *if* $s(\varphi_1) \perp s(\varphi_2)$.

PROOF:

(i) Consider the big Hilbert space $\ell^2(\mathfrak{W}_{\text{inf}})$ of all square summable functions on $\mathfrak{W}$, whose dimension is of continuum, and the tensor product $\widetilde{\mathcal{M}} = \mathcal{M} \overline{\otimes} \mathcal{L}\big(\ell^2(\mathfrak{W}_{\text{inf}})\big)$. We denote the canonical matrix unit of $\mathcal{L}\big(\ell^2(\mathfrak{W}_{\text{inf}})\big)$ by $\big\{ e_{\varphi,\psi} : \varphi, \psi \in \mathfrak{W}_{\text{inf}} \big\}$. We then consider the balanced weight $\Phi = \sum^{\oplus}_{\varphi \in \mathfrak{W}_{\text{inf}}} \varphi$ of all weights of infinite multiplicity:

$$\Phi\Bigg(\sum_{\varphi,\psi \in \mathfrak{W}_{\text{inf}}} x_{\varphi,\psi} \otimes e_{\varphi,\psi} \Bigg) = \sum_{\varphi \in \mathfrak{W}_{\text{inf}}} \varphi(x_{\varphi,\varphi}).$$

We then have $s(\Phi) = \sum_\varphi s(\varphi) \otimes e_{\varphi,\varphi}$ and $s(\varphi) \otimes e_{\varphi,\varphi} \in \widetilde{\mathcal{M}}_\Phi$. As seen in Lemma 4.3, we know that $\varphi_1 \precsim \varphi_2$ if and only if $s(\varphi_1) \otimes e_{\varphi_1,\varphi_1} \precsim s(\varphi_2) \otimes e_{\varphi_2,\varphi_2}$ in $\widetilde{\mathcal{M}}_\Phi$. Hence the comparison problem of weights are transferred to that of the corresponding projections in $\widetilde{\mathcal{M}}_\Phi$. Let $\mathcal{A}$ be the center of $\widetilde{\mathcal{M}}_\Phi$. For each $\varphi \in \mathfrak{W}_{\text{inf}}$, let $p_\mathcal{M}(\varphi)$ be the central support of $s(\varphi) \otimes e_{\varphi,\varphi}$ in $\widetilde{\mathcal{M}}_\Phi$. The reduced algebra $\widetilde{\mathcal{M}}_{\Phi,s(\varphi)\otimes e_{\varphi,\varphi}}$ is canonically isomorphic to $\mathcal{M}_\varphi$, so that the infinite multiplicity of φ means the proper infiniteness of $s(\varphi) \otimes e_{\varphi,\varphi}$ in $\widetilde{\mathcal{M}}_\Phi$. Since $\mathcal{M}$ is separable, $s(\varphi) \otimes e_{\varphi,\varphi}$ must be σ-finite, in fact $\widetilde{\mathcal{M}}_{\Phi,s(\varphi)\otimes e_{\varphi,\varphi}} \cong \mathcal{M}_\varphi$ is separable, so that its central support $p_\mathcal{M}(\varphi)$ is σ-finite as a projection in $\mathcal{A}$ (not in $\widetilde{\mathcal{M}}_\Phi$ however). The proper infiniteness and the σ-finiteness of $s(\varphi) \otimes e_{\varphi,\varphi}$ together imply that

$$\varphi_1 \precsim \varphi_2 \iff p_\mathcal{M}(\varphi_1) \leq p_\mathcal{M}(\varphi_2).$$

Let c be a σ-finite projection of $\mathcal{A}$. Let $\{\varphi_i\}$ be a maximal family of weights in $\mathfrak{W}_{\rm inf}$ such that $\{p_{\mathcal{M}}(\varphi_i)\}$ is pairwise orthogonal and $p_{\mathcal{M}}(\varphi_i) \leq c$. The σ-finiteness of c implies the countability of $\{\varphi_i\}$. Let $\{e_i\}$ be a mutually orthogonal family of projections in $\mathcal{M}$. Since $\mathcal{M}$ is a factor of type III, $e_i \sim s(\varphi_i) \sim 1$. Let u_i be a partial isometry of $\mathcal{M}$ such that $u_i^* u_i = e_i$ and $u_i u_i^* = s(\varphi_i)$. Set $\psi_i = \varphi_{i,u_i}$. Then $\psi_i \sim \varphi_i$ and so $p_{\mathcal{M}}(\psi_i) = p_{\mathcal{M}}(\varphi_i)$. Let $\psi = \sum \psi_i$. The orthogonality of $s(\psi_i) = e_i$ implies the semi-finiteness of ψ. Since $s(\psi) = \sum s(\psi_i)$ and $s(\psi_i) \otimes e_{\psi_i,\psi_i} \sim s(\varphi_i) \otimes e_{\varphi_i,\varphi_i}$ in $\widetilde{\mathcal{M}}_\Phi$, we have $p_{\mathcal{M}}(\psi) = \bigvee p_{\mathcal{M}}(\psi_i) = \bigvee p_{\mathcal{M}}(\varphi_i) \leq c$. We claim that $c = p_{\mathcal{M}}(\psi)$. If $p_{\mathcal{M}}(\psi) \neq c$, then $c - p_{\mathcal{M}}(\psi)$ is orthogonal to $p_{\mathcal{M}}(\varphi)$ for every $\varphi \in \mathfrak{W}_{\rm inf}$ by the maximality of $\{\varphi_i\}$. But $s(\Phi) = \sum s(\varphi) \otimes e_{\varphi,\varphi}$, so that $1_{\mathcal{A}} = \bigvee p_{\mathcal{M}}(\varphi)$, a contradiction. Hence $c = p_{\mathcal{M}}(\psi)$. Therefore $p_{\mathcal{M}}$ maps $\mathfrak{W}_{\rm inf}$ onto $\mathrm{Proj}_\sigma(\mathcal{A})$.

The uniqueness of $\{\mathcal{A}, p_{\mathcal{M}}\}$ follows from the facts that (i) $p_{\mathcal{M}}$ induces an order isomorphism $\dot{p}_{\mathcal{M}}$ of the quotient ordered space $\mathfrak{W}_{\rm inf}/$"$\sim$" onto $\mathrm{Proj}_\sigma(\mathcal{A})$, and (ii) any order isomorphism between the lattices of σ-finite projections in abelian von Neumann algebras extends canonically to an isomorphism.

(ii) We consider this time the tensor product $\overline{\mathcal{M}} = \mathcal{M} \overline{\otimes} \mathcal{L}(\ell^2(\mathcal{M}_*^+ \setminus \{0\}))$, and $\Omega = \sum_{\omega \in \mathcal{M}_*^+}^{\oplus} \omega$, the balanced weight of all normal non-zero positive functionals. Then Ω is a semi-finite normal weight on $\overline{\mathcal{M}}$. Let $\mathcal{F} = \overline{\mathcal{M}}_\Omega$. Since $\overline{\mathcal{M}}_{\Omega, s(\omega)\otimes e_{\omega,\omega}}$ is canonically isomorphic to $\mathcal{M}_\omega$ and ω is a finite faithful trace on $\mathcal{M}_\omega$, $\overline{\mathcal{M}}_{\Omega, s(\omega)\otimes e_{\omega,\omega}}$ is finite, so that $\mathcal{F}$ is semi-finite. Put

$$q_{\mathcal{M}}(\omega) = s(\omega) \otimes e_{\omega,\omega}, \quad \omega \in \mathcal{M}_*^+, \quad \omega \neq 0.$$

It then follows that $q_{\mathcal{M}}(\omega_1) \precsim q_{\mathcal{M}}(\omega_2)$ if and only if $\omega_1 \precsim \omega_2$.

Let $\{u_i\}$ be a family of isometries in $\mathcal{M}$ with mutually orthogonal range. For any $\omega \in \mathcal{M}_*^+$, $\omega \neq 0$, let

$$\check{\omega} = \sum_i u_i \omega u_i^*.$$

Then $\check{\omega}$ is semi-finite and of infinite multiplicity. After fixing $\{u_i\}$, the map: $\omega \in \mathcal{M}_*^+ \mapsto \check{\omega} \in \mathfrak{W}_{\rm inf}$ is well-defined. With $\mathcal{B} = \mathcal{L}(\ell^2(\mathbf{N}))$ and $\check{\Omega} = \sum^{\oplus} \check{\omega}$ we have canonically

$$\widetilde{\mathcal{M}}_{\check{\Omega}} \cong \mathcal{F} \overline{\otimes} \mathcal{B}$$

with a fixed $\{u_i\}$. Hence there exists an isomorphism of the center $\mathcal{C}$ of $\mathcal{F}$ onto the center of $\widetilde{\mathcal{M}}_{\check{\Omega}}$. Since $\check{\Omega} = \widetilde{\mathcal{M}}_{\check{\Omega}}$ can be canonically identified with a direct summand of $\mathcal{A}$. Thus we obtained an isomorphism from $\mathcal{C}$ onto the direct summand of $\mathcal{A}$. Since the equivalence class of $\check{\omega}$ does not depend on the choice of $\{u_i\}$, the final isomorphism from $\mathcal{C}$ into $\mathcal{A}$ is independent of $\{u_i\}$. Hence $\mathcal{C}$ itself may be identified with a direct summand of $\mathcal{A}$. By construction, we have

$$z(q_{\mathcal{M}}(\omega)) = p_{\mathcal{M}}(\check{\omega}).$$

(iii) On $\mathcal{F}$, Ω is a faithful semi-finite normal trace and $\Omega(q_{\mathcal{M}}(\omega)) = \omega(1) < +\infty$, $\omega \in \mathcal{M}_*^+$. The comparison of finite projections in a semi-finite von Neumann algebra is measured by a center valued trace. Let Ψ be a faithful semi-finite trace on $\mathcal{C}$ and $\mathfrak{T}$ be a faithful normal center valued trace on $\mathcal{F}$ such that $\Omega = \Psi \circ \mathfrak{T}$. Let

$$\mu_\varphi(c) = \Omega(c q_{\mathcal{M}}(\varphi)), \quad \varphi \in \mathcal{M}_*^+.$$

Since $\Omega\big(q_{\mathcal{M}}(\varphi)\big)$ is finite, μ_φ is a finite normal functional on $\mathcal{C}$. Then we have for each $c \in \mathcal{C}$

$$\Psi\Big(c\mathfrak{T}\big(q_{\mathcal{M}}(\varphi)\big)\Big) = \Psi\Big(\mathfrak{T}\big(cq_{\mathcal{M}}(\varphi)\big)\Big) = \Omega\big(cq_{\mathcal{M}}(\varphi)\big) = \mu_\varphi(c),$$

so that $\mu_{\omega_1} \leq \mu_{\omega_2}$ if and only if $\mathfrak{T}\big(q_{\mathcal{M}}(\omega_1)\big) \leq \mathfrak{T}\big(q_{\mathcal{M}}(\omega_2)\big)$, if and only if $q_{\mathcal{M}}(\omega_1) \precsim q_{\mathcal{M}}(\omega_2)$, if and only if $\omega_1 \precsim \omega_2$.

If $\omega = \omega_1 + \omega_2$ and $s(\omega_1) \perp s(\omega_2)$, then

$$\begin{aligned} q_{\mathcal{M}}(\omega) = s(\omega) \otimes e_{\omega,\omega} &= \big(s(\omega_1) + s(\omega_2)\big) \otimes e_{\omega,\omega} \\ &= s(\omega_1) \otimes e_{\omega,\omega} + s(\omega_2) \otimes e_{\omega,\omega} \\ &\sim s(\omega_1) \otimes e_{\omega_1,\omega_1} + s(\omega_2) \otimes e_{\omega_2,\omega_2} = q_{\mathcal{M}}(\omega_1) + q_{\mathcal{M}}(\omega_2), \end{aligned}$$

so that $\mu_\omega = \mu_{\omega_1} + \mu_{\omega_2}$. Q.E.D.

We now consider a one parameter group $\{\Theta_s\}$ of transformations in $\mathfrak{W}$ as follows:

$$\Theta_s \varphi = e^{-s}\varphi, \quad \varphi \in \mathfrak{W}. \tag{1}$$

Obviously, $\Theta_s\varphi \precsim \Theta_s\psi$ if and only if $\varphi \precsim \psi$. Hence $\{\Theta_s\}$ gives rise to a one parameter group of automorphisms of $\mathcal{A}$, because each order automorphism of $\mathrm{Proj}_\sigma(\mathcal{A})$ is canonically extended to an automorphism of $\mathcal{A}$, although the map: $s \in \mathbf{R} \mapsto \Theta_s(a) \in \mathcal{A}$, $a \in \mathcal{A}$, is highly discontinuous.

Definition 4.5. The flow $\{\mathcal{A}, \mathbf{R}, \Theta\}$ is called the *global flow of weights* on the factor $\mathcal{M}$.

To study the global flow of weights on a separable factor $\mathcal{M}$ of type III_λ, $0 \leq \lambda < 1$, we make a small detour to the comparison of two weights.

Definition 4.6. A *strictly* semi-finite weight on $\mathcal{M}$ is a weight φ of the form:

$$\varphi = \sum_i \omega_i,$$

where $\{\omega_i\}$ is a subset of $\mathcal{M}_*^+$ such that $\{s(\omega_i)\}$ is mutually orthogonal. This is equivalent to saying that φ is semi-finite on its centralizer $\mathcal{M}_\varphi$.

Lemma 4.7.

(i) *If $\omega \in \mathcal{M}_*^+$ is faithful and lacunary, then every maximal abelian subalgebra $\mathcal{A}$ of $\mathcal{M}_\omega$ is maximal abelian in $\mathcal{M}$. In particular, $\mathcal{M}_\omega' \cap \mathcal{M} = \mathcal{C}_\omega$.*

(ii) *If φ is a strictly semi-finite faithful normal weight on a factor $\mathcal{M}$ of type* III_λ, $0 \leq \lambda < 1$, *then $\mathcal{M}_\varphi$ contains some maximal abelian subalgebra of $\mathcal{M}$.*

PROOF:

(i) By assumption, we can choose $\delta > 0$ such that $\mathcal{M}_\omega = \mathcal{M}_\omega\big([-\delta, \delta]\big)$. Since $\mathcal{A}$ is fixed under $\{\sigma_t^\omega\}$, the relative commutant $\mathcal{A}^c = \mathcal{A}' \cap \mathcal{M}$ of $\mathcal{A}$ is globally invariant under $\{\sigma_t^\omega\}$, i.e. $\{\sigma_t^\omega\}$ acts on $\mathcal{A}^c$. Hence $\mathcal{A}^c$ is the smallest σ-weakly closed subspace of $\mathcal{M}$ which contains $\mathcal{A}^c \cap \mathcal{M}_\omega([\alpha - \delta/2, \alpha + \delta/2])$, $\alpha \in \mathbf{R}$. If $x \in \mathcal{M}_\omega([\alpha - \delta/2, \alpha + \delta/2]) \cap \mathcal{A}^c$, then x^*x and xx^* both belong to $\mathcal{M}_\omega([-\delta, \delta]) = \mathcal{M}_\omega$, so that x^*x and xx^* are both in $\mathcal{A}^c \cap \mathcal{M}_\omega = \mathcal{A}$. Hence with $x = u|x|$ the polar decomposition, the phase part u commutes with $|x|$, which means that $|x| = |x^*|$ because $u|x| = |x^*|u$. Hence $x^*x = xx^*$. We now consider the modular condition:

$$F(t) = \omega(\sigma_t^\omega(x^*)x), \qquad F(t + \mathrm{i}) = \omega(x\sigma_t^\omega(x^*))$$

with some $F \in \mathcal{A}(\mathbf{D})$. Since $\mathrm{Sp}_\omega(x^*) \subset [-\alpha - \delta/2, -\alpha + \delta/2]$ and F is a positive definite function on $\mathbf{R}$, we have a positive measure μ supported by $[-\alpha - \delta/2, -\alpha + \delta/2]$ such that

$$F(t) = \int_{-\alpha-\delta/2}^{-\alpha+\delta/2} \mathrm{e}^{\mathrm{i}st}\, \mathrm{d}\mu(s)\,.$$

We then have

$$F(0) = \mu\left(\left[-\alpha - \frac{\delta}{2},\ -\alpha + \frac{\delta}{2}\right]\right) = \omega(x^*x),$$

$$F(\mathrm{i}) = \omega(xx^*) = \int_{-\alpha-\delta/2}^{-\alpha+\delta/2} \mathrm{e}^{-s\alpha}\, \mathrm{d}\mu(s)$$

$$\geq \mathrm{e}^{\alpha-\delta/2}\mu\left(\left[-\alpha - \frac{\delta}{2},\ -\alpha + \frac{\delta}{2}\right]\right) = \mathrm{e}^{\alpha-\delta/2}\omega(x^*x).$$

If $\alpha > \delta$, then this forces $\omega(x^*x) = 0$, i.e. $x = 0$. Hence

$$\mathcal{A}^c \cap \mathcal{M}_\omega\left(\left[\alpha - \frac{\delta}{2},\ \alpha + \frac{\delta}{2}\right]\right) = \{0\}$$

for every $\alpha > \delta$, which implies $\mathcal{A}^c \subset \mathcal{M}_\omega([-\delta, \delta]) = \mathcal{M}_\omega$. Hence $\mathcal{A}^c = \mathcal{A}$.

(ii) Suppose $\varphi = \sum \omega_i$ with $\omega_i \in \mathcal{M}_*^+$ and $\{s(\omega_i)\}$ orthogonal. Let $e_i = s(\omega_i)$ and $\mathcal{M}_i = \mathcal{M}_{e_i}$. If $\mathcal{A}_i$ is a maximal abelian subalgebra of $\mathcal{M}_i$ for each i, then the von Neumann algebra $\mathcal{A}$ generated by $\{\mathcal{A}_i\}$ is in fact maximal abelian in $\mathcal{M}$. Hence the problem of finding a maximal abelian subalgebra of $\mathcal{M}$ in $\mathcal{M}_\varphi$ is reduced to the case where φ is bounded. Thus we may assume $\varphi \in \mathcal{M}_*^+$.

Type III$_\lambda$ case, $0 < \lambda < 1$: With $T = -2\pi/\mathrm{Log}\,\lambda$, σ_T^φ is inner, so we choose $h \in \mathcal{C}_\varphi^+$ such that $\mathrm{Ad}(h^{\mathrm{i}T}) = \sigma_T^\varphi$. Let $\psi = \varphi(h^{-1}\,\cdot\,)$. Then $\sigma_T^\psi = \mathrm{id}$, so that ψ is lacunary. Furthermore, we can choose an invertible h, so that ψ is bounded. Hence by (i) any maximal abelian subalgebra of $\mathcal{M}_\psi$ is maximal abelian in $\mathcal{M}$. Let $\mathcal{A}$ be a maximal abelian subalgebra of $\mathcal{M}_\psi$ which contains h. We then have, for any $a \in \mathcal{A}$,

$$\sigma_t^\varphi(a) = h^{\mathrm{i}t}\sigma_t^\psi(a)h^{-\mathrm{i}t} = h^{\mathrm{i}t}ah^{-\mathrm{i}t} = a,$$

so that $\mathcal{A} \subset \mathcal{M}_\varphi$. Hence we are done.

Type III_0 case: Let $\{e_i\}$ be a maximal orthogonal family of projections in $\mathcal{C}_\varphi$ such that $\mathcal{M}_{\varphi_{e_i}}$ contains a maximal abelian subalgebra $\mathcal{A}_i$ of $\mathcal{M}_{e_i}$. With $e = \sum e_i$, the von Neumann subalgebra $\mathcal{A}$ of $\mathcal{M}_e$ generated by $\{\mathcal{A}_i\}$ is maximal abelian in $\mathcal{M}_e$ and contained in $\mathcal{M}_{\varphi_e}$. Hence to complete the proof, we have only to show that for any faithful $\varphi \in \mathcal{M}_*^+$ there exists a non-zero projection $e \in \mathcal{C}_\varphi$ such that $\mathcal{M}_{\varphi_e}$ contains a maximal abelian subalgebra of $\mathcal{M}_e$. To this end, we shall show that there exist a projection $e \in \mathcal{C}_\varphi$ and an invertible $h \in \mathcal{C}_{\varphi_e}^+$ such that $\varphi_h = \varphi(h\,\cdot\,)$ is lacunary on $\mathcal{M}_e$. Then the arguments in the type III_λ case shows the existence of a maximal abelian subalgebra of $\mathcal{M}_e$ contained in $\mathcal{M}_{\varphi_e}$.

Since $S(\mathcal{M}) = \{0, 1\}$, there exists a non-zero $e \in \text{Proj}(\mathcal{C}_\varphi)$ such that $\text{Sp}(\sigma^{\varphi_e}) \cap \big([-2\mu, -\mu] \cup [\mu, 2\mu]\big) = \emptyset$ for some $\mu > 0$. Replacing φ by φ_e, we assume that $\text{Sp}(\sigma^{\varphi}) \cap \big([-2\mu, -\mu] \cup [\mu, 2\mu]\big) = \emptyset$. It then follows that $\mathcal{N} = \mathcal{M}_\varphi\big([-\mu, \mu]\big)$ is a von Neumann subalgebra of $\mathcal{M}$. Then we have $\text{Sp}\big(\sigma^{\varphi|\mathcal{N}}\big) \subset [-\mu, \mu]$. By Theorem XI.1.10, $\sigma_t^{\varphi|\mathcal{N}} = \text{Ad}(e^{itb})$, where

$$b = \int_{-\infty}^{\infty} \lambda \, de(\lambda) \in \mathcal{C}_\varphi$$

and $1 - e(\lambda)$ is the left support $q(\lambda)$ of $\mathcal{M}_\varphi\big([\lambda, +\infty]\big)$. Since $1 \in \mathcal{M}_\varphi\big(\{0\}\big)$, $e(\lambda) = 0$ for $\lambda \le 0$, and $e(\lambda) = 1$ for $\lambda > \mu$. Hence $0 \le b \le \mu$. Put $h = b - \mu/2$. Then we have $-\mu/2 \le h \le \mu/2$ and $\sigma_t^\varphi = \text{Ad}\big(e^{ith}\big)$ on $\mathcal{P}$. Let $k = e^{-h}$. Then $\psi = \varphi(k\,\cdot\,)$ is a faithful normal finite trace on $\mathcal{P}$. We now have

$$\text{Sp}\big(\sigma^\psi|_{\mathcal{M}_\varphi([-\mu,\mu])}\big) = \{0\};$$

$$\text{Sp}\big(\sigma^\psi|_{\mathcal{M}_\varphi(]-\infty,-2\mu])}\big) \subset \big]-\infty, \mu\big];$$

$$\text{Sp}\big(\sigma^\psi|_{\mathcal{M}_\varphi([2\mu,+\infty[)}\big) \subset \big[\mu, +\infty\big[,$$

so that $\text{Sp}(\sigma^\psi) \subset \big]-\infty, \mu\big] \cup \{0\} \cup \big[\mu, +\infty\big[$. Hence ψ is lacunary. Q.E.D.

In the above arguments, we have proved the following useful fact:

Lemma 4.8. *If $\mathcal{M}$ is a factor of type III_λ, $0 \le \lambda < 1$, then for any semi-finite normal weight φ on $\mathcal{M}$ there exists a non-zero $h \in \mathcal{C}_\varphi^+$ such that φ_h is lacunary and h is invertible on $s(h)$.*

Lemma 4.9. *If $\mathcal{M}$ is a factor of type III_λ, $0 \le \lambda < 1$, then the following two conditions for faithful strictly semi-finite normal weights φ and ψ on $\mathcal{M}$ are equivalent:*

(i) $\mathcal{M}_\varphi \subset \mathcal{M}_\psi$;

(ii) *There exists a non-singular positive self-adjoint operator h affiliated with $\mathcal{C}_\varphi$ such that*

$$\psi = \varphi(h\,\cdot\,).$$

PROOF: The implication, (ii) $\Longrightarrow$ (i), is a triviality.

(i) $\Longrightarrow$ (ii): Let $u_t = (\mathrm{D}\psi : \mathrm{D}\varphi)_t$. By Lemma 4.7, we know that $\mathcal{M}'_\varphi \cap \mathcal{M} = \mathcal{C}_\varphi$. For each $x \in \mathcal{M}_\varphi$, we have

$$x = \sigma_t^\psi(x) = u_t \sigma_t^\varphi(x) u_t^* = u_t x u_t^*,$$

so that $u_t \in \mathcal{C}_\varphi$. Thus $\{u_t\}$ is a one parameter unitary group in $\mathcal{C}_\varphi$, and so there exists h affiliated with $\mathcal{C}_\varphi$ with $u_t = h^{it}$, $t \in \mathbf{R}$. This implies that $\psi = \varphi(h\,\cdot\,)$.

Q.E.D.

We are now ready to describe the global flow of weights on a factor of type III_λ, $0 \leq \lambda < 1$, in terms of the discrete decomposition.

Let $\mathcal{M}$ be a separable factor of type III_λ, $0 \leq \lambda < 1$. We fix a faithful lacunary weight φ with infinite multiplicity. By Theorems 2.1 and 3.10, we have:

a) $\mathcal{M}_\varphi$ is of type II_∞;
b) there exists a unitary $U \in \mathcal{M}$ such that

$$\begin{gathered} U\mathcal{M}_\varphi U^* = \mathcal{M}_\varphi, \qquad \varphi \circ \mathrm{Ad}(U) \leq \lambda_0 \varphi, \quad 0 < \lambda_0 < 1; \\ \mathcal{M} \cong \mathcal{M}_\varphi \rtimes_\theta \mathbf{Z}, \quad \text{where } \theta = \mathrm{Ad}(U)\big|_{\mathcal{M}_\varphi}; \\ \varphi = (\varphi|_{\mathcal{M}_\varphi})\widehat{\ }. \end{gathered} \tag{2}$$

If $\lambda \neq 0$, then $\mathcal{M}_\varphi$ is a factor and $\varphi \circ \mathrm{Ad}(U) = \lambda\varphi$. In any case, φ is a faithful semi-finite normal trace on $\mathcal{M}_\varphi$, so that there exists $\rho \in \mathcal{C}_\varphi^+$ such that

$$\varphi \circ \mathrm{Ad}(U) = \varphi(\rho\,\cdot\,), \quad 0 \leq \rho \leq \lambda_0.$$

Let $\tau = \varphi|_\mathcal{N}$ be the restriction of φ to $\mathcal{N} = \mathcal{M}_\varphi$. With these notations, we have the following:

Theorem 4.10.

(i) *If $h \in \mathcal{N}_+$ satisfies the conditions that*
 a) $\rho s(h) \leq h < 1$,
 b) $1 - h$ *is non-singular,*
 then

$$\mathcal{M}_{\varphi_h} = \mathcal{N}_{\tau_h} = \{h\}' \cap \mathcal{N}_{s(h)}, \tag{3}$$

where φ_h and τ_h are respectively defined by:

$$\varphi_h = \varphi(h\,\cdot\,) \quad \text{and} \quad \tau_h = \tau(h\,\cdot\,). \tag{4}$$

(ii) *For each weight ψ on $\mathcal{M}$, there exists $h \in \mathcal{N}_+$ satisfying* (a) *and* (b) *in* (i) *such that $\psi \sim \varphi_h$.*
(iii) *The weight $\tau_h = \varphi_h|_\mathcal{N}$ in* (ii) *is unique within the equivalence in $\mathfrak{W}(\mathcal{N})$.*

Before the proof, we shall make a comment. This theorem reduces the comparison problem in $\mathfrak{W}(\mathcal{M})$ to that in the subclass of $\mathfrak{W}(\mathcal{N})$ determined by (a) and (b) in (i), which is surprisingly independent of the automorphism θ, but depends only on ρ.

We now proceed the proof by preparing a few lemmas. We keep the above notations.

Lemma 4.11. *If* $\psi \in \mathfrak{W}_{\text{inf}}(\mathcal{M})$, *then there exists* $h \in \mathcal{N}_+$ *such that* $\varphi_h \precsim \psi$.

PROOF: By Lemma 4.8, there exists $k \in \mathfrak{C}_\psi^+$ such that ψ_k is lacunary. Since $\mathcal{M}_{\psi,s(k)} \subset \mathcal{M}_{\psi_k}$, ψ_k is of infinite multiplicity. In the type III_λ case, $0 < \lambda < 1$, ψ_k is then a generalized trace on $\mathcal{M}_{s(k)}$ and $s(k) \sim 1$. Hence by Theorem 2.2, there exist $u \in \mathcal{M}$ and α, $\lambda < \alpha \leq 1$, such that $u^*u = 1$, $uu^* = s(k)$ and $\psi_k = \alpha\varphi_{u^*}$. Since k is invertible on $s(k)$, u^*ku is invertible in $\mathcal{N}$. Put $h = \alpha(u^*ku)^{-1}$. We then have

$$\varphi(hu^*xu) = \psi(s(k)x), \quad x \in \mathcal{M},$$

so that $\varphi_h \sim \psi_{s(k)}$. Hence $\varphi_h \precsim \psi$.

Type III_0 case: Since ψ_k is lacunary and of infinite multiplicity, Theorems 3.7 and 3.10 tell us that there exist non-zero $e \in \mathrm{Proj}(\mathfrak{C}_{\psi_k})$, $f \in \mathrm{Proj}(\mathfrak{C}_\varphi)$ and a partial isometry $v \in \mathcal{M}$ such that

$$\begin{cases} v^*v = e, & vv^* = f, \\ v\mathcal{M}_{\psi_k}v^* = \mathcal{M}_{\varphi_f} & \text{and} \quad v^*\mathcal{M}_{\varphi_f}v = \mathcal{M}_{\psi_{ke}}. \end{cases}$$

Replacing φ by φ_v, we see that φ and ψ_{ke} have the same centralizer, and that they are both lacunary weights of infinite multiplicity. Thus, by Lemma 4.8 there exists h_1 affiliated with $\mathfrak{C}_\varphi$ such that $\varphi_{h_1} = \psi_{ke}$. Reducing h_1 by a spectral projection e_1, we get a bounded operator $h = k^{-1}h_1e_1$ (recall that k is invertible on $s(k)$), so that $\psi_{e_1} = \varphi_h$. This is so after replacing φ by φ_v; hence we conclude that $\varphi_{v^*hv} \prec \psi$.

Q.E.D.

In order to simplify our discussion, we introduce a new notation for the rest of this section. For two positive operators k_1 and k_2, the inequality symbol $k_1 < k_2$ indicates the fact that $k_1 \leq k_2$ in the usual sense and $k_2 - k_1$ is non-singular on the support of k_2.

Before moving into the next formal discussion, we consider simple examples. In $\mathcal{B} = \mathcal{L}(\mathfrak{K})$, a factor of type I_∞, every weight is given by a positive self-adjoint operator. Hence the comparison of weights on $\mathcal{B}$ is precisely that of positive self-adjoint operators, which is exactly the theory of spectral multiplicity. If we consider only the infinite multiplicity case, then the comparison of h and k, $h, k \geq 0$, is exactly that of the spectral measures in the sense of absolute continuity. Hence the global flow of weights on $\mathcal{B}$ is the action of $\mathbf{R}$ on $C_\infty(\mathbf{R}_+)^{**}$ induced by $\lambda \mapsto e^{-s}\lambda$, $\lambda \in \mathbf{R}_+$, $s \in \mathbf{R}$. In the type II case, the situation is somewhat complicated. But the basic idea remains unchanged. Namely, we are considering the comparison of positive self-adjoint operators affiliated with the algebra. The spectral measure of the operator provides the first invariant for the equivalence.

Returning to our original situation, we are now going to consider the global flow of weights on the von Neumann algebra $\mathcal{N}$ of type II. Let $\mathcal{A}_{\mathcal{N}}$ denote the carrier algebra of the global flow of weights on $\mathcal{N}$.

Lemma 4.12. *For each $\psi \in \mathfrak{W}(\mathcal{N})$, let h_ψ be the positive operator affiliated with $\mathcal{N}$ such that $\psi = \tau_{h_\psi}$. If δ_1 and δ_2 are positive operators affiliated with the center $\mathcal{C}$ of $\mathcal{N}$ such that $\delta_1 < \delta_2$, then there exists a unique projection $[\delta_1, \delta_2[\in$ $\mathrm{Proj}(\mathcal{A}_{\mathcal{N}})$ such that*

$$p_{\mathcal{N}}(\psi) \leq \big[\delta_1, \delta_2\big[\iff \delta_1 s(\psi) \leq h_\psi < \delta_2 s(\psi)$$

for $\psi \in \mathfrak{W}_{\mathrm{inf}}(\mathcal{N})$.

PROOF: Let $\mathcal{E} = \big\{\psi \in \mathfrak{W}_{\mathrm{inf}} : s(h_\psi)\delta_1 \leq h_\psi < s(h_\psi)\delta_2\big\}$. Suppose that $\psi_1, \psi_2 \in \mathfrak{W}_{\mathrm{inf}}$ are equivalent and $\psi_1 \in \mathcal{E}$. Then there exists $w \in \mathcal{N}$ such that $\psi_2 = \psi_{1,w}$, $ww^* = s(\psi_1) = s(h_{\psi_1})$ and $w^*w = s(\psi_2) = s(h_{\psi_2})$. The equality $\psi_2 = \psi_{1,w}$ means that $w^* h_{\psi_1} w = h_{\psi_2}$. It then follows that

$$s(\psi_2)\delta_1 = w^* w \delta_1 = w^* \delta_1 w \leq w^* h_{\psi_1} w = h_{\psi_2};$$

$$h_{\psi_2} = w^* h_{\psi_1} w \leq w^* s(h_{\psi_1})\delta_2 w = s(h_{\psi_2})\delta_2.$$

Furthermore, we have

$$s(h_{\psi_2})\delta_2 - h_{\psi_2} = w^*\big(s(h_{\psi_1})\delta_2 - h_{\psi_1}\big)w,$$

so that $s(h_{\psi_2})\delta_2 - h_{\psi_2}$ is non-singular on the support of $s(h_{\psi_2})\delta_2$ by the assumption on h_{ψ_1}. Hence $h_{\psi_2} < s(h_{\psi_2})\delta_2$. Thus $\psi_2 \in \mathcal{E}$. Now, suppose that $\psi_1 \in \mathcal{E}$ and $\psi_2 = \psi_{1,e}$ for $e \in \mathrm{Proj}(\mathcal{M}_{\psi_1})$. It follows that

$$h_{\psi_2} = h_{\psi_1} e = e h_{\psi_1}; \qquad s(\psi_2) = s(\psi_1)e = e$$

and

$$s(\psi_2)\delta_1 = e s(\psi_1)\delta_1 \leq e h_{\psi_1} e \leq s(\psi_2)\delta_2.$$

Furthermore, $\delta_2 - h_{\psi_2} = \delta_2 - h_{\psi_1} e \geq \delta_2 - h_{\psi_1}$, so that $\delta_2 s(\psi_2) - h_{\psi_2}$ is non-singular on $\delta_2 s(\psi_2)$, so that $h_{\psi_2} < s(\psi_2)\delta$. Hence ψ_2 belongs to $\mathcal{E}$.

Let $\{\psi_n\}$ be a sequence in $\mathcal{E}$ with pairwise orthogonal supports, and put $\psi = \sum_{n=1}^{\infty} \psi_n$. We then have

$$h_\psi = \sum_{n=1}^{\infty} h_{\psi_n}, \qquad s(\psi) = \sum_{n=1}^{\infty} s(\psi_n);$$

$$s(\psi)\delta_1 = \sum_{n=1}^{\infty} s(\psi_n)\delta_1 \leq \sum_{n=1}^{\infty} h_{\psi_n} < \sum_{n=1}^{\infty} s(\psi_n)\delta_2 = s(\psi)\delta_2,$$

where the infinite sum of pairwise orthogonal self-adjoint operators $\{h_{\psi_n}\}$ should be naturally understood. Thus ψ belongs to $\mathcal{E}$. Therefore, the set $\{\, p_{\mathcal{N}}(\psi) : \psi \in \mathcal{E}\,\}$ is precisely the set of all σ-finite projections in $\mathcal{A}_{\mathcal{N}}$ majorized by $\bigvee\{\, p_{\mathcal{N}}(\psi) : \psi \in \mathcal{E}\,\} = [\delta_1, \delta_2[\;\in \mathcal{A}_{\mathcal{N}}$. Q.E.D.

Lemma 4.13. *Let $\mathcal{P}$ be a general von Neumann algebra and $\mathcal{E}$ be a faithful normal conditional expectation from $\mathcal{P}$ onto a von Neumann subalgebra $\mathcal{Q}$.*

(i) *For any weight ψ on $\mathcal{Q}$, $\psi \circ \mathcal{E}$ is a weight on $\mathcal{P}$ with the same support as ψ.*
(ii) *If ψ_1 and ψ_2 are weights on $\mathcal{Q}$ with $\psi_1 \prec \psi_2$, then $\psi_1 \circ \mathcal{E} \prec \psi_2 \circ \mathcal{E}$.*
(iii) *If $\{\psi_n\}$ is a sequence of weights on $\mathcal{Q}$ with pairwise disjoint supports, then*

$$\sum_{j=1}^{\infty}(\psi_j \circ \mathcal{E}) = \Big(\sum_{j=1}^{\infty}\psi_j\Big) \circ \mathcal{E}.$$

PROOF:
(i) Let $e = s(\psi)$. Then $\mathcal{E}(e) = e$ and $\psi\big(\mathcal{E}(1-e)\big) = 0$. If $x \in \mathcal{P}_e^+$, then $\mathcal{E}(x) \in \mathcal{Q}_e^+$, so that $\psi \circ \mathcal{E}$ is faithful on $\mathcal{P}_e$. Thus $s(\psi \circ \mathcal{E}) = e$.
(ii) Suppose $u \in \mathcal{Q}$ is a partial isometry such that $uu^* \in \mathcal{Q}_{\psi_2}$ and $\psi_{2,u} = \psi_1$. Since $\sigma_t^{\psi_2} = \sigma_t^{\psi_2\circ\mathcal{E}}|_{\mathcal{Q}}$ by Theorem IX.4.2, $\mathcal{Q}_{\psi_2} = \mathcal{P}_{\psi_2\circ\mathcal{E}} \cap \mathcal{Q}$, so that $uu^* \in \mathcal{P}_{\psi_2\circ\mathcal{E}}$. Now, we have, for each $x \in \mathcal{P}_+$, $\psi_2 \circ \mathcal{E}(uxu^*) = \psi_2\big(u\mathcal{E}(x)u^*\big) = \psi_1 \circ \mathcal{E}(x)$; thus $\psi_1 \circ \mathcal{E} \prec \psi_2 \circ \mathcal{E}$.
(iii) This is easy. We leave the proof to the reader. Q.E.D.

Lemma 4.14. *Let $\mathcal{M}$, $\mathcal{N}$, U, φ and ρ be as in Theorem* 4.10.

(i) *If $h \in \mathcal{N}_+$ satisfies the conditions of Theorem* 4.10.(i), *then with $\psi = \varphi_h$ we have $\mathcal{M}_\psi \subset \mathcal{M}_\varphi = \mathcal{N}$.*
(ii) *If ψ is the weight of* (i), *then for any $e \in \mathrm{Proj}(\mathcal{M}_\psi)$ there exists $k \in \mathcal{N}_+$ such that $\psi_e = \varphi_k$ and $\rho s(k) \le k < 1$.*
(iii) *Suppose that $\psi_i = \varphi_{h_i}$, $i = 1, 2$, with $h_1, h_2 \in \mathcal{N}_+$ satisfying the condition in* (i). *If $\psi_2 = \psi_{1,u}$ for a partial isometry $u \in \mathcal{M}$ with $uu^* = s(\psi_1)$ and $u^*u = s(\psi_2)$, then u belongs to $\mathcal{N}$.*

PROOF: The assertion (i) is nothing but Theorem 4.10.(i).
(iii) Put $k_j = \rho\big(1 - s(h_j)\big) + h_j$, $j = 1, 2$. Then $\rho \le k_j < 1$. By Lemma 4.3.(i), we have

$$\begin{aligned} uk_2^{it} &= us(\psi_2)k_2^{it} = u\,(\mathrm{D}\psi_2 : \mathrm{D}\varphi)_t = uu^*\,(\mathrm{D}\psi_1 : \mathrm{D}\varphi)_t\,\sigma_t^{\varphi}(u) \\ &= s(\psi_1)h_1^{it}\sigma_t^{\varphi}(u) = k_1^{it}s(\psi_1)\sigma_t^{\varphi}(u) = k_1^{it}\sigma_t^{\varphi}(u). \end{aligned}$$

Hence we get

$$uk_2^{it} = k_1^{it}\sigma_t^{\varphi}(u), \qquad t \in \mathbf{R}.$$

With $\theta = \mathrm{Ad}(U)$, we have $\varphi \circ \theta = \varphi_\rho$. For each $n \in \mathbf{Z}$, we define ρ_n to be the operator affiliated with $\mathcal{C}_\varphi = \mathcal{C}_{\mathcal{N}}$ such that $\varphi \circ \theta^n = \varphi(\rho_n \,\cdot\,)$. We then have $\rho_{n+m} = \rho_n \theta^{-n}(\rho_m)$. Hence $\sigma_t^\varphi(U^n) = U^n \rho_n^{\mathrm{i}t}$, and $\rho_n < \rho = \rho_1$ for $n > 0$.

Recalling that $\mathcal{M} = \mathcal{N} \rtimes_\theta \mathbf{Z}$, we choose a sequence $\{x_n\}$ in $\mathcal{N}$ so that $u = \sum_{n=-\infty}^{\infty} x_n U^n$ in the locally convex topology induced by the semi-norms: $x \mapsto \hat{\omega}(x^* x)^{1/2}$, $\omega \in \mathcal{N}_*^+$. We now have

$$k_1^{\mathrm{i}t} \sigma_t^\varphi(u) = k_1^{\mathrm{i}t} \sum_{n=-\infty}^{\infty} x_n U^n \rho_n^{\mathrm{i}t} = \sum_{n=-\infty}^{\infty} k_1^{\mathrm{i}t} x_n U^n \rho_n^{\mathrm{i}t} = \sum_{n=-\infty}^{\infty} k_1^{\mathrm{i}t} x_n \theta^n(\rho_n^{\mathrm{i}t}) U^n$$

$$u k_2^{\mathrm{i}t} = \sum_{n=-\infty}^{\infty} x_n U^n k_2^{\mathrm{i}t} = \sum_{n=-\infty}^{\infty} x_n \theta^n(k_2^{\mathrm{i}t}) U^n.$$

Hence by the uniqueness of the expansion, we have

$$k_1^{\mathrm{i}t} x_n \theta^n(\rho_n^{\mathrm{i}t}) = x_n \theta^n(k_2^{\mathrm{i}t}), \quad n \in \mathbf{Z}, \quad t \in \mathbf{R};$$

$$k_1^{\mathrm{i}t} x_n = x_n \theta^n(k_2^{\mathrm{i}t} \rho_n^{-\mathrm{i}t}).$$

We shall show that $k_1^{\mathrm{i}t} y = y \theta^n(k_2^{\mathrm{i}t} \rho_n^{-\mathrm{i}t})$, $t \in \mathbf{R}$, implies $y = 0$ for $y \in \mathcal{N}$ if $n \neq 0$. If this were done, then $u = x_0 \in \mathcal{N}$ as we want. Making use of the polar decomposition of y, we may and do assume that y is a partial isometry of $\mathcal{N}$ such that $y^* y$ commutes with $\theta^n(k_2^{\mathrm{i}t} \rho_n^{-\mathrm{i}t})$ and yy^* commutes with $k_1^{\mathrm{i}t}$. Assume $n > 0$. We have

$$k_1^{\mathrm{i}t} yy^* = y \theta^n\big(k_2^{\mathrm{i}t} \rho_n^{-\mathrm{i}t}\big) y^*, \quad t \in \mathbf{R}.$$

Since $\rho_n \le \rho \le k_2$, the right hand side of the above equality extends to a holomorphic function on the upper half plane: $z \mapsto F(z) = y\theta^n\big(k_2^{\mathrm{i}z} \rho_n^{-\mathrm{i}z}\big) y^*$ with the bound

$$\big\| F(z) \big\| \le \big\| k_2^{\mathrm{i}z} \rho_n^{-\mathrm{i}z} \big\| = \big\| (k_2^{-1}\rho)^{-\mathrm{i}z} \big\| \le 1$$

for $\mathrm{Im}\, z \ge 0$. But $k_1 < 1$, so that the left hand side $k_1^{\mathrm{i}t} yy^*$ extends to a bounded holomorphic function on the lower half plane. By the Liouville theorem, the function: $t \mapsto k_1^{\mathrm{i}t} yy^*$ must be constant, which is possible only in the case that $yy^* = 0$. Next, let $n < 0$. We then use $k_1^{\mathrm{i}t} \theta^n(\rho_n^{\mathrm{i}t}) y = y \theta^n(k_2^{\mathrm{i}t})$, $t \in \mathbf{R}$. As $y^* y$ commutes with $\theta^n(k_2^{\mathrm{i}t})$, the above arguments apply, provided that $k_1 \theta^n(p_n) \ge 1$. But the cocycle identity $1 = \rho_0 = \rho_{-n} \theta^n(\rho^n)$ implies $\theta^n(\rho_n) = \rho_{-n}^{-1}$, so that $k_1 \theta^n(\rho_n) \ge 1$ for $n < 0$ because $\rho_{-n} \le \rho_1 \le k_1$. This completes the proof for (iii).

If u is a partial isometry in $\mathcal{M}_\psi$ such that $uu^* = s(\psi) = uu^*$, then $\psi = \psi_u$, so that $u \in \mathcal{N}$ by (iii). Thus $\mathcal{M}_\psi \subset \mathcal{N} = \mathcal{M}_\varphi$.

(ii) If $e \in \mathcal{M}_\psi$, then e belongs to $\mathcal{N}$ by (i), so that e commutes with h. Hence we have

$$\psi_e = \varphi(he \,\cdot\,) \quad \text{and} \quad \rho s(eh) \le eh \le h < 1. \qquad \text{Q.E.D.}$$

Proof of Theorem 4.10: We use the notations in the theorem and others established above. We denote by $\mathcal{E}$ the conditional expectation from $\mathcal{M}$ onto $\mathcal{N} = \mathcal{M}_\varphi$.

We first prove that $\sum_{n=-\infty}^{\infty}\big[\rho_n, \rho_{n-1}\big[= 1$ in $\mathcal{A}_{\mathcal{N}}$. Let $\psi = \tau_h$ with h a self-adjoint positive operator affiliated to $\mathcal{N}$. Then all ρ_n's and h commute, and $\rho_n = \rho_{n-1}\theta^{-(n-1)}(\rho) \leq \lambda_0\rho_{n-1}$ with λ_0 of (2), so that there exists an orthogonal sequence $\{e_n\}_{n\in\mathbf{Z}}$ in $\mathrm{Proj}(\mathcal{N})$ commuting with h (in fact spectral projections of h), such that

$$\rho_n e_n \leq h e_n < \rho_{n-1} e_n, \qquad n \in \mathbf{Z}.$$

Therefore, any weight ψ on $\mathcal{N}$ is decomposed into an orthogonal sum $\psi = \sum_{n=-\infty}^{\infty}\psi_n$ such that $\psi_n = \tau_{h_n}$ with h_n satisfying $s(h_n)\rho_n \leq h_n < s(h_n)\rho_{n-1}$. Hence $p_{\mathcal{N}}(\psi) = \bigvee p_{\mathcal{N}}(\psi_n) \leq \bigvee\big[\rho_n, \rho_{n-1}\big[$ for every $\psi \in \mathfrak{W}_{\mathrm{inf}}(\mathcal{N})$. Thus $\bigvee_{n\in\mathbf{Z}}\big[\rho_n, \rho_{n-1}\big[= 1$. Suppose $n \neq m$. If $h_1, h_2 \in \mathcal{N}_+$ with $s(h_1)\rho_n \leq h_1 < \rho_{n-1}$ and $s(h_2)\rho_m \leq h_2 < \rho_{m-1}$, then no non-zero subweights[6] of $\psi_1 = \tau_{h_1}$ and $\psi_2 = \tau_{h_2}$ are equivalent. Hence $p_{\mathcal{N}}(\psi_1)$ and $p_{\mathcal{N}}(\psi_2)$ are orthogonal whenever $\psi_1, \psi_2 \in \mathfrak{W}_{\mathrm{inf}}(\mathcal{N})$. Therefore, $\big[\rho_n, \rho_{n-1}\big[$ and $\big[\rho_m, \rho_{m-1}\big[$ are orthogonal.

For any $e \in \mathrm{Proj}_\sigma(\mathcal{A}_{\mathcal{N}})$, put $I(e) = p_{\mathcal{M}}(\psi \circ \mathcal{E})$ where $\psi \in \mathfrak{W}_{\mathrm{inf}}(\mathcal{N})$ such that $p_{\mathcal{N}}(\psi) = e$.

By Lemma 4.13.(ii), $I(e)$ does not depend on the choice of ψ, and if $e_1 \leq e_2$, $e_1, e_2 \in \mathrm{Proj}_\sigma(\mathcal{A}_{\mathcal{N}})$, then $I(e_1) \leq I(e_2)$. Let $\{e_n\}$ be a sequence in $\mathrm{Proj}_\sigma(\mathcal{A}_{\mathcal{N}})$, and choose $\psi_n \in \mathfrak{W}_{\mathrm{inf}}(\mathcal{N})$ so that $p_{\mathcal{N}}(\psi_n) = e_n$ and $\{s(\psi_n)\}$ is mutually orthogonal. Then we have, by Lemma 4.13.(iii),

$$\bigvee I(e_n) = \bigvee p_{\mathcal{M}}(\psi_n \circ \mathcal{E}) = p_{\mathcal{M}}\Big(\sum \psi_n \circ \mathcal{E}\Big) = p_{\mathcal{M}}\Big(\Big(\sum \psi_n\Big) \circ \mathcal{E}\Big) = I\Big(\bigvee e_n\Big).$$

We claim that if $e \leq \big[\rho_n, \rho_{n-1}\big[$, $e \in \mathrm{Proj}_\sigma(\mathcal{A}_{\mathcal{N}})$, then there exists $f \leq [\rho_1, 1[$, $f \in \mathrm{Proj}_\sigma(\mathcal{A}_{\mathcal{N}})$ such that $I(e) = I(f)$.

By assumption, $e = p_{\mathcal{N}}(\psi)$ for some $\psi \in \mathfrak{W}_{\mathrm{inf}}(\mathcal{N})$ of the form $\psi = \tau_h$ with h satisfying $s(h)\rho_n \leq h < \rho_{n-1}$. For every $x \in \mathcal{M}_+$, we have

$$\varphi(hU^k x U^{-k}) = \varphi_{U^k}\big(\theta^{-k}(h)x\big) = \varphi\big(\rho_k\theta^{-k}(h)x\big),$$

so that $I(e) = I_{f_k}$ where $f_k = p_{\mathcal{N}}\big(\tau_{\rho_k\theta^{-k}(h)}\big)$. The inequality for h then implies

$$\theta^{-k}\big(s(h)\rho_n\big) \leq \theta^{-k}(h) < \theta^{-k}(\rho_{n-1});$$

hence

$$\rho_{n+k}\theta^{-k}\big(s(h)\big) \leq \rho_k\theta^{-k}(h) < \rho_k\theta^{-k}(\rho_{n-1}) = \rho_{n+k-1}.$$

Therefore, with $k = 1 - n$, we get $f = f_{1-k} \leq [\rho_1, 1[$ and $I(f) = I(e)$.

Applying Lemma 4.11, we now conclude that for any $g \in \mathrm{Proj}_\sigma(\mathcal{A}_{\mathcal{M}})$, there exists $e \in \mathrm{Proj}_\sigma(\mathcal{N})$ such that $I(e) = g_1 \leq g$ and $e \neq 0$. Choose $n \in \mathbf{Z}$ with $e_1 = e\big[\rho_n, \rho_{n-1}\big[\neq 0$. Then Lemma 4.13.(ii) shows that $I(e_1) \leq I(e) = g_1$, so that $g_2 = I(e_1) = I(f)$ for some $f \leq [\rho_1, 1[$. Therefore, the usual exhaustion arguments show that any $g \in \mathrm{Proj}_\sigma(\mathcal{A}_{\mathcal{M}})$ is of the form $g = \sum_{n=1}^{\infty} I(f_n)$ where $\{f_n\}$ is an orthogonal sequence in $\mathrm{Proj}_\sigma(\mathcal{A}_{\mathcal{N}})$ with $f_n \leq [\rho_1, 1[$. With $f = \sum f_n$, we have $I(f) = g$ and $f \leq [\rho_1, 1[$. This completes the proof for the existence part (ii). The rest of the claims of the theorem follows from Lemma 4.14. Q.E.D.

---

6 A *subweight* of a weight ψ on $\mathcal{P}$ means a weight ψ_e with $e \in \mathrm{Proj}(\mathcal{P}_\psi)$.

We state now a few immediate consequences of Theorem 4.10. To this end, we extend the map I in the above proof as follows:

$$I(e) = \bigvee \{ I(f) : f \in \mathrm{Proj}_\sigma(\mathcal{A}_\mathcal{N}),\ f \le e \}, \quad e \in \mathrm{Proj}(\mathcal{A}_\mathcal{N}). \tag{5}$$

It then follows that $I\big(\bigvee e_\alpha\big) = \bigvee I(e_\alpha)$ for any family $\{e_\alpha\}$ in $\mathrm{Proj}(\mathcal{A}_\mathcal{N})$.

Corollary 4.15. *With the above notations, we conclude the following:*

(i) *For each $n \in \mathbf{Z}$, the mapping I gives rise to an isomorphism from $\mathcal{A}_{\mathcal{N}[\rho_n, \rho_{n-1}[}$ onto $\mathcal{A}_\mathcal{M}$.*

(ii) *Let $\overline{\theta}$ be the automorphism of $\mathcal{A}_\mathcal{N}$ corresponding to θ, i.e., $\overline{\theta}\big(p_\mathcal{N}(\psi)\big) = p_\mathcal{N}\big(\psi \circ \theta^{-1}\big)$. Then*

$$I \circ \overline{\theta} = I;$$

$$\overline{\theta}\big([\rho_n, \rho_{n-1}[\big) = \big[\rho_{n-1}, \rho_{n-2}\big[.$$

(iii) *The map I induces an isomorphism denoted by I again from $(\mathcal{A}_\mathcal{N})^{\overline{\theta}}$ onto $\mathcal{A}_\mathcal{M}$ intertwining the flow Θ:*

$$I\Theta_t^\mathcal{N} I^{-1} = \Theta_t^\mathcal{M}, \quad t \in \mathbf{R}.$$

The proof is straightforward. So we leave it to the reader as an exercise.

Corollary 4.16. *If $\mathcal{M}$ is a separable factor of type III_λ, $0 \le \lambda < 1$, then there exists a faithful normal state ω such that $\mathcal{M}_\omega$ is maximal abelian in $\mathcal{M}$.*

PROOF: We use the above notations. Choose $e \in \mathrm{Proj}(\mathcal{N})$ with $\tau(e) = 1$, and an $h \in \mathcal{N}_+$ such that $\rho e \le h < e$ and $\{h\}''$ is maximal abelian in $\mathcal{N}_e$. Let $\omega = \varphi_h$. Then $\mathcal{M}_\omega$ is maximal abelian in $\mathcal{M}_e$ by the theorem. Furthermore, ω is faithful on $\mathcal{M}_e$, which is isomorphic to $\mathcal{M}$ because $\mathcal{M}$ is of type III. Q.E.D.

Corollary 4.17. *If $\mathcal{M}$ is a separable factor of type III_0, then for any faithful weight ψ on $\mathcal{M}$, we have*

$$\mathcal{C}'_\psi \cap \mathcal{M} = \mathcal{M}_\psi \qquad \text{and} \qquad \mathcal{M}'_\psi \cap \mathcal{M} = \mathcal{C}_\psi.$$

PROOF: The automorphism θ of $\mathcal{N}$ acts freely on the center $\mathcal{C} = \mathcal{C}_\mathcal{N}$ of $\mathcal{N}$ by the ergodicity of θ on $\mathcal{C}$ in Theorem 3.10. By Theorem 4.10, we may assume that ψ is of the form $\psi = \varphi_h$, $\rho s(h) \le h < 1$, and $\mathcal{M}_\psi \subset \mathcal{N}_{s(h)}$. Although the original weight ψ is faithful, the new weight φ_h may fail to be faithful. But with $e =$ the central support of $s(h)$ in $\mathcal{N}$, φ_e is lacunary and of infinite multiplicity. Hence we may replace φ by φ_e. Now we have $\mathcal{C}_{s(h)} \subset \mathcal{M}_\psi \subset \mathcal{N}_{s(h)}$, so that $\mathcal{C}_{s(h)} \subset \mathcal{C}_\psi$. Since $\mathcal{M}_\psi = \{h\}' \cap \mathcal{N}_{s(h)}$, our claim follows from the fact that $\mathcal{C}' \cap \mathcal{M} = \mathcal{N}$, Proposition XI.2.26. Q.E.D.

As mentioned already, the global flow of weights is highly discontinuous. We are now going to identify the continuous part of this flow.

Let $\mathcal{M}$ be a separable properly infinite von Neumann algebra. Let $\mathcal{M} = \mathcal{N} \rtimes_\theta \mathbf{R}$ be the continuous decomposition of $\mathcal{M}$ in §1, and τ be the trace scaled by θ, i.e. a faithful semi-finite normal trace on $\mathcal{N}$ such that $\tau \circ \theta_s = \mathrm{e}^{-s}\tau$. We fix $\overline{\varphi} = \hat{\tau}$, the weight dual to τ.

Theorem 4.18. *With the above notations, we conclude:*

(i) $\overline{\varphi} \sim \lambda\overline{\varphi}$ *for every* $\lambda > 0$;

(ii) *If* ψ *is a weight of infinite multiplicity such that* $\psi \sim \lambda\psi$ *for every* $\lambda > 0$, *then* ψ *is equivalent to* $\overline{\varphi}$.

PROOF:

(i) Let $\{u(s)\}$ be the one parameter unitary group associated with the crossed product $\mathcal{M} = \mathcal{N} \rtimes_\theta \mathbf{R}$. Then we have $\sigma_t^{\overline{\varphi}}(u(s)) = \mathrm{e}^{-\mathrm{i}st}u(s)$, so that

$$\overline{\varphi}(u(s)xu(s)^*) = \mathrm{e}^{-s}\overline{\varphi}(x), \quad x \in \mathcal{M}_+.$$

Hence $\mathrm{e}^{-s}\overline{\varphi} = \overline{\varphi}_{u(s)} \sim \overline{\varphi}$.

(ii) We now use the notations in the proof of Theorem 1.7. So we have $\mathcal{M} = \mathcal{P} \overline{\otimes} \mathcal{L}(L^2(\mathbf{R}))$. Suppose that $\overline{\psi}_1$ and $\overline{\psi}_2$ are faithful weights of infinite multiplicity such that $\overline{\psi}_i \sim \lambda\psi_1$, $i = 1, 2$, $\lambda > 0$. Replacing $\overline{\psi}_i$ by equivalent ones, we may assume that $\overline{\psi}_j$'s are of the form:

$$\overline{\psi}_j = \psi_j \otimes \mathrm{Tr},$$

where $\psi_j \sim \lambda\psi_j$ for every $\lambda > 0$, $j = 1, 2$. For each $s \in \mathbf{R}$, there exists a unitary $X_j(s) \in \mathcal{P}$ such that

$$\psi_{j, X_j(s)} = \mathrm{e}^s\psi_j, \qquad \text{i.e.} \quad \sigma_t^{\psi_j}(X(s)) = \mathrm{e}^{\mathrm{i}st}X_j(s).$$

By the separability of $\mathcal{P}$, we can choose a Borel map: $s \in \mathbf{R} \mapsto X_j(s) \in \mathcal{U}(\mathcal{P})$, $j = 1, 2$. Let X_j be the unitary element of

$$\mathcal{P} \overline{\otimes} U(L^\infty(\mathbf{R})) = \widetilde{U}(L^\infty(\mathbf{R}, \mathcal{P}))$$

corresponding to the function: $s \mapsto X_j(s)$. We then have

$$\sigma_t^{\psi_j \otimes \mathrm{Tr}}(X_j) = (\sigma_t^{\psi_j} \otimes \mathrm{id})(X_j) = X_j(1 \otimes U(t)).$$

Hence we get

$$X_j^*\sigma_t^{\overline{\psi}_j}(X_j) = 1 \otimes U(t) = \big(\mathrm{D}(\psi_j \otimes \omega) : \mathrm{D}\overline{\psi}_j\big)_t,$$

where ω is the weight on $\mathcal{L}(L^2(\mathbf{R}))$ such that

$$(\mathrm{D}\omega : \mathrm{D\,Tr})_t = U(t).$$

Thus, we have

$$\psi_j \otimes \omega \sim \psi_j \otimes \mathrm{Tr} = \overline{\psi}_j.$$

By the next lemma, we have $\psi_1 \otimes \omega \sim \psi_2 \otimes \omega$ for any ψ_1 and ψ_2, which completes the proof. Q.E.D.

Lemma 4.19. *For any pair ψ_1, ψ_2 of faithful weights on $\mathcal{P}$, we have $\psi_1 \otimes \omega \sim \psi_2 \otimes \omega$ on $\mathcal{P} \overline{\otimes} \mathcal{L}(L^2(\mathbf{R}))$, where ω is the weight on $\mathcal{L}(L^2(\mathbf{R}))$ such that*

$$\left(\mathrm{D}\omega : \mathrm{D}\,\mathrm{Tr}\right)_t = U(t).$$

PROOF: Let $u_t = \left(\mathrm{D}\psi_2 : \mathrm{D}\psi_1\right)_t$, $t \in \mathbf{R}$. By (X.2.37), there exists $v \in \mathcal{P} \overline{\otimes} \mathcal{L}(L^2(G))$ such that

$$u_s \otimes U(s) = v\left(1 \otimes U(s)\right)\sigma_s^{\psi_1 \otimes \mathrm{Tr}}(v^*).$$

But we have

$$u_s \otimes U(s) = \left(\mathrm{D}(\psi_2 \otimes \omega) : \mathrm{D}(\psi_1 \otimes \mathrm{Tr})\right)_s;$$

$$1 \otimes U(s) = \left(\mathrm{D}(\psi_1 \otimes \omega) : \mathrm{D}(\psi_1 \otimes \mathrm{Tr})\right)_s.$$

Hence $\psi_1 \otimes \omega \sim \psi_2 \otimes \omega$. Q.E.D.

Definition 4.20.

(i) A semi-finite normal weight φ on $\mathcal{M}$ is said to be *integrable* if the modular automorphism $\{\sigma_t^\varphi\}$ on $\mathcal{M}_{s(\varphi)}$ is integrable in the sense of §X.2.

(ii) A faithful φ on $\mathcal{M}$ is called *dominant* if it is of infinite multiplicity and satisfies the condition:

$$\varphi \sim \lambda\varphi, \qquad \lambda > 0;$$

equivalently if φ is dual to the trace τ on $\mathcal{N}$ in a continuous decomposition $\mathcal{M} = \mathcal{N} \rtimes_\theta \mathbf{R}$.

The next theorem justifies the above terminology.

Theorem 4.21. *For a semi-finite normal weight ψ on a separable properly infinite von Neumann algebra $\mathcal{M}$, the following three conditions are equivalent:*

(i) *ψ is integrable;*

(ii) *The map: $s \in \mathbf{R} \mapsto \Theta_s p_{\mathcal{M}}(\check{\psi}) \in \mathcal{A}_{\mathcal{M}}$ is σ-strongly continuous, where $\check{\psi}$ is defined by*

$$\check{\psi} = \sum_{i=1}^{\infty} u_i \psi u_i^*$$

with $\{u_i\}$ a family of isometries with orthogonal ranges.

(iii) *$\psi \prec \overline{\varphi}$ for some, and hence all, dominant weight $\overline{\varphi}$ on $\mathcal{M}$.*

PROOF:

(i) $\Longrightarrow$ (iii): Let $\{u_i\}$ be a family of coisometries[7] with $\sum_{i=1}^{\infty} u_i^* u_i = 1$. Put $\check{\psi} = \sum_{i=1}^{\infty} \psi_{u_i}$. It then follows that $\check{\psi}$ is of infinite multiplicity and integrable if ψ is. By virtue of Theorem X.2.8, $\check{\psi} \precsim \overline{\varphi}$ if ψ is integrable.

[7] A *coisometry* means an operator u such that $uu^* = 1$.

(iii) $\Longrightarrow$ (ii): Let $\overline{\varphi}$ be dominant and $\mathcal{M} = \mathcal{N} \rtimes_\theta \mathbf{R}$ be the associated continuous decomposition of $\mathcal{M}$ so that $\mathcal{M}_{\overline{\varphi}} = \mathcal{N}$. For each $e, f \in \mathrm{Proj}(\mathcal{N})$, $\overline{\varphi}_e \precsim \overline{\varphi}_f$ is equivalent to $e \precsim f$ in $\mathcal{N}$. If e is properly infinite, then $e \sim z(e)$ in $\mathcal{N}$ where $z(e)$ is the central support of e in $\mathcal{N}$. Hence $\mathcal{A} p_{\mathcal{M}}(\overline{\varphi})$ can be identified with the center $\mathcal{C}$ of $\mathcal{N}$, where $\mathcal{A}$ is the abelian von Neumann algebra of the global flow of weights. Let $\{u(s)\}$ be the one parameter unitary group in $\mathcal{M}$ associated with the continuous decomposition, so that $\overline{\varphi}_{u(s)} = \mathrm{e}^{-s}\overline{\varphi}$. If $z \in \mathrm{Proj}(\mathcal{C})$, then we get

$$\mathrm{e}^{-s}\overline{\varphi}(xz) = \overline{\varphi}\big(u(s)xzu(s)^*\big) = \overline{\varphi}_{\theta_s(z)}\big(u(s)xu(s)^*\big),$$

so that $\mathrm{e}^{-s}\overline{\varphi}_z \sim \overline{\varphi}_{\theta_s(z)}$. Hence the identification of $\mathcal{A} p_{\mathcal{M}}(\overline{\varphi})$ and $\mathcal{C}$ conjugates $\{\Theta_s\}$ and $\{\theta_s\}$. Since the flow $\{\mathcal{C}, \mathbf{R}, \theta\}$ is continuous in the sense that for each $z \in \mathcal{C}$ the map: $s \in \mathbf{R} \mapsto \theta_s(z) \in \mathcal{C}$ is σ-strongly continuous, $\{\Theta_s\}$ is σ-strongly continuous on $\mathcal{A} p_{\mathcal{M}}(\overline{\varphi})$. Thus (ii) follows.

(ii) $\Longrightarrow$ (iii): Let $e \in \mathrm{Proj}_\sigma(\mathcal{A})$ be such that the map: $s \in \mathbf{R} \mapsto \Theta_s(e) \in \mathcal{A}$ is σ-strongly continuous. Let $e_0 = \bigvee\{\Theta_s(e) : s \in \mathbf{Q}\}$. Then e_0 is σ-finite. We want to prove that $\Theta_s(e_0) = e_0$ for every $s \in \mathbf{R}$. If $\omega \in \mathcal{A}_*^+(1 - e_0)$, then $\omega\big(\Theta_s(e)\big) = 0$ for every $s \in \mathbf{Q}$, so that the continuity assumption on e implies that $\omega\big(\Theta_s(e)\big) = 0$ for every $s \in \mathbf{Q}$. This means that $\Theta_s(e) \perp (1 - e_0)$ for every $s \in \mathbf{R}$; so $\Theta_s(e) \le e_0$, which means $\Theta_s(e_0) = e_0$. Let $\overline{\psi}$ be a faithful weight on $\mathcal{M}$ of infinite multiplicity such that $p_{\mathcal{M}}(\overline{\psi}) = e_0$. By Theorem 4.18, $\overline{\psi}$ is dominant. By construction, if $p_{\mathcal{M}}(\psi) = e$, then $\psi \precsim \overline{\psi}$.

(iii) $\Longrightarrow$ (i): Suppose that $\psi = \overline{\varphi}_e$ with $e \in \mathrm{Proj}(\mathcal{C})$, where $\mathcal{C}$ is the center of $\mathcal{M}_{\overline{\varphi}} = \mathcal{N}$. Let $\mathfrak{p}$ be the hereditary subalgebra of $\sigma^{\overline{\varphi}}$-integrable elements in $\mathcal{M}$. Since $\mathfrak{p}$ is a two sided module over $\mathcal{M}_{\overline{\varphi}}$, $e\mathfrak{p}e \subset \mathfrak{p}$ and $e\mathfrak{p}e$ is σ-weakly dense in $\mathcal{M}_e$. Hence ψ is integrable. Q.E.D.

Corollary 4.22. *Let $\mathcal{M}$ be a separable factor of type* III *with continuous decomposition $\mathcal{M} = \mathcal{N} \rtimes_\theta \mathbf{R}$. The restriction of the global flow of weight $\{\mathcal{A}, \mathbf{R}, \Theta\}$ of $\mathcal{M}$ to $\{\mathcal{A} p_{\mathcal{M}}(\overline{\varphi}), \mathbf{R}, \Theta\}$, with $\overline{\varphi}$ a dominant weight, is conjugate to the flow of weights $\{\mathcal{C}, \mathbf{R}, \theta\}$ of Definition* 1.3.

In contrast to the global flow of weights, we call either $\{\mathcal{C}, \mathbf{R}, \theta\}$ or $\{\mathcal{A} p_{\mathcal{M}(\overline{\varphi})}, \mathbf{R}, \Theta\}$ the *smooth* flow of weights.

The flow of weights is by its construction a functor from the category of separable factors of type III with isomorphisms as morphisms to the category of properly ergodic flows with conjugations as morphisms. This means the automorphism group $\mathrm{Aut}(\mathcal{M})$ of a separable factor $\mathcal{M}$ of type III acts naturally on the flow of weights. In fact, the action $\widetilde{\alpha}$ of each $\alpha \in \mathrm{Aut}(\mathcal{M})$ is given simply by:

$$\widetilde{\alpha}\big(p_{\mathcal{M}}(\psi)\big) = p_{\mathcal{M}}\big(\psi \circ \alpha^{-1}\big), \quad \alpha \in \mathrm{Aut}(\mathcal{M}), \quad \psi \in \mathfrak{W}_{\mathrm{inf}}(\mathcal{M}).$$

Clearly the action $\widetilde{\alpha}$ commutes with the global flow of weights $\{\Theta_s\}$, so that its restriction to the smooth flow of weights makes sense. We call this resulted automorphism on the flow of weights $\{\mathcal{C}, \mathbf{R}, \theta\}$ the *modulus* of α and denote it by $\mathrm{mod}(\alpha)$. Since an inner automorphism does not change the equivalence class of a weight, its modulus is trivial, i.e. we conclude the following:

Corollary 4.23.

(i) *The modulus of an inner automorphism of a separable factor* $\mathcal{M}$ *of type* III *is trivial, i.e.*

$$\mathrm{Int}(\mathcal{M}) \subset \mathrm{Ker}(\mathrm{mod}).$$

(ii) *Suppose that* $\mathcal{M}$ *is a separable factor of type* III_λ *and* $\psi \in \mathfrak{W}_{\mathrm{inf}}$ *is lacunary and* $\sigma_T^\psi = \mathrm{id}$ *with* $T = -\log\lambda$. *If* $\alpha \in \mathrm{Aut}(\mathcal{M})$ *transforms* ψ *in the way that* $\psi \circ \alpha^{-1} = \mu\psi$ *with* $\lambda < \mu \leq 1$, *then we have* $\mathrm{mod}(\alpha) = \theta_{-\log\mu}$.

PROOF: We only need to prove (ii). By the assumption on $\mathcal{M}$ and ψ, there exists $U \in \mathcal{U}(\mathcal{M})$ which gives a discrete crossed product decomposition of $\mathcal{M}$, i.e. $\sigma_t^\psi(U) = \lambda^{\mathrm{i}t}U$, $t \in \mathbf{R}$, and $\mathcal{M} = \mathcal{M}_\psi \rtimes_{\mathrm{Ad}(U)} \mathbf{Z}$. Let $\theta_0 = \mathrm{Ad}(U)\big|_{\mathcal{M}_\psi}$. Then θ_0 scales the semi-finite normal trace $\tau = \psi|_{\mathcal{M}}$ by λ. Now consider the crossed product $\mathcal{N} = \mathcal{M} \rtimes_{\sigma^\psi} \mathbf{R}$ equipped with dual action θ of $\mathbf{R}$ and the extension $\widetilde{\alpha}$ of α to $\mathcal{N}$, which is given by:

$$\widetilde{\alpha}(x) = \alpha(x), \quad x \in \mathcal{M}, \qquad \widetilde{\alpha}\big(u(s)\big) = \mu^{\mathrm{i}s}u(s), \quad s \in \mathbf{R}$$

where $\{u(s)\}$ is of course the one parameter unitary group of the crossed product by σ^ψ. The center $\mathcal{C}$ of $\mathcal{N}$ is generated by $u(T)$. Therefore, $\mathrm{mod}(\alpha)\big(u(T)\big) = \mu^{\mathrm{i}T}u(T) = \theta_{-\log(\mu)}\big(u(T)\big)$. Hence the assertion follows. This completes the proof.

Q.E.D.

Exercise XII.4

1) Let $\mathcal{M}$ be a separable semi-finite factor with τ a faithful semi-finite trace. Then the flow $\{\mathcal{A}, \mathbf{R}, \theta\}$ of weights on $\mathcal{M}$ is $\{L^\infty(\mathbf{R}), \text{ translation}\}$.

(a) Show that $\mathrm{Z}_\theta^1\big(\mathbf{R}, \mathcal{U}(\mathcal{A})\big) = \mathrm{B}_\theta^1\big(\mathbf{R}, \mathcal{U}(\mathcal{A})\big)$, i.e. every $c \in \mathrm{Z}_\theta^1\big(\mathbf{R}, \mathcal{U}(\mathcal{A})\big)$ is of the form: $c(s) = b\theta_s(b^*)$ for some $b \in \mathcal{U}(\mathcal{A})$.

(b) Let h_0 be the unbounded self-adjoint operator affiliated with $\mathcal{A}$ which is represented by the identity function: $s \in \mathbf{R} \mapsto s \in \mathbf{R}$. Observe that there exists a function $f : s \in \mathbf{R}_+^* \mapsto f(s) \in \mathbf{T}$ such that $b = f(e^{h_0})$.

(c) Show that a faithful semi-finite normal weight φ of infinite multiplicity is integrable if and only if φ is of the form : $\varphi = \tau(h\,\cdot\,)$ with h a non-singular self-adjoint operator affiliated with a subfactor $\mathcal{B} \subset \mathcal{M}$ of type I_∞ with absolutely continuous spectrum.

(d) Assume that a faithful semi-finite normal weight φ of infinite multiplicity is integrable and that $\varphi = \tau(h\,\cdot\,)$ as in (c), so that $\sigma_t^\varphi = \mathrm{Ad}(h^{\mathrm{i}t})$, $t \in \mathbf{R}$. Show that $\sigma_c^\varphi = \mathrm{Ad}\big(f(h)\big)$.

2) Let $\mathcal{M}_i$, $i = 1, 2$, be an infinite factor with flow of weights $\{\mathcal{A}_i, \mathbf{R}, \theta^i\}$. Show that the flow of weights $\{\mathcal{A}, \mathbf{R}, \theta\}$ of $\mathcal{M} = \mathcal{M}_1 \mathbin{\overline{\otimes}} \mathcal{M}_2$ is conjugate to the flow obtained by restricting $\{\theta_s^1 \otimes \theta_s^2 : s \in \mathbf{R}\}$ to the fixed point subalgebra: $\{x \in \mathcal{A}_1 \mathbin{\overline{\otimes}} \mathcal{A}_2 : (\theta_t^1 \otimes \theta_{-t}^2)(x) = x,\ t \in \mathbf{R}\}$. [491]

§ 5 Action of Int($\mathcal{M}$) on $\mathfrak{W}_0$ and $\mathfrak{S}_*$

We continue to study weights and normal states. First, we discuss the density of integrable weights in $\mathfrak{W}_{\text{inf}}$. We fix a separable properly infinite von Neumann algebra $\mathcal{M}$.

Definition 5.1 (One Parameter Family of Orderings). For a positive real number $\lambda > 0$, and a pair φ_1, φ_2 of weights on $\mathcal{M}$ (of course, semi-finite and normal), we write

$$\varphi_1 \leq_\lambda \varphi_2 \tag{1}$$

if the map: $t \in \mathbf{R} \mapsto u_t = \left(\mathrm{D}\varphi_1 : \mathrm{D}\varphi_2\right)_t \in \mathcal{M}$ is extendable to a σ-strongly continuous function u_z on the horizontal strip $\overline{\mathbf{D}}_\lambda = \left\{z \in \mathbf{C} : -\lambda \leq \operatorname{Im} z \leq 0\right\}$ which is holomorphic inside $\mathbf{D}_\lambda = \left\{z \in \mathbf{C} : -\lambda < \operatorname{Im} z < 0\right\}$ and $\|u_z\| \leq 1$, $z \in \overline{\mathbf{D}}_\lambda$.

Lemma 5.2. *The relation "$\varphi_1 \leq_\lambda \varphi_2$" on $\mathfrak{W}_0$ with a fixed $\lambda > 0$ is indeed an order relation.*

PROOF: Trivially $\varphi \leq_\lambda \varphi$. Suppose $\varphi_1 \leq_\lambda \varphi_2$ and $\varphi_2 \leq_\lambda \varphi_3$. Let $u_t^{i,j} = \left(\mathrm{D}\varphi_i : \mathrm{D}\varphi_j\right)_t$, $i, j = 1, 2, 3$. By assumption, $u_t^{1,2}$ and $u_t^{2,3}$ have extensions $u_z^{1,2}$ and $u_z^{2,3}$ on $\overline{\mathbf{D}}_\lambda$. It then follows that $u_z = u_z^{1,2} u_z^{2,3}$ is a continuous function on $\overline{\mathbf{D}}_\lambda$ and holomorphic on $\mathbf{D}_\lambda$, and $u_t = u_t^{1,3}$, $t \in \mathbf{R}$, by the chain rule. Obviously, $\|u_z\| \leq \left\|u_z^{1,2}\right\| \left\|u_z^{2,3}\right\| \leq 1$, $z \in \overline{\mathbf{D}}_\lambda$; hence $\varphi_1 \leq_\lambda \varphi_3$.

Suppose $\varphi_1 \leq_\lambda \varphi_2$ and $\varphi_2 \leq_\lambda \varphi_1$. Then $u_t^{1,2}$ and $u_t^{2,1}$ have extensions $u_z^{1,2}$ and $u_z^{2,1}$ to $\overline{\mathbf{D}}_\lambda$. But, we have $u_t^{2,1} = \left(u_t^{1,2}\right)^*$. Let $\omega \in \mathfrak{S}_*(\mathcal{M})$ and put

$$f(z) = \omega\left(u_z^{2,1}\right) \qquad \text{and} \qquad g(z) = \omega\left(u_z^{1,2}\right), \quad z \in \overline{\mathbf{D}}_\lambda.$$

Since $f(t) = \overline{g(t)}$, $t \in \mathbf{R}$, f is extended to a continuous function on $\overline{\mathbf{D}}_\lambda$ which is holomorphic on $\mathbf{D}_{-\lambda}$. Hence f is holomorphic on $\mathbf{D}_{-\lambda} \cup \mathbf{R} \cup \mathbf{D}_\lambda = \mathbf{D}_\lambda$. But we have

$$f(0) = 1 \geq \left|f(z)\right|, \quad z \in \mathbf{D}_\lambda,$$

so that the maximal modulus principle implies $f(z) \equiv 1$ on $\mathbf{D}_\lambda$. Thus $u_z^{1,2} = 1$, that is $\varphi_1 = \varphi_2$. Q.E.D.

Remark 5.3. By Theorem VIII.3.17, the usual ordering $\varphi \leq \psi$ in the sense that $\varphi(x) \leq \psi(x)$, $x \in \mathcal{M}_+$, is equivalent to $\varphi \leq_{1/2} \psi$.

Lemma 5.4. *For $\varphi_1, \varphi_2 \in \mathfrak{W}_0$, consider the balanced weight $\varphi = \varphi_1 \oplus \varphi_2$ on $\mathcal{N} = \mathcal{M} \otimes M_2$, i.e.*

$$\varphi\left(\sum x_{ij} \otimes e_{ij}\right) = \varphi_1(x_{11}) + \varphi_2(x_{22}).$$

Then the inequality $\varphi_1 \leq_\infty \varphi_2$ is equivalent to the condition $1 \otimes e_{12} \in \mathcal{N}_\varphi\left([0, \infty[\right)$.

PROOF: Our assertion follows from a more general fact. Let $\{\alpha_t\}$ be an arbitrary continuous one parameter automorphism group of $\mathcal{N}$. Then $x \in \mathcal{N}^\alpha([0,\infty[)$ if and only if for every $\omega \in \mathcal{N}_*$, the Fourier transform, in the sense of tempered distribution, of the function: $t \in \mathbf{R} \mapsto \omega(\alpha_t(x))$ is supported by $[0,\infty[$; hence by the Paley-Wiener theorem if and only if the function: $t \in \mathbf{R} \mapsto \alpha_t(x) \in \mathcal{N}$ is extended to a bounded holomorphic function on the upper half plane $\mathbf{D}_{-\infty}$. With $\alpha_t = \sigma_t^\varphi$ and $x = 1 \otimes e_{21}$, our assertion follows. Q.E.D.

Definition 5.5. For a pair $\varphi_1, \varphi_2 \in \mathfrak{W}_0$, we put

$$d(\varphi_1,\varphi_2) = \inf\{\alpha > 0 : \mathrm{e}^{-\alpha}\varphi_2 \leq_\infty \varphi_1 \leq_\infty \mathrm{e}^{\alpha}\varphi_2\}. \tag{2}$$

From Lemma 5.2, it follows that d is a distance function on $\mathfrak{W}_0$ with values in $[0,+\infty]$. We note that if $d(\varphi_1,\varphi_2) < +\infty$, then the function: $t \in \mathbf{R} \mapsto (\mathrm{D}\varphi_2 : \mathrm{D}\varphi_1)_t$ extends to an entire function. The topology on $\mathfrak{W}_0$ given by d will be called the *uniform topology.*

Example 5.6. Let $\mathcal{M}$ be semifinite and τ be a faithful semifinite normal trace. Let $\varphi = \tau_h$ and $\psi = \tau_k$ where h and k are self-adjoint non-singular positive operator affiliated with $\mathcal{M}$. For $\lambda > 0$, $\varphi \leq_\lambda \psi$ is then equivalent to the operator inequality $h^\lambda \leq k^\lambda$. Hence $\varphi \leq_\infty \psi$ is equivalent to $h^n \leq k^n$, $n = 1, 2, \dots$.

(i) Let $h = \int_0^\infty \lambda\, \mathrm{d}p(\lambda)$ and $k = \int_0^\infty \lambda\, \mathrm{d}q(\lambda)$ be the spectral decompositions. Then $h^n \leq k^n$, $n = 1, 2, \dots$, is equivalent to $p(\lambda) \geq q(\lambda)$, $\lambda \geq 0$.

(ii) Let $H = \log h$ and $K = \log k$. Put

$$u_t = (\mathrm{D}\psi : \mathrm{D}\varphi)_t = k^{\mathrm{i}t}h^{-\mathrm{i}t}, \quad t \in \mathbf{R}.$$

The differentiability of $\{u_t\}$ in $\mathcal{M}$ in the σ-strong* topology is equivalent to the existence of $a \in \mathcal{M}$ such that $a = a^*$ and $K = H + a$, i.e. K is a bounded additive perturbation of H.

(iii) The distance $\alpha = d(\varphi,\psi) < +\infty$ implies that $\mathrm{Sp}_\sigma(a)$ is bounded where $\sigma = \sigma^\varphi$.

PROOF:

(i) Suppose that $h^n \leq k^n$, $n = 1, 2, \dots$. Let $\mathcal{M}$ act on $\mathfrak{H}$. A vector $\xi \in \mathfrak{H}$ belongs to $p(\lambda)\mathfrak{H}$ if and only if $\lim_{n\to\infty}\|\mu^{-n}h^n\xi\| = 0$ for every $\mu > \lambda$. Hence the inequality $h^n \leq k^n$ for every $n = 1, 2, \dots$ implies $q(\lambda)\mathfrak{H} \subset p(\lambda)\mathfrak{H}$. Conversely, suppose that $q(\lambda) \leq p(\lambda)$ for every $\lambda \geq 0$. Integrating by part, we have

$$\begin{aligned}(h\xi \mid \xi) &= \int_0^\infty \lambda\,\mathrm{d}(p(\lambda)\xi \mid \xi) = -\int_0^\infty \lambda\,\mathrm{d}(p(\lambda)^\perp\xi \mid \xi) = \int_0^\infty (p(\lambda)^\perp\xi \mid \xi)\,\mathrm{d}\lambda \\ &\leq \int_0^\infty (q(\lambda)^\perp\xi \mid \xi)\,\mathrm{d}\lambda = \int_0^\infty \lambda\,\mathrm{d}(q(\lambda)\xi \mid \xi) = (k\xi \mid \xi).\end{aligned}$$

Then the inequality $h^n \leq k^n$ follows from the fact that the spectral projection of h^n (resp. k^n) is $\{ p(\lambda^{1/n}) : \lambda > 0 \}$ (resp. $\{ q(\lambda^{1/n}) : \lambda > 0 \}$).

(ii) Suppose that $\{u_t\}$ is differentiable, and put $a = \frac{1}{\mathrm{i}} \frac{\mathrm{d}}{dt} u_t \big|_{t=0}$. Since $\mathrm{e}^{\mathrm{i}tK} = u_t \mathrm{e}^{\mathrm{i}tH}$, for every $\xi \in \mathfrak{D}(H)$ the function: $t \in \mathbf{R} \mapsto \mathrm{e}^{\mathrm{i}tK}\xi = u_t \mathrm{e}^{\mathrm{i}tH}\xi \in \mathfrak{H}$ is differentiable, so that ξ belongs to $\mathfrak{D}(K)$ and $K\xi = (H + a)\xi$. Conversely, if $\xi \in \mathfrak{D}(K)$, then the function: $t \in \mathbf{R} \mapsto \mathrm{e}^{\mathrm{i}tH}\xi = u_t^* \mathrm{e}^{\mathrm{i}tK}\xi \in \mathfrak{H}$ is differentiable; hence $\xi \in \mathfrak{D}(H)$. Thus, we get $K = H + a$. Suppose that $K = H + a$ with $a \in \mathcal{M}$ self-adjoint. If $\xi \in \mathfrak{D}(H) = \mathfrak{D}(K)$, then

$$\frac{1}{s}(u_s - 1)\xi = \frac{1}{s}(\mathrm{e}^{\mathrm{i}sK} - 1)\mathrm{e}^{-\mathrm{i}sH}\xi + \mathrm{e}^{\mathrm{i}sK}\frac{1}{s}(\mathrm{e}^{-\mathrm{i}sH} - 1)\xi$$

converges to $\mathrm{i}a\xi$ as $s \to 0$. Hence

$$\lim_{s\to 0} \frac{1}{s}(u_{s+t} - u_t)\xi = \mathrm{i}u_t\sigma_t(a)\xi, \quad \xi \in \mathfrak{D}(H). \tag{3}$$

Therefore, we get

$$u_t\xi = \xi + \mathrm{i}\int_0^t u_s\sigma_s(a)\xi \, \mathrm{d}s, \quad \xi \in \mathfrak{D}(H). \tag{4'}$$

The both sides of this equality is continuous in ξ, so that

$$u_t = 1 + \mathrm{i}\int_0^t u_s\sigma_s(a)\, \mathrm{d}s\,. \tag{4}$$

The above integral converges in the σ-strong* topology. Thus $\{u_t\}$ is differentiable in the σ-strong* topology.

Before moving on to (iii), we draw an immediate consequence of (4). Recurrent use of (4) implies that

$$u_t = 1 + \sum_{k=1}^{n} \mathrm{i}^k \int_0^t \int_0^{t_1} \cdots \int_0^{t_{n-1}} \sigma_{t_n}(a)\sigma_{t_{n-1}}(a)\cdots\sigma_{t_1}(a)\, \mathrm{d}t_n\, \mathrm{d}t_{n-1} \cdots \mathrm{d}t_1$$
$$+ \mathrm{i}^{n+1} \int_0^t \int_0^{t_1} \cdots \int_0^{t_n} u_{t_{n+1}}\sigma_{t_{n+1}}(a)\sigma_{t_n}(a)\cdots\sigma_{t_1}(a)\, \mathrm{d}t_{n+1} \cdots \mathrm{d}t_1\,.$$

Since the volume of the simplex:

$$\Delta(t) = \{ (t_1, \dots, t_n) \in \mathbf{R}^n : 0 \leq t_1 \leq t_2 \leq \cdots \leq t \}$$

is $t^n/n!$, we have an absolute convergent expansion of u_t:

$$u_t = 1 + \sum_{n=1}^{\infty} \mathrm{i}^n \int_0^t \int_0^{t_1} \cdots \int_0^{t_{n-1}} \sigma_{t_n}(a)\sigma_{t_{n-1}}(a)\cdots\sigma_{t_1}(a)\, \mathrm{d}t_n\, \mathrm{d}t_{n-1} \cdots \mathrm{d}t_1\,. \tag{5}$$

(iii) Suppose that $\alpha = d(\varphi, \psi) < +\infty$. Then the function: $t \in \mathbf{R} \mapsto u_t \in \mathcal{M}$ is extended to an entire function over $\mathbf{C}$. By (i), we know that $K = H + a$ with $a = a^* \in \mathcal{M}$. The equality $u_t^* u_t = u_t u_t^* = 1$ implies that $u_{\bar{z}}^* u_z = u_z u_{\bar{z}}^* = 1$ for every $z \in \mathbf{C}$. Hence u_z is invertible. The inequality,

$$e^{-(\alpha+\varepsilon)}\varphi \leq_\infty \psi \leq_\infty e^{(\alpha+\varepsilon)}\varphi$$

for every $\varepsilon > 0$, means that

$$e^{-(\alpha+\varepsilon)s} \leq u_{-is/2}^* u_{-is/2} \leq e^{(\alpha+\varepsilon)s}, \quad s \geq 0,$$

so that $\|u_z\| \leq e^{(\alpha+\varepsilon)|\mathrm{Im}\, z|}$. Therefore, the Fourier transform, as a tempered distribution, $\hat{u}$ of the function: $t \mapsto u_t$ is supported by $\left[-(\alpha+\varepsilon), (\alpha+\varepsilon)\right]$ for $\varepsilon > 0$ by the Paley-Wiener theorem; hence by $[-\alpha, \alpha]$. Let u_t' be the derivative of $\{u_t\}$. Then we have $\hat{u}'(p) = \mathrm{i} p\hat{u}(p)$, $p \in \mathbf{R}$, as a tempered distribution, so that $\hat{u}'$ is also supported by $[-\alpha, \alpha]$. Hence by the equality

$$\mathrm{i}\sigma_t(a) = u_t^* u_t',$$

we conclude that the Fourier transform of the function: $t \mapsto \sigma_t(a) \in \mathcal{M}$ is supported by $\left[-2\alpha, 2\alpha\right]$. Thus $\mathrm{Sp}_\sigma(a) \subset \left[-2\alpha, 2\alpha\right]$. Q.E.D.

We now return to the study of the metric space $\{\mathfrak{W}_0, d\}$ of faithful semi-finite normal weights on $\mathcal{M}$.

Proposition 5.7.

(i) *The metric space* $\{\mathfrak{W}_0, d\}$ *is complete and the function:* $\varphi \in \mathfrak{W}_0 \mapsto \varphi(x) \in [0, \infty]$ *is continuous for every* $x \in \mathcal{M}_+$.

(ii) *For any normal states* φ_1, φ_2 *on* $\mathcal{M}$,

$$\|\varphi_1 - \varphi_2\| \leq 4d(\varphi_1, \varphi_2).$$

We need the next:

Lemma 5.8. *For any* $\varepsilon > 0$ *and* $R > 0$, *there exists* $\delta = \delta(\varepsilon, R) > 0$ *such that if* f *is a Banach space valued entire function with* $\|f(z)\| \leq e^{\delta|\mathrm{Im}\, z|}$, $z \in \mathbf{C}$, *then*

$$\|f(z) - f(0)\| \leq \varepsilon, \quad |z| < R.$$

PROOF: Let H be the space of all entire functions over $\mathbf{C}$ equipped with the compact open topology. Let H_λ, $\lambda > 0$, be the set of all entire functions f with $|f(z)| \leq e^{\lambda|\mathrm{Im}\, z|}$, $z \in \mathbf{C}$. It then follows that H_λ is a compact subset of H, and $\bigcap_{\lambda>0} H_\lambda = \{\alpha 1 : |\alpha| \leq 1, \alpha \in \mathbf{C}\} = H_0$. Let

$$H_{\varepsilon,R} = \left\{ f \in H : |f(z) - f(0)| < \varepsilon, |z| \leq R \right\}.$$

Then $H_{\varepsilon,R}$ is an open subset of H and $H_{\varepsilon,R} \supset H_0$. Hence, by compactness, we have $H_\delta \subset H_{\varepsilon,R}$ for some $\delta = \delta(\varepsilon, R) > 0$.

Now, let E be a Banach space and f be an E-valued entire function on $\mathbf{C}$ with $\|f(z)\| \le e^{\delta|\mathrm{Im}\, z|}$, $z \in \mathbf{C}$. For any $\varphi \in E^*$, $\|\varphi\| \le 1$, the function: $z \in \mathbf{C} \mapsto \varphi \circ f(z) \in \mathbf{C}$ belongs to H_δ; hence to $H_{\varepsilon,R}$, so that

$$\Big|\varphi\big(f(z)\big) - \varphi\big(f(0)\big)\Big| \le \varepsilon, \quad |z| \le R.$$

Thus, we get, for any $|z| < R$,

$$\big\|f(z) - f(0)\big\| = \sup\Big\{\Big|\varphi\big(f(z)\big) - \varphi\big(f(0)\big)\Big| : \varphi \in E^*,\ \|\varphi\| \le 1\Big\} \le \varepsilon.$$

This completes the proof. Q.E.D.

Proof of Proposition 5.7:

(i) Let $\varphi_1, \varphi_2 \in \mathfrak{W}_0$, and $\delta > 0$. Suppose that $d(\varphi_1, \varphi_2) < \delta$, i.e. $e^{-\delta}\varphi_1 \le_\infty \varphi_2 \le_\infty e^{\delta}\varphi_1$. Let $\{u_z : z \in \mathbf{C}\}$ be the entire extension of $u_t = \big(\mathrm{D}\varphi_2 : \mathrm{D}\varphi_1\big)_t$. We then have $\|u_z\| \le e^{\delta|\mathrm{Im}\, z|}$, $z \in \mathbf{C}$. Hence if $\delta = \delta(\varepsilon, R)$ for any $\varepsilon > 0$, we have, by Lemma 5.8,

$$\big\|u_z - 1\big\| \le \varepsilon, \qquad |z| \le R.$$

Therefore, with a fixed $\varphi \in \mathfrak{W}_0$ such that $\alpha = d(\varphi, \varphi_1) < +\infty$, we have

$$\begin{aligned}
&\Big\|\big(\mathrm{D}\varphi_2 : \mathrm{D}\varphi\big)_z - \big(\mathrm{D}\varphi_1 : \mathrm{D}\varphi\big)_z\Big\| \\
&\quad = \Big\|\Big\{\big(\mathrm{D}\varphi_2 : \mathrm{D}\varphi\big)_z\big(\mathrm{D}\varphi_1 : \mathrm{D}\varphi\big)_z^{-1} - 1\Big\}\big(\mathrm{D}\varphi_1 : \mathrm{D}\varphi\big)_z\Big\| \\
&\quad \le \Big\|\big(\mathrm{D}\varphi_2 : \mathrm{D}\varphi_1\big)_z - 1\Big\| e^{\alpha|\mathrm{Im}\, z|} \le \varepsilon e^{\alpha|\mathrm{Im}\, z|}, \quad |z| \le R.
\end{aligned}$$

This shows that if $\{\varphi_n\}$ is a Cauchy sequence in $\mathfrak{W}_0$, then $(\mathrm{D}\varphi_n : \mathrm{D}\varphi)_z$ converges to v_z in norm uniformly on every bounded subset of $\mathbf{C}$; so that the function: $z \in \mathbf{C} \mapsto v_z \in \mathcal{M}$ is entire and $v_{s+t} = v_s\sigma_s^\varphi(v_t)$, $s, t \in \mathbf{R}$, and also each v_t is a unitary. By Theorem VIII.3.8, there exists $\psi \in \mathfrak{W}_0$ such that $v_t = (\mathrm{D}\psi : \mathrm{D}\varphi)_t$. In other words, $(\mathrm{D}\varphi_n : \mathrm{D}\varphi)_z$ converges to $(\mathrm{D}\psi : \mathrm{D}\varphi)_z$ in norm uniformly for $|z| \le R$. For any $\varepsilon > 0$, choose $m \in \mathbf{N}$ so that $\big\|(\mathrm{D}\psi : \mathrm{D}\varphi_m)_z\big\| \le e^{\varepsilon|\mathrm{Im}\, z|}$. Hence we get $d(\psi, \varphi_m) \le \varepsilon$. Thus ψ is the limit of $\{\varphi_n\}$.

The continuity: $\varphi \in \mathfrak{W}_0 \mapsto \varphi(x) \in [0, \infty]$, $x \in \mathcal{M}_+$, follows from the inequality:

$$e^{-\delta}\varphi_1(x) \le \varphi_2(x) \le e^{\delta}\varphi_1(x), \quad x \in \mathcal{M}_+,$$

if $d(\varphi_1, \varphi_2) \le \delta$.

(ii) If $d(\varphi_1, \varphi_2) \le \varepsilon$, then

$$\big(e^{-\varepsilon} - 1\big)\varphi_1(x) \le \varphi_2(x) - \varphi_1(x) \le \big(e^{\varepsilon} - 1\big)\varphi_1(x), \quad x \in \mathcal{M}_+,$$

so that for any faithful $\varphi_1, \varphi_2 \in \mathfrak{S}_*(\mathcal{M})$ we have

$$\begin{aligned}
\big\|\varphi_2 - \varphi_1\big\| &\le 2\sup\Big\{\big|\varphi_2(x) - \varphi_1(x)\big| : x \in \mathcal{M}_+,\ \|x\| \le 1\Big\} \\
&\le 2\max\big\{e^{\varepsilon} - 1,\ 1 - e^{-\varepsilon}\big\} \le 4\varepsilon.
\end{aligned}$$

Q.E.D.

In the above discussion, we have seen that the metric d on $\mathfrak{W}_0$ gives rise to a very strict topology in $\mathfrak{W}_0$. In the semifinite case, the metric d on the space of self-adjoint operators affiliated with $\mathcal{M}$ under the identification $H \longleftrightarrow \tau(\mathrm{e}^H \cdot)$ is stricter than the norm topology. The distance $d(H, K)$ measures not only the norm $\|H - K\|$ but also the degree of commutativity of H and K. The study of the metric on self-adjoint operators is interesting on its own right. But we will not dwell on this further. Instead, we prove the following density theorem for integrable weights:

Theorem 5.9. *If φ is a faithful weight of infinite multiplicity on $\mathcal{M}$, then for any $\varepsilon > 0$ there exists an integrable faithful weight ψ of infinite multiplicity commuting with φ such that $d(\varphi, \psi) < \varepsilon$.*

PROOF: Let $\mathcal{B}$ be a subfactor of type I_∞ of $\mathcal{M}_\varphi$. Choose an $h \in \mathcal{B}$, $\mathrm{e}^{-\varepsilon} \leq h \leq \mathrm{e}^{\varepsilon}$, such that (i) h has only absolutely continuous spectrum and (ii) $\{h\}' \cap \mathcal{B}$ is properly infinite. We claim first that $\mathrm{Tr}(h \cdot)$ is integrable as a weight on $\mathcal{B}$ where Tr means of course the trace on $\mathcal{B}$. The one parameter unitary group $\{u_t\} = \{h^{\mathrm{i}t}\}$ has absolutely continuous spectrum. Hence it is quasi-equivalent to a subrepresentation of the regular representation of $\mathbf{R}$ because the quasi-equivalence class of a unitary representation of an abelian group is completely determined by the absolute continuous class of the corresponding spectral measure. This means that the weight $\mathrm{Tr}(h \cdot)$ is equivalent to a subweight of a dominant weight on $\mathcal{B}$. Hence $\mathrm{Tr}(h \cdot)$ is integrable on $\mathcal{B}$.

Put $\psi = \varphi(h \cdot)$. Then the restriction of ψ on $\mathcal{B}$ is exactly $\mathrm{Tr}(h \cdot)$, which is integrable, so that ψ itself is integrable on $\mathcal{M}$ by Lemma X.2.7. Since $\mathcal{M}_\psi$ contains $\{h\}' \cap \mathcal{B}$, ψ is of infinite multiplicity. Now, we have $h^{\mathrm{i}t} = \left(\mathrm{D}\psi : \mathrm{D}\varphi\right)_t = u_t$, so that $u_z = h^{\mathrm{i}z}$, $z \in \mathbf{C}$. Hence we have for $s > 0$

$$\|u_{r+\mathrm{i}s}\| = \|h^{-s}\| \leq \mathrm{e}^{s\varepsilon}, \qquad \|u_{r+\mathrm{i}s}^{-1}\| = \|h^{s}\| \leq \mathrm{e}^{s\varepsilon}.$$

Therefore, we get $d(\varphi, \psi) \leq \varepsilon$. Q.E.D.

Corollary 5.10. *If $\mathcal{M}$ is a factor of type III_1, then for any pair φ_1, φ_2 of faithful weights of infinite multiplicity and $\varepsilon > 0$, there exists a unitary $u \in \mathcal{M}$ such that $d(\varphi_{1,u}, \varphi_2) < \varepsilon$, i.e.*

$$\mathrm{e}^{-\varepsilon}\varphi_2 \leq_\infty \varphi_{1,u} \leq_\infty \mathrm{e}^{\varepsilon}\varphi_2.$$

PROOF: Suppose that ψ_1, ψ_2 are faithful integrable weights of infinite multiplicity such that $d(\psi_i, \varphi_i) < \varepsilon/2$, $i = 1, 2$. Since $\mathcal{M}$ is of type III_1, ψ_i are both dominant, so that there exists a unitary $u \in \mathcal{M}$ with $\psi_2 = \psi_{1,u}$, and therefore we get

$$d\left(\varphi_{1,u}, \varphi_2\right) \leq d\left(\varphi_{1,u}, \psi_{1,u}\right) + d\left(\psi_2, \varphi_2\right) = d\left(\varphi_1, \psi_1\right) + d\left(\psi_2, \varphi_2\right) < \varepsilon.$$

This completes the proof. Q.E.D.

Corollary 5.11. *If $\mathcal{M}$ is a factor of type III_λ, $\lambda \neq 0$, then for any pair φ_1, φ_2 of faithful weights of infinite multiplicity on $\mathcal{M}$ ($\varphi_j(1) = +\infty$, $j = 1, 2$, is enough when $\lambda \neq 1$), there exists a unitary $u \in \mathcal{M}$ such that*

$$u^* \mathfrak{m}_{\varphi_1} u = \mathfrak{m}_{\varphi_2}$$

PROOF: The case $\lambda = 1$. Fix $\varepsilon > 0$. By the previous corollary, there exists a unitary $u \in \mathcal{M}$ such that $d(\varphi_{1,u}, \varphi_2) < \varepsilon$, which means that $e^{-\varepsilon}\varphi_2 \leq_\infty \varphi_{1,u} \leq_\infty e^{\varepsilon}\varphi_2$. In particular,

$$e^{-\varepsilon}\varphi_2(x) \leq \varphi_{1,u}(x) \leq e^{\varepsilon}\varphi_2(x), \quad x \in \mathcal{M}_+.$$

Therefore, $\mathfrak{m}_{\varphi_2} = \mathfrak{m}_{\varphi_{1,u}} = u^* \mathfrak{m}_{\varphi_1} u$.

The case $\lambda \neq 1$ and $\varphi_1(1) = \varphi_2(1) = +\infty$. Let $T = -2\pi/\log\lambda$. Choose $h_j \in \mathcal{M}_{\varphi_j}$ with $\lambda \leq h_j \leq 1$ such that

$$\sigma_T^{\varphi_j} = \mathrm{Ad}\big(h_j^{\mathrm{i}T}\big), \quad j = 1, 2.$$

Then $\psi_j = \varphi_j(h_j^{-1}\,\cdot\,)$ is a generalized trace of $\mathcal{M}$. By the inequality $\lambda\psi_j \leq \varphi_j \leq \psi_j$, we have $\mathfrak{m}_{\varphi_j} = \mathfrak{m}_{\psi_j}$, $j = 1, 2$. Thus, our assertion follows from the unitary equivalence of ψ_1 and ψ_2. Q.E.D.

So far, we have studied faithful weights of infinite multiplicity. We are now going to work on the normal state space $\mathfrak{S}_*$, in particular the action of $\operatorname{Int}(\mathcal{M})$ on it.

To motivate our further discussion, we state the following result, although we need some preparation:

Theorem 5.12. *If $\mathcal{M}$ is a separable factor of type III_λ, $0 \leq \lambda \leq 1$, then*

$$\sup\Big\{\inf\Big\{\|\omega_1 \circ \mathrm{Ad}(u) - \omega_2\| : u \in \mathcal{U}(\mathcal{M})\Big\} : \omega_1, \omega_2 \in \mathfrak{S}_*(\mathcal{M})\Big\} = 2\frac{1-\lambda^{\frac12}}{1+\lambda^{\frac12}}. \quad (6)$$

In particular, if $\mathcal{M}$ is of type III_1, then

$$\inf\Big\{\|\omega_1 \circ \mathrm{Ad}(u) - \omega_2\| : u \in \mathcal{U}(\mathcal{M})\Big\} = 0, \quad (6')$$

i.e. every orbit in $\mathfrak{S}_*$ under the actions of $\operatorname{Int}(\mathcal{M})$ is dense. If $\mathcal{M}$ is of type III_0, then the diameter of the orbit space $\mathfrak{S}_*(\mathcal{M})/\operatorname{Int}(\mathcal{M})$ is two.

To illustrate the meaning of the above theorem, we first discuss the semi-finite case.

For each $\lambda \in \mathbf{R}$, we denote by E_λ the characteristic function $\chi_{[\lambda,\infty[}$ of $[\lambda, \infty[$.

Proposition 5.13.

(i) *If τ is a faithful semi-finite normal trace τ on a von Neumann algebra $\mathcal{M}$, then for each $h, k \in L^1(\mathcal{M}, \tau)^+$ we have the inequality:*

$$\|h - k\|_1 \geq \int_{\mathbf{R}_+} \big|\tau\big(E_\lambda(h)\big) - \tau\big(E_\lambda(k)\big)\big|\, d\lambda\,. \quad (7)$$

(ii) *If* $\mathcal{M} = \mathcal{L}(\mathfrak{H})$ *and* τ *is the standard trace on it, then the inequality* (7) *becomes*

$$\|h - k\|_1 \geq \sum_{i=1}^{\infty} |\lambda_i - \mu_i|,$$

where $\lambda_1 \geq \lambda_2 \geq \cdots$ *and* $\mu_1 \geq \mu_2 \geq \cdots$ *are respectively the decreasing arrangement of eigenvalues of* h *and* k.

Lemma 5.14. *With* $\{\mathcal{M}, \tau\}$ *as above, if* $h \leq k$ *are self-adjoint positive operators affiliated with* $\mathcal{M}$, *then*

$$\tau\big(f(h)\big) \leq \tau\big(f(k)\big) \tag{8}$$

for any monotone increasing Borel function $f \geq 0$.

PROOF: We first prove

$$\tau\big(E_\lambda(h)\big) \leq \tau\big(E_\lambda(k)\big), \quad \lambda > 0. \tag{9}$$

Suppose that $\tau\big(E_\lambda(h)\big) > \tau\big(E_\lambda(k)\big)$ for some $\lambda > 0$, say λ_0. Let $e = E_{\lambda_0}(h)$ and $f = E_{\lambda_0}(k)$. Then we have, since $e - e \wedge f^\perp \sim f - e^\perp \wedge f$,

$$\tau\big(e - e \wedge f^\perp\big) = \tau\big(f - e^\perp \wedge f\big) \leq \tau(f) < \tau(e),$$

so that $e \wedge f^\perp \neq 0$. If $\xi \in (e \wedge f^\perp)\mathfrak{H}$, we have

$$\big(h\xi \mid \xi\big) \geq \lambda \|\xi\|^2 > \big(k\xi \mid \xi\big),$$

which contradicts the inequality $h \leq k$. Hence (9) holds.

With $g(\lambda) = \inf\big\{\, \mu : \mu \geq f(\lambda) \,\big\}$, we have

$$\tau\big(f(h)\big) = \int_{\mathbf{R}_+} \tau\Big(E_\lambda\big(f(h)\big)\Big)\, d\lambda = \int_{\mathbf{R}_+} \tau\big(E_{g(\lambda)}(h)\big)\, d\lambda$$

$$\leq \int_{\mathbf{R}_+} \tau\big(E_{g(\lambda)}(k)\big)\, d\lambda = \int_{\mathbf{R}_+} \tau\big(E_\lambda(f(k)\big)\, d\lambda = \tau\big(f(k)\big). \qquad \text{Q.E.D.}$$

Lemma 5.15. *If* $h, k \in L^1(\mathcal{M}, \tau)_+$, *then*

$$\|h - k\|_1 + \|h + k\|_1 = 2 \inf\Big\{\, \tau(a) : h \leq a,\ k \leq a,\ a \in L^1(\mathcal{M}, \tau) \,\Big\}. \tag{10}$$

Furthermore, this equality is attained, i.e. there exists $a_0 \in L^1(\mathcal{M}, \tau)_+$ *such that* $a_0 \geq h$, $a_0 \geq k$ *and*

$$\|h - k\|_1 + \|h + k\|_1 = 2\tau(a_0). \tag{10$'$}$$

PROOF: Suppose that $h \le a$ and $k \le a$. Then we have

$$\begin{aligned}\|h-k\|_1 &= \big\|(h-a)-(k-a)\big\|_1 \le \|a-h\|_1 + \|a-k\|_1 \\ &= \tau(a-h)+\tau(a-k) = 2\tau(a)-\tau(h+k) \\ &= 2\tau(a) - \|h+k\|_1.\end{aligned}$$

Conversely, let $a_1 = (h-k)_+$ and $a_2 = (h-k)_-$. Then we have $h-k = a_1 - a_2$ and $\|h-k\|_1 = \tau(a_1)+\tau(a_2)$. Put $a_0 = h + a_2$. Then we have $a_0 \ge h$ and $a_0 = h + a_2 = k + a_1 \ge k$, and also

$$2a_0 = h + k + a_1 + a_2 = h + k + |h-k|,$$

so that $2\tau(a_0) = \|h+k\|_1 + \|h-k\|_1$. Q.E.D.

PROOF OF PROPOSITION 5.13:

(i) With a_0 in ($10'$), we have

$$\begin{aligned}\|h-k\|_1 &= 2\tau(a_0) - \tau(h) - \tau(k) = \tau\big(2a_0 - h - k\big) \\ &= \int_{\mathbf{R}_+} \Big\{2\tau\big(E_\lambda(a_0)\big) - \tau\big(E_\lambda(h)\big) - \tau\big(E_\lambda(k)\big)\Big\}\, \mathrm{d}\lambda\,.\end{aligned}$$

By (9), we have

$$\tau\big(E_\lambda(a_0)\big) \ge \tau\big(E_\lambda(h)\big), \qquad \tau\big(E_\lambda(a_0)\big) \ge \tau\big(E_\lambda(k)\big).$$

so that for every $\lambda \ge 0$

$$\Big|\tau\big(E_\lambda(h)\big) - \tau\big(E_\lambda(k)\big)\Big| \le 2\tau\big(E_\lambda(a_0)\big) - \tau\big(E_\lambda(h)\big) - \tau\big(E_\lambda(k)\big).$$

Hence we get

$$\int_{\mathbf{R}_+} \Big|\tau\big(E_\lambda(h)\big) - \tau\big(E_\lambda(k)\big)\Big|\, \mathrm{d}\lambda \le \|h-k\|_1.$$

(ii) In the case that $\mathcal{M} = \mathcal{L}(\mathfrak{H})$, we have

$$h = \sum_{i=1}^{\infty} \lambda_i e_i \qquad \text{and} \qquad k = \sum_{j=1}^{\infty} \mu_j f_j,$$

where $\{e_i\}$ and $\{f_j\}$ are both orthogonal families of one dimensional projections. It then follows that $E_\lambda(h) = \sum\{\, e_i : \lambda_i \ge \lambda \,\}$, so that $\tau\big(E_\lambda(h)\big) = \sup\{\, n : \lambda_n \ge \lambda \,\}$, and similarly $\tau\big(E_\lambda(k)\big) = \sup\{\, m : \mu_m \ge \lambda \,\}$. Therefore, with $k' = \sum_{j=1}^{\infty} \mu_j e_j$, we have

$$\int_{\mathbf{R}_+} \Big|\tau\big(E_\lambda(h)\big) - \tau\big(E_\lambda(k)\big)\Big|\, \mathrm{d}\lambda = \|h-k'\|_1 = \sum_{i=1}^{\infty} \big|\lambda_i - \mu_i\big|.$$

Q.E.D.

Definition 5.16. We call the number of the left hand side of the equation (6), the *diameter* of *the state orbit space* of $\mathcal{M}$ and denote it by $d(\mathcal{M})$.

In fact, let $[\varphi]$ be the closure of the orbit $\varphi \circ \alpha,\ \alpha \in \mathrm{Int}(\mathcal{M})$, of φ. Then the function

$$d\big([\varphi], [\psi]\big) = \inf\Big\{ \|\varphi \circ \alpha - \psi\| : \alpha \in \mathrm{Int}(\mathcal{M}) \Big\} \tag{11}$$

is indeed a distance function on the set of $[\varphi]$'s, i.e. the set of the closures of orbits. Then $d(\mathcal{M})$ is in fact the diameter of this set, the state orbit space. It is clear that if $\mathcal{M}$ is not a factor, then $d(\mathcal{M})$ is always two. Thus $d(\mathcal{M})$ is not interesting for non-factors.

An immediate consequence of Proposition 5.13 is the following:

Corollary 5.17.
(i) *If* $\mathcal{M}$ *is a factor of type* I_n, *then* $d(\mathcal{M}) = 2\big(1 - 1/n\big)$ *for* $n < +\infty$ *and*

$$2\left(1 - \frac{1}{n}\right) = d\big([\tau], [\varphi]\big)$$

for the tracial state τ *and a pure state* φ.[8]
(ii) *If* $\mathcal{M}$ *is a factor of type* I_∞ *or of type* II, *then* $d(\mathcal{M}) = 2$.

PROOF:
(i) Identifying $\mathcal{M}$ with $M_n(\mathbf{C})$, let

$$h = \begin{pmatrix} \lambda_1 & 0 & \cdots & 0 \\ 0 & \lambda_2 & \cdots & 0 \\ 0 & \cdots & \ddots & \vdots \\ 0 & \cdots & \cdots & \lambda_n \end{pmatrix}, \qquad k = \begin{pmatrix} \mu_1 & 0 & \cdots & 0 \\ 0 & \mu_2 & \cdots & 0 \\ 0 & \cdots & \ddots & \vdots \\ 0 & \cdots & \cdots & \mu_n \end{pmatrix}$$

be diagonal matrices such that $\lambda_1 \geq \lambda_2 \geq \cdots \geq \lambda_n,\quad \mu_1 \geq \cdots \geq \mu_n$ and $\sum_{i=1}^n \lambda_i = 1 = \sum_{j=1}^n \mu_j$. By Proposition 5.13, for any unitary $u \in \mathcal{M}$, we have

$$\|uhu^* - k\|_1 \geq \sum_{i=1}^n |\lambda_i - \mu_i|.$$

Hence if φ and ψ are the states on $\mathcal{M}$ given by h and k respectively, i.e. $\varphi = \mathrm{Tr}(h\ \cdot\ )$ and $\psi = \mathrm{Tr}(k\ \cdot\ )$, then

$$d\big([\varphi], [\psi]\big) = \sum_{i=1}^n |\lambda_i - \mu_i|.$$

Then the supremum of $d\big([\varphi], [\psi]\big)$ is given by the sequences $\lambda_1 = \lambda_2 = \cdots = \lambda_n = 1/n$ and $\mu_1 = 1,\ \ \mu_2 = \cdots = \mu_n = 0$.

8 A state is called *tracial* if it is a trace, i.e., a tracial state is a trace with norm one.

(i) Let $\{e_1, e_2, \dots, e_n\}$ be orthogonal finite projections. Let τ be a faithful semi-finite normal trace on $\mathcal{M}$. For sequences $\lambda_1 \geq \lambda_2 \geq \cdots \geq \lambda_n$ and $\mu_1 \geq \mu_2 \geq \cdots \geq \mu_n$ such that $\sum_{i=1}^n \lambda_i \tau(e_i) = \sum_{j=1}^n \mu_j \tau(e_j) = 1$, we set

$$\varphi(x) = \sum_{i=1}^{u} \lambda_i \tau(xe_i), \qquad \psi(x) = \sum_{j=1}^{n} \mu_j \tau(xe_j).$$

Then we have

$$d([\varphi], [\psi]) = \sum_{i=1}^{n} |\lambda_i - \mu_i| \tau(e_i) = \|\varphi - \psi\|.$$

Therefore, if φ and ψ are of the form

$$\varphi(x) = \frac{\tau(xe)}{\tau(e)} \qquad \text{and} \qquad \psi(x) = \frac{\tau(xf)}{\tau(f)}$$

with $e, f \in \mathrm{Proj}(\mathcal{M})$ and $e \leq f$, then

$$d([\varphi], [\psi]) = 2\left(1 - \frac{\tau(e)}{\tau(f)}\right).$$

Thus, we conclude that $d(\mathcal{M}) = 2$. Q.E.D.

Since $\mathcal{M}$ is assumed to be separable the set $\mathfrak{S}_0$ of faithful normal states on $\mathcal{M}$ is norm dense in $\mathfrak{S}_*(\mathcal{M})$, so that in the computation of $d(\mathcal{M})$, we may restrict to $\mathfrak{S}_0$, i.e.

$$d(\mathcal{M}) = \sup\left\{ d([\varphi], [\psi]) : \varphi, \psi \in \mathfrak{S}_0 \right\}.$$

We first prove Theorem 5.12 for a σ-finite factor $\mathcal{M}$ of type III_1, i.e. $d(\mathcal{M}) = 0$. We consider a standard form $\{\mathcal{M}, \mathfrak{H}, \mathfrak{P}, J\}$.

Lemma 5.18. *If $\mathcal{M}$ is a σ-finite III_1-factor, and $\varphi \in \mathfrak{S}_0$, then for any non-zero $e, f \in \mathrm{Proj}(\mathcal{M}_\varphi)$ and $\varepsilon > 0$ there exists a partial isometry $u \in \mathcal{M}_\varphi$ with $u^*u = e_1 \leq e$ and $uu^* = f_1 \leq f$ such that*

a) $\|[u, \xi_\varphi]\| \leq \varepsilon \|u\|_\varphi^\sharp = \varepsilon \varphi(u^*u + uu^*)^{1/2} / \sqrt{2}$,

b) $\|[e_1, \xi_\varphi]\| \leq \varepsilon \varphi(e_1)^{1/2}, \quad \|[f_1, \xi_\varphi]\| \leq \varepsilon \varphi(f_1)^{1/2}$.

PROOF: Since $\mathcal{M}$ is of type III_1, $f\mathcal{M}_\varphi[-\delta, \delta]e \neq \{0\}$. Take non-zero $x \in f\mathcal{M}_\varphi[-\delta, \delta]e$, $\delta > 0$. Then $x\xi_\varphi$ belongs to the spectral subspace of Δ_φ corresponding to the interval $[e^{-\delta}, e^{\delta}]$, so that $\|(1 - \Delta_\varphi^{1/2})x\xi_\varphi\| \leq (e^{\delta/2} - 1)\|x\xi_\varphi\| \leq \delta \|x\xi_\varphi\|$; hence

$$\|x\xi_\varphi - \xi_\varphi x\| = \|x\xi_\varphi - Jx^*J\xi_\varphi\| = \left\|(1 - \Delta_\varphi^{\frac{1}{2}})x\xi_\varphi\right\| \leq \delta \|x\xi_\varphi\|.$$

By Corollary IX.1.23, there exists $a > 0$ such that

$$\left\|[u_a(x), \xi_\varphi]\right\| \leq 4\delta \left\|u_a(x)\right\|_\varphi^\sharp, \quad u_a(x) \neq 0.$$

Hence with $\delta < \varepsilon/4$ and $u = u_a(x)$, the partial isometry u satisfies (a). Next, we have

$$\begin{aligned}\left\|[e_1, \xi_\varphi]\right\| &= \left\|u^* u \xi_\varphi - \xi_\varphi u^* u\right\| \leq \left\|u^*(u\xi_\varphi - \xi_\varphi u)\right\| + \left\|(u^*\xi_\varphi - \xi_\varphi u^*)u\right\| \\ &\leq 2\left\|[u, \xi_\varphi]\right\| \leq 8\delta\|u\|_\varphi^\sharp;\end{aligned}$$

similarly we get $\left\|[f_1, \xi_\varphi]\right\| \leq 8\delta\|u\|_\varphi^\sharp$. Here we have $\|u\|_\varphi^\sharp = \varphi(e_1 + f_1)^{1/2}/\sqrt{2}$, and

$$\begin{aligned}\left|\varphi(e_1) - \varphi(f_1)\right| &= \left\|u\xi_\varphi\right\|^2 - \left\|u^*\xi_\varphi\right\|^2 \\ &= \left(\left\|u\xi_\varphi\right\| + \left\|u^*\xi_\varphi\right\|\right)\left(\left\|u\xi_\varphi\right\| - \left\|u^*\xi_\varphi\right\|\right) \\ &= \left(\varphi(e_1)^{\frac{1}{2}} + \varphi(f_1)^{\frac{1}{2}}\right)\left(\left\|u\xi_\varphi\right\| - \left\|\xi_\varphi u\right\|\right) \\ &\leq \left(\varphi(e_1)^{\frac{1}{2}} + \varphi(f_1)^{\frac{1}{2}}\right)\left\|[u, \xi_\varphi]\right\| \\ &\leq 4\delta\left[\varphi(e_1)^{\frac{1}{2}} + \varphi(f_1)^{\frac{1}{2}}\right]\sqrt{\frac{\varphi(e_1 + f_2)}{2}} \leq 4\delta\varphi(e_1 + f_1).\end{aligned}$$

Hence with $\delta < 1/8$, we have

$$\frac{1}{3}\varphi(e_1) \leq \varphi(f_1) \leq 3\varphi(e_1),$$

so that

$$\|u\|_\varphi^\sharp = \sqrt{\frac{\varphi(e_1 + f_1)}{2}} \leq \sqrt{2}\varphi(e_1)^{\frac{1}{2}}.$$

Thus we get

$$\left\|[e_1, \xi_\varphi]\right\| \leq 8\delta\|u\|_\varphi^\sharp \leq 12\delta\varphi(e_1)^{\frac{1}{2}}; \qquad \left\|[f_1, \xi_\varphi]\right\| \leq 12\delta\varphi(f_1)^{\frac{1}{2}}.$$

With $\delta > 0$ small enough, we get (a) and (b). Q.E.D.

Lemma 5.19. *For $\xi \in \mathfrak{P}$ and $e \in \mathrm{Proj}(\mathcal{M})$, put*

$$\xi' = e\xi e + (1 - e)\xi(1 - e).$$

Then we conclude that

(i) $\xi' \in \mathfrak{P}$ *and*

$$\omega_{\xi'}(e) = \omega_\xi(e) - \frac{1}{2}\left\|e\xi - \xi e\right\|^2;$$

(ii) *For a partial isometry $u \in \mathcal{M}$ with $u = (1 - e)ue$,*

$$\left\|u\xi' - \xi' u\right\| \leq \left\|u\xi - \xi u\right\|.$$

PROOF:

(i) By Theorem IX.1.2.(iii), we have $a\mathfrak{P}a^* \subset \mathfrak{P}$ for every $a \in \mathcal{M}$, so that ξ' belongs to $\mathfrak{P}$. Next, we have

$$\begin{aligned}\xi &= e\xi e + e\xi\,(1-e) + (1-e)\xi e + (1-e)\xi\,(1-e)\\ &= \xi' + e\xi\,(1-e) + (1-e)\xi e,\end{aligned}$$

so that

$$\big(e\xi \mid \xi\big) = \|e\xi'\|^2 + \big\|e\xi\,(1-e)\big\|^2 = \|e\xi'\|^2 + \frac{1}{2}\big\|e\xi - \xi e\big\|^2.$$

(ii) By the assumption on u, we have

$$\begin{aligned}u\xi' - \xi' u &= u\xi e - (1-e)\xi u;\\ u\xi - \xi u &= \big(u\xi e - (1-e)\xi u\big) + \big(u\xi(1-e) - e\xi u\big).\end{aligned}$$

In the last expression the first and the second terms are orthogonal, so that the inequality follows. Q.E.D.

PROOF OF THEOREM 5.12: In the type III_1 case: Assume that we have a σ-finite factor $\{\mathcal{M}, \mathfrak{H}, \mathfrak{P}, J\}$ of type III_1 in the standard form. We must prove that for a pair $\varphi_0, \psi_0 \in \mathfrak{S}_0$ of faithful states and $\varepsilon > 0$ there exists a unitary $u \in \mathcal{M}$ such that $\|\varphi_0 \circ \mathrm{Ad}(u) - \psi_0\| < \varepsilon$. Let ξ_0 and η_0 be the representing vectors in $\mathfrak{P}$ and let $\delta > 0$.

Let R be the set of all triples $r = (w, \alpha, \beta)$ such that w is a partial isometry and $\alpha, \beta \in \mathfrak{H}$:

a) with $a = w^*w$ and $b = ww^*$ we have

$$a\alpha = \alpha, \qquad b\beta = \beta;$$

b) $\|\alpha\|^2 \leq \delta\varphi_0(a)$, $\|\beta\|^2 \leq \delta\psi_0(b)$;

c) $\xi = \xi_0 - \alpha - J\alpha \in \mathfrak{P}$, $\eta = \eta_0 - \beta - J\beta \in \mathfrak{P}$ and $a\xi = \xi a$ and $b\eta = \eta b$, $\|\xi\| \leq 1$, $\|\eta\| \leq 1$;

d) with $\varphi = \omega_\xi$ and $\psi = \omega_\eta$ and $\rho = \varphi \oplus \psi$, $\bar{w} = \begin{pmatrix} 0 & 0 \\ w & 0 \end{pmatrix}$ satisfies the inequality:

$$\big\|\bar{w}\xi_\rho - \xi_\rho\bar{w}\big\|^2 \leq \delta\big(\|\bar{w}\|_\rho^\sharp\big)^2.$$

We define a partial ordering $r \leq r'$ in R by the following relation:

(i) w' is an extension of w, i.e. $w'a = w$ and $w'^*b = w^*$;

(ii) $a\,(\alpha' - \alpha) = 0$, $b\,(\beta' - \beta) = 0$;

(iii) $\|\alpha' - \alpha\|^2 \leq \delta\varphi_0(a' - a)$, $\|\beta' - \beta\|^2 \leq \delta\psi_0(b' - b)$.

We claim that this relation is transitive. In fact if $r \le r'$ and $r' \le r$, then $a \le a' \le a$ and $b \le b' \le b$ and w is an extension of w. Now $a\,(\alpha' - \alpha) = 0$ and $a'\,(\alpha - \alpha) = 0$ imply that $a\,(\alpha - a') = 0$ and then $a\,(\alpha - \alpha) = 0$ because $a \le a'$. Similarly $b\,(\beta - \beta) = 0$. Since $a'\,(\alpha' - \alpha) = \alpha' - \alpha$ and $a'\,(\alpha - \alpha') = 0$, we have

$$\|\alpha'' - \alpha\|^2 = \|\alpha'' - \alpha'\|^2 + \|\alpha' - \alpha\|^2 \le \delta\varphi_0(a'' - a') + \delta\varphi_0(a' - a)$$
$$= \delta\varphi_0(a'' - a).$$

Similarly, we have

$$\|\beta'' - \beta\|^2 \le \delta\psi_0(b'' - b).$$

Hence we get $r \le r''$. Thus it follows that the relation $\le$ is a partial ordering in R.

Suppose that $\{r_i\}$ is a totally ordered subset of R. Let $r_i = (w_i, \alpha_i, \beta_i)$ with $a_i = w_i^* w_i$ and $b_i = w_i w_i^*$. The map: $r_i \mapsto \varphi_0(a_i)$ is injective. Hence $\{r_i\}$ contains a cofinal subsequence $\{r_n\}$. Then $\{a_n\}$ and $\{b_n\}$ are both increasing sequences of projections and w_{n+1} extends w_n. Thus, $\{w_n\}$ converges to w in σ-strong* topology. Let $a = \sup a_n$ and $b = \sup b_n$. Then $a = w^* w$ and $b = ww^*$. Since we have $\|\alpha_n - \alpha_m\|^2 \le \delta\varphi(a_n - a_m)$ for $n \ge m$ by (iii), $\alpha = \lim \alpha_n$ exists, and $\beta = \lim \beta_n$ also exists similarly. It follows easily that $r = (w, \alpha, \beta) \in R$ and $r \ge r_n$ for all n.

Now, let $r = (w, \alpha, \beta) \in R$ be a maximal element of R. We claim that w is either an isometry or a coisometry. To prove this claim, assume that $a = w^* w \ne 1$ and $b = ww^* \ne 1$. We shall construct an element $r' \in R$ such that $r \le r'$ and $r \ne r'$, violating the maximality of $r \in R$.

Let $e = 1 - a$ and $f = 1 - b$. Set $\bar{e} = \left(\begin{smallmatrix} e & 0 \\ 0 & 0 \end{smallmatrix}\right)$ and $\bar{f} = \left(\begin{smallmatrix} 0 & 0 \\ 0 & f \end{smallmatrix}\right)$, and $\rho = \varphi \oplus \psi$ with φ and ψ as in (d). Since $e\alpha = 0$ by (a), we have

$$e\xi e = e\left(\xi_0 - \alpha - J\alpha\right)e = e\xi_0 e.$$

By (c), e and ξ commute, so that $e\xi = e\xi e \in \mathfrak{P}$ represents the functional φ_e. Furthermore, $\bar{e}$ and $\bar{f}$ both commute with ρ, and $\rho_{\bar{e}}$ and φ_e (resp. $\rho_{\bar{f}}$ and ψ_f) can be identified under the natural correspondence. Hence Lemma 5.18 applied to $(\mathcal{M} \otimes M_2)_{\bar{e}+\bar{f}}$ and ρ implies that for any $\varepsilon > 0$ there exists a partial isometry $\bar{u} \in \widetilde{\mathcal{M}} = \mathcal{M} \otimes M_2$ such that

$$\bar{e}_1 = \bar{u}^* \bar{u} \le \bar{e}, \qquad \bar{f}_1 = \bar{u}\bar{u}^* \le \bar{f},$$

$$\big\|[\bar{u}, \xi_\rho]\big\| \le \varepsilon \|\bar{u}\|_\rho^\sharp \ne 0, \qquad \big\|[\bar{e}_1, \xi_\rho]\big\| \le \varepsilon\rho(\bar{e}_1)^{\frac{1}{2}}, \qquad \big\|[\bar{f}_1, \xi_\rho]\big\| \le \varepsilon\rho(\bar{f}_1)^{\frac{1}{2}}.$$

Since $\bar{e} = e \otimes e_{11}$, $\bar{f} = f \otimes e_{22}$, $\bar{u}$ must be of the form $\bar{u} = \left(\begin{smallmatrix} 0 & 0 \\ u & 0 \end{smallmatrix}\right)$ for some partial isometry $u \in \mathcal{M}$, and we have $\bar{e}_1 = e_1 \otimes e_{11}$ and $\bar{f}_1 = f_1 \otimes e_{22}$ with $e_1 \in \mathrm{Proj}(\mathcal{M}_e)$ and $f_1 \in \mathrm{Proj}(\mathcal{M}_f)$. Since $e \perp a$ and $f \perp b$, $w' = w + u$ is a partial isometry such that $a' = w'^* w' = a + e_1$ and $b' = w' w'^* = b + f_1$. Let $\alpha' = \alpha + e_1\xi(1 - e_1)$ and $\beta' = \beta + f_1\eta(1 - f_1)$. We are now going to prove $r' = (w, \alpha', \beta') \in R$ and $r \le r'$, disproving the maximality of r.

Certainly, r' satisfies (a), and w' extends w so that (i) for r and r' holds. Now, we have

$$\alpha' - \alpha = e_1\xi(1 - e_1) \perp a\alpha = \alpha,$$

so that $a\,(\alpha' - \alpha) = 0$ and

$$\|\alpha'\|^2 = \|\alpha\|^2 + \big\|e_1\xi(1-e_1)\big\|^2 = \|\alpha\|^2 + \big\|e_1(e_1\xi - \xi e_1)\big\|^2$$
$$\leq \delta\varphi_0(a) + \varepsilon^2\varphi(e_1).$$

But by (c) for r, we have

$$e_1\xi = e_1 e\xi = e_1(1-a)\xi = e_1(1-a)\xi_0 = e_1\xi_0$$

because $a\xi = \xi a$ and $a\alpha = \alpha$. Hence $\varphi(e_1) = \varphi_0(e_1)$. Thus with small $\varepsilon > 0$, we have

$$\|\alpha'\|^2 \leq \delta\varphi_0(a) + \delta\varphi_0(e_1) = \delta\varphi_0(a').$$

Similarly, we have $\|\beta'\|^2 \leq \delta\varphi_0(b')$, so that (b) for r' holds. The above arguments have already proved (ii) and (iii) for $r \leq r'$.

To prove (c), let

$$\xi' = \xi_0 - \alpha' - J\alpha' \qquad \text{and} \qquad \eta' = \eta_0 - \beta' - J\beta'.$$

It then follows that

$$\xi' = \xi - e_1\xi(1-e_1) - (1-e_1)\xi e_1 = e_1\xi e_1 + (1-e_1)\xi(1-e_1) \in \mathfrak{P};$$
$$\eta' = f_1\eta f_1 + (1-f_1)\eta(1-f_1) \in \mathfrak{P}.$$

Clearly e_1 and ξ' commute. Also a and ξ commute because a and ξ' commute, and e_1 and a are orthogonal. Hence $a'\xi' = \xi' a'$. Similarly b' and η' commute. Since $\|\xi'\| \leq \|\xi\|$ and $\|\eta'\| \leq \|\eta\|$, (c) follows.

For (d), let $\rho' = \varphi' \oplus \psi'$ where $\varphi' = \omega_{\xi'}$ and $\psi' = \omega_{\eta'}$, and $\bar{w}' = w' \otimes e_{21}$. Since $[a, \varphi'] = 0$ as seen above, $\bar{a} = a \otimes e_{11}$ commutes with ρ'. The same is true for $\bar{b} = b \otimes e_{22}$ and ρ'. Now, we observe

$$\xi_{\rho'} = \begin{pmatrix} \xi' & 0 \\ 0 & \eta' \end{pmatrix} = \begin{pmatrix} e_1\xi e_1 + e_1^{\perp}\xi e_1^{\perp} & 0 \\ 0 & f_1\eta f_1 + f_1^{\perp}\eta f_1^{\perp} \end{pmatrix}$$
$$= E\xi_\varphi E + (1-E)\xi_\rho(1-E),$$

where $E = e_1 \otimes e_{11} + f_1 \otimes e_{22}$. We also know

$$s_\ell(\bar{w}) \leq (1-E), \qquad s_r(\bar{w}) \leq (1-E), \qquad s_\ell(\bar{u}) \leq E, \qquad s_r(\bar{u}) \leq E.$$

Hence, we have the orthogonal sum:

$$[\bar{w}', \xi_{\rho'}] = [\bar{w}, \xi_\rho] + [\bar{u}, \xi_\rho],$$

so that

$$\big\|[\bar{w}', \xi_{\rho'}]\big\|^2 = \big\|[\bar{w}, \xi_\rho]\big\|^2 + \big\|[\bar{u}, \xi_\rho]\big\|^2 \leq \delta\big(\|\bar{w}\|_\rho^\sharp\big)^2 + \varepsilon\big(\|\bar{u}\|_\rho^\sharp\big)^2.$$

Since $E\rho' E = E\rho E$, we have $\|\bar{w}\|_\rho^\sharp = \|\bar{w}\|_\rho^\sharp$, and $\|\bar{u}\|_\rho^\sharp = \|\bar{u}\|_{\rho'}^\sharp$. Hence the above inequality becomes, with $\varepsilon > 0$ small enough,

$$\big\|[\bar{w}', \xi_{\rho'}]\big\|^2 \leq \delta\big(\|\bar{w} + \bar{u}\|_{\rho'}^\sharp\big)^2 = \delta\big(\|\bar{w}'\|_{\rho'}^\sharp\big)^2.$$

Hence (d) for r' holds.

Thus, we obtained $r' \in R$ which dominates r. Therefore, if $r \in R$ is maximal, then either $a = 1$ or $b = 1$, which means that w is either an isometry or a coisometry. Suppose that $w^*w = 1$. Let $\varphi_1 = \varphi$ and $\psi_1 = \psi_b$ with $b = ww^*$ of course. By (b)

$$\|\varphi_0 - \varphi_1\| \leq 2\|\xi_0 - \xi\| \leq 4\delta^{\frac{1}{2}};$$

$$\|\psi_0 - \psi_1\| \leq 2\|\eta_0 - \eta\| \leq 4\delta^{\frac{1}{2}}.$$

By Theorem IX.1.2, and (d), we have

$$\begin{aligned}
\big\|w\varphi_1 w^* - \psi_1\big\|^2 &\leq \big\|\xi_{w\varphi_1 w^*} - \xi_{\psi_1}\big\|\big\|\xi_{w\varphi_1 w^*} + \xi_{\psi_1}\big\| \\
&\leq 2\big\|w\xi_\varphi w^* - \xi_{b\psi b}\big\| = 2\big\|(w\xi_\varphi - \xi_\psi w)w^*\big\| \\
&\leq 2\big\|w\xi_\varphi - \xi_\psi w\big\| = 2\big\|\bar{w}\xi_\rho - \xi_\rho \bar{w}\big\| \\
&\leq 2\sqrt{2}\delta^{\frac{1}{2}} < 4\delta^{\frac{1}{2}}.
\end{aligned}$$

Similarly, we get

$$\begin{aligned}
\big\|\varphi_1 - w^*\psi_1 w\big\|^2 &\leq 2\big\|\xi_\varphi - w^*\xi_\psi w\big\| = 2\big\|w^*w\xi_\varphi - w^*\xi_\psi w\big\| \\
&\leq 2\big\|w\xi_\varphi - \xi_\psi w\big\| \qquad (\text{as } b\xi_\psi = b\eta = \eta b \text{ by (c)}) \\
&= 2\big\|w\xi_\rho - \xi_\rho w\big\| \leq 4\delta^{\frac{1}{2}}.
\end{aligned}$$

We are going to adjust w to a unitary. To this end, we first observe

$$\psi(b) = \psi_1(1) \geq \psi_0(1) - 4\delta^{\frac{1}{2}} = 1 - 4\delta^{\frac{1}{2}},$$

so that $\psi(1-b) \leq 1 - \psi(b) \leq 4\delta^{1/2}$. Now choose a non-zero $e \in \mathrm{Proj}(\mathcal{M})$ such that $\varphi(e) < \delta$ and $\psi\big(wew^*\big) < \delta$, and set $f = 1 - b + wew^* \in \mathrm{Proj}(\mathcal{M})$. Observe that $fw(1-e) = 0$ and $\psi_0(f) < 4\delta^{1/2} + \delta$. Since $\mathcal{M}$ is a factor of type III_1, we can find a partial isometry $v \in \mathcal{M}$ such that $v^*v = e$ and $vv^* = f$. Set $u = w(1-e) + v$ and compute

$$u^*u = \big((1-e)w^* + v^*\big)\big(w(1-e) + v\big) = 1 - e + v^*v = 1;$$

$$uu^* = \big(w(1-e) + v\big)\big((1-e)w^* + v^*\big) = b - wew^* + f = 1.$$

Hence u is a unitary of $\mathcal{M}$. Also we have

$$\begin{aligned}
\big\|(u - w)\xi_\varphi\big\| &= \big\|(v - we)\xi_\varphi\big\| \leq \big\|v\xi_\varphi\big\| + \big\|we\xi_\varphi\big\| \\
&= \varphi(v^*v)^{\frac{1}{2}} + \varphi(e)^{\frac{1}{2}} = 2\varphi(e)^{\frac{1}{2}} < 2\delta^{\frac{1}{2}};
\end{aligned}$$

$$\begin{aligned}\left\|(u^* - w^*)\xi_\psi\right\| &= \left\|(v^* - ew^*)\xi_\psi\right\| \le \|v^*\xi_\psi\| + \|ew_\psi^*\xi_\psi\|\\ &= \psi(f)^{\frac12} + \psi(wew^*)^{\frac12} \le 2\psi(f)^{\frac12}\\ &< 2\big(4\delta^{\frac12} + \delta\big)^{\frac12}.\end{aligned}$$

Therefore, we get finally

$$\begin{aligned}\|u\varphi_0 u^* - \psi_0\| &\le \|u\varphi_0 u^* - u\varphi u^*\| + \|u\varphi u^* - \psi\| + \|\psi - \psi_0\|\\ &\le \|\varphi_0 - \varphi\| + 2\|\xi_{u\varphi u^*} - \xi_\psi\| + \|\psi - \psi_0\|\\ &\le 8\delta^{\frac12} + 2\|u\xi_\varphi u^* - \xi_\psi\| = 8\delta^{\frac12} + 2\|u\xi_\varphi - \xi_\psi u\|\\ &\le 8\delta^{\frac12} + 2\|(u - w)\xi_\varphi\| + \|w\xi_\varphi - \xi_\psi w\| + \|\xi_\psi(u - w)\|\\ &\le 8\delta^{\frac12} + 2\delta^{\frac12} + \|(u^* - w^*)\xi_\psi\| + \|\bar w\xi_\rho - \xi_\rho\bar w\|\\ &\le 10\delta^{\frac12} + 2\big(4\delta^{\frac12} + \delta\big)^{\frac12} + \sqrt{2\delta} < 20\delta^{\frac14} \qquad \text{for } \delta < 1.\end{aligned}$$

Thus a small enought $\delta > 0$ makes $u\varphi_0 u^*$ arbitrary close to ψ_0.

In the case that $b = ww^* = 1$ and $a = w^*w \neq 1$, we exchange the role of φ and ψ in the last arguments to complete the proof. Q.E.D.

Exercise XII.5

In this exercise, we consider only separable von Neumann algebras.

1[†]) Let $\mathcal{M}$ be an infinite factor with normal state space $\mathfrak{S}_*$ and flow of weights $\{\mathcal{A}, \mathbf{R}, \theta\}$. For each $\varphi \in \mathfrak{S}_*$, let $[\varphi]$ be the closure of the orbit $\{\varphi\circ\alpha : \alpha \in \mathrm{Int}(\mathcal{M})\}$ relative to norm in $\mathcal{M}_*$. There exists a Borel map: $\varphi \in \mathfrak{S}_* \mapsto p(\varphi) \in \mathrm{Proj}(\mathcal{A})$ such that:

(i) $p(\varphi) = p(\psi)$ if and only if $[\varphi] = [\psi]$, $\varphi, \psi \in \mathfrak{S}_*$;

(ii) $\theta_s\big(p(\varphi)\big) = p(e^{-s}\varphi)$, $\varphi \in \mathfrak{S}_*$;

(iii) The range $p(\mathfrak{S}_*)$ of p generates $\mathcal{A}$. [552]

§ 6 Structure of a von Neumann algebra of Type III, Part II – Functoriality and the Characteristic Square

We will discuss the canonical construction of a non-commutative flow of weights, to be called *the non-commutative flow of weights* of a von Neumann algebra $\mathcal{M}$ in this section. As we have seen in §1, every von Neumann algebra is uniquely decomposed as the crossed product $\mathcal{M} = \mathcal{N} \rtimes_\theta \mathbf{R}$ of a semi-finite von Neumann algebra $\mathcal{N}$ by a trace scaling one parameter automorphism group $\{\theta_t\}$. The uniqueness of $\{\mathcal{N}, \mathbf{R}, \theta\}$ was established, but the construction depended on the choice of a faithful semi-finite normal weight on $\mathcal{M}$. In fact, it was given as the crossed product $\mathcal{N} = \mathcal{M} \rtimes_{\sigma^\varphi} \mathbf{R}$ after fixing $\varphi \in \mathfrak{W}_0(\mathcal{M})$, where $\mathfrak{W}_0(\mathcal{M})$ means the set of all faithful semi-finite normal weights on $\mathcal{M}$. We will eliminate this dependence from the construction of $\mathcal{N}$.

For short, with $\mathcal{M}$ a fixed von Neumann algebra let $\mathfrak{W}_0 = \mathfrak{W}_0(\mathcal{M})$. For a fixed $t \in \mathbf{R}$, consider the relation on $\mathcal{M} \times \mathfrak{W}_0$ given by:

$$(x, \varphi) \sim (y, \psi) \quad \text{if } x = y(\mathrm{D}\varphi : \mathrm{D}\psi)_t, \quad (x, \varphi), (y, \psi) \in \mathcal{M} \times \mathfrak{W}_0.$$

By the chain rule, Theorem VIII.3.7, it is easily seen that the relation "$\sim$" is an equivalence relation in $\mathcal{M} \times \mathfrak{W}_0$. Let $\mathcal{M}(t) = (\mathcal{M} \times \mathfrak{W}_0)/\text{“} \sim \text{”}$ be the space of all equivalence classes and denote the class of (x, φ) in $\mathcal{M}(t)$ by $x\varphi^{\mathrm{i}t}$.

Lemma 6.1. *For each $t \in \mathbf{R}$ the space $\mathcal{M}(t) = (\mathcal{M} \times \mathfrak{W}_0)/\text{“} \sim \text{”}$ is a dual Banach space relative to the following structure:*

$$\left.\begin{aligned} &(x\varphi^{\mathrm{i}t}) + (y\varphi^{\mathrm{i}t}) = (x+y)\varphi^{\mathrm{i}t}; \\ &\lambda(x\varphi^{\mathrm{i}t}) = \lambda x\varphi^{\mathrm{i}t}, \qquad \lambda \in \mathbf{C}; \\ &\|x\varphi^{\mathrm{i}t}\| = \|x\|, \qquad x, y \in \mathcal{M}, \quad \varphi \in \mathfrak{W}_0. \end{aligned}\right\} \tag{1}$$

PROOF: For each fixed $\varphi \in \mathfrak{W}_0$, let $\Phi_{\varphi,t}$ be the map: $(x, \psi) \in \mathcal{M} \times \mathfrak{W}_0 \mapsto x\,(\mathrm{D}\psi : \mathrm{D}\varphi)_t \in \mathcal{M}$. Then it follows that $\Phi_{\varphi,t}(a) = \Phi_{\varphi,t}(b)$ for each pair $a, b \in \mathcal{M} \times \mathfrak{W}_0$ if and only if $a \sim b$. Thus $\Phi_{\varphi,t}$ gives rise to a bijection $\hat{\Phi}_{\varphi,t}$ of $\mathcal{M}(t)$ onto $\mathcal{M}$. Thus $\Phi_{\varphi,t}$ pulls the dual Banach space structure of $\mathcal{M}$ back to $\mathcal{M}(t)$. The only question remains to be seen is if this Banach space structure in $\mathcal{M}(t)$ is independent of the choice of $\varphi \in \mathfrak{W}_0$. So let us write $a +_\varphi b,\ a, b \in \mathcal{M}(t)$, when the addition of $a \in \mathcal{M}(t)$ and $b \in \mathcal{M}(t)$ is defined by $\hat{\Phi}_{\varphi,t}$. Suppose that $a = x\varphi^{\mathrm{i}t},\ b = y\varphi^{\mathrm{i}t}$, $x, y \in \mathcal{M}$. Then $a +_\varphi b = (x+y)\varphi^{\mathrm{i}t}$ and

$$\begin{aligned} a +_\psi b &= \big(x\,(\mathrm{D}\varphi : \mathrm{D}\psi)_t + y\,(\mathrm{D}\varphi : \mathrm{D}\psi)_t\big)\psi^{\mathrm{i}t} = (x+y)(\mathrm{D}\varphi : \mathrm{D}\psi)_t\psi^{\mathrm{i}t} \\ &= (x+y)\varphi^{\mathrm{i}t} = a +_\varphi b. \end{aligned}$$

Thus the addition in $\mathcal{M}(t)$ defined above is independent of the choice of $\varphi \in \mathfrak{W}_0$. The similar consideration ensures that the dual Banach space structure in $\mathcal{M}(t)$ is independent of φ. Q.E.D.

Define a product map:

$$(a, b) \in \mathcal{M}(s) \times \mathcal{M}(t) \mapsto a \cdot_\varphi b = \hat{\Phi}^{-1}_{\varphi,s+t}\Big(\hat{\Phi}_{\varphi,s}(a)\sigma^\varphi_s\big(\hat{\Phi}_{\varphi,t}(b)\big)\Big) \in \mathcal{M}(s+t),$$

for each pair $s, t \in \mathbf{R}$. In a more suggestive notation, we are defining the product of an element of $\mathcal{M}(s)$ and another in $\mathcal{M}(t)$ by:

$$\big(x\varphi^{\mathrm{i}s}\big) \cdot_\varphi \big(y\varphi^{\mathrm{i}t}\big) = \big(x\sigma^\varphi_s(y)\big)\varphi^{\mathrm{i}\,(s+t)}.$$

Lemma 6.2. *Set $\mathcal{F} = \dot{\bigcup}_{t\in\mathbf{R}} \mathcal{M}(t)$ the disjoint union. Then we have:*

(i) *The product defined above is independent of the choice of $\varphi \in \mathfrak{W}_0$ and satisfies the associative law and the distributive law whenever product and sum are* *possible. Thus we may write the $\mathcal{M}(s+t)$-valued product in $\mathcal{M}(s) \times \mathcal{M}(t)$ by:*

$$\big(x\varphi^{\mathrm{i}s}\big)\big(y\varphi^{\mathrm{i}t}\big) = \big(x\sigma_s^\varphi(y)\big)\varphi^{\mathrm{i}(s+t)}, \quad x, y \in \mathcal{M}, \quad s, t \in \mathbf{R}. \tag{2}$$

(ii) *The involution in $\mathcal{F}$ defined by:*

$$\big(x\varphi^{\mathrm{i}t}\big)^* = \sigma_{-t}^\varphi(x^*)\varphi^{-\mathrm{i}t}, \quad x\varphi^{\mathrm{i}t} \in \mathcal{M}(t), \tag{3}$$

is independent of $\varphi \in \mathfrak{W}_0$ and satisfies:

$$\left.\begin{aligned} (\lambda x + \mu y)^* &= \bar{\lambda} x^* + \bar{\mu} y^*, && x, y \in \mathcal{M}(t);\\ (xy)^* &= y^* x^*, && x, y \in \mathcal{F};\\ x^{**} &= x, && x \in \mathcal{F}. \end{aligned}\right\} \tag{4}$$

We leave the proof to the reader.

Let $\hat{\Phi}_\varphi(a) = \big(\hat{\Phi}_{\varphi,t}(a), t\big) \in \mathcal{M} \times \mathbf{R}$ for $a \in \mathcal{M}(t)$, so that $\hat{\Phi}_\varphi$ is a bijection of $\mathcal{F}$ onto $\mathcal{M} \times \mathbf{R}$.

Lemma 6.3.

(i) *The map $\hat{\Phi}_{\varphi,\psi} = \hat{\Phi}_\varphi \circ \hat{\Phi}_\psi^{-1} : \mathcal{M} \times \mathbf{R} \mapsto \mathcal{M} \times \mathbf{R}$ is a homeomorphism on $\mathcal{B} \times \mathbf{R}$ for each bounded subset $\mathcal{B}$ of $\mathcal{M}$ relative to the product topology of any operator topology on $\mathcal{M}$ and the usual topology on $\mathbf{R}$.*

(ii) *This map is a homeomorphism on the entire space $\mathcal{M} \times \mathbf{R}$ if we consider the Arens-Mackey $\tau(\mathcal{M}, \mathcal{M}_*)$-topology on $\mathcal{M}$.*

PROOF:

(i) Assume that $\mathcal{B}$ is contained in the ball of radius $M > 0$. Let $\{x_i\}$ be a net in $\mathcal{B}$ converging to $x \in \mathcal{M}$ in the σ-weak topology and $\{t_i\}$ a net in $\mathbf{R}$ converging to $t \in \mathbf{R}$. Then we have $\Phi_{\varphi,\psi}(x_i, t_i) = (x_i(\mathrm{D}\varphi : \mathrm{D}\psi)_{t_i}, t_i)$. With $\mathfrak{H}$ a representing Hilbert space of $\mathcal{M}$, we compute for $\xi, \eta \in \mathfrak{H}$:

$$\begin{aligned} &\Big|\big(x_i(\mathrm{D}\varphi : \mathrm{D}\psi)_{t_i}\xi \mid \eta\big) - \big(x(\mathrm{D}\varphi : \mathrm{D}\psi)_t\xi \mid \eta\big)\Big| \\ &\le \Big|\big((\mathrm{D}\varphi : \mathrm{D}\psi)_{t_i}\xi \mid x_{\mathrm{i}}^*\eta\big) - \big((\mathrm{D}\varphi : \mathrm{D}\psi)_t\xi \mid x_{\mathrm{i}}^*\eta\big)\Big| \\ &\quad + \Big|\big((\mathrm{D}\varphi : \mathrm{D}\psi)_t\xi \mid x_{\mathrm{i}}^*\eta\big) - \big((\mathrm{D}\varphi : \mathrm{D}\psi)_t\xi \mid x^*\eta\big)\Big| \\ &\le \Big\|\big((\mathrm{D}\varphi : \mathrm{D}\psi)_{t_i} - (\mathrm{D}\varphi : \mathrm{D}\psi)_t\big)\xi\Big\| \, \|x_{\mathrm{i}}^*\eta\| \\ &\quad + \Big|\big((\mathrm{D}\varphi : \mathrm{D}\psi)_t\xi \mid x_i^*\eta\big) - \big((\mathrm{D}\varphi : \mathrm{D}\psi)_t\xi \mid x^*\eta\big)\Big| \\ &\le M\Big\|\big((\mathrm{D}\varphi : \mathrm{D}\psi)_{t_i} - (\mathrm{D}\varphi : \mathrm{D}\psi)_t\big)\xi\Big\| \, \|\eta\| \\ &\quad + \Big|\big((\mathrm{D}\varphi : \mathrm{D}\psi)_t\xi \mid x_i^*\eta\big) - \big((\mathrm{D}\varphi : \mathrm{D}\psi)_t\xi \mid x^*\eta\big)\Big|. \end{aligned}$$

The strong continuity of the cocycle derivative yields the convergence of the last expression to zero along the net. Hence the map $\hat{\Phi}_{\varphi,\psi}$ is continuous relative to the product topology of the σ-weak topology of $\mathcal{B}$ and the usual topology of $\mathbf{R}$. As $\hat{\Phi}_{\varphi,\psi}^{-1} = \hat{\Phi}_{\psi,\varphi}$, we conclude the bicontinuity of $\hat{\Phi}_{\varphi,\psi}$.

As the continuity relative to other operator topologies follows similarly, we leave the rest of the proof to the reader.

(ii) To prove the continuity of $\hat{\Phi}_{\varphi,\psi}$, let K be a $\sigma(\mathcal{M}_*, \mathcal{M})$-compact convex subset of $\mathcal{M}_*$ and I a closed bounded interval in $\mathbf{R}$. Let $\{x_i\}$ be a net converging to x in the $\tau(\mathcal{M}, \mathcal{M}_*)$-topology. Observe that $L = \bigcup_{t\in I}(\mathrm{D}\varphi : \mathrm{D}\psi)_t K$ is a weakly compact subset of $\mathcal{M}_*$ as it is the image of a compact set $K \times I$ under the continuous map: $(\omega, t) \in K \times I \mapsto (\mathrm{D}\varphi : \mathrm{D}\psi)_t\, \omega \in \mathcal{M}_*$. The convex closure $\tilde{K} = \overline{\mathrm{conv}}(L)$ of L is again weakly compact by the Krein-Smulian theorem, [506, p. 434]. Let $\{t_i\}$ a net in $\mathbf{R}$ converging to a point $t \in \mathbf{R}$. Without loss of generality, we may assume that the net is contained in a closed bounded interval I. For each $\omega \in K$, we have:

$$\begin{aligned}
&\left|\left\langle \pi_\ell \circ \hat{\Phi}_{\varphi,\psi}(x_i, t_i) - \pi_\ell \circ \hat{\Phi}_{\varphi,\psi}(x, t), \omega\right\rangle\right| \\
&\quad = \left|\left\langle x_i(\mathrm{D}\varphi : \mathrm{D}\psi)_{t_i} - x(\mathrm{D}\varphi : \mathrm{D}\psi)_t, \omega\right\rangle\right| \\
&\quad \leq \left|\left\langle (x_i - x), (\mathrm{D}\varphi : \mathrm{D}\psi)_{t_i}\omega\right\rangle\right| + \left|\left\langle x, \left((\mathrm{D}\varphi : \mathrm{D}\psi)_{t_i} - (\mathrm{D}\varphi : \mathrm{D}\psi)_t\right)\omega\right\rangle\right| \\
&\quad \leq \left|\left\langle (x_i - x), (\mathrm{D}\varphi : \mathrm{D}\psi)_{t_i}\omega\right\rangle\right| + \left|\left\langle (\mathrm{D}\varphi : \mathrm{D}\psi)_{t_i} - (\mathrm{D}\varphi : \mathrm{D}\psi)_t, \omega x\right\rangle\right| \\
&\quad \leq \sup_{\rho\in\tilde{K}} \left|\langle x_i - x, \rho\rangle\right| + \sup_{\rho\in K x} \left|\left\langle (\mathrm{D}\varphi : \mathrm{D}\psi)_{t_i} - (\mathrm{D}\varphi : \mathrm{D}\psi)_t, \rho\right\rangle\right|,
\end{aligned}$$

which tends, along the net, to zero uniformly on $\omega \in K$, where π_ℓ means the projection to the first component of $\mathcal{M} \times \mathbf{R}$. Q.E.D.

Remark 6.4. As the cocycle derivative $\{(\mathrm{D}\varphi : \mathrm{D}\psi)_t\}$ need not be norm continuous, we can not expect to have a canonical norm topology in $\mathcal{F}$ independent of the choice of $\varphi \in \mathfrak{W}_0$. Namely, the map $\hat{\Phi}_{\varphi,\psi}$ is not continuous relative to the product topology of the norm on $\mathcal{M}$ and the usual topology on $\mathbf{R}$ unless the cocycle derivative $\{(\mathrm{D}\varphi : \mathrm{D}\psi)_t\}$ is norm continuous.

Definition 6.5. The τ-topology on $\mathcal{F}$ means the topology induced on $\mathcal{F}$ by $\hat{\Phi}_\varphi$ from the product topology on $\mathcal{M} \times \mathbf{R}$ of the Arens-Mackey $\tau(\mathcal{M}, \mathcal{M}_*)$-topology and the usual topology on $\mathbf{R}$.

Let $\Gamma^1(\mathcal{F})$ be the set of all L^1-cross sections of $\mathcal{F}$ in the sense that

(i) each element $x \in \Gamma^1(\mathcal{F})$ is measurable in the sense that for any $\varepsilon > 0$ and a bounded interval I of $\mathbf{R}$ there exists a compact subset $K \subset I$ such that the restriction $t \in K \mapsto x(t) \in \mathcal{F}$ is continuous relative to the τ-topology in $\mathcal{F}$ and $|I - K| < \varepsilon$;

(ii) each element $x \in \Gamma^1(\mathcal{F})$ is integrable:

$$\|x\|_1 = \int_{\mathbf{R}} \|x(t)\| \, \mathrm{d}t < \infty, \tag{5}$$

where the measurability of the function: $t \in \mathbf{R} \mapsto \|x(t)\| \in \mathbf{R}_+$ follows from the lower semi-continuity of the norm in $\mathcal{M}$ relative to the $\tau(\mathcal{M}, \mathcal{M}_*)$-topology.

Lemma 6.6. *The set $\Gamma^1(\mathcal{F})$ is an Banach involutive algebra relative to the following structure:*

$$\left.\begin{aligned} (\lambda x + \mu y)(t) &= \lambda x(t) + \mu y(t), && \lambda, \mu \in \mathbf{C}; \\ (xy)(t) &= \int_{\mathbf{R}} x(s) y(t-s) \, \mathrm{d}s, && x, y \in \Gamma^1(\mathcal{F}); \\ x^*(t) &= x(-t)^*, && x \in \Gamma^1(\mathcal{F}). \end{aligned}\right\} \tag{6}$$

The proof is routine, so we will leave it to the reader. We call $\Gamma^1(\mathcal{F})$ the L^1-*bundle algebra for* $\mathcal{M}$.

We now consider actions of von Neumann algebra $\mathcal{M}$ on a Hilbert space $\mathfrak{H}$ and study how the representation of $\Gamma^1(\mathcal{F})$ can be defined. So let $\{\mathcal{M}, \mathfrak{H}\}$ be a von Neumann algebra and $\mathcal{N} = (\mathcal{M}')^\circ$ be the opposite algebra of the commutant $\mathcal{M}'$ which acts on $\mathfrak{H}$ from the right, which makes $\mathfrak{H}$ an $\mathcal{M}$-$\mathcal{N}$-bimodule. Because we are now dealing with two different von Neumann algebras, we will use $\mathcal{F}(\mathcal{M})$ and $\mathcal{F}(\mathcal{N})$ to refer to their respective bundles, whenever there is any possibility of confusion. To construct a bundle of Hilbert spaces on which $\mathcal{F}(\mathcal{M})$ acts from the left and $\mathcal{F}(\mathcal{N})$ acts from the right, we consider the (cartesian product) $X = \mathbf{R} \times \mathfrak{W}_0(\mathcal{M}) \times \mathfrak{H} \times \mathfrak{W}_0(\mathcal{N})$, and a relation $\sim_t$ on this set, after fixing $t \in \mathbf{R}$, viz.,

$$(r_1, \varphi_1, \xi_1, \psi_1) \sim_t (r_2, \varphi_2, \xi_2, \psi_2), \tag{7}$$

whenever

$$\left(\frac{\mathrm{d}\varphi_1}{\mathrm{d}\psi_1^\circ}\right)^{ir_1} \xi_1 = \left(\frac{\mathrm{d}\varphi_2}{\mathrm{d}\psi_2^\circ}\right)^{ir_2} \xi_2 (\mathrm{D}\psi_2 : \mathrm{D}\psi_1)_t. \tag{8}$$

It is easily verified that the relation $\sim_t$ is an equivalence relation on the set X. We denote the quotient set $X/\sim_t$ by $\mathfrak{H}(t)$ and the class $[r, \varphi, \xi, \psi] \in \mathfrak{H}(t)$ of $(r, \varphi, \xi, \psi) \in X$ by $\varphi^{ir} \xi \psi^{\mathrm{i}(t-r)}$. So in $\mathfrak{H}(t)$ we have

$$\varphi^{it} \xi = \left(\frac{\mathrm{d}\varphi}{\mathrm{d}\psi^\circ}\right)^{it} \xi \psi^{it}, \quad \varphi \in \mathfrak{W}_0(\mathcal{M}), \quad \psi \in \mathfrak{W}_0(\mathcal{N}), \quad \xi \in \mathfrak{H}, \quad t \in \mathbf{R} \tag{9}$$

or equivalently

$$\varphi^{it} \xi \psi^{-it} = \left(\frac{\mathrm{d}\varphi}{\mathrm{d}\psi^\circ}\right)^{it} \xi. \tag{9'}$$

Observe that the relation $\sim_t$ is generated by subrelations: $\varphi_1^{it}\varphi_2^{-it} \sim \left(\mathrm{D}\varphi_1 : \mathrm{D}\varphi_2\right)_t$, $\varphi_1, \varphi_2 \in \mathfrak{W}_0(\mathcal{M})$, $\psi_1^{it}\psi_2^{-it} \sim \left(\mathrm{D}\psi_1 : \mathrm{D}\psi_2\right)_t$, $\psi_1, \psi_2 \in \mathfrak{W}_0(\mathcal{N})$ and the relation (9′).

In the set $\mathfrak{H}(t)$, $t \in \mathbf{R}$, the linear structure and the inner product defined by:

$$\left.\begin{aligned} \lambda\left(\varphi^{\mathrm{i}r}\xi\psi^{\mathrm{i}(t-r)}\right) + \mu\left(\varphi^{\mathrm{i}r}\eta\psi^{\mathrm{i}(t-r)}\right) &= \varphi^{\mathrm{i}r}\left(\lambda\xi + \mu\eta\right)\psi^{\mathrm{i}(t-r)}; \\ \left(\varphi^{\mathrm{i}r}\xi\psi^{\mathrm{i}(t-r)} \mid \varphi^{\mathrm{i}r}\eta\psi^{\mathrm{i}(t-r)}\right) &= (\xi \mid \eta) \end{aligned}\right\} \tag{10}$$

makes $\mathfrak{H}(t)$ a Hilbert space which does not depend on the choice of $\varphi \in \mathfrak{W}_0(\mathcal{M})$, $\psi \in \mathfrak{W}_0(\mathcal{N})$ nor on $r \in \mathbf{R}$ as seen easily. Let $\mathfrak{G} = \dot{\bigcup}_{t\in\mathbf{R}}\mathfrak{H}(t)$ be the disjoint union of Hilbert spaces $\left\{\mathfrak{H}(t) : t \in \mathbf{R}\right\}$. Observe that for each $\varphi \in \mathfrak{W}_0(\mathcal{M})$ and $\psi \in \mathfrak{W}_0(\mathcal{N})$ the maps:

$$\begin{aligned} U_\varphi(t)&\colon \xi \in \mathfrak{H} \mapsto \varphi^{\mathrm{i}t}\xi = U_\varphi(t)\xi \in \mathfrak{H}(t); \\ V_\psi(t)&\colon \eta \in \mathfrak{H} \mapsto \eta\psi^{\mathrm{i}t} = V_\psi(t)\eta \in \mathfrak{H}(t), \end{aligned} \tag{11}$$

are both unitaries satisfying

$$V_\psi(t)^*U_\varphi(t) = \left(\frac{\mathrm{d}\varphi}{\mathrm{d}\psi^\circ}\right)^{it}. \tag{12}$$

We now define a multilinear product: $\left(x\varphi^{\mathrm{i}r}, \varphi^{\mathrm{i}s}\xi, y\psi^{\mathrm{i}t}\right) \in \mathcal{M}(r) \times \mathfrak{H}(s) \times \mathcal{N}(t) \mapsto x\varphi^{\mathrm{i}r}\varphi^{\mathrm{i}s}\xi y\psi^{\mathrm{i}t} \in \mathfrak{H}\left(r+s+t\right)$ as follows:

$$\begin{aligned} x\varphi^{\mathrm{i}r}\varphi^{\mathrm{i}s}\xi y\psi^{\mathrm{i}t} &= \varphi^{\mathrm{i}(r+s)}\sigma^\varphi_{-(r+s)}(x)\xi y\psi^{\mathrm{i}t} \\ &= x\left(\frac{\mathrm{d}\varphi}{\mathrm{d}\psi^\circ}\right)^{\mathrm{i}(r+s)}\xi\sigma^\psi_{(r+s)}(y)\psi^{\mathrm{i}(r+s+t)}. \end{aligned} \tag{13}$$

It is again routine to check that the above product does not depend on the choice of $\varphi \in \mathfrak{W}_0(\mathcal{M})$ and $\psi \in \mathfrak{W}_0(\mathcal{N})$ and is associative. We omit the detail.

Lemma 6.7. *With $r, s, t \in \mathbf{R}$ fixed, we have the following statements:*

(i) *If a pair $(a', b') \in \mathcal{L}\left(\mathfrak{H}(s), \mathfrak{H}(s+t)\right) \times \mathcal{L}\left(\mathfrak{H}(r+s), \mathfrak{H}(r+s+t)\right)$ satisfies $b'(x\xi) = x(a'\xi)$ for every $x \in \mathcal{M}(r)$, i.e. if (a, b) makes the following diagram commutative:*

$$\begin{array}{ccc} \mathfrak{H}(s) & \xrightarrow{\ a'\ } & \mathfrak{H}(s+t) \\ {\scriptstyle x}\big\downarrow & & {\scriptstyle x}\big\downarrow \\ \mathfrak{H}(r+s) & \xrightarrow{\ b'\ } & \mathfrak{H}(r+s+t) \end{array}$$

then there exists $y \in \mathcal{N}(t)$ such that $a'\xi = \xi y$, $\xi \in \mathfrak{H}(s)$, and $b'\eta = \eta y$, $\eta \in \mathfrak{H}(r+s)$.

(ii) *If a pair* $(a, b) \in \mathcal{L}\big(\mathfrak{H}(s), \mathfrak{H}(s+t)\big) \times \mathcal{L}\big(\mathfrak{H}(r+s), \mathfrak{H}(r+s+t)\big)$ *satisfies* $b(\xi y) = (b\xi)y$ *for every* $y \in \mathcal{N}(r)$, *i.e. if* (a', b') *makes the following diagram commutative:*

$$\begin{array}{ccc} \mathfrak{H}(s) & \xrightarrow{\;\;a\;\;} & \mathfrak{H}(s+t) \\ \Big\downarrow y & & \Big\downarrow y \\ \mathfrak{H}(r+s) & \xrightarrow{\;\;b\;\;} & \mathfrak{H}(r+s+t) \end{array}$$

then there exists $x \in \mathcal{M}(t)$ *such that* $a\xi = x\xi$, $\xi \in \mathfrak{H}(s)$, *and* $b\eta = x\eta$, $\eta \in \mathfrak{H}(r+s)$.

PROOF: We prove only the assertion (i). The other follows by symmetry. Fix a pair $(\varphi, \psi) \in \mathfrak{W}_0(\mathcal{M}) \times \mathfrak{W}_0(\mathcal{N})$. For each $\xi \in \mathfrak{H}$ and $x \in \mathcal{M}$, we have

$$\begin{aligned} \varphi^{-\mathrm{i}(r+s)}\Big(b'\varphi^{\mathrm{i}r}\big(\varphi^{\mathrm{i}s}x\xi\big)\Big)\psi^{-\mathrm{i}t} &= \varphi^{-\mathrm{i}(r+s)}\Big(b'\big(\varphi^{\mathrm{i}(r+s)}x\xi\big)\Big)\psi^{-\mathrm{i}t} \\ &= \varphi^{-\mathrm{i}(r+s)}\Big(b'\sigma^{\varphi}_{r+s}(x)\varphi^{\mathrm{i}r}\varphi^{\mathrm{i}s}\xi\Big)\psi^{-\mathrm{i}t} \\ &= \varphi^{-\mathrm{i}(r+s)}\Big(\sigma^{\varphi}_{r+s}(x)\varphi^{\mathrm{i}r}a'\varphi^{\mathrm{i}s}\xi\Big)\psi^{-\mathrm{i}t} \\ &= \Big(x\varphi^{-\mathrm{i}s}a'\varphi^{\mathrm{i}s}\xi\Big)\psi^{-\mathrm{i}t}; \end{aligned}$$

equivalently

$$V_\psi(t)^*\, U_\varphi(r+s)^*\, b'\, U_\varphi(r+s)\, x = x\, V_\psi(t)^*\, U_\varphi(s)^*\, a'\, U_\varphi(s), \qquad x \in \mathcal{M}.$$

Taking $x = 1$, we conclude that

$$V_\psi(t)^*\, U_\varphi(r+s)^*\, b'\, U_\varphi(r+s) = V_\psi(t)^*\, U_\varphi(s)^*\, a'\, U_\varphi(s)$$

is an operator in the commutant $\mathcal{M}'$, which we denote by $y^\circ \in \mathcal{M}'$. Namely, there exists an operator $y = y^{\circ\circ} \in \mathcal{N}$ such that

$$V_\psi(t)^*\, U_\varphi(r+s)^*\, b'\, U_\varphi(r+s)\, \xi = V_\psi(t)^*\, U_\varphi(s)^*\, a'\, U_\varphi(s)\, \xi = \xi y, \qquad \xi \in \mathfrak{H}.$$

This means:

$$\begin{aligned} a'\varphi^{\mathrm{i}s}\xi &= \varphi^{\mathrm{i}s}\xi y \psi^{\mathrm{i}t}; \\ b'\varphi^{\mathrm{i}(r+s)}\xi &= \varphi^{\mathrm{i}(r+s)}\xi y \psi^{\mathrm{i}t}, \qquad \xi \in \mathfrak{H}. \end{aligned}$$

This completes the proof. Q.E.D.

Let $\widetilde{\mathfrak{H}} = \Gamma^2(\mathfrak{G})$ be the Hilbert space of square integrable cross-sections of the bundle $\mathfrak{G}$. We want to let the bundle algera $\Gamma^1\big(\mathcal{F}(\mathcal{M})\big)$ (resp. $\Gamma^1(\mathcal{F}(\mathcal{N}))$) act on $\widetilde{\mathfrak{H}}$ from the left (resp. from the right). Before doing this, we should establish the correspondence between the bundle algebras $\Gamma^1\big(\mathcal{F}(\mathcal{M}')\big)$ and $\Gamma^1\big(\mathcal{F}(\mathcal{N})\big)$.

We have defined $\mathcal{N}$ to be the opposite algebra $(\mathcal{M}')^\circ$ of $\mathcal{M}'$. Let us denote the canonical correspondence between $\mathcal{N}$ and $\mathcal{M}'$ by $y \in \mathcal{N} \longleftrightarrow y^\circ \in \mathcal{M}'$ and $\psi \in \mathfrak{W}_0(\mathcal{N}) \longleftrightarrow \psi^\circ \in \mathfrak{W}_0(\mathcal{M}')$, i.e. $\psi^\circ(y^\circ) = \psi(y)$, $y \in \mathcal{N}$. Furthermore, we write $y^{\circ\circ} = y$ and $\psi^{\circ\circ} = \psi$. Recall that $\sigma_t^\psi(y)^\circ = \sigma_{-t}^{\psi^\circ}(y^\circ)$, $y \in \mathcal{N}$. Therefore the natural extension of the $\circ$-operation from $\mathcal{N}$ and $\mathcal{M}'$ to the bundles $\mathcal{F}(\mathcal{N})$ and $\mathcal{F}(\mathcal{M}')$ is then given by:

$$(y\psi^{it})^\circ = \psi^{\circ -it} y^\circ = \sigma_{-t}^{\psi^\circ}(y^\circ)\psi^{\circ -it}, \quad y \in \mathcal{N}, \quad \psi \in \mathfrak{W}_0(\mathcal{N}). \tag{14}$$

Thus the $\circ$-operation on $\Gamma^1(\mathcal{F}(\mathcal{N}))$ and $\Gamma^1(\mathcal{F}(\mathcal{M}'))$ is given by:

$$y^\circ(t) = \big(y(-t)\big)^\circ, \quad y \in \Gamma^1(\mathcal{F}(\mathcal{N})) \cup \Gamma^1(\mathcal{F}(\mathcal{M}')), \tag{14'}$$

and we get $\Gamma^1(\mathcal{F}(\mathcal{N}))^\circ = \Gamma^1(\mathcal{F}(\mathcal{M}'))$ and $\Gamma^1(\mathcal{F}(\mathcal{M}'))^\circ = \Gamma^1(\mathcal{F}(\mathcal{N}))$. Define left and right actions of the bundle algebras $\Gamma^1(\mathcal{F}(\mathcal{M}))$ and $\Gamma^1(\mathcal{F}(\mathcal{N}))$ on $\widetilde{\mathfrak{H}}$ respectively as follows:

$$\left.\begin{aligned} (x\xi)(t) &= \int_{\mathbf{R}} x(r)\xi(t-r)\, dr; \\ (\xi y)(t) &= \int_{\mathbf{R}} \xi(s)y(t-s)\, ds\,. \end{aligned}\right\} \tag{15}$$

Also $\Gamma^1(\mathcal{F}(\mathcal{M}'))$ acts on $\widetilde{\mathfrak{H}}$ from the left:

$$y\xi = \xi y^\circ, \quad y \in \Gamma^1(\mathcal{F}(\mathcal{M}')),$$

i.e.

$$\begin{aligned} (y\xi)(t) &= \int_{\mathbf{R}} y(s)\xi(t-s)\, ds = \int_{\mathbf{R}} \xi(t-s)y^\circ(s)\, ds \\ &= \int_{\mathbf{R}} \xi(t+s)\big(y(s)\big)^\circ\, ds, \quad y \in \Gamma^1(\mathcal{F}(\mathcal{M}')). \end{aligned} \tag{15'}$$

Each $x \in \mathcal{M}(r)$ and $y \in \mathcal{N}(t)$ act on $\widetilde{\mathfrak{H}}$ from the respective side as follows:

$$\left.\begin{aligned} (x\xi)(s) &= x\xi(s-r); \\ (\xi y)(s) &= \xi(s-t)y, \\ (y^\circ\xi)(s) &= y^\circ\xi(s+t), \quad \xi \in \widetilde{\mathfrak{H}}. \end{aligned}\right\} \tag{15''}$$

In particular, $\mathcal{M} = \mathcal{M}(0)$ and $\mathcal{N} = \mathcal{N}(0)$ both act on $\widetilde{\mathfrak{H}}$ respectively. Also $\{\varphi^{it} : t \in \mathbf{R}\}$ and $\{\psi^{-it} : t \in \mathbf{R}\}$ are both one parameter unitary groups on $\widetilde{\mathfrak{H}}$ acting from their respective sides; these will be denoted by $\{u_\varphi(t)\}$ and $\{v_\psi(t)\}$ when we view them as one parameter unitary groups acting on $\widetilde{\mathfrak{H}}$ from the left.

Lemma 6.8.

(i) *With* $\varphi \in \mathfrak{W}_0(\mathcal{M})$ *fixed, the map* $U_\varphi^*: \xi \in \Gamma^2(\mathcal{G}) \mapsto \xi_\varphi \in L^2(\mathbf{R}, \mathfrak{H}) = L^2(\mathbf{R}) \otimes \mathfrak{H}$ *defined by:*

$$\big(U_\varphi^*\xi\big)(s) = \xi_\varphi(s) = \varphi^{-\mathrm{i}s}\xi(s) \in \mathfrak{H}, \quad \xi \in \widetilde{\mathfrak{H}} \tag{16}$$

is a unitary such that

a) *for each* $x \in \mathcal{M}$,

$$\big(U_\varphi^* x U_\varphi \xi\big)(s) = \sigma_{-s}^\varphi(x)\xi(s), \quad \xi \in L^2(\mathbf{R}, \mathfrak{H}); \tag{17}$$

a′) *for each* $y \in \mathcal{N}$,

$$\big(U_\varphi^* y^\circ U_\varphi \xi\big)(s) = \xi(s)y; \tag{18}$$

b) *for each* $t \in \mathbf{R}$,

$$\big(U_\varphi^* u_\varphi(t) U_\varphi \xi\big)(s) = \xi(s-t), \quad \xi \in L^2(\mathbf{R}, \mathfrak{H}); \tag{19}$$

b′) *for each* $t \in \mathbf{R}$,

$$\big(U_\varphi^* v_\psi(t) U_\varphi \xi\big)(s) = \left(\frac{\mathrm{d}\varphi}{\mathrm{d}\psi^\circ}\right)^{\mathrm{i}t} \xi(s+t). \tag{20}$$

(ii) *The unitary* U_φ^*, *with* $\varphi \in \mathfrak{W}_0(\mathcal{M})$ *fixed, carries the von Neumann algebra* $\widetilde{\mathcal{M}}$ *generated by the action of* $\Gamma^1\big(\mathcal{F}(\mathcal{M})\big)$ *onto the cross product* $\mathcal{M} \rtimes_{\sigma^\varphi} \mathbf{R}$ *isomorphically.*

(iii) *The von Neumann algebra* $\widetilde{\mathcal{M}}'$ *generated by the action of* $\Gamma^1\big(\mathcal{F}(\mathcal{M}')\big)$ *is the commutant of* $\widetilde{\mathcal{M}}$ *and is mapped isomorphically by the unitary* U_φ^* *to the crossed product* $\mathcal{M}' \rtimes_{\sigma^{\psi^\circ}} \mathbf{R}$ *with any* $\psi \in \mathfrak{W}_0(\mathcal{N})$ *fixed.*

PROOF:

(i) Let $x \in \mathcal{M}$ and $\xi \in L^2(\mathbf{R}, \mathfrak{H})$. Then

$$\big(U_\varphi^* x U_\varphi \xi\big)(s) = \varphi^{-\mathrm{i}s} x \big(U_\varphi \xi\big)(s) = \varphi^{-\mathrm{i}s} x \varphi^{\mathrm{i}s}\xi(s) = \sigma_{-\mathrm{i}s}^\varphi(x)\xi(s), \quad s \in \mathbf{R}.$$

(ii) Let us simply compute for $\xi \in L^2(\mathbf{R}, \mathfrak{H})$:

$$\begin{aligned}\big(U_\varphi^* u_\varphi(t) U_\varphi \xi\big)(s) &= \varphi^{-\mathrm{i}s}\big(u_\varphi(t) U_\varphi \xi\big)(s) = \varphi^{-\mathrm{i}s}\varphi^{\mathrm{i}t}\big(U_\varphi \xi\big)(s-t) \\ &= \varphi^{-\mathrm{i}s}\varphi^{\mathrm{i}t}\varphi^{\mathrm{i}\,(s-t)}\xi(s-t) = \xi(s-t), \quad s, t \in \mathbf{R}.\end{aligned}$$

(iii) With $\varphi \in \mathfrak{W}_0(\mathcal{M})$ fixed, for each $x \in \mathcal{A}(\mathcal{M})$ we compute:

$$\begin{aligned}\big(U_\varphi^* x U_\varphi \xi\big)(s) &= \varphi^{-\mathrm{i}s} \int_{\mathbf{R}} x(r)\big(U_\varphi \xi\big)(s-r)\,\mathrm{d}r \\ &= \varphi^{-\mathrm{i}s} \int_{\mathbf{R}} x(r)\varphi^{\mathrm{i}\,(s-r)}\xi(s-r)\,\mathrm{d}r\end{aligned}$$

$$= \varphi^{-\mathrm{i}s} \int_{\mathbf{R}} x_\varphi(r)\varphi^{\mathrm{i}r}\varphi^{\mathrm{i}(s-r)}\xi(s-r)\,\mathrm{d}r$$

$$= \int_{\mathbf{R}} \sigma^\varphi_{-s}\big(x_\varphi(r)\big)\xi(s-r)\,\mathrm{d}r$$

$$= \left(\left(\int_{\mathbf{R}} x_\varphi(r)\big(\lambda(r)\otimes 1\big)\,\mathrm{d}r\right)\xi\right)(s),$$

where $x_\varphi(r) = x(r)\varphi^{-\mathrm{i}r}$ as seen before and $\lambda(\,\cdot\,)$ is the regular representation of $\mathbf{R}$ on $L^2(\mathbf{R})$. From these calculations, the assertion follows easily. Q.E.D.

Definition 6.9. The von Neumann algebra $\widetilde{\mathcal{M}}$ generated by the action of $\Gamma^1(\mathcal{F})$ on the Hilbert space $\widetilde{\mathfrak{H}} = \Gamma^2(\mathcal{G})$ will be called the *canonical core* of $\mathcal{M}$.

On the Hilbert space $\widetilde{\mathfrak{H}} = \Gamma^2(\mathcal{G})$, consider a one parameter unitary group $\big\{U(s) : s \in \mathbf{R}\big\}$ defined by:

$$\big(U(s)\xi\big)(t) = \mathrm{e}^{-\mathrm{i}st}\xi(t), \quad s,t \in \mathbf{R}. \tag{21}$$

Theorem 6.10. *The one parameter unitary group $\big\{U(s) : s \in \mathbf{R}\big\}$ defined by (21) gives rise to an integrable one parameter automorphism group $\big\{\theta_s : s \in \mathbf{R}\big\}$ of the non-commutative flow of weights algebra $\widetilde{\mathcal{M}}$ of a von Neumann algebra $\mathcal{M}$ such that:*

(i) *The von Neumann algebra $\mathcal{M}$ is canonically identified with the fixed point algebra $\widetilde{\mathcal{M}}^\theta$;*

(ii) *For any faithful semi-finite normal weight $\varphi \in \mathfrak{W}_0$, there is a natural isomorphism Π_φ of $\widetilde{\mathcal{M}}$ onto the crossed product $\mathcal{M} \rtimes_{\sigma^\varphi} \mathbf{R}$ which conjugate the one parameter automorphism group $\big\{\theta_s : s \in \mathbf{R}\big\}$ with the action $\big\{\widehat{\sigma}^\varphi\big\}$ dual to σ^φ, hence $\Pi_\varphi(\mathcal{M}) = \mathcal{M}$ relative to the natural embedding of $\mathcal{M}$ into the respective algebras;*

(iii) *For each semi-finite normal weight $\varphi \in \mathfrak{W}_0$, the dual weight $\widetilde{\varphi}$ defined by:*

$$\widetilde{\varphi}(x) = \varphi \circ I_\theta(x), \quad x \in \widetilde{\mathcal{M}}_+,$$

where I_θ means the operator valued weight of $\widetilde{\mathcal{M}}$ onto $\mathcal{M}$ given by:

$$I_\theta(x) = \int_{\mathbf{R}} \theta_t(x)\,\mathrm{d}t, \quad x \in \widetilde{\mathcal{M}}_+,$$

has the modular automorphism group $\sigma^{\hat{\varphi}}$ given by the one parameter unitary group $\big\{\varphi^{\mathrm{i}t} : t \in \mathbf{R}\big\}$ which is in the carrier algebra $\widetilde{\mathcal{M}}$ of the non-commutative flow of weights;

(iv) *Viewing* φ *for each* $\varphi \in \mathfrak{W}_0$ *as the analytic generator of the one parameter unitary group* $\{\varphi^{it} : t \in \mathbf{R}\}$, *i.e.* $\varphi = \exp\left(\frac{1}{i}\frac{d}{dt}(\varphi^{it})\big|_{t=0}\right)$, *the weight* τ *defined by:*

$$\tau(x) = \widetilde{\varphi}(\varphi^{-1}x) = \lim_{\varepsilon \to 0} \widetilde{\varphi}\left((\varphi+\varepsilon)^{-\frac{1}{2}} x (\varphi+\varepsilon)^{-\frac{1}{2}}\right), \quad x \in \widetilde{\mathcal{M}}_+,$$

does not depend on the choice of φ *and is faithful semi-finite normal trace on* $\widetilde{\mathcal{M}}$;

(v) *The one parameter automorphism group* $\{\theta_t : t \in \mathbf{R}\}$ *scales the trace* τ, *i.e.*

$$\tau \circ \theta_s = e^{-s}\tau, \quad s \in \mathbf{R};$$

(vi)
$$\widetilde{\mathcal{M}} \rtimes_\theta \mathbf{R} \cong \mathcal{M} \mathbin{\overline{\otimes}} \mathcal{L}\big(L^2(\mathbf{R})\big).$$

PROOF:

(i)–(iii): Fix $\varphi \in \mathfrak{W}_0$. With[9] $x_\varphi(s) = \pi_\ell \circ \hat{\Phi}_\varphi^{-1}\big(x(s)\big)$, $s \in \mathbf{R}$, for each $x \in \Gamma^1(\mathcal{F})$ we have $x(s) = x_\varphi(s)\varphi^{is}$ and

$$(xy)_\varphi(t) = \int_{\mathbf{R}} x(s)\sigma_s^\varphi\big(y(t-s)\big)\, ds, \quad x, y \in \Gamma^1(\mathcal{F}).$$

For each $\xi \in \Gamma^2(\mathcal{G})$ set $\xi_\varphi(s) = U_\varphi(s)^{-1}\big(\xi(s)\big)$, i.e. $\xi(s) = \varphi^{is}\xi_\varphi(s)$. Then we have for each $x \in \Gamma^1(\mathcal{F})$

$$\varphi^{it}(x\xi)_\varphi(t) = \int_{\mathbf{R}} \big(x_\varphi(s)\varphi^{is}\big)\big(\varphi^{i(t-s)}\xi_\varphi(t-s)\big)\, ds = \int_{\mathbf{R}} x_\varphi(s)\varphi^{it}\xi_\varphi(t-s)\, ds$$
$$= \varphi^{it}\int_{\mathbf{R}} \sigma_{-t}^\varphi\big(x_\varphi(s)\big)\xi_\varphi(t-s)\, ds,$$

so that

$$(x\xi)_\varphi(t) = \int_{\mathbf{R}} \sigma_{-t}^\varphi\big(x_\varphi(s)\big)\xi_\varphi(t-s)\, ds, \quad x \in \Gamma^1(\mathcal{F}), \quad \xi \in \Gamma^2(\mathcal{G}).$$

Therefore, the map $U_\varphi^*: \xi \in \Gamma^2(\mathcal{G}) \mapsto \xi_\varphi \in L^2(\mathbf{R}, \mathfrak{H})$ is an isometry such that $U_\varphi^*\Gamma^1(\mathcal{F})U_\varphi = L^1(\mathbf{R}, \mathcal{M})$ with the twisted convolution by σ^φ on $L^1(\mathbf{R}, \mathcal{M})$, where both $\Gamma^1(\mathcal{F})$ and $L^1(\mathbf{R}, \mathcal{M})$ are viewed as the algebras of operators on the respective Hilbert spaces $\Gamma^2(\mathcal{G})$ and $L^2(\mathbf{R}, \mathfrak{H})$. Thus, $U_\varphi^*\widetilde{\mathcal{M}}U_\varphi = \mathcal{M} \rtimes_{\sigma^\varphi} \mathcal{M}$. Define $\Pi_\varphi(x) = U_\varphi^* x U_\varphi$, $x \in \widetilde{\mathcal{M}}$. It now easily seen that $\Pi_\varphi \circ \theta_s \circ \Pi_\varphi^{-1} = \widehat{\sigma}_s^\varphi$, $s \in \mathbf{R}$. Furthermore, we have $(\varphi^{is}\xi)(t) = \varphi^{is}\xi(t-s)$, $\xi \in \Gamma^2(\mathcal{G})$, $s, t \in \mathbf{R}$, so that $\big(\varphi^{is}\xi\big)_\varphi(t) = \xi_\varphi(t-s)$. Thus $U_\varphi^*\varphi^{is}U_\varphi = \lambda(s)$, $s \in \mathbf{R}$, where $\{\lambda(s) : s \in \mathbf{R}\}$ is the one parameter unitary group associated with the crossed product $\mathcal{M} \rtimes_{\sigma^\varphi} \mathbf{R}$. It is now routine to check the rest of the claims of (i) through (iii).

---

9 π_ℓ means again the projection to the first coodinate of $\mathcal{M} \times \mathbf{R}$.

(iv) From the above discussion, we see that the weight $\widehat{\varphi}$ on $\mathcal{M} \rtimes_{\sigma^\varphi} \mathbf{R}$ dual to $\varphi \in \mathfrak{W}_0$ is transformed to the weight $\widetilde{\varphi}$ on $\widetilde{\mathcal{M}}$, i.e. $\widetilde{\varphi} = \widehat{\varphi} \circ \Pi_\varphi$. Therefore, the weight defined by $\tau_\varphi(x) = \widetilde{\varphi}(\varphi^{-1}x)$, $x \in \widetilde{\mathcal{M}}$, is a trace on $\widetilde{\mathcal{M}}$ and θ scales τ_φ. Furthermore, we have $\left(\mathrm{D}\widetilde{\varphi} : \mathrm{D}\widetilde{\psi}\right)_t = (\mathrm{D}\varphi : \mathrm{D}\psi)_t = \varphi^{\mathrm{i}t}\psi^{-\mathrm{i}t}$, $t \in \mathbf{R}$. Now we have:

$$\begin{aligned}\left(\mathrm{D}\tau_\varphi : \mathrm{D}\tau_\psi\right)_t &= \left(\mathrm{D}\tau_\varphi : \mathrm{D}\widetilde{\varphi}\right)_t \left(\mathrm{D}\widetilde{\varphi} : \mathrm{D}\widetilde{\psi}\right)_t \left(\mathrm{D}\widetilde{\psi} : \mathrm{D}\tau_\psi\right)_t \\ &= \varphi^{-\mathrm{i}t}\left(\varphi^{\mathrm{i}t}\psi^{-\mathrm{i}t}\right)\psi^{\mathrm{i}t} = 1, \quad t \in \mathbf{R},\end{aligned}$$

which means that $\tau_\varphi = \tau_\psi$.

The claims in (v) and (vi) follow now easily from the above. Q.E.D.

Theorem 6.11.

(i) *The correspondence:* $\mathcal{M} \to \left\{\widetilde{\mathcal{M}}, \mathbf{R}, \theta, \tau\right\}$ *is a functor from the category of von Neumann algebras* $\mathcal{M}$ *with isomorphism as morphism to the category of semi-finite von Neumann algebra* $\widetilde{\mathcal{M}}$ *equipped with a one parameter automorphism group* $\left\{\theta_t : t \in \mathbf{R}\right\}$*, which scales a faithful semifinite normal trace* τ*, with conjugation, which preserves the traces, as morphism.*

(ii) *The correspondence:* $\{\mathcal{M}, \mathfrak{H}\} \to \left\{\widetilde{\mathcal{M}}, \mathbf{R}, \theta, \widetilde{\mathfrak{H}}\right\}$ *is a functor from the category of von Neumann algebras represented on Hilbert spaces with unitary implementing an isomorphism as morphism to the category of semi-finite von Neumann algebras, equipped with trace scaling one parameter automorphism group and trace, represented on Hilbert spaces with unitary implementing an isomorphism intertwining* θ *and* τ.

PROOF: The assertion (i) follows from (ii) by considering the canonical Hilbert space $L^2(\mathcal{M}_1)$, discussed in 1, Chapter IX, as $\mathfrak{H}_1$. Let U be a unitary of a Hilbert space $\mathfrak{H}_1$ onto $\mathfrak{H}_2$ such that $U\mathcal{M}_1U^* = \mathcal{M}_2$ where $\mathcal{M}_1$ and $\mathcal{M}_2$ are von Neumann algebras on $\mathfrak{H}_1$ and $\mathfrak{H}_2$ respectively. Let $\alpha_U(x) = UxU^*$, $x \in \mathcal{M}_1$ be the isomorphism of $\mathcal{M}_1$ onto $\mathcal{M}_2$ implemented by U. Let us drop the suffix U and write just α instead for a while until we need to indicate the dependence of α_U on U. Define $\tilde{\alpha}: \mathcal{M}_1(t) \mapsto \mathcal{M}_2(t)$, $t \in \mathbf{R}$ by:

$$\tilde{\alpha}\left(x\varphi^{\mathrm{i}t}\right) = \alpha(x)\left(\varphi \circ \alpha^{-1}\right)^{\mathrm{i}t}, \quad x\varphi^{\mathrm{i}t} \in \mathcal{M}_1(t), \quad \varphi \in \mathfrak{W}_0(\mathcal{M}_1).$$

Also define $\widetilde{U}: \mathfrak{H}_1(t) \mapsto \mathfrak{H}_2(t)$ as follows:

$$\widetilde{U}\left(\varphi^{\mathrm{i}t}\xi\right) = \left(\varphi \circ \alpha^{-1}\right)^{\mathrm{i}t}U\xi, \quad \varphi^{\mathrm{i}t}\xi \in \mathfrak{H}_1(t), \quad \varphi \in \mathfrak{W}_0(\mathcal{M}_1).$$

Observe that $\widetilde{\alpha}$ and $\widetilde{U}$ respectively map $\mathcal{F}(\mathcal{M}_1)$ and $\mathfrak{g}(\mathfrak{H}_1)$ to $\mathcal{F}(\mathcal{M}_2)$ and $\mathfrak{g}(\mathfrak{H}_2)$ consistently, which give rise to an isomorphism of $\Gamma^1\left(\mathcal{F}(\mathcal{M}_1)\right)$ onto $\Gamma^1\left(\mathcal{F}(\mathcal{M}_2)\right)$ denoted by $\widetilde{\alpha}$ again and an isometry $\Gamma^2\left(\mathfrak{g}(\mathfrak{H}_1)\right)$ onto $\Gamma^2\left(\mathfrak{g}(\mathfrak{H}_2)\right)$ denoted also by $\widetilde{U}$ such that $\widetilde{\alpha}(x) = \widetilde{U}x\widetilde{U}^*$, $x \in \Gamma^1\left(\mathcal{F}(\mathcal{M}_1)\right)$. It is straightforward to check that $\widetilde{\alpha}$ given by $\mathrm{Ad}(\widetilde{U})$ is the desired isomorphism of $\widetilde{\mathcal{M}}_1$ onto $\widetilde{\mathcal{M}}_2$. Since the correspondences $U \mapsto \alpha_U$, $\alpha_U \mapsto \widetilde{\alpha}_U$ and $U \mapsto \widetilde{U}$ are all canonical and explicitly written, they all obey the chain rule and preserve the relevant structures in question.

Q.E.D.

Definition 6.12.

(i) The covariant system $\{\widetilde{\mathcal{M}}, \mathbf{R}, \theta, \tau\}$, together with the trace, associated with a von Neumann algebra $\mathcal{M}$ in the above theorem is called *the non-commutative flow of weights* on the von Neumann algebra $\mathcal{M}$.

(ii) The group $\widetilde{\mathcal{U}}(\mathcal{M}) = \{u \in \mathcal{U}(\widetilde{\mathcal{M}}) : u\mathcal{M}u^* = \mathcal{M}\}$ of normalizers of $\mathcal{M}$ is called *the extended* unitary group of $\mathcal{M}$. For each $u \in \widetilde{\mathcal{U}}(\mathcal{M})$, we denote $\widetilde{\mathrm{Ad}}(u) = \mathrm{Ad}(u)\big|_{\mathcal{M}} \in \mathrm{Aut}(\mathcal{M})$ and $\mathrm{Cnt}_r(\mathcal{M}) = \{\widetilde{\mathrm{Ad}}(u) : u \in \widetilde{\mathcal{U}}(\mathcal{M})\} \subset \mathrm{Aut}(\mathcal{M})$.

(iii) If α is an isomorphism of a von Neumann algebra $\mathcal{M}_1$ onto $\mathcal{M}_2$, then the corresponding isomorphism $\widetilde{\alpha}$ of the non-commutative flow of weights $\widetilde{\mathcal{M}}_1$ onto $\widetilde{\mathcal{M}}_2$ is called the *quantized modulus of* α. As it will be too heavy to write $\widetilde{\mathrm{mod}}(\alpha)$ for the quantized modulus of α, we will keep writing $\widetilde{\alpha}$ instead. It should be clear, however, that the restriction of $\widetilde{\alpha}$ to the center $\mathcal{C}$ of $\widetilde{\mathcal{M}}$ is precisely the modulus of α.

For the rest of this section, we will consider a factor $\mathcal{M}$ for the sake of simplicity although this restriction is not essential.

Lemma 6.13. *Let $\mathcal{M}$ be a factor with non-commutative flow of weights $\{\widetilde{\mathcal{M}}, \mathbf{R}, \theta, \tau\}$ and $\mathcal{C} = \mathcal{C}_{\widetilde{\mathcal{M}}}$ be the center of $\widetilde{\mathcal{M}}$. Then we have*

(i) $\mathcal{M}' \cap \widetilde{\mathcal{M}} = \mathcal{C}$;

(ii) *if $\alpha \in \mathrm{Aut}(\widetilde{\mathcal{M}})$ leaves $\mathcal{M}$ elementwise fixed and commutes with $\{\theta_s : s \in \mathbf{R}\}$, then $\alpha = \theta_p$ for some $p \in \mathbf{R}$.*

PROOF:

(i) We identify, via the isomorphism Π_φ, $\widetilde{\mathcal{M}}$ with $\mathcal{M} \rtimes_\varphi \mathbf{R}$ for a fixed dominant $\varphi \in \mathfrak{W}_0(\mathcal{M})$ and prove that $\mathcal{M}' \cap (\mathcal{M} \rtimes_\varphi \mathbf{R}) =$ The center of $\mathcal{M} \rtimes_\varphi \mathbf{R}$. As φ is dominant, $\mathcal{M}$ is generated by $\mathcal{M}_\varphi$ and a one parameter unitary group $\{u(s) : s \in \mathbf{R}\}$ such that $\sigma_t^\varphi(u(s)) = e^{ist}u(s)$, $s, t \in \mathbf{R}$. Let ρ be the regular one parameter unitary group on $L^2(\mathbf{R})$ which is defined by:

$$\big(\rho(s)\xi\big)(t) = \xi(s+t), \quad \xi \in L^2(\mathbf{R}), \quad s, t \in \mathbf{R}$$

and let $\{\rho_t = \mathrm{Ad}(\rho(t)) : t \in \mathbf{R}\}$ be the one parameter automorphism group of $\mathcal{L}(L^2(\mathbf{R}))$ generated by $\rho(\mathbf{R})$. Then we have

$$\mathcal{M} \rtimes_\varphi \mathbf{R} = \Big(\mathcal{M} \mathbin{\overline{\otimes}} \mathcal{L}\big(L^2(\mathbf{R})\big)\Big)^{\sigma^\varphi \otimes \rho}.$$

Now, $\mathcal{M}_\varphi$ being the fixed point algebra of $\mathcal{M}$ under σ^φ, we get $\Pi_\varphi(\mathcal{M}_\varphi) = \mathcal{M}_\varphi \otimes \mathbf{C}$ in $\mathcal{M} \mathbin{\overline{\otimes}} \mathcal{L}\big(L^2(\mathbf{R})\big)$, so that

$$\begin{aligned}\Pi_\varphi(\mathcal{M})' \cap (\mathcal{M} \rtimes \mathbf{R}) &\subset (\mathcal{M}_\varphi \otimes \mathbf{C})' \cap \Big(\mathcal{M} \mathbin{\overline{\otimes}} \mathcal{L}\big(L^2(\mathbf{R})\big)\Big)^{\sigma^\varphi \otimes \rho} \\ &= \Big(\mathcal{C}_\varphi \otimes \mathcal{L}\big(L^2(\mathbf{R})\big)\Big)^{\sigma^\varphi \otimes \rho} = \mathcal{C}_\varphi \mathbin{\overline{\otimes}} \rho(\mathbf{R})''.\end{aligned}$$

The unitary groups $\{\Pi_\varphi(u(s))\}$ and $\{\rho(t)\}$ satisfy the Heisenberg-Weyl commutation relation, which entails:

$$\Pi_\varphi(\mathcal{M})' \cap \left(\mathcal{M} \overline{\otimes} \mathcal{L}(L^2(\mathbf{R}))\right)^{\sigma^\varphi \otimes \rho} = (\mathcal{C}_\varphi \overline{\otimes} \rho(\mathbf{R})'') \cap (\Pi_\varphi(\mathbf{R})') = \mathcal{C}_\varphi \otimes \mathbf{C}.$$

This means that $\mathcal{M}' \cap (\mathcal{M} \rtimes_{\sigma^\varphi} \mathbf{R}) = \mathcal{C} =$ the center of $\mathcal{M} \rtimes_{\sigma^\varphi} \mathbf{R}$.

(ii) Since $\theta_s(u_\varphi(t)) = e^{-ist} u_\varphi(t)$, if $\alpha \in \mathrm{Aut}(\widetilde{\mathcal{M}})$ fixes $\mathcal{M}$ pointwise, then we have $a(t) = u_\varphi(t)^* \alpha(u_\varphi(t)) \in \mathcal{M}' \cap \widetilde{\mathcal{M}}$. Therefore $\{a(t) : t \in \mathbf{R}\}$ is a one parameter unitary group in $\mathcal{C}$ with $\alpha(u_\varphi(t)) = a(t)u_\varphi(t)$. We may now compute:

$$e^{-ist} a(t) u_\varphi(t) = \alpha\Big(\theta_s\big(u_\varphi(t)\big)\Big) = \theta_s \circ \alpha\big(u_\varphi(t)\big) = e^{-ist} \theta_s\big(a(t)\big) u_\varphi(t)$$

so that $a(t) \in \mathcal{C}^\theta = \mathbf{C}$, as $\{\theta_t\}$ is ergodic on $\mathcal{C}$. This means that $a(t) = e^{-ipt}$ for some $p \in \mathbf{R}$, and hence $\alpha = \theta_p$. Q.E.D.

By construction, to every $\alpha \in \mathrm{Aut}(\mathcal{M})$ there corresponds the quantized modulus $\tilde{\alpha} \in \mathrm{Aut}(\widetilde{\mathcal{M}})$ of α such that:

$$\begin{cases} \tilde{\alpha}|_{\mathcal{M}} = \alpha; \\ \theta_s \circ \tilde{\alpha} = \tilde{\alpha} \circ \theta_s, \quad s \in \mathbf{R}; \\ \tau \circ \tilde{\alpha} = \tau \end{cases}$$

and that the map: $\alpha \in \mathrm{Aut}(\mathcal{M}) \mapsto \tilde{\alpha} \in \mathrm{Aut}(\widetilde{\mathcal{M}})$ is an injective homomorphism.

Lemma 6.14. *In the above context, we conclude:*

(i) $\tilde{\alpha} \in \mathrm{Aut}(\widetilde{\mathcal{M}})$ *is inner if and only if* α *is of the form:* $\alpha = \mathrm{Ad}(u) \circ \sigma_c^\varphi$ *for some, hence for every, dominant weight* φ *on* $\mathcal{M}$ *and a cocycle* $c \in Z_\theta(\mathbf{R}, \mathcal{U}(\mathcal{C}))$ *and some* $u \in \mathcal{U}(\mathcal{M})$;

(ii) *If* $u \in \widetilde{\mathcal{U}}(\mathcal{M})$, *then* $\widetilde{\mathrm{Ad}(u)} = \mathrm{Ad}(u)$;

(iii) *The quantized modulus:* $\alpha \in \mathrm{Aut}(\mathcal{M}) \mapsto \tilde{\alpha} \in \mathrm{Aut}(\widetilde{\mathcal{M}})$ *is a continuous homomorphism.*

PROOF: The assertion (i) follows from the Theorem 1.10. We leave the proof to the reader.

(ii) Choose arbitrarily $u \in \widetilde{\mathcal{U}}(\mathcal{M})$, and set $\alpha = \widetilde{\mathrm{Ad}}(u)$. Then we know that $\tilde{\alpha}|_{\mathcal{M}} = \widetilde{\mathrm{Ad}}(u)\big|_{\mathcal{M}}$, so that $\mathrm{Ad}(u) \circ \tilde{\alpha}^{-1}$ leaves $\mathcal{M}$ pointwise fixed. By the previous lemma, there exists $p \in \mathbf{R}$ such that $\mathrm{Ad}(u) \circ \tilde{\alpha}^{-1} = \theta_p$. As both $\mathrm{Ad}(u)$ and $\tilde{\alpha}$ leaves the trace τ invariant, $p = 0$; hence we conclude $\mathrm{Ad}(u) = \tilde{\alpha}$.

(iii) The homomorphism property of the quantized modulus is clear. Consider the standard form $\{\mathfrak{H} = \mathcal{M}, L^2(\mathcal{M}), J, \mathfrak{P}\}$ which is spanned by $\{\varphi^{1/2} : \varphi \in \mathfrak{S}_* = \mathfrak{S}_*(\mathcal{M})\}$. First each $\alpha \in \mathrm{Aut}(\mathcal{M})$ gives rise to a unitary $U(\alpha)$ on $\mathfrak{H}$ given by: $U(\alpha)\varphi^{1/2} = (\varphi \circ \alpha^{-1})^{1/2}$, $\varphi \in \mathfrak{S}_*$. Thus the map: $\alpha \in \mathrm{Aut}(\mathcal{M}) \mapsto U(\alpha) \in U(\mathfrak{H})$

is continuous by the inequality (IX.1.9) of Theorem IX.1.2. The fiber $\mathfrak{H}(t)$ is given by $\varphi^{\mathrm{i}t}\mathfrak{H}$ and the corresponding unitary

$$U_t(\alpha)\varphi^{\mathrm{i}t}\xi = (\varphi\circ\alpha)^{\mathrm{i}t}U(\alpha)\xi, \qquad \xi\in\mathfrak{H}.$$

Thus for each fixed $t\in\mathbf{R}$, the map: $\alpha\in\mathrm{Aut}(\mathcal{M})\mapsto U_t(\alpha)\in\mathcal{U}(\mathfrak{H}(t))$ is continuous. The glued unitary $\widetilde{U}(\alpha)$ of $\{U_t(\alpha) : t\in\mathbf{R}\}$ on $\Gamma^2(\mathcal{G})$ gives the quantized modulus $\widetilde{\alpha}$. The continuity of the map: $\alpha\in\mathrm{Aut}(\mathcal{M})\mapsto\widetilde{U}(\alpha)$ can easily seen as an application of the Lebesgue dominated convergence theorem. Q.E.D.

Theorem 6.15. *To each factor $\mathcal{M}$ there corresponds canonically an* $\mathrm{Aut}(\mathcal{M})\times\mathbf{R}$ *equivariant commutative square of exact sequences:*

$$\left\{\begin{array}{ccccccccc}
 & & 1 & & 1 & & 1 & & \\
 & & \downarrow & & \downarrow & & \downarrow & & \\
1 & \longrightarrow & \mathbf{T} & \longrightarrow & \mathcal{U}(\mathcal{C}) & \xrightarrow{\partial} & \mathrm{B}^1_\theta(\mathbf{R},\mathcal{U}(\mathcal{C})) & \longrightarrow & 1\\
 & & \downarrow & & \downarrow & & \downarrow & & \\
1 & \longrightarrow & \mathcal{U}(\mathcal{M}) & \longrightarrow & \widetilde{\mathcal{U}}(\mathcal{M}) & \xrightarrow{\partial} & \mathrm{Z}^1_\theta(\mathbf{R},\mathcal{U}(\mathcal{C})) & \longrightarrow & 1\\
 & & {\scriptstyle\mathrm{Ad}}\downarrow & & {\scriptstyle\widetilde{\mathrm{Ad}}}\downarrow & & \downarrow & & \\
1 & \longrightarrow & \mathrm{Int}(\mathcal{M}) & \longrightarrow & \mathrm{Cnt}_r(\mathcal{M}) & \xrightarrow{\dot{\partial}} & \mathrm{H}^1_\theta(\mathbf{R},\mathcal{U}(\mathcal{C})) & \longrightarrow & 1\\
 & & \downarrow & & \downarrow & & \downarrow & & \\
 & & 1 & & 1 & & 1 & &
\end{array}\right.$$

Here $\mathcal{C}$ is the center of the non-commutative flow of weights $\{\widetilde{\mathcal{M}},\mathbf{R},\theta\}$ of $\mathcal{M}$ and $\widetilde{\mathcal{U}}(\mathcal{M})$ is the extended unitary group of $\mathcal{M}$. If $\mathcal{M}$ is a separable factor, then except $\mathrm{H}^1_\theta(\mathbf{R},\mathcal{U}(\mathcal{C}))$ in the case of type III_0 *at the down right corner all eight groups are Polish.*

PROOF: We first note that except for those indicated, all arrows are either inclusion maps or canonical. For brevity, we write B, Z and H for the groups appearing in the far right column.

(i) First row: We define the coboundary ∂:

$$\left.\begin{aligned}(\partial b)(t) &= b^*\theta_t(b), \quad b\in\mathcal{U}(\mathcal{C});\\ \mathrm{B} &= \partial(\mathcal{U}(\mathcal{C})).\end{aligned}\right\} \tag{22}$$

(ii) Second row: We define the map ∂ by

$$(\partial(u))(t) = u^*\theta_t(u), \quad u\in\widetilde{\mathcal{U}}, \tag{22$'$}$$

which extends the ∂ of (22). For each $x\in\mathcal{M}$, we have

$$x\big(\partial(u)\big)(t) = xu^*\theta_t(u) = u^*\,\widetilde{\mathrm{Ad}}(u)(x)\theta_t(u) = u^*\theta_t\big(\widetilde{\mathrm{Ad}}(u)(x)u\big)$$
$$= u^*\theta_t(ux) = u^*\theta_t(u)x = \big(\partial(u)\big)(t)x,$$

so that $\big(\partial(u)\big)(t) \in \mathcal{M}' \cap \widetilde{\mathcal{M}} = \mathcal{C}$ by Lemma 6.14. By construction, $\partial(u)$ satisfies the cocycle identity; hence $\partial(u) \in \mathrm{Z}$.

To prove the surjectivity of ∂, take an arbitrary $c \in \mathrm{Z}$. The stability of θ on $\widetilde{\mathcal{M}}$, Theorem 1.11, implies the existence of $u \in \mathcal{U}(\widetilde{\mathcal{M}})$ with $c(t) = u^*\theta_t(u)$. We now compute for $x \in \mathcal{M}$:

$$\theta_s(uxu^*) = \theta_s(u)x\theta_s(u^*) = c(s)uxu^*c(s)^* = uxu^*$$

as $c(s)$ is in the center $\mathcal{C}$ of $\widetilde{\mathcal{M}}$, so that $uxu^* \in \widetilde{\mathcal{M}}^\theta = \mathcal{M}$. Thus u falls in $\widetilde{\mathcal{U}}$ and $c = \partial(u)$. Thus, ∂ is surjective. Now the exactness of the second row can be seen as follows: for any $u \in \widetilde{\mathcal{U}}$, $\partial(u) = 1$ if and only if $u^*\theta_s(u) = 1$ for every $s \in \mathbf{R}$; if and only if $\theta_s(u) = u$, $s \in \mathbf{R}$; if and only if $u \in \mathcal{U}(\mathcal{M})$.

(iii) Third row: For any $\sigma \in \mathrm{Cnt}_r(\mathcal{M})$, choose $u \in \widetilde{\mathcal{U}}$ with $\sigma = \widetilde{\mathrm{Ad}}(u)$ and set $c = \partial(u)$. Suppose $\sigma \in \mathrm{Int}(\mathcal{M})$, and hence $\sigma = \mathrm{Ad}(a)$ for some $a \in \mathcal{U}(\mathcal{M})$. Then we have $b = a^*u \in \mathcal{M}' \cap \widetilde{\mathcal{M}} = \mathcal{C}$. As $\theta_s(a) = a$, $s \in \mathbf{R}$, we obtain:

$$(\partial b)(t) = b^*\theta_t(b) = u^*a\theta_t(a^*u) = u^*\theta_t(u) = \big(\partial(u)\big)(t) = c(t).$$

Hence c falls in B. Conversely, suppose that $c = \partial b \in \mathrm{B}$ with some $b \in \mathcal{U}(\mathcal{C})$. Let $a = ub^*$. Then we have:

$$\widetilde{\mathrm{Ad}}(a) = \widetilde{\mathrm{Ad}}(ub^*) = \widetilde{\mathrm{Ad}}(u);$$

$$\theta_t(a) = \theta_t(u)\theta_t(b^*) = c(t)ub^*(\partial b)(t)^* = ub^* = a,$$

hence a belongs to $\mathcal{M}$ and $\sigma = \widetilde{\mathrm{Ad}}(u) = \mathrm{Ad}(a) \in \mathrm{Int}(\mathcal{M})$. Thus, we conclude that $\sigma \in \mathrm{Int}(\mathcal{M})$ if and only if $c \in \mathrm{B}$. Therefore, the map:

$$\dot{\partial} : \sigma \in \mathrm{Cnt}_r(\mathcal{M}) \to [c] \in \mathrm{H} \tag{23}$$

is well-defined, and its kernel is precisely $\mathrm{Int}(\mathcal{M})$.

(iv) First column: Trivial.

(v) Second column: As $\mathcal{M}' \cap \widetilde{\mathcal{M}} = \mathcal{C}$, we have exactness by the definition of $\widetilde{\mathrm{Ad}}$: $\widetilde{\mathrm{Ad}}(u) = \mathrm{Ad}(u)\big|_{\mathcal{M}}$, $u \in \widetilde{\mathcal{U}}(\mathcal{M})$.

(vi) Third column: This is precisely the definition of H.

(vii) $\mathrm{Aut}(\mathcal{M}) \times \mathbf{R}$-action: Each $\alpha \in \mathrm{Aut}(\mathcal{M})$ gives rise to $\tilde{\alpha} \in \mathrm{Aut}(\widetilde{\mathcal{M}})$ which induces an action on $\widetilde{\mathcal{U}}$. As $\alpha\,\widetilde{\mathrm{Ad}}(u)\alpha^{-1} = \widetilde{\mathrm{Ad}}\big(\tilde{\alpha}(u)\big)$, $\alpha \in \mathrm{Aut}(\mathcal{M})$ and $u \in \widetilde{\mathcal{U}}$, the $\mathrm{Aut}(\mathcal{M})$-action on $\mathrm{Cnt}_r(\mathcal{M})$ by conjugation is consistent with the one on $\widetilde{\mathcal{U}}$. The restriction of $\tilde{\alpha}$ to $\mathcal{C}$ is precisely the modulus, $\mathrm{mod}(\alpha)$, of α. Since $\{\theta_t : t \in \mathbf{R}\}$ acts trivially on $\mathcal{M}$, it induces an $\mathbf{R}$-action on $\widetilde{\mathcal{U}}$ which is trivial in the first column and is the flow of weights on the third column. Since $\{\theta_t : t \in \mathbf{R}\}$ commutes with $\{\tilde{\alpha} : \alpha \in \mathrm{Aut}(\mathcal{M})\}$, we conclude that the exact square is $\mathrm{Aut}(\mathcal{M}) \times \mathbf{R}$-equivariant.

(viii) Polishness of the eight groups for a separable factor $\mathcal{M}$: Suppose that $\mathcal{M}$ is separable. First, $\mathbf{T}$ is the compact group of complex numbers of modulus one, which is of course Polish. The groups $\mathcal{U}(\mathcal{C})$, $\mathcal{U}(\mathcal{M})$ and $\widetilde{\mathcal{U}}(\mathcal{M})$ are all closed subgroup of a Polish group $\mathcal{U}(\widetilde{\mathcal{M}})$, so that they are all Polish. The groups B, Z, $\mathrm{Int}(\mathcal{M})$ and $\mathrm{Cnt}_r(\mathcal{M})$ are all quotient groups of Polish groups by closed subgroups, hence they are all Polish. One should, however, observe that the topologies in the groups $\mathrm{Int}(\mathcal{M})$ and $\mathrm{Cnt}_r(\mathcal{M})$ are stronger than the topologies inherited from the group $\mathrm{Aut}(\mathcal{M})$. Q.E.D.

Definition 6.16. The square of exact sequences of Theorem 6.15 is called the *characteristic square* for $\mathcal{M}$. An element $\alpha \in \mathrm{Cnt}_r(\mathcal{M})$ is called an *extended inner* automorphism of $\mathcal{M}$.

In order to understand the characteristic square better, we put it in an abstract setting. Let E be a Polish group equipped with an action γ of a Polish group $\widetilde{\Gamma}$ which is a cartesian product $\Gamma \times \mathbf{R}$ of another Polish group Γ and $\mathbf{R}$. Let A be a $\widetilde{\Gamma}$-invariant subgroup of the center of E such that

$$\mathrm{B}^1(\mathbf{R}, E) = \mathrm{Z}^1(\mathbf{R}, A) \tag{24}$$

in the sense that $c(t) = x^{-1}\gamma_t(x) \in A$ whenever $x \in E$ and $t \in \mathbf{R}$, so that

$$\gamma_t(x) = c(t)x, \quad c(t) \in A, \quad x \in A, \quad t \in \mathbf{R}, \tag{25}$$

and every $c \in \mathrm{Z}^1(\mathbf{R}, A)$ is of the above form. Then the $\widetilde{\Gamma}$-equivariant exact sequence:

$$1 \longrightarrow A \longrightarrow E \longrightarrow N \longrightarrow 1,$$

(with trivial $\mathbf{R}$-action on N) gives rise to the following exact square:

$$\begin{array}{ccccccccc}
 & & 1 & & 1 & & 1 & & \\
 & & \downarrow & & \downarrow & & \downarrow & & \\
1 & \longrightarrow & A^{\mathbf{R}} & \longrightarrow & A & \xrightarrow{f} & \mathrm{B} & \longrightarrow & 1 \\
 & & \downarrow & & \downarrow & & \downarrow & & \\
1 & \longrightarrow & E^{\mathbf{R}} & \longrightarrow & E & \xrightarrow{g} & \mathrm{Z} & \longrightarrow & 1 \\
 & & \downarrow & & \downarrow & & \downarrow & & \\
1 & \longrightarrow & D & \longrightarrow & N & \xrightarrow{h} & \mathrm{H} & \longrightarrow & 1 \\
 & & \downarrow & & \downarrow & & \downarrow & & \\
 & & 1 & & 1 & & 1 & &
\end{array} \tag{26}$$

where

$$\left.\begin{array}{l}
\mathrm{B} = \mathrm{B}^1(\mathbf{R}, A), \qquad \mathrm{Z} = \mathrm{Z}^1(\mathbf{R}, A), \qquad \mathrm{H} = \mathrm{H}^1(\mathbf{R}, A); \\
\qquad \big(f(a)\big)(t) = a^{-1}\gamma_t(a), \quad a \in A; \\
\qquad \big(g(x)\big)(t) = x^{-1}\gamma_t(x), \quad x \in E.
\end{array}\right\} \tag{27}$$

To each Borel cross-section $x : m \in N \to x(m) \in E$ with $x(1) = 1$ of the quotient map: $E \to N = E/A$, there correspond two Borel maps $\lambda_x : N \times \widetilde{\Gamma} \to A$ and $\mu_x : N \times N \to A$ such that

$$\left.\begin{aligned} \lambda_x(m,\gamma) &= (\partial_1 x)(m,\gamma) = \gamma\Big(x\big(\gamma^{-1}(m)\big)\Big)x(m)^{-1}, \quad \gamma \in \widetilde{\Gamma}, \quad m \in N;\\ \mu_x(m,n) &= (\partial_2 x)(m,n) = x(m)x(n)x(mn)^{-1}, \qquad n \in N. \end{aligned}\right\} \tag{28}$$

The pair (λ_x, μ_x) satisfies the following cocycle identities:

$$\left.\begin{aligned} \mu(\ell,m)\mu(\ell m,n) &= \mu(\ell,mn)\mu(m,n), \qquad \ell,m,n \in N;\\ \lambda(m,\alpha\beta) &= \lambda(m,\alpha)\alpha\Big(\lambda\big(\alpha^{-1}(m),\beta\big)\Big), \quad \alpha,\beta \in \widetilde{\Gamma};\\ \lambda(m,\alpha)\lambda(m,\alpha)\lambda(mn,\alpha)^{-1} &= \alpha\Big(\mu\big(\alpha^{-1}(m),\alpha^{-1}(n)\big)\Big)\mu(m,n)^{-1};\\ \lambda(m,\alpha) = \mu(m,n) = 1 &\quad \text{if any of } m,n \in N, \text{ or } \alpha \in \widetilde{\Gamma} \text{ is the identity.} \end{aligned}\right\} \tag{29}$$

Let $\mathrm{Z}\big(\widetilde{\Gamma}, N, A\big)$ be the set of all Borel maps (λ, μ) satisfying the cocycle identities (29). It follows that $\mathrm{Z}\big(\widetilde{\Gamma}, N, A\big)$ is a group relative to pointwise multiplication. Let $\mathrm{Map}(N, A)$ denote the group of all A-valued Borel functions ξ on N with $\xi(1) = 1$. Changing the cross-section x to ξx, $\xi \in \mathrm{Map}(N, A)$, results in a change in (λ_x, μ_x) by multiplying by $\partial\xi = (\partial_1\xi, \partial_2\xi)$ where

$$\left.\begin{aligned} (\partial_1\xi)(m,\gamma) &= \gamma\Big(\xi\big(\gamma^{-1}(m)\big)\Big)\xi(m)^{-1}, \quad m \in N, \quad \gamma \in \widetilde{\Gamma};\\ (\partial_2\xi)(m,n) &= \xi(m)\xi(n)\xi(mn)^{-1}, \qquad n \in N. \end{aligned}\right\} \tag{30}$$

Let

$$\left.\begin{aligned} \mathrm{B}\big(\widetilde{\Gamma}, N, A\big) &= \partial\,\mathrm{Map}(N, A);\\ \Lambda\big(\widetilde{\Gamma}, N, A\big) &= \mathrm{Z}\big(\widetilde{\Gamma}, N, A\big)\big/\mathrm{B}\big(\widetilde{\Gamma}, N, A\big). \end{aligned}\right\} \tag{31}$$

The class $\tilde{\chi} = [\lambda, \mu]$ of (λ_x, μ_x) in the quotient group $\Lambda(\widetilde{\Gamma}, N, A)$ is now independent of the choice of cross-section x.

Proposition 6.17. *The equivariant square* (26) *is completely determined by the corresponding cohomology class* $\tilde{\chi}$ *in* $\Lambda\big(\widetilde{\Gamma}, N, A\big)$. *Conversely, if* $\tilde{\chi} = \big[\lambda, \mu\big] \in \Lambda\big(\widetilde{\Gamma}, N, A\big)$ *has the property that* $\lambda_{\mathbf{R}}(N, \cdot)\mathrm{B}^1(\mathbf{R}, A) = \mathrm{Z}^1(\mathbf{R}, A)$, *where* $\lambda_{\mathbf{R}}(m, \cdot)$ *means the element of* $\mathrm{Z}^1(\mathbf{R}, A)$ *corresponding to* $\big\{\lambda(m, \gamma_t) : t \in \mathbf{R}\big\}$, *then* $\tilde{\chi}$ *determines a* $\widetilde{\Gamma}$*-equivariant square of the form* (26) *uniquely up to isomorphism.*

We leave the proof to the reader. One should note, however, that the surjectivity of g and h follow from the assumption that $\lambda_{\mathbf{R}}(N, \cdot)\mathrm{B}^1(\mathbf{R}, A) = \mathrm{Z}^1(\mathbf{R}, A)$. As in [700, Theorem 2.2], each element $\tilde{\chi} \in \Lambda\big(\widetilde{\Gamma}, N, A\big)$ determines a pair $(\chi, \nu) \in \Lambda\big(\Gamma, N, A\big)^{\mathbf{R}} \times \mathrm{Hom}_{\Gamma}(N, \mathrm{H})$. As to the characterisation of pairs which occur in this fashion, we refer the reader to the paper cited above.

Definition 6.18. We call the cohomology class $\tilde{\chi} \in \Lambda\big(\mathrm{Aut}(\mathcal{M}) \times \mathbf{R}, \mathrm{Cnt}_r(\mathcal{M}), \mathcal{U}(\mathcal{C})\big)$, which determines the characteristic square of Theorem 6.15, the *intrinsic invariant* of $\mathcal{M}$, and write $\tilde{\chi}_{\mathcal{M}}$ or $\Theta(\mathcal{M})$ for $\tilde{\chi}$ sometimes.

Exercise XII.6

1) Let $\mathcal{M}$ be a separable factor of type III_λ. Show that the characteristic square of $\mathcal{M}$ in Theorem 6.15 takes the following form with $T = -2\pi/\log\lambda$, $T' = -\log\lambda$ and $\{\mathcal{C}, \mathbf{R}, \theta\} = \{L^\infty(\mathbf{R}/T'\mathbf{Z}), \mathbf{R}, \text{translation})\}$:

$$\left\{\begin{array}{ccccccccc}
 & & 1 & & 1 & & 1 & & \\
 & & \downarrow & & \downarrow & & \downarrow & & \\
1 & \longrightarrow & \mathbf{T} & \longrightarrow & \mathcal{U}(\mathcal{C}) & \xrightarrow{\partial} & \mathrm{B}^1_\theta\big(\mathbf{R}, \mathcal{U}(\mathcal{C})\big) & \longrightarrow & 1 \\
 & & \downarrow & & \downarrow & & \downarrow & & \\
1 & \longrightarrow & \mathcal{U}(\mathcal{M}) & \longrightarrow & \widetilde{\mathcal{U}}(\mathcal{M}) & \xrightarrow{\partial} & \mathrm{Z}^1_\theta\big(\mathbf{R}, \mathcal{U}(\mathcal{C})\big) & \longrightarrow & 1 \\
 & & \downarrow & & \downarrow{\scriptstyle \widetilde{\mathrm{Ad}}} & & \downarrow{\scriptstyle \delta} & & \\
1 & \longrightarrow & \mathrm{Int}(\mathcal{M}) & \longrightarrow & \mathrm{Cnt}_r(\mathcal{M}) & \xrightarrow{\nu} & \mathbf{R}/T\mathbf{Z} & \longrightarrow & 1 \\
 & & \downarrow & & \downarrow & & \downarrow & & \\
 & & 1 & & 1 & & 1 & &
\end{array}\right.$$

2) For each $\alpha \in \mathrm{Aut}(\mathcal{M})$ set $\mathrm{mod}(\alpha) = \tilde{\alpha}\big|_{\mathcal{C}}$ where $\mathcal{C}$ means the center of $\widetilde{\mathcal{M}}$ and let $\mathrm{Aut}_0(\mathcal{M}) = \mathrm{Ker}(\mathrm{mod})$. Let $\widetilde{\mathcal{M}_{\varphi,0}} = \mathcal{M} \rtimes_{\sigma^\varphi} \mathbf{R}/T\mathbf{Z}$ with φ a faithful semi-finite normal weight such that $\sigma^\varphi_T = \mathrm{id}$.

(a) Show that the exact square of $\mathcal{M}$ takes the following form:

$$\left\{\begin{array}{ccccccccc}
 & & 1 & & 1 & & 1 & & \\
 & & \downarrow & & \downarrow & & \downarrow & & \\
1 & \longrightarrow & \mathbf{T} & \longrightarrow & \mathbf{T} & \longrightarrow & 1 & \longrightarrow & 1 \\
 & & \downarrow & & \downarrow & & \downarrow & & \\
1 & \longrightarrow & \mathcal{U}(\mathcal{M}) & \longrightarrow & \widetilde{\mathcal{U}}_0(\mathcal{M}) & \xrightarrow{\partial} & \mathbf{T} & \longrightarrow & 1 \\
 & & \downarrow & & \downarrow{\scriptstyle \widetilde{\mathrm{Ad}}} & & \downarrow{\scriptstyle \delta} & & \\
1 & \longrightarrow & \mathrm{Int}(\mathcal{M}) & \longrightarrow & \mathrm{Cnt}_r(\mathcal{M}) & \xrightarrow{\nu} & \mathbf{R}/T\mathbf{Z} & \longrightarrow & 1 \\
 & & \downarrow & & \downarrow & & \downarrow & & \\
 & & 1 & & 1 & & 1 & &
\end{array}\right.$$

where $\widetilde{\mathcal{U}}_0(\mathcal{M}) = \big\{u \in \mathcal{U}(\widetilde{\mathcal{M}}_0) : u\mathcal{M}u^* = \mathcal{M}\big\}$ with $\widetilde{\mathcal{M}}_0 \cong \mathcal{M} \rtimes_{\sigma^\varphi} (\mathbf{R}/T\mathbf{Z})$.

(b) Show that $\mathrm{Aut}_0(\mathcal{M}) \times \mathbf{Z}$ acts on the exact square. Is it possible to make this square equivariant under the larger group $\mathrm{Aut}(\mathcal{M}) \times \mathbf{Z}$? Examine if there is any obstruction to do this.

In the rest of this Exercise, assume that $\mathcal{M}$ is a general separable factor.

3) Set $\{\mathcal{A}, \mathbf{R}, \rho\}$ to be the covariant system $\{L^\infty(\mathbf{R}),\ \text{translation}\}$, i.e.

$$\big(\rho_t(f)\big)(s) = f(s+t), \quad f \in \mathcal{A}, \quad s, t \in \mathbf{R}.$$

In $\{\mathcal{A}, \mathbf{R}, \rho\}$, we fix the following one parameter unitary group $\{\, V(t) : t \in \mathbf{R}\,\}$ which generates $\mathcal{A}$ as a von Neumann algebra and its analytic generator H affiliated with $\mathcal{A}$:

$$\begin{cases} \big(V(t)\big)(s) = \mathrm{e}^{-ist}, & s, t \in \mathbf{R}; \\ H(t) = \mathrm{e}^{-t}. \end{cases}$$

It then follows that:

$$\rho_t\big(V(s)\big) = \mathrm{e}^{-ist}\, V(t), \quad s, t \in \mathbf{R}.$$

Fix $\varphi \in \mathfrak{W}(\mathcal{M})$ and $e = s(\varphi)$ the support of φ.

(a) Show that there exists uniquely an equivariant isomorphsim π^φ from $\{\mathcal{A}, \mathbf{R}, \rho\}$ into $\{\widetilde{\mathcal{M}}_{s(\varphi)}, \mathbf{R}, \theta\}$ such that

$$\pi^\varphi\big(V(t)\big) = \varphi^{\mathrm{i}t}, \quad t \in \mathbf{R}.$$

(b) Show that if π is an equivariant isomorphism of $\{\mathcal{A}, \mathbf{R}, \theta\}$ into $\{\widetilde{\mathcal{M}}_e, \mathbf{R}, \theta\}$ with $e \in \mathrm{Proj}(\mathcal{M})$, then there exists uniquely $\varphi = \varphi_\pi \in \mathfrak{W}(\mathcal{M})$ such that $\pi = \pi^\varphi$.

4) Suppose $\pi = \pi^\varphi$ for a fixed $\varphi \in \mathfrak{W}(\mathcal{M})$. Let $h = \pi(H)$ be the analytic generator of the one parameter unitary group $\{v(t)\} = \{\pi\big(V(t)\big)\}$ in $\widetilde{\mathcal{M}}_{s(\varphi)}$. Prove that if $I_\theta(a) = \int_{\mathbf{R}} \theta_t(a)\, dt = 1,\ a \in \widetilde{\mathcal{M}}_+$, then

$$\varphi(x) = \lim_{\varepsilon \searrow 0} \tau\Big(h^{\frac12}\big(1+\varepsilon h^{\frac12}\big)^{-\frac12} x^{\frac12} a x^{\frac12} h^{\frac12}\big(1+\varepsilon h^{\frac12}\big)^{-\frac12}\Big), \quad x \in \widetilde{\mathcal{M}}_+.$$

Symbolically, we can write:

$$\varphi(x) = \tau\Big(h^{\frac12} x^{\frac12} a x^{\frac12} h^{\frac12}\Big) = \tau\Big(a^{\frac12} x^{\frac12} h x^{\frac12} a^{\frac12}\Big), \quad x \in \widetilde{\mathcal{M}}_+.$$

Observe that the result of the right hand does not depend on the choice of $a \in \widetilde{\mathcal{M}}_+$ as long as $I_\theta(a) = 1$ holds.

5) Prove that if μ is a normal weight on $\mathcal{A}$ such that $\mu \circ \rho_s = \mathrm{e}^{-s}\mu$ and $0 < \mu(f_0) < \infty$ for some $f_0 \in \mathcal{A}_+$, then the weight μ is of the following form:

$$\mu(f) = \int_{\mathbf{R}} \mathrm{e}^s f(s)\, \mathrm{d}s, \quad f \in \mathcal{A} = L^\infty(\mathbf{R}),$$

and it is therefore faithful and semi-finite.

6) Prove the equivalence of the following statements for $\varphi \in \mathfrak{W}(\mathcal{M})$:

(i) The weight φ is finite, i.e. $\varphi(1) < \infty$;

(ii) $\pi^{\varphi}(\mathcal{A}) \cap \mathfrak{m}_{\tau} \neq \{0\}$;

(iii) The operator φ in $\widetilde{\mathcal{M}}$ is τ-measurable in the sense of §2, Chapter IX.

7) For a densely defined closed operator T affiliated with $\widetilde{\mathcal{M}}$, we say that T has *grade* $\alpha \in \mathbf{C}$ if $\theta_s(T) = \mathrm{e}^{-\alpha s}T$ and write $\alpha = \mathrm{grad}(T)$. Let $\mathfrak{M}$ be the set of all τ-measurable operators affiliated with $\widetilde{\mathcal{M}}$ and $\mathfrak{M}(\alpha) = \left\{T \in \mathfrak{M} : \mathrm{grad}(T) = \alpha\right\}$ for $\alpha \in \mathbf{C}$.

(a) Show that $\mathfrak{M}(\alpha) = \{0\}$ for $\mathrm{Re}(\alpha) < 0$ and that $\mathfrak{M}(0) = \mathcal{M}$.

(b) Show that if $T \in \mathfrak{M}$ has $\mathrm{grad}(T) = \alpha \in \mathbf{D}$ with $\mathbf{D}$ the right half plane in $\mathbf{C}$ and $p = \mathrm{Re}(\alpha)$, then $|T|^{\frac{1}{p}}$ is given by an element $\varphi \in \mathcal{M}_*^+$.

(c) Show that if $T \in \mathfrak{M}(1)$, then for any $a \in \widetilde{\mathcal{M}}_+$ with $I_\theta(a) = 1$ the value $\tau\left(a^{1/2}Ta^{1/2}\right)$ is finite and independent of the choice of such a. Write this value $\int T$ for $T \in \mathfrak{M}(1)$.

(d) Show that $\mathfrak{M}(1)$ is a Banach space under the norm:

$$\|T\| = \int |T|, \quad T \in \mathfrak{M}(1).$$

(e) Show that if $h, k \in \mathfrak{M}(1)_+$, then on the vertical strip $\mathbf{S} = \left\{\alpha \in \mathbf{C} : 0 \leq \mathrm{Re}(\alpha) \leq 1\right\}$ we have $h^\alpha k^{1-\alpha} \in \mathfrak{M}(1)$ and the multiplicative convexity:

$$\left\|h^\alpha k^{1-\alpha}\right\| \leq \|h\|^{\mathrm{Re}(\alpha)}\|k\|^{1-\mathrm{Re}(\alpha)}, \quad \alpha \in \mathbf{S}.$$

(f) Show that for fixed $\alpha \in \mathbf{D}+1$ with $p = \mathrm{Re}(\alpha)$ the quantity:

$$\|T\| = \left(\int |T|^{\frac{1}{p}}\right)^p, \quad T \in \mathfrak{M}(\alpha),$$

makes $\mathfrak{M}(\alpha)$ a Banach space. Note that unlike the usual integration theory, the intersection of different $\mathfrak{M}(\alpha)$'s is $\{0\}$ so that the norm $\|T\|$ is defined for only one $\alpha \in \mathbf{D}+1$.

(g) Show that if $\left\{T_1, T_2, \dots, T_n\right\} \subset \mathfrak{M}$ and $\mathrm{grad}(T_1)+\mathrm{grad}(T_2)+\cdots+\mathrm{grad}(T_n)= 1$, then $T_1T_2\cdots T_n \in \mathfrak{M}(1)$ and

$$\int T_1T_2\cdots T_n = \int T_nT_1T_2\cdots T_{n-1};$$

$$\left\|T_1T_2\cdots T_n\right\| \leq \|T_1\|\|T_2\|\cdots\|T_n\|.$$

(h) Show that if $\alpha \in \mathbf{D}$ and $1-\alpha \in \mathbf{D}$, then the Banach spaces $\mathfrak{M}(\alpha)$ and $\mathfrak{M}(1-\alpha)$ is in duality under the pairing:

$$\langle S, T\rangle = \int ST, \quad S \in \mathfrak{M}(\alpha), \quad T \in \mathfrak{M}(1-\alpha).$$

[543, 563, 567, 596, 719]

8) Fix a $\varphi \in \mathfrak{W}(\mathcal{M})$ and set $\mathcal{A}^\varphi = \pi^\varphi(\mathcal{A})$ and $\mathcal{D}^\varphi = \mathcal{C} \vee \mathcal{A}^\varphi = \mathcal{C} \vee \pi^\varphi\big(L^\infty(\mathbf{R})\big)$. Observe the following facts:

$$\begin{cases} (\mathcal{A}^\varphi)' \cap \mathcal{M} = \mathcal{M}_\varphi; \\ \mathcal{D}^\varphi \cap \mathcal{M} \subset \mathcal{C}_\varphi, \end{cases}$$

where $\mathcal{M}_\varphi$ is the centralizer of φ in $\mathcal{M}$ and $\mathcal{C}_\varphi$ is the center of $\mathcal{M}_\varphi$.

9) Recall the middle row for the extended unitary group $\widetilde{\mathcal{U}}(\mathcal{M})$:

$$1 \longrightarrow \mathcal{U}(\mathcal{M}) \longrightarrow \widetilde{\mathcal{U}}(\mathcal{M}) \xrightarrow{\partial} Z^1_\theta\big(\mathbf{R}, \mathcal{U}(\mathcal{C})\big) \longrightarrow 1,$$

where $(\partial b)_t = b\theta_t(b^*),\ t \in \mathbf{R},\ b \in \widetilde{\mathcal{U}}(\mathcal{M})$.

10) Following the steps suggested below, show that to each $\varphi \in \mathfrak{W}_0(\mathcal{M})$ there corresponds a right inverse $b_\varphi \colon c \in Z^1_\theta\big(\mathbf{R}, \mathcal{U}(\mathcal{C})\big) \mapsto b_\varphi(c) \in \widetilde{\mathcal{U}}(\mathcal{M}) \cap \mathcal{D}^\varphi$ of the coboundary map ∂ such that

(i) $$\partial\big(b_\varphi(c)\big) = c, \quad c \in Z^1_\theta\big(\mathbf{R}, \mathcal{U}(\mathcal{C})\big);$$

(ii) b_φ is a continuous homomorphism of $Z^1_\theta\big(\mathbf{R}, \mathcal{U}(\mathcal{C})\big)$ into $\widetilde{\mathcal{U}}(\mathcal{M}) \cap \mathcal{D}^\varphi$;

(iii) $$b_{\varphi\circ\alpha^{-1}} = \tilde{\alpha} \circ b_\varphi \circ \tilde{\alpha}^{-1}, \quad \alpha \in \mathrm{Aut}(\mathcal{M});$$

(iv) If $\varphi \in \mathfrak{W}_0(\mathcal{M})$ is dominant, then

$$\widetilde{\mathrm{Ad}}\big(b_\varphi(c^*)\big) = \sigma^\varphi_c, \quad c \in Z^1_\theta\big(\mathbf{R}, \mathcal{U}(\mathcal{C})\big);$$

(v) For every pair $\varphi, \psi \in \mathfrak{W}_0(\mathcal{M})$ of dominant weights

$$b_\varphi(c^*)b_\psi(c^*)^* = (\mathrm{D}\varphi : \mathrm{D}\psi)_c, \quad c \in Z^1_\theta\big(\mathbf{R}, \mathcal{U}(\mathcal{C})\big);$$

(vi) For each $c^1, c^2 \in Z^1_\theta\big(\mathbf{R}, \mathcal{U}(\mathcal{C})\big)$,

$$(\mathrm{D}\varphi : \mathrm{D}\psi)_{c^1c^2} = (\mathrm{D}\varphi : \mathrm{D}\psi)_{c^1}\sigma^\psi_{c^1}\big((\mathrm{D}\varphi : \mathrm{D}\psi)_{c^2}\big). \qquad [520]$$

(**a**) Fix $\varphi \in \mathfrak{W}_0(\mathcal{M})$ and identify $\widetilde{\mathcal{M}}$ and the crossed product $\mathcal{M} \rtimes_{\sigma^\varphi} \mathbf{R}$ where the one parameter unitary group $\{\varphi^{\mathrm{i}t}\}$ corresponds to the modular automorphism group σ^φ.

(**b**) We want to choose the diagonalization of $\{\varphi^{\mathrm{i}t}\}$ rather than the dual action $\{\theta_t\}$. So let F be the Fourier transform and the Fourier inverse transform on $\widetilde{\mathfrak{H}} = L^2(\mathbf{R}, \mathfrak{H})$ given by:

$$\begin{aligned} (F\xi)(p) &= \sqrt{\frac{1}{2\pi}} \int_{\mathbf{R}} \mathrm{e}^{-\mathrm{i}sp}\xi(s)\,\mathrm{d}s, \\ (F^*\xi)(s) &= \sqrt{\frac{1}{2\pi}} \int_{\mathbf{R}} \mathrm{e}^{\mathrm{i}sp}\xi(p)\,\mathrm{d}p, \end{aligned} \qquad \xi \in L^2(\mathbf{R}, \mathfrak{H}),$$

where the above integral should be understood as the improper integral:

$$\int_{\mathbf{R}} f(t)\,\mathrm{d}t = \lim_{n\to\infty} \int_{-n}^{n} f(t)\,\mathrm{d}t$$

for those functions f such that the above limit exists. We then work with the von Neumann algebra $F(\mathcal{M} \rtimes_{\sigma^\varphi} \mathbf{R})F^*$ instead of the original crossed product. We have then:

$$\begin{cases} \big(F\varphi^{\mathrm{i}t}F^*\xi\big)(s) = \mathrm{e}^{-\mathrm{i}st}\xi(s) = \big(V(t)\xi\big)(s); \\ \big(F\mu(p)F^*\xi\big)(s) = \xi(s-t) = \big(\bar\lambda(t)\xi\big)(s). \end{cases}$$

The von Neumann algebra $F(\mathcal{M} \rtimes_{\sigma^\varphi} \mathbf{R})F^* = F\big(\mathcal{M} \mathbin{\overline{\otimes}} \mathcal{L}(L^2(\mathbf{R}))\big)^{(\sigma^\varphi\otimes\rho)}F^*$ is then identified with the fixed point algebra $\big(\mathcal{M} \mathbin{\overline{\otimes}} \mathcal{L}(L^2(\mathbf{R}))\big)^{\sigma^\varphi\otimes\mathrm{Ad}(\mu)}$. In order to keep track of the identification, let us denote by $\tilde\pi^\varphi$ the isomorphism of $\widetilde{\mathcal{M}}$ onto $F(\mathcal{M} \rtimes_{\sigma^\varphi} \mathbf{R})F^*$.

(**c**) Identify $\widetilde{\mathcal{M}}$ with $F(\mathcal{M} \rtimes_{\sigma^\varphi} \mathbf{R})F^*$ and $\varphi^{\mathrm{i}t}$ with $V(t)$. Observe that the non-commutative flow $\{\theta_t\}$ is identified with the one parameter unitary group $\big\{\bar\rho(s) : s \in \mathbf{R}\big\} = \big\{1 \otimes \rho(s) : s \in \mathbf{R}\big\}$.

(**d**) Set

$$\big(\tilde b_\varphi(c)\xi\big)(s) = \big(\tilde\pi^\varphi(c_{-s})\xi\big)(s), \quad c \in \mathrm{Z}^1_\theta\big(\mathcal{U}(\mathcal{C})\big), \quad \xi \in \widetilde{\mathfrak{H}}, \quad s \in \mathbf{R}.$$

Put

$$b_\varphi(c) = (\tilde\pi^\varphi)^{-1}\big(\tilde b_\varphi(c)\big).$$

(**e**) Prove that

$$b_\varphi(c) \in \mathcal{D}^\varphi = \mathcal{C} \vee \mathcal{A}^\varphi$$

and that

$$b_\varphi(c^1c^2) = b_\varphi(c^1)b_\varphi(c^2), \quad c^1, c^2 \in \mathrm{Z}^1_\theta\big(\mathcal{U}(\mathcal{C})\big).$$

(**f**) Show that the map $b_\varphi \colon c \in \mathrm{Z}^1_\theta\big(\mathcal{U}(\mathcal{C})\big) \mapsto b_\varphi(c) \in \mathcal{D}^\varphi$ is continuous, i.e. (ii).

(**g**) Show that with $b = b_\varphi(c)$

$$\partial(b)_t = b\theta_t(b^*) = c_t, \quad t \in \mathbf{R}.$$

Thus (i) follows.

(**h**) Representing $\mathcal{M}$ in the standard form $\{\mathcal{M}, \mathfrak{H}, \mathfrak{H}_+\}$ and writing the unitary corresponding to $\alpha \in \mathrm{Aut}(\mathcal{M})$ by $U(\alpha)$, prove that

$$\begin{cases} \pi^{\varphi\circ\alpha^{-1}} = \tilde\alpha \circ \pi^\varphi; \\ \tilde\pi^{\varphi\circ\alpha^{-1}}(x) = \tilde U(\alpha)\tilde\pi^\varphi(x)U(\alpha)^*, \quad x \in \widetilde{\mathcal{M}}; \\ (\varphi\circ\alpha^{-1})^{\mathrm{i}t} = \tilde\alpha(\varphi^{\mathrm{i}t}); \\ \mathcal{A}^{\varphi\circ\alpha^{-1}} = \tilde\alpha(\mathcal{A}^\varphi); \\ \mathcal{D}^{\varphi\circ\alpha^{-1}} = \tilde\alpha(\mathcal{D}^\varphi) \end{cases}$$

where $\widetilde U(\alpha) = U(\alpha) \otimes 1$ on $\widetilde{\mathfrak{H}} = L^2(\mathbf{R}, \mathfrak{H}) = \mathfrak{H} \otimes L^2(\mathbf{R})$.

(i) Prove that

$$\tilde{\pi}^{\varphi}\Big(\tilde{\alpha}\big(b_{\varphi}(\tilde{\alpha}^{-1}(c))\big)\Big) = \tilde{\pi}^{\varphi\circ\alpha^{-1}}\big(b_{\varphi\circ\alpha^{-1}}(c)\big)$$

and observe that the equivariance (iii) of the maps b_{φ}, $\varphi \in \mathfrak{W}_0(\mathcal{M})$, follows.

(j) Prove directly that $\widetilde{\sigma_c^{\varphi}} = \widetilde{\mathrm{Ad}}\big(b_{\varphi}(c^*)\big)$ for a dominant $\varphi \in \mathfrak{W}_0(\mathcal{M})$. Thus (iv) follows.

(k) Extend the above arguments to the balanced weight on $\mathcal{M}_2 = M_2(\mathbf{C}) \otimes \mathcal{M}$ and show (v) and (vi).

(l) Show that if $\{\varphi_n\}$ is a sequence in $\mathfrak{W}$ converging to φ in the strong resolvent sense as closed operator affiliated with $\widetilde{\mathcal{M}}$, then for each $f \in L^{\infty}(\mathbf{R})$ the sequence $\{\tilde{\pi}^{\varphi_n}(f)\}$ converges to $\tilde{\pi}^{\varphi}(f)$ in the σ^*-strongly; hence $\{b_{\varphi_n}(c)\}$ also converges to $b_{\varphi}(c)$ σ^*-strongly for every $c \in \mathrm{Z}_{\theta}^1\big(\mathcal{U}(\mathcal{C})\big)$.

Notes on Chapter XII

The general structure theorem, Theorem 1.1, for a von Neumann algebra of type III was a result of long struggle of many mathematicians ever since the pioneering work of Murray and von Neumann. At their time, the mere existence of such factors was a big issue, [242]. A von Neumann algebra of type III had stayed in mystery for many many years. In the midst of these years, Araki and Woods showed in 1963 that most of von Neumann algebras appearing in quantum physics are indeed of type III, [430]. This discovery shocked many specialists as they hoped to apply the theory of operator algebras to quantum physics, but alas one had to understand von Neumann algebras of type III first to fulfill their dream. Well, the study of von Neumann algebras of type III was intensified in the sixties: R. T. Powers showed the existence of continuously many non-isomorphic factors of type III, [670]. In the today's language, what Powers proved was the existence of a factor of type III_{λ} for each λ, $0 < \lambda < 1$. Powers' discovery prompted specialists to further intensify their efforts in the study of the structure of factors in two directions: one toward the construction of non-isomorphic factor of type II_1 and the other toward the classification theory of concretely constructed, (infinite tensor product of type I factors), factors of type III, which appears frequently in quantum physics. The first line of effort resulted the construction of continuously many non-isomorphic factors of McDuff, [635], and of Sakai, [686], in 1969. The second line of research was taken up by Araki and Woods and produced their classification theory of factors of infinite tensor product of type I factors, (abbreviated ITPFI), [431], in 1969. They introduced two algebraic invariants for factors of type III, the asymptotic ratio set $r_{\infty}(\mathcal{M})$ and ρ-set for a factor $\mathcal{M}$. Those invariants were driven from ideas hidden in an unconfirmed unofficial preprint of Tomita distributed in 1967. Their theory was further sharpened by W. Krieger in his study of ergodic transformation and associated factors, the subject taken up in the next chapter. It was A. Connes who unveiled the true nature of these two invariants in terms of the Tomita-Takesaki modular theory. It turned out that the asymptotic ratio set was the modular spectrum, $S(\mathcal{M})$, and the

ρ-set was the modular period for an ITPFI factor $\mathcal{M}$. In his thesis, [458], Connes obtained the discrete decomposition, with cocycle conjugacy uniqueness, of a factor of type III_λ, $0 \leq \lambda < 1$, in 1972, published in 1973. At the same time, H. Araki obtained a similar decomposition for a factor which admits a almost periodic state without uniqueness, [426]. Takesaki also proved such a decomposition for a factor with periodic homogeneous states with a weaker uniqueness of cocycle conjugacy of the associated covariant system, [713]. These three works were all completed in 1972 and appeared in 1973. Then the duality theorem, Theorem X.2.3, was proved to yield the general structure theorem, Theorem 1.1, with cocycle conjugacy uniqueness. The stronger uniqueness of conjugacy was then proven in the joint work of Connes and Takesaki, [491], in 1974 and published in 1977.

The materials of this chapter were taken mostly from the works of Connes, [458], Takesaki, [714, 362] and the joint work of Connes and Takesaki, [491]. The results of §5 were from the joint work of Connes and Størmer, [490]. The non-commutative flow of weights on a factor in §6 is taken from a joint work of Falcone and Takesaki, [520], whilst the characteristic square was discovered by Y. Katayama, C. E. Sutherland and Takesaki, [591].

Appendix

A.1 Analyticity in Banach Spaces. Let E be a Banach space and G a complex domain. For an E-valued function f on G, the following conditions are equivalent:

(i) For each $\alpha_0 \in G$ and a sufficiently small $\delta > 0$, there exists a sequence $\{a_n\}$ in E such that

$$f(\alpha) = \sum_{n=0}^{\infty} (\alpha - \alpha_0)^n a_n, \quad |\alpha - \alpha_0| < \delta, \tag{1}$$

converges in norm;

(ii) There exists a norm closed subspace F of E^* such that

$$\|x\| = \sup\Big\{ \big|\langle x, x'\rangle\big| : x' \in F, \ \|x'\| \le 1 \Big\}, \quad x \in E,$$

and each $x' \in F$ gives rise to a holomorphic function: $\alpha \in G \mapsto \big\langle f(\alpha), x'\big\rangle$.

If this is the case, then the sequence $\{a_n\}$ is given by

$$a_n = \frac{1}{n!} f^{(n)}(\alpha_0), \quad n \ge 0. \tag{2}$$

(iii) The function f is locally bounded in the sense that f is bounded on any compact subset of G and there exists a subspace F, not necessarily norm closed, of E^* such that

$$\|x\| = \sup\Big\{ \big|\langle x, x'\rangle\big| : x' \in F, \ \|x'\| \le 1 \Big\}, \quad x \in E,$$

and each $x' \in F$ gives rise to a holomorphic function: $\alpha \in G \mapsto \big\langle f(\alpha), x'\big\rangle$.

PROOF:

(i) $\Longrightarrow$ (ii): The conclusion holds with $F = E^*$.

(iii) $\Longrightarrow$ (ii): By assumption, we may regard E as a norm closed subspace of F^*. Let $\alpha_0 \in G$ and a closed disk $D \subset G$ with center α_0. By the assumption, f is bounded on D so that there exists $M > 0$ such that $\big\| f(\alpha)\big\| \le M$, $\alpha \in D$. Therefore, if a sequence $\{x'_n\}$ in F converges to $x' \in E^*$ in norm, then the inequality: $\big\{\big|\langle f(\alpha), x'_n\rangle - \langle f(\alpha), x'\rangle\big| \le M\|x'_n - x'\|\big\}$ yields the uniform convergence of the functions: $\big\{\big\langle f(\,\cdot\,), x'_n\big\rangle\big\}$ to $\big\langle f(\,\cdot\,), x'\big\rangle$. Thus the limit function: $\big\langle f(\,\cdot\,), x'\big\rangle$ is holomorphic on G. Hence, the closure $\bar{F}$ of F has the same property of F which means the property (ii).

(ii) $\Longrightarrow$ (i): Assume (ii). Then F is a Banach space and E is a norm-closed subspace of F^*. For each fixed $x' \in F$, the function: $\langle f(\,\cdot\,), x'\rangle$ is analytic, hence it is bounded on a closed disk, say $D \subset G$. The uniform boundedness theorem implies that f is bounded on D. Since the problem is local, we may assume that G is the unit disk and α_0 is the origin 0 of G. For a fixed $\lambda \in G$, choose $0 < r < 1$ with $|\lambda| < r$. For each $n \in \mathbf{N}$ and $x' \in F$, consider the contour integral:

$$\frac{1}{2\pi \mathrm{i}} \oint_{|\alpha|=r} \frac{\langle f(\alpha), x'\rangle}{\alpha^{n+1}}\, d\alpha,$$

which gives rise to an element $a_n \in F^*$ such that

$$\langle a_n, x'\rangle = \frac{1}{2\pi \mathrm{i}} \oint_{|\alpha|=r} \frac{\langle f(\alpha), x'\rangle}{\alpha^{n+1}}\, d\alpha\,.$$

The Cauchy integral formula yields for every $\lambda \in \mathbf{D}_r = \{\lambda \in \mathbf{C} : |\lambda| < r\}$ that:

$$\begin{aligned}\langle f(\lambda), x'\rangle &= \frac{1}{2\pi \mathrm{i}} \oint_C \frac{\langle f(\alpha), x'\rangle}{\alpha - \lambda}\, d\alpha = \frac{1}{2\pi \mathrm{i}} \oint_C \left(\sum_{n=0}^{\infty} \frac{\langle f(\alpha), x'\rangle \lambda^n}{\alpha^{n+1}} \right) d\alpha \\ &= \sum_{n=0}^{\infty} \frac{1}{2\pi \mathrm{i}} \left(\oint_{|\alpha|=r} \frac{\langle f(\alpha), x'\rangle}{\alpha^{n+1}}\, d\alpha \right) \lambda^n = \sum_{n=0}^{\infty} \langle a_n, x'\rangle \lambda^n.\end{aligned}$$

With $M = \sup\{\, \|f(\alpha)\| : |\alpha| = r \,\}$, we estimate

$$\left|\langle a_n, x'\rangle\right| \le \frac{1}{2\pi} \left| \oint_{|\alpha|=r} \frac{\langle f(\alpha), x'\rangle}{\alpha^{n+1}}\, d\alpha \right| \le M \frac{\|x'\|}{r^n},$$

so that the power series $\sum \lambda^n a_n$ converges in norm and uniformly on any closed disk contained in $\mathbf{D}_r$ to the element $f(\lambda) \in E$. Therefore, the function $\lambda \in \mathbf{D}_r \mapsto f(\lambda)$ is norm continuous. Thus the integral

$$a_n = \frac{1}{2\pi \mathrm{i}} \oint \frac{f(\alpha)}{\alpha^{n+1}}\, d\alpha$$

can be obtained as a norm limit of Riemann sum, which guarantees that a_n belongs to E and the sum of the power series converges in norm. Thus we get the assertion (i).

(ii) $\Longrightarrow$ (iii): This is a triviality. Q.E.D.

A.2. Any one of the conditions (A.1.i), (A.1.ii) and (A.1.iii) implies also the norm differentiability of the function f.

A.3 Stone's Theorem. Let $\{U(t) : t \in \mathbf{R}\}$ be a family of unitary operators on a Hilbert space $\mathfrak{H}$ such that

$$U(s)U(t) = U(s+t), \quad s, t \in \mathbf{R}, \tag{1}$$

$$\lim_{s \to 0} \| U(s)\xi - \xi \| = 0, \quad \xi \in \mathfrak{H}. \tag{2}$$

Then $\{U(t)\}$ is called a one parameter unitary group on $\mathfrak{H}$. In this case, there exists a self-adjoint operator H, possibly unbounded, on $\mathfrak{H}$ such that:

(i) The domain $\mathfrak{D}(H)$ is precisely the set of those vectors ξ for which the limit:

$$\lim_{t \to 0} \frac{1}{t}\big(U(t) - 1\big)\xi = \mathrm{i}H\xi \tag{3}$$

exists;

(ii) The one parameter unitary group $U(t)$ is given by the formula:

$$U(t) = \exp(\mathrm{i}tH) \tag{4}$$

in the sense that

$$U(t) = \int_{\mathbf{R}} \mathrm{e}^{\mathrm{i}\lambda t}\, \mathrm{d}E(\lambda), \tag{5}$$

where $\{E(\lambda)\}$ gives the spectral decomposition of H:

$$H = \int_{\mathbf{R}} \lambda \, \mathrm{d}E(\lambda)\,. \tag{6}$$

PROOF: For each non-zero s, we set

$$H_s = \frac{1}{\mathrm{i}s}\big(U(s) - 1\big), \quad s \neq 0, \tag{7}$$

and define

$$H\xi = \lim_{s \to 0} H_s \xi \tag{3'}$$

for those ξ such that the limit (3′) exists. Clearly, H is a linear operator with domain $\mathfrak{D}(H)$ given in (3). For each $\varepsilon \neq 0$, we set $T_\varepsilon = \frac{1}{\varepsilon}\int_0^\varepsilon U(t)\,\mathrm{d}t \in \mathcal{L}(\mathfrak{H})$. We then have

$$\| T_\varepsilon \xi - \xi \| \le \frac{1}{\varepsilon}\int_0^\varepsilon \| U(t)\xi - \xi \| \,\mathrm{d}t \le \max_{0 \le t \le \varepsilon} \| U(t)\xi - \xi \| \to 0 \quad \text{as } \varepsilon \to 0.$$

Making use of the group property (1), we have

$$H_s T_\varepsilon = H_\varepsilon T_s, \quad s \neq 0, \quad \varepsilon \neq 0, \tag{8}$$

which implies that $T_\varepsilon \mathfrak{H} \subset \mathfrak{D}(H)$ and

$$\lim_{s \to 0} H_s T_\varepsilon \xi = H_\varepsilon \xi. \tag{9}$$

This shows that $\mathfrak{D}(H)$ is dense in $\mathfrak{H}$. If $\xi \in \mathfrak{D}(H)$, then (8) implies that

$$T_\varepsilon H\xi = T_\varepsilon \lim_{s\to 0} H_s\xi = \lim_{s\to 0} T_\varepsilon H_s\xi = \lim_{s\to 0} H_\varepsilon T_s\xi = H_\varepsilon\xi, \quad \varepsilon \neq 0. \tag{10}$$

If $\{\xi_n\} \subset \mathfrak{D}(H)$, $\xi_n \to \xi$ and $H\xi_n \to \eta$ as $n \to \infty$, then the above calculation shows that $H_\varepsilon\xi = T_\varepsilon H\xi_n$, so that $H_\xi\xi = \lim_{n\to\infty} T_\varepsilon H\xi_n = T_\varepsilon\eta$. Hence as $\varepsilon \to 0$, $H_\varepsilon\xi$ converges to η; thus $\xi \in \mathfrak{D}(H)$ and $H\xi = \eta$. Therefore, H is a closed operator.

Now, suppose that $\eta \in \mathfrak{D}(H^*)$. Since $H_s^* = H_{-s}$ and $T_\varepsilon^* = -T_{-\varepsilon}$ by the unitary property of $\{U(t)\}$, we have

$$\begin{aligned} H^*\eta &= \lim_{\varepsilon\to 0} T_\varepsilon H^*\eta = -\lim_{\varepsilon\to 0} T_{-\varepsilon}^* H^*\eta = \lim_{\varepsilon\to 0}(HT_\varepsilon)^*\eta \\ &= \lim_{\varepsilon\to 0} H_\varepsilon^*\eta = \lim_{\varepsilon\to 0} H_{-\varepsilon}\eta\,, \end{aligned}$$

so that $\eta \in \mathfrak{D}(H)$ and $H^*\eta = H\eta$. Therefore, H is self-adjoint. Since $\{U(t)\}$ and $\{H_s\}$ commute, $\{U(t)\}$ leaves $\mathfrak{D}(H)$ invariant, and $\{U(t)\}$ and H commute also. With the spectral decomposition (6) of H, we define a new one parameter unitary group $V(t)$ by the right hand side of (5). If $\mathfrak{M} = \bigcup_{n=1}^{\infty}[E(n) - E(-n)]\mathfrak{H}$, then each vector $\xi \in \mathfrak{M}$ has the property, by the Paley-Winner Theorem, that the $\mathfrak{H}$-valued function: $t \in \mathbf{R} \to V(t)\xi = \int_{-n}^{n} e^{i\lambda t}\, dE(\lambda)\,\xi$ has an entire extension and

$$\frac{d}{dt}V(t)\xi = iH\xi = \frac{d}{dt}U(t)\xi.$$

Hence $V(t)\xi - U(t)\xi$ must be constant for $\xi \in \mathfrak{M}$, namely zero by setting $t = 0$. Thus we obtain (4) or equivalently (5) since $\mathfrak{M}$ is dense. Q.E.D.

The operator H is called the *infinitesimal generator* of $\{V(t)\}$.

A.4 Core of the Infinitesimal Generator. A subset $\mathfrak{D}$ of the domain $\mathfrak{D}(T)$ of a closed operator T on a Hilbert space $\mathfrak{H}$ is called a *core* if T is the closure of its restriction to $\mathfrak{D}$, i.e. the minimal closed extension of $T|_{\mathfrak{D}}$. Let H be the generator of a one parameter unitary group $\{V(t)\}$ on $\mathfrak{H}$. A subspace $\mathfrak{D}$ of the definition domain $\mathfrak{D}(H)$ is a core of H if $\mathfrak{D}$ is dense in $\mathfrak{H}$ and invariant under $\{V(t)\}$.

PROOF: Considering the closure of $H|_{\mathfrak{D}}$, we may assume that $H|_{\mathfrak{D}}$ is already closed. We then want to prove that $\mathfrak{D} = \mathfrak{D}(H)$. In $\mathfrak{D}$, we define a new inner product:

$$(\xi \mid \eta)_H = (\xi \mid \eta) + \big(H\xi \mid H\eta\big), \quad \xi, \eta \in \mathfrak{D}.$$

By the closedness of $H|_{\mathfrak{D}}$, $\mathfrak{D}$ is a Hilbert space with respect to the norm $\|\cdot\|_H$. Since $V(t)$ and H commute, the restriction $V_{\mathfrak{D}}(t)$ of $V(t)$ to $\mathfrak{D}$ is a one parameter unitary group in $\mathfrak{D}$. Therefore, the operator T_ε defined in A.3 leaves $\mathfrak{D}$ invariant, since the integration used to define T_ε makes sense in $\mathfrak{D}$. Suppose $\xi \in \mathfrak{D}(H)$. Let $\{\xi_n\}$ be a sequence in $\mathfrak{D}$ such that $\|\xi - \xi_n\| \to 0$ as $n \to \infty$. Then $T_\varepsilon\xi_n \to T_\varepsilon\xi$ as $n \to \infty$ since T_ε is bounded. Furthermore, we have, as $n \to \infty$,

$$HT_\varepsilon\xi_n = H_\varepsilon\xi_n \longrightarrow H_\varepsilon\xi = HT_\varepsilon\xi,$$

so that $T_\varepsilon\xi$ belongs to $\mathfrak{D}$ and $T_\varepsilon\xi_n$ converges to $T_\varepsilon\xi$ in the $\|\cdot\|_H$-norm. Therefore, we get

$$\lim_{\varepsilon\to 0}\left\|T_\varepsilon\xi-\xi\right\|=0 \qquad \text{and} \qquad \lim_{\varepsilon\to 0}\left\|HT_\varepsilon\xi-\xi\right\|=\lim_{\varepsilon\to 0}\left\|H_\varepsilon\xi-\xi\right\|=0,$$

so that $\xi\in\mathfrak{D}$. Hence $\mathfrak{D}=\mathfrak{D}(H)$. Q.E.D.

A.5 Analytic Vectors. Let H be a fixed densely defined closed operator on a Hilbert space $\mathfrak{H}$ with domain $\mathfrak{D}$. A vector $\xi\in\mathfrak{D}$ is said to be an *analytic vector* of H if

there exists $r>0$ such that $\sum_{n=0}^{\infty}\frac{r^n}{n!}\|H^n\xi\|$ converges. It follows that analytic vectors form a linear subspace, not necessarily closed, of $\mathfrak{H}$. If H is symmetric and the set $\mathfrak{D}_0$ of analytic vectors is dense in $\mathfrak{H}$, then H is *essentially self-adjoint* on $\mathfrak{D}_0$ in the sense that the closure of the restriction $H|_{\mathfrak{D}_0}$ of H to $\mathfrak{D}_0$ is self-adjoint.

PROOF: For each $\xi\in\mathfrak{D}_0$, let $r(\xi)$ denote the convergence radius of the power series $\sum\frac{\alpha^n}{n!}\|H^n\xi\|$ and set

$$\xi(\alpha)=\sum_{n=0}^{\infty}\frac{\alpha^n}{n!}H^n\xi, \qquad |\alpha|<r(\xi). \tag{1}$$

It follows that $\xi(\,\cdot\,)$ is an $\mathfrak{H}$-valued holomorphic function on the disk, $|\alpha|<r(\xi)$. Furthermore, H leaves $\mathfrak{D}_0$ globally invariant and

$$\xi'(\alpha)=(H\xi)(\alpha)=H\big[\xi(\alpha)\big], \qquad |\alpha|<r(\xi). \tag{2}$$

For each $\xi\in\mathfrak{D}_0$ and $\alpha\in\mathbf{C}$ with $|\alpha|<r(\xi)$, we set

$$V(\alpha)\xi=\xi(\mathrm{i}\alpha)=\sum_{n=0}^{\infty}\frac{(\mathrm{i}\alpha)^n}{n!}H^n\xi, \qquad |\alpha|<r(\xi). \tag{3}$$

Since the power series (1) is the Taylor series expansion of $\xi(\,\cdot\,)$ around the origin, we have $V(\alpha)\xi\in\mathfrak{D}_0$ if $|\alpha|<r(\xi)$ and

$$V(\alpha)V(\beta)\xi=V(\alpha+\beta)\xi, \qquad |\alpha|+|\beta|<r(\xi). \tag{4}$$

Denoting the real variable by t, we have for $\xi\in\mathfrak{D}_0$

$$\begin{aligned}\frac{\mathrm{d}}{\mathrm{d}t}\|V(t)\xi\|^2&=\left(\frac{\mathrm{d}}{\mathrm{d}t}V(t)\xi\,\middle|\,V(t)\xi\right)+\left(V(t)\xi\,\middle|\,\frac{\mathrm{d}}{\mathrm{d}t}V(t)\xi\right)\\&=\left(\mathrm{i}HV(t)\xi\,\middle|\,V(t)\right)+\left(V(t)\xi\,\middle|\,\mathrm{i}HV(t)\xi\right)\\&=0.\end{aligned}$$

provided $|t|<r(\xi)$. Hence we have

$$\left\|V(t)\xi\right\|=\|\xi\|, \qquad \xi\in\mathfrak{D}_0, \qquad |t|<r(\xi), \qquad t\in\mathbf{R}. \tag{5}$$

Since the differentiation and the translation commute on holomorphic functions, $HV(t)\xi = V(t)H\xi$, $|t| < r(\xi)$. Hence we have $r(V(t)\xi) = r(\xi)$ by (5), so that we can iterate $V(t)$, i.e. we can definite $V(nt)\xi$ by

$$V(nt)\xi = \underbrace{V(t)\cdots V(t)}_{n \text{ times}}\xi, \quad |t| < r(\xi). \tag{6}$$

By equality (4), the expression (6) defines $V(nt)$ uniquely, so that we get a one parameter group $\{V(t) : t \in \mathbf{R}\}$ on $\mathfrak{D}_0$ which acts as a group of isometries by (5). Furthermore, $t \in \mathbf{R} \to V(t)\xi$, $\xi \in \mathfrak{D}_0$, is strongly continuous. Therefore, we can extend $\{V(t)\}$ to the whole space $\mathfrak{H}$ as a continuous one parameter unitary group, which will be denoted by $V(t)$ again. Therefore, Stone's Theorem, A.3, applies to $V(t)$ so that we have the spectral decomposition:

$$V(t) = \int_{\mathbf{R}} e^{i\lambda t}\, dE(\lambda) \tag{7}$$

and if we set

$$K = \int_{\mathbf{R}} \lambda\, dE(\lambda), \tag{8}$$

then $\mathfrak{D}(K)$ is precisely the set of all $V(t)$-differentiable vectors and

$$K\xi = \frac{1}{i}\frac{d}{dt}V(t)\xi\Big|_{t=0}, \quad \xi \in \mathfrak{D}(K). \tag{9}$$

Since $\{V(t)\}$ leaves $\mathfrak{D}_0$ invariant, $\mathfrak{D}_0$ is a core of K. But the restriction of K to $\mathfrak{D}_0$ is precisely that of H. Therefore H is essentially self-adjoint in $\mathfrak{D}_0$. Q.E.D.

A.6 Resolvent Convergence. A sequence $\{H_n\}$ of self-adjoint operators on $\mathfrak{H}$ is said to converge to a self-adjoint operator H in the *strong resolvent sense* if $(H_n - \lambda)^{-1}$ to $(H-\lambda)^{-1}$ strongly for every $\lambda \in \mathbf{C}$ with $\operatorname{Im}\lambda \neq 0$. By Theorem II.4.7, this is equivalent to the strong convergence of $\{(H_n - \lambda_0)^{-1}\}$ to $(H - \lambda_0)^{-1}$ for some λ_0 with $\operatorname{Im}\lambda_0 \neq 0$ because $(x-\lambda)^{-1} = (x-\lambda_0)^{-1}\big(1+(\lambda_0-\lambda)(x-\lambda_0)^{-1}\big)^{-1}$ and the function: $\mu \in \mathbf{R} \mapsto (\mu-\lambda_0)^{-1}\big(1+(\lambda_0-\lambda)(\mu-\lambda_0)^{-1}\big)^{-1} \in \mathbf{C}$ maps $\mathbf{R}$ onto a circle in the complex plane tangent to the real line at the origin and the points $\pm\infty$ to the origin if $\operatorname{Im}\lambda \neq 0$. By virtue of Theorem II.4.7, if $\{H_n\}$ converges to H in the strong resolvent sense, then $f(H_n)$ converges to $f(H)$ for every continuous function f with compact support on $\mathbf{R}$ because $f(H_n) = g\big((H_n-\lambda)^{-1}\big)$ and $f(H) = g\big((H-\lambda)^{-1}\big)$ with $g(\mu) = f(\lambda + \mu^{-1})$ for any $\lambda \in \mathbf{C}$, $\operatorname{Im}\lambda \neq 0$, because the function g on the circle is continuous as f has a compact support on $\mathbf{R}$.

Theorem. *Let $\{H_n\}$ be a sequence of self-adjoint operators on $\mathfrak{H}$.*

(i) *Let H be another self-adjoint operator on $\mathfrak{H}$. The sequence $\{H_n\}$ converges to H in the strong resolvent sense if and only if $\{\exp(itH_n)\}$ converges to*

$\exp(\mathrm{i}tH)$ *strongly for each* $t \in \mathbf{R}$. *If this is the case, then the strong convergence of* $\exp(\mathrm{i}tH_n)$ *to* $\exp(\mathrm{i}tH)$ *is uniform on every finite interval of* t.

(ii) *If there exist two points* λ_0 *and* μ_0 *with* $\operatorname{Im}\lambda_0 > 0$ *and* $\operatorname{Im}\mu_0 < 0$ *such that* $\{(H_n - \lambda_0)^{-1}\}$ *and* $\{(H_n - \mu_0)^{-1}\}$ *both converge strongly, then* $\{H_n\}$ *converges to a self-adjoint operator* H *in the strong resolvent sense.*

(iii) *Let* $\mathfrak{D}$ *be a core of a self-adjoint operator* H *which is contained in* $\bigcap \mathfrak{D}(H_n)$. *If* $\{H_n\xi\}$ *converges to* $H\xi$ *in norm for every* $\xi \in \mathfrak{D}$, *then* $\{H_n\}$ *converges to* H *in the strong resolvent sense.*

(iv) *If* $\{H_n\}$ *is a sequence of non-singular positive self-adjoint operators which converges to a non-singular positive self-adjoint operator* H *in the strong resolvent sense, then the sequence* $\{\log H_n\}$ *of logarithm converges to* $\log H$ *in the strong resolvent sense as well and therefore the sequence* $\{H_n^{\mathrm{i}t}\}$ *of one parameter unitary groups converges strongly to* $H^{\mathrm{i}t}$ *and uniformly for* t *in any bounded interval.*

PROOF:

(i) Assume that $\{H_n\}$ converges to H in the strong resolvent sense. We claim that for any bounded continuous function f on $\mathbf{R}$ the sequence $\{f(H_n)\}$ converges to $f(H)$ strongly, which implies our assertion. Fix $\xi \in \mathfrak{H}$. Then for any $\varepsilon > 0$ there exists a continuous function φ with compact support on $\mathbf{R}$ such that $0 \le \varphi(x) \le 1$, $x \in \mathbf{R}$, and $\|\xi - \varphi(H)\xi\| < \varepsilon$. Let $\lambda_0 \in \mathbf{C}$, $\operatorname{Im}\lambda_0 \neq 0$, be such that $\{(\lambda_0 - H_n)^{-1}\}$ converges to $(\lambda_0 - H)^{-1}$ strongly. Let C be the circle: $\{(\lambda_0 - x)^{-1} : x \in \mathbf{R}\} \cup \{0\}$. Set $g(\mu) = f\big(\frac{1}{\mu} + \lambda_0\big)\varphi\big(\frac{1}{\mu} + \lambda_0\big)$, $\mu \in C$, and view g as a continuous function on the circle C. Then Theorem II.4.7 applied to C, g and φ yields that $\{g\big((\lambda_0 - H_n)^{-1}\big)\} = \{f(H_n)\varphi(H_n)\}$ converges to $\{g\big((\lambda_0 - H)^{-1}\big)\} = \{f(H)\varphi(H)\}$ and $\{\varphi(H_n)\}$ converges to $\varphi(H)$ strongly. Therefore there exists an $n_0 \in \mathbf{N}$ such that for every $n \ge n_0$ we have

$$
\begin{aligned}
&\|f(H_n)\xi - f(H)\xi\| \\
&\le \Big\|f(H_n)\big(\xi - \varphi(H)\big)\xi\Big\| + \Big\|f(H_n)\big(\varphi(H) - \varphi(H_n)\big)\xi\Big\| \\
&\quad + \Big\|f(H_n)\varphi(H_n)\xi - f(H)\varphi(H)\xi\Big\| + \Big\|f(H)\varphi(H)\xi - f(H)\xi\Big\| \\
&\le \|f\|_\infty\Big(\|\xi - \varphi(H)\xi\| + \Big\|\big(\varphi(H) - \varphi(H_n)\big)\xi\Big\|\Big) \\
&\quad + \Big\|f(H_n)\varphi(H_n)\xi - f(H)\varphi(H)\xi\Big\| + \|f\|_\infty\|\varphi(H)\xi - \xi\| \\
&\le 4\varepsilon.
\end{aligned}
$$

This means that $\{f(H_n)\}$ converges to $f(H)$ strongly. In particular, $\{\exp(\mathrm{i}tH_n)\}$ converges to $\exp(\mathrm{i}tH)$ strongly.

To prove the converse assertion, we first show the formula:

$$(H-\lambda)^{-1} = \frac{1}{\mathrm{i}} \int_0^a \mathrm{e}^{-\mathrm{i}\lambda t} U(t)\,\mathrm{d}t\,, \qquad \operatorname{Im}\lambda < 0, \tag{1}$$

where $U(t) = \exp(\mathrm{i}tH)$. Let $H = \int_{\mathbf{R}} \lambda\,\mathrm{d}E(\lambda)$ be the spectral decomposition. Then we have

$$\begin{aligned}(H-\lambda)^{-1} &= \int_{\mathbf{R}} \frac{1}{\mu-\lambda}\,\mathrm{d}E(\mu) = \int_{\mathbf{R}} \frac{1}{\mathrm{i}}\left(\int_0^\infty \mathrm{e}^{\mathrm{i}(\mu-\lambda)t}\,\mathrm{d}t\right)\mathrm{d}E(\mu)\\ &= \frac{1}{\mathrm{i}}\int_0^\infty \left[\int_{\mathbf{R}} \mathrm{e}^{\mathrm{i}(\mu-\lambda)t}\,\mathrm{d}E(\mu)\right]\mathrm{d}t = \frac{1}{\mathrm{i}}\int_0^\infty \mathrm{e}^{-\mathrm{i}\lambda t} U(t)\,\mathrm{d}t\,,\end{aligned}$$

so that (1) follows. Therefore, we have, for each $\psi \in \mathfrak{H}$,

$$\left\|(H_n-\lambda)^{-1}\xi - (H-\lambda)^{-1}\xi\right\| \le \int_0^\infty \mathrm{e}^{(\operatorname{Im}\lambda)t} \left\|U_n(t)\xi - U(t)\xi\right\|\mathrm{d}t\,, \quad \operatorname{Im}\lambda < 0,$$

with $U_n(t) = \exp(\mathrm{i}tH_n)$. Hence the Lebesgue dominated convergence theorem yields the assertion if $\operatorname{Im}\lambda < 0$. Making use of the similar formula:

$$(H-\lambda)^{-1} = \frac{1}{\mathrm{i}} \int_{-\infty}^0 \mathrm{e}^{-\mathrm{i}\lambda t} U(t)\,\mathrm{d}t\,, \qquad \operatorname{Im}\lambda > 0, \tag{2}$$

we can prove the assertion for $\operatorname{Im}\lambda > 0$.

To prove the uniform convergence for t in a finite interval, say $[a, b]$, we set

$$U_n(t) = \exp(\mathrm{i}tH_n), \qquad U(t) = \exp(\mathrm{i}tH),$$

and

$$U_n(f) = \int_{\mathbf{R}} f(s)U_n(s)\,\mathrm{d}s\,, \qquad U(f) = \int_{\mathbf{R}} f(s)U(s)\,\mathrm{d}s\,, \qquad f \in L^1(\mathbf{R}).$$

We prove first that $U_n(t)U(f)\xi$ converges to $U(t)U(f)\xi$ in norm uniformly in $t \in [a, b]$ for any $f \in L^1(\mathbf{R})$ and $\xi \in \mathfrak{H}$. If $f \in L^1(\mathbf{R})$, then

$$\begin{aligned}U(t)U(f) &= \int_{\mathbf{R}} f(s)U(s+t)\,\mathrm{d}s = \int_{\mathbf{R}} f(s-t)U(s)\,\mathrm{d}s\\ &= \int_{\mathbf{R}} \big(\lambda(t)f\big)(s)U(s)\,\mathrm{d}s = U\big(\lambda(t)f\big).\end{aligned} \tag{3}$$

For a given $\varepsilon > 0$, choose $t_1, \ldots, t_m \in [a, b]$ so that

$$\inf_j \left\| \lambda(t)f - \lambda(t_j)f \right\|_1 < \varepsilon, \quad t \in [a, b]. \tag{4}$$

Take $N \in \mathbf{N}$ large enough so that

$$\left\| \left[U_n(t_j)U(f) - U(t_j)U(f) \right] \xi \right\| < \varepsilon, \quad 1 \le j \le m, \quad n \ge N. \tag{5}$$

We now estimate:

$$\begin{aligned}
&\left\| \left[U_n(t) - U(t) \right] U(f)\xi \right\| \\
&\quad \le \left\| \left[U_n(t)U(f) - U_n(t)U_n(f) \right] \xi \right\| \\
&\qquad + \left\| \left[U_n(t)U_n(f) - U(t)U(f) \right] \xi \right\| \\
&\quad = \left\| \left(U(f) - U_n(f) \right) \xi \right\| + \left\| \left[U_n\left(\lambda(t)f \right) - U\left(\lambda(t)f \right) \right] \xi \right\| \\
&\quad \le \int_{\mathbf{R}} |f(s)| \left\| \left[U(s) - U_n(s) \right] \xi \right\| \, \mathrm{d}s + \left\| \left[U_n\left(\lambda(t)f \right) - U_n\left(\lambda(t_j)f \right) \right] \xi \right\| \\
&\qquad + \left\| U_n\left(\lambda(t_j)f \right)\xi - U\left(\lambda(t_j)f \right)\xi \right\| + \left\| \left[U\left(\lambda(t_j)f \right) - U\left(\lambda(t)f \right) \right] \xi \right\|.
\end{aligned}$$

The first term of the above expression tends to zero as $n \to \infty$ by the Lebesgue convergence theorem. The second and the last terms are both majorized by $\varepsilon\|\xi\|$ for some j by (4). The third term is majorized by ε due to (5). Hence we conclude that $\left\| \left[U_n(t) - U(t) \right] U(f)\xi \right\|$ is uniformly small for large n if $t \in [a, b]$. Thus, $U_n(t)U(f)\xi$ converges to $U(t)U(f)\xi$ uniformly in $t \in [a, b]$. For an arbitrary $\xi \in \mathfrak{H}$, choose an $f \in L^1(\mathbf{R})$ so that $\left\| U(f)\xi - \xi \right\| < \varepsilon$ for a preassigned $\varepsilon > 0$. Then we have

$$\begin{aligned}
&\left\| \left[U_n(t) - U(t) \right] \xi \right\| \\
&\quad \le \left\| U_n(t)\left(1 - U(f) \right)\xi \right\| + \left\| \left[U_n(t) - U(t) \right] U(f)\xi \right\| + \left\| U(t)\left(1 - U(f) \right)\xi \right\| \\
&\quad = 2\left\| \left(1 - U(f) \right)\xi \right\| + \left\| \left[U_n(t) - U(t) \right] U(f)\xi \right\|,
\end{aligned}$$

so that the uniform convergence of $U_n(t)\xi$ to $U(t)\xi$ for $t \in [a, b]$ follows.

(iii) Let $\eta = (H + \mathrm{i})\xi$ with $\xi \in \mathfrak{D}$. Then

$$\begin{aligned}
\left\| \left[(H_n + \mathrm{i})^{-1} - (H + \mathrm{i})^{-1} \right] \eta \right\| &= \left\| (H_n + \mathrm{i})^{-1} \left(H - H_n \right) (H + \mathrm{i})^{-1} \eta \right\| \\
&\le \left\| (H - H_n)\xi \right\| \to 0 \quad \text{as } n \to \infty.
\end{aligned}$$

Since $\mathfrak{D}$ is a core of H, $(H + \mathrm{i})\mathfrak{D}$ is dense in $\mathfrak{H}$. Hence the boundedness of $(H_n + \mathrm{i})^{-1}$, in fact $\left\| (H_n + \mathrm{i})^{-1} \right\| \le 1$, entails the strong resolvent convergence of $\left\{ H_n \right\}$ to H.

(iv) Set $k_n = (1 + H_n)^{-1}$ and $k = (1 + H)^{-1}$. Then the sequence $\{k_n\}$ of bounded positive operators converges strongly to k. Define continuous functions $f_\pm$ on the closed unit interval $[0, 1]$ by:

$$f_\pm(\lambda) = \begin{cases} \dfrac{1}{\log(\frac{1}{\lambda} - 1) \pm \mathrm{i}} & \text{for } 0 < \lambda < 1; \\ 0 & \text{for } \lambda = 0, 1. \end{cases}$$

Then we have $(\log H_n \pm \mathrm{i})^{-1} = f_\pm(k_n)$ and $(\log H \pm \mathrm{i})^{-1} = f_\pm(k)$. So the strong resolvent convergence of $\{\log H_n\}$ to $\log H$ follows from Theorem II.4.7 or rather Lemma II.4.3. Q.E.D.

A.7 Sesquilinear Form and Quadratic Form. Let $\mathfrak{D}$ be a complex vector space. If s is a positive sesquilinear form on $\mathfrak{D}$, then the function $q_s : \xi \in \mathfrak{D} \to s(\xi, \xi)$ satisfies the following:

$$q(\lambda\xi) = |\lambda|^2 q(\xi), \quad \lambda \in C, \quad \xi \in \mathfrak{D}, \tag{1}$$

$$q(\xi + \eta) + q(\xi - \eta) = 2q(\xi) + 2q(\eta), \quad \xi, \eta \in \mathfrak{D}. \tag{2}$$

Conversely, a function q on $\mathfrak{D}$ satisfying (1) and (2) is called a *quadratic form.* If we consider only positive valued quadratic forms, then we can extend the formulas (1) and (2) to $[0, \infty]$-valued quadratic form with convention $0\,(+\infty) = 0$. In this case, $\mathfrak{D}(q) = \{\xi \in \mathfrak{D} : q(\xi) < +\infty\}$ is a subspace of $\mathfrak{D}$, called the *domain* of q. In $\mathfrak{D}(q)$, we consider the $\mathbf{C}$-valued function:

$$s_q(\xi, \eta) = \frac{1}{4}\sum_{k=0}^{3} \mathrm{i}^k q(\xi + \mathrm{i}^k\eta), \quad \xi, \eta \in \mathfrak{D}(q). \tag{3}$$

We claim that s_q is sesquilinear on $\mathfrak{D}(q)$ and $q(\xi) = s_q(\xi, \xi)$. To prove this, let S_R and S_I be the real and the imaginary parts of s_q, i.e.

$$\left.\begin{aligned} S_\mathrm{R}(\xi, \eta) &= \frac{1}{4}\big(q(\xi + \eta) - q(\xi - \eta)\big) \\ S_\mathrm{I}(\xi, \eta) &= \frac{1}{4}\big(q(\xi + \mathrm{i}\eta) - q(\xi - \mathrm{i}\eta)\big). \end{aligned}\right\} \tag{3'}$$

It is clear that $S_\mathrm{R}(\xi, \eta) = S_\mathrm{R}(\eta, \xi) = S_\mathrm{I}(\xi, -\mathrm{i}\eta)$. Hence it suffices to prove that S_R is real linear. We observe first

$$\begin{aligned} S_\mathrm{R}(\xi, 2\eta) &= \frac{1}{4}\big[q(\xi + 2\eta) - q(\xi - 2\eta)\big] \\ &= \frac{1}{4}\Big[2q(\xi + \eta) + 2q(\eta) - q(\xi) - 2q(\xi - \eta) - 2q(\eta) + q(\xi)\Big] \\ &= \frac{1}{2}\big[q(\xi + \eta) - q(\xi - \eta)\big] = 2S_\mathrm{R}(\xi, \eta). \end{aligned}$$

For each $\xi, \eta, \zeta \in \mathfrak{D}(q)$, we have

$$\begin{aligned} S_{\mathrm{R}}(\xi+\eta, \zeta) &= \frac{1}{4}[q(\xi+\eta+\zeta) - q(\xi+\eta-\zeta)] \\ &= \frac{1}{4}\left[q\left(\xi+\frac{1}{2}\zeta+\eta+\frac{1}{2}\zeta\right) - q\left(\xi-\frac{1}{2}\zeta+\eta-\frac{1}{2}\zeta\right)\right] \\ &= \frac{1}{4}\left[2q\left(\xi+\frac{1}{2}\zeta\right) + 2q\left(\eta+\frac{1}{2}\zeta\right) - q(\xi-\eta)\right. \\ &\quad \left. - 2q\left(\xi-\frac{1}{2}\zeta\right) - 2q\left(\eta-\frac{1}{2}\zeta\right) + q(\xi-\eta)\right] \\ &= 2S_{\mathrm{R}}\left(\xi, \frac{1}{2}\zeta\right) + 2S_{\mathrm{R}}\left(\eta, \frac{1}{2}\zeta\right) = S_{\mathrm{R}}(\xi,\zeta) + S_{\mathrm{R}}(\eta,\zeta). \end{aligned}$$

Since $S_{\mathrm{R}}(-\xi, \eta) = -S_{\mathrm{R}}(\xi, \eta)$, the above calculation shows that S_{R} is biadditive. Now, we observe that

$$\begin{aligned} S_{\mathrm{R}}(\lambda\xi, \xi) &= \frac{1}{4}\Big[q\big((\lambda+1)\xi\big) - q\big((\lambda-1)\xi\big)\Big] = \frac{1}{4}\Big[(\lambda+1)^2 - (\lambda-1)^2\Big]q(\xi) \\ &= \lambda q(\xi) = \lambda S_{\mathrm{R}}(\xi, \xi). \end{aligned}$$

Next, we get, for each $\xi, \eta \in \mathfrak{D}(q)$, and $\lambda \in \mathbf{R}$,

$$\begin{aligned} S_{\mathrm{R}}\big(\lambda(\xi+\eta), \xi+\eta\big) &= \lambda q(\xi) + S_{\mathrm{R}}(\lambda\xi, \eta) + S_{\mathrm{R}}(\lambda\eta, \xi) + \lambda q(\eta) \\ &= \lambda S_{\mathrm{R}}\big(\xi+\eta, \xi+\eta\big) = \lambda\Big[q(\xi) + 2S_{\mathrm{R}}(\xi,\eta) + q(\eta)\Big], \end{aligned}$$

so that

$$\lambda S_{\mathrm{R}}(\xi, \eta) = \frac{1}{2}\big[S_{\mathrm{R}}(\lambda\xi, \eta) + S_{\mathrm{R}}(\xi, \lambda\eta)\big]. \tag{4}$$

We now use the fact that $S_{\mathrm{R}}(\lambda\xi, \mathrm{i}\xi) = 0$ for $\lambda \in \mathbf{R}$. So

$$\begin{aligned} 0 = S_{\mathrm{R}}\big(\lambda(\xi+\eta), \mathrm{i}\,(\xi+\eta)\big) &= S_{\mathrm{R}}\big(\lambda\xi, \mathrm{i}\eta\big) + S_{\mathrm{R}}\big(\lambda\eta, \mathrm{i}\xi\big) \\ &= S_{\mathrm{R}}\big(\lambda\xi, \mathrm{i}\eta\big) + S_{\mathrm{R}}\big(\mathrm{i}\lambda\eta, -\xi\big), \end{aligned}$$

because $S_{\mathrm{R}}(\mathrm{i}\xi, \mathrm{i}\eta) = S_{\mathrm{R}}(\xi, \eta)$. Hence we get

$$S_{\mathrm{R}}\big(\xi, \mathrm{i}\lambda\eta\big) = S_{\mathrm{R}}\big(\lambda\xi, \mathrm{i}\eta\big).$$

Replacing $\mathrm{i}\eta$ by η, we get $S_{\mathrm{R}}(\xi, \lambda\eta) = S_{\mathrm{R}}(\lambda\xi, \eta)$, which together with (4) implies that

$$S_{\mathrm{R}}(\lambda\xi, \eta) = \lambda S_{\mathrm{R}}(\xi, \eta). \tag{4$'$}$$

Thus, we proved that S_{R} is real linear, so that s_q is linear on $\mathbf{C}$.

A.8 Closable Quadratic Forms. Let $\mathfrak{D}$ be a not necessarily closed vector subspace of $\mathfrak{H}$. We consider a positive quadratic form q on $\mathfrak{D}$ and the associated sesquilinear form s_q on $\mathfrak{D}(q)$. The form q is called *closed* if $\mathfrak{D}(q)$ is complete under the norm:

$$\|\xi\|_q = \sqrt{\|\xi\|^2 + q(\xi)}, \quad \xi \in \mathfrak{D}(q). \tag{1}$$

A subspace $\mathfrak{M}$ of $\mathfrak{D}(q)$ is called a *core* of q if $\mathfrak{M}$ is dense in $\{\mathfrak{D}(q), \|\cdot\|_q\}$.

If H is a self-adjoint positive operator on $\mathfrak{H}$, then the form q_H on $\mathfrak{D}(H^{1/2})$ defined by:

$$q_H(\xi) = \|H^{\frac{1}{2}}\xi\|^2, \quad \xi \in \mathfrak{D}(H^{\frac{1}{2}}) \tag{2}$$

is a closed positive quadratic form.

A positive quadratic form q with domain $\mathfrak{D}(q)$ is called *closable* if it has a closed extension.

Theorem.

(i) *Every densely defined closed positive quadratic form is associated with a unique self-adjoint positive operator on* $\mathfrak{H}$.

(ii) *The closability of a positive quadratic form* q *is equivalent to the lower semi-continuity.*

PROOF:

(i) Let q be a closed positive form with domain $\mathfrak{D}(q)$, and let s_q be the associated sesquilinear form on $\mathfrak{D}(q)$. It follows that $\mathfrak{D}(q)$ is a Hilbert space with the inner product:

$$(\xi \mid \eta)_q = (\xi \mid \eta) + s_q(\xi, \eta), \quad \xi, \eta \in \mathfrak{D}(q). \tag{3}$$

We then observe that the embedding i of $\mathfrak{D}(q)$ into $\mathfrak{H}$ is norm decreasing and has a dense range. So the adjoint i^* maps $\mathfrak{H}$ injectively into $\mathfrak{D}(q)$ and its range is dense in $\mathfrak{D}(q)$ with respect to the q-norm. Let $i = ah$ and $i^* = bk$ be the polar decomposition of i and i^* with $b = a^*$ and $k = aha^*$. Since the range of i is dense, a is an isometry of $\mathfrak{D}(q)$ onto $\mathfrak{H}$ and b maps $\mathfrak{H}$ onto $\mathfrak{D}(q)$. We then have, for each $\xi, \eta \in \mathfrak{H}$,

$$(\xi \mid \eta) = (b\xi \mid b\eta)_q = (b\xi \mid b\eta) + s_q(b\xi, b\eta).$$

We should note here that a and b are isometries with respect to the norms in $\mathfrak{H}$ and $\mathfrak{D}(q)$. Now, we have $\|b\xi\|^2 \leq \|b\xi\|_q^2 = \|\xi\|^2$, $\xi \in \mathfrak{H}$, so that b is a contraction. Hence $a = b^{-1}$ is a closed operator with domain $\mathfrak{D}(q)$ such that

$$(a\xi \mid a\eta) = (\xi \mid \eta) + s_q(\xi, \eta), \quad \xi, \eta \in \mathfrak{D}(q).$$

Let $H = a^*a - 1 \geq 0$. Then we have $(H\xi \mid \eta) = s_q(\xi, \eta)$ for every $\xi, \eta \in \mathfrak{D}(a^*a) = \mathfrak{D}(H)$. Since $\mathfrak{D}(a^*a)$ is a core of $\mathfrak{D}(a)$, we conclude that $\mathfrak{D}(H^{1/2}) = \mathfrak{D}(q)$ and q is associated with H.

The uniqueness of H follows from that of the polar decomposition for closed operators.

(ii) Suppose that $q = q_H$ for some self-adjoint operator $H \geq 0$. Then we have for each $\xi \in \mathfrak{D}(H^{1/2})$

$$q(\xi) = \|H^{\frac{1}{2}}\xi\|^2 = \sup\Big\{ |(H^{\frac{1}{2}}\xi \mid \eta)|^2 : \eta \in \mathfrak{D}(H^{\frac{1}{2}}),\ \|\eta\| \leq 1 \Big\}$$
$$= \sup\Big\{ |(\xi \mid H^{\frac{1}{2}}\eta)|^2 : \eta \in \mathfrak{D}(H^{\frac{1}{2}}),\ \|\eta\| \leq 1 \Big\},$$

so that q is lower semi-continuous.

Conversely, suppose that q is lower semi-continuous. In $\mathfrak{D}(q)$, we consider the inner product given by (3), and the completion $\mathfrak{D}$ of $\mathfrak{D}(q)$. Since the embedding i is a contraction, it can be extended to a contraction of $\mathfrak{D}$ into $\mathfrak{H}$, which we denote by i again. We prove that i is injective. To this end, let $\{\xi_n\}$ be a sequence in $\mathfrak{D}(q)$ such that $\lim_{n\to\infty}\|\xi_n\| = 0$ and $\lim_{n,m\to+\infty} q(\xi_n - \xi_m) = 0$. Let $\xi = \lim \xi_n \in \mathfrak{D}$. By assumption, for any $\varepsilon > 0$ there exists an $n_0 \in \mathbf{N}$ such that $q(\xi_n - \xi_m) < \varepsilon$ for every $n, m \geq n_0$. By the lower semi-continuity, we have

$$q(\xi_n) \leq \liminf_{m\to\infty} q\big(\xi_n - \xi_m\big) \leq \varepsilon, \quad n \geq n_0.$$

Hence we get $\lim_{n\to\infty} q(\xi_n) = 0$, which means $\xi = 0$. Therefore, $\mathfrak{D}$ can be viewed as a subspace of $\mathfrak{H}$ and q is extended to a closed form $\bar{q}$ defined on $\mathfrak{D}$. Q.E.D.

A.9 Sum of Positive Self-Adjoint Operators. Let $H_1, H_2, \dots, H_n$ be positive self-adjoint operators on $\mathfrak{H}$. Let $q_1, \dots, q_n$ be their associated positive quadratic forms. If $\mathfrak{D} = \mathfrak{D}(q_1) \cap \cdots \cap \mathfrak{D}(q_n)$ is dense, then the sum $q = q_1 + \cdots + q_n$ is a closed positive quadratic form with $\mathfrak{D}(q) = \mathfrak{D}$. Hence there exists uniquely a positive self-adjoint operator H such that

$$\|H^{\frac{1}{2}}\xi\|^2 = \|H_1^{\frac{1}{2}}\xi\|^2 + \cdots + \|H_n^{\frac{1}{2}}\xi\|^2, \quad \xi \in \mathfrak{D}.$$

The operator H is called the *form sum* of $H_1, \dots, H_n$.

PROOF: Suppose that $\{\xi_n\}$ is a Cauchy sequence in $\mathfrak{D}$ with respect to q-norm. Since $q_i \leq q$, $\{\xi_n\}$ is also a Cauchy sequence with respect to each q_i-norm, $1 \leq i \leq n$. Since each q_i is closed, $\{\xi_n\}$ converges to some $\xi \in \mathfrak{H}$ in the q_i-norm. The limit ξ does not depend on q_i's, because ξ is also the limit of $\{\xi_n\}$ in the norm of $\mathfrak{H}$. By the closedness of q_i, ξ belongs to $\mathfrak{D}(q_i)$; hence to $\mathfrak{D} = \bigcap \mathfrak{D}(q_i)$, and $\{\xi_n\}$ converges to ξ in q-norm. Therefore $\mathfrak{D}$ is complete with respect to the q-norm. Q.E.D.

When we have two positive self-adjoint operators H_1 and H_2, the intersection $\mathfrak{D}(H_1) \cap \mathfrak{D}(H_2)$ need not be dense even if $\mathfrak{D}\big(H_1^{1/2}\big) \cap \mathfrak{D}\big(H_2^{1/2}\big)$ is dense. It is a simple exercise to see that if $H_1 + H_2$ is essentially self-adjoint on $\mathfrak{D}(H_1)\cap\mathfrak{D}(H_2)$, then the form sum of H_1 and H_2 agrees with the strong sum $H_1 \dot{+} H_2$.

A.10 Order of Positive Self-Adjoint Operators. For two positive self-adjoint operators H_1 and H_2 on $\mathfrak{H}$, the order $H_1 \le H_2$ is defined by the fact that

$$\mathfrak{D}\big(H_1^{\frac{1}{2}}\big) \supset \mathfrak{D}\big(H_2^{\frac{1}{2}}\big) \qquad \text{and} \qquad \big\|H_1^{\frac{1}{2}}\xi\big\| \le \big\|H_2^{\frac{1}{2}}\xi\big\|, \quad \xi \in \mathfrak{D}\big(H_2^{\frac{1}{2}}\big). \tag{1}$$

This relation is equivalent to the bounded operator inequality:

$$(1 + H_1)^{-1} \ge (1 + H_2)^{-1}. \tag{2}$$

PROOF: Suppose (1) holds. For any $\xi \in \mathfrak{H}$, let $\eta_1 = (1 + H_1)^{-1}\xi$ and $\eta_2 = (1 + H_2)^{-1}\xi$. We then have

$$\begin{aligned}
\big((1 + H_2)^{-1}\xi \mid \xi\big)^2 &= \big(\eta_2 \mid (1 + H_1)\eta_1\big)^2 \\
&= \big((1 + H_1)^{\frac{1}{2}}\eta_2 \mid (1 + H_1)^{\frac{1}{2}}\eta_1\big)^2 \\
&\le \big\|(1 + H_1)^{\frac{1}{2}}\eta_2\big\|^2 \big\|(1 + H_1)^{\frac{1}{2}}\eta_1\big\|^2 \\
&= \Big(\|\eta_2\|^2 + \big\|H_1^{\frac{1}{2}}\eta_2\big\|^2\Big)\big((1 + H_1)\eta_1 \mid \eta_1\big) \\
&\le \Big(\|\eta_2\|^2 + \big\|H_2^{\frac{1}{2}}\eta_2\big\|^2\Big)\big(\xi \mid (1 + H_1)^{-1}\xi\big) \\
&= \big((1 + H_2)\eta_2 \mid \eta_2\big)\big((1 + H_1)^{-1}\xi \mid \xi\big) \\
&= \big(\xi \mid (1 + H_2)^{-1}\xi\big)\big((1 + H_1)^{-1}\xi \mid \xi\big),
\end{aligned}$$

so that $\big((1 + H_2)^{-1}\xi \mid \xi\big) \le \big((1 + H_1)^{-1}\xi \mid \xi\big)$, $\xi \in \mathfrak{H}$.

Conversely, suppose (2) holds. Set $K_1 = (1 + H_1)^{-1}$ and $K_2 = (1 + H_2)^{-1}$. For $\lambda > 0$, put $\mu = 1/(1 + \lambda)$. Then we have

$$\frac{1}{1 + \varepsilon\lambda} = \frac{1}{1 - \varepsilon}\left(1 - \frac{\varepsilon}{\varepsilon + (1 - \varepsilon)\mu}\right), \qquad 0 < \varepsilon < 1.$$

By the spectral decomposition, we have

$$\begin{aligned}
\big(1 + \varepsilon H_1\big)^{-1} &= \frac{1}{1 - \varepsilon}\left(1 - \varepsilon\Big(\varepsilon + (1 - \varepsilon)K_1\Big)^{-1}\right) \\
&\ge \frac{1}{1 - \varepsilon}\left(1 - \varepsilon\Big(\varepsilon + (1 - \varepsilon)K_2\Big)^{-1}\right) = \big(1 + \varepsilon H_2\big)^{-1}.
\end{aligned} \tag{3}$$

Thus, we get

$$\begin{aligned}
H_{1,\varepsilon} = H_1\big(1 + \varepsilon H_1\big)^{-1} &= \frac{1}{\varepsilon}\left(1 - \Big(1 + \varepsilon H_1\Big)^{-1}\right) \\
&\le \frac{1}{\varepsilon}\left(1 - \Big(1 + \varepsilon H_2\Big)^{-1}\right) = H_2\big(1 + \varepsilon H_2\big)^{-1} = H_{2,\varepsilon}.
\end{aligned} \tag{4}$$

For every $\xi \in \mathfrak{D}(H_1^{1/2})$, we have

$$\| H_2^{\frac{1}{2}} \xi \|^2 = \lim_{\varepsilon \to 0} \| H_{2,\varepsilon}^{\frac{1}{2}} \xi \|^2 \geq \lim_{\varepsilon \to 0} \| H_{1,\varepsilon}^{\frac{1}{2}} \xi \|^2 = \| H_1^{\frac{1}{2}} \xi \|^2,$$

so that ξ belongs to $\mathfrak{D}(H_1^{1/2})$ and (1) holds. Q.E.D.

A.11 Increasing Limit of Positive Self-Adjoint Operators. Let $\{H_n\}$ be an increasing sequence of positive self-adjoint operators on $\mathfrak{H}$. If $\bigcap_{n=1}^{\infty} \mathfrak{D}(H_n^{1/2}) = \mathfrak{D}$ is dense in $\mathfrak{H}$, then there exists a positive self-adjoint operator H such that

$$\mathfrak{D}(H^{\frac{1}{2}}) = \mathfrak{D} \quad \text{and} \quad \| H^{\frac{1}{2}} \xi \|^2 = \lim_{n \to \infty} \| H_n^{\frac{1}{2}} \xi \|^2, \quad \xi \in \mathfrak{D}. \tag{1}$$

Furthermore, $\{H_n\}$ converges to H in the strong resolvent case. If one of $\{H_n\}$ is non-singular, then $\log H_n$ converges to $\log H$ in the strong resolvent sense. Therefore, $\{H_n^{it}\}$ converges strongly to H^{it} uniformly in t of any finite interval.

PROOF: Let $q_k(\xi) = \| H_k^{1/2} \xi \|^2$, $\xi \in \mathfrak{D}(H_k^{1/2})$ and $q(\xi) = \lim_{k \to \infty} q_k(\xi)$, $\xi \in \mathfrak{D}$. Suppose that $\{\xi_n\}$ is a Cauchy sequence of $\mathfrak{D}$ in q-norm, and let $\xi = \lim \xi_n$ in $\mathfrak{H}$. For each k, we have $q_k(\xi_n - \xi_m) \leq q(\xi_n - \xi_m)$, so that $\{\xi_n\}$ converges to ξ in the q_k-norm. Hence $\xi \in \mathfrak{D}(H_k^{1/2})$ for each k, so that $\xi \in \mathfrak{D}$. For any $\varepsilon > 0$, choose n_0 so that $q(\xi_n - \xi_m) \leq \varepsilon$ for every $n, m \geq n_0$. Now, the lower semi-continuity of q inherited from $\{q_k\}$ implies

$$q(\xi - \xi_m) \leq \lim_{n \to \infty} \inf q(\xi_n - \xi_m) \leq \varepsilon.$$

Hence q is closed. Therefore, there exists a positive self-adjoint operator H such that $\mathfrak{D} = \mathfrak{D}(H^{1/2})$ and $q(\xi) = \| H^{1/2} \xi \|^2$, $\xi \in \mathfrak{D}$.

Since $\{H_n\}$ is increasing and bounded by H from above, $\{(1 + H_n)^{-1}\}$ is decreasing and bounded by $(1 + H)^{-1}$ from below, so that it converges strongly to a bounded positive operator K. By the inequality $K \geq (1 + H)^{-1}$, K is non-singular and $H_n \leq K^{-1} - 1 \leq H$, so that $K = (1 + H)^{-1}$. Hence $\{(1 + H_n)^{-1}\}$ converges to $(1 + H)^{-1}$ strongly. By A.6, $\{H_n\}$ converges to H in the strong resolvent sense.

If any one of $\{H_n\}$ is non-singular, then H is non-singular too. The rest of the assertion follows from Theorem II.4.7. Q.E.D.

A.12 Complex Interpolation Method. Let X_0 and X_1 be a pair of Banach spaces which are continuously embedded in a Hausdorff topological vector space $\tilde{X}$. We say that such a pair is *compatible*. In this case, the intersection $\Delta(X_0, X_1) = X_0 \cap X_1$ and the algebraic sum $\Sigma(X_0, X_1) = X_0 + X_1$ can be made into Banach spaces respectively by the following norms:

$$\|x\|_{\Delta} = \max\{\|x\|_0, \|x\|_1\}, \quad x \in \Delta, \tag{1}$$

$$\|x\|_{\Sigma} = \inf\left\{ \|x_0\|_0 + \|x_1\|_1 : x = x_0 + x_1,\ x_i \in X_i,\ i = 0, 1 \right\}, \tag{2}$$

where $\| \cdot \|_0$ and $\| \cdot \|_1$ mean the norms of X_0 and X_1 respectively.

Given a compatible pair $X = (X_0, X_1)$, let $\mathcal{F}(X_0, X_1)$ be the space of all bounded continuous $\Sigma(X_0 - X_1)$-valued functions f on the closed vertical strip $\mathbf{S} = \{ z \in \mathbf{C} : 0 \le \operatorname{Re} z \le 1 \}$ which is holomorphic in the interior $\mathbf{S}_0$ such that

(i) $f(it) \in X_0,\ f(1+it) \in X_1,\ t \in \mathbf{R}$;

(ii) For $j = 0, 1$, the map: $t \in \mathbf{R} \to f(j+it) \in X_j$ is continuous and $\lim_{|t|\to\infty} \| f(j+it) \|_j = 0,\ j = 0, 1$.

Since $\| \cdot \|_\Sigma$ is majorized by $\| \cdot \|_j$ on X_j for $j = 0, 1$, the Phragmen-Lindelöf theorem implies that $\mathcal{F}(X)$ is a Banach space under the norm:

$$\|f\| = \max\Big\{ \sup_t \| f(it) \|_0,\ \sup_t \| f(1+it) \|_2 \Big\}. \tag{3}$$

For each $0 < \theta < 1$, the *complex interpolation space* between X_0 and X_1, C_θ $(= C_\theta(X) = C_\theta(X_0, X_1))$, is the space of all $f(\theta),\ f \in \mathcal{F}(X)$, equipped with the *complex interpolation norm*:

$$\|x\|_\theta = \inf\{ \|f\| : f \in \mathcal{F}(X),\ x = f(\theta) \}. \tag{4}$$

It then follows that

$$\| f(\theta) \|_\theta \le \|f\|, \quad f \in \mathcal{F}(X). \tag{5}$$

Furthermore, if $K_\theta = \{ f \in \mathcal{F}(X) : f(\theta) = 0 \}$, then C_θ is naturally identified with the quotient Banach space $\mathcal{F}(X)/K_\theta$, which then implies the completeness of C_θ.

Theorem. *Let $X = (X_0, X_1)$ and $Y = (Y_0, Y_1)$ be two compatible pairs of Banach spaces, and T be a linear operator from $\Sigma(X)$ into $\Sigma(Y)$. If T maps X_0 into Y_0 and X_1 into Y_1 and if there exist constants M_0 and $M_1 > 0$ such that*

$$\begin{aligned} \| T x_0 \|_0 &\le M_0 \|x_0\|_0, \quad x_0 \in X_0, \\ \| T x_1 \|_1 &\le M_1 \|x_1\|_1, \quad x_1 \in X_1, \end{aligned} \tag{6}$$

then T maps $C_\theta(X)$ into $C_\theta(Y)$ for each $0 < \theta < 1$ and

$$\|Tx\|_\theta \le M_0^{1-\theta} M_1^\theta \|x\|_\theta, \quad x \in C_\theta(X). \tag{7}$$

PROOF: It is easy to check that $\|Tx\|_\Sigma \le M\|x\|_\Sigma$ for every $x \in \Sigma(X)$ with $M = \max\{M_0, M_1\}$. Hence $(Tf)(z) = Tf(z),\ f \in \mathcal{F}(X)$, gives a bounded linear operator T from $\mathcal{F}(X)$ into $\mathcal{F}(Y)$. For each $f \in \mathcal{F}(X)$, we put

$$(Sf)(z) = M_0^{z-1} M_1^{-z} Tf(z), \quad z \in \mathbf{S}.$$

Fix an $x \in C_\theta$ and $\varepsilon > 0$. Choose an $f \in \mathcal{F}(X)$ such that $x = f(\theta)$ and $\|f\| < \|x\|_\theta + \varepsilon$, and set $g = Sf$. Then g belongs to $\mathcal{F}(Y)$ and $g(\theta) = M_0^{\theta-1} M_1^{-\theta} Tx$. By assumption on T, we have $\|g\| \le \|f\| \le \|x\|_\theta + \varepsilon$. Hence

$$\|Tx\|_\theta \le M_0^{1-\theta} M_1^\theta \|g\| \le M_0^{1-\theta} M_1^\theta \big(\|x\|_\theta + \varepsilon \big).$$

Thus (7) follows. Q.E.D.

A.13 Numerical Range. Let A be a unital Banach algebra. Of course, we assume $\|1\| = 1$. A *unital* functional means an element $\omega \in A^*$ such that

$$\omega(1) = \|\omega\| = 1.$$

If A is a C^*-algebra, then a unital functional is precisely a state. The set $V(A)$ of all unital functionals on A is a weakly* compact convex subsets of A^*. The *numerical range* of an element $a \in A$ is defined to be the set:

$$W_A(a) = \left\{ \omega(a) : \omega \in V(A) \right\}. \tag{1}$$

For each $a \in A$, let $E(a)$ be the subspace of A spanned by 1 and a, which is at most of dimension two. In view of the Hahn-Banach extension theorem, we have

$$W_A(a) = \left\{ f(a) : f \in E(a)^*, \ \|f\| = f(1) = 1 \right\}, \tag{2}$$

which is also given by:

$$W_A(a) = \left\{ \alpha \in \mathbf{C} : |\lambda + \mu\alpha| \leq \|\lambda + \mu a\|, \ \lambda, \mu \in \mathbf{C} \right\}. \tag{2'}$$

But the inequality $\left|\lambda + \mu\alpha\right| \leq \left\|\lambda + \mu a\right\|$ for all $\lambda, \mu \in \mathbf{C}$ is equivalent to the following:

$$\left|\lambda + \alpha\right| \leq \left\|\lambda + a\right\|, \quad \lambda \in \mathbf{C},$$

i.e.

$$W_A(a) = \left\{ \alpha \in \mathbf{C} : |\lambda + \alpha| \leq \|\lambda + a\|, \ \lambda \in \mathbf{C} \right\}. \tag{2''}$$

Proposition. *In a unital Banach algebra A, we have, for each $a \in A$,*

$$\begin{aligned} \sup \operatorname{Re} W_A(a) &= \inf\left\{ \frac{1}{t}\big(\|1 + ta\| - 1\big) : t > 0 \right\} \\ &= \lim_{t \to 0+} \frac{1}{t}\big(\|1 + ta\| - 1\big); \end{aligned} \tag{3}$$

$$\begin{aligned} \sup \operatorname{Re} W_A(a) &= \sup\left\{ \frac{1}{t} \log\left\|\exp ta\right\| : t > 0 \right\} \\ &= \lim_{t \to 0+} \frac{1}{t} \log\left\|\exp ta\right\|. \end{aligned} \tag{4}$$

PROOF: We observe that

$$\operatorname{Re} W_A(a) = \left\{ \operatorname{Re} \varphi(a) : \varphi \in V(A) \right\}.$$

For each $\varphi \in A^*$, put $\varphi_R(x) = \operatorname{Re} \varphi(x)$. Then the map $\varphi \in A^* \to \varphi_R \in A^*_{\mathbf{R}}$ is an isometry, where $A_{\mathbf{R}}$ is the real Banach space obtained from A by restricting the scalar field to $\mathbf{R}$, and $A^*_{\mathbf{R}}$ is its dual as a real Banach space.

Let $E_{\mathbf{R}}(a)$ be the real subspace of $A_{\mathbf{R}}$ spanned by 1 and a, and $V(a)$ be the set of all real linear functionals φ on $E_{\mathbf{R}}(a)$ such that $1 = \|\varphi\| = \varphi(1)$. Then the restriction map: $\varphi \in V(A) \to \varphi_R\big|_{E_{\mathbf{R}}(a)} \in V(a)$ is a surjection. Hence we have

$$\operatorname{Re} W_A(a) = \big\{ \varphi(a) : \varphi \in V(a) \big\}.$$

Since $\varphi(\lambda + \mu a) = \lambda + \mu\varphi(a),\ \lambda, \mu \in \mathbf{R},\ \varphi \in V(a)$ is determined by the value $\varphi(a)$, say α, so that

$$\operatorname{Re} W_A(a) = \big\{ \alpha \in \mathbf{R} : |\lambda + \mu\alpha| \le \|\lambda + \mu a\|,\ \lambda, \mu \in \mathbf{R} \big\}.$$

But the inequality $\big|\lambda + \mu\alpha\big| \le \big\|\lambda + \mu a\big\|$ is equivalent to $\big|\lambda + \alpha\big| \le \big\|\lambda + a\big\|$, which is further equivalent to

$$-\big\|\lambda + a\big\| - \lambda \le \alpha \le \big\|\lambda + a\big\| - \lambda.$$

Therefore, we get

$$\operatorname{Re} W_A(a) = \Big[\sup_s\big(-\|s + a\| - s\big),\ \inf_t\|t + a\| - t\Big].$$

The function $f(t) = \|t + a\| - t$ on $\mathbf{R}$ is decreasing, thus

$$\begin{aligned}\inf f(t) &= \lim_{t \to +\infty}\big(\|t + a\| - t\big) = \lim_{t \to 0^+}\left(\left\|\frac{1}{t} + a\right\| - \frac{1}{t}\right)\\ &= \lim_{t \to 0^+}\frac{1}{t}\big(\|1 + ta\| - 1\big).\end{aligned}$$

This shows the formula (3).

Now, we have

$$\Big|\big\|\exp ta\big\| - 1\Big| \le \big\|\exp ta - 1\big\| \le \sum_{n=1}^{\infty}\frac{|t|^n}{n!}\|a^n\|,$$

so that $\frac{1}{t}\big(\|\exp ta\| - 1\big) = g(t)$ is bounded for small t, and $\big\|\exp ta\big\| = 1 + tg(t)$. With this remark, we compute

$$\begin{aligned}\lim_{t \to 0^+}\frac{1}{t}\log\big\|\exp ta\big\| &= \lim_{t \to 0^+}\left(\frac{tg(t)}{t}\frac{\log\big(1 + tg(t)\big)}{tg(t)}\right)\\ &= \lim_{t \to 0^+}\frac{\big\|\exp ta\big\| - 1}{t} \cdot \lim_{t \to 0^+}\frac{\log\big(1 + tg(t)\big)}{tg(t)}\\ &= \lim_{t \to 0^+}\frac{\big\|\exp ta\big\| - 1}{t}\end{aligned}$$

since $\lim_{x \to 0}\frac{\log(1+x)}{x} = 1$. Since $\exp ta = 1 + ta + O(t^2)$ as $t \to 0$, we have

$$\lim_{t \to 0^+}\frac{\big\|\exp ta\big\| - 1}{t} = \lim_{t \to 0^+}\frac{\big\|1 + ta\big\| - 1}{t}.$$

Therefore, the second half of (4) follows from (3).

To prove the first part of (4), consider $f(t) = \log\|\exp ta\|$. Then we have $f(s+t) \leq f(s) + f(t)$. Put $\alpha = \sup_{t>0} f(t)/t$. For any $\varepsilon > 0$, choose $t > 0$ so that $\alpha - \varepsilon \leq f(t)/t$. For any $s > 0$, choose $n \in \mathbf{N}$ and δ with $0 \leq \delta < s$ so that $t = ns + \delta$. Then we have

$$\alpha - \varepsilon \leq \frac{f(t)}{t} = \frac{f(ns+\delta)}{t} \leq \frac{nf(s) + f(\delta)}{t} = \frac{ns}{t}\frac{f(s)}{s} + \frac{f(\delta)}{t}$$
$$\leq \frac{ns}{t}\alpha + \frac{f(\delta)}{t}.$$

As $s \to 0$, we have $\frac{ns}{t} \to 1$ and $\frac{f(\delta)}{t} \to 0$, so that

$$\alpha - \varepsilon \leq \liminf_{s\to 0+} \frac{f(s)}{s} \leq \limsup_{s\to 0+} \frac{f(s)}{s} \leq \alpha;$$

hence we get

$$\alpha = \lim_{s\to 0+} \frac{f(s)}{s} = \sup \operatorname{Re} W_A(a).$$ Q.E.D.

A.14 Hermitian Elements. In a unital Banach algebra A, an element $a \in A$ is called *hermitian* if $W_A(a) \subset \mathbf{R}$. In the case of an involutive Banach algebra, a self-adjoint element, an element a with $a = a^*$, may fail to be hermitian.

(i) An element $a \in A$ is hermitian if and only if $\|\exp ita\| = 1$ for every $t \in \mathbf{R}$.

(ii) if $a \in A$ is hermitian, then

$$\|a\|_{\mathrm{sp}} = \|a\|,$$

where $\|a\|_{\mathrm{sp}}$ is the spectral radius of a, i.e. $\|a\|_{\mathrm{sp}} = \lim_{n\to\infty} \|a^n\|^{1/n}$.

PROOF:

(i) This equivalence follows immediately from (A.12.ii).

(ii) Suppose $a \in A$ is hermitian. Then the one parameter group: $\exp ita = \cos ta + \mathrm{i}\sin ta$, $t \in \mathbf{R}$, lies in the unit sphere of A, so that

$$\|\sin ta\| \leq \frac{1}{2}\Big(\|\exp(ita)\| + \|\exp(-ita)\|\Big) = 1.$$

By the spectral mapping theorem, if $\|ta\|_{\mathrm{sp}} = |t|\|a\|_{\mathrm{sp}} < \pi/2$, then we have $ta = \sin^{-1}(\sin ta)$. Now consider the Taylor series expansion of arc sin around zero:

$$\sin^{-1} x = \sum_{n=0}^{\infty} \frac{1\cdot 3\cdot 5\cdots(2n-1)}{2^n n!(2n+1)} x^{2n+1} = \sum_{n=0}^{\infty} \alpha_n x^{2n+1}.$$

Since $\alpha_n > 0$ for all n and $|\sin^{-1} x| < \pi/2$ for $|x| < 1$, we get $\sum_{n=0}^{\infty} \alpha_n = \pi/2$, so that if $\|ta\|_{\mathrm{sp}} < \pi/2$,

$$\|ta\| \leq \sum_{n=0}^{\infty} \alpha_n \|\sin ta\|^{2n+1} \leq \sum_{n=0}^{\infty} \alpha_n = \pi/2.$$

Thus we have $\|a\| \leq \|a\|_{\mathrm{sp}}$. But we know in general $\|a\|_{\mathrm{sp}} \leq \|a\|$. Hence $\|a\| = \|a\|_{\mathrm{sp}}$. Q.E.D.

A.15 Perron-Frobenius Eigenvalue.

Theorem. *Let* $T = \left[a_{i,j}\right] \in M_n(\mathbf{R}_+)$ *be an* $n \times n$ *matrix with strictly positive entries. Then the largest eigenvalue, say* ρ*, is positive and simple, i.e. there exits a unique vector* $x \in \mathbf{R}^n_+$*, up to positive scalar multiple, such that* $Tx = \rho x$*.*

PROOF: First set $\delta = \inf_{1 \le i,j \le n} a_{i,j} > 0$ and consider the commutative C^*-algebra $A = \ell^\infty_n(\mathbf{C})$ of dimension n and its positive cone $A_+ = \mathbf{R}^n_+$. Then T is a positive linear map of A into itself. The positivity of the coefficient of T means that T is positive also as a map of A. The strict positivity of all the coefficients of T means that if $x \in A_+$ is non-zero, then Tx is an invertible positive element of A. For each $x \in A_+$, $x \neq 0$, set $r(x) = \sup\left\{ r \ge 0 : rx \le Tx \right\}$. Since $r(\lambda x) = r(x),\ x \in A_+$ and $\lambda > 0$ we consider the function r on the unit sphere S_+ of A_+. As T is a bounded linear map of A, r is a bounded function on S_+. Set $\rho = \sup\left\{ \lambda > 0 : \lambda x \le Tx \right\}$. Let $\{x_k\}$ be a sequence in S_+ such that $\rho = \lim_{k\to\infty} r(x_k)$. Passing to a subsequence if necessary, we may and do assume that the sequence $\{x_k\}$ converges to a point $x \in S_+$. Then the inequality: $Tx_k - r(x_k)x_k \ge 0$ implies that $Tx - \rho x \ge 0$ as the positive cone A_+ is closed. Hence $\rho \le r(x)$. But by definition we have $\rho \ge r(x)$. Thus $\rho = r(x)$. We claim that $Tx = \rho x$. Suppose $Tx \neq \rho x$. Then $T\,(Tx - \rho x)$ is invertible and hence there exists $\varepsilon > 0$ such that $T\left(Tx - (\rho + \varepsilon)x\right) \ge 0$ so that $r(Tx) \ge \rho + \varepsilon$ which contradicts the definition of ρ.

We claim that ρ dominates the absolute value of all the eigenvalues of T. Suppose that λ is an eigenvalue of T and v is a corresponding eigenvector in the unit sphere S of A, so that $Tv = \lambda v$. In terms of coordinate we have

$$\lambda v_i = \sum_{j=1}^{n} a_{i,j} v_j, \quad i = 1, 2, \ldots, n,$$

which yields

$$|\lambda||v_i| \le \sum_{j=1} a_{i,j}|v_j|, \quad i = 1, 2, \ldots, n.$$

Hence $|\lambda| \le r\left(|v|\right) \le \rho$. Hence the absolute value of every eigenvalue of T is majorized by ρ.

Finally, suppose that $y \in A$ satisfies $Ty = \rho y$. Then $\rho \le r\left(|y|\right) \le \rho$. Hence $r\left(|y|\right) = \rho$. This means that $T|y| = \rho|y|$ from the above discussion. Hence

$$\left|\rho y_i\right| = \sum_{j=1}^{n} a_{i,j}|y_j| = \left|\sum_{j=1}^{n} a_{i,j} y_j\right|,$$

which is possible only when y and $|y|$ are proportional. So we may and do assume that $y \ge 0$. If x and y are linearly independent, then there are scalars α and β such that $z = \alpha x + \beta y \neq 0$ is not invertible and $Tz = \rho z$. Then the above discussion shows that $T\left(|z|\right) = \rho|z|$ is invertible, which is a contradiction. Q.E.D.

The eigenvalue ρ and vector x are called respectively *Perron-Frobenius eigenvalue* and *Perron-Frobenius eigenvector* of the matrix T.

A.16 Countable Selection Theorem.

Theorem. *Suppose that X is a Polish space and let $\mathfrak{C}_0(X)$ be the standard Borel space of all non-empty closed subsets of X defined in* Volume I, A.17.

(i) *For any open subset $U \subset X$ there exists a Borel choice function f from the standard Borel space $\mathfrak{C}_0(X)$ into X such that $f(F) \in F \cap U$ for every $F \in \mathfrak{C}_0(X)$ with $F \cap U \neq \emptyset$.*

(ii) *There exists a sequence $\{f_n\}$ of Borel choice functions from $\mathfrak{C}_0(X)$ into X such that $\{ f_n(F) : n \in \mathbf{N} \}$ is dense in F.*

PROOF:

(i) Fix a point $x_0 \in X$. As one can set $f(F) = x_0$ whenever $F \cap U = \emptyset$ and U is a Polish space on its own right with respect to the relative topology. So the proof reduces to the case that $X = U$. If X is a countable space, then the proof is trivial. So we may and do assume that X is uncountable.

Let $\{ x_n : n \in \mathbf{N} \}$ be a dense subset of X. For each $n \in \mathbf{N}$, let $X(n)$ be the open ball of radius $\frac{1}{2}$ with center x_n. Clearly $\{ X(n) : n \in \mathbf{N} \}$ is an open covering of X and each member $X(n)$ carries a metric d_n with which $X(n)$ is complete. Thus we obtained an open covering $\{ X(n) : n \in \mathbf{N} \}$ of X such that each member has diameter at most $\frac{1}{2}$.

Fix $n \in \mathbf{N}$. Repeating the above procedure, we obtain an open covering $\{ X(n,k) : k \in \mathbf{N} \}$ of $X(n)$ such that each member has diameter at most $\frac{1}{2^2}$ and $\overline{X(n,k)} \subset X(n)$. Let $d_{n,k}$ be a complete metric of $X(n,k)$ for each $k \in \mathbf{N}$. By induction, we obtain a series of open covering $\{ X(n_1, n_2, \dots, n_k) : (n_1, \dots, n_k) \in \mathbf{N}^k \}$ such that $X(n_1, \dots, n_{k-1}) = \bigcup_{j=1}^{\infty} X(n_1, \dots, n_{k-1}, j)$ and the diameter $\delta(X(n_1, \dots, n_{k-1}, j)) \leq \frac{1}{2^k}$ and $\overline{X(n_1, \dots, n_{k-1}, j)} \subset X(n_1, \dots, n_{k-1})$.

With this construction, for each $\bar{n} = (n_1, \dots, n_k, \dots) \in \Lambda = \mathbf{N}^{\mathbf{N}}$ we have $\bigcap_{k=1}^{\infty} X(n_1, \dots, n_k)$ is a singleton and set $\varphi(\bar{n})$ to be the element of the above intersection. This gives a continuous open mapping φ from Λ onto X such that $\varphi(\Lambda(n_1, \dots, n_k)) = X(n_1, \dots, n_k)$ where $\Lambda(n_1, \dots, n_k) = \{ \bar{m} \in \Lambda : m_1 = n_1, \dots, m_k = n_k \}$.

Now for each $F \in \mathfrak{C}_0(X)$, set $g(F)$ be the least element of $\varphi^{-1}(F)$ which is a closed subset of Λ, so that it admits the least element relative to the lexicographic ordering. Put $f(F) = \varphi(g(F)) \in F$. It remains to prove that f is a Borel map from $\mathfrak{C}_0(X)$ to X. Every non-empty open set U of X is a countable union of $X(\,\cdot\,)$'s, so that it is enough to prove that $f^{-1}(X(n_1, \dots, n_k))$ is a Borel subset of $\mathfrak{C}_0(X)$ for every $(n_1, \dots, n_k) \in \mathbf{N}^k$, $k \in \mathbf{N}$. But now we have

$f(F) \in X(n_1, \ldots, n_k)$

$\iff$ The least element of $\varphi^{-1}(F)$ belongs to $\Lambda(n_1, \ldots, n_k)$.

$$\iff \varphi^{-1}(F) \cap \left\{\bar{m} < (n_1, \ldots, n_k, 1, 1, \ldots)\right\} = \emptyset$$

and $\varphi^{-1}(F) \cap \Lambda(n_1, \ldots, n_k) \neq \emptyset$;

$$\iff F \cap X(m_1, m_2, \ldots, m_k) = \emptyset \quad \text{for any } (m_1, \ldots, m_k) < (n_1, \ldots, n_k)$$

and $F \cap X(n_1, \ldots, n_k) \neq \emptyset$.

Thus it is enough to prove that $\{F \cap X(n_1, \ldots, n_k) \neq \emptyset\}$ is a Borel subset of $\mathfrak{C}_0(X)$. But $\{F \cap X(n_1, \ldots, n_k) \neq \emptyset\}$ is precisely a member of the generating family of the Borel structure of $\mathfrak{C}_0(X)$. Hence it is Borel in $\mathfrak{C}_0(X)$.

(ii) Let $\{U_n : n \in \mathbf{N}\}$ be a countable basis of open sets of X. To each $n \in \mathbf{N}$, we associate a Borel choice function f_n of $\mathfrak{C}_0(X)$ by (i) such that $f_n(F) \in F \cap U_n$. Clearly $\{f_n(F) : n \in \mathbf{N}\}$ is dense in F. Q.E.D.

A.17 Second Separation Theorem.

(i) *If A and B are two Souslin sets in a standard Borel space X, then there exists a pair C and D of disjoint co-Souslin sets such that*

$$A \setminus B \subset C, \qquad B \setminus A \subset D,$$

where a subset of a standard Borel space is said to be co-Souslin if it is the complement of a Souslin set.

(ii) *If $\{A_n\}$ is a sequence of Souslin sets in a standard Borel space X, then there exists a sequence $\{C_n\}$ of mutually disjoint co-Souslin sets such that*

$$A_n \setminus \left(\bigcup_{m \neq n} A_m\right) \subset C_n.$$

Co-Souslin sets are often called *coanalytic* in the literature.

The proof requires the preparations spanning from A.18 through A.22 and will be completed in A.23.

A.18 Trees. For any set X, we denote the set of finite sequences in X by $X^{<\mathbf{N}}$, which may be viewed as the free semi-group generated by X. For each $s = a_0 a_1 \cdots a_{n-1} \in X^{<\mathbf{N}}$ the number n is called the *length* of s and denoted by $|s|$. For $m < n = |s|$ we put $s|m = a_0 a_1 \cdots a_{m-1}$ and call it the *initial segment* of s at m. If $|s| < |t|$, $s, t \in X^{<\mathbf{N}}$ and $s = t\big||s|$, then we write $s \prec t$ and $s \perp t$ otherwise. A *tree* on X is a subset $T \subset X^{<\mathbf{N}}$ such that if $t \in T$ and $s \prec t$, $s \in X^{<\mathbf{N}}$, then $s \in T$. If a tree T on X has no indefinitely increasing chain: $s_1 \prec s_2 \prec s_3 \prec \cdots$, then we say that the tree T is *finitely terminated.* It is easy to see the equivalence of the following two statements for a tree T on $\mathbf{N}$:

(i) T is finitely terminated;
(ii) The order $\leq_{\mathrm{KB}}$ in $\mathbf{N}^{<\mathbf{N}}$ defined by:

$$s = a_0 a_1 \cdots a_{m-1} \leq_{\mathrm{KB}} t = b_0 b_1 \cdots b_{n-1}$$
$$\Longleftrightarrow$$
$$t \prec s \quad \text{or} \quad \exists j < \min\{m, n\}: \ a_0 = b_0, \ a_1 = b_1, \ldots, a_{j-1} = b_{j-1}, \ a_j < b_j$$

is well-ordered on T.

Each $x = a_0 a_1 \cdots a_n \cdots \in X^{\mathbf{N}}$ gives a sequence $\{s_n\}$ in $X^{<\mathbf{N}}$ by $s_n = x|n$. Relative to the cartesian product topology on $X^{\mathbf{N}}$ of the discrete topology on X, the system $\Sigma_X(s) = \{y \in X^{\mathbf{N}} : s \prec y\}$, $s \prec x$, is the neighborhood system of $x \in X^{\mathbf{N}}$. If $X = \mathbf{N}$, then we write simply $\Sigma(s)$.

Lemma. *Suppose X is a countable discrete space.*

(i) *The set* $\mathrm{Tree}(X)$ *of trees on X is a Borel subspace of the Polish space* $\mathcal{C}_0(X^{<\mathbf{N}})$.

(ii) *The set* $\mathrm{Tree}_{\mathrm{F}}(X)$ *of finitely terminated trees on X is a co-Souslin set in* $\mathcal{C}_0(X^{<\mathbf{N}})$.

PROOF:

(i) As $X^{<\mathbf{N}}$ is a countable discrete space, $\mathcal{C}_0(X^{<\mathbf{N}})$ can be identified with the compact metrizable space $\{0, 1\}^{X^{<\mathbf{N}}}$, where the topology is given by the family of functions: $\{A \in \mathcal{C}_0(X^{<\mathbf{N}}) \mapsto \chi_A(x) \in \{0, 1\} : x \in X^{<\mathbf{N}}\}$. Now $T \in \mathcal{C}_0(X^{<\mathbf{N}})$ is a tree if and only if $\chi_T(s) \geq \chi_T(x)$ for every pair $x, s \in X^{<\mathbf{N}}$ with $s \prec x$. Thus $\mathrm{Tree}(X)$ is a closed subset of the compact metrizable space $\mathcal{C}_0(X^{<\mathbf{N}})$.

(ii) Observe

$$\{T \in \mathrm{Tree}(X) \setminus \mathrm{Tree}_{\mathrm{F}}(X)\}$$
$$= \pi_r\left(\left\{(y, S) \in X^{\mathbf{N}} \times \mathrm{Tree}(X) : y|n \in S \text{ for all } n \in \mathbf{N}\right\}\right)$$

where π_r means the projection to the second component of the product Polish space $X^{\mathbf{N}} \times \mathcal{C}_0(X^{<\mathbf{N}})$. Thus, the complement of $\mathrm{Tree}_{\mathrm{F}}(X)$ in the standard Borel space $\mathrm{Tree}(X)$ is a Souslin set. Q.E.D.

A.19 Binary Relations on a Countable Set. Fix a countable discrete space X. A subset $R \subset X \times X$ is identified with a binary relation on X, i.e. $xRy \Longleftrightarrow (x, y) \in R$, $x, y \in X$. For a pair $R, S \subset X \times X$ we get the third set $R * S$ in $X \times X \times X$ defined by

$$R * S = \{(x, y, z) \in X \times X \times X : (x, y) \in R, \ (y, z) \in S\}.$$

We then set $R \circ S = (\pi_\ell \times \pi_r)(R * S) \subset X \times X$ where π_ℓ and π_r mean the projections respectively to the first and the third components. Since $\chi_{R*S}(x, y, z) = \chi_R(x, y)\chi_S(y, z)$, the map: $(R, S) \in \mathcal{C}_0(X \times X) \times \mathcal{C}_0(X \times X) \longmapsto R * S \in \mathcal{C}_0(X \times X \times X)$ is continuous. With $f \in \ell^1(X)$ such that $f(y) > 0$, $y \in X$, we have

$$R \circ S = \left\{(x, z) \in X \times X : f_{R,S}(x, z) = \sum_{y \in Z} f(y)\chi_R(x, y)\chi_S(y, z) > 0\right\}.$$

The function: $(R, S) \in \mathcal{C}_0(X) \times \mathcal{C}_0(X) \mapsto f_{R,S}(x, z) \in \mathbf{R}_+$ is a continuous function for each $(x, z) \in X \times X$. Therefore, the map $(R, S) \in \mathcal{C}_0(X \times X) \times \mathcal{C}(X \times X) \mapsto R \circ S \in \mathcal{C}_0(X \times X)$ is Borel. The relation $R \subset X \times X$ is an order relation on the domain $D(R) = \{x \in X : (x, x) \in R\}$ if and only if

$$R \cap R^{-1} = \Delta_{D(R)}, \qquad \pi_\ell(R) = \pi_r(R) = D(R), \qquad R \circ R \subset R$$

where $(x, y)^{-1} = (y, x)$ and Δ_A means the diagonal set $\{(x, x) : x \in A\}$ for every $A \subset X$. Let $\mathrm{Ord}(X)$ denote the set of all order relations $R \subset X \times X$ on the domain $D(R) \subset X$ and $\mathrm{Ord}_{\mathrm{L}}(X)$ be the set of all linear order relations on X. Finally let $\mathrm{Ord}_{\mathrm{W}}(X)$ denote the set of all well order relations on X.

Lemma. *For a countable discrete set X, we have the following two statements:*

(i) $\mathrm{Ord}_{\mathrm{L}}(X)$ *is a standard Borel space;*

(ii) $\mathrm{Ord}_{\mathrm{W}}(X)$ *is a co-Souslin subset of* $\mathrm{Ord}_{\mathrm{L}}(X)$.

PROOF:

(i) From the preceding arguments, it follows that $\mathrm{Ord}(X)$ is a Borel subset of $\mathcal{C}_0(X \times X)$. Now $R \in \mathrm{Ord}(X)$ is a linear order if and only if $R \cup R^{-1} = D(R) \times D(R)$. Hence it is a Borel subset of $\mathrm{Ord}(X)$.

(ii) Each $R \subset \mathcal{C}_0(X \times X)$ gives rise uniquely a tree $T_R \in \mathrm{Tree}(X)$ on X by $T_R = \{x = x_0 x_1 \cdots x_{n-1} \in X^{<\mathbf{N}} : (x_{i-1}, x_i) \in R \text{ for every } i,\ 1 \le i \le n-1\}$. The order R is a well-order if and only if the tree T_R is finitely terminated. The map: $R \in \mathcal{C}_0(X \times X) \mapsto T_R = \bigcup_{n=0}^{\infty} R^{*n} \in \mathcal{C}_0(X^{<\mathbf{N}})$, where $R^{*0} = X$ and $R^{*n} = \underbrace{R * R * \cdots * R}_{n \text{ times}}$, is an injective Borel map.

The set $\mathrm{Ord}_{\mathrm{W}}(X)$ is the preimage of the co-Souslin set $\mathrm{Tree}_{\mathrm{F}}(X)$ of finitely terminated trees on X under this injective Borel map. Thus the assertion follow. Q.E.D.

A.20 Trees and Closed Subsets of $X^{\mathbf{N}}$. Keep the assumption that X is a countable discrete space. If $T \in \mathrm{Tree}(X)$, then the *body* $[T]$ of T is defined to be the set of all those $x \in X^{\mathbf{N}}$ such that $x|n \in T$ for all $n \in \mathbf{N}$. It is clear that T is finitely terminated if and only if $[T] = \emptyset$. It is also clear that the body $[T]$ of every tree T is a closed subset of $X^{\mathbf{N}}$. For each closed subset $A \subset X^{\mathbf{N}}$, we set $T_A = \{x|n \in X^{<\mathbf{N}} : x \in A,\ n \in \mathbf{N}\}$. Then $A = [T_A]$.

Lemma. *The map: $A \in \mathcal{C}_0(X^{\mathbf{N}}) \mapsto T_A \in \mathrm{Tree}(X)$ is an injective Borel map.*

PROOF: With $\pi_n : x \in X^{\mathbf{N}} \mapsto \pi_n(x) = x|n \in X^n \subset X^{<\mathbf{N}}$, we have $T_A = \bigcup_{n=0}^{\infty} \pi_n(A)$. For each $n \in \mathbf{N}$ and $s \in X^n$, we have $x \in T_A$ if and only if $\pi_n^{-1}(s) \cap A \neq \emptyset$, i.e. $\Sigma_X(s) \cap A \neq \emptyset$. Thus for each $s \in X^n$, we have

$$\{A \in \mathcal{C}_0(X^{\mathbf{N}}) : s \in T_A\} = \{A \in \mathcal{C}_0(X^{\mathbf{N}}) : A \cap \Sigma_X(s) \neq \emptyset\}.$$

The latter set is an open set in $\mathcal{C}_0(X^{\mathbf{N}})$. Therefore, the map: $A \in \mathcal{C}_0(X^{\mathbf{N}}) \mapsto T_A \in \mathcal{C}_0(X^{<\mathbf{N}})$ is a Borel map. The injectivity of the map is obvious. Q.E.D.

A.21 Trees and Closed Sets in $X^{\mathbf{N}} \times Y^{\mathbf{N}}$. We consider two countable discrete spaces X and Y. For each $T \in \mathrm{Tree}(X \times Y)$ and a point $x \in X^{\mathbf{N}}$, set $T(x) = \{ y \in Y^{<\mathbf{N}} : (x||y|, y) \in T \} \in \mathrm{Tree}(Y)$ and call it the *branch* of T at x.

Lemma. *For every $C \in \mathfrak{C}_0(X^{\mathbf{N}} \times Y^{\mathbf{N}})$ the map: $x \in X^{\mathbf{N}} \mapsto T_C(x) \in \mathrm{Tree}(Y)$ is a Borel map such that $x \notin \pi_\ell(C)$ if and only if $T_C(x) \in \mathrm{Tree}_{\mathrm{F}}(Y)$.*

PROOF: It is easy to see that $x \in X^{\mathbf{N}} \setminus \pi_\ell(C)$ if and only if $T_C(x) \in \mathrm{Tree}_{\mathrm{F}}(Y)$. Thus, it suffices to show that for each $T \in \mathrm{Tree}(X \times Y)$ the map: $x \in X^{\mathbf{N}} \mapsto T(x) \in \mathrm{Tree}(Y)$ is a Borel map. For each $y \in Y^n$, $n \in \mathbf{N}$, we have that $y \in T(x)$ if and only if $(x|n, y) \in T$ so that $\chi_{T(x)}(y) = \chi_T(x|n, y)$. But the function: $x \in X^{\mathbf{N}} \mapsto (x|n, y) \in X^n \times Y^n$ is continuous, so that $x \in X^{\mathbf{N}} \mapsto \chi_{T(x)}(y) \in \{0, 1\}$ is continuous. Thus, the map: $x \in X^{\mathbf{N}} \mapsto T(x) \in \mathrm{Tree}(Y)$ is described by the countable family $\{ \chi_{T(\cdot)}(y) : y \in Y^{<\mathbf{N}} \}$ of continuous functions, which implies the assertion. Q.E.D.

A.22 The Universality of $\mathrm{Ord}_{\mathrm{L}}$ and $\mathrm{Ord}_{\mathrm{W}}$. We write Ord for $\mathrm{Ord}(\mathbf{N})$, $\mathrm{Ord}_{\mathrm{L}}$ for $\mathrm{Ord}_L(\mathbf{N})$ and $\mathrm{Ord}_{\mathrm{W}}$ for $\mathrm{Ord}_{\mathrm{W}}(\mathbf{N})$. We also omit $\mathbf{N}$ from Tree's: Tree $= \mathrm{Tree}(\mathbf{N})$, $\mathrm{Tree}_{\mathrm{F}} = \mathrm{Tree}_{\mathrm{F}}(\mathbf{N})$.

Theorem. *If A is a co-Souslin set in a standard Borel space X, then there exists a Borel map f from X into $\mathrm{Ord}_{\mathrm{L}}$ such that $A = f^{-1}(\mathrm{Ord}_{\mathrm{W}})$.*

PROOF: We may and do assume that $X = \mathbf{N}^{\mathbf{N}}$ and that there exists a continuous map g from $\mathbf{N}^{\mathbf{N}}$ to X and a closed subset $D \subset \mathbf{N}^{\mathbf{N}}$ such that $\complement A = g(D)$. Setting $C = \{ (x, g(x)) : x \in \mathbf{N}^{\mathbf{N}} \}$, we obtain a closed set C in $\mathbf{N}^{\mathbf{N}} \times \mathbf{N}^{\mathbf{N}}$ such that $\complement A = \pi_\ell(C)$. We then get a tree T_C on $\mathbf{N} \times \mathbf{N}$ such that $x \in A$ if and only if $T_C(x) \in \mathrm{Tree}_{\mathrm{F}}$ if and only if $T_C(x)$ is well-ordered relative to the ordering $\leq_{\mathrm{KB}}$. The map: $x \in \mathbf{N}^{\mathbf{N}} \mapsto T_C(x) \in \mathrm{Tree}$ is a Borel map as seen in the last lemma.

Now let $x \colon n \in \mathbf{N} \mapsto x(n) \in \mathbf{N}^{<\mathbf{N}}$ be a fixed bijection. We know that $\mathbf{N}^{<\mathbf{N}}$ is linearly ordered by $\leq_{\mathrm{KB}}$. Thus every $T \in \mathrm{Tree}$ is linearly ordered. For each $T \in \mathrm{Tree}$, we define a linear order R_T by the following:

$$(m, n) \in R_T, \ (m, n) \in \mathbf{N} \times \mathbf{N},$$

$$\iff \begin{cases} m \leq n & \text{if } m \notin x^{-1}(T) \text{ and } n \notin x^{-1}(T), \\ m \in x^{-1}(T) \text{ and } n \notin x^{-1}(T), & \\ x(m) \leq_{\mathrm{KB}} x(n) & \text{if } m \in x^{-1}(T) \text{ and } n \in x^{-1}(T). \end{cases}$$

It is easy to see that $R_T \in \mathrm{Ord}_{\mathrm{L}}$ and that $R_T \in \mathrm{Ord}_{\mathrm{W}}$ if and only if $T \in \mathrm{Tree}_{\mathrm{F}}$. With R_0 the natural order in $\mathbf{N}$, the order R_T is given by

$$R_T = \Big(R_0 \cap \big(\complement x(T) \times \complement x(T) \big) \Big)$$
$$\cup \big(x(T) \times \complement x(T) \big) \cup \Big((x \times x) \big(R_{\mathrm{KB}} \cap (T \times T) \big) \Big) \subset \mathbf{N} \times \mathbf{N}.$$

Hence we have

$$\chi_{R_T}(m,n) = \max\left\{\begin{array}{l} \chi_{R_0}(m,n)\chi_{\complement T}\big(x(m)\big)\chi_{\complement T}\big(x(n)\big), \\ \chi_T\big(x(m)\big)\chi_{\complement T}\big(x(n)\big), \\ \chi_{R_{\mathrm{KB}}}\big(x(m),x(n)\big)\chi_T\big(x(m)\big)\chi_T\big(x(n)\big) \end{array}\right\}$$

for every $(m,n) \in \mathbf{N} \times \mathbf{N}$, which shows that the map: $T \in \mathrm{Tree} \mapsto R_T \in \mathrm{Ord_L}$ is Borel. Thus the map $f\colon x \in X \mapsto f(x) = R_{T_C(x)} \in \mathrm{Ord_L}$ is Borel and $A = f^{-1}(\mathrm{Ord_W})$. Q.E.D.

A.23 Proof of Theorem A.17.

(i) First we define an order relation $\preceq$ on Ord by

$$R_1 \preceq R_2 \iff \exists \varphi \in \mathbf{N}^{\mathbf{N}}:$$

$$\left\{\begin{array}{l} \varphi \text{ maps } D(R_1) \text{ injectively into } D(R_2) \text{ and} \\ \text{for } m,n \in D(R_1),\ (m,n) \in R_1 \text{ if and only if } \big(\varphi(m),\varphi(n)\big) \in R_2. \end{array}\right.$$

Consider

$$E = \left\{\begin{array}{l} \big(\varphi, R_1, R_2\big) \in \mathbf{N}^{\mathbf{N}} \times \mathrm{Ord} \times \mathrm{Ord}: \\ \quad \left\{\begin{array}{l} (\varphi \times \varphi)\Big(\big(D(R_1) \times D(R_1)\big) \setminus \Delta_{D(R_1)}\Big) \cap \Delta = \emptyset, \\ R_1 = (\varphi \times \varphi)^{-1}(R_2) \cap \big(D(R_1) \times D(R_1)\big). \end{array}\right. \end{array}\right\},$$

where Δ is the diagonal set in $\mathbf{N} \times \mathbf{N}$ and $\Delta_{D(R_1)} = \Delta \cap \big(D(R_1) \times D(R_1)\big)$. Then E is a Borel subset of $\mathbf{N}^{\mathbf{N}} \times \mathrm{Ord} \times \mathrm{Ord}$ and

$$R_{\preceq} = \pi(E) \subset \mathrm{Ord} \times \mathrm{Ord},$$

where π is the projection to $\mathrm{Ord} \times \mathrm{Ord}$ from $\mathbf{N}^{\mathbf{N}} \times \mathrm{Ord} \times \mathrm{Ord}$. Thus the order relation $\preceq$ is a Souslin set in $\mathrm{Ord} \times \mathrm{Ord}$.

Now let f and g be Borel maps from X to $\mathrm{Ord_L}$ such that $A = f^{-1}(\mathrm{Ord_W})$ and $B = g^{-1}(\mathrm{Ord_W})$ whose existences are guaranteed by Theorem A.22. Set

$$C = A \cap \complement\{\, g(x) \preceq f(x) : x \in X \,\}; \qquad D = B \cap \complement\{\, f(x) \preceq g(x) : x \in X \,\}.$$

Let

$$F = \left\{\begin{array}{l} (\varphi, x) \in \mathbf{N}^{\mathbf{N}} \times X: \\ \quad \left\{\begin{array}{l} \Big((\varphi \times \varphi)\big(D(f(x)) \times D(f(x)\big) \setminus \Delta_{D(f(x))}\Big) \cap \Delta = \emptyset, \\ f(x) = (\varphi \times \varphi)^{-1}\big(g(x)\big) \cap \big(D(f(x)) \times D(f(x)\big). \end{array}\right. \end{array}\right\}.$$

Then F is a Borel set in $\mathbf{N}^{\mathbf{N}} \times X$ and therefore $\{x \in X : f(x) \preceq g(x)\} = \pi_r(F)$ is a Souslin set in X. Thus C is co-Souslin and similarly D as well. Suppose $x \in A \setminus B$. Then $f(x) \in \mathrm{Ord}_{\mathrm{W}}$ and $g(x) \notin \mathrm{Ord}_{\mathrm{W}}$. Since every subset of a well-ordered set is well-ordered, $g(x) \npreceq f(x)$, so that $x \in C$. Similarly if $x \in B \setminus A$, then $f(x) \npreceq g(x)$ and $x \in D$.

Suppose that $x \in C \cap D$. Then $f(x) \in \mathrm{Ord}_{\mathrm{W}}$ and $g(x) \in \mathrm{Ord}_{\mathrm{W}}$. The comparability theorem for well-ordered sets implies that $f(x) \preceq g(x)$ or $g(x) \preceq f(x)$. In the first case, x cannot be in D and in the second case x cannot be in C. This shows that $C \cap D = \emptyset$.

(ii) For each $n \in \mathbf{N}$, by (i) there exist disjoint co-Souslin sets C'_n and D'_n such that $A_n \setminus \big(\bigcup_{m \neq n} A_m\big) \subset C'_n$ and $\big(\bigcup_{m \neq n} A_m\big) \setminus A_n \subset D'_n$. Put $C_n = C'_n \cap \big(\bigcap_{m \neq n} D'_n\big)$. Then the family $\{C_n\}$ has the desired property. Q.E.D.

A.24 Set of Uniqueness.

Theorem. *Let X and Y be two standard Borel spaces and $f : X \mapsto Y$ is a Borel map. Then the set of uniqueness: $Z_f = \{y \in Y : f^{-1}(y)$ is a singleton$\}$ is a co-Souslin set in Y.*

PROOF: We may and do assume that X is a closed subset of $\mathbf{N}^{\mathbf{N}}$, Y is Polish and $f : X \mapsto Y$ is continuous. For each $s \in \mathbf{N}^{<\mathbf{N}}$, set

$$A_s = f\big(\Sigma(s) \cap X\big) \qquad \text{and} \qquad B_s = f\big(X \setminus \Sigma(s)\big)$$

where $\Sigma(s) = \{x \in \mathbf{N}^{\mathbf{N}} : x \succ s\}$. Observe that

$$A_s = \bigcup_{n \in \mathbf{N}} A_{sn}, \qquad B_{sn} \supset B_s, \qquad B_s = \bigcup \{ A_t : |t| = |s| \text{ and } t \neq s \}$$

and

$$A_s \setminus B_s \supset A_{sn} \setminus B_{sn}, \qquad n \in \mathbf{N}. \tag{$*$}$$

Since f is continuous, we have $\{f(x)\} = \bigcap_{n=1}^{\infty} A_{x|n} = \bigcap_{n=1}^{\infty} \overline{A_{x|n}}$ and therefore

$$\begin{aligned} Z_f &= \bigcup_{x \in X} \Big(\{f(x)\} \setminus f\big(X \setminus \{x\}\big)\Big) \\ &= \bigcup_{x \in X} \left(\bigcap_{k=1}^{\infty} f(X \cap \Sigma(x|k)) \setminus f\Big(\bigcup_{k=1}^{\infty} (X \setminus \Sigma(x|k))\Big)\right) \\ &= \bigcup_{x \in X} \Big(\bigcap_{k=1}^{\infty} (A_{x|k} \setminus B_{x|k})\Big) = \bigcup_{x \in X} \Big(\bigcap_{k=1}^{\infty} \overline{A_{x|k}} \setminus B_{x|k}\Big). \end{aligned}$$

By Theorem A.17, for each $k \in \mathbf{N}$, there exists a disjoint family $\{C_s : |s| = k,\ s \in \mathbf{N}^{<\mathbf{N}}\}$ of co-Souslin sets in Y such that $A_s \setminus B_s \subset C_s$. Replacing

C_s by $\bigcap_{t \prec s} C_t$, we may choose the system $\left\{ C_s : s \in \mathbf{N}^{<\mathbf{N}} \right\}$ so that $C_s \supset C_{sn}$, $n \in \mathbf{N}$. Set $C_s^* = C_s \cap \left(\overline{A_s} \setminus B_s \right) = C_s \cap \overline{A_s} \cap \complement B_s$, which is co-Souslin as the intersection of the three co-Souslin sets, C_s, $\overline{A_s}$ and $\complement B_s$. Then we have

$$A_s \setminus B_s \subset C_s^* \subset \overline{A_s} \setminus B_s$$

and

$$C_{sm}^* = \left(\overline{A_{sm}} \setminus B_{sm} \right) \cap C_{sm} \subset \left(\overline{A_s} \setminus B_s \right) \cap C_s = C_s^*. \tag{$**$}$$

Therefore, we have $C_s^* \cap C_t^* = \emptyset$ if $|s| = |t|$ and $s \neq t$, $s, t \in \mathbf{N}^{<\mathbf{N}}$. Now we have

$$Z_f \subset \bigcup_{x \in X} \left(\bigcap_{k=1}^{\infty} C_{x|k}^* \right) \subset \bigcup_{x \in X} \left(\bigcap_{k=1}^{\infty} \overline{A_{x|k}} \setminus B_{x|k} \right) = Z_f,$$

so that $Z_f = \bigcup_{x \in X} \left(\bigcap_{k=1}^{\infty} C_{x|k}^* \right)$. If $y \in \bigcup_{x \in X} \left(\bigcap_{k=1}^{\infty} C_{x|k}^* \right)$, there exists $x \in X$ such that $y \in \bigcap_{k=1}^{\infty} C_{x|k}^*$, so that $y \in \bigcup_{|s|=k,\ s \in \mathbf{N}^{<\mathbf{N}}} C_s^*$ for every $k \in \mathbf{N}$, i.e. $y \in \bigcap_{k=1}^{\infty} \left(\bigcup_{|s|=k,\ s \in \mathbf{N}^{<\mathbf{N}}} C_s^* \right)$. Conversely, if y belongs to $\bigcap_{k=1}^{\infty} \left(\bigcup_{|s|=k,\ s \in \mathbf{N}^{<\mathbf{N}}} C_s^* \right)$, then for every $k \in \mathbf{N}$ there exists a unique, as $\left\{ C_s^* : |s| = k \right\}$ is disjoint, $s_k \in \mathbf{N}^{<\mathbf{N}}$ with $|s_k| = k$ such that $y \in C_{s_k}^*$. The disjointness of $\left\{ C_s^* : |s| = k,\ s \in \mathbf{N}^{<\mathbf{N}} \right\}$ and the fact that $C_{sn}^* \subset C_s^*$ together imply that $s_1 \prec s_2 \prec \cdots \prec s_k \cdots$. Hence there exists $x \in \mathbf{N}^{\mathbf{N}}$ such that $x|k = s_k$. The inclusion of $(**)$ yields the existence of $x_k \in X \cap \Sigma(s_k)$ such that $\lim_{k \to \infty} f(x_k) = y$. Since $\left\{ \Sigma(s_k) : k \in \mathbf{N} \right\}$ is a decreasing sequence of open sets in $\mathbf{N}^{\mathbf{N}}$ converging to x, we have $\lim x_k = x \in X$ and $x|k = s_k$. Therefore, we conclude that $y \in \bigcup_{x \in X} \left(\bigcap_{k=1}^{\infty} C_{x|k}^* \right) = Z_f$. This implies that Z_f is co-Souslin as claimed. Q.E.D.

A.25 Theorem (Lusin).

If f is a Borel map from a standard Borel space X into another Y such that $f^{-1}(y)$ is at most countable, then the image $f(A)$ of a Borel set A in X is a Borel set in Y and there exists a Borel map $g\colon y \in f(X) \mapsto g(y) \in X$ such that $f \circ g(y) = y$, $y \in f(X)$.

PROOF: We may and do assume that X and Y are both Polish and f is continuous. Then the countability and the closedness of $f^{-1}(y)$ imply that either $f^{-1}(y) = \emptyset$ or $f^{-1}(y)$ contains isolated points. Let $\left\{ U_n : n \in \mathbf{N} \right\}$ be a countable basis of the open sets in X. For every Borel set $A \subset X$ and every point $y \in f(A)$ there exists $n \in \mathbf{N}$ such that $\varphi^{-1}(y) \cap \left(U_n \cap A \right)$ is a singleton. Hence $B_n = \left\{ y \in f(A) : f^{-1}(y) \cap (A \cap U_n) \text{ is a singleton} \right\}$ is the set of uniqueness of f restricted to $A \cap U_n$ and hence co-Souslin. Thus $f(A) = \bigcup_{n=1} B_n$ is Souslin and co-Souslin; hence Borel.

The second assertion on the existence of a right inverse g of f now follows from Theorem A.15, Volume I. Q.E.D.

Bibliography

[418] C. A. Akemann, G. Elliott, G. K. Pedersen and J. Tomiyama, *Derivations and multipliers of C^*-algebras,* Amer. J. Math., **98** (1976), 679–708.

[419] C. A. Akemann and P. Østrand, *On a tensor product C^*-algebra associated with the free group on two generators,* J. Math. Soc. Japan, **27** (1975), 585–599.

[420] W. Ambrose, *The L^2-system of a unimodular group* I, Trans. Amer. Math. Soc., **65** (1949), 27–48

[421] W. Ambrose and S. Kakutani, *Structure and continuity of measurable flow,* Duke Math. J., **9** (1942), 25–42.

[422] H. Anzai, *Ergodic skew product transformations on the torus,* Osaka Math. J., **3** (1951), 83–99.

[423] H. Araki, *Asymptotic ratio set and property L'_λ,* Publ. Res. Inst. Math. Sci., **6** (1971), 443–460.

[424] H. Araki, *One parameter family of Radon-Nikodym theorems for states of a von Neumann algebra,* Publ. Res. Inst. Math. Sci., **10** (1974), 1–10.

[425] H. Araki, *Some properties of the modular conjugation operator of von Neumann algebras and a non-commutative Radon-Nikodym theorem with a chain rule,* Pacific J. Math., **50,** 309–354.

[426] H. Araki, *Structure of some von Neumann algebras with isolated discrete modular spectrum,* Publ. Res. Inst. Math. Sci., **9** (1973), 1–44.

[427] H. Araki, *Expansional in Banach algebras,* Ann. Sci. École Norm. Sup., **6** (1973), 67–84.

[428] H. Araki and T. Masuda, *Positive cones and L_p-spaces for von Neumann algebras,* Publ. Res. Inst. Math. Sci., **18** (1982), 759–831.

[429] H. Araki, R. Haag, D. Kastler and M. Takesaki, *Extension of KMS states and chemical potential,* Comm. Math. Phys., **53** (1977), 97–134.

[430] H. Araki and E. J. Woods, *Representations of the canonical commutation relations describing a non-relativistic infinite free Bose gas,* J. Math. Phys., **4** (1963), 637–662.

[431] H. Araki and E. J. Woods, *A classification of factors,* Publ. Res. Inst. Math. Sci., **3** (1968), 51–130.

[432] W. Arveson, *Subalgebras of C^*-algebras.* I, Acta Math., **123** (1969), 141–244. II, Acta Math., **128** (1972), 271–308. III, *Multivariable operator theory,* Acta Math., **181** (1998), 159–228.

[433] W. Arveson, *On groups of automorphisms of operator algebras,* J. Funct. Anal., **15** (1974), 217–243.

[434] W. Arveson, *The harmonic analysis of automorphism groups,* Proc. Sympos. Pure Math., Amer. Math. Soc., **38** (1982), 199–269.

[435] W. Arveson, *Ten lectures on operator algebras,* CBMS, **55,** Amer. Math. Soc. (1984), viii + 93.

[436] D. Avitzour, *Free products of C^*-algebras,* Trans. Amer. Math. Soc., **271** (1982), 423–435.

[437] S. I. Bezuglyi and V. Y. Golodets, *Outer conjugacy for actions of countable amenable groups on a measure space,* Izvestiya Acad. Sci. USSR ser Math., **50** (1986), 643–660.

[438] B. Blackadar, *K-theory for operator algebras,* MSRI. Publ., **5,** Springer-Verlag (1986), viii + 338.

[439] R. Blattner, *Automorphic group representations,* Pacific J. Math., **8** (1958), 665–677.

[440] H. Borchers, *Energy and momentum as observables in quantum field theory,* Comm. Math. Phys., **2** (1966), 49–54.

[441] N. Bourbaki, *Intégration, Éléments de Mathématique, Livre* VI, Hermann, Paris (1963).

[442] O. Bratteli and D. W. Robinson, *Operator algebras and quantum statistical mechanics* I, Springer-Verlag (1979), xiv + 519. II, ibid. (1981), xiii + 505.

[443] L. Brown and H. Kosaki, *Jensen's inequality in semi-finite von Neumann algebras,* J. Operator Theory, **23** (1990), 3–19.

[444] H. Choda, *A Galois correspondence in a von Neumann algebra,* Tôhoku Math.J., **30** (1978), 491–504.

[445] M. D. Choi and E. G. Effros, *The completely positive lifting problem for C^*-algebras,* Ann. Math., **104** (1976), 585–609.

[446] M. D. Choi and E. G. Effros, *Separable nuclear C^*-algebras and injectivity,* Duke Math. J., **43** (1976), 309–322.

[447] M. D. Choi and E. G. Effros, *Injectivity and operator sapces,* J. Funct. Anal., **24** (1977), 156–209.

[448] M. D. Choi and E. G. Effros, *Nuclear C^*-algebras and injectivity, The general case,* Ind. Univ. Math. J., **26** (1977), 443–446.

[449] M. D. Choi and E. G. Effros, *Nuclear C^*-algebras and the approximation property,* Amer. J. Math., **100** (1978), 61–79.

[450] F. Combes, *Poids sur une C^*-algèbre,* J. Math. Pure et Appl., **47** (1968), 57–100.

[451] F. Combes, *Poids associés a une algèbre hilbertienne à gauche,* Compos. Math., **23** (1971), 49–77.

[452] F. Combes and C. Delaroche, *Groupe modulaire d'une espérance conditionelle dans une algèbre de von Neumann,* Bull. Soc. Math. France, **103** (1975), 385–426.

[453] A. Connes, *Un nouvel invariant pour les algèbres de von Neumann,* C. R. Acad. Sc. Paris, **273, Série A,** (1971), 900–903.

[454] A. Connes, *Calcul des deux invariants d'Araki-Woods par la théorie de Tomita-Takesaki,* C. R. Acad. Sc. Paris, **274, Série A,** (1972), 175–178.

[455] A. Connes, *Etats presque périodiques sur une algèbre de von Neumann,* C. R. Acad. Sc. Paris, **274,** Série A, (1972), 1402–1405.

[456] A. Connes, *Groupe modulaire d'une algèbre de von Neumann,* C. R. Acad. Sc. Paris, **274,** Série A, (1972), 1923–1926.

[457] A. Connes, *Une classification des facteurs de type* III, C. R. Acad. Sc. Paris, **275,** Série A, (1972), 523–525.

[458] A. Connes, *Une classification des facteurs de type* III, Ann. Scient. Ecole Norm. Sup., 4ème Série, **6** (1973), 133–252.

[459] A. Connes, *Caractérisation des espaces vectoriels ordonnés sous-jacents aux algèbres de von Neumann,* Ann. Inst. Fourier, **24** (1974), 121–155.

[460] A. Connes, *Almost periodic states and factors of type* III_1, J. Funct. Anal., **16** (1974), 415–445.

[461] A. Connes, *Sur le théorème de Radon Nikodym,* Bull. Soc. Math. de France, 2ème Série, **97** (1973), 253–258.

[462] A. Connes, *On hyperfinite factors of type* III_0 *and Krieger's factors,* J. Funct. Anal., **18** (1975), 318–327.

[463] A. Connes, *On the hierarchy of W. Krieger,* Illinois J. Math., **19** (1975), 428–432.

[464] A. Connes, *A factor not antiisomorphic to itself,* Bull. London Math. Soc., **7** (1975), 171–174.

[465] A. Connes, *A factor not antiisomorphic to itself,* Ann. of Math., **101** (1975), 536–554.

[466] A. Connes, *Periodic automorphisms of the hyperfinite factor of type* II_1, Acta Math. Szeged., **39** (1977), 39–66.

[467] A. Connes, *Outer conjugacy of automorphisms of factors,* Symposia Mathematica, **20** , 149–159.

[468] A. Connes, *Classification of injective factors,* Ann. of Math., **104** (1976), 73–115.

[469] A. Connes, *Outer conjugacy classes of automorphisms of factors,* Ann. Sci. École Norm. Sup., 4ème Série, **8** (1975), 383–419.

[470] A. Connes, *Sur la classification des facteurs de type* II, C. R. Acad. Sci. Paris, Ser. A-B **281** (1975), 13–15.

[471] A. Connes, *On the cohomology of operator algebras, J. Funct. Anal.,* **28** (1978), 248–253.

[472] A. Connes, *On the spatial theory of von Neumann algebras,* J. Funct. Anal., **35** (1980), 153–164.

[473] A. Connes, *Sur la théorie non commutative de l'intégration,* Lecture Notes in Math., Springer-Verlag, **725** (1979), 19–143.

[474] A. Connes, *The von Neumann algebra of a foliation,* Lecture Notes in Physics, **80** (1977), 145–151.

[475] A. Connes, *Von Neumann algebras,* Proceedings of the International Congress of Mathematicians Helsinki, (1978), 97–109.

[476] A. Connes, *A factor of type* II_1 *with countable fundamental group,* J. Operator Theory, **4** (1980), 151–153.

[477] A. Connes, *A survey of foliations and operator algebras, Operator algebras and applications,* Proc. Symposia Pure Math., **38, Part I,** (1982), 521–628.

[478] A. Connes, *Property T for von Neumann Algebras,* Proc. London Math. Soc., **17** (1985), 57–62.

[479] A. Connes, *Factors of type* III_1 *and property* L'_λ, J. Operator theory, **14** (1985), 189–211.

[480] A. Connes, *Noncommutative geometry,* Academic Press (1994), xiv + 661.

[481] A. Connes and V. Daele, *The group property of invariant S of von Neumann algebras,* Math. Scand., **32** (1973), 187–192.

[482] A. Connes, J. Feldman and B. Weiss, *An Amenable equivalence relation is generated by a single transformation,* Ergodic Theory Dynamical Systems, **1** no. 4 (1981), 431–450.

[483] A. Connes, U. Haagerup and E. Størmer, *Diameters of state spaces of type* III *factors, Operator algebras and their connections with topology and ergodic theory,* Lecture Notes in Math., Springer-Verlag, **1132** (1985), 91–116.

[484] A. Connes and V. Jones, *A* II_1 *factor with two nonconjugate Cartan subalgebras,* Bull. Amer. Math. Soc., **6** (1982), 211–212.

[485] A. Connes and V. F. R. Jones, *Property T for von Neumann algebras,* Bull. London Math. Soc., **17** (1985), 57–62.

[486] A. Connes and W. Krieger, *Measure space automorphism groups, the normalizer of their full groups, and approximate finiteness,* J. Func. Anal., **24** (1977), 336–352.

[487] A. Connes, P. Chez Lima, Testard and J. Woods, *Review of a paper of Golodets,* Mimeographed Notes, (1972).

[488] A. Connes and C. Rovelli, *Von Neumann algebra automorphisms and time-thermodynamics relation in generally covariant quantum theories,* Class. Quantum Grav., **11** (1994), 2899–2917.

[489] A. Connes and E. Størmer, *Entropy for automorphisms of finite von Neumann algebras,* Acta Math., **134** (1975), 289–306.

[490] A. Connes and E. Stømer, *Homogeneity of the state space of factors of type* III_1, J. Funct. Anal., **28** (1978), 187–196.

[491] A. Connes and M. Takesaki, *The flow of weights on factors of type* III, Tôhoku Math. J., **29** (1977), 473–575.

[492] A. Connes and B. Weiss, *Property T and asymptotically invariant sequences,* Israel Journal of Math., **37** (1980), 209–210.

[493] A. Connes and E. J. Woods, *Approximately transitive flows and* ITPFI *factors,* Ergodic Theory and Dynamical System, **5** (1985), 203–236.

[494] A. Connes and E. J. Woods, *Hyperfinite von Neumann algebras and Poisson boundaries of time dependent random walks,* Pacific J. Math., **137** (1989), 225–243.

[495] A. van Daele, *A new approach to the Tomita-Takesaki theory of generalized Hilbert algebras,* J. Funct. Anal., **15** (1974), 378–393.

[496] A. van Daele, *Continuous crossed products and type* III *von Neumann algebras,* London Math. Soc., Lecture Notes Series, **31** (1978), vii + 68.

[497] A. van Daele, *Two weights with orthogonal supports but equal on a dense *-subalgebra,* Bull. Amer. Math. Soc., **5** (1981), 179–181.

[498] A. van Daele and M. Rieffel, *The commutation theorem for tensor products of von Neumann algebras,* Bull. London Math. Soc., **7** (1975), 257–260.

[499] A. Van Daele and M. Rieffel, *A bounded operator approach to Tomita-Takesaki theory,* Pacific J. Math., **69** (1977), 187–221.

[500] T. Digernes, *Poids duals sur un produit croise,* C. R. Acad. Sci. Paris,, **278** (1974), 937–940.

[501] T. Digernes, *Dual weights and the commutant theorem for crossed products of* W^**-algebras,* Thesis, UCLA, (1975).

[502] J. Dixmier, *Algèbres quasi-unitaires,* Comment. Math. Helv., **26** (1952), 275–322.

[503] J. Dixmier and C. Lance, *Deux nouveaux facteurs de type* II, Invent. Math., **7** (1969), 226–234.

[504] J. Dixmier and O. Marechal, *Vecteurs totalisateurs d'une algèbre de von Neumann,* Comm. Math. Phys., **22** (1971), 44–50.

[505] S. Doplicher, D. Kastler and D. Robinson, *Covariance algebras in field theory and statistical mechanics,* Comm. Math. Phys., **3** (1966), 1–28.

[506] N. Dunford and J. T. Schwartz, *Linear Operators, Part I: General Theory,* Wiley, New York (1988).

[507] H. Dye, *On groups of measure preserving transformations* I, Amer. Math. J., **81** (1959), 119–159. II, ibid., **85** (1963), 551–576.

[508] E. G. Effros, *Transformation groups and* C^**-algebras,* Ann. Math., **81** (1965), 38–55.

[509] E. G. Effros, *Dimensions and* C^**-algebras,* CBMS 46, Amer. Math. Soc. (1981), v + 74.

[510] E. G. Effros and C. Lance, *Tensor products of operator algebras,* Adv. Math., **25** (1977), 1–34.

[511] E. G. Effros and Z. J. Ruan, *Operator Spaces,* London Math. Soc. Monographs, New Series, **23**, Oxford University Press, New York (2000), xvi + 363.

[512] M. Enock and J. M. Schwartz, *Une dualité dans les algèbres de von Neumann,* Bull. Soc. Math. France Mem., **no. 44.** Supplement au Bull. Soc. Math. France, **103** (1975), 1–144.

[513] M. Enock and J. M. Schwartz, *Produit croise d'une algèbre de von Neumann par une algèbre de Kac,* J. Funct. Anal., (1977), 16–47. II, Publ. Res. Inst. Math. Sci., **16** (1980), 189–232.

[514] M. Enock and J. M. Schwartz, *Kac alagebras and duality of locally compact groups,* Springer-Verlag (1992), x + 257.

[515] D. E. Evans and Y. Kawahigashi, *Quantum symmetries on operator algebras,* Oxford Mathematical Monographs. Oxford Science Publications. Oxford University Press, New York (1998), xvi + 829.

[516] P. Eymard, *L'algèbre de Fourier d'un groupe localement compact,* Bull. Math. Soc. France, **92** (1964), 181–236.

[517] T. Fack and H. Kosaki, *Generalized s-numbers of τ-measurable operators,* Pacific J. Math., **123** (1986), 269–300.

[518] A. J. Falcone, *Operator valued weights, L^2-von Neumann modules and their relative tensor products,* Thesis, UCLA, (1996), viii + 94.

[519] A. J. Falcone and M. Takesaki, *Operator valued weights without structure theory,* Trans. Amer. Math. Soc., **351** (1999), 323–341.

[520] A. J. Falcone and M. Takesaki, *Non-commutative flow of weights on a von Neumann algebra,* J. Funct. Anal., **182** (2001), 170–206.

[521] J. Feldman, P. Hahn and C. C. Moore, *Orbit Structure and countable sections for actions of continuous groups,* Adv. Math., **28** (1982), 186–230.

[522] J. Feldman and C. C. Moore, *Ergodic equvialence relations cohomology, and von Neumann algebras* I, Trans. Amer. Math., **234** (1977), 289–324. II, ibid., 325–359.

[523] J. M. G. Fell, *An extension of Mackey's method to Banach *-algebraic bundles,* Mem. Amer. Math. Soc., **90** (1969), 168.

[524] J. M. G. Fell, *Induced representations and Banach *-algebraic bundles,* Lecture Notes in Math., Springer-Verlag, **582** (1977), iii + 349.

[525] F. Forelli, *Analytic and quasi-invariant measures,* Acta Math., **118** (1967), 33–59.

[526] J. Glimm, *Locally compact transformation groups,* Trans. Amer. Math. Soc., **101** (1961), 124–138.

[527] R. Godement, *Théorèmes tauberiens et théorie spectrale,* Ann. Sci. Ecole Norm. Sup., **(3) 64** (1947), 119–138.

[528] R. Godement, *Mémoire sur le théorie des caractères dans les groupes localement compacts unimodulaires,* J. Math. Pure et Appl., **30** (1951), 1–110.

[529] R. Godement, *Théorie des caractères,* I *Algèbres unitaires,* Ann. Math., **59** (1954), 47–62.

[530] F. Goodman, P. de la Harpe and V. F. R. Jones, *Coxeter graphs and towers of algebras,* MSRI. Publ., **14** Springer Verlag, New York (1989), x + 288.

[531] F. P. Greenleaf, *Invariant means on topological groups and their applications,* Van Nostrand Math. Studies, **16,** Van Nostrand Reinhold Co., New York – Toronto – London (1969), ix + 113.

[532] A. Grothendieck, *Résumé de la théorie métrique des produits tensoriels topologiques,* Bol. Soc. Mat. São Paulo, **8** (1953), 1–79.

[533] A. Guichardet, *Systemes dynamiques non-commutatifs,* Asterisque, Soc. Math. France, **13–14** (1974).

[534] R. Haag, *Local Quantum Physics. Fields, particles, algebras,* Second edition, Texts and Monographs in Phys., Springer Verlag, Berlin (1996), xvi + 390.

[535] R. Haag, N. M. Hugenholtz and M. Winnink, *On the equilibrium states in quantum mechanics,* Comm. Math. Phys., **5** , 215–236.

[536] R. Haag and D. Kastler, *An algebraic approach to quantum field theory,* J. Math. Phys., **5** (1964), 848–861.

[537] U. Haagerup, *The standard form of von Neumann algebras,* Math. Scand., **37** (1975), 271–283.

[538] U. Haagerup, *Normal weights on W^*-algebras,* J. Functional Analysis, **19** (1975), 302–317.

[539] U. Haagerup, *An example of a weight with type* III *centralizer,* Proc. Amer. Math. Soc., **62** , 278–280.

[540] U. Haagerup, *On the dual weight for crossed products of von Neumann algebras* I, Math. Scand., **43** (1978), 99–118. II, ibid., 119–140.

[541] U. Haagerup, *Operator valued weights in von Neumann algebras* I, J. Funct. Anal., **32** (1979), 175–206. II, ibid., **33** (1979), 339–361.

[542] U. Haagerup, *An example of non-nuclear C^*-algebra, which has the metric approximation property,* Invent. Math., **50** (1979), 279–293.

[543] U. Haagerup, *L^p-spaces associated with an arbitrary von Neumann algebra,* Colloques Internationaux C.N.R.S., Algèbres d'opérateurs et leurs applications en physique mathématique, **274** (1979), 175–184.

[544] U. Haagerup, *A density theorem for left Hilbert algebras,* Algebrès d'operateurs (1978), 170–179. Lecture Notes in Math., Springer-Verlag, **725** (1979).

[545] U. Haagerup, *Solution of the similarity problem for cyclic representations of C^*-algebras,* Ann. Math., **118** (1983), 215–240.

[546] U. Haagerup, *All nuclear C^*-algebras are amenable,* Invent. Math., **74** (1983), 305–319.

[547] U. Haagerup, *A new proof of the equivalence of injectivity and hyperfiniteness for factors on a separable Hilbert space,* J. Funct. Anal., **62** (1985), 160–201.

[548] U. Haagerup, *The Grothendieck inequality for bilinear forms on C^*-algebras,* Adv. Math., **56** (1985), 93–116.

[549] U. Haagerup, *The injective factors of type* III_λ, $0 < \lambda < 1$, Pacific J. Math., **137** (1989), 265–310.

[550] U. Haagerup, *Connes bicentralizer problem and uniqueness of the injective factor of type* III_1, Acta Math., **158** (1987), 95–147.

[551] U. Haagerup and E. Størmer, *Pointwise inner automorphisms of von Neumann algebras with an appendix by C. Sutherland,* J. Funct. Anal., **92** (1990), 177–201.

[552] U. Haagerup and E. Størmer, *Equivalence of normal states of von Neumann algebras and the flow of weights,* Adv. Math., **83** (1990), 180–262.

[553] U. Haagerup and L. Zsido, *Sur la properiété de Dixmier pour les C^*-algebras,* C. R. Acad. Sci., Paris, Série I, **298** no. 8 (1984), 173–176.

[554] Y. Haga, *Crossed products of von Neumann algebras by compact groups,* Tôhoku Math. J., **28** (1976), 511–522.

[555] Y. Haga and Z. Takeda, *Correspondence between subgroups and subalgebras in a cross product von Neumann algebra,* Tôhoku Math. J., **24** (1972), 167–190.

[556] P. Hahn, *Haar measure for measure groupoids,* Trans. Amer. Math. Soc., **242** (1978), 1–33. *The regular representations of measure groupoids,* ibid., 35–72.

[557] J. Hakeda and J. Tomiyama, *On some extension properties of von Neumann algebras,* Tôhoku Math. J., **19** (1967), 315–323.

[558] T. Hamachi, Y. Oka and M. Osikawa, *A classification of ergodic non-singular transformation groups,* Mem. Fac. Sci. Kyushu Univ., Ser. A **28** (1974), 113–133.

[559] T. Hamachi, Y. Oka and M. Oshikawa, *Flows associated with ergodic non-singular transformation groups,* Publ. Res. Inst. Math. Sci., **11** (1975), 31–50.

[560] T. Hamachi and M. Oshikawa, *Ergodic groups of automorphisms and Krieger's theorem,* Seminar on Math. Sci. Keio Univ., (1981), 113.

[561] R. Herman and M. Takesaki, *States and automorphism groups of operator algebras,* Comm. Math. Phys., **19** (1970), 142–160.

[562] F. Hiai, *Minimizing indices of conditional expectations onto a subfactor,* Publ. Res. Inst. Math. Sci., **24** (1988), 673–678.

[563] M. Hilsum, *Les espaces L^p d'une algèbre de von Neumann définies par la derive spatiale,* J. Funct. Anal., **40** (1981), 151–169.

[564] N. M. Hugenholtz, *On the factor type of equilibrium states in quantum mechanics,* Comm. Math. Phys., **6** (1967), 189–193.

[565] U. K. Hui, *Cocycle conjugacy of one parameter automorphism groups of* AFD *factors of type* III, Internat. J. Math., **13** no. 6 (2002), 579–603.

[566] B. Iochum and H. Kosaki, *Linear Radon-Nikodým theorems for states on* JBW *and W^* algebras,* Publ. Res. Inst. Math. Sci., **21** (1985), 1205–1222.

[567] H. Izumi, *Construction of non-commutative L^p-spaces with a complex parameter,* Internat. J. Math., **8** (1997), 1029–1066.

[568] B. E. Johnson, *Cohomology in Banach algebras,* Mem. Amer. Math. Soc., **127** (1972), iii + 96.

[569] B. E. Johnson, R. V. Kadison and J. Ringrose, *Cohomology of operator algebras* III, Bull. Soc. Math. France, **100** (1972), 73–96.

[570] V. F. R. Jones, *Actions of finite groups on the hyperfinite type* II_1 *factor,* Mem. Amer. Math. Soc., **237** (1980), v + 70.

[571] V. F. R. Jones, *Index for subfactors,* Invent. Math., **72** (1983), 1–25.

[572] V. F. R. Jones, *A polynomial invariant for knots via von Neumann algebras,* Bull. Amer. Math. Soc., **12** (1985), 103–111.

[573] V. F. R. Jones, *Heck algebra representations of braid groups and link polynomials,* Ann. Math., **126** (1987), 335–388.

[574] V. F. R. Jones and M. Takesaki, *Actions of compact abelian groups on semifinite injective factors,* Acta Math., **153** (1984), 213–258.

[575] G. I. Kac, *Ring groups and the duality principle,* Trudy Moskov. Mat. Obšč., **12** (1963), 259–301. *Annular groups and the principle of duality,* II, ibid., **13** (1965), 84–113.

[576] R. V. Kadison, *Derivations of operator algebras,* Ann. Math., **83** (1966), 280–293.

[577] R. V. Kadison, *Strong continuity of operator functions,* Pacific J. Math., **26** (1968), 121–129.

[578] R. V. Kadison, *Limits of states,* Comm. Math. Phys., **85** (1982), 143–154.

[579] R. V. Kadison and J. Ringrose, *Derivations and automorphisms of operator algebras,* Comm. Math. Phys., **4** (1967), 32–63.

[580] R. V. Kadison and J. Ringrose, *Cohomology of operator algebras* I. *Type* I *von Neumann algebras,* Acta Math., **126** (1971), 227–243. II. *Extended cobounding and the hyperfinite case,* Arkiv für Math., **9** (1971), 55–63.

[581] R. V. Kadison and J. Ringrose, *Fundamentals of the theory of operator algebras* I, Academic Press (1983), xv + 398. II, Academic Press (1986), xiv + 399–1074.

[582] S. Kakutani, *Induced measure preserving transformations,* Proc. Japan Acad., **19** (1943), 635–641.

[583] S. Kakutani, *On equivalence of infinite product measures,* Ann. Math., **49** (1948), 214–224.

[584] R. R. Kallman, *A generalization of free action,* Duke Math. J., **36** (1969), 781–789.

[585] R. R. Kallman, *Unitary groups and automorphisms of operator algebras,* Amer. J. Math., **91** (1969), 785–806.

[586] R. R. Kallman, *Groups of inner automorphisms of von Neumann algebras,* J. Funct. Anal., **7** (1971), 43–60.

[587] R. R. Kallman, *One-parameter groups of* **-automorphisms of* II_1 *von Neumann algebras,* Proc. Amer. Math. Soc., **24** (1970), 336–340.

[588] D. Kastler, *The* C^**-algebras of a free Boson field,* Comm. Math. Phys., **1** (1965), 14–48.

[589] D. Kastler and M. Takesaki, *Group duality and the Kubo-Martin-Schwinger condition,* Comm. Math. Phys., **70** (1979), 193–212. II, Comm. Math. Phys., **85** (1982), 155–176.

[590] Y. Katayama, C. E. Sutherland and M. Takesaki, *The intrinsic invariant of an approximately finite-dimensional factor and the cocycle conjugacy of discrete amenable group actions,* Electron. Res. Announc. Amer. Math. Soc., **1** (1995), 43–47.

[591] Y. Katayama, C. E. Sutherland and M. Takesaki, *The characteristic square of a factor and the cocycle conjugacy of discrete group actions on factors,* Invent. Math., **132** (1998), 331–380.

[592] Y. Kawahigashi, *One parameter automorphism groups of the hyperfinite type* II_1 *factor,* J. Operator Theory, **25** (1991), 37–59.

[593] Y. Kawahigashi, C. E. Sutherland and M. Takesaki, *The structure of the automorphism group of an injective factor and the cocycle conjugacy of discrete abelian group actions,* Acta Math., **169** (1992), 105–130.

[594] Y. Kawahigashi and M. Takesaki, *Compact abelian group actions on injective factors,* J. Funct. Anal., **105** (1992), 112–128.

[595] A. Kishimoto, *Remarks on compact automorphism groups on a certain von Neumann algebra,* Publ. Res. Inst. Math. Sci., **13** (1977), 573–581.

[596] H. Kosaki, *Canonical* L^p*-spaces associated with an arbitrary abstract von Neumann algebra,* Thesis, UCLA, (1980).

[597] H. Kosaki, *Positive cones associated with a von Neumann algebra,* Math. Scand., **47** (1980), 295–307.

[598] H. Kosaki, *Positive cones and* L^p*-spaces associated with a von Neumann algebra,* J. Operator Theory, **6** (1981), 13–23.

[599] H. Kosaki, *Remarks on positive cones associated with a von Neumann algebra,* Tôhoku Math. J., **33** (1981), 587–591.

[600] H. Kosaki, *Non-commutative Lorentz spaces associated with a semi-finite von Neumann algebra and applications,* Proc. Japan Acad., **57** (1981), 303–306.

[601] H. Kosaki, *T-theorem for* L^p*-spaces associated with a von Neumann algebra,* J. Operator Theory, **7** (1982), 267–277.

[602] H. Kosaki, *A Radon-Nikodym theorem for natural cones associated with von Neumann algebras,* Proc. Amer. Math. Soc., **84** (1982), 207–211.

[603] H. Kosaki, *Linear Radon-Nikodym theorem for states on a von Neumann algebra,* Publ. Res. Inst. Math. Sci., **18** (1982), 379–386. II, Proc. Amer. Math. Soc., **87** (1983), 283–288.

[604] H. Kosaki, *Applications of the complex interpolation method to a von Neumann algebra: Non-commutative* L^p*-spaces,* J. Funct. Anal., **56** (1984), 29–78.

[605] H. Kosaki, *Interpolation theory and the Wigner-Yanase-Dyson-Lieb concavity,* Comm. Math. Phys., **87** (1982), 315–329.

[606] H. Kosaki, *Remarks on Lebesgue-type decomposition of positive operators,* J. Operator Theory, **11** (1984), 137–143.

[607] H. Kosaki, *Applications of uniform convexity of non-commutative* L^p*-spaces,* Trans. Amer. Math. Soc., **283** (1984), 265–282.

[608] H. Kosaki, *On the Bures distance and the Uhlmann transition probability of states on a von Neumann algebra,* Proc. Amer. Math. Soc., **89** (1983), 285–288.

[609] H. Kosaki, *Lebesgue decomposition of states on a von Neumann algebra,* Amer. J. Math., **107** (1985), 697–735.

[610] H. Kosaki, *On the continuity of the map* $\varphi \to |\varphi|$ *from the predual of a* W^**-algebra,* J. Funct. Anal., 59(1984), 123-131.

[611] H. Kosaki, *Extension of Jones' theory on index to arbitrary factors,* J. Funct. Anal., **66** (1986), 123–140.

[612] H. Kosaki, *Relative entropy of states: a variational expression,* J. Operator Theory, **16** (1986), 335–348.

[613] H. Kosaki, *Characterization of crossed product (properly infinite case),* Pacific J. Math., **137** (1989), 159–167.

[614] H. Kosaki, *An inequality of Araki-Lieb-Thirring (von Neumann algebra case),* Proc. Amer. Math.Soc., **114** (1992), 477–481.

[615] H. Kosaki, *Unitarily invariant norms under which the map $A \to |A|$ is Lipschitz continuous,* Publ. Res. Inst. Math. Sci., **28** (1992), 299–313.

[616] H. Kosaki, *A remark on Sakai's quadratic Radon-Nikodym theorem,* Proc. Amer. Math. Soc., **116** (1992), 783–786.

[617] H. Kosaki and R. Longo, *A remark on the miminal index of subfactors,* J. Funct. Anal., **107** (1992), 458–470.

[618] U. Krengel, *Darstellungssätze für Strömungen und Halbströmungen* II, Math. Ann., **182** (1969), 1–39.

[619] W. Krieger, *On non-singular transormations of a measure space* I, Z. Wahrscheinlichkeitstheorie verw. Geb., **11** (1969), 83–97. II, **ibid.,** 98–119.

[620] W. Krieger, *On the Araki-Woods asymptotic ratio set and non-singular transformations of a measure space,* Lecture Notes in Math., Springer-Verlag, **160** (1970), 158–177.

[621] W. Krieger, *On ergodic flows and the isomorphism of factors,* Math. Ann., **223** (1976), 19–70.

[622] I. Kubo, *Quasi-flows,* Nagoya Math. J., **35** (1969), 1–30.

[623] C. Lance, *Direct integrals of left Hilbert algebras,* Math. Ann., **216** (1975), 11–28.

[624] M. Landstad, *Duality for dual covariant algebras,* Comm. Math. Phys., **52** (1977), 191–202.

[625] H. Leptin, *Verallgemeinerte L^1-Algebren und projektive Darstellungen lokalkompakter Gruppen,* I, Invent. Math., **3** (1967), 257–281. II, ibid., **4** (1967), 68–86.

[626] P. H. Loi, *On the theory of index and type* III *factors,* Pennsylvania State University, Thesis (1988).

[627] R. Longo, *Index of subfactors and statistics of quantum fields* I, Comm. Math. Phys., **126** (1989), 217–247. II, ibid., **130** (1990), 285–309.

[628] R. Longo, *Minimal index and braided subfactors,* J. Funct. Anal., **109** (1992), 98–112.

[629] G. Loupias and S. Miracle-Sole, *C^*-algèbres des systèmes canoniques.* I, Comm. Math. Phys., **2** (1966), 31–48. II, Ann. Inst. H. Poincar Sect. A (N.S.), **6** (1967), 39–58.

[630] G. W. Mackey, *Induced representations of locally compact groups* I, Ann. Math., **55** (1952), 101–139. *Induced representation of locally compact groups* II, *The Frobenius reciprocity theorem,* ibid., **58** (1953), 193–221.

[631] G. W. Mackey, *Borel structure in groups and their duals,* Trans. Amer. Math. Soc., **85** (1957), 134–165.

[632] G. W. Mackey, *Point realizations of transformation groups,* Illinois J. Math., **6** (1962), 327–335.

[633] G. W. Mackey, *Ergodic theory and virtual groups,* Math. Ann., **166** (1966), 187–207.

[634] T. Masuda, *A note on a theorem of A. Connes on Radon-Nikodym cocycles,* Publ. Res. Inst. Math. Sci., **20** (1984), 131–136.

[635] D. McDuff, *Uncountably many* II_1 *factors,* Ann. Math., **90** (1969), 372–377.

[636] D. McDuff, *Central sequences and the hyperfinte factor,* Proc. London Math. Soc., **21** (1970), 443–461.

[637] C. C. Moore, *Invariant measues on product spaces,* Proc. Fifth Berkeley Sympos. Math. Stat. and Prob., II , (1967), 447–459.

[638] P. Muhly, J. Renault and D. P. Williams, *Equivalence and isomorphism for groupoid* C^**-algebras,* J. Operator theory, **17** (1987), 3–22.

[639] Y. Nakagami, *Duality for crossed products of von Neumann algebras by locally compact groups,* Bull. Amer. Math. Soc., **81** (1975), 1106–1108.

[640] Y. Nakagami, *Dual action on a von Neumann algebra and Takesaki's duality for a locally compact group,* Publ. Res. Inst. Math. Sci., **12** (1977), 727–775.

[641] Y. Nakagami, *Essential spectrum* $\Gamma(\beta)$ *of a dual action on a von Neumann algebra,* Pacific J. Math., **70** (1977), 437–479.

[642] Y. Nakagami and C. E. Sutherland, *Takesaki's duality for regular extensions of von Neumann algebras,* Pacific J. Math., **83** (1979), 221–229.

[643] Y. Nakagami and M. Takesaki, *Duality for crossed products of von Neumannm algebras,* Lecture Notes in Math., Springer-Verlag **731,** (1979), ix + 139.

[644] M. Nakamura and Z. Takeda, *On some elementary properties of the crossed products of von Neumann algebras,* Proc. Japan Acad., **34** (1958), 489–494. *A Galois theory for finite factors,* Proc. Japan Acad., **36** (1960), 258–260.

[645] M. Nakamura and Z. Takeda, *On the fundamental theorem of the Galois theory for finite factors,* Proc. Japan Acad., **36** (1960), 313–318.

[646] H. Nakano, *Hilbert algebras,* Tôhoku Math. J., **2** (1950), 4–23.

[647] E. Nelson, *Notes on non-commutative integration,* J. Funct. Anal., **15** (1974), 103–116.

[648] A. Ocneanu, *Actions of discrete amenable groups on factors,* Lecture Notes in Math., Springer-Verlag, Berlin – New York **1138,** (1985), iv + 115.

[649] D. Olesen, G. K. Pedersen and M. Takesaki, *Ergodic actions of compact abelian groups,* J. Operator Theory, **3** (1980), 237–269.

[650] V. I. Paulsen, *Completely bounded maps and dilations,* Pitman Research Notes in Math. Sci., **146** (1986), 187.

[651] G. K. Pedersen, *Measure theory for C^*-algebras,* I, Math. Scand., **19** (1966), 131–145. II, ibid., **22** (1968), 63–74. III, ibid., **25** (1969), 71–93. IV, ibid., **25** (1969), 121–127.

[652] G. K. Pedersen, *C^*-algebras and their automorphism groups,* Academic Press, London Mathematical Society Monographs, 14 London, New York (1979), ix + 416.

[653] G. K. Pedersen, *Analysis Now,* Graduate Text in Math., 118, Springer-Verlag, Berlin, Heidelberg, London, New York, Paris and Tokyo (1988), xiv + 277.

[654] G. K. Pedersen and M. Takesaki, *The Radon-Nikodym theorem for von Neumann algebras,* Acta Math., **130** (1973), 53–87.

[655] M. Pimsner and S. Popa, *Entropy and index for subfactor,* Ann. Sci. Ec. Norm. Sup., **19** (1986), 57–106.

[656] M. Pimsner and S. Popa, *Iterating the basic contruction,* Trans. Amer. Math. Soc., **310** (1988), 127–133.

[657] M. Pimsner and D. Voiculescu, *Exact sequences for K-groups and Ext-groups of certain cross-product C^*-algebras,* J. Operator Thoery, **4** (1980), 93–118.

[658] G. Pisier, *Grothendieck's theorem for noncommutative C^*-algebras with an appendix on Grothendieck's constant,* J. Funct. Anal., **29** (1978), 397–415.

[659] S. Popa, *On a problem of R. V. Kadison on maximal abelian *-subalgebras in factors,* Invent. Math., **65** (1981), 269–281.

[660] S. Popa, *Orthogonal pairs of *-subalgebras in finite von Neumann algebras,* J. Operator Theory, **9** (1983), 253–268.

[661] S. Popa, *Singular maximal abelian *-subalgebras in continuous von Neumann algebras,* J. Funct. Anal., **50** (1983), 151–166.

[662] S. Popa, *Maximal injective subalgebras in factors associated with free groups,* Adv. Math., **50** (1983), 27–48.

[663] S. Popa, *Semiregular maximal abelian *-subalgebras and the solution to the factor state Stone-Weierstrass problem,* Invent. Math., **76** (1984), 157–161.

[664] S. Popa, *Notes on Cartan subalgebras in type* II_1 *factors,* Math. Scand., **57** (1985), 171–188.

[665] S. Popa, *A short proof that injectivity implies hyperfiniteness for finite von Newmann algebras,* J. Operator Theory, **16** (1986), 261–272.

[666] S. Popa, *Classification of subfactors: the reduction to commuting squares,* Invent. Math., **101** (1990), 19–43.

[667] S. Popa, *On the classification of actions of amenable groups on subfactors,* C. R. Acad. Sci. Paris, **315** (1992), 295–299.

[668] S. Popa, *Classification of amenable subfactors of type* II, Acta Math., **172** (1994), 163–255.

[669] S. Popa and M. Takesaki, *The topological structure of the unitary and authormorphism groups of a factor,* Comm. Math. Phys., **155** (1993), 93–101.

[670] R. T. Powers, *Representations of uniformly hyperfinite algebras and their associated von Neumann rings,* Ann. Math., **86** (1967), 138–171.

[671] R. T. Powers and E. Størmer, *Free states of the canonical anticommutation relations,* Comm. Math. Phys., **16** (1970), 1–33.

[672] L. Pukanszky, *On the theory of quasi-unitary algebras,* Acta Sci. Math., **16** (1955), 103–121.

[673] F. Rădulescu, *Random matrices, amalgamated free products and subfactors of the von Neumann algebra of a free group, of noninteger index,* Invent. Math., **115** (1994), 347–389.

[674] A. Ramsay, *Virtual groups and group actions,* Adv. Math., **6** (1971), 253–322.

[675] A. Ramsay, *Topologies on measured groupoids,* J. Funct. Anal., **47** (1982), 314–343.

[676] J. Renault, *A groupoid approach to C^*-algebras,* Lecture Notes Series in Math., Springer-Verlag, **793** (1980), ii + 160.

[677] J. E. Roberts, *Cross products of von Neumann algebras by group duals,* Symposia Mathematica, **XX,** Academic Press, London (1976), 335–363.

[678] D. W. Robinson, E. Størmer and M. Takesaki, *Derivations of simple C^*-algebras tangential to compact automorphism groups,* J. Operator Theory, **13** (1985), 189–200.

[679] Z.-J. Ruan, *Subspaces of C^*-algebras,* J. Funct. Anal., **76** (1988), 217–230.

[680] W. Rudin, *Functional Analysis,* McGraw-Hill, New York (1973), xiii + 397.

[681] K. Saito, *On a dulality for locally compact groups,* Tôhoku Math. J., **20** (1968), 355–367.

[682] T. Saitô, *On incomplete infinite direct product of W^*-algebras,* Tôhoku Math. J., **10** (1958), 165–171.

[683] S. Sakai, *On a conjecture of Kaplansky,* Tôhoku Math.J., **12** (1960), 31–33.

[684] S. Sakai, *Derivations of W^*-algebras,* Ann. Math., **83** (1966), 273–279.

[685] S. Sakai, *Derivations of simple C^*-algebras,* J. Funct. Anal., **2** (1968), 202–206. II, Bull. Soc. Math. France, **99** (1971), 259–263.

[686] S. Sakai, *An uncountable number of* II_1 *and* II_∞ *factors,* J. Funct. Anal., **5** (1970), 236–246.

[687] S. Sakai, *Operator algebras in dynamical systems,* Encyclopedia of Mathematics and its applications, **41,** Cambridge Univ. Press (1991), xi + 219.

[688] J. L. Sauvageot, *Sur le produit tensoriel relatif d'éspace de Hilbert,* J. Operator Theory, **9** (1983), 237–252.

[689] I. E. Segal, *An extension of Plancherel's formula to separable unimodular groups,* Ann. Math., **52** (1950), 272–292.

[690] C. F. Skau, *Finite subalgebras of a von Neumann algebra,* J. Funct. Anal., **25** (1977), 211-235.

[691] S. M. Srivastava, *A course on Borel sets, Graduate Texts in Math.,* Springer-Verlag, **180** (1998), xvi + 261.

[692] W. Stinespring, *Integration theorems for gages and duality for unimodular groups,* Trans. Amer. Math. Soc., **90** (1959), 15–56.

[693] E. Størmer, *On infinite tensor products of von Neumann algebras,* Amer. J. Math., **93** (1971), 810–818.

[694] E. Størmer, *Hyperfinite product factors,* Ark. Mat., **9** (1971), 165–170. II, J. Funct. Anal., **10** (1972), 471–481. III, Amer. J. Math., **97** (1975), 589–595.

[695] E. Størmer, *Spectra of states, and asymptotically abelian C^*-algebras,* Comm. Math. Phys., **28** (1972), 279–294. *Corrections,* ibid., **38** (1974), 341–343.

[696] C. E. Sutherland, *Direct integral theory for weights and the Plancherel formula,* Bull. Amer. Math. Soc., **20** (1974), 456–461.

[697] C. E. Sutherland, *Notes on orbit equivalence: Krieger's Theorem,* Lecture Notes Series, **No. 23** (1976), Oslo.

[698] C. E. Sutherland, *A Borel parametrization of Polish groups,* Publ. Res. Inst. Math. Sci., **21** (1985), 1067–1086.

[699] C. E. Sutherland and M. Takesaki, *Actions of discrete amenable groups and groupoids on von Neumann algebras,* Publ. Res. Inst. Math. Sci., **21** (1985), 1087–1120.

[700] C. E. Sutherland and M. Takesaki, *Actions of discrete amenable groups on injective factors of type* III_λ, $\lambda \neq 1$, Pacific J. Math., **137** (1989), 405–444.

[701] N. Suzuki, *A Linear representation of a countably infinite group,* Proc. Japan Acad., **34** (1958), 575–579.

[702] N. Suzuki, *Crossed products of rings of operators,* Tôhoku Math. J., **11** (1959), 113–124.

[703] H. Takai, *On a duality for crossed products of C^*-algebras,* J. Funct. Anal., **19** (1975), 25–39.

[704] Z. Takeda, *Inductive limit and infinite direct product of operator algebras,* Tôhoku Math. J., **7** (1955), 67–86.

[705] O. Takenouchi, *On type classification of factors constructed as infinite tensor products,* Publ. Res. Inst. Math. Sci., **4** (1968), 467–482.

[706] M. Takesaki, *Covariant representations of C^*-algebras and their automorphism groups,* Acta Math., **119** (1967), 273–302.

[707] M. Takesaki, *A characterization of group algebras as a converse of Tannaka-Stinespring-Tatsuuma duality,* Amer. J. Math., **91** (1969), 529–564.

[708] M. Takesaki, *Tomita's theory of modular Hilbert algebras and its applications,* Lecture Notes in Math., Springer-Verlag, **128** (1970), 123.

[709] M. Takesaki, *Theory of operator algebras,* UCLA Lecture Notes, (1970).

[710] M. Takesaki, *Disjointness of the KMS states of different temperatures,* Comm. Math. Phys., **17** (1970), 33–41.

[711] M. Takesaki, *Duality and von Neumann algebras,* Lectures on operator algebras, Tulane Univ. Ring and Operator Theory Year (1970–1971), Vol. II. Lecture Notes in Math., Springer-Verlag, **247** (1972), 665–786.

[712] M. Takesaki, *Conditional expectation on von Neumann algebras,* J. Funct. Anal., **9** (1972), 306–321.

[713] M. Takesaki, *The structure of a von Neumann algebra with a homogenous periodic state,* Acta Math., **131** (1973), 249–310.

[714] M. Takesaki, *Structure of factors and automorphism groups,* CBMS, **51** (1983), 107.

[715] M. Takesaki, *A classification theory based on groupoids,* Proc. of the US-Japan Seminar, Kyoto, 1983, Pitman Research Notes in Math. Ser., **123** (1986), 400–410.

[716] M. Takesaki and N. Tatsuuma, *Duality and subgroups,* Ann. Math., **98** (1971), 344–364.

[717] T. Tannaka, *Über den Dualität der nichtkommutativen topologischen Gruppen,* Tôhoku Math. J., **45** (1938), 1–12.

[718] N. Tatsuuma, *A duality theorem for locally compact groups,* J. Math. Kyoto Univ., **6** (1967), 187–293.

[719] M. Terp, *Interpolation spaces between a von Neumann algebra and its predual,* J. Operator Theory, **8** (1982), 327–360.

[720] M. Tomita, *Spectral theory of operator algebras.* I, Math. J. Okayama Univ., **9** (1959), 63–98. II, ibid., **10** (1960), 19–60.

[721] J. Tomiyama, *Tensor products and projections of norm one in von Neumann algebras,* Lecture Notes, Univ. Copenhagen (1970), 71.

[722] J. Tomiyama, *Tensor products and approximation problems of C^*-algebras,* Publ. Res. Inst. Math. Sci., **11** (1975), 163–183.

[723] J. Tomiyama, *Fubini algebras and the commutation theorem for tensor products of C^*-algebras,* Symposia Math., **20** (1976), 27–37.

[724] J. Tomiyama, *Invitation to C^*-algebras and topological dynamics,* World Scientific Pub. Co., Singapore (1987), x + 167.

[725] T. Turumaru, *Crossed products of operator algebras,* Tôhku J. Math., **10** (1958), 355–365.

[726] H. Umegaki, *Conditional expectation in an operator algebra* II, Tôhoku Math. J., **8** (1956), 86–100. III, Kōdai Math. Sem. Rep., **11** (1959), 51–64. IV, *Entropy and Information,* Kōdai Math. Sem. Rep., **14** (1962), 59–85.

[727] D. Voiculescu, *A non-commutative Weyl-von Neumann theorem,* Rev. Rounamine Math. Pures and Appl., **21** (1976), 97–113.

[728] D. Voiculescu, *Addition of certain noncommuting random variables,* J. Funct. Anal., **66** (1986), 323–346.

[729] D. Voiculescu, *On the existence of quasicentral approximate units relative to normed ideals,* I, J. Funct. Anal., **91** (1990), 1–36.

[730] D. Voiculescu, *Free probability theory: random matrices and von Neumann algebras,* Proc. ICM 94, Zürich, I , 227–241.

[731] D. Voiculescu, *The analogues of entropy and of Fisher's information measure in free probability theory.* I, Comm. Math. Phys., **155** (1993), 71–92. II, Invent. Math., **118** (1994), 411–440. III, *The absence of Cartan subalgebras,* Geometric Functional Analysis, **6** (1996), 172–199. IV, *Maximum entropy and freeness. Free probability theory,* Fields Inst. Comm., **12,** Amer. Math. Soc., Providence, RI (1997), 293–302. V, *Noncommutative Hilbert transforms,* Invent. Math., **132** (1998), 189–227.

[732] D. V. Voiculescu, K. J. Dykema and A. Nica, *Free random variables,* CRM Monograph Series, I , (1992), v + 70.

[733] S. Wassermann, *Exact C^*-algebras and related topics,* Lecture Notes Series, Seoul National Univ., Korea, **19** (1994), 92.

[734] H. Wenzl, *Hecke algebras of type* A_n *and subfactors,* Invent. Math., **92** (1988), 349–383.

[735] H. W. Wiesbrock, *Half-sided modular inclusions of von Neumann algebras,* Comm. Math. Phys., **157** (1993), 83–92 , 683–685.

[736] H. W. Wiesbrock, *Symmetries and half-sided modular inclusions of von Neumann algebras,* Letter in Math. Phys., **28** (1993), 107–114.

[737] C. Winsløw, *Strongly free actions on subfactors,* Internat. J. Math., **4** (1993), 675–688.

[738] C. Winsløw, *Classification of strongly amenable subfactors of type* III_0, J. Funct. Anal., **148** (1997), 296–313.

[739] C. Winsløw, *The flow of weights in subfactor theory,* Publ. Res. Inst. Math. Sci., **31** (1995), 519–532.

[740] P. Wojtaszczyk, *Banach spaces for analysts,* Cambridge Univ. Press, Cambridge Studies in Adv. Math., **25** (1991), xiv + 382.

[741] S. L. Woronowicz, *Operator systems and their application to the Tomita-Takesaki theory,* J. Operator Theory, **2** (1979), 169–209.

[742] S. L. Woronowicz, *Compact matrix pseudogroups,* Comm. Math. Phys., **111** (1987), 613–665.

[743] S. L. Woronowicz, *Tannaka-Krein duality for compact matrix pseudogroups. Twisted* $\mathrm{SU}(N)$ *groups,* Invent. Math., **93** (1988), 35–76.

[744] S. L. Woronowicz, *Differential calculus on compact matrix pseudogroups (quantum groups),* Comm. Math. Phys., **122** (1989), 125–170.

[745] S. L. Woronowicz, *A remark on compact matrix quantum groups,* Letter Math. Phys., **21** (1991), 35–39.

[746] S. Yamagami, *Modular theory for bimodules,* J. Funct. Anal., **125** (1994), 327–357.

[747] S. Yamagami, *Algebraic aspects in modular theory,* Publ. Res. Inst. Math. Sci., **28** (1992), 1075–1106.

[748] T. Yamanuouchi, *Duality for actions and coactions of measured groupoids on von Neumann algebras,* Mem. Amer. Math. Soc., **101** no. 484 (1993), vi + 109.

[749] G. Zeller-Meier, *Produit croisé d'une C^*-algèbre par un groupe d'automorphismes,* J. Math. Pure et Appl., **47** (1968), 102–239.

[750] R. J. Zimmer, *Ergodic theory and semisimple groups,* Birkhäuser, Monograph in Math., **81** (1984), viii + 209.

[751] A. Zygmund, *Trigonometric series,* **I, II,** Cambridge Univ. Press (1959).

Notation Index

Subject Index

GPSR Compliance

The European Union's (EU) General Product Safety Regulation (GPSR) is a set of rules that requires consumer products to be safe and our obligations to ensure this.

If you have any concerns about our products, you can contact us on ProductSafety@springernature.com

In case Publisher is established outside the EU, the EU authorized representative is:

Springer Nature Customer Service Center GmbH
Europaplatz 3
69115 Heidelberg, Germany

Batch number: 10755310

Printed by Printforce, the Netherlands